The chemistry of
phenols

Patai Series: The Chemistry of Functional Groups

A series of advanced treatises founded by Professor Saul Patai and under the general editorship of Professor Zvi Rappoport

The **Patai Series** publishes comprehensive reviews on all aspects of specific functional groups. Each volume contains outstanding surveys on theoretical and computational aspects, NMR, MS, other spectroscopical methods and analytical chemistry, structural aspects, thermochemistry, photochemistry, synthetic approaches and strategies, synthetic uses and applications in chemical and pharmaceutical industries, biological, biochemical and environmental aspects.
To date, over 100 volumes have been published in the series.

Recently Published Titles

The chemistry of the Cyclopropyl Group (2 volumes, 3 parts)
The chemistry of the Hydrazo Azo and Azoxy Groups (2 volumes, 3 parts)
The chemistry of Double-Bonded Functional Groups — (3 volumes, 6 parts)
The chemistry of Organophosphorus Compounds (4 volumes)
The chemistry of Halides, Pseudo-Halides and Azides (2 volumes, 4 parts)
The chemistry of the Amino, Nitro and Nitroso Groups (2 volumes, 4 parts)
The chemistry of Dienes and Polyenes (2 volumes)
The chemistry of Organic Derivatives of Gold and Silver
The chemistry of Organic Silicon Compounds (3 volumes, 6 parts)
The chemistry of Organic Germanium, Tin and Lead Compounds (2 volumes, 3 parts)
The chemistry of Phenols

Forthcoming Titles

The chemistry of Organolithium Compounds
The chemistry of Cyclobutanes
The chemistry of Peroxides (Volume 2)

The Patai Series Online

Starting in 2003 the **Patai Series** will be available in electronic format on Wiley InterScience. All new titles will be published online and a growing list of older titles will be added every year. It is the ultimate goal that all titles published in the **Patai Series** will be available in electronic format.
For more information see the **Patai Series Online** website:

www.interscience.wiley.com/bookfinder.html

The chemistry of
phenols
Part 2

Edited by

ZVI RAPPOPORT

The Hebrew University, Jerusalem

2003

An Interscience® Publication

Copyright © 2003   John Wiley & Sons Ltd, The Atrium, Southern Gate, Chichester,
West Sussex PO19 8SQ, England

Telephone (+44) 1243 779777

Email (for orders and customer service enquiries): cs-books@wiley.co.uk
Visit our Home Page on www.wileyeurope.com or www.wiley.com

All Rights Reserved. No part of this publication may be reproduced, stored in a retrieval system or transmitted in any form or by any means, electronic, mechanical, photocopying, recording, scanning or otherwise, except under the terms of the Copyright, Designs and Patents Act 1988 or under the terms of a licence issued by the Copyright Licensing Agency Ltd, 90 Tottenham Court Road, London W1T 4LP, UK, without the permission in writing of the Publisher. Requests to the Publisher should be addressed to the Permissions Department, John Wiley & Sons Ltd, The Atrium, Southern Gate, Chichester, West Sussex PO19 8SQ, England, or emailed to permreq@wiley.co.uk, or faxed to (+44) 1243 770620.

This publication is designed to provide accurate and authoritative information in regard to the subject matter covered. It is sold on the understanding that the Publisher is not engaged in rendering professional services. If professional advice or other expert assistance is required, the services of a competent professional should be sought.

Other Wiley Editorial Offices

John Wiley & Sons Inc., 111 River Street, Hoboken, NJ 07030, USA

Jossey-Bass, 989 Market Street, San Francisco, CA 94103-1741, USA

Wiley-VCH Verlag GmbH, Boschstr. 12, D-69469 Weinheim, Germany

John Wiley & Sons Australia Ltd, 33 Park Road, Milton, Queensland 4064, Australia

John Wiley & Sons (Asia) Pte Ltd, 2 Clementi Loop #02-01, Jin Xing Distripark, Singapore 129809

John Wiley & Sons Canada Ltd, 22 Worcester Road, Etobicoke, Ontario, Canada M9W 1L1

Wiley also publishes its books in a variety of electronic formats. Some content that appears in print may not be available in electronic books.

Library of Congress Cataloging-in-Publication Data

The chemistry of phenols / edited by Zvi Rappoport.
    p. cm.—(The chemistry of functional groups)
  Includes bibliographical references and indexes.
  ISBN 0-471-49737-1 (set: acid-free paper)
    1. Phenols. I. Rappoport, Zvi. II. Series.

QD341.P5C524 2003
547'.632–dc21

2003045075

British Library Cataloguing in Publication Data

A catalogue record for this book is available from the British Library

ISBN 0-471-49737-1

Typeset in 9/10pt Times by Laserwords Private Limited, Chennai, India
Printed and bound in Great Britain by Biddles Ltd, Guildford, Surrey
This book is printed on acid-free paper responsibly manufactured from sustainable forestry in which at least two trees are planted for each one used for paper production.

Dedicated to

Gadi, Adina,

Sharon and **Michael**

Contributing authors

L. Ross C. Barclay	Department of Chemistry, Mount Allison University, Sackville, New Brunswick, Canada, E4L 1G8
M. Berthelot	Laboratoire de Spectrochimie, University of Nantes, 2, rue de la Houssiniere BP 92208, F-44322 Nantes Cedex 3, France
Volker Böhmer	Johannes Gutenberg-Universität, Fachbereich Chemie und Pharmazie, Abteilung Lehramt Chemie, Duesbergweg 10–14, D-55099 Mainz, Germany
Luis Castedo	Departamento de Química Orgánica y Unidad Asociada al C.S.I.C., Facultad de Química, Universidad de Santiago de Compostela, 15782 Santiago de Compostela, Spain
M. J. Caulfield	Polymer Science Group, Department of Chemical Engineering, The University of Melbourne, Victoria 3010, Australia
Victor Glezer	National Public Health Laboratory, Ministry of Health, 69 Ben Zvi Rd., Tel Aviv, Israel
J. Graton	Laboratoire de Spectrochimie, University of Nantes, 2, rue de la Houssiniere BP 92208, F-44322 Nantes Cedex 3, France
Concepción González	Departamento de Química Orgánica, Facultad de Ciencias, Universidad de Santiago de Compostela, 27002 Lugo, Spain
Poul Erik Hansen	Department of Life Sciences and Chemistry, Roskilde University, P.O. Box 260, DK-4000 Roskilde, Denmark
William M. Horspool	Department of Chemistry, The University of Dundee, Dundee DD1 4HN, Scotland, UK
Menahem Kaftory	Department of Chemistry, Technion—Israel Institute of Technology, Haifa 32000, Israel
Alla V. Koblik	ChemBridge Corporation, Malaya Pirogovskaya str., 1, 119435 Moscow, Russia
Eugene S. Kryachko	Department of Chemistry, University of Leuven, B-3001, Belgium, and Departement SBG, Limburgs Universitaire Centrum, B-3590 Diepenbeek, Belgium
Dietmar Kuck	Fakultät für Chemie, Universität Bielefeld, Universitätsstraße 25, D-33615 Bielefeld, Germany

Contributing authors

C. Laurence	Laboratoire de Spectrochimie, University of Nantes, 2, rue de la Houssiniere BP 92208, F-44322 Nantes Cedex 3, France
Joel F. Liebman	Department of Chemistry and Biochemistry, University of Maryland, Baltimore County, 1000 Hilltop Circle, Baltimore, Maryland 21250, USA
Sergei M. Lukyanov	ChemBridge Corporation, Malaya Pirogovskaya str., 1, 119435 Moscow, Russia
P. Neta	National Institute of Standards and Technology, Gaithersburg, Maryland 20899, USA
Minh T. Nguyen	Department of Chemistry, University of Leuven, B-3001 Leuven, Belgium
Ehud Pines	Chemistry Department, Ben-Gurion University of the Negev, P.O.B. 653, Beer-Sheva 84105, Israel
G. K. Surya Prakash	Loker Hydrocarbon Research Institute and Department of Chemistry, University of Southern California, Los Angeles, California 90089-1661, USA
G. G. Qiao	Polymer Science Group, Department of Chemical Engineering, The University of Melbourne, Victoria 3010, Australia
V. Prakash Reddy	Department of Chemistry, University of Missouri-Rolla, Rolla, Missouri 65409, USA
Suzanne W. Slayden	Department of Chemistry, George Mason University, 4400 University Drive, Fairfax, Virginia 22030, USA
D. H. Solomon	Polymer Science Group, Department of Chemical Engineering, The University of Melbourne, Victoria 3010, Australia
Jens Spanget-Larsen	Department of Life Sciences and Chemistry, Roskilde University, P.O. Box 260, DK-4000 Roskilde, Denmark
S. Steenken	Max-Planck-Institut für Strahlenchemie, D-45413 Mülheim, Germany
Luc G. Vanquickenborne	Department of Chemistry, University of Leuven, B-3001 Leuven, Belgium
Melinda R. Vinqvist	Department of Chemistry, Mount Allison University, Sackville, New Brunswick, Canada, E4L 1G8
Masahiko Yamaguchi	Department of Organic Chemistry, Graduate School of Pharmaceutical Sciences, Tohoku University, Aoba, Sendai, 980-8578 Japan
Shosuke Yamamura	Department of Chemistry, Faculty of Science and Technology, Keio University, Hiyoshi, Yokohama 223-8522, Japan
Jacob Zabicky	Institutes for Applied Research, Ben-Gurion University of the Negev, P. O. Box 653, Beer-Sheva 84105, Israel

Foreword

This is the first volume in The 'Chemistry of Functional Groups' series which deals with an aromatic functional group. The combination of the hydroxyl group and the aromatic ring modifies the properties of both groups and creates a functional group which differs significantly in many of its properties and reactions from its two constituents. Phenols are important industrially, in agriculture, in medicine, in chemical synthesis, in polymer chemistry and in the study of physical organic aspects, e.g. hydrogen bonding. These and other topics are treated in the book.

The two parts of the present volume contain 20 chapters written by experts from 11 countries. They include an extensive treatment of the theoretical aspects, chapters on various spectroscopies of phenols such as NMR, IR and UV, on their mass spectra, on the structural chemistry and thermochemistry, on the photochemical and radiation chemistry of phenols and on their synthesis and synthetic uses and on reactions involving the aromatic ring such as electrophilic substitution or rearrangements. There are also chapters dealing with the properties of the hydroxyl group, such as hydrogen bonding or photoacidity, and with the derived phenoxy radicals which are related to the biologically important antioxidant behavior of phenols. There is a chapter dealing with polymers of phenol and a specific chapter on calixarenes — a unique family of monocyclic compounds including several phenol rings.

Three originally promised chapters on organometallic derivatives, on acidity and on the biochemistry of phenols were not delivered. Although the chapters on toxicity and on analytical chemistry deal with biochemistry related topics and the chapter on photoacidity is related to the ground state acidity of phenols, we hope that the missing chapters will appear in a future volume.

The literature coverage in the various chapters is mostly up to 2002.

I will be grateful to readers who draw my attention to any mistakes in the present volume.

Jerusalem
February 2003

ZVI RAPPOPORT

The Chemistry of Functional Groups
Preface to the series

The series 'The Chemistry of Functional Groups' was originally planned to cover in each volume all aspects of the chemistry of one of the important functional groups in organic chemistry. The emphasis is laid on the preparation, properties and reactions of the functional group treated and on the effects which it exerts both in the immediate vicinity of the group in question and in the whole molecule.

A voluntary restriction on the treatment of the various functional groups in these volumes is that material included in easily and generally available secondary or tertiary sources, such as Chemical Reviews, Quarterly Reviews, Organic Reactions, various 'Advances' and 'Progress' series and in textbooks (i.e. in books which are usually found in the chemical libraries of most universities and research institutes), should not, as a rule, be repeated in detail, unless it is necessary for the balanced treatment of the topic. Therefore each of the authors is asked not to give an encyclopaedic coverage of his subject, but to concentrate on the most important recent developments and mainly on material that has not been adequately covered by reviews or other secondary sources by the time of writing of the chapter, and to address himself to a reader who is assumed to be at a fairly advanced postgraduate level.

It is realized that no plan can be devised for a volume that would give a complete coverage of the field with no overlap between chapters, while at the same time preserving the readability of the text. The Editors set themselves the goal of attaining reasonable coverage with moderate overlap, with a minimum of cross-references between the chapters. In this manner, sufficient freedom is given to the authors to produce readable quasi-monographic chapters.

The general plan of each volume includes the following main sections:

(a) An introductory chapter deals with the general and theoretical aspects of the group.

(b) Chapters discuss the characterization and characteristics of the functional groups, i.e. qualitative and quantitative methods of determination including chemical and physical methods, MS, UV, IR, NMR, ESR and PES—as well as activating and directive effects exerted by the group, and its basicity, acidity and complex-forming ability.

(c) One or more chapters deal with the formation of the functional group in question, either from other groups already present in the molecule or by introducing the new group directly or indirectly. This is usually followed by a description of the synthetic uses of the group, including its reactions, transformations and rearrangements.

(d) Additional chapters deal with special topics such as electrochemistry, photochemistry, radiation chemistry, thermochemistry, syntheses and uses of isotopically labelled compounds, as well as with biochemistry, pharmacology and toxicology. Whenever applicable, unique chapters relevant only to single functional groups are also included (e.g. 'Polyethers', 'Tetraaminoethylenes' or 'Siloxanes').

This plan entails that the breadth, depth and thought-provoking nature of each chapter will differ with the views and inclinations of the authors and the presentation will necessarily be somewhat uneven. Moreover, a serious problem is caused by authors who deliver their manuscript late or not at all. In order to overcome this problem at least to some extent, some volumes may be published without giving consideration to the originally planned logical order of the chapters.

Since the beginning of the Series in 1964, two main developments have occurred. The first of these is the publication of supplementary volumes which contain material relating to several kindred functional groups (Supplements A, B, C, D, E, F and S). The second ramification is the publication of a series of 'Updates', which contain in each volume selected and related chapters, reprinted in the original form in which they were published, together with an extensive updating of the subjects, if possible, by the authors of the original chapters. A complete list of all above mentioned volumes published to date will be found on the page opposite the inner title page of this book. Unfortunately, the publication of the 'Updates' has been discontinued for economic reasons.

Advice or criticism regarding the plan and execution of this series will be welcomed by the Editors.

The publication of this series would never have been started, let alone continued, without the support of many persons in Israel and overseas, including colleagues, friends and family. The efficient and patient co-operation of staff-members of the publisher also rendered us invaluable aid. Our sincere thanks are due to all of them.

The Hebrew University SAUL PATAI
Jerusalem, Israel ZVI RAPPOPORT

Sadly, Saul Patai who founded 'The Chemistry of Functional Groups' series died in 1998, just after we started to work on the 100th volume of the series. As a long-term collaborator and co-editor of many volumes of the series, I undertook the editorship and I plan to continue editing the series along the same lines that served for the preceeding volumes. I hope that the continuing series will be a living memorial to its founder.

The Hebrew University ZVI RAPPOPORT
Jerusalem, Israel
June 2002

Contents

1 General and theoretical aspects of phenols — 1
   Minh Tho Nguyen, Eugene S. Kryachko
   and Luc G. Vanquickenborne

2 The structural chemistry of phenols — 199
   Menahem Kaftory

3 Thermochemistry of phenols and related arenols — 223
   Suzanne W. Slayden and Joel F. Liebman

4 Mass spectrometry and gas-phase ion chemistry of phenols — 259
   Dietmar Kuck

5 NMR and IR spectroscopy of phenols — 333
   Poul Erik Hansen and Jens Spanget-Larsen

6 Synthesis of phenols — 395
   Concepción González and Luis Castedo

7 UV-visible spectra and photoacidity of phenols, naphthols and pyrenols — 491
   Ehud Pines

8 Hydrogen-bonded complexes of phenols — 529
   C. Laurence, M. Berthelot and J. Graton

9 Electrophilic reactions of phenols — 605
   V. Prakash Reddy and G. K. Surya Prakash

10 Synthetic uses of phenols — 661
   Masahiko Yamaguchi

11 Tautomeric equilibria and rearrangements involving phenols — 713
   Sergei M. Lukyanov and Alla V. Koblik

12 Phenols as antioxidants — 839
   L. Ross C. Barclay and Melinda R. Vinqvist

13 Analytical aspects of phenolic compounds — 909
   Jacob Zabicky

14 Photochemistry of phenols — 1015

William M. Horspool

15	Radiation chemistry of phenols **P. Neta**	1097
16	Transient phenoxyl radicals: Formation and properties in aqueous solutions **S. Steenken and P. Neta**	1107
17	Oxidation of phenols **Shosuke Yamamura**	1153
18	Environmental effects of substituted phenols **Victor Glezer**	1347
19	Calixarenes **Volker Böhmer**	1369
20	Polymers based on phenols **D. H. Solomon, G. G. Qiao and M. J. Caulfield**	1455
	Author index	1507
	Subject index	1629

List of abbreviations used

Ac	acetyl (MeCO)
acac	acetylacetone
Ad	adamantyl
AIBN	azoisobutyronitrile
Alk	alkyl
All	allyl
An	anisyl
Ar	aryl
Bn	benzyl
Bz	benzoyl (C_6H_5CO)
Bu	butyl (C_4H_9)
CD	circular dichroism
CI	chemical ionization
CIDNP	chemically induced dynamic nuclear polarization
CNDO	complete neglect of differential overlap
Cp	η^5-cyclopentadienyl
Cp*	η^5-pentamethylcyclopentadienyl
DABCO	1,4-diazabicyclo[2.2.2]octane
DBN	1,5-diazabicyclo[4.3.0]non-5-ene
DBU	1,8-diazabicyclo[5.4.0]undec-7-ene
DIBAH	diisobutylaluminium hydride
DME	1,2-dimethoxyethane
DMF	N,N-dimethylformamide
DMSO	dimethyl sulphoxide
ee	enantiomeric excess
EI	electron impact
ESCA	electron spectroscopy for chemical analysis
ESR	electron spin resonance
Et	ethyl
eV	electron volt

List of abbreviations used

Fc	ferrocenyl
FD	field desorption
FI	field ionization
FT	Fourier transform
Fu	furyl(OC_4H_3)
GLC	gas liquid chromatography
Hex	hexyl(C_6H_{13})
c-Hex	cyclohexyl(c-C_6H_{11})
HMPA	hexamethylphosphortriamide
HOMO	highest occupied molecular orbital
HPLC	high performance liquid chromatography
i-	iso
ICR	ion cyclotron resonance
Ip	ionization potential
IR	infrared
LAH	lithium aluminium hydride
LCAO	linear combination of atomic orbitals
LDA	lithium diisopropylamide
LUMO	lowest unoccupied molecular orbital
M	metal
M	parent molecule
MCPBA	m-chloroperbenzoic acid
Me	methyl
MNDO	modified neglect of diatomic overlap
MS	mass spectrum
n	normal
Naph	naphthyl
NBS	N-bromosuccinimide
NCS	N-chlorosuccinimide
NMR	nuclear magnetic resonance
Pen	pentyl(C_5H_{11})
Ph	phenyl
Pip	piperidyl($C_5H_{10}N$)
ppm	parts per million
Pr	propyl (C_3H_7)
PTC	phase transfer catalysis or phase transfer conditions
Py, Pyr	pyridyl (C_5H_4N)

R	any radical
RT	room temperature
s-	secondary
SET	single electron transfer
SOMO	singly occupied molecular orbital
t-	tertiary
TCNE	tetracyanoethylene
TFA	trifluoroacetic acid
THF	tetrahydrofuran
Thi	thienyl(SC_4H_3)
TLC	thin layer chromatography
TMEDA	tetramethylethylene diamine
TMS	trimethylsilyl or tetramethylsilane
Tol	tolyl(MeC_6H_4)
Tos or Ts	tosyl(*p*-toluenesulphonyl)
Trityl	triphenylmethyl(Ph_3C)
Xyl	xylyl($Me_2C_6H_3$)

In addition, entries in the 'List of Radical Names' in *IUPAC Nomenclature of Organic Chemistry*, 1979 Edition, Pergamon Press, Oxford, 1979, p. 305–322, will also be used in their unabbreviated forms, both in the text and in formulae instead of explicitly drawn structures.

CHAPTER **12**

Phenols as antioxidants

L. ROSS C. BARCLAY and MELINDA R. VINQVIST

Department of Chemistry, Mount Allison University, Sackville, New Brunswick, Canada, E4L 1G8
Fax: (506) 364-2313; e-mail: rbarclay@mta.ca

> 'Pure' air might be very useful in medicine, but... as a candle burns out much faster in it,
> so a man might *live out too fast*.
> Joseph Priestley, 1775

> *We dedicate this chapter to Keith Ingold*

I. INTRODUCTION	840
II. KINETICS AND MECHANISM	841
A. Autoxidation	841
B. Inhibition by Phenols	842
1. Antioxidant activity and stoichiometric factor—H-atom transfer and electron transfer mechanisms	842
2. Reaction products of antioxidants: α-Toc	845
III. EFFICIENCIES OF PHENOLIC ANTIOXIDANTS	850
A. Some Experimental Methods	850
1. Reaction of phenolic antioxidants with peroxyl radicals	850
a. Inhibited oxygen uptake (IOU) measurement techniques	850
b. Product studies—Hydroperoxide products	851
2. Reaction of phenolic antioxidants with other radical sites	855
3. Overall assessment of strategies to determine antioxidant activities	858
B. Structural Effects on Efficiencies of Antioxidants	859
1. Monohydroxy phenols: Substituent effects	859
2. Dihydroxy phenols: Catechols and 1,4-hydroquinones—Intramolecular hydrogen bonding revisited	871
C. Media Effects	876
1. Solvent effects	876
a. Solvent interactions with the attacking radicals	877
b. Solvent interactions with phenolic antioxidants—Effects on antioxidant mechanisms	880

The Chemistry of Phenols. Edited by Z. Rappoport
© 2003 John Wiley & Sons, Ltd ISBN: 0-471-49737-1

       2. Antioxidants in heterogeneous systems: Lipid peroxidation and
          inhibition in micelles and lipid membranes 884
          a. Monohydroxy phenols. Factors controlling antioxidant activities in
             membranes. The unique behavior of α-Toc 886
          b. Di- and polyhydroxy phenols in membranes. Ubiquinols and
             flavonoids 892
IV. CHEMICAL CALCULATIONS ON PHENOLS 895
    A. Introduction 895
    B. Application to Antioxidants 896
       1. Antioxidant mechanisms by phenols: Hydrogen atom transfer (HAT)
          and single electron transfer (SET) 896
       2. Calculations of substituent effects for monophenols 897
       3. Calculations of more complex polyhydroxy phenols 898
V. FUTURE PROSPECTS FOR ANTIOXIDANTS 899
VI. REFERENCES 902

I. INTRODUCTION

In general terms, an antioxidant can be defined as 'any substance, when present at low concentrations compared with those of an oxidizable substrate, significantly delays or prevents oxidation of that substrate'[1]. There are two general classes of antioxidants. *Preventative antioxidants* are those that prevent the attack of reactive oxygen species (ROS) on a substrate. For example, in biological systems superoxide dismutase (SOD) catalyzes the deactivation of the superoxide anion, $O_2^{-\bullet}$, by converting it to hydrogen peroxide, which is subsequently reduced by catalase. *Chain-breaking antioxidants* reduce or delay the attack of ROS, usually by trapping chain-propagating, oxygen-centered free radicals. Phenolic antioxidants accomplish this by hydrogen-atom transfer to peroxyl radicals, converting them to hydroperoxides (equation 1).

$$ROO^{\bullet} + ArOH \xrightarrow{k_{inh}} ROOH + ArO^{\bullet} \qquad (1)$$

Our current knowledge on phenolic antioxidants developed after the discovery of vitamin E, and its role as antioxidant, since its existence was first reported in 1922 by Evans and Bishop[2]. Six decades later Ingold and coworkers reported that vitamin E is the main chain-breaking, lipid-soluble antioxidant in human blood[3]. This outstanding discovery helped spark an increased interest in antioxidants. Uninhibited free radical peroxidation *in vivo* is implicated in a wide variety of degenerative diseases such as cancer, heart disease, inflammation and even ageing; consequently, it is not surprising to find that during the last two decades the large volumes of literature reports on phenols as antioxidants are concentrated on their role in biochemical or biological systems. This is especially evident in publications from international symposia[4–6], other reviews in books[7–9] and even a new journal founded in 1999[10].

While we are not unaware of the biological significance of antioxidants, our review will concentrate on the more basic *chemical* aspects of their function, since this has lagged behind the attention to their practical use. Accordingly, this chapter begins with a brief outline of the kinetics and mechanism of autoxidation and its inhibition by phenolic antioxidants. There are many methods to study antioxidants and report their activities. Unfortunately, some of them give quite unreliable data so we will attempt to point out advantages and disadvantages of some of the common methods. Most of this chapter will be devoted to the structural effects on the activities of antioxidants, since this may present possibilities for selecting or even designing more active ones. Solvation phenomena, especially hydrogen bonding, can have a profound effect on the activity of phenols as antioxidants, so a separate section is included on media effects in solution and in heterogeneous phases. Calculations of the hydrogen–oxygen bond strengths and ionization energies of the phenolic hydroxyl groups on various phenols allow for predictions of their potential as antioxidants, and some typical examples will be cited of this theoretical approach.

II. KINETICS AND MECHANISM

A. Autoxidation

The reaction of organic compounds with oxygen, known as *autoxidation*, is the most common of all organic reactions. The reaction is a free radical chain process involving peroxyl radicals which includes initiation, propagation and termination steps and is the subject of earlier reviews[11–13]. For control of these reactions under laboratory conditions, the reaction is usually initiated by azo initiators. The reactions are outlined briefly in equations 2–4.

Initiation:

$$R-N=N-R \xrightarrow{k_i} 2R^{\bullet} + N_2 \quad (2)$$

$$R^{\bullet} + O_2 \text{ (fast)} \longrightarrow ROO^{\bullet} \text{ (peroxyl radical)}$$

Propagation:

$$ROO^{\bullet} + R_s-H \text{ (substrate)} \xrightarrow{k_p} ROOH + R_s^{\bullet} \quad (3)$$

Termination:

$$2 ROO^{\bullet} \xrightarrow{2k_t} \text{non-radical products} + O_2 \quad (4)$$

The kinetic expressions for these reactions are given in equations 5–8. Since the reaction of oxygen with carbon-centered radicals is fast and essentially diffusion controlled, the rate of oxygen uptake is given by equation 5.

$$\frac{-d[O_2]}{dt} = k_p[RO_2^{\bullet}][R_s-H] \quad (5)$$

The rate of chain initiation, R_i, can be controlled and calculated by using an initiator with a known rate of decomposition, k_i, and known initiator efficiency, e. This correction, e, is needed since only those radicals which 'escape' the solvent cage in which they are formed can react with oxygen to initiate reaction on the substrate. At steady state, the rate of chain initiation = the rate of termination, as shown in equation 6 for an azo-initiator.

$$R_i = 2k_i e[R-N=N-R] = 2k_t \times [RO_2^{\bullet}]^2 \quad (6)$$

Substituting for the reactive intermediate, $[RO_2^{\bullet}]$, into equation 5 gives equation 7, the general expression for uninhibited oxygen uptake.

$$\frac{-d[O_2]}{dt} = \frac{k_p}{2k_t^{\frac{1}{2}}} (2k_i \times e[R-N=N-R])^{\frac{1}{2}} \times [R_s-H] = \frac{k_p}{2k_t^{\frac{1}{2}}} \times [R_s-H] \times R_i^{\frac{1}{2}} \quad (7)$$

The susceptibility of a substrate to undergo autoxidation, known as its *oxidizability*, is given by equation 8, a very useful concept in free radical oxidation of different substrates.

$$\text{Oxidizability} = \frac{k_p}{2k_t^{\frac{1}{2}}} = \frac{-d[O_2]/dt}{[R_s - H] \times R_i^{\frac{1}{2}}} \tag{8}$$

For quantitative kinetic determinations, the R_i must be controlled and it can be measured. This is usually done by adding a phenolic inhibitor, known to trap two peroxyl radicals (see Section II.B), and measuring the induction period, τ, during which oxidation is suppressed (equation 9).

$$R_i = \frac{2[\text{Ar} - \text{OH}]}{\tau} \tag{9}$$

B. Inhibition by Phenols

1. Antioxidant activity and stoichiometric factor — H-atom transfer and electron transfer mechanisms

The kinetics and mechanism of inhibition (inh) of free radical oxidation has been the subject of several earlier reviews[13–16]. The main reactions for inhibited oxidation by phenols are outlined below. When a phenolic antioxidant is present, peroxyl radicals are 'trapped' by H-atom abstraction from a phenolic hydroxyl group, followed by rapid recombination of peroxyl and resulting aryloxyl radicals (equations 10 and 11).

$$\text{ROO}^\bullet + \text{ArOH} \xrightarrow{k_{inh}} \text{ROOH} + \text{ArO}^\bullet \tag{10}$$

$$\text{ROO}^\bullet + \text{ArO}^\bullet \xrightarrow{\text{fast}} \text{non-radical products} \tag{11}$$

In the presence of an 'efficient' antioxidant, most of the peroxyl radicals are trapped so that a new steady-state approximation applies, where the rate of peroxyls formed in initiation equals the rate of peroxyls trapped in the process of equation 10 (equation 12).

$$R_i = 2k_i e[\text{initiator}] = k_{inh} \times n[\text{ArOH}] \times [\text{ROO}^\bullet] \tag{12}$$

Now the reactive intermediate is redefined by equation 13.

$$[\text{ROO}^\bullet] = \frac{R_i}{k_{inh} \times n[\text{ArOH}]} \tag{13}$$

Substituting for [ROO$^\bullet$] in equation 5, for the rate-limiting reaction of peroxyl radicals, gives the basic expression for suppressed oxygen uptake in the presence of the antioxidant (equation 14).

$$\frac{-d[O_2]}{dt} = \frac{k_p}{k_{inh}} \times [R_s - H] \times \frac{R_i}{n[\text{ArOH}]} \tag{14}$$

The factor 'n' in equations 12–14 represents the number of peroxyl radicals trapped by the antioxidant in reactions 10 and 11 the stoichiometric factor. This value is expected to approximate 2 for those phenols, which are efficient antioxidants.

This simple kinetic treatment of inhibited autoxidation provides for a useful semiquantitative explanation of what is meant by *antioxidant* and *antioxidant activity* under known and controlled R_i. The ability of a known amount of 'potential' antioxidant to suppress the oxygen uptake depends on the value of the absolute rate constant for inhibition, k_{inh}, compared to the propagation rate constant, k_p, for reaction of the substrate *with peroxyl radicals*, e.g. the ratio of the rate constants in equation 14. Unsaturated organic compounds such as alkenes, arylalkenes and unsaturated fatty esters readily undergo initiated autoxidation and their k_p values are in the range of about 1.0 for an alkene to 200 M^{-1} s^{-1} for a polyunsaturated ester (triene)[11]. Consequently, for a compound to be an *effective antioxidant* its *antioxidant activity*, k_{inh}, must be several orders of magnitude greater than k_p, or $k_{inh} \geqslant 10^4$ M^{-1} s^{-1}. An antioxidant can also be defined graphically, as illustrated in Figure 1, which compares the typical profile of uninhibited oxygen uptake with the suppressed profiles in the presence of antioxidants. By definition, the oxygen uptake in the presence of the antioxidant is significantly suppressed, when equation 14 applies, until all of the antioxidant is consumed, and then the oxidation returns to its uninhibited rate and the kinetic equation 7 applies. By determination of the length of the induction period, τ, for an antioxidant where the stoichiometric factor, n, is known (e.g. $n = 2$, Figure 1), the rate of chain initiation is calculated (equation 15).

$$R_i = \frac{2[Ar-OH]}{\tau} \tag{15}$$

There are very many organic compounds that can have an effect on oxygen uptake during free radical oxidation but which do not possess sufficiently high *antioxidant activities* to suppress the oxygen uptake significantly. Such compounds do not rapidly trap peroxyl radicals, so peroxyls still undergo self-recombination. As a result, such compounds do not give measurable induction periods (Figure 1). Such compounds are NOT by definition antioxidants but are classed as *retarders*. The kinetics in this situation become quite

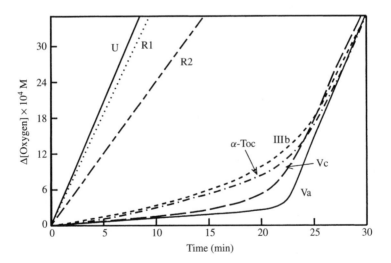

FIGURE 1. Oxygen uptake profiles for oxidation of 0.12 M methyl linoleate in 0.5 M SDS micelles, initiated by 0.03 M of the thermal azo initiator di-*tert*-butylhyponitrite, comparing the effects of the retarder melatonin (**R**) with phenolic antioxidants: **U**—uninhibited oxidation, **R1**—8.72 × 10^{-5} M melatonin, **R2**—87.2 × 10^{-5} M melatonin, α-**Toc**—8.72 × 10^{-5} M α-Toc, **IIIb**—8.72 × 10^{-5} M BHT [butylated hydroxytoluene (2,6-di-*t*-butyl-4-methylphenol)], **Vc**—8.72 × 10^{-5} M Trolox (2,5,6,7-tetramethyl-2-carboxy-5-hydroxychroman), **Va**—8.72 × 10^{-5} M PMHC (2,2,5,6,7-pentamethyl-5-hydroxychroman). Reproduced by permission of Elsevier Science from Reference 283

complex as discussed before[17]. The reaction in equation 4 will occur simultaneously with that in equation 10 and, in addition, a retarder, XH, may react with peroxyl radicals to give an X$^\bullet$ radical, which will in turn abstract hydrogen from the substrate and continue the oxidation chain (equations 16 and 17).

$$ROO^\bullet + X-H \longrightarrow ROOH + X^\bullet \quad (16)$$

$$X^\bullet + R_s-H \longrightarrow X-H + R_s^\bullet \quad (17)$$

Consequently, the retarder may be consumed slowly while oxygen uptake is only reduced slightly, but the effect occurs well past the time at which two peroxyl radicals have been generated from the initiator for every molecule of retarder. Under these conditions, a retarder may appear to react with more than two peroxyl radicals. This situation is quite often observed and causes misinterpretation of results concerning inhibition efficiency, unless a reliable method is used to determine the stoichiometric factor and antioxidant activity (See Section II.A.)

The detailed pathway involved in the antioxidant mechanism by phenols has been the subject of considerable debate. The main question is whether the pathway is a *direct*, concerted mechanism for H-atom transfer (HAT) from the phenolic hydroxyl, or, alternately, if the process involves a stepwise mechanism whereby a rate-determining *single electron transfer* (SET) precedes the hydrogen transfer. In a general manner, one could consider a range of possible 'structures', **1a–1d**, along the pathway where concerted H-atom abstraction is at one end, while at the other extreme electron transfer is complete before the hydrogen (proton) moves over giving ion pairs **1c** or **1d**. In addition, a hydrogen bonded complex, ArOH····O—O—R, in a pre-equilibrium followed by the rate-controlling atom transfer may be involved.

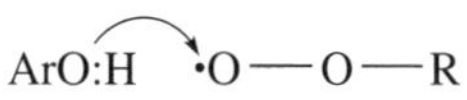

ArO:H $^\bullet$O—O—R   Ar—O·H···O—O—R
(1a) 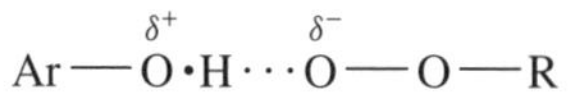   (1b)

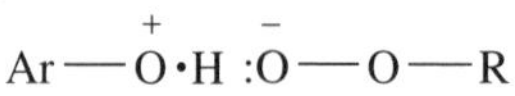

Ar—O·H :O—O—R   Ar—O·H   :O—O—R
(1c) 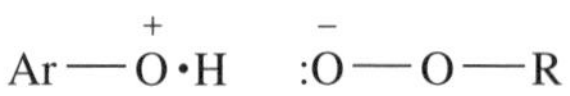   (1d)

Ingold and coworkers found substantial deuterium kinetic isotope effects for phenol-inhibited autoxidations in non-polar media, including results with more reactive phenols, and concluded that 'H-atom transfer is rate-controlling in all cases'[18]. A transition state with partial charge transfer, **1b**, was also considered in their HAT mechanism, since rate constants for *para*- and *meta*-substituted phenols correlated with σ^+, $\rho = 2.2$[19]. More recently Bisby and Parker reported[20] that α-Toc reduces duroquinone triplet by direct H-atom transfer even in polar media such as acetonitrile or SDS micelles, media that would be expected to favor electron transfer. Since excited triplet ketones are well known to be effective H-atom abstractors, this supported direct H-atom transfer as the usual mechanism. On the other hand, Nagaoka and Mukai and coworkers interpreted large deuterium kinetic isotope effects for the α-Toc reaction with an aryloxyl radical in terms of electron transfer followed by proton tunneling, through a complex such as **1b** or **1c**[21–23]. Nagaoka and Ishihara[24] interpreted their femtosecond spectroscopic evidence on the lifetime of the singlet state of a tethered vitamin E-duroquinone in terms of 'an initial electron transfer', the opposite conclusion to that reached by Bisby and Parker[20] for the similar kind of process. The different result may be due to a more restricted spatial relationship between the excited carbonyl and the hydroxyl group in the tethered system[24] which prevents direct H-atom abstraction, so that electron transfer takes over. As Neta and coworkers showed, the *reactivity* of the attacking oxygen-centered radical as well as *solvent effects*[25] can influence the mechanism of the antioxidant mechanism by phenols and, for example, the effect of polar solvents support the electron transfer mechanism[26,27] for the reaction of reactive halogenated peroxyl radicals. It is clear that one must be very careful when applying an interpretation using results obtained from different kinds of reactive species and different solvents to reactions propagated by peroxyl radicals. We will return to this question of the mechanism for hydrogen transfer under substituent effects in Section III.B.1.

2. Reaction products of antioxidants: α-Toc

The aryloxyl radicals formed in the initial antioxidant reaction of phenols (equation 1) may undergo several different kinds of secondary reactions, including: Type (1), rapid combination (termination) with the initiating oxygen-centered radicals (equation 11); Type (2), self-reactions; Type (3), initiation of new oxidation chains by H-atom abstraction from the substrate, the so-called *prooxidant effect*; and Type (4), reduction or regeneration by other H-atom donors resulting in synergistic inhibition. The relative importance of these secondary reactions will be considered briefly here, since they may affect the overall efficiency of the antioxidant, which includes the antioxidant activity, as measured by the rate constant, k_{inh} (equation 10), and the number of radicals trapped, n.

Type (1) reaction is the usual fast reaction on initiation by oxygen-centered radicals. The primary products of this reaction depend on the nature of the initiating radical and the structure of the antioxidant. For example, early product studies on a trialkyl phenol, **2**, on reaction with *tert*-butylperoxyl radical, yielded the 4-*t*-butylperoxy-2,5-cyclohexadienone, **4**, by recombination with **3** (*cf.* **3′**) at the *para* position[28], whereas the more reactive *tert*-butoxyl radical gives dimeric products of the type shown in Scheme 1. This could occur through 'hydrogen migration' from the methyl group, as proposed, however dimerization via a quinone methide, **3a**, is more likely[29]. Alternately, the reactive *tert*-butoxyl can abstract hydrogen directly from the *para* methyl, a known reaction[30]. Some general trends were recognized and indicate how the product distribution depends on steric hindrance in the antioxidant[31]. If the antioxidant has bulky R groups (e.g. *tert*-butyls) at positions 2 and 6 and a substituent at position 4, self-reaction [Type (2)] is slow, and Type (1) will predominate so that the principal product is the 4-alkylperoxy adduct, **4** (Scheme 1), which can be isolated in high yield. With *ortho*-substituted phenols, *ortho*-coupling or disproportionation may also occur. Oxidation of compounds containing a *para*-methyl can, under certain conditions, lead to the formation of stilbenequinones, **6**, possibly by disproportionation of **3** to **5**, dimer formation and continued oxidation of the dimer[32], or more likely by formation of a quinone methide followed by dimer formation. Oxidation of phenols with a 'free' *ortho* position gave (unexpectedly) a complex array of at least ten products[32], classified into three types, as shown in Scheme 2: (A) Peroxyl adduct at the free *ortho* site, followed by decomposition to an *ortho* quinone, **7** → **8** → **9**; (B) carbon–carbon radical recombination at the *ortho* positions through **11** yielding a bis-*ortho*-phenol, **12**, which in turn adds peroxyl to form the peroxycyclohexadienone dimer, **13**; and (C) phenoxyl radical self-addition at the *ortho* position followed by the reaction sequence **14** → **15** → **16**, the latter being the isolated product[33].

Products from reaction Types (1)–(2) are of particular interest with chromanol antioxidants of the vitamin E class. They have been studied by various researchers[29,31,34–40] and are the subject of a detailed review[41]. The main features of these reactions as applicable

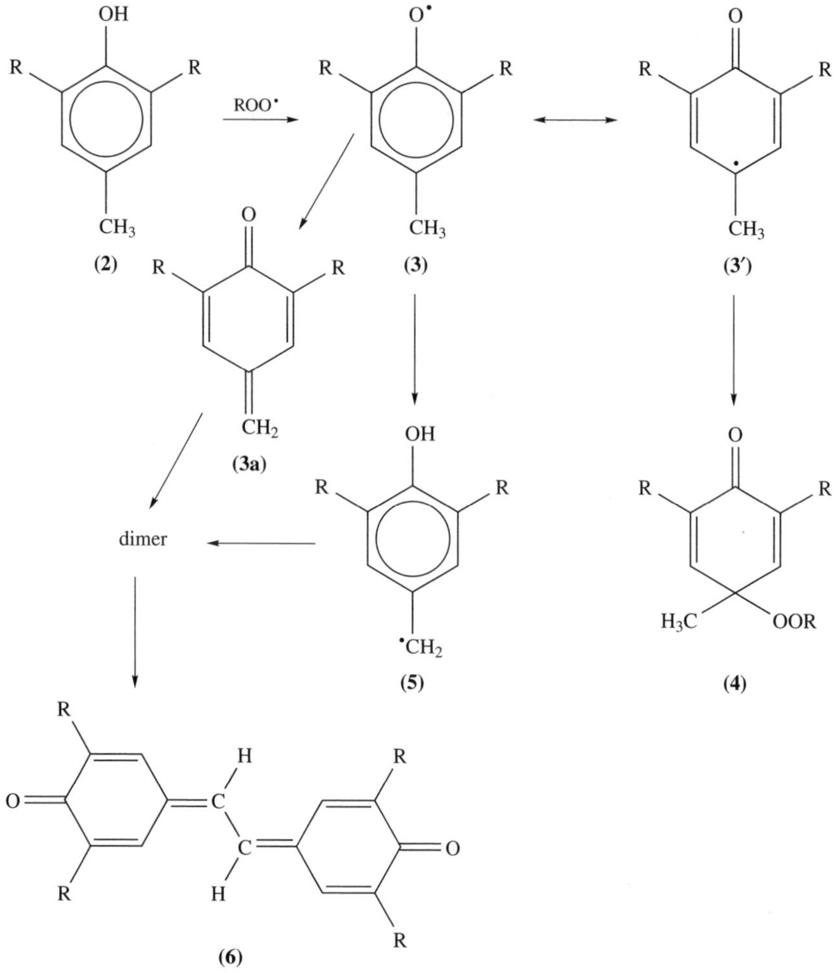

SCHEME 1. Oxidation products of 2,4,6-trialkylphenols with peroxyl radicals, R = *tert*-butyl

to α-Toc (α-Toc) are summarized in Scheme 3. The α-Toc quinone, **19**, formed by *para* coupling and rapid reaction of the adduct, **18**, is a major product of oxidation by alkylperoxyl radicals in organic solvents and determination of the consumption of peroxyls gave a stoichiometric value of 2^{31}. Products of reaction at the 5-*ortho* methyl position are also observed. In particular, disproportionation of the α-Toc radical, **17**, could form the reactive quinone methide, **22**, which dimerizes to **21** or the spirodimer **23**. A pathway via the benzyl-type radical **20**[41] is considered less probable. The dimer **21** has been found as a 'natural' impurity in vitamin E and it caused an 'extraordinary' kinetic behavior by accelerating the rate of decay of the α-Toc radical[42]. Epoxides are also oxidation products, apparently formed from oxidation of the radical **17**[38]. Oxidation of α-Toc in polar protic solvents including water can lead to an additional array of products resulting from polar addition of solvent to the intermediate quinone methide as outlined before[41]. Product distribution using chemical oxidizing agents, such as metal ions, will therefore give different products and one must be careful in assuming that these are also typical of the products from free radical oxidation of these chromanols in non-polar solvents.

As has been emphasized before[18], α-Toc and related phenols owe their effective antioxidant activity to the rapid reaction with peroxyl radicals (equations 10 and 11), and the ArO• 'wasting' reactions (equations 18–21) are all relatively slow reactions.

$$\text{ArO}^\bullet + \text{ArO}^\bullet \longrightarrow \text{non-radical products} \qquad (18)$$

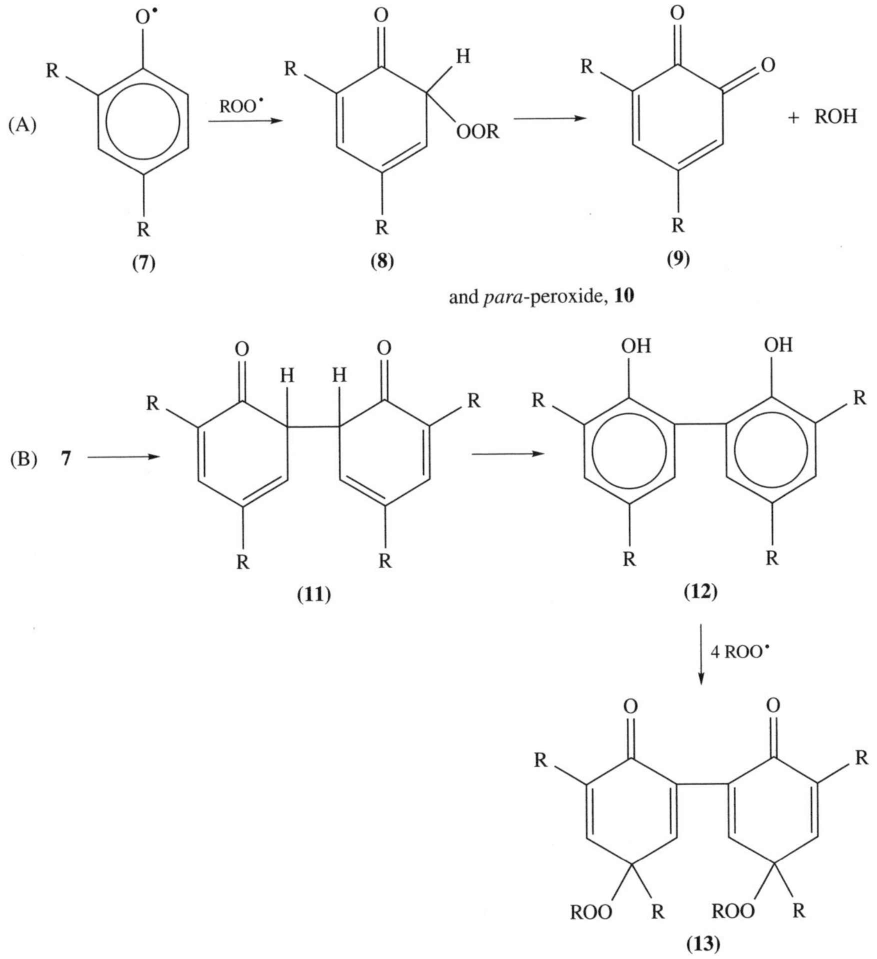

SCHEME 2. Oxidation products of 2,4-di-*tert*-butylphenoxyl radical with *tert*-butylperoxyl radicals, $R = tert$-butyl

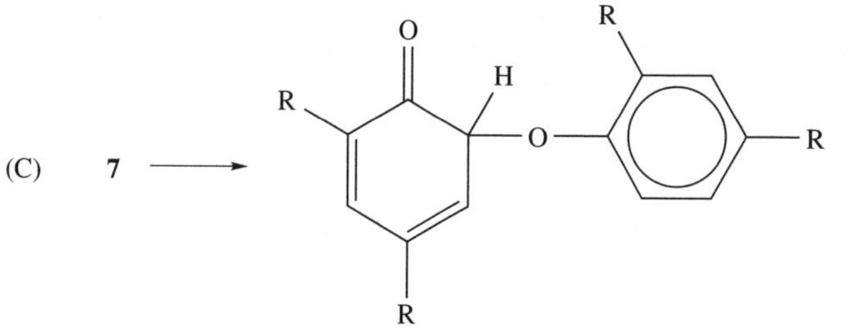

(C) 7 ⟶ **(14)**

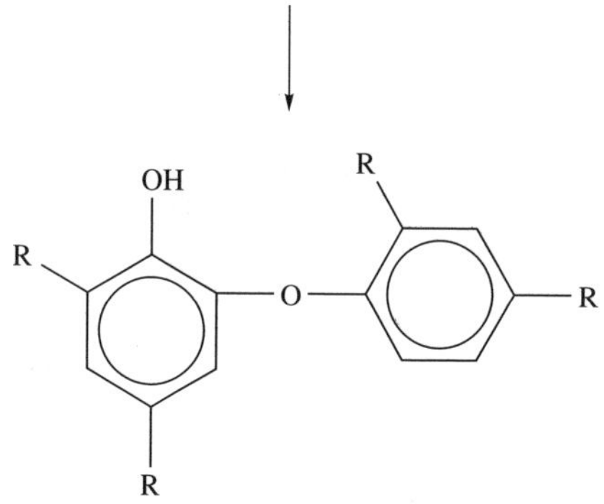

↓

(15)

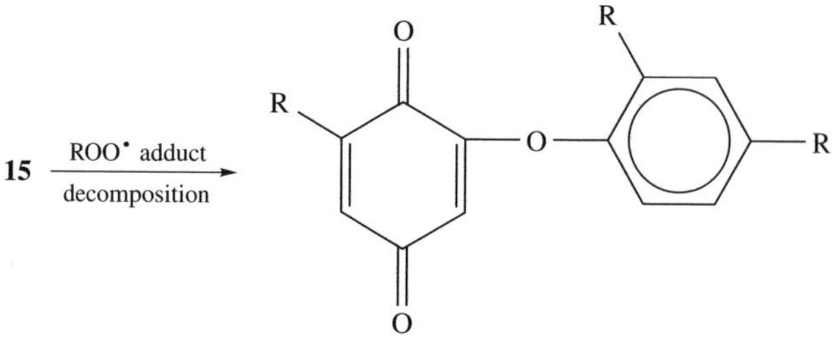

15 $\xrightarrow{\text{ROO}^\bullet \text{ adduct decomposition}}$ **(16)**

R = t-Bu

SCHEME 2. (*continued*)

$$\text{ArO}^\bullet + \text{O}_2 \longrightarrow \text{no reaction} \tag{19}$$

$$\text{ArO}^\bullet + \text{R}_s\text{—H} \longrightarrow \text{ArOH} + \text{R}_s^\bullet \tag{20}$$

$$\text{ArO}^\bullet + \text{ROOH} \longrightarrow \text{ArOH} + \text{ROO}^\bullet \tag{21}$$

Reaction 20, the pro-oxidant effect (Type 3), can become significant during high local concentrations of α-Toc in heterogeneous systems of lipids and will be discussed in Section III.C.2. Similarly, synergism (Type 4) is of particular interest in the inhibition of lipid peroxidation and will be reviewed in that section.

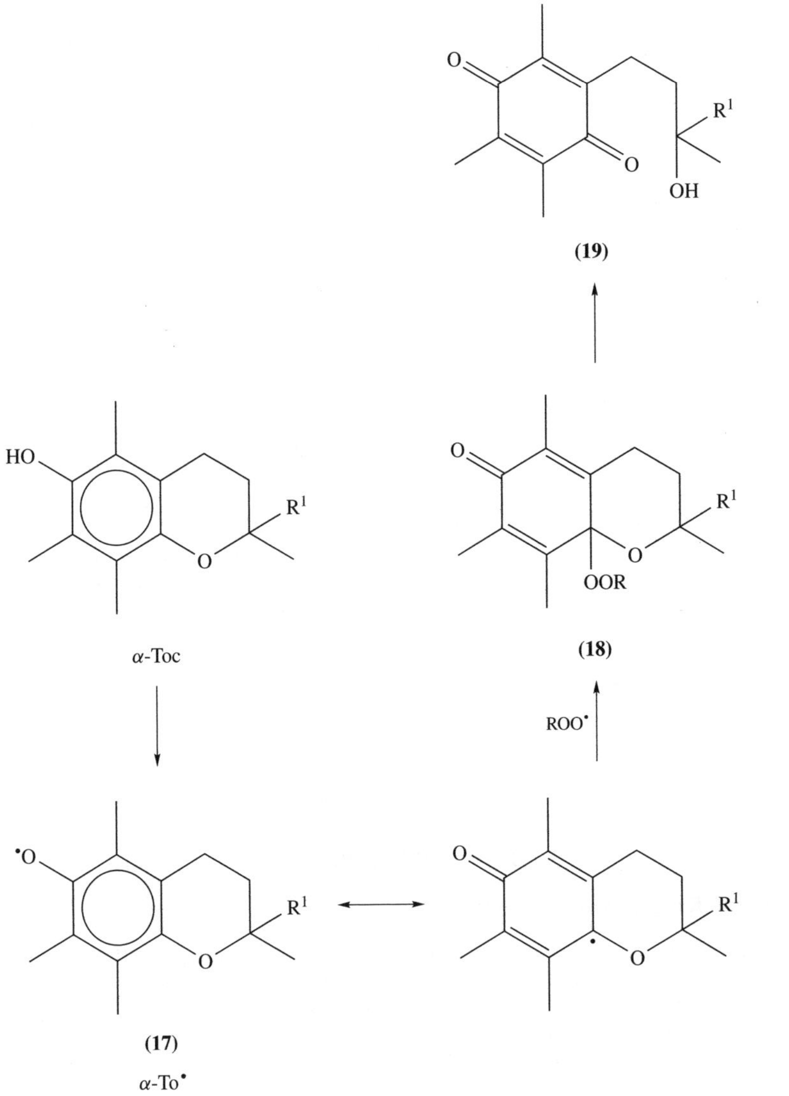

SCHEME 3. Oxidation products of α-tocopheroxyl radical with peroxyl radicals, $R^1 = C_{16}H_{33}$

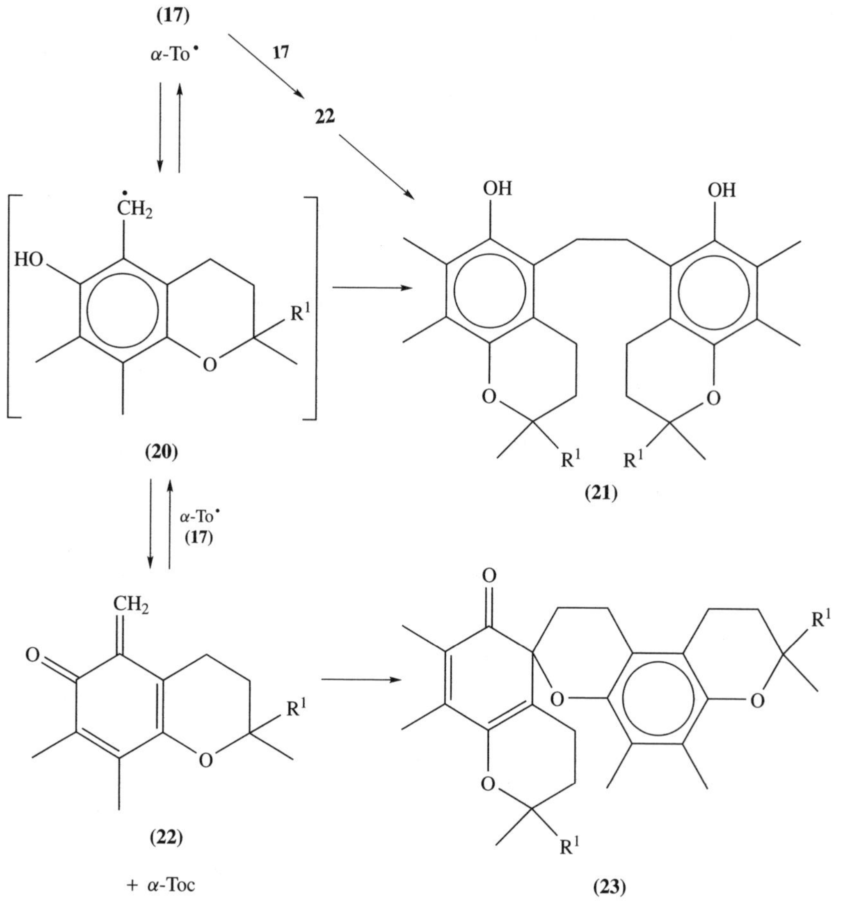

SCHEME 3. (*continued*)

III. EFFICIENCIES OF PHENOLIC ANTIOXIDANTS

A. Some Experimental Methods

There are numerous techniques available to measure antioxidant effectiveness, which generally involve monitoring the suppression of oxygen uptake, the loss of antioxidant or substrate or else the formation of reaction products over a time period, in order to compare differences when antioxidant is present or absent. A variety of the more commonly used techniques, and some of their advantages and disadvantages, are outlined in this section.

1. Reaction of phenolic antioxidants with peroxyl radicals

  a. Inhibited oxygen uptake (IOU) measurement techniques. As shown in equations 2–5, peroxidation of substrates including fatty biological material (lipids) involves the consumption of oxygen; consequently, one technique to study antioxidant activity is to monitor the effect of the phenolic antioxidant on the rate of oxygen uptake over time (Inhibited Oxygen Uptake or IOU technique). It is possible to monitor oxygen consumption using either a pressure transducer system, or using an oxygen electrode system.

Pressure Transducer. This is an extremely sensitive method that measures minute changes in gas pressure in a sample cell containing an oxidizing sample, compared to a reference cell, using a pressure transducer[43,44]. Thermal control of the experiment is possible by immersing the equipment in a thermostated water bath. It is not specific to oxygen, consequently one corrects the oxygen uptake profiles obtained for release of nitrogen from decomposition of azo-initiators, oxygen consumption by the initiator and oxygen evolution during the termination step. Using a thermal azo-initiator with known efficiency and rate of decomposition (such as azo-bis-isobutyrylnitrile, AIBN, 2,2-azo-bis-2,4-dimethylvaleronitrile, AMVN, or azo-bis-amidinopropane 2HCl, ABAP or AAPH) means that one can control the rate of chain initiation, and thus determine the antioxidant activity quantitatively (i.e. its k_{inh} value, equation 10), provided the propagation constant, k_p, of the substrate is known (equations 12–14). The stoichiometry of the reaction, its n value (equation 9), is also obtained. This technique also differentiates between antioxidants and retarders because of the distinctly different oxygen uptake profiles (see Figure 1). This technique can be used for studies in homogeneous solution with lipid[44–47] or simple organic substrates[48–51], and for studies in aqueous model systems (micelles or liposomes)[44–48,52–60].

Oxygen Electrode. Oxygen electrodes are used mainly in aqueous systems to measure absorption of oxygen across a gas-permeable membrane[53,61,62]. It is possible to obtain quantitative kinetic data from this technique by relating the oxygen consumption to the moles of oxidizing substrate, and one does not have to correct for nitrogen evolution by azo-initiators since the electrode is oxygen specific. Also, the profiles of oxygen uptake will differentiate between antioxidants and retarders. The oxygen electrode is useful for studies in aqueous systems such as micelles and phosphatidylcholine liposomes, however it is limited in the number of organic solvents that can be used[63]. Oxygen depletion in the reaction cell occurs fairly rapidly, requiring frequent oxygen purging, and there are other problems associated with routine use of the electrodes[64].

b. Product studies—Hydroperoxide products

Direct Measurements—UV/VIS Techniques. The conjugated diene (CD) formed among the polyene hydroperoxide products that are formed as a result of oxidation of polyunsaturated fatty acids (PUFAs) have a UV absorbance that can be monitored to follow the progress of the oxidation. The effect of antioxidants on the suppressed rate of product formation can be followed with time. For example, conjugated dienes from oxidation of linoleate lipid molecules absorb at 234 nm and can be monitored directly[65–69], or else after HPLC separation (via normal phase[59,70,71] or reverse phase[72–74]) of the individual isomers. In order to use these findings to calculate the antioxidant activity of phenols and relate it to oxygen uptake studies (equations 7 and 14), one also has to make a correction to account for loss of absorbance due to loss (from decomposition) of hydroperoxides (equation 22)[67,68].

$$\frac{-d[O_2]}{dt} = K \frac{-d[CD]}{dt} \tag{22}$$

The rate of oxygen uptake, $-d[O_2]/dt$, is therefore directly related to the rate of conjugated diene formation, $-d[CD]/dt$, corrected for product decomposition by the proportionality constant, K, which was found to equal 1.19, in other words 19% of the conjugated dienes decompose, so the absorbance of products has to be corrected to that degree to equal the oxygen consumption. Using the molar absorptivity of the oxidation products[68,75] or else individual molar absorptivity values for the 4 main product isomers separated by HPLC[76], it is possible to relate the changing absorbance to moles of products formed (equation 23)[67], where A_t is the absorbance at 70% of the induction period (after which point the rate of oxygen consumption is no longer increasing in a steady manner due to decreased antioxidant concentrations), A_0 is the absorbance when the antioxidant is added, t_t is the time at 70% of the induction period, t_0 is the time that the antioxidant is added, ε is the molar absorptivity of the conjugated dienes and L is the path length.

$$\frac{-d[CD]}{dt} = \frac{-(A_t - A_0)}{(\varepsilon L)(t_t - t_0)} \tag{23}$$

To calculate the antioxidant activity for the antioxidant used with this technique, one first determines the slope, S, from the plot of $-d[CD]_{inh}/dt(t_0)$ vs $[Inh]^{-1}$. The k_{inh} can be represented as the k_{inh}/k_p ratio, or calculated if the propagation rate constant, k_p, is known for the substrate LH = lipids (equation 24)[67].

$$\text{Antioxidant Efficiency} = AE = \frac{k_{inh}}{k_p} = \frac{[LH]R_i}{nKS} \tag{24}$$

The concentration of the oxidizable substrate must be low enough so that the absorbance from the products during the course of the oxidation does not exceed maximum reliable absorbance readings. Direct UV examination of the oxidizing material is possible when conducting studies on homogeneous systems in organic solvents[77], or studies in heterogeneous systems like micelles[67,69] and unilamellar liposomes[78]. Extraction of the lipids before analysis is often required for purposes of HPLC analyses[79], or, in the case of direct UV studies on multilamellar liposome systems or biological samples, lipid extraction may be necessary due to opacity of the material and/or UV interference from other compounds. The sensitivity of direct UV analyses can be improved by monitoring more than one wavelength or using tandem cuvettes[78]. If one uses UV study on lipid peroxidation products post-HPLC, then using product ratios of the cis–trans/trans–trans isomers provides information on the mechanism of initiation (e.g. free radical vs singlet oxygen, Section III.C.2) and peroxidation[80,81] and also on the antioxidant behavior[60,71,75,82]. The separation of products on HPLC also can reduce interference due to absorbance from non-hydroperoxide conjugated dienes from other sources in biological samples. In order to obtain relevant, semiquantitative results, one has to be able to stop any further oxidation of the lipid during extraction and work-up, and have an appropriate control sample for comparison purposes, especially in studies of tissue samples[65]. Conjugated trienes absorbance can be monitored at 288 nm[69,83] or 235 nm (methyl linolenate oxidation products, $\varepsilon = 24,400$)[75]. However, with oxidation of triene fats one can get non-conjugated products, making the UV analysis less useful for quantitative results unless an appropriate correction factor can be determined, and conjugated trienes can be formed from diene lipids during the course of photoinitiation[69].

The application of UV alone to monitor substrate oxidation product formation is limited to studies on polyunsaturated fats with conjugated oxidation products. Reverse-phase HPLC elution of conjugated and non-conjugated lipid hydroperoxides (LOOH) can be followed electrochemically (concurrently with UV detection) so long as the eluent contains a supporting electrolyte (such as sodium chloride[84] or tetraethylammonium perchlorate[72]).

UV analysis can also be used to monitor loss of antioxidants or formation of antioxidant radicals or their product quinones[85,86], although there may be products other than just quinones (vide supra). Absolute second-order rate constants for hydrogen atom transfer from antioxidants ($k_{ROO^\bullet/ArOH}$) can be determined by following the rate of radical formation k_{obs}, and plotting k_{obs} vs [ArOH] (equation 25)[86].

$$k_{obs} = k_0 + k_{ROO^\bullet/ArOH}[ArOH] \tag{25}$$

In general, equation 25 applies to most spectrophotometric procedures where rate of product formation or rate of loss of a colored indicator is used to monitor antioxidant behavior.

Direct Measurements—Fluorescence/Chemiluminescence Techniques. The progress of the peroxidation of lipids can be followed by chemiluminescence, specifically by measuring the chemiluminescence of lipid peroxyl recombination products like singlet oxygen and triplet carbonyl products. It is a technique that has been found to correlate well with the UV method of monitoring conjugated diene formation, while avoiding the problem in UV studies of absorbance by other compounds at the relevant wavelength[64,87]. The degree of chemiluminescence is very low however, so detection may require the use of chemiluminescent enhancers, and also singlet oxygen can be generated by other means, which may cause interference. The polyunsaturated fatty acid *cis*-parinaric acid has been used as a marker to follow lipid oxidation[61,88,89] due to its fluorescence, since the rate of oxidation can be monitored by following the rate of loss of fluorescence, which would be influenced by antioxidants according to their efficiency. This marker has the potential for application in a variety of media, as it can be used embedded in low density lipoprotein LDL[61] or added in solution to liposomes[88]. Another way to measure antioxidant potential directly using luminescence is to heat di-*tert*-butyl peroxyoxalate with *o*-dichlorobenzene and ethylbenzene, and monitor the resultant luminescence of the di-*tert*-butyl peroxyoxalate and its suppression by the addition of an antioxidant[90]. β-Phycoerythrins can also be used as fluorescent markers of peroxyl radical damage by monitoring the loss of their fluorescence in the presence of peroxyls and the protection of the β-phycoerythrins by antioxidants[89], although the β-phycoerythrins method does not actually distinguish between antioxidants or retarders[64].

Indirect Measurement of Hydroperoxide Formation. The rate of substrate oxidation in the presence and absence of antioxidants can be followed by measuring reactions of the hydroperoxide products formed with various compounds. Some of these techniques are outlined here.

(A) Fluorescence/Chemiluminescence Techniques. Lipid hydroperoxides can be reacted with chemiluminescent indicators such as luminol or diphenyl-1-pyrenylphosphine post-HPLC[72], which allows separation and identification of phospholipid and cholesterol ester peroxides[64]. This technique is applicable to both conjugated and non-conjugated lipids, however it tends to involve a relatively long delay between injection and final fluorescent analysis, and it probably provides inaccurate assessments of total levels of peroxide[64].

(B) UV/VIS Techniques. (i) The absorbance of β-carotene at 455–465 nm decreases in the presence of oxidizing linoleic acid in micelles, and the addition of antioxidants slows down the rate of this decrease, depending upon the effectiveness of the antioxidant[91]. There has been some work to quantify the protective effect of antioxidants using this technique, and Rosas-Romero and coworkers[91] defined a parameter for this purpose, ω, based upon the change of β-carotene absorbance with time and concentration of the antioxidant, which they then correlated with the estimated ionization potential and ^{13}C NMR chemical shift, δ, of the *ipso*-carbon of the OH group. The technique has, so far, been used in micellar systems[91]. It may be difficult to use β-carotene as an indicator in this technique for analyses of tissue samples or plant extracts because the materials for study would have to be corrected for natural levels of β-carotene, which is itself an antioxidant[92] and singlet oxygen quencher[93,94].

(ii) The reaction of 2,2′-azinobis-(3-ethylbenzothiazoline-6-sulfonic acid, ABTS) with lipid peroxyl radicals forms ABTS$^{+\bullet}$, which in the presence of a peroxidase (metmyoglobin) and ferryl myoglobin produces absorbance at 650, 734 and 820 nm, a high enough wavelength to avoid interference from most other compounds. Addition of antioxidants or hydrogen atom donating compounds results in a loss of absorbance, and the net loss is related to that from a standard antioxidant, a water-soluble vitamin E analogue,

Trolox (2,5,7,8-tetramethyl-2-carboxy-6-hydroxychroman)[95,96]. This test is referred to as the TEAC or Trolox-Equivalent Antioxidant Capacity assay.

(iii) For the Fox Assay, Fe(II) is oxidized to Fe(III) under acidic aqueous conditions in the presence of LOOH, and the Fe(III) subsequently forms a complex with xylenol orange which is measured at 560–580 nm; however, lengthy incubation periods make this procedure less reliable if oxidation of the substrate continues during work-up[97]. This technique is a useful indicator of hydroperoxide in both liposome preparations and in assays of biological samples such as plasma[64].

(C) Titration for Hydroperoxides. The iodometric assay technique is useful in both aqueous[65,98] and organic media[99] and involves the measurement of the iodine produced from the reaction between LOOH and potassium iodide by sodium thiosulfate titration with a starch indicator. The two main problems with iodine measurements in aqueous systems are that free iodine can be absorbed by lipid material, and iodine can be produced in the absence of substrate due to reaction between oxygen and potassium iodide[98]. In organic systems, water can interfere with measurements, however treatment with sodium bicarbonate can eliminate the error in measurements when oxygen is present[99]. It is also possible to measure the amount of iodine formed via other techniques, such as potentiometrically using a platinum electrode[98], or spectrophotometrically measuring I_3^- absorbance at 350 or 290 nm (although this technique has solvent restrictions, and requires that nothing else present interferes at those wavelengths).

Analysis of the Decomposition Products of Hydroperoxides. Some authors have monitored formation of some of the decomposition products of the lipid hydroperoxides. Direct spectrophotometric measurements of the formation of oxo-octadecadienoic acids at 280 nm are possible[74], as are measurements of secondary oxidation products like α-diketones and unsaturated ketones at 268 nm. The formation of various aldehyde products of lipid peroxide decomposition can be monitored by reacting them with 2,4-dinitrophenylhydrazine and, after HPLC separation, measuring at 360–380 nm the DNPH derivatives formed[100], although the sensitivity of this particular technique makes it very susceptible to interference.

A commonly used technique to follow lipid oxidation is to monitor the formation of malondialdehyde (MDA), a water-soluble compound produced from lipid hydroperoxide and endoperoxide decomposition. The formation of the MDA can be followed by measuring it directly via HPLC[101,102], however most often it is reacted with thiobarbituric acid (TBA) with a 1 : 2 stoichiometry (equation 26), and formation of the product is followed spectrophotometrically at 535 nm[83,98] or followed by fluorescence[103].

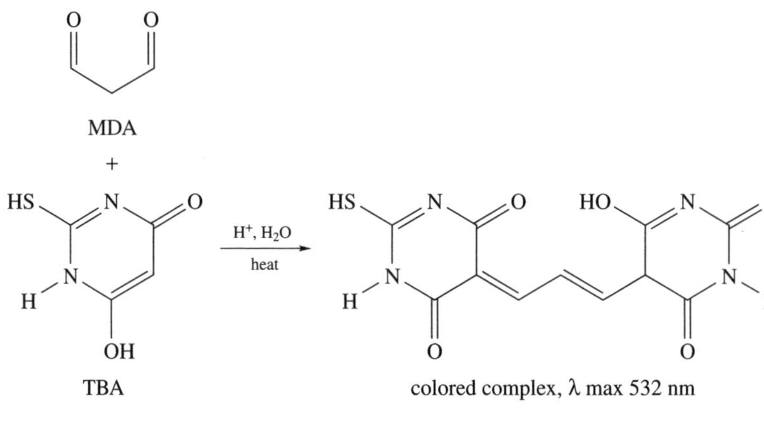

It should be obvious that this method is limited to those lipids that form MDA on oxidation. In biological samples, the work-up for sample preparation involves addition of trichloroacetic acid to precipitate proteins, acid to change the pH and heating to dissolve the TBA. The technique is widely used because it is both simple and sensitive. However, there are a number of factors to consider when interpreting results using this technique, particularly when studying biological samples: (a) the presence of oxygen, Fe^{+2} and acid during sample work-up results in further substrate oxidation which can distort the results (although the addition of an inhibitor like BHT can help protect against this)[65,66,101]; (b) TBA will react with other compounds, such as other aldehydes[65] or even sucrose[101]; (c) MDA can be generated from other processes than just lipid peroxidation, and can be lost due to dimerization, reaction with other biological compounds or metabolized by mitochondrial enzymes[65,83,101]; (d) as with other spectrophotometric techniques, TBA absorbance can reach a maximum level, limiting the upper range of MDA concentration that can be analyzed[104]. The acid, pH, and heating temperatures and times should ideally be optimized for each type of analysis, rather than just following a standard procedure[103].

2. Reaction of phenolic antioxidants with other radical sites

Studies have been conducted to measure antioxidant effectiveness in reactions with radical species other than peroxyl radicals, generally in terms of hydrogen atom donating ability. It must be kept in mind that different radical species may well differ in their hydrogen atom abstracting ability, and so measurements using different radical centers might lead to different relative reactivities of antioxidants. There are several different radical species, generated in a variety of ways, used in the study of antioxidants. Hydroxyl radicals are reported to be generated by the xanthine–xanthine oxidase system[105–107], but this only occurs when traces of metal ions are present. They are also formed by the ascorbate/Fe^{+2} combination[108], and by irradiation of hydroperoxyl species[109]. Pulse radiolysis is also a common method to generate hydroxyl radicals. However, it is very unlikely that reaction with hydroxyl radicals provides a reliable method to determine an antioxidant's H-atom donating ability. As pointed out before[110,111], a hydroxyl radical reacts rapidly with almost any organic molecule in its proximity and addition to unsaturated systems is a common reaction[111]. Stable radical species like diphenylpicrylhydrazyl radical DPPH•, galvinoxyl and a phenoxyl species (*vide infra*) can be used with spectrophotometric or ESR analyses. Alkoxyl radicals can be generated by irradiation of hydroperoxyl species[109], and with NADPH/ADP/Fe^{+3} [73]. Aryloxyl radicals can be generated using time-resolved laser flash photolysis, and the phenoxyl radical generated in this way is a much more active hydrogen atom abstractor than peroxyl radicals[112].

Use of Electron Spin Resonance Techniques. Electron spin resonance (ESR) studies have been used to examine both 'activity' of antioxidants[18,113–115] and their location within the liposome[113]. Studies of antioxidant radicals via ESR provide data on the electron delocalization within the antioxidant, which can be correlated with antioxidant activity, although not always with very good agreement with inhibition studies[18]. Spin traps have been themselves examined as potential antioxidants, and have been used to attempt to trap peroxyl species for study[116]. However, trapped peroxyl species are not very stable and carbon-centered radicals have been preferentially trapped, even though in some studies other techniques (e.g. malondialdehyde/thiobarbituric acid, MDA/TBARS-technique) indicate the presence of peroxide species in the sample[117]. Fremy's salt (($K^+SO_3^-)_2NO$•) has been used in micellar systems to determine rate constants quantitatively for the antioxidants α-Toc and ascorbic acid and their derivatives, because it reacts with them in a way similar to peroxyl radicals and can be used as a spin probe in stop-flow ESR studies[114,115]. ESR has also been used to monitor the loss of DPPH•[90,105,118] and galvinoxyl[119,120] signal intensity to follow the rate at which the radicals abstract the phenolic hydrogen of the antioxidant (thus becoming ESR silent), and to follow the formation of the antioxidant radicals[90].

UV/VIS Techniques. The UV/VIS method can be utilized to measure the hydrogen atom donating ability of antioxidants to alkoxyl and nitrogen-centered radicals, and some techniques are outlined here:

(i) In one technique, alkoxyl radicals (RO•) are generated by irradiating *tert*-butyl hydroperoxide or 13-hydroperoxylinoleic acid in the presence of *tert*-butyl alcohol (to scavenge hydroxyl radicals). The rate at which the alkoxyl radicals bleached the carotenoids crocin, monitored at 440 nm in aqueous systems, or canthaxanthin, monitored at 450 nm in hexane, was determined in the presence and absence of antioxidants (which would preferentially react with the alkoxyl radicals, protecting the carotenoid)[109]. From these rates it is possible to calculate relative rate constants; however, in this study the rates were calculated based upon the change in absorbance measured before and after 5 minutes of photolysis, rather than continually monitoring changes in absorbance over longer periods of time. The technique is limited to non-thiol antioxidants that do not have absorbance in the same region as the carotenoids. In addition, the bleaching reaction studied must be faster than the known rapid unimolecular decay, 1.5×10^6 s^{-1}, of t-butoxyl in water[121], and the stoichiometic factors for antioxidants cannot be determined from this technique[109].

(ii) The chemiluminescent indicators luminol and lucigenin can also be used to determine the reaction of phenols with superoxide radicals and hydrogen peroxide by monitoring the degree to which the phenols protected the indicators from oxidation, since the oxidation of luminol causes it to emit a blue light which can be measured[122].

(iii) Aryloxyl radicals (in this example, phenoxyl) can be generated through laser flash photolysis, and can be monitored by their absorbance at 400 nm. The rate at which phenolic antioxidants donate a hydrogen to the aryloxyl radical can then be followed not only by measuring the loss of the aryloxyl radical absorbance, but also by the growth in absorbance for the antioxidant radical (so long as the species do not absorb at similar wavelengths)[112].

(iv) A common technique to measure a phenol's hydrogen atom donating ability is to measure the reaction rate between the phenol and a colored, stable radical species such as: (a) DPPH•, which in solution is purple with an absorbance at approximately 515 nm[50,88,123–125], (b) the galvinoxyl radical, which in solution is orange with an absorbance at approximately 424 nm[119,126], or (c) with a phenoxyl radical species (2,6-di-*tert*-butyl-4-(4′-methoxyphenyl)phenoxyl radical, ArO•), which is generated by oxidizing a solution of the starting phenol (a white solid) with lead dioxide, yielding a purple solution which has a strong absorbance at 370 nm[50,127–129]. With these techniques, the progress of the hydrogen abstraction can be monitored by the loss of absorbance for the indicator radical. Galvinoxyl has even been incorporated into liposomes to study antioxidant mobility and action[130]. Different researchers using the DPPH• technique to measure antioxidant activity may monitor the loss of DPPH• signal until a steady state is reached[124,125,131], or may monitor it for only a short time period to determine the initial rate of signal loss[85,88,105]. The calculation to determine a quantitative rate constant (k_2 for the second-order reaction with the phenol) requires the initial and final concentrations of the DPPH[125,132] (equation 27)[132].

$$k_2 = \frac{2.303}{t} \log \frac{[\text{DPPH}^\bullet]_0}{[\text{DPPH}^\bullet]_t} \times \frac{1}{[\text{ArOH}]} \qquad (27)$$

It also does not take into consideration the rate of the reaction, although one group has addressed this by defining a new parameter to characterize antioxidant compounds that does incorporate time, T_{EC50}, to reach the 50% reaction of the DPPH• (EC_{50}) value[124] (equation 28).

$$\text{Antiradical efficiency} = \text{AE} = \frac{1}{EC_{50}} T_{EC50} \quad (28)$$

In summary, these are some of the common techniques used to measure lipid peroxidation in the presence and absence of antioxidants, and to study antioxidants directly. There are other techniques, ranging from simple gas chromatographic analysis and measurement of amounts of unoxidized polyunsaturated fatty acids (PUFA) before and after oxidation[83], to enzyme-mediated assay techniques such as the cyclooxygenase or glutathione peroxidase-based systems[64]. There is even a spectrophotometric technique to measure the ability of a phenol to bind with the iron reagent, in which the phenol is incubated with ferrozine (monosodium 3-(2-pyridyl)-5,6-diphenyl-1,2,4-triazine-p,p'-disulfonate) and ammonium ferrous sulfate with ammonium acetate. The phenol will compete with the ferrozine to complex with iron, and this is followed by monitoring the absorbance of the iron(II) ferrozine complex at 562 nm and comparing that to an untreated sample[88,131]. This technique, however, is a measure of a phenol's ability to act as a preventative antioxidant (since metal ions like copper and iron have a significant role in initiating lipid peroxidation *in vivo*[133,134], and are also used to initiate lipid peroxidation in *in vitro* studies[108,134,135]) as opposed to a chain-breaking antioxidant.

These techniques to study the effectiveness of phenolic antioxidants tend to be used on individual antioxidants. There are also investigations conducted on plant/tissue extracts that contain many phenolic compounds, and thus any technique used will only be able to screen overall antioxidant capacity of the mixed system. They will not be quantitative for any single phenolic species, and will provide little useful information on interactions (synergistic or otherwise) of the phenolic antioxidants present with other compounds. Studies on mixed systems also will not take into consideration varying reaction rates of the antioxidants, and may not differentiate among the roles of free radical scavenging (chain breaking) or metal chelating (preventative) actions, or the ability of some compounds to repair oxidative damage[61,134]. Results of such studies may provide a relative numerical value for 'antioxidant activity' within the system. For example, one paper using the chemiluminescence technique with luminol on plant extracts reported results in terms of percent inhibition of luminol intensity, and the authors compared their results to two standard phenolic antioxidants[122]. Other papers also were interested in total antioxidant capacity of plant or tissue/plasma preparations[64,89], but assays in such complex systems may be measuring protein peroxides as well as lipid peroxides[64], and there are non-phenolic compounds which play a role in protection against peroxidative damage[134]. One technique used for this kind of combined antioxidant assay is called the FRAP (ferric reducing ability of plasma) assay, which follows spectrophotometrically (at 593 nm) the reduction of ferric ion in a complex with tripyridyltriazine to ferrous ions[64,96]. Another technique, called the TRAP (total radical antioxidant parameter) assay, is used on plasma samples to assess total antioxidant capacity, and uses the oxygen uptake technique to determine the induction period (length of inhibition) provided by a plasma sample when initiation is induced and controlled by an azo-initiator, and results are reported relative to the water-soluble phenolic antioxidant Trolox[62,64,96]. A variation of this technique, called ORAC (oxygen radical absorbance capacity) follows the progress of the reactions via a decrease in the fluorescence of phycoerythrin[96]. Techniques which examine total antioxidant capacity of plant/tissue preparations directly (for example, looking at plasma directly via UV/VIS) often require substantial dilution of the material, which can cause misleading results, because although the concentration of water-soluble antioxidants decreases with the dilution, the concentration of lipid particles with LDL is not affected by the dilution[64]. Also, direct UV analysis, particularly for complex biological samples, is susceptible to interference due to the low wavelength required to observe conjugated diene formation, and thus overall has a tendency to overestimate lipid peroxidation[64].

3. Overall assessment of strategies to determine antioxidant activities

A quote from Halliwell sums up the overall problem of assessing antioxidant activities[134]: 'Many substances have been suggested to act as antioxidants *in vivo*, but few have been proved to do so'. Given the wide variety of techniques available to the investigator, it is important to clarify what makes an optimal strategy for the determination of antioxidant activity. The main problems associated with the variety of methods available to determine antioxidant 'activity' are summarized very well in the review by Frankel and Meyer[133]. Antioxidant studies are conducted using a wide array of substrates (e.g. simple organic substrates like styrene or cumene, or lipids, which vary in type, charge and degree of unsaturation), types of initiators (e.g. thermal azo-initiators, photoinitiators, enzyme mediated initiation, metal ion mediated initiation), system compositions (e.g. homogeneous solution, bulk lipids, oil/aqueous emulsions, micelles, liposomes, biological samples such as plasma, tissue and plant extracts), pH, temperatures and assay techniques. Studies to determine if natural 'potential' antioxidant compounds show reasonable activity should be conducted at antioxidant concentrations that would be relevant to those in biological systems[134]. Studies on biological samples need to take into consideration that there may be more than one biological role for the antioxidant in question[133,134], that the antioxidant may be regenerated through the action of other compounds (e.g. the regeneration of vitamin E by vitamin C, see Section III.C.2) and that there may be partitioning[58,136] and/or charge factors[56] affecting its activity, and that other components in the samples may respond to the assay chemicals to varying degrees depending upon the assay used[96]. Often, determinations of antioxidant activity are conducted using only one type of technique, even though there can be extreme inconsistencies in results found when comparing different techniques[96,133]. Even comparing results that use the same assay technique can be difficult, depending on what the different researchers use as an end-point in their technique, or if there are slight modifications to the experimental conditions[96,133]. This makes it very difficult to compare results from different researchers.

A standardized testing system is needed that provides not only a quantitative kinetic rate constant for the activity of the antioxidant, but also indicates the stoichiometry of the reaction, and which is relevant to the systems of interest (for example, reaction with peroxyl radicals, which is relevant to biological systems). To obtain kinetic data in a manner that applies the principle of autoxidation and inhibition as outlined in equations 2–15 (see Section II) requires consideration of the following factors:

(a) The rate of chain initiation must be controlled and measurable, for example by using azo-initiators with known rate constants of decomposition and efficiencies.

(b) The concentration of oxidizable substrate must be known, and the propagation rate constant (k_p) for the substrate should be known or measurable. If k_p is not known, the relative values of k_{inh}/k_p may be used.

To give a specific example, the advantages of styrene as a substrate for peroxyl radical trapping antioxidants are well known[43]: (i) Its rate constant, k_p, for chain propagation is comparatively large (41 $M^{-1} s^{-1}$ at 30 °C) so that oxidation occurs at a measurable, suppressed rate during the inhibition period and the inhibition relationship (equation 14) is applicable; (ii) styrene contains no easily abstractable H-atom so it forms a polyperoxyl radical instead of a hydroperoxide, so that the reverse reaction (equation 21), which complicates kinetic studies with many substrates, is avoided; and (iii) the chain transfer reaction (pro-oxidant effect, equation 20) is not important with styrene since the mechanism is one involving radical addition of peroxyls to styrene.

For very weak antioxidants ($k_{inh} \leqslant 5 \times 10^4$ M^{-1} s^{-1}), in order to measure the stoichiometric factor, which requires determination of the inhibition period, one should select a substrate with a relatively low k_p. For this purpose, cumene ($k_p = 0.18$ M^{-1} s^{-1} at 30 °C[137]) has proven to be useful and with this substrate the inhibition periods of a variety of inhibitors were measured[138]. Overall, the relative efficiencies of different phenolic antioxidants vary markedly with substrate and in different solvents (see Section III.C). These factors must be carefully controlled for quantitative studies of activities. Qualitative screening methods are widely used, without regard to the controls outlined above, to determine the relative effectiveness of antioxidants, as reviewed in detail in this section. These methods do provide some useful data on the relative 'potential' of compounds as antioxidants, but they do not determine the actual 'activity' of individual compounds.

B. Structural Effects on Efficiencies of Antioxidants

1. Monohydroxy phenols: Substituent effects

It is well known that substituents have a profound effect on the hydrogen atom donating ability of phenols. Indeed, only those phenols bearing electron donating substituents, particularly at the *ortho* and/or *para* positions, are active as antioxidants. In general, this is as expected since such groups are expected to lower the phenolic O−H bond dissociation enthalpy and increase the reaction rates with peroxyl radicals.

In general, the effects of alkyl and alkoxyl (e.g. CH_3−O) groups are well known and understood. For example, *ortho* and *para* alkyls (at positions 2,4,6) stabilize the phenoxyl radical by inductive and hyperconjugative effects and, in addition, *ortho* groups provide steric hindrance to minimize undesirable 'wasting' reactions such as pro-oxidation (equation 21). In addition, the conjugative effect of a heteroatom, for example at the *para*-position, provides stabilization through resonance (Scheme 4).

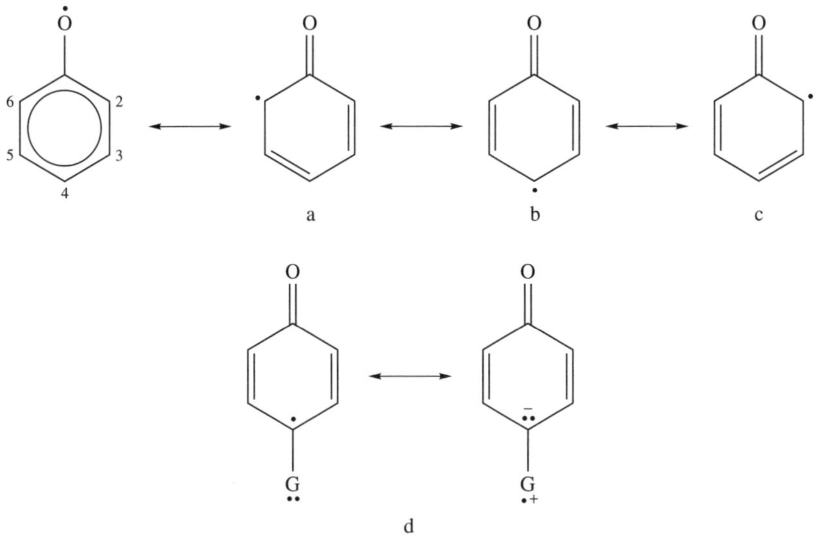

G = resonatively electron donating group

SCHEME 4. Electron delocalization in phenoxyl radicals

Quantitative kinetic studies of absolute rate constants for hydrogen atom transfer from substituted phenols to polystyrene peroxyl radicals by Howard and Ingold in the 1960s provided the first reliable data on substituent effects[19,139,140] on antioxidant activities of phenols. Later, a very detailed report appeared providing data on substituent and structural effects on various classes of monohydroxy phenols[18]. In addition, detailed reviews were given of substituent effects[14,141]. These reports provide the basis for understanding how substituent and structural effects control the antioxidant activities of phenols and will be summarized in part below.

Substituent Effects of Alkyls and para-Methoxy in Simple Phenols. Table 1 gives some data on substituent effects of alkyls, especially methyls, and *para*-methoxy on antioxidant activities, k_{inh}, of three classes, I, II and III, of monophenols. The data were interpreted for the most part in terms of the relative inductive or resonance effects on stabilizing the aryloxyl radical intermediates[18]. For example, the large increase in activity of Ib over Ia, and IIb over IIa, can be attributed to increased stabilization of the radical by conjugative interaction with the *para*-ether oxygen (Scheme 4d). In general, the *para*-methoxy exerts this effect in most of these structures; however, when it is flanked by two *ortho*-methyl groups, as in IId, this increased 'activity' is lost, apparently due to lack of coplanarity which is required for the conjugative, resonance effect (see the following paragraph).

TABLE 1. Effects of alkyl and methoxy (*para*) substituents on antioxidant activities of monohydroxy phenols

Structure	No.	R^1	R^2	R^3	k_{inh} ($M^{-1} s^{-1} \times 10^{-4}$)
I^a	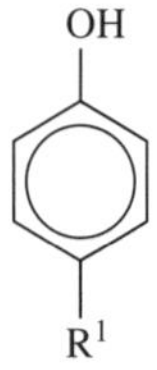 Ia	CH_3			0.917
	Ib	OCH_3			4.78
II^b	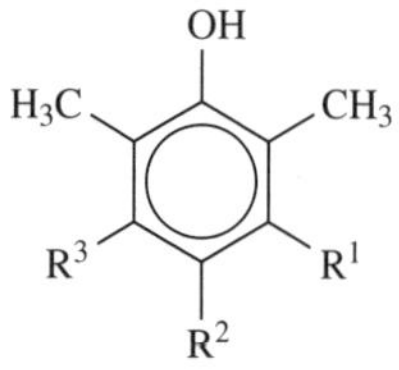 IIa	H	CH_3	H	8.5
	IIb	H	OCH_3	H	94
	IIc	H	OCH_3	CH_3	130
	IId	CH_3	OCH_3	CH_3	39
	IIe	CH_3	CH_3	CH_3	36
	IIf	CH_3	CH_3	H	11
	IIg	CH_3	H	CH_3	7.5
	IIh	H	H	H	2.5
III^b	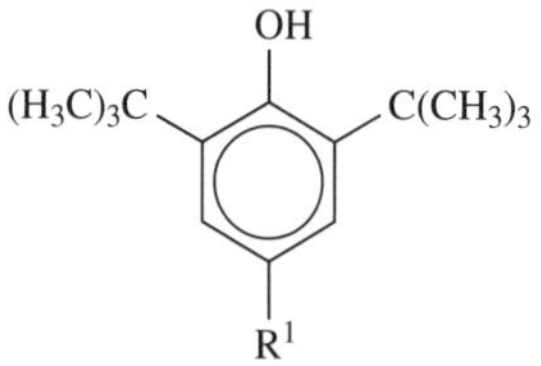 IIIa	OCH_3			11
	IIIb	CH_3			1.4
	IIIc	$(CH_3)_3C$			0.31

[a] Taken from Reference 19. At 65 °C.
[b] Taken from Reference 18.

When the phenolic hydroxyl is flanked by two *ortho-tert*-butyl groups (*cf.* III), the lower activity is attributed to steric hindrance to abstraction of the hydrogen atom by peroxyl radicals.

Stereoelectronic Effects. Stereoelectronic effects of *para*-methoxyl (as well as inductive effects of methyls) are important in controlling the antioxidant activities of methoxy phenols of Class II[18]. As noted above, a *para*-methoxy stabilizes a phenoxyl radical by conjugative electron delocalization with the oxygen. For stabilization, the oxygen p-type lone-pair orbital must overlap with the semi-occupied orbital (SOMO) of the radical. The extent of overlap depends on the dihedral angle, θ, between the oxygen lone pair and the SOMO which is perpendicular to the atoms of the aromatic plane, and the angle θ should be the same as the angle θ' between the O_1-C_2 bond and this plane (see Figure 2). Stabilization of the radical will be at a maximum when $\theta = 0°$ and at a minimum when $\theta = 90°$. In fact, the angle for the solid IId (89°) is almost that of perpendicular arrangement, whereas it was estimated to be only 8° for IIc, in agreement with the markedly higher activity of IIc as an antioxidant. However, as pointed out, the radical activity of the 'twisted' IId ($k_{inh} = 39 \times 10^4$ M^{-1} s^{-1}) is higher than expected in comparison with the *para*-methyl compound, IIg, or the unsubstituted, IIh[18]. In fact, a perpendicular *para*-methoxyl in IId might be expected to reduce its activity by the $-I$ (inductive) effect of oxygen. That this is *not* the case might be due to the possibility that the 'effective' θ for IId in solution is less than 90° or, as suggested, the $-I$ effect of a perpendicular methoxy group is outweighed by a residual $+M$ (mesomeric) effect attributed to '... a resonance contribution from the other lone pair on the oxygen'[18].

Effect of a Heterocyclic Ether Ring: Vitamin E Class. The antioxidant activities of four classes, IV, V, VI and VII, of chromans are summarized in Table 2. α-Toc was determined to be the most active antioxidant of the vitamin E tocopherols[18]. Differences in k_{inh} among the tocopherols appear to be due to inductive effects based upon the number and positions of methyl groups, the maximum effect being attained with completely substituted α-Toc. It is interesting to note that β-tocopherol has the same activity as the 'planar' acyclic compound, IIc. Thus the stereoelectronic effect enforced by the ring system together with inductive effects of methyls contribute to the overall high reactivity of the tocopherols. Minor differences observed in the activity of α-Toc compared to Va and Vb were attributed to different 1,3-interactions within the half-chair conformation of the

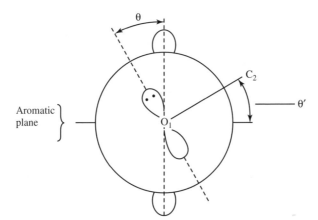

FIGURE 2. The angles involved in interpreting the stereoelectronic effect. Reprinted with permission from Reference 18. Copyright 1985 American Chemical Society

TABLE 2. Effect of heterocyclic ether ring and substituents on antioxidant activities of monohydroxy phenols

Structure	Name/No.	R^1	R^2	R^3	k_{inh} ($M^{-1} s^{-1} \times 10^{-4}$)
IV[a] 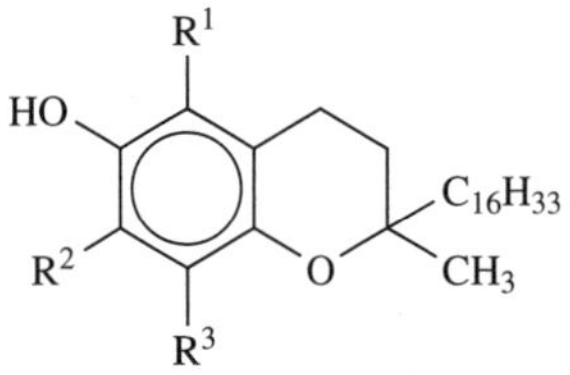	α-Toc	CH_3	CH_3	CH_3	320
	DMT	CH_3	CH_3	H	180
	β-Toc	CH_3	H	CH_3	130
	γ-Toc	H	CH_3	CH_3	140
	δ-Toc	H	H	CH_3	44
V[a] 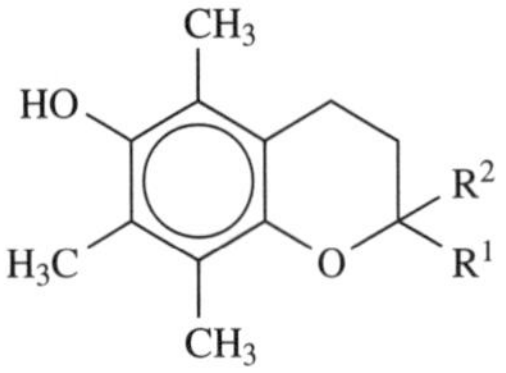	Va	CH_3	CH_3		380
	Vb	H	H		270
	Vc	CH_3	COOH		110
	Vd	CH_3	$COOCH_3$		180
	Ve	CH_3	CH_2COOH		190
	Vf	CH_3	CH_2COOCH_3		270
	Vg	CH_3	$(CH_2)_2COOH$		370
	Vh	CH_3	$(CH_2)_2COOCH_3$		330
	Vi	CH_3	CH_2OH		270
	Vj	CH_3	OCH_3		150
VI 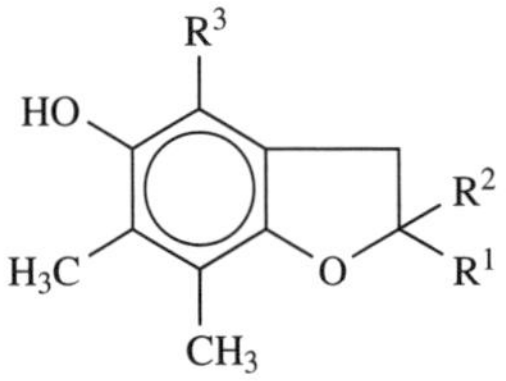	VIa[a]	CH_3	CH_3	CH_3	570
	VIb[a]	H	CH_3	CH_3	540
	VIc[b]	$R^1R^2 = CH_2CH_2$ (spiro)			379
VII[c] 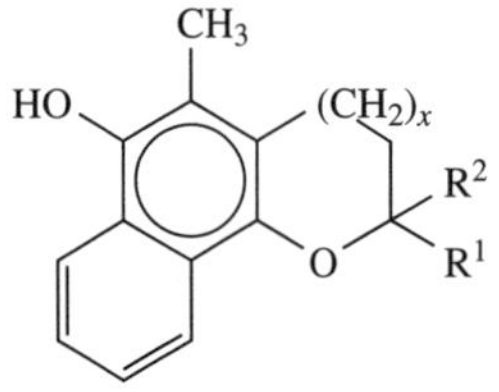	VIIa	$C_{16}H_{33}$	CH_3	$x = 1$	1140
	VIIb	CH_3	CH_3	$x = 0$	2870[d]

[a]Taken from Reference 18.
[b]Taken from Reference 142.
[c]Taken from Reference 49.
[d]k_{inh} for α-Toc was 290×10^4 $M^{-1} s^{-1}$ under the same conditions.

heterocyclic ring. In contrast, significant differences are observed when electron attracting groups replace one of the methyls on this ring, so that a carboxyl group in the commercial, water-soluble antioxidant Trolox (Vc) reduces the reactivity to one-third that of α-Toc. In water, however, where the group is ionized, $-COO^-$, it is expected to be more reactive. Two types of convincing evidence indicate that the reduced reactivities of the type V series is due to $-I$ effects of the substituents[18]. First, the relative reactivities show a regular (approximately) linear trend d > g > h > i ~ f > e > j > c with the σ_1 substituent constants. Second, interesting electron spin resonance data confirmed this trend. The relative spin densities in the various ArO• radicals were determined by measuring the hyperfine splittings (hfs) of the 2- and 6-CH$_3$ groups of IIb, c, d and the 5- and 7-CH$_3$ of type V compounds. The sum of the hfs splittings exhibited a remarkable linear trend with log k_{inh} of the antioxidant activities.

The search for compounds more reactive than α-Toc led to the discovery of the 5-hydroxy-6,7-dimethyl-2,3-dihydrofurans and derivatives, class VI (Table 2). The very significant increase in reactivity for VIa (1.78 times that of α-Toc) was, at least in part, attributed to the increased planarity of the 5-membered heterocyclic ring which reduced the θ torsion angle to 6°, compared to 17° for Va. Following the lead of Ingold and coworkers, there were some interesting attempts to produce even more reactive phenolic antioxidants bearing heterocyclic ether rings. Incorporation of a spiro ring (cf. VIc) did not cause an increase in reactivity, although the calculated θ angle was less than 1°[142]. However, incorporation of a second benzene ring in 6-hydroxy-2,5-dimethyl-2-phytyl-7,8-benzochroman, vitamin E derivative VIIa, and a corresponding chromene (3,4-double bond, not shown) raised the reactivity to four times that of α-Toc and the compound with both a second ring and a 5-membered heterocyclic ether ring, 2,3-dihydro-5-hydroxy-2,2,4-trimethylnaphtho[1,2-b]furan, VIIb, has a reactivity about nine times that of α-Toc, and is undoubtedly the most reactive phenolic antioxidant known[49]. However, with an increase in reactivity of compounds of class VIIa, b, there is a drop in the stoichiometric factor, n, to around 1.5–1.6 compared to $n = 2$ for α-Toc. The lower n values could be due to ArO• 'wasting' reactions resulting from chain transfer reactions with styrene and/or terminating self-reaction of ArO•.

Some phenols containing large *ortho*-alkyl groups and heterocyclic oxygen rings were synthesized to determine if this combination would provide efficient antioxidants. Two examples of the chromanol class, VIIIa, b, shown in Table 3, having *iso*-propyl groups or methyl and *tert*-butyl groups, have antioxidant activities less than α-Toc. The decreased reactivity here is attributed to a combination of steric hindrance to attack by peroxyl radicals and the lack of a stabilizing +I effect of a *meta*-methyl[55].

TABLE 3. Effect of large *ortho*-alkyl groups on antioxidant activities of chromanols

Structure	No.	R^1	R^2	R^3	R^4	k_{inh} (M^{-1}s^{-1} ×10^{-4})
VIII^a 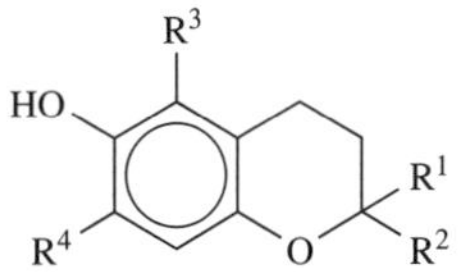	VIIIa	CH$_3$	CH$_3$	(CH$_3$)$_2$CH	(CH$_3$)$_2$CH	238[a]
	VIIIb	CH$_3$	CH$_3$	CH$_3$	(CH$_3$)$_3$C	199[a]
IX^b 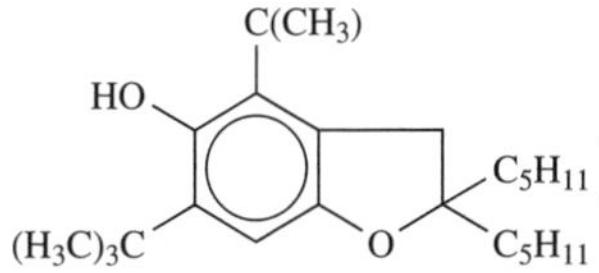						$k_{inh}/k_p(IX) = 2.24 \times 10^3$[b], methyl linoleate in acetonitrile; $k_{inh}/k_p(\alpha\text{-Toc}) = 4.55 \times 10^3$[b], methyl linoleate in acetonitrile

[a]Taken from Reference 55.
[b]Taken from Reference 126.

The hindered phenol IX was 'designed' as an antioxidant, especially as an inhibitor of lipid peroxidation (see Section III.C). In acetonitrile, its antioxidant activity was about one-half that of α-Toc against azo-initiated peroxidation of methyl linoleate (Table 3). This reduced reactivity was attributed to steric hindrance to attack by peroxyl radicals at the phenolic hydroxyl[126].

Effect of a Heterocyclic Nitrogen or Sulfur Ring and Substituents. There is considerable interest and some data on antioxidant activities of nitrogen and sulfur analogs at the heterocyclic center. It was anticipated that X[143] (Table 4) might be a better antioxidant than its analog, Vb (Table 2), because nitrogen should provide better conjugative delocalization of its lone pair. That this was *not* observed (k_{inh}^s are about the same) was attributed to a conformational interaction between the N−CH$_2$CH$_3$ and the 8-CH$_3$ alkyl, which in turn forces the N−CH$_2$CH$_3$ to be axial and the nitrogen lone pair to be coplanar with the benzene ring, an unfavorable position for maximum stabilization of the radical[18]. It

TABLE 4. Selected examples of the effect of hetero-nitrogen and sulfur on the antioxidant activities, k_{inh}, and H-atom donating ability, k_{-H}, of phenols

Structure	No.	R^1	R^2	k_{inh}^a (M^{-1} s^{-1} × 10^{-4})
X	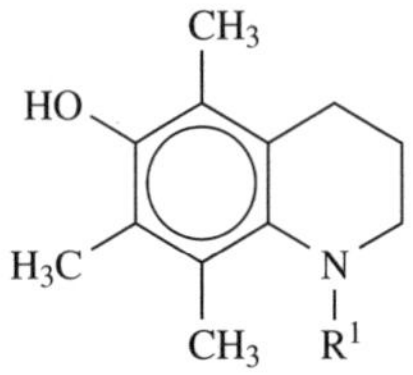 Xa	C$_2$H$_5$		200
	Xb	COCH$_3$		12
XI	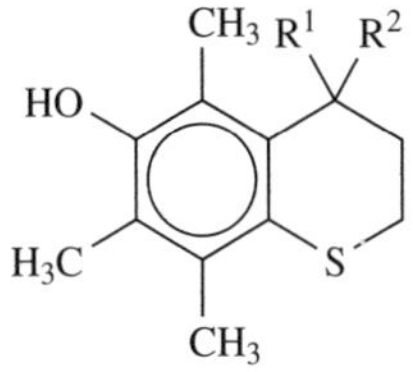 XI	CH$_3$	CH$_3$	280

	No.	R^1	R^2	$n\, k_{inh}^b$ (M^{-1} s^{-1} × 10^{-6})	
	XIIa	C$_{16}$H$_{33}$	CH$_3$	2.6	α-Toc^c
					6.4
	XIIb	CH$_3$	CH$_3$	2.7	Va
					7.6
XII	XIIc	CH$_3$	H	2.0	—
	XIId	H	H	2.1	Vb
					5.4
	XIIe	CH$_3$	COOCH$_3$	1.2	Vd
					3.6

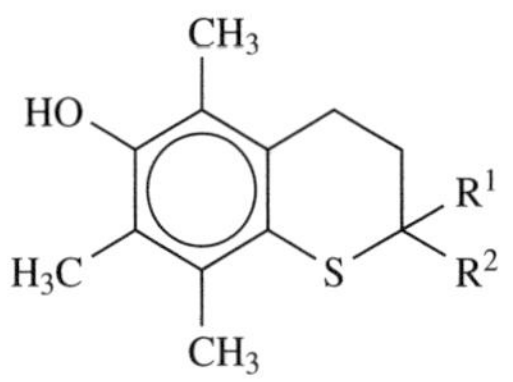

^{a}Taken from References 18 and 143.
^{b}Taken from Reference 144.
^{c}Presented in the same format for comparison.

would be of interest to have results for the N–CH$_3$ derivative; the N–H compound was found to be unstable.

Several sulfur analogs (XI, XII) were also synthesized and their reactivities measured during inhibition of styrene autoxidation[144]. The stoichiometric factors, n, were less than 2 for these compounds, so their antioxidant activities were reported as $n \times k_{inh}$ values. Compounds XII are compared with α-Toc and hydroxychromans in Table 4. It is seen that in all cases the activities of the sulfur analogs are lower than the vitamin E class.

Recently, an interesting report appeared on the 'antioxidant profiles' of some 2,3-dihydrobenzo[b]furan-5-ol and 1-thio, 1-seleno and 1-telluro analogs[145]. Redox properties and rate constants with the *tert*-butoxyl radical were measured in acetonitrile. The values found for the oxygen and sulfur analogs were the same, 2×10^8 M^{-1} s^{-1}, less than the value for α-Toc, 6×10^8 M^{-1} s^{-1}, under the same conditions.

Hydrogen Atom Donating Ability of Antioxidants. Phenolic antioxidants can transfer hydrogen to radicals other than peroxyls in the absence of oxygen and various methods have been employed to measure the hydrogen donating ability of antioxidants (see Section III.A). We propose the term 'antioxidant ability', k_{ab}, for the ability of phenols to react in this way, because the attacking radicals are quite different from peroxyl radicals, the chain-carrying radicals in autoxidation, and the methods employed and kinetic data differ from 'antioxidant activities' determined with peroxyl radicals.

Mukai and coworkers[148,149] developed the use of the stable, colored 2,6-di-*tert*-butyl-4-(4'-methoxyphenyl)phenoxyl radical (ArO$^\bullet$) and other *para* derivatives in stopped-flow measurements of hydrogen bond donating ability of a wide variety of chromanol-type antioxidants, and some of their results are reviewed below.

The relative k_{ab} values of α, β, γ, δ tocopherols (1.00 : 0.44 : 0.47 : 0.20) in Table 5 are in good agreement with the relative values given in Table 2 of k_{inh} (α, β, γ, δ = 1.00 : 0.41 : 0.44 : 0.14), although the stopped-flow k_{ab} values are 600 times smaller. These stopped-flow measurements were made using a hindered reactant in a protic solvent (ethanol) which could account for low reactivities of the phenols as hydrogen atom donors (see Section III.C.1). However, this agreement does not hold for the more reactive antioxidants of the class VI (Table 2) and XIV (Table 5). Only with a *tertiary* butyl group at position 8 (*meta* to the phenolic OH, XIIIa) does the k_{ab} value approach that of k_{inh} (relative) for VI and the explanation involved interaction with the ether oxygen to 'increase the orbital overlap between the 2 p type lone pair...and the aromatic π electron system'[146]. As shown in Table 5, compounds XV and XVI, with a second benzene ring, were found to be very active hydrogen bond donors.

These researchers present a number of arguments and evidence, including large deuterium kinetic isotope effects, in support of a mechanism involving 'proton-tunneling' in a charge transfer complex (equation 29)[127], as the rate-determining step for the reaction of the hindered aryloxyl radical, ArO$^\bullet$, with phenolic antioxidants and they propose that the mechanism applies equally well to attack by peroxyl radicals, R–O–O$^\bullet$, on phenols.

$$Y^\bullet + HX \longrightarrow \{[Y^{\bullet\delta-} \cdots HX^{\delta+}] \longrightarrow \text{proton tunneling}$$
$$\downarrow \qquad (29)$$
$$YH + X^\bullet \longleftarrow [YH \cdots X^\bullet]\}$$

However, the evidence for their interpretation, and in particular extension to the reaction with peroxyl radicals, is far from convincing, especially since Mukai and coworkers

TABLE 5. Hydrogen donating ability of tocopherol and related antioxidants to a substituted phenoxyl radical (PhO$^\bullet$)[a]

Structure	Name/No.	R^1	R^2	R^3	k_{ab}[a] $(M^{-1} s^{-1} \times 10^{-3})$	
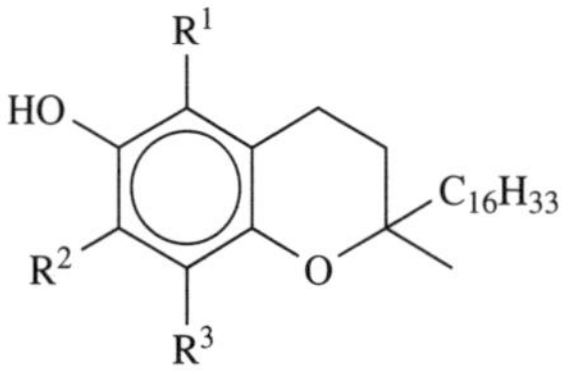	α-Toc	CH_3	CH_3	CH_3	5.12	
	β-Toc	CH_3	H	CH_3	2.24	
	γ-Toc	H	CH_3	CH_3	2.42	
	δ-Toc	H	H	CH_3	1.00	
	Tocol	H	H	H	0.56	
	XIIIa	CH_3	H	$(CH_3)_3C$	3.62	
	XIIIb	CH_3	$(CH_3)_3C$	H	2.97	
	XIIIc	$(CH_3)_2CH$	$(CH_3)_2CH$	H	2.51	
	XIIId	CH_3	CH_3	H	2.39	
	XIIIe	CH_3CH_2	CH_3CH_2	H	1.97	

						k_{ab}/k_{ab}^b α-Toc
XIVa	CH_3	H	$(CH_3)_3C$	9.10	1.77	
XIVb	CH_3	CH_3	CH_3	6.99	1.36	
XIVc	$(CH_3)_2CH$	$(CH_3)_2CH$	H	5.40	1.05	
XIVd	CH_3	CH_3	H	3.49	0.68	
XIVe	H	H	H	0.88	0.17	
XV	$C_{16}H_{33}$	—	—	35.4	6.91	
XVI	$C_{16}H_{33}$	—	—	24.8	4.84	

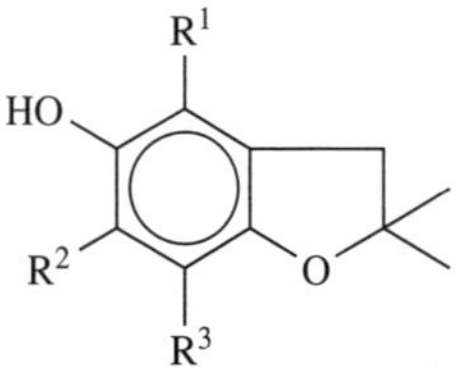

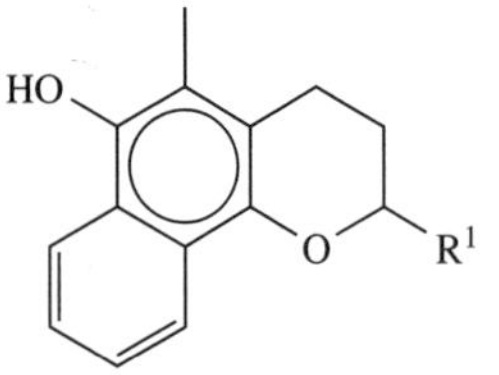

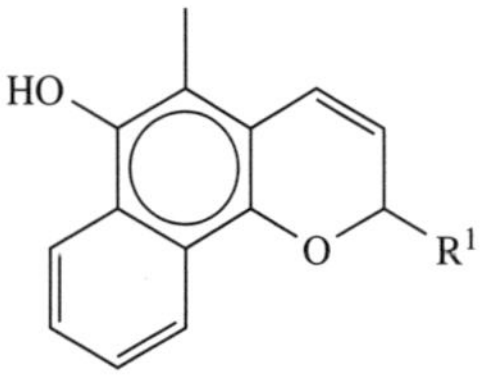

[a]Taken from Reference 147.
[b]Taken from Reference 146.

reported that the 'reverse reaction'[21] (equation 30)

$$R\text{—}O\text{—}O\text{—}H + \text{di-}i\text{-propyl-Toc}^\bullet \underset{k_{inh}}{\rightleftarrows} R\text{—}O\text{—}O^\bullet + 5,7\text{-di-}i\text{-propyl-Toc}$$

(30)

does not exhibit an unusual kinetic deuterium isotope effect; that is, tunneling does not play a role in this reaction. Now, of course the back reaction (i.e. equation 30) is simply the reaction involved in the rate-determining step of the mechanism of antioxidation. From the Principle of Microscopic Reversibility, the mechanism of the back reaction and its forward one must be the same, since they should follow the same potential energy surface. This means that the evidence from reactions of a hindered aryloxyl radical (ArO$^\bullet$) may not be applicable in detail to those involving peroxyl radicals (R−O−O$^\bullet$). Alkyl peroxyl radicals, the main chain-carrying radicals in autoxidation, are highly polarized due to stabilization through resonance[150] and through inductive effects from the alkyl group[151]:

$$R-\ddot{\underset{..}{O}}-\ddot{\underset{..}{O}}\cdot \longleftrightarrow R-\overset{+}{\underset{..}{O}}-\ddot{\underset{..}{O}}:^{-}$$

The hindered aryloxyl radicals may not 'model' exactly the antioxidant mechanism of phenols with phenoxyl radicals.

Ortho-Methoxyphenols: Effect of Intramolecular Hydrogen Bonding. Current interest in *ortho*-methoxyphenols as antioxidants is driven by their frequent occurrence and importance in various natural products including ubiquinols, curcumin, lignin model compounds and others. An *ortho*-methoxy group could provide stabilization of the phenoxyl radical formed by resonance of the type shown in structure **24**[152]. The parent methoxyphenol is intramolecularly hydrogen bonded as shown in structure **25**, so that in non-polar solvents less than 0.1% exists as the free phenol[153]. This hydrogen bond is estimated to stabilize the parent compound by 4 kcal mol^{-1}, which opposes the electronic effect of the methoxy group[154], to decrease the reactivity compared to the *para*-methoxyphenol. As suggested before, the non-linearity of the intramolecular hydrogen bond in the *ortho*-methoxy isomer 'leaves the phenolic hydrogen atom available for abstraction'[155]. The net result of these opposing effects, the activating effect of the *ortho*-methoxy versus the stabilizing effect of H-bonding, is a decreased reactivity of the 2-methoxy isomer

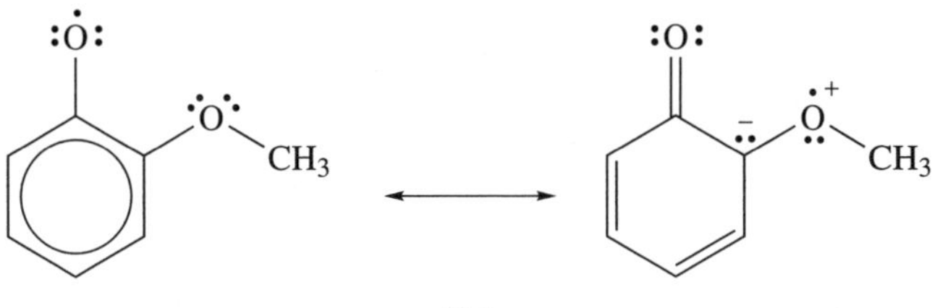

(**24**)

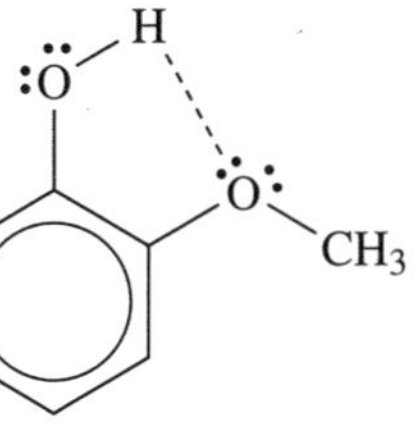

(**25**)

compared to the 4-methoxy one. For example, in non-polar solvents (tetrachloromethane, benzene) the relative rate constants for hydrogen atom abstraction by alkoxyl radicals are in the range $k_{4\text{MeO}}/k_{2\text{MeO}} = 25-32$ and these differences drop remarkably in more polar solvents. The effects of hydrogen bonding and polar solvents must be kept in mind when evaluating the antioxidant activities of all *ortho*-methoxyphenols (see Section III.C).

The antioxidant activities of a series of *para*-substituted *ortho*-methoxyphenols, and related lignin model compounds, were determined by Barclay and coworkers[152] in styrene/chlorobenzene initiated by AIBN. Their data are summarized in Table 6, relative to the commercial antioxidant 2,6-di-*tert*-butyl-4-methylphenol (BHT)[152]. It is interesting to note that 2-methoxyphenol itself was not sufficiently reactive with *peroxyl* radicals to measure its activity since it acted only as a retarder under these conditions. The lignin model compounds isoeugenol, XVIIc, and coniferyl alcohol, XVIId, show twice the activity of XVIIa and XVIIb, and this was attributed to the conjugated double bond in XVIIc and XVIId which provides additional stabilization of the phenoxyl radical through extended delocalization. A strong electron-attracting carbonyl group in XVIIe appears to reduce this effect. It is interesting to note that the overall efficiencies of the dimers, XVIIg and XVIIh, and the tetramer, XVIII, as determined by the product $n \times k_{\text{rel}}$ is greater than those of the monomeric compounds, even when corrected by the number of phenolic hydroxyls. The increased reactivities of these compounds are probably due to the conjugative effect of the *ortho*-phenyl linkage in the case of XVIIg and the inductive effect of the *ortho* $-\text{CH}_2-$ linkages in XVIII.

Curcumin (**26a**) is an important example of a natural *ortho*-methoxyphenol which is reported to have many health benefits[51,156,157]. The structure is a dimeric phenol linked through a β-diketone system. Solutions of curcumin in non-polar media consist mainly of the enol form due to extended conjugation and hydrogen bonding. Jovanovic and coworkers[157] proposed a novel mechanism for the antioxidant mechanism of curcumin. From its reactions with reactive radicals such as methyl and *tert*-butoxyl generated by pulse radiolysis or laser photolysis, an absorption band appeared at 490 nm assigned to the carbon-centered radical from hydrogen atom transfer from the $-\text{CH}_2-$ group. However, other researches assigned this transient absorption to phenoxyl radicals derived from curcumin[111,158]. The antioxidant mechanism of curcumin was re-examined in our laboratory[51] using chemical kinetic methods of autoxidation, namely inhibition of the azo-initiated oxidation of styrene or of the lipid methyl linoleate by curcumin and some methylated, non-phenolic curcumin derivatives.

Typical oxygen uptake method results, employing methyl linoleate as a substrate in chlorobenzene and initiation by AIBN, are shown in Figure 3. These results clearly show that curcumin, **26a**, is a moderately active phenolic antioxidant against lipid peroxidation since $k_{\text{inh}} = 3.9 \times 10^4$ M^{-1} s^{-1}, $n = 4$, compared to 2,6-di-*tert*-butyl-4-methoxyphenol (DBHA), $k_{\text{inh}} = 11 \times 10^4$ M^{-1} s^{-1}, $n = 2$. The effect of the non-phenolic derivative, **26b**, is even more striking; there is no effect at all on the oxygen uptake (see Figure 3 for **26b**), consequently the antioxidant mechanism does not operate by hydrogen atom transfer from the $-\text{CH}_2-$ group. The inhibiting effect of curcumin and dehydrozingerone, **27**, was also examined during AIBN-initiated oxidation of styrene in chlorobenzene. Here, curcumin was a somewhat better antioxidant, $k_{\text{inh}} = 34 \times 10^4$ M^{-1} s^{-1}, $n = 4$. It was of interest to find that dehydrozingerone gave $k_{\text{inh}} = 17 \times 10^4$ M^{-1} s^{-1}, $n = 2$, one half the activity of curcumin, which is not unexpected considering its structure. Again the non-phenolic derivative **26b** gave no inhibition of oxygen uptake under these conditions. Finally, the somewhat suppressed activity of curcumin in the presence of methyl linoleate ($k_{\text{inh}} = 3.9 \times 10^4$ M^{-1} s^{-1}) compared to that in non-polar styrene ($k_{\text{inh}} = 34 \times 10^4$ M^{-1} s^{-1}) is expected for a phenolic antioxidant. Antioxidant activities of phenols are known to be reduced in the presence of esters, which are strong hydrogen bond acceptors for the phenolic hydroxyl[86].

TABLE 6. Antioxidant activities of *ortho*-methoxyphenols

Structure	Name/No.	R^1	R^2	k_{rel}^a	n^b	$n \times k_{rel}$
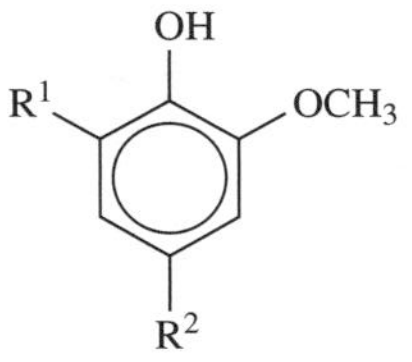 (R^1 — phenol with OH, OCH$_3$, R^2)	XVIIa	H	CH$_2$CH$_2$CH$_3$	1.69	1.7	2.9
	XVIIb	H	CH$_2$CH=CH$_2$	1.83	1.6	2.9
	XVIIc	H	CH=CH–CH$_3$	3.95	1.6	6.3
	XVIId	H	CH=CH–CH$_2$OH	4.25	1.7	7.2
	XVIIe	H	CH=CH–CHO	2.00	1.7	3.4
	XVIIf	CH$_3$O	CH$_2$CH=CH$_2$	4.09	1.7	7.0
(bridged bis-phenol structure)	XVIIg	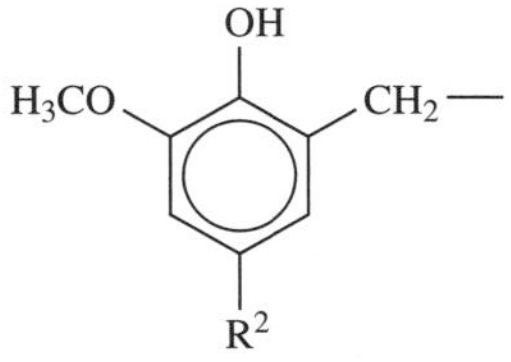		5.36	3.2	17/2 = 8.5
(phenol with CH$_2$OCH$_3$)	XVIIh	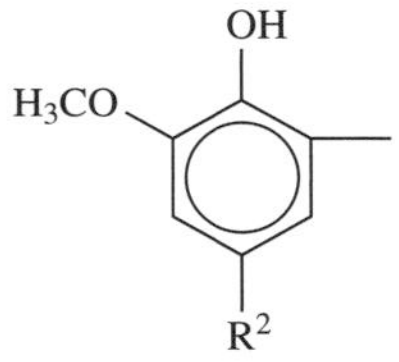	CH$_2$OCH$_3$	4.48	3.3	27/2 = 13.5
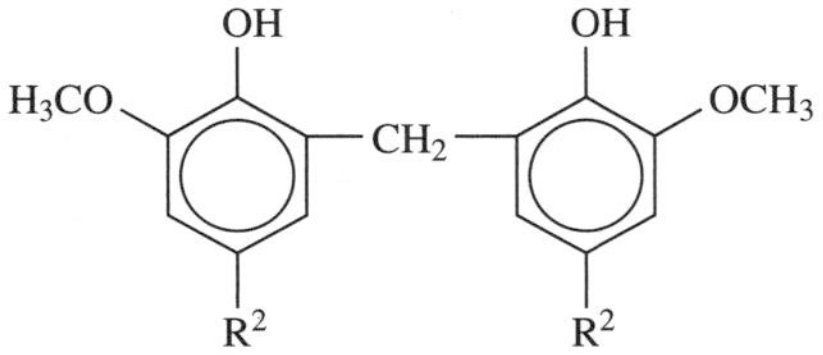 (bis-phenol structure)	XVIII		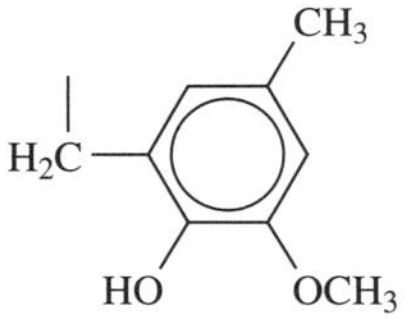	7.75	6.4	50/2 = 25

^{a}Relative k_{inh} values to BHT = 1.82×10^4 M^{-1} s^{-1} determined under the same conditions; taken from Reference 152.
^{b}The stoichiometric factor, n, determined compared to 2,6-di-*tert*-butyl-4-methoxyphenol, equals 2.

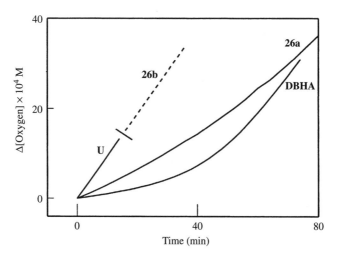

(26) (a) $R^1 = R^3 = OH, R^2 = R^4 = OCH_3$
(b) $R^1 = R^2 = OCH_3, R^3 = R^4 = H$

(27)

FIGURE 3. Oxidation of 0.74 M methyl linoleate in chlorobenzene, initiated with 0.04 M AIBN: **U** = uninhibited rate of oxidation; **26a** = 5.7 μM curcumin (**26a**), $\tau = 64$ min, $R_i = 5.61 \times 10^{-9}$ M s^{-1}, $n = 4.0$, $k_{inh} = 3.85 \times 10^4$ M^{-1} s^{-1}; **26b** = 19.3 μM of **26b**, which shows no inhibition; **DBHA** = 5.9 μM DBHA, $\tau = 46$ min, $R_i = 4.00 \times 10^{-9}$ M s^{-1}, $n = 2.0$, $k_{inh} = 11.1 \times 10^4$ M^{-1} s^{-1}. Reprinted in part with permission from Reference 51. Copyright 2000 American Chemical Society

Recently, another antioxidant mechanism was proposed for curcumin which involved an initial carbon-centered radical at the β-diketone moiety that subsequently undergoes rapid intramolecular hydrogen shift to a phenoxyl radical[157]. Obviously, this mechanism does not account for the antioxidant activity results with peroxyl radicals[51].

2. Dihydroxy phenols: Catechols and 1,4-hydroquinones — Intramolecular hydrogen bonding revisited

Catechols, 1,2-dihydroxybenzene and derivatives are remarkably active antioxidants compared to most *ortho*-methoxyphenols and this structure is very widely distributed in nature, especially as the flavonoids, as well as various flavonal compounds. When one considers the similarity in basic structure of catechol with *ortho*-methoxyphenol, the interesting question arises: What is the origin of the increased activity of catechol? The answer lies in increased stabilization of the semiquinone radical formed from catechol, and of the corresponding transition state, through strong hydrogen bonding in resonance canonical structures such as **28a** and **28b**, as suggested before[68]. Increased stabilization of the radical, **28**, over that of the parent catechol (or, of course, the *ortho*-methoxyphenol, **25**) provided by hydrogen bonding was confirmed by calculations. The parent catechol is stabilized by a moderately strong hydrogen bond of 4 kcal mol^{-1} while the radical has a much stronger hydrogen bond of about 8 kcal mol^{-1} [154]. Thus overall, an *ortho*-methoxyphenol is somewhat deactivated as an antioxidant by intramolecular hydrogen bonding whereas a catechol is activated. A few selected examples shown in Table 7 indicate the effects of hydrogen bonding on antioxidant activities of some simple catechols[68,159] compared to an *ortho*-methoxyphenol. The 4-*tert*-butylcatechol, XIXb, has nearly 30 times the activity of the *ortho*-methoxyphenol, XVIIa[152], and 3,5-di-*tert*-butylcatechol, XIXc, has about half the activity of α-Toc under the same conditions. The methyl catechols, XIXd and XIXe, showed similar increases in k_{inh} values.

(28a) ↔ **(28)** ↔ **(28b)**

Yamamura and coworkers used an oxygen absorption method to study the effects of a series of 46 dihydric phenols on inhibition of azo-initiated oxidation of tetralin[160]. They reported activities in terms of the stoichiometric factor, n, and the rate of oxygen absorption, R_{inh}, during induction periods. The 13 catechols studied all showed higher n factors ($n = 2.0–2.3$) and lower R_{inh} values than any other of the diols. Unfortunately, they were not able to obtain k_{inh} values.

Flavonoids are widely distributed in fruits and vegetables and are very common nutritional supplements as antioxidants. The results on antioxidant activities of simple catechols provide a useful basis for evaluating results for the many, more complex natural compounds containing the catechol structure, such as the flavonoids, steroidal catechols and hormonal catecholamines. There are several reviews on the antioxidant properties of flavonoids[8,9,161] and several reports on experimental[156,162–168] and theoretical evidence[154,169] linking their antioxidant properties to the catechol moiety usually found in their structure. The basic flavonoid structure (**29**) is shown in Chart 1, with a few selected examples (**30–36**) from different groups to illustrate some of the relationships between their detailed structures and related antioxidant properties. Efforts to elucidate these relationships are hampered by their very low solubility in non-polar solvents, and the tendency of some researchers to employ metal ions as initiators of oxidation in aqueous media so that one cannot distinguish between their action as chain-breaking

TABLE 7. Comparison of antioxidant activities of simple catechols with an *ortho*-methoxyphenol

Structure	Name/No.	R^1	R^2	$k_{inh}{}^a$ $M^{-1}s^{-1} \times 10^{-4}$	n^b
(structure with OH, R^1, OCH$_3$, R^2)	XVIIa (Table 6)	H	CH$_2$CH$_2$CH$_3$	3.07	1.7
(catechol structure with OH, OH, R^1, R^2)	XIXa	H	H	55^a	2.3^b
	XIXb	H	(CH$_3$)$_3$C	88^a	2.1^b
	XIXc	(CH$_3$)$_3$C	(CH$_3$)$_3$C	149^a	2.3^b
	XIXd	CH$_3$	H	85^c	—
	XIXe	H	CH$_3$	150^c	—

^{a}Results for these catechols taken from Reference 139, using inhibited oxidation of styrene initiated by AIBN.
^{b}Earlier n factors for catechols of 2–3 were attributed to reactions of the initial catechol-oxidation products with peroxyl radical[138].
^{c}These results from Reference 68 were determined from inhibited oxidation of linoleic acid in cyclohexane.

antioxidants and their metal ion complexing ability. Non-kinetic methods, such as electron spin resonance of the radicals and ionization energies, are sometimes used to evaluate their 'antioxidant potential'.

It is a 'risky' business to attempt an evaluation of literature reports on antioxidant properties of flavonoids, since it is not surprising that these polyphenols give different results from different experimental methods. Nevertheless, a report on their rate constants for reactions with peroxyl radicals generated by azo-initiated reaction on diphenylmethane (DPM) in chlorobenzene provided a quantitative measure of the relative antioxidant activities[168]. Although the chemiluminescence kinetic method did not follow 'classical' theory, the relative reactivity obtained for α-Toc and 2,6-di-*tert*-butyl-4-methylphenol (BHT) was the same (*ca* 230) as that reported by the IOU method (compare Tables 1 and 2). The rate constants for reaction with these DPM peroxyls were similar for quercetin (**31**), dihydroquercetin (**34**), luteolin (**30**) and 3,5-di-*tert*-butylcatechol: 21×10^6, 19×10^6, 22×10^6 and 19×10^6 M^{-1} s^{-1}, respectively (see Chart 1, examples **30–36**), which indicates that the main contribution to the activity is the catechol structure in ring B of **29**. This is confirmed by the value for kaempferol (**33**), which dropped to 1.0×10^6 M^{-1} s^{-1}. This remaining activity in **33** could be due to the phenolic hydroxyl in ring B which is activated by the *para* conjugated enol group, since the flavonoid naringenin (**35**) has only very slight activity according to their chemiluminescent method (3.4×10^3 M^{-1} s^{-1})[168]. Polar solvents exert a very large effect on the hydrogen atom donating ability of catechols[50]. This is also true, as expected, for the flavonoids, since the reactivity is mainly in the catechol ring. In chlorobenzene, the antioxidant activities for quercetin and epicatechin during inhibited peroxidation of methyl linoleate determined by a spin probe method were determined to be 4.3×10^5 and 4.2×10^5 M^{-1} s^{-1}, respectively[166]. In *tert*-butyl alcohol, these values dropped to 2.1×10^4 and 1.7×10^4 M^{-1} s^{-1}, respectively.

A quantum mechanical explanation for the antioxidant activity of flavonoids found a planar structure for quercetin and compared the torsion angles of other flavonoids with spin densities and oxidation potentials with $\Delta\Delta H_f$, the heat of formation of the radicals

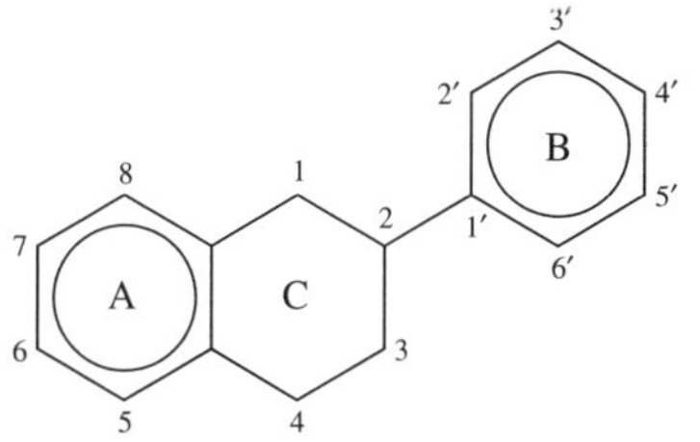

12. Phenols as antioxidants

(29)

Flavone group

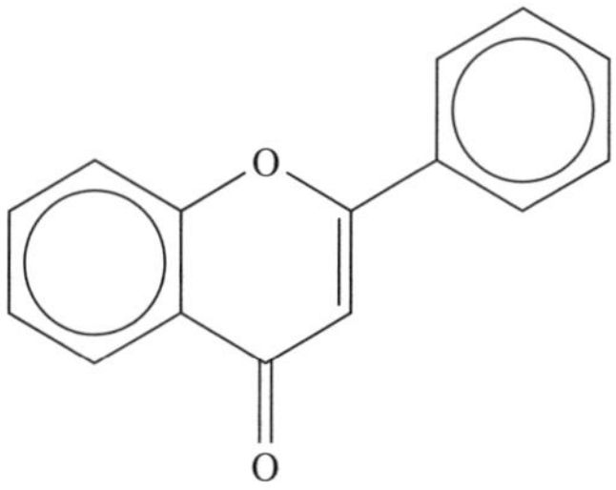

(30) luteolin: OH^S at 3′, 4′, 5, 7

Flavanonol group

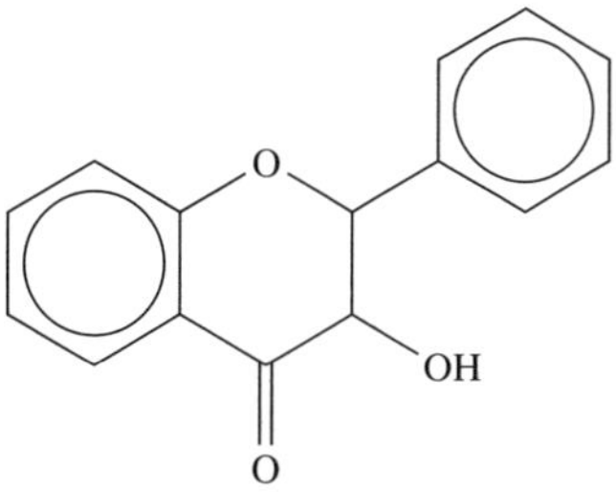

(34) dihydroquercetin: OH^S at 3′, 4′, 5, 7
(35) naringenin: OH^S at 4′, 5, 7

Flavonol group

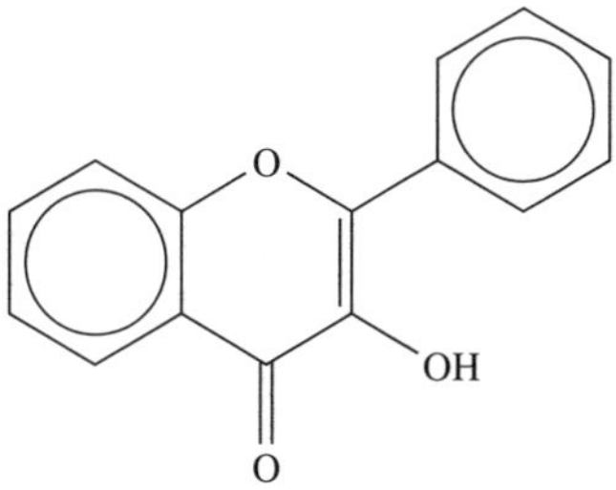

(31) quercetin: OH^S at 3′, 4′, 5, 7
(32) fisetin: OH^S at 3′, 4′, 7
(33) kaempferol: OH^S at 4′, 5, 7

Flavanol group

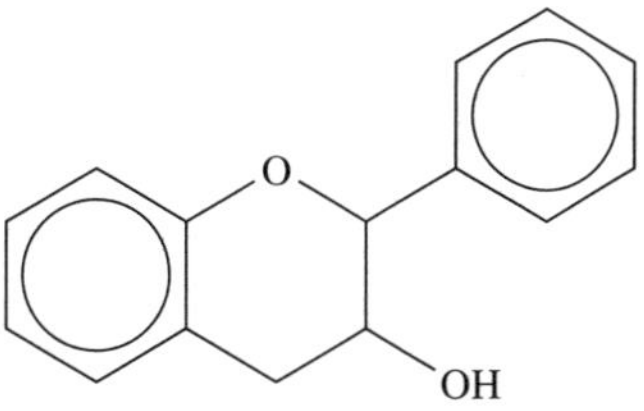

(36) catechin: OH^S at 3′, 4′, 5, 7

CHART 1. Some flavonoid structural groups

relative to that of the parents[169]. Van Acker and coworkers correlated the torsion angle of ring B with the rest of the molecule to the scavenging activity, the latter apparently improving with conjugation. An interesting conclusion was reached which attributed the 'good' antioxidant activity of the flavonols, quercetin, fisetin (**32**) and kaempferol to an intramolecular hydrogen-bond-like interaction between the protons on carbons 2′ and 6′ and the 3-OH moiety. However, this interpretation is not supported experimentally, since evidence so far indicates that luteolin, which lacks the 3-OH, is equally as active as quercetin while kaempferol is relatively inactive[168]. Electron transfer reactions on model catechols and flavonoids initiated by pulse radiolysis support the general conclusions that the catechol ring B is responsible for the antioxidant activity of flavonoids[163,165], and similarly for the activity of gallocatechins from green tea[170].

The catechol structure is also present in several steroids and in natural hormonal amines such as adrenalin, L-dopa and dopamine, and their effects as antioxidants in natural biological systems are of interest. Some catechol steroids were reported to have effective antioxidant properties in lipoproteins[171] and rat liver microsomes[172]. The hormonal catecholamines are of particular interest. They are known to have both antioxidant and toxic effects[1,173–175]. Both catechols and 1,4-hydroquinones have associated toxic properties in biological systems. The cytotoxicity is attributed to two processes. In one, redox cycling[1,176,177] between a semi-quinone radical, formed in an initial hydrogen atom transfer to attacking radicals, and a quinone results in the formation of superoxide and subsequently the reactive conjugate acid, $^{\bullet}$O−O−H. In another process, quinone methides cause damage through alkylation of cellular proteins or DNA[178].

There are many examples in the literature on applications of 1,4-hydroquinones as polymer stabilizers and as antioxidants. The natural ubiquinols are 2,3-dimethoxy dialkyl derivatives of these hydroquinones and these natural compounds are now known to be of great importance in biological systems. We select a few examples of 1,4-hydroquinones as antioxidants to illustrate the effect of structure (e.g. substituents) on their reactivity, but especially to emphasize the role that hydrogen bonding plays in the reactivity of catechols, 1,4-hydroquinones and methoxy derivatives.

In the 1970s Pospíšil and coworkers reported on hydroquinones as polymer stabilizers and antioxidants[179–181]. For the latter studies they used tetralin as substrate, initiated by AIBN. Their usual method of reporting antioxidant properties, the 'relative activities' from the induction periods in the presence of antioxidants, A_{IP}, and the time for absorption of a measured volume of oxygen, A_τ, do not permit actual evaluation of quantitative antioxidant activities[179]. Nevertheless, some interesting results were recorded. They found that the oxidation rates did not return to the uninhibited rates after the 'end' of the induction period, especially with the catechols. In other words, the *ortho*-quinones formed were acting as retarders. We think this could be due to addition of peroxyl radicals to the conjugate quinone chromophore. They also reported that the 2-alkoxyalkyl-substituted hydroquinones were the most 'active' and attributed this to hydrogen bonding between the phenolic group and the alkoxy group[180]. Actually, this conclusion is quite ambiguous because the longer induction periods for these derivatives, shown in larger A_{IP} and A_T values, are probably due to *decreased* reactivity resulting from hydrogen bonding in these derivatives. They also reported that the stoichiometric factors depended markedly on experimental conditions and proposed pathways to account for this, such as reaction of the semi-quinone radicals with hydroperoxides[181], an explanation that has been invoked by others later (*vide infra*). The antioxidant properties of 1,4-hydroquinones has been reexamined by several groups more recently. Yamamura and coworkers[160] found that these hydroquinones were less effective in reducing oxygen uptake than catechols (by about one-half) and the stoichiometric factors of hydroquinones ranged (0.6–1.1) to about half that of the catechols. It is interesting to note that they attributed the higher activity of catechols, compared to hydroquinones, to the increased stability of the derived phenoxyl radicals due to the intramolecular hydrogen bond.

Last year Roginsky and coworkers determined the chain-breaking activity of thirteen hydroquinones, p-QH$_2$, during azo-initiated oxidation of styrene, '...known as the most suitable oxidation substrate for testing chain-breaking antioxidants'[182]. Their results provide a useful comparison with antioxidant activities of monophenols studied under similar conditions (e.g. Section III.B.1), therefore their results are reproduced in summary form for compounds XX in Table 8. The following conclusions can be drawn from the results:

TABLE 8. Antioxidant activities of 1,4-hydroquinones during azo-initiated oxidation of styrene at 37 °C[182]

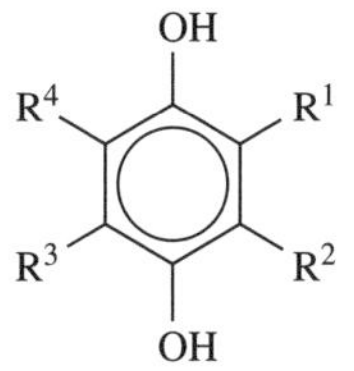

OH, R⁴, R¹, R³, R², OH substitution pattern on benzene ring.

QH$_2$	R^1	R^2	R^3	R^4	k_{inh} (M^{-1} s^{-1} × 10^5)	n
XXa	H	H	H	H	5.54	2.09
XXb	CH$_3$	H	H	H	7.13	1.94
XXc	CH$_3$CH$_2$	H	H	H	13.1	1.00–1.63
XXd	(CH$_3$)$_3$C	H	H	H	12.0	0.76–1.79
XXe	CH$_3$	H	H	CH$_3$	15.6	0.83–2.00
XXf	CH$_3$	H	CH$_3$	H	11.9	0.35–0.99
XXg	CH$_3$	CH$_3$	H	H	18.8	0.26–1.35
XXh	CH$_3$	H	CH$_3$CH(CH$_3$)CH$_2$	H	17.1	0.27–1.07
XXi	CH$_3$O	H	H	CH$_3$O	13.7	0.21–0.50
XXj	C$_6$H$_5$	H	H	C$_6$H$_5$	4.7	0.29–0.48
XXk	Cl	H	Cl	H	0.9	1.99
XXl	CH$_3$	CH$_3$	CH$_3$	H	23.2	0.09–0.31
XXm	CH$_3$O	CH$_3$O	CH$_3$	H	4.4	1.59

(1) The antioxidant activities of 1,4-hydroquinones are greater than those of the mono phenols of similar structure. For example, the k_{inh} value of the 2,6-dimethyl derivative, XXe (1.56 × 10^6 M^{-1} s^{-1}), is nearly two orders of magnitude greater than that of 2,6-dimethylphenol (IIh, Table 1, 2.5 × 10^4 M^{-1} s^{-1}). (2) Alkyl groups increase the reactivity which increases in the order mono-, di-, and tri-substituted QH$_2$, whereas electron attracting chlorines reduce the activity (cf. XXk). (3) Methoxy groups (cf. XXi) cause a drop in activity, especially in the ubiquinol XXm. This was attributed to a decrease in oxygen p-type and aromatic π overlap if two adjacent methoxy groups are forced out-of-plane. However, a much more feasible explanation is that the lower reactivity of XXm is due to strong intramolecular hydrogen bonding of each phenolic hydroxy group to an *ortho*-methoxy; that is, the same phenomena encountered with *ortho*-methoxyphenol (*vide supra*) applies here. (4) The stoichiometric factors, n, of the substituted 1,4-hydroquinones are less than two and depend on the experimental conditions; e.g. n decreases with oxygen concentration, and increases somewhat at high rate of initiation. Low and variable n values are characteristic of 1,4-hydroquinones. Of the two probable 'wasting' reactions (equations 31 and 32) involving the semi-quinone radical, $^\bullet$QH, the authors give arguments favoring the disproportionation reaction (equation 31) as the main factor reducing n.

$$QH^\bullet + O_2(O_2, LH) \longrightarrow Q + LO_2^\bullet + H_2O_2 \qquad (31)$$

$$QH_2 + O_2 \longrightarrow QH^\bullet + HO_2^\bullet \qquad (32)$$

The antioxidant properties of the ubiquinols have been reviewed recently[183] and this material will not be discussed here. Nevertheless, we emphasize the importance of intramolecular hydrogen bonding again here. The 2,3-dimethoxy-1,4-hydroquinones are strongly hydrogen bonded at both sites as shown in structure **37**. Transition state

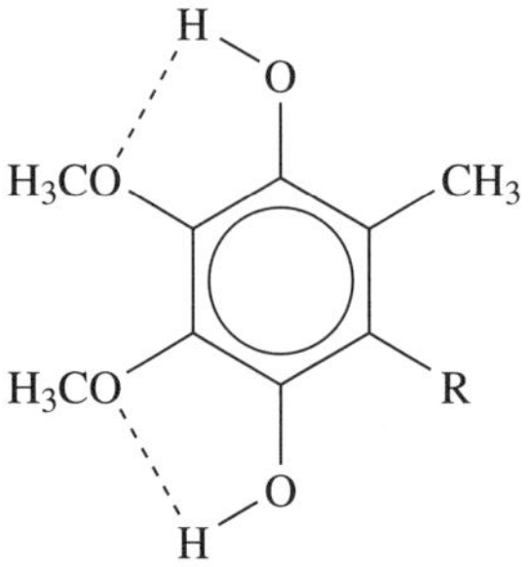

(**37**) R = H, or a C_5-isoprenoid

calculations as well as rate constant data show that intramolecularly hydrogen bonded phenolic hydrogens in *ortho*-methoxyphenols are abstracted surprisingly readily by oxygen-centered radicals[155]. The internal hydrogen bonds in **37** 'protect' the molecule from the strong solvent interactions shown for mono-hydroxy phenols. As a consequence, although ubiquinol has only one-tenth the activity of α-Toc in the styrene–AIBN system[184], the activities are the same in aqueous dispersions because the mono-phenol is more susceptible to strong external hydrogen bonding[185]. Thus ubiquinol becomes very important in biomembranes (see Section III.C.2).

C. Media Effects

1. Solvent effects

Studies on antioxidant activity have been conducted in homogeneous systems in organic solvents, and in heterogeneous systems like micelles and liposomes. In biological systems, peroxyl radical reactions can take place in hydrophilic (e.g. in plasma, cytosol, serum) or hydrophobic (e.g. within lipid bilayers) environments. The nature of solvent interactions with phenolic antioxidants and their effect on antioxidant activity are of considerable interest when attempting to understand antioxidant behavior in biological systems and the diverse solvent environments there. Many have conducted studies on antioxidant activity in homogeneous solution and attempted to apply those findings to what occurs in natural systems. However, before one can do that, it is important to consider the solvent in which antioxidant activity is studied, and determine whether the solvent itself can interact with the reactants to increase or decrease the reaction rate. For convenience, and to overcome problems with solubility, antioxidants or initiators have often been dissolved and added in solvents different from the rest of the reaction mixture, and because the effects of these solvents on reaction rates are not taken into consideration, the results are often difficult to interpret. The effect the solvent has on the rate constant for the reaction of antioxidants with the radical species is dependent upon how the solvent interacts with reactants, and on the mechanism of the antioxidant action. It should be noted that in order to predict the effect of solvents on antioxidant activities, it is important to be able to utilize solvent parameters that are established for a large number of solvents. Abraham and coworkers[186] have developed a scale of *solute* hydrogen bond basicity, β_2^H, for a very large number of solutes by measuring their log K_B^H values against several reference acids and using equation 33 to calculate the β_2^H value. The K_B^H values are the equilibrium constants for hydrogen bond complexation of bases with various reference acids.

$$\beta_2^H = \frac{(\log K_B^H + 1.1)}{4.636} \tag{33}$$

They clearly indicate that the β_2^H is *not* the same as *solvent* hydrogen bond basicity, β_1, because the β_2^H value treats the solvent as a solute in the chemical interactions, and that the β_2^H and β_1 scales are relatively collinear but not interchangeable[186]. (The latter scale is based on the comparison of the indicators *p*-nitroaniline and *p*-nitro-*N*, *N*-dimethylaniline.) Neither is the β_2^H value connected to solute proton-transfer basicity[187]. They have established that this β_2^H value is relatively constant for homologous series of solvents, and that substituents on the parent structure of the solvent do not overly influence the β_2^H value in terms of inductive or polar effects, unless the substituent is halogenated, in which case the β_2^H will decrease. Chain branching of the parent also has little effect on the β_2^H value. This makes it possible to predict 'average' β_2^H values for solutes whose K_B^H values are not known. Correlations of kinetic data with β_2^H are not always accurate because the β_2^H parameter does not take into consideration solvent size, which can lead to steric hindrance of hydrogen bond formation[188].

a. Solvent interactions with the attacking radicals. Ingold and coworkers[189,190] have reported that an increase in solvent polarity also increases the likelihood that an alkoxyl radical, in this case the cumyloxyl radical, undergoes β-scission (equation 34) in preference to hydrogen abstraction (equation 35) from a hydrocarbon substrate, due to improved solvation of the late transition state for the β-scission reaction.

β-scission:

$$C_6H_5-(CH_3)_2C-O^{\bullet} \xrightarrow{k_{\beta}^{CumO}} C_6H_5-(CH_3)C=O + CH_3^{\bullet} \quad (34)$$

H-abstraction:

$$C_6H_5-(CH_3)_2C-O^{\bullet} + R-H \xrightarrow{k_{\alpha}^{CumO}} C_6H_5-(CH_3)_2C-OH + R^{\bullet} \quad (35)$$

They conducted these studies using several techniques to monitor product formation in six different solvents. They concluded under the conditions of these experiments that the decrease in the ratio for the rate constants $k_{\alpha}^{CumO}/k_{\beta}^{CumO}$ with increase in solvent polarity was due to increase in the k_{β}^{CumO} while the k_{α}^{CumO} was solvent independent of abstraction from a hydrocarbon (although there is a significant solvent effect for hydrogen atom abstraction from phenols, *vide infra*[188,191,192]). Although hydrogen bonding between the cumyloxyl radical and a polar solvent can occur, it has been speculated that the hydrogen bonding does not involve the unpaired electron on the cumyloxyl oxygen, and thus its reactivity is not affected[191].

Franchi and coworkers reported on the effect of solvents on hydrogen atom abstraction from phenolic antioxidants by primary alkyl radicals[193], which could be useful when studying oxidations at low oxygen partial pressures because under those conditions the addition of oxygen to substrate alkyl radicals can be reversible (equation 36) and alkyl radicals could play a role in termination reactions (equation 37)[116]

$$R^{\bullet} + O_2 \rightleftarrows R-O-O^{\bullet} \quad (36)$$

$$R^{\bullet} + ROO^{\bullet} \rightleftarrows R-O-O-R \quad (37)$$

These authors demonstrated that a kinetic solvent effect (KSE) was observed; in other words, that the solvent affects the reaction rate between attacking radical and antioxidant, and some of their results are presented in Table 9. Table 9 also includes data on KSEs observed when different radical species are used to abstract hydrogen from phenolic antioxidants (*vide supra*).

The effect of solvent on the hydrogen atom abstracting ability of the nitrogen-centered radical DPPH• can be significant (see Table 9). However, it seems to be due only slightly to hydrogen bonding interactions between the DPPH• and the solvent. Polar solvents could have two influences on the DPPH•: (a) polar solvents could stabilize the charged resonance

TABLE 9. Summary of the effects of solvents on antioxidant effectiveness, reported in terms of their rate constant, k, and the ratio of rate constants, k_{alkane}/k_s

Solvent $\beta_2^{\text{H}a}$	Alkane[a] 0	Cl-Ph 0.09	CH$_3$OPh 0.26	CH$_3$CN 0.44	1°ROH[a] 0.45	3°ROH[a] 0.49	(CH$_3$)$_2$CO 0.50
α-Toc + DPPH•							
α-Toc[b], $k \times 10^{-2}$ M^{-1}s^{-1}	74	27	14	4.9		5.7	
k_{alkane}/k_s		2.7	5.3	15		13	
α-Toc + RO•							
α-Toc[b], $k \times 10^{-8}$ M^{-1}s^{-1}	99	36	20	9.4		1.8	
k_{alkane}/k_s		2.8	4.9	10.3		55	
α-Toc + ROO•							
α-Toc[c], $k \times 10^{-5}$ M^{-1}s^{-1}	68	27	15	3.8		5.6	
k_{alkane}/k_s		2.5	4.5	18		12	
α-Toc + R•							
α-Toc[d], $k \times 10^{-4}$ M^{-1}s^{-1}	115		44	23		3.2	
k_{alkane}/k_s			2.6	5		35	
phenol + DPPH•							
phenol[b], $k \times 10^{-3}$ M^{-1}s^{-1}	160	59	7.2			2.9	
k_{alkane}/k_s		2.7	22			31	
phenol + RO•							
phenol[b], $k \times 10^{-7}$ M^{-1}s^{-1}	110	48	5.6[e]	0.58[e]		0.36	
k_{alkane}/k_s		2.3	20	189		306	
Va + DPPH•							
Va[f], $k \times 10^{-2}$ M^{-1}s^{-1}	68				2.3	2.2	2.0
k_{alkane}/k_s					29	31	34
Va + RO•							
Va[f], $k \times 10^{-3}$ M^{-1}s^{-1}	93.9				4.4	5.6	3.3
k_{alkane}/k_s					21	17	28
IIIa + DPPH•							
IIIa[f], k M^{-1}s^{-1}	37.6				5.23	4.27	1.14
k_{alkane}/k_s					7.2	8.8	33
IIIa + RO•							
IIIa[f], $k \times 10^{-2}$ M^{-1}s^{-1}	44.3				5.8	6.4	2.8
k_{alkane}/k_s					7.6	6.9	16

[a] The different alkanes and cycloalkanes used were octane, isooctane, hexane and cyclohexane. 1° ROH was n-propyl alcohol, 3° ROH was tert-butyl alcohol. All β_2^{H} values are from Reference 186.
[b] Results are selected from Reference 191. RO• is tert-butoxyl radical with α-Toc, and cumyloxyl radical with phenol.
[c] Results are selected from Reference 86. ROO• is the cumylperoxyl radical. α-Toc rate in alkane is for the rate measured with cyclohexane.
[d] Results are selected from Reference 193.
[e] Results are selected from Reference 188. RO• is the cumyloxyl radical.
[f] Results are selected from Reference 50. RO• is the 2,6-di-tert-butyl-4-(4'-methoxyphenyl)phenoxy radical.

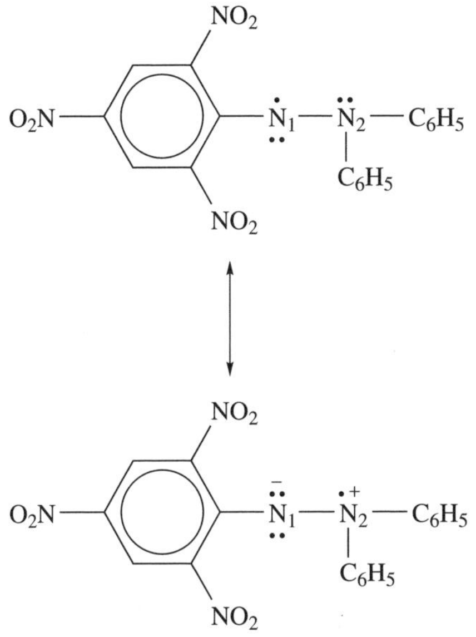

structure of the radical (see above), and (b) hydrogen bonding at the N_2 nitrogen (since the picryl group is very electron-withdrawing) could increase the localization of the unpaired electron on N_1, increasing DPPH• reactivity with increasing solvent polarity[191]. A more important consideration, however, may be due to solvent structure[194]. The 'bulky' alcohols 2-propanol and especially *tert*-butyl alcohol have a dramatic effect on the reactivity of DPPH•. It was speculated that this effect may be due to these solvents forming a 'solvation shell' around the radical, causing steric crowding between solvent molecules that compete for sites on the DPPH•. This causes increased spin density on both N_1 and N_2, resulting in increased reactivity[194]. In order to ensure that the effect observed was not due to interactions between solvent and substrate, Ingold and coworkers[194] used the hydrocarbon cyclohexadiene as the hydrogen atom donor.

Solvent effects on peroxyl radical reactions have been conducted by a variety of researchers. In particular, the effect of solvent dielectric constant and polarity on the propagation and termination rate constants ($k_p/2k_t$ ratio) of substrates were studied[195–197]. Hendry and Russell attempted to determine if an increase in solvent polarity would increase the value of k_p, which could have been attributed to increased solvation of the polar peroxyl radical structure and polar transition state for hydrogen atom abstraction; however, there was insufficient evidence to state this absolutely[197]. Instead, they concluded that the increased solvent polarity had a more significant effect on the $2k_t$ value, since the polar character of the peroxyl is lost upon termination, a theory which has been supported by other groups[198]. Other researchers also examined the effects of solvent polarity on k_p for hydrogen abstraction by cumylperoxyl radical and found the value of k_p to be relatively constant for five solvents of varying β_1 values; thus variations in the oxidizability ($k_p/2k_t^{1/2}$) due to solvent could be attributed to changes in $2k_t$[199]. The $2k_t$ value also appears to decrease as the size of the peroxyl species increases within the same solvent system[198]. The independence of k_p with respect to solvent indicates that the kinetic solvent effects (KSE) observed with hydrogen atom abstraction from phenols are due to interactions of the solvent with the substrate rather than the attacking radical[191,192,198]. If propagation by peroxyl species is unaffected by solvent polarity, it seems reasonable that hydrogen atom abstraction by DPPH•, due to its similar electronic structure (e.g. its charged resonance structure) to peroxyl radicals[199], should also be relatively unaffected by solvent polarity.

Alpincourt and coworkers[151] used theoretical calculations to predict the effects of polar and non-polar solvation on the structure of and electron distribution in peroxyl radical species. Structurally, bond angles of peroxyl species are not affected significantly by polar solvents. However, there is a strong increase in dipole moment for the peroxyl (in terms of the charge distribution), even though there is a slight decrease in spin density on the outer oxygen, due to electrostatic interactions. Furthermore, hydrogen bonding with the solvent by the peroxyl allows for intermolecular charge transfer from the peroxyl to water, the polar medium used in these calculations. Their results seem to confirm that the stabilization of the peroxyl radical by a polar solvent is significant. On the other hand, Sumarno and coworkers[200] state that small amounts of a polar solvent, ethanol, can promote decomposition of a hydroperoxide species, producing reactive radical products (alkoxyl and peroxyl radicals) which would promote further oxidation of substrate.

b. Solvent interactions with phenolic antioxidants—Effects on antioxidant mechanisms

Electron Transfer. Neta and coworkers[25,26,201] have worked extensively with halogen-substituted methyl peroxyl radicals ($X_n H_m COO^•$, where X = Cl, Br or F) in aqueous and non-aqueous media, using combinations of solvents in different ratios to change the polarity of the mixture. They describe the mechanism for the reaction of the water-soluble antioxidant Trolox with their peroxyl radicals as 'H-mediated electron transfer', having determined that the rate of the reaction increases with an increase in solvent polarity. They examined solvent polarity in terms of the dielectric constant of the solvent, ε, and solvent basicity, reported as either the coordinate covalency parameter, ξ, which is a measure of solvent proton-transfer basicity, or the β_1 value, which is a measure of solvent hydrogen bond basicity[25].

If the antioxidant reaction proceeds via electron transfer from antioxidant to radical, then the increased polarity of the solvent (Solv) will help to stabilize the polar transition state. If the solvent basicity increases, then the reaction may proceed first via deprotonation of the antioxidant by the solvent to form an antioxidant anion, which would be much more reactive in terms of electron transfer to the radical than the protonated phenol. The reaction pathway may be shown to be three steps (equations 38–40), especially in strongly basic solvents[201].

$$ArO\text{—}H + Solv \longrightarrow ArO^- + Solv\text{—}H^+ \qquad (38)$$

$$ArO^- + Cl_3COO^• \longrightarrow ArO^• + Cl_3COO^- \qquad (39)$$

$$Cl_3COO^- + Solv\text{—}H^+ \longrightarrow Cl_3COOH + Solv \qquad (40)$$

The other possible mechanism is the removal of a proton from the transition state concerted with the electron transfer. Overall, the equations can be combined and expressed as a hydrogen atom transfer reaction (equation 41)[201].

$$ArO\text{—}H + Cl_3COO^• + Solv \longrightarrow ArO^• + Cl_3COOH + Solv \qquad (41)$$

However, this simplified reaction does not indicate the involvement of solvent or electron transfer. They supported their theory using kinetic isotope studies, which showed the involvement of solvent protons in the electron transfer step, via hydrogen bonding by the solvent to the hydroperoxide anion formed (equation 39)[25]. They suggested that in environments of low polarity and decreased proton donating ability (e.g. in the hydrophobic phase of lipid membranes), electron transfer would be slowed down enough that hydrogen atom transfer would become the predominant mechanism[25].

Maki and coworkers designed a group of hindered phenolic antioxidants that showed reversible and non-reversible electron transfer depending on the location (*ortho* or *para*) of an α-alkylamino group, which would hydrogen-bond with the phenolic hydrogen[202]. These compounds were designed to model what happens biologically in protein systems to allow the formation of a persistent tyrosyl radical with histidine residues, and show that hydrogen bonding can have a significant effect on redox potential for an antioxidant[202].

Hydrogen Atom Transfer. In the presence of a hydrogen bond accepting (HBA) solvent, the phenolic hydrogen can form a hydrogen bond with the solvent that interferes with hydrogen atom abstraction by attacking radicals[192]. If the attacking radical is, for example, an alkoxyl such as cumyloxyl radical, it is unable to approach the hydrogen-bonded complex to abstract the phenolic hydrogen for steric reasons[192].

Looking at this in terms of the Gibbs free energy, as expressed by Ingold and coworkers[191], when hydrogen atom abstraction reactions take place in a strongly HBA solvent (ii), the overall ΔG_{ii} to reach the transition state is the sum of the ΔG_i for the reaction in a poor HBA solvent (i) and the extra ΔG_{i-ii} required to overcome the energy barrier caused by increased solvation of the reactants (specifically, the hydrogen bonded phenolic antioxidant). The amount of this difference can then be calculated from the measured rate constants for the reactions of the antioxidant with the attacking radical in the two solvents (equation 42)[191].

$$\log\left(\frac{k^i}{k^{ii}}\right) = \frac{\Delta G_{i-ii}}{2.3RT} \tag{42}$$

Ingold and coworkers have also expressed a kinetic equation (equation 43) to determine the equilibrium constant ($K^{A/nA}$) for hydrogen bonding between the phenolic antioxidant and dilute HBA solvent, A, if rate constants in neat HBA and non-HBA solvent, nA, CCl_4 in this example, are known. For this study they monitored hydrogen-atom transfer from the antioxidant phenol to cumyloxyl radicals[188,203].

$$k^{nA} = k^A \left(1 + K^{A/nA}[A]\right) \tag{43}$$

Ingold and coworkers have concluded that the kinetic solvent effect, KSE, is independent of the nature of the attacking radical, in other words the radical species used would not influence the *trend* of hydrogen bonding interactions of solvent with antioxidant[191]. That means that if one examined rate constants for the antioxidant activity of a phenolic antioxidant over a series of solvents with a particular radical, one would then be able to predict the rate constants for that antioxidant with a new radical for the same series of solvents, based on a measurement in just one of the solvents. This can be shown by plotting the log of the rate constants for the same antioxidant with two different radical species over a series of solvents, and if the plot is linear with a slope of one, then this indicates that the KSE is independent of the nature of the radical[193]. The KSE on antioxidant activities can also be examined in terms of the *ratio* of rate constants in an alkane (non-H-bonding) solvent versus different solvents, k_{alkane}/k_s, which should in general be similar for an antioxidant independent of the nature of the attacking radical, even though absolute rate constants will differ. Table 9 shows some rate constants and k_{alkane}/k_s ratios for some phenolic antioxidants in various solvents, with various attacking radical species. For example, for α-Toc, the ratio $k_{alkane}/k_s = 2.7$(DPPH•) *vs* 2.8 (RO•) *vs* 2.5 (ROO•) when S = chlorobenzene; $k_{alkane}/k_s = 5.3$ (DPPH•) *vs* 4.9 (RO•) *vs* 4.5

(ROO$^\bullet$) when S = anisole and $k_\text{alkane}/k_\text{s}$ = 15 (DPPH$^\bullet$) vs 10 (RO$^\bullet$) vs 18 (ROO$^\bullet$) when S = acetonitrile[191,86]. There are sometimes discrepancies in using the rate ratios to examine the KSE, for example when using DPPH in *t*-butyl alcohol as discussed earlier[191,194], and it generally is used to examine only a few points, whereas plotting the log k values for antioxidant reactions with two different radicals provides more reliable information using more data points. That the KSE *is* independent of the attacking radical is the most important single discovery about solvent effects on hydrogen atom transfer reaction, and it has important implications for antioxidant activities determined in solution.

Recently, Snelgrove and coworkers derived an empirical relationship to describe quantitatively KSEs for hydrogen atom abstraction[195]. They found a linear correlation between the KSEs for a range of solvents and the hydrogen bond basicity values, β_2^H, of Abraham and coworkers[186], and also a simple linear correlation between the magnitude of the KSE for substrates, XH, and the α_2^H parameter of Abraham and coworkers[196], the ability of a substrate to act as a hydrogen bond donor, HBD. A combination of the linear relationships gave a general, empirical equation which can be used to predict KSEs for hydrogen atom donors (equation 44).

$$\log(k^\text{s}_{\text{XH/Y}^\bullet}/\text{M}^{-1}\,\text{s}^{-1}) = \log(k^\text{o}_{\text{XH/Y}^\bullet}/\text{M}^{-1}\,\text{s}^{-1}) - 8.3\alpha_2^\text{H}\beta_2^\text{H} \qquad (44)$$

So if the rate constant is measured in a non-hydrogen-bonding solvent, k°, one can predict the rate constant, k^s, in any other solvent by the use of equation 44.

Hydrogen bonding between the phenolic antioxidant and solvent is not straightforward when the structure of the antioxidant allows for internal hydrogen bonding, as outlined in Scheme 5[155]. In HB1, there is a linear hydrogen bond formed between the phenolic hydrogen and the solvent, and the OH group is twisted out of the plane of the aromatic ring by approximately 25° in the transition state. In HB2, the bond between solvent and phenolic hydrogen is no longer linear because the hydrogen is also involved in an internal hydrogen bond with an *ortho* HBA oxygen (in this case another hydroxyl group, although it could also be an alkoxy substituent). In HB3, only an internal hydrogen bond is illustrated. With an internal hydrogen bond, the OH is closer to the plane of the aromatic ring, with a dihedral angle of about 14.6° in the transition state. It is hydrogen bonding with the solvent that has the most influence on the magnitude of the KSE, because such an *inter*molecular hydrogen bond interferes with hydrogen abstraction[155]. If one tries to predict rate constants for hydrogen atom abstraction from a phenol capable of internal hydrogen bonding, based on the assumption that *intra*molecular hydrogen bonds also interfere with hydrogen abstraction, one discovers that the predicted rate is lower than the experimental rate. In other words, internal hydrogen bonds do not appear to prevent hydrogen atom abstraction in the way that hydrogen bonds to solvents do, and Ingold and coworkers[155] speculated it is due to the non-linearity of the hydrogen bond, which leaves the hydrogen still open to radical attack. The fact that phenols capable of *intra*molecular hydrogen bonds do still show KSE effects is evidence that the 'doubly' hydrogen-bonded complex, HB2, in Scheme 5 also exists, although for steric reasons hydrogen atom abstraction from this complex is unlikely.

Another potential site of hydrogen bonding interactions between solvent and antioxidant is at the *para*-ether oxygen of α-Toc analogs and 2,6-di-*tert*-butyl-4-methoxyphenol. Electrons from *para*-ether oxygen can assist through resonance stabilization of phenoxyl radicals formed after hydrogen atom abstraction (Scheme 4, part d). We had speculated that hydrogen bond formation between the solvent and the ether oxygen on the antioxidant could also affect the rate constant, by tying up electrons that would otherwise have stabilized the phenoxyl radical[55], however it has since been shown that hydrogen bonding at that site is *not* important[204]. Iwatsuki and coworkers[204] compared KSE for α-Toc and for a derivative of α-Toc in which the *para*-ether oxygen of the chroman ring was replaced

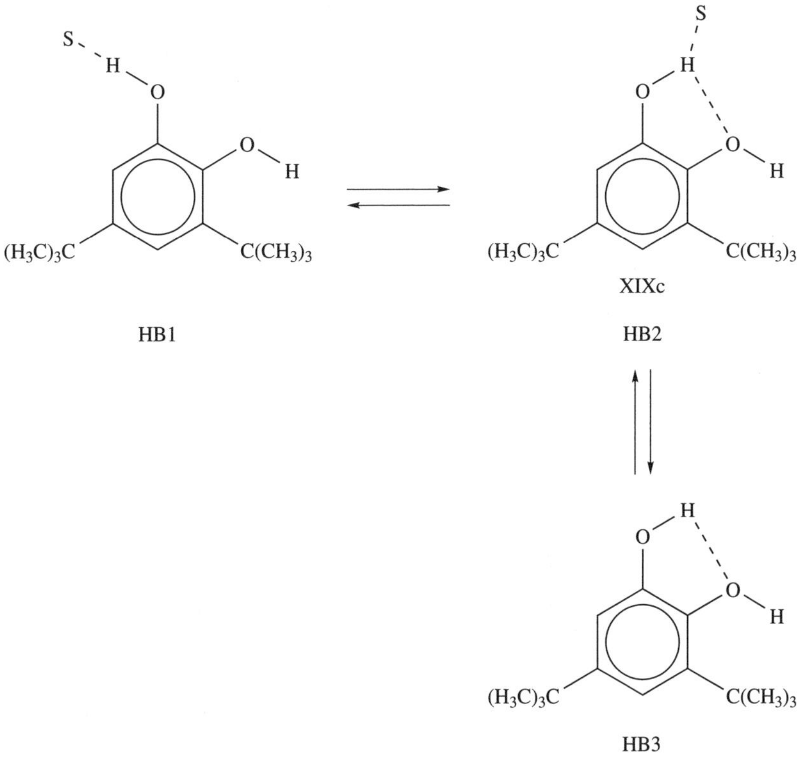

SCHEME 5. Hydrogen bonding interactions with 3,5-di-*tert*-butylcatechol, XIXc, and an HBA solvent[155]

with a CH$_2$ group (and the ring itself is five- rather than six-membered), and the addition of methanol decreased the rate constants for the two antioxidants to the same degree, indicating that the *para*-ether oxygen had no significant effect on the KSE.

Aside from the considerations of hydrogen bonding, solvent can have a 'physical' effect on the reaction rate, generally by steric or viscosity influences. As far as steric effects are concerned, as mentioned earlier in this section, if the approach of a peroxyl radical towards the hydrogen atom of the substrate (antioxidant or hydrogen bond donor, HBD) is sterically hindered either by solvent or antioxidant substituents, then the KSE will differ from the predicted solvent effect. To briefly summarize some of the influences: (a) a steric effect on rate was observed with hydrogen abstraction by DPPH•, where the bulky *tert*-butyl alcohol solvent actually enhanced the reaction rate by enclosing the DPPH• in a 'solvent cage'[191,194], (b) hydrogen bonding between solvent and substrate sterically interferes with the approach of the attacking radical species, cumyloxyl[192], to decrease the reaction rate, and (c) attempts to relate rate constants to β_2^H values are not always successful because the β_2^H values ignore steric considerations[188].

Changes in solvent viscosity can reduce reaction rates by reducing the rate at which reactants can diffuse towards each other. This was seen with α-Toc in homogeneous solution, where viscosity affected the α-Toc diffusion, resulting in $k_{\text{Toc/RO}}$• values which deviated from those predicted based on solvent polarity alone[191,205]. Solvent viscosity can also change the efficiency of radical production from the decomposition of thermal azo-initiators, by affecting the rate of solvent cage escape. For example, the efficiency at 65 °C for azo-bis-isobutyronitrile can vary from 0.60 (*tert*-butyl alcohol) to 0.80 (chlorobenzene) to 1.33 (acetonitrile)[206]. Consequently, changes in solvent will have a significant effect on the R_i, as seen from equation 6. Even high concentrations of unhindered phenols (so that they can be considered to be part of the solvent) can enhance the rate of peroxyl radical formation from the thermal azo-initiator AIBN[206], and can decrease the efficiency of the antioxidant because of complex formation between the antioxidant molecules[193,206].

2. Antioxidants in heterogeneous systems: Lipid peroxidation and inhibition in micelles and lipid membranes

Inhibition of peroxidation of unsaturated lipid chains in biomembranes is of particular significance and interest, because uncontrolled oxidation disrupts the protective layer around cells provided by the membranes. Furthermore, radical chain transfer reactions can also initiate damage of associated proteins, enzymes and DNA. The volume of literature is immense and expanding in the field of antioxidants. We will select certain *milestones* of advances where micelles and lipid bilayers, as mimics of biomembranes, provided media for quantitative studies on the activities of phenolic antioxidants. One of us, L. R. C. Barclay, was fortunate to be able to spend a sabbatical in Dr. Keith Ingold's laboratory in 1979–1980 when we carried out the first *controlled initiation* of peroxidation in lipid bilayers of egg lecithin and its inhibition by the natural antioxidant α-Toc[45]. A typical example of the early results is shown in Figure 4. The oxidizability of the bilayer membrane was determined in these studies, but we were not aware that phosphatidyl cholines aggregate into reverse micelles in non-protic solvents like chlorobenzene, so this determination was not correct in solution. This was later corrected by detailed kinetic and ^{31}P NMR studies, which concluded that the oxidizability of a lipid chain in a bilayer is very similar to that in homogeneous solution[207,208].

A second *milestone* of quantitative studies of lipid peroxidation in 1980 came from Porter and coworkers. They provided quantitative studies of lipid hydroperoxides found

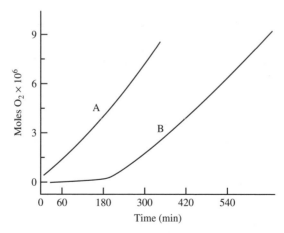

FIGURE 4. The first reported antioxidant profile for Vitamin E inhibited oxidation of egg lecithin in water, DBHN initiated showing (A) uninhibited and (B) inhibited oxygen consumption. Reprinted with permission from Reference 44. Copyright 1981 American Chemical Society

during controlled initiation of linoleate in solution and dilinoleoylphosphatidyl choline (DLPC) bilayers which showed how the *cis, trans* to *trans, trans* product ratios (kinetic to thermodynamic ratios) depended directly on the hydrogen atom donating ability of the medium, such as provided by an antioxidant like α-Toc[81,209-211]. These pioneering studies were the basis for others to use such product studies together with kinetic studies to examine the effects of antioxidants during peroxidation of linoleate chains in micelles and bilayers (*vide infra*).

Another *milestone* in the 1980s was the discovery that either water-soluble or lipid-soluble initiators with water-soluble or lipid-soluble phenolic antioxidants can be used for quantitative kinetic studies in micelles and lipid membranes[207,212]. This made measurements in these systems less difficult than before when the initiators were included in high concentrations in lipid membranes due to low initiator efficiency[45].

A fourth major advance towards quantitative studies of antioxidant activities in heterogeneous phases was the determination that the classical rate law of autoxidation (equation 7) is applicable to micelles and membranes. We showed that the kinetic order in substrate was unity and half order in R_i for varying linoleate concentrations and initiator in SDS micelles[47,213,214]. Similarly, the classical rate law was discovered to apply to phospholipid bilayers by using mixtures of unsaturated and saturated lipid systems[46,54,215]. Through the use of rotating sector experiments for micelles[47] and lipid bilayers[54], the absolute rate constants for propagation, k_p, and termination, $2k_t$, were determined in these systems. These advances set the stage for determinations of the absolute rate constants for antioxidant activities, k_{inh} of equation 15, in heterogeneous phase, in particular in the laboratories of Niki and Pryor. Examples of antioxidant activities are given in Table 10, so that comparisons can be made between homogeneous solutions, SDS micelles and lipid bilayers in terms of the effects influencing the antioxidant activities of phenols in solutions to heterogeneous systems.

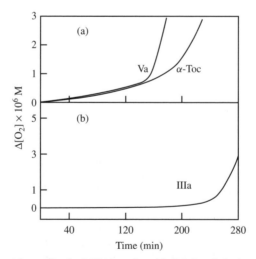

FIGURE 5. Oxygen uptake profiles for inhibition of azo-bis-2,4-dimethylvaleronitrile (ADVN) initiated oxidation of 5.7×10^{-5} mol DLPC at 37 °C. Va is 1.20×10^{-8} mol PMHC (0.95 µmol ADVN), an analog of α-Toc without the long phytyl tail, which, compared to 5.04×10^{-9} mol α-Toc (3.17 µmol ADVN), shows a much sharper break and more rapid return to uninhibited rate. IIIa is 9.74×10^{-9} mol DBHA (3.05 µmol ADVN). Reproduced by permission of NRC Research Press from Reference 55

TABLE 10. Antioxidant activities of phenolic antioxidants in different media

		k_{inh} (M^{-1} s^{-1} × 10^{-4})			
		Medium			
Structure	No.	Styrene/ $C_6H_5Cl^a$	$(CH_3)_3COH^b$	0.5 M SDS^c	Liposomes/ DLPC^b
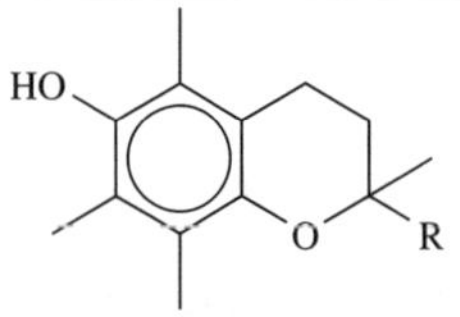 [HO-chromanol, R]	α-Toc, R = $C_{16}H_{33}$	320	23 (51)^d	3.7	0.58
	Va, R = CH_3	380	21	15	1.78
	Vc, R = COOH	110	15	11	0.58
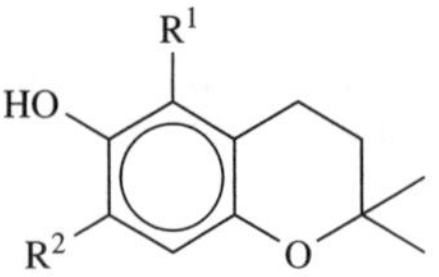 [HO-chromanol with R^1, R^2]	VIIIa, $R^1 = R^2 =$ $(CH_3)_2CH$	238			5.5
	VIIIb, $R^1 = CH_3$, $R^2 = (CH_3)_3C$	199			6.1
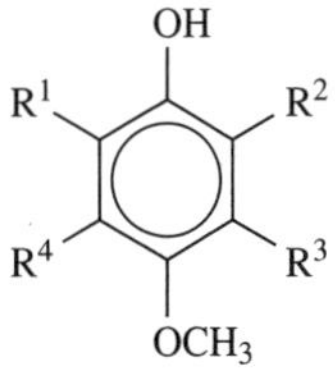 [phenol with OH, OCH$_3$, R^1-R^4]	IIc, R^1–R^3 = CH_3, R^4 = H	130			1.04
	IId, R^1–R^4 = CH_3	39			0.21
	IIIa, $R^1 = R^2 = (CH_3)_3C$, $R^3 = R^4 = H$	11			2.75
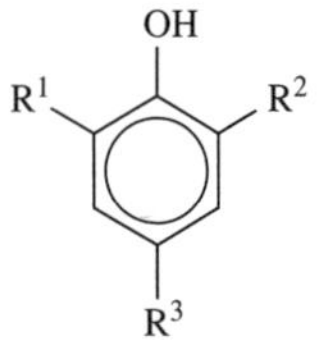 [phenol with OH, R^1, R^2, R^3]	IIIb, $R^1 = R^2 = (CH_3)_3C$, $R^3 = CH_3$	1.4		1.1	0.37
	IIa, R^1–R^3 = CH_3	8.5			0.056

^{a} Data taken from Tables 1, 2 and 3.
^{b} Data taken from Reference 55.
^{c} Data taken from Reference 216.
^{d} From the oxidation of methyl linoleate[233].

a. Monohydroxy phenols. Factors controlling antioxidant activities in membranes. The unique behavior of α-Toc. A qualitative comparison of the inhibition period provided by α-Toc with that of simple mono-phenols such as 2,6-di-*tert*-butyl-4-methoxyphenol (DBHA) in DLPC membranes (Figure 5) indicated at the outset that very important differences influence their activities in membranes compared to solution. That is, while DBHA is a relatively weak antioxidant in styrene oxidation (k_{inh} DBHA/α-Toc = 0.034, styrene), it appears to be far superior compared to α-Toc in aqueous DLPC (k_{inh} DBHA/α-Toc = 4.7, DLPC). This earlier data of k_{inh} in DLPC for the typical classes of monohydroxy phenols together with results in styrene/chlorobenzene, and available data in *tert*-butyl alcohol and SDS micelles are given in Table 10. For α-Toc, PMHC (Va) and Troloc (Vc), very significant decreases in activity were observed from styrene → *t*-butyl alcohol → SDS micelles → DLPC bilayers. Attempts were made to explain the drop in activity by hydrogen bonding of the phenolic hydroxyl by the polar protic media in alcohol or in the aqueous systems[55,216]. This was a reasonable interpretation at the time, since α-Toc was found to be located in bilayers of egg lecithin with its polar chromanol head group near the aqueous–lipid interface where the phenolic hydroxyl would be in contact with the aqueous phase[217]. Also, spin labeling studies show that α-Toc scavenges lipophilic radicals close to the membrane–aqueous surface[218]. The reduced activity of lipophilic phenols like α-Toc in aqueous SDS micelles was attributed to a limiting diffusion between micelles[136].

The lower antioxidant activity of α-Toc in bilayers cannot be due to hydrogen bonding by water alone because the hydrogen bond accepting ability of water, as measured by the value $\beta = 0.31$[205], is less than the value for *tert*-butyl alcohol, $\beta = 1.01$[219], and the kinetic data for α-Toc in *tert*-butyl alcohol and aqueous DLPC are *not* at all in agreement with this β parameter. In order to obtain measurable rates in bilayers, rather high concentrations of antioxidants were used (Table 10, about 10^{-4} M). This means that local concentrations of aryloxy radicals could initiate chain transfer reactions or pro-oxidant effects. In order to avoid such effects we determined the antioxidant activity of α-Toc in palmitoyl, linoleoyl phosphatidyl choline (PLPC) bilayers by an independent method of product analyses[220]. This method involved (1) determination of the *cis/trans* to *trans/trans* (*c,t/t,t*) ratio of the 9- and 13-linoleate hydroperoxides as the membrane concentration of PLPC varied, and (2) the variation of this *c,t/t,t* ratio for various α-Toc concentrations in the presence of excess ascorbate and homocysteine to keep the antioxidant in its reduced form. The k_{inh} value for α-Toc in PLPC bilayers by this method was 4.7×10^4 M^{-1} s^{-1}, which better reflects its 'intrinsic' activity in biomembranes. This value represents a factor of 68 times less active in lipid membranes than in styrene, while it was estimated that the actual effect due to hydrogen bonding by water should reduce the activity of α-Toc by only 3.9 times[205]. The larger drop in activity is probably due to some 'unique behaviors' of α-Tocs in lipid membranes. For example, it was suggested that in mixed water–lipid systems, the small magnitude of $k_{\alpha\text{-Toc/ROO}}$• is due to non-uniform distribution of the α-Toc so that much of it was 'physically inaccessible' to the lipid peroxyl radicals[205]. This would explain why induction periods found for α-Toc in bilayers typically do not give sharp breaks like those observed for other chromanols like PMHC (Va in Figure 5) or even DBHA. Some of the lipid particles may not contain α-Toc, so they undergo rapid oxidation while oxidation is completely suppressed in others. The result is that the rate does not return quickly to the uninhibited rate. There is other evidence that diffusion of α-Toc between and within lipid bilayers is limited. Earlier, Niki and coworkers discovered that the phytyl side tail enhances the retainment of vitamin E in liposomes so that it did not transfer, as other smaller molecules do (e.g. PMHC), between liposomes[221]. Later, Kagan and coworkers observed intermembrane transfer of α-Toc, but transfer was incomplete and it did not transfer to give a homogeneous distribution[222]. Barclay and coworkers found that α-Toc transferred only very slowly from a water-soluble protein complex into liposomes[52], and quantitative studies showed that it took nearly ten hours for it to transfer completely from saturated liposomes into DLPC liposomes[60]. In contrast, chromanols like PMHC transferred readily between liposomes and this was a very efficient method to incorporate such antioxidants into liposomes for determination of antioxidant activities compared to the more conventional co-evaporation from solvents (see Figure 6)[60].

There is independent physical evidence for non-uniform distribution and restriction from transmembrane diffusion of α-Toc in lipid membranes. Differential scanning calorimetry results indicated that it partitioned into the most fluid domains in lipid vesicles[223]. Fluorescence studies showed that α-Toc has a very high lateral diffusion rate in egg lecithin[224] but it does not take part in transbilayer (flip-flop) migration even over 'many hours'[225]. It is not known if this behavior of α-Toc extends to natural biomembranes where actual structures and conditions may dramatically change migration phenomena.

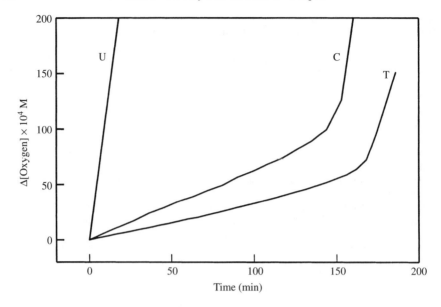

FIGURE 6. Comparison of oxygen uptake profiles for co-evaporated and transferred PMHC during oxidation of DLPC bilayers at 37 °C, pH 7.0, initiated with 2.8–3.0 μmol ADVN: U = uninhibited oxidation of DLPC, C = 15.0 nmol PMHC co-evaporated with 5.76×10^{-5} mol DLPC, T = 9.44 nmol PMHC transferred into liposomes containing 7.43×10^{-5} mol DLPC. Reproduced by permission of Elsevier Press from Reference 60

For example, flip-flop transfer of phospholipids in membranes is usually also very slow, from hours to days. However, there are exceptions: for example, phosphatidylethanol undergoes rapid and reversible transbilayer distribution in unilamellar PC vesicles in the presence of multivalent cations, including calcium[226].

Electrostatic Effects: Membranes and Antioxidants. Phospholipid bilayers bearing surface charges, such as negatively charged phosphatidyl acids and phosphatidyl glycerol, are significant mimics of charged natural membranes. We found that charged water-soluble antioxidants like ionized Trolox (COO⁻), **38**, and 2,5,7,8-tetramethyl-2-(β-trimethylammoniummethyl)-chromanol, **39**, exhibit some unique behavior as antioxidants on charged bilayers compared to zwitterionic ones.

(38)   (39)

Ordinary Trolox is an effective antioxidant for inhibition of peroxidation in micelles and lipid bilayers. This was attributed in part to its partitioning between the aqueous and lipid phases of PC membranes, according to ^{14}C tracer studies[58]. It was proposed that Trolox traps peroxyl radicals near the lipid–water interface because peroxyl radicals may diffuse towards the aqueous phase due to their high polarity, as illustrated in Figure 7[45]. It is not surprising to find that Trolox does not function as an antioxidant at pH = 7 and when the bilayer contains a surface with negatively charged groups such as phosphatidyl glycerols, whereas the positively charged antioxidant, **39**, is very effective under these conditions[56]. Natural biomembranes having charged head groups exhibit important interactions with other constituents, such as proteins[56] and it is important to elucidate the efficiency of water-soluble antioxidants in these systems.

Synergistic Effects between Antioxidants. A synergistic effect operates between antioxidants when the total inhibition period observed when two (or more) antioxidants are present is greater than the sum of the inhibition periods when they act singly. Synergism between two phenolic antioxidants during hydrocarbon oxidation was observed by Mahoney and DaRooge[227]. The conditions and the magnitude of synergism with two phenols, such as a hindered, BH, phenol and a non-hindered one, AH, are reviewed by Mahoney[13]. In particular, the most important factor is the rate of regeneration of the non-hindered phenol (equation 45) compared to chain transfer reactions that may be started by reaction of the non-hindered A$^{\bullet}$ with the substrate or hydroperoxides that are formed.

$$A^{\bullet} + B-H \longrightarrow A-H + B^{\bullet} \qquad (45)$$

The 'reinforcing action' of ascorbic acid with α-Toc during inhibition of oxidation of fats was observed by Golumbic and Mattill as early as 1941[228]. The regeneration of α-Toc (α-Toc) from the α-To$^{\bullet}$ by reduction with vitamin C (ascorbate) has attracted a great

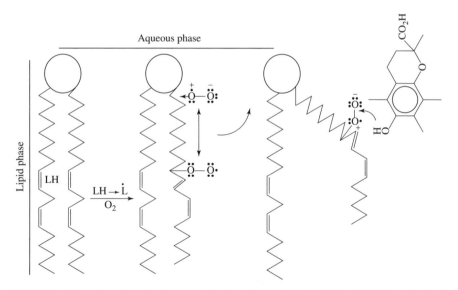

FIGURE 7. Illustration of the floating peroxyl radical theory, where the lipid chain (LH = dilineloyl phosphatidyl choline) bearing the polar peroxyl radical migrates towards the polar surface of the liposome, where it can interact with water-soluble antioxidants like Trolox

deal of interest since Tappel proposed in 1968 that the nutritional relationship between these vitamins could be explained if vitamin C reduced the oxidized form of vitamin E *in vivo*[229]. Since 1968 there have been kinetic[230,231] and spectroscopic evidence[232] to show that vitamin C does in fact regenerate α-Toc from the α-To$^{\bullet}$ radical in solution, and during inhibited oxidation of methyl linoleate by combinations of α-Toc and vitamin C[233]. In 1983 we reported synergism between α-Toc and vitamin C during inhibited peroxidation of linoleic acid in the biphasic system of SDS micelles[213] and quantitative studies in micelles showed that vitamin C regenerates a mole of α-Toc (or Trolox) per mole of vitamin C introduced[214]. The next year ESR results showed that ascorbate recycled α-Toc from α-To$^{\bullet}$ in DLPC liposomes[234]. In 1985, two independent reports appeared to demonstrate that vitamin C acts synergistically with α-Toc during peroxidation of phosphatidylcholine membranes in aqueous dispersions[212,235]. These reports are of particular interest because they showed that ascorbate, which resides in the aqueous phase, is able to regenerate α-Toc from the α-To$^{\bullet}$ radical across the interface in the hydrophobic phase of membranes. In addition to vitamin C, other natural hydrogen atom donors are known to act synergistically with α-Toc, such as cysteine[236,237], although glutathione appears to react cooperatively, not synergistically, during inhibited peroxidation of DLPC liposomes[53]. Natural thiols, such as homocysteine or glutathione, are known to regenerate ascorbic acid from dehydroascorbic acid and it was found that combinations of the two inhibitors, thiols and ascorbate, interact with a phenolic antioxidant during inhibited peroxidation of linoleate in micelles, to extend the inhibition further than any two combined[238]. These results provided evidence for a 'cascade' of antioxidant effects as illustrated in Scheme 6. It is possible that interactions observed between endogenous antioxidants in human blood plasma[239] or in rat hepatocytes[240] involve cascades of this type.

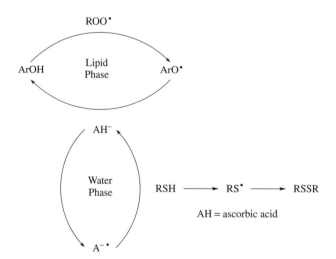

SCHEME 6. Cooperative cascade of three antioxidants during lipid peroxidation

These various *in vitro* synergistic interactions between vitamins E and C can be summed up in equations 46–48.

$$ROO^{\bullet} + \alpha\text{-Toc} \longrightarrow ROOH + \alpha\text{-To}^{\bullet} \qquad (46)$$

$$\alpha\text{-To}^\bullet + \text{AH}^- \text{ (ascorbate)} \longrightarrow \alpha\text{-Toc} + \text{A}^{-\bullet} \quad (47)$$

$$\text{A}^{-\bullet} + \text{A}^{-\bullet} \xrightarrow{\text{H}^+} \text{A} + \text{AH}^- \quad (48)$$

While there is much clear and convincing evidence for this synergistic interaction between these two vitamins *in vitro*, such compelling evidence is lacking to date *in vivo*. Indeed, at least one detailed study using guinea pigs showed that vitamin C does 'not' spare vitamin E *in vivo*[241]. So one cannot immediately assume that the laboratory *in vitro* findings on interactions between antioxidants are applicable to living systems.

Pro-oxidant Effects of Antioxidants. Our review to this point has concentrated on the beneficial effects of phenolic antioxidants through their efficiency in trapping damaging oxygen-centered radicals. However, there are limitations in these beneficial effects. Under certain conditions, phenols which normally act as antioxidants can display pro-oxidant activity. This pro-oxidant effect can be attributed to two quite different phenomena:

(1) Phenoxyl radicals formed in the inhibition step (equation 10) are normally terminated by rapid reaction with peroxyl radicals (equation 11). However, phenoxyl radicals, particularly unhindered ones, are also able to participate in chain transfer reactions by hydrogen atom abstraction from hydroperoxides which build up (equation 21), which is the reverse of equation 10, or initiate new reaction chains by hydrogen atom abstraction from substrate ($R_s H$) (equation 20).

(2) In certain 'media', even normally unreactive aryloxyl radicals participate in these so-called pro-oxidant reactions due to local high concentrations of ArO$^\bullet$ or certain physical restrictions which prevent their termination by radical–radical reactions.

It has been known for some decades that phenoxyl radicals will initiate hydrogen atom abstraction in solution from hydrocarbons and they exhibit high selectivity; e.g. the 4-methoxyphenoxyl radical is more selective than phenoxyl[242]. More recently, the unsubstituted phenoxyl radical, $C_6H_5O^\bullet$, was discovered to possess 'surprisingly high reactivity', being approximately 100–300 times more reactive than peroxyl radicals, on hydrogen atom abstraction from phenols[112]. Storozhok and coworkers[243] used pulse radiolysis methods to estimate rate constants for hydrogen atom abstraction by the α-tocopheroxyl radical (Ar-O$^\bullet$) from lipids and reported that rate constants $k_{\text{effective}}$ varied with the degree of unsaturation but their '$k_{\text{effective}}$' values do not appear reliable by this method. Nagaoka, Mukai and coworkers[244,245] reported rate constants for hydrogen atom abstraction from fatty acid esters by several 5,7-dialkyltocopheroxyl radicals using stopped flow methods, that also depended on the degree of unsaturation, with varying $k_{\text{abstr.}}$ being 1.04×10^{-5}, 1.82×10^{-2}, 3.84×10^{-2} and 4.83×10^{-2} M^{-1} s^{-1} for oleate, linoleate, linolenate and arachidonate, respectively. The $k_{\text{abstr.}}$ rate constants per active H were lower in *t*-butyl alcohol compared to benzene, but large in Triton X-100 compared to *t*-butyl alcohol. The $k_{\text{abstr.}/\text{H}}$ values were approximately the same in benzene and in *t*-butyl alcohol for lipids containing 2, 3, 4 and 6 double bonds. However, $k_{\text{abstr.}/\text{H}}$ actually decreased along this series. This interesting effect in the micelles was attributed to local restriction of motion between the attacking radicals and the 'tail' of a polyunsaturated lipid chain. They reported that rate constants for hydrogen atom abstraction from alkyl hydroperoxides (equation 21) by *these* α-To$^\bullet$ radicals were approximately an order of magnitude larger[246] and the value for abstraction from linoleate hydroperoxide was 2.5×10^{-1} M^{-1} s^{-1}[247]. It is expected that *the* α-To$^\bullet$ radical would be less reactive than $C_6H_5O^\bullet$ towards hydrogen abstraction. There is some evidence to support this; hydrogen atom abstraction from ubiquinol by 5,7-diethyl-To$^{\bullet}$[21] in hexane is estimated to be at least two orders of magnitude less than the value ($8-9 \times 10^7$ M^{-1} s^{-1}) using $C_6H_5O^\bullet$ in benzene.

During the inhibited self-initiated autoxidation of methyl linoleate by α-Toc in solution, Niki and coworkers[71] made the interesting observation that α-Toc acts as an antioxidant at low concentrations, but high concentrations (up to 18.3 mM) actually increased hydroperoxide formation due to a pro-oxidant effect. The pro-oxidant effect of α-Toc was observed earlier by Cillard and coworkers[248] in aqueous micellar systems and they found that the presence of co-antioxidants such as cysteine, BHT, hydroquinone or ascorbyl palmitate 'inverted' the reaction into antioxidant activity, apparently by reduction of α-To$^•$ to α-Toc[249]. Liu and coworkers[250] found that a mixture of linoleic acid and linoleate hydroperoxides and α-Toc in SDS micelles exhibited oxygen uptake after the addition of α-Toc. The typical ESR spectrum of the α-To$^•$ radical was observed from the mixture. They attributed the rapid oxidation to decomposition of linoleate hydroperoxides, resulting in the formation of linoleate oxy radicals which initiated reactions on the lipid in the high concentration of the micellar micro-environment. Niki and coworkers reported pro-oxidant activity of α-Toc when it was added with metal ions, Fe^{3+} [251] or Cu^{2+}, in the oxidation of phosphatidyl choline liposomes. α-Toc was found to reduce the metal ions to their more reactive valence states. These in turn reacted with hydroperoxides to give reactive alkoxyl radicals which accelerate the oxidation (equations 49–51).

$$Cu^{2+} + \alpha\text{-Toc} \longrightarrow Cu^{1+} + \alpha\text{-Toc}^{•+} \qquad (49)$$

$$\alpha\text{-Toc}^{•+} \longrightarrow \alpha\text{-To}^{•} + H^+ \qquad (50)$$

$$ROOH + Cu^{1+} \longrightarrow RO^{•} + Cu^{2+} + OH^- \qquad (51)$$

These observations on the pro-oxidant behavior of the antioxidant (normally) α-Toc in micelles and lipid membranes provide some insight into the remarkable pro-oxidant activity of α-Toc in low density lipoprotein, LDL, under *in vitro* conditions. The radical initiated oxidation of LDL is of great interest because it is implicated in heart disease. The tocopherol-mediated peroxidation (TMP) of LDL was reported on in detail by Bowry and Stocker[252] and is also the subject of timely, detailed reviews[185,253]. This interesting story will not be reviewed again here, but these articles are highly recommended since they provide important insight into the 'unique behavior' of α-Toc.

b. Di- and polyhydroxy phenols in membranes. Ubiquinols and flavonoids. The ubiquinols (**40**, UQH_2) consist of a series of 2,3-dimethoxyhydroquinones which differ in the number of isoprenoid (C5) units in a side chain. As their name implies, the ubiquinols are widely distributed in nature, especially the $n = 6-10$ types, as are the corresponding *para*-quinones which are referred to as Coenzyme Q, CoQ. Ubiquinol is an effective scavenger of peroxyl radicals during lipid peroxidation and can regenerate α-Toc in lipoproteins and other lipid membranes[1,155]. Thus the ubiquinols are of great interest as antioxidants in LDL[185,252-254]. As pointed out in Section III.C.1, the internal hydrogen bond in *ortho*-methoxyphenols is an important factor in their antioxidant activity. In particular, internal hydrogen bonding from the phenolic hydrogen to an *ortho*-methoxy does not prevent hydrogen atom abstraction by oxygen-centered radicals in the same way that external hydrogen bonds to solvents do[155]. To illustrate the difference, de Heer and coworkers reported that the rate constant for hydrogen atom abstraction by *tert*-butoxyl from 4-methoxyphenol, which is susceptible to external hydrogen bonding only, in *tert*-butyl alcohol is only about 2 percent of the value in hexane, whereas for ubiquinol-0 the drop is to 20 percent[155]. Consequently, ubiquinol has eight to nine times the activity in *tert*-butyl alcohol compared to the externally hydrogen-bonded 4-methoxyphenol.

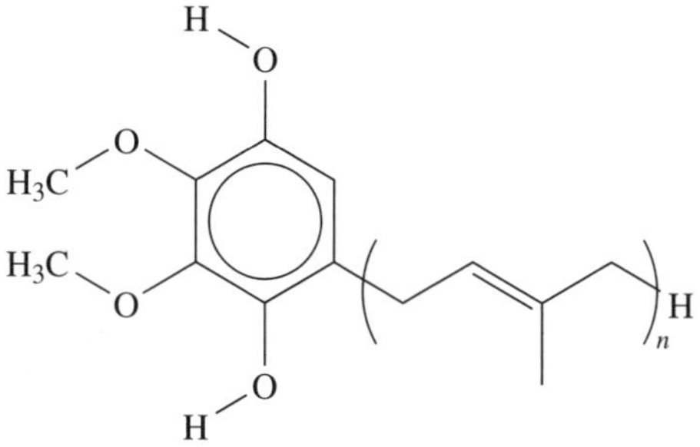

(40)

The antioxidant properties of the ubiquinols in lipid membranes as well as in solution and of other 'biological hydroquinones' (e.g. of the α-Toc hydroquinone class) is the subject of a recent review[183]. We will add only two points by way of emphasis at this time, concerning the relationship between experimental conditions, structure (hydrogen bonding) and the antioxidant efficiencies of the ubiquinols. First, Niki and coworkers[255] re-examined the antioxidant properties of ubiquinol-10 compared to α-Toc during peroxidation of linoleate in solution and during oxidation of liposomes. The results showed that α-Toc was more effective as an antioxidant than ubiquinols in acetonitrile solution, but their antioxidant properties were similar in membranes and micelles. However, in liposomes the stoichiometric factor (n) for ubiquinol was typically less than 1, being around 0.61, whereas on hydrogen atom donation to the galvinoxyl radical in solution the n factor for ubiquinols UBH_2 was 2 (while for α-Toc n was 1). The low n factor for ubiquinol in oxidation by lipid peroxyl radicals LOO$^\bullet$ was interpreted in terms of a competition with its autoxidation. When the attacking radicals are in comparative low concentration, as is usual for the steady-state concentration of peroxyl radicals during autoxidation, oxygen may compete with the reactive semiquinone radical, UQH$\bullet$, forming ubiquinone, UB according to the sequence outlined in equations 52–55.

$$LO_2^\bullet + UQH_2 \longrightarrow LOOH + UQH^\bullet \qquad (52)$$

$$UQH^\bullet + O_2 \longrightarrow UQ + HO_2^\bullet \qquad (53)$$

$$HO_2^\bullet + LH \longrightarrow H_2O_2 + L^\bullet \qquad (54)$$

$$L^\bullet + O_2 \longrightarrow LO_2^\bullet \qquad (55)$$

Secondly, we consider again the effect of structure, in particular hydrogen bonding with adjacent methoxy groups. It has been suggested by others that the effect of the *ortho*-methoxy groups is to decrease the antioxidant activity of the 1,4-hydroquinone system due to a decrease in the stereoelectronic effect of these groups because they are expected to become non-planar with the aromatic ring[182,183]. Calculations by de Heer and coworkers[153] do show that the methyls of the two *ortho*-methoxyl groups are tilted out of the phenyl plane, but the non-planarity of the methoxy groups has little if any impact on the strength of the hydrogen bonds. This conclusion was recently confirmed by observing the FTIR spectrum of UQ-0 in CCl_4 which showed only one absorption band at 3554 cm^{-1}, indicating that both phenolic groups are hydrogen-bonded to methoxyls, since this absorption appeared in this region for *ortho*-methoxyphenol at 3558 cm^{-1}[256]. Consequently, we propose again that it is intramolecular hydrogen bonding that lowers the reactivity of ubiquinols in organic solvents but provides protection against intermolecular hydrogen bonding in aqueous dispersions, and thus antioxidant activity is significant in lipid membranes.

Flavonoids as antioxidants have been reviewed several times[161,257,258], including an outline of many claims to their beneficial health effects[259]. Due to their complex structures and different classes (eight thousand different compounds are known[258]), researchers often resorted to qualitative screening methods to evaluate their antioxidant potentials in mixed aqueous/lipid phases. For example, the so-called Trolox equivalent antioxidant capacity (TEAC), the concentration of Trolox with 'equivalent antioxidant activity' of a 1 mM concentration of the substrate, is frequently used in heterogeneous systems. Unfortunately, this can be an unreliable measure of the activity of the substance, especially if initiation is also carried out in the aqueous phase. Nevertheless, there have been some efforts made to evaluate antioxidant activities of specific flavonoids using more quantitative methods in heterogeneous systems in order to mimic natural environments. A few examples are cited below to illustrate some approaches to determine flavonoid activities in micelles or lipid membranes.

Several groups have used aqueous micelles as the heterogeneous media for determining the activity of flavonoids[260–262]. Mukai and coworkers[260] determined the effect of pH on the hydrogen atom donating ability of quercetin and rutin (see Chart 1, rutin is the rutinose derivative of quercetin at position 3) to their hindered ArO• radical (k_s) and the 5,7-diisopropyltocopheroxyl, Toc• (k_r) in Triton X-100 micelles. The values of both k_s and k_r increased with increasing pH 7–10, $k_s = 2.28 \times 10^2 – 3.89 \times 10^3$ and $k_r = 5.48 \times 10^2 – 3.38 \times 10^5$ $M^{-1} s^{-1}$ for rutin, and $k_s = 3.73 \times 10^4 – 3.38 \times 10^5$ $M^{-1} s^{-1}$ for quercetin at pH 8–10. Calculations were made of k_s values for the different ionic species of rutin at different pK_a levels. From the dependence on pH, they concluded that the reaction rates increased with the electron-donating ability of the flavonoids. Roginsky and coworkers[261] reported that typical flavonoids like quercetin did not behave as classical phenolic antioxidants during azo-initiated peroxidation of methyl linoleate in chlorobenzene nor in 0.2 M SDS micelles, and the classical rate law for autoxidation was not followed. Contrary to other investigators, they found that flavonoids showed only 'moderate' chain-breaking activity. For example, in chlorobenzene, quercetin was less active even than BHT. However, in SDS micelles they reported higher relative activities, where that for quercetin was about 38 percent of the value for α-Toc. Various explanations were offered for the non-classical behavior of flavonoids as antioxidants, including their pro-oxidant effects. Foti and coworkers[262] used the spectral method reported by Pryor and coworkers[67] to determine the 'relative antioxidant efficiencies (RAE)' of ten flavonoids in 0.1 M SDS micelles containing linoleic acid. Quercetin, the most active by this method, gave an RAE 90% of α-Toc, whereas Pryor and coworkers[67] reported 19% by this method. The lower activity of other flavonoids (e.g. catechin, **36**, RAE = 22) was attributed to the lack of conjugation with ring C (see Chart 1) as suggested for similar structural effects on flavonoid activities observed in solution (*vide supra*)[262].

Although phospholipid bilayers are better mimics of biomembranes than are micelles, there are few reliable quantitative data on flavonoid antioxidant activities in lipid bilayers. Terao and coworkers[263] compared the antioxidant efficiency of quercetin and catechins (epicatechin and epicatechin gallate) with that of α-Toc in egg yolk PC liposomes using initiation by the water-soluble initiator, ABAP, and analysis of hydroperoxide formation and antioxidant consumption by HPLC. Based on the length of the induction periods and the profile of suppressed hydroperoxide formation, they concluded that quercetin and the catechins were more efficient antioxidants than α-Toc in these bilayers. Apparently the 'unique behavior' of α-Toc in bilayers is responsible for these results (*vide supra*). In hexane and alcohols solution during suppressed peroxidation of methyl linoleate, the relative antioxidant activities reversed so that the flavonoids were 5–20 times less active than α-Toc. Arora and coworkers[264] used a fluorescent probe to determine the antioxidant efficiencies of flavonoids in 1-stearoyl-2-linoleoyl PC vesicles during initiation by ABAP. They found *t*-butylhydroquinone to be more effective as an antioxidant and quercetin to be more active among the six flavonoids examined. A number of groups used ferrous or ferric ion as initiator in aqueous/bilayer systems[265–267]. However, it is not clear whether the effects observed with flavonoids are due to radical scavenging or iron chelating properties and we have not reviewed these in detail. The water-soluble flavonoid gluconide, isoorientin-6″-*O*-glucoside, inhibited the copper ion initiated peroxidation of LDL and may be useful in 'antioxidant therapy'[268].

To sum up the state of antioxidant efficiencies of the flavonoids, a few general conclusions can be reached.

(1) The structural features responsible for their antioxidant properties: (a) the catechol structure in ring B is most important, (b) the 2,3-double bond in conjunction with the 4-oxo function provides additional, beneficial electron delocalization and (c) coplanarity of the system is beneficial and the 3-hydroxyl group may help lock a coplanar configuration.

(2) Reliable quantitative studies of flavonoid antioxidant activities in model membranes are lacking.

(3) Flavonoids can exhibit pro-oxidant activities. This could be due to redox cycling of semiquinones, which is well known. Also, we point out that isolated phenolic groups in ring A could form reactive phenoxyl radicals by chain transfer processes and contribute to pro-oxidant effects, especially during local high concentrations. In any event, the enthusiasm for the incorporation of large quantities of flavonoids in the diet of humans should be tempered with the knowledge that they can have mutagenic effects[269].

IV. CHEMICAL CALCULATIONS ON PHENOLS

A. Introduction

In Sections I–III our review of antioxidant activities is based entirely on *experimental* observations with interpretations of relative activities based upon classical concepts of electronic and steric effects operating on phenoxyl radicals. In the last decade there have been some applications of chemical calculations on phenols as antioxidants, with applications to interpretation on known phenols and extension of this to predictions of activities of novel molecules. In general, the thrust of the theoretical approaches endeavor to: (1) clarify the antioxidant mechanism of phenols, (2) calculate antioxidant activities and (3) make predictions on potentially new antioxidants.

Four kinds of quantitative information are useful for evaluating (or predicting) antioxidant activities of phenols: Bond dissociation energies (BDE) of phenolic −O−H bonds, ionization potentials (IP) of phenols or one-electron reduction potentials, $E°$, and overall molecular geometry. The antioxidant activities of substituted phenols can be related to the bond dissociation enthalpies of the O−H bonds since the weaker the O−H bond, the more rapidly it will donate the hydrogen atom to an attacking radical. For different phenols, the BDE is influenced by electron-donating and electron-withdrawing substituent effects, steric effects and hydrogen bonding of the OH group. Various strategies are used to obtain useful BDE values, including theoretical calculations using 'full basis methodology', and locally dense basis sets (LDBS) as described by Wright and coworkers[154]. Others have been able to use simple empirical correlations of known BDE values with Brown–Okamoto σ^+ Hammett values for substituents to calculate BDE for a molecule with 'unknown' BDE (see Section IV.B.2).

The ionization potential (IP) of phenols is a measure of how readily an electron can be donated from the OH group to yield the phenolic cation. As pointed out before[154], the IP is related to the energy of the HOMO and the global molecular geometry, therefore a full-basis calculation is used for both the phenol and cation. The IP values are important, since they may indicate how readily a phenol will enter into the single electron transfer (SET) mechanism (*vide infra*).

The one-electron reduction potential ($E°$) also provides redox information to predict the direction of free radical processes since the change, $\Delta E°$, indicates the position of the equilibrium[270]. For example, if one considers hydrogen atom abstraction from polyunsaturated lipids (PUFA-H) by peroxyl radicals (PUFA-OO$^{\bullet}$), information on the two redox couples can be used, where for PUFA-OO$^{\bullet}$, H$^+$/PUFA-OOH, $E° = 1000$ mV and for PUFA$^{\bullet}$, H$^+$/PUFA-H, $E° = 600$ mV, so the reaction in equation 56

$$\text{PUFA—OO}^{\bullet} + \text{PUFA—H} \longrightarrow \text{PUFA—OOH} + \text{PUFA}^{\bullet}, \Delta E° = +400 \text{ mV}$$

(56)

is favorable and (of course) so is the reaction in equation 57.

$$\text{PUFA—OO}^{\bullet} + \alpha\text{-Toc} \longrightarrow \text{PUFA—OOH} + \alpha\text{-To}^{\bullet}, \Delta E° = +500 \text{ mV}$$

(57)

The $\Delta E°$ do not provide data on the reaction rates which will depend on the free energy of activation. It is well known that the rate constant for reaction of peroxyl radicals with α-Toc (equation 57) is much larger than the chain propagation rate constant (equation 56).

Calculations of molecular geometry are important when intramolecular hydrogen bonding is involved, and more complex, polycyclic molecules are being considered for calculations, such as the flavonoids. Some specific examples will be reviewed briefly to illustrate how the various methods have been applied to elucidate the mechanisms and in particular predict the effects of substituents and overall structure on the antioxidant activities of phenols.

B. Application to Antioxidants

1. Antioxidant mechanisms by phenols: Hydrogen atom transfer (HAT) and single electron transfer (SET)

It would be very significant if theoretical methods could resolve the question of the antioxidant mechanism, HAT or SET, for a given antioxidant under known conditions. An attempt to do this for a different reaction, that of formation of substituted benzylic radicals from *para*-substituted toluenes, concluded that radical cation formation (SET) is subject to strong substituent effects whereas hydrogen atom transfer is 'mainly independent' of the nature of the substituent[271]. A decision on the mechanism might be made on this basis. It is true that calculated ΔIP values relative to phenol for α, β, γ and δ tocopherols, of -36.1, -33.6, -32.9 and -30.5 kcal mol^{-1}, are significantly higher than the corresponding ΔBDE values of -11.3, -9.4, -8.9 and -7.3 kcal mol^{-1}[154]; however, these trends can be used to support either mechanism! It has been suggested as cut-off values, that up to ΔIP of 36 kcal mol^{-1} and for ΔBDE of -10 kcal mol^{-1}, the mechanism is *dominated* by hydrogen atom transfer in aqueous solution, whereas for ΔIP > -45 kcal mol^{-1} the mechanism is predominantly SET[154]. As already pointed out, solvent polarity (see Section III.C.1) may be the deciding factor about the determination of the predominant pathway. As we pointed out in Section III.B.1, Mukai and coworkers[146-148] interpreted experimental results from deuterium isotope effects and correlation of rate constants with ionization and activation energies in terms of a antioxidant mechanism involving a charge transfer complex and proton tunneling.

When considering effects which may promote either the HAT or SET mechanisms, attention should also be paid to the attacking radicals, especially peroxyl radicals. The reactivities of peroxyls are strongly influenced by substituents on the alkyl or aryl group[150]. Recent calculations on solvated peroxyl radicals by water showed a strong increase of the dipole moment of alkyl peroxyls in water, indicative of quite high polarizability[151]. We are not aware of such studies on other oxygen-centered radicals such as hindered aryloxyls, but speculate that polar solvent effects should not be as significant as with peroxyls due to their polarity.

2. Calculations of substituent effects for monophenols

Some empirical methods have provided useful correlations concerning the effects of substituents on thermochemical properties of phenols. Griller and coworkers[272] developed a photoacoustic method for measuring bond dissociation energies (BDE) of phenols and showed for the first time a linear relationship between the Hammett σ^+ *para*-substituent constant and BDEs. Wayner and coworkers[273] found a correlation between experimental ΔBDE values for a series of substituted phenols compared with phenol and the Hammett σ^+ constants (equation 58).

$$\Delta\text{BDE(O} - \text{H) kcal mol}^{-1} = 7.32[\Sigma(\sigma_o^+ + \sigma_m^+ + \sigma_p^+)] - 0.64 \tag{58}$$

This relationship was then used to calculate the BDE for α-Toc, giving a value of 77.2 kcal mol^{-1}, in excellent agreement with the experimental value of 77.3 kcal mol^{-1}. This empirical method does depend on the electronic effects of groups (methyls and *para*-ether) around the phenyl ring and provides some confirmation of the role these play in weakening the O—H bond and thus raising the antioxidant activity. In a similar manner, Jovanovic and coworkers obtained a correlation between the measured reduction potentials and the σ^+ constants for twenty-one substituted phenols at pH 7 (equation 59) and pH 0 (equation 60)[274].

$$E_7 = 0.95 + 0.31\sigma^+ \tag{59}$$

$$E_0 = 1.34 + 0.32\sigma^+ \tag{60}$$

From equation 59, the derived reduction potential of the phenoxyl radical is 0.95 V and $\rho = 0.31$. This calculated value E_7 was in good agreement with the experimental value of 0.97 V. They noted in particular that strong electron-donating substituents (having negative σ^+ values) reduced the redox potential of the phenols and increased their efficacy as antioxidants. Strong electron-withdrawing substituents (high positive σ^+ values) increased the redox potential, 'disqualifying' such phenols as antioxidants.

Actual *theoretical* calculations of the O—H bond strengths of a group of 35 phenols using density functional theory (DFT) were reported in 1997 by Wright and coworkers[275]. More recently, the calculations were extended to other phenols and to calculations of ionization potentials (IP)[154]. In the earlier report, an additivity scheme was found for most methyl and methoxy phenols but not for 3,5-dimethyl-4-methoxyphenol nor for 2,3,5,6-tetramethyl-4-methoxyphenol. However, this was readily explained by the effect of two *meta* methyls which forced the *para*-methoxy group out of plane, which almost eliminates the normal substituent effect of this group. Thus the BDE results were in agreement with antioxidant activities reported earlier[18] (see also Section III.B.1). It is also interesting to note that the BDE values for the dihydrobenzofuranols were 1.0 kcal mol^{-1} smaller than for the corresponding chromanols (for structures see Table 2). This added further support to the stereoelectronic explanation for the high antioxidant activities of these furanols and chromanols of the vitamin E class[18].

It was found that calculations of BDEs for substituted phenols using the less rigorous locally dense basis sets (LDBS) gave results similar to those with full basis set (FB), with good agreement with experimental values for phenol and 12 alkyl and methoxy derivatives[154]. The exception was compounds bearing *ortho*-di-*t*-butyl groups, where large errors appeared due to 'excessive destabilization' as a result of strain in the parent compound. Wright and coworkers[154] calculated the ΔBDEs compared with phenol for a series of substituted phenols bearing substituents at the *ortho, meta* or *para* positions ranging from the very strong electron-supplying (NH_2) to the strongest electron-attracting (NO_2). They proposed 'additivity values' for combinations of 12 types of substituents. As already noted in Section IV.B.1, the calculations for the tocopherols were of particular interest in that the calculated order of ΔBDE (compared to phenol, 87.1) for α, β, γ and δ-tocopherol was -11.3, -9.4, -8.9 and -7.3 kcal mol^{-1}, so that the predicted order of antioxidant activity in a non-polar solvent, $\alpha > \beta \approx \gamma > \delta$, is the same as that found by experiment[18]. This can be taken as evidence in support of the HAT mechanism for reaction with peroxyl radicals. However, the ΔIP values showed the same trend: α, β, γ and δ values were -36.1, -33.6, -32.9 and -30.5 kcal mol^{-1} or a drop of 3 kcal mol^{-1} per methyl group, and this could be taken as support of the SET pathway! Their calculations for *ortho*-substituted phenols subject to hydrogen bonding (e.g. *o*-methoxy, *o*-hydroxy) were of particular application to our interpretation of the antioxidant activities of these compounds in terms of stabilization of the resulting radicals, compared to the parent compounds (Section III.B.2).

3. Calculations of more complex polyhydroxy phenols

Owing to their very common occurrence in nature and widespread use as dietary supplements, there have been several approaches to calculate relative antioxidant properties of the flavonoids. Lein and coworkers[164] used an empirical relationship based on calculated parameters such as heat of formation and the number of OH groups to estimate antioxidant properties of a large number (42) of flavonoids. General agreement was found with the Trolox equivalent parameter (TEAC), but unfortunately this parameter is not a reliable measure of antioxidant efficiency. Jovanovic and coworkers approached this (rather complex) problem by selecting simpler structural models for rings A and B of the flavonoids[163]. Reduction potentials were then obtained for these models (e.g. substituted catechols or derivatives for ring B and 5,7-dihydroxy compounds or derivatives for ring A) compared to some typical flavonoids. As expected, the flavonoid ring whose radical had the lower reduction potential was ring B for the catechol group including hesperidin, rutin, dihydroquercetin and quercetin, whereas in galangin, a 5,7-dihydroxy compound (modeled by 2,4-dihydroxyacetophenone), ring A has the lower reduction potential and takes over the antioxidant property. Van Acker and coworkers[169] carried out *ab initio* quantum mechanical calculations using heats of formation and the geometry of the parent flavonoids and their corresponding radicals. In addition, calculated spin densities were compared with ESR data. They concluded that oxidation of flavonoids takes place in ring B in those containing the catechol structure and that ring B is the site of antioxidant activity for the catechol flavonoids. Also, they concluded that the 'extremely good' antioxidant activity of the flavonols was due to an intramolecular hydrogen bond between the hydroxyl at position 3 on ring C and ring B. As pointed out in Section III.B.2, this interesting conclusion is not always supported by experimental results. Russo and coworkers[276] carried out semiempirical calculations at the AM1 and PM3 levels on quercetin and the radical species. Their AM1 optimized structure gave a non-planar structure with ring B out of plane. They reported that two radicals derived by hydrogen atom transfer from the 3−OH (ring C) and 4′−OH (catechol ring B) were almost isoenergetic. This result implies that the enolic hydrogen (3−OH) is more readily abstracted than the second hydrogen in the catechol ring B. This (3′−OH) is now strongly hydrogen-bonded to the adjacent radical site, which could possibly make abstraction more difficult. However, a radical site at oxygen on carbon 3 would be very unfavorable due to the adjacent carbonyl. It would be interesting to observe high level calculations comparing the two proposed isoenergetic sites.

Wright and coworkers[154] applied their 'additivity of substituent effects' to calculate relative BDEs of the catechin flavonoids, and consequently their order of antioxidant activity, by selecting model structures representing rings A, B and C separately. Then, by additivity of BDEs they applied these calculations to the more complex tricyclic flavonoids in order to establish their expected order of antioxidant activity. The reactivity order was in agreement with experimental results on the reactivities with superoxide radical observed by Jovanovic and coworkers[170].

V. FUTURE PROSPECTS FOR ANTIOXIDANTS

Most of our review to this point has focused on well-defined quantitative aspects of the mechanism and efficiency of phenolic antioxidants. In this final section, we will attempt to raise some long-term qualitative questions on future prospects or expectations for antioxidants. Some of the questions may be worthy of future pursuits while others will be of a more provocative nature.

(1) Is there a practical limit to the antioxidant activity of a phenolic antioxidant?

We have already commented on examples of the search for antioxidants more active than Vitamin E and actually encountered one exhibiting an order of magnitude higher activity than α-Toc in solution (see Table 2). However, is there a practical limit? To restate the problem in terms of ΔBDE: Is there a limit in the magnitude of ΔBDE above which the antioxidant itself undergoes autoxidation directly with oxygen? This can be answered in part by the example of the weak O−H bond in the semi-*para*-quinone radical, which is known to react with oxygen, giving rise to toxicity of such compounds (Section III.B.2). The enthalpy for this reaction (*cf.* equation 61) has recently been calculated by Johnson[277] using density functional theory. This gave a BDE for the O−H bond in the semi-quinone radical QH$^{\bullet}$ of 59.0 kcal mol^{-1} and a BDE of 52.4 kcal mol^{-1} for the O−H bond in H−O−O$^{\bullet}$.

$$Q^{\bullet}\!-\!H + O_2 \longrightarrow Q + H\!-\!O\!-\!O^{\bullet}, \Delta H = 6.6 \text{ kcal mol}^{-1} \quad (61)$$

Consequently, this reaction is endothermic in the gas phase. Also, ΔG was calculated to be +5.3 kcal mol^{-1} in the gas phase. However, in the aqueous phase, solvation by water of the H−O$_2^{\bullet}$ radical provides some driving force for this reaction and, using a known solvent model[278], it is estimated that $\Delta G_{\text{aqueous}}$ will drop to 1.6 kcal mol^{-1} due to solvation effects. Substituent effects on the Q$^{\bullet}$−H could make ΔG negative and accelerate the reaction even further. Consequently, a very weak phenolic O−H bond (BDE < 60 kcal mol^{-1}) can cause the phenolic antioxidant to turn into a pro-oxidant, since the HO$_2^{\bullet}$ formed will start new oxidation chains.

Despite a practical limit on antioxidant activity, as determined by the strength of the O−H bond, the search will continue for more efficient antioxidants, especially those that are active but non-toxic. With this in mind we are currently investigating the 1,8-naphthalenediol, **41**, and derivatives[279]. The derived radical **42**, is stabilized by a strong

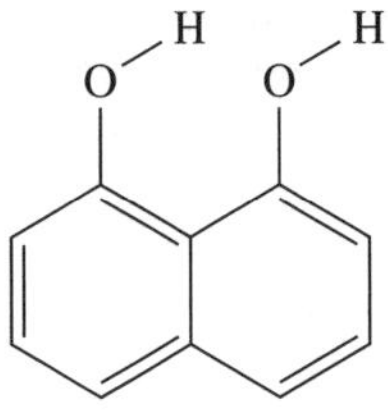

(41) 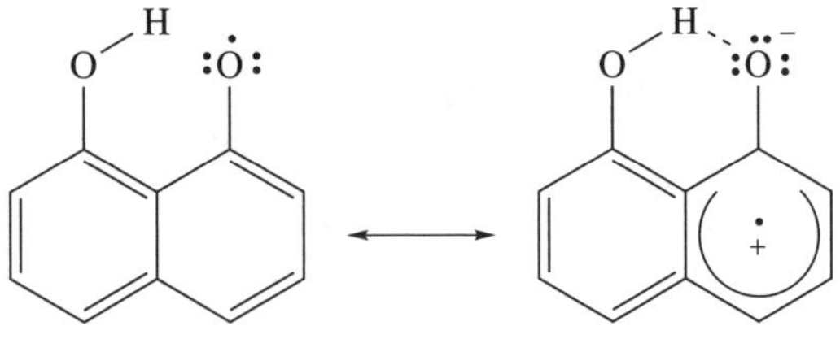   (42)

intramolecular hydrogen bond, like that in the *ortho*-semiquinone radical, but formation of a quinone and the associated toxicity are not possible for the radical.

Pratt and coworkers[280] recently used both calculations and syntheses to develop a novel and promising group of antioxidants by incorporating nitrogens into the aromatic ring of hydroxy aromatics. For example, one of their compounds bearing a *para*-dimethylamino group and two nitrogens is **43**, possessing a *higher* ionization potential but a *lower* O−H BDE than α-Toc. As a result, it is more stable in air than α-Toc, but reacts about twice as fast with peroxyl radicals than α-Toc.

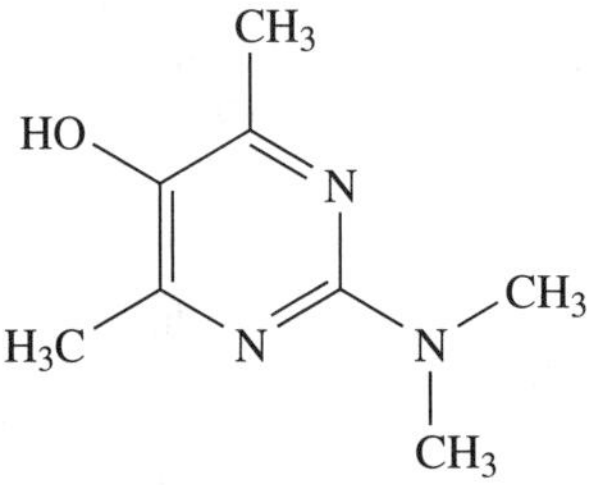

(43)

It should be realized that there are factors other than activity that determine the efficacy of an antioxidant. As outlined recently by Noguchi and Niki[281], 'the potency of antioxidants...is determined by many factors...

1. the chemical reactivity towards radicals,
2. localization of antioxidants,
3. concentration and mobility at the microenvironment,
4. fate of antioxidant-derived radical,
5. interaction with other antioxidants, and
6. absorption, distribution, retention, metabolism, and safety.'

The natural RRR-α-Toc isomer meets the above criteria[14], with the possible exception of 'mobility', and more than this it has been shown that α-tocopheryl quinone, the common oxidation product of α-Toc, is converted back into vitamin E in man[282].

(2) Is there a preferred method to determine antioxidant activity?

We have reviewed briefly the many different methods to evaluate antioxidants indicating advantages or limitations where appropriate. It is emphasized again that in order to determine the *antioxidant activity*, one must control the rate of free radical initiation. A preferred way to do this is by using azo initiators which decompose to form peroxyl radicals at a known, controlled rate. The advantages of the simple oxygen uptake method were given. While this method is not suitable for rapid, qualitative 'screening' for antioxidants, in fact these popular screening methods can give completely unreliable results. For example, the so-called Trolox Equivalent Antioxidant Capacity (TEAC) method measures the concentration of Trolox with the same antioxidant capacity as a 1 mM concentration of unknown antioxidant. However, Trolox is water-soluble, so if peroxyl radicals are generated in an aqueous phase for this test, Trolox will trap them there for the most part and one may not know the real function of the unknown. In addition, this method, and one like it called ORAC, the oxygen radical absorbing capacity versus Trolox, will not usually distinguish between an antioxidant and a retarder. This can lead to erroneous conclusions about the efficiency of a compound as antioxidant compared to Trolox or vitamin E, for example in the case of melatonin[283].

(3) How are radical reactions initiated in vivo?

The many methods to initiate lipid peroxidation *in vitro*, such as azo initiators, metal ions, pulse radiolysis, photoinitiation (Type I), enzymes (oxidases), to mention a few, have been reviewed[279]. However, as Bucala emphasized in a review[111], 'oxidation initiation is a pivotal first step and there is little understanding of how initiation proceeds *in vivo*.' Transition metal ions, iron or copper, are frequently used to initiate lipid oxidation, but free (unchelated) redox-active transition metals are virtually absent from biological systems[110] and appear to have little bearing on known pathological processes[111].

We have found that the DNA/RNA bases, purine and pyrimidine, will photoinitiate the radical peroxidation of lipids in a model heterogeneous system (e.g. micelles)[284] but the relevance of this *in vivo* is not known. As Pryor pointed out some years ago[285], superoxide is found in all aerobically metabolizing cells, but at physiological pH only a small portion, 1%, will exist as the conjugate acid, the hydroperoxyl radical HOO•, which can initiate lipid peroxidation. Of course, it would not seem to be possible to directly test the overall significance of initiation of radical reactions in the living cell by HOO•. In a different approach to this problem, Salvador, Antunes and coworkers developed a mathematical kinetic modeling procedure as applied to the mitochondrial inner membranes[286,287]. Their model was based on certain known rate constants and possible relevant reactions in mitochondria. Their results included the importance of the HOO• radical compared with HO• in the initiation, with an order of magnitude higher rate of initiation by HOO• of 10^{-7} M^{-1} s^{-1} in mitochondrial membranes.

(4) Are there specific benefits (or dangers) from nutritional additives like flavonoids?

Flavonoids that possess the catechol structure (ring B) are active antioxidants. The intake level of the human diet is high, ranging from about 50–500 mg per day, compared to vitamins C and E[258]. However, it appears that little is known about the bioavailability or efficiency *in vivo*. The percentage absorbed according to blood levels is only a few percent of flavonoids ingested[258]. It was suggested that their *in vivo* antioxidant effect may be tested by measuring the increase in the total antioxidant potential of blood plasma after a single, large intake of flavonoid-containing food or beverages. The antioxidant capacity of plasma can be determined by measuring TRAP, the total radical trapping antioxidant parameter and the contribution of each antioxidant to TRAP evaluated by analysis[288]. It is reported that long-term consumption of green tea improves the levels of α-Toc in red blood cells and LDL; that is, the flavonoids apparently have a sparing effect on other antioxidants[258].

The modern media hype on nutritional additives has put their use very much into the general public domain. In 1995 the NIH set up an Office of Dietary Supplements and the first Director, Dr. Bernadette Marriott, on the occasion of launching a new journal, *Antioxidants and Redox Signaling*[10], wrote in the Introduction: 'For the public, antioxidants embody a solution to most health problems and to living a long life without looking old!'.

VI. REFERENCES

1. B. Halliwell and J. M. C. Gutteridge (Eds.), *Free Radicals in Biological Medicine*, 3rd edn., Oxford Science Publications, 1999.
2. K. E. Mason, in *Vitamin E. Basic and Clinical Nutrition*, Vol. 1 (Ed. L. J. Machlin), M. Dekker, Inc., New York, 1980, pp. 1–6; H. M. Evans and K. S. Bishop, *Science*, **56**, 650 (1922).
3. G. W. Burton, A. Joyce and K. U. Ingold, *Arch. Biochem. Biophys.*, **221**, 281 (1983).
4. C. de Duve and O. Hayaishi (Eds.), *Tocopherol, Oxygen and Biomembranes. Proceedings of the International Symposium at Lake Yamanaka, Japan*, Elsevier/North-Holland Biomed. Press, Amsterdam, 1977.
5. *Biology of Vitamin E. Ciba Foundation Symposium 101*, (Eds. R. Porter and J. Whelan) Pitman, London, 1983.
6. A. T. Diplock and L. J. Machlin, in *Vitamin E. Biochemistry and Health Effects Implications* (Eds. L. Packer and W. A. Pryor), Ann. N. Y. Acad. Sci., Vol. 570, 1989.
7. H. F. DeLuca (Ed.), *The Fat-Soluble Vitamins. Handbook of Lipid Research*, Plenum Press, New York, 1978.
8. F. Shahidi (Ed.), *Natural Antioxidants. Chemistry, Health Effects, and Applications*, AOCS Press, Champaign, Illinois, 1997.
9. L. Packer and A. S. H. Ong (Eds.), *Biological Oxidants and Antioxidants. Molecular Mechanisms and Health Effects*, AOCS Press, Champaign, Illinois, 1998.
10. *Antioxidants and Redox Signaling* (Eds. C. K. Sen and D. K. Das), Vol. **1**, M. A. Liebert, Inc, New York, 1999.
11. J. A. Howard, *Adv. Free Radical Chem.*, **4**, 49 (1972).
12. K. U. Ingold, *Adv. Chem. Ser.*, **75**, 207 (1968).
13. L. R. Mahoney, *Angew. Chem., Int. Ed. Engl.*, **8**, 547 (1969).
14. G. W. Burton and K. U. Ingold, *Acc. Chem. Res.*, **19**, 194 (1986).
15. L. R. C. Barclay, *Can. J. Chem.*, **71**, 1 (1993).
16. K. U. Ingold, *Chem. Rev.*, **61**, 563 (1961).
17. J. A. Howard and K. U. Ingold, *Can. J. Chem.*, **42**, 2324 (1964).
18. G. W. Burton, T. Doba, E. J. Gabe, L. Hughes, F. L. Lee, L. Prasad and K. U. Ingold, *J. Am. Chem. Soc.*, **107**, 7053 (1985).
19. J. A. Howard and K. U. Ingold, *Can. J. Chem.*, **41**, 1744 (1963).
20. R. H. Bisby and A. W. Parker, *J. Am. Chem. Soc.*, **117**, 5664 (1995).
21. S-I. Nagaoka, M. Inoue, C. Nishioka, Y. Nishioku, S. Tsunoda, C. Ohguchi, K. Ohara, K. Mukai and U. Nagashimi, *J. Phys. Chem. B*, **104**, 856 (2000).
22. S-I. Nagaoka, A. Kuranaka, H. Tsuboi, U. Nagashima and K. Mukai, *J. Phys. Chem.*, **96**, 2754 (1992).
23. S-I. Nagaoka, K. Mukai, T. Itoh and S. Katsumata, *J. Phys. Chem.*, **96**, 8184 (1992).
24. S-I. Nagaoka and K. Ishihara, *J. Am. Chem. Soc.*, **118**, 7361 (1996).
25. P. Neta, R. E. Huie, P. Maruthamuthu and S. Steenken, *J. Phys. Chem.*, **93**, 7654 (1989).
26. Z. B. Alfassi, R. E. Huie, M. Kumar and P. Neta, *J. Phys. Chem.*, **96**, 767 (1992).
27. Z. B. Alfassi, S. Marguet and P. Neta, *J. Phys. Chem.*, **98**, 8019 (1994).
28. T. W. Campbell and G. M. Coppinger, *J. Am. Chem. Soc.*, **74**, 1469 (1952).
29. K. U. Ingold, *Can. J. Chem.*, **41**, 2807 (1963).
30. D. Griller, L. R. C. Barclay and K. U. Ingold, *J. Am. Chem. Soc.*, **97**, 6151 (1975).
31. J. Winterle, D. Dulin and T. Mill, *J. Org. Chem.*, **49**, 491 (1984).
32. E. C. Horswill and K. U. Ingold, *Can. J. Chem.*, **44**, 263 (1966) and references cited therein.
33. E. C. Horswill and K. U. Ingold, *Can. J. Chem.*, **44**, 269 (1966).
34. W. A. Skinner and R. M. Parkhurst, *Lipids*, **6**, 240 (1971).
35. W. A. Skinner, *Biochem. Biophys. Res. Commun.*, **15**, 469 (1964).
36. R. Yamauchi, K. Kato and Y. Ueno, *Lipids*, **23**, 779 (1988).
37. M. J. Thomas and B. H. J. Bielski, *J. Am. Chem. Soc.*, **111**, 3315 (1989).
38. D. C. Liebler, P. F. Baker and K. L. Kaysen, *J. Am. Chem. Soc.*, **112**, 6995 (1990).
39. D. C. Liebler and J. A. Burr, *Lipids*, **30**, 789 (1995).
40. R. Yamauchi, Y. Yagi and K. Kato, *Biosci. Biotech. Biochem.*, **60**, 616 (1996).
41. A. Kamal-Eldin and L-A. Appelqvist, *Lipids*, **31**, 671 (1996).
42. V. W. Bowry and K. U. Ingold, *J. Org. Chem*, **60**, 5456 (1995).

43. G. W. Burton and K. U. Ingold, *J. Am. Chem. Soc.*, **103**, 6472 (1981).
44. L. R. C. Barclay and K. U. Ingold, *J. Am. Chem. Soc.*, **103**, 6478 (1981).
45. L. R. C. Barclay and K. U. Ingold, *J. Am. Chem. Soc.*, **102**, 7792 (1980).
46. L. R. C. Barclay, D. Kong and J. van Kessel, *Can. J. Chem.*, **64**, 2103 (1986).
47. L. R. C. Barclay, K. A. Baskin, S. J. Locke and T. D. Schaefer, *Can. J. Chem.*, **65**, 2529 (1987).
48. F. Xi and L. R. C. Barclay, *Can. J. Chem.*, **76**, 171 (1998).
49. L. R. C. Barclay, M. R. Vinqvist, K. Mukai, S. Itoh and H. Morimoto, *J. Org. Chem.*, **58**, 7416 (1993).
50. L. R. C. Barclay, C. D. Edwards and M. R. Vinqvist, *J. Am. Chem. Soc.*, **121**, 6226 (1999).
51. L. R. C. Barclay, M. R. Vinqvist, K. Mukai, H. Goto, Y. Hashimoto, A. Tokunaga and H. Uno, *Org. Lett.*, **2**, 2841 (2000).
52. L. R. C. Barclay, A. M. H. Bailey and D. Kong, *J. Biol. Chem.*, **260**, 15809 (1985).
53. L. R. C. Barclay, *J. Biol. Chem.*, **263**, 16138 (1988).
54. L. R. C. Barclay, K. A. Baskin, S. J. Locke and M. R. Vinqvist, *Can. J. Chem.*, **67**, 1366 (1989).
55. L. R. C. Barclay, K. A. Baskin, K. A. Dakin, S. J. Locke and M. R. Vinqvist, *Can. J. Chem.*, **68**, 2258 (1990).
56. L. R. C. Barclay and M. R. Vinqvist, *Free Radical Biol. Med.*, **16**, 779 (1994) and references cited therein.
57. L. R. C. Barclay, C. D. Edwards, K. Mukai, Y. Egawa and T. Nishi, *J. Org. Chem.*, **60**, 2739 (1995).
58. L. R. C. Barclay, J. D. Artz and J. J. Mowat, *Biochem. Biophys. Acta*, **1237**, 77 (1995).
59. L. R. C. Barclay, K. A. McLaughlin and M. R. Vinqvist, *J. Liposome Res.*, **5**, 955 (1995).
60. L. R. C. Barclay, F. Antunes, Y. Egawa, K. L. McAllister, K. Mukai, T. Nishi and M. R. Vinqvist, *Biochem. Biophys. Acta*, **1328**, 1 (1997).
61. G. Hsiao, C-M. Teng, C-L. Wu and F-N. Ko, *Arch. Biochem. Biophys.*, **334**, 18 (1996).
62. D. D. M. Wayner, G. W. Burton, K. U. Ingold, L. R. C. Barclay and S. J. Locke, *Biochim. Biophys. Acta*, **924**, 408 (1987).
63. L. Escobar, C. Salvador, M. Contreras and J. E. Escamilla, *Anal. Biochem*, **184**, 139 (1990).
64. P. M. Abuja and R. Albertini, *Clin. Chim. Acta*, **306**, 1 (2001).
65. H. Kappus, in *Oxidative Stress* (Ed. H. Sies) Chap. 12, Academic Press, London, 1985, pp. 273–309.
66. J. M. C. Gutteridge, *Free Radical Res. Commun.*, **1**, 173 (1986).
67. W. A. Pryor, J. A. Cornicelli, L. J. Devall, B. Tait, B. K. Trivedi, D. T. Witiak and M. Wu, *J. Org. Chem.*, **58**, 3521 (1993).
68. M. Foti and G. Ruberto, *J. Agric. Food Chem.*, **49**, 342 (2001).
69. K. E. Fygle and T. B. Melo, *Chem. Phys. Lipids*, **79**, 39 (1996).
70. L. R. C. Barclay, M. R. Vinqvist, F. Antunes and R. E. Pinto, *J. Am. Chem. Soc.*, **119**, 5764 (1997).
71. M. Takahashi, Y. Yoshikawa and E. Niki, *Bull. Chem. Soc. Jpn.*, **62**, 1885 (1989).
72. D. E. Henderson, A. M. Slickman and S. K. Henderson, *J. Agric. Food Chem.*, **47**, 2563 (1999).
73. P. Palozza, S. Moualla and N. I. Krinsky, *Free Radical Biol. Med.*, **13**, 127 (1992).
74. S. Banni, M. S. Contini, E. Angioni, M. Deiana, M. A. Dessi, M. R. Melis, G. Carta and F. P. Corongui, *Free Radical Res.*, **25**, 43 (1996).
75. K. E. Peers, D. T. Coxon and H. W-S. Chan, *J. Sci. Food Agric.*, **32**, 898 (1981).
76. H. W.-S. Chan and G. Levett, *Lipids*, **12**, 99 (1977).
77. D. Fiorentini, L. Cabrini and L. Landi, *Free Radical Res. Commun.*, **18**, 201 (1993).
78. R. C. R. M. Vossen, M. C. E. vanDam-Mieras, G. Hornstra and R. F. A. Zwaal, *Lipids*, **28**, 857 (1993).
79. D. C. Liebler, K. L. Kaysen and J. A. Burr, *Chem. Res. Toxicol.*, **4**, 89 (1991).
80. N. A. Porter and D. G. Wujek, *J. Am. Chem. Soc.*, **106**, 2626 (1984).
81. N. A. Porter, *Acc. Chem. Res.*, **19**, 262 (1986).
82. K. E. Peers, D. T. Coxon and H. W-S. Chan, *J. Sci. Food Agric.*, **32**, 898 (1981).
83. M. K. Logani and R. E. Davies, *Lipids*, **15**, 485 (1980).
84. W. Korytowski, T. Sarna and M. Zaręba, *Arch. Biochem. Biophys.*, **319**, 142 (1995).
85. A. Mellors and A. L. Tappel, *J. Biol. Chem.*, **241**, 4353 (1966).

86. L. Valgimigli, J. T. Banks, J. Lusztyk and K. U. Ingold, *J. Org. Chem.*, **64**, 3381 (1999).
87. R. Albertini, P. M. Abuja and B. Halliwell, *Free Radical Res.*, **26**, 75 (1998).
88. T. C. P. Dinis, V. M. C. Madeira and L. M. Almedia, *Arch. Biochem. Biophys.*, **315**, 161 (1994).
89. H. Zieliński and H. Kozlowska, *Pol. J. Food Nutr. Sci.*, **8**, 147 (1999).
90. E. Niki, J. Tsuchiya, R. Tanimura and Y. Kamiya, *Oxid. Commun.*, **4**, 261 (1983).
91. A. J. Rosas-Romero, B. Rojano, C. A. Hernández, C. M. Manchado, J. Silva and J. C. Herrera, *Ciencia*, **7**, 78 (1999).
92. G. W. Burton and K. U. Ingold, *Science*, **224**, 569 (1984).
93. C. S. Foote and R. W. Denny, *J. Am. Chem. Soc.*, **90**, 6233 (1968).
94. D. R. Kearns, *Chem. Rev.*, **71**, 395 (1971).
95. N. J. Miller, C. Rice-Evans, M. J. Davies, V. Gopinathan and A. Milner, *Clin. Sci.*, **84**, 407 (1993).
96. C. A. Rice-Evans, *Free Radical Res.*, **33**, 559 (2000).
97. M. Hermes-Lima, W. G. Willmore and K. B. Storey, *Free Radical Biol. Med.*, **19**, 271 (1995).
98. J. I. Gray, *J. Am. Oil Chem. Soc.*, **55**, 539 (1978).
99. V. R. Kokatnur and M. Jelling, *J. Am. Chem. Soc.*, **63**, 1432 (1941).
100. H. Esterbauer and H. Zollner, *Free Radical Biol. Med.*, **7**, 197 (1989).
101. S. D. Aust, in *Methods in Toxicology*, Vol. 1B (Eds. C. A. Tyson and J. M. Frazier), Chap. 30, Academic Press, San Diego, 1994, pp. 367–376.
102. G. Lepage, G. Munoz, J. Champagne and C. C. Roy, *Anal. Biochem.*, **197**, 277 (1991).
103. K. Yagi, In *Lipid Peroxides in Biology and Medicine* (Ed. K. Yagi), Academic Press, New York, 1982, pp. 223–242.
104. E. Kishida, A. Kamura, S. Tokumaru, M. Oribe, H. Iguchi and S. Kojo, *J. Agric. Food Chem.*, **41**, 1 (1993).
105. M. Hiramatsu, J. Liu, R. Edamatsu, S. Ohba, D. Kadowaki and A. Mori, *Free Radical Biol. Med.*, **16**, 201 (1994).
106. J. Aikens and T. A. Dix, *J. Biol. Chem.*, **266**, 15091 (1991).
107. A. W. Girotti and J. P. Thomas, *J. Biol. Chem.*, **259**, 1744 (1984).
108. T. Ohyashiki, Y. Yabunaka and K. Matsui, *Chem. Pharm. Bull.*, **39**, 976 (1991).
109. W. Bors, C. Michel and M. Saran, *Biochim. Biophys. Acta*, **796**, 312 (1984).
110. T. A. Dix and J. Aikens, *Chem. Res. Toxicol.*, **6**, 2 (1993).
111. R. Bucala, *Redox Report*, **2**, 291 (1996).
112. M. Foti, K. U. Ingold and J. Lusztyk, *J. Am. Chem. Soc.*, **116**, 9440 (1994).
113. M. Takahashi, J. Tsuchiya, E. Niki and S. Urano, *J. Nutr. Sci. Vitaminol.*, **34**, 25 (1988).
114. Z.-L. Liu, Z-X. Han, K-C. Yu, Y-L. Zhang and Y-C. Liu, *J. Phys. Org. Chem.*, **5**, 33 (1992).
115. Z.-L. Liu, Z-X. Han, P. Chen and Y-C. Liu, *Chem. Phys. Lipids*, **56**, 73 (1990).
116. L. R. C. Barclay and M. R. Vinqvist, *Free Radical Biol. Med.*, **28**, 1079 (2000).
117. T. Yoshida, H. Otake, Y. Aramaki, T. Hara, S. Tsuchiya, A. Hamada and H. Utsumi, *Biol. Pharm. Bull.*, **19**, 779 (1996).
118. E. Popa, T. Constantinescu and S. Rădulescu, *Rev. Roum. Chim.*, **49**, 75 (1995).
119. J. Tsuchiya, T. Yamada, E. Niki and Y. Kamiya, *Bull. Chem. Soc. Jpn.*, **58**, 326 (1985).
120. M. Lucarini, G. F. Pedulli and M. Cipollone, *J. Org. Chem.*, **59**, 5063 (1994).
121. W. Bors, C. Nichel and K. Stettmaier, *J. Chem. Soc., Perkin Trans. 2*, 1513 (1992).
122. A. Lugasi, E. Dworschak and J. Hovari, *Curr. Status Future Trends Anal. Food Chem., Proc. Eur. Conf. Food Chem.*, 8th, **3**, 639 (1995).
123. K. E. Malterud, O. H. Diep and R. B. Sund, *Pharmacol. Toxicol.*, **78**, 111 (1996).
124. C. Sánchez-Moreno, J. A. Larrauri and F. Saura-Calixto, *J. Sci. Food Agric.*, **76**, 270 (1998).
125. V. Bondet, W. Brand-Williams and C. Berset, *Food Sci. Technol. (London)*, **30**, 609 (1997).
126. N. Noguchi, Y. Iwaki, M. Takahashi, E. Komura, Y. Kato, K. Tamura, O. Cynshi, T. Kodama and E. Niki, *Arch. Biochem. Biophys.*, **342**, 236 (1997).
127. S-I. Nagaoka, A. Kuranaka, H. Tsuboi, U. Nagashima and K. Mukai, *J. Phys. Chem.*, **96**, 2754 (1992).
128. K. Mukai, M. Nishimura and S. Kikuchi, *J. Biol. Chem.*, **266**, 274 (1991).
129. K. Mukai, Y. Watanabe and K. Ishizu, *Bull. Chem. Soc. Jpn.*, **59**, 2899 (1986).
130. E. Niki, in *Clinical and Nutritional Aspects of Vitamin E* (Eds. O. Hayaishi and M. Mino), Elsevier Science Publications, Amsterdam, 1987, pp. 3–13.
131. A. C. S. Woolard, S. P. Wolff and Z. A. Bascal, *Free Radical Biol. Med.*, **9**, 299 (1990).

12. Phenols as antioxidants

132. S. Singh, K. R. Bhaskar and C. N. R. Rao, *Can. J. Chem.*, **44**, 2657 (1966).
133. E. N. Frankel and A. S. Meyer, *J. Sci. Food. Agric.*, **80**, 1925 (2000).
134. B. Halliwell, *Free Radical Res. Commun.*, **9**, 1 (1990).
135. J. M. Braughler, L. A. Duncan and R. L. Chase, *J. Biol. Chem.*, **261**, 10282 (1986).
136. L. Castle and M. J. Perkin, *J. Am. Chem. Soc.*, **108**, 6382 (1986).
137. J. A. Howard, K. U. Ingold and M. Symonds, *Can. J. Chem.*, **46**, 1017 (1968).
138. E. C. Horswill and K. U. Ingold, *Can. J. Chem.*, **44**, 985 (1966).
139. J. A. Howard and K. U. Ingold, *Can. J. Chem.*, **40**, 1851 (1962).
140. J. A. Howard and K. U. Ingold, *Can. J. Chem.*, **41**, 2800 (1963).
141. G. W. Burton, L. J. Johnston, J. C. Walton and K. U. Ingold, in *Substituent Effects in Radical Chemistry* (Eds. H. G. Viehe, Z. Janousek and R. Merényi), D. Reidel Publishing Company, Boston, 1986, pp. 107–122.
142. J. C. Gilbert and M. Pinto, *J. Org. Chem.*, **57**, 5271 (1992).
143. G. W. Burton, L. Hughes and K. U. Ingold, *J. Am. Chem. Soc.*, **105**, 5950 (1983).
144. H. Zahalka, B. Robillard, L. Hughes, J. Lusztyk, G. W. Burton, E. G. Janzen, Y. Kotake and K. U. Ingold, *J. Org. Chem.*, **53**, 3739 (1988).
145. J. Malmström, M. Jonsson, I. A. Cotgreave, L. Hammarström, M. Sjödin and L. Engman, *J. Am. Chem. Soc.*, **123**, 3434 (2001).
146. K. Mukai, K. Okabe and H. Hosose, *J. Org. Chem.*, **54**, 557 (1989).
147. K. Mukai, Y. Kageyama, T. Ishida and K. Fukuda, *J. Org. Chem.*, **54**, 552 (1989).
148. K. Mukai, K. Fukuda, K. Tajima and K. Ishizu, *J. Org. Chem.*, **53**, 430 (1988).
149. K. Mukai, K. Daifuku, K. Okabe, T. Tanigaki and K. Inoue, *J. Org. Chem.*, **56**, 4187 (1991).
150. L. R. C. Barclay, in *Peroxyl Radicals* (Ed. Z. B. Alfassi), Chap. 3, Wiley, Chichester, 1997, pp. 27–48.
151. P. Alpincourt, M. F. Ruiz-López, X. Assfeld and F. Bohr, *J. Comput. Chem.*, **20**, 1039 (1999).
152. L. R. C. Barclay, F. Xi and J. Q. Norris, *J. Wood. Chem. Tech.*, **17**, 73 (1997).
153. M. I. de Heer, H-G. Korth and P. Mulder, *J. Org. Chem.*, **64**, 6969 (1999).
154. J. S. Wright, E. R. Johnson and G. A. DiLabio, *J. Am. Chem. Soc.*, **123**, 1173 (2001).
155. M. I. de Heer, P. Mulder, H-G. Korth, K. U. Ingold and J. Lusztyk, *J. Am. Chem. Soc.*, **122**, 2355 (2000).
156. S. V. Jovanovic, S. Steenken, C. W. Boone and M. G. Simic, *J. Am. Chem. Soc.*, **121**, 9677 (1999).
157. S. V. Jovanovic, C. W. Boone, S. Steenken, M. Trinoga and R. B. Kasey, *J. Am. Chem. Soc.*, **123**, 3064 (2001).
158. S. M. Khopde, K. I. Priyadarsini, P. Venkatesan and M. N. A. Rao, *Biophys. Chem.*, **80**, 85 (1999).
159. F. Xi and L. R. C. Barclay, *Can. J. Chem.*, **76**, 171 (1998).
160. T. Yamamura, K. Nishiwaki, Y. Tanigaki, S. Terauchi, S. Tomiyama and T. Nishiyama, *Bull. Chem. Soc. Jpn.*, **68**, 2955 (1995).
161. C. A. Rice-Evans, N. J. Miller and G. Paganga, *Trends Plant Sci.*, **2**, 152 (1997).
162. G. Cao, E. Sofic and R. L. Prior, *Free Radical Biol. Med.*, **22**, 749 (1997).
163. S. V. Jovanovic, S. Steenken, Y. Hara and M. G. Simic, *J. Chem. Soc., Perkin Trans. 2*, 2497 (1996).
164. E. J. Lein, S. Ren, H-H. Bui and R. Wang, *Free Radical Biol. Med.*, **26**, 285 (1999).
165. S. V. Jovanovic, S. Steenken, M. Tosic, B. Marjanovic and M. G. Simic, *J. Am. Chem. Soc.*, **116**, 4846 (1994).
166. P. Pedrielli, G. F. Pedulli and L. H. Skibsted, *J. Agric. Food Chem.*, **49**, 3034 (2001).
167. O. Dangles, G. Fargeix and C. Dufour, *J. Chem. Soc., Perkin Trans. 2*, 1387 (1999).
168. V. A. Belyakov, V. A. Roginsky and W. Bors, *J. Chem. Soc., Perkin Trans. 2*, 2319 (1995).
169. S. A. B. E. van Acker, M. J. de Groot, D-J. van den Berg, M. N. J. L. Tromp, G. D-O. den Kelder, W. J. F. van der Vijgh and A. Bast, *Chem. Res. Toxicol.*, **9**, 1305 (1996).
170. S. V. Jovanovic, Y. Hara, S. Steenken and M. G. Simic, *J. Am. Chem. Soc.*, **117**, 9881 (1995).
171. S. Taniguchi, T. Yanase, K. Kobayashi, R. Takayanagi, M. Haji, F. Umeda and H. Nawata, *Endocrin J.*, **41**, 605 (1994).
172. K. Takanashi, K. Watanabe and I. Yoshizawa, *Biol. Pharm. Bull.*, **18**, 1120 (1995).
173. C. Mytilineou, S-K. Han and G. Cohen, *J. Neurochem.*, **61**, 1470 (1993).
174. H. Ishii, D. B. Stanimirovic, C. J. Chang, B. B. Mrsulja and M. Spatz, *Neurochem. Res.*, **18**, 1193 (1993).

175. T. G. Hastings, D. A. Lewis and M. J. Zigmond, *Proc. Natl. Acad. Sci. USA*, **93**, 1956 (1996).
176. J. J. Inbaraj, R. Gandhidasan and R. Murugesan, *Free Radical Biol. Med.*, **26**, 1072 (1999).
177. J. J. Inbaraj, M. C. Krishna, R. Gandhidasan and R. Murugesan, *Biochim. Biophys. Acta*, **1472**, 462 (1999).
178. E. S. Krol and J. L. Bolton, *Chem.-Biol. Interact.*, **104**, 11 (1997).
179. J. Pospíšil, E. Lisá and I. Buben, *Eur. Polym. J.*, **6**, 1347 (1970).
180. E. Lisá, L. Kotulák, J. Petránek and J. Pospíšil, *Eur. Polym. J.*, **8**, 501 (1972).
181. E. Lisá and J. Pospíšil, *J. Polym. Sci., Polym. Symp*, **40**, 209 (1973).
182. D. Loshadkin, V. Roginsky and E. Pliss, submitted to *Int. J. Chem. Kinetics*, **34**, 162 (2002).
183. K. Mukai, in *Free Radical Chemistry of Coenzyme Q* (Eds. V. F. Kagan and P. J. Quinn), Chap. 3, CRC Press, Boca Raton, Florida, 2001, pp. 43–61.
184. L. R. C. Barclay, C. D. Edwards, K. Mukai, Y. Egawa and T. Nishi, *J. Org. Chem.*, **60**, 2739 (1995).
185. K. U. Ingold, V. W. Bowry, R. Stocker and C. Walling, *Proc. Natl. Acad. Sci. USA*, **90**, 45 (1993).
186. M. H. Abraham, P. L. Grellier, D. V. Prior, J. J. Morris and P. J. Taylor, *J. Chem. Soc., Perkin Trans. 2*, 521 (1990).
187. R. W. Taft, D. Gurka, L. Joris, P. v. R. Schleyer and J. W. Rakshys, *J. Am. Chem. Soc.*, **91**, 4801 (1969).
188. P. A. MacFaul, K. U. Ingold and J. Lusztyk, *J. Org. Chem.*, **61**, 1316 (1996).
189. D. V. Avila, C. E. Brown, K. U. Ingold and J. Lusztyk, *J. Am. Chem. Soc.*, **115**, 466 (1993).
190. J. A. Howard, W. J. Schwalm and K. U. Ingold, *Adv. Chem. Ser.*, **75**, 6 (1968).
191. L. Valgimigli, J. T. Banks, K. U. Ingold and J. Lusztyk, *J. Am. Chem. Soc.*, **117**, 9966 (1995).
192. D. V. Avila, K. U. Ingold, J. Lusztyk, W. H. Green and D. R. Procopio, *J. Am. Chem. Soc.*, **117**, 2929 (1995).
193. P. Franchi, M. Lucarini, G. F. Pedulli, L. Valgimigli and B. Lunelli, *J. Am. Chem. Soc.*, **121**, 507 (1999).
194. L. Valgimigli, K. U. Ingold and J. Lusztyk, *J. Org. Chem.*, **61**, 7947 (1996).
195. D. W. Snelgrove, J. Lusztyk, J. T. Banks, P. Mulder and K. U. Ingold, *J. Am. Chem. Soc.*, **123**, 469 (2001).
196. M. H. Abraham, P. L. Grellier, D. V. Prior, P. P. Duce, J. J. Morris and P. J. Taylor, *J. Chem. Soc., Perkin Trans. 2*, 699 (1989).
197. D. G. Hendry and G. A. Russell, *J. Am. Chem. Soc.*, **86**, 2368 (1964).
198. J. A. Howard and K. U. Ingold, *Can. J. Chem.*, **44**, 1119 (1966).
199. M. Lucarini, G. F. Pedulli and L. Valgimigli, *J. Org. Chem.*, **63**, 4497 (1998).
200. M. Sumarno, E. Atkinson, C. Suarna, J. K. Saunders, E. R. Cole and P. T. Southwell-Keely, *Biochim. Biophys. Acta*, **920**, 247 (1987).
201. Z. B. Alfassi, R. E. Huie and P. Neta, *J. Phys. Chem.*, **97**, 7253 (1993).
202. T. Maki, Y. Araki, Y. Ishida, O. Onomura and Y. Matsumura, *J. Am. Chem. Soc.*, **123**, 3371 (2001).
203. J. T. Banks, K. U. Ingold and J. Lusztyk, *J. Am. Chem. Soc.*, **118**, 6790 (1996).
204. M. Iwatsuki, J. Tsuchiya, E. Komuro, Y. Yamamoto and E. Niki, *Biochim. Biophys. Acta*, **1200**, 19 (1994).
205. L. Valgimigli, K. U. Ingold and J. Lusztyk, *J. Am. Chem. Soc.*, **118**, 3545 (1996).
206. J. A. Howard and K. U. Ingold, *Can. J. Chem.*, **42**, 1044 (1964).
207. L. R. C. Barclay, S. J. Locke, J. M. MacNeil, J. van Kessel, G. W. Burton and K. U. Ingold, *J. Am. Chem. Soc.*, **106**, 2479 (1984).
208. L. R. C. Barclay, J. M. MacNeil, J. van Kessel, B. J. Forrest, N. A. Porter, L. S. Lehman, K. J. Smith and J. C. Ellington, *J. Am. Chem. Soc.*, **106**, 6740 (1984).
209. N. A. Porter, B. A. Weber, H. Weenen and J. A. Khan, *J. Am. Chem. Soc.*, **102**, 5597 (1980).
210. N. A. Porter, L. S. Lehman, B. A. Weber and K. J. Smith, *J. Am. Chem. Soc.*, **103**, 6447 (1981).
211. H. Weenen and N. A. Porter, *J. Am. Chem. Soc.*, **104**, 5216 (1982).
212. T. Doba, G. W. Burton and K. U. Ingold, *Biochim. Biophys. Acta*, **835**, 298 (1984).
213. L. R. C. Barclay, S. J. Locke and J. M. MacNeil, *Can. J. Chem.*, **61**, 1288 (1983).
214. L. R. C. Barclay, S. J. Locke and J. M. MacNeil, *Can. J. Chem.*, **63**, 366 (1985).
215. L. R. C. Barclay, K. A. Baskin, D. Kong and S. J. Locke, *Can. J. Chem.*, **65**, 2541 (1987).
216. W. A. Pryor, T. Strickland and D. F. Church, *J. Am. Chem. Soc.*, **110**, 2224 (1988).

217. I. H. Ekiel, L. Hughes, G. W. Burton, P. A. Jovall, K. U. Ingold and I. C. P. Smith, *Biochemistry*, **27**, 1432 (1988).
218. M. Takahashi, J. Tsuchiya and E. Niki, *J. Am. Chem. Soc.*, **111**, 6350 (1989).
219. M. J. Kamlet, J. L. Abboud, M. H. Abraham and R. W. Taft, *J. Org. Chem.*, **48**, 2877 (1983).
220. L. R. C. Barclay, M. R. Vinqvist, F. Antunes and R. E. Pinto, *J. Am. Chem. Soc.*, **119**, 5764 (1997).
221. E. Niki, A. Kawakami, M. Saito, Y. Yamamoto, J. Tsuchiya and Y. Kamiya, *J. Biol. Chem.*, **260**, 2191 (1985).
222. V. E. Kagan, R. A. Bakalova, Zh. Zh. Zhelev, D. S. Rangelova, E. A. Serbinova, V. A. Tyurin, N. K. Denisova and L. Packer, *Arch. Biochem. Biophys.*, **280**, 147 (1990).
223. J. C. Gomez-Fernandez, J. Villaiain and F. J. Aranda, in *Vitamin E Health Disease*, (Eds. L. Packer and J. Fuchs), Dekker, New York, 1993, p. 223.
224. F. J. Aranda, A. Coutinho, M. N. Berberan-Santos, M. J. E. Prieto and J. C. Gómez-Fernández, *Biochim. Biophys. Acta*, **985**, 26 (1989).
225. V. A. Tyurin, V. E. Kagan, E. A. Serbinova, N. V. Gorbunov, A. N. Erin, L. L. Prilipko and Ts. S. Stoytchev, *Bull. Exp. Biol. Med. USSR*, **12**, 689 (1986).
226. A. V. Victorov, N. Janes, G. Moehren, E. Rubin, T. F. Taraschi and J. B. Hoek, *J. Am. Chem. Soc.*, **116**, 4050 (1994).
227. L. R. Mahoney and M. A. DaRooge, *J. Am. Chem. Soc.*, **89**, 5619 (1967).
228. C. Golumbic and H. A. Mattill, *J. Am. Chem. Soc.*, **63**, 1279 (1941).
229. A. L. Tappel, *Geriatrics*, **23**, 97 (1968).
230. J. E. Packer, T. F. Slater and R. L. Wilson, *Nature*, **275**, 737 (1979).
231. K. Mukai, K. Fukuda, K. Ishizu and Y. Kitamura, *Biochem. Biophys. Res. Commun.*, **146**, 134 (1987).
232. E. Niki, J. Tsuchiya, R. Tanimura and Y. Kamiya, *Chem. Lett.*, 789 (1982).
233. E. Niki, T. Saito, A. Kawakami and Y. Kamiya, *J. Biol. Chem.*, **259**, 4177 (1984).
234. M. Scarpa, A. Rigo, M. Maiorino, F. Ursini and C. Gregolin, *Biochim. Biophys. Acta*, **801**, 215 (1984).
235. E. Niki, A. Kawakami, Y. Yamamoto and Y. Kamiya, *Bull. Chem. Soc. Jpn.*, **58**, 1971 (1985).
236. T. Motoyama, M. Miki, M. Mino, M. Takahashi and E. Niki, *Arch. Biochem. Biophys.*, **270**, 655 (1989).
237. C. Rousseau, C. Richard and R. Martin, *J. Chim. Phys. Phys.-Chim. Biol.*, **85**, 145 (1988).
238. L. R. C. Barclay, K. A. Dakin and J. A. Y. Khor, *Res. Chem. Intermed.*, **21**, 467 (1995).
239. M. K. Sharma and G. R. Buettner, *Free Radical Biol. Med.*, **14**, 649 (1993).
240. Y. Nakagawa, I. A. Cotgreave and P. Moldéus, *Biochem. Pharmacol.*, **442**, 883 (1991).
241. G. W. Burton, U. Wronska, L. Stone, D. O. Foster and K. U. Ingold, *Lipids*, **25**, 199 (1990).
242. L. R. Mahoney, *J. Am. Chem. Soc.*, **89**, 1895 (1967).
243. N. M. Storozhok, N. O. Pirogov, S. A. Krashakov, N. G. Khrapova and E. B. Burlakova, *Kinet. Catal. (Engl. Transl.)*, **36**, 751 (1995).
244. S-I. Nagaoka, Y. Okauchi, S. Urano, U. Nagashima and K. Mukai, *J. Am. Chem. Soc.*, **112**, 8921 (1990).
245. K. Mukai, H. Morimoto, Y. Okauchi and S.-I. Nagaoka, *Lipids*, **28**, 753 (1993).
246. S-I. Nagaoka, K. Sawada, Y. Fukumoto, U. Nagashima, S. Katsumata and K. Mukai, *J. Phys. Chem.*, **96**, 6663 (1992).
247. K. Mukai, K. Sawada, Y. Kohno and J. Terao, *Lipids*, **28**, 747 (1993).
248. J. Cillard, P. Cillard and M. Cormier, *J. Am. Oil Chem. Soc.*, **57**, 255 (1980).
249. J. Cillard and P. Cillard, *J. Am. Oil Chem. Soc.*, **63**, 1165 (1986).
250. Z-L. Liu, L-J. Wang and Y-C. Liu, *Science in China*, **34**, 787 (1991).
251. K. Yamamoto and E. Niki, *Biochim. Biophys. Acta*, **958**, 19 (1988).
252. V. W. Bowry and R. Stocker, *J. Am. Chem. Soc.*, **115**, 6029 (1993).
253. V. W. Bowry and K. U. Ingold, *Acc. Chem. Res.*, **32**, 27 (1999).
254. S. R. Thomas, J. I. Neuzil and R. Stocker, *Arteriosclerosis, Thrombosis and Vascular Biology*, **16**, 687 (1996).
255. H. Shi, N. Noguchi and E. Niki, *Free Radical Biol. Med.*, **27**, 334 (1999).
256. L. R. C. Barclay, Unpublished results.
257. C. Rice-Evans, N. J. Miller and G. Paganga, *Free Radical Biol. Med.*, **20**, 933 (1996).
258. P.-G. Pietta, *J. Nat. Prod.*, **63**, 1035 (2000).

259. C. Kandaswami and E. Middleton, in *Natural Antioxidants: Chemistry, Health Effects and Applications* (Ed. F. Shahidi), Chap. 10, AOCS Press, Champaign, Illinois, 1997, pp. 174–203.
260. K. Mukai, W. Oka, K. Watanabe, Y. Egawa and S-I. Nagaoka, *J. Phys. Chem. A*, **101**, 3746 (1997).
261. V. A. Roginsky, T. K. Barsukova, A. A. Remorova and W. Bors, *J. Am. Oil Chem. Soc.*, **73**, 777 (1996).
262. M. Foti, M. Piatteli, M. T. Baratta and G. Ruberto, *J. Agric. Food Chem.*, **44**, 497 (1996).
263. J. Terao, M. Piskula and Q. Yao, *Arch. Biochem. Biophys.*, **308**, 278 (1994).
264. A. Arora, M. G. Nair and G. M. Strasburg, *Free Radical Biol. Med.*, **24**, 1355 (1998).
265. P. Sestili, A. Guidarelli, M. Dachà and O. Cantoni, *Free Radical Biol. Med.*, **25**, 196 (1998).
266. A. C. Santos, S. A. Uyemura, J. L. C. Lopes, J. N. Bazon, F. E. Mingatto and C. Curti, *Free Radical Biol. Med.*, **24**, 1455 (1998).
267. G. W. Plumb, K. R. Price, M. J. C. Rhodes and G. Williamson, *Free Radical Res.*, **27**, 429 (1997).
268. F. N. Ko, C. C. Chu, C. N. Lin, C. C. Chang and C-M. Teng, *Biochim. Biophys. Acta*, **1389**, 81 (1998).
269. W. Bors, C. Michel, W. Heller and H. Sandermann, in *Free Radicals, Oxidative Stress, and Antioxidants* (Ed. Özber), Chap. 8, Plenum Press, New York, 1998, pp. 85–92.
270. G. R. Buettner, *Arch. Biochem. Biophys.*, **300**, 535 (1993).
271. T. Fox and P. A. Kollman, *J. Phys. Chem.*, **100**, 2950 (1996).
272. P. Mulder, O. W. Saasted and D. Griller, *J. Am. Chem. Soc.*, **110**, 4090 (1988).
273. D. D. M. Wayner, E. Lusztyk and K. U. Ingold, *J. Org. Chem.*, **61**, 6430 (1996).
274. S. V. Jovanovic, M. Tosic and M. G. Simic, *J. Phys. Chem.*, **95**, 10824 (1991).
275. J. S. Wright, D. J. Carpenter, D. J. McKay and K. U. Ingold, *J. Am. Chem. Soc.*, **119**, 4245 (1997).
276. N. Russo, M. Toscano and N. Uccella, *J. Agric. Food Chem.*, **48**, 3232 (2000).
277. E. Johnson, Unpublished results.
278. G. A. DiLabio and J. S. Wright, *Free Radical Biol. Med.*, **29**, 480 (2000).
279. L. R. C. Barclay, M. Foti and M. R. Vinqvist, Unpublished results.
280. D. A. Pratt, G. A. DiLabio, G. Brigati, G. F. Pedulli and L. Valgimigli, *J. Am. Chem. Soc.*, **123**, 4625 (2001).
281. N. Noguchi and E. Niki, *Free Radical Biol. Med.*, **28**, 1538 (2000).
282. A. N. J. Moore and K. U. Ingold, *Free Radical Biol. Med.*, **22**, 931 (1997).
283. F. Antunes, L. R. C. Barclay, K. U. Ingold, M. King, J. Q. Norris, J. C. Scaiano and F. Xi, *Free Radical Biol. Med.*, **26**, 117 (1999).
284. L. R. C. Barclay, E. Crowe and C. D. Edwards, *Lipids*, **32**, 237 (1997).
285. W. A. Pryor, *Ann. Rev. Physiol.*, **48**, 657 (1986).
286. F. Antunes, A. Salvador, H. S. Marinho, R. Alves and R. E. Pinto, *Free Radical Biol. Med.*, **21**, 917 (1996).
287. A. Salvador, F. Antunes and R. E. Pinto, *Free Radical Res.*, **23**, 151 (1995).
288. D. D. M. Wayner, G. W. Burton, K. U. Ingold and S. Locke, *FEBS*, **187**, 33 (1985).

CHAPTER 13

Analytical aspects of phenolic compounds

JACOB ZABICKY

Institutes for Applied Research, Ben-Gurion University of the Negev, P. O. Box 653, Beer-Sheva 84105, Israel
Fax: +972-8-6472969; e-mail: zabicky@bgumail.bgu.ac.il

I. ACRONYMS	912
II. INTRODUCTION	913
A. Phenols in Nature and the Technological World	913
B. Some Considerations in Modern Analysis	928
C. Scope of the Chapter	929
III. GAS CHROMATOGRAPHY	930
A. Sample Preparation	930
1. General	930
2. Solid phase extraction vs. liquid–liquid extraction	930
3. Simultaneous distillation and extraction	931
4. Solid phase microextraction	932
5. Supercritical fluid extraction	932
B. Derivatization	933
1. General	933
2. Acylation	933
3. Silylation	934
4. Alkylation	935
5. Bromination	937
C. End Analysis	937
1. General	938
2. Environmental samples	938
3. Foodstuff samples	938
4. Miscellaneous industrial samples	941
IV. LIQUID CHROMATOGRAPHY	942
A. Sample Preparation	942
1. General	942

The Chemistry of Phenols. Edited by Z. Rappoport
© 2003 John Wiley & Sons, Ltd   ISBN: 0-471-49737-1

2. Solid phase extraction vs liquid–liquid extraction 944
   a. General aspects 944
   b. Environmental samples 945
   c. Foodstuffs 947
   d. Miscellaneous industrial products 948
   e. Biological and biomedical samples 948
3. Other preconcentration methods 949
B. Derivatization 949
  1. General 949
  2. Formation of azo dyes 950
  3. Formation of imino dyes 950
  4. Fluorescent tags and fluorescence enhancement 950
C. End Analysis 952
  1. General 954
  2. Foodstuffs 957
   a. General 957
   b. Fruit juice 957
   c. Wine and liquors 960
   d. Oils and fats 961
   e. Miscellaneous 961
  3. Environmental samples 962
   a. General 962
   b. Water 962
   c. Soil 965
  4. Biological and biomedical samples 965
   a. Plant extracts 965
   b. Effects of food intake 968
   c. Physiological and toxicological monitoring 968
  5. Miscellaneous industrial products 969
   a. Polymeric materials 969
   b. Disinfectants 970
   c. Liquid fuels 970
   d. Dyestuffs 970
  6. Structural and functional characterization 970
V. ELECTROPHORESIS 971
A. Environmental Samples 971
  1. Water 971
   a. Sensitivity enhancement 971
   b. Miscellaneous samples 971
  2. Soil 972
  3. Air 972
B. Foodstuffs 972
  1. General 972
  2. Wine and beer 972
  3. Honey 973
C. Biological Samples 973
D. Miscellaneous Industrial Samples 974
  1. Wood and paper 974
  2. Fuels 974

	E. Structural and Functional Characterization	974
VI.	BIOSENSORS	974
	A. Electrochemical Detection	974
	1. Working principles	974
	2. Biosensors based on tyrosinase	974
	3. Biosensors based on peroxidases and other enzymes	977
	4. Biosensors incorporating tissues and microorganisms	978
	5. Amplification processes	979
	B. Spectrophotometric and Colorimetric Detection	980
VII.	ELECTROCHEMICAL METHODS	981
	A. Voltammetric Detection	981
	1. Quantitative analysis	981
	2. Structural and functional characterization	982
	B. Amperometric Detection	983
	1. Quantitative analysis	983
	2. Structural and functional characterization	984
	C. Polarography	984
	D. Potentiometric Titrations	984
	1. Quantitative analysis	984
	2. Structural and functional characterization	984
VIII.	ULTRAVIOLET-VISIBLE DETECTION METHODS	984
	A. Spectrophotometry and Colorimetry	984
	1. Direct determination	984
	2. Derivatization	985
	a. Halogenation	985
	b. Oxidative coupling	985
	c. Coupling with diazonium ions	987
	d. Complex formation	988
	e. Miscellaneous color-developing methods	990
	3. Structural and functional characterization	991
	B. Fluorescence Detection	993
	1. Direct determination	993
	2. Fluorescent labelling	995
	C. Chemiluminescence Detection	995
IX.	INFRARED AND RAMAN SPECTRAL METHODS	997
	A. Quantitative Analysis	997
	B. Structural and Functional Characterization	997
X.	NUCLEAR MAGNETIC RESONANCE	997
	A. Quantitative Analysis	997
	B. Structural and Functional Characterization	998
XI.	MASS SPECTROMETRY	999
XII.	MISCELLANEOUS METHODS	1000
	A. Surface Plasmon Resonance	1000
	B. Miscellaneous Titrations	1000
	C. Piezoelectric Detection	1001
	D. Thermal Analysis	1002
	E. Phenol as a Measuring Stick	1002
XIII.	REFERENCES	1002

I. ACRONYMS

AED	atomic emission detector
AMD	amperometric detector/detection
APCI-MS	atmospheric pressure chemical ionization MS
BET	Brunauer–Emmett–Teller method
CE	capillary electrophoresis
CLD	chemiluminescence detector/detection
CPE	carbon paste electrode
CZE	capillary zone electrophoresis
DA-UVD	diode array UVD
DPV	differential pulse voltammetry
DRD	differential refractometric detection
ECD	electron capture detector/detection
ELD	electrochemical detector/detection
EPA	U.S. Environmental Protection Agency
ESI-MS	electrospray ionization MS
FAB-MS	fast atom bombardment MS
FIA	flow injection analysis
FID	flame ionization detector/detection
FLD	fluorimetric detector/detection
GCB	graphitized carbon black
GCE	glassy carbon electrode
GPC	gel permeation chromatography
IEC	ion exchange chromatography
ISP-MS	ion spray MS
ITD-MS	ion trap detector MS
LLE	liquid–liquid extraction
LOD	limits of detection
LOQ	limits of quantation
MALDI	matrix assisted desorption-ionization
MAP	microwave assisted process
MEC	micellar electrokinetic chromatography
MLC	micellar liquid chromatography
MS	mass spectrum/spectra/spectrometry
PBEI-MS	particle-beam electron-impact MS
PCR	principal components regression
PLS	partial least squares
RP-	reversed phase, for example RP-HPLC
SAX	strong anion exchanger
SDE	simultaneous distillation and extraction
SFC	supercritical fluid chromatography
SFE	supercritical fluid extraction
SIA	sequential injection analysis
SIM	selected ion monitoring
SNR	signal-to-noise ratio
SPE	solid phase extraction
SPME	solid phase microextraction
SPR	surface plasmon resonance
TSP-MS	thermospray-MS
UVD	ultraviolet-visible detector/detection
UVMA	ultraviolet multiwavelengths absorptiometry
UVV	ultraviolet-visible

II. INTRODUCTION
A. Phenols in Nature and the Technological World

Phenolic compounds are extensively distributed in living organisms. Only a small sampling of the intense research activity involving these compounds will be mentioned here. L-Tyrosine (**1**) is a protein-building amino acid present in all cells. Some of the simple phenolic plant constituents yield polymeric materials such as lignins and procyanidins. Lignin is a macromolecular substance derived from n-propylbenzene building blocks. Together with cellulose and other polysaccharides lignin forms the woody tissues that constitute the mechanical support of higher plants. Humic acids are phenolic degradation products of plant debris found in the soil. Flavonoids are polyphenolic compounds derived from 1,3-diphenylpropane, where the aliphatic chain is part of a six-membered heterocyclic ring. Various classes are distinguished, depending on the structure of the heterocyclic ring: Catechins, such as (−)-epicatechin (**2**) and (+)-catechin (**3**), flavones, such as apigenin (**4**), flavanones, such as naringenin (**5**) and flavonols, such as kaempferol (**6**); many of these compounds appear as glycosides. Flavonoids and other phenolic constituents may contribute to leaf or seed resistance to insect, or pathogenic fungal attack[1–4]. Tannins are polyhydic phenolic compounds of complex structure, found in extracts from many parts of the plants; for example, corilagin (**7**) is formed from gallic acid (**8**) and glucose blocks. GC-MS identification of gallic acid and inositol in extracts from an Egyptian mummy pointed to the use of tannins in the embalming process[5].

A plethora of simpler phenolcarboxaldehydes and phenolcarboxylic acids are found in plant extracts that contribute to the organoleptic properties of derived foodstuffs, such as fruit juices, wine and oil. The state of ripening of a cultivar may also affect these properties, by changing the nature and concentration of the relevant phenolic compounds[6,7]. The distribution of phenolic compounds usually varies from tissue to tissue in the same individual, from variety to variety for the same species[8–15] and from species to species for the same genus[16,17]. A close correlation between the phenolic compound patterns and the botanical origin of plants was found[18]. An individual hybrid plant tends to reproduce the characteristics of the parental taxa[19], thus, for example, the genetic composition of *Equisetum* hybrids in the British Isles could be determined not only from their morphological characteristics, but also based on HPLC and TLC determination of their phenolic constituents, such as caffeic acid conjugates, flavonoids and styrylpyrones. The involved analytical profiles shown by the phenolic compounds present in certain tissues may serve for forensic identification. Thus, reversed phase HPLC (RP-HPLC) analysis with diode-array ultraviolet-visible detection (DA-UVD) of the phenolic extracts from samples of Portuguese quince jams showed that certain specimens contained arbutin (**9**), suggesting adulteration with pear puree[20].

Important phenolic compounds that can be found in the animal kingdom are the catecholamines (e.g. dopamine, **10a**; dopa, **10b**) that are essential to the physiology of the nervous system, steroidal hormones such as estrone (**11**), amino acid hormones such as thyroxine (**12**) and polypeptidic hormones containing tyrosine (**1**) residues such as the drugs shown in Table 1 carrying note *f*.

Development of analytical methods for certain classes of phenolic compounds is necessary in support of food related clinical investigations. A monograph appeared on biologically active oxidants and antioxidants[21] and the effects of the latter on the food intake[22]. Interest in the flavonoids, isoflavonoids and other phenolic constituents arose for their varied potential pharmacological action, such as anticarcinogenic properties[23–25]. Tea catechins and flavonoids have been reported as antioxygenic[26,27], antimutagenic[28] and possessing prophylactic activity against hypertension[29]. Caffeic (**25**), coumaric (**26**) and protocatechuic (**27**) acids were investigated *in vitro* for their inhibitory action on the oxidation of human low-density lipoprotein in serum[30]. The antioxidant action of **25**, **26**[30]

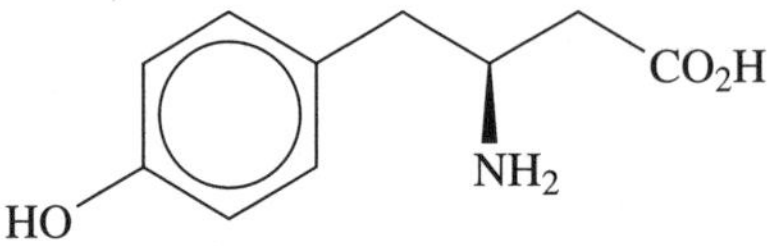

(1)

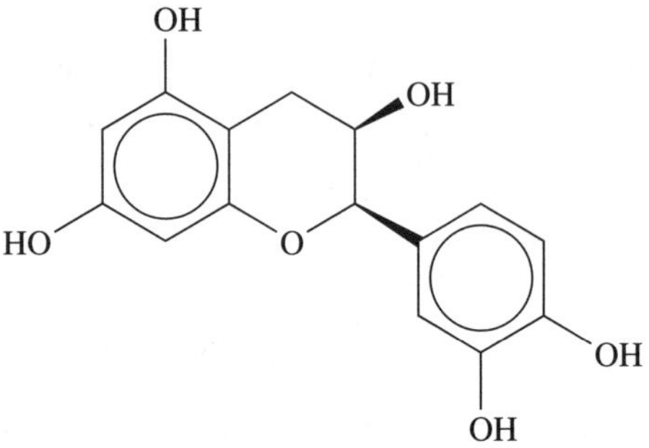

(2)

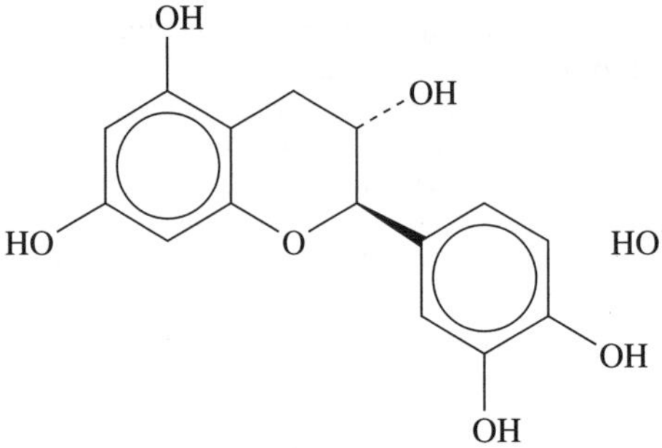

(3)

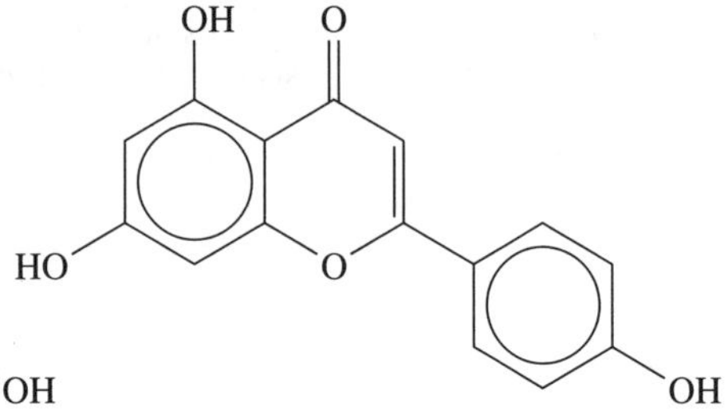

(4)

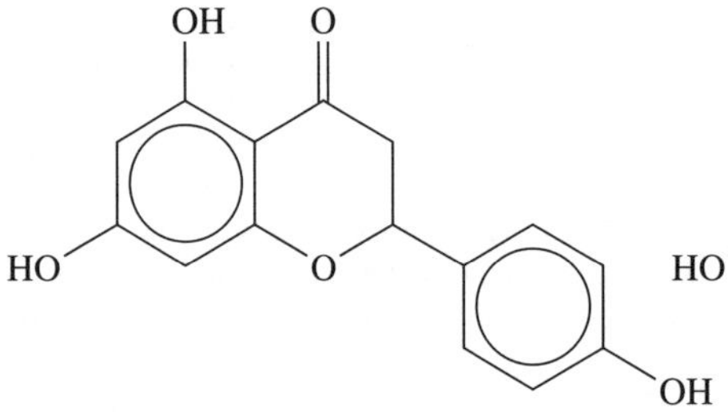

(5)

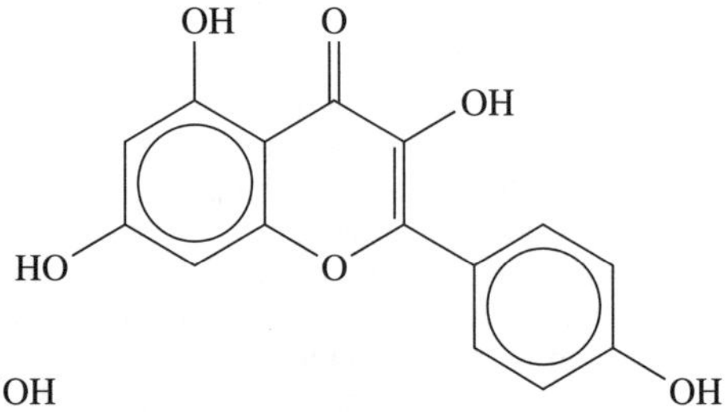

(6)

and flavonoids[27] has been linked to the lower incidence of coronary disease in populations with high red wine intake. Resveratrol (**28a**) has been attributed many properties of clinical relevance. This compound and its 3β-glycoside (picein, **28b**) are found in groundnuts (*Arachis hypogaea*) and grape products[31].

Phenol is a heavy chemical used to manufacture phenolic resins and organic intermediates such as bisphenol A (**29**), salicylic acid (**21**), alkylphenols, aniline, xylenols and cyclohexanone (as a precursor of adipic acid, **30**). Natural and synthetic phenolic

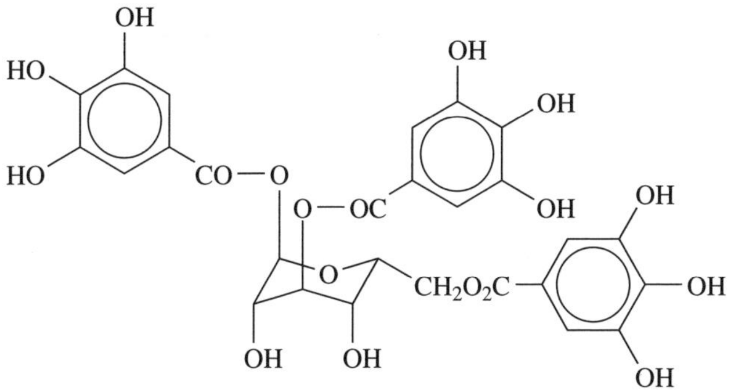

(7)

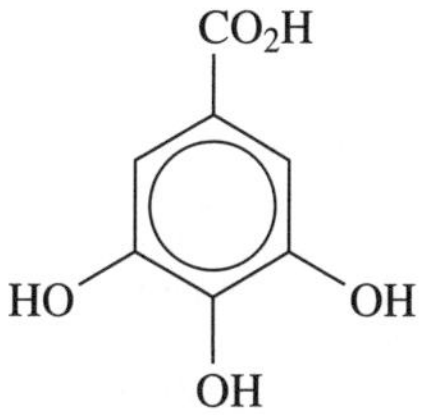

(8)

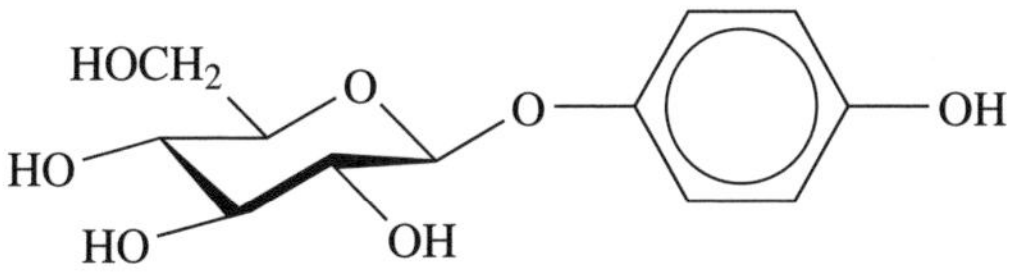

(9)

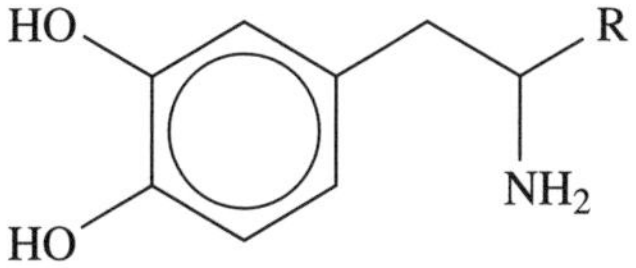

(10) (a) R = H
(b) R = CO$_2$H

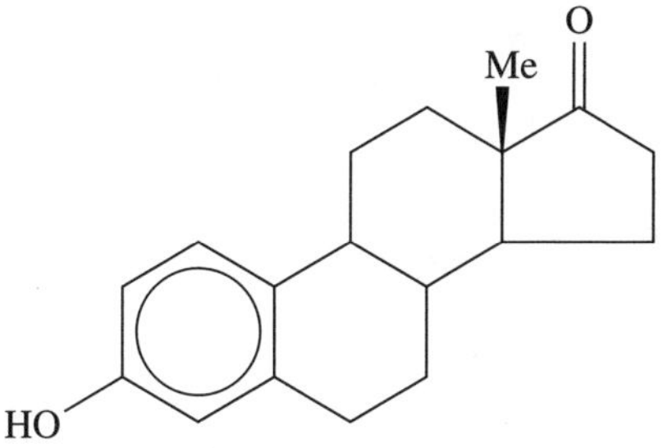

(11)

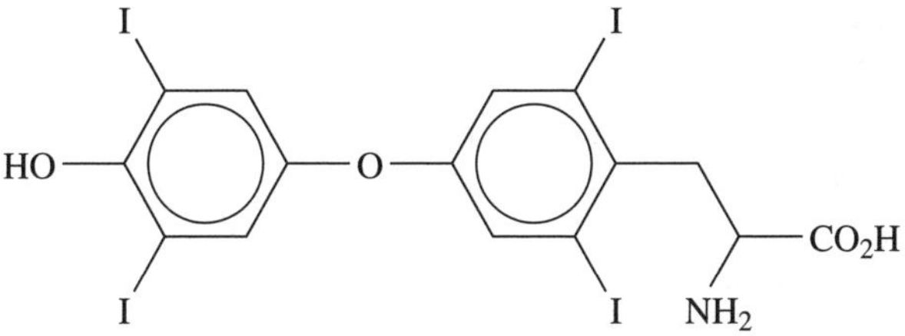

(12)

TABLE 1. Phenolic compounds listed in the USP[a]

Compound [CAS registry number]	Notes
Acetaminophen [103-90-2]	
Albuterol [18558-94-9]	b
Apomorphine hydrochloride [41372-20-7]	c
Bismuth subgallate [99-26-3]	
Buprenorphine hydrochloride [53152-21-9]	c
Butorphanol tartrate [58786-99-5]	c
Carbidopa [28860-95-9]	b
Cefadroxil (**13**) [119922-85-9]	
Cefoperazone sodium (**14**) [62893-20-3]	
Cefpiramide [70797-11-4]	
Cefprozil [121123-17-9]	
Chlorotetracycline hydrochloride [64-72-2]	d
Chloroxylenol [88-04-0]	
Clioquinol [130-26-7]	
Demeclocycline [127-33-3]	d
Dienestrol [84-17-3; 13029-44-2]	
Diethylstilbestrol [56-53-1]	
Dioxybenzone [131-53-3]	
Dobutamine hydrochloride [49745-95-1]	b
Dopamine (**10a**) hydrochloride [62-31-7]	b
Doxycycline [17086-28-1]	c
Doxylamine succinate [562-10-7]	
Dronabinol [1972-08-3]	
Edrophonium chloride [116-38-1]	
Epinephrine (**15a**) [51-43-4]	b
Epitetracycline hydrochloride [23313-80-6]	d
Equilin [474-86-2]	e
Estradiol [50-28-2]	e
Estriol [50-27-1]	e
Estrone [53-16-7]	e
Ethinyl estradiol [57-63-6]	e
Eugenol [97-53-0]	
Fluorescein (**16**) [2321-07-5]	
Glucagon [16941-32-5]	f
Hexylresorcinol [136-77-6]	
Homosalate [118-56-9]	g
Hydroxymorphone hydrochloride [71-68-1]	c
Hydroxyamphetamine hydrochloride [306-21-8]	
Hydroxyzine pamoate [10246-75-0]	h
Insulin [11070-73-8, 11061-68-0]	f
Iodoquinol [83-73-8]	
Isoetharine hydrochloride [2576-92-3]	b
Isoproterenol hydrochloride [51-30-9]	b
Isoxsuprine hydrochloride [579-56-6, 34331-89-0]	
Labetalol hydrochloride [32780-64-6]	
Levodopa [59-92-7]	b
Levonordefrin [18829-78-2, 829-74-3]	b
Levorphanol tartrate [125-72-4]	c
Levothyroxin sodium [25416-65-3, 55-03-8]	i
Liothyronine sodium [55-06-1]	i
Meclocycline sulfosalicylate [73816-42-9]	d, g
Mesalamine [89-57-6]	g
Metacresol (**17**) [108-39-4]	
Metaproterenol sulfate [5874-97-5]	
Metanminol bitartrate [33402-03-8]	
Methacycline hydrochloride [3963-95-9]	d

13. Analytical aspects of phenolic compounds

TABLE 1. (*continued*)

Compound [CAS registry number]	Notes
Methyldopa [41372-08-1, 555-30-6]	*b*
Methyldopate hydrochloride [5123-53-5, 2509-79-4]	*b*
Methyrosine [672-87-7]	*i*
Minocycline hydrochloride [13614-98-7]	*d*
Morphine (**18**) sulfate [6211-15-0, 64-51-3]	
Nalorphine hydrochloride [57-29-4]	*c*
Naloxone hydrochloride [357-08-4]	*c*
Naltrexone hydrochloride [16676-29-2]	*c*
Norepinephrine (**15b**) bitartrate [69815-49-2]	*b*
Octyl salicylate [118-60-5]	*g*
Oxybenzone [131-57-7]	*g*
Oxymetazoline hydrochloride [2315-02-8]	
Oxymorphine hydrochloride [357-07-3]	*c*
Oxytetracycline [6153-64-6]	*d*
Oxytocin [50-56-6]	*f*
Parachlorophenol [106-48-9]	
Pentazocine [359-83-1]	*c*
Phenol [108-95-2]	
Phenolphthalein (**19**) [77-09-8]	*j*
Phenylephrine hydrochloride [61-76-7]	
Potassium guaiacolsulfonate [78247-49-1]	
Probucol [23288-49-5]	
Pyrantel pamoate [22204-24-6]	*h*
Pyridoxine hydrochloride [58-56-0]	*k*
Pyrvinium pamoate [3546-41-6]	*h*
Raclopride C-11	*l*
Recepinephrine [329-65-7]	*b*
Resorcinol (**20**) [108-46-3]	
Rifampin [13292-46-1]	*m*
Ritodrine hydrochloride [23239-51-2]	
Rosebengal sodium I-131	*l*
Roxarsone [121-19-7]	
Salicylamide [65-45-2]	*g*
Salicylic acid (**21**) [69-72-7]	
Salsalate [552-94-3]	*g*
Sargamostim [123774-72-1]	*f*
Sulfasalazine [599-79-1]	*g*
Sulisobenzone [4065-45-6]	*g*
Terbutaline sulfate [23031-32-5]	*b*
Tetracycline (**22**) [60-54-8]	
Tubocurarine chloride [41354-45-4]	
Tyrosine (**1**) [60-18-4]	
Vancomicin [1404-93-9]	
Vasopressin [50-57-7]	*f*

[a] The United States Pharmacopeia[42]. Entries are presented with quality control analytical procedures.
[b] A catecholamine. Model compound: Dopamine (**10a**).
[c] A morphine (**18**) alkaloid analogue.
[d] A tetracycline (**22**) antibiotic analogue.
[e] A steroidal hormone. Model compound: Estrone (**11**).
[f] A polypeptide hormone with at least one tyrosine (**1**) residue.
[g] A derivative of salicylic acid (**21**).
[h] Pamoic acid (**23**) is a phenolic compound.
[i] A phenolic α-amino acid. Model compound: Tyrosine (**1**).
[j] This pharmaceutical was removed from the USP[42] but it is still listed in its British counterpart[43].
[k] A phenolic derivative of pyridine.
[l] A radioactive isotope carrier.
[m] A rifamycine (**24**) antibiotic analogue.

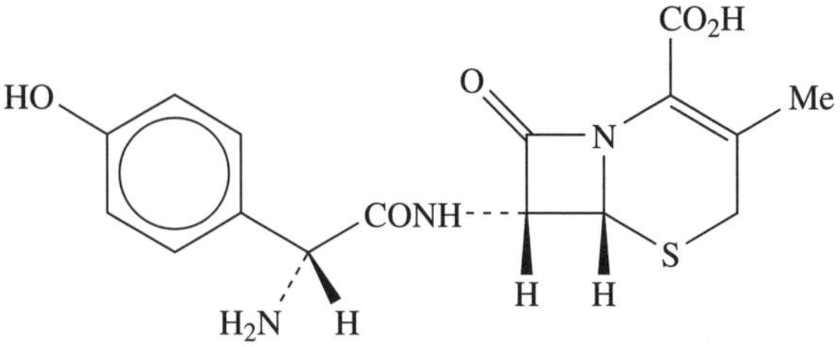

(13)

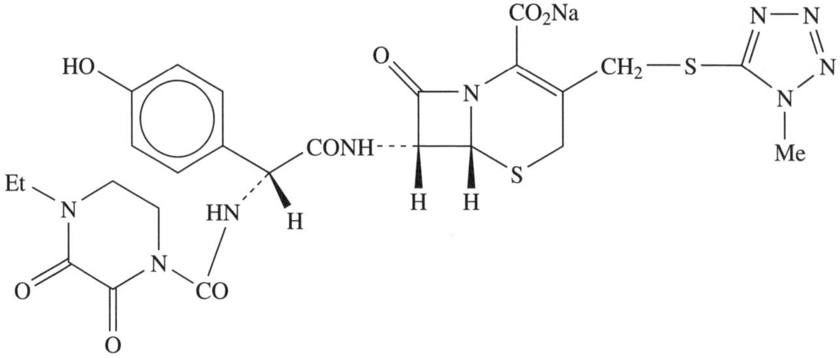

(14)

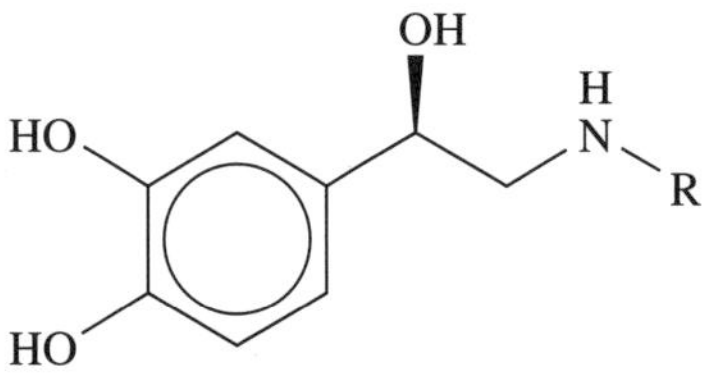

(15) (a) R = Me
(b) R = H

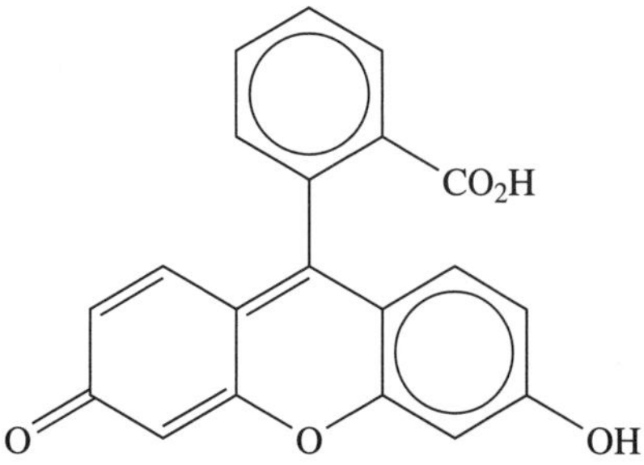

(16)

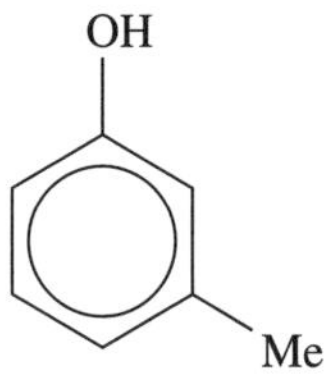

(17)

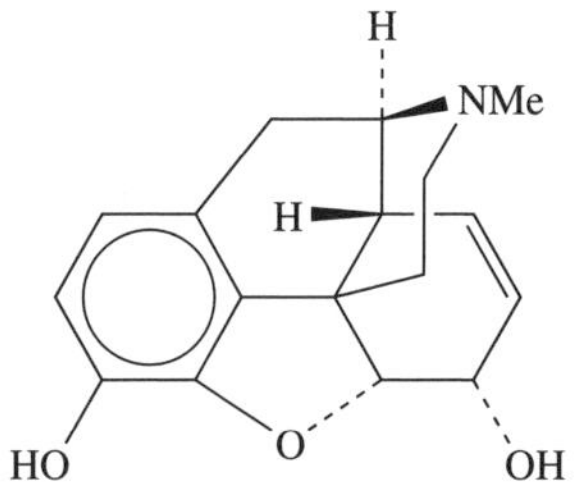

13. Analytical aspects of phenolic compounds

(18)

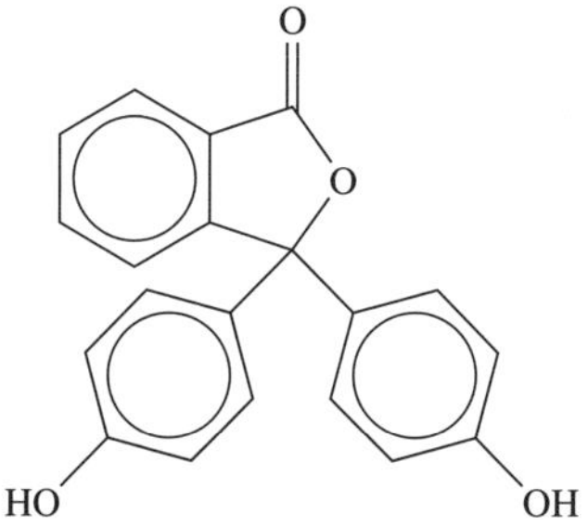

(19)

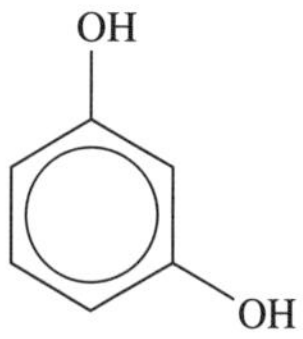

(20)

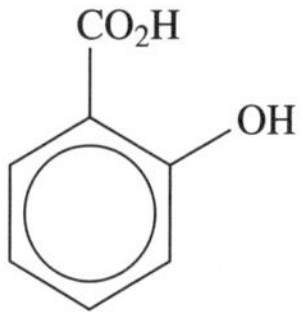

(21)

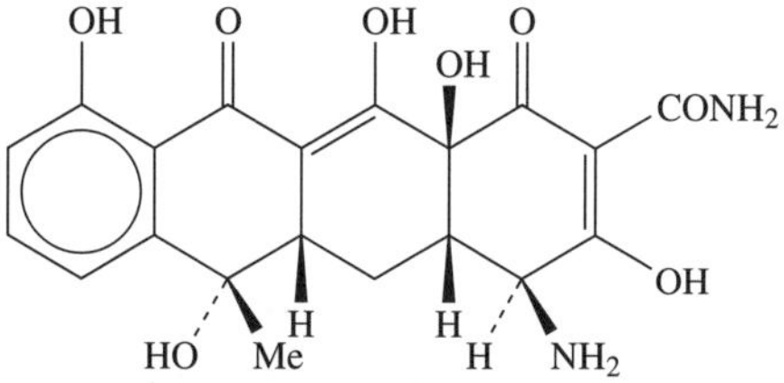

(22)

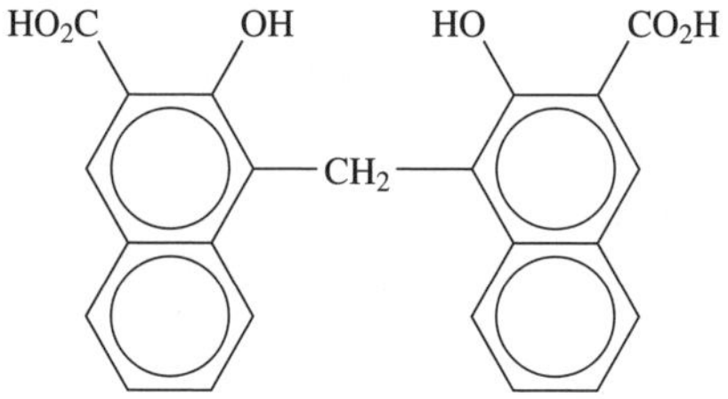

(23)

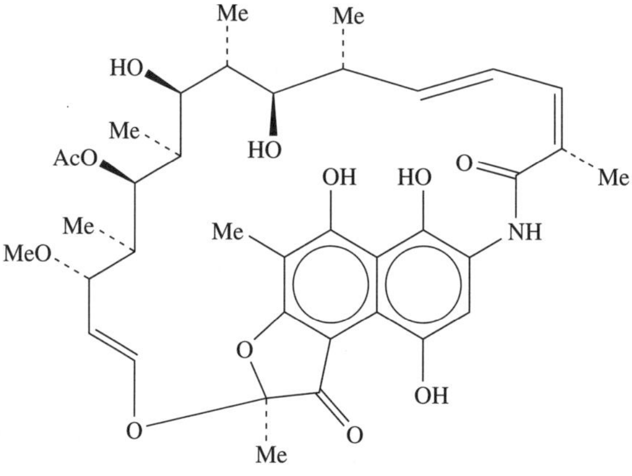

(24)

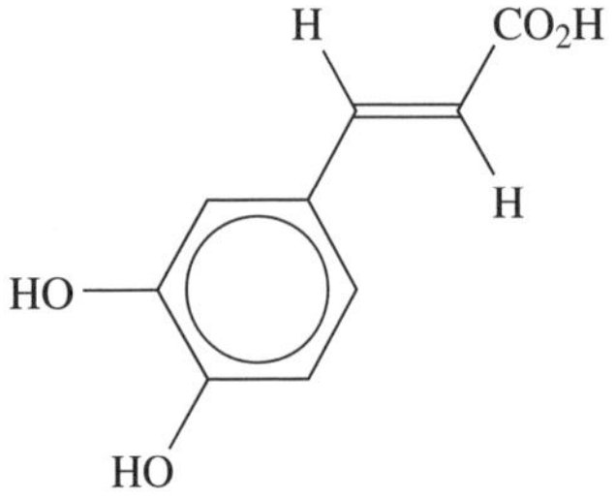

compounds that have found pharmaceutical application are listed in Table 1. The dyes are an important class of industrial products, including hundreds of organic compounds of varied structure, many of which contain phenolic moieties. In Table 2 appear commercially available dyes listed in the Color Index. Although compounds containing the azo group are predominant in this list, several other classes of dyes are represented. Synthetic antioxidants such as those appearing in Table 3 are added to foods, drugs and other manufactured products to inhibit autooxidation. As these additives are somewhat toxic, it is necessary to control the amount added to any food or drug. A plethora of computational methods have been developed to correlate structure and properties of compounds, including many aspects of biological behavior (toxicity, pharmacological activity, growth promotion and inhibition, etc.)[32]. The presence of phenolic compounds in urine points to exposure to aromatic hydrocarbons, such as benzene and condensed polycyclic hydrocarbons, frequent in gasoline station operators and tar-related industries[33–35].

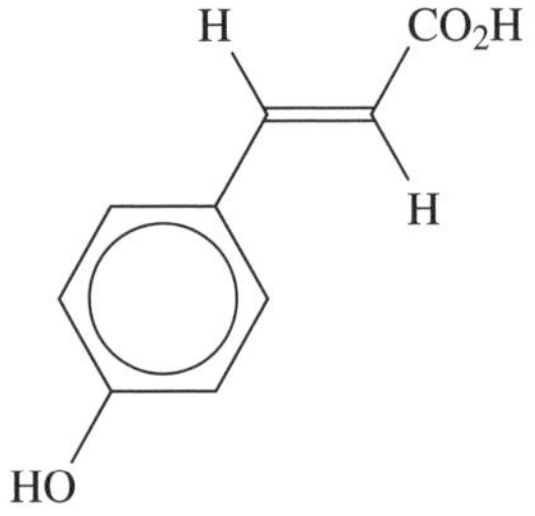

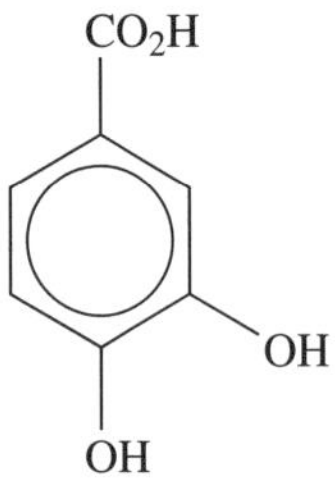

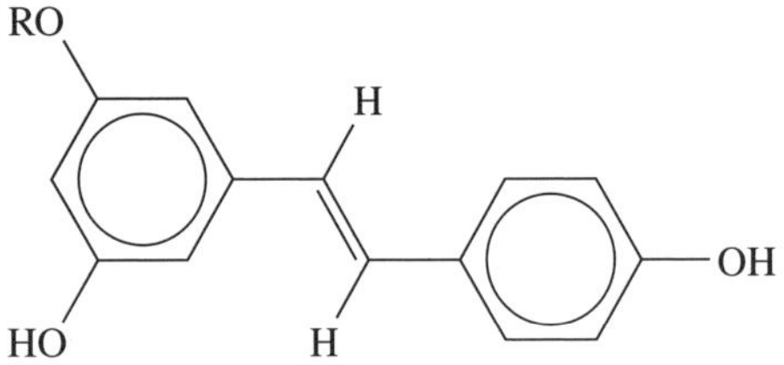

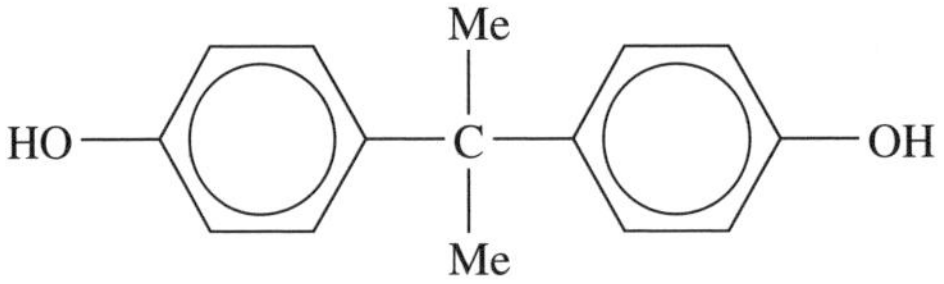

The appearance of simple phenolic compounds in water points to pollution stemming from industrial sources, such as manufacturers of dyes, drugs, antioxidants, pulp and paper, or may be the result of pesticide application. The presence of certain phenols in

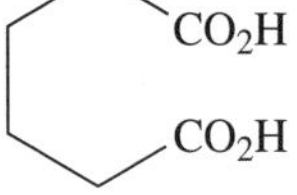

13. Analytical aspects of phenolic compounds

TABLE 2. Commercially available phenolic dyes listed in the Color Index (CI)

Dyes [CAS registry number] and Color Index number	Properties[a]	Notes
Acid Alizarin Violet N [2092-66-9] CI 15670	λ_{max} 663 nm, I(2)997C, U1, SR(2)2747B, DB7012000	b
Acid Blue 45 [2861-02-1] CI 63010	λ_{max} 595 nm, I(2)1017D, U8, SR(2)2773F, CB0548500	b
Acid Orange 8 [5850-86-2] CI 15575	λ_{max} 490 nm, I(2)981D, U25, SR(2)2747A,	b
Acid Red 88 [1658-56-6] CI 15620	λ_{max} 505 nm, I(2)986D, U35, SR(2)2751A, QK2420000	b
Acid Red 97 [10169-02-5] CI 22890	λ_{max} 498 nm, I(2)993D, U36, SR(2)2765C, GC5787480	b
Acid Red 114 [6459-94-5] CI 23635	λ_{max} 514(365) nm, I(2)993B, U38, QJ6475500	b
Acid Red 151 [6406-56-0] CI 26900	λ_{max} 512(356) nm, I(2)989C, U39, SR(2)2761O, DB7084500	b, c
Acid Red 183 [6408-31-7] CI 18800	λ_{max} 494 nm, I(2)1001B, U40, SR(2)2753K	b
Acid Violet 7 [1658-56-6] CI 18055	λ_{max} 520 nm, I(2)984D, U42, SR(2)2749E, QJ6000000	b
Acid Yellow 99 [10343-58-5] CI 13900	λ_{max} 445 nm, I(2)978B, U56, SR(2)2743I	b, d
Alizarin [72-48-0] CI 58000	λ_{max} 609(567) nm, I(2)86D, U75, SH93C, SR(2)1689E, CB6580000	e
Alizarin Blue Black B [1324-21-6] CI 63615	λ_{max} 548 nm, I(2)511C, U77, SR(2)2217H	e
Alizarin Red S [130-22-3] CI 58005	λ_{max} 556(596) nm, I(2)510D, U80, SR(2)2217D, CB1095300	e
Alizarin Yellow GG [584-42-9] CI 14025	λ_{max} 362 nm, I(2)972A, U84, SR(2)2741H, DH2528550	b
Allura Red AC [25956-17-6] CI 16035	λ_{max} 504 nm, QK2260000	b
Amaranth [915-67-3] CI 16185	λ_{max} 521 nm, I(2)988B, U92, SH118A, SR(2)2751K, QJ6550000	b
Aurintricarboxylic acid trisodium salt [13186-45-3] CI 43810	λ_{max} 525 nm, I(2)1031B, U106, SR(2)2775D, DG4975325	f
Biebrich Scarlet [4196-99-0] CI 26905	λ_{max} 505 nm, I(2)986B, U137, SH413A, SR(2)2763D	b
Bordeaux R [5858-33-3] CI 16180	λ_{max} 518 nm, I(2)999D, U148, SR(2)2751I, QJ6479500	b
Brilliant Black BN [2519-30-4] CI 28440	λ_{max} 570(407) nm, I(2)998A, U150, SR(2)2763M, QJ5950000	b
Brilliant Crocein MOO [5413-75-2] CI 27290	λ_{max} 510 nm, I(2)989D, U158, SR(2)2763C	b
Brilliant Yellow [3051-11-4] CI 24890	λ_{max} 397 nm, I(2)980A, U164, SR(2)2759H	b
Carmine [1390-65-4] CI 75470	λ_{max} 531(563) nm, U190, FH8891000	d, e
Carminic acid [1260-17-9] CI 75470	λ_{max} 495 nm, I(2)246D, U193, SR(2)1835E	e
Celestine Blue [1562-90-9] CI 51050	λ_{max} 642 nm, I(2)1041B, U196, SR(2)2811J	g
Chicago Sky Blue 6B [2610-05-1] CI 24410	λ_{max} 618 nm, I(2)996D, U198, SR(2)2765H, QJ6430000	b
Chrome Azurol S [1667-99-8] CI 43825	λ_{max} 458 nm, I(2)1032B, U203, SR(2)2775E	f
Chromotrope FB [3567-69-9] CI 14720	λ_{max} 515(383) nm, I(2)988C, U204, SR(2)2751J, QK1925000	b

(*continued overleaf*)

TABLE 2. (*continued*)

Dyes [CAS registry number] and Color Index number	Properties[a]	Notes
Chromotrope 2B [548-80-1] CI 16575	λ_{max} 514 nm, I(2)986A, U206, SR(2)2747N	b
Chromotrope 2R [4197-07-3] CI 16570	λ_{max} 510(530) nm, I(2)982D, U207, SR(2)2747K, QJ6418000.	b
Chromoxane Cyanine R [3564-18-9] CI 43820	λ_{max} 512 nm, I(2)1031C, U209, SR(2)2791C.	f
Cibacron Brilliant Red 3B-A [17681-50-4] CI 18105	λ_{max} 517 nm, I(2)1005C, U215, SH3030D, SR(2)2757H	b, h
Crocein Orange G [1934-20-9] CI 15970	λ_{max} 482 nm, I(2)981B, U238, SR(2)2745M	b
Crystal Scarlet [2766-77-0] CI 16250	λ_{max} 510 nm, I(2)987D, N(3)550B, SR(2)2751H	b
Diazine Black [4443-99-6] CI 11815	λ_{max} 584 nm	
4′,5′-Dibromofluorescein [596-03-2] CI 45370.1	λ_{max} 450 nm, I(2)1011A, U249, SR(2)2793N, LM5200000	i
Diiodofluorescein [31395-16-1] CI 45425.1	λ_{max} 522 nm, I(2)1011B, U257, SR(2)2793O	i
Direct Blue 71 [4399-55-7] CI 34140	λ_{max} 594 nm, I(2)992C, U268, SR(2)2767F	b
Direct Red 23 [3441-14-3] CI 29160	λ_{max} 507 nm, I(2)996B, U270, SR(2)2763O	b
Direct Red 75 [2829-43-8] CI 25380	λ_{max} 522 nm, I(2)995D, U272, SR(2)2765B	b
Direct Red 80 [2610-10-8] CI 35780	λ_{max} 528 nm, SR(2)2767G	b
Direct Red 81 [2610-11-9] CI 28160	λ_{max} 508(397) nm, I(2)994D, U274, SR(2)2763G, QK1370000	b
Direct Violet 51 [5489-77-0] CI 27905	λ_{max} 549 nm, I(2)995B, U276, SR(2)2763F	b
Disperse Orange 13 [6253-10-7] CI 26080	λ_{max} 427 nm, SR(2)2759k	b
Disperse Yellow 3 [2832-40-8] CI 11855	λ_{max} 357 nm, I(2)969A, U293, SH1477D, SR(2)2741J	b
Disperse Yellow 7 [6300-37-4] CI 26090	λ_{max} 385 nm, I(2)969A, U293, SR(2)2759C, SM1140030	b
Eosin B [548-24-3] CI 45400	λ_{max} 514(395) nm, I(2)1011D, U300, SR(2)2797L	i
Eosin Y [17372-87-1] CI 45380	λ_{max} 517 nm, I(2)1011C, U304, SR(2)2795C, LM5850000	i
Eriochrome Black T [1787-61-7] CI 14645	λ_{max} 503 nm, I(2)997D, U308, SR(2)2751F, QK2197000	b
Eriochrome Blue Black B [3564-14-5] CI 14640	λ_{max} 528 nm, I(2)987C, U310, SR(2)2751D, QK2195000	b
Erythrosin B [16423-68-0] CI 45430	λ_{max} 525 nm, I(2)1014D, U314, SR(2)2795I, LM5950000	i
Ethyl Eosin [6359-05-3] CI 45386	λ_{max} 532 nm, I(2)1012B, U320, SR(2)2795G	i
Evans Blue [314-13-6] CI 23860	λ_{max} 611 nm, I(2)996C, U327, SH1673A, SR(2)2765G, QJ6440000	b
Fast Green FCF [253-45-9] CI 42053	λ_{max} 622(427) nm, I(2)1033C, U345, SH1678A, SR(2)2779I, BQ4425000	f
Fluorescein sodium salt [518-47-8] CI 43350	λ_{max} 491 nm, I(2)1010B, U375, SH1688C, SR(2)2793B, LM5425000	i
Gallocyanine [1562-85-2] CI 51030	λ_{max} 601 nm, I(2)888A, U386, SR(2)2811I, SP7692000	g

13. Analytical aspects of phenolic compounds 923

TABLE 2. (continued)

Dyes [CAS registry number] and Color Index number	Properties[a]	Notes
8-Hydroxy-1,3,6-pyrenetrisulfonic acid, trisodium salt [6358-59-6] CI 59040	λ_{max} 403 nm, UR2700000	
Indoine Blue [4569-88-4] CI 12210	λ_{max} 589 nm, U409, SH1981B, SR(2)2809E, SG1608000	b, j
Methylene Violet [2516-05-4] CI 52041	λ_{max} 580 nm, I(2)1036D, U453, SR(2)2815A	b, k
Methyl Eosin [23391-49-3] CI 45385	λ_{max} 520 nm, I(2)1012A, U456, SR(2)2795F	i
Mordant Blue 9 [3624-68-8] CI 14855	λ_{max} 622(427) nm, I(2)1033C, U473, SR(2)2747L	b
Mordant Brown 1 [3564-15-6] CI 20110	λ_{max} 373(487) nm, I(2)990B, U474, SR(2)2763A	b
Mordant Brown 4 [6247-27-4] CI 11335	λ_{max} 500(374) nm, I(2)967D, U475, SR(2)2739L	b
Mordant Brown 24 [5370-46-3] CI 11880	λ_{max} 373(487) nm, I(2)978A, U477, SR(2)2741K	b
Mordant Brown 33 [3618-62-0] CI 13250	λ_{max} 442 nm, I(2)977D, U478, SR(2)2743F	b
Mordant Brown 48 [6232-63-7] CI 11300	λ_{max} 492 nm, I(2)968A, U479, SR(2)2739M	b
Mordant Orange 1 [2243-76-7] CI 14030	λ_{max} 385 nm, I(2)972B, U480, SR(2)2741I, VO5310000	b
Mordant Orange 10 [6406-37-7] CI 26560	λ_{max} 386 nm, I(2)978D, U483, SR(2)2759G	b
Mordant Red 19 [1934-24-3] CI 18735	λ_{max} 413 nm, I(2)1000C, U484, SR(2)2753I	b, c
Mordant Yellow 10 [6054-99-5] CI 14010	λ_{max} 354 nm, I(2)977A, U486, SR(2)2743G, GB4450000	b
Mordant Yellow 12 [6470-98-0] CI 14045	λ_{max} 380 nm, I(2)968D, U487, SR(2)2741G	b
Naphthochrome Green [5715-76-4] CI 44530	λ_{max} 362 nm, I(2)1031D, U490, SR(2)2775F	f
Naphthol Blue Black [1064-48-8] CI 20470	λ_{max} 618 nm, I(2)990C, U496, SH2505A, SR(2)2759, QJ6196000	b
Naphthol Green B [19381-50-1] CI 10020	λ_{max} 714 nm, I(2)954B, U498, SR(2)2833N	d
Naphthol Yellow S [846-70-8] CI 10316	λ_{max} 428 nm, I(2)954A, U500, SH1685A, SR(2)2833O, QK1813000	
New Coccine [2611-82-7] CI 16255	λ_{max} 506(350) nm, I(2)988D, U506, SH2534A, SR(2)2751L, QJ6530000	b
Nitrazine Yellow [5423-07-4] CI 14890	λ_{max} 586 nm, I(2)985A, U500, SH2446A, SR(2)2749A	b
Nuclear Fast Red [6409-77-4] CI 60760	λ_{max} 518 nm, I(2)511A, U532, SR(2)2217E	e
Oil Red EGN [4477-79-6] CI 26120	λ_{max} 521 nm, I(2)975C, U536, SH2651C, SR(2)2761L	b
Oil Red O [1320-06-5] CI 26125	λ_{max} 518(359) nm, I(2)976B, U537, SH2651D, SR(2)2761M	b
Orange G [1936-15-9] CI 16230	λ_{max} 475 nm, I(2)982C, U539, SH2655D, SR(2)2747H, QJ6500000	b
Orange II [633-96-5] CI 15510	λ_{max} 483 nm, U541, SH2656A, SR(2)2745N, DB7084000	b

(continued overleaf)

TABLE 2. (continued)

Dyes [CAS registry number] and Color Index number	Properties[a]	Notes
Orange OT [2646-17-5] CI 12100	λ_{max} 505 nm, I(2)949C, SR(2)2745C, QL5425000	b
Palatine Chrome Black 6BN [2538-85-4] CI 15705	λ_{max} 569 nm, I(2)987A, U545 SR(2)2751B, QK2200000	b
Palatine Fast Black WAN [5610-64-0] CI 15711	λ_{max} 588 nm, I(2)989A, U547, SR(2)2751E	b, d
4-Phenylazophenol [1689-82-3] CI 11800	λ_{max} 347 nm, I(2)965B, U574, SH2763C, SR(2)2737C, SM8300000	b
Phloxine B [18472-87-2] CI 45410	λ_{max} 515(383) nm, I(2)1013A, U577, SR(2)2795J, LM5900000	i
Plasmocorinth B [1058-92-0] CI 16680	λ_{max} 527 nm, I(2)983A, U581, SR(2)2747M	b
Ponceau SS [6226-78-4] CI 27190	λ_{max} 514(351) nm, I(2)990A, U585, SR(2)2763B	b
Purpurin [81-54-9] CI 58205	λ_{max} 515(521) nm, I(2)913C, U592, SH2985D, SR(2)1689K, CB8200000	e
Quinizarin [81-64-1] CI 58050	I(2)87A, U606, SH3024A, SR(2)1689F, CB6600000	e
Reactive Orange 16 [12225-83-1] CI 17757	λ_{max} 494(388) nm, U616, SR(2)2749J	b
Rosolic Acid [603-45-2] CI 43800	λ_{max} 482 nm, I(2)1028D, U640, SH3055B, SR(2)2775A	f
Sudan I [842-07-9] CI 12055	λ_{max} 476(418) nm, I(2)969B, U653, SH3193C, SR(2)2745B, QL4900000	b
Sudan II [3118-97-6] CI 12140	λ_{max} 493(420) nm, I(2)976A, U654, SH3193D, SR(2)2745D, QL5850000	b
Sudan III [85-86-9] CI 26100	λ_{max} 507(354) nm, I(2)975B, U656, SH3194A, SR(2)2761I, QK4250000	b
Sudan IV [65-53-6] CI 26105	λ_{max} 520(357) nm, I(2)976C, U658, SH3194B, SR(2)2761J, QL5775000	b
Sudan Orange G [2051-85-6] CI 11920	λ_{max} 388 nm, I(2)965C, U662, SR(2)2737D, CZ9027500	b
Sudan Red B [3176-79-2] CI 25110	λ_{max} 521 nm, SR(2)2761K	b
Sunset Yellow FCF [2783-94-0] CI 15985	λ_{max} 482 nm, QK2450000	b
Toluidine Red [2425-85-6] CI 12120	λ_{max} 507(398) nm, I(2)974D, U716, SR(2)2745K, QK4247000	b
Tropaeolin O [547-57-9] CI 14270	λ_{max} 490 nm, I(2)977B, U 719, SR(2)2741N	b
Trypan Blue [72-57-1] CI 23850	λ_{max} 520(357) nm, U721, SH3552D, SR(2)2765F, QJ6475000	b
Xylidine Ponceau 2R [3761-53-3] CI 16150	λ_{max} 503(388) nm, U742, SH3622B, SR(2)2747I, QJ6825000	b

[a] Codes beginning with I, N and U denote FTIR spectra in Reference 36, NMR spectra in Reference 37 and UVV spectra in Reference 38, respectively. Codes beginning with SH denote data for safety handling in Reference 39, while those beginning with SR are entries on safety regulations in Reference 40. A code of two letters followed by seven digits is a reference to a protocol in *Registry of Toxic Effects of Chemical Substances* (RTECS) of the National Institute for Occupational Safety and Health/Occupational Safety and Health Administration (NIOSH/OSHA). See also Reference 41 for toxicological data.
[b] Azo dye. [c] Pyrazole dye.
[d] Complex with a metal ion. [e] Alizarin dye.
[f] Triphenylmethane dye. [g] Phenoxazinium dye.
[h] s-Triazine reactive dye. [i] Xanthene dye.
[j] Phenazinium dye. [k] Phenothiazinium dye.

drinking water may have untoward effects even at ppb levels, because on chlorine disinfection they yield chlorophenols that confer bad odor and taste[46]. Alkylphenols and their derivatives containing one or more $-CH_2CH_2O-$residues are water pollutants, related to the use of nonionic surfactants, recognized as estrogenic endocrine-disrupting chemicals[47]. Over two score nonestrogenic anthropogenic compounds that mimic the action of 17β-estradiol (**37**) have been recently found in wastewaters even after treatment[48–51]. These are the so-called xenoestrogens, many of which are simple phenol derivatives such as bisphenol A (**29**), BHA (**31**, Table 3), 4-nonylphenol, 4-t-octylphenol, 2-t-butyl-4-methylphenol, 2-hydroxybiphenyl, 4-hydroxybiphenyl, 4-chloro-3-methylphenol and 4-chloro-2-methylphenol[52]. The condition for estrogenic activity of a pollutant seems to be the presence of an unhindered phenolic OH group in a *para* position and a molecular mass of 140 to 250 Da[53].

Quantitative structure–toxicity models were developed that directly link the molecular structures of a set of 50 alkylated and/or halogenated phenols with their polar narcosis toxicity, expressed as the negative logarithm of the 50% growth inhibitory concentration (IGC50) value in mM units. Regression analysis and fully connected, feed-forward neural networks were used to develop the models. The best model was a quasi-Newton neural network that had a root-mean-square error of 0.070 log units for the 45 training set phenols and 0.069 log units for the five cross-validation set of phenols[54]. The toxicity and untoward organoleptic properties conferred by phenols has induced governmental agencies to limit their concentration in water for human consumption. The phenolic compounds listed in Table 4 are among the so-called priority pollutants defined by the U.S. Environmental Protection Agency (EPA) and the European Community, and should be of main concern in the detection of water and soil pollution. In the European Community the maximum admissible concentration of all phenols present in drinking water was set to 0.5 μg L^{-1}, excluding those that do not react with chlorine, or to 0.1 μg L^{-1} for

TABLE 3. Some antioxidants used for foodstuffs[44,45]

Antioxidant	Properties[a]
Butylated hydroxyanisole[b] [25013-16-5] (BHA, **31**)	SK1575000
Butylated hydroxytoluene [128-37-0] (BHT, **32a**)	I(1)1094D, N(2)285A, SH1101C, SR(1)1285L, GO7875000
2,6-Di-t-butyl-4-(hydroxymethyl)phenol [88-26-6] (Lonox 100, **32b**)	DO0750000
n-Propyl gallate [121-79-9] (PG, **33a**)	I(2)301A, N(2)1261B, SH2967D, SR(2)1909K, LW8400000
n-Octyl gallate [1034-01-1] (OG, **33b**)	SH2648D, SR(2)1909L, LW8225000
n-Dodecyl gallate [1166-52-5] (DG, **33c**)	SH2068D, SR(2)1909M, DH9100000
2,4,5-Trihydroxybutyrophenone [1421-63-2] (THBP, **34**)	EU5425000
t-Butylhydroquinone [1848-33-0] (TBHQ, **35**)	I(1)1108B, N(2)305B, SR(1)1301D, MX4375000
Nordihydroguaiaretic acid [500-38-9] (NDGA, **36**)	I(1)1119D, N(2)325A, SR(1)1311H, UX1750000

[a]Codes beginning with I, N and U denote FTIR spectra in Reference 36, NMR spectra in Reference 37 and UVV spectra in Reference 38, respectively. Codes beginning with SH denote data for safety handling in Reference 39, while those beginning with SR are entries on safety regulations in Reference 40. A code of two letters followed by seven digits is a reference to a protocol in *Registry of Toxic Effects of Chemical Substances* (RTECS) of the National Institute for Occupational Safety and Health/Occupational Safety and Health Administration (NIOSH/OSHA). See also Reference 41 for toxicological data.
[b]Mixed isomers of **31**.

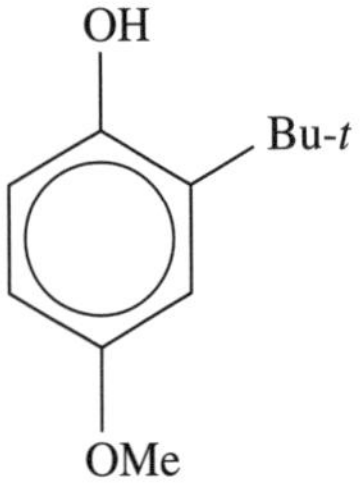

(31) 2-tert-butyl-4-methoxyphenol

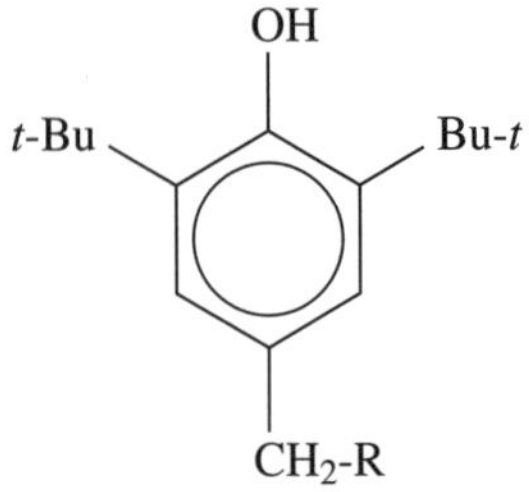

(32) (a) R = H
     (b) R = OH

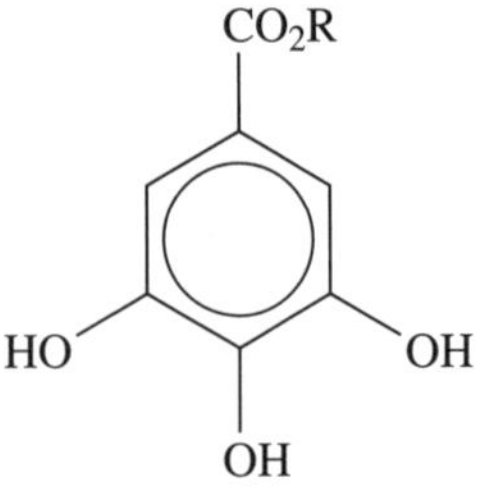

(33) (a) R = n-Pr
     (b) R = n-C$_8$H$_{17}$
     (c) R = n-C$_{12}$H$_{25}$

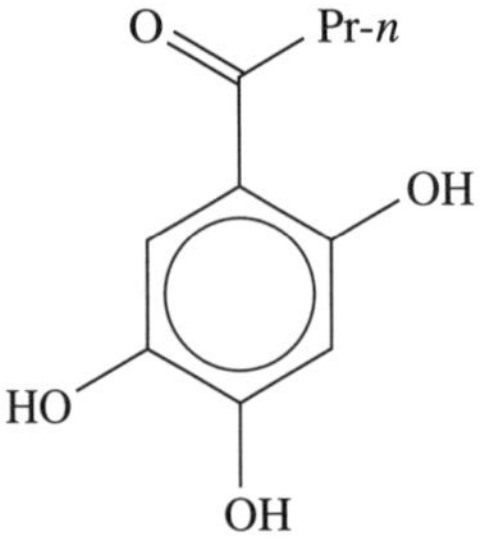

(34)

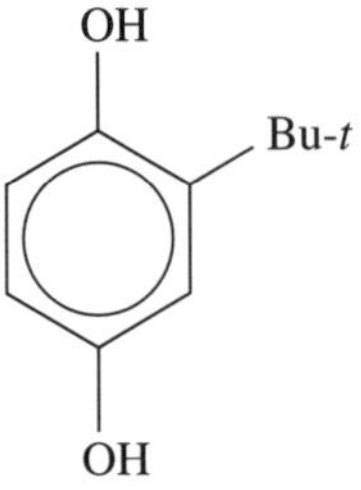

(35)

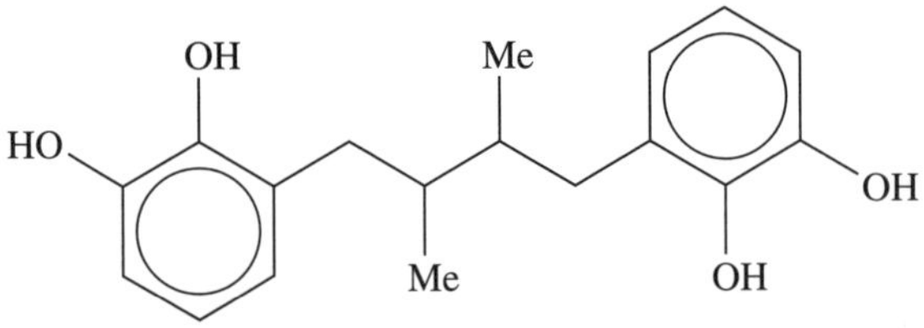

(36)

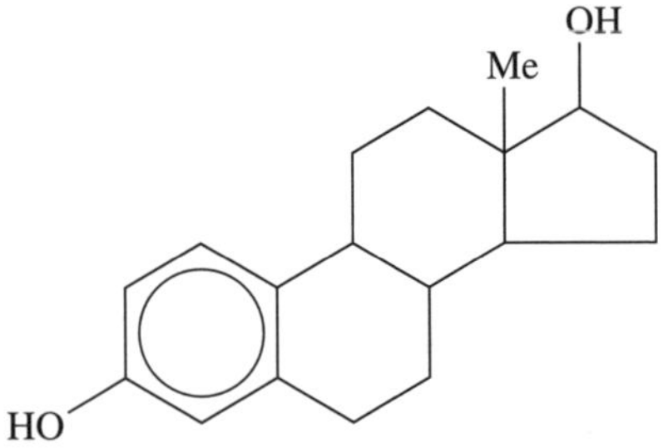

(37)

TABLE 4. The priority phenol pollutants according to US-EPA[57,58] and the European Community directives[59,60]

Compound [CAS No.]	Properties[a]	Notes
Phenol [108-95-2]	I(1)1069A, N(2)243A, SH2745A, SR(1)1265A, SJ3325000	b
2-Methylphenol [95-48-7]	I(1)1069B, N(2)243B, SH923B, SR(1)1265B, GO6300000	
3-Methylphenol [108-39-4]	I(1)1073B, N(2)248B, SH923D, SR(1)1267E, GO6125000	
4-Methylphenol [106-44-5]	I(1)1075D, SH924B, SR(1)1269F, GO6475000	
2,4-Dimethylphenol [105-67-9]	I(1)1087B, SH1403C, SR(1)1277I, ZE5600000	b
2-Nitrophenol [88-75-5]	I(1)1331C, N(2)682C, SH2581B, SR(1)1553D, SM2100000	b
4-Nitrophenol [100-02-7]	I(1)1341B, N(2)696C, SH2582B, SR(1)1559J, SM2275000	b
2,4-Dinitrophenol [51-28-5]	I(1)1370C, N(2)750C, SH1439D, SR(1)1587J, SL2800000	b
2,4-Dinitro-6-methylphenol [534-52-1]	I(1)1375D, N(2)763A, SH1436C, SR(1)1595B, GO9625000	b
2-s-Butyl-4,6-dinitrophenol (Dinoseb, DNBP) [88-85-7]		
2-Cyclohexyl-4,6-dinitrophenol [131-89-5]		
2-Chlorophenol [95-57-8]	I(1)1072A, N(2)246C, SH832C, SR(1)1256N, SK2625000	b
3-Chlorophenol [108-43-0]	I(1)1075A, N(2)249C, SH833A, SR(1)1267M, SK2450000	
4-Chlorophenol [106-48-9]	I(1)1078B, N(2)253B, SH833B, SR(1)1271G, SK2900000	
4-Chloro-3-methylphenol [59-50-7]	I(1)1086D, N(2)265B, SH807D, SR(1)1277H, GO7100000	b
2,4-Dichlorophenol [120-83-2]	I(1)1089C, N(2)274B, SH1150B, SR(1)1281J, SK8575000	b
2,6-Dichlorophenol [87-65-0]	I(1)1083B, N(2)260C, SH1151A, SR(1)1275F, SK8750000	
2,4,5-Trichlorophenol [95-95-4]	I(1)1097C, N(2)292C, SH3417A, SR(1)1289K, SN1400000	
2,4,6-Trichlorophenol [88-06-2]	I(1)1095C, N(2)287C, SH3417C, SR(1)1287I, SN1575000	b
2,3,4,5-Tetrachlorophenol [879-39-0]		
2,3,4,6-Tetrachlorophenol [58-90-2]		
2,3,5,6-Tetrachlorophenol [935-95-5]		
Pentachlorophenol [87-86-5]	I(1)1100D, SH2696B, SR(1)1291E, SM6300000	b
2-Amino-4-chlorophenol [95-85-2]	I(1)1228A, N(2)517B, SH789A, SR(1)1417F, SJ5700000	

[a] Codes beginning with I and N denote FTIR spectra in Reference 36 and NMR spectra in Reference 37, respectively. Codes beginning with SH denote data for safety handling in Reference 39, while those beginning with SR are entries on safety regulations in Reference 40. A code of two letters followed by seven digits is a reference to a protocol in *Registry of Toxic Effects of Chemical Substances* (RTECS) of the National Institute for Occupational Safety and Health/Occupational Safety and Health Administration (NIOSH/OSHA). See also Reference 41 for toxicological data.
[b] Belongs to the eleven priority phenols defined by EPA.

individual compounds. An evaluation of the odor threshold concentrations of the iodine derivatives of phenol, obtained on iodine disinfection of water, was carried out for the USA space program[55]. A review appeared dealing with the determination of phenolic pollutants in water and wastewaters, where an evaluation was given of sample preparation methods such as liquid–liquid extraction (LLE), solid phase extraction (SPE) and solid phase microextraction (SPME), and end analysis by well established analytical methods such as LC with various detectors, as well as emerging techniques such as capillary zone electrophoresis (CZE), ELISA and biosensors[56].

Phenols derived from lignin degradation were used as markers to determine the origin of waters of the Seine estuary in France. Thus, fluvial run-off contains syringic, hydroxybenzoic and vanillic phenols, whereas upstream penetrating marine waters contain cinnamic phenols derived from the estuarine herbs. In the maximum turbidity zone the vanillic acid (**38**) to vanillin (**39**) ratio increases due to aerobic degradation of lignin[61].

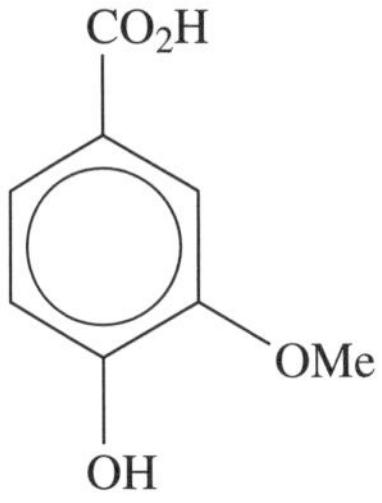

B. Some Considerations in Modern Analysis

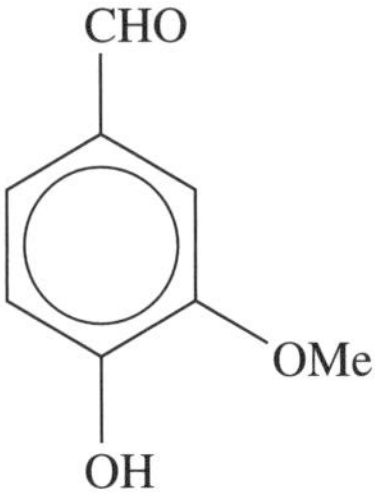

The analytical procedures in the chemical industry may include part or all of the following steps: Sampling, sample reduction, sample preparation, end analysis, disposal of analytical wastes, data processing and feedback into the control system. Automatization of all stages of the analytical process is a trend that can be discerned in the development of modern analytical methods for chemical manufacture; however, the extent to which this has been applied varies from case to case, depending on reliability and cost-benefit considerations. Among the elements of reliability one counts conformity of the accuracy and precision of a method to the specifications of the manufacturing process, stability of the analytical system and closeness to real-time analysis. The latter is a requirement for feedback into automatic process-control systems. The investment in equipment for automatic on-line analysis of specific components may be high. Thus, this is frequently replaced by monitoring an overall property of the chemical mixture that is easy and inexpensive to measure, and correlating that property with the analyte of interest. Such compromise is usually supplemented by collection of samples that are sent to the analytical laboratory for determination, possibly at a lower cost.

A different approach is required to solve analytical problems related to fields such as biological research, pharmacology, forensic investigations, occupational hygiene and environmental protection. Often one confronts samples that are difficult to deal with because of their small size, instability, the low concentration of analyte or the nature of the matrix.

Many advances of modern analysis are concerned with pushing down the limits of detection and quantation (LOD, LOQ) to lower and lower concentrations, using smaller and smaller samples. Modern methods frequently deal with concentrations in the μM or nM range, or require only picomoles or even femtomoles of analyte for an accurate response. These advancements are the result of improved selectivity of chemical reagents and supporting media, development of sensors with increased sensitivity that are backed-up by reliable electronic systems, and optimization of the analytical methodology. Many publications show some concern for the efficiency and analytical throughput, and less frequently advances deal with making the analytical equipment cheaper, or easier to handle.

Application of the methods of chemometrics may help solving difficult analytical problems, and provide an alternative to methods based on separation and quantation of individual components of a mixture. Avoiding separation, when feasible, may afford considerable savings in labor and instrumental investment costs. The following computational techniques have been applied in problems involving analysis of phenolic compounds: Experimental design[62], information theory, evaluation of discriminating power, cluster formation, dendrograms[63], artificial neural networks[54,64,65], least squares, partial least squares (PLS), principal components regression (PCR)[66], multilinear regression[67], derivative spectrometry[67,68], double Fourier transform filtering[69], the Kalman filter algorithm[70] etc. Multivariate analysis and pattern recognition were applied to phenolic compounds to find correlations between physical and spectral properties[71].

International trade and regulations agencies have introduced a healthy tendency toward continuous revision of standards. Even well-established analytical fields are undergoing modification and shaping-up. Thus, the 1993 IUPAC nomenclature recommendations for chromatography[72] have been revised, introducing definitions for *chromatographic system* and *chromatographic process*, modifying that of *hold-up volume* and discouraging the use of the terms *corrected retention time*, *net retention time*, *total retention volume*, *total retention time*, *specific retention volume at* $0\,°C$ and *relative pressure*[73,74].

C. Scope of the Chapter

The largest part of the literature included in this chapter belongs to the last decade of the 20th century and carries on until the first quarter of 2001. Analytical methods are heavily oriented toward those involving separation of individual components, such as chromatography, and especially LC, in accord with the intense research effort invested in this area. Electrophoresis of phenolic compounds is gaining popularity, and in some cases advantage over conventional chromatographic methods has been claimed. Biosensors are now in an intense phase of development, and some commercial applications for ultratrace analysis are emerging. Most biosensors involve electrochemical detection (ELD); however, design involving UVD, fluorometric detection (FLD) and other principles begin to appear. Of the other analytical methods, those based on ultraviolet-visible (UVV) spectrophotometry, including fluorescence effects and chemiluminescence detection (CLD), are probably the most useful, both for the spectral response of phenolic analytes themselves in the UVV region, and the easy production of intensely colored derivatives. Only brief accounts are given of the structural and functional characterization of phenolic compounds based on various analytical techniques, as this subject is more amply discussed elsewhere in this book. The LOD and LOQ concepts are used rather loosely in the literature, where LOD are given in extensive as well as intensive terms (e.g. μmol *vs.* nmol mL^{-1}). Except for cases where sample size was reported and a lower limit concentration could be discerned, extensive LOD values appear as reported in the original paper.

III. GAS CHROMATOGRAPHY

A. Sample Preparation

1. General

Many investigations have been carried out of procedures for improving the analytical quality of GC methods by changing the matrix, increasing the concentration of the pertinent analytes and reducing the interference of other compounds present in the sample. Preconcentration by LLE, before or after derivatization, is most frequently applied in GC trace analysis; however, other techniques, such as SPE, sample stacking (see Section V.A.1) and some of their modifications, such as simultaneous distillation and extraction (SDE) and SPME, are also mentioned. Application of microwave-assisted processes (MAP) during sample preparation seems to improve recoveries.

2. Solid phase extraction vs. liquid–liquid extraction

Twenty seven phenols including mono-, di-, tri-, tetra- and pentachlorophenols, mono- and dinitrophenols, mono- and dinitrocresols and dimethylnitrophenols have been extracted from aqueous samples by solid phase extraction using both modified silica gel (C_{18}) and XAD resin-adsorbents. When a 1 L sample of spiked water was used, a considerable breakthrough was observed with phenol itself, while all other phenols were almost quantitatively extracted. The recovery of phenol itself can be improved by employing smaller sample volumes. End analysis was by GC using capillary columns with specially deactivated weakly polar phases[75]. The extraction efficiency of hydrophobic solvents in LLE of phenols and other organic compounds may be rather poor. For example, the recoveries of phenol and aniline from water by LLE with *n*-octanol were 75.5% and 46%, respectively. The efficiency of hydrophilic solvents such as *n*-butyl and isobutyl alcohols was greatly improved by salting out with sodium sulfate or sodium chloride, attaining extraction efficiencies about 95% for phenol and nitroanilines[76]. SPE on a C_{18} sorbent phase followed by silylation showed better recoveries for the GC determination of phenolic compounds in olive oil than the usual LLE procedures; however, some interference was observed in the determination of oleuropein (**40**)[77]. Despite these findings, total recovery after LLE of phenolic compounds was attained when DMF was used as solvent[78]. Simultaneous LLE and derivatization of phenol and methylphenols in soil were much improved by MAP. Thus, soil samples immersed in hexane containing acetic anhydride and pyridine showed much higher recoveries and shorter extraction times when subjected to MAP as compared with ultrasonic treatment. End analysis by GC-MS was carried out without preliminary clean-up or concentration. LOD was in the lower ppb range[79].

Preconcentration of analytes in aqueous solution may be performed by a miscible organic phase followed by salting out. Thus, microextraction of anionic solutes such as phenol, cresols and xylenols in industrial effluents can be carried out with a small amount of isopropyl alcohol, followed by demixing of the phases with ammonium sulfate. End analysis of the extract by GC-MS in the selected ion monitoring (SIM) mode allowed a LOD of 1 ppb for 50 mL samples[80]. The best conditions for eliminating petroleum products from the concentrate were found for the GC determination of volatile phenols in natural waters. Losses of volatile phenols due to preconcentration were insignificant and caused no increase in the relative error of determination by the internal-standard method[81]. The concentration of phenol in the atmosphere can be determined by sorption on Chromosorb 102, desorption with benzene and 0.1 M NaOH and GC using a capillary column. LOD was about 1 $\mu g\,m^{-3}$, with accuracy within 15%[82].

A method for detection of exposure to aromatic hydrocarbons was based on simultaneous detection of metabolites such as phenol, and isomers of cresol, xylenol and naphthol

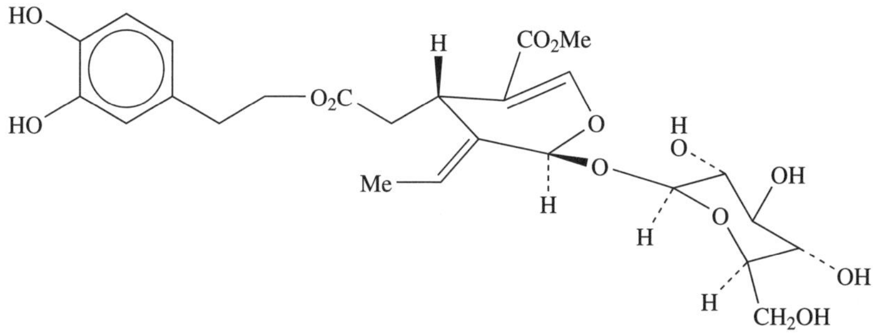

(40)

in hydrolyzed urine by SPE preconcentration, followed by capillary GC on cross-linked 5% phenylmethylsilicone. For all the phenols tested LOD was 0.1 to 0.2 ppm, with RSD 2.6 to 16.6% and linearity from 5 to 100 ppm; recoveries were generally over 80%[34]. Determination of phenolic flame-retardants in human plasma involved SPE with styrene–divinylbenzene copolymer, treatment of the SPE column with concentrated sulfuric acid to decompose the plasma lipids and GC end analysis with electron capture MS detection. The method was validated for 2,4,6-tribromophenol, pentabromophenol, tetrachlorobisphenol-A and tetrabromobisphenol-A in the concentration range 1.2–25, 0.4–40, 4–200 and 4–200 pg(g plasma)$^{-1}$, respectively. Analyte recovery was 51 to 85%, repeatability had RSD 4 to 39% and LOD was 0.3 to 0.8 pg(g plasma)$^{-1}$. A positive detection of these analytes points to potential occupational exposure[83].

Phenolic pollutants in the effluent from tertiary sewage treatment plants were preconcentrated by SPE on a styrene–divinylbenzene copolymer. The performance was superior to that of graphitized carbon black (GCB). Recoveries were good in spite of the wide polarity range of the phenols[84].

Determination of Irgasan DP 300 (**41**) in slaughterhouse wastewater involved alkalinization to pH 11, removal of fats and oils by LLE with petroleum ether, acidification to pH 1, LLE with benzene, further purification by sodium sulfate/silica gel adsorption, desorption, derivatization with diazomethane and end analysis by GC with electron capture detection (ECD). LOD was 8.2 ng L^{-1}; recovery was better than 88% regardless of concentration[85].

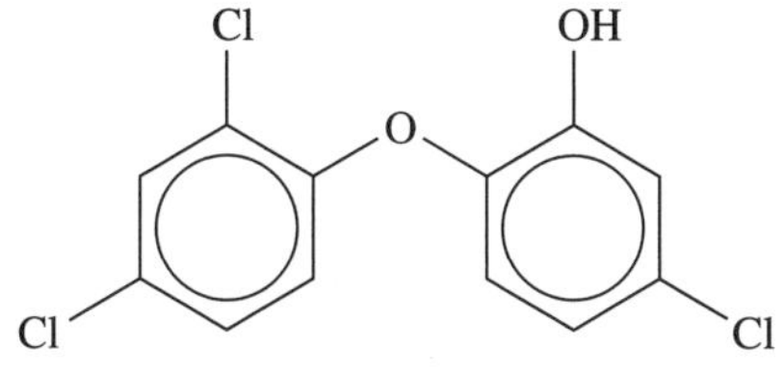

(41)

3. Simultaneous distillation and extraction

This sample preparation method involves steam distillation of the volatile organic components of a sample followed by preconcentration by LLE using a water-insoluble solvent. SDE served as unique clean-up and preconcentration step before derivatization, in the GC-MS determination of polycyclic aromatic hydrocarbons, phenols and aromatic amines in particulate phase mainstream cigarette smoke[86]. Preconcentration by the SDE technique was proposed for environmental water and soil samples, with end analysis by GC with flame ionization detection (FID). LOD was 0.01 mg(L water)$^{-1}$ and 0.1 mg(kg soil)$^{-1}$ [87].

4. Solid phase microextraction

A way of avoiding the use of solvents, either for LLE of analytes from a matrix or for their elution after SPE, is by SPME. In this technique, the analytes become adsorbed on suitably coated silica fibers, which are placed directly in the injector of a GC, where the analytes become thermally desorbed. SPME with poly(acrylate)-coated fibers was applied for preconcentration of phenols regulated by EPA wastewater methods 604 and 625 and Ontario MISA Group 20 regulations. LOD were in the sub-ppb range with RSD 5–12%, depending on the compound. Low pH levels and saturated salt conditions significantly increase the sensitivity of the method. SPME of phenolics from the headspace over water has also been investigated[88–90]. SPME was applied to the detection of phenolic compounds in mainstream smoke of tobacco cigarettes. End analysis was by GC-MS in the SIM mode. The following compounds were detected: Phenol, cresols, xylenols, methoxyphenols, ethylphenols, 2,4,6-trimethylphenol, vanillin and the naphthols. Recoveries were excellent, except for the naphthols in the 50% range[91]. SPME with various sorbents was investigated for trace analysis of phenols in water. End analysis was by GC with FID. LOD was 0.3 to 2 μg L^{-1} for 100 mL of water at acid pH, with RSD 2.3–4.5% ($n = 11$)[92]. Conditions for the optimization of SPME of phenol and chlorophenol soil contaminants were investigated; end analysis was by GC-FID. The method was applied to soil analysis after acute contamination in industrial sites. The method was validated by comparison with an EPA certified extraction method[93]. Soil samples were suspended in water and the extracted phenols were acetylated *in situ* with acetic anhydride in the presence of potassium bicarbonate. Acid was added after the end of the derivatization and SPME was performed by placing a poly(dimethylsiloxane) fiber in the headspace. End analysis was performed by introducing the fiber into the injector of a GC-MS apparatus. LOD was in the sub-ppm range, with good precision, sensitivity and linearity[94].

5. Supercritical fluid extraction

Chlorinated phenolic compounds in air-dried sediments collected downstream of chlorine-bleaching mills were treated with acetic anhydride in the presence of triethylamine. The acetylated derivatives were removed from the matrix by supercritical fluid extraction (SFE) using carbon dioxide. The best overall recovery for the phenolics was obtained at 110 °C and 37 MPa pressure. Two SFE steps had to be carried out on the same sample for quantitative recovery of the phenolics in weathered sediments. The SFE unit was coupled downstream with a GC for end analysis[95]. Off-line SFE followed by capillary GC was applied in the determination of phenol in polymeric matrices[96]. The sonication method recommended by EPA for extraction of pollutants from soil is inferior to both MAP and SFE techniques in the case of phenol, *o*-cresol, *m*-cresol and *p*-cresol spiked on soil containing various proportions of activated charcoal. MAP afforded the highest recoveries (>80%), except for *o*-cresol in a soil containing more than 5% of activated carbon. The SFE method was inefficient for the four phenols tested; however, *in situ* derivatization of the analytes significantly improved the performance[97].

B. Derivatization

1. General

Two main objectives are pursued when analytes are derivatized before GC analysis: Increasing volatility and attaining enhanced sensitivity when certain detection methods are used. Derivatization methods of phenolic compounds for GC analysis have been reviewed[98].

2. Acylation

In situ acetylation of phenolic compounds using acetic anhydride in the presence of a base is frequently applied as a precolumn derivatizing technique, and some applications were mentioned above[95,99] (see also Sections III.B.2, 4, 5). Trace concentrations of mono-, di- and trihydroxybenzenes in water were directly acetylated and the acetates were concentrated on C_{18}-silica SPE columns. End analysis was by GC-MS. Detection of phenols in the ng L^{-1} range can be obtained with 500 mL water samples. The method is unsuitable for nitrophenols[100]. Phenolic compounds in bleached pulp and wastewater treatment plant sludges were subjected to Soxhlet extraction with ethanol–toluene. After concentration the phenols were acetylated, cleaned up with silica gel and determined by GC-MS; LOD was 0.5 ppm on dry basis[101]. Determination of polychlorinated biphenyls, chlorinated pesticides, chlorinated phenols, polycyclic aromatic-hydrocarbons and chlorophenoxyalkanecarboxylic acids in water began with *in situ* acetylation and preconcentration on a SPE cartridge packed with 500 mg of Separon SGX C_{18}. Recovery of pollutants at concentrations in the $\mu g L^{-1}$ to ng L^{-1} level generally ranged from 54 to 109% (RSD 3–16%). End analysis was carried out by GC with ECD or HPLC with UVD or FLD, after elution from the cartridge with 2 mL of *n*-hexane and 2 mL of 1% NH_3/EtOH[99]. Direct acetylation of phenol, alkylphenols, chlorophenols and nitrophenols in environmental waters was followed by SPE of the acetates with a C_{18} disk. End analysis was by GC with ion trap detector MS (ITD-MS). In most cases recoveries were better than 80% at concentrations of 0.1 and 1.0 $\mu g L^{-1}$. LOD was in the 2–15 ng L^{-1} range for phenol, alkylphenols and halogenated phenols, and in the 25–50 ng L^{-1} range for nitrophenols[102]. The analysis of wastewaters of a coal gasification plant involved direct acetylation, SPE with a polystyrene resin, elution with *n*-hexane, concentration of the extract and end analysis by GC-MS. The following compounds were identified: Phenol and alkylated derivatives up to C_3-phenol, catechol (**42**) and alkylated derivatives up to C_2-catechol, vanillin (**39**) and a biphenyldiol[103]. Wetted soil samples were directly acetylated in vials, and the phenol and cresol acetates were determined by GC headspace analysis. LOD was 0.03–0.08 $\mu g g^{-1}$. The method is suitable for soils with carbon content below 5%[104]. Pressurized LLE using acetic anhydride for simultaneous acylation was applied to the analysis of phenolic pollutants, sterols and carboxylic acids in environmental and microbial samples[105].

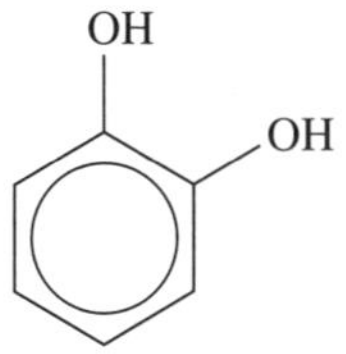

(**42**)

The main GC advantages from analysis of trifluoroacetate esters as compared to plain phenols are enhanced volatility and improved resolution. The elution temperature of a given phenol is typically 50 °C greater than that of the corresponding trifluoroacetate ester. The retention of compounds with two trifluoroacetate groups is only moderately greater than that of the monoesters, whereas underivatized dihydroxy compounds are very difficult to elute from any GC column. The GC-MS characteristics of trifluoroacetate esters of phenolic compounds were discussed. Linear temperature programmed retention indices and total ion current MS response factors of over 120 phenolic esters are reported. Complete resolution of isomeric C_0-, C_1- and C_2-alkylphenol esters is readily achieved on conventional fused silica GC columns; resolution of the corresponding underivatized compounds requires specialized GC columns with low temperature limits. In general, MS of trifluoroacetate esters are more characteristic of a given structure than those of the corresponding phenols and may be more rigorously interpreted toward structural elucidation. Some of the more important spectral features used in compound identification were summarized. Example applications in analysis of coal-, shale- and petroleum-derived materials were presented; SIM was used to determine individual phenolic components in whole distillates; reconstructed ion chromatograms were used to illustrate distributions of selected species as a function of fuel storage and thermal stress[106].

Parameters such as solvent, basic medium and reaction time, affecting the derivatization of alcohols and phenols with benzoyl chloride, were investigated. End analysis was by GC with UVD[107]. A sensitive method proposed for trace determination of phenols in water consists of preconcentration by SPE with a commercial styrene–divinylbenzene copolymer, acylation with pentafluorobenzoyl chloride in the presence of tetrabutylammonium bromide and end analysis by GC with either ECD or ITD-MS. LOD was 3 to 20 ng L^{-1} for ECD and 10 to 60 ng L^{-1} for ITD-MS, with 500 mL samples[108]. Acylation with the fluorinated glutaric acid derivative **43** was proposed for determination of urinary phenols, as indicative of exposure to benzene and other aromatic hydrocarbons. End analysis by GC-MS shows strong molecular ions of the derivatives by electron ionization. The protonated ions are the base peaks obtained by chemical ionization. LOD was 0.5 mg L^{-1} and the linearity range 0–100 mg L^{-1} for phenol[109].

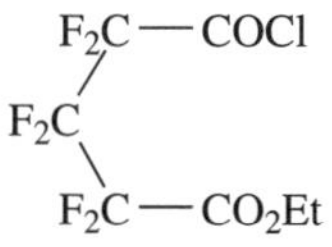

$$F_2C-COCl$$
$$F_2C$$
$$F_2C-CO_2Et$$
(43)

3. Silylation

Silylation is an affective means for aiding volatilization of phenolic compounds. More than fifty substituted phenols of various types were determined by GC-MS, after derivatizing with *N*-(*t*-butyldimethylsilyl)-*N*-methyltrifluoroacetamide (**44**). The MS of all the examined analytes was dominated by the M−57 peak, resulting from the loss of *t*-butyl from the molecular ion. LOD of 5 pg could be achieved with electron ionization in the SIM mode[110]. A method for separation and determination of flavonoids is based on preparation of the aglycons by acid hydrolysis in methanol, solvent evaporation, pH adjustment, SPE with a C_{18} cartridge, elution with AcOEt, solvent evaporation under a nitrogen stream and derivatization with a mixture of *N*,*O*-bis(trimethylsilyl)trifluoroacetamide (**45**) and chlorotrimethylsilane (**46**). End analysis was by GC with FID. The method was applied to the analysis of flavonoids in tea leaves extract, where the following aglycons were identified: The catechins epicatechin (**2**) and catechin (**3**) and the flavonols kaempferol (**6**), quercetin (**47**) and myricetin (**48**)[111]. GC-MS analysis of polyphenols in wine, including the flavonoids, was carried out after a similar derivatization procedure[112]. Pressurized LLE using **45** for simultaneous silylation was applied to the analysis of phenolic pollutants, sterols and carboxylic acids in environmental and microbial samples[113].

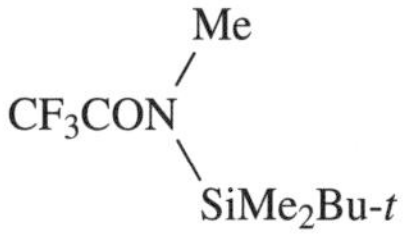

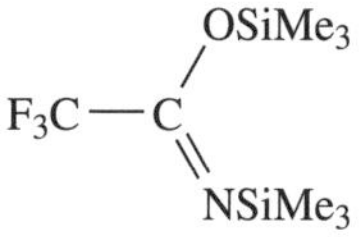

4. Alkylation

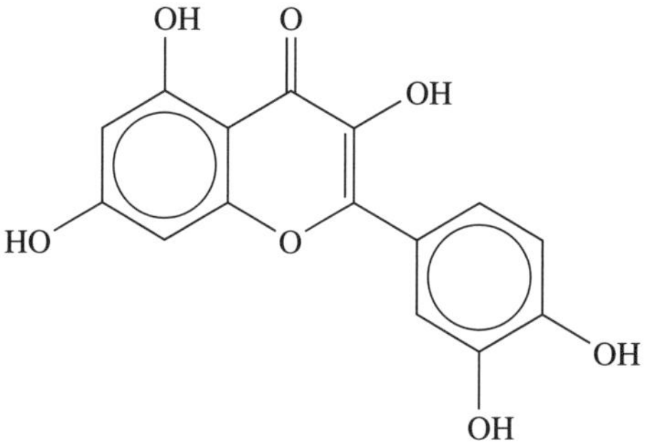

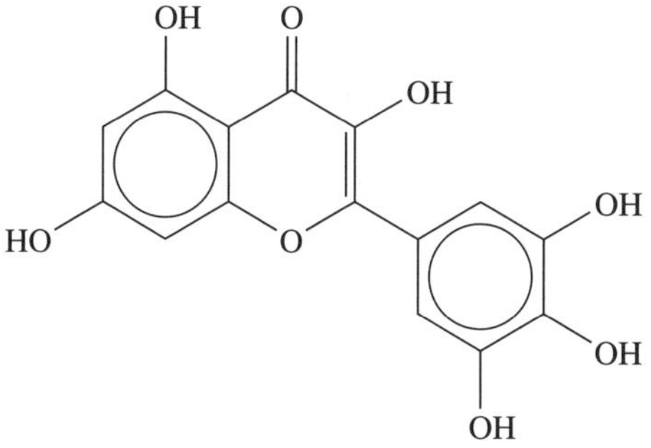

Bupivacaine (**49a**) and its phenolic metabolites (**49b–c**) were detected in urine after the samples underwent hydrolysis, LLE with ether, concentration by evaporation and derivatization with diazomethane, to obtain the methyl ethers (**49d**). End analysis was by GC-MS in the SIM mode[114]. Another methylation reaction with this reagent was mentioned in Section III.A.2[85]. On-site methylation with tetramethylammonium hydroxide was proposed for the analysis of phenolic additives in polymeric materials, by the pyrolysis-GC method[115]. Pressurized LLE using phenytrimethylammonium hydroxide (**50**) or trimethylsulfonium hydroxide (**51**) for simultaneous methylation was applied to the analysis of phenolic pollutants, sterols and carboxylic acids in environmental and microbial samples[113]. Methylation of various phenolic xenoestrogens in MeOH solution was achieved at room temperature with **50**[51]. The phenolic herbicides bromoxynil (**52a**) and ioxynil (**52b**) and the 2,4-dinitrophenol derivatives DNOC (**53a**), dinoseb (**53b**), dinoseb acetate (**53c**), binapacryl (**53d**), dinobuton (**53e**), dinoterb (**53f**) and dinoterb acetate (**53g**), become strongly adsorbed on GCB; however, the compounds with a free phenolic group cannot be eluted to any practical extent. Thus, after SPE preconcentration with a GCB, from spiked water samples and elution of esterified derivatives, the phenolic pesticides were treated *in situ* with diazomethane or trimethylsulfonium hydroxide (**51**), eluted with ethyl acetate and determined by GC-MS[116].

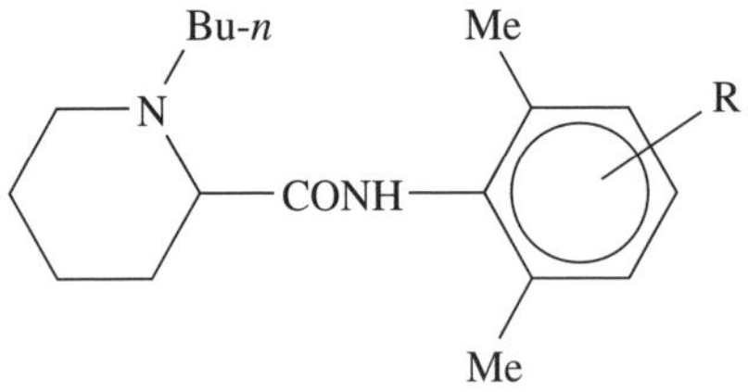

(**49**) (a) R = H
(b) R = 3-OH
(c) R = 4-OH
(d) R = OMe

(**50**) [PhNMe₃]⁺OH⁻

(**51**) [Me₃S]⁺OH⁻

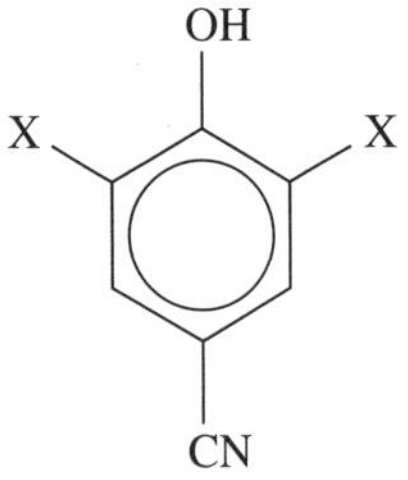

(**52**) (a) X = Br
(b) X = OH

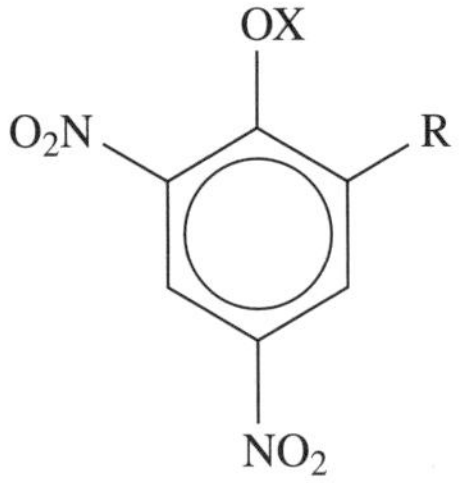

(**53**) (a) R = Me, X = H
(b) R = s-Bu, X = H
(c) R = s-Bu, X = Ac
(d) R = s-Bu, X = CH₂=CHC(=O)
(e) R = s-Bu, X = i-PrOC(=O)
(f) R = t-Bu, X = H
(g) R = t-Bu, X = Ac

After LLE of phenols and carboxylic acids in water, on-line methylation with **51** was applied together with large volume injection (100 μL). The solvent was removed before the analytes were transferred into the GC column with MS detection in full scan mode. Volatile fatty acids, dicarboxylic acids, benzoic acids and phenols in water, at concentrations of 0.4 to 0.1 μM, could be determined in 5 mL samples. Lactic, pyruvic and malonic acids required higher concentrations due to their higher water solubility and lower methylation rates[117]. Samples of particulate matter were subjected to LLE with THF, and hydrolysis/methylation of the extract with tetramethylammonium hydroxide, followed by GC. A linear correlation was found between the amount of methoxybenzene measured in the chromatogram and the concentration of phenolic resin in the particulate matter[118].

On column benzylation of phenols was carried out with 3,5-bis(trifluoromethyl)benzyl-dimethylphenylammonium fluoride (**54**). Fluorinated benzyl derivatives allow very sensitive detection of phenols at ppt levels, by GC-MS in the negative ion chemical ionization mode[84]. Derivatizing with pentafluorobenzyl bromide ($C_6F_5CH_2Br$) was proposed for GC-MS detection of airborne carboxylic acids and phenols[119].

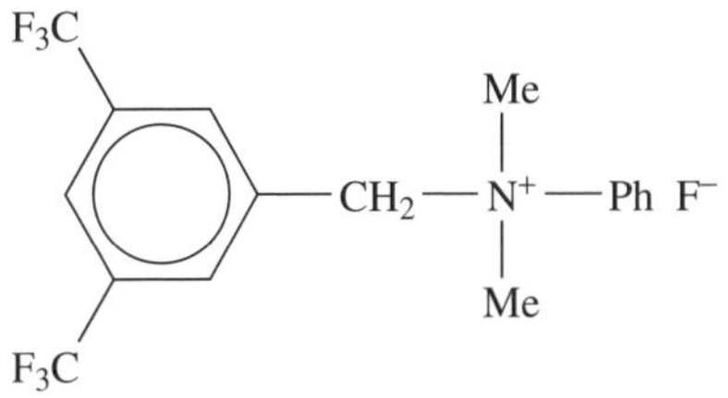

F_3C—[benzene ring with second F_3C]—CH_2—$N^+(Me)(Me)$—Ph F^-

(**54**)

5. Bromination

A very sensitive method for determination of phenol, methylated phenols and resorcinol is based on bromination in acidic solution, LLE with benzene and GC-ECD, with or without previous silylation. For phenol–cresol mixtures, RSD was 4.9–8.5%, using 10 mL of 0.1 μM aqueous solution and 2 mL benzene for extraction. The method was applied for determination of phenols in cigarette smoke and human urine[120]. Conditions were investigated for precolumn quantitative bromination of phenols in water solution for subsequent determination by GC-ECD. Analytical errors of 5 to 25% were found for concentrations in the 0.5 to 100 $\mu g\,L^{-1}$ range[121].

C. End Analysis

The present section is organized mainly according to the different matrix types related to the origin of the samples undergoing GC analysis. In Table 5 are summarized detection methods applied after the chromatographic separations mentioned in Sections III.A–C.

TABLE 5. Post-column detection methods for GC mentioned in Section III

Detection method	Subsection and references
MS	**A.2** 79, 80, 83[a]; **A.3** 86; **A.4** 91, 94; **B.2** 100–103, 106, 108, 109; **B.3** 110, 112; **B.4** 84, 114, 116, 117; **C.2** 125, 126, 128; **C.3** 129, 130; **C.4** 131.
ECD	**A.2** 85; **B.2** 99, 108; **B.5** 120, 121.
FID	**A.3** 87; **A.4** 92, 93; **B.3** 111; **C.4** 132.
UVD[b]	**B.2** 107.
AED	**C.1** 123.
Hyperthermal[c]	**C.1** 124.

[a]Electron capture mass spectrometry.
[b]Gas-phase molecular absorption spectrometry.
[c]Hyperthermal negative surface ionization operating principle.

1. General

Besides MS detection, identification of unknown peaks in GC routine analysis of environmental samples can be aided by the use of correlations between physicochemical parameters and structure of the analytes to predict the retention times. The correlation between the boiling points and the retention times of chloro- and bromo-benzenes and of some chloro- and nitro-substituted phenols was investigated for nonpolar capillary columns and allowed tentative identification of many compounds belonging to these analogous series[122].

The use of an atomic emission detector (AED) coupled to a GC may provide under ideal conditions information about the empirical formula of the analyte corresponding to a GC peak. However, it was found that the AED responses of C, Cl and O of a series of phenols is related to the working condition of the AED. The elemental response of Cl is independent of molecular structure, but those of C and O are not, probably due to formation of CO in the plasma. The O response is also affected in nitrophenols, probably due to NO_2 formation[123]. A novel detector, based upon hyperthermal negative surface ionization, shows up to 100-fold higher sensitivity than that of the FID for alcohols and phenolic compounds[124].

2. Environmental samples

An automatic method for the analysis of trace phenolic pollutants in water consists of off-line acetylation with acetic anhydride followed by sampling (10 mL), SPE on a polydimethylsiloxane cartridge, loading, drying to remove water and derivatizing agent excess, thermal desorption and GC-MS in the SIM mode. LOD was about $1-5$ ng L^{-1}[125]. A fast method was proposed for field screening of phenolic pollutants in soil. The method is based on thermal desorption GC-MS in the SIM mode, aided by a compound-specific data analysis algorithm[126]. The phenolic resin content of particulate matter collected on roads can be correlated with the amount of asbestos present, as this type of particulate matter originates in brakes abrasion. The resin was extracted with tetrahydrofuran, and was estimated from the GC determination of the amount of phenol generated in a Curie-point pyroliser coupled to the chromatograph. In Figure 1 is shown a schematic representation of the pyrolysis of a phenol-formaldehyde resin, leading to the formation of various products, in their order of appearance in the chromatogram; the peak intensities of phenol and the cresols are preponderant[127].

Fly ash as obtained from the incineration of municipal waste was subjected to clean up, and extractions designed to isolate the phenolic compounds from more acidic and from neutral or basic components. More than sixty phenolic compounds belonging to various structural systems (**55–61**) were identified in the concentrated extracts by GC-MS, most of them containing chloro and bromo substituents, at three levels of confidence based on the MS of each peak in the chromatogram: Positive identification, presumed and tentatively presumed. Compounds of structures **59–61** belong to the latter two classes[128].

3. Foodstuff samples

Dried vanilla bean sections were subjected to ballistic heating in a short-path thermal desorber, and the volatiles so obtained were analyzed by GC and GC-MS. Over 60 flavor compounds, 18 of them phenolics, were detected by this technique. The method is

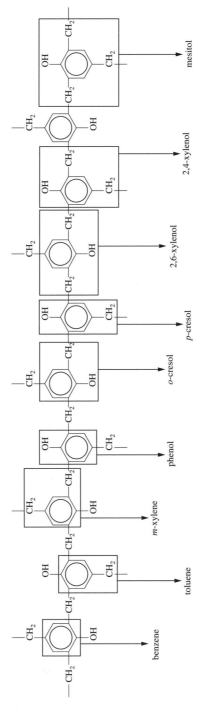

FIGURE 1. Schematic representation of the pyrolysis of a phenol-formaldehyde resin, showing possible fragmentation sites leading to simple aromatic products

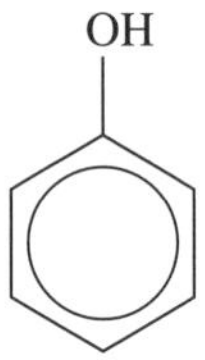

(55) 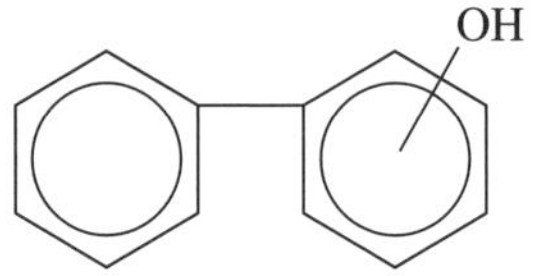 (56) 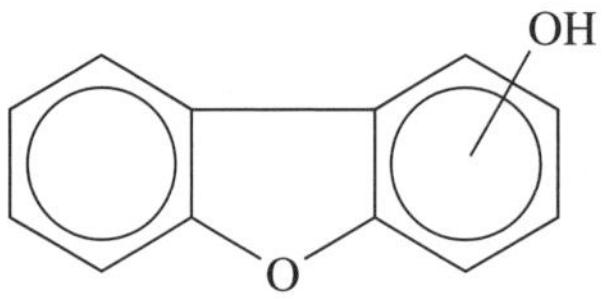 (57) 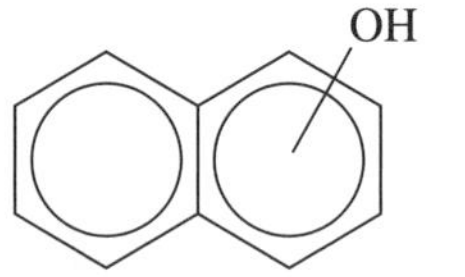 (58) 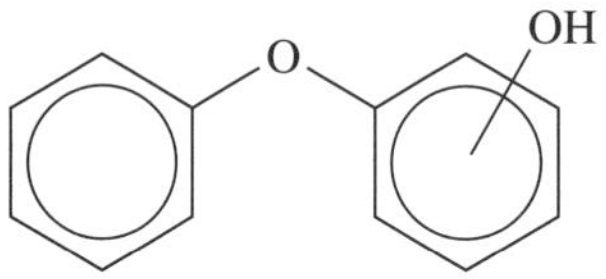 (59) 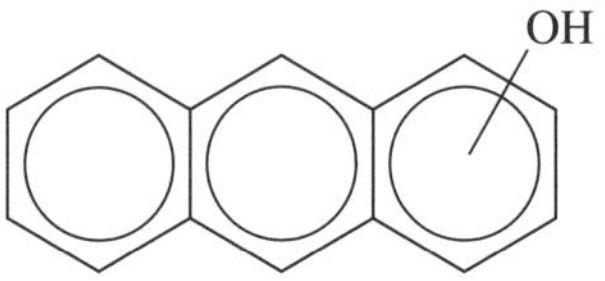 (60) 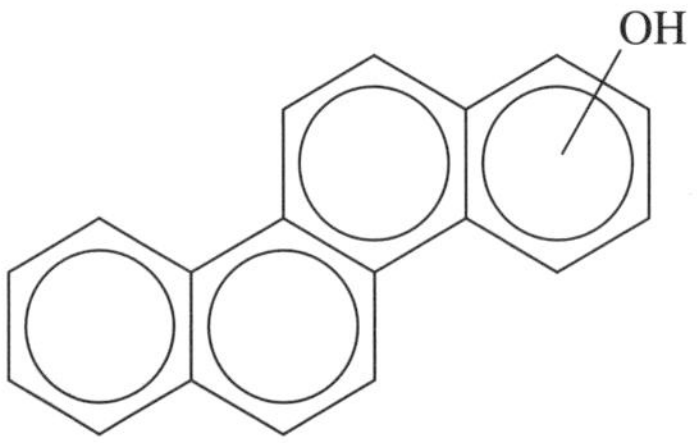 (61)

quantitative and reproducible. Determination of vanillin (**39**) and other compounds could be achieved by spiking with 2,6-dimethoxyphenol (**62**) as internal standard[129].

A profiling protocol for the aliphatic carboxylic acid and phenolic compounds in distilled alcoholic beverages consisted of SPE on an ion exchange disk, simultaneous elution and silylation of the analytes and direct analysis of the extract by GC-MS. The profile consisted of fourteen open-chain mono- and dicarboxylic acids, up to C_{12}, some carrying hydroxy or keto substituents, and vanillin (**39**), syringaldehyde (**63**), coniferaldehyde (**64**), vanillic acid (**38**) and gallic acid (**8**). Recovery of individual analytes was affected by the level of tannins in the spirit. For a given brand the contents of these analytes increased with aging[130].

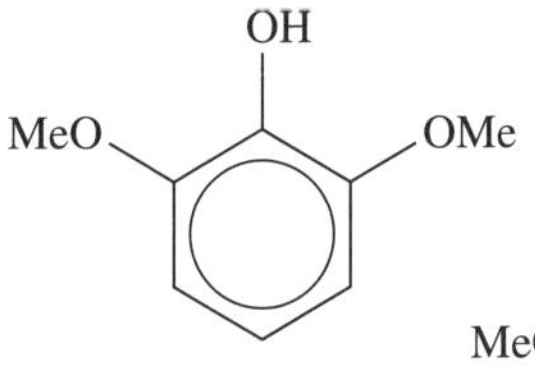

(62) 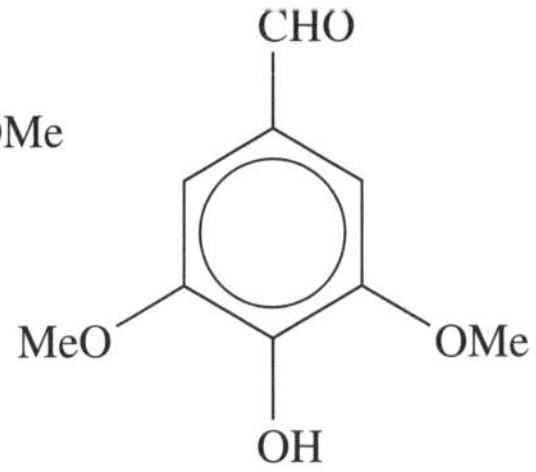   (63) 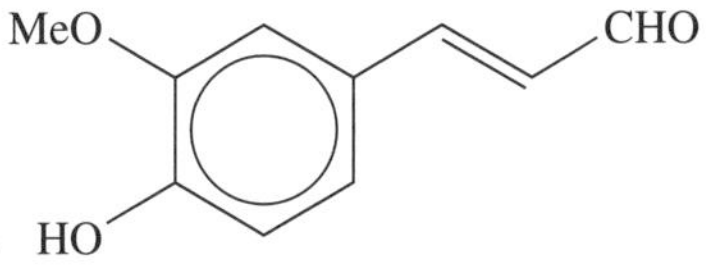   (64)

4. Miscellaneous industrial samples

Pyrolysis-GC with MS detection in the SIM mode was applied to the characterization of natural and industrial lignins of various species, as for the presence of *p*-hydroxyphenyl (**65a**), guaiacyl (**65b**) and syringyl (**65c**) moieties. This objective was achieved after permethylation of the lignin samples. Such moieties are characteristic of ligin in straw, softwoods and hardwoods, respectively[131].

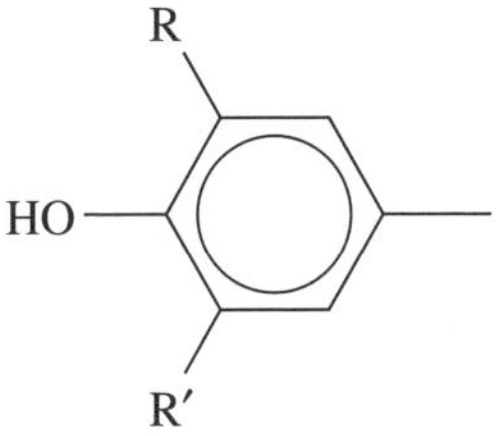

(**65**) (a) R = R′ = H
(b) R = OMe, R′ = H
(c) R = R′ = OMe

Mannich base hardeners for curing epoxy resins may contain residual formaldehyde, phenol and benzyl alcohol, that can have undesirable effects when present in the final product. Determination of these compounds in the hardener was carried out by GC-FID of a 2% solution in chloroform–EtOH, using amyl alcohol as internal standard. LOD was 18, 30 and 26 ppm and LOQ was 75, 86 and 79 ppm of formaldehyde, phenol and benzyl alcohol, respectively[132].

Reaction mixtures containing phenol and hydrogen peroxide show high concentrations of *p*-quinone (**67**) when analyzed by GC, whereas only small concentrations of **67** are observed by HPLC analysis. The reason for this may be the reaction shown in equation 1 taking place in the gas phase, where phenol undergoes stepwise oxidation to hydroquinone (**66**) and **67**. It is therefore proposed that such systems be analyzed by LC as long as hydrogen peroxide is present in the sample[133].

Isobutylene (2-methyl-1-propene) is used for catalytic alkylation of phenol, to produce *t*-butylated phenolic antioxidants to improve the shelf life of fuels and lubricants. Some of the alkyl groups found in these phenolics are dimeric (octyl) or trimeric (dodecyl) derivatives of isobutylene. A procedure was developed based on high resolution capillary GC for analysis of these antioxidants, using an SE-30 stationary phase[134].

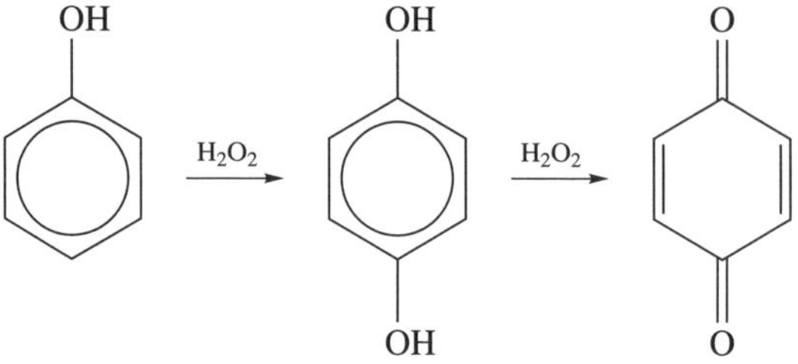

$$\text{benzene} \xrightarrow{H_2O_2} \text{(66) hydroquinone} \xrightarrow{H_2O_2} \text{(67) quinone} \quad (1)$$

IV. LIQUID CHROMATOGRAPHY

A. Sample Preparation

1. General

Many investigations have been carried out dealing with improvement of the analytical quality of LC methods by changing the matrix, increasing the concentration of the pertinent analytes (phenolic compounds) and reducing the interference by other compounds present in the sample. This section deals mainly with preconcentration by SPE, although the alternative LLE method is also mentioned, and sometimes its efficiency compared with that of SPE. This part of the analytical process may become very laborious due to factors such as complexity of the analyte, complexity of the matrix and required analytical quality. A critical review appeared on HPLC analysis of phenol and its chloro, methyl and nitro derivatives in biological samples, with special emphasis on sample preparation[135]. The sample preparation methods adequate for determination of phenolic compounds in fruits have been reviewed[136]. Certain phenolic compounds frequently serve as internal markers for the analysis of botanical extracts. Various factors affecting the concentration of such markers in the sample need consideration to improve the quality of analysis[137].

The sample preparation scheme shown in Figure 2 for the chromatographic determination of phenolic acids in wine may serve as illustrative example of the involved procedures sometimes applied. It should be pointed out that the four extracts obtained by this procedure have each a volume of 50 μL. The *alkaline* extract contains weakly acidic and basic analytes, while stronger acids, such as carboxylic acids, are retained in the aqueous solution; the *acid* extract contains all the neutral and acidic components that are soluble in ether, while the basic components are retained in the aqueous solution; the *anionic* extract was prepared for further refining the chromatograpic analysis, as it is supposed to contain mainly compounds that bear the carboxyl group; the *hydrolysate* extract is used for further characterization of the depsides (**68**), that are derivatives of tartaric acid esterified by a phenolic carboxylic acid (group RCO_2 in **68**), such as caffeic acid (**25**), coumaric acid (**26**) and ferulic acid (**69**). It should be noted that caffeic acid is totally destroyed at pH 10 or in more alkaline solutions. End analysis of the acid extract was by RP-HPLC with UVD, on a Spherosorb ODS_2 microbore column, using a gradient of phosphate buffer (pH 2.4) and methanol. The presence of the following carboxylic acids was detected, in increasing order of retention time: the depside of caffeic acid (**25**), gallic acid (**8**), the depsides of coumaric or ferulic acids, and the following carboxylic acids: 3,4-dihydroxybenzoic, vanillic (**38**), syringic (**70**), caffeic (**25**), coumaric (**26**) and ferulic (**69**). Appearance in wine of the *cis* forms of **26** and **69** was attributed to isomerization of the *trans* form caused by exposure to air and light[138].

Of course, no scheme for sample preparation is universal, and the operations have to fit the nature of the samples and the analytes in hand. Many separation schemes

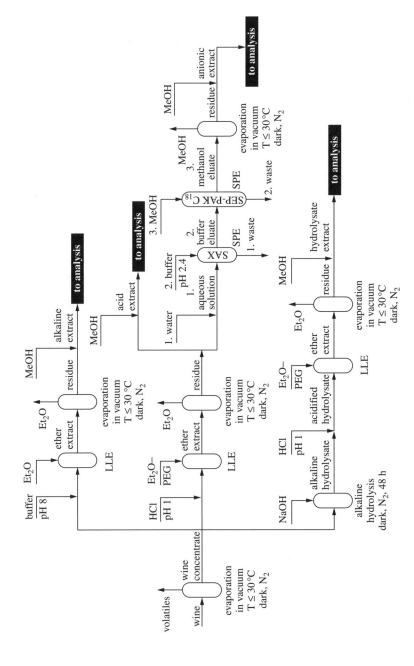

FIGURE 2. Sample preparation scheme for the chromatographic analysis of nonvolatile phenolic acids and depsides in wine[138]

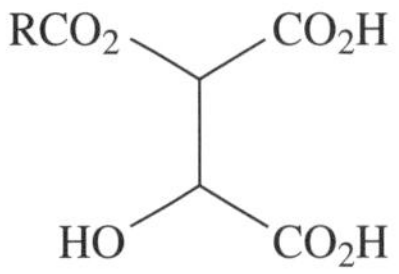

(68) RCO₂–CH(OH)–... structures shown

(68), 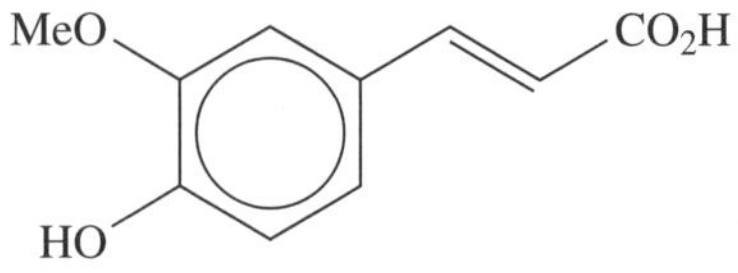 **(69)**, 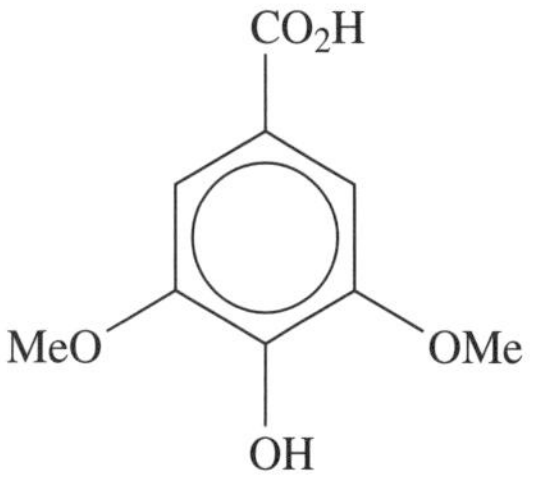 **(70)**

have been reported in the recent literature[139−141], e.g. four different extraction methods have been investigated for the determination of 4-hydroxybenzoic, protocatechuic (**27**), coumaric (**26**), caffeic (**25**) and vanillic (**38**) acids in the flowers of *Delphinium formosum*. End analysis was by RP-HPLC with DA-UVD[142]. However, in contrast with the scheme of Figure 2, sample separation has been found to be superfluous in some cases, e.g. determination of phenolic compounds in white wine[143].

2. Solid phase extraction vs liquid–liquid extraction

a. General aspects. The mechanism of phenol adsorption on activated charcoal was investigated by controlled transformation rate thermal analysis and high resolution argon adsorption at 77 K, processed by the derivative isotherm summation procedure. The most energetic sites for phenol adsorption were identified as micropores that are filled by argon at $-12 \leqslant \ln(P/P_s) \leqslant -7$. Larger and smaller pores are less energetic. This method may be applied to other adsorbents used in SPE[144]. Activated carbons were prepared by carbonization of oxidized or unoxidized coals followed by activation in CO_2 to various degrees of burnoff. Both Brunauer–Emmett–Teller (BET) specific surface areas and pore volume affected the adsorption capacity of the activated carbons. Adsorption of phenols closely followed the Langmuir isotherm, pointing to monolayer formation. The amounts adsorbed on reaching surface saturation decreased with the burnoff extent and with the carbon particle size; the latter effect can be attributed to an increase of diffusion path. The adsorption capacity decreased with the carbonization temperature of unoxidized coals, while it increased for the oxidized coals; this is probably related to different populations of oxygen functional groups on the carbon surfaces[145]. A detailed investigation of the adsorption mechanism was carried out for phenol and pentachlorophenol on carbonized slash pine bark[146]. Analytes differing much in nature from the precolumns on which they are sorbed may give on elution broad chromatographic peaks. To avoid peak broadening, elution from the precolumn should be carried out only with the organic solvent used in the mobile phase[147].

A comparison study of C_{18}-bonded silica cartridges and polystyrene–divinylbenzene copolymer membrane absorption disks showed that the latter were the more effective for SPE of phenols at the 0.5 ppb concentration levels (70–98% recoveries), whereas the C_{18} cartridges were preferable for higher concentration levels (10 ppb) because smaller sample and solvent volumes were required and analysis time was therefore shorter. End analysis was by LC-ELD, with a phosphate buffer–acetonitrile–methanol mixture as mobile phase and coulometric detection at $+750$ mV[148]. A study was carried on the preconcentration step of phenol, *o*-, *m*-, *p*-methylphenol, *o*-, *m*-, *p*-chlorophenol, 2,5-, 2,6-dichlorophenol, catechol (**42**), resorcinol (**20**) etc., at 0.5 and 5 μg L^{-1} concentrations. SPE utilizing a divinylbenzene–hydrophilic methacrylate copolymer gel showed recoveries better than

90%, except for catechol and resorcinol. The performance of this gel was better than that of C_{18}-bonded silica[149].

b. Environmental samples. The main disadvantage of using SPE with certain environmental samples is the presence of suspended matter that may clog the preconcentration devices. Filtration of such suspensions is to be avoided, lest part of the trace analytes be lost in the manipulations. Water samples can be preserved for a long time after adjustment at pH 5 with phosphoric acid and addition of copper sulfate to avoid bacterial and chemical degradation. Improvement of end analytical quality may be achieved by performing the desorption of the analytes preconcentrated on a precolumn using only the organic solvent that serves to modify the mobile phase. This modification allows determination of phenols in water at low ppb levels; LOD was $0.1-2$ µg L^{-1} in tap water, for a 10 mL sample[147]. Operating variables such as concentration, pH and ionic strength of the influent, presence of concurrent solutes, fluid flow-rate and column length were investigated for their effect on the frontal analysis of phenols in water, undergoing SPE on Amberlite XAD-2 or XAD-4[150]. An investigation was carried out of the conditions for selective SPE of aromatic amines and phenols in environmental samples, prior to LC with amperometric detection (AMD)[151].

Off-line SPE with a styrene–divinylbenzene copolymer gave better results than activated carbon, for the preconcentration of phenol, chloro- and nitrophenols, 2,6-dimethylphenol and 2,4,6-trimethylphenol in water, at 100 ppb concentration levels. Except for the last two, these are EPA priority phenols (Table 4). End analysis was by RP-HPLC with DA-UVD. Recoveries were better than 90% and the RSD for real samples was lower than 10%[152]. The breakthrough volumes and selectivity were studied for the SPE performance in the preconcentration of several phenolic water pollutans. An acetylated polystyrene resin and commercial sorbents such as PLRP-S, Amberchrom, Envi-Chrom P and LiChrolut EN were used. End analysis was by HPLC. Retention times increased for phenolic compounds adsorbed on the acetylated resin[153]. A styrene–divinylbenzene copolymer, derivatized with keto groups, was described as a selective SPE medium for phenols in environmental water samples. Various preconcentration techniques were discussed. End analysis was by LC with ELD[154,155]. Comparison between various materials for SPE preconcentration of the eleven EPA priority phenols showed better performance of functionalized polymer resins over carbon black. LOD was less than 35 ng L^{-1} for most analytes in tap water, with linearity range from 0.05 to 20 µg L^{-1}; RSD for repeatability was lower than 8% and for reproducibility between days was lower than 10%, for samples spiked at 0.1 µg L^{-1} [156]. On-line PTFE membranes incorporating a cation exchange resin, based on cross-linked poly-(*endo,endo*-norborn-2-ene-5,6-dicarboxylic acid) (**71**), were investigated as SPE devices for the EPA priority phenolic pollutants. The efficiency of these membranes was better than that of C_{18} or carboxypropylsilica[157].

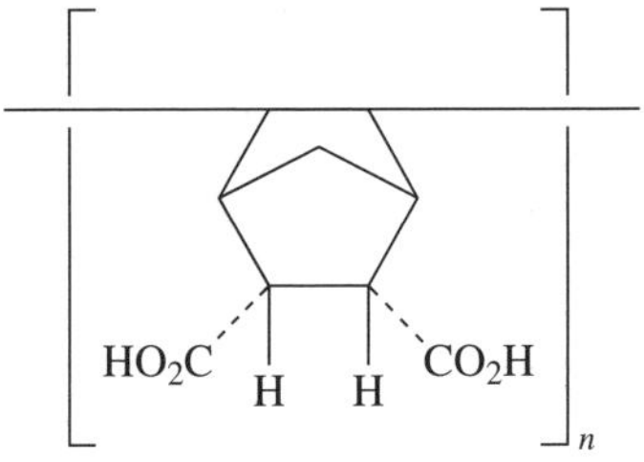

(**71**)

An SPE precolumn made of eight different sorbents was coupled on-line to LC with UVV detection, using 50–100 mL samples of ground water. The performance of this system was compared with that of an off-line method using Empore extraction disks and 1 L water samples. Recovery of phenols varied from <20 to 100% for concentrations in the range 0.1–10 μg L^{-1} at an acid pH. The system was validated by interlaboratory exercises with samples containing 0.1 to 0.5 μg L^{-1} of 2,4,6-trichlorophenol and pentachlorophenol[158].

The stability and recovery of phenolic pollutants in water after SPE was investigated. Three types of polymeric materials were used. Long-term storage of the phenol-loaded sorbants showed losses up to 70% at room temperature while recovery was complete after storing for two months at $-20\,°C$. Stability depends on the water matrix, storage temperature, and the properties of each analyte such as water solubility and vapor pressure. End analysis was by LC with UVD[159].

A semiautomatic module was devised for alkaline extraction of phenols from soil samples, followed by SPE preconcentration on XDA-2. Average recoveries above 60% were obtained for 0.1 to 10 g soil samples, containing 50 to 5000 ppb of phenols, except for 2-t-butyl-4-methylphenol that showed a poor recovery. Soil composition affects in different ways the recovery of alkyl-, chloro- and nitrophenols[160]. Microwave-assisted recovery of SPE-preconcentrated phenols on Empore C_{18} disc membranes was carried out with water in a closed vessel. Under optimal conditions recoveries for eleven priority phenols were above 85%, except for phenol and 4-nitrophenol. Results were similar to those obtained by LLE or SPE on C_{18} cartridge techniques. End analysis was by LC with UVD[161]. An automated SPE method for determination of phenol, o-chlorophenol, 2-amino-4-chlorophenol, 2,4,6-trichlorophenol and pentachlorophenol was developed using a tandem of styrene–divinylbenzene copolymer and C_{18} cartridges. The analytes were recovered with 1N NaOH solution, evaporated under N_2 at room temperature, acidified with glacial acetic acid and subjected to end analysis by HPLC with UVD. Recovery rates were from 54 to 78%; LOQ was less than 50 μg L^{-1} for a signal-to-noise ratio (SNR) of 10[162]. Excellent recoveries were reported for the same analytes in water, after SPE with Amberlite XAD-4 mixed with 10% Norit CN-1 (active carbon)[163]. Automated trace enrichment of phenolic compounds was achieved using a 10 × 2 mm ID precolumn packed with Polysphere RP-S, coupled on-line with RP-HPLC and ELD. Instead of gradient elution, that may be problematic with ELD, two different eluents were used to account for the different polarities of the phenols. When analyzing waters the sample volumes varied according to the origin. Thus, with 4 mL of tap water the LOD for phenolic compounds were between 1 and 10 ng L^{-1}, except for 2,4-dinitrophenol (75 ng L^{-1}) and 2,4-dinitro-6-methylphenol (50 ng L^{-1}); when river waters were analyzed only 1 mL samples could be used due to the interference of humic and fulvic acids, and the LOD were about four times higher[164].

Samples of drinking (2 L), ground (1 L) and river (0.5 L) water, containing eleven EPA priority pollutant phenols, were passed through a GCB cartridge (1 g), at ca 70 mL min^{-1}. After drying with MeOH (1.5 mL), the phenols were eluted with acidic CH_2Cl_2-MeOH and the solvent was partially removed. End analysis was by RP-HPLC with UVD. Recovery of phenols from drinking water, at 0.05 to 4 ppb levels, was higher than 90%. The extraction efficiency of GCB was better than that of C_{18}-bonded silica for the more water-soluble phenols. Interference of the presence of fulvic acids in the SPE of phenols was investigated[165]. Preconcentration by SPE using styrene–divinylbenzene copolymer disks followed by LC-AMD at +1100 mV allowed recoveries of 80–100%, except for the more polar phenolic pollutants. LOD was 0.01 to 0.1 ppb for tap water and 0.1 to 1.0 ppb for river water[166].

The ability of a two-trap tandem system to extract trace amounts of phenols from environmental waters and isolate them from base-neutral species was evaluated. The first trap contains 300 mg of GCB and the second one 50 mg of a strong anion exchanger (SAX), Sephadex QAE A-25. After the water sample had passed through the GCB cartridge, the latter was connected to the SAX cartridge and the base-neutral species were removed from the GCB surface by a neutral eluent. The very weakly acidic phenols were eluted and selectively readsorbed on the SAX surface. Still maintaining the two cartridges in series, an acidified eluent was allowed to flow through both cartridges to recover the most acidic phenols from the GCB cartridge and the least acidic phenols from the SAX cartridge. After partial removal of the solvent, the final extract was submitted to RP-HPLC with UVD. Recoveries of 17 phenols of environmental concern added to 21 of drinking water at levels between 0.2 and 2 $\mu g\,L^{-1}$ were higher than 90%. The effect of the presence of fulvic acids in water on the efficiency of the extraction device was assessed. The recovery efficiency of the GCB-SAX tandem system was compared to that of single extraction cartridges, one containing a chemically bonded siliceous material (C_{18}) and the other SAX material. The LOD of the analytes considered were well below 0.1 $\mu g\,L^{-1}$[167].

A porous membrane impregnated with organic solvent forming a barrier between two aqueous phases can be used for selective LLE of chlorophenols, that are transferred to the second phase for end analysis by LC with ELD. LOD was ca 25 $ng\,L^{-1}$ for 30 min extraction[168].

Various preconcentration methods were evaluated to monitor phenol and monochlorophenols in drinking and river waters. LLE showed large losses during solvent removal. SPE with Amberlite XAD-2, XAD-4, C_{18} Si 100, Tenax and Polysphere RP-18 showed the best results with the latter solid phase. End analysis was by RP-HPLC with LiChrospher RP-18e[169]. An extensive study was performed on the factors affecting the analytical quality of the HPLC determination of phenolic pollutants in water at the 1 ppm level (44 compounds). A preconcentration step was carried out by LLE with n-C_6H_{14}, Et_2O, AcOEt, $CHCl_3$ and CH_2Cl_2. The latter solvent was found to give the best overall recoveries (55–99%)[170].

A study was carried out for LLE by the Soxhlet method and microwave-assisted extraction for the determination of the priority phenols in soil samples. Recoveries varied from 67 to 97% with RSD between 8 and 14% for LLE, and >70% for the MAP, except for nitrophenols that underwent degradation when the latter method was applied. LOD was from 20 $ng\,g^{-1}$ for 2,4-dimethylphenol to 100 $ng\,g^{-1}$ for pentachlorophenol. The best detection method for LC was atmospheric pressure chemical ionization MS (APCI-MS)[171]. The most abundant ions obtained by this detection method were $[M - H]^-$ for the lowly chlorinated phenols and $[M - H - HCl]^-$ for tri-, tetra- and pentachlorophenols[172].

To determine phenolic acids in soil, samples were subjected to LLE with 0.1 M NaOH for 16 h, centrifugation, filtration and pH adjustment. End analysis was by RP-HPLC on a C_{18} column with UVD at 280 nm. Recoveries were as follows: *p*-hydroxybenzoic acid 123%, vanillic acid (**38**) 83%, syringic acid (**70**) 66%, coumaric acid (**26**) 100%, ferulic acid (**69**) 58% and caffeic acid (**25**) 0%. LOD was 0.5 ppm for the derivatives of benzoic acid and 1 ppm for those of cinnamic acid, excepting **25** that could not be detected by this method[173].

c. Foodstuffs. The HPLC determination of synthetic phenolic antioxidant additives (see Table 3) in food has been reviewed[174]. On-line SPE was proposed where the samples of wine were injected and adsorbed onto polystyrene–divinylbenzene cartridges in a flow injection analysis (FIA) system. End analysis was by RP-HPLC with DA-UVD[175]. Application of SPE was studied instead of the well established LLE for the volatile phenols in wine. Thus, percolation of clarified wine at pH 9 on the anion exchange resin AG 2-X8 permits adsorption of derivatives of phenol (e.g. **72a–e**) and guaiacol (e.g. **73a,b**). This left the organic acids in solution; the basic compounds were rinsed out with 1 N HCl, and the adsorbed phenols were eluted with methanol, diluted with water and directly determined by RP-HPLC with UVD at 280 nm, with high sensitivity (e.g. 20–40 ppb for compounds **72a–d**) and good recoveries (91%) and repeatability. No interference from other compounds was noted in various wines[176].

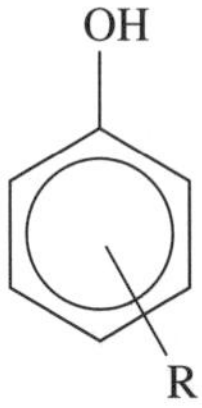

(72) (a) R = H
(b) R = 2-Me
(c) R = 3-Me
(d) R = 4-Me
(e) R = 4-Et

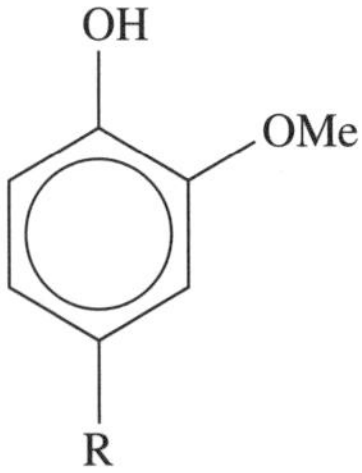

(73) (a) R = Et
(b) R = Vi

d. Miscellaneous industrial products. The SPE preconcentration step was simplified for a series of 34 phenolic compounds used in plastic manufacture, that could contaminate water which came into contact with plastic utensils. These compounds included phenol, its alkyl and chloro derivatives, dihydroxybenzenes, their alkyl derivatives and other phenolic compounds. After a single extraction of the SPE cartridges with MeOH, end analysis was by LC with DA-UVD, at the 1–5 ppm level RSD 1–6% ($n = 3$) with recoveries of 50–100%[177,178].

A simplified method for determination of phenolic compounds in crude oils, gasoline and diesel fuel consists of on-line SPE with a silicone membrane followed by LC with ELD and UVD[179,180]. On-line coupling of a preconcentration device to an HPLC analyzer with ELD or UVD, in an overall automatic operation, is claimed to substantially improve analytical performance. A silicone membrane device has been used for SPE in the determination of phenols dissolved in complex organic matrices such as gasoline and kerosene[181].

p-Nonylphenol is a surfactant used in commercial sprays and aminocarb insecticide formulations. After removing the insecticide by alkaline hydrolysis the surfactant was extracted with *n*-heptane and determined with good reproducibility by LC using a Partisil(R) ODS-2 column, 95% MeOH/water as mobile phase and UVD at 278 nm. LOQ was 30 ppm with 10 μL injection[182].

e. Biological and biomedical samples. Proanthocyanidins or condensed tannins consist of chains of epicatechin (**2**) or epigallocatechin (**74**) of varying degree of polymerization and mode of linking. Methanol or ethanol can be used to extract low molecular phenolics and oligomeric proanthocyanidins from fresh tissue, while aqueous acetone is required for larger polymeric units. The tannin fraction was separated by SPE on Toyopearl HW-40 (F), and recovered with aqueous acetone. Size separation of the condensed tannins was performed by HPLC on a Lichrospher Si 100 column. Although the retention times

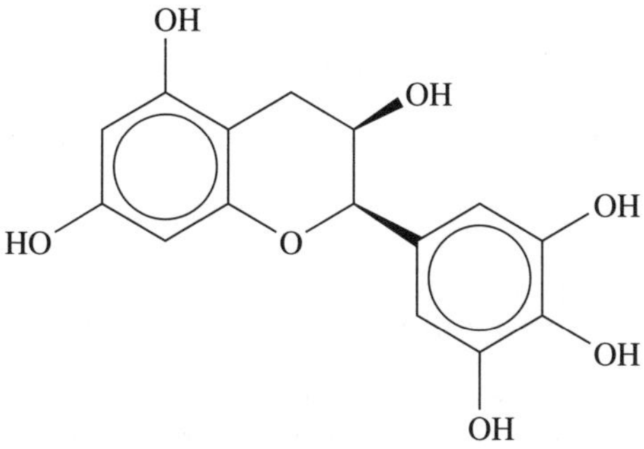

(74)

increased with the degree of polymerization, no functional correlation could be developed for these parameters[183].

Methods involving LLE were developed for determination of phenol[184] and other phenolic compounds. For example, for simultaneous determination of phenol, hydroquinone (**66**) and catechol (**42**) in urine, the samples were subjected to acid hydrolysis, saturation with sodium sulfate and LLE with diethyl ether. End analysis was by RP-HPLC on a C_{18} column, elution with sodium acetate–acetic acid buffer–acetonitrile gradients, and FLD. The recovery and reproducibility were generally over 90%. The method appears to be more sensitive than GC or HPLC with UVD. It is proposed for cigarette smokers and refinery workers exposed to low benzene concentrations. Good recoveries of these metabolites was attained at 0.1 to 50 mg L^{-1} concentrations, with coefficients of variation of a few percent, both for within a day and between day determinations[33].

3. Other preconcentration methods

A preconcentration method that bears some resemblance to SDE consists of isolating the volatile phenols by steam distillation, followed by freeze-drying of the distillate. End analysis was by HPLC with ELD. The method was applied for determination of such phenolic components in foodstuffs and packing materials[185]. Determination of phenolic antioxidants in polyolefins was carried out by dissolving the polymer sample in a heptane–isopropanol mixture (1000/5, v/v), at 160–170 °C, in an autoclave. The polymer precipitated on cooling the solution, and the dissolved antioxidant could be determined by LC with UVD. The advantage of the method is the relatively short time of analysis (about 2 h) and its reproducibility (RSD 3–5%)[186].

B. Derivatization

1. General

Various objectives are sought when preparing derivatives of phenolic compounds prior to performing a LC separation: Facilitating separation of the analytes from the matrix, modifying the chromatographic on-column behavior of the analytes and improving the sensitivity toward the analyte during detection. The most important modifications aiming at the latter objective consist of introducing chromophoric or fluorophoric groups that will enhance the response of UVDs and FLDs to the analyte.

Derivatization methods of phenolic compounds for LC analysis have been reviewed[98].

2. Formation of azo dyes

Phenols in trace concentrations were derivatized by coupling with a 4-sulfobenzene-diazonium salt at pH 10.5. The azo dyes were combined with tetradecyldimethylbenzy-lammonium ions at pH 5.0 and, after SPE on a PTFE membrane filter, the end analysis was carried out by RP-HPLC with UVD at 352 nm. LOD was between 40 ppt (phenol) and 2 ppb (2,5-xylenol) for eight phenols tested. The method was used for determination of phenols in river water[187].

A macroporous reactive polymer was prepared by copolymerization of the methacrylate esters **75** and **76**, using the methacrylate ester **77** as crosslinking agent. After removal of the benzylidene protecting groups the polymer could be diazotized and used for immobilizing phenolic analytes by a coupling reaction. The azo dye formed on the polymer was split by hydrolysis of the ester and quantitatively determined by LC[188].

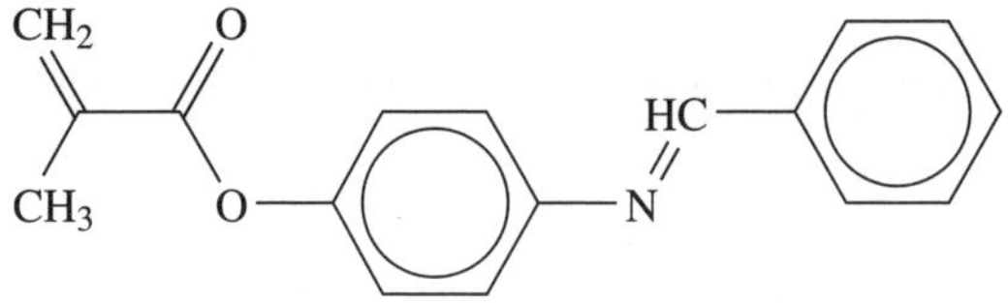

(75) 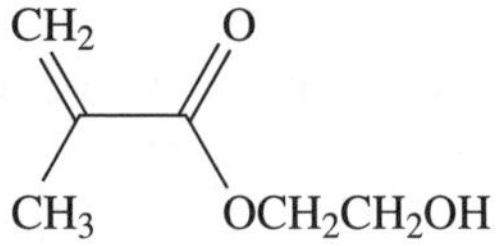 (76)

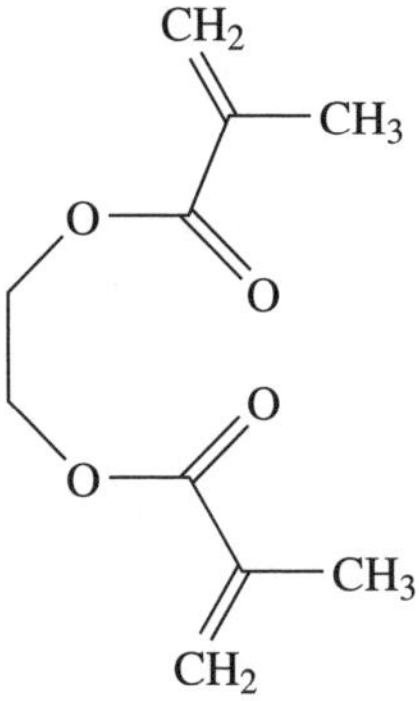

(77)

3. Formation of imino dyes

Phenols combine with 4-aminoantipyrine (**78**) in the presence of an oxidant to yield imino dyes, for example, according to equation 2 for phenol[189]. The dye product can be concentrated by LLE with chloroform or SPE on a C_{18}-silica column, followed by LC-UVD. LOD was in the ppb range. The process can be carried out in a FIA system. Phenolics possessing additional acidic functional groups are in the anionic form and cannot be extracted and determined by this method[190].

4. Fluorescent tags and fluorescence enhancement

Reaction of 2-(9-anthrylethyl) chloroformate (**79**) with phenols (phenol, 4-methylphenol, 3,4-dimethylphenol and 4-t-butylphenol), to yield the corresponding carbonates (**80**) (equation 3), was investigated as derivatizing method before RP-HPLC with FLD. LOD was 7 to 10 nM[191].

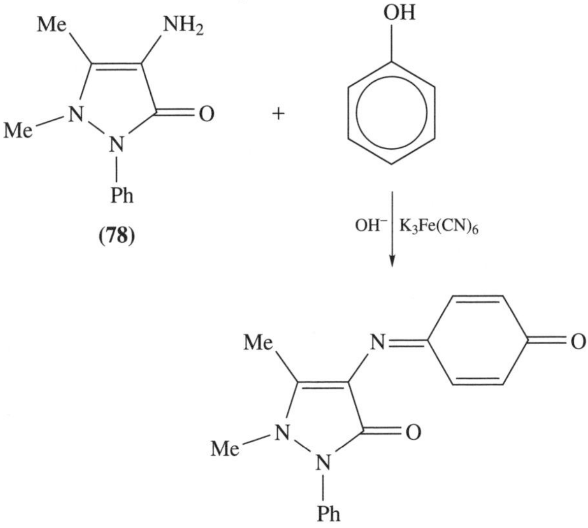

4-(N-Chloroformylmethyl-N-methyl)amino-7-(N,N-dimethylaminosulfonyl)-2,1,3-benzoxadiazole (**81**) was proposed as a precolumn derivatizing reagent for alcohols, phenols, amines and thiols, conferring a fluorescent tag (λ_{ex} 437–445 nm, λ_{fl} 543–555 nm). The presence of quinuclidine (**82**) was required to ensure complete reaction of analytes other than amines. End analysis was by RP-HPLC with FLD. LOD was in the femtomol range for the derivatives on column[192]. 4-(4,5-Diphenyl-1H-imidazol-2-yl)benzoyl chloride (**83**) was proposed as a precolumn fluorescent label for acylation of phenols. LOD for phenol and various chlorophenols was below 0.1 μM for 20 μL injections. The average concentration of free and total phenols in human urine is 4.3 ± 2.5 and 29.5 ± 14.0 μM, respectively[193,194]. A method for simultaneous determination of phenol and p-cresol (**72d**) in urine was based on acid hydrolysis, LLE with isopropyl ether and reaction with labeling reagents such as **84**, to give fluorescent sulfonic acid esters (λ_{ex} 300–308 nm, λ_{fl} 410 nm) and RP-HPLC. LOD for the analytes (SNR 3) was about 0.2 pmol per injection for **84a** and about 15 fmol per injection for **84b–d**. The content of phenol and p-cresol in human urine is 12–294 and 8–246 nmol(mg creatinine)$^{-1}$, respectively[195,196].

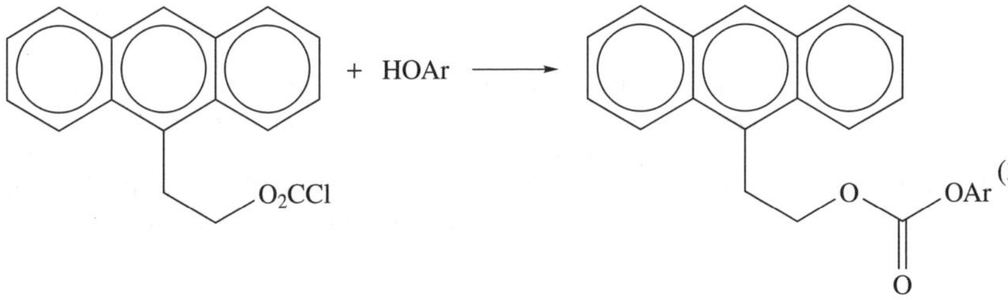

Phenolic compounds form inclusion complexes with α-cyclodextrin (**85**), enhancing the fluorescent properties of the aromatic analytes. For example, p-hydroxybenzoic acid (**86a**),

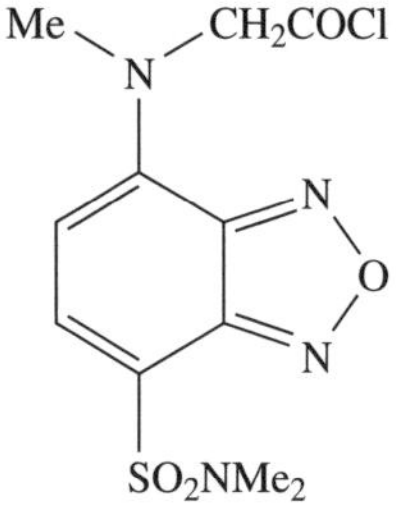

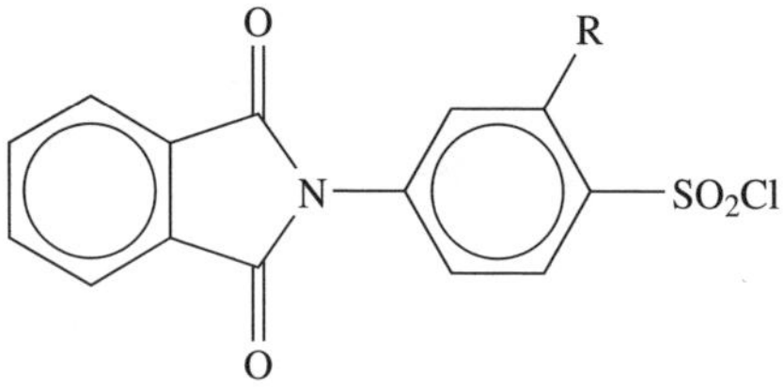

(84) (a) R = H
(b) R = OMe
(c) R = OEt
(d) R = OPr

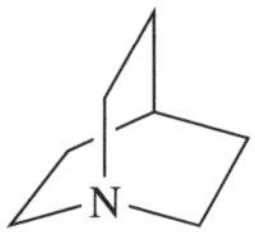

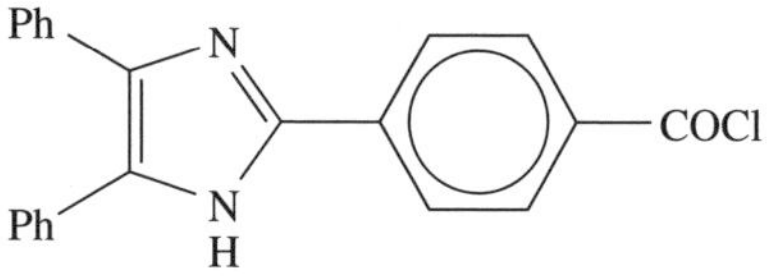

methylparaben (**86b**), ferulic acid (**69**) and vanillic acid (**38**) were determined by RP-HPLC with FLD, using as mobile phase a 10^{-2} M concentration of **85** in acetate buffer of pH 4.6. Formation constants of the complexes were calculated from retention parameters. LOD was in the 1–5 µg L^{-1} levels. The method was applied to the analysis of phenolics in beer[197]; see also a method for determination of β-cyclodextrin in Section VIII.A.2.d.

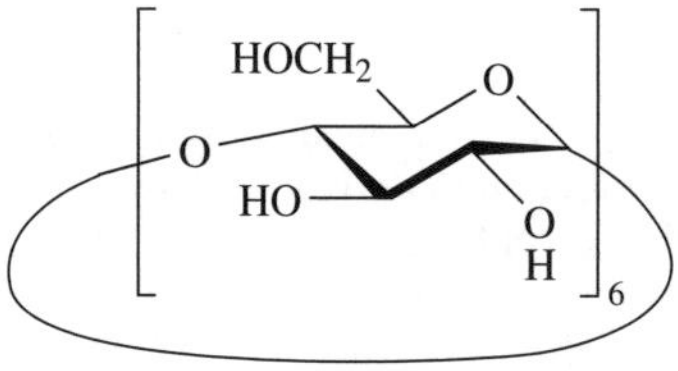

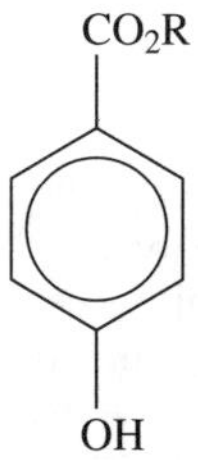

(86) (a) R = H
(b) R = Me

C. End Analysis

The present section is organized according to the different matrix types related to the origin of the samples undergoing LC analysis. In Table 6 are summarized the chromatographic techniques other than RP-HPLC, and in Table 7 the detection methods applied after the chromatographic separation mentioned in Sections IV.A–C.

13. Analytical aspects of phenolic compounds

TABLE 6. Liquid chromatography techniques mentioned in Section IV, other than conventional RP-HPLC

Chromatographic technique	Subsection and references
Capillary electrochromatography (CEC)	**C.3.c** 265.
Gel permeation chromatography (GPC)	**C.5.c** 289.
Ion chromatography	**C.1** 214.
Ion exchange chromatography (IEC)	**C.2.a** 215; **C.3.b** 257.
Micellar liquid chromatography (MLC)[a]	**C.2.d** 237, 238; **C.6** 292.
Microbore column	**A.1** 138; **C.1** 210.
Normal phase HPLC	**C.2.a** 215.
Reversed phase (RP) microcolumn	**C.3.b** 261; **C.3.c** 266.
Size exclusion chromatography	**C.5.a** 286.
Supercritical fluid chromatography (SFC)	**C.1** 207–209; **C.3.b** 256; **C.5.a** 285.
Thin layer chromatography (TLC)	**C.3.a** 250, 251; **C.3.b** 264; **C.4.a** 63, 271–273

[a] Applications of the micellar electrokinetic chromatography (MEC) technique appear in Section V, dealing with electrophoresis.

TABLE 7. Post-column detection methods for liquid chromatography mentioned in Section IV

Detection method	Subsection and references
UVD[a]	**A.1** 138; **A.2.b** 158, 159, 161, 162, 165, 167, 173; **A.2.c** 176; **A.2.d** 180–182; **A.3** 186; **B.2** 187; **B.3** 190; **C.1** 201–203; **C.2.b** 223; **C.2.d** 237, 238; **C.3.a** 247; **C.3.b** 253, 254, 257–260, 264; **C.4.a** 268, 273[b,c] **C.4.c** 281, 282; **C.5.d** 290.
DA-UVD	**A.1** 142; **A.2.b** 152; **A.2.c** 175; **A.2.d** 179, 180; **C.1** 202; **C.2.a** 217; **C.2.b** 218, 219, 221, 222, 224, 225; **C.2.c** 229–234; **C.2.e** 241, 242; **C.3.a** 249; **C.3.b** 256; **C.4.a** 267, 269; **C.5.b** 287.
FLD	**A.2.e** 33; **B.4** 191–197; **C.1** 198; **C.2.c** 234; **C.2.d** 240; **C.3.b** 258; **C.4.c** 279, 232.
ELD	**A.2.a** 148[d]; **A.2.b** 151[e], 154, 155, 164, 166[e], 168; **A.2.d** 179–181; **A.3** 185; **C.1** 201, 202, 203[d], 204[f], 207[g], 208, 210, 211, 212[e], 214; **C.2.a** 216[d], **C.2.b** 221, 222, 227[h]; **C.2.e** 244[h]; **C.3.a** 246, 248[h]; **C.3.b** 254[e], 255[e], 261, 262[h], 263[h]; **C.3.c** 266; **C.4.b** 275; **C.4.c** 280[i], 282[e], 283, 284[f]; **C.5.c** 288.
MS	**A.2.b** 171[j], 172[j]; **C.1** 203[k], 206[j,k,l]; **C.2.b** 224; **C.2.c** 236[m]; **C.2.d** 239[l,n], 240[m]; **C.4.a** 269[j], 270[n]; **C.4.c** 277[m], 278[j]; **C.5.a** 285[j].
DRD	**C.5.a** 286.

[a] See also Reference 99 in Section III.B.2.
[b] Dual wave-length spot densitometer for TLC.
[c] Video image analysis for TLC.
[d] Coulometric detection.
[e] Amperometric detection.
[f] Tyrosinase-based biosensor. See also Section VI.A.2.
[g] Voltammetric detection.
[h] Multiple electrode coulometric array.
[i] Phenoloxidase-based biosensor. See also Section VI.A.3.
[j] APCI-MS.
[k] PBEI-MS.
[l] ISP-MS.
[m] ESI-MS.
[n] TSP-MS.

1. General

Experimental design methodology was applied for the optimization of the elution program for the RP-HPLC resolution of a mixture of nine phenols, with a ternary solvent system (water–AcOH–MeCN). Important factors were the initial isocratic elution, the gradient running time and the gradient curvature[62].

Peaks with a large degree of overlapping can be resolved by the H-point standard additions method. The method was applied to the LC-FLD determination of phenol and some monosubstituted phenols (**72b–d**) in water, using as analytical signals the heights or the areas obtained at two selected emission wavelengths. Good results are obtained for highly overlapping peaks with highly overlapping fluorescence spectra. The principal benefits of the method are the ease of finding the required wavelengths and its insensitivity to changes in the retention time of the peak from one injection to another[198]. The pK_a values of polychlorinated phenols (**87–89**) were computed by a nonlinear regression algorithm, and were applied to estimation of the capacity factors at various pH values, using a standard reversed-phase column and acetonitrile as organic modifier. The method can also be applied to mobile phase optimization. A resolution map is presented based on pK_a values obtained either from the literature or from chromatographic data[199].

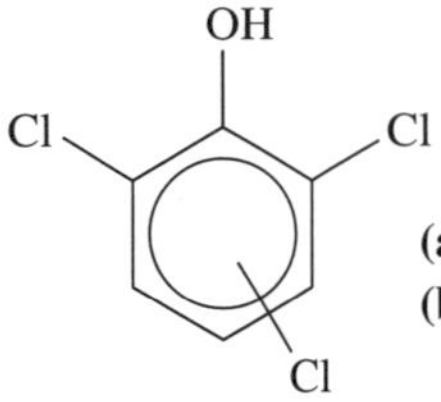

(**87**)
(a) 2,3,6-Cl_3
(b) 2,4,6-Cl_3

(**88**)
(a) 2,3,4,5-Cl_4
(b) 2,3,4,6-Cl_4
(c) 2,3,5,6-Cl_4

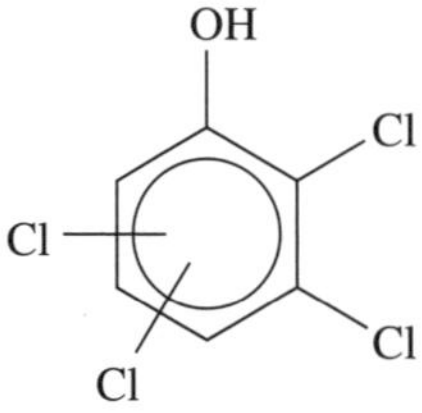

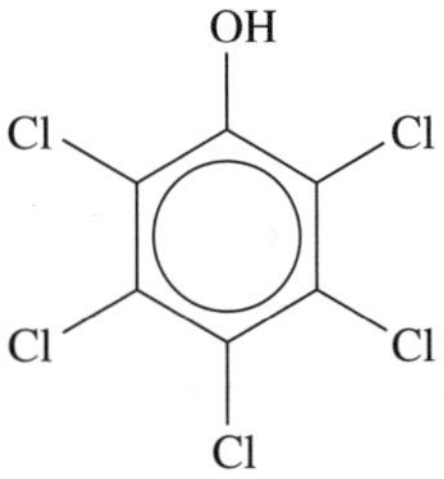

(**89**)

Several physical-chemical properties of alkanes, polyaromatic hydrocarbons, alkylbenzenes, polychlorobenzenes, polymethylphenols and polychlorophenols were determined using various software packages. The ionization potentials calculated by the MOPAC program was the most suitable property with which to adjust the capacity ratios of polychlorobenzene, polymethylphenol and polychlorophenol isomers[200].

Comparative studies were carried out to determine the efficiency of the various detection methods in the analysis of phenolic compounds. On-line SPE of sixteen priority phenol pollutants in water on polystyrene was followed by HPLC separation and detection. The sensitivity of ELD was higher than that of UVD. LOD down to the ppt level was attained by ELD on 100 mL samples for all the chorinated phenols; however, nitrophenols could not be equally determined because they require working potentials different from those chosen for the chlorinated phenols in this study. Phenol could be detected at 0.02 ppb on reducing the sample volume to 10 mL[201].

An extensive study was performed on the factors affecting the analytical quality of the HPLC determination of phenolic pollutants in water at the 1 ppm level (44 compounds). Various types of column were considered, taking advantage of certain modes of analyte–stationary phase interaction: for C_{18} dispersion forces, for diphenyl π-electron interactions and for propylnitrile dipole moment interactions. Although optimization of the operating conditions improved the resolution of the column, no single column was capable of separating the complete set. Single components in very complex mixtures could be analyzed without MS detectors applying multidimensional chromatography and PCR. LOD was in the ppb range at SNR 2[170].

A comparison of performance was carried out of DA-UVD and ELD for the determination of trace amounts of phenolic antioxidants, such as BHA (**31**), 2-*tert*-butylphenol, 2-*tert*-butyl-4-methylphenol, PG (**33a**) and OG (**33b**). The LOD were lower and the linearity ranges wider for ELD than UVD[202].

UVD, ELD by controlled-potential coulometry and particle-beam electron-impact MS (PBEI-MS) in the SIM mode were applied to the HPLC analysis of fifteen benzoic and cinnamic acid derivatives. LOD for ELD was in the range from 1 to 5 pg injected, RSD was 0.6–3.0% at the 0.1 ng level ($n = 4$), with linear dynamic range of at least 10^3; LOD for UVD was in the 5–50 ng range with linearity up to at least 15 µg for most analytes, RSD was from 1.2 to 3.1% at the 500 ng level ($n = 4$); LOD for PBEI-MS was 2–5 ng, with nonlinear behavior over the entire range investigated (from 10 ng to 10 µg), RSD was 0–1.8% at the 100 ng level ($n = 4$) except for caffeic acid (**25**, RSD 75% at the 50 µg level, $n = 4$)[203].

The use of a carbon paste electrode (CPE) incorporating tyrosinase (see Section VI.A.2) as ELD for LC determination of phenols was investigated. The enzyme-modified electrode showed higher stability than the unmodified one[204].

Chlorinated phenols in solid matrices, at concentrations down to sub-ppm levels, could be determined without any sample clean up, by placing a SFE device in tandem with a LC instrument. The speed of analysis and selectivity of the system compared favorably with conventional methods[205].

Three LC-MS interfacing techniques were compared. When using the thermospray (TSP) interface, $[M - H]^-$ or $[M + CH_3COO]^-$ were obtained as the main ions. APCI and ion spray (ISP) interfaces gave $[M - H]^-$ at 20–30 V as the main ion. Calibration graphs were linear from 1 to 100 ng for each compound with repeatability values of 15–20%. Instrumental LOD for APCI were 3–180 ng in full scan and from 0.001–0.085 ng in SIM mode. Instrumental LOD for ISP and TSP were larger by approximately one order of magnitude[206].

Supercritical fluid chromatography (SFC) with ELD, using CO_2 or CO_2–MeOH as mobile phase, was applied to simultaneous determination of 11 priority phenols and 13 polycyclic aromatic hydrocarbons. Voltammetric measurements allow low-nanogram detection limits of reducible and oxidizable analytes, even if they elute simultaneously from the chromatographic column[207]. SFC with MeOH-modified CO_2 was performed under isobaric and pressure-programmed conditions, combined with ELD. LOD was 250 µg of 2,6-dimethylphenol for oxidative ELD and 100 pg of 1,3-dinitrobenzene for reductive ELD[208]. Various sorbents were investigated for SPE preconcentration prior to SFC[209].

Microbore columns are of advantage due to the low mobile phase volumetric flow rates involved, the reduced on-column samples and the reduced chromatographic dilution, conferring high efficiency. Microbore columns with ELD were applied to the analysis of antioxidants, which are usually electroactive compounds. This combination led to highly selective and sensitive analyses. A micro ELD was designed and tested with catecholamines. LOD for noradrenaline (**15b**) was 1 pg per 0.2 μL injection (3 nM), on a 0.7 mm bore column. The tested antioxidants were gallic acid (**8**), its propyl ester (**33a**) and three dihydroxybenzenes (**20, 42, 66**). The dynamic range was of four orders of magnitude and LOD was down to 0.1 fmol (20 fg injected) with a 0.3 mm bore column[210].

A great enhancement of sensitivity for phenols analyzed by LC with ELD was attained using a glassy carbon electrode (GCE) chemically modified with polymerized Ni-protoporphyrin IX (**90**). This modification can also suppress oxidation of substrates more polar than phenols, such as ascorbic acid (**91**) and potassium hexacyanoferrate(II) (**92**). LOD was 13 μg L^{-1} of *p*-nitrophenol with linearity up to 1.3 ppm[211].

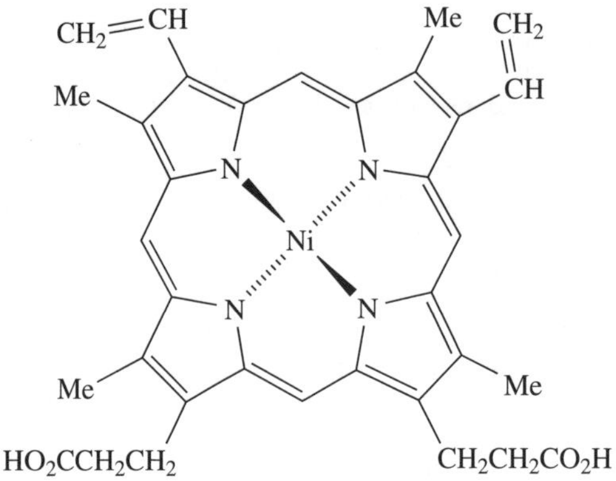

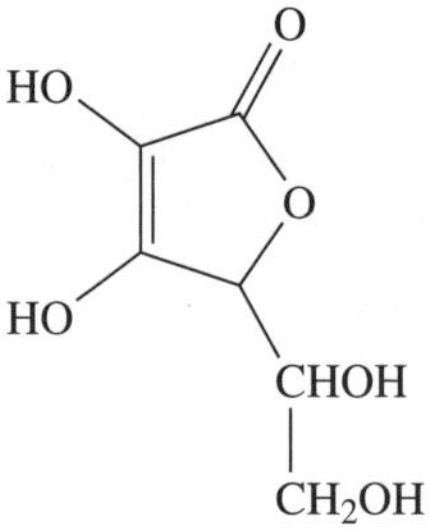

A nickel phthalocyanine (**93**) polymer-coated GCE, working at an applied potential of +0.70 V *vs.* Ag/AgCl, was used for AMD of phenolic antioxidants. LOD was 0.11, 0.60 and 0.15 mg L^{-1} for BHA (**31**), BHT (**32a**) and PG (**33a**), respectively, using 50 μL injections with TBHO (**35**) as internal standard[212].

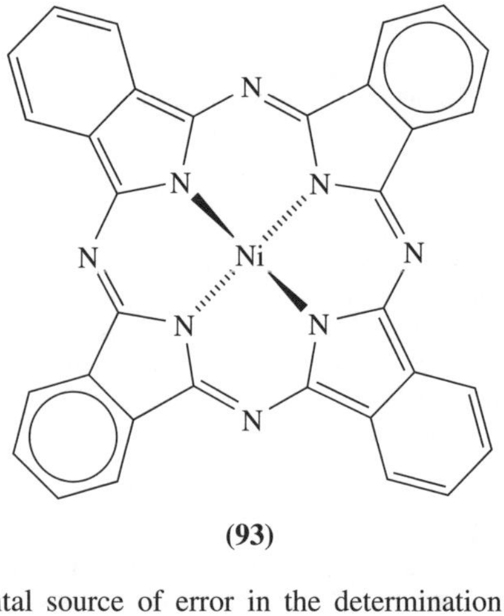

A possible instrumental source of error in the determination of organic analytes by LC methods or in FIA setups is adsorption on the tubing and ducts of the instrument.

Thus, for example, the deviations from the expected behavior for the higher homologs in the determination of the diffusion coefficients of the m-alkoxyphenol and alkyl p-hydroxybenzoate homologous series, in alkaline aqueous ethanol solution, was attributed in part to solute adsorption on the walls of the Teflon dispersion tube[213].

Pentachlorophenol, 4-chlorophenol, 2-nitrophenol and 4-nitrophenol, in 0.9 to 3.6 mM concentrations, were investigated as modifiers of the mobile phase ($NaHCO_3$, Na_2CO_3 and NaOH solutions) in the ion chromatographic determination of various anions. These included species that usually are determined by this technique, such as F^-, Cl^-, NO_3^- and PO_4^{3-}, and also ions that are strongly retained on the column, such as I^-, SCN^-, CrO_4^{2-}, MoO_4^{2-} and ClO_4^-. The phenols were effective in substantially reducing the retention times of the strongly adsorbed anions; e.g. the retention time of ClO_4^- changed from 93.0 to 15.2 min in the presence of 3.6 mM of 4-nitrophenol. NO_3^- and PO_4^- both had a retention time of 9.10 min with the ordinary mobile phase, but could be resolved in the presence of pentachlorophenol and 4-nitrophenol[214].

2. Foodstuffs

a. General. A review appeared on normal phase HPLC, RP-HPLC and ion exchange chromatography (IEC) separation of phenolic compounds in food, including anthocyanins, flavones, carotenoids, beet pigments, curcumins, mangiferin, gingerol and phenolic components produced from degradation of natural products during food processing[215].

A general method for the evaluation of phenolic compounds in fermented beverages, fruit juices and plant extracts was developed using gradient HPLC and coulometric detection. In a 10 μL injection it was possible to identify and determine 36 different flavonoids and simple and complex phenols, without sample extraction, purification or concentration, in several kinds of beers, red and white wines, lemon juice and soya, forsythia and tobacco extracts. This may also be useful for the characterization of beverages and extracts[216].

An optimization strategy was presented for the validation of a unique LC method, including the use of a single solvent gradient, for the LC analysis with DA-UVD of the most representative phenolic compounds from different food sources[217].

b. Fruit juice. A method was described involving SPE and isocratic LC with DA-UVD, for rapid determination of five phenolic acids, namely gallic (**8**), caffeic (**25**), ferulic (**69**), ellagic (**94**) and chlorogenic (**95**), in fruit juices[218,219]. A single-gradient RP-HPLC run was recommended for the initial investigation of phenolic compounds in plants, whereas multiple runs after optimization for individual components are recommended when chromatographic resolution is required. This approach was applied to the analysis of apple juice, where the principal components were chlorogenic acid (**95**), phloridzin (**96**), caffeic acid (**25**) and coumaric acid (**26**), and tomato juice, containing **95**, **25**, **26**, naringenin (**5**) and rutin (**97**). Similar quantitative estimates were obtained for these components by both chromatographic approaches[220].

Phenolic and furfural compounds in apple juice were determined by HPLC using a combination of ELD and DA-UVD. LOD for ELD were 4 to 500 times greater than those for spectrophotometric detection. The content of phenolics varied from 30 to 115 mg L^{-1}, including major phenolic components such as chlorogenic acid (**95**), p-coumaroylquinic acid (**98**) and phloridzin (**96**) and minor ones such as caffeic acid (**25**), p-coumaric acid (**26**), ferulic acid (**69**), gallic acid (**8**), protocatechuic acid (**27**) and catechin (**3**)[221]. The same methods were also applied for the analysis of maple products. The phenolic content was dependent on the source of the product. Application of reverse osmosis to maple sap caused a relative decrease of aldehydes and alcohols and an increase of phenolic acids.

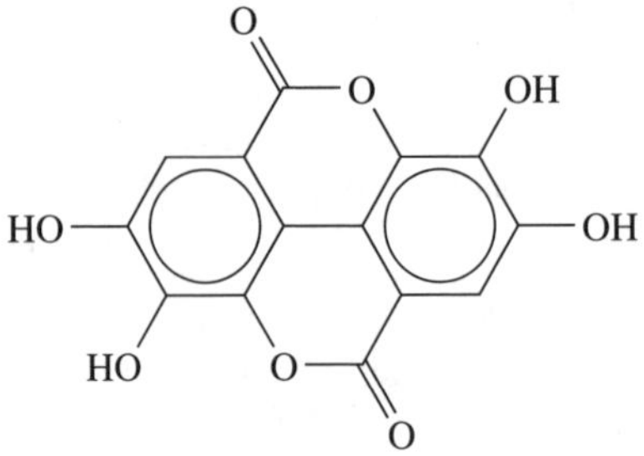

(94)

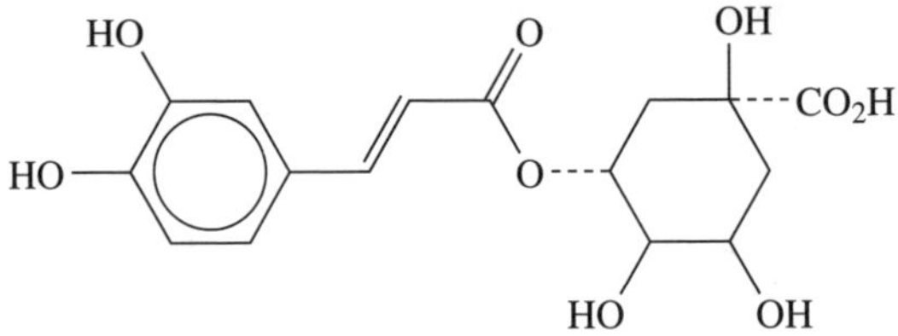

(95)

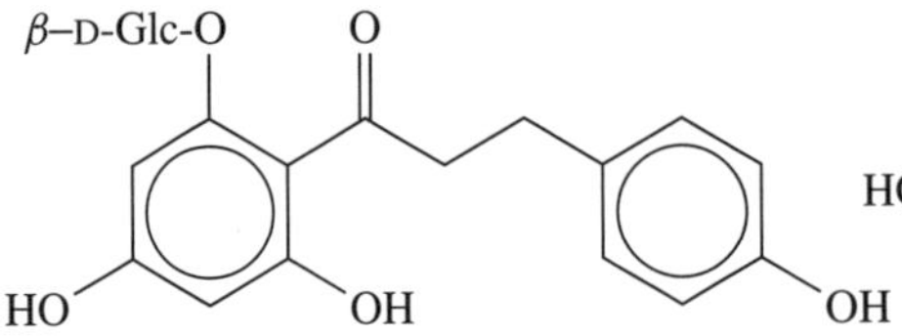

(96)

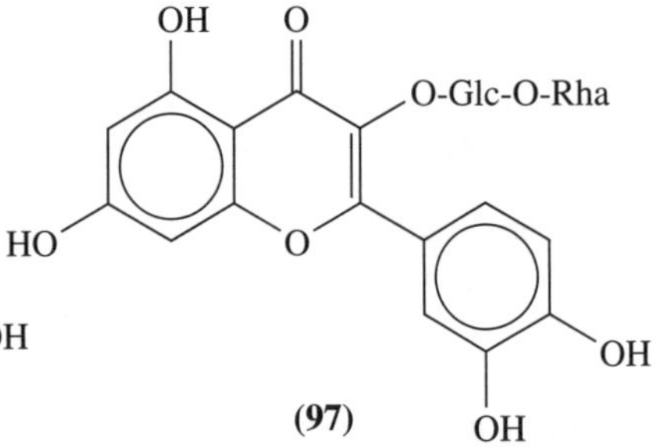

(97)

Thermal evaporation brought about an increase of ferulic acid (**69**), vanillin (**39**) and syringaldehyde (**63**) with an attendant drastic decrease in sinapic acid (**99**)[222].

SPE on a CLX cartridge was applied to separate 'acidic' phenols such as chlorogenic acid (**95**) from 'neutral' phenols such as (−)-epicatechin (**2**), (+)-catechin (**3**), phloridzin (**96**) and quercitrin (**100**). The neutral phenols were determined in apple juice by capillary LC with UVD at 280 nm, as an alternative to conventional HPLC. LOD were from 9 pg for **96** to 97 pg for **3**[223]. HPLC analysis with MS and DA-UVD showed that apple pomace is a good potential source for phenolics. The usefulness of arbutin (**9**) as specific marker for pear products was placed in doubt[224] (see Section II.A[20]).

The following profile of phenolics in pear fruit was established by HPLC with DA-UVD: Quinic acid (**101**) esterified in various positions by caffeic (**25**), coumaric (**26**) and malic (**102**) acids, and a mixture of flavonols that included three quercetin (**47**) 3-*O*-glycosides (rutinoside, glucoside and malonyl glucoside) and five isorhamnetin (**103**) 3-*O*-glycosides (rutinoside, galactorhamnoside, glucoside, malonylgalactoside and malonylglucoside). Identification was aided by chemical and spectral methods, such as FAB-MS[225].

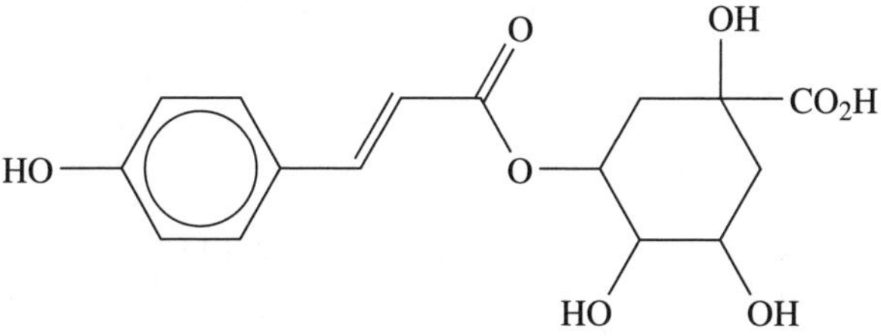

(98)

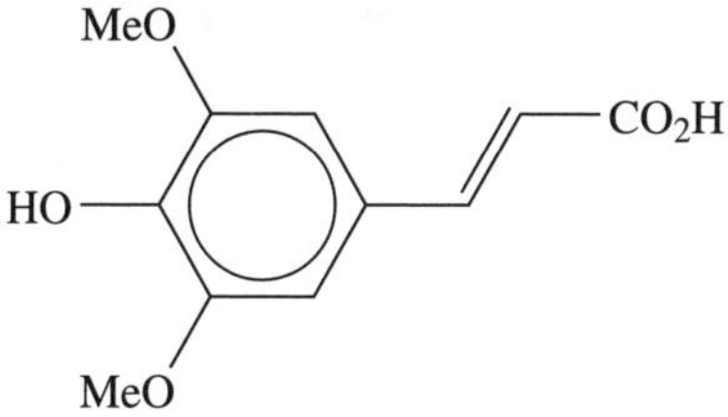

(99)

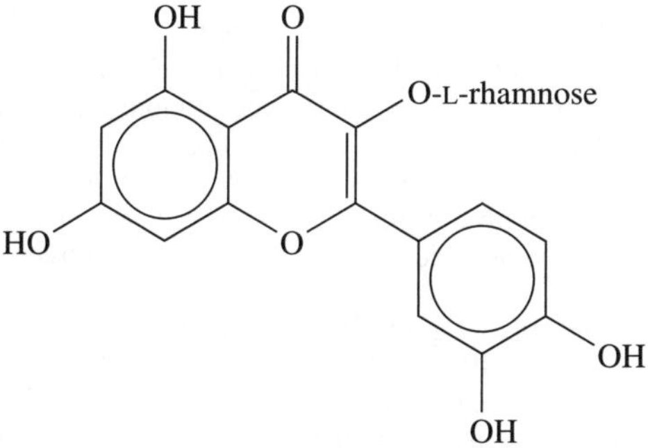

(100)

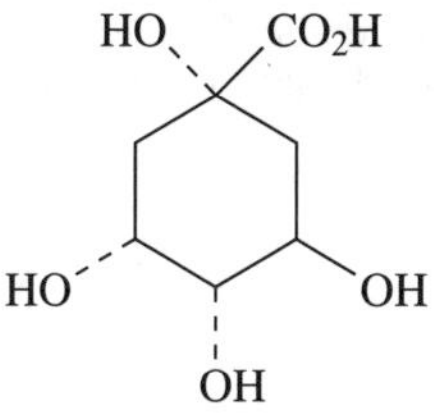

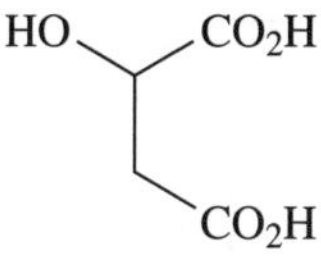

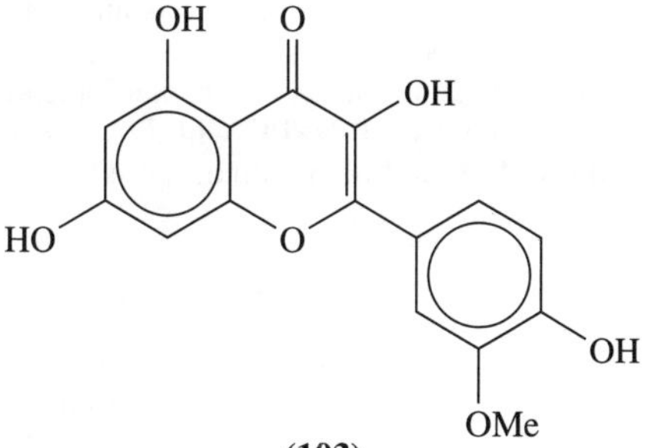

(101) **(102)** **(103)**

After alkaline hydrolysis and acidification to pH 3.4, samples of juices (green grape, black grape and cherry) were subjected to LLE with AcOEt. End analysis was by isocratic RP-HPLC. Gallic (**8**), chlorogenic (**95**), caffeic (**25**) and ferulic (**69**) acids were separated and determined. Determination of ellagic acid (**94**) required a modification of the elution regime. The phenolic acids of cherry juice were shown to have anticancergenic properties[218,226].

A coulometric array consisting of sixteen detectors was set up to generate voltammetric data for ELD after RP-HPLC of phenolics and flavonoids in juice beverages. Such detection could be used for on-line resolution of compounds with similar retention times. Within each class of compounds (phenolics and flavonoids), the oxidation potential changed with the substitution pattern as depicted in equation 4. A mixture of twenty-seven reference compounds was resolved in a run of 45 min duration. LOD was in the low $\mu g\,L^{-1}$ range with a linear response range of at least three orders of magnitude[227].

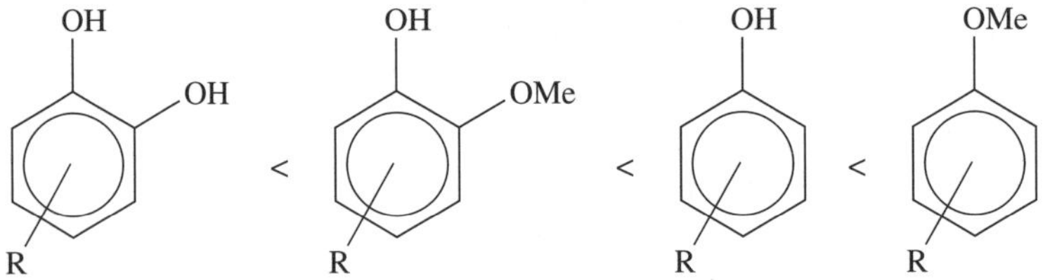

$$\underset{R}{\text{OH, OH}} < \underset{R}{\text{OH, OMe}} < \underset{R}{\text{OH, OMe}} < \underset{R}{\text{OMe}} \qquad (4)$$

c. Wine and liquors. Direct phase HPLC of an SPE preconcentrate of phenolic compounds in red wine has been attempted[228]. The RP-HPLC analysis with DA-UVD of the polyhydroxy phenols in wine was carried out, after suitable preparation, using a series of solvent gradients and columns; detection was at 280, 313, 365 and 520 nm[229]. Similar analyses were also carried out by direct injection of the wine in the column[230]; more than fifteen phenolic compounds with antioxidant properties were detected, including flavan-3-ols, anthocyanins, cinnamic acid derivatives, flavonol derivatives and *trans*-resveratrol (**28a**)[231].

After optimization of the solvent gradient, analysis of wine samples could be carried out by direct injection into the RP-HPLC column with DA-UVD in the 240–390 nm region. The following low molecular mass components could be detected, in increasing order of retention times: Gallic acid (**8**), furfural (**104a**), 5-hydroxymethylfurfural (**104b**), *p*-hydroxybenzaldehyde, vanillic acid (**38**), syringic acid (**70**), vanillin (**39**), syringaldehyde (**63**) and ellagic acid (**94**)[232]. Direct injection RP-HPLC with DA-UVD was applied to the detection of phenolic compounds and furans in fortified wines that underwent extended periods of wood ageing[233]. A method was developed for optimal separation of *trans*-resveratrol (**28a**) and its *cis*-isomer, epicatechin (**2**), catechin (**3**), quercetin (**47**) and rutin (**97**) in wine, by RP-HPLC with DA-UVD. Application of FLD considerably lowered the LOD of **2**, **3**, and both isomers of **47**[234].

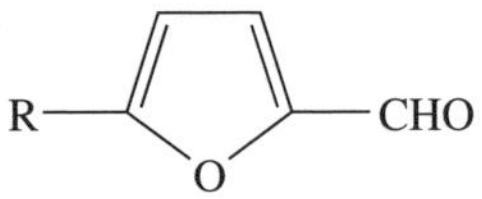

(**104**) (**a**) R = H
(**b**) R = CH$_2$OH

Various HPLC-MS methods were examined to establish the optimal ion source and detector operating conditions for the detection or determination of low molecular mass phenols and flavan-3-ols. Atmospheric pressure electrospray ionization MS (ESI-MS) in the negative ion mode for the low molecular mass phenols and both negative and positive ion modes for the flavan-3-ols were found most suitable. This was applied to the analysis of phenolic compounds in a complex matrix such as wine[235].

A capillary-scale particle beam interface was used for the analysis of phenols in red wine by LC with MS detection. The interface allows very low mobile phase flows and sensitive detection of the analytes in complex matrices[236].

d. Oils and fats. A cooperative study involving many laboratories was carried out for the determination of the phenolic antioxidants listed in Table 3, used in foodstuffs to stabilize animal fat. The limitations imposed by the standards on the concentration of antioxidant in the fats and the relative amounts of fats present in various types of foodstuffs require detection limits for the analytical methods in the ppm range. This made HPLC the method of choice, for AOAC LC method 983.15[44,45].

A simple method for determination of antioxidants (**31, 32a, 33a–c, 35**) in oils and fats consists of dissolving the sample in *n*-propanol, filtering and analyzing by micellar liquid chromatography (MLC) with UVD at 290 nm. LOD was 0.2 to 1.3 ng, corresponding to concentrations well below those allowed in food; with RSD 2% for samples spiked at 200 ppm[237,238].

HPLC with TSP-MS and ISP-MS detection methods was used to identify phenolic glycoside components of olive leaves, such as oleuropein (**40**), directly from crude extracts[239]. Phenolic compounds in extracts from freeze-dried olives were cleaned up by SPE and subsequently analyzed by HPLC with both fluorescence and ESI-MS detection. Oleuropein (**40**) was the major phenolic component in the fruit[240].

e. Miscellaneous. The HPLC analysis of catechins in green tea leaves using DA-UVD has been described[241,242]. A rapid HPLC method for determination of phenolic antioxidants (**31, 32a, 33a–c**) in bakery products has been described[243]. The following phenolic esters were determined simultaneously by RP-HPLC with ELD: Methyl 4-hydroxybenzoate (**105a**), methyl vanillate (**105b**), methyl syringate (**105c**), methyl *p*-coumarate (**106a**) and methyl *trans*-ferulate (**106b**). An array of sixteen coulometric electrodes was used, with potentials increasing from 300 to 900 mV. These compounds were found in honey, in concentrations between 1.31 and 5044 ppb; LOD was 0.1 to 1.0 µg(kg honey)$^{-1}$ (SNR 3). The method was sensitive enough to discriminate between rape honey and other varieties[244]. HPLC analysis showed that honeys originating from heather contain ellagic (**94**), *p*-hydroxybenzoic, syringic (**70**) and the *ortho* isomer of coumaric (**106c**) acids while those originating from lavender contain gallic acid (**8**)[245].

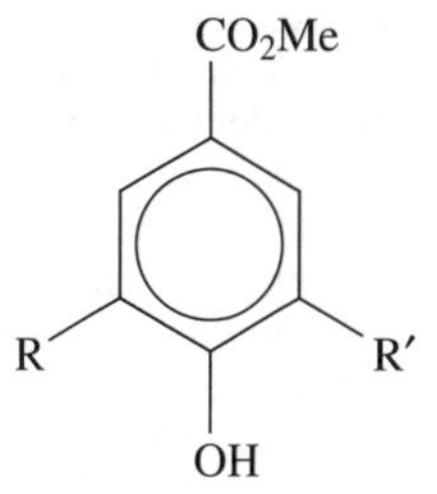

(**105**) (a) R = R' = H
(b) R = OMe, R' = H
(c) R = R' = OMe

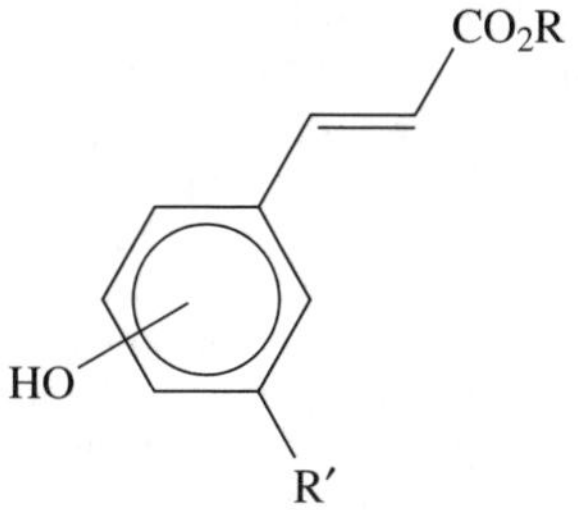

(**106**) (a) 4-OH, R = Me, R' = H
(b) 4-OH, R = Me, R' = OMe
(c) 2-OH, R = R' = H

3. Environmental samples

a. General. After on-line enrichment on a styrene–divinylbenzene copolymer column end analysis followed RP-HPLC with ELD was applied to the determination of phenols in seawater and marine sediments, at $ng L^{-1}$ levels[246]. The effect of octylammonium phosphate as modifier of the water–MeCN mobile phase was investigated for the RP-HPLC analysis of the EPA priority pollutant phenols. LOD was lower than 30 ppb without preconcentration, for UVD at 285 nm[247]. Modifications of the ELD cell were reported for coulometric detection of phenolic pollutants, including methyl, chloro and nitro derivatives. An array of four GCE set at 550, 700, 750 and 800 mV vs. Ag/AgCl, respectively, was proposed. This type of ELD allowed the determination of 2,4-dinitrophenol and 4,6-dinitro-2-methylphenol[248]. Phenols derived from lignin present in environmental samples were determined by HPLC with DA-UVD. The use of the diode array allowed detection of impurities within individual chromatographic peaks[249].

Separation of eleven chlorophenols was attempted on RP-TLC plates. Best resolution was obtained for a 7 : 3 mixture of MeCN and water, although 3- and 4-chlorophenol were not resolved; other solvents separated this pair[250]. Various stationary and mobile phases were investigated for TLC separation of phenol and its derivatives[251].

b. Water. A new type of polystyrene resin has been proposed as stationary phase for determination of ppb levels of priority water pollutants[252]. After LLE of thirteen water pollutants, they were determined by RP-capillary chromatography with gradient elution and UVD. LOD was 0.10 to 0.81 ppm (100 nL injections). The effect of temperature programming was also investigated[253]. The priority phenols listed in Table 4 were determined in drinking water at the concentration levels allowed in the European Union[59,60]; after on-line enrichment on a styrene–divinylbenzene copolymer column, end analysis was carried out by RP-HPLC with both AMD and UVD. The former detection method, using a GCE at 1.0 V, was more sensitive than UVD, except for the nitrophenols, of which 4-nitrophenol was detected at 380 nm and the others at 280 nm. LOD was 20 to 50 $ng L^{-1}$, RSD was lower than 10% for all tested analytes, with linearity from 0.1 to 5 $\mu g L^{-1}$ for most of them[254]. Ultratrace concentrations of phenol in river or wastewaters were determined by a preconcentration step on a guard C_{18} column, followed by RP-HPLC using a methanol in water solvent gradient, from 40% to 100%, with AMD at +1.1 V; LOD was 30 ppt[255].

The priority phenols (Table 4) in tap and river waters were determined by SPE on line with SFC with DA-UVD. Tetrabutylammonium bromide was used in the extraction process to increase breakthrough volumes. The mobile phase was CO_2 at 40 °C, modified by a gradient of MeOH. LOD was 0.4 to 2 $\mu g L^{-1}$, for 20 mL samples, with good repeatability and reproducibility between days ($n = 3$) for real samples spiked with 10 $\mu g L^{-1}$ [256]. Seven pollutant phenols, **107a–f** and pentachlorophenol, were determined by IEC with a basic SAX resin (styrene–divinylbenzene copolymer with quaternary ammonium groups) and single channel UVD. Resolution of overlapping peaks was carried out by inverse least-squares multivariate calibration. LOD was 0.6 to 6.6 ng, with better than 90% recovery from spiked pure water and 83% from river water. No extensive clean-up was necessary[257].

An ultrasensitive method for phenols in water consisted of several steps: LLE with dichloromethane, evaporation of the solvent, RP-HPLC on a C_{18} column with a water–acetonitrile gradient, and tandem UVD-FLD. Dinitrophenols are first determined by UVD; then an oxidation of phenols with Ce(IV) takes place in the FLD cell, where the fluorescence of the reduced Ce(III) ions is measured. LOD in the lower ppt range can be achieved. Quantation can be improved using internal standards[258]. Phenolic constituents of industrial wastewaters could be detected by a post-column reaction after RP-HPLC. A color reaction with maximum absorbance at *ca* 500 nm takes place with many phenols in

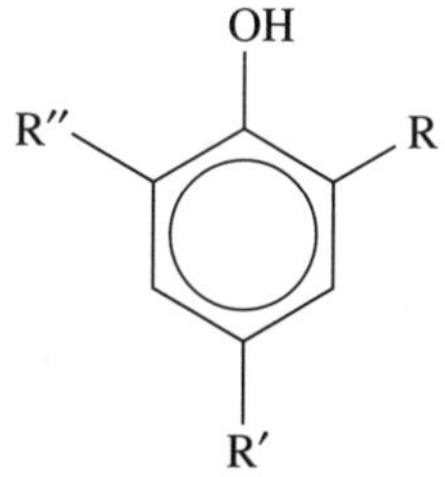

(107) (a) R = Cl, R′ = R″ = H
(b) R = R′ = Me, R″ = H
(c) R = Me, R′ = Cl, R″ = H
(d) R = R′ = NO_2, R″ = H
(e) R = R′ = R″ = NO_2
(f) R = R′ = NO_2, R″ = Me

the presence of 3-methyl-2-benzothiazolidinone hydrazone (108) and the cerium complex $Ce(NH_4)_2(SO_4)_3$ in highly acidic media. The spectra were independent of the eluents and the matrix complexity. Except for nitrophenols, LOD were about 1 to 20 ng of phenol per injection. Aldehydes are passive under these conditions. Interference of thiophenols can be eliminated under neutral to basic conditions. Aromatic amines show a large hypsochromic shift accompanied by a decrease of absorbance intensity[259].

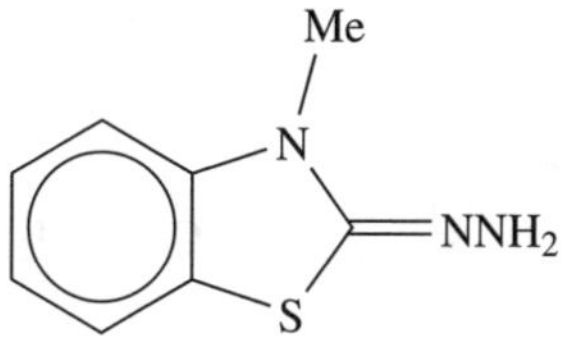

(108)

The presence in drinking water of phenol, cresol and various antioxidants (109–113) used for synthetic rubber preservation was tested by SPE with a C_{18} cartridge followed by RP-HPLC on a C_{18} column with UVD at 280 nm[260]. An automatic LC system was devised for determination of trace amounts of phenols in water, based on SPE preconcentration, RP microcolumns, isocratic methanol–water mobile phase and ELD in the autoincrement mode. LOD was 40–600 ng L^{-1}, for eleven priority phenols[261].

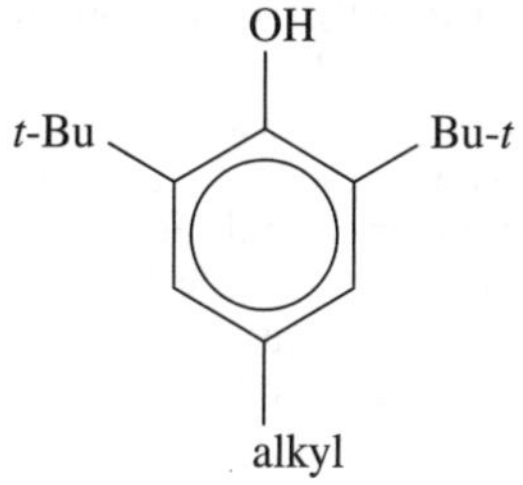

(109)

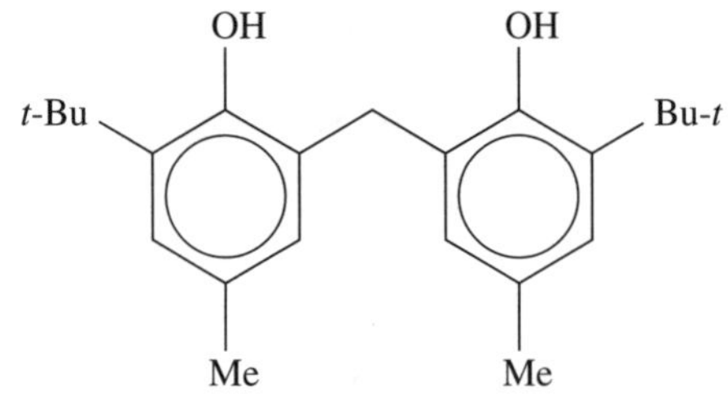

An array of two electrodes was set up with the first one at a low potential (250 mV) for sample clean up, while the second electrode served for measurements. This array allowed LOD in the ppt range for 5 mL samples of water, after applying an SPE preconcentration step[262]. Simultaneous determination of phenol, 26 substituted phenols and herbicides was carried out by SPE followed by RP-HPLC using a gradient of a solvent modifier and a counter-ion with an array of ELDs. The identity of each chromatographic peak was based on its retention time and the peak height ratio across the electrode array, as compared with those of an authentic standard. The method was applied to determination of phenylurea herbicide residuals, phenol, chlorophenols and nitrophenols in waters of various origins. LOD for the less sensitive analyte, the herbicide Linuron (**114**), was 0.5 ng L^{-1} at SNR 3, much lower than the European Community specification[263].

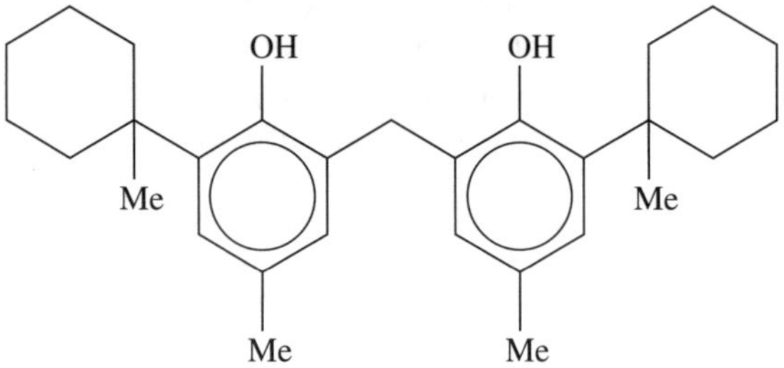

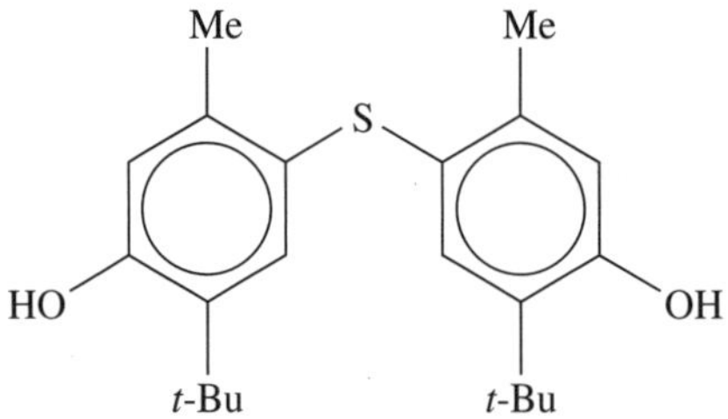

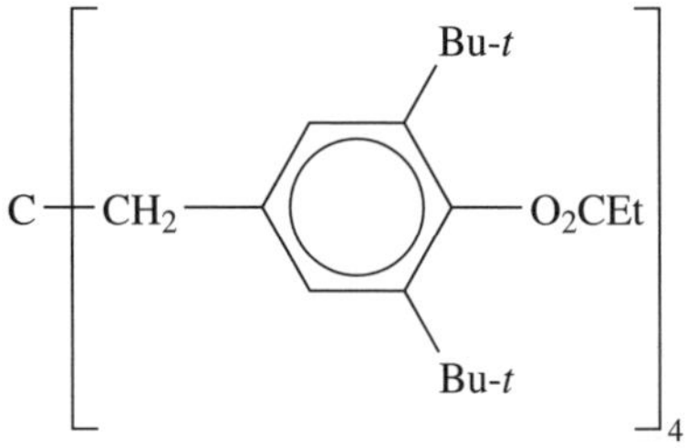

The possibility of determining trace phenolic pollutants in water by TLC was investigated. The analytes were preconcentrated by SPE and subjected to both classical and multiple gradient development TLC. *In situ* quantation was performed by UV absorption or by visible light absorption after treatment with Wuster's reagent (**115**). LOD was

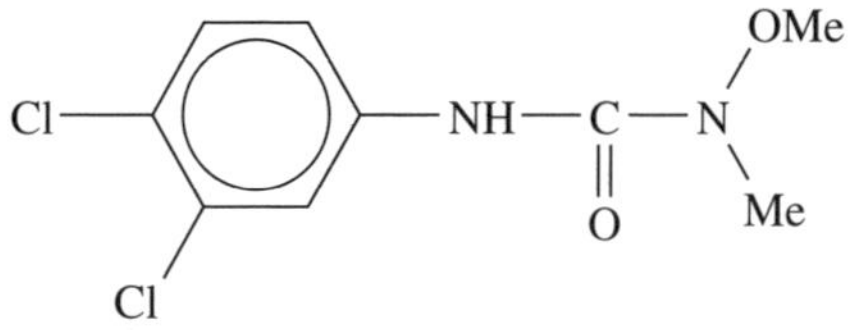

$$\text{(114)} \quad \text{Cl-C}_6\text{H}_3(\text{Cl})\text{-NH-C(=O)-N(OMe)(Me)}$$

(114)

10–100 ng per band, with 60–80% recovery, and RSD 1.3–2.8% ($n = 3$) at 10 µg L^{-1}, for seven priority phenols[264].

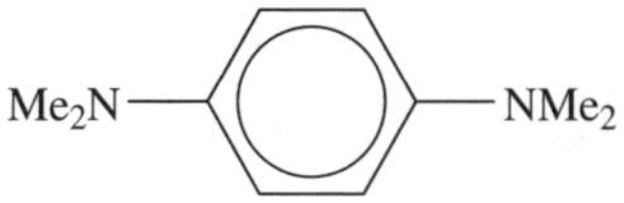

$$\text{(115)} \quad \text{Me}_2\text{N-C}_6\text{H}_4\text{-NMe}_2$$

(115)

c. Soil. A combination of SFE with capillary electrochromatography (CEC) was proposed for determination of phenolic contaminants of soil. At optimal operation conditions baseline resolution was achieved for a mixture of ten compounds, including phenol, cresols, xylenols and other alkylated phenols[265]. A method for determination of total phenols in soil samples is based on extraction with sodium hydroxide solution. After pH adjustment the phenols present in solution were collected by SPE on a C$_{18}$ cartridge. End analysis of the phenols was on a short RP-column using ELD. This allows fast elution of the phenolic species into a single peak, which can be integrated, while ELD provides both sensitivity and functional selectivity[266].

4. Biological and biomedical samples

a. Plant extracts. In contrast to the sample simplification strategy illustrated in Figure 2, direct analysis of complex samples has also been attempted, based on finding the most adequate columns and operating conditions. For example, wood, bark and leaf extracts of *Eucalyptus* sp. were analyzed for phenolic acids, phenolic aldehydes and flavonoids by RP-HPLC with DA-UVD. The analytes were identified by retention time and spectrophotometric response against a set of 46 standards, including phenolic acids, phenolic aldehydes, catechins, isoflavones, flavones, flavanones, flavonols, dihydroflavonols and the glycosides of some flavonoids[267]. The methods for HPLC determination of resveratrol (**28a**) and piceid (**28b**) have been critically reviewed[31].

The anthocyanins and colorless phenolics in eleven cultivars and hybrids of sweet cherries were characterized and quantified by HPLC and GC. All of the dark-colored cherry genotypes were found to contain the 3-rutinoside and the 3-glucoside of cyanidin (**116b**) as the major anthocyanins and the same glycosides of peonidin (**116c**) as minor anthocyanins. Another minor anthocyanin, pelargonidin (**116a**) 3-rutinoside, was identified in sweet cherries for the first time. The major colorless phenolics were characterized as neochlorogenic acid (enantiomer of **95**) and *p*-coumaroylquinic acid (presumably **98**). The total anthocyanin content ranged from 82 to 297 mg per 100 g of pitted cherry for the dark cherries and from 2 to 41 mg per 100 g of pitted fruit for the light-colored cherries. **98** and **95** ranged from 24 to 128 mg and from 23 to 131 mg per 100 g of pitted cherry,

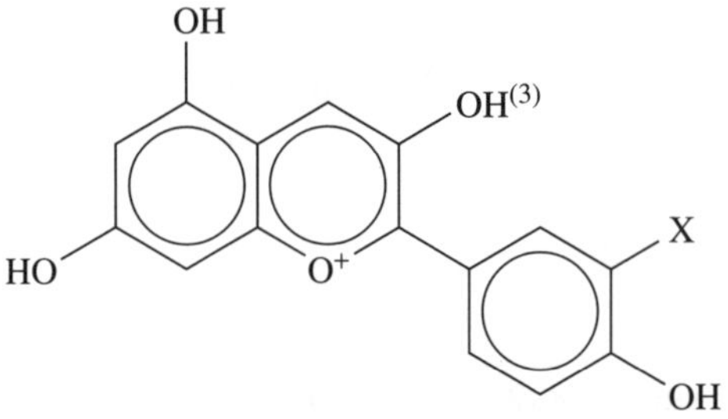

(116) (a) X = H
(b) X = OH
(c) X = OMe

respectively. The relative amounts of the two phenolic acids varied widely across the cherry cultivars examined in this study[8].

Aloesin (**117**), aloenin (**118**) and aloe-emodin (**119**) are among the phenolic constituents that were identified in MeOH extracts of various aloe species, by RP-HPLC with UVD at 290 nm, using a linear gradient of MeCN–water. LOD was 8 to 70 ppb for these and other compounds in the extract[268]. The natural antioxidants present in crude aqueous mate leave extracts (*Ilex paraguayensis*) were analyzed by LC with APCI-MS in the negative ion mode and DA-UVD. Among the identified polyphenolic compounds were isomers of quinic acid (**101**) mono- and diesterified with caffeic acid (**25**), such as chlorogenic acid (**95**), rutin (**97**) and various glucosides[269].

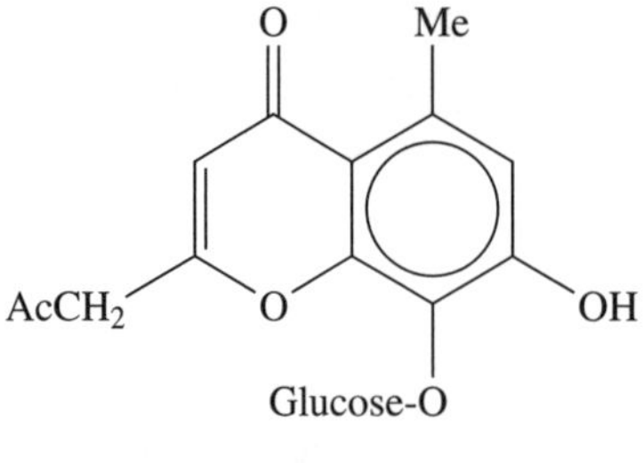

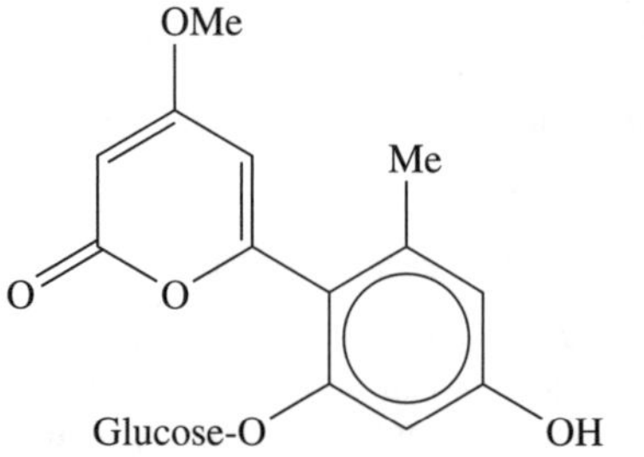

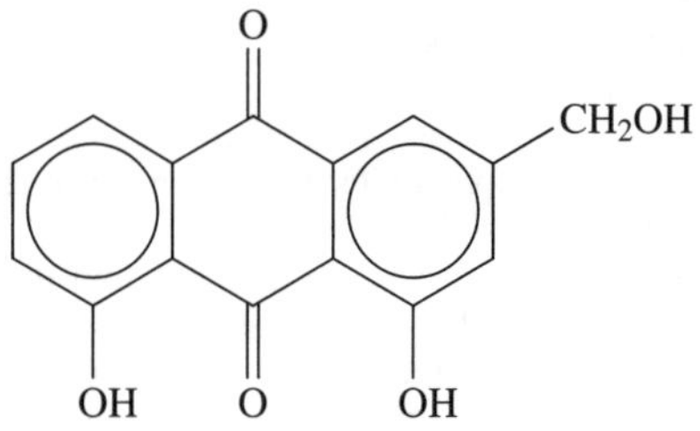

Plant extracts often contain compounds of biological and pharmaceutical interest as glycosides. MS investigation of these metabolites requires soft ionization techniques such as

13. Analytical aspects of phenolic compounds

desorption chemical ionization or fast atom bombardment (FAB) if information on molecular mass or sugar sequence is desired. Thermospray (TSP) provides MS results similar to those obtained with positive-ion desorption chemical ionization MS, using NH_3, and thus is potentially applicable to on-line analyses of these compounds and can be applied to plant extract analysis. Extracts of Gentianaceae (containing secoiridoids and xanthone mono- and diglycosides), Polygalaceae (containing flavonol di- and triglycosides), Pedaliaceae (containing iridoids, phenylpropanoid glycosides) and Leguminosae (containing triterpene glycosides) were analyzed by RP-HPLC with TSP-MS detection, using methanol–water or acetonitrile–water gradients. Good optimization of the temperature of the source and the vaporizer was crucial for the observation of pseudomolecular ions of glycosides. For example, in Figure 3 rutin (**97**) shows an $[M + H^+]$ ion (m/e 611) that loses rhamnose and glucose residues, giving strong peaks at 465 and 303, respectively[270].

Information theory and numerical taxonomy methods were applied in the evaluation of the efficiency of mobile phases for the TLC separation of flavonoids and phenolic acids identified in a MeOH extract of *Rosmarini folium*. The optimal mobile phase was an $AcOEt-HCO_2H-AcOH-H_2O$ mixture of 100 : 11 : 11 : 27 volumetric ratio[271]. Multiple gradient development TLC was applied to the analysis of phenolic acids from *Lycopus europaeus L.* (Lamiaceae). For the first time 2,4-, 2,5-, 3,4- and 3,5-dihydroxybenzoic acids were detected in this genus[272]. Thirteen TLC analyses by various methods were performed, of a methanolic extract of leaves of *Helleborus atrorubens Waldst. et Kit.* that contained fifteen flavonoids and phenolic acids. The results were subjected to numerical methods, including calculation of the information content, determination of discriminating power and formation of clusters and dendrogram. This allowed evaluation of the separating power of the various methods, and pointed to the $AcOEt-HCO_2H-H_2O$ (65 : 15 : 20 v/v/v) mixture as the optimal one for the separation of the given compounds[63]. The settings of a dual-wavelength spot densitometer and a video image analyzing system were investigated for their effect on the repeatability of these detection methods for TLC. The results of both methods at the optimal settings were equivalent for the analysis of

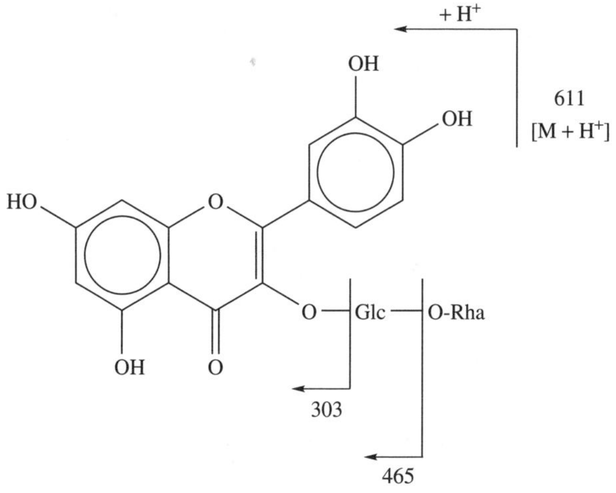

(**97**)

FIGURE 3. Main fragmentation pattern in the MS of rutin (**97**)

phenolic compounds in the leaves of *Phyllantus emblica L.*, separated by normal and RP-TLC, and determined at 254 and 366 nm[273].

b. Effects of food intake. Seven flavonoid and two anthraquinone phenolic derivatives were found in human urine by RP-HPLC, after administration of the traditional Chinese herbal medicines Dachaihu-tang and Xiaochaihu-tang[274]. The accumulation of three antioxidants, PG (**33a**), BHA (**31**) and BHT (**32a**), in the omentum originating in dietary intake was demonstrated by HPLC with ELD. Evidence for peroxidase-catalyzed oxidation of **31** was obtained by detection of the dimeric oxidized metabolite **120**. The method was sensitive to 0.1 to 1 µg L^{-1} antioxidant in plasma or tissue homogenate[275].

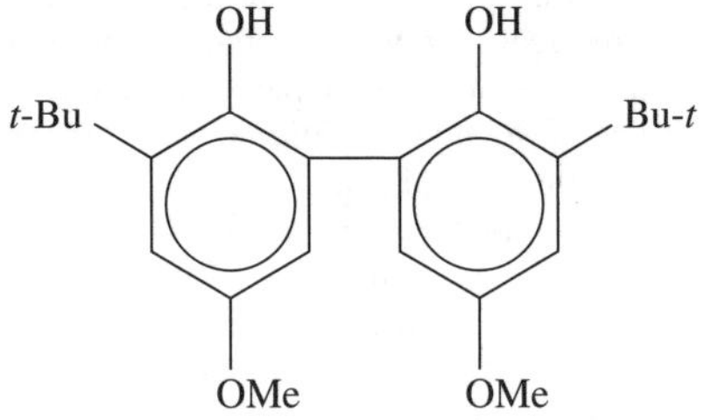

(**120**)

The determinaton of the human bioavailability of flavonoids and hydroxy derivatives of cinnamic acid, such as coumaric (**26**), caffeic (**25**), chlorogenic (**95**) and ferulic (**69**) acids, was carried out by HPLC analysis of urine[276].

c. Physiological and toxicological monitoring. Antipyrine (**121**) was used as an exogenous marker for the damage caused in the organism by free radicals and oxidative stress. Antipyrine and its phenol derivatives were determined by RP-HPLC with ESI-MS detection, with the spectrometer operating in the multiple reaction mode. LOD for antipyrine was 25 ng L^{-1} for 20 µL injections[277]. The APCI-MS detection method was also investigated, in both the single ion and selective reaction modes. LOD for antipyrine was 300 ng L^{-1} for 20 µL injections[278].

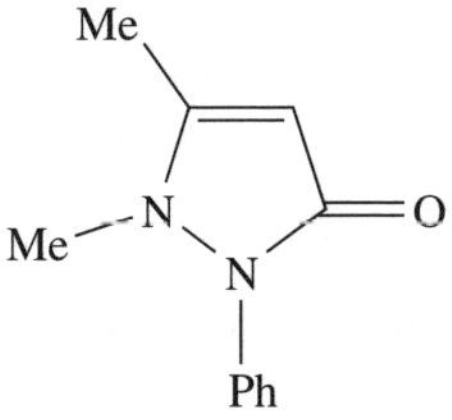

(**121**)

A sensitive method was developed for determination of phenol and 4-methylphenol in serum, based on LLE with ethyl acetate and HPLC with FLD. This was simpler than GC, as no derivatizing was required. This was applied to control hemodialysis of uremic patients, who have significantly higher concentration of these analytes than in normal blood[279]. A method for determination of phenolic compounds in plasma involved centrifugation, SPE with a polytetrafluoroethene membrane impregnated with a water immiscible organic solvent (Hex-O-Hex), redissolution in an alkaline phase, transfer to LC analysis via an ion exchange phase and detection with a phenoloxidase-based biosensor. LOD was below 50 µg(L plasma)$^{-1}$ [280]. Phenol and p-cresol were determined in urine and feces by acid hydrolysis, LLE with ether and aqueous NaOH, evaporation under nitrogen and redissolution in water. Hydrolysis is necessary because phenols are usually conjugated by the liver and colonic epithelial cells as sulfates or glucuronates. End analysis was by RP-HPLC with UVD at 270 nm[281]. Determination of phenolic metabolites of benzo[a]pyrene (**122**) in water and urine was performed by SPE followed by HPLC with AMD. This detection method is 2 to 12 times more sensitive than UVD and FLD[282].

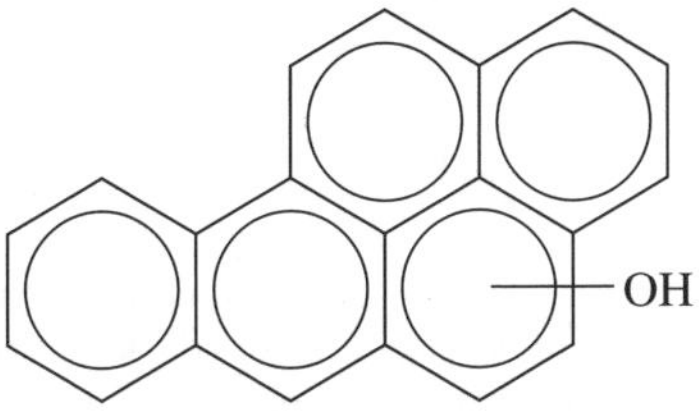

(**122**)

Immobilized β-glucosidase served for enzymatically catalyzed hydrolysis of benzene metabolites in urine. End analysis of phenol was by RP-HPLC with ELD at 0.85 V *vs.* Ag/AgCl electrode. ELD avoids interference from other compounds present in urine. LOD was 10 µg L^{-1} (20 µL injection, 0.2 ng), with RSD 1.16% and 3.38% for 1.2 ng and 2.0 ng, respectively[283]. A study was carried out of two FIA systems for enzymatically catalyzed determination of dopamine (**10a**). Thus, a combination of a packed bed reactor containing immobilized tyrosinase followed by photometric detection was compared with ELD based on a graphite electrode with its surface covered by immobilized tyrosinase. The former configuration was linear up to 0.75 mM while the latter reached 1 mM. LC separation and post-column detection with the bioelectrode was applied to analysis of spiked serum samples[284].

5. Miscellaneous industrial products

a. Polymeric materials. Low molecular weight species present in resol prepolymers were analyzed by SFC with APCI-MS detection, without derivatization. Thirty-four components were identified, including the initial phenol and cresol reagents and their oligomerization products, ranging from dimers to pentamers with varying amounts of methylol substitution[285]. Size exclusion chromatography with differential refractometric detection (DRD) apparently failed to yield monodisperse fractions of polybisphenol A carbonate. This is due to self-association by hydrogen bonding of phenol-terminated polymer chains, leading to formation of macromolecular aggregates of higher hydrodynamic volume (see also Section XI)[286].

b. Disinfectants. Compounds **123–125** are used in combination as active ingredients in hospital disinfectant formulations. Their concentration can be determined by RP-HPLC with a C_{18} column, using an isocratic mobile phase consisting of methanol and phosphate buffer, and DA-UVD to help identification of the eluted fractions[287].

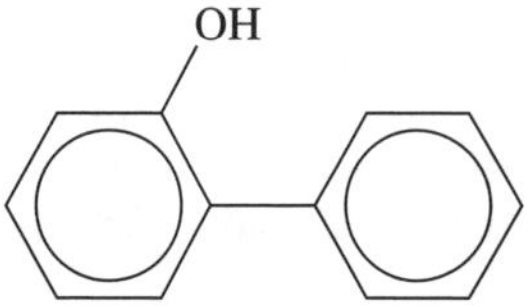

(123) 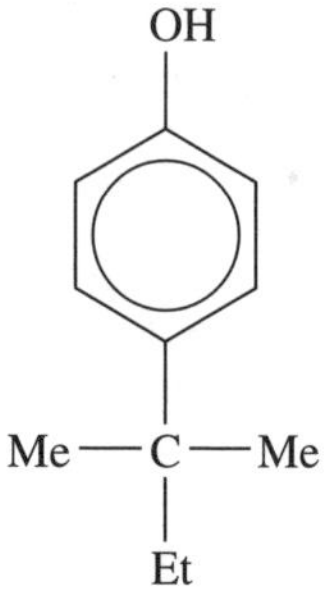   **(124)** 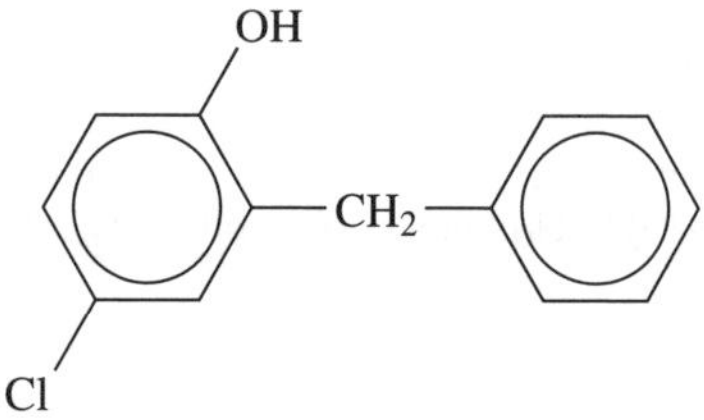   **(125)**

c. Liquid fuels. Kerosene type fuels for jet aircraft contain phenolic antioxidants at 10 to 20 $mg\,L^{-1}$ levels. An LC-ELD method was developed for determination of these additives. LOD was 0.1 $mg\,L^{-1}$ [288]. Phenols in pyrolysis oils were determined by a combination of gel permeation chromatography (GPC) and multidimensional LC. GPC served to separate the high molecular mass 'lignins' from the phenolic fraction that remained adsorbed on the column. Subsequent elution from this precolumn followed by introduction into the LC analytical column completed the analysis[289].

d. Dyestuffs. The presence of phenols and aromatic amines in dyestuffs was determined by dissolving the sample in water cleaning with a SAX cartridge, and HPLC with on-line preconcentration and UVD[290].

6. Structural and functional characterization

The pK_a values of six polychlorinated phenolic compounds (**87–89**) were estimated by the Marquard–Levenberg algorithm, from the capacity factors of the compounds obtained on varying the pH of the mobile phase[199]. The pK_a values of 64 phenolic and 50 nitrogen-containing compounds were determined from their RP-HPLC behavior and by a computational method on the basis of the pK_a values of reference compounds and the Hammett equation. Good correspondence was found in general for the phenolic compounds, whereas the computational estimates were higher than the chromatographic values for the nitrogen-containing compounds. Substitution in the *ortho* position and a nitro group may disturb the computational method[291].

MLC has been applied for the determination of partition properties and the hydrophobicity of monosubstituted phenols. The enthalpy and entropy of partition were estimated from the temperature dependence of the partition properties; these values were interpreted in terms of molecular size and the ability of the solute to establish a hydrogen bond. The π, $\pi(H)$ and $\pi(S)$ constants were determined from the experimental partition properties and applied to quantitative structure–activity relationship (QSAR) analysis[292].

V. ELECTROPHORESIS

A. Environmental Samples

1. Water

a. Sensitivity enhancement. Preconcentration of very dilute phenolic analytes was achieved by on-line flow sample stacking. Thus, the sample is continuously delivered over the opening of a capillary containing a suitable electrolyte. On applying a high potential, phenols are stacked in the interface between the sample and the electrolyte. For example, up to 2000-fold preconcentration was attained from a sample of deionized water spiked with low concentrations of the eleven EPA priority phenols, on filling the capillary with a buffer made of 20 mM phosphate, 8% 2-butanol and 0.001% N,N,N,N',N',N'-hexamethyl-1,10-decanediammonium bromide at pH 11.95 and applying 2 kV for 240 s. No matrix removal was necessary to carry out the capillary electrophoresis (CE) analysis[293]. Determination of the priority phenols and others in water, at the concentrations levels stated by the international regulations for the public supply, was carried out by off-line preconcentration followed by CE using the stacking procedure for matrix removal[294]. A method involving field-amplified injections was proposed instead of off-line preconcentration, because of its simplicity, speed and the high enrichment factors achieved. This technique was applied to the CZE analysis of the eleven EPA priority phenols. LOD was in the ppb range, SNR between 1.4 and 8.8% for the tested phenols[295]. A significant sensitivity enhancement was reported when using a nonaqueous running buffer for determination of the priority phenols by CE with AMD[296,297]. A triacetylated derivative of β-cyclodextrin (**126**) served for selectivity enhancement in the analysis of phenolic water priority pollutants. CE with AMD (Pt vs. Ag/AgCl, poised at $+1.6$ V) using a nonaqueous solvent achieved 3- to 8-fold LOD diminution factors as compared to aqueous buffers[298]. CZE with laser-induced fluorescence (LIF) for indirect FLD, using a sodium borate ($Na_2B_4O_7$) buffer containing fluorescein (**16**), achieved LOD in the 10^{-7} to 10^{-6} M range for eleven priority phenols[299].

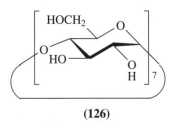

(**126**)

b. Miscellaneous samples. Application of CZE to the determination of chlorophenols in water samples was investigated, using UVD at 214 nm[300]. The combination of a micromachined CE chip with a thick-film AMD was proposed for simultaneous determination of phenolic pollutants in water. At pH 8, LOD was $1-2$ μM with linearity up to 0.2 mM and RSD was 3.7% ($n = 3$) for seven priority chlorophenolic pollutants. Additional phenols could be determined on raising the pH to 10.5[301]. A CZE method combined with AMD was developed for phenols in industrial wastewaters. The column had ID 50 μm and 62.5 cm length, operating at 9 kV, with a 20 mM buffer solution of N-cyclohexyltaurine

(**127**) at pH 10.1. The detector was made of a 9 μm diameter carbon fiber microelectrode inserted at the end of the detection capillary, operating at +1.10 V vs. Ag/AgCl. LOD was at the μM level for eleven priority pollutant phenols, with linearity over two orders of magnitude[302].

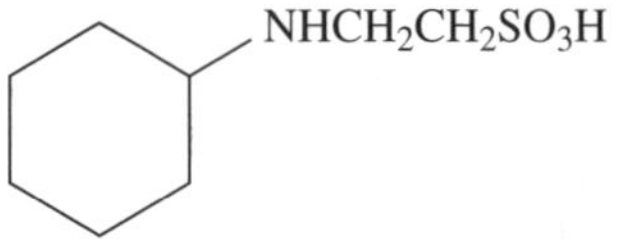

$$\text{C}_6\text{H}_{11}\text{-NHCH}_2\text{CH}_2\text{SO}_3\text{H}$$

(**127**)

Double-chain surfactants with two sulfonate groups were proposed for micellar electrokinetic chromatography (MEC) analysis of phenolic pollutants in water[303]. CE with ESI-MS detection was applied to the analysis of phenolic compounds in olive mill wastewaters. Quantitative analysis was performed in the negative SIM mode, using *p*-chlorophenol as internal standard. LOD ranged from 1 pg for 4-hydroxybenzaldehyde and protocatechuic acid (**27**) to 386 pg for vanillic acid (**38**)[304]. A modified montmorillonite served for SPE preconcentration of phenols, followed by EtOH desorption and CE end analysis[305].

2. Soil

Soil samples were extracted with 0.6 M NaOH in 95% MeOH and subjected to CZE for determination of chlorophenols. LOD were usually sub-ppm and recoveries were usually fair; however, in some cases they were very small and in others they were in excess of 200%[300].

3. Air

The analysis of phenolic pollutants in the atmosphere involves collection of the pollutants in a liquid film. End analysis is by MEC with direct UVD of the analytes[306].

B. Foodstuffs

1. General

MEC was investigated as an alternative to HPLC for the determination of simple phenolic constituents, e.g. vanillin (**39**), vanillic acid (**38**) and ferulic acid (**69**) in spirituous beverages[307] and those of nutritional or pharmacological significance, such as catechol (**42**), hydroquinone (**66**), caffeic acid (**25**), catechin (**3**), chlorogenic acid (**95**) and vanillin (**39**)[308].

2. Wine and beer

A simple CZE method, using a borate buffer at pH 9.5 and UVD at 280 nm, was applied for analysis of Spanish red wines. Although the electrophoretic profile was similar for different wines, the quantitative analysis varied much between them. The following phenolic components were identified: (−)-epicatechin (**2**), (+)-catechin (**3**), (−)-epigallocatechin (**74**), syringic acid (**70**), vanillic acid (**38**), gallic acid (**8**), protocatechuic

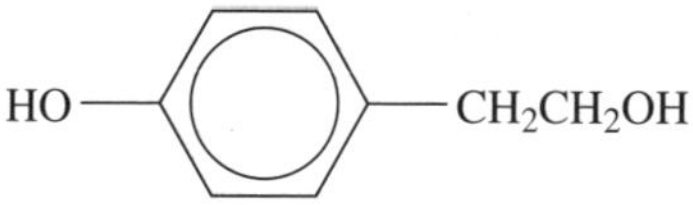

13. Analytical aspects of phenolic compounds

HO—⟨benzene⟩—CH$_2$CH$_2$OH

(128)

acid (**27**), coumaric acid (**26**), caffeic acid (**25**) and the depsides (**68**) derived from *cis*- and *trans*-coumaric acid and *cis*-caffeic acid[309]. A comparative study of HPLC and CZE for noncolored phenolic components in wine showed good agreement between both methods, but CZE was less sensitive for detection of flavonoids. However, *p*-hydroxyphenethyl alcohol (**128**) was detected by CZE for the first time in wine[310].

The CE analysis of phenolic acids in complex matrices such as beer was investigated. The voltammetric end determination required separation of interfering components and optimization of pH at the various stages of the procedure[311]. Application of CZE and MEC with DA-UVD to the analysis of antioxidants was investigated. Gallic acid (**8**) and some of its derivatives (**33a–c**, the amide of **8** and its trimethyl ether), BHA (**31**) and BHT (**32a**) were only partially resolved by CZE, whereas full resolution was achieved by MEC[312].

3. Honey

More than twenty phenolic compounds were found in honey extracts from various floral species, by CZE with DA-UVD. Individual compounds that were identified by total spectrum recording included, in order of increasing migration time, naringenin (**5**), clorogenic acid (**95**), *m*-coumaric (**129a**) acid, quercetin (**47**), syringic acid (**70**), ferulic acid (**69**), *o*-coumaric acid (**129b**), kaempferol (**6**), *p*-coumaric acid (**26**), apigenin (**4**), vanillic acid (**38**), ellagic acid (**94**), *p*-hydroxybenzoic acid, caffeic acid (**25**), gallic acid (**8**) and 2,4-dihydroxybenzoic acid[313].

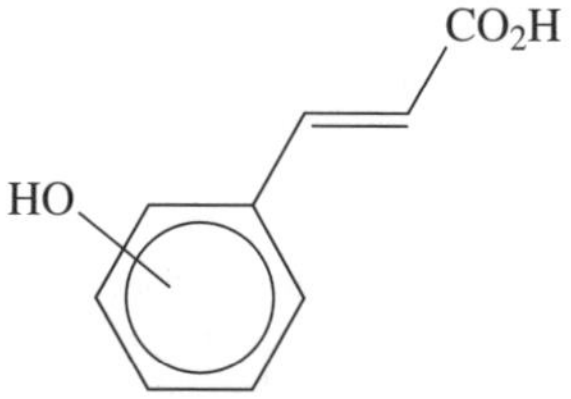

HO—⟨benzene⟩—CH=CH—CO$_2$H

(129) (a) *m*-OH
(b) *o*-OH

C. Biological Samples

A direct injection method was proposed for phenolic acid extracts from plant tissue or soil, based on CZE at pH higher than the pK_a of the acids. Tetradecyltrimethylammonium bromide was added to reverse the electroosmotic flow. LOD was 1–7 μM for eight phenolic acids at pH 7.20[314].

D. Miscellaneous Industrial Samples

1. Wood and paper

Phenolic degradation products of lignin in Kraft black liquors were extracted with chloroform after acidification and separated by CE with UVD at 214 nm[315]. Simple CZE was insufficient for the separation of low molecular mass phenolic and neutral degradation products of lignin. Enhanced separation was attained on turning to the MEC technique, where the analytes interact with micelles present in the carrier buffer solution[316].

2. Fuels

Biomass carbonization oils constitute an important source of chemicals and, more recently, an alternative to fossil oils as energy source. Phenol derivatives (alkyl- and methoxyphenols, alkyldihydroxybenzenes, hydroxybenzaldehydes) and naphthol derivatives were determined by the MEC method[317].

E. Structural and Functional Characterization

The change in mobility as a function of pH, observed for phenolphthalein (**19**) during CZE, was used to estimate the pK_a values of this compound (8.64 and 9.40)[318]; see also Section VIII.A.3.

VI. BIOSENSORS

A. Electrochemical Detection

1. Working principles

A review appeared on determination of phenolic compounds by amperometric measurement, taking advantage of the catalytic properties of certain immobilized enzymes[319]. The operation principles of the most popular biosensors for phenol analysis are shown schematically in Figure 4. The reactive form E_{ox} of an oxidase is reduced to E_{red} by a phenol molecule (Ph). The enzyme is regenerated by oxygen or by hydrogen peroxide, as the case might be (e.g. tyrosinase or horseradish peroxidase, respectively). The measurements can be carried out electrochemically, following the consumption of the regenerating agent or the appearance of the phenol oxidation product, such as a reactive free radical (Ph*) or a quinone (Q).

The process depicted for phenol in equations 5 consists of an enzyme-catalyzed oxidation to a quinone, and a reduction process taking place at the electrode; these reactions may serve for electrode calibration. The development of AMD biosensors for detection of phenols in environmental waters has been described for phenoloxidases such as tyrosinases and laccases and less specific oxidases such as peroxidases. Such biosensors may be part of a FIA system for direct determination of phenols or may serve as detectors for LC[320].

2. Biosensors based on tyrosinase

Various aspects of the kinetic behavior of the tyrosinase biosensor were investigated, including parameters affecting the enzyme activity and the rate of oxygen consumption. The Michaelis–Menten constant was determined for tyrosinase using several substrates and different experimental conditions. Performance parameters of the biosensor in the

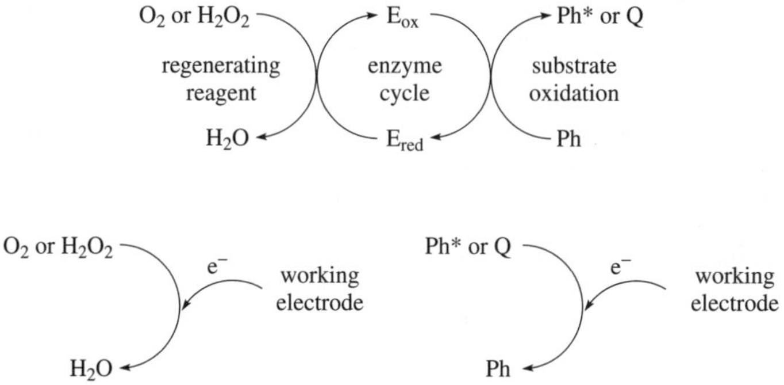

13. Analytical aspects of phenolic compounds

FIGURE 4. Operation principles of a biosensor based on enzymatic oxidation of a phenol (top) and electrochemical detection by determining oxygen or hydrogen peroxide (bottom left) or the oxidation products derived from the phenol (bottom right). Ph denotes a phenol, Ph* an activated form of a phenol and Q a quinone

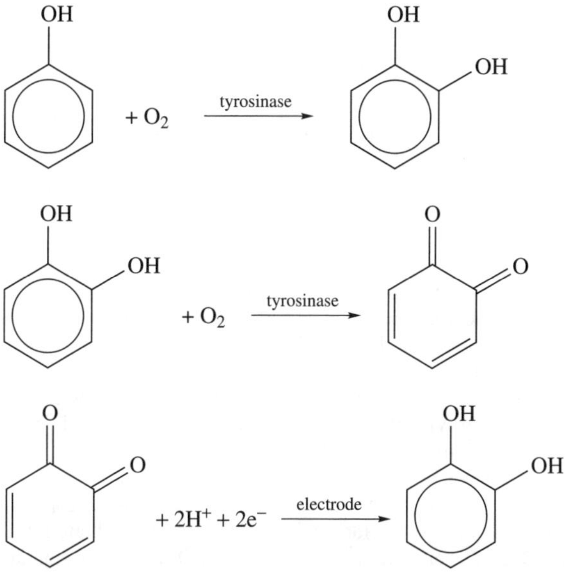

(5)

analysis of phenols were evaluated, such as sensitivity, linearity range, optimal temperature and pH operative conditions, and some interference effects[321]. The performance was evaluated of a graphite–epoxy electrode incorporating tyrosinase, working in an AMD flow cell. LOD was 1.0 μM of phenol and 0.04 μM of catechol (**42**) (SNR 3 and RSD <2%)[322]; the sensitivities of the electrode were 1.53, 1.28, 1.05, 0.687, 0 and 0 for catechol, phenol, p-cresol, m-cresol, o-cresol and 2-chlorophenol, respectively[323]. A comparative study was carried out of the efficiency of tyrosinase-modified CPEs, using lyophilized powder of the enzyme purchased from different companies. Cyclic voltammetry and FIA measurements indicated that the response of the modified electrodes was limited by the rate of the enzymatic oxidation of the catechols. The highest sensitivity for the studied phenols and catechols was obtained when the enzyme was directly mixed into the graphite powder doped with an osmium-based mediator. The best selectivity, on the other hand, was dependent on the source of enzyme used for electrode preparation[324]. Tyrosinase combined with a CPE was found to be a more effective biosensor than the combination with immobilized laccase or coconut tissue; LOD were 72, 37 and 32 µg L^{-1} for hydroquinone, phenol and catechol, respectively[325]. A tyrosinase–modified electrode showed advantage over a GCE with or without modification with Ni-protoporphyrin IX methyl ester **130** for determination of oleuropein (**40**) in olive oil[326]. Tyrosinase was immobilized on a zeolite modified with N-methylphenazonium ion (**131**) and spread over a strip detector with a polyurethane hydrogel. This biosensor achieved subnanomolar LOD for priority phenolic pollutants (0.25 nM for phenol)[327].

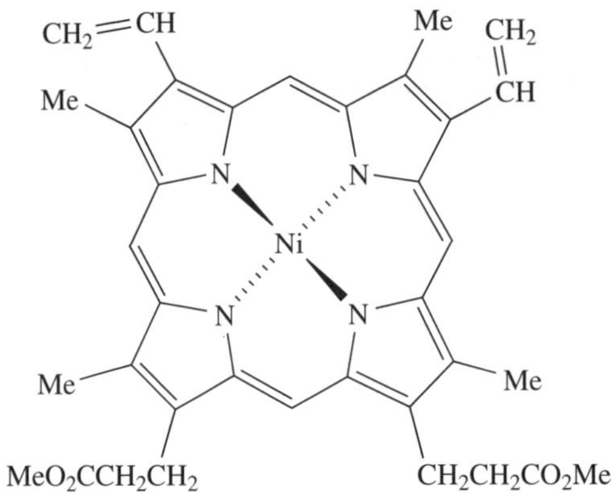

(130)

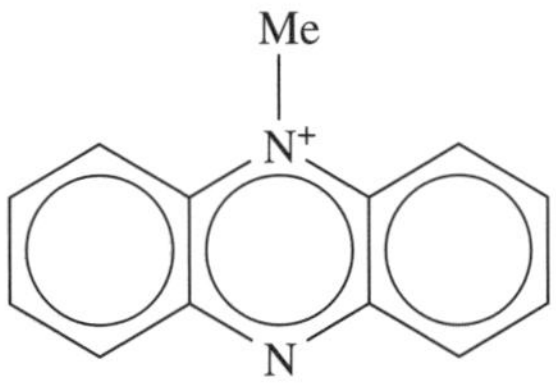

(131)

A study was made on the optimal immobilizing phase for tyrosinase in combination with a GCE and the experimental conditions for AMD of phenols. The apparent Michaelis–Menten constants and the stability of the biosensor were discussed[328]. A study was made on the operational and storage stability of phenol-sensitive AMD electrodes, based on immobilized tyrosinase, varying the electrode material and mode of deposition of the enzyme. The electrode with the best performance was obtained for tyrosinase immobilized on Nafion, sensitivity being 11.51 nA µM^{-1}; LOD was 0.015 µM of catechol, with a throughput of 36 per hour. After 90 consecutive measurements of extremely contaminated wastewaters this electrode retained 70% of its initial response[329]. Tyrosinase, covalently immobilized on the surface of a carbodiimide-activated graphite electrode, serves for the AMD of the enzymatic products at −50 mV vs. a standard calomel electrode. The biosensor responds to phenolic substrates with different conversion efficiencies in a FIA system. LOD for phenol is 3 nM (SNR = 3), LOQ 10 nM, RSD 3.7%, dynamic range up to 5 µM, with a throughput of 110 samples per hour[330]. A specially designed electrode included tyrosinase immobilized on hydrophobic porous carbon, with a supply of gaseous oxygen. This afforded enhanced AMD signals and linear ranges (1 nM to 50 µM), as compared to dissolved oxygen. The gas-diffusion electrode may also be applied for determination of phenols in the gas phase[331]. The efficiency dependence on the fabrication method of bulk-modified epoxy-graphite tyrosinase biosensors was investigated by cyclic voltammetry. On introducing Au/Pd into the epoxy-graphite body, current densities as high as 27.70 and 4.90 µA cm^{-2} were achieved for catechol (**42**) and phenol, respectively[332]. Tyrosinase and laccase were immobilized on a GCE that was used for

13. Analytical aspects of phenolic compounds 977

AMD of the enzymatic products derived from phenols in a FIA system. Measurements were carried out at 0.05 V vs. Ag/AgCl. The combination of the two enzymes allows analysis of many phenolic compounds[333].

Biosensors based on a Clark oxygen electrode, coupled to tyrosinase immobilized by three different methods, were investigated for the determination of phenol in real matrices, such as water of various natural sources, industrial wastes and oil press. The feasibility study included direct use of the biosensors and *in situ* analysis[334]. An integrated system, incorporating SPE, desorption, fractionation and biosensor detection, was validated for screening phenolic compounds in water. Two types of electrode were tested, solid graphite and CPE incorporating tyrosinase. Correct analyses were found for river water samples spiked with phenol (10 $\mu g\,L^{-1}$), *p*-cresol (25 $\mu g\,L^{-1}$) and catechol (1 $\mu g\,L^{-1}$)[335]. A multimembrane AMD biosensor based on immobilized tyrosinase on a Pt disk electrode was proposed for determination of multiple phenol mixtures in a FIA system. Simultaneous measurements with various biosensors of different selectivity were applied for determination of a mixture of phenol, catechol and *m*-cresol. Data processing was carried out by a three-layer artificial neural network with feed-forward connections, sigmoidal transfer function and back propagation learning algorithm. Best results were obtained for a network with 5 inputs, 3 neurons in the hidden layer and 10,000 learning cycles. Correlation coefficients for 36 analyzed samples are: catechol 0.96, phenol 0.88 and *m*-cresol 0.67. The latter result is only semiquantitative, due to the weak amperometric signals obtained with all the tested biosensors[64].

3. Biosensors based on peroxidases and other enzymes

A calorimetric study pointed to peroxidase as a catalyst faster than tyrosinase, being therefore more suitable for biosensor applications[336]. A study of the electrochemical determination of phenols in a FIA system, using solid graphite electrodes modified with peroxidases of various types, showed that, excepting the chloroperoxidase electrode, all the electrodes were sensitive to all the tested phenols[337]. The sensitivity of tyrosinase-based biosensors for AMD of phenol at −0.2 V vs. Ag/AgCl can be improved by horseradish peroxidase in the presence of hydrogen peroxide[338]. A GCE was developed, coated with horseradish peroxidase and a redox osmium polymer. The biosensor had low operating potential (0 V vs. Ag/AgCl) and high sensitivity in the determination of phenols. LOD was in the μM range[339]. Horseradish peroxidase-catalyzed hydroxylation of phenol in the presence of dihydroxyfumaric acid (**132**) and oxygen should not be used as measuring process because introduction of hydroxy in the phenol groups is independent of the catalytic cycle of the enzyme, as indicated by a thermodynamic analysis of the process[340]. The effect of the presence of phenols on the peroxidase activity toward *o*-dianisidine (**133**) can be used to estimate their concentration. Phenol and resorcinol (**20**) are inhibitors, whereas pyrogallol (**134**) and hydroquinone (**66**) produce a lag period on the kinetic curve, the duration of which depends on their concentration. The fungal peroxidase from *Phellinus igniarius* exhibited the highest sensitivity toward phenols, at concentration levels in the 10^{-7} to 10^{-6} M range[341]. Also, peanuts were a good source of peroxidase for this method[342]. A GCE modified by polyphenol oxidase immobilized on a pyrrole amphiphilic monomer served for the direct AMD of phenol, 3-chlorophenol and 4-chlorophenol. Determination of 2-chlorophenol and polychlorinated phenols could be carried out based on inhibitory effects of the analytes on the bioelectrode[343].

Quinoprotein glucose dehydrogenase and recombinant tyrosinase from *Streptomyces antibioticus* were immobilized on polyvinyl alcohol and coupled to a Clark oxygen electrode. LOD was 5 nM for dopamine (**10a**), L-dopa (**10b**) and adrenaline (epinephrine, **15a**)[344]. An electroimmunological biosensor for *p*-cresol was developed, based on the

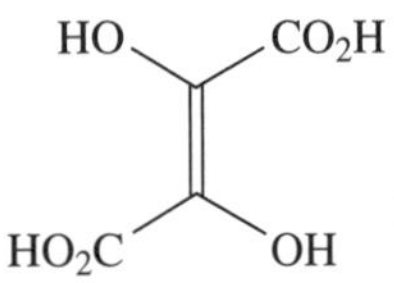

(132) 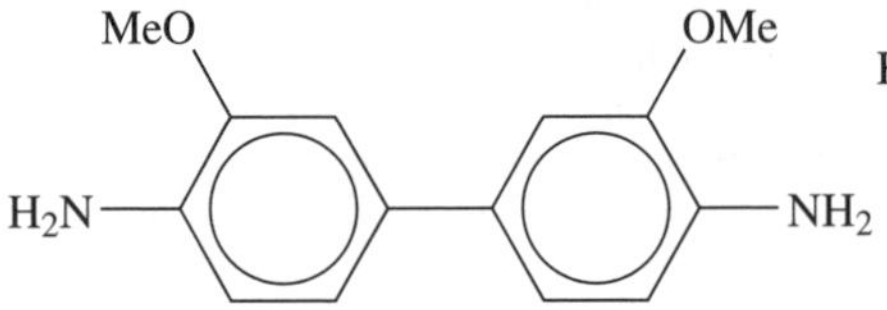   (133) 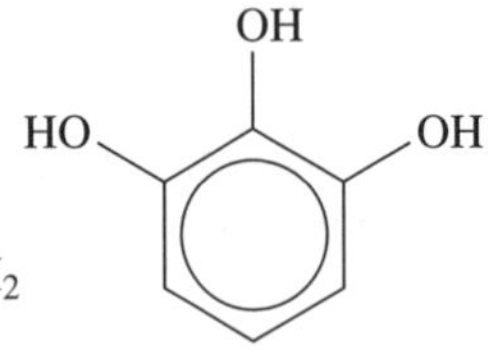   (134)

production of antibodies to a *p*-cresol bovine serum albumin conjugate and their incorporation into a conducting polymer. Fast, sensitive and reproducible analysis of *p*-cresol and other phenols could be obtained in a FIA system by pulsed ELD. The sensor was reusable[345]. An AMD biosensor was developed by incorporating quinoprotein glucose dehydrogenase into a CPE. The oxidation of glucose was coupled to the regeneration of the enzyme by the oxidation product of a phenol at the electrode set at 500 mV (vs. a Ag/AgCl electrode). The presence of the enzyme allows very sensitive AMD measurements of redox species such as hydroquinone, *p*-aminophenol and catecholamines such as epinephrine (**15a**), norepinephrine (**15b**) and dopamine (**10a**). The highest sensitivity was observed for *p*-aminophenol and could be determined at sub-nM levels[346]. A comparative study of the response of peroxidases of various origins was carried out for the determination of phenol and its derivatives. The most sensitive enzyme was obtained from a fungus, *Phellinius igniarius*, followed by those obtained from horseradish roots and a lucerne cell culture. LOD of various phenols were in the 10^{-7} to 10^{-6} M range[347].

4. Biosensors incorporating tissues and microorganisms

Fruit tissues of a palm tree, *Latania* sp., were used as immobilized polyphenol oxidase enzymes, for phenol oxidation, followed by AMD. Various modes of action were tested for the tissues: On-line fresh or dried tissue-based reactor in a FIA system and incorporation of the fresh or dried tissues in CPEs. Determinations of catechol (**42**) and dopamine (**10a**) showed that using these tissues endowed the biosensor with high sensitivity, reproducibility and long-term stability. This seems to be the first time dry tissues were used as enzyme source in biosensors[348]. A biosensor was designed based on mushroom tissue, as a source of polyphenol oxidase, and cobalt(II) phthalocyanine (**135**) dispersed in a CPE. Electrodes containing **135** give shorter response times and require a lower applied potential, as compared to conventional tissue biosensors[349]. Crude extract of sweet potato (*Ipomoea-batatas* (L.) Lam.) was used as a source of phenol oxidases (polyphenoloxidase, tyrosinase, catecholoxidase, EC 1.14.18.1). A biosensor was produced by immobilizing the crude extract with glutaraldehyde and bovine serum albumin onto an oxygen membrane. A linear response in the 20 to 430 μM range was observed for phenol, *p*-cresol, catechol and pyrogallol. This biosensor was proposed for determination of phenols in industrial wastewaters[350].

Various biosensors have been developed, incorporating microorganisms instead of specific enzymes. An AMD biosensor was proposed that is more sensitive to chlorophenols, especially 3- and 4-chlorophenol, than to phenol, and does not respond to their benzoates. The sensor incorporates *Trichosporon beigelii* (cutaneum). LOD was 2 ppb for all studied compounds, with RSD 5.5% and linearity up to 40 ppb for 4-chlorophenol[351]. An AMD biosensor incorporating *Rhodococcus* was investigated for the determination of phenol and its three monochloro derivatives. A linear relationship between the current and the concentration of these compounds was observed up to 20 μM; LOD was 4 μM

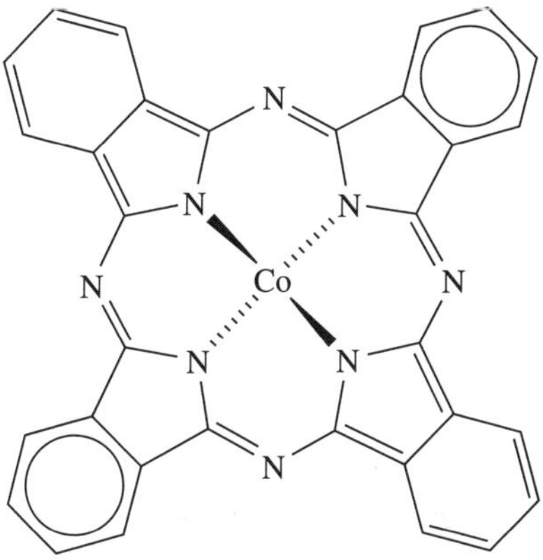

(135)

for all studied substrates. The current difference was reproducible within 5.5% for 40 μM phenol[352]. *Pseudomonas putida* GFS-8 immobilized in poly(vinyl alcohol) cryogel was used as a biological transducer due to its capacity to oxidize phenol, pyrocatechol, mesityl oxide and aniline, but it does not react with a number of xenobiotics, sugars and alcohol. The relationship between phenol concentration in the activating medium and endogenic cell respiration is linearly dependent in the 0.1–1.0 mg L^{-1} range. A Clark membrane electrode was used as physiochemical transducer. The assay may be completed within 5 min. With the exception of aniline, most components found in wastewaters from phenol production do affect the cell ability to use phenol as exogenous respiratory substrate. The immobilized cells retained their activity for up to 1 month[353]. A membrane incorporating living *Bacillus stearothermophilus* cells coupled to a dissolved oxygen electrode resulted in a biosensor for AMD of phenols over the 35–55 °C temperature range, at pH 4.5–8.0, in matrices containing compounds that are toxic to most enzymes and microorganism used. Optimal performance was observed at 55 °C and pH 7.2. Response was very fast and stable for months. This biosensor was proposed for on-line monitoring of phenols in industrial waste effluents[354].

5. Amplification processes

Amplification factors of 8 to 12 were claimed for the determination of phenol in a FIA system by a cyclic process depicted in equations 5 (Section VI.A.1). Phenol is converted to *o*-benzoquinone in contact with immobilized tyrosinase held in a fixed bed reactor; the quinone reacts with ascorbic acid (**91**) to yield catechol and dehydroascorbic acid (**136**); catechol can be enzymatically oxidized again to *o*-benzoquinone and so forth. The accumulated dehydroascorbic acid forms with *o*-phenylenediamine (**137**) a highly fluorescent product (λ_{ex} 345 nm, λ_{fl} 410 nm). LOD was *ca* 0.02 μM for phenol and catechol; the linear range for phenol was 0.1 to 2 μM and for catechol 0.02 to 2 μM[355].

An analogous amplification process for determination of phenols was proposed based on the kinetics of disappearance of β-NADH reacting with quinone, which is derived from a phenol in a tyrosinase-catalyzed oxidation. LOD was as low as 50 nM in a 10 min assay[356]. Amplification cycles were also achieved by combining a Pt electrode where phenols are oxidized with a polyurethane layer embedding pyrroloquinoline quinone-dependent glucose dehydrogenase, to catalyze the reduction of the oxidation products[357].

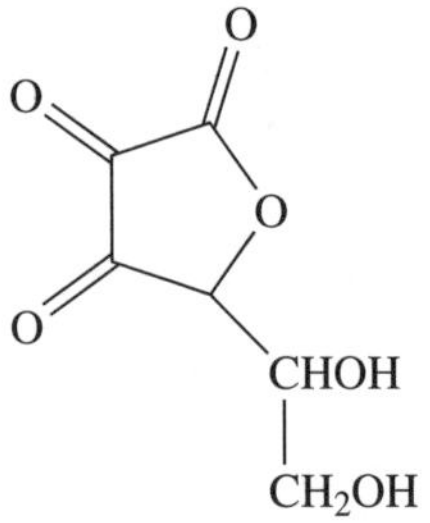

(136) 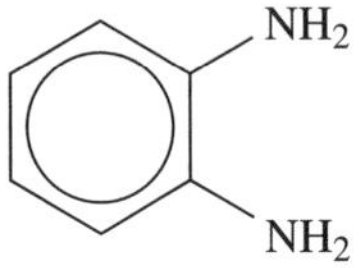 **(137)**

B. Spectrophotometric and Colorimetric Detection

A portable disposable bioprobe for detection and semiquantitative determination of phenols consists of a mushroom polyphenol oxidase immobilized on a nylon membrane, acting in the presence of 3-methyl-2-benzothiazolinone hydrazone. Maroon to orange colored dyes of (**138**) are developed, as illustrated for phenol (equation 6), of intensity proportional to the concentration of the substrate, down to 0.05 mg L^{-1}. Enzyme activity remained unscathed in the pH range 4 to 10, in the presence of various concentrations of salt and metal ions and at temperatures from 5 to 25 °C[358].

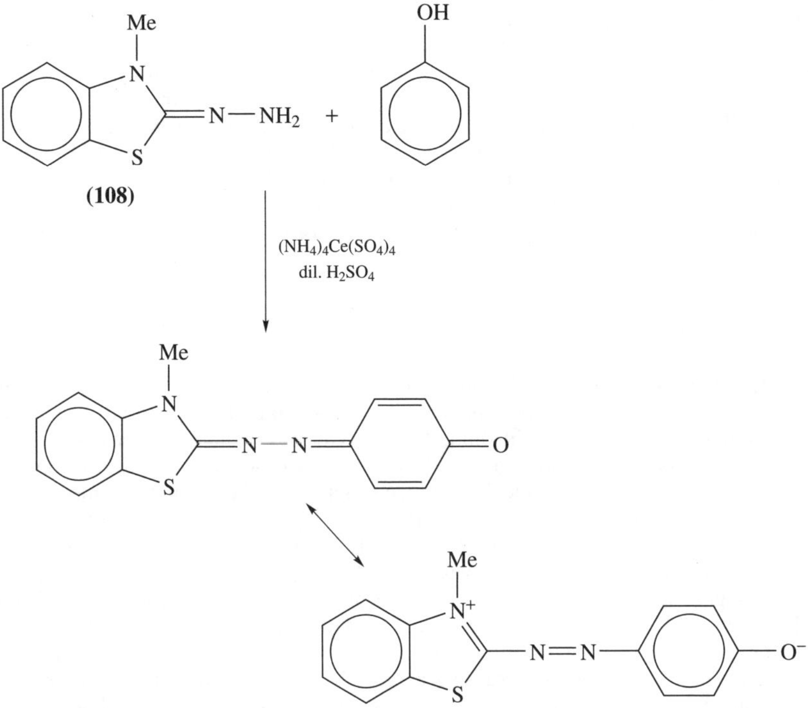

(6)

(**108**)

(**138**)

A crude extract of sweet potato (*Ipomoea-batatas* (L.) Lam.) was used as a source of phenol oxidases (polyphenoloxidase, tyrosinase, catecholoxidase, EC 1.14.18.1). The extract was directly placed in the carrier of a FIA system with UVD, to promote oxidation of phenolic compounds to *o*-quinones that condense to form melanin-like pigments with a strong absorption at 410 nm. The determination of phenols in industrial waste waters showed good agreement with conventional methods (correlation coefficient 0.9954); LOD was 10 μM, with RSD <2.7% ($n = 6$). Under optimal storage conditions the enzymatic activity did not vary for at least five months[359].

The enhanced chemiluminescense obtained with the horseradish peroxidase-H_2O_2-luminol (**139**) system was applied to the development of a CLD biosensor for *p*-iodophenol, coumaric acid (**26**), 2-naphthol and hydrogen peroxide. The enzyme was immobilized by microencapsulation in a sol-gel matrix. LOD for the phenolic compounds were 0.83 μM, 15 nM and 48 nM, respectively. A remote version of the enhanced biosensor was designed by directly immobilizing the enzyme on the tip of an optical fiber. This model was used for H_2O_2 assay. LOD was 52.2 μM, with RSD 4.7% ($n = 4$)[360]. A bioluminescent response was obtained for phenols with $pK_a > 7$ in the presence of a recombinant *Escherichia coli* strain, DPD2540, containing a fabA::luxCDABE fusion[361]; this behavior may have analytical applications.

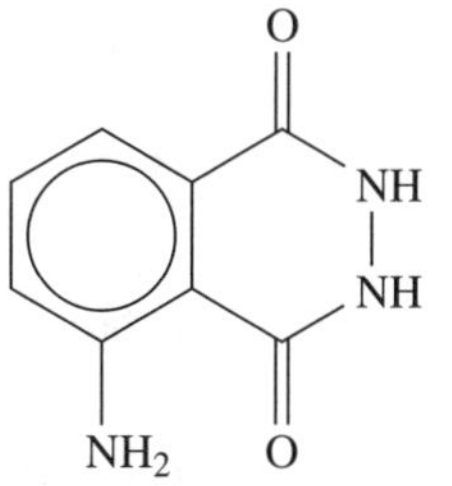

(**139**)

VII. ELECTROCHEMICAL METHODS

A. Voltammetric Detection

1. Quantitative analysis

The results of the simultaneous differential pulse voltammetry (DPV) determination of an aqueous solution containing nitrobenzene, *o*-, *m*-, *p*-nitrophenol and 2,4-dinitrophenol were subjected to data processing by three chemometric methods: PLS, PCR and classical least squares, to resolve overlapping peaks. The relative prediction error of the former two was acceptable (*ca* 10%) whereas that of the latter was not (38%). The method was applied to the analysis of field samples[66]. Binary mixtures of phenols were determined by DPV using a carbon fiber electrode with titania. The records of overlapping signals were processed using a Fourier transform filter and PCA for noise reduction and data compression and then as a neural network. Results of such calculations were better for hydroquinone than those obtained by PLS methods; however, for catechol errors were similar by both procedures[65]. Phenol in the concentration range from 50 nM to 60 μM was determined by 2.5th order differential voltammetry, using a CPE/polyamide electrode. LOD was 5.6 nM, with RSD 4.5%. The method was applied for determination of phenol in cola drinks[313]. A simultaneous voltammetric determination of the phenolic antioxidants BHA (**31**) and BHT (**32a**) was carried out in acetonitrile medium, using a carbon fiber microelectrode (8 μm × 8 mm). LOD for DPV of both analytes was about 70 ppb. Square-wave voltammetry with the microelectrode showed much higher current densities than with the conventional GCE. The background current observed for the microelectrode was several orders of magnitude lower. The voltammograms of **31** and **32a** mixtures showed well-defined oxidation peaks, with a difference in potential of about 300 mV, allowing good simultaneous determination[362]. The behavior of CPEs modified with bentonite was investigated for the DPV determination of phenols in seawater in a FIA system. Good electrode stability and recoveries were obtained for seawater spiked with EPA priority phenols in the 0.5 to 2.5 ppm range[363].

Using solid paraffin as binder for CPEs was claimed to improve electrode performance in the analysis of phenols. LOD was 50 nM of phenol, with RSD <3.5% ($n = 6$) and linear range from 0.25 to 5 μM[364]. Modification of a GCE with Co(II) phthalocyanine (**135**) increased the oxidation currents and the electrode stability in the cyclic voltammetric determination of phenolic compounds. Analogs of **135** with other metal(II) species were less effective[365]. A polypyrrole electrode modified with nickel phthalocyanine (**93**) was investigated for cyclic voltammetry and DPV determinations of the phenolic antioxidants TBHQ (**35**) and BHA (**31**), used as food preservatives. LOD was 2.1 ppm for both, using cyclic voltammetry[366]. A CPE modified with β-cyclodextrin (**126**) was applied to the cyclic voltammetric determination of phenol and its derivatives. A complex was formed before the measurement by immersing the electrode in the sample for a few minutes. Regeneration of the electrode was achieved by immersion in 1 M nitric acid for a few seconds. LOD was 5×10^{-7} M for 25 min deposition, by DPV, with RSD 5.2% ($n = 4$). The presence of benzoic acid, hippuric acid (**140a**) and the isomers of methylhippuric acid (**140b–d**) interferes with the determination[367]. Gradual passivation of GCEs takes place under flow conditions when a polymeric layer is formed on the electrode. This can be avoided by means of laser ablation of the surface[368].

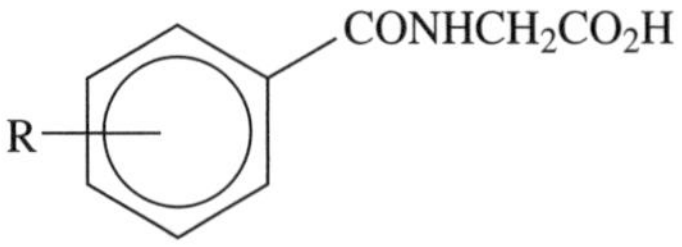

(**140**) (a) R = H
(b) R = *o*-Me
(c) R = *m*-Me
(d) R = *p*-Me

2. Structural and functional characterization

The antioxidant efficiency of phenolic acids, as determined by the accelerated autooxidation of methyl linoleate[369] and scavenging of the free radical 2,2-diphenyl-1-picrylhydrazyl (**141**)[370] methods, was found to be inversely proportional to the maximal detector response potential in the voltammetric determination of these compounds. No similar correlation was found for the flavonoids[371]. A good correlation was found between the O−H bond dissociation energy of a phenolic compound and its effectiveness as antioxidant, expressed as the rate constant of free radical scavenging[372]. The bond dissociation energy of the phenol O−H bond was estimated by a three-dimensional quantitative structure–activity relationship method incorporating electron densities computed using the Austin Method 1 (AM1) followed by correlation of the

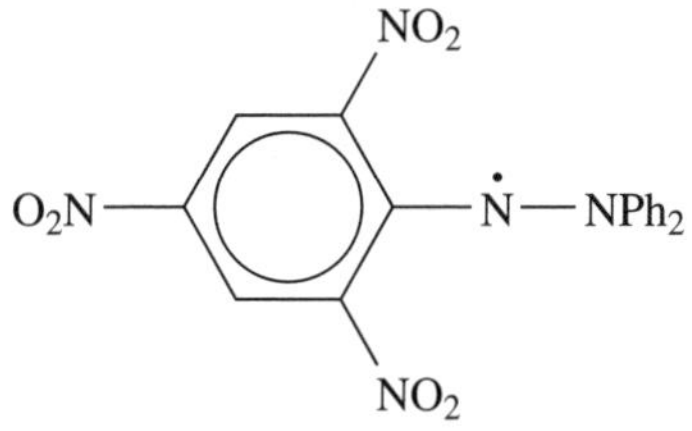

(141)

electron density with the relative bond dissociation energies. Such information is important in medicinal chemistry[373].

B. Amperometric Detection

1. Quantitative analysis

Phenol and the three dihydroxybenzenes (**20, 42, 66**) in water were determined by LLE with a hydrophilic solvent followed by amperometric titration. LOD was in the ppm range[374]. A dual electrode in a FIA system has been used as detector for total phenols in waste water. The upstream coulometric electrode has a large surface area and is used to eliminate compounds that cause interference and the second one is an amperometric electrode for oxidative detection of all phenols. Optimal results were found working with a phosphate buffer at pH 6.8, at potentials of +0.35 V and +0.78 V for the coulometric and amperometric electrodes, respectively. A high sample throughput of 60 per hour can be attained with RSD of 0.1–4%. This method is more reliable than the colorimetric method[375]. The concentration of fenobucarb (**142**) in drinking water was determined after a short alkaline hydrolysis, and oxidation of the resulting 2-s-butylphenol with a GCE at 750 mV, pH 3.5; LOD was 3.6×10^{-6} M, RSD 3.74% for 1×10^{-5} M ($n = 11$, $p = 0.05$)[376].

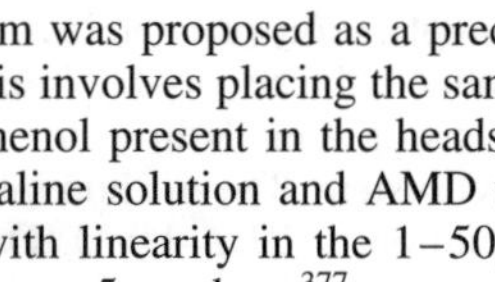

(142)

Pervaporation in a FIA system was proposed as a preconcentration step for the determination of phenol in water. This involves placing the sample in concentrated brine at pH 2, diffusion of the salted out phenol present in the headspace through the pervaporation membrane into a collecting alkaline solution and AMD using a GCE set at +0.6 V. At 20 °C, LOD was 0.9 mg L^{-1}, with linearity in the 1–50 mg L^{-1} range and RSD 1–4% ($n = 3$). The sample thoughput was 5 per hour[377].

The current response of a GCE used for AMD was greatly improved after modification with polyhistidine. LOD was 6 nM for dopamine (**10a**), 8 nM for epinephrine (**15a**) and 20 nM for catechol (**42**). The modified electrode has also been applied for AMD after CE[378]. PVC membranes were designed to serve as selective barriers for the amperometric detection of phenols and elimination of thiocyanate interference[379].

2. Structural and functional characterization

Henry's law constants of phenols were determined dynamically by a nonequilibrium method based on pervaporation in a FIA system. Good agreement was found between these values and those determined by the single equilibrium static technique for 2-methylphenol, 3-methylphenol and 2,4,6-trichlorophenol[380].

C. Polarography

A method for determination of phenols in air consisted of absorption on a membrane loaded with 2.0 M NaOH, coupling with p-bromobenzenediazonium ion and polarographic end analysis of the azo dye. Peak currents were proportional to concentration in the 2.0×10^{-8} to 2.0×10^{-5} M range; LOD was 5.0×10^{-9} M[381].

D. Potentiometric Titrations

1. Quantitative analysis

Acid–base potentiometric titration of phenol in aqueous solution is precluded because of its high pK_a value (9.98), while 4-nitrophenol (7.41) and 2,4,6-trinitrophenol (0.71) can be directly titrated in that solvent. Nonaqueous titrations of phenol are possible; however, difficulties are met when nitrophenols are also present in the system[382]. The determination of carboxylic and phenolic groups in humic acids was carried out by acid–base potentiometric titrations in NaCl solutions up to 1 M. Titration data were processed by linear and nonlinear calculation techniques[383].

2. Structural and functional characterization

An automated system was used for the potentiometric determination of the protonation constants of phenol, 2-chlorophenol, 2-nitrophenol, 2,4-dichlorophenol and 2-methylphenol in 1.0 mol L^{-1}NaCl at 25 °C. The estimation of the constants has been carried out using both graphical and numerical methods[384].

VIII. ULTRAVIOLET-VISIBLE DETECTION METHODS

A. Spectrophotometry and Colorimetry

1. Direct determination

Application of UVV spectrophotometric methods to the analysis of waters and wastewaters has great practical interest. However, interference of certain species has to be eliminated, either by actual application of chemical or physical separation methods, or, alternatively, by computational balance of the interferences, based on reasonable assumptions. Simultaneous analysis of phenols in waters was carried out in an automatic sequential injection analysis (SIA) system. The method involved preconcentration by LLE, back extraction into a NaOH solution and DA-UVD. Data processing using multilinear regression and first derivative spectroscopic techniques yielded the concentrations of the various components[67]. Derivative spectrometry using the zero-crossing technique was applied for the simultaneous determination of binary mixtures of a series of phenols and herbicides at ppm levels. The method was extended to the resolution of overlapping peaks obtained in LC with DA-UVD[68]. A data processing method was proposed for simultaneous determination of a mixture of analytes, based on double Fourier transform filtering and second ratio UVV spectra derivatives. This was applied to determination of phenol, catechol and hydroquinone in solution, in the concentration range of 10 to 50 mg L^{-1}, with RSD from 0.07 to 5.4%[69].

The principles of ultraviolet multiwavelengths absorptiometry (UVMA) with computational balance of interferences, including turbidity, have been discussed and applied[385,386]. An application of UVMA for the determination of phenols has been proposed using the PLS algorithm. A simplification of practical importance was introduced, consisting of selecting three model compounds for the phenolic pollutants, based on their preponderance in actual cases: The catechol group (including resorcinol), the phenol group and the hydroquinone group. It is possible to analyze phenols selectively within three groups. The UV spectrum of a water sample polluted by phenols is resolved into the contribution of these three tracers instead of the more difficult analysis of individual components. Moreover, two methods of background correction have been explored, UVMA and the turbid standard solutions method. The described procedure provides advantages in the determination of polyhydric and *para*-substituted phenols. It can be used preferably for the analysis of phenolic wastewaters of the brown coal conversion industry. Furthermore, sample preparation is not required because turbidity does not interfere with the analysis. The method was used as an alternative to the definition of a phenolic index according to the German standard method DIN 38 409 H16, which is based on application of equation 2[387].

The spectrophotometric method for determination of phenolphthalein (**19**, see Section VIII.A.3) as raw material and in pharmaceutical formulations recommended by the British Pharmacopeia[43] was compared with the HPLC method recommended by older editions of the US Pharmacopeia[42] (see Table 1). The former method was better for the raw materials, whereas the latter one was found to be better for routine analysis of formulations from the point of view of the linearity, sensitivity, reproducibility and lack of interference by other components present in the sample[388].

Preconcentration by SPE of trace phenolic pollutants in water was recommended, prior to UVD, FLD or ELD[389]. The optimum extraction procedure was established for the spectrophotometric determination of phenol and aniline in water. LOD were in the approximate range of the maximum permissible concentrations (about 5 ppm for phenol)[390].

2. Derivatization

a. Halogenation. The precision and sensitivity of the UVV spectrophotometric determination of microgram amounts of phenols monosubstituted with methyl, ethyl, chloro and nitro groups, catechol (**42**), resorcinol (**20**), guaiacol (**143a**), 4-ethylguaiacol (**143b**), dimethylphenols, dichlorophenols, trichlorophenols and pentachlorophenol, after treatment with iodine monobromide, was improved by using iron(III) sulfate as catalyst. The interference of reducing compounds was eliminated by addition of a bromate solution, and that of certain organic acids was reduced by LLE of the analytes into cyclohexane. However, the interference of phenylamine compounds could not be removed. If the organic solution showed emulsification, this was eliminated by anhydrous sodium sulfate. It was proposed to prepare standard mixtures of phenols as comparison standards in the determination of total phenol content of wastewaters and whisky samples[391].

b. Oxidative coupling. Sub-μg L^{-1} levels of phenols in water and soil extracts were determined in a FIA system by preconcentration in an Amberlite XAD-4 column at pH 2.0 that did not retain interfering aromatic amines, followed by elution at pH 13.0 and spectrophotometric measurement of the analytes by the 4-aminoantipyrine (**78**) method, according to equation 2. LOD was 0.2 μg L^{-1}, with linearity over the 0.5–60 μg L^{-1}

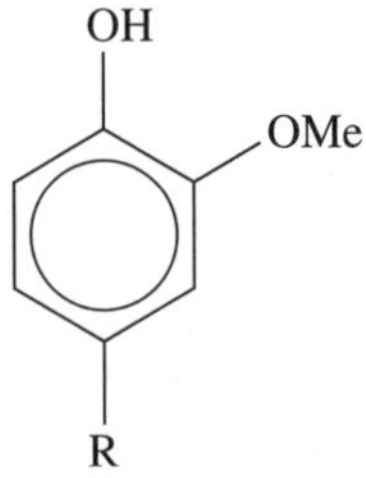

OH
OMe
R

(143) (a) R = H
    (b) R = Et

range. A throughput of 8 samples per hour was achieved, including 5 min preconcentration periods[392]. A standard method for the determination of total phenols in oil can be improved by on-line SPE preconcentration followed by absorbance measurement at 500 nm of the color developed according to equation 2 in the presence of potassium persulfate ($K_2S_2O_8$) as oxidant, in a FIA system. LOD was 0.09 mg L^{-1} of phenol, 0.18 mg L^{-1} of o-cresol and 0.02 mg L^{-1} of m-cresol[393]. The dyes derived from trace phenolic pollutants in water according to equation 2 were concentrated by SPE on a finely divided ion-exchange resin. The color intensity of the dye was compared with a calibration curve to determine the phenol concentration in the sample[394]. An SIA scheme for the simultaneous determination of nitrite, nitrate, sulfate and phenolic compounds in wastewaters was proposed, with equation 2 as part of the analytical scheme[395]. The dye produced by 4-aminoantipyrine (78) at pH 9.0 with phenols in water was concentrated by SPE on a nitrocellulose filter, eluted with 2-methoxyethanol and determined at 480 nm. The linear range was from 0.25 to 6 mg of phenol in the final eluate[396]. Phenolic compounds in wastewaters were determined in a fully automatic SIA system, by oxidative coupling with 4-aminoantipyrine (78), and UVD at 510 nm. The linear range was from 0.05 to 25 ppm, and the sample throughput was 24 per hour with RSD <0.6%[397]. Modifications of 4-aminoantipyrine (78) were proposed as various combinations of substituents in formula 144. Phenol derivatives of the tested reagents had λ_{max} around 480 nm and good stability. No great advantage over 78 was observed in general[398]. The effect of adding a poly(ethylene glycol) phase on the enhancement of equation 2 was investigated[399].

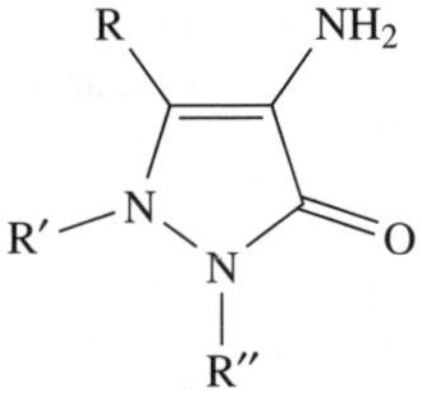

(144) R = Me, Ph, p-ClC$_6$H$_4$, p-H$_2$NC$_6$H$_4$
    R' = Ph, Me, Et
    R'' = Ph, p-H$_2$NC$_6$H$_4$

The dependence on the structure of the phenols of analytically useful color development by processes such as equation 6 was investigated[400]. A fast method for monitoring phenols in water and wastewaters consisted of on-line SPE preconcentration at pH 2, followed by elution at pH 12 and spectrophotometric determination of the color developed in equation 6, using potassium hexacyanoferrate(III), $K_3[Fe(CN)_6]$, in a FIA system. For phenol, the linear calibration range was 0.01 to 1 mg L^{-1}, LOD 0.004 mg L^{-1} (SNR 3), with RSD 2.4% for 0.2 mg L^{-1}. The throughput was 12 samples per hour[401]. A comparative study of determination of phenolic compounds was carried out for the oxidative coupling of phenols with 4-aminoantipyrine (**78**), according to equation 2, and 3-methyl-2-benzothiazolinone hydrazone (**108**), according to equation 6. Both methods were found to be readily applicable in FIA systems, with an output of 40 to 60 analyses per hour. However, the sensitivity of reagent **108** may be significantly higher for phenol. Furthermore, some p-substituted phenols are nearly insensitive to reagent **78** but give good results with **108**[402–404]. Equation 6 was used to develop a method for determination of chlorine dioxide in water. Possible interference from metal ions and other oxychlorinated moieties, such as hypochlorite, chlorite and chlorate, can be avoided[405].

A kinetic method was applied for simultaneous determination of phenol, o-cresol, m-cresol, resorcinol and m-aminophenol at ppm levels, by reaction with p-aminophenol in basic solution and in the presence of potassium periodate. As color developed, UVV scans were recorded every few seconds between 400 and 700 nm for 600 s. The data were processed by the PLS method, using the UNSCRAMBLER program[406]. A method for determination of phenol in water at the ng L^{-1} level consists of a preconcentration step of the pollutants at the top of a solid probe, achieved by controlled freezing of the water sample. The upper end of the probe is collected by partial melting, and a dye is developed on addition of potassium iodate and N,N-diethyl-p-phenylenediamine (**145**)[407]. A method for simultaneous determination of phenolic compounds was proposed, based on kinetic measurement of the oxidative coupling of these analytes to reagent **145** in the presence of hexacyanoferrate(III), $K_3[Fe(CN)_6]$, following the appearance of dyes by changes in the absorbance at 660 nm. The kinetic data are processed by the Kalman filter algorithm. Phenols can be determined individually over the concentration range of 1.25 to 25 μM with RSD of ca 0.6–0.8%. Differences in the kinetic behavior of various phenolic species can be applied to analyze mixtures at the μM level, in a wide variety of concentration ratios with errors less than 10%[70].

$$H_2N-\underset{(145)}{\langle\bigcirc\rangle}-NEt_2$$

c. Coupling with diazonium ions. A scheme was proposed for determination of total phenols based on derivatization with a 4-nitrobenzenediazonium salt, SPE of the diazophenolate of cetyltrimethylammonium on polyurethane foam and UVV determination of the azo dyes mixture. Individual phenols can be determined by HPLC of the mixture with UVD[408]. Spectrophotometric determination of phenols by measuring the diazo dye developed by coupling with a diazonium ion reagent may be accompanied by absorbance instability and large blank values. This is due to decomposition of the reagent into a phenolic byproduct that can also undergo a coupling reaction according to equation 7. Using 2,4,6-trimethylaniline (**146**) as source for the diazonium ion reagent avoids these analytic problems, because its byproduct is incapable of undergoing the coupling reaction with excess reagent. The method was applied in a FIA system to the UVV determination of thymol (**147**), guaiacol (**143a**), dopamine (**10a**), epinephrine (**15a**) and paracetamol (**148**) in pharmaceutical preparations, using a sodium dodecyl sulfate micellar medium to prepare the solutions of **146** and to catalyze the coupling reaction. LOD was in the sub-μM range, linear range from 15 to 170 μM, with RSD

<1% ($n = 3$)[409,410]. Coupling with diazotized *p*-aminoacetophenone (**149**) and measuring at 475 nm ($\varepsilon\, 3.13 \times 10^5$ L mol^{-1} cm^{-1}) was proposed for the spectrophotometric determination of phenol liberated from certain pesticides, present in environmental and biological samples[411]. Similarly, diazotized benzocaine (**150**) couples with phenolic compounds to yield azo dyes. Thus, this reaction was applied to determination of phenolic antibiotics such as amoxicillin (**151**), cefadroxil (**13**) and vancomycin (**152**) in pharmaceutical preparations, yielding an orange yellow coloration that could be measured spectrophotometrically[412].

$$\text{Ar}-\text{N}_2^+ \xrightarrow{\text{H}_2\text{O}} \text{Ar}-\text{OH} \xrightarrow{\text{Ar}-\text{N}_2^+} \text{Ar}-\text{N}=\text{N}-\text{Ar}-\text{OH} \quad (7)$$

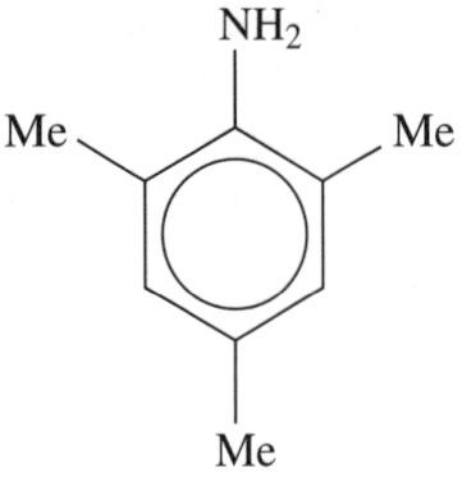

(**146**) 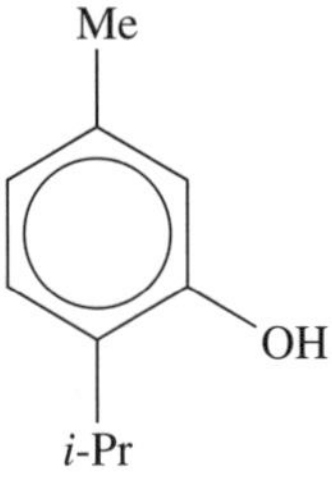 (**147**) 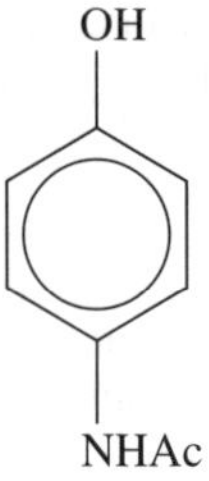 (**148**) 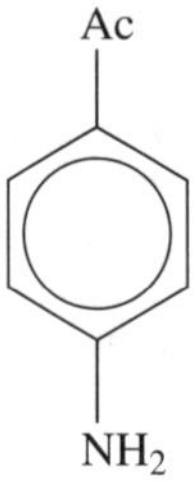 (**149**)

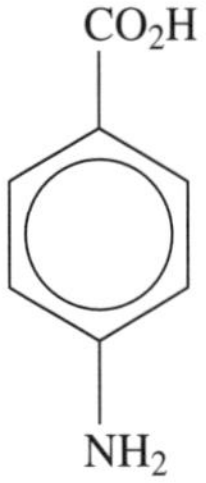

(**150**) 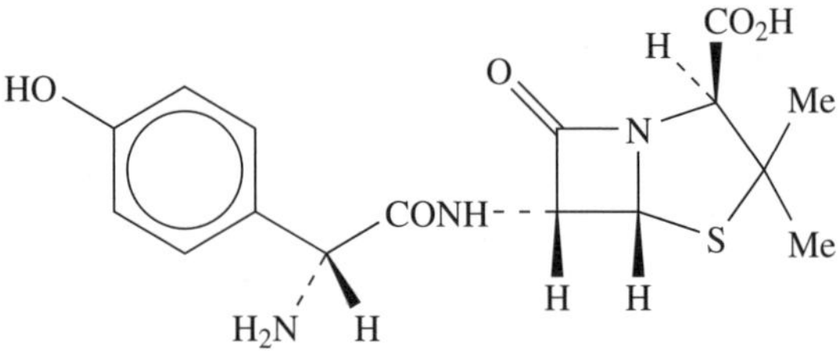 (**151**)

Phenols and naphthols were derivatized by coupling with 4-nitrobenzenediazonium (**153a**) tetrafluoroborate. The diazo dyes were adsorbed on a polyurethane foam and measured photometrically. LOD are low[413]. Coupling of phenol with **153b**, produced by diazotization of 2-cyano-4-nitroaniline, led to formation of a reddish dye that was extracted with 2-methyl-1-butanol and measured at 580 nm. Beer's law held in the 0.05 to 0.4 ppm range[414].

d. Complex formation. Blue to violet complexes are formed at pH 4.0–6.5, between Fe(III) ions, oxalate ions ($C_2O_4^{2-}$) and phenolic compounds carrying two hydroxyl groups on the same ring, with 1 : 2 : 1 stoichiometric ratio. The complex could be used for direct spectrophotometric quantation or as indicator for EDTA titration of the analytes. The method was applied for determination of catechol (**42**), pyrogallol (**134**), dopamine (**10a**), adrenaline (**15a**) and sulbutamol (**154**)[415].

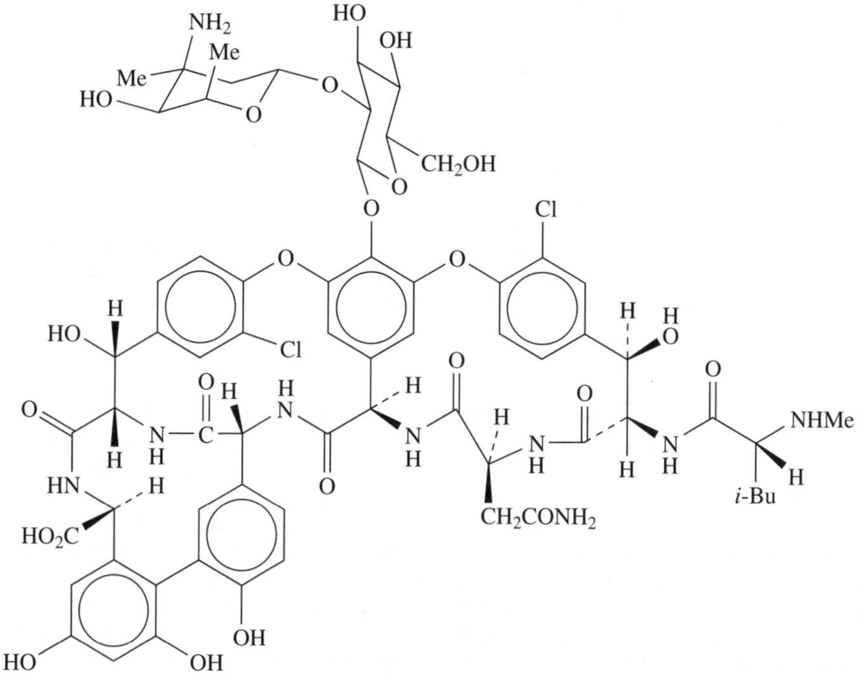

(152)

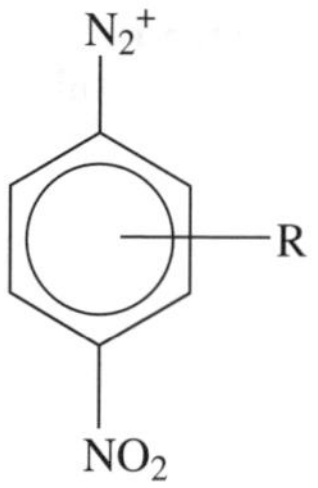

(153) (a) R = H
(b) R = 2-CN

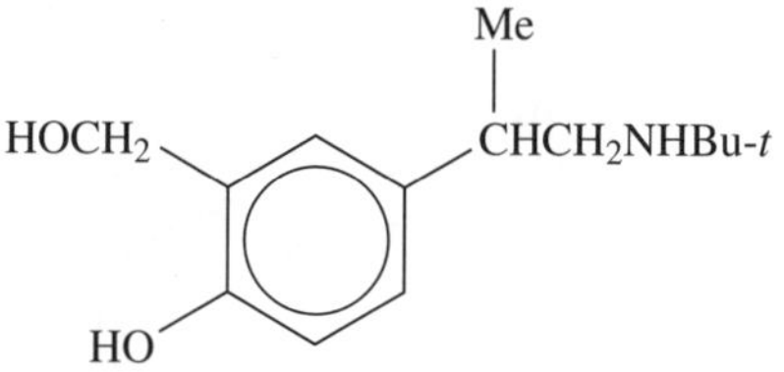

(154)

After LLE into ethanolic KOH, the antioxidant BHT (**32a**) used in aircraft fuel was determined in the presence of Cu(II) ions, by UVV spectrophotometry at 368 nm. Linearity was observed in the 0 to 30 ppm range, RSD $\leqslant 2\%$[416]. A UVV spectrophotometric method for determination of β-cyclodextrin (**126**) is based on the formation of a complex with phenolphthalein (**19**, Section VIII.A.3). Both the intensity and linear range are affected by the pH and the concentration of **19** in solution. The method was considered to be inadequate for precise determinations of **126** of purity higher than 98%[417]; see also application of α-cyclodextrin (**85**) for analysis of phenolics in Section IV.B.4[197]. Phenol in the presence of sodium nitroprusside, $Na_2[Fe(CN)_5NO]$, and hydroxylamine, at pH 10.26–11.46, developed a blue coloration that could be applied for quantitative analysis (λ_{max} 700 nm, ε 1.68 × 10^4 L mol^{-1} cm^{-1}, Sandell sensitivity 0.0052 μg phenol cm^{-2}). Beer's law was found to be valid from 0.1 to 6.5 ppm[418].

e. Miscellaneous color-developing methods. A modification of the Folin–Ciocalteu method has been proposed for the colorimetric determination of total phenolic content of complex samples, such as wine. The main components of the Folin–Ciocalteu reagent are sodium tungstate (Na_2WO_4) and sodium molybdate (Na_2MoO_4) in acid solution; this reagent in the presence of phenolic compounds develops a measurable color[419,420]. The Folin–Ciocalteu method was found to be inadequate for determination of phenolic compounds in citrate extracts of soil samples, due to strong interference by dissolved organic matter[421]. The analogous Folin–Denis reagent was also frequently used for the colorimetric determination of total phenols, for example, in canola oil[422]. Etilefrine (**155**), prenalterol (**156**) and ritodrine (**157**) were determined spectrophotometrically in their formulations using the Gibbs reagent (**158a**). The maximum absorbance is in the 610–650 nm region, obeying Beer's law. LOD was 0.2–0.4 mg L^{-1}. The method was considered to be fast and simple to apply[423]. The possible course of this process is shown in equation 8, using the Gibbs reagent (**158a**) or its bromo analogue (**158b**) to yield an indophenol dye (**159**) with certain labile *para*-substituted phenols in alkaline solution[424].

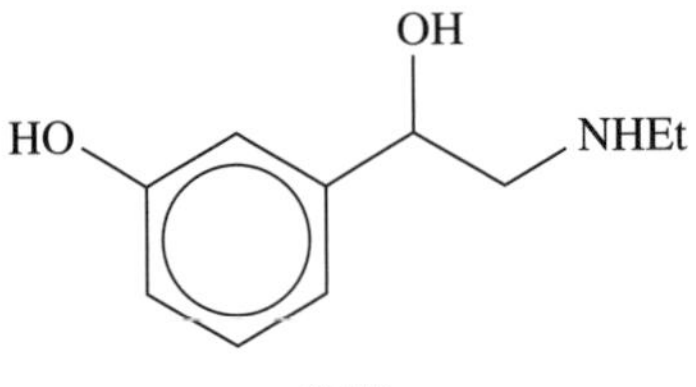

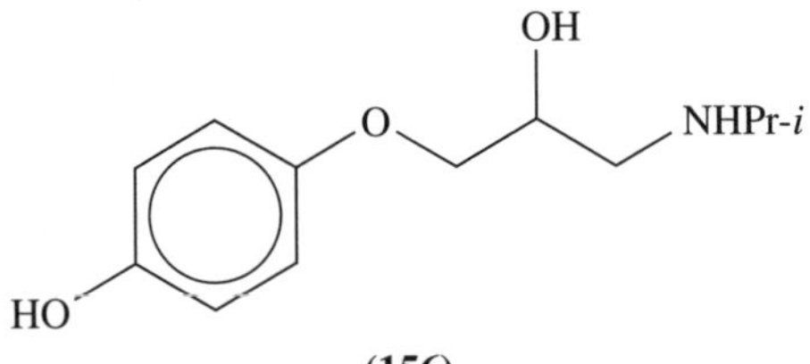

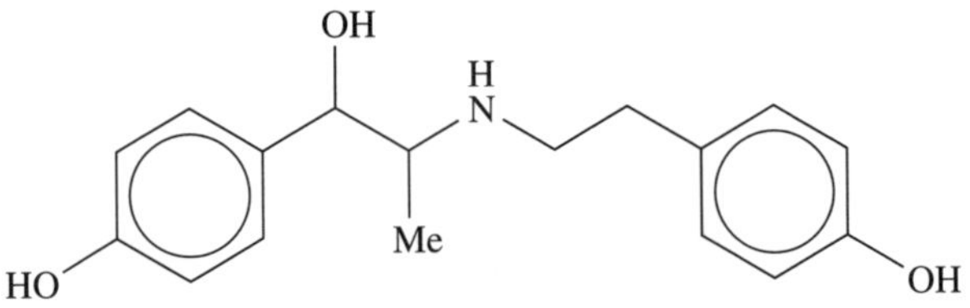

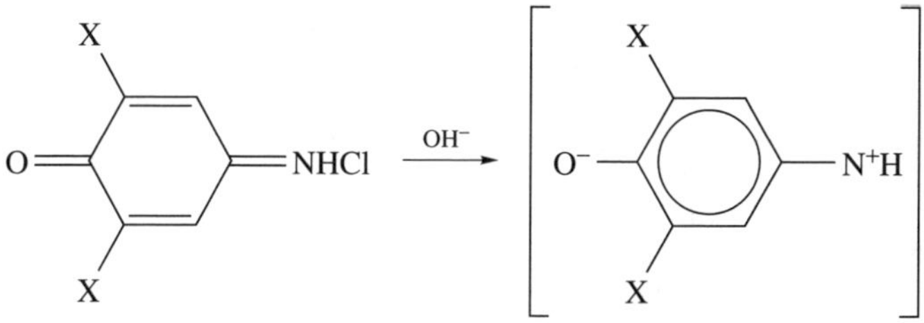

(158) (a) X = Cl
(b) X = Br

(8a)

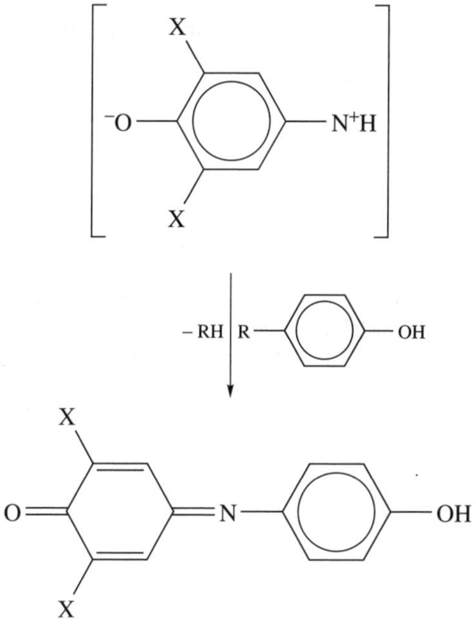

(8b)

(159)

3. Structural and functional characterization

The UVV spectra of free, esterified and insoluble-bound fractions of phenolic acids isolated from *Triton canola* were recorded between 250 and 520 nm. These spectra were analyzed as linear combinations of Gaussian bands using the CHAOS-B computer program. The spectra of the free and esterified fractions were derived from three separate component bands at approximately 280, 300 and 328 nm, and that of the insoluble-bound phenolic acid fraction from four bands at 254, 282, 319 and 384 nm. All three fractions displayed a shorter wavelength component that could be represented by a Gaussian band located between 217 and 235 nm. The second and fourth theoretical derivative spectra yielded a very good fit to the corresponding numerical derivatives of the experimental data; this analysis was applied to a model system consisting of mixtures of protocatechuic (**27**) and sinapic (**99**) acids. The content of **99** could be estimated with an accuracy of

6%[425]. The difference UVV spectra between phenols, naphthols, quinolines, aniline and its derivatives and pyridine and its derivatives, measured at the same concentrations at pH values from 8 to 13 and 1 to 2, presented similar features among analogous compounds. The pH dependence of the spectra was attributed to changes in the conjugated bond system related to acid–base equilibria[426]. The torsional splitting caused by hindered rotation of water and methanol hydrogen-bonded to phenol was investigated by high resolution UV spectroscopy[427,428].

The effects of polar and nonpolar solvents on the peroxyl-radical-trapping antioxidant activity of some flavonoids, catechol derivatives, hydroquinone and monophenols have been studied. The inhibition rate constants k_{inh} of the antioxidants have been determined by following the increase in absorbance at 234 nm of a dilute solution of linoleic acid at 50 °C containing small amounts of antioxidant and radical initiator. Phenols with two *ortho*-hydroxyl groups are the most effective antioxidants in nonpolar solvents (k_{inh} up to 15×10^5 L mol^{-1} s^{-1} in cyclohexane); however, this rate constant significantly declines in strongly hydrogen-bonding acceptor solvents (e.g. *t*-BuOH); in polar solvents that are not strong hydrogen-bonding acceptors (e.g. MeCN) the peroxy radical scavenging efficiency of *ortho*-dihydroxy phenols approaches that of these phenols in nonpolar solvents[429] (see also Sections VII.A.2 and IX.B).

The structure of the various dissociation stages of phenolphthalein (H$_2$PP, **19**) in phosphate buffers of pH 5 to 13, as depicted in equation 9, was correlated with the UVV spectrum. Thus, the aqueous solution of H$_2$PP is colorless; HPP$^-$, preserving the lactone structure, is colorless too; PP^{2-} (**160**), where the lactone is opened, is red; at higher pH colorless PP(OH)$^{3-}$ (**161**) is formed, where incorporation of the hydroxy group disturbs the conjugated structure of PP^{2-} (**160**). The dissociation of sulfonaphthalein (H$_2$PS, **162**) in aqueous solution is shown in equation 10; at low pH, H$_2$PS has a zwitterionic structure and no lactone moiety, and the solution is orange-red; it dissociates to yellow HPS$^-$, and then to red PS^{2-} (**163**), of absorption spectrum similar to that of PP^{2-} (**160**)[430].

$$H_2PP \underset{pK_1=9.05}{\overset{+OH^-}{\rightleftharpoons}} HPP^- \underset{pK_2=9.50}{\overset{+OH^-}{\rightleftharpoons}} PP^{2-} \underset{pK_3=12}{\overset{+OH^-}{\rightleftharpoons}} PP(OH)^{3-} \quad (9)$$

$$H_2PS \overset{+OH^-}{\rightleftharpoons} HPS^- \overset{+OH^-}{\rightleftharpoons} SPP^{2-} \quad (10)$$

The host–guest structural relation in the complexes of β-cyclodextrin (**126**) and phenol or 2,4,6-trimethylphenol were studied by correlating simulated complexation trajectories with the induced circular dichroism measured for the solutions. The relative importance of various contributions to the solvation energy is discussed and it is shown that those terms arising from the interaction of hydrophobic groups with the aqueous environment are essential for the dynamic simulation model; the sign and strength of the calculated rotatory strength are in perfect agreement with induced circular dichroism obtained from experimentally determined averaged spectra[431]. The equilibrium constants for the formation of 1 : 1 and 2 : 1 inclusion complexes of phenols with β-cyclodextrin (**126**) and γ-cyclodextrin (**164**), respectively, were correlated with the molecular polarizability of the guest molecules[432]. Quantitative structure–affinity relationships have been established for the formation of inclusion complexes between *para*-substituted phenols and β-cyclodextrin (**126**) and formation constants of the complexes have been estimated. Experimental results came from potentiometry, circular dichroism, ^{1}H NMR and UVV spectrophotometry. The contribution of van der Waals interactions is a significant factor, provided the *para*-substituent causes no large dipole moment difference[433].

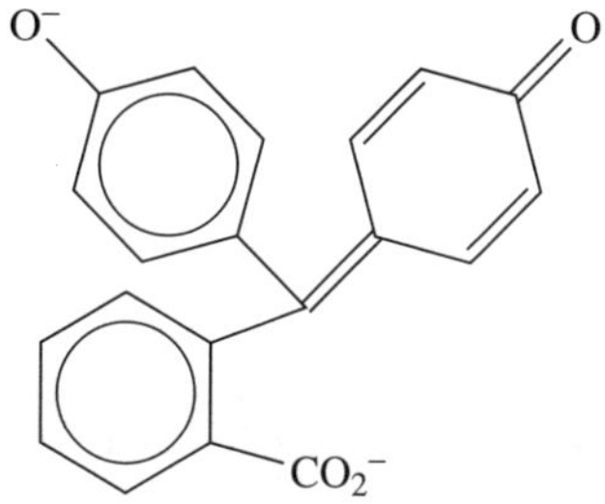

(160) 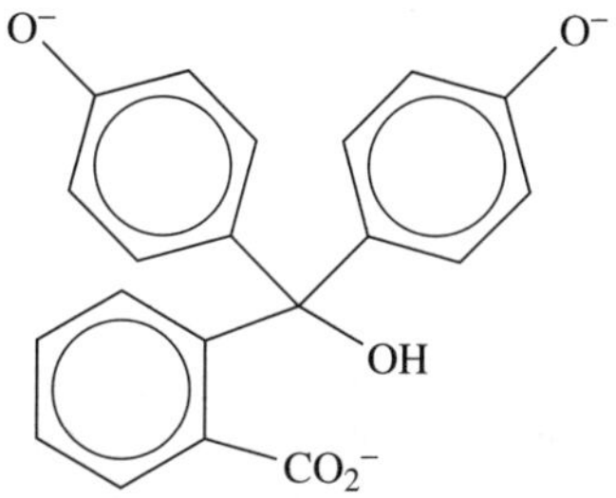 (161) 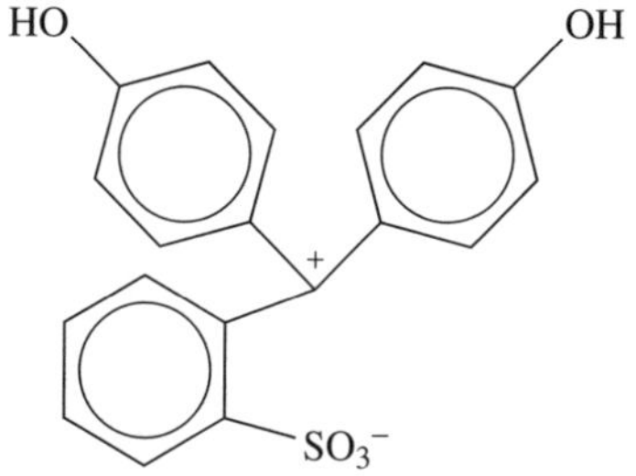 (162) 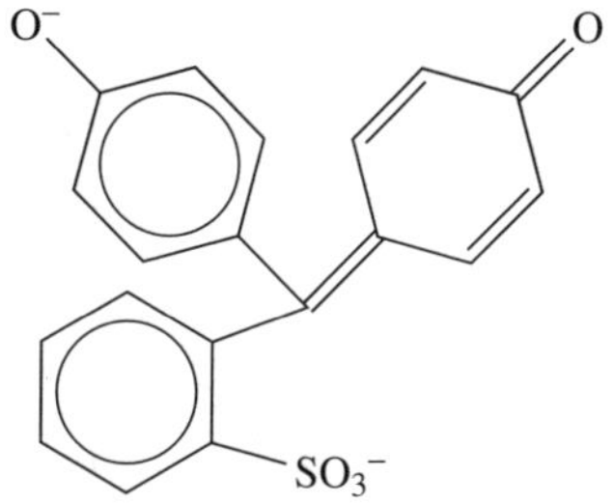 (163)

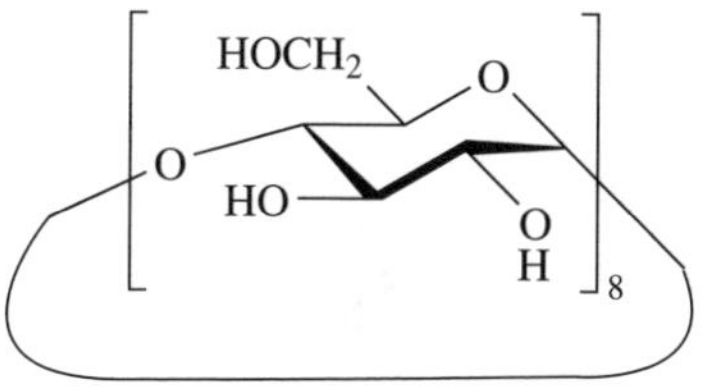

(164)

B. Fluorescence Detection

1. Direct determination

Determination of phenolic and oil product contaminants in water using a FIA system was carried out with an intermittent water sample flow regime. The method involved LLE of the oily constituents with tributyl phosphate-hexane, SPE on a chromatographic absorption column and a PTFE membrane. In the case of natural waters the humic acids had to be eliminated before end analysis of the polluting phenols by LC with FLD ($\lambda_{ex} 270 \pm 10$ nm, $\lambda_{fl} 310 \pm 10$ nm)[434]. The excitation fluorescence spectrum in the 245–290 nm range with emission at 306 nm was used for the simultaneous determination of phenol, bisphenol A (**29**) and its diglycidyl ether (**165**) at ppb levels, after micro-LLE. As the spectra of the analytes considerably overlapped, a full-spectrum multivariate calibration method combined with a PLS calculation algorithm were applied[435].

A fluorescein derivative (**166**) immobilized on a PVC membrane showed fluorescence enhancement in the presence of carboxylic acids and fluorescence quenching in the presence of phenols. This property was applied for development of a fluorescence sensor for

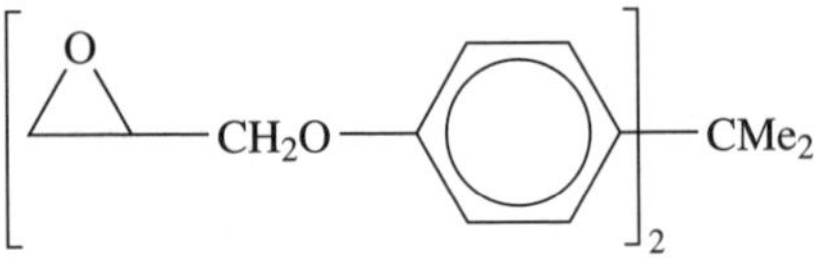

$$\left[\overset{O}{\underset{\triangle}{\diagup\!\!\!\diagdown}}\!\!-\!CH_2O-\!\!\underset{}{\bigcirc}\!\!-\!CMe_2 \right]_2$$

(165)

direct measurement of concentration of phenolic compounds in the μM to dM range[436]. A detection method for phenolic compounds was based on the strong fluorescence quenching caused by these compounds on a poly(ethylene glycol methacrylate) macroporous resin, crosslinked with the fluorescent monomer **167** (λ_{ex} 310 nm, λ_{fl} 395 nm). The quenching effect of phenols and anilines is much stronger than that of aliphatic alcohols and amines[437].

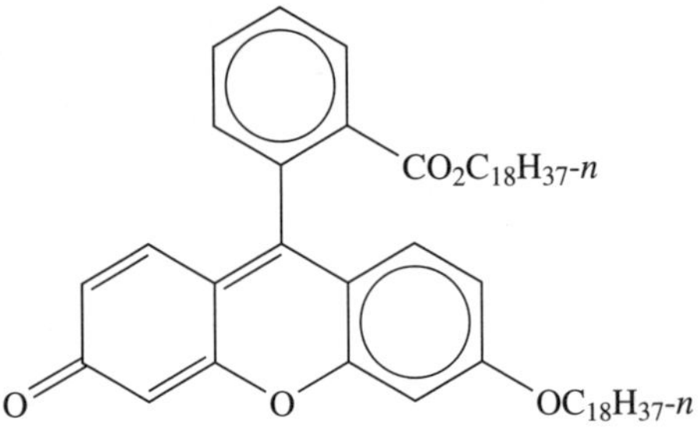

(166)

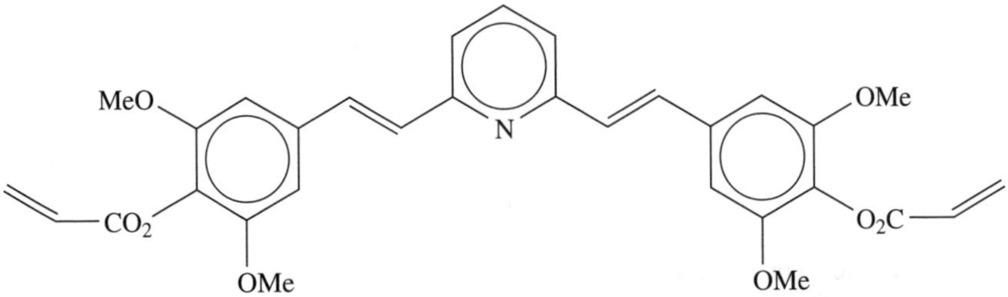

(167)

Microspectrofluorometry was employed for mapping the location of phenolic substances in maize kernels. Autofluorescence due to phenolic acids was detected mainly in the embryo, aleurone and pericarp of maize kernel cross sections. Boric acid (H_3BO_3) reagent enhanced the fluorescence due to flavonoids in the aleurone layer. The amides of phenolic acids required derivatization with Ehrlich's reagent (**168**) to reveal fluorescence in the embryo and aleurone. The localization of phenolic amines was confirmed by HPLC analysis. Phenolic compounds are important in the resistance of maize kernels to pests. Resistant maize types showed higher intensities of phenolic fluorescence but no unusual distributions of these compounds[438].

$$p\text{-}Et_2N-C_6H_4-CHO$$

(168)

2. Fluorescent labelling

The von Pechman–Duisberg condensation, illustrated for phenol in equation 11, was applied to the β-lactam phenolic antibiotics amoxicillin (**151**), cefadroxil (**13**) and cefoperazone (acid form of **14**) to yield the corresponding coumarin derivatives. The determination was spectrofluorometric, with λ_{ex} at 401 to 467 nm and λ_{fl} at 465 to 503 nm. The method is of advantage as compared to established procedures[439].

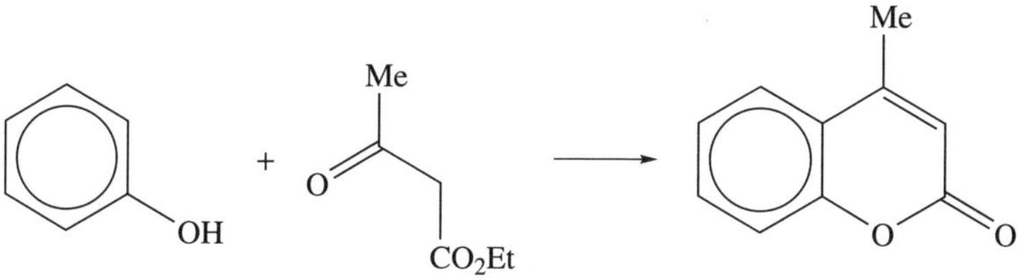

Sympathomimetic drugs can be determined by various procedures. Optimal reaction conditions have been developed for a FIA system with FLD, based on the reaction with 4-aminoantipyrine (**78**) in the presence of potassium hexacyanoferrate (equation 2). Pure samples or pharmaceutical formulations of etilefrine (**155**), orciprenaline (**169**), fenoterol (**170**), hexoprenaline (**171**) and reproterol (**172**) were determined, after dilution to the 2 to 50 ppm range. Results agreed with the official or the referee methods[440].

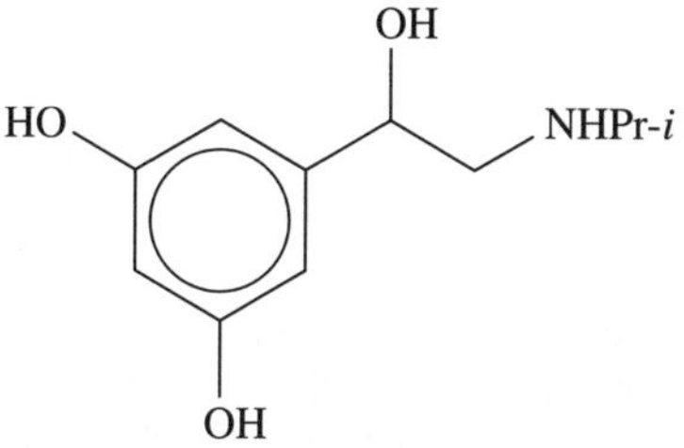

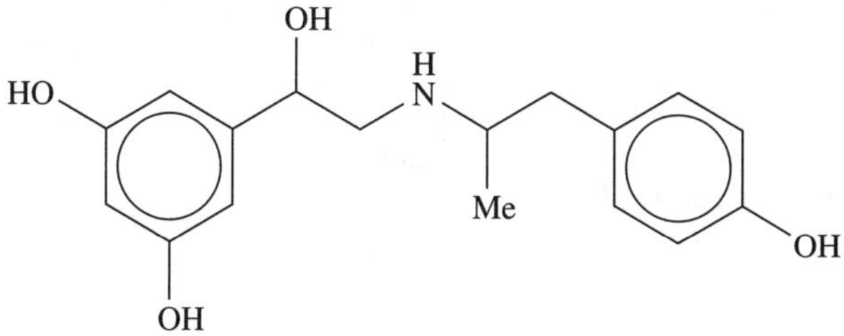

C. Chemiluminescence Detection

A FIA system with CLD was proposed for determination of phenols in natural waters, based on the reaction with potassium permanganate in the presence of sulfuric acid. Preconcentration by SPE on XAD-4 resin lowers the LOD to 5 μg L^{-1}. The method has

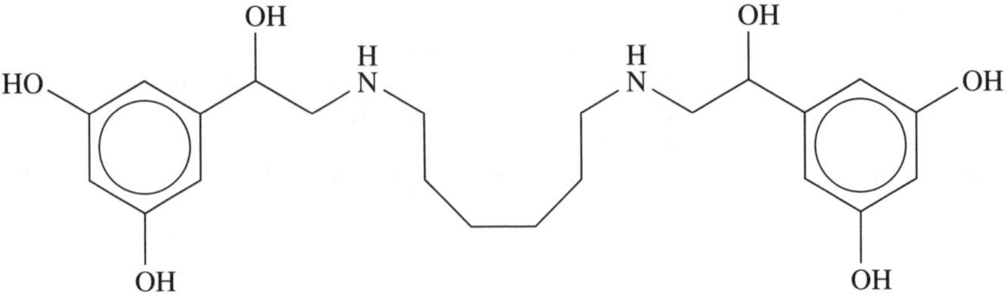

(171)

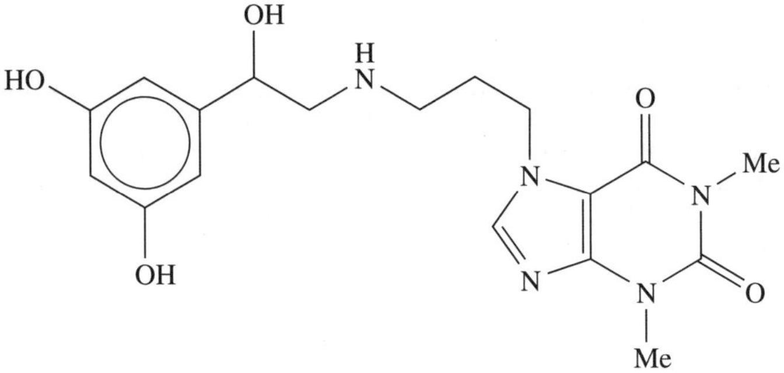

(172)

very low consumption of reactives. The sample throughput was 60 per hour for water as received and 12 per hour when preconcentration was applied[441]. Phenols cause quenching of chemiluminescence of 4-chlorobenzenediazonium fluoroborate (173) in alkaline solution, in the presence of hydrogen peroxide. This is more sensitive than UVD of the analytes alone or in the presence of 4-aminoantipyrine (78). The following LOD and linearity ranges were determined: Phenol 15 ppb, 0–6.0 ppm (RSD 3.0% for 1 ppm solution); 2-nitrophenol 20 ppb, 0–5.0 ppm; p-cresol 25 ppb, 0–4.5 ppm; 2,4-xylenol 30 ppb, 0–8.0 ppm[442]. The sympathomimetic drugs etilefrine (155), isoxsuprine (174) and prenalterol (156) were determined by CLD in a FIA system, where a reaction with $KMnO_4$ in the presence of formic acid was induced. Linearity ranges were 0.2–9, 0.2–12.5 and 0.025–1.25 mg L^{-1}, respectively[443].

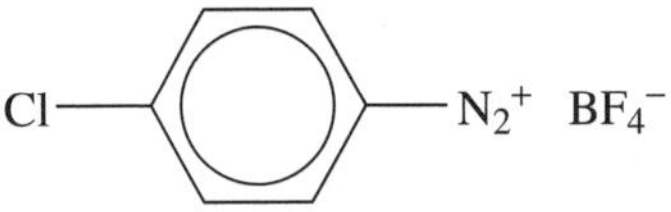

(173) 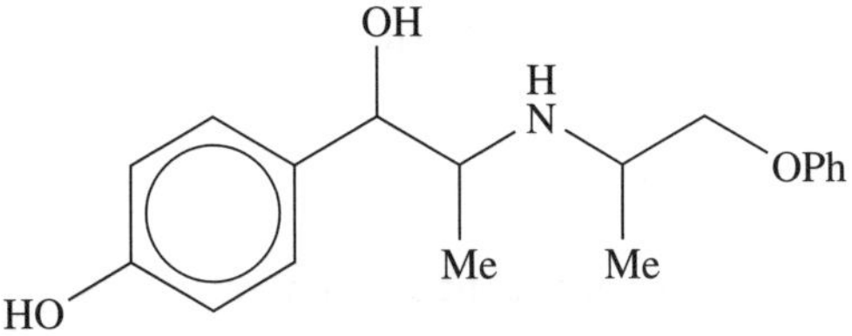   (174)

IX. INFRARED AND RAMAN SPECTRAL METHODS

A. Quantitative Analysis

The phenolic hydroxyl group content in acetylated milled wood lignins was determined by selective aminolysis of the aromatic acetoxy groups. FTIR spectra of the lignins and their acetylated derivatives were recorded. PCR and PLS calibrations were carried out to correlate between the aminolysis results and the FTIR spectra. Spectra of acetylated lignins and a PLS regression gave the best correlation between predicted and observed values; the standard error (SE) ±0.06% (abs.) was about one sixth of the best SE obtained by a simple regression. PCR was slightly inferior to PLS. Calibration with nonacetylated lignins also gave satisfactory results[444]. IR spectroscopy was applied for the determination of free phenol and the formaldehyde-to-phenol ratio in formaldehyde–phenol resol resins. Results were also correlated with ^{13}C NMR spectroscopy data[445,446]. The concentration of antioxidants of type **109** in low-density polyethylene, with the alkyl group varying from C_0 to C_{17}, was determined by FTIR of the polymer without extraction[447].

B. Structural and Functional Characterization

The structure of self assembled monolayers terminated with phenol and 2-chlorophenol moieties was studied by reflectance FTIR, X-ray reflectometry, solid state ^{13}C NMR spectroscopy and measurements of contact angle with water. The pH values at half dissociation ($pH_{1/2}$) of the monolayers were $\geqslant 12.5$ and $\geqslant 12$, respectively, which is at least 2.5 pH units higher than those of half dissociation of the corresponding phenols in solution, as denoted by their pK_a values. The pH at which a certain contact angle was achieved was lower for the 2-chlorophenol moieties, in accordance with their higher acidity[448].

The H-bond complexes formed between phenol derivatives and bis-1,8-(dimethylamino)-naphthalene (**175**) in 1,2-dichloroethane and tetrachloroethylene solution were characterized by FTIR spectroscopy. Compound **175** acts as an effective 'proton sponge' for its ability to form a six-membered chelate-type structure including a N $\cdots$ H $\cdots$ N moiety. The stability constants of the 1 : 1 and 2 : 1 complexes are strongly dependent on the pK_a value of the phenols and increase also with the polarity of the solvent. No complex formation was detected in tetrachloroethylene when H was replaced by D[449].

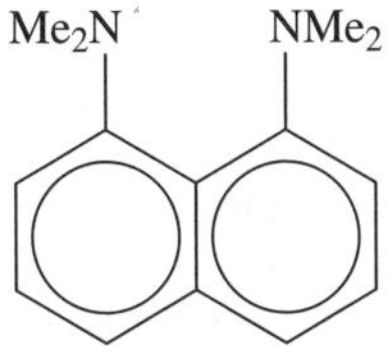

(**175**)

The ultraviolet resonance Raman spectra of the phenolate anion and phenoxy radical in aqueous solution indicate that the C–O bond has a substantial double bond character and the carbon frame has substantial quinonoid character[450].

X. NUCLEAR MAGNETIC RESONANCE

A. Quantitative Analysis

The phenolic hydroxy groups of lignin were determined by two independent spectroscopic methods. The UVV method was based on the difference between the absorption maxima near 300 and 350 nm of samples dissolved in alkaline and neutral solutions. The ^{1}H NMR method was based on the integrated OH proton intensities of the sample dissolved in DMSO, before and after addition of D_2O[451]. One- and two-dimensional ^{1}H NMR was used in the analysis of olive oil phenolic constituents[452,453]. Two-dimensional ^{1}H NMR spectroscopy techniques were applied for the analysis of the phenolic acids in MeOH extracts of two oregano species[454].

Labile hydrogen (phenolic OH and moisture) of coal liquefaction resids was determined using the ^{31}P NMR tagging agent 2-chloro-4,4,5,5-tetramethyl-1,3,2-dioxaphospholane (**176**). Although the presence of organic free radicals in the resids contributed to the breadth of the derivatized phenolic ^{31}P resonances, the results were in excellent agreement with the phenolic contents obtained by FTIR spectroscopy. The best results were obtained by processing the ^{31}P NMR spectra with an NMR-matched filter apodization program[455]. Coal liquefaction products were separated into a nonpolar and a polar fraction. The latter was analyzed for phenols after derivatizing with **176**, separating by RP-HPLC and determining the emerging fractions by ^{31}P NMR. The quantitative analysis of phenols was in accord with independent FTIR determinations[456]. The use of **176** as a phosphitylation reagent in quantitative ^{31}P NMR analysis of the hydroxyl groups in lignins has been thoroughly examined, and an experimental protocol recommended for spectra acquisition has been developed. Quantitative analysis of diverse lignin samples gave results comparable to those obtained by other analytical methods. Excellent resolution was obtained for the various phenolic hydroxyl environments including those present in condensed moieties. However, resolution in the aliphatic hydroxyl region was poor and no distinction could be made between primary, secondary, and the *erythro* and *threo* forms of the secondary hydroxyls of the β-O-4 bonds[457].

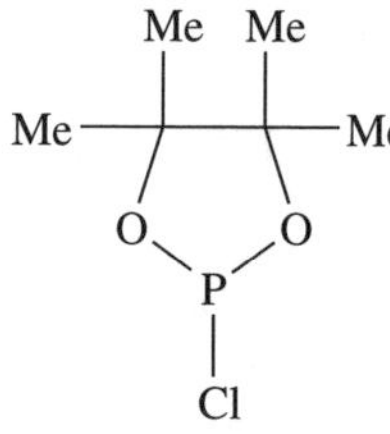

(**176**)

Determination of phenol by ^{13}C NMR spectroscopy has the advantage that each determination affords three independent results that can be averaged, allowing rejection of results with too large RSD. The method was applied to the determination of phenol in tars of the cumene process, and were correlated with those of ^{1}H NMR, UVV spectroscopy and titration with bromine. RSD for single results was 0.8%[458].

B. Structural and Functional Characterization

A ^{1}H NMR study was carried out of the equilibrium 12. If P represents a phenol molecule, with the OH proton at frequency ν_1, and P_n is the preferred oligomer formed of n hydrogen-bonded phenol molecules, with the OH proton at frequency ν_n, in CCl_4 solution, then the equilibrium constant for oligomer formation is given by equation 13. If the rate of exchange between P and P_n is large in comparison with the frequency difference $|\nu_1 - \nu_n|$, as is indeed the case at room temperature, a single peak will be observed at the averaged frequency ν, as shown in equation 14. An experimental procedure was devised for the estimation of the preferred association number n and the equilibrium constant

K_n from the measured average frequencies. At room temperature and concentrations in the 1 M range, oligomerization with $n = 3$ is predominant[459]. At concentrations in the 10^{-3} M range, formation of P_2 species appears to be the preferred oligomerization, as determined by IR measurements[460]. In the solid phase the phenol hydroxy groups form extended linear chains[461,462].

$$n\text{-}C_6H_5OH \rightleftharpoons [C_6H_5OH]_n \qquad (12)$$

$$K_n = \frac{[P_n]}{[P]^n} \qquad (13)$$

$$v = \frac{[P]}{[P] + n[P_n]} v_1 + \frac{n[P_n]}{[P] + n[P_n]} v_n = \frac{[P]v_1 + nK_n[P]^n v_n}{[P] + nK_n[P]^n} \qquad (14)$$

A graphical method was proposed for the assessment of dimerization from the chemical shifts of the monomer and the dimer. The enthalpy and entropy of dimerization could be estimated from the effect of temperature on the dimerization constant[463].

The ^{31}P NMR spectra were investigated after carrying out phosphitylation of lignin-related model compounds, using 2-chloro-1,3,2-dioxaphospholane (**177**) or its tetramethylated analogue **176**. The chemical shifts of phosphitylated carboxylic acids, phenols and aliphatic alcohols were clearly distinguished. A Hammett σ-ρ linear relationship was obtained for the phosphorus chemical shifts of lignin-related phenols. In addition, a correlation between ^{31}P NMR chemical shifts for *ortho*- and *para*-substituted phosphitylated phenols was obtained. A set of empirical parameters was proposed for the accurate prediction of ^{31}P NMR chemical shifts of lignin-related phenolic compounds derivatized with reagent **176**[464].

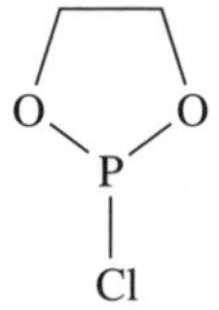

(**177**)

One- and two-dimensional ^{1}H NMR spectral analysis at 500 MHz showed that the site of hydroxy substitution in two metabolites previously reported as 3-nitrofluoranthen-8-ol (**178b**) and 3-nitrofluoranthen-9-ol (**178c**) had to be revised. A third and previously unidentified metabolite was shown to be 3-nitrofluoranthen-7-ol (**178a**). Analysis of NMR spectral data on 2- and 3-nitrofluoranthenes enabled confirmation of the previously reported structures of 2-nitrofluoranthen-8-ol (**178d**) and 2-nitrofluoranthen-9-ol (**178e**) from derived chemical shift substituent effects. Chemical shift data suggest that the nitro group is not strictly coplanar with the aromatic ring system in solution and that metabolism at a distant site can alter the conformation about the C-N bond of the nitro group. A correlation was attempted between reported mutagenicity data and various factors, such as imine quinone formation, chemical shift substituent effects, electronegativity effects and conformation[465].

XI. MASS SPECTROMETRY

In the application of ESI-MS for the analysis of phenols the use of negative and positive ion modes is complementary of each other. Thus, phenols are detected with greater sensitivity in the negative mode; however, the positive mode shows fragmentation that can

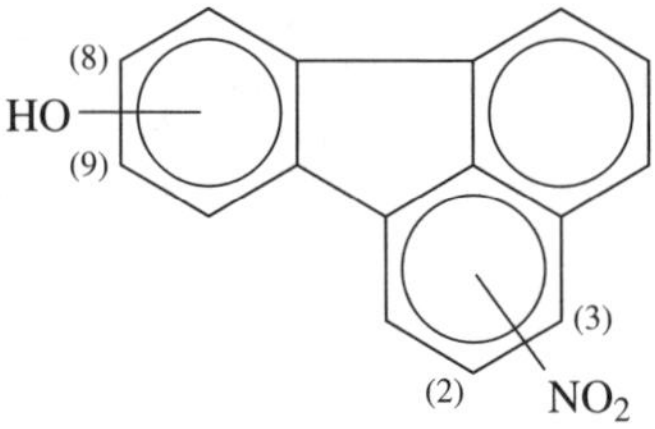

(**178**) (**a**) 3-NO$_2$-7-OH
(**b**) 3-NO$_2$-8-OH
(**c**) 3-NO$_2$-9-OH
(**d**) 2-NO$_2$-8-OH
(**e**) 2-NO$_2$-9-OH

be correlated with the structure of the analyte. The latter feature allowed identification of phenolic components in olive oil that were not previously reported[466]. A method for establishing the profile of phenolic components of edible oils, and especiallly crude olive oil, is based on APCI-MS of the methanolic extract[467]. Water pollutants can be determined by CO$_2$ laser ablation of the frozen sample, followed by resonance-enhanced multiphoton ionization technique coupled with reflection time of flight MS. For phenol LOD was 0.1 pg L^{-1}, with linearity from 0.1 ppb to 10 ppm[468].

The self-association by hydrogen bonding of phenol-terminated polybisphenol A carbonate chains, leading to formation of macromolecular aggregates of higher hydrodynamic volume, was confirmed by MS, applying the matrix-assisted desorption-ionization (MALDI) method. MALDI is a sensitive method for detection of polymer association in dilute solution (see also Section IV.C.5.a)[286].

XII. MISCELLANEOUS METHODS

A. Surface Plasmon Resonance

An optical sensing device for surface plasmon resonance (SPR) was proposed for determination of the concentration of phenolic compounds in water. The phenols become adsorbed on a thin gold or silver film that has been spin-coated with a sol-gel layer containing receptor molecules. Best SPR signals for phenolic compounds were obtained when the receptors were viologen-type polymers with polymeric counterions (**179**). The SPR signal intensity was concentration dependent and had to be calibrated for individual phenolic compounds[469].

B. Miscellaneous Titrations

Thermometric titrations and back-titrations of gallic acid (**8**) and tannic acids with various oxidants were investigated for their possible application in the analysis of polyhydric phenols in wine. Consistent results were obtained using as titrants potassium permanganate (**180**) (to the first equivalence point) and potassium hexacyanoferrate(III) (**181**), with the latter providing sharper end points at higher analyte concentrations. However, use of excess **180** and back titration with Mohr's salt, Fe(NH$_4$)$_2$(SO$_4$)$_2$, is precluded. Titration of tannins with **181** or cerium(IV) sulfate (**182**) were in agreement with the results obtained by the classical volumetric titrations with **180**, using indigo carmine (**183**) as indicator (Löwenthal's method[470]) and **181** (the Folin–Ciocalteu method[471]). As

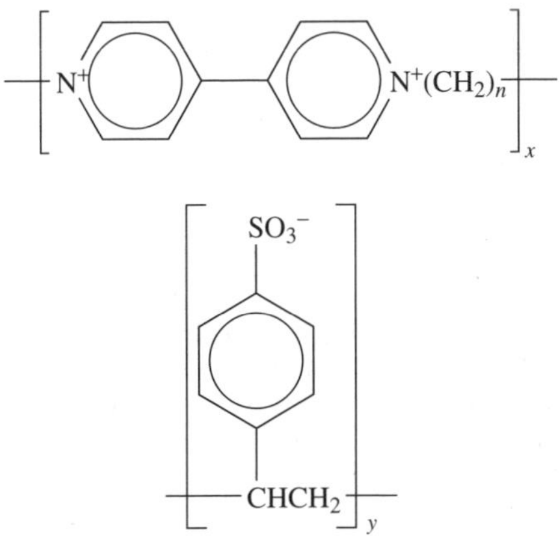

$$\left[-N^+\underset{}{\bigcirc}-\underset{}{\bigcirc}N^+(CH_2)_n-\right]_x$$

$$\left[\underset{-CHCH_2-}{\overset{SO_3^-}{\bigcirc}}\right]_y$$

(179)

opposed to Löwenthal's method, thermometric titrations with **181** and **182** do not require oxidation restrictors or matrix correction. The faster titrations make the method more selective for tannin, as proteins and reducing sugars have slower oxidation kinetics. The presence of sulfur dioxide in wine has a much lower interference effect in thermometric titrations than with the Folin–Ciocalteu method; however, the presence of ascorbic acid is undesirable[472]. The Folin–Ciocalteu method for determination of total phenolic content is still being investigated for various applications[473].

$KMnO_4$	$K_3[Fe(CN)_6]$	$Ce(SO_4)_2$
(180)	**(181)**	**(182)**

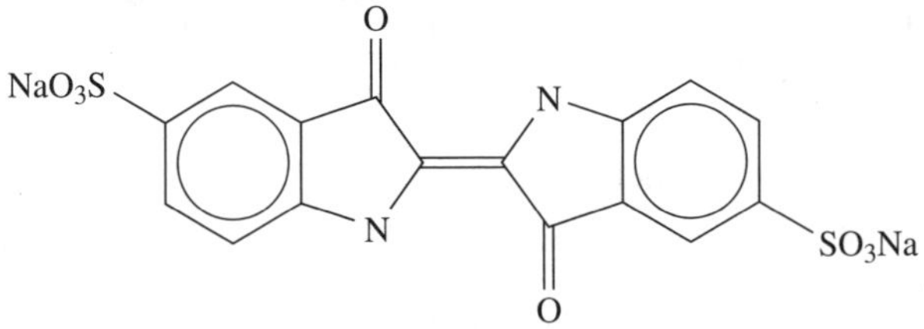

(183)

C. Piezoelectric Detection

A review appeared on piezoelectric quartz crystals used as detectors for phenols in air, after coating with Triton X-100 and 4-aminoantipyrine (**78**), or with activated carbon cloth impregnated with various compounds, such as poly(vinyl pyrrolidone)[474]. A piezoelectric sensor was proposed for determination of trace amounts of phenol and alkylphenols in air. The problems attaining selectivity of the adsoption membranes and operating conditions were addressed[475,476]. An AT-cut quartz crystal, coated with a hydrophobic PVC layer and operating in the thickness shear mode, has been used to detect 4-aminophenol, after conversion to a hydrophobic indophenol dye and adsorption on the polymer layer. The mode of preparation of the PVC coating affects the sensitivity of the detector[477]. A bulk acoustic wave device sensor oscillating in a thickness shear mode was developed for detecting phenols in the atmosphere, based on a piezoelectric quartz crystal coated with various materials for selective binding of the analytes. The highest sensitivity was achieved for a 4-aminoantipyrine (**78**) coating, and the maximal frequency response was about 100 Hz for 20 μg of coating and for a phenol concentration of 0.05 mg L^{-1} in air[478].

D. Thermal Analysis

The constituents of binary phenol mixtures can be identified by differential thermal analysis of a sample to which any of the aroyl chlorides **184–186** has been added. The thermogram is compared with a bank of differential thermograms of phenols, binary phenol mixtures and binary phenol derivatives. Most such systems show well-resolved endotherms corresponding to the melting points of the phenols and their acylated derivatives. The method is proposed for rapid identification of phenols in the solid state[479].

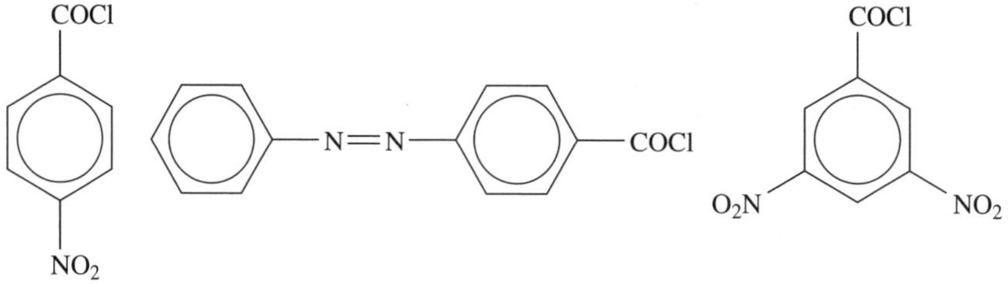

(184)   **(185)**   **(186)**

Differential scanning calorimetry was applied to investigate the kinetic behavior and to evaluate the effect of lignin addition on the curing behavior of phenolic resins. Heat evolution was increased when methylolated lignins were used instead of lignin in the formation of lignin–phenol–formaldehyde thermosets. The curing process followed the Borchardt and Daniel's nth-order kinetic model and showed a 50% to 100% order increase when using methylolated lignin instead of lignin[480].

E. Phenol as a Measuring Stick

Adsorption of phenol in aqueous solution has been applied to the estimation of the specific surface area of granulated activated carbon. The values obtained according to the Langmuir or the BET methods are in agreement with estimations made by other methods. More than 97% of the surface in the activated carbon samples used can be assigned to the micropores of diameter below 7 nm[481]. Phenol adsorption on inorganic carbon-supported microfiltration membranes followed Langmuir and BET isotherm equations, and therefore formed unimolecular adsorption layers. This characteristic could be applied to the determination of specific surface area of porous materials. The results obtained by this method were in close agreement with those derived from mercury porosimetry measurements[482].

XIII. REFERENCES

1. D. Treutter and W. Fencht. *J. Hort. Sci.*, **65**, 511 (1990).
2. J. A. Serratos, A. Blanco-Labra, J. A. Mihm, L. Pietrzak and J. T. Arnason, *Can. J. Bot.*, **71**, 1176 (1993).
3. E. von Röpenack, A. Parr and P. Schulze-Lefert, *J. Biol. Chem.*, **273**, 9013 (1998).

13. Analytical aspects of phenolic compounds   1003

4. N. Zouiten, Y. Ougass, A. Hilal, N. Ferriere, J. J. Macheix and I. El Hadrami, *Agrochimica*, **44**, 1 (2000).
5. P. Mejanelle, J. Bleton, S. Goursaud and A. Tchapla, *J. Chromatogr. A*, **767**, 177 (1997).
6. G. Beltrán, A. Jiménez, M. P. Aguilera and M. Uceda, *Grasas Aceites (Sevilla)*, **51**, 320 (2000); *Chem. Abstr.*, **134**, 309912 (2001).
7. F. Caponio, T. Gomes and A, Pasqualone, *Eur. Food Res. Technol.*, **212**, 329 (2001).
8. L. Gao and G. Mazza, *J. Agric. Food Chem.*, **43**, 343 (1995).
9. A. Escarpa and M. C. González, *J. Chromatogr. A*, **823**, 331 (1998).
10. S. Guyot, N. Marnet, D. Laraba, P. Sanoner and J. F. Drilleau, *J. Agric. Food Chem.*, **46**, 1698 (1998).
11. C. E. Lewis, J. R. L. Walker, J. E. Lancaster and K. H. Sutton, *J. Sci. Food Agric.*, **77**, 45 (1998).
12. C. E. Lewis, J. R. L. Walker, J. E. Lancaster and K. H. Sutton, *J. Sci. Food Agric.*, **77**, 58 (1998).
13. E. Kozukue, N. Kozukue and H. Tsuchida, *J. Jpn. Soc. Hort. Sci.*, **67**, 805 (1998).
14. T. Beta, L. W. Rooney, L. T. Marovatsanga and J. R. N. Taylor, *J. Sci. Food Agric.*, **79**, 1003 (1999).
15. A. Escarpa and M. C. González, *Chromatographia*, **51**, 33 (2000).
16. H. D. Smolarz, *Acta Soc. Bot. Pol.*, **69**, 21 (2000).
17. J. A. Pedersen, *Biochem. Syst. Ecol.*, **28**, 229 (2000).
18. P. B. Andrade, R. M. Seabra, P. Valentao and F. Areias, *J. Liq. Chromatogr. Relat. Technol.*, **21**, 2813 (1998).
19. M. Veit, K. Bauer, C. Beckert, B. Kast, H. Geiger and F. C. Czygan, *Biochem. Syst. Ecol.*, **23**, 79 (1995).
20. B. M. Silva, P. B. Andrade, G. C. Mendes, P. Valentao, R. M. Seabra and M. A Ferreira, *J. Agric. Food Chem.*, **48**, 2853 (2000).
21. L. Packer (Ed.), *Oxidants and Antioxidants—Part A*, Academic Press, San Diego, 1999.
22. T. Yoshikawa, *Antioxidant Food Supplements in Human Health*, Academic Press, San Diego, 1999.
23. M. G. L. Hertog, P. C. H. Hollman and D. P. Venema, *J. Agric. Food Chem.*, **40**, 1591 (1992).
24. A. A. Franke, L. J. Custer, W. Wang and C. Y. Shi, *Proc. Soc. Exp. Biol. Med.*, **217**, 263 (1998).
25. E. A. Hudson, P. A. Dinh, T. Kokubun, M. S. J. Simmonds and A. Gescher, *Cancer Epidemiol. Biomarkers Prev.*, **9**, 1163 (2000).
26. T. Matsuzaki and Y. Hara, *Nippon Nougei Kagaku Kaishi*, **59**, 129 (1985); *Chem. Abstr.*, **102**, 165534 (1985).
27. J. A. Vinson, Y. A. Dabbagh, M. M. Serry and J. Jang. *J. Agric. Food Chem.*, **43**, 2800 (1995).
28. K. Shimoi, Y. Nakamura, I. Tomita, Y. Hara and T. Kada, *Mutat. Res.*, **173**, 239 (1986).
29. Y. Hara, T. Matsuzaki and T. Suzuki, *Nippon Nougei Kagaku Kaishi*, **61**, 803 (1987); *Chem. Abstr.*, **107**, 193927 (1987).
30. R. Abu-Amsha, K. D. Croft, I. B. Puddey, J. M. Proudfoot and L. J. Beilin, *Clin. Sci.*, **91**, 449 (1996).
31. R. M. Lamuela-Raventós and A. L. Waterhouse, *Methods Enzymol.*, **299**, 184 (1999).
32. S. Bumble, *Computer Generated Physical Properties*, Lewis Publishers, Boca Raton, 1999.
33. B. L. Lee, H. Y. Ong, C. Y. Shi and C. N. Ong, *J. Chromatogr.-Biomed. Appl.*, **619**, 259 (1993).
34. G. Bieniek, *J. Chromatogr. B*, **682**, 167 (1996).
35. G. Khoschsorur and W. Petek, *Anal. Sci.*, **16**, 589 (2000).
36. C. J. Pouchert, *The Aldrich Library of FT-IR Spectra*, Aldrich Chemical, Milwaukee, WI, 1997.
37. C. J. Pouchert, *The Aldrich Library of NMR Spectra*, 2nd edn., Aldrich Chemical, Milwaukee, WI, 1983.
38. F. J. Green, *The Sigma-Aldrich Handbook of Stains, Dyes and Indicators*, Aldrich Chemical, Milwaukee, WI, 1990.
39. R. E. Lenga (Ed.), *The Sigma-Aldrich Library of Chemical Safety Data*, 2nd edn., Aldrich Chemical, Milwaukee, WI, 1988.

40. R. E. Lenga and K. L. Votoupal (Eds.), *The Sigma-Aldrich Library of Regulatory and Safety Data*, Aldrich Chemical, Milwaukee, WI, 1992.
41. R. J. Lewis, Sr., *Sax's Dangerous Properties of Industrial Materials*, 8th edn., Van Nostrand Reinhold, New York, 1992.
42. *The United States Pharmacopeia*, 24th edn., United States Pharmacopeial Convention, Inc., Rockville, MD, USA, 2000.
43. *The British Pharmacopoeia* (2000), The Stationery Office, London, 2000.
44. B. D. Page, *J. AOAC Int.*, **76**, 765 (1993).
45. A. Dieffenbacher, *Deut. Lebensmit.-Rundsch.*, **94**, 381 (1998).
46. S. Rauzy and J. Danjou, *J. Fr. Hydrol.*, **24**, 233 (1993).
47. R. A. Rudel, S. J. Melly, P. W. Geno, G. Sun and J. G. Brody, *Environ. Sci. Technol.*, **32**, 861 (1998).
48. A. M. Soto, C. Sonnenschein, K. L. Chung, M. F. Fernandez, M. Olea and M. F. Olea-Serrano, *Environ. Health Perspect.*, **103**, 113 (1995).
49. S. Jobling, T. Reinolds, R. White, M. G. Parker and J. P. Sumpter, *Environ. Health Perspect.*, **103**, 582 (1995).
50. D. M. Klotz, B. S. Beckman, S. M. Hill, J. A. McLachlan, M. R. Walters and S. F. Arnold, *Environ. Health Perspect.*, **104**, 1084 (1996).
51. W. Körner, U. Bolz, W. Süßmuth, G. Hiller, W. Schuller, V. Hanf and H. Hagenmaier, *Chemosphere*, **40**, 1131 (2000).
52. U. Bolz, W. Körner and H. Hagenmaier, *Chemosphere*, **40**, 929 (2000).
53. D. Miller, B. B. Wheals, N. Beresford and J. P. Sumpter, *Environ. Health Perspect.*, **109**, 133 (2001).
54. L. Xu, J. W. Ball, S. L. Dixon and P. C. Jurs, *Environ. Toxicol. Chem.*, **13**, 841 (1994).
55. A. M. Dietrich, S. Mirlohi, W. F. DaCosta, J. P. Dodd, R. Sauer, M. Homan and J. Schultz, *Water Sci. Technol.*, **40**, 45 (1999).
56. D. Puig and D. Barceló, *Trends Anal. Chem.*, **15**, 362 (1996).
57. *EPA Method 604, Phenols*, Environmental Protection Agency, Part VIII, 40 CFR Part 136, *Fed. Regist.*, Oct. 26 (1984), p. 58.
58. *EPA Method 625, Bases, Neutrals and Acids*, Environmental Protection Agency, Part VIII, 40 CFR Part 136, *Fed. Regist.*, Oct. 26 (1984), p. 153.
59. *Directive 76/160/EEC*, Commission of the European Communities, Brussels, 1975.
60. *Drinking Water Directive 80/778/EEC*, Commission of the European Communities, Brussels, 1980.
61. B. Motamed and H. Texier, *Oceanol. Acta*, **23**, 167 (2000).
62. B. Motamed, J. L. Bohm, D. Hennequin, H. Texier, R. Mosrati and D. Barillier, *Analusis*, **28**, 592 (2000).
63. Z. Males and M. Medic-Saric, *J. Pharm. Biomed. Anal.*, **24**, 353 (2001).
64. M. Trojanowicz, A. Jagielska, P. Rotkiewicz and A. Kierzek, *Chem. Anal. (Warsaw)*, **44**, 865 (1999).
65. R. M. de Carvalho, C. Mello and L. T. Kubota, *Anal. Chim. Acta*, **420**, 109 (2000).
66. Y. Ni, L. Wang and S. Kokot, *Anal. Chim. Acta*, **431**, 101 (2001).
67. A. Cladera, M. Miró, J. M. Estela and V. Cerda, *Anal. Chim. Acta*, **421**, 155 (2000).
68. I. Baranowska and C. Pieszko, *Analyst*, **125**, 2335 (2000).
69. J. Liu and Z. Lin, *Spectrosc. Spectr. Anal.*, **20**, 480 (2000).
70. A. Velasco, X. Rui, M. Silva and D. Pérez-Bendito, *Talanta*, **40**, 1505 (1993).
71. S. R. Salman, *Spectrosc. Lett.*, **30**, 173 (1997).
72. L. S. Ettre, *Pure Appl. Chem.*, **65**, 819 (1993).
73. http://www.iupac.org/reports/provisional/abstract00/jagd_280201.html
74. http://www.iupac.org/reports/provisional/abstract00/davankov_280201.html
75. P. Mussmann, K. Levsen and W. Radeck, *Fresenius J. Anal. Chem.*, **348**, 654 (1994).
76. Ya. I. Korenman and T. A. Kuchmenko, *J. Anal. Chem.*, **50**, 723 (1995).
77. L. Liberatore, G. Procida, N. d'Alessandro and A. Cichelli, *Food Chem.*, **73**, 119 (2001).
78. M. Brenes, A. García, P. García and A. Garrido, *J. Agric. Food Chem.*, **48**, 5178 (2000).
79. M. P. Llompart, R. A. Lorenzo, R. Cela, J. R. Jocelyn Paré, J. M. R. Bélanger and K. Li, *J. Chromatogr. A*, **757**, 153 (1997).
80. F. I. Onuska and K. A. Terry, *J. High Resolut. Chromatogr.*, **18**, 564 (1995).
81. Ya. I. Korenman, K. I. Zhilinskaya and V. N. Fokin, *J. Anal. Chem.*, **51**, 1038 (1996).

13. Analytical aspects of phenolic compounds

82. J. Bartulewicz, E. Bartulewicz, J. Gawlowski and J. Niedzielski, *Chem. Anal. (Warsaw)*, **41**, 939 (1996).
83. C. Thomsen, K. Janak, E. Lundanes and G. Becher, *J. Chromatogr. B*, **750**, 1 (2001).
84. J. Cheung and R. J. Wells, *J. Chromatogr. A*, **771**, 203 (1997).
85. M. Graovac, M. Todorovic, M. I. Trtanj, M. M. Kopecni and J. J. Comor, *J. Chromatogr. A*, **705**, 313 (1995).
86. J. B. Forehand, G. L. Dooly and S. C. Moldoveanu, *J. Chromatogr. A*, **898**, 111 (2000).
87. P. Bartak, P. Frnkova and L. Cap, *J. Chromatogr. A*, **867**, 281 (2000).
88. K. D. Buchholz and J. Pawliszyn, *Environ. Sci. Technol.*, **27**, 2844 (1993).
89. K. D. Buchholz and J. Pawliszyn, *Anal. Chem.*, **66**, 160 (1994).
90. P. Bartak and L. Cap, *J. Chromatogr. A*, **767**, 171 (1997).
91. T. J. Clark and J. E. Bunch, *J. Chromatogr. Sci.*, **34**, 272 (1996).
92. M. A. Crespín, E. Ballesteros, M. Gallego and M. Valcárcel, *J. Chromatogr. A*, **757**, 165 (1997).
93. R. Baciocchi, M. Attina, G. Lombardi and M. R. Boni, *J. Chromatogr. A*, **911**, 135 (2001).
94. M. Llompart, B. Blanco and R. Cela, *J. Microcol. Sep.*, **12**, 25 (2000).
95. H. B. Lee, T. E. Peart and R. L. Hongyou, *J. Chromatogr.*, **636**, 263 (1993).
96. M. L. Marín, A. Jiménez, J. Vilaplana, J. López and V. Berenguer, *J. Chromatogr. A*, **819**, 289 (1998).
97. M. P. Llompart, R. A. Lorenzo, R. Cela, K. Li, J. M. R. Bélanger and J. R. Jocelyn Paré, *J. Chromatogr. A*, **774**, 243 (1997).
98. P. I. Demyanov, *J. Anal. Chem.*, **47**, 1407 (1992).
99. J. Lenicek, J. Holoubek, M. Sekyra and S. Kocianova, *Chem. Listy*, **87**, 852 (1993).
100. T. J. Boyd, *J. Chromatogr. A*, **662**, 281 (1994).
101. J. R. Louch, L. E. Lafleur, G. Wilson, D. Bautz, D. Woodrow, H. Teitzel, J. Jones and M. Mark, *TAPPI J.*, **76**, 71 (1993).
102. M. L. Bao, F. Pantani, K. Barbieri, D. Burrini and O. Griffini, *Chromatographia*, **42**, 227 (1996).
103. T. M. Pissolatto, P. Schossler, A. M. Geller, E. B. Caramão and A. F. Martins, *J. High Resolut. Chromatogr.*, **19**, 577 (1996).
104. M. P. Llompart-Vizoso, R. A. Lorenzo-Ferreira and R. Cela-Torrijos, *J. High Resolut. Chromatogr.*, **19**, 207 (1996).
105. J. Porschmann, J. Plugge and R. Toth, *J. Chromatogr. A*, **909**, 95 (2001).
106. J. B. Green, S. K.-T. Yu and R. P. Vrana, *J. High Resolut. Chromatogr.*, **17**, 439 (1994).
107. I. Sanz-Vicente, S. Cabredo and J. Galbán, *Chromatographia*, **48**, 542 (1998).
108. M. L. Bao, K. Barbieri, D. Burrini, O. Griffini and F. Pantani, *Ann. Chim. (Rome)*, **86**, 343 (1996).
109. A. Dasgupta, W. Blackwell and E. Burns, *J. Chromatogr. B*, **689**, 415 (1997).
110. T. Heberer and H. J. Stan, *Anal. Chim. Acta*, **341**, 21 (1997).
111. P. Stremple, *J. High Resolut. Chromatogr.*, **19**, 581 (1996).
112. G. J. Soleas and D. M. Goldberg, *Methods Enzymol.*, **299**, 137 (1999).
113. J. Porschmann, J. Plugge and R. Toth, *J. Chromatogr. A*, **909**, 95 (2001).
114. A. Q. Zhang, S. C. Mitchell and J. Caldwell, *J. Pharm. Biomed. Anal.*, **17**, 1139 (1998).
115. N. Manabe, T. Toyoda and Y. Yokota, *Bunseki Kagaku*, **48**, 449 (1999); *Chem. Abstr.*, **131** 32344 (1999).
116. J. Nolte, B. Graß, F. Heimlich and D. Klockow, *Fresenius J. Anal. Chem.*, **357**, 763 (1997).
117. A. Zapf and H. J. Stan, *J. High Resolut. Chromatogr.*, **22**, 83 (1999).
118. R, Nishino and T. Saito, *Bunseki Kagaku*, **45**, 915 (1996); *Chem. Abstr.*, **126**, 2298 (1997).
119. C. J. Chien, M. J. Charles, K. G. Sexton and H. E. Jeffries, *Environ. Sci. Technol.*, **32**, 299 (1998).
120. L. Maros and S. Igaz, *Mag. Kem. Fol.*, **103**, 347 (1997).
121. Ya. I. Korenman, I. V. Gruzdev, B. M. Kondratenok and V. N. Fokin, *J. Anal. Chem.*, **54**, 1134 (1999).
122. T. C. Gerbino and G. Castello, *J. High Resolut. Chromatogr.*, **19**, 377 (1996).
123. C. Webster and M. Cooke, *J. High Resolut. Chromatogr.*, **18**, 319 (1995).
124. H. Kishi, H. Arimoto and T. Fujii, *Anal. Chem.*, **70**, 3488 (1998).
125. E. Baltussen, F. David, P. Sandra, H. G. Janssen and C. Cramers, *J. Microcol. Separation*, **11**, 471 (1999).

126. K. S. Jiao and A. Robbat, *J. AOAC Int.*, **79**, 131 (1996).
127. T. Saito, *Anal. Chim. Acta*, **276**, 295 (1993).
128. S. Nito and R. Takeshita, *Chemosphere*, **33**, 2239 (1996).
129. T. G. Hartman, K. Karmas, J. Chen, A. Shevade, M. Deagro and H. I. Hwang, in *Phenolic Compounds in Food and Their Effects on Health* (Eds. C.-T. Ho, C. Y. Lee and M.-T. Huang), American Chemical Society, Washington, *Am. Chem. Soc. Symp. Ser.*, **506**, 60 (1992).
130. L.-K. Ng, P. Lafontaine and J. Harnois, *J. Chromatogr. A*, **873**, 29 (2000).
131. S. Camarero, P. Bocchini, G. C. Galletti and A. T. Martínez, *Rapid Comm. Mass Spectrom.*, **13**, 630 (1999).
132. C. Pérez Lamela, J. Simal Lozano, P. Paseiro Losada, S. Paz Abuín and J. Simal Gandara, *Analusis*, **21**, 367 (1993).
133. A. J. H. P. van der Pol, A. J. Verduyn and J. H. C. van Hooff, *Appl. Cat. A*, **96**, L13 (1993).
134. S. K. Chopra, V. B. Kapoor, S. C. Vishnoi and S. D. Bhagat, *Erdol Kohle Erdgas Petrochem.*, **47**, 230 (1994); *Chem. Abstr.*, **121**, 112971 (1994).
135. F. Bosch Reig, F. Campíns Falcó and J. Verdú Andrés, *J, Liq. Chromatogr.*, **18**, 2229 (1995).
136. M. Antolovich, P. Prenzler, K. Robards and D. Ryan, *Analyst*, **125**, 989 (2000).
137. S. Baugh and S. Ignelzi, *J. AOAC Int.*, **83**, 1135 (2000).
138. F. Buiarelli, G. Cartoni, F. Coccioli and Z. Levetsovitou, *J. Chromatogr. A*, **695**, 229 (1995).
139. F. Kader, B. Rovel, M. Girardin and M. Metche, *Food Chem.*, **55**, 35 (1996).
140. L. E. Vera-Avila, J. Reza and R. Covarrubias, *Int. J. Environ. Anal. Chem.*, **63**, 301 (1996).
141. C.-K. Wang and W.-H. Lee, *J. Agric. Food Chem.*, **44**, 2014 (1996).
142. N. Durust, B. Bozan, S. Ozden, Y. Durust and K. H. C. Baser, *Anal. Lett.*, **32**, 2841 (1999).
143. M. T. Benassi and H. M. Cecchi, *J. Liq. Chromatogr. Relat. Technol.*, **21**, 491 (1998).
144. L. J. Michot, F. Didier, F. Villieras and J. M. Cases, *Pol. J. Chem.*, **71**, 665 (1997).
145. H. Teng and C.-T. Hsieh, *Ind. Eng. Chem. Res.*, **37**, 3618 (1998).
146. R. U. Edgehill and G. Q. Lu, *J. Chem. Technol. Biotechnol.*, **71**, 27 (1998).
147. E. Pocurull, R. M. Marcé and F. Borrull, *Chromatographia*, **41**, 521 (1995).
148. M. T. Galceran and O. Jauregui, *Anal. Chim. Acta*, **304**, 75 (1995).
149. N. Shimizu and Y. Inoue, *Bunseki Kagaku*, **43**, 1107 (1994); *Chem. Abstr.*, **122**, 305546 (1995).
150. L. E. Vera-Avila, J. L. Gallegos-Perez and E. Camacho-Frias, *Talanta*, **50**, 509 (1999).
151. V. Piangerelli, F. Nerini and S. Cavalli, *Ann. Chim. (Rome)*, **87**, 571 (1997).
152. E. Pocurull, M. Calull, R. M. Marcé and F. Borrull, *J. Chromatogr. A*, **719**, 105 (1996).
153. N. Masqué, M. Galià, R. M. Marcé and F. Borrull, *J. Chromatogr. A*, **771**, 55 (1997).
154. V. Piangerelli, F. Nerini and S. Cavalli, *Ann. Chim. (Rome)*, **83**, 331 (1993).
155. G. Carrieri, S. Cavalli, P. P. Ragone and N. Cardellicchio, *Ann. Chim.*, **88**, 619 (1998).
156. N. Masqué, E. Pocurull, R. M. Marcé and F. Borrull, *Chromatographia*, **47**, 176 (1998).
157. K. Eder, M. R. Buchmeiser and G. K. Bonn, *J. Chromatogr. A*, **810**, 43 (1998).
158. D. Puig and D. Barceló, *Chromatographia*, **40**, 435 (1995).
159. M. Castillo, D. Puig and D. Barceló, *J. Chromatogr. A*, **778**, 301 (1997).
160. M. A. Crespín, M. Gallego and M. Valcárcel, *Anal. Chem. (Rome)*, **71**, 2687 (1999).
161. K. K. Chee, M. K. Wong and H. K. Lee, *Mikrochim. Acta*, **126**, 97 (1997).
162. S. Dupeyron, M. Astruc and M. Marbach, *Analusis*, **23**, 470 (1995).
163. M.-V. Jung, D. W. Lee, J.-S. Rhee and K.-J. Paeng, *Anal. Sci.*, **12**, 981 (1996).
164. E. Pocurull, G. Sánchez, F. Borrull and R. M. Marcé, *J. Chromatogr. A*, **696**, 31 (1995).
165. A. Di Corcia, S. Marchese, R. Samperi, G. Cecchini and L. Cirilli, *J. AOAC Int.*, **77**, 446 (1994).
166. O. Jáuregui and M. T. Galceran, *Anal. Chim. Acta*, **340**, 191 (1997).
167. A. Di Corcia, S. Marchese and R. Samperi, *J. Chromatogr.*, **642**, 175 (1993).
168. M. Knutsson, L. Mathiasson and J. Å. Jonsson, *Chromatographia*, **42**, 165 (1996).
169. B. Makuch, K. Gazda and M. Kaminski, *Anal. Chim. Acta*, **284**, 53 (1993).
170. C. Baciocchi, M. A. Roggero, D. Giacosa and E. Marengo, *J. Chromatogr. Sci.*, **33**, 338 (1995).
171. M. C. Alonso, D. Puig, I. Silgoner, M. Grasserbauer and D. Barceló, *J. Chromatogr. A*, **823**, 231 (1998).
172. O. Jáuregui, E. Moyano and M. T. Galceran, *J. Chromatogr. A*, **823**, 241 (1998).
173. W. T. Kelley, D. L. Coffey and T. C. Mueller, *J. AOAC Int.*, **77**, 805 (1994).
174. J. Karovicova and P. Simko, *J. Chromatogr. A*, **882**, 271 (2000).

13. Analytical aspects of phenolic compounds   1007

175. C. Chilla, D. A. Guillén, C. G. Barroso and J. A. Pérez-Bustamante, *J. Chromatogr. A*, **750**, 209 (1996).
176. J. Siegrist, C. Salles and P. Etievant, *Chromatographia*, **35**, 50 (1993).
177. B. Brauer and T. Funke, *Deut. Lebensmit.-Rundsch.*, **88**, 243 (1992).
178. B. Brauer and T. Funke, *Deut. Lebensmit.-Rundsch.*, **91**, 146 (1995).
179. M. T. García Sanchez, J. L. Pérez Pavón and B. Moreno Cordero, *J. Chromatogr. A*, **766**, 61 (1997).
180. M. E. Fernández Laespada, J. L. Pérez Pavón and B. Moreno Cordero, *J. Chromatogr. A*, **852**, 395 (1999).
181. M. E. Fernández Laespada, J. L. Pérez Pavón and B. Moreno Cordero, *J. Chromatogr. A*, **823**, 537 (1998).
182. K. M. S. Sundaram, *J. Liq. Chromatogr.*, **18**, 1787 (1995).
183. V. Cheynier, J. M. Souquet, E. Le Roux, S. Guyot and J. Rigaud, *Methods Enzymol.*, **299**, 158 (1999).
184. M. L. Menezes and A. C. C. O. Demarchi, *J. Liq. Chromatogr. Relat. Technol.*, **21**, 2355 (1998).
185. I. Barlow, G. T. Lloyd and E. H. Ramshaw, *Aust. J. Dairy Technol.*, **48**, 113 (1993).
186. T. Macko, R. Siegl and K. Lederer, *Angew. Makromol. Chem.*, **227**, 179 (1995).
187. J. Itoh, Y. Hirosawa and M. Komata, *Bunseki Kagaku*, **43**, 959 (1994); *Chem. Abstr.*, **122**, 16690 (1995).
188. G. Kossmehl and K. Buche, *J. Macromol. Sci.*, **A30**, 331 (1993).
189. S. Gottlieb and P. B. Marsh, *Ind. Eng. Chem. Anal. Ed.*, **18**, 16 (1946).
190. W. Frenzel and S. Krekler, *Anal. Chim. Acta*, **310**, 437 (1995).
191. W. J. Landzettel, K. J. Hargis, J. B. Caboot, K. L. Adkins, T. G. Strein, H. Veening and H. D. Becker, *J. Chromatogr. A*, **718**, 45 (1995).
192. K. Imai, T. Fukushima and H. Yokosu, *Biomed. Chromatogr.*, **8**, 107 (1994).
193. K. Nakashima, S. Kinoshita, M. Wada, N. Kuroda and W. R. G. Baeyens, *Analyst*, **123**, 2281 (1998).
194. M. Wada, S. Kinoshita, Y. Itayama, N. Kuroda and K. Nakashima, *J. Chromatogr. B*, **721**, 179 (1999).
195. Y. Tsuruta, S. Watanabe and H. Inoue, *Anal. Biochem.*, **243**, 86 (1996).
196. Y. Tsuruta, S. Kitai, S. Watanabe and H. Inoue, *Anal. Biochem.*, **280**, 36 (2000).
197. F. García Sánchez, A. Navas Díaz, J. Lovillo and L. S. Feria, *Anal. Chim. Acta*, **328**, 73 (1996).
198. J. Verdú-Andrés, F. Bosch-Reig and P. Campíns-Falcó, *Chromatographia*, **42**, 283 (1996).
199. P. Chaminade, A. Baillet, D. Ferrier, B. Bourguignon and D. L. Massart, *Anal. Chim. Acta*, **280**, 93 (1993).
200. T. Hanai, H. Hatano, N. Nimura and T. Kinoshita, *Analyst*, **119**, 1167 (1994).
201. D. Puig and D. Barceló, *Anal. Chim. Acta*, **311**, 63 (1995).
202. N. Yagoubi, A. Baillet, C. Mur and D. Baylocq-Ferrier, *Chromatographia*, **35**, 455 (1993).
203. C. Bocchi, M. Careri, F. Groppi, A. Mangia, P. Manini and G. Mori, *J. Chromatogr. A*, **753**, 157 (1996).
204. K. R. Rogers, J. Y. Becker, J. Wang and F. Lu, *Field Anal. Chem. Technol.*, **3**, 161 (1999).
205. M. H. Liu, S. Kapila, K. S. Nam and A. A. Elseewi, *J. Chromatogr.*, **639**, 151 (1993).
206. D. Puig, D. Barcelo, I. Silgoner and M. Grasserbauer, *J. Mass Spectrom.*, **31**, 1297 (1996).
207. S. F. Dressman, A. M. Simeone and A. C. Michael, *Anal. Chem.*, **68**, 3121 (1996).
208. S. R. Wallenborg, K. E. Markides and L. Nyholm, *J. Chromatogr. A*, **785**, 121 (1997).
209. E. Pocurull, R. M. Marcé, F. Borrull, J. L. Bernal, L. Toribio and M. L. Serna, *J. Chromatogr. A*, **755**, 67 (1996).
210. R. Boussenadji, M. Porthault and A. Berthod, *J. Pharm. Biomed. Anal.*, **11**, 71 (1993).
211. V. Campo Dall'Orto, C. Danilowicz, S. Sobral, A. Lo Balbo and I. Rezzano, *Anal. Chim. Acta*, **336**, 195 (1996).
212. M. A. Ruiz, E. García-Moreno, C. Barbas and J. M. Pingarrón, *Electroanalysis*, **11**, 470 (1999).
213. W. Loh, C. A. Tonegutti and P. L. Q. Volpe, *J. Chem. Soc., Faraday Trans.*, **89**, 113 (1993).
214. A. D. Karpyuk, N. V. Bragina, Z. N. Ignatova and I. L. Poletaeva, *J. Anal. Chem.*, **50**, 1149 (1995).

215. A. L. Khurana, in *Phenolic Compounds in Food and Their Effects on Health* (Eds. C.-T. Ho, C. Y. Lee and M.-T. Huang), American Chemical Society, Washington, *Am. Chem. Soc. Symp. Ser.*, **506**, 77 (1992).
216. G. Achilli, G. P. Cellerino, P. H. Gamache and G. V. M. Deril, *J. Chromatogr.*, **632**, 111 (1993).
217. A. Escarpa and M. C. González, *J. Chromatogr. A*, **897**, 161 (2000).
218. S. Shahrzad and I. Bitsch, *J. Chromatogr. A*, **741**, 223 (1996).
219. Y. Amakura, M. Okada, S. Tsuji and Y. Tonogai, *J. Chromatogr. A*, **891**, 183 (2000).
220. P. D. Bremner, C. J. Blacklock, G. Paganga, W. Mullen, C. A. Rice-Evans and A. Crozier, *Free Radical Res.*, **32**, 549 (2000).
221. S. Kermasha, M. Goetghebeur, J. Dumont and R. Couture, *Food Res. Int.*, **28**, 245 (1995).
222. S. Kermasha, M. Goetghebeur and J. Dumont, *J. Agric. Food Chem.*, **43**, 708 (1995).
223. D. Blanco Gomis, N. Fraga Palomino and J. J. Mangas Alonso, *Anal. Chim. Acta*, **426**, 111 (2001).
224. A. Schieber, P. Keller and R. Carle, *J. Chromatogr. A*, **910**, 265 (2001).
225. W. Oleszek, M. J. Amiot and S. Y. Aubert, *J. Agric. Food Chem.*, **42**, 1261 (1994).
226. S. Shahrzad and I. Bitsch, in *Vitamine und Zusatzstoffe in der Ernährung von Mensch und Tiere* (Eds. R. Schubert, G. Flachowsky and R. Bitsch), 5th Symposium, Jena, Sep. 28–29 (1995); p. 567.
227. P. Gamache, E. Ryan and I. N. Acworth, *J. Chromatogr.*, **635**, 143 (1993).
228. J. A. Kennedy and A. L. Waterhouse, *J. Chromatogr. A*, **866**, 25 (2000).
229. A. L. Waterhouse, S. P. Price and J. D. McCord, in Reference 21, p. 113.
230. D. M. Goldberg and G. J. Soleas, *Methods Enzymol.*, **299**, 122 (1999).
231. E. Revill and J. M. Ryan, *J. Chromatogr. A*, **881**, 461 (2000).
232. C. G. Barroso, M. C. Rodríguez, D. A. Guillén and J. A. Pérez-Bustamante, *J. Chromatogr. A*, **724**, 125 (1996).
233. P. Ho, T. A. Hogg and M. C. M. Silva, *Food Chem.*, **64**, 115 (1999).
234. P. Viñas, C. López-Erroz, J. J. Marín-Hernández and M. Hernández-Córdoba, *J. Chromatogr. A*, **871**, 85 (2000).
235. S. Pérez-Magariño, I. Revilla, M. L. González-SanJosé and S. Beltrán, *J. Chromatogr. A*, **847**, 75 (1999).
236. A. Cappiello, G. Famiglini, F. Mangani, M. Careri, P. Lombardi and C. Mucchino, *J. Chromatogr. A*, **855**, 515 (1999).
237. J. F. Noguera-Ortí, R. M. Villanueva-Camañas and G. Ramis-Ramos, *Anal. Chim. Acta*, **402**, 81 (1999).
238. A. Aparicio, M. P. San Andrés and S. Vera, *J. High Resolut. Chromatogr.*, **23**, 324 (2000).
239. A. De Nino, N. Lombardo, E. Perri, A. Procopio, A. Raffaelli and G. Sindona, *J. Mass Spectrom.*, **32**, 533 (1997).
240. D. Ryan, K. Robards and S. Lavee, *J. Chromatogr. A*, **832**, 87 (1999).
241. T. Goto and Y. Yoshida, *Methods Enzymol.*, **299**, 107 (1999).
242. P. C. H. Hollman, D. P. Venema, S. Khokhar and I. C. W. Arts, *Methods Enzymol.*, **299**, 202 (1999).
243. M. Rafecas, F. Guardiola, M. Illera, R. Codony and J. Boatella, *J. Chromatogr. A*, **822**, 305 (1998).
244. E. Joerg and G. Sontag, *J. Chromatogr.*, **635**, 137 (1993).
245. P. Andrade, F. Ferreres and M. T. Amaral, *J. Liq. Chromatogr. Relat. Technol.*, **20**, 2281 (1997).
246. N. Cardellicchio, S. Cavalli, V. Piangerelli, S. Giandomenico and P. Ragone, *Fresenius J. Anal. Chem.*, **358**, 749 (1997).
247. S. Angelino and M. C. Gennaro, *Anal. Chim. Acta*, **346**, 61 (1997).
248. M. Kuribayashi, N. Shinozuka, N. Takai, Y. Shima and F. Mashige, *Bunseki Kagaku*, **43**, 1001 (1994); *Chem. Abstr.*, **122**, 305541 (1995).
249. J. M. Lobbes, H. P. Fitznar and G. Kattner, *Anal. Chem.*, **71**, 3008 (1999).
250. W. Fischer, O. Bund and H. E. Hauck, *Fresenius J. Anal. Chem.*, **354**, 889 (1996).
251. I. Baranowska and C. Pieszko, *J. Planar Chromatogr.*, **11**, 119 (1998).
252. N. A. Penner, P. N. Nesterenko, A. V. Khryaschevsky, T. N. Stranadko and O. A. Shpigun, *Mendeleev Commmun.*, 24 (1998).

253. M. P. San Andrés, M. E. León-González, L. V. Pérez-Arribas and L. M. Polo-Díez, *J. High Resolut. Chromatogr.*, **23**, 367 (2000).
254. E. Pocurull, R. M. Marcé and F. Borrull, *J. Chromatogr. A*, **738**, 1 (1996).
255. J. Lehotay, M. Baloghova and S. Hatrik, *J. Liq. Chromatogr.*, **16**, 999 (1993).
256. J. L. Bernal, M. J. Nozal, L. Toribio, M. L. Serna, F. Borrull, R. M. Marcé and E. Pocurull, *Chromatographia*, **46**, 295 (1997).
257. L. V. Pérez-Arribas, M. E. León-González and L. M. Polo-Díez, *Analusis*, **22**, 369 (1994).
258. G. Lamprecht and J. F. K. Huber, *J. Chromatogr. A*, **667**, 47 (1994).
259. O. Fiehn and M. Jekel, *J. Chromatogr. A*, **769**, 189 (1997).
260. H. J. Kretzschmar, V. Neyen and B. Fritz, *Deut. Lebensmit.-Rundsch.*, **91**, 273 (1995).
261. J. Ruana, I. Urbe and F. Borrull, *J. Chromatogr. A*, **655**, 217 (1993).
262. D. Puig and D. Barceló, *J. Chromatogr. A*, **778**, 313 (1997).
263. G. Achilli, G. P. Cellerino, G. Melzi d'Eril and S. Bird, *J. Chromatogr. A*, **697**, 357 (1995).
264. J. Bladek, A. Rostkowski and S. Neffe, *J. Planar Chromatogr.*, **11**, 330 (1998).
265. Y. S. Fung and Y. H. Long, *J. Chromatogr. A*, **907**, 301 (2001).
266. C. Webster, M. Smith, P. Wilson and M. Cooke, *J. High Resolut. Chromatogr.*, **16**, 549 (1993).
267. E. Conde, E. Cadahía and M. C. García-Vallejo, *Chromatographia*, **41**, 657 (1995).
268. N. Okamura, M. Asai, N. Hine and A. Yagi, *J. Chromatogr. A*, **746**, 225 (1996).
269. M. Carini, R. M. Facino, G. Aldini, M. Calloni and L. Colombo, *Rapid Comm. Mass Spectrom.*, **12**, 1813 (1998).
270. J. L. Wolfender, M. Maillard and K. Hostettmann, *J. Chromatogr.*, **647**, 183 (1993).
271. M. Medicsaric and Z. Males, *J. Planar Chromatogr.*, **10**, 182 (1997).
272. I. Fecka and W. Cisowski, *Chromatographia*, **49**, 256 (1999).
273. D. Summanen, T. Yrjonen, R. Hiltunen and H. Vuorela, *J. Planar Chromatogr.*, **11**, 421 (1998).
274. C. Li, M. Homma and K. Oka, *J. Chromatogr. B*, **693**, 191 (1997).
275. L. Bianchi, M. A. Colivicchi, L. Della Corte, M. Valoti, G. P. Sgaragli and P. Bechi, *J. Chromatogr. B*, **694**, 359 (1997).
276. L. C. Bourne and C. A. Rice-Evans, *Methods Enzymol.*, **299**, 91 (1999).
277. S. A. J. Coolen, M. Van Lieshout, J. C. Reijenga and F. A. Huf, *J. Microcol. Separation*, **11**, 701 (1999).
278. S. A. J. Coolen, T. Ligor, M. van Lieshout and F. A. Huf, *J. Chromatogr. B*, **732**, 103 (1999).
279. T. Niwa, *Clin. Chem.*, **39**, 108 (1993).
280. J. Norberg, J. Emnéus, J. Å. Jönsson, L. Mathiasson, E. Burestedt, M. Knutsson and G. Marko-Varga, *J. Chromatogr. B*, **701**, 39 (1997).
281. A. M. Birkett, G. P. Jones and J. G. Muir, *J. Chromatogr. B*, **674**, 187 (1995).
282. E. Fischer, G. Henze and K. L. Platt, *Fresenius J. Anal. Chem.*, **360**, 95 (1998).
283. J.-F. Jen, J.-H. Zen, F.-C. Cheng and G.-Y. Yang, *Anal. Chim. Acta*, **292**, 23 (1994).
284. F. Ortega, J. L. Cuevas, J. I. Centenera and E. Domínguez, *J. Pharm. Biomed. Anal.*, **10**, 789 (1992).
285. M. J. Carrott and G. Davidson, *Analyst*, **124**, 993 (1999).
286. C. Puglisi, F. Samperi, S. Carroccio and G. Montaudo, *Rapid Comm. Mass Spectrom.*, **13**, 2268 (1999).
287. L. A. Ohlemeier and W. K. Gavlick, *J. Liq. Chromatogr.*, **18**, 1833 (1995).
288. M. Bernabei, G. Bocchinfuso, P. Carrozzo and C. De Angelis, *J. Chromatogr. A*, **871**, 235 (2000).
289. T. Andersson, T. Hyötylainen and M. L. Riekkola, *J. Chromatogr. A*, **896**, 343 (2000).
290. C.-S. Lu and S.-D. Huang, *J. Chromatogr. A*, **696**, 201 (1995).
291. T. Hanai, K. Koizumi, T. Kinoshita, R. Arora and F. Ahmed, *J. Chromatogr. A*, **762**, 55 (1997).
292. K. Nakamura, K. Hayashi, I. Ueda and H. Fujiwara, *Chem. Pharm. Bull.*, **43**, 369 (1995).
293. P. Kuban, M. Berg, C. García and B. Karlberg, *J. Chromatogr. A*, **912**, 163 (2001).
294. I. Rodríguez, M. I. Turnes, H. M. Bollain, M. C. Mejuto and R. Cela, *J. Chromatogr. A*, **778**, 279 (1997).
295. D. Martínez, F. Borrull and M. Calull, *J. Chromatogr. A*, **788**, 185 (1997).
296. S. Morales and R. Cela, *J. Chromatogr. A*, **846**, 401 (1999).
297. S. Morales and R. Cela, *J. Chromatogr. A*, **896**, 95 (2000).

298. J. H. T. Luong, A. Hilmi and A.-L. Nguyen, *J. Chromatogr. A*, **864**, 323 (1999).
299. Y.-C. Chao and C.-W. Whang, *J. Chromatogr. A*, **663**, 229 (1994).
300. W. E. Rae and C. A. Lucy, *J. AOAC Int.*, **80**, 1308 (1997).
301. J. Wang, M. P. Chatrathi and B. M. Tian, *Anal. Chim. Acta*, **416**, 9 (2000).
302. I.-C. Chen and C.-W. Whang, *J. Chin. Chem. Soc.*, **41**, 419 (1994).
303. H. Harino, S. Tsunoi, T. Sato and M. Tanaka, *Anal. Sci.*, **16**, 1349 (2000).
304. F. Lafont, M. A. Aramendia, I. García, V. Borau, C. Jiménez, J. M. Marinas and F. J. Urbano, *Rapid Comm. Mass Spectrom.*, **13**, 562 (1999).
305. F. B. Erim and I. Alemdar, *Fresenius J. Anal. Chem.*, **361**, 455 (1998).
306. S. Kar and P. K. Dasgupta, *J. Chromatogr. A*, **739**, 379 (1996).
307. T. Watanabe, A. Yamamoto, S. Nagai and S. Terabe, *J. Chromatogr. A*, **793**, 409 (1998).
308. A. Versari, G. P. Parpinello and S. Galassi, *Ann. Chim. (Rome)*, **89**, 901 (1999).
309. M. I. Gil, C. García-Viguera, P. Bridle and F. A. Thomas-Barberán, *Z. Lebensmit.-Unters. Forsch.*, **200**, 278 (1995).
310. C. García-Viguera and P Bridle, *Food Chem.*, **54**, 349 (1995).
311. S. Moane, S. Park, C. E. Lunte and M. R. Smyth, *Analyst*, **123**, 1931 (1998).
312. J. Summanen, H. Vuorela, R. Hiltunen, H. Sirén and M. L. Riekkola, *J. Chromatogr. Sci.*, **33**, 704 (1995).
313. P. Andrade, F. Ferreres, M. I. Gil and F. A. Tomás-Barberán, *Food Chem.*, **60**, 79 (1997).
314. Z. L. Chen, G. S. R. Krishnamurti and R. Naidu, *Chromatographia*, **53**, 179 (2001).
315. D. Volgger, A. Zemann and G. Bonn, *J. High Resolut. Chromatogr.*, **21**, 3 (1998).
316. T. Javor, W. Buchberger and I. Tanzcos, *Mikrochim. Acta*, **135**, 45 (2000).
317. C. Rony, J. C. Jacquier and P. L. Desbene, *J. Chromatogr. A*, **669**, 195 (1994).
318. T. Takayanagi and S. Motomizu, *Chem. Lett.*, 14 (2001).
319. S. S. Rosatto, R. Sanches Freire, N. Duran and L. T. Kubota, *Quim. Nova*, **24**, 77 (2001); *Chem. Abstr.*, **135**, 338544 (2001).
320. G. Marko-Varga, J. Emnéus and T. Ruzgas, *Trends Anal. Chem.*, **14**, 319 (1995).
321. L. Campanella, Y. Su, M. Tomassetti, G. Crescentini and M. P. Sammartino, *Analusis*, **22**, 58 (1994).
322. P. Önnerfjord, J. Emnéus, G. Marko-Varga, L. Gorton, F. Ortega and E. Domínguez, *Biosensors Bioelectronics*, **10**, 607 (1995).
323. J. Li, L. S. Chia, N. K. Goh and S. N. Tan, *Anal. Chim. Acta*, **362**, 203 (1998).
324. A. Lindgren, T. Ruzgas, J. Emnéus, E. Csöregi, L. Gorton and G. Marko-Varga, *Anal. Lett.*, **29**, 1055 (1996).
325. M. Szewczynska and M. Trojanowicz, *Chem. Anal. (Warsaw)*, **45**, 667 (2000).
326. V. C. Dall'Orto, C. Danilowicz, I. Rezzano, M. Del Carlo and M. Mascini, *Anal. Lett.*, **32**, 1981 (1999).
327. H. Kotte, B. Grundig, K. D. Vorlop, B. Strehlitz and U. Stottmeister, *Anal. Chem.*, **67**, 65 (1995).
328. B. Wang and S. Dong, *J. Electroanal. Chem.*, **487**, 45 (2000).
329. C. Nistor, J. Emnéus, L. Gorton and A. Ciucu, *Anal. Chim. Acta*, **387**, 309 (1999).
330. F. Ortega, E. Domínguez, G. Jönsson-Pettersson and L. Gorton, *J. Biotechnol.*, **31**, 289 (1993).
331. A. Kaisheva, I. Iliev, R. Kazareva, S. Christov, U. Wollenberger and F. Scheller, *Sensors Actuators, B*, **33**, 39 (1996).
332. E. S. M. Lutz and E. Domínguez, *Electroanalysis*, **8**, 117 (1996).
333. A. I. Yaropolov, A. N. Kharybin, J. Emnéus, G. Marko-Varga and L. Gorton, *Anal. Chim. Acta*, **308**, 137 (1995).
334. L. Campanella, T. Beone, M. P. Sammartino and M. Tomassetti, *Analyst*, **118**, 979 (1993).
335. E. Burestedt, J. Emnéus, L. Gorton, G. Marko-Varga, E. Domínguez, F. Ortega, A. Narváez, H. Irth, M. Lutz, D. Puig and D. Barceló, *Chromatographia*, **41**, 207 (1995).
336. A. Wolf, R. Huttl and G. Wolf, *J. Therm. Anal.*, **61**, 37 (2000).
337. A. Lindgren, J. Emnéus, T. Ruzgas, L. Gorton and G. Marko-Varga, *Anal. Chim. Acta*, **347**, 51 (1997).
338. S. Cosnier and I. C. Popescu, *Anal. Chim. Acta*, **319**, 145 (1996).
339. S. A. Kane, E. I. Iwuoba and M. R. Smyth, *Analyst*, **123**, 2001 (1998).
340. A. Courteix and A. Bergel, *Enzyme Microb. Technol.*, **17**, 1087 (1995).

341. I. G. Gazaryan, D. B. Loginov, A. L. Lialulin and T. N. Shekhovtsova, *Anal. Lett.*, **27**, 2917 (1994).
342. N. A. Bagirova, T. N. Shekhovtsova, N. V. Tabatchikova and R. B. van Huystee, *J. Anal. Chem.*, **55**, 82 (2000).
343. J. L. Besombes, S. Cosnier, P. Labbe and G. Reverdy, *Anal. Lett.*, **28**, 405 (1995).
344. K. Streffer, E. Vijgenboom, A. W. J. W. Tepper, A. Makower, F. W. Scheller, G. W. Canters and U. Wollenberger, *Anal. Chim. Acta*, **427**, 201 (2001).
345. D. Barnett, D. G. Laing, S. Skopec, O. Sadik and G. G. Wallace, *Anal. Lett.*, **27**, 2417 (1994).
346. U. Wollenberger and B. Neumann, *Electroanalysis*, **9**, 366 (1997).
347. T. N. Shekhovtsova, A. L. Lyalyulin, E. I. Kondrateva, I. G. Gazaryan and I. F. Dolmanova, *J. Anal. Chem.*, **49**, 1191 (1994).
348. A. W. O. Lima, E. K. Vidsiunas, V. B. Nascimento and L. Angnes, *Analyst*, **123**, 2377 (1998).
349. M. Ozsoz, A. Erdem, E. Kilinc and L. Gokgunnec, *Electroanalysis*, **8**, 147 (1996).
350. I. da Cruz Vieira and O. Fatibello-Filho, *Anal. Lett.*, **30**, 895 (1997).
351. K. Riedel, B. Beyersdorf-Radeck, B. Neumann, F. Schaller and F. Scheller, *Appl. Microbiol. Biotechnol.*, **43**, 7 (1995).
352. K. Riedel, J. Hensel, S. Rothe, B. Neumann and F. Scheller, *Appl. Microbiol. Biotechnol.*, **38**, 556 (1993).
353. E. I. Rainina, I. E. Badalian, O. V. Ignatov, A. Yu. Fedorov, A. L. Simonian and S. D. Varfolomeyev, *Appl. Biochem. Biotechnol.*, **56**, 117 (1996).
354. R. Rella, D. Ferrara, G. Barison, L. Doretti and S. Lora, *Biotech. Appl. Biochem.*, **24**, 83 (1996).
355. B. Fuhrmann and U. Spohn, *Biosensors Bioelectronics*, **13**, 895 (1998).
356. E. Valero and F. García-Carmona, *J. Mol. Catal. B*, **6**, 429 (1999).
357. A. Rose, F. W. Scheller, U. Wollenberger and D. Pfeiffer, *Fresenius J. Anal. Chem.*, **369**, 145 (2001).
358. I. M. Russell and S. G. Burton, *Anal. Chim. Acta*, **389**, 161 (1999).
359. I. D. Vieira and O. Fatibello, *Anal. Chim. Acta*, **366**, 111 (1998).
360. M. C. Ramos, M. C. Torijas and A. Navas Díaz, *Sensors Actuators, B*, **73**, 71 (2001).
361. S. H. Choi and M. B. Gu, *Environ. Toxicol. Chem.*, **20**, 248 (2001).
362. M. L. Agüí, A. J. Reviejo, P. Yáñez-Sedeno and J. M. Pingarrón, *Anal. Chem.*, **67**, 2195 (1995).
363. I. Naranjo Rodríguez, J. A. Muñoz Leyva and J. L. Hidalgo Hidalgo de Cisneros. *Analyst*, **122**, 601 (1997).
364. H. Wang, A. Zhang, H. Cui, D. Liu and R. Liu, *Microchem. J.*, **59**, 448 (1998).
365. T. Mafatle and T. Nyokong, *Anal. Chim. Acta*, **354**, 307 (1997).
366. C. de la Fuente, J. A. Acuña, M. D. Vázquez, M. L. Tascón and P. Sánchez Batanero, *Talanta*, **49**, 441 (1999).
367. S. H. Kim, M. S. Won and Y. B. Shim, *Bull. Korean Chem. Soc.*, **17**, 342 (1996).
368. Q. Fulian and R. G. Compton, *Analyst*, **125**, 531 (2000).
369. M. E. Cuvelier, H. Richard and C. Berset. *Sci. Aliment.*, **10**, 797 (1990).
370. W. Brand-Williams, M. E. Cuvelier and C. Berset. *Lebensm.-Wiss. Technol.*, **28**, 25 (1995).
371. M. N. Peyrat-Maillard, S. Bonnely and C. Berset, *Talanta*, **51**, 709 (2000).
372. H.-Y. Zhang, Y.-M. Sun, G.-Q. Zhang and D.-Z. Chen, *Quant. Struct.-Act. Relat.*, **19**, 375 (2000).
373. R. J. Vaz, M. Edwards, J. Shen, R. Pearlstein and D. Kominos, *Int. J. Quantum Chem.*, **75**, 187 (1999).
374. Ya. I. Korenman, T. N. Ermolaeva, T. A. Kuchmenko and A. V. Mishina, *J. Anal. Chem.*, **49**, 1069 (1994).
375. M. J. Christophersen and T. J. Cardwell, *Anal. Chim. Acta*, **323**, 39 (1996).
376. A. Guiberteau Cabanillas, T. Galeano Díaz, F. Salinas, J. M. Ortiz and J. M. Kauffmann, *Electroanalysis*, **9**, 952 (1997).
377. S. Y. Sheikheldin, T. J. Cardwell, R. W. Cattrall, M. D. Luque de Castro and S. D. Kolev, *Anal. Chim. Acta*, **419**, 9 (2000).
378. H.-Y. Chen, A.-M. Yu and D. K. Xu, *Fresenius J. Anal. Chem.*, **359**, 542 (1997).
379. I. M. Christie, P. Treloar, S. Reddy, C. Hepburn, J. Hulme and P. Vadgama, *Electroanalysis*, **9**, 1078 (1997).

380. S. Y. Sheikheldin, T. J. Cardwell, R. W. Cattrall, M. D. Luque de Castro and S. D. Kolev, *Environ. Sci. Technol.*, **35**, 178 (2001).
381. Z.-Q. Zhang, H. Zhang and Q. Deng, *Talanta*, **53**, 517 (2000).
382. Ya. I. Korenman, T. N. Ermolaeva and T. A. Kuchmenko, *J. Anal. Chem.*, **51**, 468 (1996).
383. J. C. Masini, G. Abate, E. C. Lima, L. C. Hahn, M. S. Nakamura, J. Lichtig and H. R. Nagatomy, *Anal. Chim. Acta*, **364**, 223 (1998).
384. G. Arana, N. Etxebarria and L. A. Fernández, *Fresenius J. Anal. Chem.*, **349**, 703 (1994).
385. O. Thomas and S. Gallot, *Fresenius J. Anal. Chem.*, **338**, 234 (1990).
386. O. Thomas, S. Gallot and N. Mazas, *Fresenius J. Anal. Chem.*, **338**, 238 (1990).
387. F. Martin and M. Otto, *Fresenius J. Anal. Chem.*, **352**, 451 (1995).
388. S. Torrado, S. Fraile, J. J. Torrado and E. Selles, *J. Pharm. Biomed. Anal.*, **13**, 1167 (1995).
389. A. G. Dedov, V. V. Nekrasova, N. K. Zaitsev, V. A. Khlebalkin, S. G. Suslov and L. A. Travnikova, *Pet. Chem.*, **40**, 55 (2000).
390. G. M. Smolskii and T. A. Kuchmenko, *J. Anal. Chem.*, **52**, 85 (1997).
391. P. Zhang and D. Littlejohn, *Analyst*, **118**, 1065 (1993).
392. Z. L. Zhi, A. Ríos and M. Valcárcel, *Analyst*, **121**, 1 (1996).
393. J. F. van Staden and H. E. Britz, *Fresenius J. Anal. Chem.*, **357**, 1066 (1997).
394. I. Nukatsuka, S. Nakamura, K. Watanabe and K. Ohzeki, *Anal. Sci.*, **16**, 269 (2000).
395. R. A. S. Lapa, J. L. F. C. Lima and I. V. O. S. Pinto, *Analusis*, **28**, 295 (2000).
396. Y. Chen, S. Wang and T. Zhou, *Anal. Lett.*, **31**, 1233 (1998).
397. R. A. S. Lapa, J. L. F. C. Lima and I. V. O. S. Pinto, *Int. J. Environ. Anal. Chem.*, **76**, 69 (2000).
398. Y. C. Fiamegos, C. D. Stalikas, G. A. Pilidis and M. I. Karayannis, *Anal. Chim. Acta*, **403**, 315 (2000).
399. Ya. I. Korenman, T. A. Kuchmenko and S. A. Karavaev, *J. Anal. Chem.*, **53**, 258 (1998).
400. M. Pospisilova, M. Polasek and D. Svobodova, *Mikrochim. Acta*, **129**, 201 (1998).
401. W.-L. Song, Z. L. Zhi and L.-S. Wang, *Talanta*, **44**, 1423 (1997).
402. S. Hünig and K. H. Fritsch, *Ann. Chem.*, **609**, 143 (1959).
403. H. O. Friestad, D. E. Ott and F. A. Gunther, *Anal. Chem.*, **41**, 1750 (1969).
404. W. Frenzel, J. Olesky-Frenzel and J. Möller, *Anal. Chim. Acta*, **261**, 253 (1992).
405. T. Watanabe, T. Ishii, Y. Yoshimura and H. Nakazawa, *Anal. Chim. Acta*, **341**, 257 (1997).
406. M. de la Guardia, K. D. Khalaf, B. A, Hasan, A. Morales-Rubio, J. J. Arias, J. M. Garcia-Fraga, A. I. Jimenez and F. Jemenez, *Analyst*, **121**, 1321 (1996).
407. L. P. Eksperiandova, I. I. Fokina, A. B. Blank, T. I. Ivkova and B. P. Soukhomlinov, *Anal. Chim. Acta*, **396**, 317 (1999).
408. E. N. Myshak, S. G. Dmitrienko, E. N. Shapovalova, A. V. Zhigulev, O. A. Shpigun and Yu. A. Zolotov, *J. Anal. Chem.*, **52**, 939 (1997).
409. J. S. Esteve Romero, L. Alvarez Rodríguez, M. C. García Alvarez-Coque and G. Ramis-Ramos, *Analyst*, **119**, 1381 (1994).
410. L. Alvarez-Rodriguez, J. Esteve-Romero, I. Escrig-Tena and M. C. Garcia Alvarez-Coque, *J. AOAC Int.*, **82**, 937 (1999).
411. O. Agrawal and V. K. Gupta, *Microchem. J.*, **62**, 147 (1999).
412. S. M. El-Ashry, F. Belal, M. M. El-Kerdawy and D. R. El Wasseef, *Mikrochim. Acta*, **135**, 191 (2000).
413. S. G. Dmitrienko, E. N. Myshak, V. K. Runov and Yu. A. Zolotov, *Chem. Anal. (Warsaw)*, **40**, 291 (1995).
414. M. Bhattacharjee and V. K. Gupta, *Chem. Anal. (Warsaw)*, **39**, 687 (1994).
415. F. A. Nour El-Dien, *Spectrosc. Lett.*, **33**, 347 (2000).
416. M. Solar and J. Pesakova, *Chem. Listy*, **90**, 392 (1996).
417. W. Li, J. Zhou, L. Zhang, H. Corke and L. Yang, *Anal. Lett.*, **29**, 1201 (1996).
418. P. Nagaraj, J. M. Bhandari and B. N. Achar, *Indian J. Chem., Sect. A*, **32**, 641 (1993).
419. V. L. Singleton, R. Orthofer and R. M. Lamuela-Raventós, *Methods Enzymol.*, **299**, 152 (1999).
420. J. G. Ritchey and A. L. Waterhouse, *Am. J. Enol. Viticult.*, **50**, 91 (1999).
421. T. Ohno and P. R. First, *J. Environ. Qual.*, **27**, 776 (1998).
422. L. Xu and L. L. Diosady, *Food Res. Int.*, **30**, 571 (1997).
423. R. S. Bakry, A. F. M. El Walily and S. F. Belal, *Mikrochim. Acta*, **127**, 89 (1997).
424. P. David Josephy and A. Van Damme, *Anal. Chem.*, **56**, 813 (1984).

425. D. Pink, M. Naczk, K. Baskin and F. Shahidi, *J. Agric. Food Chem.*, **42**, 1317 (1994).
426. A. Labudzinska and K. Gorczynska, *J. Mol. Struct.*, **349**, 469 (1995).
427. R. M. Helm, H. P. Vogel and H. J. Neusser, *J. Chem. Phys.*, **108**, 4496 (1998).
428. M. Schmitt, J. Kupper, D Spangenberg and A. Westphal, *Chem. Phys.*, **254**, 349 (2000).
429. M. Foti and G. Ruberto, *J. Agric. Food Chem.*, **49**, 342 (2001).
430. Z. Tamura, S. Abe, K. Ito and M. Maeda, *Anal. Sci.*, **12**, 927 (1996).
431. B. Mayer, G. Marconi, C. Klein, G. Kohler and P. Wolschann, *J. Incl. Phenom. Mol. Recognition Chem.*, **29**, 79 (1997).
432. Y. Sueishi and T. Miyakawa, *J. Phys. Org. Chem.*, **12**, 541 (1999).
433. D. Landy, S. Fourmentin, M. Salome and G. Surpateanu, *J. Incl. Phenom. Macrocycl. Chem.*, **38**, 187 (2000).
434. L. Moskvin, L. N. Moskvin, A. V. Moszhuchin and V. V. Fomin, *Talanta*, **50**, 113 (1999).
435. M. del Olmo, A. Zafra, N. A. Navas and J. L. Vílchez, *Analyst*, **124**, 385 (1999).
436. W. H. Chan, A. W. M. Lee, Y. S. Lam and K. M. Wang, *Anal. Chim. Acta*, **351**, 197 (1997).
437. M. A. Reppy, M. E. Cooper, J. L. Smithers and D. L. Gin, *J. Org. Chem.*, **64**, 4191 (1999).
438. A. Sen, D. Bergvinson, S. S. Miller, J. Atkinson, R. G. Fulcher and J. T. Arnason, *J. Agric. Food Chem.*, **42**, 1879 (1994).
439. A. F. M. El Walili, A. A. K. Gazy, S. F. Belal and E. F. Khamis, *J. Pharm. Biomed. Anal.*, **20**, 643 (1999).
440. A. E. El-Gendy, *Anal. Lett.*, **33**, 2927 (2000).
441. J. Michalowski, P. Halaburda and A. Kojlo, *Anal. Lett.*, **33**, 1373 (2000).
442. H.-S. Zhuang, F. Zhang and Q.-E. Wang, *Analyst*, **120**, 121 (1995).
443. F. A. Aly, S. A. Al-Tamimi and A. A. Alwarthan, *J. AOAC Int.*, **83**, 1299 (2000).
444. O. Faix and J. H. Böttcher, *Holzforschung*, **47**, 45 (1993); *Chem. Abstr.*, **118**, 215140 (1993).
445. P. Luukko, L. Alvila, T. Holopainen, J. Rainio and T. T. Pakkanen, *J. Appl. Polym. Sci.*, **69**, 1805 (1998).
446. T. Holopainen, L. Alvila, J. Rainio and T. T. Pakkanen, *J. Appl. Polym. Sci.*, **69**, 2175 (1998).
447. K. Möller and T. Gevert, *J. Appl. Polym. Sci.*, **51**, 895 (1994).
448. B. Zhao, D. Mulkey, W. J. Brittain, Z. H. Chen and M. D. Foster, *Langmuir*, **15**, 6856 (1999).
449. K. Platteborze-Stienlet and T. Zeegers-Huyskens, *J. Mol. Struct.*, **378**, 29 (1996).
450. A. Mukherjee, M. L. McGlashen and T. G. Spiro, *J. Phys. Chem.*, **99**, 4912 (1995).
451. E. Tiainen, T. Drakenberg, T. Tamminen, K. Kataja and A. Hase, *Holzforschung*, **53**, 529 (1999); *Chem. Abstr.*, **131**, 287890 (1999).
452. M. Servili, M. Baldioli, R. Selvaggini, A. Macchioni and G. F. Montedoro, *J. Agric. Food Chem.*, **47**, 12 (1999).
453. M. Servili, M. Baldioli, R. Selvaggini, E. Miniati, A. Macchioni and G. Montedoro, *J. Am. Oil Chem. Soc.*, **76**, 873 (1999).
454. I. P. Gerothanassis, V. Exarchou, V. Lagouri, A. Troganis, M. Tsimidou and A. Boskou, *J. Agric. Food Chem.*, **46**, 4185 (1998).
455. T. Mohan and J. G. Verkade, *Energy Fuels*, **7**, 222 (1993).
456. K. Erdmann, T. Mohan and J. G. Verkade, *Energy Fuels*, **10**, 378 (1996).
457. A. Granata and D. S. Argyropoulos, *J. Agric. Food Chem.*, **43**, 1538 (1995).
458. J. Skarzynski, K. Gorczynska and I. Leszczynska, *Analyst*, **124**, 1823 (1999).
459. L. Lessinger, *J. Chem. Educ.*, **72**, 85 (1995).
460. H. Frohlich, *J. Chem. Educ.*, **70**, A3 (1993).
461. C. Scheringer, *Z. Krist.*, **119**, 273 (1963).
462. H. Gillier-Pandraud, *Bull. Soc. Chim. Fr.*, 1988 (1967).
463. J.-S. Chen, J.-C. Shiau and C.-Y. Fang, *J. Chin. Chem. Soc.*, **42**, 499 (1995).
464. Z. H. Jiang, D. S. Argyropoulos and A. Granata, *Magn. Res. Chem.*, **33**, 375 (1995).
465. F. E. Evans, J. Deck and P. C. Howard, *Chem. Res. Toxicol.*, **7**, 352 (1994).
466. D. Ryan, K. Robards, P. Prenzler, D. Jardine, T. Herlt and M. Antolovich, *J. Chromatogr. A*, **855**, 529 (1999).
467. D. Caruso, R. Colombo, R. Patelli, F. Giavarini and G. Galli, *J. Agric. Food Chem.*, **48**, 1182 (2000).
468. S. S. Alimpiev, V. V. Mlynski, M. E. Belov and S. M. Nikiforov, *Anal. Chem.*, **67**, 181 (1995).

469. J. D. Wright, J. V. Oliver, R. J. M. Nolte, S. J. Holder, N. A. J. M. Sommerdijk and P. I. Nikitin, *Sensors Actuators, B*, **51**, 305 (1998).
470. S. S. Deshpande, M. Cheyran and D. K. Salunkhe, *Crit. Rev. Food Sci. Nutr.*, **24**, 401 (1986).
471. V. L. Singleton and J. A. Rosi, *Am. J. Enol. Viticult.*, **16**, 144 (1965).
472. M. Celeste, C. Tomás, A. Cladera, J. M. Estela and V. Cerdá, *Analyst*, **118**, 895 (1993).
473. V. L. Singleton, R. Orthofer and R. M. Lamuela-Raventos, *Methods Enzymol.*, **299**, 152 (1999).
474. L. V. Rajakovic and A. E. Onjia, in *Polymers in Sensors* (Eds. N. Akmal and A. M. Usmani), American Chemical Society, Washington, *Am. Chem. Soc. Symp. Ser.*, **690**, 168 (1998).
475. Ya. I. Korenman, S. A. Tunikova, N. V. Belskikh, M. Bastic and L. Rajakovic, *J. Anal. Chem.*, **52**, 278 (1997).
476. T. A. Kuchmenko, K. V. Trivunac, L. V. Rajakovic, M. B. Bastic and Ya. I. Korenman, *J. Anal. Chem.*, **54**, 161 (1999).
477. S. M. Reddy, *Mater. Sci. Eng. C*, **12**, 23 (2000).
478. L. V. Rajakovic, M. B. Bastic, Ya. I. Korenman, S. A. Tunikova and N. V. Belskikh, *Anal. Chim. Acta*, **318**, 77 (1995).
479. J. O. Hill, N. G. Buckman and R. J. Magee, *J. Therm. Anal.*, **39**, 795 (1993).
480. A. O. Barry, W. Peng and B. Riedl, *Holzforschung*, **47**, 247 (1993); *Chem. Abstr.*, **119**, 204557 (1993).
481. M. K. N. Yenkie and G. S. Natarajan, *Separation Sci. Technol.*, **28**, 1177 (1993).
482. T. Bialopiotrowicz, P. Blanpain-Avet and M. Lalande, *Separation Sci. Technol.*, **34**, 1803 (1999).

CHAPTER **14**

Photochemistry of phenols

WILLIAM M. HORSPOOL

Department of Chemistry, The University of Dundee, Dundee DD1 4HN, Scotland, UK
Telephone: 44(0)1382 360478;   e-mail: william.horspool@tesco.net

I. INTRODUCTION	1016
II. GENERAL OBSERVATIONS	1016
A. Spectral and Luminescent Properties	1016
B. Photooxidation and Phenoxyl Radicals	1017
C. Electron Transfer Processes	1019
III. HYDROGEN TRANSFER REACTIONS	1020
A. Hydrogen Transfer in Acyl and Related Phenols	1020
B. Addition Reactions of Alkenyl Phenols	1022
C. Hydrogen Transfer in Salicylidene Derivatives	1024
D. Hydrogen Transfer in Heterocyclic Systems	1025
E. Hydrogen Transfer and Cyclization	1027
IV. CHALCONES	1027
A. Hydrogen Transfer Reactions	1027
B. Cyclizations	1028
C. Other Processes	1030
V. QUINONEMETHIDES FROM PHENOL DERIVATIVES	1032
A. *o*-Quinonemethides	1032
B. *m*- and *p*-Quinonemethides	1035
C. Quinonemethides from Biphenyl Derivatives	1035
D. Quinonemethides from Fluorenols	1038
VI. CYCLIZATIONS WITHIN *o*-ALLYLPHENOLS	1038
A. Miscellaneous Reactions	1042
B. Photolabile Protecting Groups	1043
C. Other Hydrogen Transfers	1045
VII. REARRANGEMENTS	1048
A. Skeletal Rearrangements	1048
B. Side-chain Rearrangement	1048
C. Side-chain Migration	1055
VIII. CARBENE FORMATION	1057

The Chemistry of Phenols.   Edited by Z. Rappoport
© 2003 John Wiley & Sons, Ltd   ISBN: 0-471-49737-1

	A. Elimination of Hydrogen Halides from Phenols	1057
	B. Other Carbenes	1059
IX.	CYCLOADDITIONS	1059
	A. Intermolecular Addition Reactions	1059
	B. Intramolecular Addition Reactions	1061
	C. Cyclizations Involving Aryl Radicals	1063
	D. Other Cyclizations	1067
X.	MISCELLANEOUS ADDITIONS AND ELIMINATIONS	1071
	A. Reimer–Tiemann Reaction and Related Processes	1071
	B. Reactions with Tetranitromethane and Nitrate Ion	1071
	C. Loss of Halide	1072
	D. Other Bond Fission Processes	1075
	E. Reactions of Hydroxyanthraquinones	1075
	F. Miscellaneous Processes	1077
XI.	HYDROXYCARBOXYLIC ACIDS	1079
	A. Decarboxylation	1079
	B. Reactions in the Presence of TiO_2	1080
XII.	OXIDATIONS	1080
	A. Phenol	1080
	B. Alkylphenols	1084
	C. Miscellaneous Phenols	1084
	D. Dihydroxyphenols	1084
	E. Chlorophenols	1085
XIII.	REFERENCES	1087

I. INTRODUCTION

This chapter deals with some of the photochemical processes undergone by phenols. The original article in this series dealing with these compounds was published in 1971[1]. One of the principal sources of reference material is the useful annual compendia of photochemical results published by the Royal Society of Chemistry[2]. These were used as the starting point to assemble the key areas dealing with this subject area. Much has been reported since 1971 and there is insufficient space here to record all of it. Thus the references cited are usually in the period 1980–2001. In addition a sifting process was used. The decision of whether to include an article or not was based on the reactivity exhibited by the phenol. If in the author's judgement the reaction type did not involve the phenolic system in the prime photochemical event, then usually it was excluded. Hopefully this treatment will give a flavour for the work that has been carried out and what is going on at the present time.

II. GENERAL OBSERVATIONS

A. Spectral and Luminescent Properties

Several studies have been reported that have examined the spectral and luminescent properties of phenol. These have examined the photoinduced OH bond cleavage processes occurring on excitation and the neutral, anionic (PhO$^-$) and cationic (PhOH$_2^+$ and PhOH$_3^{2+}$) forms of phenol were observed in the pH range 4–9[3]. The photoionization of phenols, such as the parent molecule and p-cresol, in alkaline aqueous solution occurs from the singlet-excited state of the phenolates[4]. Other researchers have reported that following irradiation of phenol in water, the fluorescence quantum yield decreases with increasing excitation energy and this is attributable to an enhancement of the OH cleavage reaction[5]. The effect of chlorine substitution on the spectral and luminescent properties and photolysis of phenol and its 1 : 1, 1 : 2 and 1 : 3 complexes with water was studied by quantum chemical methods. Substitution of chlorine in the *para* position of phenol decreases the fluorescence quantum yield and makes it dependent on excitation energy, in the absence of any phototransformations. Photodissociative states do exist and these lead to photocleavage of OH and CCl bonds[6]. The adsorption and photochemistry of phenol on Ag(111) has also been investigated and irradiation brings about photochemical transformations on the surface. It is likely that this is a charge transfer induced dissociation of the OH bond of phenol[7]. Furthermore, two-photon processes permit the population of highly excited states of phenol[8]. Photoelectron spectra have also been recorded. In these, after the primary excitation, a second photon excites the species to what is described as a set of superexcited molecular states[9,10].

B. Photooxidation and Phenoxyl Radicals

The transients formed from phenol (irradiation at 266 nm in ethanol) have been identified as solvated electrons, phenoxyl radicals (an absorption around 400 nm) and the triplet state of phenol (450 nm)[11]. The formation of phenoxyl radicals and hydrated electrons display a low-frequency/high-field absorption and a high-frequency (low-field) emission polarization pattern generated by a radical pair mechanism. Phenoxyl radicals have also been observed following electron transfer from phenols (as solutes) to molecular radical cations of some non-polar solvents (cyclohexane, *n*-dodecane, 1,2-dichloroethane, *n*-butyl chloride)[12]. This study used pulsed radiolysis and the formation of the phenoxyl radicals is thought to involve Scheme 1.

$$c\text{-}C_6H_{12}^{+\bullet} + ArOH \longrightarrow c\text{-}C_6H_{12} + ArOH^{+\bullet}$$

$$ArOH^{+\bullet} \longrightarrow ArO^{\bullet} + H\text{-solv}^+$$

SCHEME 1

A CIDEP study of the photooxidation of a range of phenols by benzophenone has concluded that the reaction proceeds by abstraction of a hydrogen atom to give the corresponding phenoxyl radicals[13]. Others[14] have reported that aromatic ketones such as 3-methoxyacetophenone and 2-acetonaphthone mediate efficiently the photooxidative degradation of phenols by a one-electron process producing the radical cations of the phenols in aerated aqueous solution. A possible reaction sequence is shown in equation 1.

$$PhOH + [Ar_2CO]^3 \longrightarrow PhOH^{+\bullet} + Ar_2CO^{-\bullet} \qquad (1)$$

Rates of quenching of excited state triplets have been measured and the influence of substituents on the phenols studied has shown that electron-donating substituents enhance the degradation process ($\phi > 0.5$) while phenol itself has a quantum yield for disappearance of only 0.1.

The photooxidation of 2,6-dimethylphenol[15] with UO_2^{2+} and with the oxidant[16] [Co $(NH_3)_5N_3]^{2+}$ has been investigated and the first step has been shown to be the formation of the phenoxyl radical. 2,6-Dimethylphenol gives the corresponding *p*-quinone and the dimer **1c**. In degassed solutions only the dimer is formed[15]. Dimerization of *o*-phenylphenol can also be brought about by irradiation in the presence of $[Co(NH_3)_5N_3]^{2+}$ with concomitant reduction of Co^{3+} to its Co^{2+} state[17]. Similar dimerization has been reported for *m*- and *p*-phenylphenols. The irradiation affords the corresponding phenylphenoxyl radicals that react efficiently to give phenolic dimers as the major product[18]. *o*-, *m*- and *p*-phenylphenol all undergo oxidation with UO_2^{2+}. The *o*-phenylphenol yields two dimers and a *p*-quinone. In degassed solutions, however, only the dimer is formed[19]. Another study has shown that isoeugenol also undergoes dimerization and is converted into a 7,7'-linked lignan[20]. Phenoxyl radicals also arise as key intermediates in photosensitization of hindered phenols (2,6-di-*t*-butyl, 2,6-di-*i*-propyl, 2,6-dimethyl and 2-*t*-butyl) using acridine as the sensitizer. A triplet excited-state radical pair is formed following transfer of hydrogen to acridine. An electron transfer does not occur in this system and it is proposed that the presence of *ortho* substituents promote the dimerization to afford (**1a–d**) by inhibiting electron transfer[21]. As can be seen from the yields quoted, the dimer can be produced in 75% yield when the reaction is carried out in acetonitrile. When *o*-substituents are absent, e.g. with phenol, the dimerization fails. The oxidation of 2,6-di-*t*-butylphenol in the crystalline phase produces triplet phenoxyl radical pairs[22]. Radical pairs produced by the photolysis of 4-bromo-2,6-di-*t*-butylphenol single crystals doped with 2,6-di-*t*-butyl-*p*-quinone have been studied by EPR spectroscopy. The mechanism of radical pair generation changes from hydrogen-atom transfer to electron transfer (without proton transfer)[23]. Lead dioxide will oxidize 4,4'-(trimethylene)bis(2,6-di-*t*-butylphenol) (**2**) leading to formation of a dispiro-compound (**3**) by intramolecular cyclization at the 4,4'-positions. The spiro compound **3** is photochemically reactive and, on irradiation in a methylcyclohexane matrix at −150 °C, gives 4,4'-(trimethylene)bis(2,6-di-*t*-butylphenoxy) diradical as a stable triplet species[24].

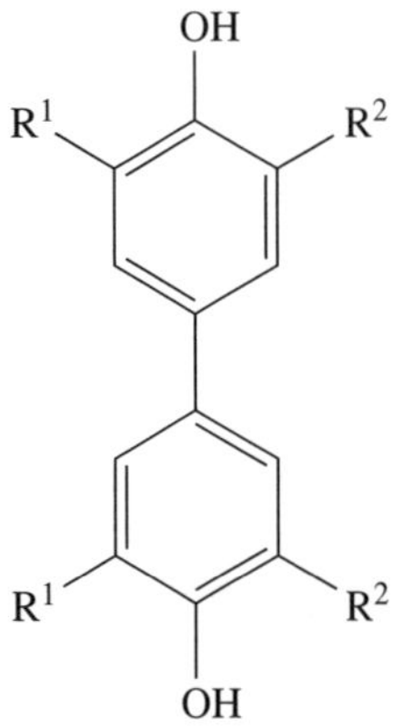

	R^1	R^2	yield (%)
(a)	*t*-Bu	*t*-Bu	75
(b)	*i*-Pr	*i*-Pr	64
(c)	Me	Me	42
(d)	H	*t*-Bu	47

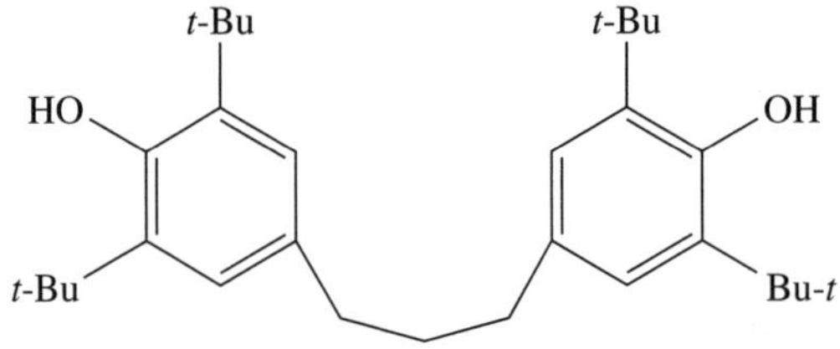

(**1**)

(**2**)

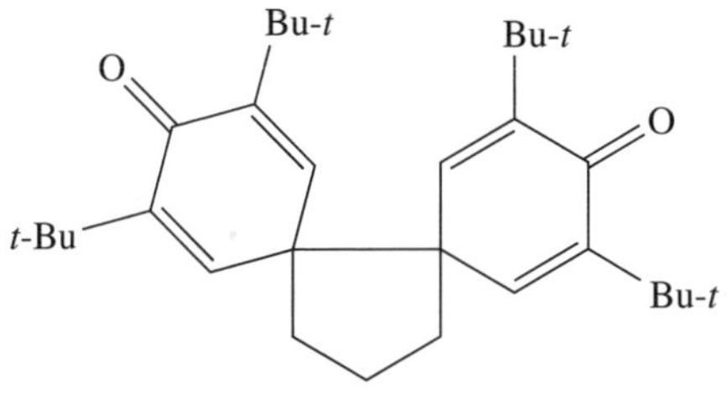

[Structure (3): spiro bis-cyclohexadienone with t-Bu substituents]

(3)

C. Electron Transfer Processes

Photoelectron transfer oxidation of phenols, 3,5-dimethyl and 2,6-dimethylphenol, takes place using 2-nitrofluorene as the electron-accepting sensitizer in both acetonitrile and cyclohexane solution. In acetonitrile the anion radical of 2,6-dimethylphenol is observed as the final product[25]. Other phenols such as the 2,4,6-trimethyl derivative also undergo electron transfer reactions with 1,1'-, 1,2'- and 2,2'-dinaphthyl ketones[26]. Other sensitizers such as 1,4-dicyanonaphthalene with biphenyl as a co-sensitizer in acetonitrile have also been used. The resultant phenol radical cations (**4a–h**) have absorption maxima in the 410–460 nm region with the exception of **4i** that absorbs at 580 nm[27]. When the reactions are carried out in the presence of a trace of water, the radical cations are not observed. Instead, phenoxyl radicals are detected. This presumably is due to the reaction shown in equation 2.

$$PhOH^{+\bullet} + H_2O \longrightarrow PhO^{\bullet} + H_3O^+ \qquad (2)$$

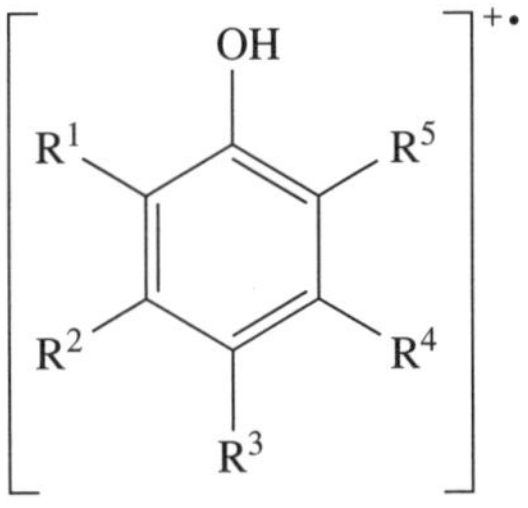

	R^1	R^2	R^3	R^4	R^5
(a)	MeO	H	H	H	H
(b)	H	MeO	H	H	H
(c)	MeO	H	H	H	MeO
(d)	H	H	MeO	H	H
(e)	H	MeO	MeO	H	H
(f)	MeO	H	Me	H	H
(g)	Me	H	Me	H	Me
(h)	H	MeO	MeO	MeO	H
(i)	H	MeO	H	MeO	H

(4)

Electron transfer oxidation of 4-methoxyphenol using *meso*-tetraphenylporphyrin as the electron acceptor brings about dehydrodimerization of the phenol to yield **5**. The presence of the radical cation of the phenol has been detected by CIDNP techniques[28]. The same product is obtained by irradiation of the tetraphenylporphyrin/benzoquinone/*p*-methoxyphenol system[29]. Pyrimidinopteridine *N*-oxide has been used as a sensitizer to effect the hydroxylation of phenols, also involving the radical cation of the phenol. Thus phenol can be converted to catechol and hydroquinone while cresol yields 4-methylcatechol[30]. Hydroquinone can itself be oxidized by the cobalt azide complex in aqueous acidic solution to generate the semiquinone radical[31]. There is also some interest in electron transfer between sterically hindered quinones and quinhydrones. In these cases the outcome can often result in either a proton transfer from the phenol to the quinone or an electron transfer in the same direction. Prokof'ev[32] has shown that in glassy media the formation of radical pairs, two paramagnetic species in the triplet state, results. In other studies in the crystalline phase with quinhydrones formed between, for example p-quinone and 2-phenylhydroquinone, the mechanism of hydrogen transfer and the involvement of charge transfer has been investigated[33]. Rate constants (8.0×10^8 s^{-1}) have been measured for photoinduced electron transfer between the hydroxyl groups of a non-covalent assembly of a calix[4]-arene (**6**)-substituted Zn(II) metalloporphyrin and benzoquinone in methylene chloride solution[34].

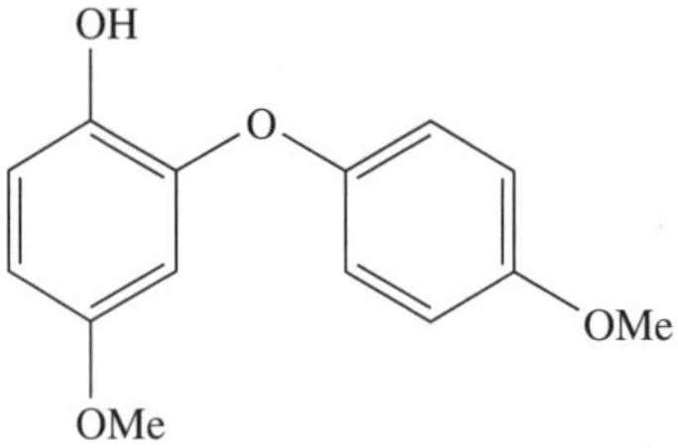

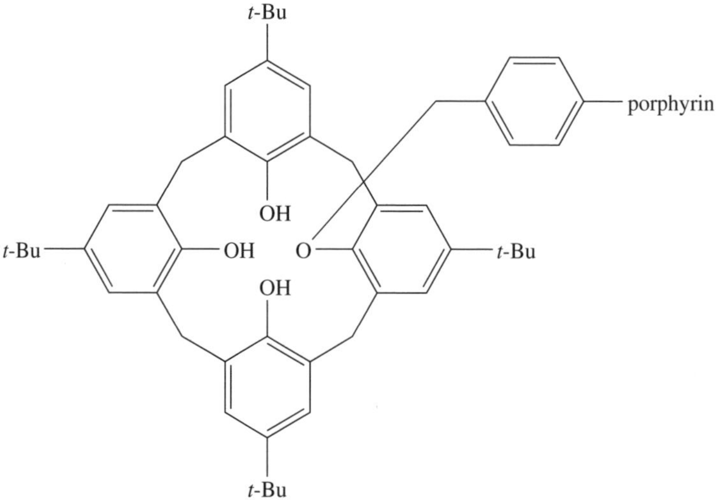

III. HYDROGEN TRANSFER REACTIONS

A. Hydrogen Transfer in Acyl and Related Phenols

Intramolecular hydrogen transfer in phenol systems has been studied in considerable detail over the years[1]. A review dealing with this subject area has also been published[35]. In the earlier studies the mechanistic details were not worked out in great detail. However, in the last decade or so considerable advances have been made in our understanding of such processes. Much of the work has been associated with photochromicity associated with the intramolecular hydrogen transfer. A laser flash study examined the process in o-hydroxyacetophenone and methyl salicylate where it is clear that a triplet state is involved[36]. Other o-acylphenols also undergo excited state hydrogen transfer and a theoretical investigation of this has been published[37]. Interest has also been shown in the photophysics of such systems and the influence that the position of the phenolic OH group can exercise on the overall processes. In this regard the o-, m- and p-derivatives (7) have been studied[38–41]. The proton transfer has been shown to be solvent sensitive[38] and there is a tendency for the formation of CT complexes in protic solvents. This involves the S^1 state of the carbonyl function. In this regard earlier work has examined the intermolecular hydrogen abstraction from phenolic hydroxy groups by photoexcited ketones[42,43], such as the reaction between benzophenone and p-cresol[43]. Another study related to this has examined geometrical effects on intramolecular quenching of aromatic ketone $\pi\pi^*$ triplets using alkoxyacetophenone derivatives **8** and **9** with remote phenolic groups. The triplet lifetimes of the phenolic ketones vary with the positions of attachment (*meta* or *para*) of the oxyethyl spacer with respect to the carbonyl and phenolic moieties. This indicates a very strong dependence of the rate of intramolecular H-abstraction on geometric factors. In these cases hydrogen-bonded triplet exciplexes are thought to be involved. A hydrogen transfer is the key chemical step in this quenching process. In the intermolecular processes the proton transfer must involve a transition state with a cyclophane-like geometry[44].

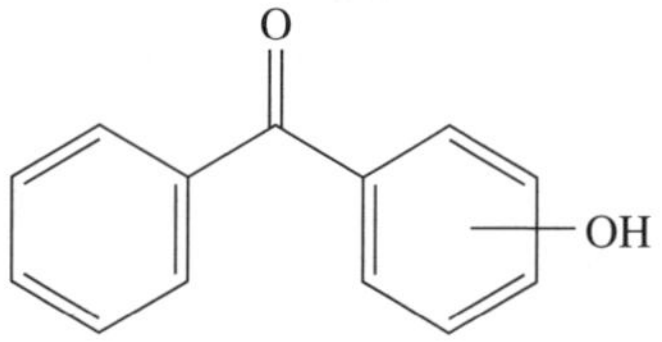

(7)

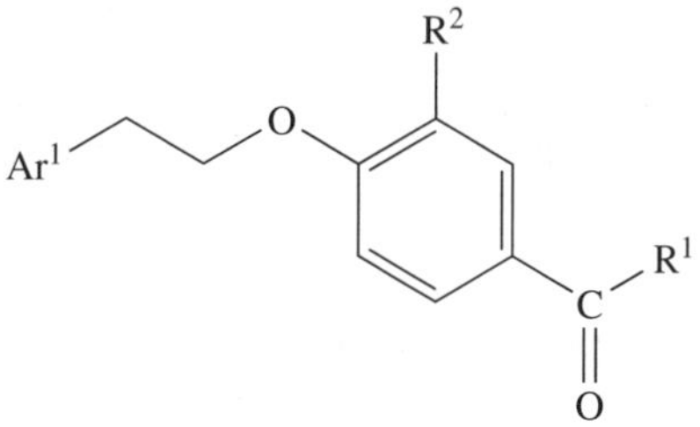

(8)

$R^1 = Me, R^2 = H, Ar^1 = p\text{-HOC}_6H_4$
$R^1 = Me, R^2 = OMe, Ar^1 = p\text{-HOC}_6H_4$
$R^1 = Ph, R^2 = H, Ar^1 = p\text{-HOC}_6H_4$

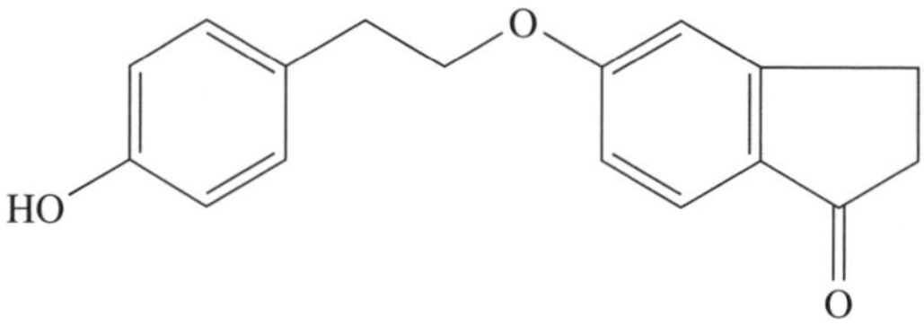

(9)

The ground and excited state proton transfer processes of 4-methyl-2,6-diacetylphenol[45–47] and the influence of polar solvents on the outcome of the photochemical transformation of 2,6-diformyl-4-methylphenol (**10**) has been evaluated[48,49]. The results obtained from the study of **10** indicate that the CO···HO hydrogen bond is stronger in the diacetyl derivative than in the diformyl compound. The hydrogen transfer occurs from both the S^1 and T^1 states. The analogous process in the more constricted environment of hydroxyindanone (**11**) in its triplet state has also been studied[50]. Excited state proton transfer from 4-methyl-2,6-diamidophenol has been studied in alcoholic solvents, using steady-state and nanosecond spectroscopy, at room temperature[51].

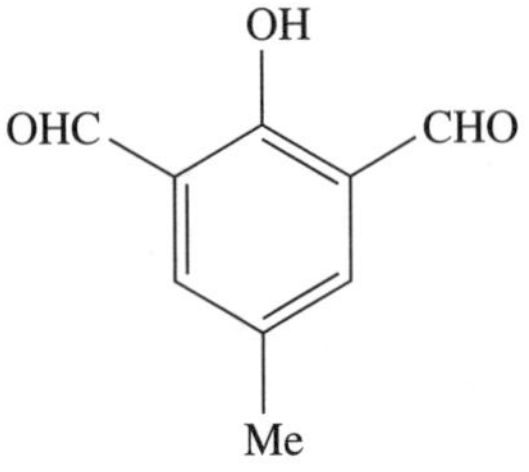

B. Addition Reactions of Alkenyl Phenols

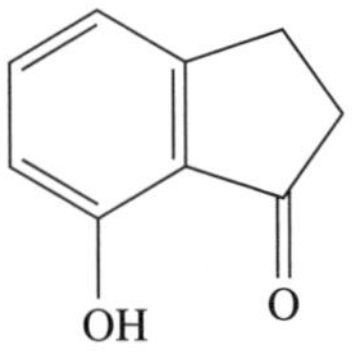

One of the interesting synthetic applications of proton transfer is the ability to bring about photochemical addition of solvents (water, alcohols and amines etc.) to the double bond adjacent to the hydroxy group. One of the early examples of this is the photohydration of 2-hydroxyphenylacetylene and 2-hydroxystyrene[52,53]. The study showed that the quantum yield for the formation of the product from the alkyne was at its highest at pH 7. It is clear that the hydration to yield o-hydroxyacetophenone from the alkyne and 2-(1-hydroxyethyl)phenol from the styrene arises by an intramolecular proton transfer from the phenolic OH under these conditions. Interestingly, the hydration of the styrene is more pH sensitive and the detailed studies have shown that the reaction can also be brought about by an intermolecular proton transfer[53]. A similar addition reaction occurs with the o-alkenyl phenols (**12, 14** and **16**). These undergo amination in good yields as shown below the structures when irradiated in the presence of alkylamines. The products formed were identified as **13, 15** and **17**, respectively. The formation of the products involves the S^1 state of the phenol resulting in transfer of the phenolic proton to the amine to afford the ion pair. Proton transfer to the alkenyl group then yields the corresponding benzylic cation that is trapped by the amine, as illustrated in Scheme 2[54,55].

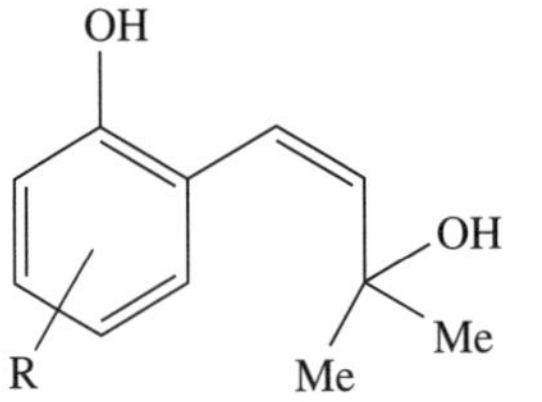

(**12**) R = H, 4-Me, 5-Me

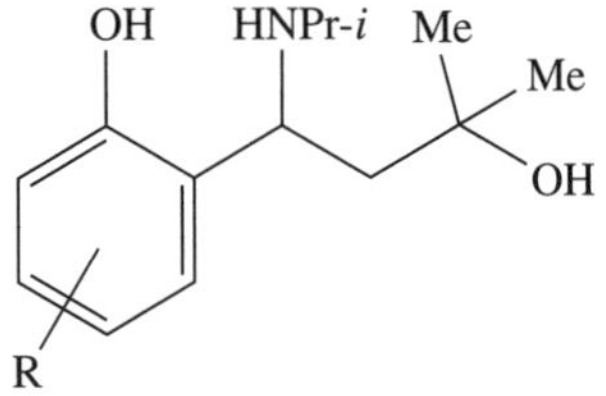

(**13**) R = H; 69% yield
R = 4-Me; 35% yield
R = 5-Me; 60% yield

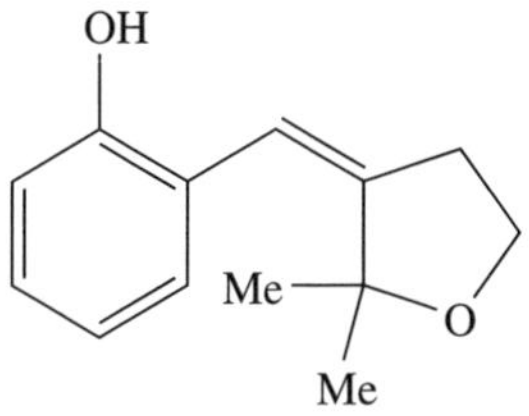

14. Photochemistry of phenols

(14)

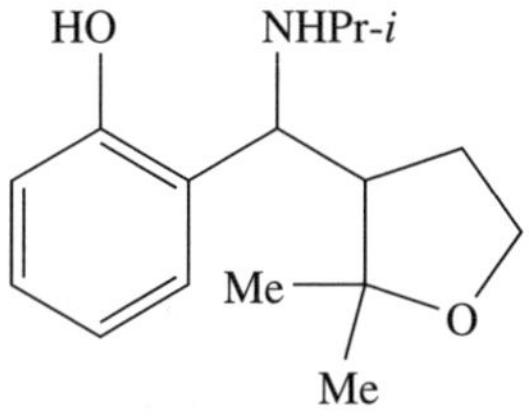

(15) 90% yield

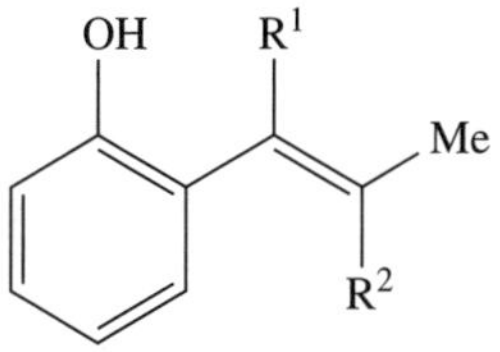

(16)

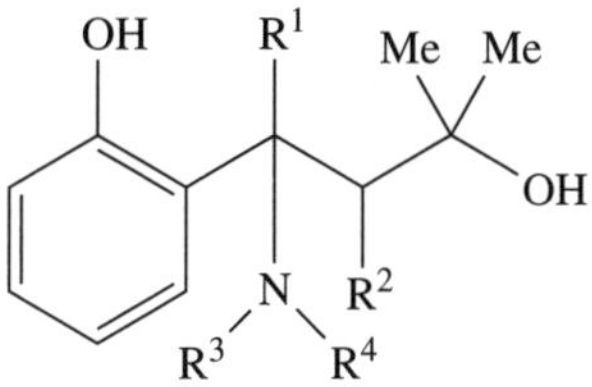

(17)

R^1	R^2	R^3	R^4	yield (%)
H	Me	Et	Et	64
H	Me	Me	H	62
H	Me	i-Pr	H	71
H	Me	H	H	35
H	Me	H	$CH_2=CH-CH_2$~	88
H	H	Et	Et	71
H	H	i-Pr	H	92
Me	H	i-Pr	H	58

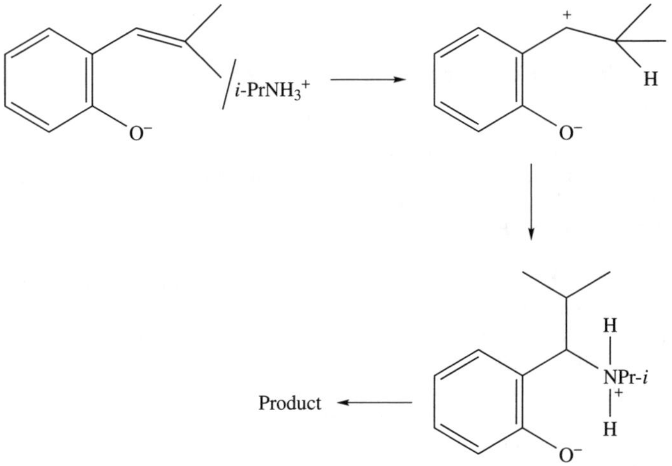

Product

SCHEME 2

Hydration has also been recorded by Fischer and Wan[56,57], who reported that the phenol derivatives **18**, **19** and **20** undergo addition of water to the double bond when they are irradiated in acetonitrile/water. The study has shown that the proposed mechanism of *m*-quinonemethide (see later for further discussion of quinonemethides) formation probably involves a solvent-mediated proton transfer of the phenolic hydrogen to the β-carbon of the alkene moiety. This must occur with the participation of a so-called water trimer. This yields the zwitterion **21** that is responsible for the formation of the products, e.g. **22** from **18**. The reactions are efficient with quantum yield values of 0.1–0.24.

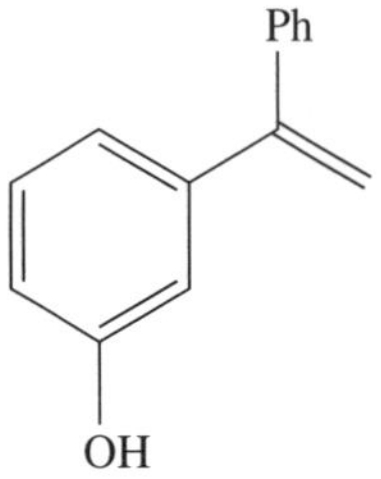

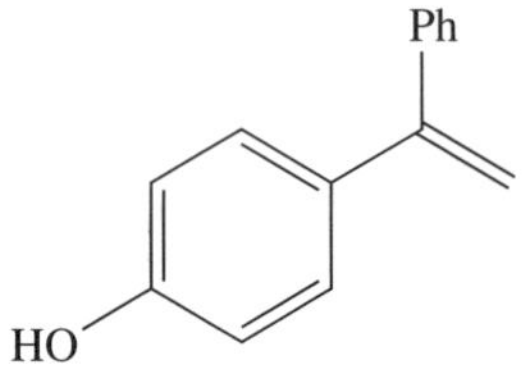

C. Hydrogen Transfer in Salicylidene Derivatives

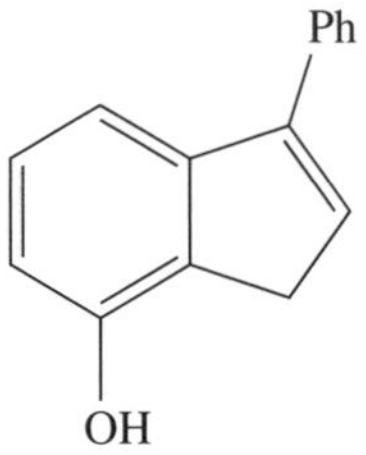

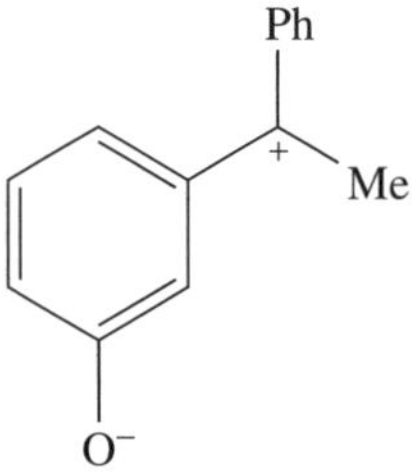

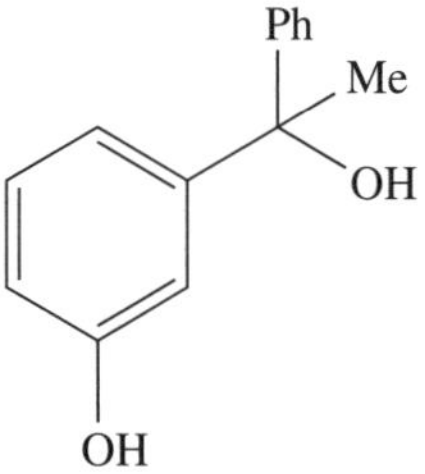

Studies associated with proton transfer in the salicylidenes and related systems have been carried out over the years. Fundamentally, the process involves the migration of the phenol proton to a neighbouring heteroatom. This simple process, that often leads to photochromism, is illustrated schematically in Scheme 3. In specific terms a solid-state study[58] and semiempirical PM3 calculations[59,60] have been carried out on the light-induced transformations of *N*-salicylideneaniline. Such isomerism has also been investigated in *N*-salicylidene-(4-*N*,*N*-dimethylamino)aniline[61] and also for the hydrogen transfer in *N*-salicylidene-(2-methyl-5-chloro)aniline in the solid state[62]. Intramolecular hydrogen transfer is reported for *N*-(*R*-salicylidene)alkylamines where the process was studied using UV-visible absorption spectroscopy[63]. Others[64] have used ^{15}N NMR to examine this proton transfer. Intramolecular proton transfer reactions in internally hydrogen-bonded Schiff bases such as *N*,*N'*-bis(salicylidene)-*p*-phenylenediamine and *N*,*N*-1-bis(2-hydroxy-1-naphthylmethylene)-*p*-phenylenediamine were studied by *ab initio* and semiempirical methods[65]. The photochromic properties and the influence that substituents have on such processes have been studied for the bis imines (**23**)[66].

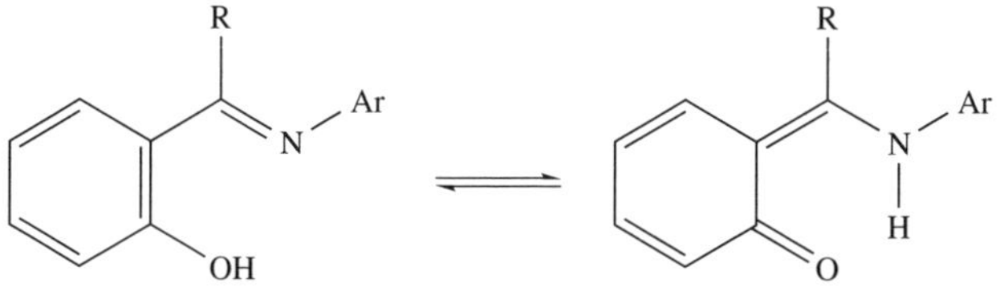

SCHEME 3

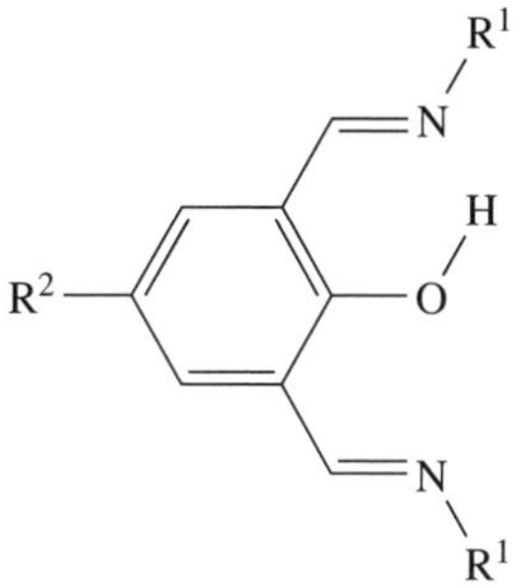

$R^1 = p\text{-MeC}_6H_4$, $R^2 = $ Me or MeO
$R^1 = $ Ph, $R^2 = $ Br
$R^1 = $ Ph, $R^2 = $ NO$_2$
$R^1 = c\text{-}C_6H_{11}$, $R^2 = $ Me
$R^1 = c\text{-}C_6H_{11}$, $R^2 = $ Br
$R^1 = c\text{-}C_6H_{11}$, $R^2 = $ NO$_2$

(23)

D. Hydrogen Transfer in Heterocyclic Systems

Proton transfer also arises from phenolic groups to the nitrogen of several heterocyclic compounds. Thus the pyridine derivative **24** shows photochemical proton transfer and the influence of restricted rotation on the process was assessed using the locked derivatives **25** and **26**[67]. Proton transfer is also observed in [2,2'-bipyridyl]-3,3'-diol[68] and within the anil of hydroxyindanone[69]. Benzimidazole derivatives such as **27**, X = CH[70–73] also undergo proton transfer in ethanol solution. Such a process had been suggested from earlier work that had detected the enhanced acidity of the phenolic hydrogen. The resultant enol is in the S^1 state and deactivation results in fluorescence. Enhanced acidity has also been observed with **27**, X = N[74,75]. The analogous processes in the imidazoles **28**[70,76] have also been studied and this yields the keto tautomers **29**. Solvent effects in the proton transfer processes in **27** have been examined by Monte Carlo simulations[77]. Apparently, polar solvents stabilize the keto forms. The benzoxazole **30**[71,78–84] undergoes facile conversion into the tautomer **31**. An analogous process occurs for the corresponding benzothiazole derivatives[85]. The quantum yield for the process is unity at 280 K, but falls with deceasing temperature to a value of 0.01 at 170 K. The influence of substituents on the phototransformation was assessed in the derivatives **32**[86]. Calculations concerning these molecules have also been reported[87]. Ground and excited state pK data have also been determined for such molecules[88]. Interestingly, photochemically induced proton transfer in the related

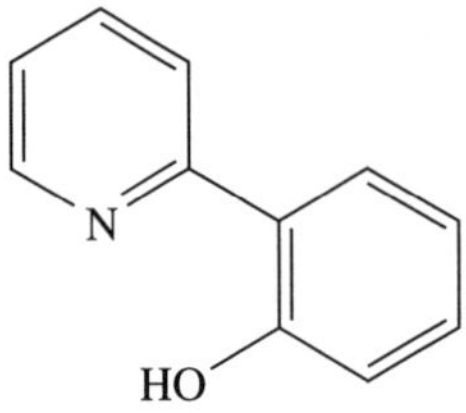

(24)

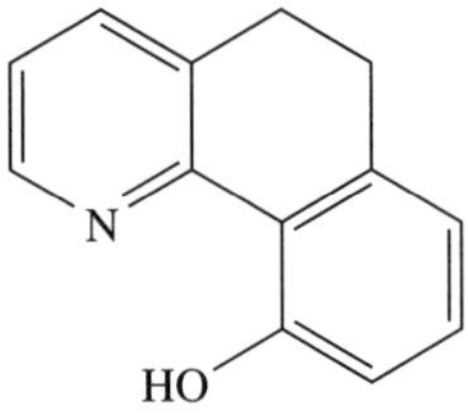

(25)

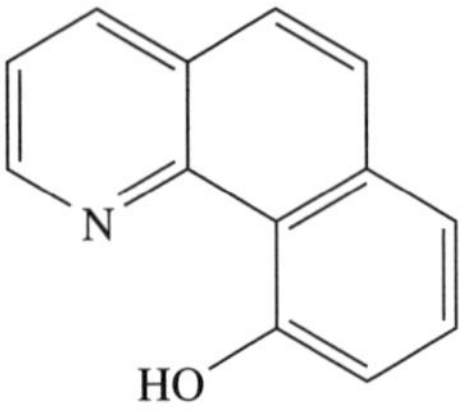

(26)

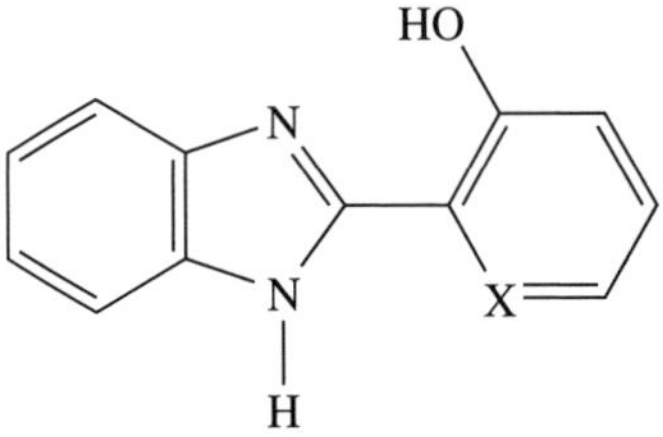

(27) X = CH or N

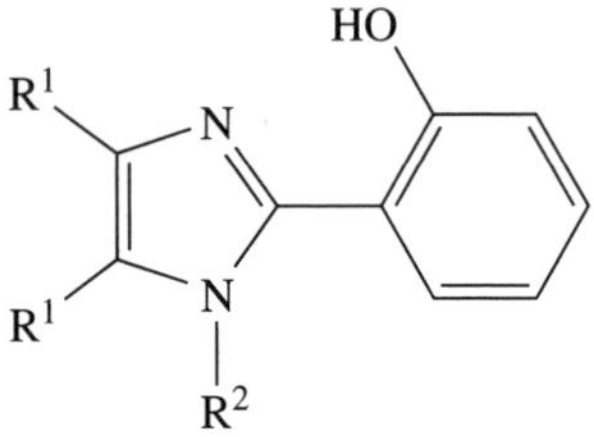

(28) R^1 = Me, R^2 = H
R^1 = H, R^2 = Me

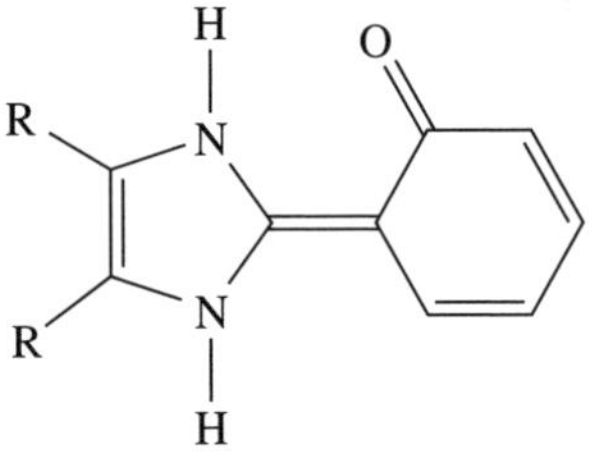

(29) R = H or Me

2-hydroxyphenyllapazole occurs from the singlet state, but the efficiency of the process is an order of magnitude greater than for 2-hydroxyphenylbenzoxazole[89]. Intramolecular hydrogen bonding is also shown in the 2-(2′-hydroxyphenyl)-4,6-diaryl-1,3,5-triazines (**33**). These compounds are phosphorescent in polar solvents at 77 K. It is likely that the phosphorescence emission arises from open conformers that have intermolecular hydrogen bonds[90].

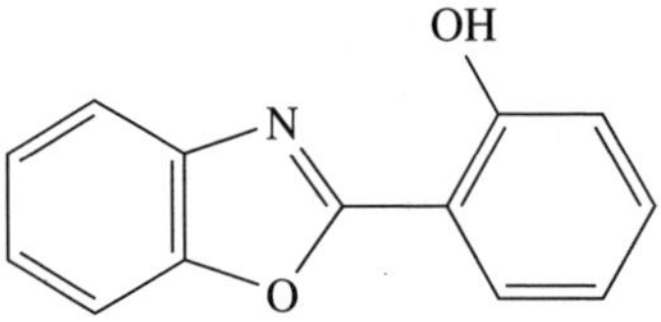

(30)

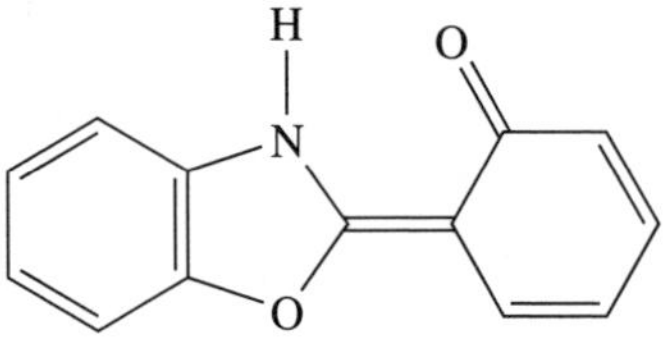

(31)

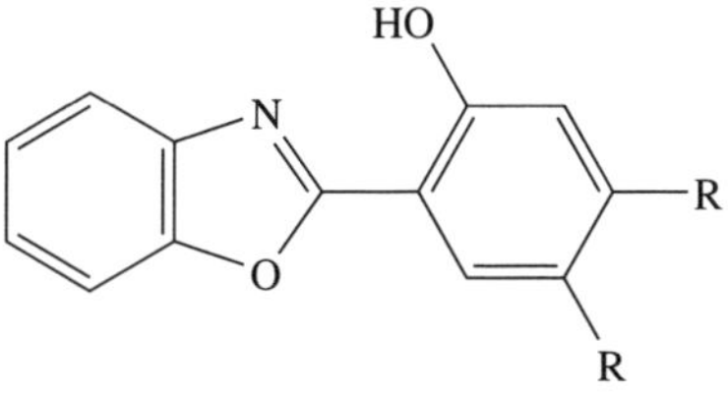

14. Photochemistry of phenols

(32) R = H or RR = −(CH=CH)$_2$−

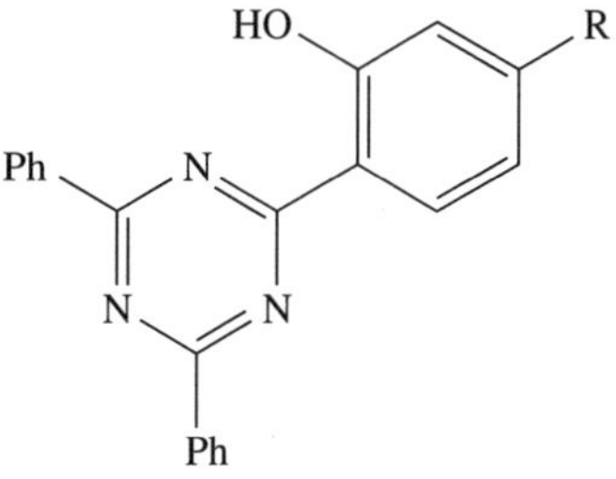

(33) R = H, OMe

E. Hydrogen Transfer and Cyclization

In some rigid planar systems such as **34** photochemically induced proton transfer occurs in benzene as solvent, but this is followed by cyclization resulting in the formation of the acridine **35**[91,92]. The cyclization is a common oxidative process in *cis*-stilbenoid systems. The proton transfer is an essential feature in the cyclization, since it was demonstrated that the reaction fails with the methoxylated analogue. With the bis hydroxy compound **34**, X^1 = X^2 = OH, a second cyclization affords **36** albeit in lower yield (14%).

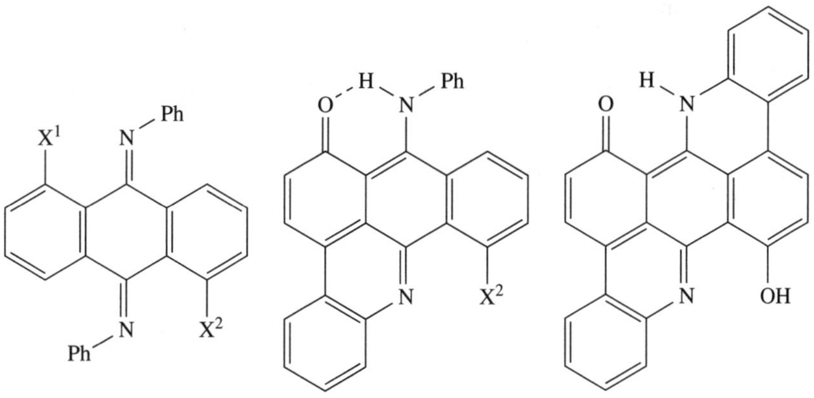

(34) (35) (36)

X^1	X^2	yield (%)
HO	HO	64
HO	H	18

IV. CHALCONES

A. Hydrogen Transfer Reactions

Irradiation of **37** at 366 nm is reported to give no observable reaction. The failure to react at this wavelength is thought to be due to the intramolecular hydrogen bonding, since the corresponding 4-hydroxy derivative does undergo facile *trans,cis*-isomerism around the double bond. However, the compound **37** is reactive using 308 nm light from an excimer laser[93] or laser flash photolysis in *n*-hexane[94]. This treatment brings about proton transfer from the phenolic OH with the formation of the keto-enol **38**[93]. A later study of **37** suggests that irradiation brings about irreversible *cis,trans*-isomerism of the double bond[95].

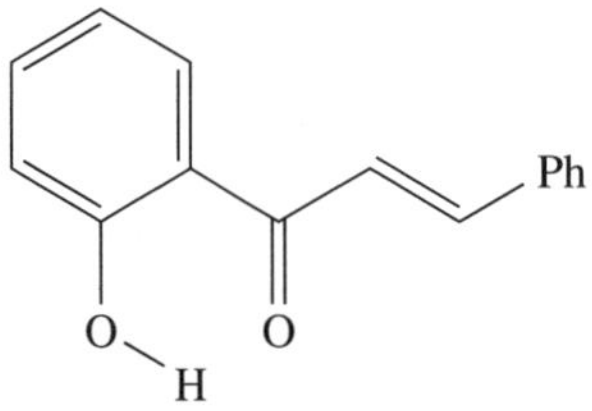

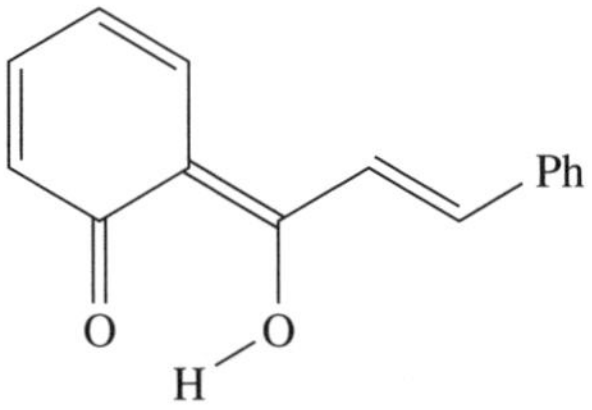

B. Cyclizations

Others[96] have reported that there is a definite effect of aryl substituents and that the derivative **39** undergoes cyclization in the presence of dissolved oxygen to yield the hydroxyflavone **40**. The cyclization involves the formation of a biradical **41** that cyclizes in the presence of oxygen to yield the hydroxyflavone **42**. The cyclization of such chalcones has been known for many years and studied in some detail[97-99]. Research showed that the derivatives **43** undergo efficient cyclization to **44** (Scheme 4) on irradiation at wavelengths >365 nm in dioxan or ethyl acetate solution[97]. The reaction is solvent-dependent and poorer yields are obtained in benzene or chloroform solution[97]. Further studies demonstrated, for the conversions shown in Scheme 5, that the cyclizations probably arose from a $\pi\pi^*$ transition[98].

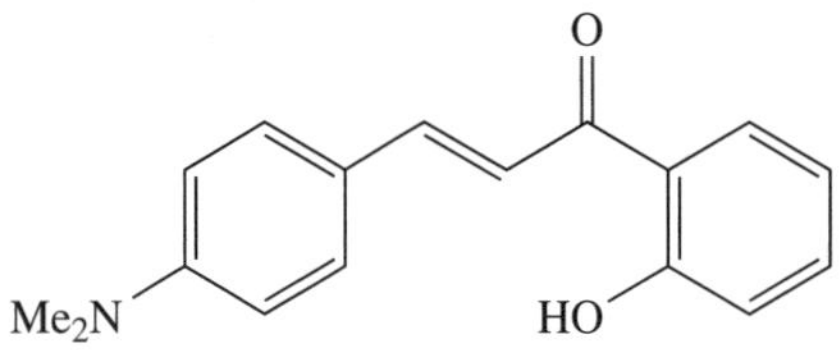

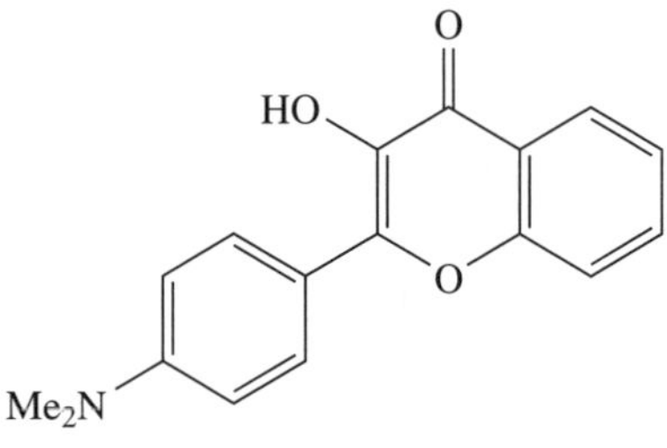

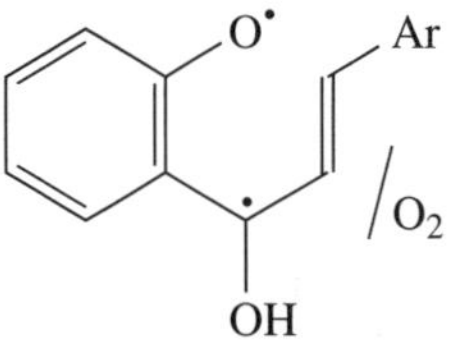

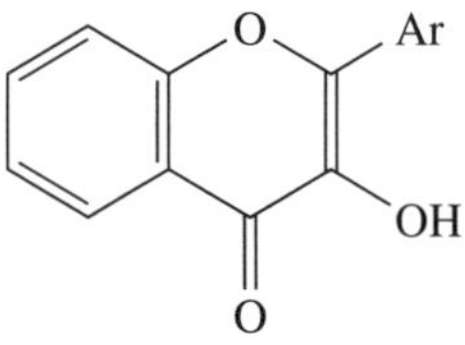

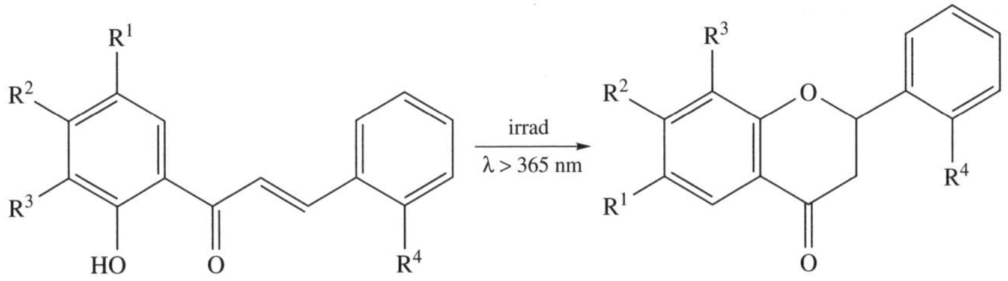

14. Photochemistry of phenols

(43) → irrad, λ > 365 nm → **(44)**

R¹	R²	R³	R⁴
H	H	H	H
H	H	H	MeO
H	OMe	H	H
MeO	H	H	H
(CH=CH$_2$)$_2$		H	H

SCHEME 4

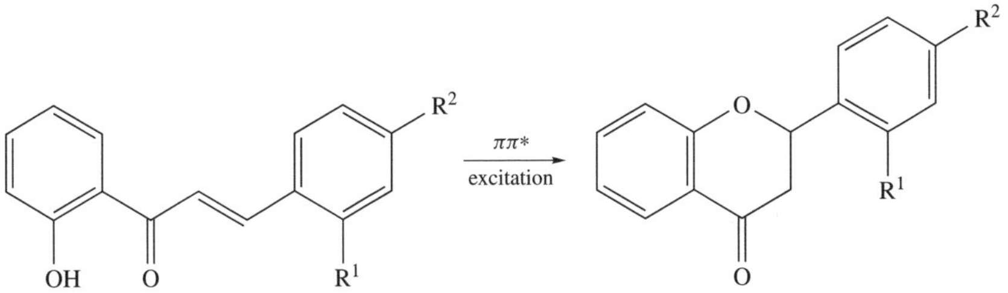

ππ* excitation

R¹	R²
H	Me, MeO, F, Cl, Br or NO$_2$
MeO	MeO

SCHEME 5

With the double bond of the chalcone systems adjacent to the hydroxy-substituted ring, cyclization is often the outcome of irradiation. This is demonstrated for the chalcone **45** that cyclizes to the flavylium salt **46** in acidic medium[100,101]. A more recent study of the cyclization of **45** has established that the precursor to the cyclic species is the ground state enol **47**[102]. The cyclization brings about a marked colour change both in solution and in plastic films. This photochromicity is substitution-dependent, as can be seen from the influence of methoxy substitution (Scheme 6) where the quantum yields vary from 0.02–0.12 depending on the position of the substituent[100]. A variety of substituents have been examined and the presence of a *p*-dimethylamino group appears to give the best results. The examples tested for photochromism are shown as **48**. Any variations on the aryl group were demonstrated to be effective. The influence of substituents on the photoreactions of the related photochromic chalcone **49** in both neutral and acidic solution has been investigated. The quantum yields for the cyclizations in both acid and neutral solution were determined and these are shown below the structure. The influence of the substituents on the process can be seen from these results[103].

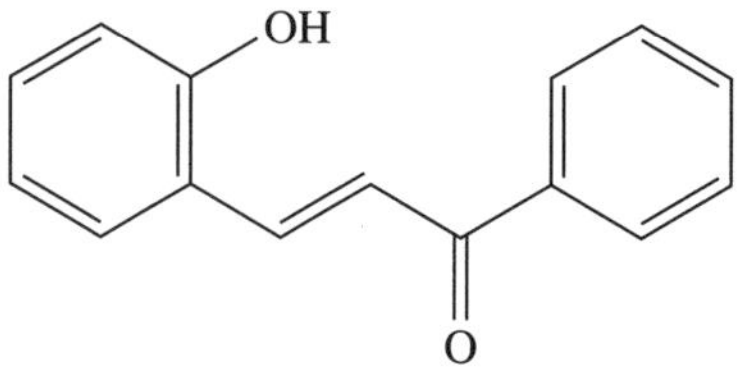

(45)

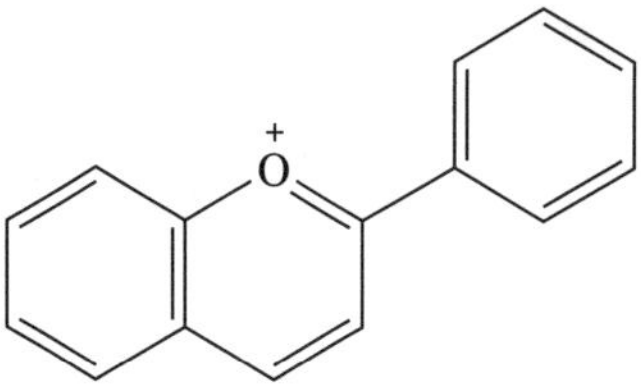

(46) 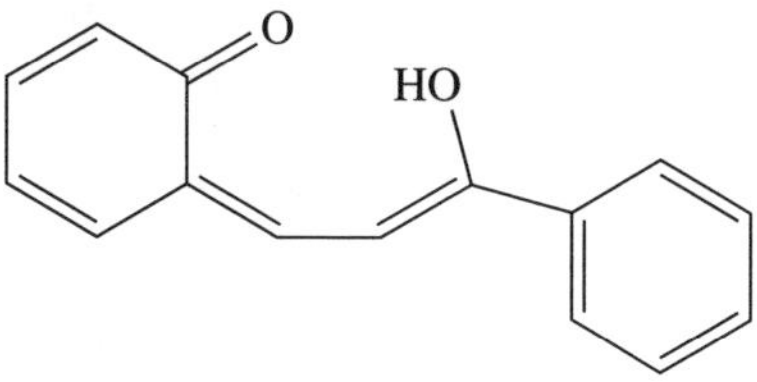   (47)

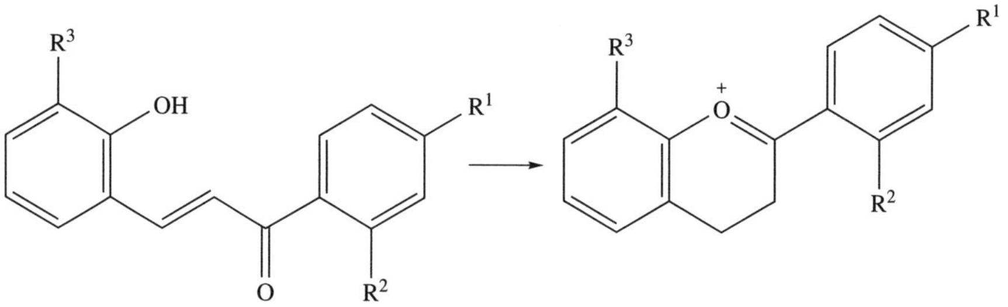

R¹	R²	R³	$\phi_{cyclization}$
MeO	MeO	H	0.08
MeO	H	MeO	0.02
MeO	H	H	0.12

SCHEME 6

C. Other Processes

In some instances double-bond isomerism is the principal event on irradiation, as with the chalcones **50**. The quantum yields for this process are in the range 0.2–0.4 in neutral aprotic solvents[104]. A study of the photochemistry of some chalcone derivatives using a variety of wavelengths (313, 334, 366 and 406 nm) has been reported[105]. Other reactivity

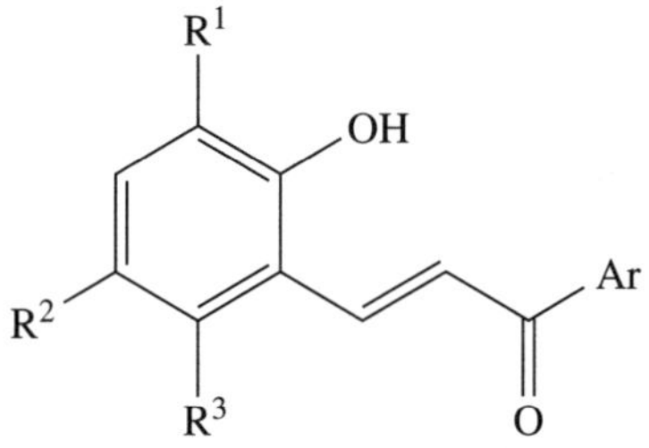

14. Photochemistry of phenols

(48)

R¹	R²	R³	Ar
H	H	H	Ph
H	H	H	o-MeOC$_6$H$_4$
H	H	H	m-MeOC$_6$H$_4$
H	H	H	p-MeOC$_6$H$_4$
MeO	H	H	p-MeOC$_6$H$_4$
H	H	H	p-Me$_2$NC$_6$H$_4$
H	benzo		p-Me$_2$NC$_6$H$_4$
H	H	H	3(2-cyano-dimepyrrolyl)
H	H	H	3(2,5-dimethylthienyl)
H	H	H	2-thienyl
H	H	H	2-furyl
H	H	H	3(2-methylbenzo[b]thienyl)

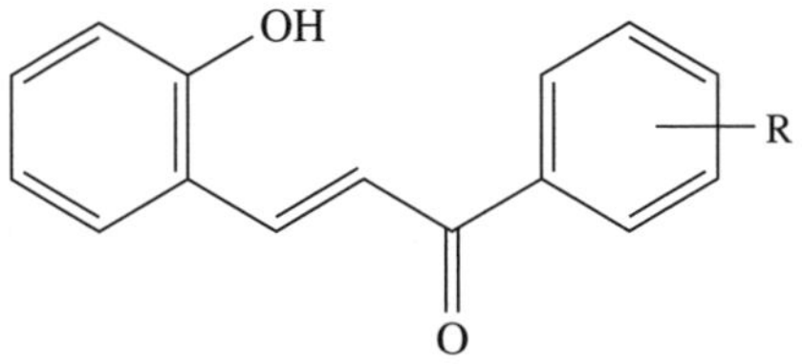

(49)

R	ϕ_{cycl} in H$^+$	ϕ_{cycl}, no acid
H	0.34	0.36
4-Cl	0.35	0.38
4-Me	0.36	0.34
4-OMe	0.36	0.31
4-NMe$_2$	0.33	0.072

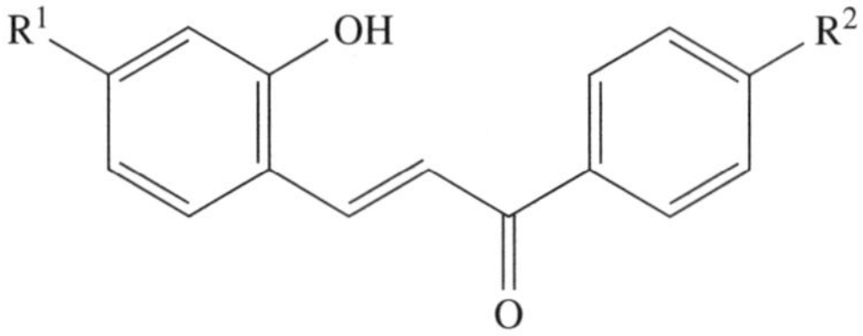

(50)

R¹	R²
NEt₂	H
NEt₂	OMe
NEt₂	NEt₂
NMe₂	NMe₂

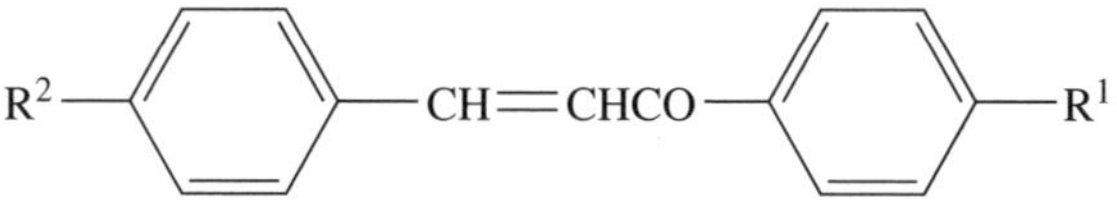

(51) R¹ = R² = H; R¹ = H, R² = Cl; R¹ = OH, R² = H

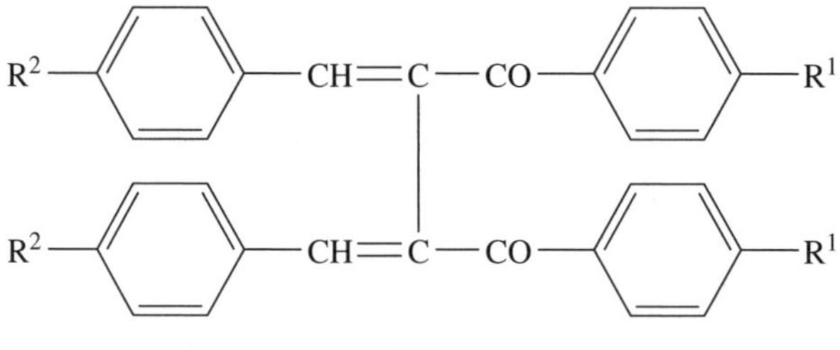

(52)

has also been observed, as with the chalcone **51** that undergoes oxidative dimerization to afford **52** on irradiation[106].

V. QUINONEMETHIDES FROM PHENOL DERIVATIVES

A. *o*-Quinonemethides

Earlier, a reference was made to the hydration of *o*-hydroxy-α-phenylstyrene and the amination of alkenes. The mechanism of these reactions has been probed in some depth. It is clear that proton transfer takes place on the irradiation of such systems and the transfer takes place to the alkenyl carbon and results in the formation of a quinonemethide such as **53**. Early work on the results of irradiation of *o*-hydroxybenzyl alcohol showed that a quinonemethide was formed. In the absence of other trapping agents phenol/formaldehyde resin-like materials were formed. Minor products such as **54** and **55** were also produced

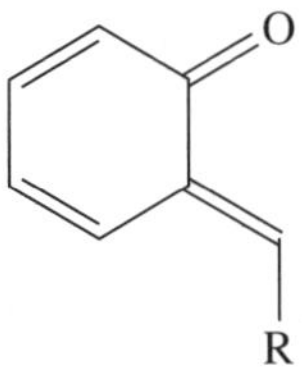

14. Photochemistry of phenols

(53)

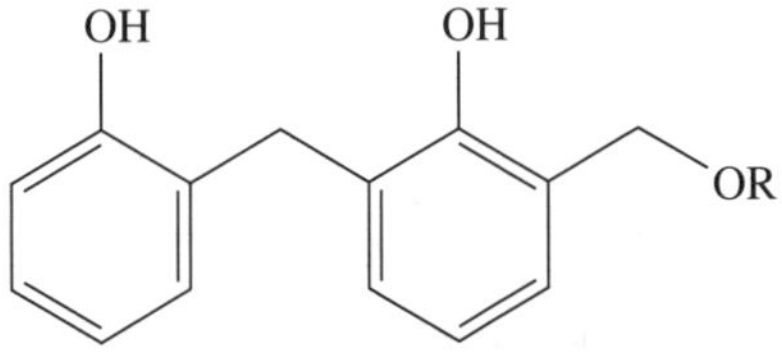

(54) R = H or Me

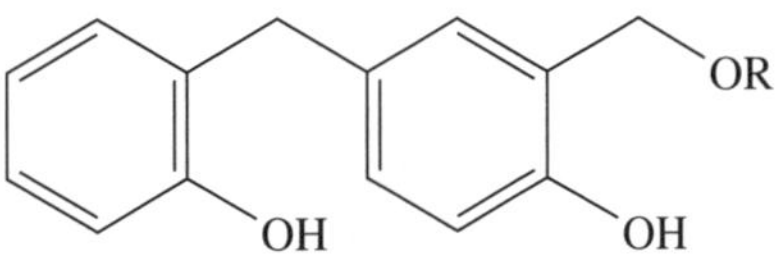

(55) R = H or Me

that did give some justification for the intermediacy of the quinonemethide[107]. The ultimate proof for the formation of an intermediate of this type comes from laser-flash studies and fluorescence measurements[108]. The quinonemethide **53** is formed from *o*-hydroxybenzyl alcohol and some α-substituted derivatives on irradiation at 254 nm and is the result of elimination of a molecule of water. When these reactive species are formed in methanol the ethers **56** (Scheme 7) are produced with reasonable photochemical efficiency. With change of solvent to water/acetonitrile and with added methyl vinyl ether the Diels–Alder adducts **57** are obtained almost quantitatively. Again this is good evidence for the involvement of quinonemethide intermediates[109,110]. A further example of photoelimination of water, this time from diol **58**, affords the quinonemethide **59** that undergoes

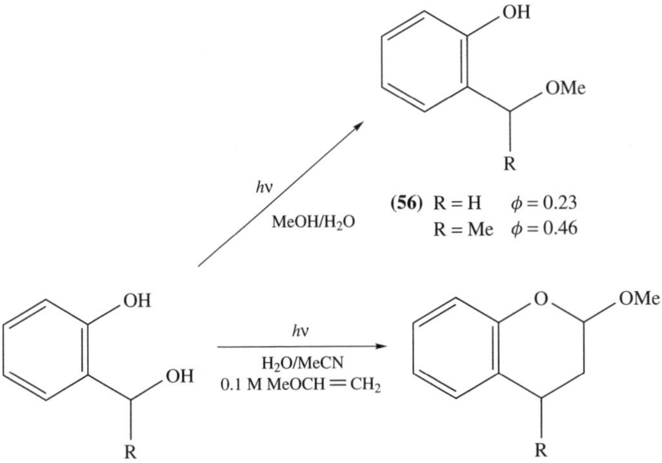

(56) R = H $\phi = 0.23$
R = Me $\phi = 0.46$

(57)

SCHEME 7

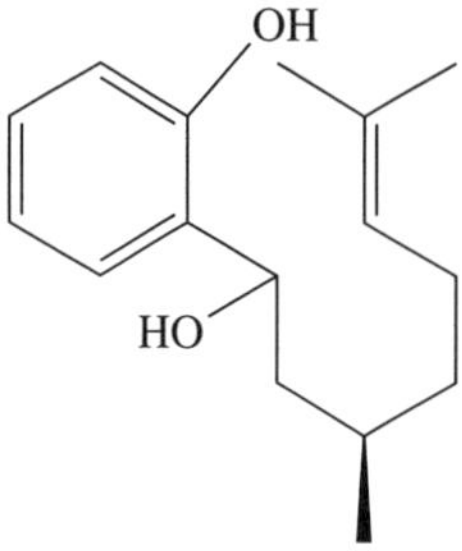

(58) 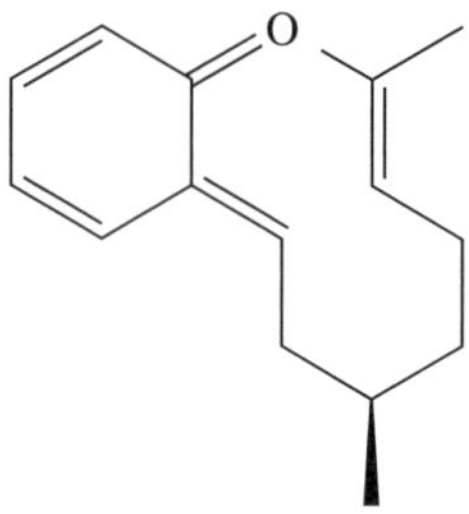   (59) 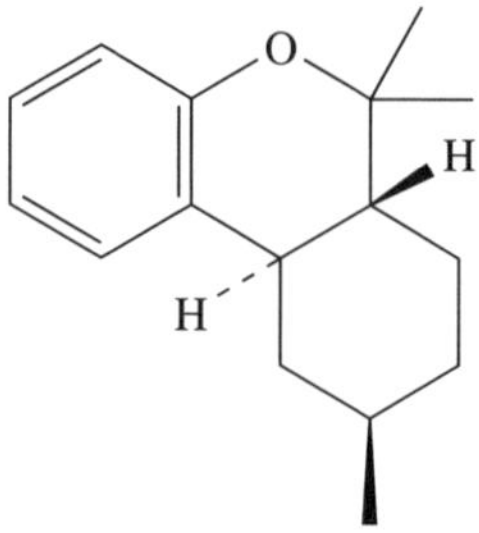   (60)

intramolecular cycloaddition to yield the hexahydrocannabinol **60**. The intermediate **59**, with an absorption at λ_{max}. ca 400 nm, was detected by laser-flash studies[111].

Elimination of simple amines from Mannich bases such as **61** also brings about the formation of o-quinonemethide intermediate such as **62** (from **61**, Ar = 4-Ph, X = NMe$_2$) using irradiation at λ > 300 nm in acetonitrile/water. Interestingly, the position of the aryl substituent in **61** is important and the best yields (the details are shown below the structure) were obtained when the phenyl group was p- to the OH. In the specific case, the formation of **62**, the quinonemethide can be trapped readily in a Diels–Alder reaction with ethoxyethene to yield the adduct **63** in 71%[112]. Other systems **64** and **65** were also studied and these again undergo elimination with the formation of the corresponding quinonemethide that also afford Diels–Alder adducts in 38% and 17% yields, respectively[113]. Other laser-flash studies have also reported the generation of o-quinonemethide from the phenol derivatives **66**. In these examples elimination of water, p-cyanophenol or an ammonium salt afforded the quinonemethide[113].

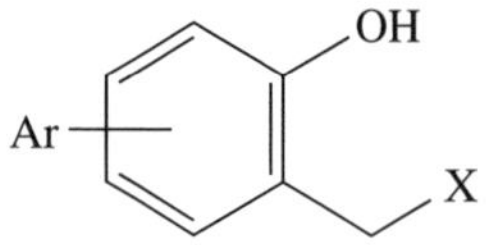

(61) 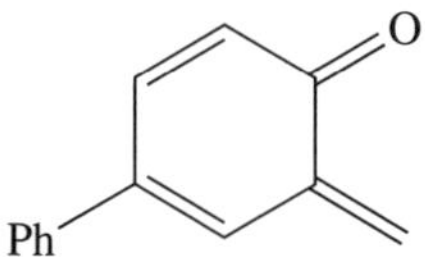   (62) 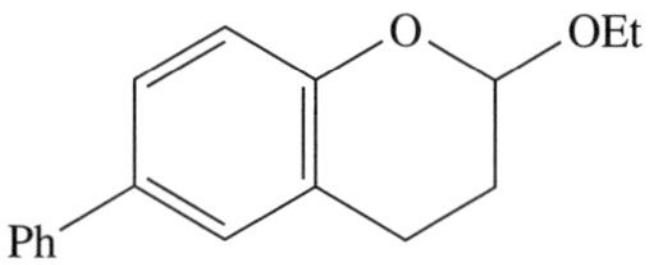   (63)

Ar	X	yield (%)
6-Ph	NMe$_2$	76
5-Ph	NMe$_2$	36
4-Ph	NMe$_2$	71
4-Ph	NEt$_2$	64
4-Ph	 N(piperidine)	57
4-Ph	HO	<2
4-Ph	MeO	<2

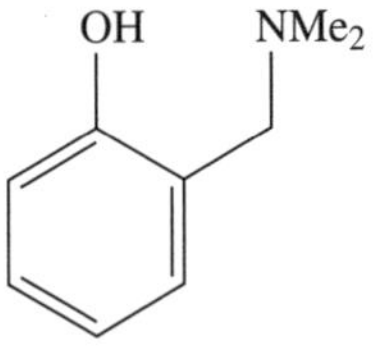

(64) 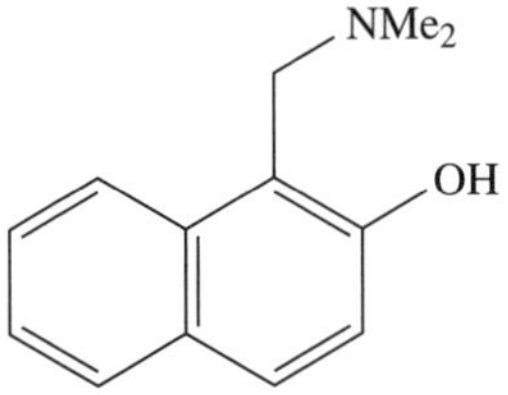   **(65)** 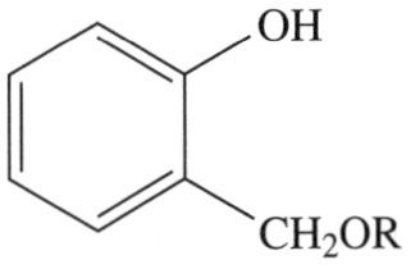   **(66)**

R = H or p-CNC$_6$H$_4$
OR = Me$_3$N$^+$I$^-$

B. *m*- and *p*-Quinonemethides

As mentioned earlier, quinonemethides other than the *ortho*-isomers can also be formed, such as **21** from **18**. In the cases cited previously, elimination of water from an appropriate hydroxy-substituted benzyl alcohol was the path followed. Other research has demonstrated that the irradiation of *p*-hydroxyphenyl ketones in water/acetonitrile mixtures brings about singlet excited-state proton transfer to afford the quinonemethide **67**. Apparently this proton transfer occurs in competition with intersystem crossing[114].

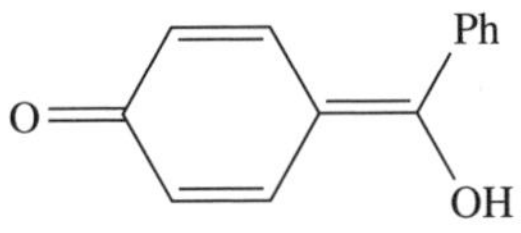

(67)

C. Quinonemethides from Biphenyl Derivatives

It is also possible to form quinonemethides involving the phenyl groups of biphenyl derivatives. The simplest of these has been shown for 2-hydroxybiphenyl. This undergoes excited state intramolecular proton transfer from the phenol moiety to the 2′-carbon position of the phenyl ring (not containing the phenol hydroxy group), to generate the corresponding quinonemethide **68**[115]. The transfer of hydrogen, in some respects, resembles the formation of the two keto-tautomers cyclohexa-2,4-dienone (**69**) and cyclohexa-2,5-dienone (**70**) of phenol that can be generated by flash photolysis[116]. In earlier work Shi and Wan[117] reported that the biphenyls **71** underwent deuterium exchange at the *ortho*-position of the ring distant from the oxygen substituent. In this case they proposed that the S^1 state of the biphenyl was strongly polarized. Related to this study is the report that laser

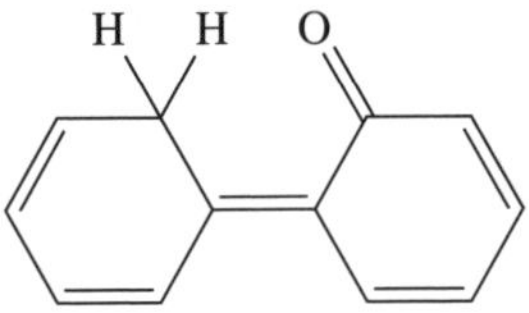

(68) 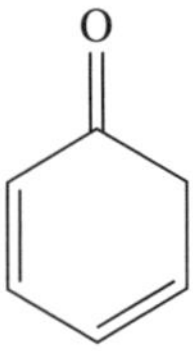   **(69)** 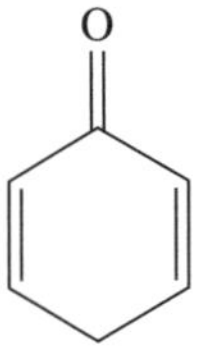   **(70)**

flash photolysis of the biphenyls **72** and **73** brings about their transformation into the quinonemethides **74** and **75**, respectively. This postulate is substantiated by preparative irradiation of the biphenyls in methanol/water when the ethers **76** and **77** are obtained with quantum yields of 0.24 and 0.03, respectively[117]. Quinonemethides are also involved in the photoconversion of the three biarylmethyl alcohols **78**, **79** and **80** into the corresponding pyrans **81**, **82** and **83** on irradiation at 254 nm in acetonitrile or acetonitrile/water[117]. The singlet excited state is thought to be involved in these transformations. Irradiation brings about elimination of water and the formation of **84** takes place from **78**. Cyclization of this quinonemethide intermediate affords the final product. The same is so for the other examples. Interestingly, compound **80** is highly twisted from planarity. Even so the quantum yield for the conversion to the pyran **83** is reasonable at 0.17. The biphenylmethanol **85** is also twisted from planarity and the X-ray crystal structure shows it to have a dihedral angle of 80° between the rings of the biphenyl ring system. Again, this compound cyclizes efficiently when it is irradiated in acetonitrile solution or in the solid state to afford the corresponding pyran **86**. In solution, the mechanism of the reaction involves intramolecular proton transfer from the phenolic OH to the benzyl alcohol function. In the solid state the proton transfer is thought to occur intermolecularly[118]. Earlier work had shown that such cyclizations were feasible with the conversion of the less heavily substituted derivative **87** into **88** again by way of the quinonemethide[119].

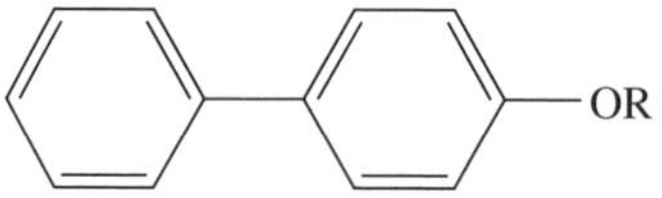

(71) R = H or Me

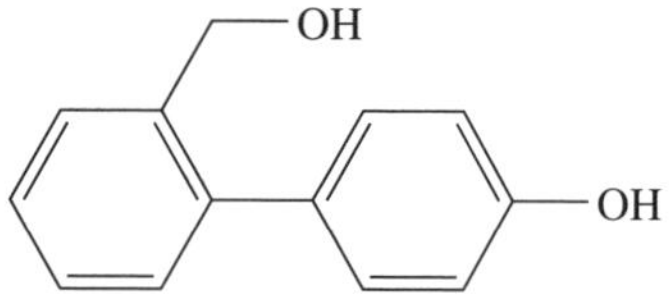

(72)

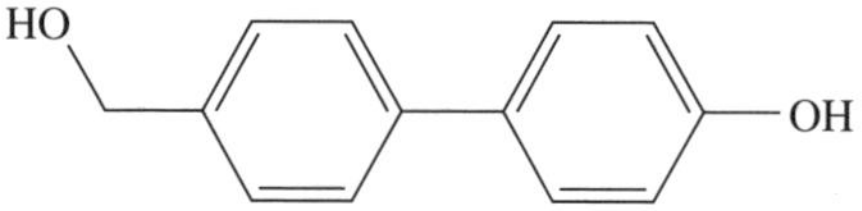

(73)

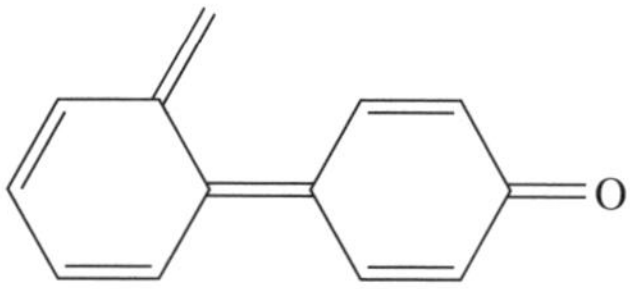

(74)

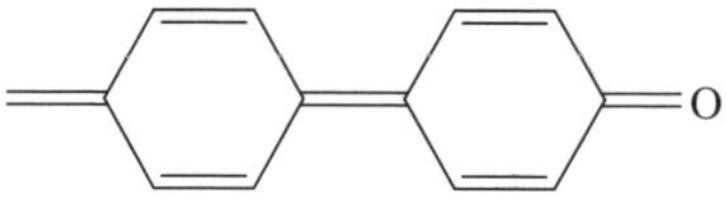

(75)

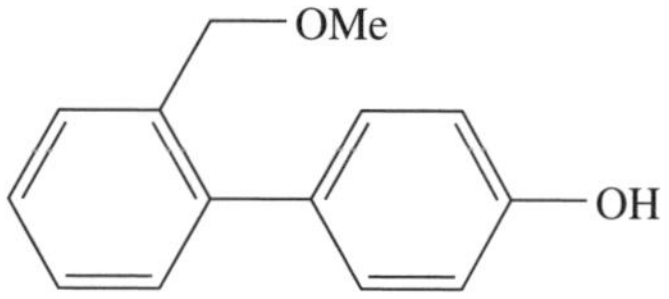

(76)

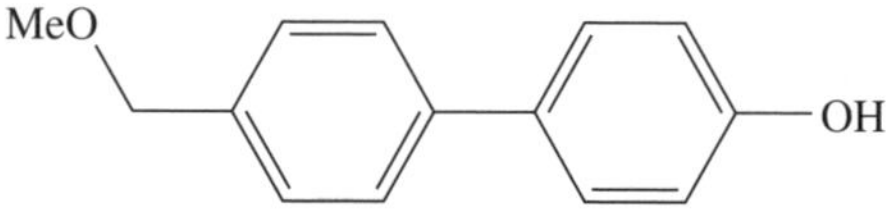

(77)

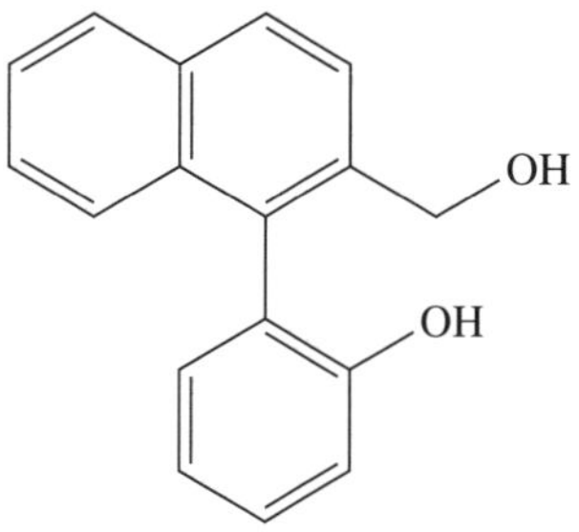

(78) 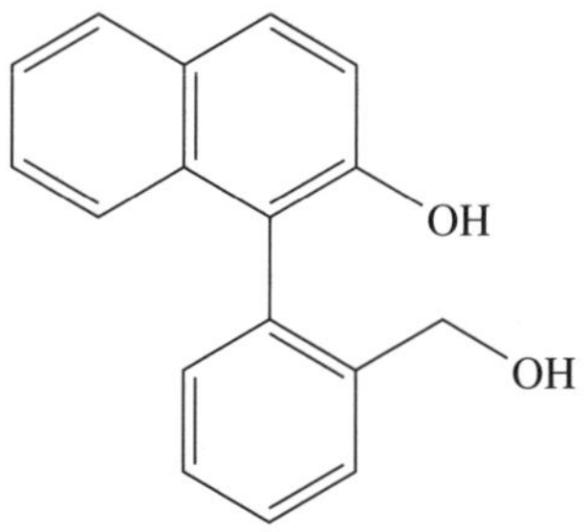 (79)

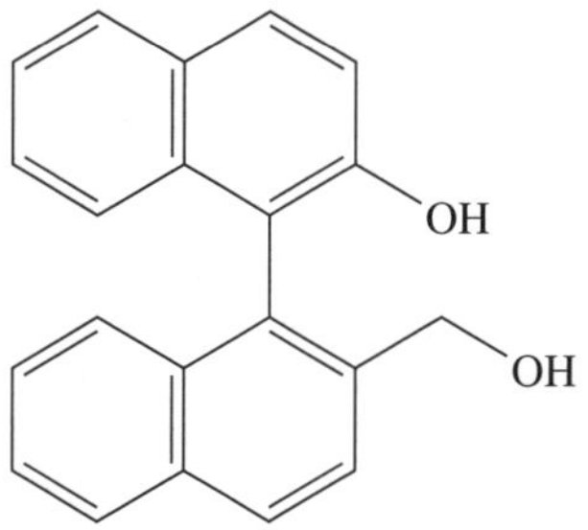

(80) 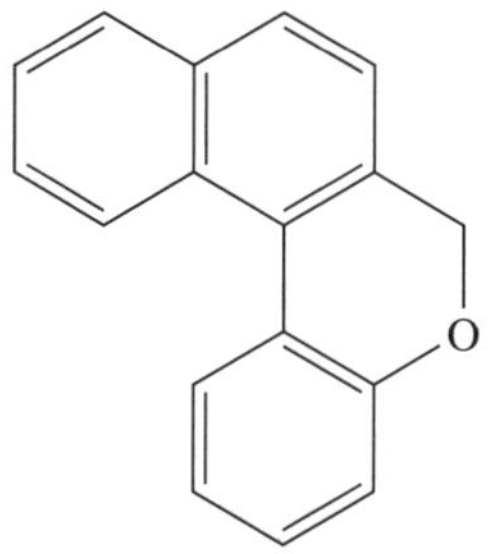 (81)

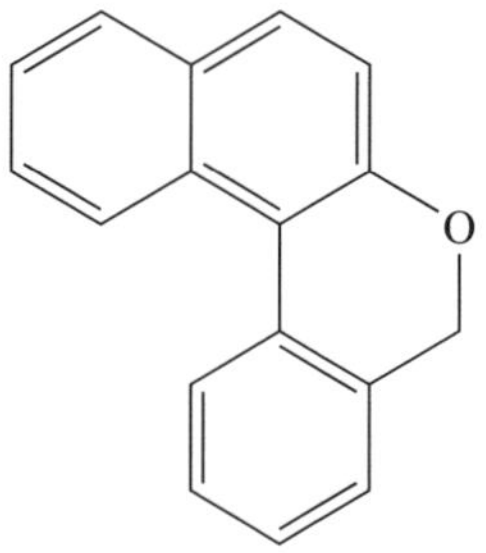

(82) 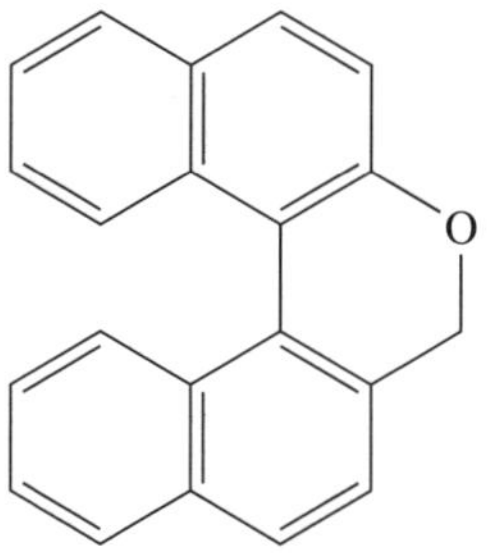 (83) 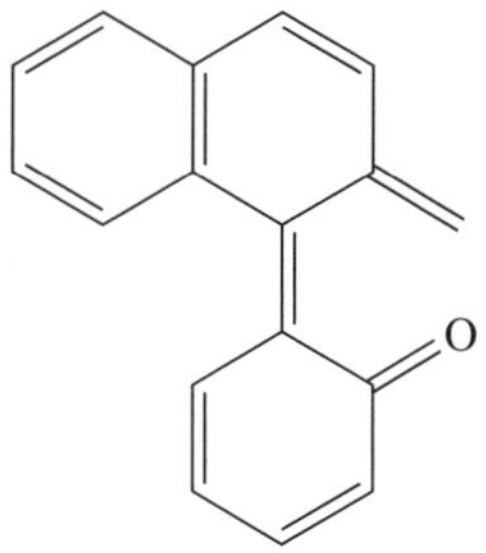 (84)

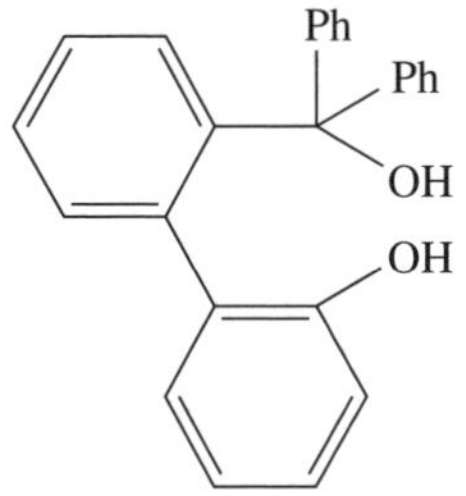

(85) 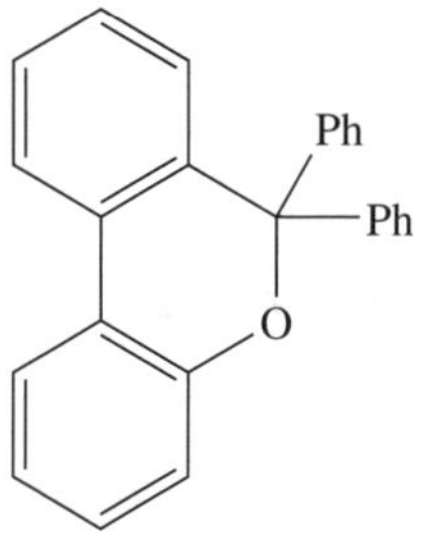 (86)

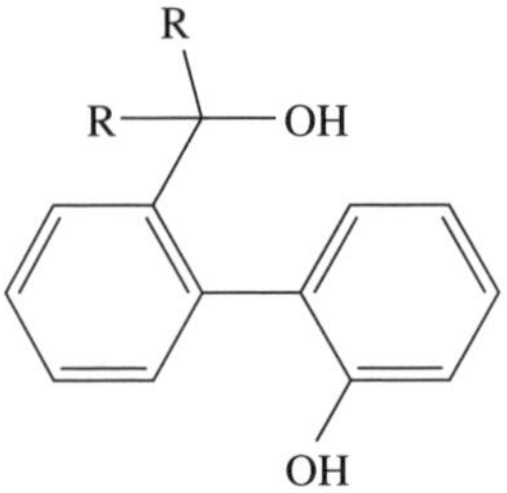

(87) R = H, Me

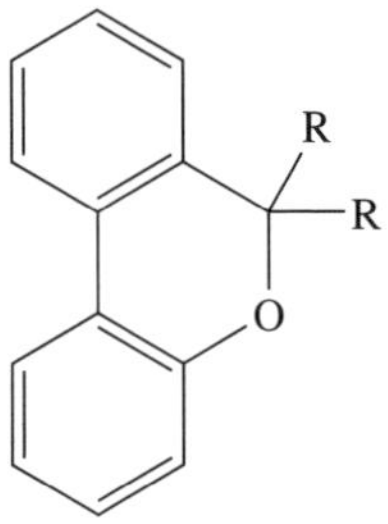

(88) R = H, Me

D. Quinonemethides from Fluorenols

The elimination of water from the biphenyl systems has been extended to include the hydroxyfluorenols **89**. Irradiation in 1 : 1 water/methanol results in conversion to the ether **90**. While the ether-forming reaction is thought to involve the generation of the fluorenyl cation by heterolysis of the CO bond the production of the quinonemethide **91** also takes place. This intermediate can be trapped by ethyl vinyl ether as the *cis*-adduct **92**[120]. Triplet state characteristics of 2,2'- and 4,4'-biphenyldiols have been investigated in different organic solvents using 248 nm nanosecond laser flash photolysis technique. The differences observed in the two diols is explained on the basis of the presence and the absence of intramolecular hydrogen bonding[121].

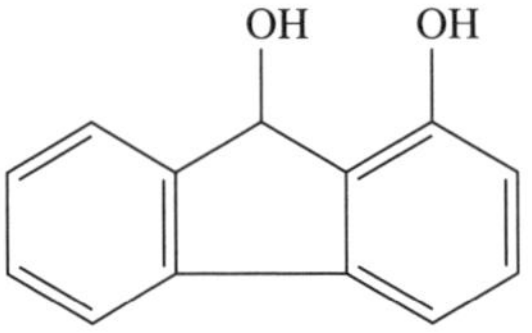

(89)

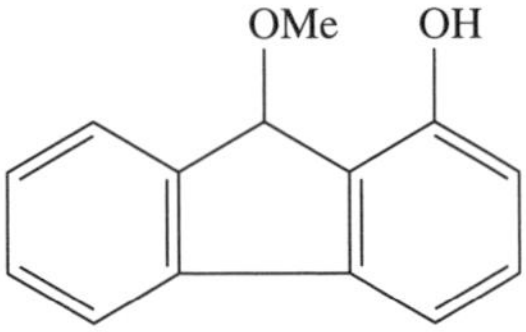

(90)

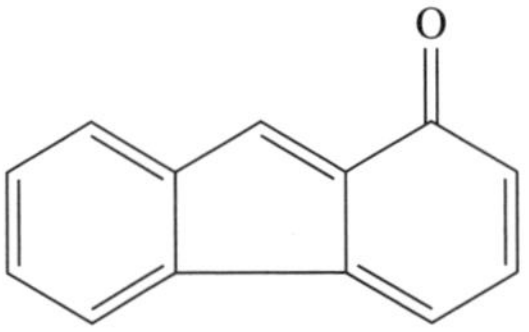

(91)

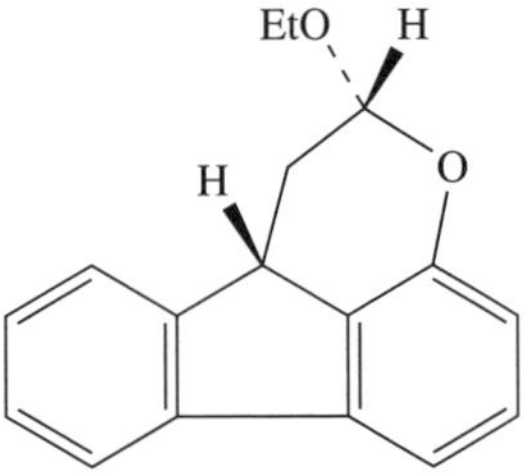

(92)

VI. CYCLIZATIONS WITHIN o-ALLYLPHENOLS

The earlier work on the photochemical cyclizations of *o*-allylphenol (**93a**) were commented upon in the original article in this series[1]. Some further studies have examined the influence of aryl substituents on the reaction and the ionic nature of the process[122a]. The photochemical cyclization of the corresponding phenoxides has also been examined[122b]. Others[122c] have examined the *trans,cis*-isomerism of **93b** and its subsequent cyclization into **94** and **95**. The whole area has also been the subject of a review[122d]. The cyclization exhibited by **93b** is the result of phenolic proton transfer to the alkene followed by cyclization within the resultant zwitterion. The yields for the formation of the two products are shown under the illustrations and it can be seen that the best yields are obtained in benzene in the presence of oxygen. The influence of aryl substituents on the reaction of such systems was studied using the phenol **96a**. Here again photochemical cyclization into a mixture of the dihydrofuran (**97a**) and the dihydropyran (**98a**) occurs. The corresponding *cis*-alkene is also formed. Interestingly, the outcome of the reaction is substituent-dependent and **96b** affords the furan **97b** as the main product (see yields). The compound **96b** apparently reacts via an SET process. With an acetyl substituent the cyclization process of **96c** fails and the reaction is diverted along the *trans,cis*-isomerism path[123].

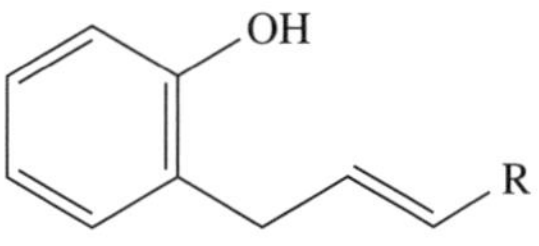

(**93**) (a) R = H
(b) R = Ph

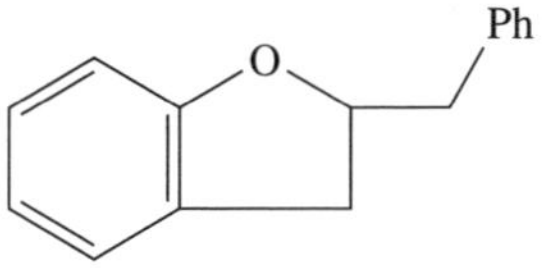

(**94**)

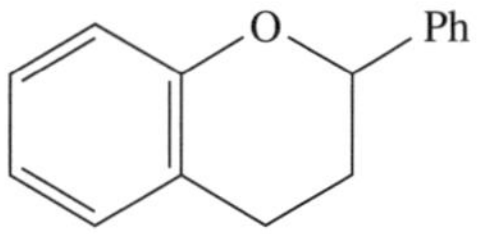

(**95**)

yield (%)		conditions
94	**95**	
22	30	Ar, λ > 200 nm, hexane
32	48	O_2, λ > 200 nm, hexane
62	23	O_2, λ > 200 nm, benzene

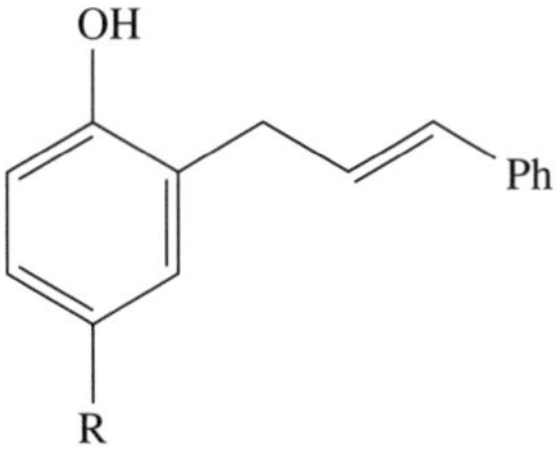

(**96**) (a) R = Ph
(b) R = MeO
(c) R = COMe

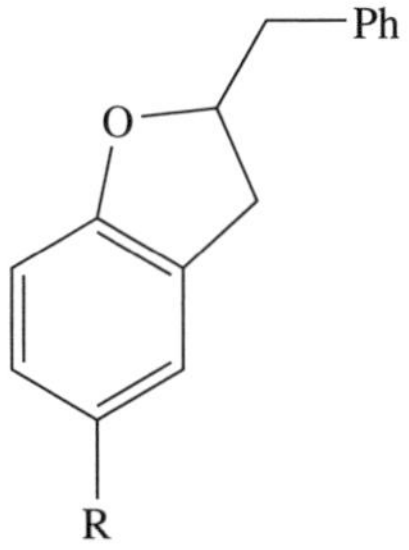

(**97**) (a) R = Ph
(b) R = OMe

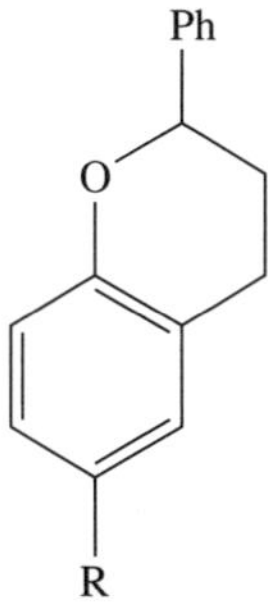

(**98**) (a) R = Ph
(b) R = OMe

	yield (%)	
	97	98
a	20	40
b	10	86
c	—	—

Direct irradiation of the arylalkenes (**99**) again results in their conversion into the cyclic ethers (**100**) and (**101**) (see also the conversions of **93** and **96**). Under different photochemical conditions, using 2,4,6-triphenylpyrylium tetrafluoroborate as an electron-accepting sensitiser, the compounds (99, R = CH$_3$) and (99, R = Meo) undergo oxidative cleavage of the double bond with the formation of the corresponding aldehyde. This occurs even though the reactions are carried out under argon[124]. In addition to *cis,trans*-isomerism, irradiation of the allylnaphthols **102a, b** brings about photochemical cyclization to afford the two products **103a** and **104a**, and **103b** and **104b**, respectively[125]. Cyclization also occurs with phenyl derivatives such as **105**. Again the cyclizations observed follow two paths to yield a mixture of the cyclized derivatives **106** and **107**. The influence on the photochemistry of substituent groups attached to the styryl moiety has been evaluated. It is clear that a charge transfer (CT) is involved and that the outcome of the reactions is solvent-dependent. The CT in the excited state brings about a proton transfer followed by cyclization to yield the products **106** and **107**, where the latter is predominant. When acetone is used as sensitizer no reaction is observed[126].

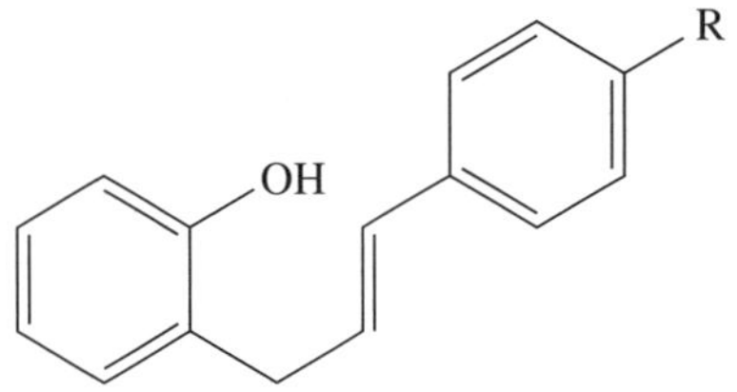

(**99**) R = H, Me, MeO

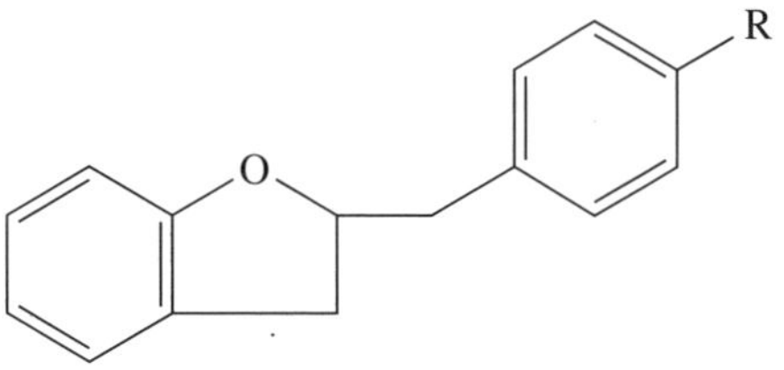

(**100**)

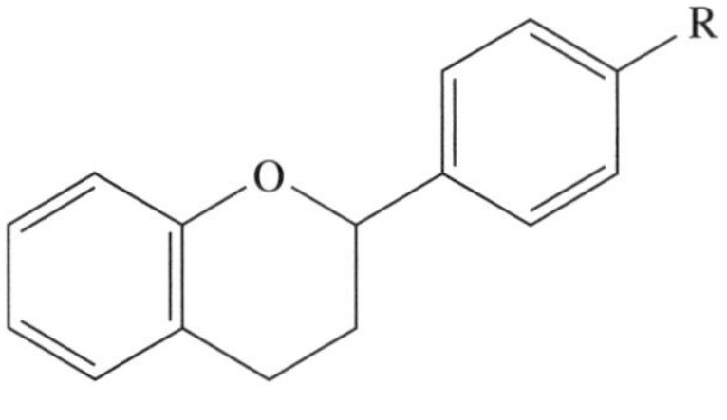

(**101**)

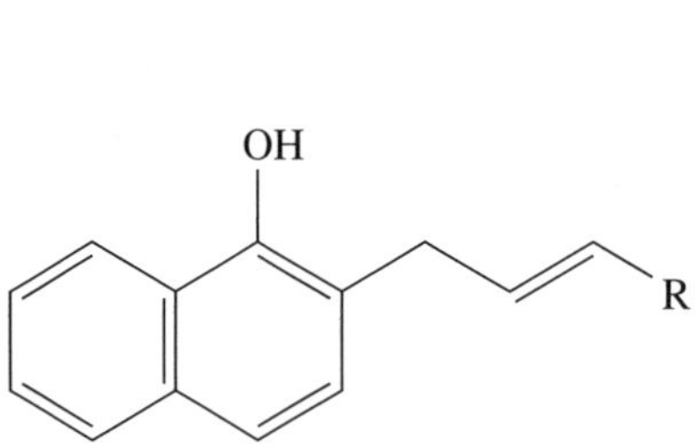

(**102a**) R = H, Ph

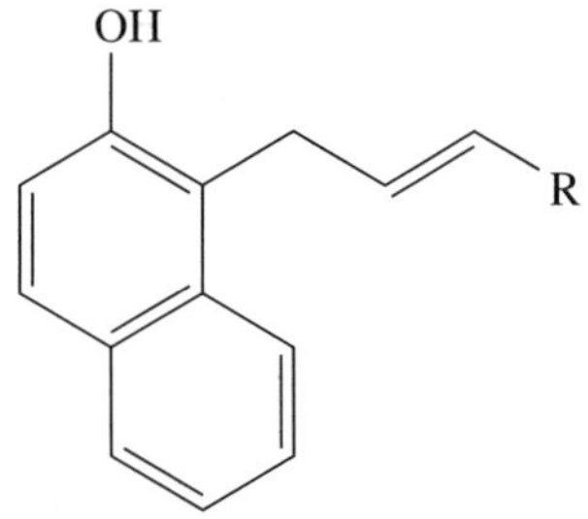

(**102b**) R = H, Ph

14. Photochemistry of phenols

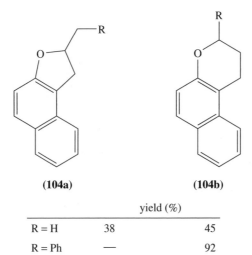

(103a) / **(103b)**

	yield (%)	
R = H	81	11
R = Ph	55	44

(104a) / **(104b)**

	yield (%)	
R = H	38	45
R = Ph	—	92

(105) R = Ph, Me, OMe

(106)

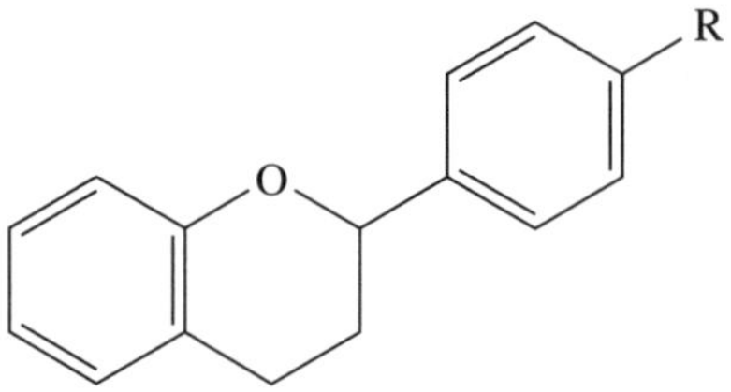

(107)

	yield (%)	
	106	107
R = Ph	21	64
R = Me	20	40
R = MeO	22	59

A. Miscellaneous Reactions

The competition between dehalogenation and cyclization within the derivatives **108** has been studied. Here, fission of C—Cl and C—Br bonds occurs and addition of solvent to the aromatic ring takes place[127]. Direct irradiation through a quartz filter of the *trans*-cyclopropane **109** in cyclohexane populates the singlet state. Within this excited state the acidity of the phenolic hydrogen is enhanced. This leads to the formation of the two tight zwitterions **110** and **111**. Reaction within these affords the principal products **112** and **113**. Other products **114–116** are also formed in low yield and the authors[128] suggest that an electron transfer mechanism is involved. The irradiation of *o*-allylphenol has also been reinvestigated in cyclohexane as solvent. A di-π-methane reaction was observed with the formation of a photolabile cyclopropyl derivative **117**. This fragments on excitation to afford a carbene **118** which inserts into a solvent molecule to yield **119** (Scheme 8)[129].

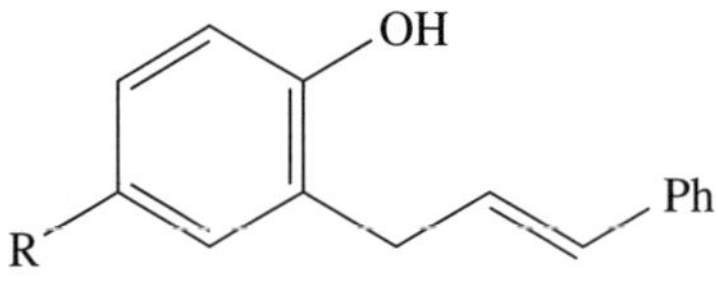

(108) R = Cl or Br

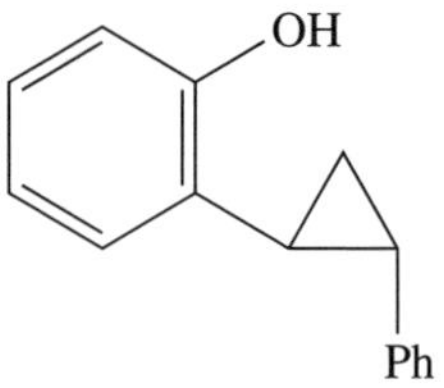

(109)

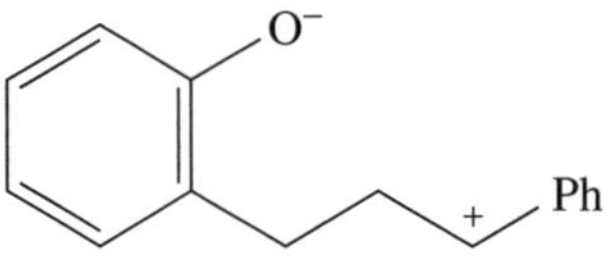

(110)

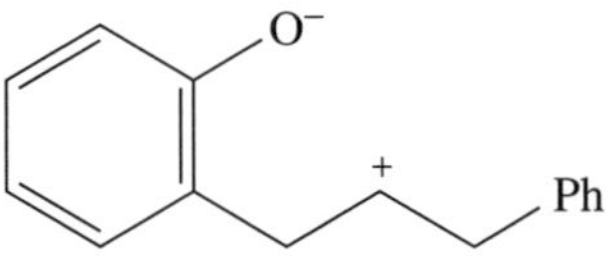

(111)

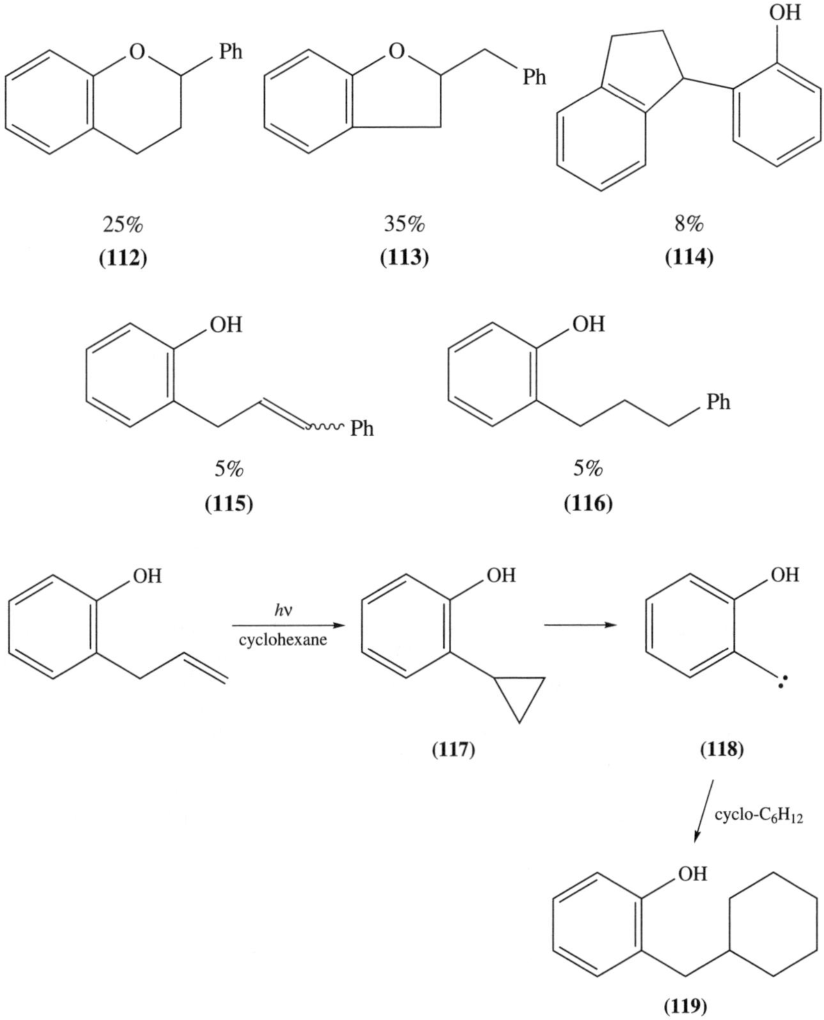

SCHEME 8

B. Photolabile Protecting Groups

The passing of the years has not diminished the interest in photolabile protecting groups. Many such systems are available[130] and have been described in the literature. A new method has been proposed for the protection of amino acids. This involves the conversion of the amino acid into the phenacyl derivatives **120**. Irradiation of these derivatives in a buffered aqueous solution results in the release of the amino acid and the transformation of the phenacyl group into a phenylacetic acid. This occurs via the triplet state within which there is intramolecular displacement of the amino acid moiety as represented in **121**. The resultant intermediate **122** undergoes ring opening by attack by water to afford the p-hydroxyphenylacetic acid as the by-product of the deprotection[131,132]. The photolabile silyl-based protecting group **123** has also been described[133]. The photochemical reaction involves the *trans,cis*-isomerism of the double bond at 254 nm followed by interaction between the phenolic OH group and the double bond. Some of the alcohol derivatives used are shown below structure **123**. This results in photochemical proton transfer and the formation of the isomer **124** (Scheme 9). Transfer of the silyl group and subsequent hydrolysis releases the protected alcohol. Additional study has demonstrated the feasibility of the hydrogen transfer by experiments using **125**. In this case deuterium transfer is the outcome yielding **126**[134]. A detailed account of photochemical reactions of alcohol

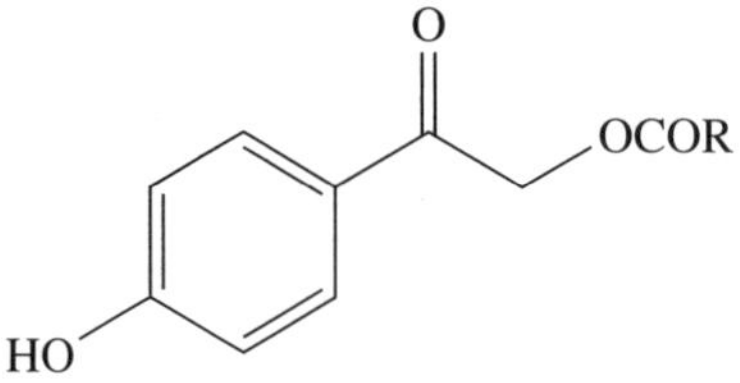

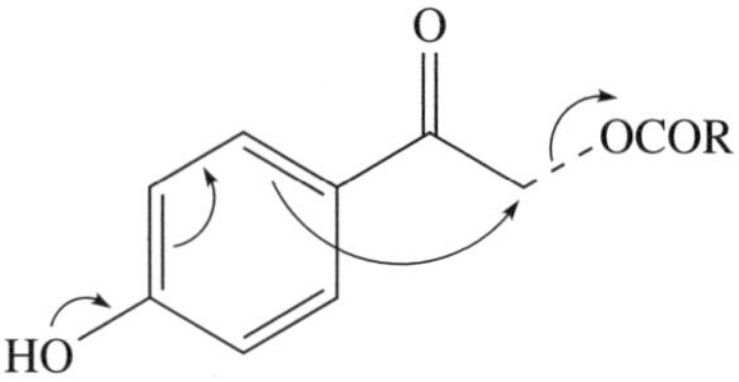

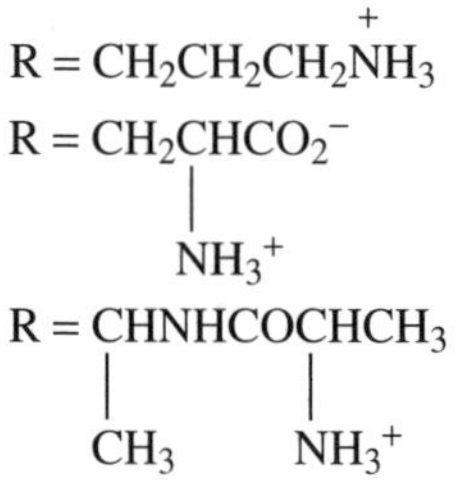

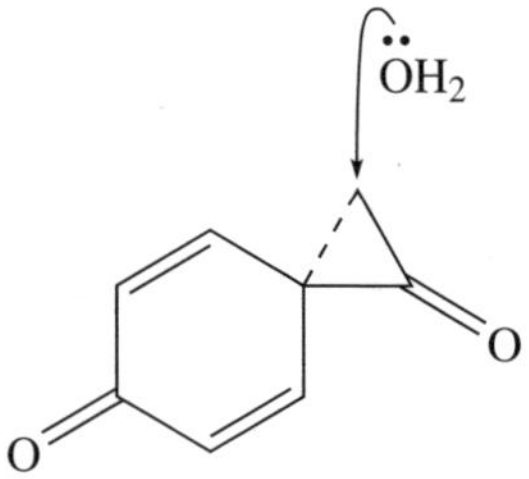

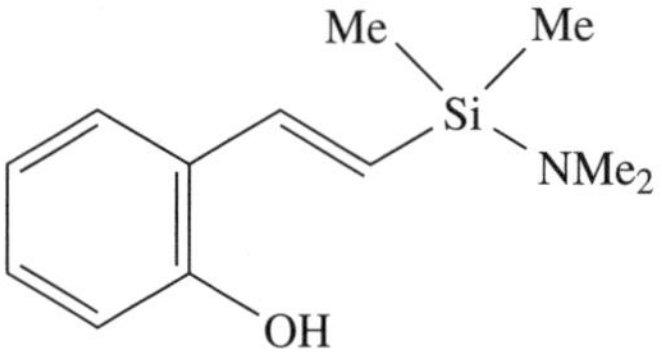

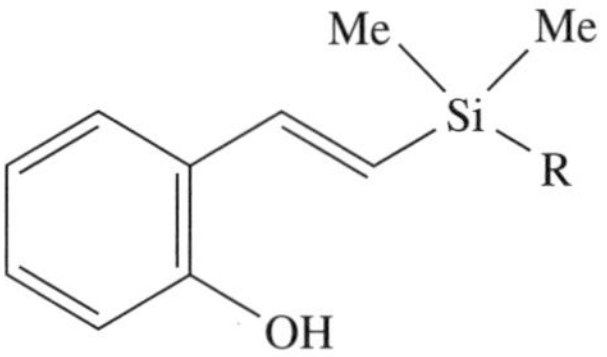

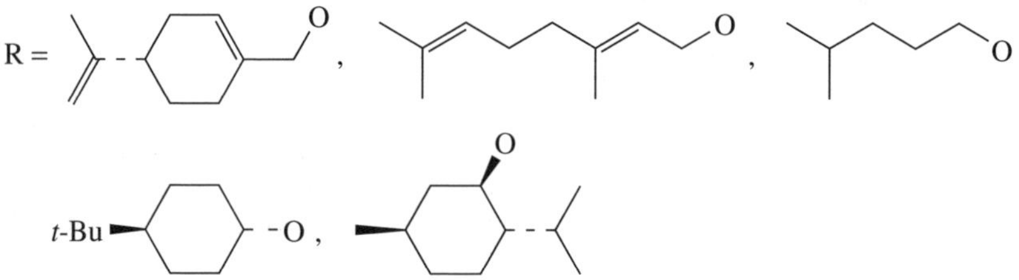

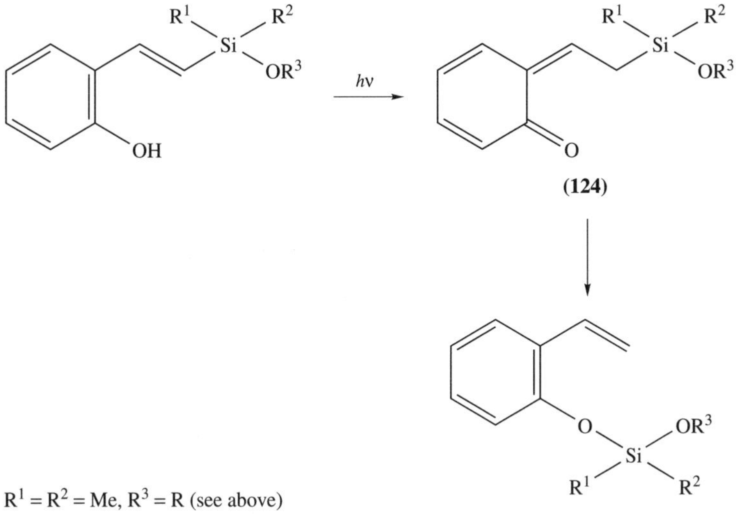

SCHEME 9

$R^1 = R^2 = Me, R^3 = R$ (see above)

(124)

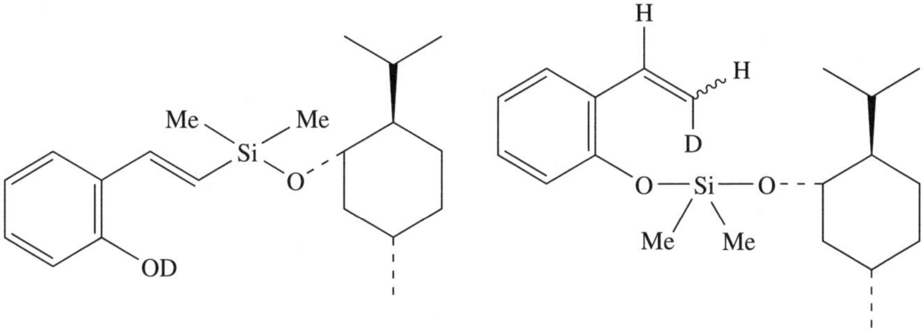

(125) **(126)**

protecting groups **127** has been reported. The deprotection of the alcohol is dependent on a primary *trans,cis*-isomerization path on irradiation at 254 nm. Some of the classes of alcohol used and the yields obtained (the percentage yields given refer to the free alcohol obtained) are shown[135].

C. Other Hydrogen Transfers

An intramolecular excited state proton transfer occurs on irradiation of hypericin **128**[136–139]. Excitation of hypericin in lipid vesicles results in excited state regioselective transfer of a proton to the substrate from one of the *peri*-hydroxyl groups[140]. Hypericin in its triplet state reacts with reducing agents to afford a long-lived transient presumed to be the resultant radical anion[141]. Both electron donors and acceptors can quench the fluorescence of hypericin[142]. A detailed review of the reactions of the photosensitizer 'hypericin' has been published. Some of the work described dealt with its photochemical deprotonation in the excited state[143].

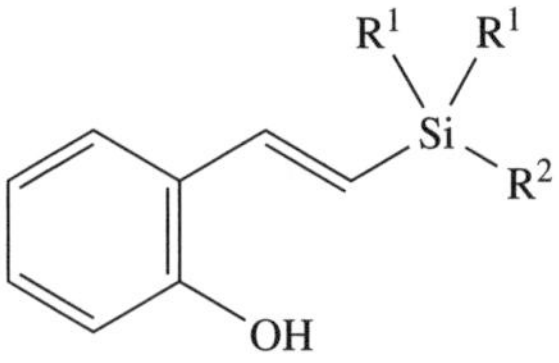

(127) R¹ = Me, i-Pr

R²		R²	
4-isopropenylcyclohex-1-enylmethyl-O	83%	3'-OTBS-thymidine-5'-O	92%
geranyl-O	91%	5'-OTBS-thymidine-3'-O	91%
isohexyl-O	89%		
4-t-Bu-cyclohexyl-O	84%	menthyl-O	87%
CH₃O	87%		

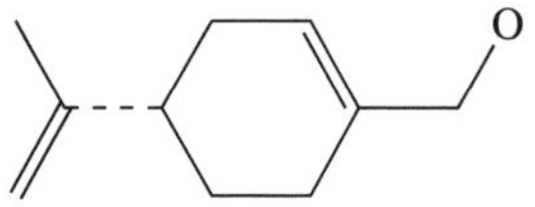

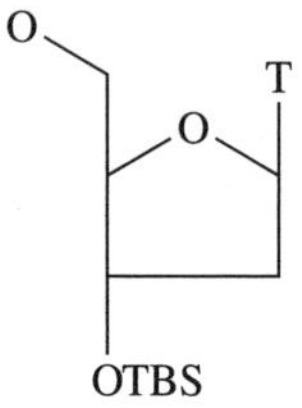

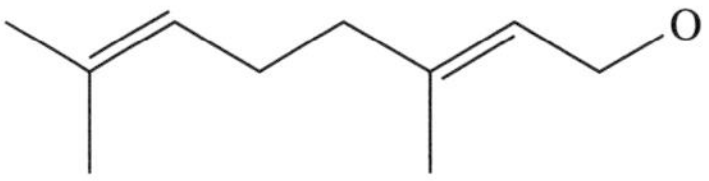

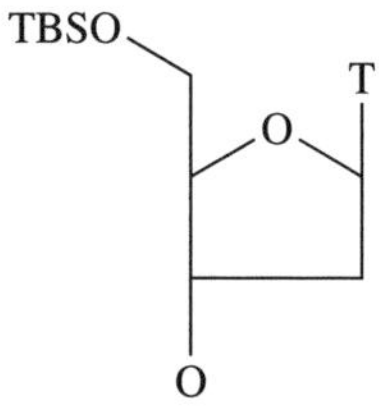

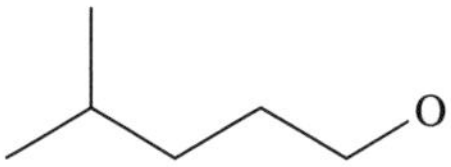

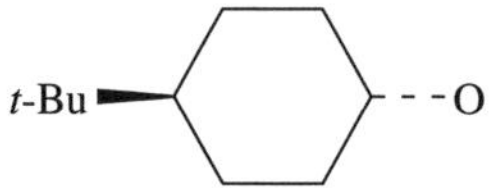

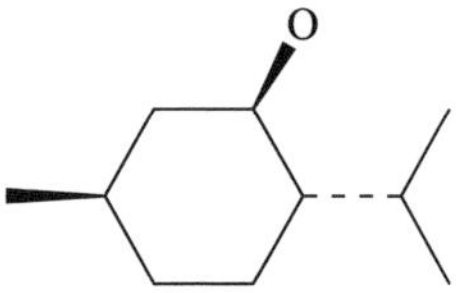

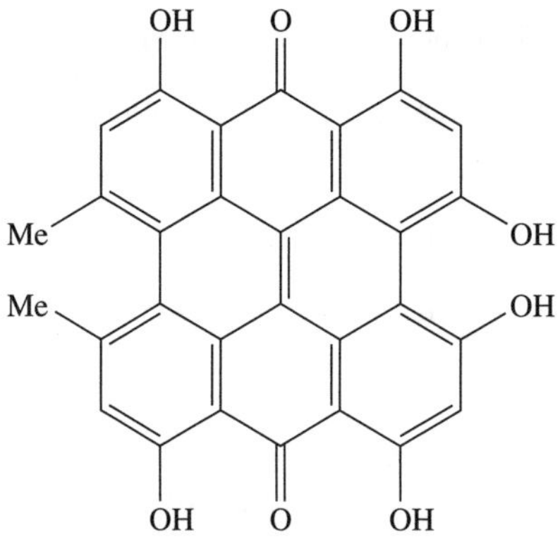

(128)

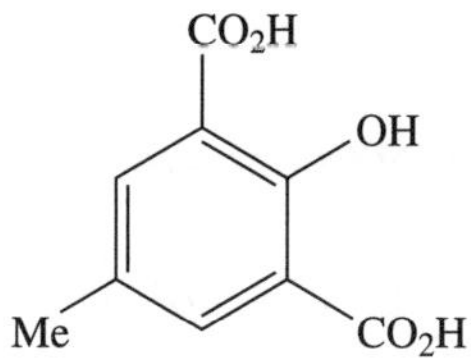

(129)

14. Photochemistry of phenols

The photochemically induced proton transfer in 2-hydroxy-6-methyl-*m*-phthalic acid has been studied[144]. Guha and his coworkers have also investigated the proton transfer processes in the isomeric diacid **129** as well as in the corresponding diester and diamide[145]. The photochemically induced hydrogen transfer reactivity in the salicylate derivatives **130** has been studied[146,147] as has the photoinduced proton transfer within 3-hydroxy-2-naphthoic acid (**131**). In this latter case a large Stokes-shifted emission is observed. This shift is dependent upon pH, solvent, temperature and excitation wavelength. The large Stokes shift is the result of intramolecular hydrogen transfer[148]. A detailed study of the photoinduced proton transfer within the acetonaphthol **132** has been carried out. The work investigated the internal twisting processes within the molecule. Interestingly, the H-bonded structure in the S^1 state is stabilized by about 2 kcal mol^{-1} [149]. Photoketonization of the hydroxyquinoline derivative **133** occurs on irradiation[150]. Excited-state intermolecular hydrogen bonding has been observed (emission at 400 nm) for aqueous solutions of *p-N,N*-dimethylaminosalicylic acid[151].

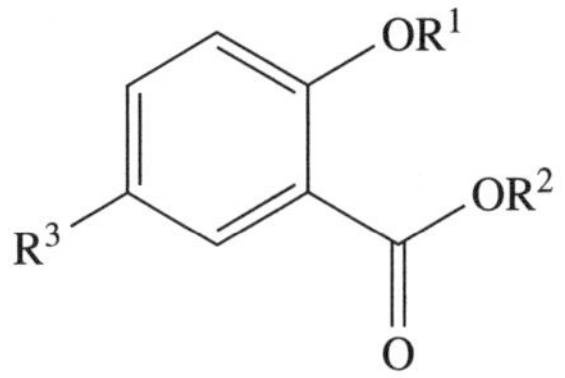

(**130**)

R^1	R^2	R^3
H	H	H
Me	H	H
H	H	Me
H	H	MeO
Me	Me	H
H	Me	Me
H	Me	MeO

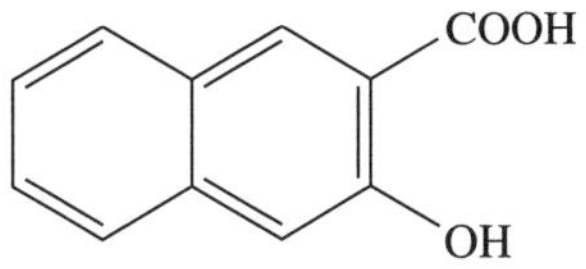

(**131**) 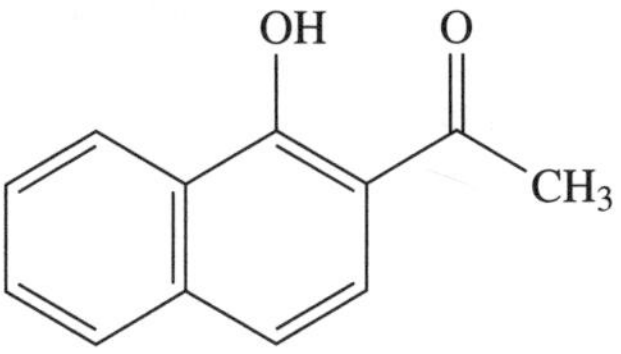 (**132**) 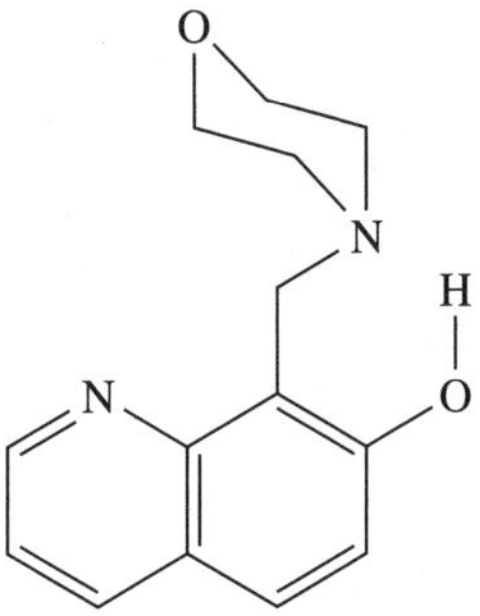 (**133**)

VII. REARRANGEMENTS

A. Skeletal Rearrangements

The structural rearrangement of the phenol skeleton can be brought about photochemically. Childs and coworkers[152] were among the earliest to report the low-yield photoisomerization of phenols using FSO_3H at low temperatures. The process involves the protonation of the phenol at the *para*-position. A better reaction system was found that made use of CF_3SO_3H[153]. Under these acidic conditions and ambient temperatures irradiation gives good yields of the bicyclic enones[154]. The wavelengths required to bring this about depend on the substitution on the phenol. Thus, for the parent phenol 254 nm light is used while for 2,3,5,6-tetramethylphenol 300 nm light is sufficient (Scheme 10). Others also demonstrated the formation of umbellulone **134** by irradiation of 2-isopropyl-5-methylphenol. The yield of **134** is low at 9.5% and this product is accompanied by 2-methylbicyclo[4.1.0]hex-2-enone in 5% yield. Other processes were reported, notably the group migration reactions to yield the isomeric phenols **135** that are formed by intermolecular alkylation processes[155]. The initial reports of these photochemical transformations demonstrated that there was a wavelength dependence upon the isomerization. Thus, the irradiation at λ > 320 nm of alkylphenols **136** in the presence of FSO_3H at −78 °C leads to structural isomerism with the formation of isomeric phenols shown in Scheme 11[152]. The reaction is wavelength-dependent and, for example, irradiation of 2,4,6-trimethylphenol **137** under the same conditions as above but using 360 nm leads to the formation of the bicyclic ketone **138** as well as the alkyl migrated phenol **139**. Indeed, the reaction path to these bicyclic ketones is quite general and an example is shown in Scheme 12. From these investigations it is clear that the reaction path involves protonation at C4 of the phenol and irradiation converts this into a bicyclic ion that rearranges to ion **140** by migration of C4. These ions can be quenched to afford the bicyclohexenones or can undergo photochemical rearrangement to the isomeric phenols (Scheme 13). Quantum yields for the rearrangements have been measured and are in the 0.65 to 0.018 range[156]. Some idea of the scope of the process and the regioselectivity exhibited of the rearrangement of the protonated phenol into the bicyclic cation can be seen in Scheme 14. Chadda and Childs[157] also noted that phenols underwent photochemical isomerism in the presence of $AlBr_3$. Fundamentally, the outcome is the same as the use of acids described above. There is the involvement of the *p*-protonated ion **141** and irradiation converts this to the bicyclic ion **142** that can be isolated as the enone **143** or can undergo further photochemical reaction to yield the isomeric phenols (Scheme 15). Kakiuchi and coworkers[158] also examined the reactivity of differently substituted phenols in the presence of $AlBr_3$. This, like the earlier work, involves the formation of cation **144** from the phenol **145**. This cation undergoes photochemical conversion into the bridged ion **146** and it is from this that the bicyclohexenones **147** are formed. The reaction is substitution-pattern-dependent and only **145a** and **145b** undergo the rearrangement[158]. They also examined some alkylated phenols, the 3-, 4- and 5-methyl derivatives that are also photochemically reactive under the same conditions. Thus independent irradiation of the three phenols **148**, **149** and **150** affords a mixture of all three. The reason for this is that the ions undergo methyl migrations and undergo transformation via the three species shown in Scheme 16. Earlier work by the same group[159] demonstrated that rearrangement of this type took place with 2-naphthols.

B. Side-chain Rearrangement

Rearrangement within side-chains has also been observed. Eugenol is photochemically reactive and irradiation in methanol brings about conversion of the side-chain into a cyclopropyl moiety (Scheme 17). The path to this product is a di-π-methane process that brings

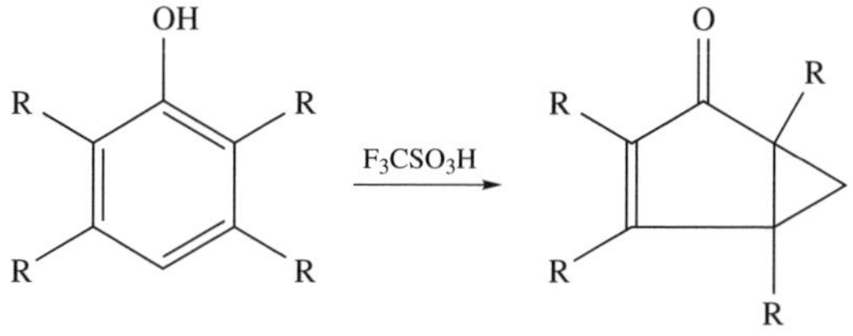

14. Photochemistry of phenols

SCHEME 10

R = H   75%
R = Me   50%

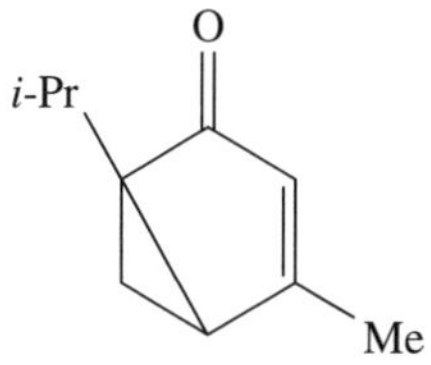

(134) 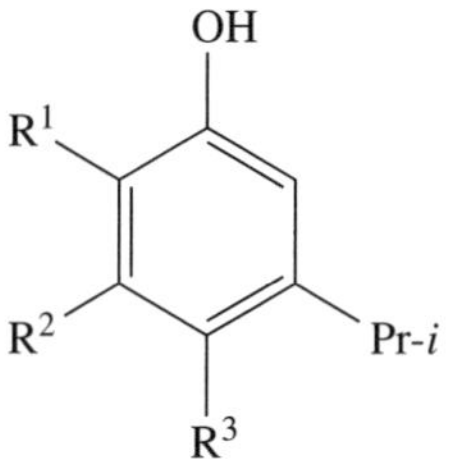   **(135)**

R^1	R^2	R^3
H	i-Pr	Me
i-Pr	Me	H
i-Pr	H	Me

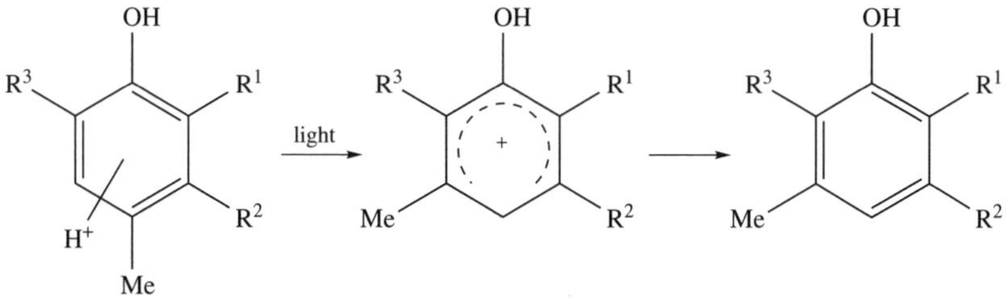

(136)

R^1	R^2	R^3
H	Me	H
Me	H	H
Me	Me	H
H	Me	Me
Me	H	Me
Me	Me	Me

SCHEME 11

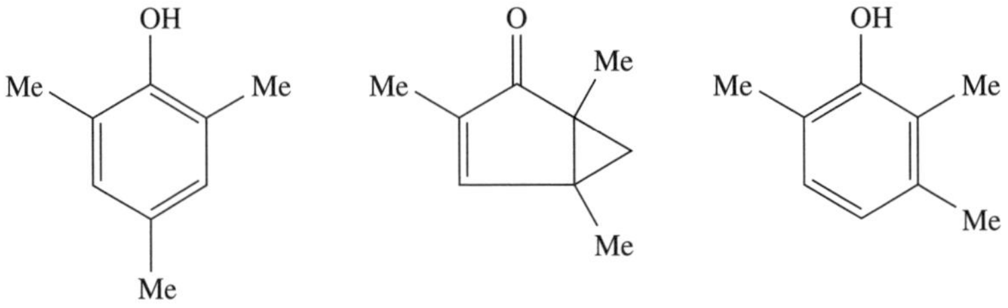

(137) **(138)** **(139)**

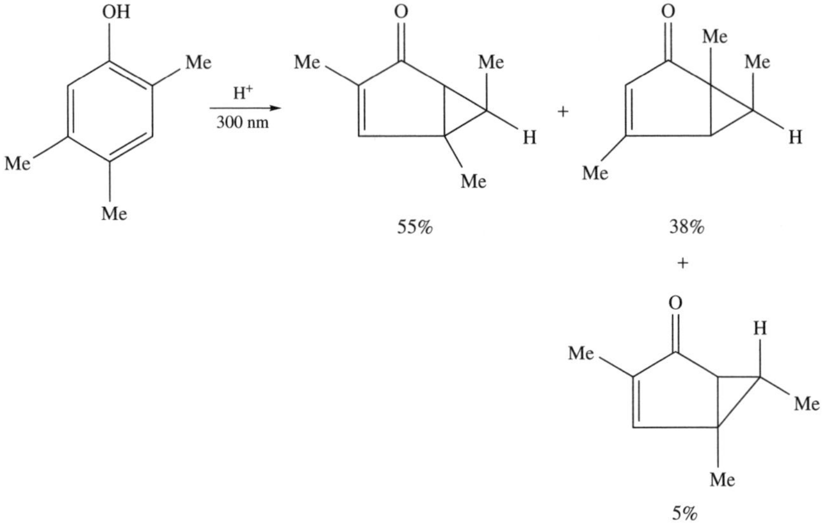

SCHEME 12

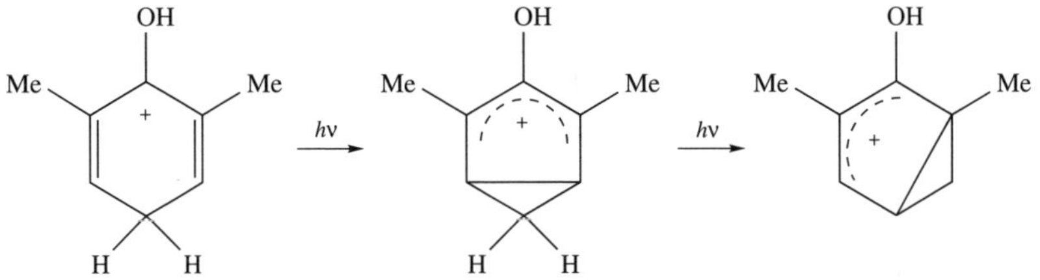

SCHEME 13

about the rearrangement via the biradicals **151** and **152**. Photoaddition of alcohol to the double bond takes place when the reaction is carried out in methanol[160]. The alkene Latiofolin (**153**) is also photochemically active by a di-π-methane reaction mode and converts on irradiation in CCl_4 into the cyclopropane derivative **154**, 58%[161]. Conversion of a side-chain to a three-membered ring is also reported for the irradiation at wavelengths >300 nm of 1,2-diaryl-2-bromoalkenes **155** along with NaH in a 18-crown-6 ether. This reaction

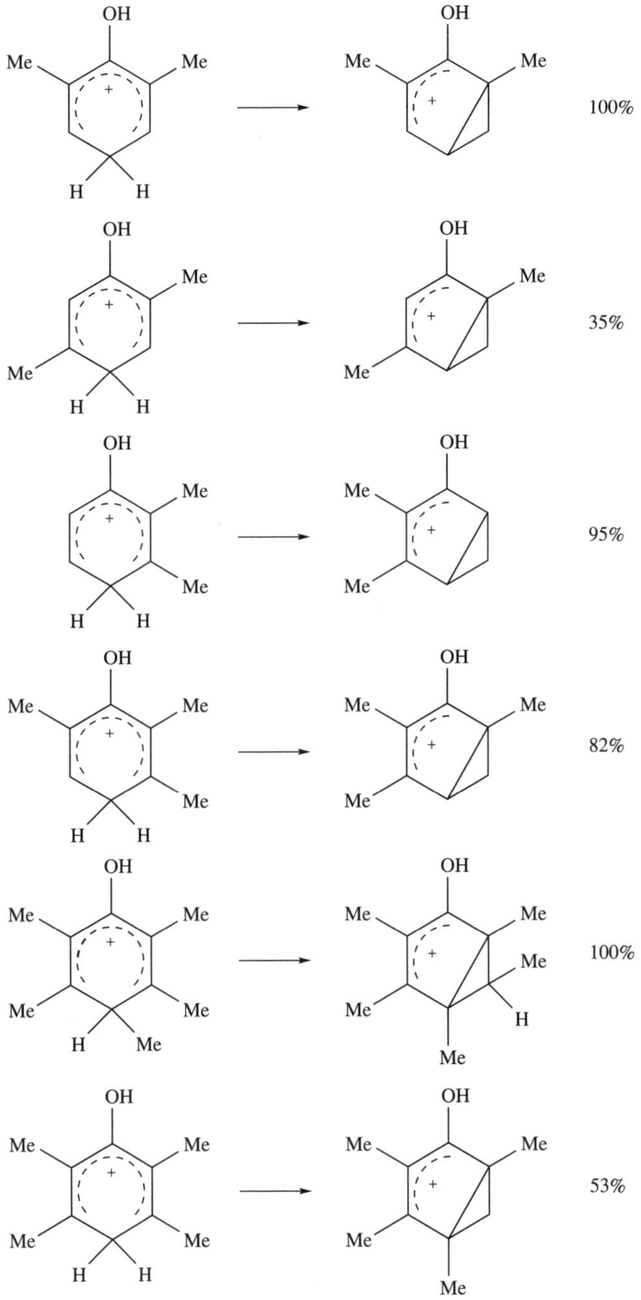

SCHEME 14

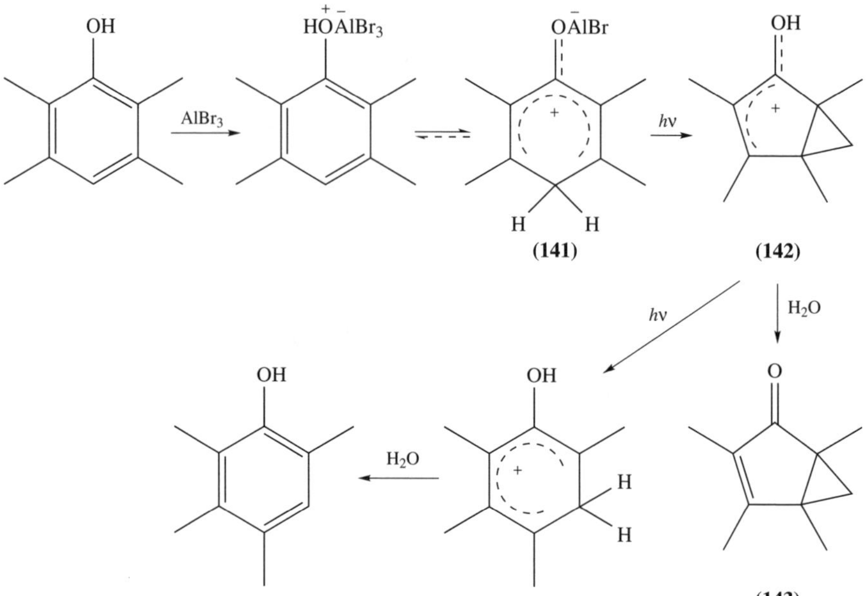

SCHEME 15

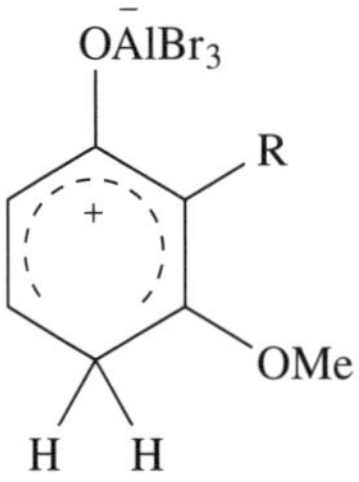

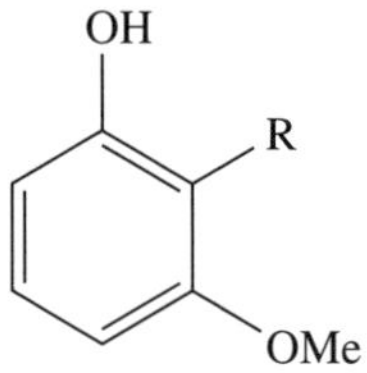

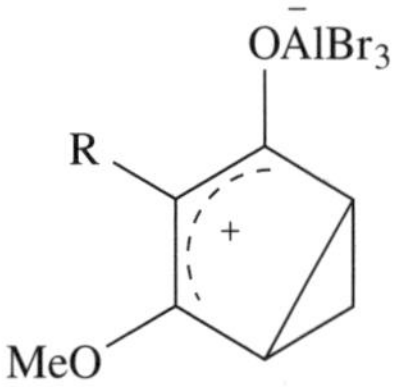

(141) (142) (143)

(144) (145)
(a) R = H
(b) R = Me

(146)

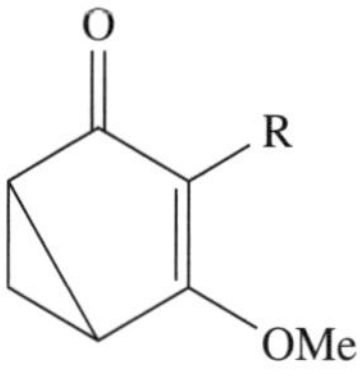

(147)

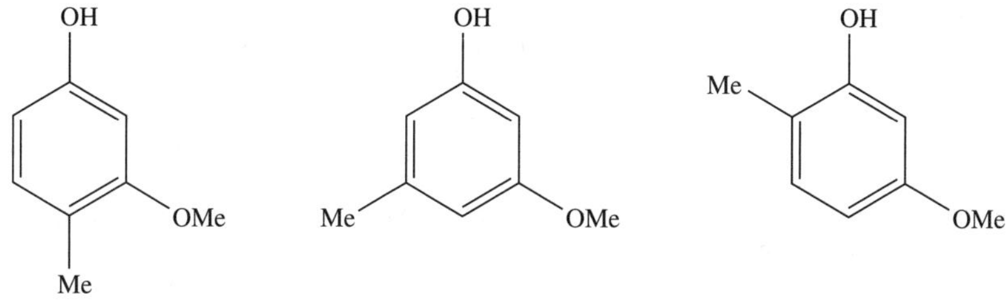

14. Photochemistry of phenols

(148), **(149)**, **(150)**

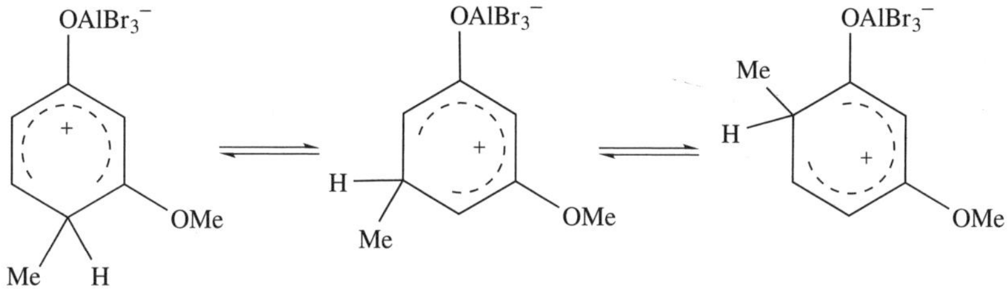

SCHEME 16

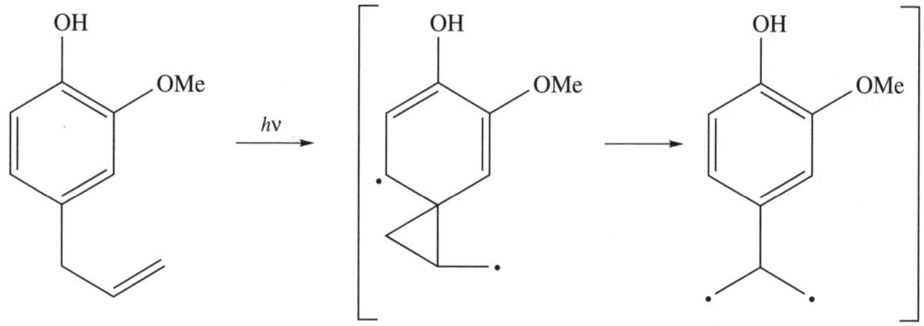

(151), **(152)**

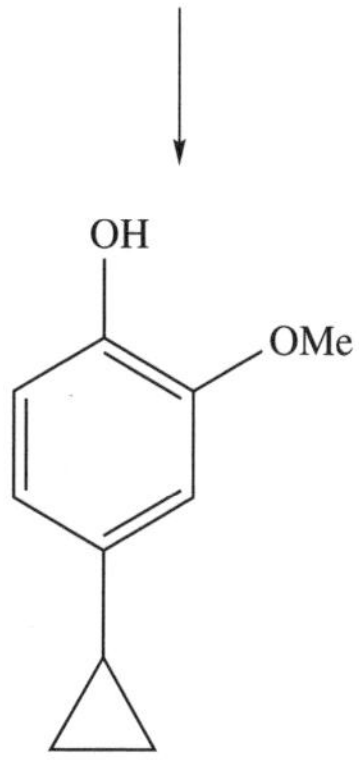

SCHEME 17

involves excitation of the corresponding phenoxide with intramolecular displacement of the bromo substituent. This reaction path affords the products **156**[162].

4-Hydroxybenzonitrile is converted on irradiation in deoxygenated water, methanol or ethanol into 4-hydroxybenzoisonitrile in high chemical yield. A two-photon process is involved and the intermediate **157** is the key to the reaction[163]. An analogous process is observed on irradiation of 4-nitrosophenol[164]. This yields *p*-benzoquinone as the final product. The rearrangement is thought to follow the path shown in Scheme 18. Again the rearrangement of the side-chain involves a three-membered ring.

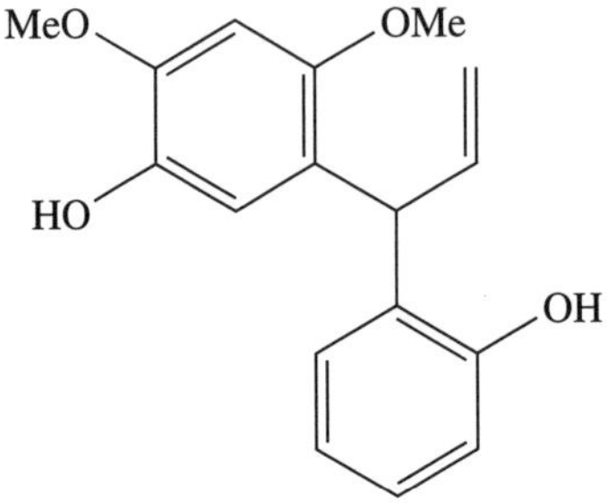

(153)

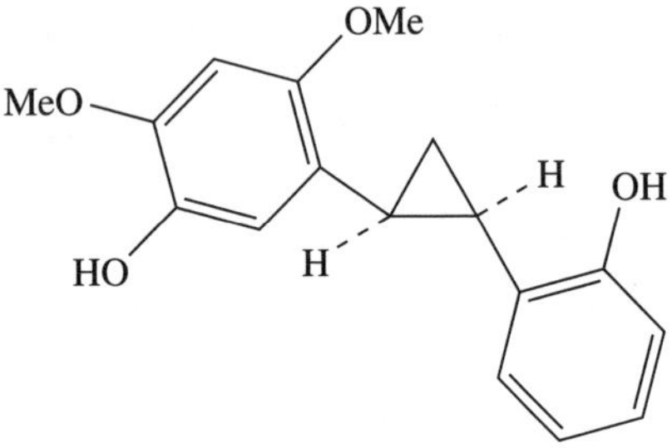

(154)

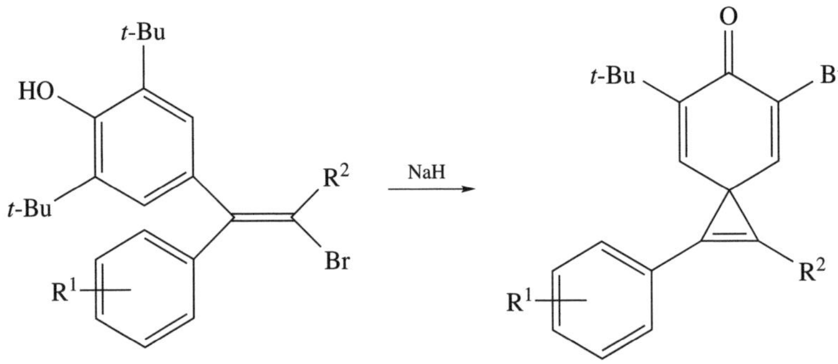

(155)

(156)

R¹	R²	yield (%)
4-MeO	Ph	74
4-Me	Ph	38
H	Ph	89
2-Me	Ph	51
4-MeO	2-MeOC$_6$H$_4$	83
4-Me	2-MeOC$_6$H$_4$	54
H	2-MeOC$_6$H$_4$	69
4-Br	2-MeOC$_6$H$_4$	33
4-MeO	Me	20

(157)

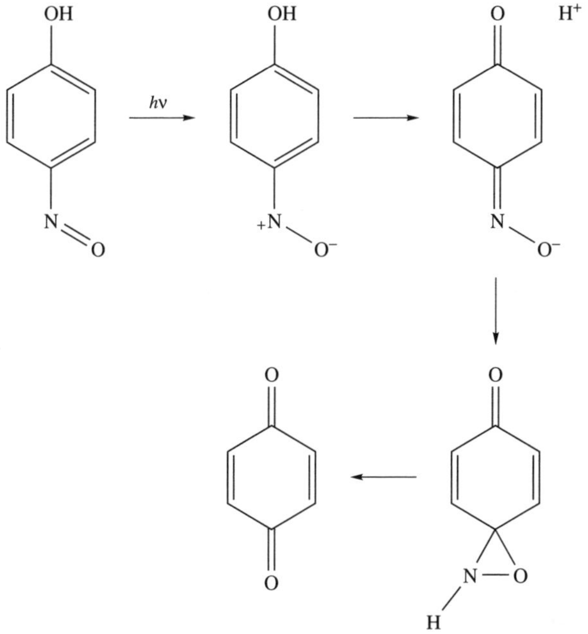

SCHEME 18

C. Side-chain Migration

One of the ubiquitous reactions of phenol derivatives is the photo-Fries process. This has been studied in great detail over the years since it was first uncovered in the 1960s[165,166]. Examples of this process are the photochemical conversion of the salicylate **158** into the 2,2′-dihydroxyketone **159** in a low yield of 8%[167] and the chlorosalicylate **160** into **161**[168]. The mechanistic details have demonstrated that the reaction is basically an intramolecular process. If radical pairs are involved, there appears to be little escape from the cages in which they are formed and little or no intermolecular products are formed. In more recent times the reaction of phenols **162** with free radicals has been investigated. The radicals are formed by the irradiation ($\lambda > 280$ nm) of benzene solutions of pinacolone. The authors[169] suggest that the products obtained (Scheme 19) are the result of Norrish Type I fission of the ketone to afford a *t*-butyl radical. This then abstracts hydrogen from the phenol to yield a phenoxyl radical. Coupling between this and the acetyl radical forms

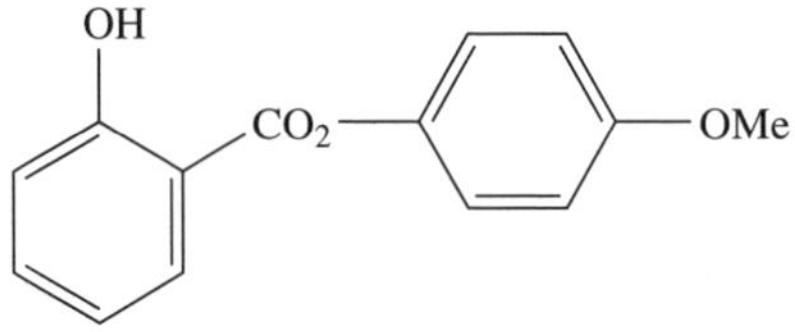

(158)

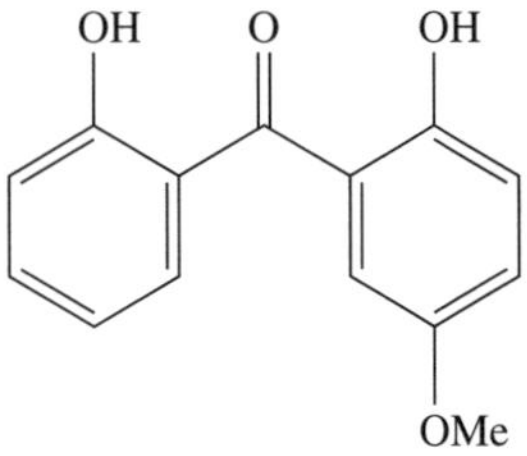

(159)

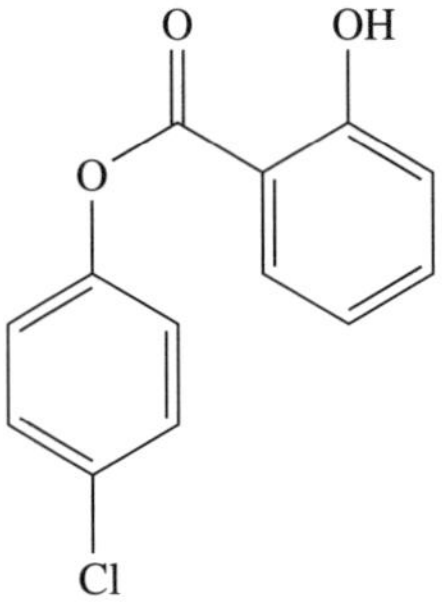

(160)

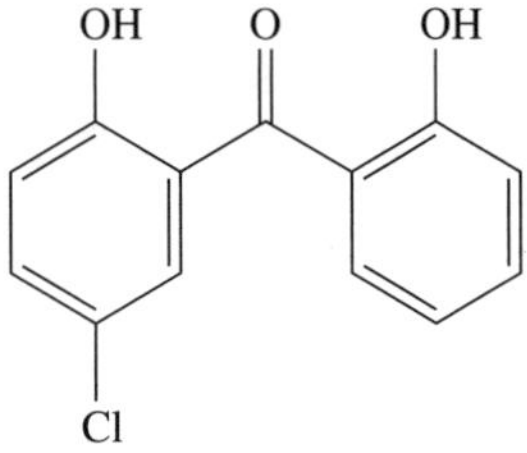

(161)

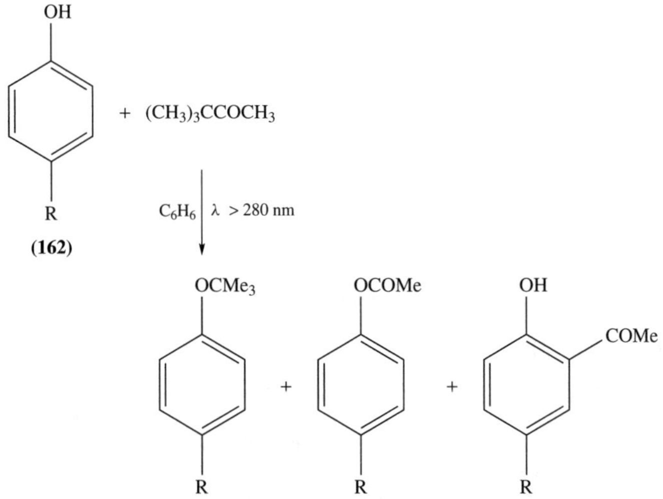

(162)

R	yield (%)		
t-Bu	22	36	42
Ph	27	31	42
CF_3	—	37	63
OMe	11	15	30

SCHEME 19

14. Photochemistry of phenols

the final products. The reaction is interpreted as an intermolecular photo-Fries process. An *ab initio* MO study on twisting around the carbons of the double bond and around the aryl-alkene bond of coumaric acid (*p*-hydroxycinnamic acid) has calculated the potential energy surfaces for such a process[170].

VIII. CARBENE FORMATION

A. Elimination of Hydrogen Halides from Phenols

Elimination of hydrogen halides from *p*-substituted halophenols has provided a path to the triplet carbene 4-oxocyclohexa-2,5-dienylidene (**163**). Initially, 4-chlorophenol in aqueous solution was subjected to nanosecond laser-flash photolysis[171]. Other studies using FT-EPR on this species indicate that a mechanism involving free radicals does not operate and most likely the elimination of HCl is a concerted process[172]. Other halophenols, i.e. 4-fluoro, 4-bromo and 4-iodophenol, have been studied using both steady-state and time-resolved photolysis and again the carbene **163** is formed by loss of HX. Reaction between the carbene and oxygen produces benzoquinone O-oxide that ultimately rearranges to 1,4-benzoquinone. This path is the same as that described for 4-chlorophenol[173]. Substituted carbenes can also be formed from this reaction mode and the irradiation of 5-chloro-2-hydroxybenzonitrile (**164**) in aqueous solutions results in the formation of the triplet carbene **165** by loss of HCl[174]. In oxygenated solutions the oxide **166** is formed by trapping of the carbene and it is this compound that leads to the quinone. In the absence of oxygen the main products observed are the isomeric biphenyls **167** and **168** and the hydroquinone **169**. Carbenes are presumably also involved in the photochemical conversion of 2-chlorophenolate[175,176] or 2-chlorophenol[177] to cyclopentadiene 5-carboxylic acid as illustrated in Scheme 20. Substituted phenolate derivatives behave similarly as do di- and tri-chlorophenolates[175]. This process is reminiscent of the Wolff rearrangement of α-diazoketones.

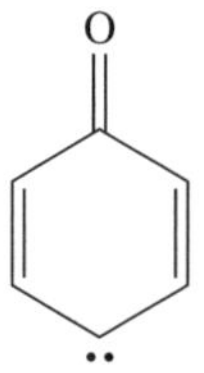

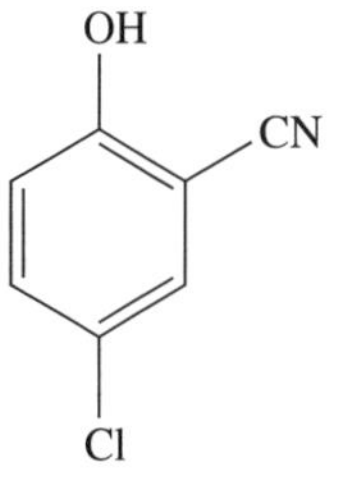

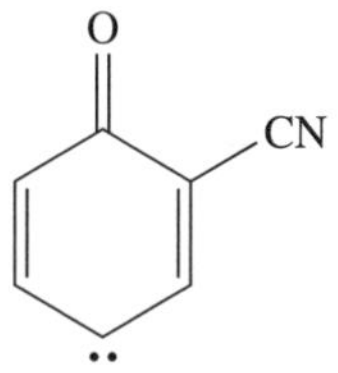

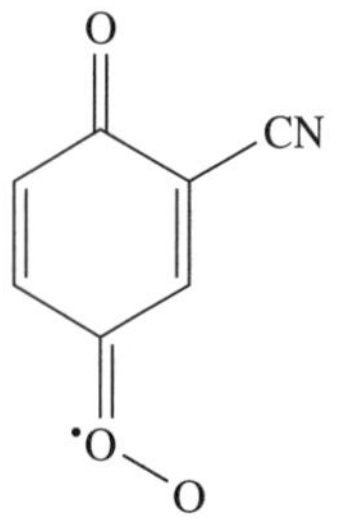

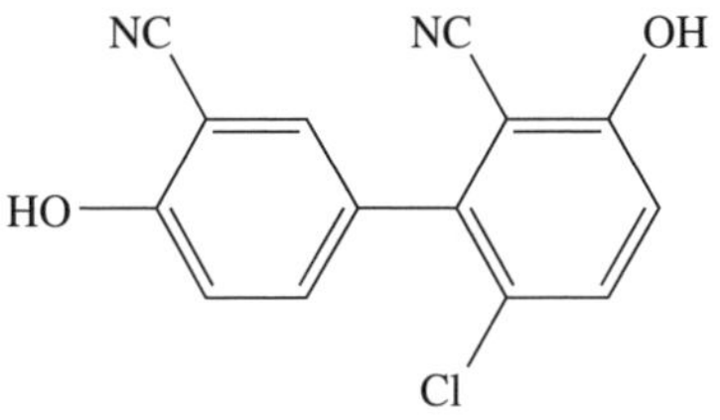

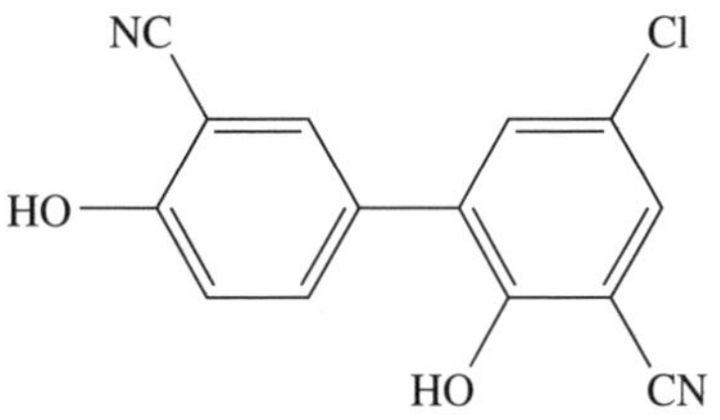

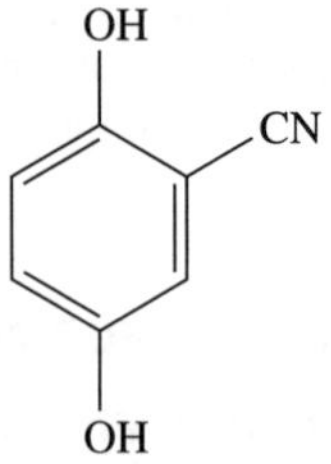

(169)

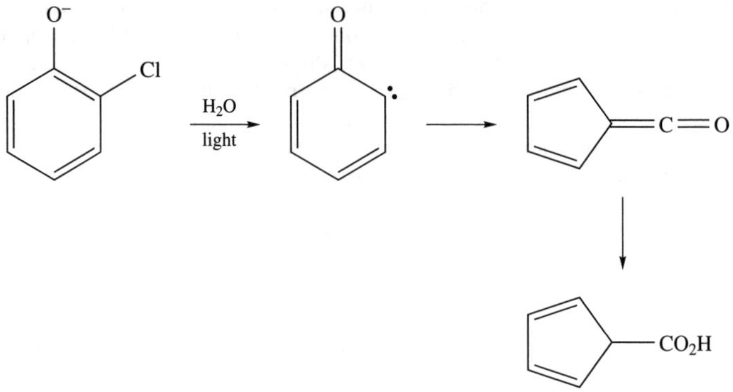

SCHEME 20

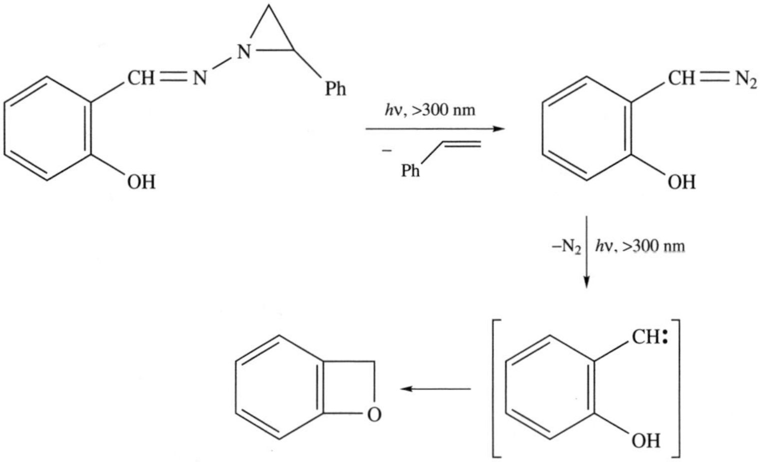

SCHEME 21

B. Other Carbenes

Carbenes on sites adjacent to the phenolic group have also been generated. Thus, 2-hydroxyphenyl carbene **170** has been obtained by the pathway shown in Scheme 21. The carbene reacts with the hydroxy group to afford the oxetene (**171**). This itself is photochemically labile and undergoes ring opening to a quinonemethide[178,179].

IX. CYCLOADDITIONS
A. Intermolecular Addition Reactions

Gilbert and his coworkers[180] have demonstrated the intermolecular addition of alkenes to phenols. The example shown in Scheme 22 is an example of the [3 + 2]-addition of *trans*-1,2-dichloroethene to phenol. The reaction is efficient and yields the principal adduct **172** in 70% yield. The reaction follows the path shown in Scheme 22. The cresols (*o-*, *m-* and *p-*) also undergo this mode of addition and from this study the orientation of the alkene to the cresol is as shown in **173**. This mode of addition yields the adduct **174**[180].

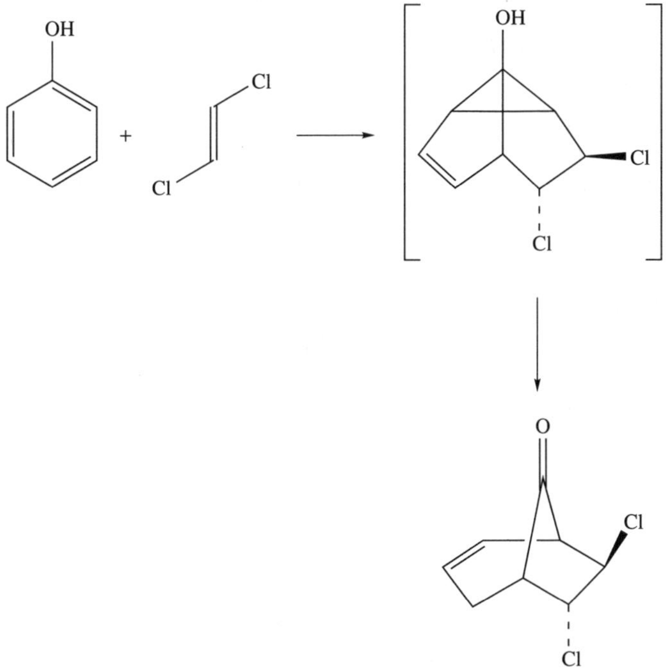

SCHEME 22

(2 + 2)-Intermolecular photocycloaddition also occurs between alkenes and simple phenols. The swing from *meta* addition illustrated above in the [3 + 2]-mode to *ortho*-addition is a result of charge-transfer interactions between the alkene and the phenol and a greater charge transfer favours the *ortho*-addition mode. These aspects have been the subjects of reviews[181,182]. This reaction mode is exemplified by the addition of acrylonitrile

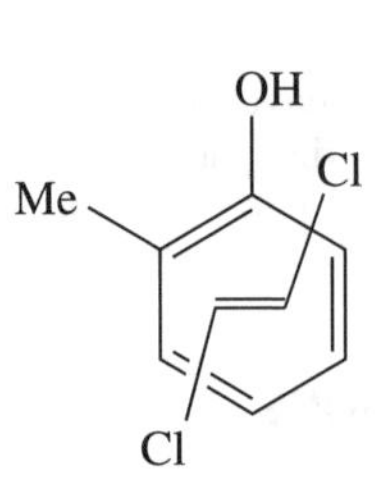

(173) 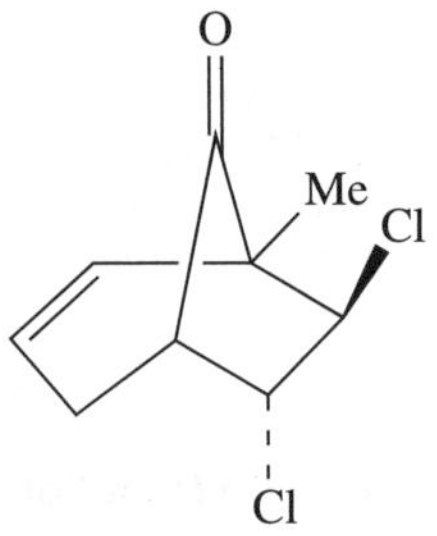 (174)

to *p*-cyanophenol and *p*-carboxymethylphenol. The product from the first addition is the cyclooctadienone **175** that arises from ring opening of the (2 + 2)-adduct **176**. The addition of *p*-carboxymethylphenol affords the bicyclic adduct **177**. An increase in the electron-donating ability of the phenol changes the reaction path between the phenol and acrylonitrile and substitution results. Thus, with hydroquinone, **178** is formed while *p*-methoxyphenol affords **179** and **180**[183].

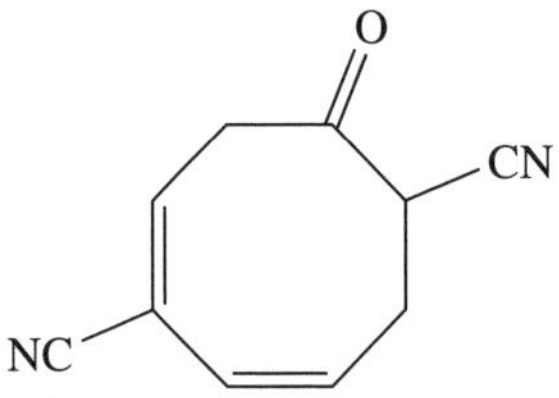

(175) 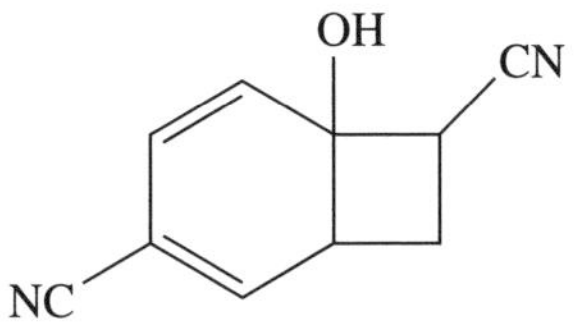 (176) 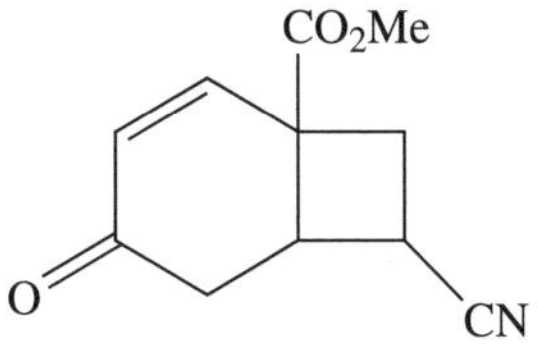 (177)

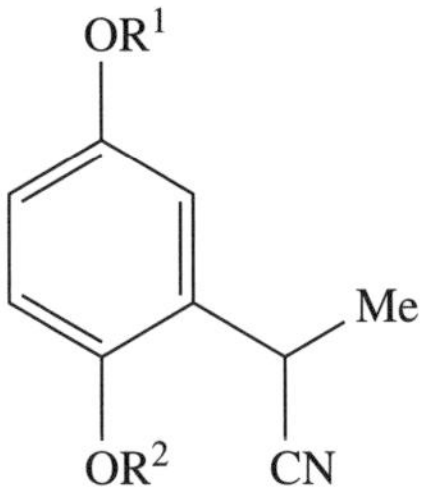

(178) $R^1 = R^2 = H$
(179) $R^1 = H, R^2 = OMe$
(180) $R^1 = OMe, R^2 = H$

Intermolecular addition accounts for the formation of the products **181** in Scheme 23. Here, irradiation brings about addition of the cyano group of the naphthalene derivative to the phenol immediately adjacent to the hydroxy group. The resultant (2 + 2)-cycloadduct is unstable and ring-opens readily to yield the azocines[184,185].

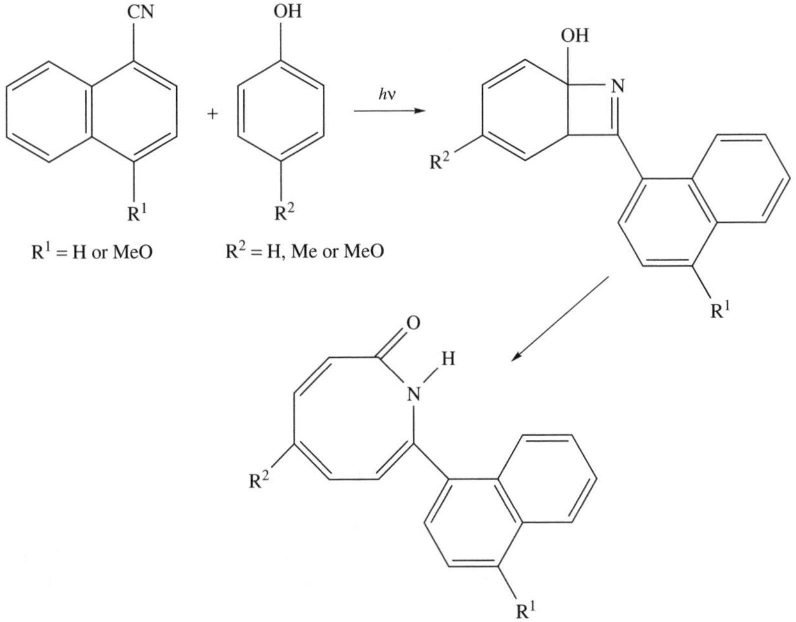

SCHEME 23

B. Intramolecular Addition Reactions

Intramolecular addition is also reported for the quinhydrone derivative **182**. This cyclophane apparently can exist in two conformations **182** and **183** but the intramolecular addition involves **182** only. The addition product formed is the *meta*-product **185** that arises

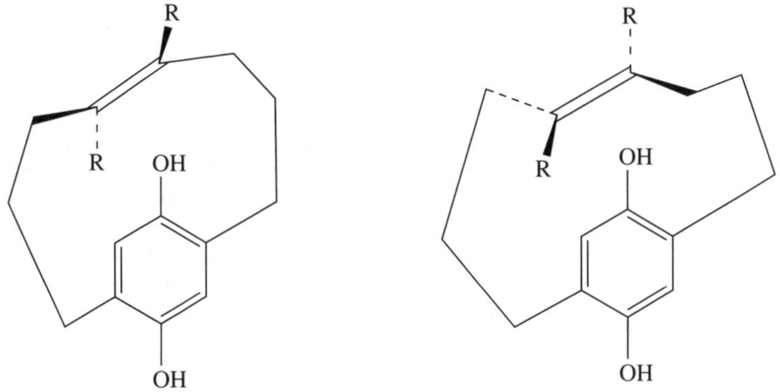

(**182**) R = H
(**186**) R = MeO or Me

(**183**) R = H

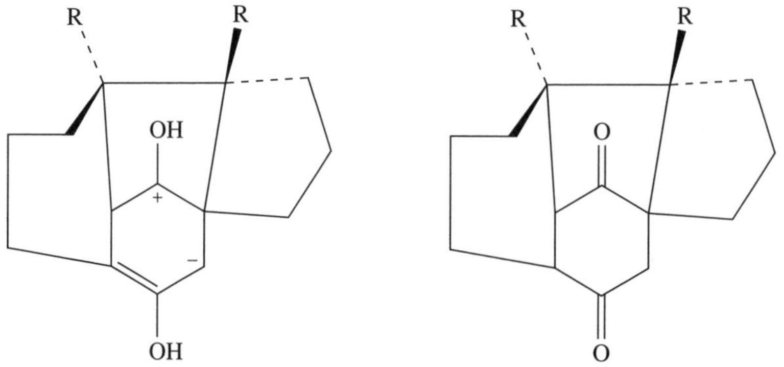

(184) R = H
(187) R = MeO or Me

(185) R = H
(188) R = MeO or Me

via the zwitterionic intermediate **184**. When the alkene group is more heavily substituted, as in **186** R = MeO or Me, the interconversion between the conformational arrangements is slowed down. However, the addition mode is the same with these derivatives and irradiation affords the adducts **188**, R = MeO or Me via the corresponding zwitterions **187**, R = MeO or Me[186]. Intramolecular cycloaddition is also exhibited by the irradiation of the phenol derivative **189** in benzene. This treatment affords the crystalline product **190** in 25% yield[187].

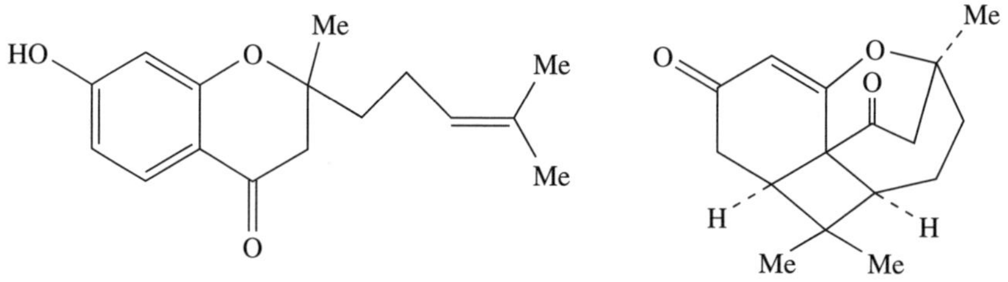

(189)

(190)

The reaction has been developed further since these earlier observations. This reaction mode of phenols has provided a useful synthetic path for the synthesis of complex molecules. The UV irradiation of the phenols **191**, **192** in the presence of acid affords the adducts **193** where addition has taken place at C3—C4 of the phenol. This can be readily ring-opened to yield **194**. To a lesser extent addition also follows the C2—C3 addition path that yields **195**, which can be ring-opened to afford **196**[188,189]. Interestingly, the addition only occurs with either a 2 or 3 carbon chain separating the alkene from the phenol. With a longer chain the addition fails. The study also examined disubstitution on the terminal carbon of the alkene moiety. Addition does occur with methyl and ethyl substituents, but fails with isopropyl groups (Scheme 24)[189]. Additional studies (Scheme 25) have demonstrated that the addition can be quite general. Thus irradiation of the phenol derivative **197** in acetonitrile with dilute sulphuric acid affords the enone **198**. This can be treated further to bring about cleavage of the cyclic enol ether moiety. This undergoes

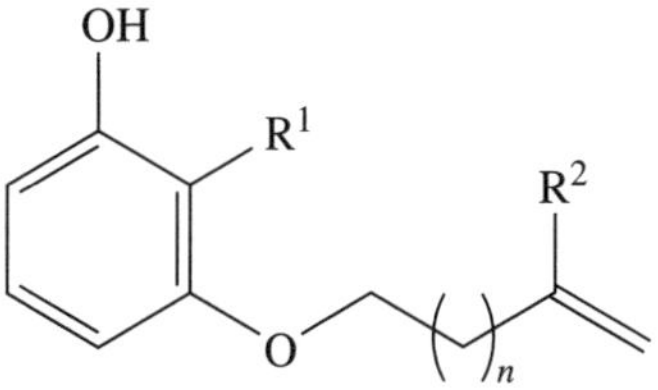

(191) n = 1, R¹ = R² = H
n = 1, R¹ = Me, R² = H
n = 1, R¹ = H, R² = Me
(192) n = 2, R¹ = R² = H

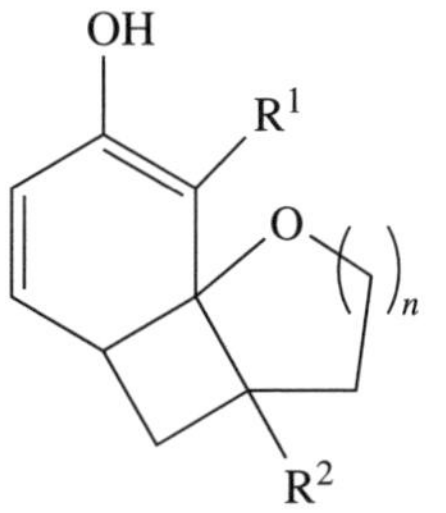

(193)

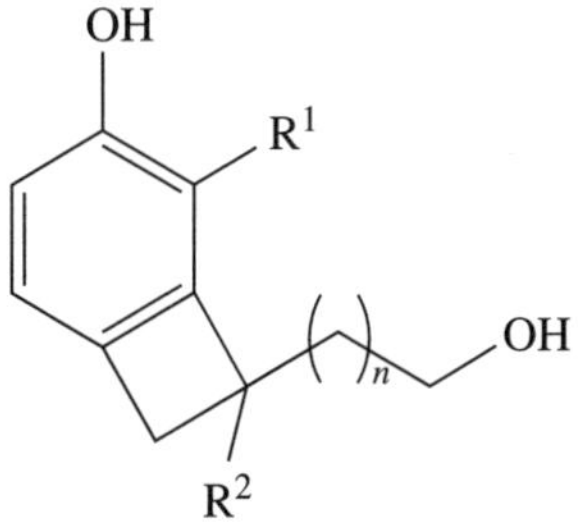

(194) 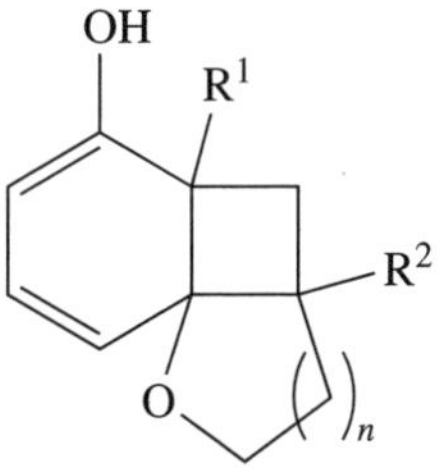   (195) 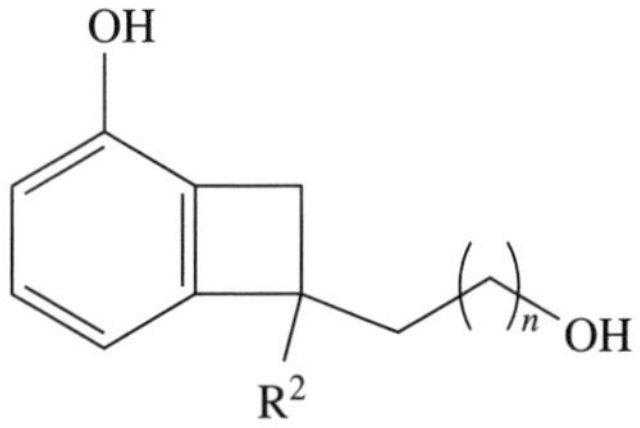   (196)

R¹	R²	n	yield (%)	194:196 ratio
H	H	1	50	9:1
H	H	2	59	2:1
H	H	3	—	—
H	Me	1	60	5:1
Me	H	1	49	1:0

acid-catalysed ring opening in methanol to yield the two products **199** and **200**[190]. Other examples of this elegant approach to such molecules are illustrated by the conversion of **201** into **202** and **203**. The yields of **203** are given. Intramolecular addition also occurs on irradiation of the cyanophenols **204a, b**. This yields the enones **205a, b** in 62% and 39%, respectively. Interestingly, another product is obtained in 11% from the derivative **204b** and this was identified as the phenol **206**. This arises from the biradical **207**, which either cyclizes to yield **205b** or undergoes attack on the cyano function. This on hydrolysis affords the phenol **206**[191].

C. Cyclizations Involving Aryl Radicals

One of the more common cyclization processes of phenols is the formation of the phenanthrene skeleton. These processes utilize the fission of a C—X bond to yield aryl radicals that then attack a neighbouring aryl group. Typical of this is the bromo or iodo derivative of **208** that cyclizes to afford the lactam **209** in moderate yields[192]. The

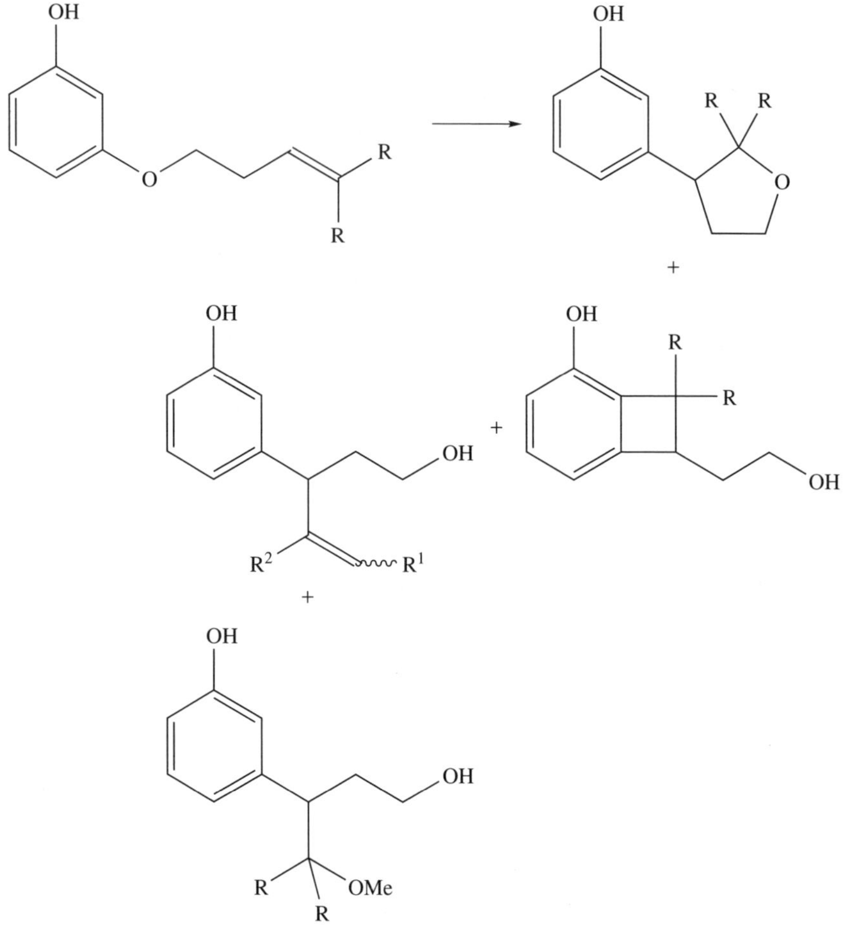

R				
Me	10	13	10[a]	37
Et	20	11	40[b]	15

[a] R^1 = H, R^2 = Me, [b] R^1 = Me, R^2 = Et.

SCHEME 24

bromophenol derivative **210** is also reactive and the cyclization of this has been used as an approach to the phenanthrene skeleton **211** of the alkaloid bulbocapnine. Cyclization is also observed with the isomeric compound **212** that yields the aporphine alkaloid domesticine **213**[193]. Interestingly, the irradiation of the stilbene system **214**, X = Br also follows the same reaction path, i.e. loss of a bromine atom. However, the resultant radical does not cyclize but merely undergoes reduction to **214**, X = H. Further irradiation of **214**, X = H does bring about cyclization via the stilbene-type reaction mode and affords

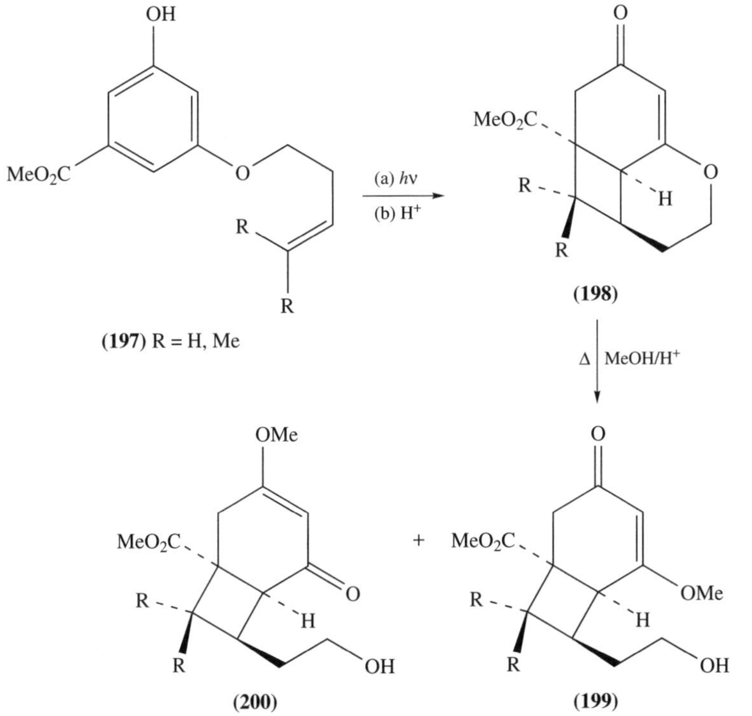

14. Photochemistry of phenols

SCHEME 25

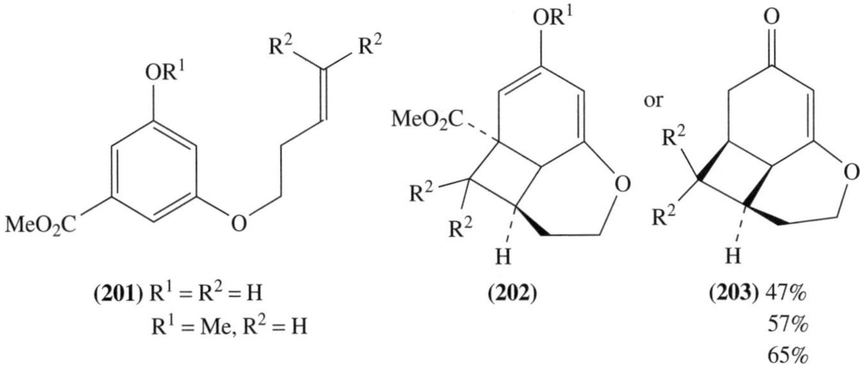

(197) R = H, Me

(198)

(200) + **(199)**

(201) $R^1 = R^2 = H$
$R^1 = Me, R^2 = H$

(202)

(203) 47%
57%
65%

215[194]. The free radical obtained on irradiation of **216** undergoes both modes of addition to the phenol moiety and affords the two cyclized products **217** and **218**. The first of these (**217**) arises by radical formation at the site *para* to the OH group while **218** arises by attack at the methoxy-bearing carbon. The intermediate obtained loses the substituents to afford **218**[195]. Two reports have been made concerning the use to which such cyclisations

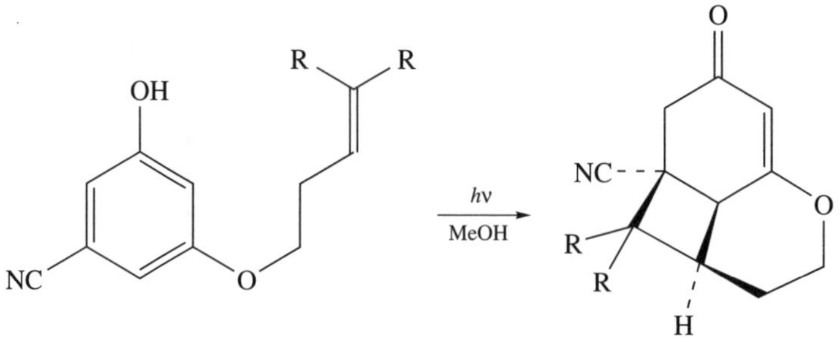

(204) (a) R = H
(b) R = Me

(205) 62%
39%

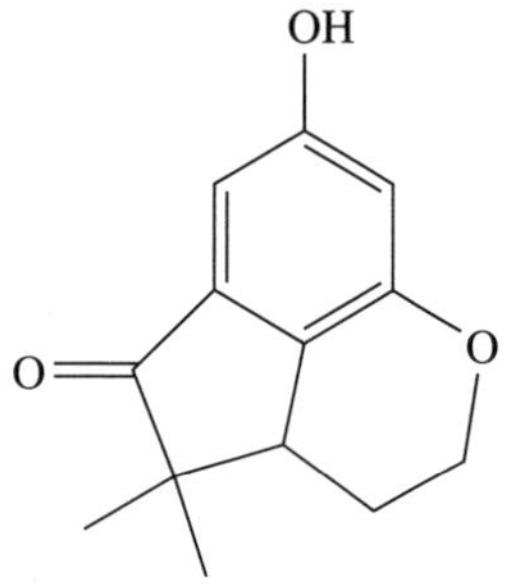

(206) —
11%

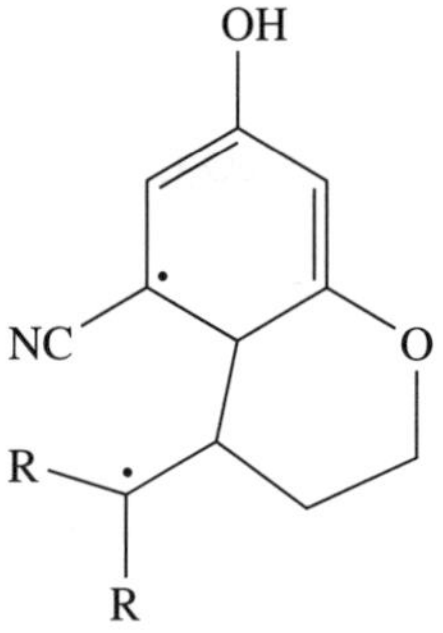

(207)

can be put to synthesize 11-membered ring lactams[196,197]. Specifically, the irradiation of the amide **219** follows the CBr fission path and the resultant radical cyclises to yield the two products (**220**) and (**221**) by both possible addition modes[197]. The major product (**221**) can be transformed into the alkaloid dysazecine.

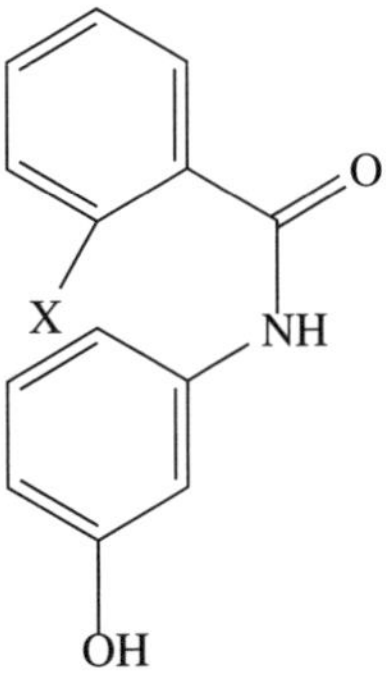

(208) X = Br
X = I

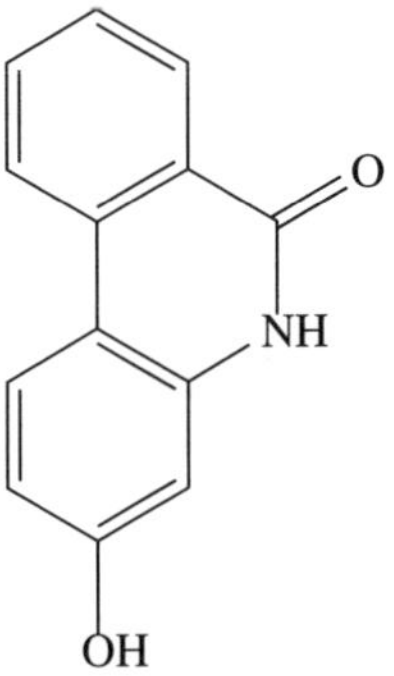

(209) 27%
49%

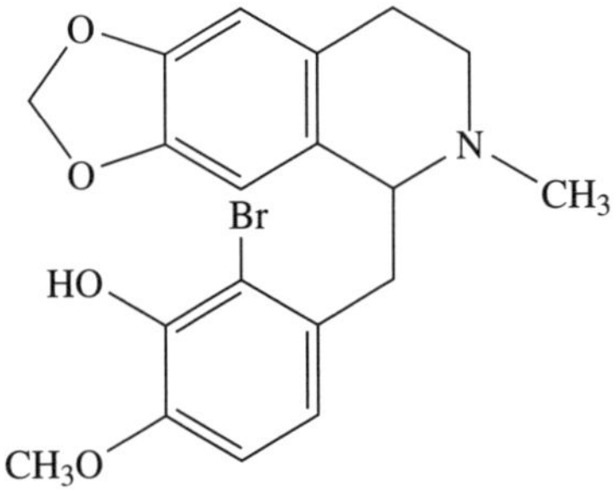

(210)

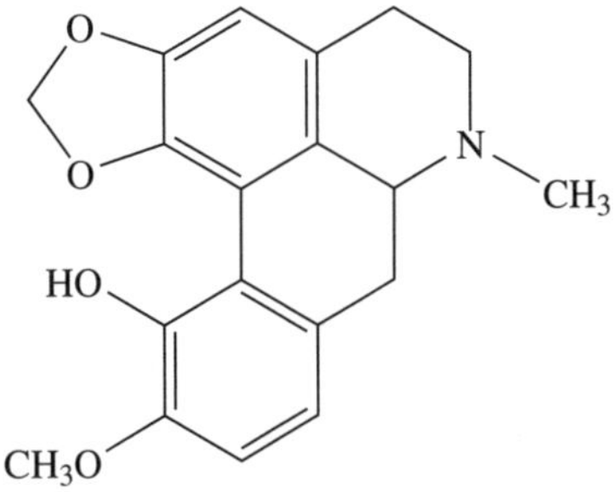

(211)

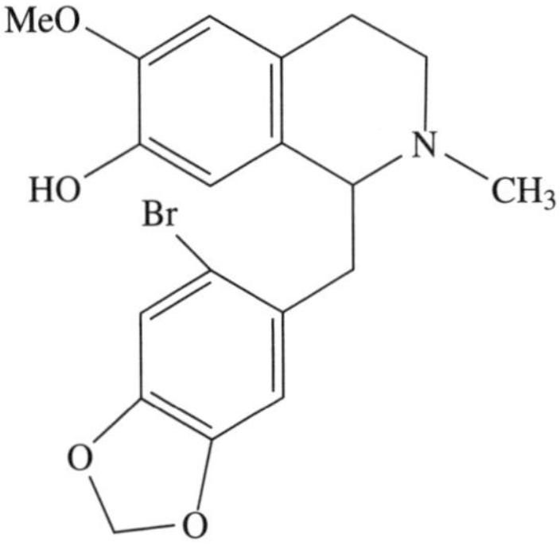

(212)

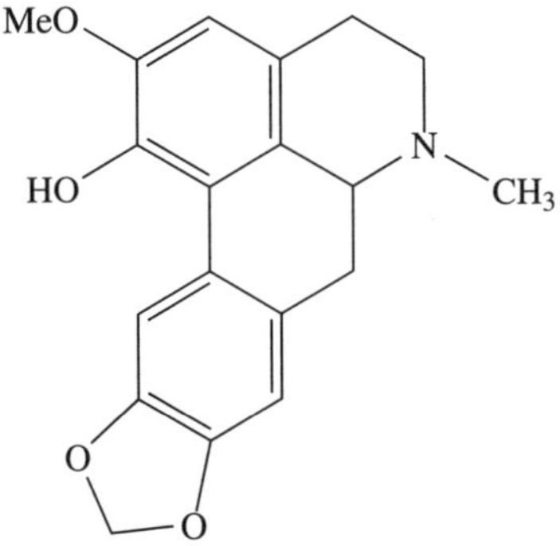

(213)

D. Other Cyclizations

Other cyclizations, this time of the stilbene type, have been reported for the naphthalene derivatives **222**. This provides a route to the highly fluorescent azaphenanthrenoid **223**[198]. The phenol derivative **224** undergoes cyclization to afford **225**. The other cyclized derivative **226** is also formed. Nitrene intermediates have also been suggested and these give

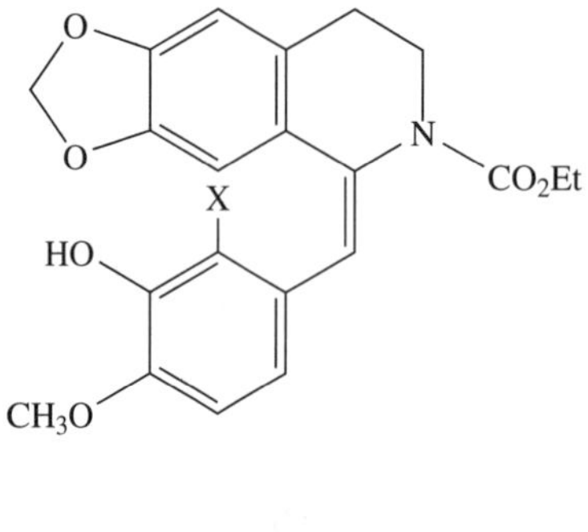

(**214**) X = Br
X = H

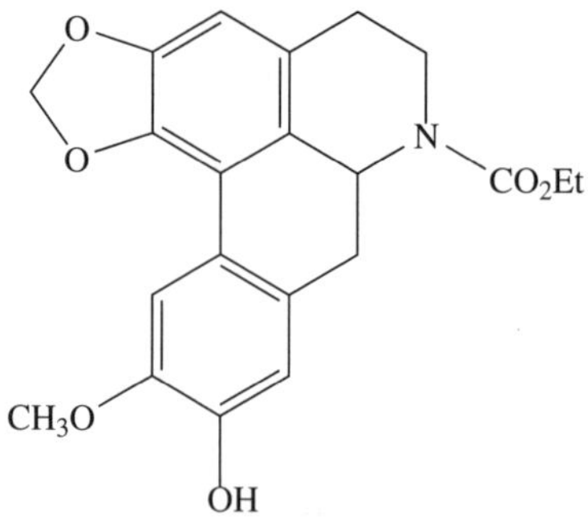

(**215**)

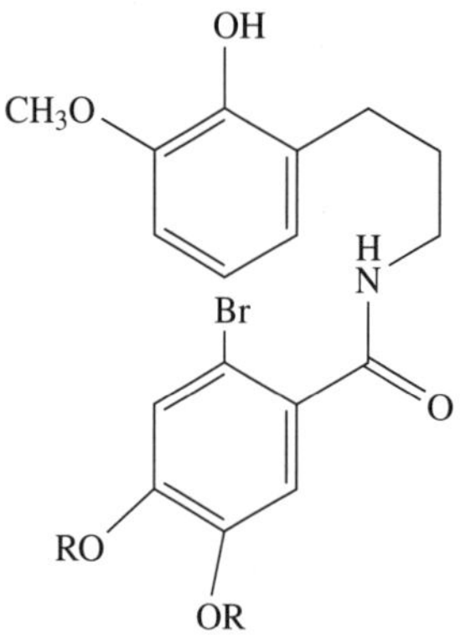

(**216**)
R = Me
RR = CH$_2$

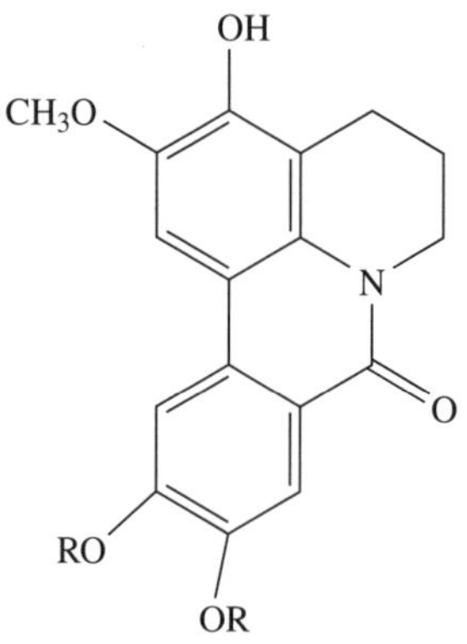

(**217**)
22.6%
9.6%

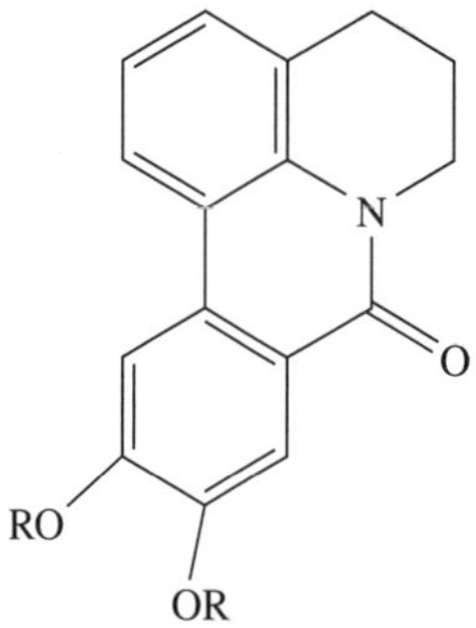

(**218**)
6.4%
7.4%

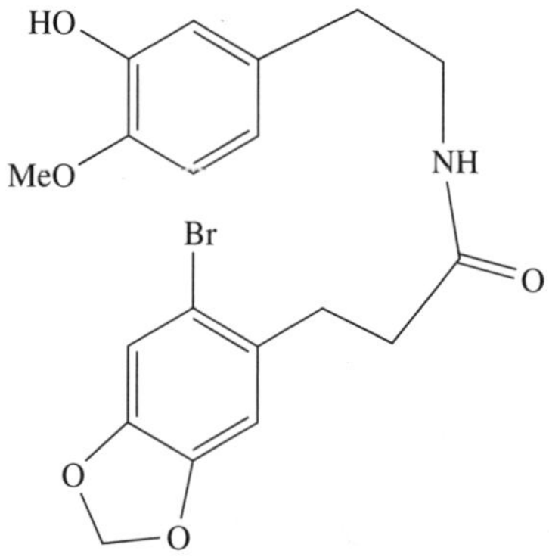

(**219**)

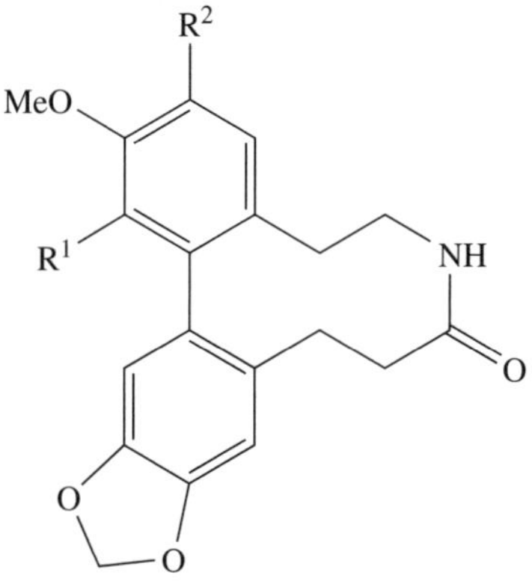

14. Photochemistry of phenols

(**220**) R^1 = OH, R^2 = H
(**221**) R^1 = H, R^2 = OH

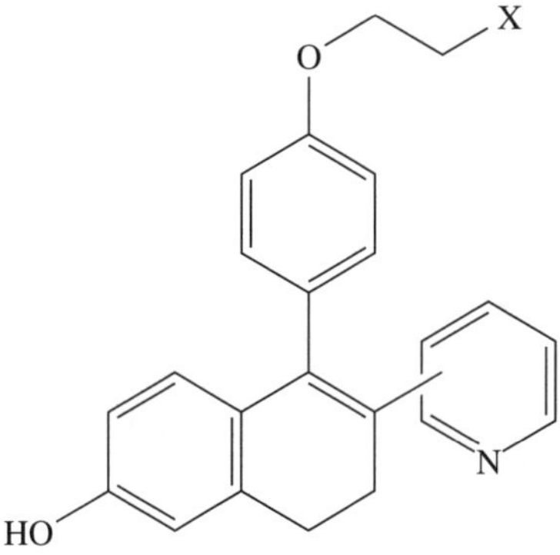

(**222**)
2-, 3- or 4-pyridyl
$X = N{\triangleleft}$
$X = OCH_2Ph$

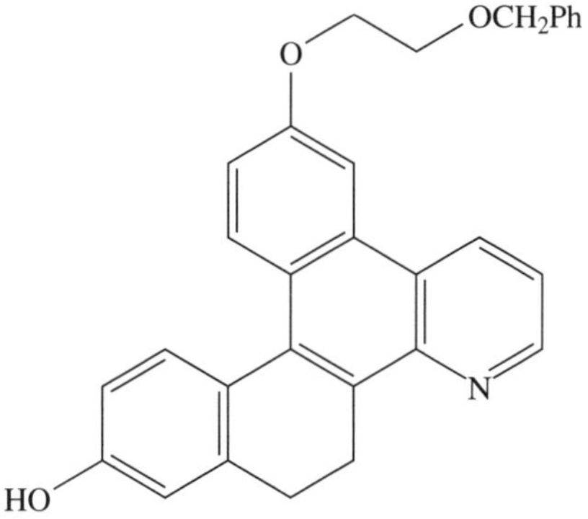

(**223**)

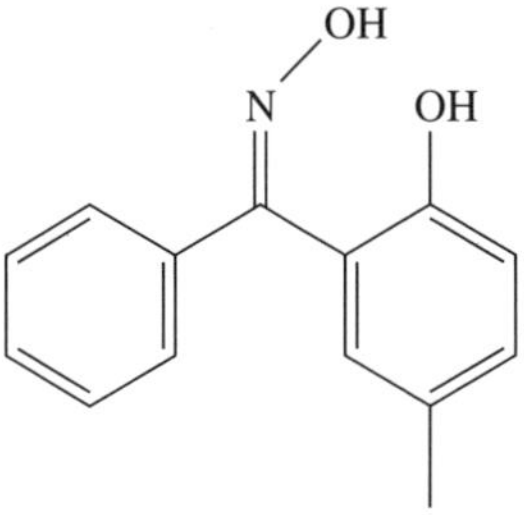

(**224**)

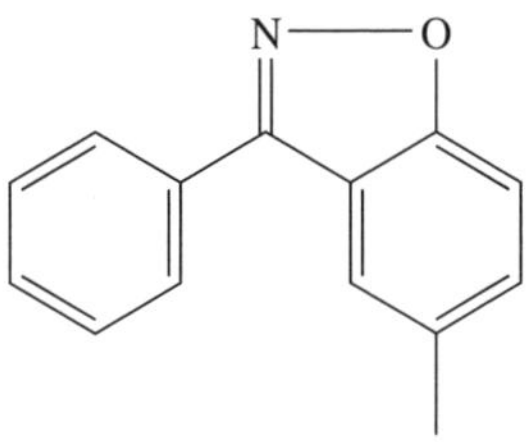

(**225**)

rise to the products identified as **227** and **228**[199]. The irradiation of the phenolic enone orientalinone **229** yields the two products isoboldine (**230**) and isothebaine (**231**) in low yields of 9% and 3%, respectively[200].

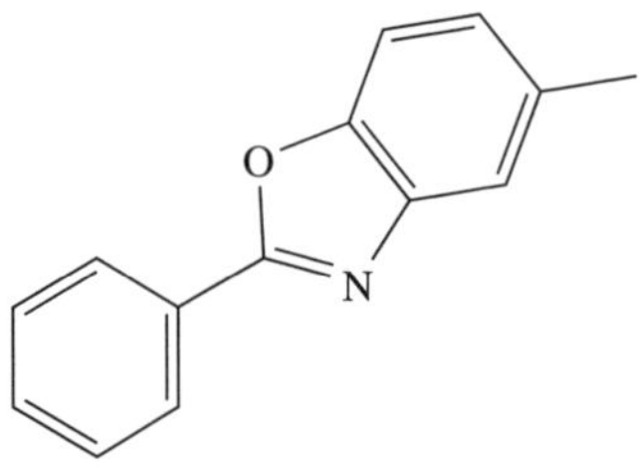

(226)

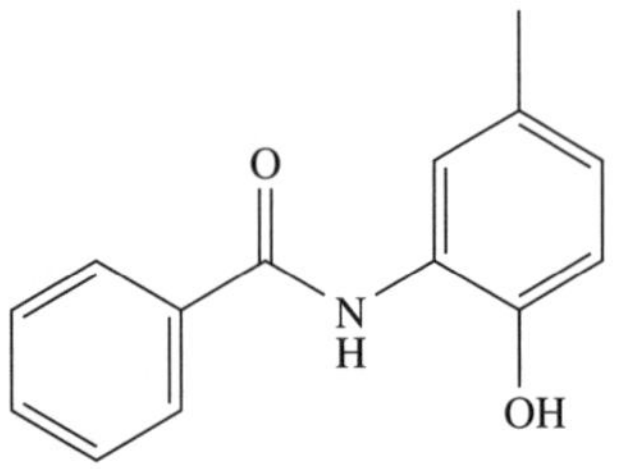

(227)

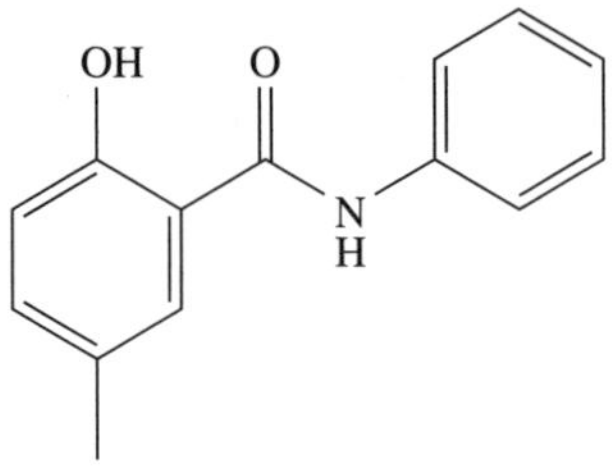

(228)

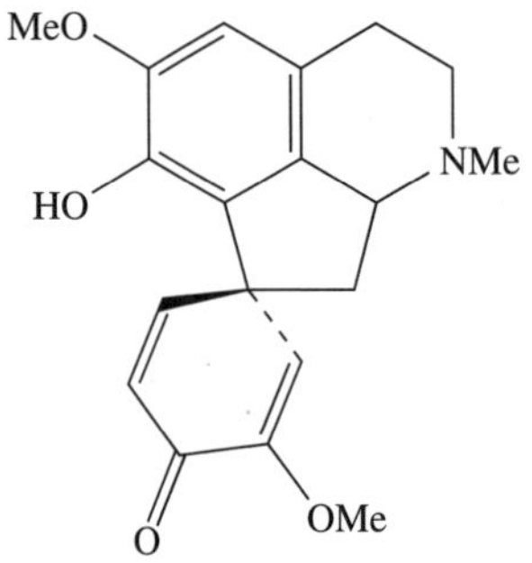

(229)

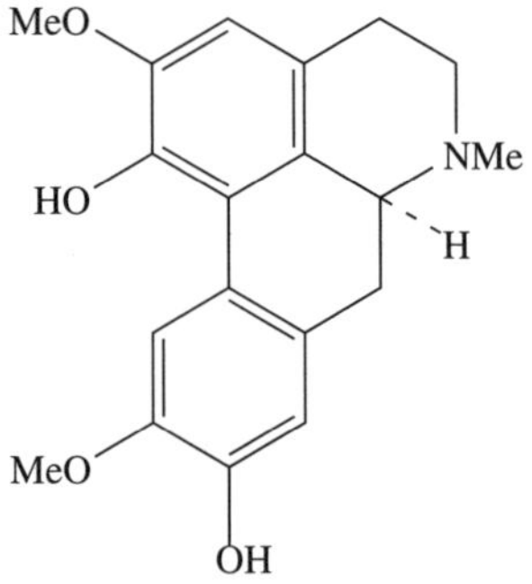

(230)

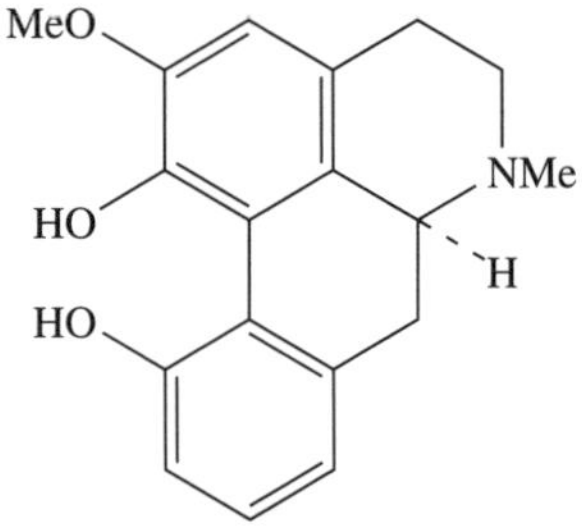

(231)

X. MISCELLANEOUS ADDITIONS AND ELIMINATIONS

A. Reimer–Tiemann Reaction and Related Processes

Photoformylation (the photo-Riemer–Tiemann reaction) of phenols (phenol, 2-methyl, 3-methyl, 4-methyl, 4-chloro, 4-bromo, 4-nitro and 4-phenyl phenol) has also been studied by irradiation in chloroform with KOH and pyridine. The yields reported are variable but formylation is reported only to occur in the 2-position[201–203]. This process involves the addition to phenol of radicals produced from chloroform. An electron transfer mechanism (transfer from excited state phenol to chloroform) is thought to be involved. The radical ion pair eliminates HCl and combination affords the products **232–234** (Scheme 26). The principal product is the ether and this undergoes partial conversion to the formate. The other products formed in low yield are the aldehydes[204]. In another application of the photo-Riemer–Tiemann reaction, this time in cyclodextrin, the phenols can be converted into 4-hydroxybenzaldehydes with high selectivity[205].

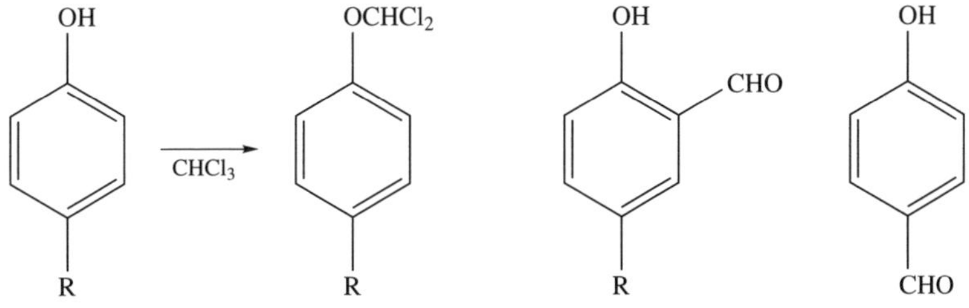

R	yield (%)		
H	75	15	10
Cl	78	15	—
CH$_3$	79	19	—
CN	87	5	—

SCHEME 26

B. Reactions with Tetranitromethane and Nitrate Ion

A charge transfer complex is involved in the photochemical reaction between 4-cresol and tetranitromethane. Irradiation at 350 nm yields the *o*-nitrated product **235**[206]. Other phenols such as phenol, 2- and 4-chlorophenol and 2- and 4-cresol behave in a similar manner and irradiation yields 2- and 4-nitrated products (**236, 237**)[207]. The quantum yields for product formation are in the range 0.12–0.31. Only the formation of 3-nitrophenol from phenol is inhibited, as might be expected from attack at the 3-position, and shows a low quantum yield. It has been reported that 2-hydroxy- or 4-hydroxybiphenyl and 4,4′-dihydroxybiphenyl are the primary products formed from the photochemical reaction of biphenyl with sodium nitrate in aqueous methanol[208]. Apparently the hydroxybiphenyls are prone to undergo photochemical nitration as a secondary process and yield the biphenyls **238** and **239** as well as 4,4′-dihydroxy-3,3′-dinitrobiphenyl, originally reported by Suzuki and coworkers[209] under heterogeneous conditions.

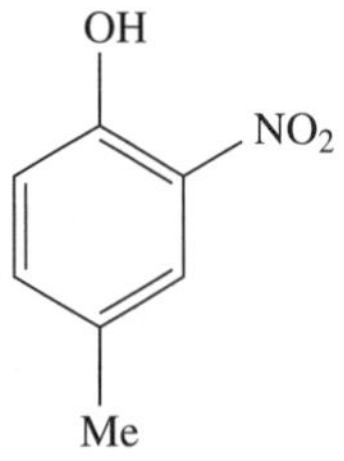

(235) — 2-nitro-4-methylphenol structure

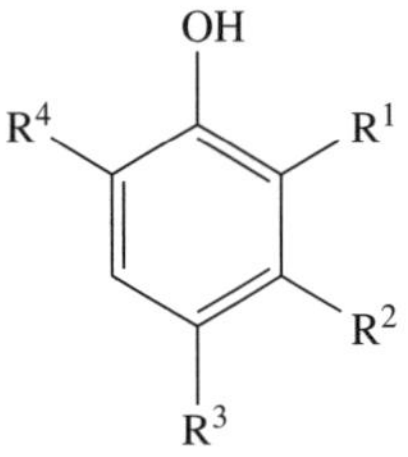

(236)

R^1	R^2	R^3	R^4	ϕ
NO_2	H	H	H	0.17
H	NO_2	H	H	1.2×10^{-3}
H	H	NO_2	H	0.12
Cl	H	NO_2	H	0.15
Cl	H	H	NO_2	0.15
Me	H	NO_2	H	0.31
Me	H	H	NO_2	0.19

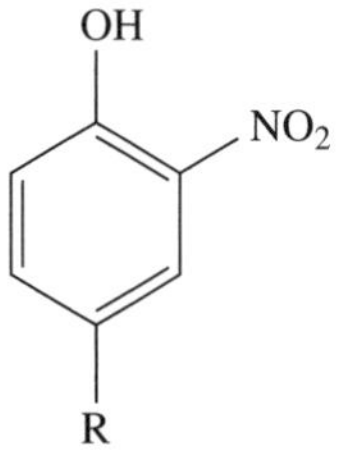

(237)

R	ϕ
Me	0.15
Cl	0.15

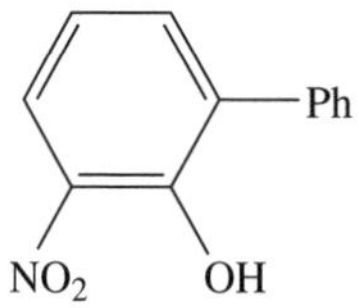

(238) 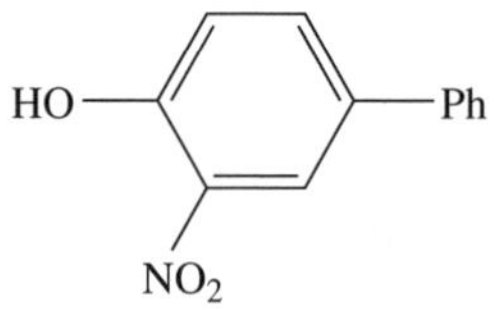 (239)

C. Loss of Halide

Carbene formation was mentioned in an earlier section. This elimination of HCl from 4-chlorophenol or elimination of other hydrogen halides from halophenols could have been inferred from earlier photochemical studies on this and other derivatives. Boule and his coworkers irradiated 4-chlorophenol under deoxygenated conditions and obtained the corresponding quinhydrone and the 2,4′-dihydroxy-5-chlorobiphenyl[177,210]. Other research demonstrated that its irradiation in neutral aqueous solutions gave the corresponding quinone[211] and also that de-aeration did not seem to affect the reaction[212].

14. Photochemistry of phenols

The same reactivity was shown for 4-bromophenol[213,214]. The herbicide bromoxynil **240** is photochemically reactive in the presence of chloride ion undergoing loss of the bromo substituents or else substitution of the bromo substituents. It is converted into 4-hydroxybenzonitrile, 3-bromo, 3-chloro and 3-bromo-5-chloro derivatives[215]. Others have shown that this herbicide can be oxidized using TiO_2[216].

3-Chlorophenol is also reactive and irradiation in water leads to its conversion into resorcinol[211,217] or in methanol to yield 3-methoxyphenol in 94% yield[218]. Photoamidation with N-methylacetamide of 3-chlorophenol is also efficient and results in the formation of the phenol **241** in a yield of 77%. Intramolecular amidation arises on irradiation of **242** in basic methanol. This results in the formation of the indole derivative **243** as well as the methoxylated product **244**[218]. More complex halophenols such as **245** are also photochemically reactive, but this yields a complex mixture of products including a benzofuran. The formation of this must be similar to the cyclizations described earlier and involves the attack of a radical, produced by the C—I bond fission, on the other ring[219]. 3-Nitrophenol is converted on irradiation in aqueous solution into a variety of products such as nitrocatechols, nitroresorcinol and resorcinol itself[220].

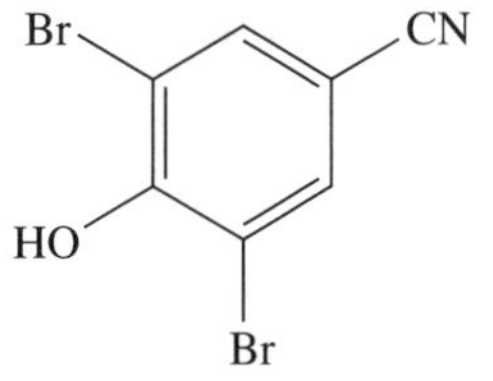

(240) 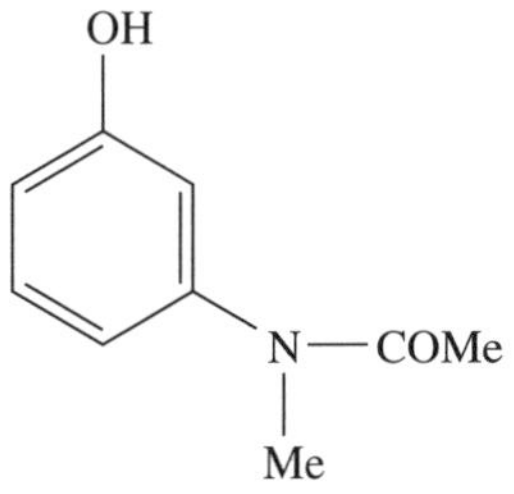 **(241)**

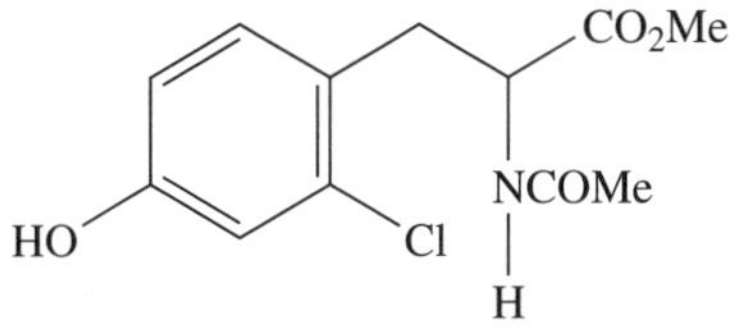

(242) 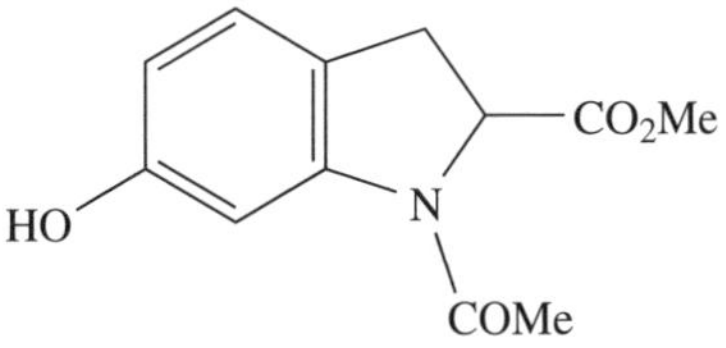 **(243)**

4-Chloro-1-hydroxynaphthalene is converted into the sulphonate **246** on eosin-sensitized irradiation in the presence of sodium sulphite[221–223]. A study of the chain substitution of the chloro group in 4-chloro-1-hydroxynaphthalene by aqueous sodium sulphite has shown that two mechanisms for the photoinitiation have been identified and two intermediates have been detected: a radical anion of 4-chloro-1-naphthoxide and the sulphite radical anion. Thus, an $S_{RN}1$ mechanism is suggested and is one that involves reaction with the radical anion of sulphite[224]. An example of the $S_{RN}1$ process between a phenol and the (2-cyanoaryl)azo-t-butylsulphides[225] has been reported. The $S_{RN}1$ reactivity of several compounds (Scheme 27) have demonstrated that **247** is a product; however, this is also photochemically reactive and is converted into the cyclic ether **248**[226].

Intramolecular loss of halide is observed when the phenoxide **249** is irradiated either directly or in the presence of triethylamine where an electron transfer mechanism

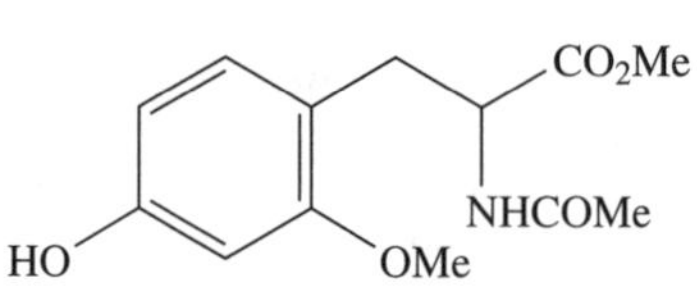

(244)

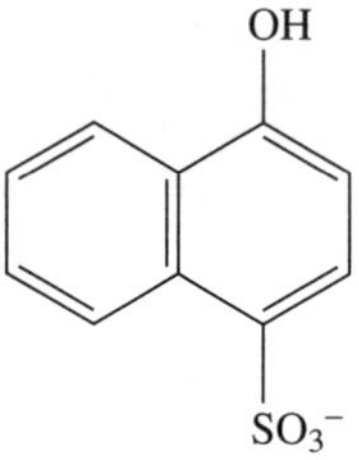

(246)

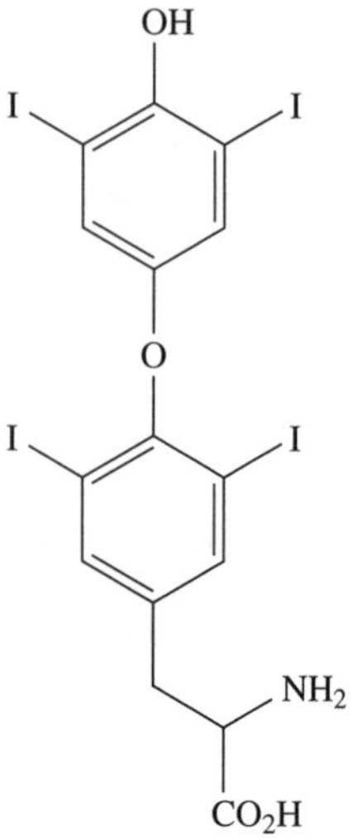

(245)

(ArX)* + Nu⁻ ⟶ (ArX)⁻• + Nu•
(ArX)⁻• ⟶ Ar• + X⁻
Ar• + Nu⁻ ⟶ (ArNu)⁻•
(ArNu)⁻• + ArX ⟶ ArNu + (ArX)⁻•

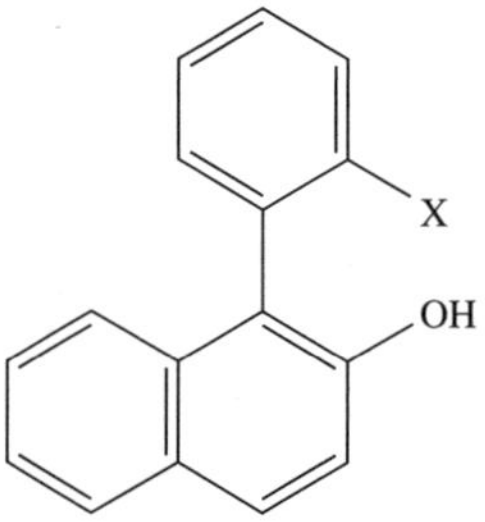

X = halogen
(247)   (248)

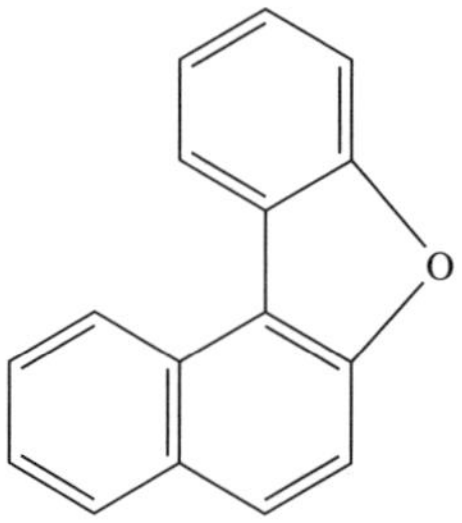

SCHEME 27

is involved. This affords the octachlorodioxin **250**. Sensitized irradiation using *m*-nitroacetophenone follows a different path and brings about the formation of ether cleavage products such as penta- and tetrachlorophenol[227]. Octachlorodioxin is also formed by irradiation of chlorophenoxyphenol **251**[228,229]. The *meta*-isomer **252** is also reactive and undergoes dechlorination and cyclization to yield[230] a chlorinated dibenzofuran. The isomeric 2,3,5,6-, 2,3,4,5- and 2,3,4,6-tetrachlorophenols also undergo photoreactions to yield a series of chlorinated biphenyls such as hexachloro, heptachloro and octachlorodihydroxybiphenyls[231].

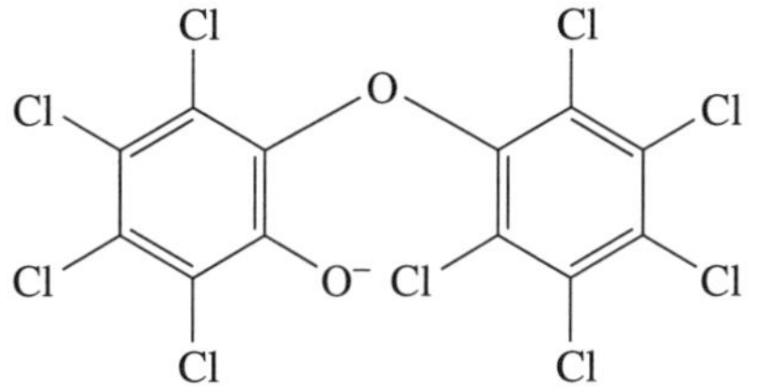

(249) 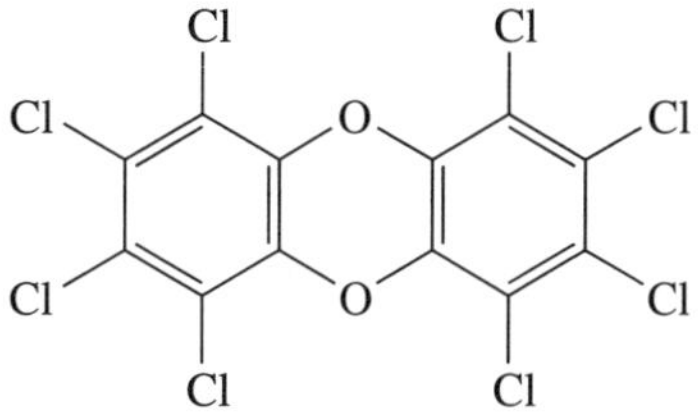 **(250)** 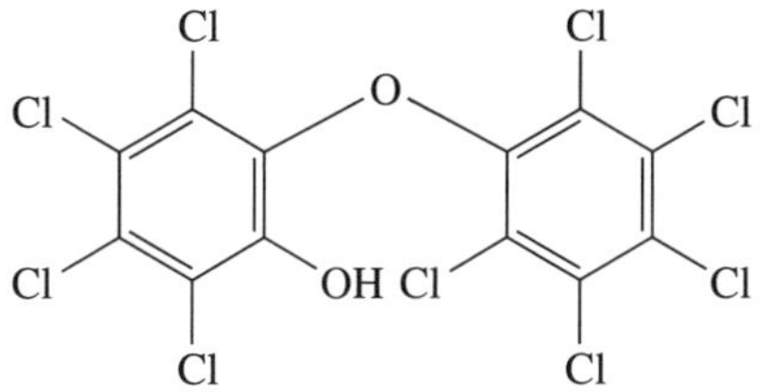 **(251)** 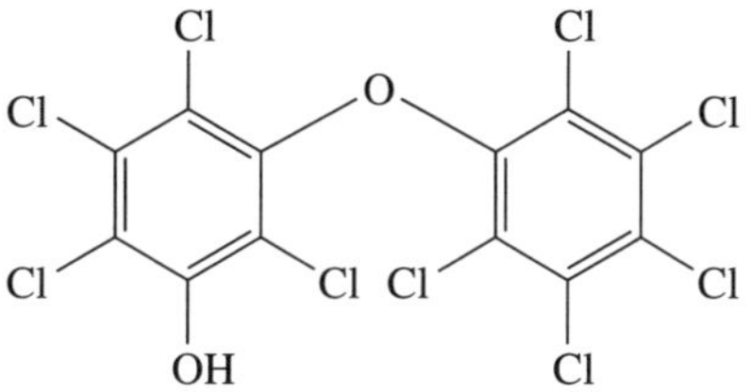 **(252)**

D. Other Bond Fission Processes

The zwitterionic iodonium phenolate **253** is photochemically reactive and a variety of products can be obtained depending upon the substrate in which the reactions are carried out. The mechanism of formation of these products could be an electrophilic reaction of the iodonium species or could involve fission of the I−C bond to yield the phenolate zwitterion **254**, which itself could undergo electrophilic reactions. Regardless of the route, addition of **254** to alkenes yields **255**, to alkynes gives **256** and to arenes produces the arylated products **257**[232]. Pyridine, thiourea, phenyliminobenzoxathiole and phenyl isocyanate also act as addends[233].

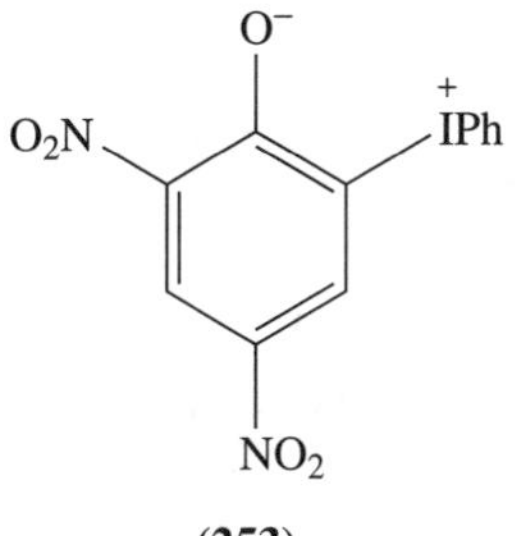

(253) 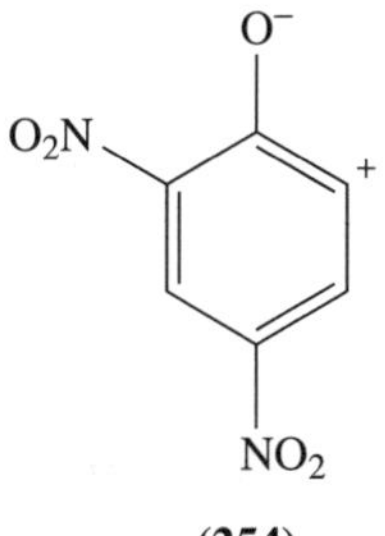 **(254)**

E. Reactions of Hydroxyanthraquinones

1-Hydroxyanthraquinone (**258a**) undergoes photochemical amination on irradiation in the presence of *n*-butylamine. Two products, **258b** and **258c**, are formed in a ratio that is dependent on the reaction conditions used. In acetonitrile under an atmosphere of air the ratio of **258b** : **258c** is 5 : 1. This changes to 0.3 : 1 when the reaction is run under nitrogen. Interestingly, the corresponding 1-aminoanthraquinone does not undergo

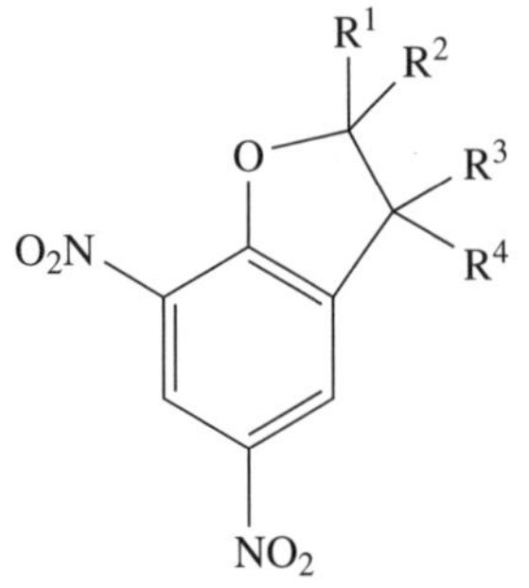

(255)

R¹	R²	R³	R⁴	yield (%)
H	—(CH$_2$)$_4$—		H	9
Me	Me	Me	Me	30
Ph	Ph	Ph	Ph	17
H	H	H	Ph	22
H	H	H	AcO	38
H	H	H	OCH$_2$Pr-i	34

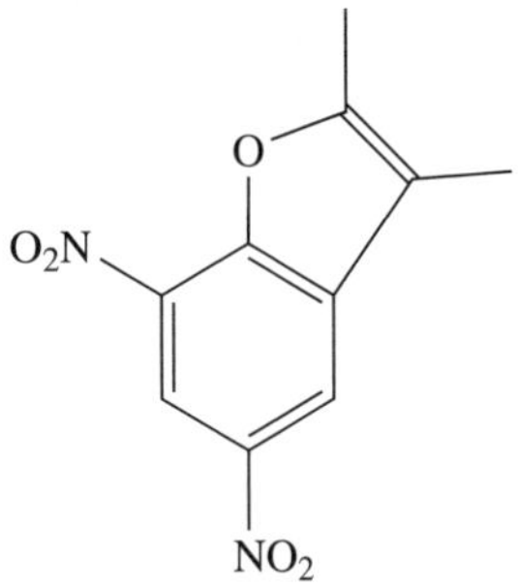

(256)

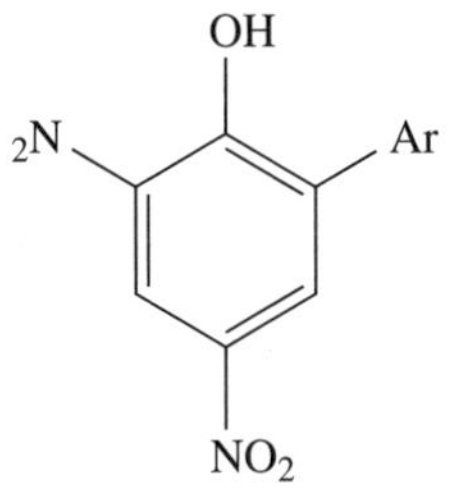

(257) Ar = Ph, 2-furyl, 2-thienyl, 2-(5-methylfuryl)

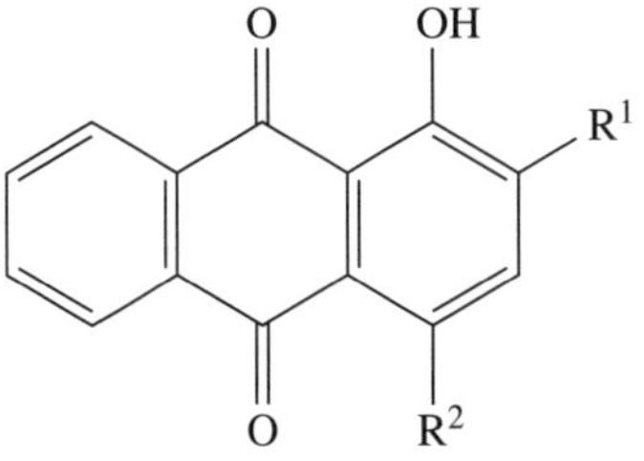

(258) (a) R¹ = R² = H
(b) R¹ = H, R² = n-BuNH
(c) R¹ = n-BuNH, R² = H

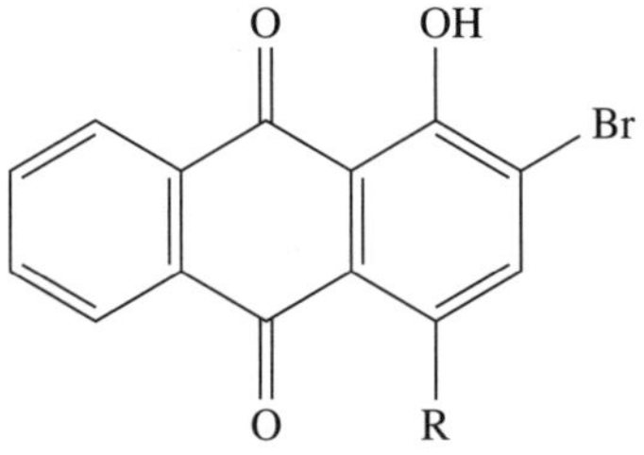

(259) R = H or Br

amination. The quinones **259** also undergo amination with the same amine to yield the 4-butylaminoquinone **259**, R = NHBu-n[234]. The same quinone also undergoes amination at the 4-position on irradiation in the presence of ammonia or methylamine[235]. 1-Hydroxyanthraquinone can also undergo sulphonation with sodium sulphite on irradiation. The products obtained are the 2- and 4-sulphonates and 2,4-disulphonates, which are obtained in 34%, 18% and 24% yield, respectively[236]. From these results it can be seen that the selectivity is poor, perhaps as a result of ionization of the phenol group.

F. Miscellaneous Processes

The calix[4]arene-based 2-naphthoate **260** undergoes photochemical cyclization to afford **261**[237]. Hydrogen bonding controls the cyclization of **262** into **263**. If the hydrogen bonding is broken by carrying out the reaction in methanol, the cyclization follows the path where attack occurs at the phenolic carbon[238]. The stilbene derivatives **264** have also been investigated. This study was associated with work to establish why some phenolic

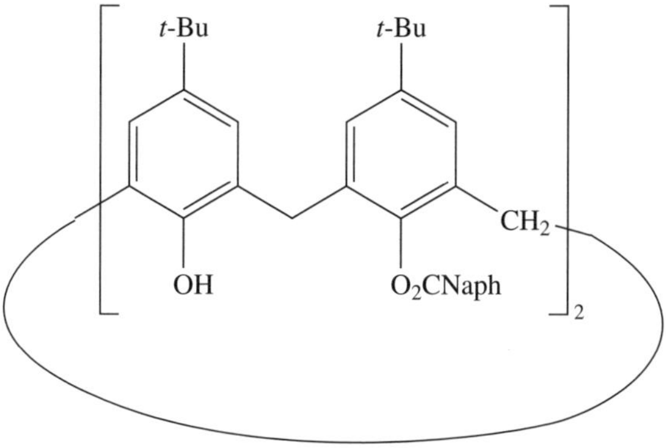

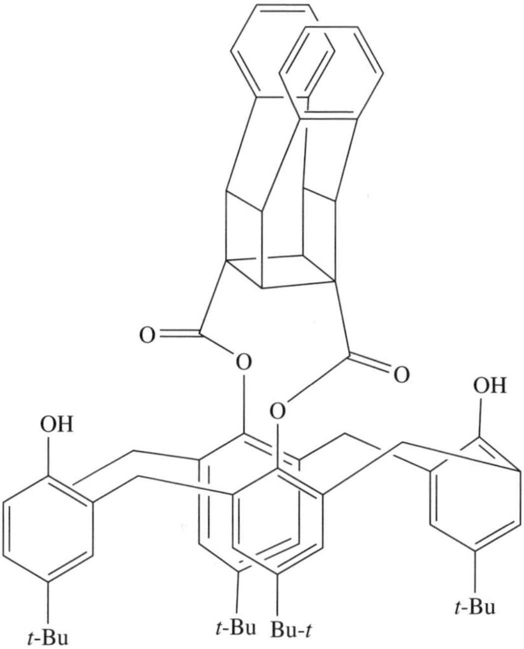

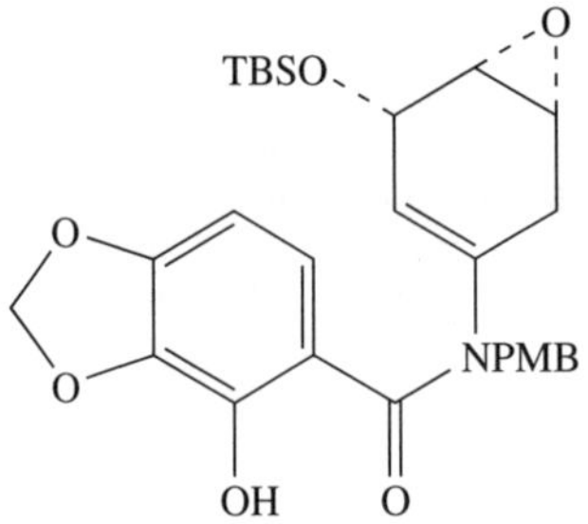

(262)

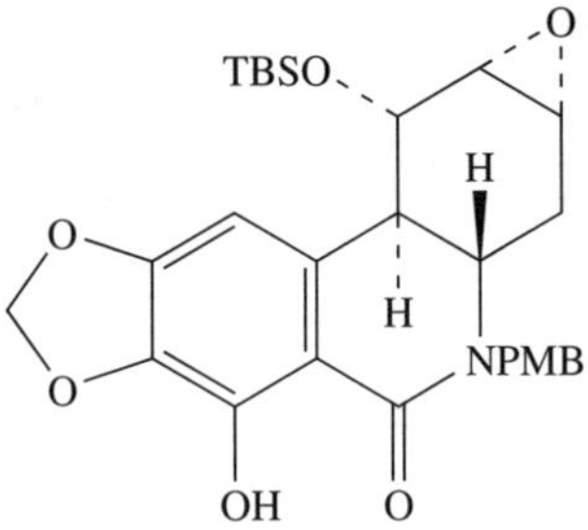

(263)

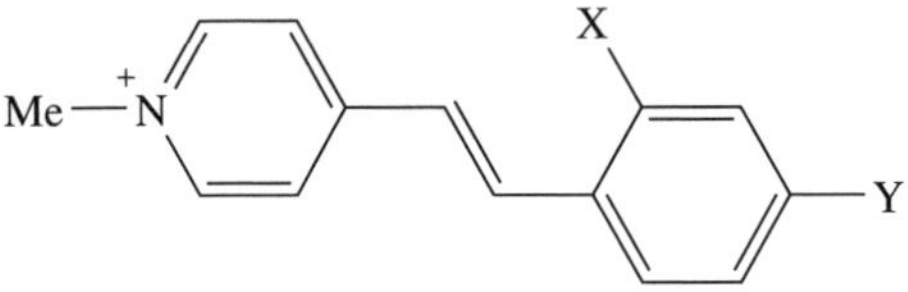

X	Y
H	O⁻ or OH
O⁻ or OH	MeO

(264)

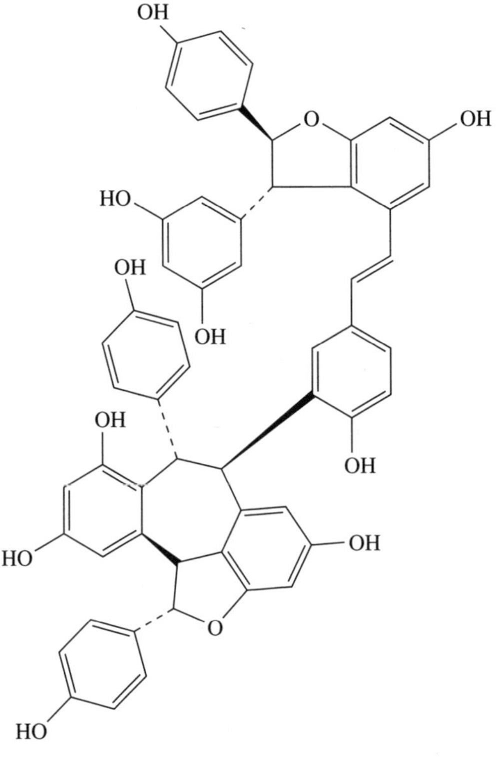

(265)

stilbenes do not cyclize[239]. Irradiation of the stilbene **265** at 254 nm in methanol transforms it into the corresponding *cis*-isomer[240]. The dimerization of the enone **266** in the crystalline phase has been described[241].

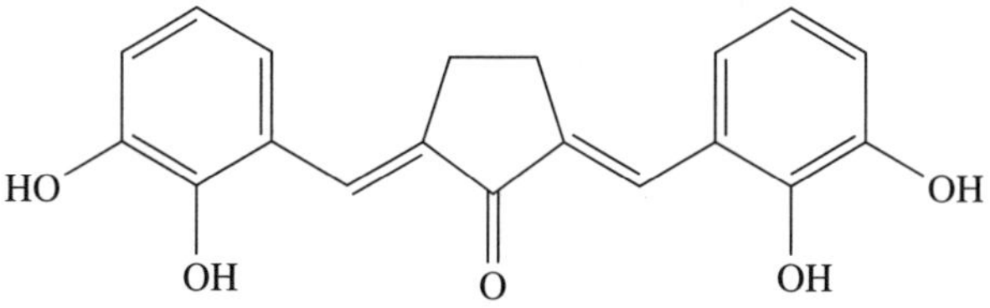

(**266**)

Homolytic C−S bond fission occurs in compounds such as **267**. This process yields the 1,4-dihydroxybenzene in yields as high as 60%. Desulphurization of the thio ethers **268** results on irradiation. This only occurs when hydroxy groups or the corresponding methoxy-substituted compounds are used[242].

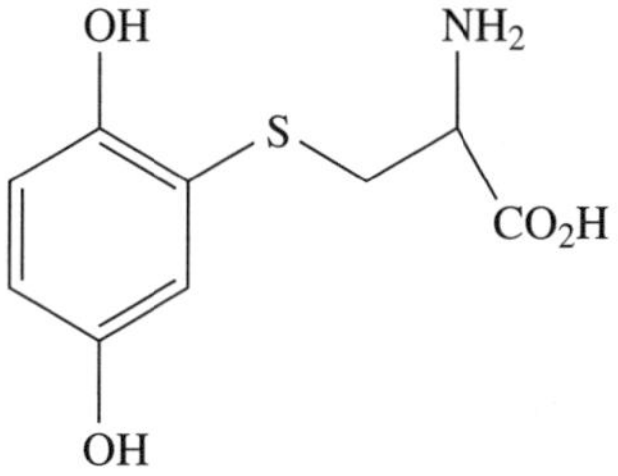

(**267**)

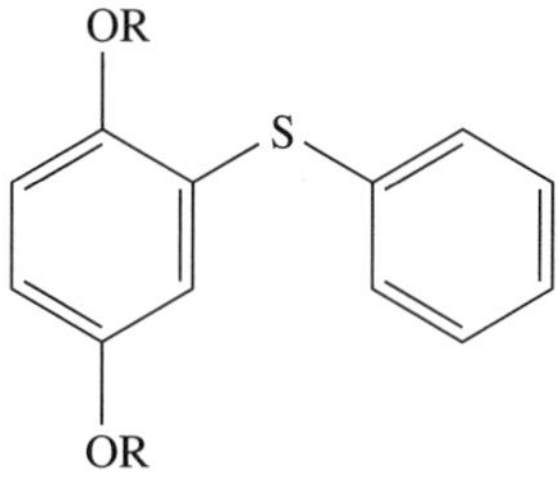

(**268**) R = H
R = Me

XI. HYDROXYCARBOXYLIC ACIDS

A. Decarboxylation

2-Hydroxy-4-trifluoromethylbenzoic acid is pharmacologically active and has been shown to be photolabile under various conditions. Its major photodegradation pathway is nucleophilic attack on the trifluoromethyl moiety. The triplet state is involved in the photodegradation[243]. The use of photochemically induced degradation of benzoic acid derivatives (syringic, gallic, veratric, vanillic, protocatechuic and *p*-hydroxybenzoic) using electron transfer to pyrylium salts has been reported. The degradation observed was significant (20–40%), even though this was contra-indicated by the presence of an electron-withdrawing carboxyl group attached to the aromatic ring[244]. The photochemical decomposition of 4-chlorophenoxyacetic acid and 2,4-dichlorophenoxyacetic acid has been studied, and in the presence of the sensitizer anthraquinonesulfonate, the degradation is accompanied with chloride ion release. In addition, decarboxylation is also observed[245]. Other studies have reported the retarded decomposition of 2,4-dichlorophenoxyacetic acid in mixed industrial effluent in the presence of copper[246].

B. Reactions in the Presence of TiO$_2$

The photochemically induced reaction of 2,4-dichlorophenoxyacetic acid in the presence of a suspension of TiO$_2$ follows an apparent first-order rate process[247]. Others have demonstrated that TiO$_2$ is a powerful oxidant for carboxylic acids and have shown that salicylic acid undergoes ready decomposition[248,249]. Salicylic acid can also be oxidized by ferric oxalate and molecular oxygen. Under these conditions hydroxylation occurs, perhaps involving hydroxy radicals with the formation of the two isomeric dihydroxyacids **269** and **270**[250]. Others have also studied the photooxidation of salicylic acid and have observed that the formation of **270** is not proof of the involvement of singlet oxygen[251]. Furthermore, the rate of decomposition of this acid in TiO$_2$ can be enhanced by the use of ultrasound during photolysis[252]. 2,4-Dihydroxybenzoic acid also undergoes decomposition in water catalysed by titanium dioxide[253]. Other carboxylic acids can also be decomposed in wastewaters using other oxidants in conjunction with photolysis. For example, hydroxyl radicals from Fenton's reagent can bring about the decomposition of *p*-hydroxyphenylacetic acid[254] while protocatechuic acid will also undergo photooxidation[255] as will vanillic acid when it is irradiated in the presence of ozone[256].

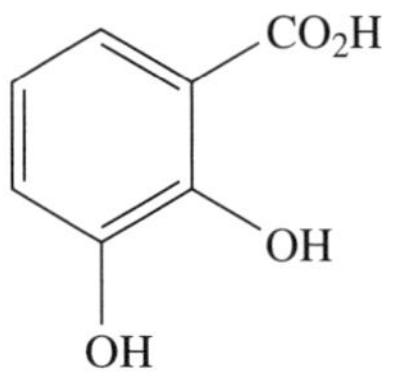

(269) 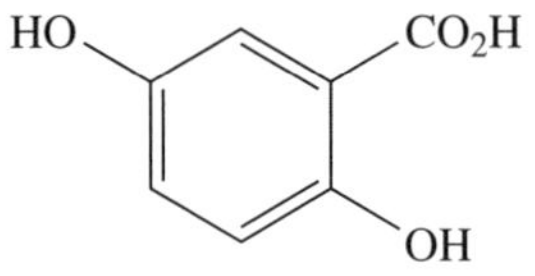   (270)

XII. OXIDATIONS

Over the last decade there is little doubt that the oxidation of phenols has been an area of considerable interest. While this chapter does not deal in detail with this subject area it is nevertheless of considerable importance. Thus, some of the material from the last ten or so years has been included. This will give the reader a flavour of what has and is going on in this area.

A. Phenol

The phenol derivative **271** undergoes oxidation to the quinone **272** by constant current electrolysis. Concomitant irradiation of this quinone transforms it into the novel (2 + 2)-cycloadduct **273**. This was the key intermediate in a synthetic approach to racemic isoitalicene **274**[257]. The oxidation of α-tocopherol **275** can be brought about using Methylene Blue as the sensitizer for the production of $O_2(^1\Delta_g)$. This converts the compound into the previously unknown enedione **276** as well as the usual quinone[258]. The reaction of phenols with $O_2(^1\Delta_g)$ is a common process. This involves attack on the phenol, usually at the *p*-position. A typical example of this is the conversion of the phenols **277** and **278** into the corresponding cyclohexadienones **279** and **280**[259]. The oxidation of phenols has been the subject of reviews[260,261].

Perhaps the largest area of research on the chemistry of phenols relates to methods for their photodegradation. The methods used are many and varied, such as the combination of ozone and UV irradiation. This is an effective method for degradation and is more

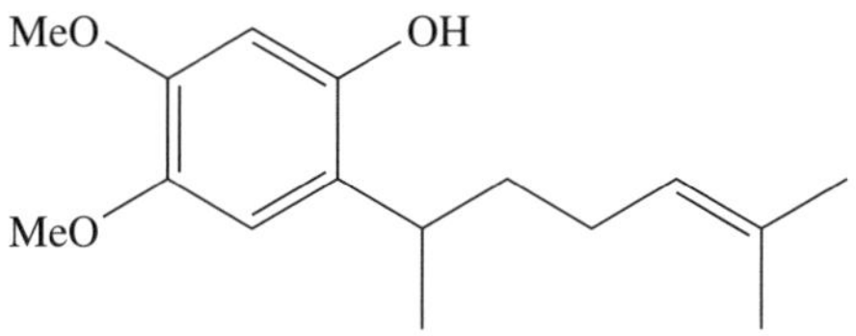

(271)

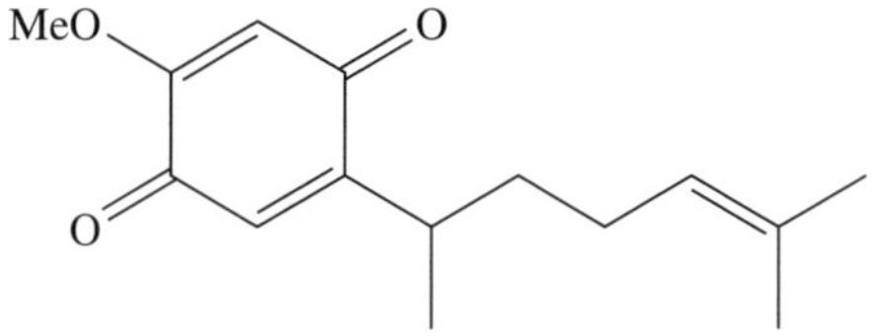

(272)

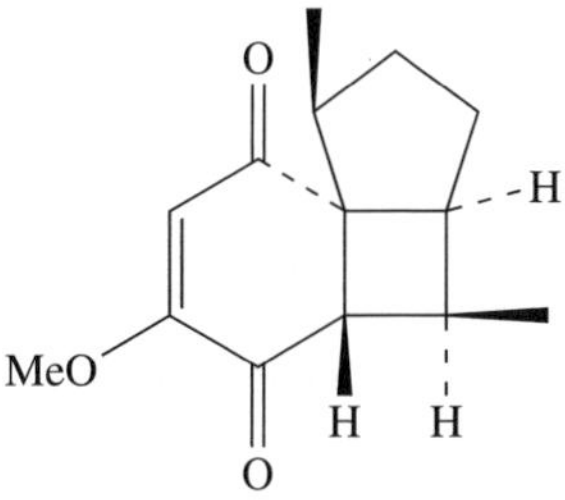

(273)

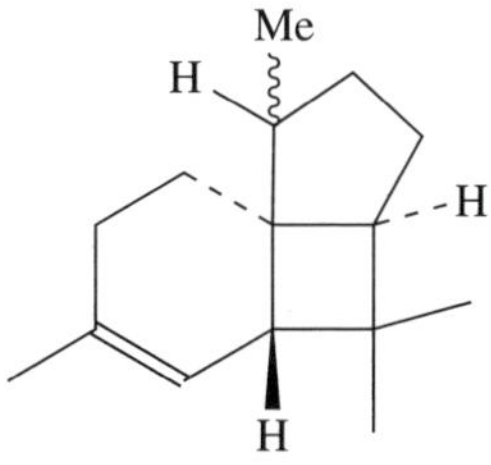

(274)

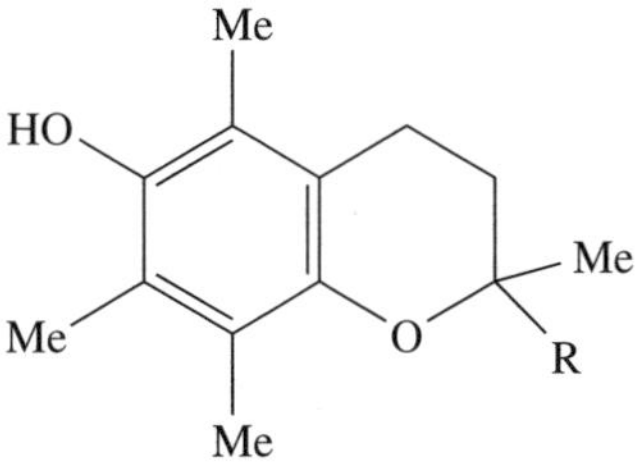

(275)

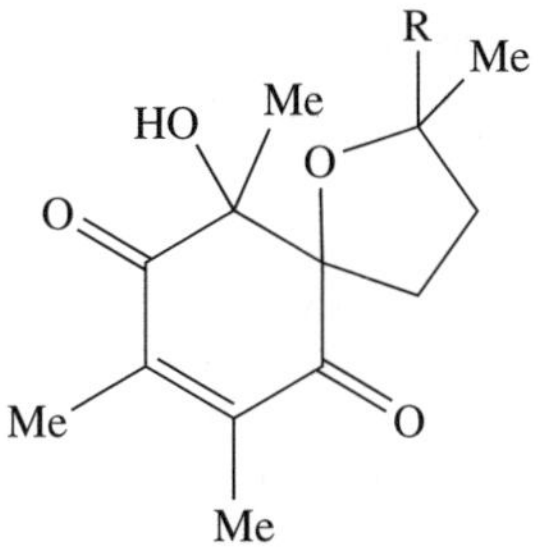

(276)

R = phytyl

efficient than the use of peroxide[262]. That aside, there are several reports[263–266] on the use of UV irradiation and peroxide as a means for the removal of phenol from wastewaters. The only by-product from this treatment was carbon dioxide. A kinetic model for the photooxidative degradation of phenol by hydrogen peroxide has been derived[267]. A heterogeneous copper-based catalyst has been developed for the removal of phenol using peroxide as the oxidant[268]. The photo-Fenton oxidation of phenol is also useful for the degradation of phenol in wastewaters[269].

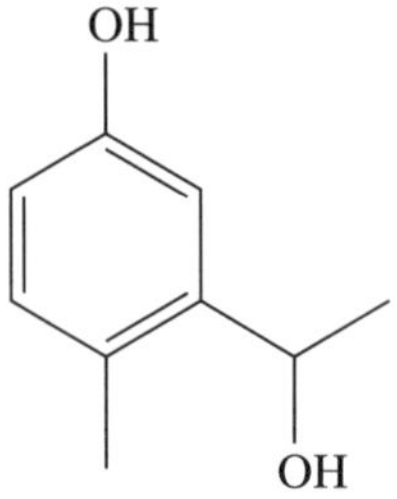

(277)

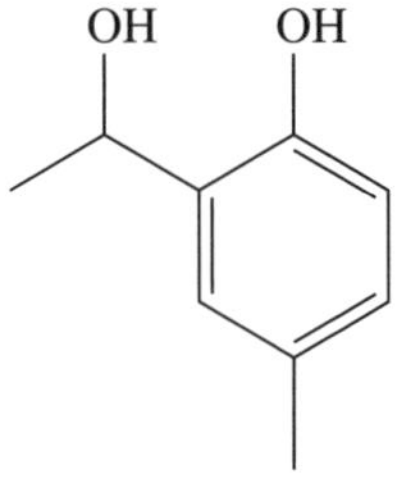

(278)

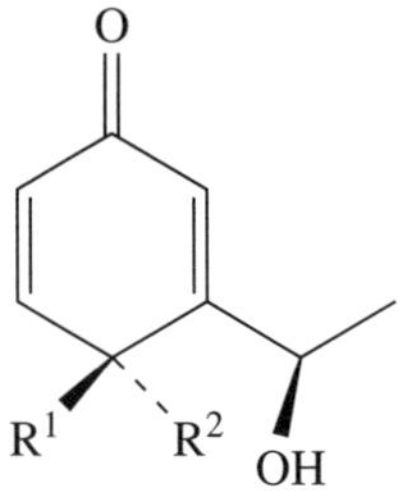

(279)

R¹ = HOO, R² = Me
R¹ = Me, R² = HOO

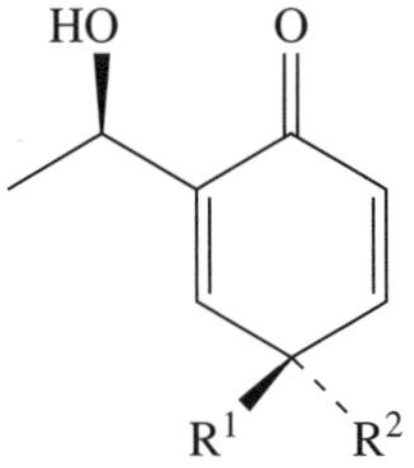

(280)

R¹ = HOO, R² = Me
R¹ = Me, R² = HOO

Oxygen and suitable catalysts can also be used for the conversion of phenol to benzoquinone. Thus, irradiation of phenol in the presence of $[Cu(bpy)_2]^{2+}$ or $[Cu(1,10\text{-}phenanthroline)]^{2+}$ brings about degradation by a path that shows both pH and solvent dependency[270,271]. Thus in acetonitrile benzoquinone predominates, but in water carbon dioxide is the sole product[272]. Benzoquinone can also be formed from phenol by continuous irradiation in the presence of the catalysts $[Cr(bpy)_3]^{3+}$ or less effectively with $[Ru(bpz)_3]^{2+}$ and $[Ru(bpy)_3]^{2+}$. The reaction path involves $O_2(^1\Delta_g)$ as the oxidant[273–275]. Porphyrins such as 5,10,15,20- tetrakis(2,6-dichlorophenyl)porphyrin and chlorins can also be used to convert naphthols and phenols to the corresponding quinones (Scheme 28)[276]. Phthalocyanines immobilized on polymers have also been used as the catalyst system to effect photooxidation[277].

Much research has been carried out to establish the efficiency of TiO_2[278,279] systems for catalysing decomposition of phenol. An examination of the product distribution and reaction pathway in the photocatalytic oxidation of phenol on TiO_2 particles reveals that the reaction proceeds in three stages, namely hydroxylation, carboxylation and mineralization[280]. Titania colloids and particles for the sol-gel technique have been assessed as a suitable method for photooxidation[281,282]. The sol-gel technique has been used to prepare TiO_2 films in a variety of substrates, such as a fibreglass cloth[283] or glass fibre[284] on γ-Al_2O_3[285] or silica gel[286] stoichiometric membranes[287,288], and evidence has been collected that suggests that such sol-gel preparations show higher activity than commercial TiO_2 catalysts[289]. In general, sol-gel methods of preparation provide a catalyst with a high surface area[290]. Modified catalysts have also been developed, such as a $ZnFe_2O_4$/TiO_2 nanocomposite[291].

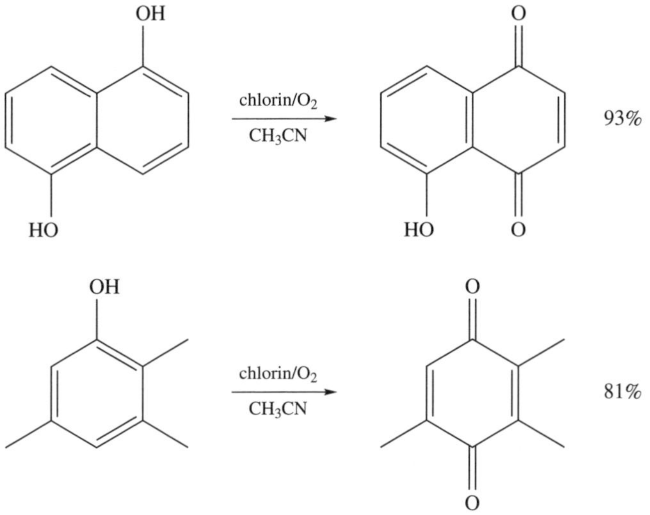

SCHEME 28

A kinetic model for the oxidation of phenol on titania has been published[292]. An FT-IR study has examined the photocatalytic decomposition of phenol on TiO_2 powders in an effort to demonstrate the existence of intermediates[293]. A study of 3-aminophenol has suggested that 1,4-dihydroxy-3-aminocyclohexa-2,5-dienyl is the most likely intermediate[294]. Pulsed-laser oxidation of phenol on TiO_2 in a variety of aqueous acidic media has endeavoured to identify the transients involved. In HCl, $Cl_2^{•-}$ is the main oxidant arising from surface-absorbed Cl^- on the titania, or from $TiCl_4$ if that has been used to produce the colloidal TiO_2[295]. Oxygen radicals are formed on irradiation of hydrated TiO_2 at 77 K and their interaction with adsorbed phenol has been monitored using ESR spectroscopy. The results imply that irradiation generated $HO_2^•$ and O^- that reacts with adsorbed phenol[296]. A study of the photoreactivity ($\lambda = 350$ nm) of phenols in a suspension of TiO_2 as a function of pH shows that the initial rate and the Langmuir–Hinshelwood kinetic parameters are comparable from pH 3–9 and at pH 13.7, but change at pH 1 and pH 11. This has been interpreted in terms of speciation of the reactants and changes in the TiO_2 surface as a function of pH. At pH >12, the oxide radical anion is thought to participate[297]. The formation of a hydroxylated intermediate is in line with a study that has shown that the photochemical oxidation of phenol at low concentration on TiO_2 surfaces involves hydroxylation by transfer from the immobile surface oxidant[298]. Kinetic data have also been obtained for the TiO_2 photocatalytic degradation of phenol. Again, oxidation is thought to occur by direct hole-oxidation of the substrate on the pre-absorbed catalyst[299]. The photocatalytic oxidation of phenol and p-substituted phenols also takes place at a positive hole and this leads to the loss of the p-substituent[300]. The influence of additives to the oxidizing media has also been evaluated and the presence of aromatics tends to have a profound effect on the oxidation[301]. Interestingly, the photooxidation of phenol using TiO_2 is greatly accelerated by the presence of Fe^{3+}/Fe^{2+}[302]. Others have reported that both Fe^{3+} and Cu^{2+} affect the rate of oxidation of phenol using peroxide and TiO_2[303].

Several reactors based on TiO_2 catalytic systems have been described for the phototreatment of wastewater, such as Pt/TiO_2-coated ceramics pipes[304], systems using potassium modified TiO_2[305] and a batch reactor using either TiO_2 and air or peroxide and solar irradiation[306]. The configuration for such systems has also been discussed[307]. Flat plate photoreactors have also been described[308] as has a study in shallow ponds[309]. The decomposition rate for the oxidation of phenol using a Pt/TiO$_2$ catalyst and by solar radiation (λ > 360 nm) has been studied[310]. Other batch reactors have also been described using Pyrex or Quartz jackets depending upon the wavelength being used for the study[311]. Another apparatus has been described that can be used to determine total carbon in wastewaters using photooxidation of phenol on titania[312].

B. Alkylphenols

All six isomeric xylenols undergo degradation when irradiated in air using TiO$_2$ dispersions as the catalyst[313]. The aqueous photolysis of trifluoromethylphenols such as 3-trifluoromethyl-4-nitrophenol has indicated that it undergoes photohydrolytic degradation under actinic radiation to yield trifluoroacetic acid[314]. 2,4,6-Trimethylphenol can be oxidized using $O_2(^1\Delta_g)$[315]. An FT-IR study has shown the presence of intermediates in the oxidation of this phenol on TiO$_2$ powders[316]. The rate of sensitized oxidation of 2,4,6-trimethylphenol in aqueous humic acid has been determined[317]. Other polyalkylated phenols also undergo oxidation[318].

C. Miscellaneous Phenols

The degradation of 2-phenylphenol, commonly used as fungicide, can be photocatalysed by TiO$_2$, although the oxidation is more efficient using ZnO. The principal photoproducts identified are hydroquinone, p-benzoquinone, phenylhydroquinone, phenylbenzoquinone, 2,2'- and 2,3-dihydroxybiphenyls. A minor product, 2-hydroxydibenzofuran, is also formed and this arises by the photocyclization of phenylbenzoquinone[319]. Other p-substituted phenols (4-methoxyphenol, 4-methylphenol, 4-chlorophenol, 4-bromophenol, 4-fluorophenol, 4-acetylphenol, 4-trifluoromethylphenol and 4-cyanophenol) undergo photooxidation catalysed by TiO$_2$. The results indicate that a number of mechanistic paths may be involved[320].

Various 4-(arylazo)phenols and naphthols have been photooxidized using $O_2(^1\Delta_g)$ involving a type II mechanism[321]. 1,1'-Binaphthol undergoes enantioselective oxidation (5.2% ee) when the chiral complex Δ-[Ru(4,4'-dimenthoxycarbonyl-2,2'-bipyridine)$_3$]$^{2+}$ is used as the photocatalyst[322].

Efficient photocatalytic degradation of p-nitrophenol can be brought about using ZnS or, less efficiently, in TiO$_2$[323]. A laboratory experiment has been devised to demonstrate TiO$_2$ catalytic decomposition of p-nitrophenol[324]. Experiments have also been described detailing the decomposition of nitrophenol on TiO$_2$ doped with Fe(III)[325]. Others have reported on the photooxidation of nitrophenols using tungsten oxide/TiO$_2$ catalysts[326,327].

Oxidation of 2,6-dichloroindophenol (sodium salt) can be carried out on TiO$_2$ with unit quantum efficiency. This particular reaction has been suggested as a method for testing for photocatalytic activity of semiconductor powders[328].

D. Dihydroxyphenols

Photocatalytic degradation of 1,3-dihydroxy-5-methoxybenzene in the presence of TiO$_2$ follows zero-order kinetics. The product formed from this process is CO_2 with the best results obtained at pH 9[329]. Other dihydroxybenzenes can also be photooxidized using dye-mediated oxidation involving $O_2(^1\Delta_g)$. A charge transfer mechanism is thought to be operative. The suggestion has been made that this process could be used for degradation of phenols in the environment[330]. The photooxidation of polyhydroxylated phenols[331] and methoxylated phenols[332] has also been studied.

E. Chlorophenols

The oxidation efficiency of hydroxy radicals towards chlorophenol has been assessed. The reaction has been shown to proceed via hydroquinone, catechol and resorcinol intermediates[333]. Oxidation of chlorophenol can also be brought about using a Xe-excimer laser irradiating at 172 nm. Hydroxy radicals formed from the photolysis of water bring about the degradation[334]. Chlorophenols have been removed from water by TiO_2-catalysed oxidation[335]. Texier and coworkers[336] have investigated the solar-induced photodecomposition of aqueous solutions of 2-chlorophenol in the presence of both titanium dioxide and sodium decatungstate $Na_4W_{10}O_{32}$. The influence of pH and cadmium sulphide on the action of TiO_2 on 2-chlorophenol has been assessed. Apparently, the addition of the semiconductor to the reaction mixture diminishes the efficiency of the photodegradation. Chlorocatechol, hydroquinone, benzoquinone and phenol were identified as the predominant products from the degradation[337]. Interestingly, research has also shown that the pseudo-first-order rate constant falls as the pH rises[338]. The same chlorophenol has been subjected to oxidation using a variety of methods, such as sonication and photocatalysis[339]. Sonication has also been used to bring about oxidation of 2-chlorophenol in water using Fenton's reagent[340]. Others have also reported the study of the photodegradation of monochlorophenols[341–343]. Hydroquinone and phenol have been shown to promote the dehalogenation of halophenols in aqueous solutions[344].

Oxidation of 4-chlorophenol can be brought about by single photodecomposition by hydroxy radicals generated from Fenton's reagent (H_2O_2 plus Fe^{2+} ions)[345]. Irradiation in the 320–400 nm range with Fenton's reagent is also effective in the oxidation of 4-chlorophenol[346]. Continuous irradiation at 365 nm has identified two different reaction pathways with formation of the 4-chlorodihydroxycyclohexadienyl radical and also of the chlorophenoxyl radical. The quantum yields of these processes have been determined to be 0.056 and 0.015, respectively[347]. Reaction of 4-chlorophenol with ozone leads to the formation of 4-chloro-1,3-dihydroxybenzene and 4-chloro-1,2-dihydroxybenzene. The latter product is produced in quantity in the presence of hydroxyl radicals[348].

Titania catalysts on a metallic support are useful for the photodegradation of 4-chlorophenol. The immobilized titania was about twice as efficient as UV photolysis but was less efficient than suspensions of TiO_2[349]. Other photocatalyst systems are also effective, such as modified sol-gel preparations of TiO_2 using different alkoxide precursors. The catalysts prepared in this fashion have good stability[350]. Generally, degradation of 4-chlorophenol on TiO_2 slurries using oxidative γ-radiolysis occurs by combination of $HO^{\bullet}$ oxidation and surface oxidation by valence-band holes. The usual products detected were 4-chlorocatechol and hydroquinone[351,352]. Solar photodegradation of 4-chlorophenol can be brought about by oxidation in the presence of TiO_2 or sodium decatungstate $Na_4W_{10}O_{32}$[353]. Interestingly, the decatungstate anion becomes as efficient or even more efficient than TiO_2. Online monitoring of the photocatalytic degradation of 4-chlorophenol can be carried out using cyclic voltammetry and UV-Vis spectrometry[354]. Riboflavin can be used as a catalyst for the degradation of 4-chlorophenol in the pH range 4–10. It is suggested that riboflavin-modified electrodes could be used for photodegradation[355]. The complexes zinc and aluminium tetrasulphophthalocyanines are also effective as catalysts[356]. The phototransformation of 4-chloro-2-methylphenol was studied in distilled, natural or water containing humic substances under a variety of irradiation conditions and wavelengths (monochromatic light at 280 nm, polychromatic light with lamps emitting within the wavelength ranges 290–350 nm and 300–450 nm and solar light). When 4-chloro-2-methylphenol is irradiated in pure water, dechlorination occurs with a good efficiency ($\phi = 0.66$). Methylbenzoquinone is the main primary photoproduct in oxygenated solution while other products, methylhydroquinone and methylhydroxybenzoquinone, are produced via a second photochemical step[357].

Photooxidation of 2,4-dichlorophenol can also be carried out using CdS in the presence or absence of thioacetamide. There is a marked pH dependence observed in the oxidation. Thus, at pH < 6, oxidation is favoured by positive holes in the semiconductor while with pH > 6, negative holes are involved[358]. Electron-transfer oxidation of chlorophenols is also reported using uranyl ion as the electron acceptor. The presence of oxygen is important to ensure that the quantum yield for the disappearance of the phenol is high[359]. As with the oxidation of other phenols, chlorophenols can be readily oxidized by the photo-assisted Fenton system. In the case of the oxidation of 2,4-dichlorophenol, the process can be brought about using a low-energy (36 watt) black fluorescent mercury light[360]. The same phenol can be readily oxidized using near-visible light with polyoxometallate catalysts. As with the Fenton system the main oxidant is the hydroxy radical. There are several principal products formed[361,362]. Other polyoxymetalates such as $[W_{10}O_{32}]^{4-}$ and $[SiW_{12}O_{40}]^{4-}$ are also effective[362]. 2,4-Dichlorophenol can also be completely mineralized using a combination of photolysis and ozonation[363]. Other dichlorophenols have also been subjected to study[364,365].

2,4-Dichlorophenol also undergoes oxidation in the presence of a TiO_2 suspension. The influence of additional reagents such as hydrogen peroxide on the efficiency of TiO_2 oxidation has been assessed[366]. The yield of oxidation product is dependent on irradiation time while pH and temperature have little effect[367]. It is reported that there is a good relationship between the disappearance of dichlorophenols and the Hammett σ-constants using titania in aqueous suspensions[368]. Heterogeneous photocatalysis with TiO_2 nanoparticles also brings about degradation of 2,4-dichlorophenol in an oxygen-free system. However, the degradation is inhibited by the presence of electron donors such as polyethyleneimine, triethylamine or 2-propanol. In the presence of EDTA, degradation of 2,4-dichorophenol still takes place by dechlorination[369]. Photocatalytic degradation using TiO_2 with Fenton's reagent of 2,4-dichlorophenol takes place more slowly than by direct photolysis[370]. The decomposition of 2,4-dichlorophenol in an aqueous system using TiO_2 supported on a film has been analysed using an electrochemical method[371]. A study using solar irradiation has examined the decomposition of 2,4-dichlorophenol in the presence of TiO_2[336]. Others have investigated the solar irradiation process at the pilot plant scale[372].

Photolysis of pentachlorophenol in water can be brought about using a high-pressure mercury arc lamp. This treatment results in several photodegradation products such as less-chlorinated phenols, catechol and trihydroxylated products. The formation of the hydroxylated products is the result of attack by hydroxyl radicals. Other minor products were also observed such as polychlorinated biphenyl ethers, hydroxylated polychlorobiphenyl ethers and polychlorinated dibenzo-p-dioxins[373]. The phenol can also be degraded in artificial fresh water streams[374]. Hydroxyl radicals react with the pentachloro-, pentafluoro- and pentabromophenolates either by addition of hydroxy radicals or by an electron transfer process[375]. Hydroxyl radical addition to the pentachlorophenolate is followed by rapid halide elimination, giving rise to hydroxytetrachlorophenoxyl radical anions. The use of the photo-Fenton system in peroxide solution with pentachlorophenol leads to the formation of octachlorodibenzo-p-dioxin and its precursor 2-hydroxynonachlorodiphenyl ether[376]. Polychlorinated phenols can be dechlorinated using poly(sodium styrenesulphonate-co-N-vinylcarbazole) as the sensitizer[377]. Trichlorophenols have also been subjected to degradation studies[378–380]. For example, the rate of decomposition of 2,4,6-trichlorophenol on TiO_2 increases with rising pH values up to 7.0^{381}. The photodegradation of sodium pentachlorophenolate has been studied using TiO_2 prepared from tetra-Bu titanate hydrolysis[382] and on TiO_2 prepared by a sol-gel method[383].

XIII. REFERENCES

1. H. D. Becker, in *The Chemistry of the Hydroxyl Group, Part 2*, (Ed. S. Patai), Wiley, New York, 1971, p. 894.
2. *Photochemistry Volumes 15–32*, Royal Society of Chemistry, London (1978–present time).
3. O. N. Tchaikovskaya, R. T. Kuznetsova, I. V. Sokolova and N. B. Sul'timova, *Zh. Fiz. Khim.*, **74**, 1806 (2000); *Chem. Abstr.*, **134**, 243297 (2001).
4. A. Bussandri and H. van Willigen, *J. Phys. Chem. A*, **105**, 4669 (2001).
5. O. K. Bazyl', V. Y. Artyukhov, G. V. Maier and I. V. Sokolova, *High Energy Chem.*, **34**, 30 (2000); *Chem. Abstr.*, **132**, 315700 (2000).
6. O. K. Bazyl', V. Ya. Artyukhov, G. V. Maier and I. V. Sokolova, *High Energy Chem.*, **35**, 33 (2001); *Chem. Abstr.*, **134**, 273412 (2001).
7. J. Lee, S. Ryu and S. K. Kim, *Surf. Sci.*, **481**, 163 (2001).
8. O. N. Tchaikovskaya, I. V. Sokolova, R. T. Kuznetsova, V. A. Swetlitchnyi, T. N. Kopylova and G. V. Mayer, *J. Fluoresc.*, **10**, 403 (2000).
9. C. P. Schick and P. M. Weber, *J. Phys. Chem. A*, **105**, 3725 (2001).
10. C. P. Schick and P. M. Weber, *J. Phys. Chem. A*, **105**, 3735 (2001).
11. Y. Kajii and R. W. Fessenden, *Res. Chem. Intermed.*, **25**, 567 (1999).
12. O. Brede, M. R. Ganapathi, S. Naumov, W. Naumann and R. Hermann, *J. Phys. Chem. A*, **105**, 3757 (2001).
13. R. Das, R. Bhatnagar and B. Venkataraman, *Proc. Indian Acad. Sci., Chem. Sci.*, **106**, 1681 (1994).
14. S. Canonica, B. Hellrung and J. Wirz, *J. Phys. Chem. A*, **104**, 1226 (2000).
15. M. Sarakha, M. Bolte and H. D. Burrows, *J. Photochem. Photobiol. A*, **107**, 101 (1997).
16. M. Sarakha and M. Bolte, *J. Photochem. Photobiol. A*, **97**, 87 (1996).
17. M. Sarakha, A. Rossi and M. Bolte, *J. Photochem. Photobiol. A*, **85**, 231 (1995).
18. M. Sarakha, H. Burrows and M. Bolte, *J. Photochem. Photobiol. A*, **97**, 81 (1996).
19. M. Sarakha, M. Bolte and H. D. Burrows, *J. Photochem. Photobiol. A*, **107**, 101 (1997).
20. Y. H. Kuo, L. H. Chen and L. M. Wang, *Chem. Pharm. Bull.*, **39**, 2196 (1991).
21. K. Okada, K. Okubo, M. Odo and H. Murai, *Chem. Lett.*, 845 (1995).
22. M. C. Depew, S. Emori and J. K. S. Wan, *Res. Chem. Intermed.*, **12**, 275 (1989).
23. D. S. Tipikin, G. G. Lazarev and A. Rieker, *Kinet. Catal.*, **42**, 246 (2001); *Chem. Abstr.*, **135**, 137161 (2001).
24. K. Nakatuji, M. Oda, M. Kozaki, Y. Morimoto and K. Okada, *Chem. Lett.*, 845 (1998).
25. S. Sinha, R. De and T. Ganguly, *J. Phys. Chem. A*, **101**, 2852 (1997).
26. S. V. Jovanovic, D. G. Morris, C. N. Piva and J. C. Scaiano, *J. Photochem. Photobiol. A*, **107**, 153 (1997).
27. T. A. Gadosy, D. Shukla and L. J. Johnston, *J. Phys. Chem. A*, **103**, 8834 (1999).
    (a) M. Sako, S. Ohara, K. Hirota and Y. Maki, *J. Chem. Soc., Perkin Trans. 1*, 3339 (1990).
28. K. Maruyama, H. Furuta and A. Osuka, *Tetrahedron*, **42**, 6149 (1986).
29. K. Maruyama and H. Furuta, *Chem. Lett.*, 243 (1986).
30. M. Sako, S. Ohara, K. Hirota and Y. Maki, *J. Chem. Soc., Perkin Trans. 1*, 3339 (1990).
31. F. Bonnichon, M. Sarakha and M. Bolte, *J. Photochem. Photobiol. A*, **109**, 217 (1997).
32. A. I. Prokof'ev, *Russ. Chem. Rev.*, **68**, 727 (1999).
33. K. Kalninsh, R. Gadonas, V. Krasauskas and A. Pugzhlys, *Chem. Phys. Lett.*, **209**, 63 (1993).
34. T. Arimura, C. T. Brown, S. L. Springs and J. L. Sessier, *J. Chem. Soc., Chem. Commun.*, 2293 (1996).
35. (a) W. Klopffer, *Adv. Photochem.*, **10**, 311 (1997).
    (b) D. M. Rentzepis and P. F. Barbara, *Adv. Chem. Phys.*, **47**, 627 (1981).
    (c) D. Huppert, M. Gutman and K. J. Kaufmann, *Adv. Chem. Phys.*, **47**, 627 (1981).
    (d) I. Y. Martynov, A. B. Demyashkevich, B. M. Uzhinov and M. G. Kuzmin, *Russ. Chem. Rev. (Engl. Transl.)*, **46**, 1 (1977).

36. E. Hoshimoto, S. Yamauchi, N. Hirota and S. Nagaoka, *J. Phys. Chem.*, **95**, 10229 (1991).
37. S. J. Formosinho and L. G. Arnaut, *J. Photochem. Photobiol. A*, **75**, 21 (1993).
38. A. C. Bhasikuttan, A. K. Singh, D. K. Palit, A. V. Sapre and J. P. Mittal, *J. Phys. Chem. A*, **102**, 3470 (1998).
39. M. Hoshino, *J. Phys. Chem.*, **91**, 6385 (1987).
40. J. C. Scaiano, *Chem. Phys. Lett.*, **92**, 97 (1982).
41. H. Miyasaka, K. Morita, K. Kamada and N. Mataga, *Bull. Chem. Soc. Jpn.*, **63**, 3385 (1990).
42. P. K. Das, M. V. Encinas and J. C. Scaiano, *J. Am. Chem. Soc.*, **103**, 4154 (1981).
43. W. J. Leigh, E. C. Lathioor and M. J. St. Pierre, *J. Am. Chem. Soc.*, **118**, 12339 (1996).
44. E. C. Lathioor, W. J. Leigh and M. J. St. Pierre, *J. Am. Chem. Soc.*, **121**, 11984 (1999).
45. A. Mandal, D. Guha, R. Das, S. Mitra and S. J. Mukherjee, *J. Chem. Phys.*, **114**, 1336 (2001).
46. S. Mitra, R. Das and S. Mukherjee, *Chem. Phys. Lett.*, **228**, 393 (1993).
47. R. Das, S. Mitra, D. N. Nath and S. Mukherjee, *J. Chim. Phys.*, **93**, 458 (1996).
48. R. Das, S. Mitra and S. Mukherjee, *J. Photochem. Photobiol. A*, **76**, 33 (1993).
49. R. Das, S. Mitra and S. Mukherjee, *Bull. Chem. Soc. Jpn.*, **66**, 2492 (1993).
50. P.-T. Chou, M. L. Martinez and S. L. Studer, *J. Phys. Chem.*, **95**, 10306 (1991).
51. D. Guha, A. Mandal, D. Nath, S. Mitra, N. Chattopadhyay and S. Mukherjee, *J. Mol. Struct. (THEOCHEM)*, **542**, 33 (2001).
52. M. Isaks, K. Yates and P. Kalanderopoulos, *J. Am. Chem. Soc.*, **106**, 2728 (1984).
53. P. Kalanderopoulos and K. Yates, *J. Am. Chem. Soc.*, **108**, 6290 (1986).
54. M. Yasuda, T. Sone, K. Tanabe and K. Shima, *Chem. Lett.*, 453 (1994).
55. M. Yasuda, T. Sone, K. Tanabe and K. Shima, *J. Chem. Soc., Perkin Trans. 1*, 459 (1995).
56. M. Fischer and P. Wan, *J. Am. Chem. Soc.*, **120**, 2680 (1998).
57. M. Fischer and P. Wan, *J. Am. Chem. Soc.*, **121**, 4555 (1999).
58. T. Goto and Y. Tashiro, *J. Lumin.*, **72–74**, 921 (1997).
59. M. E. Kletskii, A. A. Millov, A. V. Metelista and M. L. Knyazhansky, *J. Photochem. Photobiol. A*, **110**, 267 (1997).
60. S. M. Ormson and R. G. Brown, *Prog. React. Kinet.*, **19**, 45 (1994).
61. F. Wang, Y. Li, X. Li and J. Zhang, *Res. Chem. Intermed.*, **24**, 67 (1998).
62. A. Elmali and Y. Elermand, *J. Mol. Struct.*, **442**, 31 (1998).
63. W. Schilf, B. Kamienski, T. Dziembowska, Z. Rozwadowski and A. Szady-Chelmieniecka, *J. Mol. Struct.*, **552**, 33 (2000).
64. I. Krol-Starzomska, M. Rospenk, Z. Rozwadowski and T. Dziembowska, *Pol. J. Chem.*, **74**, 1441 (2000); *Chem. Abstr.*, **132**, 334977 (2000).
65. V. Enchev, A. Ugrinov and G. A. Neykov, *J. Mol. Struct. (THEOCHEM)*, **530**, 223 (2000).
66. A. V. Metelista, M. S. Korobov, L. E. Nivorozhkin, V. I. Minkin and W. E. Smith, *Russ. J. Org. Chem.*, **334**, 1149 (1998).
67. D. LeGourrierec, V. Kharlanov, R. G. Brown and W. Rettig, *J. Photochem. Photobiol. A*, **117**, 209 (1998).
68. P. Borowicz, A. Gabowska, A. Les, L. Kaczmarek and B. Zagrodzki, *Chem. Phys. Lett.*, **291**, 351 (1999).
69. M. L. Knyazhansky, S. M. Aldoshin, A. V. Metelista, A. Y. Bushkov, O. S. Filipenko and L. O. Atovmyan, *Khim. Fiz.*, **10**, 964 (1991); *Chem. Abstr.*, **115**, 207585 (1991).
70. (a) M. Brauer, M. Mosquera, J. L. Perez-Lustres and F. Rodriguez-Prieto, *J. Phys. Chem. A*, **102**, 10736 (1998).
    (b) A. Douchal, F. Amat-Guerri, M. P. Lillo and A. U. Acuna, *J. Photochem. Photobiol. A*, **78**, 127 (1994).
71. (a) K. Das, N. Sarkar, D. J. Majumdar and K. Bhattcharyya, *Chem. Phys. Lett.*, **198**, 463 (1992).
    (b) K. Das, N. Sarkar, A. K. Ghosh, D. Majumdar, D. N. Nath and K. Bhattcharyya, *J. Phys. Chem.*, **98**, 9126 (1994).
72. M. Mosquera, J. C. Penedo, M. C. Rios-Rodriguez and F. Rodriguez-Prieto, *J. Phys. Chem.*, **100**, 5398 (1996).
73. E. L. Roberts, J. Day and J. M. Warner, *J. Phys. Chem. A*, **101**, 5296 (1997).
74. F. Rodriguez-Prieto, M. C. Rios-Rodriguez, M. Mosquera and M. A. Rios-Fernandez, *J. Phys. Chem.*, **98**, 8666 (1994).
75. M. Mosquera, M. C. Rios-Rodriguez and F. Rodriguez-Prieto, *J. Phys. Chem. A*, **101**, 2766 (1997).

76. M. C. Rois-Rodriguez, F. Rodriguez-Prieto and M. Mosquera, *Phys. Chem. Chem. Phys.*, **1**, 253 (1999).
77. M. Fores, M. Duran, M. Sola, M. Orozco and F. J. Luque, *J. Phys. Chem. A*, **103**, 4525 (1999).
78. M. Itoh and Y. Fujiwara, *J. Am. Chem. Soc.*, **107**, 1561 (1985).
79. A. Mordzinski and K. H. Grellmann, *J. Phys. Chem.*, **90**, 5503 (1986).
80. K. H. Grellmann, A. Mordzinski and A. Heinrich, *Chem. Phys.*, **136**, 201 (1989).
81. J. S. Stephan and K. H. Grellmann, *J. Phys. Chem.*, **99**, 10066 (1995).
82. G. J. Woolfe, M. Melzig, S. Schneider and F. Dorr, *Chem. Phys.*, **77**, 213 (1983).
83. A. Mordzinski and A. Grabowski, *Chem. Phys. Lett.*, **90**, 122 (1982).
84. T. Arthen-Engeland, T. Bultmann, N. P. Erusting, M. A. Rodriguez and W. Thiel, *Chem. Phys.*, **163**, 43 (1992).
85. (a) P. F. Barbara, L. E. Brus and P. M. Rentzepis, *J. Am. Chem. Soc.*, **101**, 5631 (1989).
    (b) W. E. Brewer, M. L. Martinez and P.-T. Chou, *J. Phys. Chem.*, **94**, 1915 (1990).
86. S. Nagaoka, J. Kusunoki, T. Fujibuchi, S. Hatakenaka, K. Mukai and U. Nagashima, *J. Photochem. Photobiol. A*, **122**, 151 (1999).
87. P. Purkayastha and N. Chattopadhyay, *Phys. Chem. Chem. Phys.*, **2**, 203 (2000).
88. D. LeGourrierec, V. Kharlanov, R. G. Brown and W. Rettig, *J. Photochem. Photobiol. A*, **130**, 101 (2000).
89. C. E. M. Carvalho, I. M. Brinn, A. V. Pinto and M. C. F. R. Pinto, *J. Photochem. Photobiol. A*, **123**, 61 (1999).
90. F. Waiblinger, A. P. Fluegge, J. Keck, M. Stein, H. E. A. Kramer and D. Leppard, *Res. Chem. Intermed.*, **27**, 5 (2001).
91. K. Kobayashi, M. Iguchi, T. Imakubo, K. Iwata and H. Hamaguchi, *J. Chem. Soc., Chem. Commun.*, 763 (1998).
92. K. Kobayashi, M. Iguchi, T. Imakubo, K. Iwata and H. Hamaguchi, *J. Chem. Soc., Perkin Trans. 2*, 1993 (1998).
93. T. Arai and Y. Norikane, *Chem. Lett.*, 339 (1997).
94. P.-T. Chou, M. L. Martinez and W. C. Cooper, *J. Am. Chem. Soc.*, **114**, 4943 (1992).
95. Y. Norikane, H. Itoh and T. Arai, *Chem. Lett.*, 1094 (2000).
96. T. L. Brack, S. Conti, C. Radu and N. Wachter-Jurcsak, *Tetrahedron Lett.*, **40**, 3995 (1999).
97. R. Matsushima and I. Hirao, *Bull. Chem. Soc. Jpn.*, **53**, 518 (1980).
98. R. Matsushima and H. Kageyama, *J. Chem. Soc., Perkin Trans. 1*, 743 (1985).
99. A. Kasakara, T. Izumi and M. Ooshima, *Bull. Chem. Soc. Jpn.*, **47**, 2526 (1974).
100. R. Matsushima, K. Miyakawa and N. Nishihata, *Chem. Lett.*, 1915 (1988).
101. R. Matsushima and M. Suzuki, *Bull. Chem. Soc. Jpn.*, **65**, 39 (1992).
102. H. Horiuchi, A. Yokawa, T. Okutsu and H. Hiratsuka, *Bull. Chem. Soc. Jpn.*, **72**, 2429 (1999).
103. R. Matsushima and T. Murakami, *Bull. Chem. Soc. Jpn.*, **73**, 2215 (2000).
104. R. Matsushima, S. Fujimoto and K. Tokumura, *Bull. Chem. Soc. Jpn.*, **74**, 827 (2001).
105. V. V. Kudryavtsev, N. I. Rtishchev, G. I. Nosova, N. A. Solovskaya, V. A. Luk'yashina and A. V. Dobrodumov, *Russ. J. Gen. Chem.*, **70**, 1272 (2000); *Chem. Abstr.*, **134**, 359376 (2001).
106. S. Jain, *Asian J. Chem.*, **5**, 659 (1993); *Chem. Abstr.*, **119**, 249647 (1993).
107. P. Wan and D. Hennig, *J. Chem. Soc., Chem. Commun.*, 939 (1987).
108. K. L. Foster, S. Baker, D. W. Brousmiche and P. Wan, *J. Photochem. Photobiol. A*, **129**, 157 (1999).
109. L. Diao, C. Yang and P. Wan, *J. Am. Chem. Soc.*, **117**, 5369 (1995).
110. P. Wan, B. Barker, L. Diao, M. Fischer, Y. J. Shi and C. Yang, *Can. J. Chem.*, **74**, 465 (1996).
111. B. Barker, L. Diao and P. Wan, *J. Photochem. Photobiol. A*, **104**, 91 (1997).
112. K. Nakatani, N. Higashita and I. Saito, *Tetrahedron Lett.*, **38**, 5005 (1997).
113. Y. Chiang, A. J. Kresge and Y. Zhu, *J. Am. Chem. Soc.*, **123**, 8089 (2001).
114. D. W. Brousmiche and P. Wan, *J. Photochem. Photobiol. A*, **130**, 113 (2000).
115. M. Lukeman and P. Wan, *J. Chem. Soc., Chem. Commun.*, 1004 (2001).
116. M. Capponi, I. G. Gut, B. Hellrung, G. Persy and J. Wirz, *Can. J. Chem.*, **77**, 605 (1999).
117. Y. Shi and P. Wan, *J. Chem. Soc., Chem. Commun.*, 1217 (1995).
118. Y. Shi, A. MacKinnon, J. A. K. Howard and P. Wan, *J. Photochem. Photobiol. A*, **113**, 271 (1998).
119. C. G. Huang, K. A. Beveridge and P. Wan, *J. Am. Chem. Soc.*, **113**, 7676 (1991).
120. M. Fischer, Y. J. Shi, B. P. Zhao, V. Snieckus and P. Wan, *Can. J. Chem.*, **77**, 868 (1999).

121. J. Mohanty, H. Pal, R. D. Saini and A. V. Sapre, *Chem. Phys. Lett.*, **342**, 328 (2001).
122. (a) S. Geresh, O. Levy, Y. Markovits and A. Shani, *Tetrahedron*, **31**, 2803 (1975).
    (b) T. Kitamura, T. Imagawa and M. Kawanisi, *Tetrahedron*, **34**, 3451 (1978).
    (c) M. C. Jiminez, F. Marquez, M. A. Miranda and R. Tormos, *J. Org. Chem.*, **59**, 197 (1994).
    (d) H. Morrison, *Org. Photochem.*, **4**, 143 (1979).
123. M. C. Jiminez, P. Leal, M. A. Miranda and R. Tormos, *J. Org. Chem.*, **60**, 3243 (1995).
124. J. Delgado, A. Espinos, M. C. Jiminez, M. A. Miranda and R. Tormos, *Tetrahedron*, **53**, 681 (1997).
125. M. C. Jiminez, P. Leal, M. A. Miranda, J. C. Scaiano and R. Tormos, *Tetrahedron*, **54**, 4337 (1998).
126. M. C. Jiminez, M. A. Miranda and R. Tormos, *Tetrahedron*, **53**, 14729 (1997).
127. M. C. Jiminez, M. A. Miranda and R. Tormos, *J. Org. Chem.*, **63**, 1323 (1998).
128. J. Delgado, A. Espinos, M. C. Jiminez, M. A. Miranda, H. D. Roth and R. Tormos, *J. Org. Chem.*, **64**, 6541 (1999).
129. M. A. Miranda and R. Tormos, *J. Org. Chem.*, **58**, 3304 (1993).
130. (a) D. Dopp, in *Handbook of Organic Photochemistry and Photobiology* (Eds. W. M. Horspool and P.-S. Song), CRC Press, Boca Raton, 1994, p. 1019.
    (b) R. W. Binkley and T. W. Flechtner, in *Synthetic Organic Photochemistry* (Ed. W. M. Horspool), Plenum Press, New York, 1984, p. 375.
131. R. S. Givens, A. Jung, C. H. Park, J. Weber and W. Bartlett, *J. Am. Chem. Soc.*, **119**, 8369 (1997).
132. K. Khang, J. E. T. Corrie, V. R. N. Munasinghe and P. Wan, *J. Am. Chem. Soc.*, **121**, 5625 (1999).
133. M. C. Pirrung and Y. R. Lee, *J. Org. Chem.*, **58**, 6961 (1993).
134. M. C. Pirrung and Y. R. Lee, *Tetrahedron Lett.*, **34**, 8217 (1993).
135. M. C. Pirrung, L. Fallon, J. Zhu and Y. R. Lee, *J. Am. Chem. Soc.*, **123**, 3638 (2001).
136. D. S. English, W. Zang, G. A. Kraus and J. W. Petrich, *J. Am. Chem. Soc.*, **119**, 2980 (1997).
137. D. S. English, K. Das, K. D. Ashby, J. Park, J. W. Petrich and E. W. Castner, *J. Am. Chem. Soc.*, **119**, 11585 (1997).
138. A. V. Smirnov, K. Das, D. S. English, Z. Wan, G. A. Kraus and J. W. Petrich, *J. Phys. Chem. A*, **103**, 7949 (1999).
139. K. Das, A. V. Smirnov, M. D. Snyder and J. W. Petrich, *J. Phys. Chem. B*, **102**, 6098 (1998).
140. R. A. Obermueller, G. J. Schuetz, H. J. Gruber and H. Falk, *Monatsh. Chem.*, **130**, 275 (1999).
141. J. Malkin and Y. Mazur, *Photochem. Photobiol.*, **57**, 929 (1993).
142. G. Xia, X. He, Y. Zhou, M. Zhang and T. Shen, *Gangguang Kexue Yu Guang Huaxue*, **16**, 15 (1998); *Chem. Abstr.*, **128**, 301129 (1998).
143. H. Falk, *Angew. Chem., Int. Ed. Engl.*, **38**, 3117 (1999).
144. R. Das, S. Mitra, D. Guha and S. Mukherjee, *J. Lumin.*, **81**, 61 (1999).
145. D. Guha, A. Mandal and S. Mukherjee, *J. Lumin.*, **85**, 79 (1999).
146. H. C. Ludemann, F. Hillenkamp and R. W. Redmond, *J. Phys. Chem. A*, **104**, 3884 (2000).
147. D. LeGourrierec, S. M. Ormson and R. G. Brown, *Prog. React. Kinet.*, **19**, 211 (1994).
148. H. Mishra, H. C. Joshi, H. B. Tripathi, S. Maheshwary, N. Sathyamurthy, M. Panda and J. Chandrasekhar, *J. Photochem. Photobiol. A*, **139**, 23 (2001).
149. J. A. Organero, M. Moreno, L. Santos, J. M. Lluch and A. Douhal, *J. Phys. Chem. A*, **104**, 8424 (2000).
150. J. D. Geerlings, C. A. G. O. Varma and M. C. van Hemert, *J. Phys. Chem. A*, **104**, 7409 (2000).
151. Y. Kim, M. Yoon and D. Kim, *J. Photochem. Photobiol. A*, **138**, 167 (2001).
152. R. F. Childs, B. D. Parrington and M. Zeya, *J. Org. Chem.*, **44**, 4912 (1979).
153. R. F. Childs and B. D. Parrington, *Can. J. Chem.*, **52**, 3703 (1974).
154. R. F. Childs, G. S. Shaw and A. Varadarajan, *Synthesis*, 198 (1982).
155. P. Baeckstrom, U. Jacobsson, B. Koutek and T. Norin, *J. Org. Chem.*, **50**, 3728 (1985).
156. R. F. Childs and B. E. George, *Can. J. Chem.*, **66**, 1343 (1988).
157. S. K. Chadda and R. F. Childs, *Can. J. Chem.*, **63**, 3449 (1985).
158. K. Kakiuchi, M. Ue, B. Yamaguchi, A. Nishimoto and Y. Tobe, *Bull. Chem. Soc. Jpn.*, **64**, 3468 (1991).

159. M. Ue, M. Kinugawa, K. Kakiuchi, Y. Tobe and Y. Odaira, *Tetrahedron Lett.*, **30**, 6193 (1989).
160. S. Mihara and T. Shibamoto, *J. Agric. Food Chem.*, **30**, 11215 (1982); *Chem. Abstr.*, **97**, 182084 (1982).
161. S. Walia, S. K. Kulshrestha and S. K. Mukerjee, *Tetrahedron*, **42**, 4817 (1986).
162. I. Ikeda, S. Kobayashi and H. Taniguchi, *Synthesis*, 393 (1982).
163. F. Scavarda, F. Bonnichon, C. Richard and G. Grabner, *New J. Chem.*, **21**, 1119 (1997).
164. J.-F. Pilichowski, P. Boule and J.-P. Billard, *Can. J. Chem.*, **73**, 2143 (1995).
165. M. A. Miranda and H. Garcia, 'Rearrangements in acid derivatives', Supplement B2 *The Chemistry of Acid Derivatives* (Ed. S. Patai), Wiley, Chichester, 1992.
166. M. A. Miranda, in *Handbook of Organic Photochemistry and Photobiology*, (Eds. W. M. Horspool and P.-S. Song), CRC Press, Boca Raton, 1994, p. 570.
167. M. R. Diaz-Mondejar and M. A. Miranda, *Tetrahedron*, **38**, 1523 (1982).
168. H.-C. Chiang and C.-H. Chi'en, *Hua Hsueh*, 7 (1979); *Chem. Abstr.*, **94**, 46445 (1981).
169. M. C. Jiminez, P. Leal, M. A. Miranda and R. Tormos, *J. Chem. Soc., Chem. Commun.*, 2009 (1995).
170. A. Yamada, S. Yamamoto, T. Yamato and T. Kakitani, *J. Mol. Struct. (THEOCHEM)*, **536**, 195 (2001).
171. G. Grabner, C. Richard and G. Kohler, *J. Am. Chem. Soc.*, **116**, 11470 (1994).
172. A. Quardaoui, C. A. Steren, H. Vanwilligen and C. Q. Yang, *J. Am. Chem. Soc.*, **117**, 6803 (1995).
173. A.-P. Y. Durand, R. G. Brown, D. Worrall and F. Wilkinson, *J. Chem. Soc., Perkin Trans. 2*, 365 (1998).
174. F. Bonnichon, G. Grabner, G. Guyot and C. Richard, *J. Chem. Soc., Perkin Trans. 2*, 1203 (1999).
175. C. Guyon, P. Boule and J. Lemaire, *Nouv. J. Chem.*, **8**, 685 (1984).
176. P. Boule, C. Guyon and J. Lemaire, *Chemosphere*, **11**, 1179 (1982).
177. P. Boule and C. Guyon, *Toxicol. Environ. Chem.*, **7**, 97 (1984).
178. H. Tomioka and T. Matsushita, *Chem. Lett.*, 399 (1997).
179. H. Tomioka, *Pure. Appl. Chem.*, **69**, 837 (1997).
180. A. Gilbert, P. Heath and P. W. Rodwell, *J. Chem. Soc., Perkin Trans. 1*, 1867 (1989).
181. J. Mattay, *Angew. Chem., Int. Ed. Engl.*, **26**, 825 (1987).
182. J. Mattay, *J. Photochem.*, **37**, 167 (1987).
183. N. A. Al-Jalal, *Gazz. Chim. Ital.*, **119**, 569 (1989).
184. N. A. Al-Jalal, *J. Heterocycl. Chem.*, **27**, 1323 (1990).
185. N. A. Al-Jalal, *J. Photochem. Photobiol. A*, **54**, 1323 (1990).
186. M. Miyake, T. Tsuji and S. Nishida, *Chem. Lett.*, 1395 (1988).
187. G. P. Kalena, P. Pradham and A. Banerji, *Tetrahedron Lett.*, **33**, 7775 (1992).
188. N. Hoffmann and J.-P. Pete, *Tetrahedron Lett.*, **37**, 2027 (1996).
189. N. Hoffmann and J.-P. Pete, *J. Org. Chem.*, **62**, 6952 (1997).
190. N. Hoffmann and J.-P. Pete, *Tetrahedron Lett.*, **39**, 5027 (1998).
191. N. Hoffmann and J.-P. Pete, *Synthesis*, 1236–(2001).
192. B. R. Pai, H. Suguna, B. Geetha and K. Sarada, *Indian J. Chem., Sect. B*, **17**, 503 (1979).
193. B. R. Pai, H. Suguna, B. Geetha and K. Sarada, *Indian J. Chem., Sect. B*, **17**, 525 (1979).
194. H. Soicke, G. Al-Hassan, U. Frenzel and K. Gorler, *Arch. Pharm.*, **321**, 149 (1988).
195. O. Hoshino, H. Ogasawara, A. Hirokawa and B. Umezawa, *Chem. Lett.*, 1767 (1988).
196. O. Hoshino, H. Ogasawara, A. Takahashi and B. Umezawa, *Heterocycles*, **23**, 1943 (1985).
197. H. Tanaka, Y. Takamura, K. Ito, K. Ohira and M. Shibata, *Chem. Pharm. Bull.*, **32**, 2063 (1984).
198. A. Olszanowski and E. Kryzanowski, *Hydrometallurgy*, **35**, 79 (1994); *Chem. Abstr.*, **120**, 322542 (1994).
199. B. Gozler, H. Guinaudeau, M. Shamma and G. Sariyar, *Tetrahedron Lett.*, **27**, 1899 (1986).
200. A. W. Scribner, S. A. Haroutounian, K. E. Karlson and J. A. Katzenellenbogen, *J. Org. Chem.*, **62**, 1043 (1997).
201. S. A. Magoub, M. M. Aly and M. Z. A. Badr, *Rev. Roum. Chim.*, **30**, 749 (1985).
202. A. M. Fahmy, S. A. Magoub, M. M. Aly and M. Z. A. Badr, *Indian J. Chem., Sect. B*, **23B**, 474 (1984).

203. A. M. Fahmy, S. A. Magoub, M. M. Aly and M. Z. A. Badr, *Bull. Fac. Sci., Assiut Univ.*, **11**, 17 (1982); *Chem. Abstr.*, **98**, 53304 (1983).
204. M. C. Jiminez, M. A. Miranda and R. Tormos, *Tetrahedron*, **51**, 5825 (1995).
205. R. Ravichandran, *J. Mol. Catal. A: Chem.*, **130**, L205 (1998).
206. N. Tanaka, *Fukuoka Kyoiku Daigaku Kiyo*, **43**, 35 (1994); *Chem. Abstr.*, **122**, 20193 (1985).
207. S. Seltzer, E. Lam and L. Packer, *J. Am. Chem. Soc.*, **104**, 6470 (1982).
208. N. J. Bunce, S. R. Cater and J. M. Willson, *J. Chem. Soc., Perkin Trans 2*, 2013 (1985).
209. (a) J. Suzuki, H. Okazaki, T. Sato and S. Suzuki, *Chemosphere*, **11**, 437 (1982).
    (b) J. Suzuki, H. Okazaki, Y. Nishio and S. Suzuki, *Bull. Environ. Contam. Toxicol.*, **29**, 511 (1982); *Chem. Abstr.*, **98**, 7992 (1983).
210. K. Oudjehani and P. Boule, *J. Photochem. Photobiol. A*, **68**, 363 (1992).
211. E. Lipczynska-Kochany and J. R. Bolton, *J. Photochem. Photobiol. A*, **58**, 315 (1991).
212. A.-P. Y. Durand, D. Brattan and R. G. Brown, *Chemosphere*, **25**, 783 (1992).
213. E. Lipczynska-Kochany and J. R. Bolton, *Chemosphere*, **24**, 911 (1992).
214. E. Lipczynska-Kochany and J. Kochany, *J. Photochem. Photobiol. A*, **73**, 27 (1993).
215. J. Kochany, G. G. Choudry and G. R. B. Webster, in *Chemistry for the Protection of the Environment; Proc. 7th Conf. Environmental Sci. Res.*, Vol. 42 (Eds. L. Pawlovski, W. J. Lacy and J. J. Dlugosz), Plenum Press, New York, 1991, p. 259.
216. I. Texier, C. Giannotti, S. Malato, C. Richter and J. Delaire, *Catal. Today*, Vol. 54 297 (2000).
217. P. Boule, C. Guyan, A. Tissot and J. Lemaire, *J. Chim. Phys.*, **82**, 513 (1985).
218. B. Zhang, J. Zhang, D.-D. H. Yang and N. C. Yang, *J. Org. Chem.*, **61**, 3236 (1996).
219. C. J. Chandler, D. J. Craik and K. J. Waterman, *Aust. J. Chem.*, **42**, 1407 (1989).
220. A. Alif, P. Boule and J. Lemaire, *J. Photochem. Photobiol. A*, **50**, 331 (1990).
221. A. B. Artynkhin, A. V. Ivanov and V. L. Ivanov, *Vestn. Mosk. Univ. Ser. 2 Khim.*, **35**, 234 (1994); *Chem. Abstr.*, **121**, 178906 (1994).
222. V. L. Ivanov and L. Eggert, *Zh. Org. Khim.*, **19**, 2372 (1983); *Chem. Abstr.*, **100**, 84994 (1984).
223. V. L. Ivanov, J. Aurich, L. Eggert and M. G. Kuzmin, *J. Photochem. Photobiol. A*, **50**, 275 (1990).
224. V. L. Ivanov, S. Yu. Lyashkevich and H. Lemmetyinen, *J. Photochem. Photobiol. A*, **109**, 21 (1997).
225. G. Petrillo, M. Novi, C. Dell'Erba and C. Tavani, *Tetrahedron*, **47**, 9297 (1991).
226. M. T. Baumgartner, A. B. Pierini and R. A. Rossi, *J. Org. Chem.*, **58**, 2593 (1993).
227. P. K. Freeman, *J. Org. Chem.*, **51**, 3939 (1986).
228. P. K. Freeman and R. Srinivasa, *J. Agric. Food Chem.*, **32**, 1313 (1984).
229. P. K. Freeman and R. Srinivasa, *J. Agric. Food Chem.*, **31**, 775 (1983).
230. P. K. Freeman and V. Jonas, *J. Agric. Food Chem.*, **32**, 1307 (1984).
231. C. G. Choudry, V. W. M. Van der Wieln, G. R. B. Webster and O. Hutzinger, *Can. J. Chem.*, **63**, 469 (1985).
232. (a) S. P. Spyroudis, *J. Org. Chem.*, **51**, 3453 (1986).
    (b) S. Spyroudis and A. Varvoglis, *J. Chem. Soc., Perkin Trans. 1*, 135 (1984).
    (c) M. Papadopoulos, S. Spyroudis and A. Varvoglis, *J. Org. Chem.*, **50**, 1509 (1985).
233. S. P. Spyroudis, *Liebig's Ann. Chem.*, 947 (1986).
234. M. Tajima, K. Kato, K. Matsunga and H. Inoue, *J. Photochem. Photobiol. A*, **140**, 127 (2001).
235. O. P. Studzinskii, R. P. Ponomareva and V. N. Seleznev, *Izv. Vysshh. Uchebn. Zaved. Khim. Khim Technol.*, **23**, 511 (1980); *Chem. Abstr.*, **93**, 168004 (1980).
236. K. Hamilton, J. A. Hunter, P. N. Preston and J. O. Morley, *J. Chem. Soc., Perkin Trans. 1*, 1544 (1980).
237. H. Ji, Z. Tong and C. Tung, *Ganguang Kexue Yu Guang Huaxue*, **14**, 289 (1996); *Chem. Abstr.*, **127**, 148896 (1997).
238. J. H. Rigby and M. E. Mateo, *J. Am. Chem. Soc.*, **119**, 12655 (1997).
239. G. L. Laines, *Angew. Chem., Int. Ed. Engl.*, **26**, 341 (1987).
240. J. Ito, K. Gobaru, T. Shimamura, M. Niwa, Y. Takaya and Y. Oshima, *Tetrahedron*, **54**, 6651 (1998).
241. S. Kato, M. Nakatani and H. Harashina, *Chem. Lett.*, 847 (1986).
242. M. D'Ischia, G. Testa, D. Mascagna and A. Napolitano, *Gazz. Chim. Ital.*, **125**, 315 (1995).
243. F. Bosca, M. C. Cuquerella, M. L. Marin and M. A. Miranda, *Photochem. Photobiol.*, **73**, 463 (2001).

244. M. A. Miranda, F. Galindo, A. M. Amat and A. Arques, *Appl. Catal., B*, **30**, 437 (2001); *Chem. Abstr.*, **134**, 285138 (2001).
245. S. Klementova and J. Matouskova, *Res. J. Chem. Environ.*, **4**, 25 (2000); *Chem. Abstr.*, **134**, 245102 (2001).
246. A. J. Chaudhary, S. M. Grimes and Mukhtar-ul-Hassan, *Chemosphere*, **44**, 1223 (2001); *Chem. Abstr.*, **135**, 33468 (2001).
247. J. Matos, J. Laine and J.-M. Herrmann, *J. Catal.*, **200**, 10 (2001); *Chem. Abstr.*, **135**, 96852 (2001).
248. G. Colon, M. C. Hidalgo and J. A. Navio, *J. Photochem. Photobiol. A*, **138**, 79 (2001).
249. S. M. Ould-Mame, O. Zahraa and M. Bouchy, *Int. J. Photoenergy*, **2**, 59 (2000); *Chem. Abstr.*, **135**, 215328 (2001).
250. K. Lang, D. M. Wagnerova and J. Brodilova, *Collect. Czech. Chem. Commun.*, **59**, 2447 (1994).
251. C. Richard and P. Boule, *J. Photochem. Photobiol. A*, **84**, 151 (1994).
252. I. Davydov, E. P. Reddy, P. France and P. G. Smirniotis, *Appl. Catal., B*, **32**, 95 (2001); *Chem. Abstr.*, **135**, 126572 (2001).
253. F. Benoit-Marquie, E. Puech-Costes, A. M. Braun, E. Oliveros and M.-T. Maurette, *J. Photochem. Photobiol. A*, **108**, 73 (1997); *Chem. Abstr.*, **127**, 285784 (1997).
254. F. J. Benitez, J. L. Acero, F. J. Real, F. J. Rubio and A. I. Leal, *Water Res.*, **35**, 1338 (2001).
255. F. J. Benitez, J. Beltran-Heredia, T. Gonzalez and J. L. Acero, *Water Res.*, **28**, 2095 (1994).
256. F. J. Benitez, J. Beltran-Heredia, J. L. Acero and T. Gonzalez, *Toxicol. Environ. Chem.*, **47**, 141 (1995).
257. Y. Harada, S. Maki, H. Niwa, T. Hirano and S. Yamamura, *Synlett*, 1313 (1998).
258. W. Yu, L. Yang and Z. L. Liu, *Chin. Chem. Lett.*, **9**, 823 (1998).
259. (a) M. Prein, M. Mauer, E. M. Peters, K. Peters, H. G. von Schnering and W. Adam, *Chem.- Eur. J.*, **1**, 89 (1995).
 (b) H. A. Wasserman and J. E. Picket, *Tetrahedron*, **41**, 2155 (1985).
 (c) W. Adam and P. Lupon, *Chem. Ber.*, **121**, 21 (1988).
 (d) K. Endo, K. Seya and H. Hikino, *Tetrahedron*, **45**, 3673 (1989).
260. T. Koepp, M. Koether, B. Brueckner and K. H. Radeke, *Chem. Tech. (Leipzig)*, **45**, 401 (1993).
261. I. Saito and T. Matsuura, in *Singlet Oxygen* (Eds. H. H. Wasserman and R. W. Murray), Academic Press, New York, 1979, p. 511.
262. A. Albini and M. Freccero, in *Handbook of Organic Photochemistry and Photobiology* (Eds. W. M. Horspool and P.-S. Song), CRC Press, Boca Raton, 1994, p. 346.
263. M. B. Arkhipova, L. Ya. Tereshchenko, I. A. Martynova and Yu. M. Arkhipoz, *Zh. Prikl. Khim.*, **67**, 598 (1994); *Chem. Abstr.*, **121**, 307486 (1994).
264. M. B. Arkhipova, L. Ya. Tereshchenko and Yu. M. Arkhipov, *Zh. Prikl. Khim.*, **68**, 1563 (1995); *Chem. Abstr.*, **124**, 135453 (1996).
265. X. Kang, M. Liu and S. Wang, *Huaxue Gongye Yu Gongcheng*, **11**, 24 (1994); *Chem. Abstr.*, **122**, 273108 (1995).
266. H. Kawaguchi, *Chemosphere*, **24**, 1707 (1992).
267. A. De, S. Bhattachatarjee and B. K. Dutta, *Ind. Eng. Chem. Res.*, **36**, 3607 (1997).
268. X. Hu, F. L. Y. Lam, L. M. Cheung, K. F. Chan, X. S. Zhao and G. Q. Lu, in *Sustainable Energy Environ. Technol., Proc. Asia-Pac. Conf., 3rd* (Eds. X. Hu and P. L. Yue), 2001, pp. 51–55; *Chem. Abstr.*, **134**, 184587 (2001).
269. W. Spacek, R. Bauer and G. Heilser, *Chemosphere*, **30**, 477 (1995).
270. J. Sykora, C. Kutal and R. G. Zepp, *Conf. Coord Chem.*, 15, 263 (1995); *Chem. Abstr.*, **124**, 131135 (1996).
271. J. Sykora, E. Brandsteterova and A. Jabconova, *Bull. Soc. Chim. Belg.*, **101**, 821 (1992).
272. J. Kovacova, J. Sykora, E. Brandsteterova and I. Kentosova, *Conf. Coord. Chem.*, 14, 391 (1993); *Chem. Abstr.*, **120**, 257152 (1994).
273. C. Pizzocaro, M. Bolte, H. Sun and M. Z. Hoffman, *New J. Chem.*, **18**, 737 (1994).
274. C. Li and M. Z. Hoffman, *J. Phys. Chem. A*, **104**, 5998 (2000).
275. C. Pizzocaro, M. Bolte and M. Z. Hoffman, *J. Photochem. Photobiol. A*, **68**, 115 (1992).
276. D. Murtinho, M. Pineiro, M. M. Pereira, A. M. D. Gonsalves, L. G. Arnaut, M. D. Miguel and H. D. Burrows, *J. Chem. Soc., Perkin Trans. 2*, 2441 (2000).

277. R. Gerdes, O. Bartels, G. Schneider, D. Wohrle and G. Schulz-Ekloff, *Polym. Adv. Technol.*, **12**, 152 (2001); *Chem. Abstr.*, **134**, 340142 (2001).
278. R. W. Matthews and S. R. McEvoy, *J. Photochem. Photobiol. A*, **64**, 231 (1992).
279. T. Y. Wei and C. C. Wan, *Ind. Eng. Chem. Res.*, **30**, 1293 (1991).
280. Y. Wang, C. Hu and H. Tong, *Huanjing Kexue Xuebao*, **15**, 472 (1995); *Chem. Abstr.*, **124**, 240943 (1996).
281. E. Pelizzetti, C. Minero, E. Borgarello, L. Tinucci and N. Serpone, *Langmuir*, **9**, 2995 (1993).
282. G. Li, G. Grabner, R. M. Quint, R. Quint and N. Getoff, *Proc. Indian Acad. Sci., Chem. Sci.*, **103**, 505 (1991).
283. H. Wei and X. Yan, *Shanghai Huanjing Kexue*, **14**, 31 (1995); *Chem. Abstr.*, **124**, 154630 (1996).
284. V. Brezova, A. Blazkova, M. Breznan, P. Kottas and M. Ceppan, *Collect. Czech. Chem. Commun.*, **60**, 788 (1995).
285. G.-Y. Li, J.-Q Ma, H.-D. Li and F. Zhu, *Gaoxiao Huaxue Gongcheng Xuebao*, **15**, 187 (2001); *Chem. Abstr.*, **134**, 356946 (2001).
286. S. Cheng, S. J. Tsai and Y. F. Lee, *Catal. Today*, **26**, 87 (1995).
287. B. Barni, A. Cavicchioli, E. Riva, L. Zanoni, F. Bignoli, I. R. Bellobono, F. Gianturco, A. De Giorgi and H. Muntau, *Chemosphere*, **30**, 1847 (1995).
288. R. Molinari, M. Borgese, E. Drioli, L. Palmisano and M. Schiavello, *Ann. Chim.*, **91**, 197 (2001); *Chem. Abstr.*, **135**, 246895 (2001).
289. S.-S. Hong, C.-S. Ju, C.-J. Lim, B-H. Ahn, K.-T. Lim and G.-D. Lee, *J. Ind. Eng. Chem. (Seoul, Repub. Korea)*, **7**, 99 (2001); *Chem. Abstr.*, **134**, 371248 (2001).
290. R. Campostrini, G. Carturan, L. Palmisano, M. Schiavello and A. Sclafini. *Mater. Chem. Phys.*, **38**, 277 (1994).
291. Z.-H. Yuan and L.-D. Zhang, *J. Mater. Chem.*, **11**, 1265 (2001); *Chem. Abstr.*, **135**, 84129 (2001).
292. T. Y. Wei and C. Wan, *J. Photochem. Photobiol. A*, **69**, 241 (1992).
293. L. Palmisano, M. Schiavello, A. Sclafani, G. Marta, E. Borello and S. Coluccia, *Appl. Catal., B*, **3**, 117 (1994).
294. N. San, A. Hatipoglu, G. Kocturk and Z. Cinar, *J. Photochem. Photobiol. A*, **139**, 225 (2001).
295. G. Grabner, G. Li, R. M. Quint, R. Quint and N. Getoff, *J. Chem. Soc., Faraday Trans.*, **87**, 1097 (1991).
296. M. J. Lopez-Munoz, J. Soria, J. C. Conesa and V. Augugliaro, *Stud. Surf. Sci. Catal.*, **82**, 693 (1994).
297. K. E. O'Shea and C. Cardona, *J. Photochem. Photobiol., A*, **91**, 67 (1995).
298. S. Goldstein, G. Czapski and J. Rabani, *J. Phys. Chem.*, **98**, 6586 (1994).
299. T. R. N. Kutty and S. Auhuja, *Mater. Res. Bull.*, **30**, 233 (1995).
300. C. Richard and P. Boule, *New J. Chem.*, **18**, 547 (1994).
301. G. Marci, A. Sclafani, V. Augugliaro, L. Palmisano and M. Schiavello, *J. Photochem. Photobiol. A*, **89**, 69 (1995).
302. A. Sclafini and L. Palmisano, *Gazz. Chim. Ital.*, **120**, 599 (1990).
303. T. Y. Wei, Y. Y. Wang and C. C. Wan, *J. Photochem. Photobiol.*, **55**, 115 (1990).
304. T. Kanki, N. Sano, Z. Lianfeng and A. Toyoda, *Kemikaru Enjiniyaringu*, **46**, 269 (2001); *Chem. Abstr.*, **135**, 36855 (2001).
305. A. W. Morawski, J. Grzechulska, M. Tomaszewska, K. Karakulski and K. Kalucki, *Gazz. Woda Tech. Sanit.*, **69**, 89 (1995); *Chem. Abstr.*, **123**, 92686 (1995).
306. C. Hu, Y. Z. Wang and H. X. Tang, *Chemosphere*, **41**, 1205 (2000).
307. K. Kobayakawa, Y. Sato and A. Fujishima, *Kino Zairyo*, **15**, 1847 (1995); *Chem. Abstr.*, **123**, 92714 (1995).
308. P. Wyness, J. F. Klausner, D. Y. Goswami and K. S. Schanze, *J. Sol. Energ. Eng.*, **116**, 2 (1994).
309. P. Wyness, J. F. Klausner, D. Y. Goswami and K. S. Schanze, *J. Sol. Energ. Eng.*, **116**, 8 (1994).
310. C. Gao, J. Shi, Q. Dong and H. Liu, *Huaxue Shijie*, **35**, 97 (1994); *Chem. Abstr.*, **121**, 212061 (1994).
311. A. A. Yawalkar, D. S. Bhatkhande, V. G. Pangarkar and A. A. C. M. Beenackers, *J. Chem. Technol. Biotechnol.*, **76**, 363 (2001); *Chem. Abstr.*, **134**, 343923 (2001).
312. P. A. Bennett and S. Beadles, *Am. Lab.*, **26**, 29 (1994).

313. R. Terzian and N. Serpone, *J. Photochem. Photobiol. A*, **89**, 163 (1995).
314. D. A. Ellis and S. A. Mabury, *Environ. Sci. Technol.*, **4**, 632 (2000).
315. K. E. O'Shea and C. Cardona, *J. Org. Chem.*, **59**, 5005 (1994).
316. P. G. Tratnyek and J. Hoigne, *J. Photochem. Photobiol, A*, **84**, 153 (1994).
317. H. Kawaguchi, *Chemosphere*, **27**, 2177 (1993).
318. R. Terzian, N. Serpone and H. Hidaka, *Catal. Lett.*, **32**, 227 (1995).
319. A. A. Khodja, T. Sehili, J. F. Pilichowski and P. Boule, *J. Photochem. Photobiol. A*, **141**, 231 (2001).
320. L. Palmisano, M. Schiavello, A. Sclafani, G. Matras, E. Borello and S. Coluccia, *Appl. Catal., B*, **3**, 117 (1994).
321. K. Itoho, *Daiichi Yakka Daigaku Kenkyu Nenpo*, **22**, 1 (1991); *Chem. Abstr.*, **117**, 25724 (1992).
322. T. Hamada, H. Ishida, S. Usui, K. Tsumura and K. Ohkubo, *J. Mol. Catal.*, **88**, L1 (1994).
323. C. L. Torres-Martinez, R. Kho, O. I. Mian and R. K. Mehra, *J. Colloid Interface Sci.*, **240**, 525 (2001); *Chem. Abstr.*, **135**, 361911 (2001).
324. J. A. Herrera-Melian, J. M. Dona-Rodriguez, E. T. Rendon, A. S. Vila, M. B. Quetglas, A. A. Azcarate and L. P. Pariente, *J. Chem. Educ.*, **78**, 775 (2001).
325. L. Palmisano, M. Schiavello, A. Sclafani, C. Martin, I. Martin and V. Rives, *Catal. Lett.*, **24**, 303 (1994).
326. G. Marci, L. Palmisano, A. Selafani, M. Venezia, R. Campostrini, G. Carturan, C. Martin, V. Rives and G. Solana, *J. Chem. Soc., Faraday Trans.*, **92**, 819 (1996).
327. J. Bandara, J. Kiwi, C. Pulgarin, P. Peringer, G.-M. Pajonk, A. Elaloui and P. Albers, *Environ. Sci. Technol.*, **30**, 1261 (1996).
328. V. Brezova, M. Ceppan, M. Vesely and L. Lapcik, *Chem. Pap.*, **45**, 233 (1991).
329. A. N. Okte, M. S. Resat and Y. Inel, *J. Catal.*, **198**, 172 (2001); *Chem. Abstr.*, **135**, 11970 (2001).
330. D. O. Martire, S. E. Braslavsky and N. A. Garcia, *J. Photochem. Photobiol.*, **61**, 113 (1991).
331. S. Preis and J. Kallas, *I. Chem. E. Res. Event-Eur. Conf. Young Res. Chem. Eng., 1st*, **1**, 550 (1995).
332. S. Canonica and J. Hoigne, *Chemosphere*, **30**, 2365 (1995).
333. H. Kawaguchi, *Environ. Technol.*, **14**, 289 (1993).
334. L. Jakob, T. M. Hashem, S. Buerki, N. M. Guindy and A. M. Braun, *J. Photochem. Photobiol. A*, **75**, 97 (1993).
335. J. M. Tseng and C. P. Huang, *Water Sci. Technol.*, **23**, 377 (1991).
336. I. Texier, C. Giannotti, S. Malato, C. Richter and J. Delaire, *J. Phys. IV*, **9**, 289 (1999).
337. R.-A. Doong, C.-H. Chen, R. A. Maithreepala and S.-M. Chang, *Water Res.*, **35**, 2873 (2001); *Chem. Abstr.*, **135**, 246817 (2001).
338. Y. H. Hsieh and K. H. Wang, *Xingda Goncheng Xuebao*, **4**, 85 (1993); *Chem. Abstr.*, **121**, 69231 (1994).
339. V. Ragaini, E. Selli, C. L. Bianchi and C. Pirola, *Ultrason. Sonochem.*, **8**, 251 (2001); *Chem. Abstr.*, **135**, 261915 (2001).
340. J. G. Lin and Y. S. Ma, *J. Environ.*, **126**, 137 (2000).
341. K.-H. Wang, Y.-H. Hsieh and K.-S Huang, *Zhongguo Huanjing Gongcheng Xuekan*, **4**, 111 (1994); *Chem. Abstr.*, **122**, 273671 (1995).
342. Y. H. Hsieh, K.-S. Huang and K.-H. Wang, *Organohalogen Compd.*, **19**, 503 (1994).
343. C. Dong, C.-W. Chen and S.-S. Wu, *Hazard Ind. Wastes*, **27**, 361 (1995); *Chem. Abstr.*, **124**, 63455 (1996).
344. K. David-Oudjehani and P. Boule, *New J. Chem.*, **19**, 199 (1995).
345. F. J. Benitez, J. Beltran-Heredia, J. L. Acero and F. J. Rubio, *J. Chem. Technol. Biotechnol.*, **76**, 312 (2001); *Chem. Abstr.*, **134**, 241942 (2001).
346. J. Yoon, S. Kim, D. S. Lee and J. Huh, *Water Sci. Technol.*, **42**, 219 (2000).
347. P. Mazellier, M. Sarakha and M. Bolte, *New J. Chem.*, **23**, 133 (1999).
348. R. Sauleda and E. Brillas, *Appl. Catal., B*, **29**, 135 (2001); *Chem. Abstr.*, **134**, 182777 (2001).
349. W. S. Kuo, *J. Environ. Sci. Health Part A-Toxic/Hazard. Subst. Environ.*, **35**, 419 (2000).
350. C. Lettmann, K. Hildenbrand, H. Kisch, W. Macyk and W. F. Maier, *Appl. Catal., B,*, **32**, 215 (2001); *Chem. Abstr.*, **135**, 350386 (2001).
351. U. Stafford, K. A. Gray and P. V. Kamat, *J. Phys. Chem.*, **98**, 6343 (1994).
352. U. Stafford, K. A. Gray, P. V. Kamat and A. Varma, *Chem. Phys. Lett.*, **205**, 55 (1993).

353. I. Texier, C. Giannotti, S. Malato, C. Richter and J. Delaire, *Catal. Today*, **54**, 297 (1999).
354. F. He, W. Shen, W. Zhao, C. Fang and Y. Fang, *Cuihua Xuebao*, **22**, 168 (2001); *Chem. Abstr.*, **134**, 356242 (2001).
355. Y. J. Fu and C. M. Wang, *Jpn. J. Appl. Phys., Part 2*, **40**, L160 (2001); *Chem. Abstr.*, **134**, 256236 (2001).
356. K. Ozoemena, N. Kuznetsova and T. Nyokong, *J. Photochem. Photobiol. A*, **139**, 217 (2001).
357. D. Vialaton, C. Richard, D. Baglio and A. B. Paya-Perez, *J. Photochem. Photobiol. A*, **119**, 39 (1998).
358. W. Z. Tang and C. P. Hguang, *Water Res.*, **29**, 745 (1995).
359. M. Sarakha, M. Bolte and H. D. Burrows, *J. Phys. Chem. A*, **104**, 3142 (2000).
360. M. P. Ormad, J. L. Ovelleiro and J. Kiwi, *Appl. Catal., B*, **32**, 157 (2001); *Chem. Abstr.*, **135**, 215326 (2001).
361. A. Hiskia, E. Androulaki, A. Mylonas, S. Boyatzis, D. Dimoticali, C. Minero, E. Pelizzetti and E. Papaconstantinou, *Res. Chem. Intermed.*, **26**, 235 (2000).
362. A. Mylonas and E. Papaconstantinou, *J. Mol. Catal.*, **92**, 261 (1994).
363. W. S. Kuo, *Chemosphere*, **39**, 1853 (1999).
364. W. Z. Tang and C. P. Huang, *Chemosphere*, **30**, 1385 (1995).
365. C. Minero, E. Pelizzetti, P. Pichat, M. Sega and M. Vincenti, *Environ. Sci. Technol.*, **29**, 2226 (1995).
366. H. Al-Ekabi, B. Butters, D. Delany, J. Ireland, N. Lewis, T. Powell and J. Sntorey, *Trace Met. Environ.*, **3**, 321 (1993).
367. F. Serra, M. Trillas, J. Garcia and X. Domenech, *J. Environ. Sci. Health, Part A*, **A29**, 1409 (1994).
368. J. C. D'Oliviera, C. Minero, E. Pelizzetti and P. Pichat, *J. Photochem. Photobiol. A*, **72**, 261 (1993).
369. N. Serpone, I. Texier, A. V. Emeline, P. Pichat, H. Hidaka and J. Zhao, *J. Photochem. Photobiol. A*, **136**, 145 (2000).
370. R. A. Doong, R. A. Maithreepala and S. M. Chang, *Water Sci. Technol.*, **42**, 253 (2000).
371. S. Ahmed, T. J. Kemp and P. R. Unwin, *J. Photochem. Photobiol. A*, **141**, 69 (2001).
372. S. Malato, J. Blanco, P. Fernandez-Ibanez and J. Caceres, *J. Sol. Energy Eng.*, **23**, 138 (2001); *Chem. Abstr.*, **135**, 126559 (2001).
373. J. Hong, D. G. Kim, C. Cheong, S. Y. Jung, M. R. Yoo, K. J. Kim, T. K. Kim and Y. C. Park, *Anal. Sci.*, **16**, 621 (2000).
374. J. J. Pignatello, M. M. Martinson, J. G. Steirert and R. E. Carlson, *Appl. Environ. Microbiol.*, **46**, 1024 (1983).
375. X. W. Fang, H.-P. Schuchmann and C. von Sonntag, *J. Chem. Soc., Perkin Trans. 2*, 1391 (2000).
376. M. Fukushima and K. Tatsumi, *Environ. Sci. Technol.*, **35**, 1771 (2001); *Chem. Abstr.*, **134**, 285147 (2001).
377. M. Nowakowska, K. Szczubialka and S. Zapotoczny, *J. Photochem. Photobiol.*, **97**, 93 (1996).
378. Y. M. Artem'ev, M. A. Artem'eva, M. G. Vinogradov and T. I. Ilika, *Zh. Prikl. Khim.*, **67**, 1542 (1994); *Chem. Abstr.*, **123**, 40603 (1995).
379. S. Tanaka and U. K. Saha, *Water Sci. Technol.*, **30**, 47 (1994).
380. X. Rao, J. Wang and S. Dai, *Zhongguo Huanjing Kexue*, **15**, 107 (1995); *Chem. Abstr.*, **123**, 722446 (1995).
381. S. Tanaka and U. K. Saha, *Yosui to Haisui*, **36**, 883 (1994); *Chem. Abstr.*, **122**, 226444 (1995).
382. B. Xi and H. Liu, *Huanjing Kexue Xuebao*, **21**, 144 (2001); *Chem. Abstr.*, **134**, 285152 (2001).
383. G. Pecchi, P. Reyes, P. Sanhueza and J. Villasenor, *Chemosphere*, **43**, 141 (2001); *Chem. Abstr.*, **134**, 285139 (2001).

CHAPTER 15

Radiation chemistry of phenols

P. NETA

National Institute of Standards and Technology, Gaithersburg, Maryland 20899, USA
Fax: 301-975-3672;   e-mail: pedatsur.neta@nist.gov

and

S. STEENKEN

Max-Planck-Institut für Strahlenchemie, D-45413 Mülheim, Germany
Fax: 208-304-3951;   e-mail: steenken@mpi-muelheim.mpg.de

I. INTRODUCTION	1097
II. RADIATION CHEMISTRY OF PHENOLS IN AQUEOUS SOLUTIONS	1098
III. RADIATION CHEMISTRY OF PHENOLS IN ORGANIC SOLVENTS	1100
A. Halogenated Solvents	1100
B. Alkane Solvents	1102
IV. RADIATION CHEMISTRY OF NEAT PHENOLS (AND IN SOLID MATRICES)	1103
V. RADIATION CHEMISTRY OF PHENOLS IN THE GAS PHASE	1104
VI. REFERENCES	1105

I. INTRODUCTION

Ionizing radiation (γ- and X-rays, high energy electrons and other particles) is absorbed by molecules rather indiscriminately, so that most of the energy is absorbed by the solvent and not by the solutes that are present at low concentrations. Thus radiation chemistry involves in most cases the reactions of solvent radicals with the solutes. Deposition of ionizing radiation leads, as the name implies, to ionization of the solvent, i.e. formation of electrons and radical cations. These undergo subsequent processes to form a complex mixture of species. In many solvents, however, the primary events are followed by solvent-specific reactions, which result in the formation of one or two main radicals that can undergo simple reactions with the solute. Thus, despite the complexity of the early events, radiation chemistry may provide a means to study reactions of simple radicals or reduction

The Chemistry of Phenols. Edited by Z. Rappoport
© 2003 John Wiley & Sons, Ltd   ISBN: 0-471-49737-1

and oxidation reactions in relatively simple systems. The field has been summarized in several books and we recommend the excellent book by Spinks and Woods[1] as an introductory text.

The effect of ionizing radiation on phenols has been studied mainly in aqueous solutions under oxidizing conditions, where the phenols are reacted with hydroxyl radicals or with transient one-electron oxidants to yield, indirectly or directly, phenoxyl radicals. The reactions leading to formation of phenoxyl radicals, as well as the properties and reactions of phenoxyl radicals in aqueous solutions, are discussed in the chapter on transient phenoxyl radicals. In this chapter, other aspects of the radiation chemistry of phenols are summarized. These include studies with phenols in organic solvents and in the solid state, reactions leading to reduction of substituted phenols in various media and radiation treatment of phenols for detoxification purposes.

II. RADIATION CHEMISTRY OF PHENOLS IN AQUEOUS SOLUTIONS

Radiolysis of water produces hydrogen atoms, hydroxyl radicals and hydrated electrons, along with the molecular products hydrogen and hydrogen peroxide (equation 1).

$$H_2O \longrightarrow H^\bullet, {}^\bullet OH, e_{aq}^-, H^+, H_2, H_2O_2 \tag{1}$$

All phenols react very rapidly with $^\bullet OH$ radicals via addition and the adducts undergo water elimination to form phenoxyl radicals (as discussed in the chapter on transient phenoxyl radicals). All phenols also react very rapidly with H atoms via addition to the ring to form hydroxycyclohexadienyl radicals (equation 2).

$$C_6H_5OH + H^\bullet \longrightarrow {}^\bullet C_6H_6OH \tag{2}$$

The rate constants for such reactions are generally of the order of 10^9 M^{-1} s^{-1} [2]. The adducts of the simple phenols exhibit absorption maxima near 300 to 350 nm[3-8] and decay via second order reactions to form various isomeric dimers (equation 3).

$$2\ {}^\bullet C_6H_6OH \longrightarrow HOC_6H_6 \!-\! C_6H_6OH \longrightarrow HOC_6H_5C_6H_5OH \tag{3}$$

The initial dimers generally undergo aromatization by oxidation or water elimination to form substituted biphenyls.

Most phenols do not react rapidly with solvated electrons unless another substituent enhances such reaction by serving as the electron sink. The rate constant for reaction of phenol with e_{aq}^- is 2×10^7 M^{-1} s^{-1} [2] and the adduct undergoes very rapid protonation to yield the hydroxycyclohexadienyl radical[3], the same radical produced by addition of hydrogen atoms. At the other extreme, *ortho*-, *meta*- and *para*-nitrophenols react with e_{aq}^- with very high rate constants, *ca* 2×10^{10} M^{-1} s^{-1} [2] and cyanophenols react only slightly more slowly. These reactions produce the radical anions, which are similar to those derived from nitrobenzene and cyanobenzene (equation 4).

$$HOC_6H_4NO_2 + e_{aq}^- \longrightarrow HOC_6H_4NO_2^{\bullet -} \tag{4}$$

Another example that has been studied in detail is that of the halogenated phenols. Chloro-, bromo- and iodo-phenols, like their benzene analogues, react rapidly with solvated electrons to undergo dehalogenation and produce hydroxyphenyl radicals (equation 5).

$$XC_6H_4OH + e_{aq}^- \longrightarrow X^- + {}^\bullet C_6H_4OH \tag{5}$$

Hydroxyphenyl radicals are very different from phenoxyl radicals in that the electron is localized on the ring carbon where the halogen was located and the radical is a σ- rather than a π-radical. As a result, phenyl and hydroxyphenyl radicals are very reactive in hydrogen atom abstraction and addition reactions but not in electron transfer reactions. Hydrogen abstraction is favored in many cases because the aromatic C—H bond is stronger than most aliphatic C—H bonds and phenolic O—H bonds. This high reactivity along with the fact that phenyl radicals absorb only in the UV region made it more difficult to detect and characterize the hydroxyphenyl radicals by pulse radiolysis, as compared with phenoxyl radicals.

Early γ-radiolysis experiments with p-bromophenol have shown that reaction with solvated electrons yields Br^- ions quantitatively[9,10] and that the organic products include hydroxylated biphenyl and terphenyl[11–14]. From pulse radiolysis experiments[15] it was concluded that the hydroxyphenyl radical, formed by reaction of e_{aq}^- with p-bromophenol, adds rapidly ($k = 7 \times 10^7$ M^{-1} s^{-1}) to the parent compound (equation 6).

$$^{\bullet}C_6H_4OH + BrC_6H_4OH \longrightarrow HOC_6H_4C_6H_4(Br)(OH)^{\bullet} \qquad (6)$$

This adduct is the precursor of the polyphenyl products. The hydroxyphenyl radical also can abstract hydrogen atoms from alcohols (equation 7).

$$^{\bullet}C_6H_4OH + (CH_3)_2CHOH \longrightarrow C_6H_5OH + (CH_3)_2\overset{\bullet}{C}OH \qquad (7)$$

The rate constant for 2-propanol was determined[16] by competition kinetics based on the addition rate constant and found to be $k = 3 \times 10^7$ M^{-1} s^{-1}. The competition was determined by quantifying the yields of hydroxylated biphenyl vs the yield of phenol as a function of the relative concentrations of the p-bromophenol and the alcohol.

Reaction of e_{aq}^-, produced by pulse radiolysis, with bromophenols in alkaline solutions exhibited completely different pathways[17]. When the hydroxyl group of the hydroxyphenyl radical is dissociated, the negative charge is partly delocalized from O^- to the site of the radical on the aromatic ring and this site then undergoes very rapid protonation by water to form a phenoxyl radical (equation 8).

$$^{\bullet}C_6H_4OH \xrightarrow{-H^+} {}^{\bullet}C_6H_4O^- \xrightarrow{+H^+} C_6H_5O^{\bullet} \qquad (8)$$

This first order protonation was rapid with p-hydroxyphenyl ($k = 1.7 \times 10^5$ s^{-1}) and o-hydroxyphenyl ($k = 5 \times 10^4$ s^{-1}) radicals at pH 11.5 but was much slower for the m-hydroxyphenyl radical and was not observed in neutral solutions. As a result, the reductive radiation chemistry of bromophenols in neutral and acid solutions becomes the chemistry of phenoxyl radicals in alkaline solutions. Moreover, it was observed that the phenoxyl radical thus produced oxidizes another molecule of bromophenol to produce the bromophenoxyl radical (equation 9)[17].

$$C_6H_5O^{\bullet} + BrC_6H_4O^- \longrightarrow C_6H_5O^- + BrC_6H_4O^{\bullet} \qquad (9)$$

The reaction of e_{aq}^- with diiodotyrosine was found to lead to elimination of I^- as well as NH_3[18]. Iodide ions were formed at all pH values studied (pH 4 to pH 12) whereas NH_3 was produced only at pH $\leqslant 7$, i.e. when diiodotyrosine is in the protonated (NH_3^+) form.

Monofluoro aromatic compounds do not react rapidly with e_{aq}^- and do not undergo dehalogenation. Polyfluorinated derivatives, however, react very rapidly. Thus, pentafluorophenol was found to react with e_{aq}^- with a diffusion-controlled rate constant[19] and to undergo defluorination. The hydroxytetrafluorophenyl radical formed by this reaction undergoes rapid protonation at the ring carbon to yield the tetrafluorophenoxyl radical. Unlike the case of *p*-bromophenol, where such protonation occurs only at high pH, protonation of the perfluoro analogue takes place even in neutral solution because the pK_a of the phenolic OH group in pentafluorophenol is much lower than that in bromophenol. In alkaline solutions, the tetrafluorophenoxyl radical undergoes replacement of one fluoride with a hydroxide group ($k = 3 \times 10^4$ M^{-1} s^{-1}) to yield the trifluorobenzosemiquinone radical. Reaction of pentafluorophenol with hydrogen atoms also produces the tetrafluorophenoxyl radical, but the mechanism in this case was suggested to be different than that of the e_{aq}^- reaction; it involves hydrogen addition to the ring and HF elimination (H from the OH group and F from the *ortho* or *para* addition sites).

All the phenyl radicals, phenoxyl radicals and hydroxycyclohexadienyl radicals produced from phenols by various reactions react with each other and with other radicals to form, at least in part, new C—C or C—O bonds. As a result of these reactions, irradiation of phenols can lead to dimeric and polymeric products and irradiation of phenols in mixtures with other compounds can lead to crosslinking of the two materials. For example, irradiation of tyrosine or dopa with albumin in aqueous solutions leads to binding of these phenols to the protein[20]. Similarly, irradiation of tyrosine and its peptides[21,22] or mixtures of tyrosine and thymine[23] led to various dimerization products. The latter case was studied as a model for radiation-induced crosslinking between proteins and DNA.

Chlorinated phenols are common environmental pollutants, introduced as pesticides and herbicides. Studies have been carried out on the potential use of radiation to destroy these compounds as a means of environmental cleanup[8,24-32]. While these studies were concerned with mechanisms (and are discussed in the chapter on transient phenoxyl radicals), other studies involved large-scale irradiation to demonstrate the decomposition of phenol in polluted water[33,34]. Continuous irradiation led to conversion of phenol into various degradation products (formaldehyde, acetaldehyde, glyoxal, formic acid) and then to decomposition of these products. At high phenol concentrations, however, polymeric products were also formed.

III. RADIATION CHEMISTRY OF PHENOLS IN ORGANIC SOLVENTS

Radiolysis of organic solvents can lead to reducing and/or oxidizing radicals, as is the case with water. Water is inert to its radiolytic species and thus it is necessary to use additives to create purely reducing or oxidizing conditions. Many organic solvents, however, are reactive toward some of their radicals and thus lead to reducing or oxidizing conditions without added solutes. For example, radiolysis of alcohol solutions generally results in the reduction of added solutes while radiolysis of halogenated alkanes leads to oxidation of the solutes. Since phenols are difficult to reduce, their radiolysis in alcohol solutions is ineffective and has not been studied in detail. In contrast, their radiolysis in halogenated alkanes has been thoroughly examined and is known to lead to oxidation. These studies are summarized in the following section. In a subsequent section the radiolysis of phenols in alkane solutions will be discussed.

A. Halogenated Solvents

Radiolysis of CCl$_4$ solutions has been shown to lead to one-electron oxidation of many solutes. While the detailed mechanisms of the radiolysis of this solvent have been under study by several groups and some contradictory conclusions have been drawn, it is certain that many compounds are readily oxidized in this solvent. Oxidation may be effected by solvent or fragment cations, by chlorine atoms or chlorine complexes and, in the presence of oxygen, by chlorinated peroxyl radicals. In the case of phenols, it has been shown that 2,4,6-tri-*tert*-butylphenol is oxidized in irradiated CCl_4 solutions to form the phenoxyl radical with a radiolytic yield of 0.20 μmol J^{-1} [35]. In a later study, phenol and *p*-methoxyphenol have been shown to form the respective phenoxyl radicals when irradiated in CCl_4 solutions[36]. In deoxygenated solutions, the yield of phenoxyl radical increased with phenol concentration and reached a value of 0.35 μmol J^{-1} at 1 mol L^{-1} phenol. In oxygen-saturated solutions the maximum yield was higher by a factor of two and was more strongly dependent on concentration. Also, the phenoxyl radicals in oxygenated solutions were produced in two steps, a rapid step due to oxidation by solvent cations and Cl atoms, and a slower step due to oxidation by the $CCl_3O_2^•$ peroxyl radicals. The rate constant for oxidation of *p*-methoxyphenol by the $CCl_3O_2^•$ radical in this solvent was only $\leqslant 8 \times 10^6$ M^{-1} s^{-1}; the rate constant for phenol was 100 times lower[36]. It should be noted, however, that these reactions take place much more rapidly in aqueous solutions, as discussed in the chapter on transient phenoxyl radicals.

Similar radiolytic yields of phenoxyl radicals have been found in CH_2Cl_2 solutions[37]. The radiolysis of this solvent appeared to be simpler than that of CCl_4 and permitted determination of rate constants for reactions of Cl atoms. Both phenol and *p*-methoxyphenol react with diffusion-controlled rate constants (2.5×10^{10} and 5×10^9 M^{-1} s^{-1}). The slower oxidation steps were interpreted as reactions of two types of peroxyl radicals that can be formed in this solvent, i.e. $CHCl_2O_2^•$ and $CH_2ClO_2^•$. These radicals oxidize *p*-methoxyphenol with rate constants of 6×10^5 and 2×10^5 M^{-1} s^{-1}, respectively. Phenol was oxidized more slowly by these peroxyl radicals and its rate constants could not be measured under the experimental conditions used.

The very rapid oxidation of phenols by solvent radical cations can be expected to yield phenol radical cations as the first products. These species are short-lived, except in highly acidic solutions, and were not observed in the microsecond pulse radiolysis experiments described above. They were detected, however, in frozen matrices and with nanosecond pulse radiolysis[38-40]. Gamma irradiation of phenols in *n*-butyl chloride or in 1,1,2-trichloro-1,2,2-trifluoroethane (Freon 113) at 77 K produced phenol radical cations, which were detected by their optical absorption and ESR spectra[38]. Annealing to 133 K resulted in deprotonation of the radical cations to yield phenoxyl radicals. Pulse radiolysis of *p*-methoxyphenol and its 2,6-di-*tert*-butyl derivative in *n*-butyl chloride at room temperature produced both the phenol radical cations and the phenoxyl radicals. The phenol radical cations were formed very rapidly ($k = 1.5 \times 10^{10}$ M^{-1} s^{-1}) and decayed in a first-order process ($k = 2.2 \times 10^6$ s^{-1}) to yield the phenoxyl radicals. The phenoxyl radicals were partially formed in this slower process and partially in a fast process. The fast process of phenoxyl formation probably involves proton transfer to the solvent along with the electron transfer. When the *p*-methoxy group was replaced with alkyl or H, the stability of the phenol radical cation was lower and the species observed at short times were more predominantly phenoxyl radicals.

Similar results were obtained with naphthols and hydroxybiphenyls[39]. However, the extended π-system of these compounds, as compared with the simple phenols, led to a red shift of the absorption peaks of the radical cations (from 400–450 nm to 550–650 nm) and increased their lifetime (from 0.2–0.7 μs to 1.5–2.5 μs). The radical cations of naphthols and hydroxybiphenyl were found to oxidize triethylamine rapidly ($k = 4 \times 10^9$ to 1.2×10^{10} M^{-1} s^{-1}) and to transfer a proton to ethanol ($k = 3 \times 10^8$ to 6×10^8 M^{-1} s^{-1}).

Irradiation of bromoalkanes leads to formation of Br atoms, which can form complexes with the solvent, e.g. $Br^•CH_2Br_2$. This complex was found to oxidize *p*-methoxyphenol very rapidly to form the corresponding phenoxyl radical[41]. In a later study[42], rate constants were determined for the oxidation of a series of *p*-substituted phenols and found to vary from 5×10^8 to 6×10^9 M^{-1} s^{-1}. A Hammett correlation between log k and σ_p

gave a reasonably straight line with a slope of $\rho = -1.9$. Similar measurements were carried out in bromoethane and bromoform solutions; the Hammett plots gave a higher slope for $Br^{\bullet}C_2H_5Br$, $\rho = -3.1$, and a lower one for $Br^{\bullet}CHBr_3$, $\rho = -1.3$. The Br atom complex with benzene was produced by pulse radiolysis of benzene solutions containing CBr_4. The rate constants for oxidation of phenols by this complex were determined for a more extended series of p-substituted phenols, which included the less reactive acetyl- and cyanophenol. The values varied from 3×10^5 to 6×10^9 M^{-1} s^{-1} and the slope of the Hammett plot ($\rho = -4.2$) was larger than those of the aliphatic complexes. The Br complexes are clearly dipolar and bear partial negative charge at the Br atom. The extent of this negative charge was related to the variations in the ρ values.

In the same study, Br atom complexes with a series of substituted benzenes were prepared and the rate constants for their reactions with phenol were determined. The rate constants for $^{\bullet}BrC_6H_6$, $^{\bullet}BrC_6H_5F$, $^{\bullet}BrC_6H_5Br$, $^{\bullet}BrC_6H_5CF_3$ and $^{\bullet}BrC_6H_5CN$ increased gradually from 3.5×10^7 to 2×10^8 M^{-1} s^{-1}. In the same manner, the rate constants for oxidation of phenol by $^{\bullet}ClC_6H_6$, $^{\bullet}ClC_6H_5Cl$ and $^{\bullet}ClC_6H_5CCl_3$ were determined to be 1×10^9 M^{-1} s^{-1}, with variations of only 10%. The rate constant for oxidation of phenol by $^{\bullet}IC_6H_6$ was found to be only ca 10^5 M^{-1} s^{-1}. The extreme variations between the different halogens are of course due to the differences in electron affinity, and the reactivity is further modified by the electron-withdrawing effect of substituents on the benzene.

While the above studies concentrated on kinetics and mechanisms, other studies were aimed at measuring the yield of final products following γ-radiolysis of nitrophenols and other nitro compounds in CCl_4 solutions[43,44]. The gaseous products derived from the solvent included mainly HCl, $COCl_2$, $CHCl_3$, $Cl_2C=CCl_2$ (ca 0.01 μmol J^{-1}) and C_2Cl_6 (ca 0.05 μmol J^{-1}). The products derived from o-nitrophenol included mainly chloronitrophenol (by ring chlorination), dichloro- and trichlorophenols (by $ipso$ and other chlorination), and dichloroisocyanatobenzene (via attack of carbene on the nitro group).

B. Alkane Solvents

Pulse radiolysis of 2,6-di-*tert*-butyl-4-methylphenol (BHT) in n-heptadecane led to production of the phenoxyl radical[40,45]. The rate constant for the formation reaction was 8×10^8 M^{-1} s^{-1} and the process was ascribed to hydrogen abstraction from the phenol by alkyl radicals. The reaction of the phenol with alkylperoxyl radicals was too slow to be observed in this system. Another reaction observed in deoxygenated solutions was that of the phenol with hydrogen atoms, leading to formation of both the phenoxyl radical and the hydrogen adduct. This reaction was suppressed in the presence of O_2 because of the very fast reaction of hydrogen atoms with O_2. The same phenoxyl and hydrogen-adduct radicals were also observed in the pulse radiolysis of the same phenol in n-hexadecane[46] and the assignment of the optical spectra was further confirmed. When the phenol contained a carboxylic ester or a phenylthio group at the p-position, an additional reaction was observed in the pulse radiolysis and was ascribed to reaction of these phenols with solvated electrons to produce phenolate anions and hydrogen atoms (equation 10). The rapid decay of the anions was ascribed to charge neutralization with a solvent radical cation to produce phenoxyl radicals (equation 11).

$$R'C_6H_4OH + e_{solv}^- \longrightarrow R'C_6H_4O^- + H^{\bullet} \qquad (10)$$

$$R'C_6H_4O^- + RH^{\bullet +} \longrightarrow R'C_6H_4O^{\bullet} + RH \qquad (11)$$

The same results on the radiolytic oxidation of BHT were also obtained in cyclohexane solutions[47]. In this case, the rate constant for oxidation of this phenol by the cyclohexylperoxyl radical was estimated to be around 10^4 M^{-1} s^{-1}. Furthermore, the final radiolytic products were analyzed and found to include products of dimerization as well as a product formed by coupling of the phenol with the cyclohexyl radical. The yield of the latter product was considerable in the absence of O_2 but very low in the presence of O_2, clearly due to the fast reaction of the alkyl radical with oxygen.

The rate constants for hydrogen abstraction from BHT by alkyl radicals, as suggested in the above studies, are much higher than expected on the basis of related literature values. In fact, later studies by the same authors and by other authors demonstrated that the fast reactions discussed above are due to a different process and that the reactions of the alkyl radicals with BHT are quite slow ($k < 10^5$ M^{-1} s^{-1})[48]. The main reactions leading to phenoxyl radical formation are hydrogen abstraction by the hydrogen atoms and scavenging of electrons (equation 10) followed by reaction of the phenolate with a solvent radical cation. Scavenging of electrons by BHT was determined to have a rate constant of 3×10^8 M^{-1} s^{-1} and the reaction of the resulting phenolate anion with the solvent radical cation must be diffusion-controlled. Phenoxyl radicals may be produced also via direct electron transfer from the phenol to the solvent radical cation, followed by deprotonation of the phenol radical cation.

It should be pointed out in this context that deprotonation of phenols to phenolate anions upon irradiation was also detected in aqueous solutions[49]. Pulse radiolysis of N_2O-saturated neutral solutions containing α-naphthol produced a high initial concentration of OH^-, which reacted rapidly with the naphthol to form the anion (observed through increased UV absorption). The anion decayed back to the neutral naphthol by reacting with H^+ with a rate constant of 5.9×10^{10} M^{-1} s^{-1}.

IV. RADIATION CHEMISTRY OF NEAT PHENOLS (AND IN SOLID MATRICES)

Radiolysis of liquid cresols under vacuum was found[50] to produce H_2 as the main gaseous product; the radiolytic yield varied from 0.019 for m-cresol to 0.031 μmol J^{-1} for the o-cresol. Small amounts of CH_4 were also detected. Radiolysis of cyanophenols produced less H_2, only ca 0.003 μmol J^{-1}, various yields of CO and CO_2, mainly from the *ortho* isomer, and minute amounts of N_2. The difference in the yield of H_2 may be due to reaction of hydrogen atoms with the methyl group of the cresols to form H_2 as compared with addition to the CN group and to the ring, which do not produce H_2. No mechanistic details were derived from these studies.

In other studies, the phenols were irradiated in the solid state and the radicals were identified by ESR. Several aromatic compounds, including resorcinol, hydroquinone and hydroxybenzoic acids, were found to produce the hydrogen adducts upon irradiation[51]. Other phenols (amino, nitro, chloro) did not exhibit the expected ESR spectra upon irradiation. In a subsequent study, resorcinol was γ-irradiated at 77 K as a powder and as a single crystal and the ESR spectra were interpreted in terms of two types of radical pairs in which the m-hydroxyphenoxyl radical is the main component[52]. The mechanism of radical formation involves ionization of a resorcinol molecule and capture of the electron by another resorcinol molecule, followed by proton transfer from the cation to the anion to form phenoxyl radicals and H_2. The difference between the two pairs was suggested to be related to their position in the lattice relative to other molecules but could not be determined with certainty. Upon warming the solid to room temperature, the ESR spectra disappeared. However, irradiating the solid at room temperature was found to produce cyclohexadienyl-type radicals[53]. These were suggested to be formed not by addition of hydrogen atoms, since exposure of the crystal to external hydrogen atoms did not yield the same radical. Possibly, they were formed by protonation of electron adducts. In fact, addition of photochemically produced electrons to phenol and tyrosine in glassy NaOH or LiCl concentrated aqueous solutions produced radicals, which were identified as the hydrogen adducts by their ESR spectra[54]. γ-Irradiation of p-bromophenol in aqueous or methanolic glass at 77 K produced a radical, which exhibited a large hyperfine interaction with Br and was suggested to be the hydrogen adduct[55], although the exact structure remained in doubt.

Gamma irradiation of single crystals of 2-*tert*-butyl-4-methylphenol and 2,6-di-*tert*-butyl-4-methylphenol at room temperature produced the corresponding phenoxyl radicals, which were identified by their ESR spectra[56]. Similar irradiation of 2-amino-4-methylphenol did not give a resolved ESR spectrum, but after warming the crystal until it melted a resolved spectrum of the corresponding phenoxyl radical was observed. ESR spectra of phenoxyl radicals were observed also after X-ray irradiation of tyrosine and thyroxine and their iodo derivatives as compressed pellets at 100–300 K[57]. Gamma irradiation of nitrophenols at 77 K produced two types of radicals, the nitrophenoxyl radical and a nitroxide radical[58]. The mechanism of formation was suggested to involve initial ionization, electron and proton transfer from nitrophenol to an adjacent radical cation and finally rearrangement and recapture of the electron by the latter product to yield the nitroxide radical.

ESR spectra in frozen matrices have been used also to monitor the reactions of lipid-derived alkyl and alkylperoxyl radicals with antioxidants[59]. Gamma irradiation of the lipids at 100 K produces alkyl radicals and annealing to about 137 K permits migration of O_2 within the matrix and formation of peroxyl radicals. Further warming to 170 K permits reactions of these radicals with the phenols as well as self-reactions of the peroxyl radicals and warming to higher temperatures leads to decay of the phenoxyl radicals. The rates of formation of phenoxyl radicals in this system were found to decrease in the order BHT > *tert*-butylhydroquinone > α-tocopherol > propyl gallate > BHA; the extremes differ by a factor of 10. This order does not necessarily reflect the antioxidant activity, since these rates depend on the rate of migration of the molecule within the viscous lipid.

V. RADIATION CHEMISTRY OF PHENOLS IN THE GAS PHASE

Since phenols are solid under ambient conditions, few studies were concerned with the radiation chemistry of phenols in the gas phase. An early study demonstrated the acetylation of phenols when irradiated in a specific gaseous mixture. Gas-phase γ-irradiation of a mixture of CH_3F and CO was found to form the acetyl cation, CH_3CO^+, and to lead to acetylation of substrates[60]. Gaseous phenol, cresols and xylenols present in such a mixture were acetylated mainly at the OH group to form 80–97% aryl acetate. The remaining products, hydroxyacetophenones, were mainly the *ortho* and *para* derivatives.

In a more recent study[61] pulse radiolysis was utilized to produce the phenoxyl radical in the gas phase and to measure some reaction kinetics. The irradiated gas mixture contained mainly SF_6 (at 980 to 1000 mbar), which served as a source of F atoms. Phenol was present at 0.1 mbar. The rate constant for reaction of phenol with F atoms was determined to be 1.9×10^{11} M^{-1} s^{-1}. This reaction led to formation of the phenoxyl radical (45%) and other products, probably fluorine-adducts to the ring. When HCl (20 mbar) was added to the mixture, most fluorine atoms reacted with HCl to produce chlorine atoms and these reacted with phenol to produce the phenoxyl radical as the predominant product. The rate constant for reaction of chlorine atoms with phenol, derived from several competition kinetic experiments, was 1.2×10^{11} M^{-1} s^{-1}, slightly lower than the value for fluorine atoms. The spectrum of the phenoxyl radical in the gas phase was very similar to that recorded in aqueous solutions. It exhibits several peaks between 350 nm and 400 nm and much more intense absorptions in the UV, the main peak being at 235 nm (molar absorption coefficient 2.3×10^4 M^{-1} cm^{-1}). By following the decay of the phenoxyl radical absorption, rate constants were determined for the reaction of phenoxyl with NO (1.1×10^9 M^{-1} s^{-1}) and NO$_2$ (1.3×10^9 M^{-1} s^{-1}). No reaction was detected between phenoxyl and O$_2$. This is similar to the reactivity of phenoxyl radicals in aqueous solutions.

Rate constants for reactions of phenols with several other radicals in the gas phase were determined by techniques not involving ionizing radiation and the results are relevant for understanding the behavior of phenols in irradiated gaseous systems. For example, the rate constants for reactions of H• atoms and •OH radicals with phenol at high temperatures were determined by single-pulse shock tube experiments[62]. The rate constant for •OH radicals was found to be 6×10^9 M^{-1} s^{-1}, close to that determined in aqueous solutions, and independent of temperature. The reaction produces phenoxyl radical. On the other hand, the reaction of H• atoms with phenol was suggested to take place via two paths, each one with a significant activation energy: (a) hydrogen abstraction from the phenolic group to form phenoxyl, with $k = 1.15 \times 10^{11}\ e^{-51.9/RT}$ M^{-1} s^{-1}, and (b) displacement of the OH group to form benzene, with $k = 2.21 \times 10^{10}\ e^{-33.1/RT}$ M^{-1} s^{-1}. The rate constant for reaction of NO$_3$• radicals, formed by thermal decomposition of N$_2$O$_5$, with phenol in the gas phase at 298 K was determined by competition kinetics and found to be 2.3×10^9 M^{-1} s^{-1} [63], several orders of magnitude higher than the values for p-methoxyphenol, benzaldehyde and toluene. The reaction was suggested to involve mainly the functional group or side chain, not addition to the aromatic ring. These and other rate constants for reactions of phenols with radicals (including phenyl and methyl radicals) in the gas phase are summarized in recent compilations[64,65] and in the NIST kinetics database (http://kinetics.nist.gov/index.php).

VI. REFERENCES

1. J. W. T. Spinks and R. J. Woods, *Introduction to Radiation Chemistry*, third edn., Wiley, New York, 1990.
2. G. V. Buxton, C. L. Greenstock, W. P. Helman and A. B. Ross, *J. Phys. Chem. Ref. Data*, **17**, 513 (1988).
3. E. J. Land and M. Ebert, *Trans. Faraday Soc.*, **63**, 1181 (1967).
4. P. Neta and L. M. Dorfman, *J. Phys. Chem.*, **73**, 413 (1969).
5. J. Chrysochoos, *J. Phys. Chem.*, **73**, 4188 (1969).
6. J. Feitelson and E. Hayon, *J. Phys. Chem.*, **77**, 10 (1973).
7. M. Ye, K. P. Madden, R. W. Fessenden and R. H. Schuler, *J. Phys. Chem.*, **90**, 5397 (1986).
8. N. Getoff and S. Solar, *Radiat. Phys. Chem.*, **28**, 443 (1986).
9. M. Anbar, Z. B. Alfassi and H. Bregman-Reisler, *J. Am. Chem. Soc.*, **89**, 1263 (1967).
10. F. A. Peter and P. Neta, *J. Phys. Chem.*, **76**, 630 (1972).
11. M. Namiki, T. Komiya, S. Kawakishi and H. Aoki, *Chem. Commun.*, 311 (1970).
12. T. Komiya, S. Kawakishi, H. Aoki and M. Namiki, *Agric. Biol. Chem.*, **34**, 349 (1970).
13. T. Komiya, S. Kawakishi, H. Aoki and M. Namiki, *Agric. Biol. Chem.*, **35**, 1558 (1971).
14. T. Komiya, S. Kawakishi and M. Namiki, *Agric. Biol. Chem.*, **38**, 1589 (1974).
15. B. Cercek and M. Kongshaug, *J. Phys. Chem.*, **74**, 4319 (1970).
16. K. Bhatia and R. H. Schuler, *J. Phys. Chem.*, **77**, 1356 (1973).
17. R. H. Schuler, P. Neta, H. Zemel and R. W. Fessenden, *J. Am. Chem. Soc.*, **98**, 3825 (1976).
18. T. N. Das, *J. Phys. Chem.*, **98**, 11109 (1994).
19. L. C. T. Shoute, J. P. Mittal and P. Neta, *J. Phys. Chem.*, **100**, 3016 (1996).
20. O. Yamamoto and A. Okuda, *Radiat. Res.*, **61**, 251 (1975).
21. W. A. Pruetz, J. Butler and E. J. Land, *Int. J. Radiat. Biol.*, **44**, 183 (1983).
22. L. R. Karam, M. Dizdaroglu and M. G. Simic, *Int. J. Radiat. Biol.*, **46**, 715 (1984).
23. M. G. Simic and M. Dizdaroglu, *Biochemistry*, **24**, 233 (1985).
24. N. Getoff and S. Solar, *Radiat. Phys. Chem.*, **31**, 121 (1988).
25. R. B. Draper, M. A. Fox, E. Pelizzetti and N. Serpone, *J. Phys. Chem.*, **93**, 1938 (1989).
26. R. Terzian, N. Serpone, R. B. Draper, M. A. Fox and E. Pelizzetti, *Langmuir*, **7**, 3081 (1991).

27. X. X. Huang, H. Arai, S. Matsuhasi and T. Miyata, *Chem. Lett.*, 273 (1996).
28. S. Schmid, P. Krajnik, R. M. Quint and S. Solar, *Radiat. Phys. Chem.*, **50**, 493 (1997).
29. M. Trojanowicz, A. Chudziak and T. Bryl-Sandelewska, *J. Radioanal. Nucl. Chem.*, **224**, 131 (1997).
30. A. Kovacs, K. Gonter, G. Foldiak, I. Gyorgy and L. Wojnarovits, *ACH-Models Chem.*, **134**, 453 (1997).
31. X. W. Fang, Y. K. He, J. Liu and J. Wu, *Radiat. Phys. Chem.*, **53**, 411 (1998).
32. X. Fang, H.-P. Schuchmann and C. von Sonntag, *J. Chem. Soc., Perkin Trans. 2*, 1391 (2000).
33. K. J. Lin, W. J. Cooper, M. G. Nickelsen, C. N. Kurucz and T. D. Waite, *Appl. Radiat. Isotopes*, **46**, 1307 (1995).
34. M. H. O. Sampa, C. L. Duarte, P. R. Rela, E. S. R. Somassari, C. G. Silveira and A. L. Azevedo, *Radiat. Phys. Chem.*, **52**, 365 (1998).
35. H. D. Burrows, D. Greatorex and T. J. Kemp, *J. Phys. Chem.*, **76**, 20 (1972).
36. J. Grodkowski and P. Neta, *J. Phys. Chem.*, **88**, 1205 (1984).
37. Z. B. Alfassi, S. Mosseri and P. Neta, *J. Phys. Chem.*, **93**, 1380 (1989).
38. O. Brede, H. Orthner, V. Zubarev and R. Hermann, *J. Phys. Chem.*, **100**, 7097 (1996).
39. H. Mohan, R. Hermann, S. Naumov, J. P. Mittal and O. Brede, *J. Phys. Chem. A*, **102**, 5754 (1998).
40. O. Brede, M. R. Ganapathi, S. Naumov, W. Naumann and R. Hermann, *J. Phys. Chem. A*, **105**, 3757 (2001).
41. L. C. T. Shoute and P. Neta, *J. Phys. Chem.*, **94**, 2447 (1990).
42. L. C. T. Shoute and P. Neta, *J. Phys. Chem.*, **94**, 7181 (1990).
43. J. Kuruc, M. K. Sahoo and P. Kuran, *J. Radioanal. Nucl. Chem.*, **174**, 103 (1993).
44. J. Kuruc, M. K. Sahoo, P. Kuran and R. Kubinec, *J. Radioanal. Nucl. Chem.*, **175**, 359 (1993).
45. O. Brede, R. Hermann and R. Mehnert, *J. Chem. Soc., Faraday Trans. 1*, **83**, 2365 (1987).
46. O. Brede, *ZfI-Mitt.*, **132**, 197 (1987).
47. O. Brede and L. Wojnarovits, *Radiat. Phys. Chem.*, **37**, 537 (1991).
48. I. A. Shkrob and A. D. Trifunac, *Radiat. Phys. Chem.*, **46**, 97 (1995).
49. R. H. Schuler, *J. Phys. Chem.*, **96**, 7169 (1992).
50. J. Weiss, *Radiat. Res.*, **45**, 252 (1971).
51. D. Campbell and D. T. Turner, *Can. J. Chem.*, **45**, 881 (1967).
52. D. Campbell and M. C. R. Symons, *J. Chem. Soc., A*, 1494 (1969).
53. D. Campbell and M. C. R. Symons, *J. Chem. Soc., A*, 2977 (1969).
54. C. V. Paemel, H. Frumin, V. L. Brooks, R. Failor and M. D. Sevilla, *J. Phys. Chem.*, **79**, 839 (1975).
55. J. Edwards, D. J. Hills, S. P. Mishra and M. C. R. Symons, *J. Chem. Soc., Chem. Commun.*, 556 (1974).
56. F. Koksal, S. Osmanoglu and R. Tapramaz, *Z. Naturforsch., A, Phys. Sci.*, **48**, 560 (1993).
57. P. R. Crippa, E. Loh and S. Isotani, *Radiat. Eff.*, **25**, 13 (1975).
58. O. E. Yakimchenko, N. I. Boguslavskaya, F. I. Dubovitskii and Y. S. Lebedev, *High Energy Chem. (Engl. Transl.)*, **10**, 332 (1976).
59. J. Zhu, W. J. Johnson, C. L. Sevilla, J. W. Herrington and M. D. Sevilla, *J. Phys. Chem.*, **94**, 7185 (1990).
60. M. Speranza and C. Sparapani, *Radiochim. Acta*, **28**, 87 (1981).
61. J. Platz, O. J. Nielsen, T. J. Wallington, J. C. Ball, M. D. Hurley, A. M. Straccia, W. F. Schneider and J. Sehested, *J. Phys. Chem. A*, **102**, 7964 (1998).
62. Y. Z. He, W. G. Mallard and W. Tsang, *J. Phys. Chem.*, **92**, 2196 (1988).
63. R. Atkinson, S. M. Aschmann and A. M. Winer, *Environ. Sci. Technol.*, **21**, 1123 (1987).
64. D. L. Baulch, C. J. Cobos, R. A. Cox, C. Esser, P. Frank, T. Just, J. A. Kerr, M. J. Pilling, J. Troe, R. W. Walker and J. Warnatz, *J. Phys. Chem. Ref. Data*, **21**, 411 (1992).
65. D. L. Baulch, C. J. Cobos, R. A. Cox, P. Frank, G. Hayman, T. Just, J. A. Kerr, T. Murrells, M. J. Pilling, J. Troe, R. W. Walker and J. Warnatz, *J. Phys. Chem. Ref. Data*, **23**, 847 (1994).

CHAPTER 16

Transient phenoxyl radicals: Formation and properties in aqueous solutions

S. STEENKEN

Max-Planck-Institut für Strahlenchemie, D-45413 Mülheim, Germany
Fax: 208-304-3951; e-mail: steenken@mpi-muelheim.mpg.de

and

P. NETA

National Institute of Standards and Technology, Gaithersburg, Maryland 20899, USA
Fax: 301-975-3672; e-mail: pedatsur.neta@nist.gov

I. INTRODUCTION	1108
II. FORMATION OF PHENOXYL RADICALS	1108
A. Oxidation of Phenols by Metal Ions	1109
B. Oxidation of Phenols by Free Radicals	1110
C. Oxidation of Phenols by Radical Cations	1113
D. Reaction of Phenols with Hydroxyl Radicals. The Addition/Elimination Mechanism	1113
E. Formation of Phenoxyl Radicals by Oxidative Replacement of Substituents	1116
F. Formation of Phenoxyl Radicals by Intramolecular Electron Transfer	1117
G. Formation of Phenoxyl Radicals from Phenols and Hydroxyphenols by Reaction with O_2 and/or $O_2^{\bullet-}$ (Autoxidation)	1118
III. PROPERTIES OF PHENOXYL RADICALS	1120
A. Electron Spin Resonance Spectra of Phenoxyl Radicals	1120
1. Phenoxyl and monosubstituted phenoxyl radicals	1120
2. Phenoxyl radicals with extended π-systems	1124
3. Kinetic ESR measurements	1125
4. Comparison with isoelectronic radicals	1127

The Chemistry of Phenols. Edited by Z. Rappoport
© 2003 John Wiley & Sons, Ltd   ISBN: 0-471-49737-1

B. Optical Spectra of Phenoxyl Radicals 1127
C. Acid–Base Equilibria of Phenoxyl Radicals 1132
D. Reactions of Phenoxyl Radicals 1135
E. Reduction Potentials of Phenoxyl Radicals 1140
IV. REFERENCES 1143

I. INTRODUCTION

Phenoxyl radicals are the intermediate products from a large variety of thermochemical, photochemical, radiation chemical and biochemical processes which involve the oxidation of phenols or the reduction of quinones. Phenolic and quinonoidic compounds are found among the groups of hormones, vitamins, antibiotics and antioxidants (including natural and synthetic food antioxidants). The mechanism by which phenols function as antioxidants has to do with their ability to scavenge reactive radicals by the transfer of an electron or a hydrogen atom. By such processes the phenol is converted into a phenoxyl radical[1–10]. The phenol may be regenerated by reaction of the phenoxyl radical via hydrogen or electron transfer (followed by proton transfer) from another molecule, which again may be a phenol. In this case one is dealing with the phenomenon of synergism. An additional example is the involvement of phenols and phenol-like substances, such as flavins, in biological redox processes. Furthermore, phenoxyl radicals are involved in the biosynthesis of numerous natural products (predominantly by oxidative coupling)[11–13], including many alkaloids. Oxidative coupling of phenols has been explored with respect to its general preparative value[7]. The phenoxyl radical from tyrosine serves a very important (catalytic) function in the enzyme ribonucleotide reductase (RNR)[14] and also in photosystem II[14–17].

For these reasons, there has been a large and still growing interest in the chemistry of phenoxyl radicals and the earlier results have been summarized in a series of excellent reviews[1–6,8,18–20]. In this review the emphasis will be placed on results obtained by direct and fast detection techniques for phenoxyl radicals, mainly in aqueous solutions. Results for other solvents, however, will be included if they appear relevant to the aqueous phase chemistry of phenoxyl or if they are of general interest.

II. FORMATION OF PHENOXYL RADICALS

Oxidation of phenols may proceed by hydrogen atom abstraction from the phenolic OH group (equation 1),

$$A + ArOH \underset{k_{-1}}{\overset{k_1}{\rightleftharpoons}} AH^\bullet + ArO^\bullet \qquad (1)$$

or by electron transfer to an acceptor with a sufficiently high electron affinity (equation 2),

$$A + ArOH \underset{k_{-2}}{\overset{k_2}{\rightleftharpoons}} A^{\bullet -} + ArOH^{\bullet +} \qquad (2)$$

where A is any atom or molecule able to accept an electron or a hydrogen atom. The former process is thermodynamically feasible when the bond dissociation energy for ArO-H (2.6–3.7 eV)[21–25] (see also Section III.E) is lower than that for A-H. The electron transfer reaction is possible with many electron acceptors due to the low gas-phase ionization

16. Transient phenoxyl radicals: Formation and properties in aqueous solutions

potentials of phenols, which are normally ≤ 9 eV. The ease of oxidation of phenols is considerably increased upon deprotonation to give the phenoxide anion (equation 3).

$$A + ArO^- \underset{k_{-3}}{\overset{k_3}{\rightleftharpoons}} A^{\bullet -} + ArO^{\bullet} \quad (3)$$

From a product point of view, hydrogen and electron transfer are not always easy to distinguish due to the acid–base equilibria of equations 4 and 5.

$$ArOH^{\bullet +} \rightleftharpoons ArO^{\bullet} + H^+ \quad (4)$$

$$AH \rightleftharpoons A^- + H^+ \quad (5)$$

However, with respect to oxidation *mechanism* a distinction can be made between hydrogen and electron transfer on the basis of the kinetic isotope effect for the rate of oxidation[26], which is expected to be large ($k_H/k_D \geq 1.5$) for k_1 and small ($k_H/k_D \leq 1.5$) for k_2.

The redox reactions 1–3 may be reversible or proceed predominantly in one direction. Equilibrium reactions will be discussed in Section III.E. In what follows, reactions will be discussed that proceed essentially to completion.

A. Oxidation of Phenols by Metal Ions

PbO_2 has been used as an oxidant in some of the earliest ESR studies on sterically hindered phenoxyl radicals[27–29]. Later, the method found wider application to include anilinyl[30,31] and semiquinone radicals[32]. PbO_2 can be replaced by Ag_2O[33,34] or MnO_2[35]. The unsubstituted phenoxyl radical was first produced by oxidation with Ce(IV) in aqueous solution at low pH[36], and this oxidant was subsequently used to produce substituted phenoxyl radicals including semiquinones[37–39], phenol radical cations[40–44] and oxypyrones[45]. In basic solution also $[Fe^{III}(CN)_6]^{3-}$ can be used to oxidize phenols[29,37]. The reaction probably proceeds from the phenolates via equation 3. Various other oxidants have been used in studies related to oxidative coupling of phenols. This area has been reviewed by Musso[18].

The bimolecular rate constants for oxidation of phenols by metal ions in high oxidation states can readily be determined, since the production of the oxidized metal ions can be carried out in ≤ 1 µs using the pulse radiolysis method and the rate of formation of the phenoxyl radical can then be measured as a function of phenol concentration. Metal ions in unusual (and unstable) oxidation states can be produced by reaction with OH in aqueous solution[46] and can react with phenols[47].

Tl^{2+} and Ag^{2+} were found to react with 4-methoxyphenol and 3,5-dimethoxyphenol by 100% electron transfer (equation 2), whereas with $TlOH^+$ the efficiency of electron transfer is only ca 75%. The ease of oxidation increases considerably in going from the neutral phenols to the phenolates: even the weak oxidants[46] $Tl(OH)_2$ and $Ag(OH)_2$ are able to oxidize the phenolates with 100% yield to give the corresponding phenoxyl radicals[48]. In going from phenol to the dihydroxybenzenes the oxidizability increases: hydroquinone and resorcinol are oxidized with 100% yield not only by Tl^{2+} but also by the weaker oxidant $TlOH^{+}$[46]. Catechol forms a complex with Tl(II), which has the same structure as that[46] produced by reaction of *ortho*-semiquinone radical with Tl^+ or by reaction of *ortho*-benzoquinone with Tl°[49,50]. The rate constants for reaction of the Tl(II) and Ag(II) species are between $10^8–10^9$ M^{-1} s^{-1} (see Table 1).

Phenol is also oxidized by ferrate(V) ions and ferrate(VI) ions[51]. It has been suggested that ferrate(VI) ions oxidize phenol by a one-electron transfer mechanism

($k = 10^7$ M^{-1} s^{-1}) whereas ferrate(V) ions oxidize it by a two-electron transfer mechanism ($k = 3.8 \times 10^5$ M^{-1} s^{-1}).

B. Oxidation of Phenols by Free Radicals

The halogen and pseudohalogen dimer radical anions, $X_2^{\bullet-}$ (X = Cl, Br, I, SCN), react efficiently with phenols and phenolate ions to give the corresponding phenoxyl radicals (equation 6).

$$X_2^{\bullet-} + \text{ArOH} \longrightarrow \text{ArO}^{\bullet} + \text{H}^+ + 2\text{X}^- \tag{6}$$

For the case of $Cl_2^{\bullet-}$, the rate constants have been measured at pH 1 for a series of *p*-substituted phenols, the value for phenol being 2.5×10^8 M^{-1} s^{-1}[52]. The rate constants increase with increasing electron-donating power of the substituent. A plot of the rate constants *vs* the Hammett σ values yields $\rho = -1.5$, indicating an electron transfer mechanism for the formation of the phenoxyl radicals[52]. The weaker oxidant $Br_2^{\bullet-}$ reacts with phenol more slowly, $k = 6 \times 10^6$ M^{-1} s^{-1}[53]. However, upon increasing the reducing power by going from phenol to phenolate, the rate constant increases to *ca* 4×10^8 M^{-1} s^{-1}[53]. $(SCN)_2^{\bullet-}$ and $I_2^{\bullet-}$ are even weaker oxidants than $Br_2^{\bullet-}$ and thus oxidation of phenol was not observed ($k < 10^7$ M^{-1} s^{-1})[53,54]. However, phen*olate* reacts with $I_2^{\bullet-}$ with $k = 5.7 \times 10^7$ M^{-1} s^{-1}[53] and with $(SCN)_2^{\bullet-}$ the rate constant is *ca* 3×10^8 M^{-1} s^{-1}[53,54]. The rate constants for the reactions of $Br_2^{\bullet-}$ and $(SCN)_2^{\bullet-}$ with *p*-substituted phenolates follow a Hammett relationship with $\rho = -1.1$ for $Br_2^{\bullet-}$ and -1.2 for $(SCN)_2^{\bullet-}$[53], demonstrating the electrophilic nature of these radicals.

The aminyl radical $^{\bullet}NH_2$ is also able to produce phenoxyl radicals from substituted phenolates[55]. The rate constants for this reaction ($k = 3.0 \times 10^6$ M^{-1} s^{-1} for phenolate) increase strongly with increasing electron-donating power of the substituent to give a Hammett $\rho = -3.3$, from which it was concluded that the reaction proceeds by electron transfer[55]. The value $\rho = -3.3$ is more than twice that determined for the oxidation of substituted phenols by $Cl_2^{\bullet-}$[52] or of phenolates by $Br_2^{\bullet-}$ or $(SCN)_2^{\bullet-}$[53]. This increased selectivity of the $^{\bullet}NH_2$ radical is in line with its lower reactivity.

The radical $N_3^{\bullet}$, produced in the reaction of N_3^- with $^{\bullet}OH$, gives the phenoxyl-type radical on reaction with phenols and phenolate ions[56,57], whereby $N_3^{\bullet}$ shows very little tendency to perform hydrogen-abstraction reactions from C—H bonds additionally present.

A large number of oxygen-centered radicals react with phenols to yield phenoxyl radicals (Table 1). Whereas $^{\bullet}OH$, the simplest of the oxygen-centered radicals, reacts mainly by an addition/elimination mechanism (see Section II.D), most of the other radicals seem to produce phenoxyl via hydrogen or electron transfer. An example for this is $SO_4^{\bullet-}$. In its reaction with tyrosine, a delayed formation of tyrosinoxyl radical was not found and it was concluded that the reaction proceeds by electron transfer[58]. The oxide radical, $O^{\bullet-}$, the conjugate base of the $^{\bullet}OH$ radical, also has been proposed to react with phenolate by electron transfer[59].

In the reaction of $(CH_3)_3CO^{\bullet}$ with *p*-substituted phenols to yield phenoxyl radicals[26] (for rate constants see Table 1) a large kinetic isotope effect was observed, i.e. $k_H/k_D = 3-5$, which means that in the transition state an O—H bond is broken as in a hydrogen-abstraction reaction. However, there is also an increase in the rate constant with increasing electron-donating power of the substituent[26] (Hammett $\rho = -0.9$), which indicates that there is charge separation in the transition state with partial positive and negative charge on the aromatic ring and on *t*-butoxyl, respectively.

As compared to *tert*-butoxyl radical, peroxyl radicals $RO_2^{\bullet}$ react more slowly with phenols (equation 7). These reactivity differences can be related to differences in the

16. Transient phenoxyl radicals: Formation and properties in aqueous solutions

O—H bond dissociation energies which are *ca* 440 kJ mol^{-1} for RO—H but only *ca* 360 kJ mol^{-1} for ROO—H[60]. The rate constants for reaction 7,

$$ROO^{\bullet} + ArOH \longrightarrow ROOH + ArO^{\bullet} \tag{7}$$

which describes the antioxidant action of phenols, do not depend on the nature of ROO$^{\bullet}$ (for a particular phenol)[3,61–63] but are quite sensitive to the nature of the phenol[7]. For example, for reaction of a peroxyl radical from styrene with phenol $k_7 = 5 \times 10^3$ M^{-1} s^{-1}[64], for reaction with 2,5-di-*tert*-butyl-4-methylphenol $k_7 = 1.2 \times 10^4$ M^{-1} s^{-1}, for 2,3,5,6-tetramethyl-4-methoxyphenol $k_7 = 2.1 \times 10^5$ M^{-1} s^{-1} and for α-tocopherol (Vitamin E) $k_7 = 2.35 \times 10^6$ M^{-1} s^{-1}[7]. An explanation for this trend has been given in terms of ArO—H bond strengths[65] and of stereoelectronic factors that determine the stabilization of the phenoxyl radical[7]. As expected for a hydrogen-abstraction mechanism, reaction 7 exhibits a large kinetic isotope effect ($k_H/k_D = 4-11$)[7].

Alkylperoxyl radicals substituted at the α-position by halogens show a higher reactivity with respect to oxidation of phenolates (equation 8).

$$CH_{3-n}Cl_nO_2^{\bullet} + C_6H_5O^- + H_2O \longrightarrow CH_{3-n}Cl_nO_2H + C_6H_5O^{\bullet} + OH^- \tag{8}$$

The rate constant increases from 1.1×10^7 M^{-1} s^{-1} for $n = 1$ to 2.3×10^8 M^{-1} s^{-1} for $n = 3$, due to the withdrawal of electron density from the reaction center by the electronegative halogens[66]. Halogenated peroxyl radicals have been suggested[67,68] as intermediates involved in the toxic effects of CCl$_4$ on the liver. In general, rate constants for oxidation by methylperoxyl radicals substituted at the α-position with various groups are greatly dependent on the electron-withdrawing power of these groups[69]. The rate constants for halogenated peroxyl radicals are highly influenced also by the solvent[70,71]; variations of nearly two orders of magnitude have been observed for the reaction with Trolox C (6-hydroxy-2,5,7,8-tetramethylchroman-2-carboxylic acid, a Vitamin E analogue), the rate constant generally increasing with solvent polarity. Rate constants for the reactions of chlorinated methylperoxyl radicals with Trolox C in aqueous solutions have been measured as a function of temperature and the activation energies were found to be 6.4 kJ mol^{-1} for CH$_2$ClO$_2^{\bullet}$ and 17 kJ mol^{-1} for the dichloro- and trichloromethylperoxyl radicals[72].

From the reactivity point of view, 2-alkanonyl radicals may be considered as oxygen-centered (vinoxyl) radicals (cf hybrid **b**)[73].

(a)   (b)

For the case of the 2-cyclohexanonyl radical the contribution of the mesomeric structure **b** has been estimated[74] to be *ca* 15%. 2-Alkanonyl radicals react with substituted phenolates to yield the corresponding phenoxyl radicals[73,75,76] (equation 9).

$$^{\bullet}CH_2CHO + ArO^- + H_2O \longrightarrow CH_3CHO + ArO^{\bullet} + OH^- \tag{9}$$

The rate constant of this reaction ($k = 4.3 \times 10^6$ M^{-1} s^{-1} for $^{\bullet}CH_2CHO +$ phenolate) increases strongly with increasing electron-donating power of the substituent to give Hammett $\rho = -7.9$[73]. This value indicates that the reaction proceeds by electron transfer.

The value is more than twice that $(-3.3)^{55}$ for oxidation of phenolates by $^{\bullet}NH_2$. The rate constants for oxidation of (unsubstituted) phenolate by $^{\bullet}NH_2$ and $^{\bullet}CH_2CHO$ are, however, approximately equal (Table 1). This shows that conclusions relating to mechanism cannot be based solely on reaction rate constants.

The rate constants for oxidation of the hydroquinone anion by 2-alkanonyl radicals ($R^1C^{\bullet}HCOR^2$) decrease from 2.2×10^9 to 5.6×10^8 M^{-1} s^{-1} on going from $R^1 = R^2 = $ H to $R^1 = R^2 = CH_3{}^{73}$. This effect is a result of the decrease of the electron deficiency in the 2-alkanonyl radical, as expected for an electron transfer mechanism. Steric hindrance may reduce the rate constant but its effect is not decisive. This is shown by the fact that the rate constant is 1.2×10^9 M^{-1} s^{-1} for $R^1 = R^2 = CH_2OH$, which is even more bulky than CH_3, but which, in contrast, is electron-withdrawing ($-I$ effect).

TABLE 1. Rate constants and ρ values (in parentheses) for reactions of radicals with phenols and phenolate ions[a]

Radical	C_6H_5OH	$C_6H_5O^-$	4-$CH_3OC_6H_4OH$	4-$CH_3OC_6H_4O^-$
e_{aq}^-	2×10^7	4×10^6		
H$^{\bullet}$	2×10^9			
$^{\bullet}$OH	1×10^{10}	1×10^{10}	3×10^{10}	
O$^{\bullet -}$		7×10^8		
$O_2^{\bullet -}$	6×10^2			2×10^4
$^1O_2^*$	2×10^6	2×10^8	1×10^7	7×10^8
O_3	1×10^3	1×10^9		
$^{\bullet}NH_2$		3×10^6 $(-3.3)^{55}$		9×10^6
$Cl_2^{\bullet -}$	3×10^8 $(-1.5)^{52}$		1×10^9	
$Br_2^{\bullet -}$	6×10^6	5×10^8 $(-1.1)^{53}$	8×10^7	1×10^9
$I_2^{\bullet -}$		3×10^7		1×10^8
$(SCN)_2^{\bullet -}$	1×10^6	3×10^8 $(-1.2)^{53}$	5×10^7	
$N_3^{\bullet}$	4×10^7	4×10^9	4×10^9	4×10^9
Cl$^{\bullet}$	2×10^{10}			
I$^{\bullet}$	2×10^7			
$ClO_2^{\bullet}$	0.2	4×10^7	3×10^4	1×10^9
$BrO_2^{\bullet}$	3×10^5	3×10^9		
$NO_2^{\bullet}$		2×10^7		2×10^8
$CO_3^{\bullet -}$	1×10^7	3×10^8 $(-1.0)^{77}$		1×10^9
$PO_4^{\bullet 2-}$		6×10^8 $(-0.7)^{78}$		8×10^8
$SO_4^{\bullet -}$	3.0×10^9			
$SO_3^{\bullet -}$		6×10^5		4×10^7
Tl^{2+}	10^9	10^9		
$Tl(OH)^+$	10^9	10^9		
Ag^{2+}	10^8	10^8		
$^{\bullet}CH_2CHO$		4×10^6 $(-7.9)^{73}$		1×10^9
$CH_3OO^{\bullet}$		$<1 \times 10^6$		9×10^5
$CF_3OO^{\bullet}$	2×10^6		5×10^7	
$CCl_3OO^{\bullet}$	$<1 \times 10^5$	2×10^8	3×10^6	8×10^8
$C_6H_5OO^{\bullet}$				2×10^8
$(CH_3)_3CO^{\bullet}$	2.2×10^7 $(-0.9, -1.2)^b$		$1.1 \times 10^{8\,b}$	

[a]The rate constants, in M^{-1} s^{-1}, are taken from References 79 and 80 and the NDRL-NIST Solution Kinetics Database. The values in parentheses are the Hammett ρ values derived from substituent effects, given with their respective references.
[b]The rate constants for this radical were measured in methanol and the ρ values were measured in benzene/di-t-butylperoxide, $(-0.9)^{26}$ and in CCl$_4$ $(-1.2)^{81}$.

16. Transient phenoxyl radicals: Formation and properties in aqueous solutions 1113

C. Oxidation of Phenols by Radical Cations

Radical cations of methoxybenzenes efficiently oxidize phenols and other reductants. For example, the radical cations of anisole, 1,3-dimethoxybenzene (DMB), and 1,3,5-trimethoxybenzene (TMB), produced in $\geq 90\%$ yield by reaction of OH with the methoxybenzenes at pH 1, can oxidize phenols and other reductants[48]. The product radicals were identified in most cases by their known absorption spectra and extinction coefficients. The rate constants, determined by monitoring the buildup of the product radical and/or the decay of the radical cation as a function of the concentration of reductant, are summarized in Table 2. The rate constants are high for phenols bearing electron-donating substituents and much lower for phenols bearing strong electron-withdrawing substituents.

D. Reaction of Phenols with Hydroxyl Radicals. The Addition/Elimination Mechanism

Early work[82] on the radiation chemistry of aqueous phenol solutions indicated that dihydroxycyclohexadienyl (OH adduct) and phenoxyl radicals were formed. In the first pulse radiolysis investigation concerning phenol[83] it was concluded that only dihydroxycyclohexadienyl radicals were produced. In contrast, from the first ESR study[84] on reactions of phenol with •OH radicals, generated by the Ti(III)/H_2O_2 method[85], it appeared that the phenoxyl radical was the only radical formed in that reaction. These apparently conflicting observations were reconciled by pulse radiolysis[86,87] and later by ESR[88] studies. These studies showed that the •OH radical reacts by addition to yield dihydroxycyclohexadienyl radicals, which then may undergo a 'spontaneous', an acid-catalyzed or a base-catalyzed dehydration to yield phenoxyl radical[86] (equation 10). From the kinetics of formation of phenoxyl radical at pH 3–5 there was evidence for more than one OH adduct isomer responsible for phenoxyl production[86]. The isomer distribution from the reaction of •OH with phenol was later determined[89] by a combination of product analysis and pulse radiolysis methods using specific scavengers for the isomeric dihydroxycyclohexadienyl radicals. On this basis, the reactions leading to the formation and decay by dehydration of the OH adducts are as shown in equation 10.

TABLE 2. Rate constants for oxidation of phenols by methoxybenzene radical cations[a]

Compound	Anisole•+	1,3-DMB•+	1,3,5-TMB•+
4-Hydroxyphenol		3.6×10^9	4.8×10^9
4-Methylphenol		4.6×10^9	
4-Carboxyphenol		4.0×10^9	1.1×10^9
Phenol	4.9×10^9	2.4×10^9	4.8×10^9
4-Chlorophenol		3.7×10^9	4.4×10^9
4-Bromophenol		4.3×10^9	3.7×10^9
4-Formylphenol		$<10^7$	
4-Acetylphenol		$<10^7$	
4-Cyanophenol		$<10^7$	1.7×10^8
Tyrosine	3.4×10^9	1.8×10^9	2.4×10^9
Tryptophan	2.8×10^9	2.4×10^9	
Ascorbic acid	2.1×10^9	2.3×10^9	2.5×10^9
3,5-Dimethoxyphenol	3.5×10^9	4.2×10^9	4.4×10^9

[a]The rate constants are given in $M^{-1} s^{-1}$.

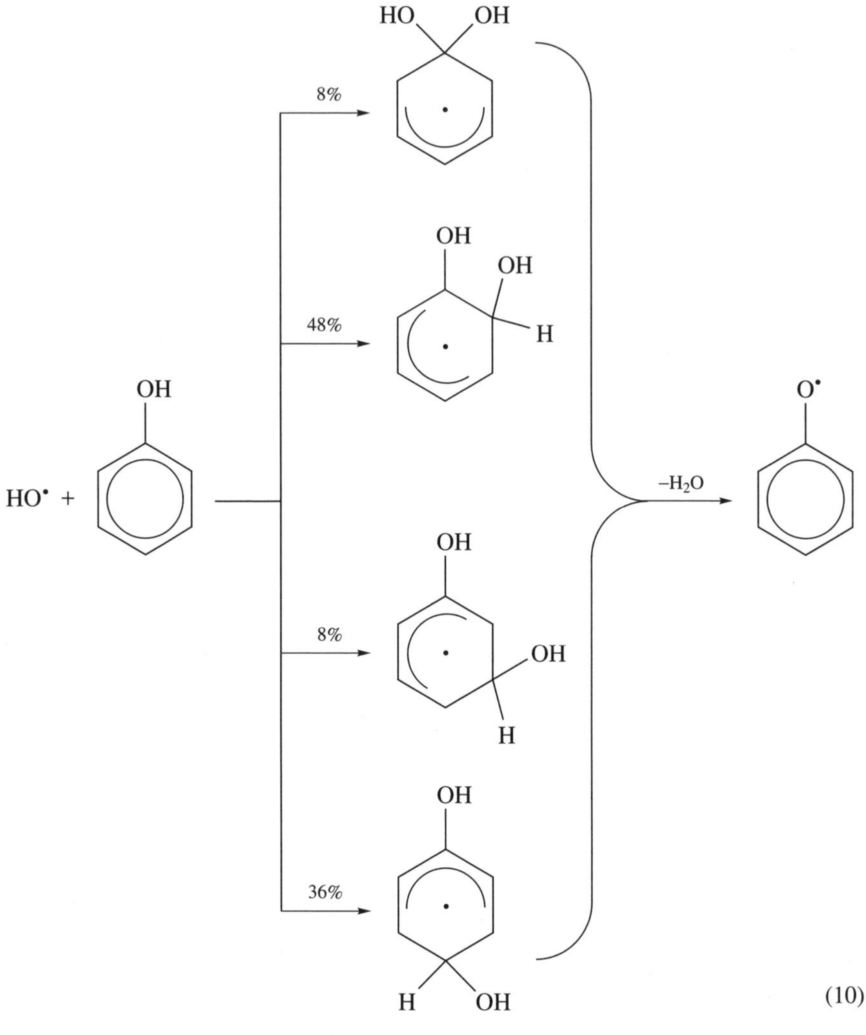

From the yields of the isomeric OH adducts, the probabilities p for attachment of OH to one ring position are calculated to be p(*para*): p(*ortho*): p(*meta*) = 9 : 6 : 1[89], which shows the pronounced preference of the electrophilic •OH radical for addition at the positions activated by the phenolic OH group. This selectivity is remarkable in view of the fact that the rate constant for reaction of •OH with phenol is very high, i.e. 1.4×10^{10} M^{-1} s^{-1} [86].

The *para*-isomer undergoes H$^+$ catalyzed dehydration a factor of 10 more rapidly than does the *ortho*-isomer[89,90]. This explains the observation by ESR[88] of only the less reactive (with respect to dehydration) *ortho*-isomer in slightly acid solutions. The mechanism for the 'spontaneous' (k_{sp}) and the H$^+$ (k_a)[47] and OH$^-$ catalyzed (k_b) dehydration steps may be formulated as in equation 11, taking the *para*-isomer as an example.

16. Transient phenoxyl radicals: Formation and properties in aqueous solutions   1115

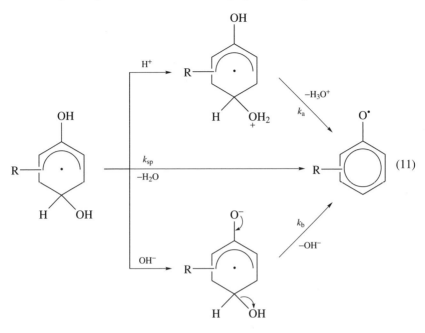

For R = H, k_a is 1.6×10^6 s^{-1} [91], in agreement with the earlier estimate of $\geqslant 5 \times 10^5$ s^{-1} [86]. As expected by the heterolytic mechanism of formation of the phenoxyl radical from the OH adduct, there is a strong influence of R on the rates of the individual steps. For example, as compared to $k_{sp} = 4.7 \times 10^3$ s^{-1} for R = H[86], for R = OH (hydroquinone) and R = OH (resorcinol) $k_{sp} = 4.6 \times 10^4$ s^{-1} and 4.3×10^4 s^{-1}, respectively[92], and $k_{sp} = 2.4 \times 10^3$ s^{-1} if R = CH$_3$ (p-cresol)[86], and 1×10^4 s^{-1} if the ring is substituted by two methoxy groups[47]. Concerning the k_b values, substitution of the phenol by the electron-withdrawing groups CN, CHO and COCH$_3$ results in values of 3×10^5 s^{-1}, 4×10^5 s^{-1} and 7×10^5 s^{-1}, respectively, as compared to $\geqslant 10^7$ s^{-1} for unsubstituted phenol[48]. This dependence of k_b on substituent is part of a more general reaction mechanism of formation of oxidized species by elimination of OH$^-$ from the corresponding (ionized) OH adducts. In this connection it may be mentioned that the rate constants for OH$^-$ elimination increase in a systematic way with decreasing ionization potential of the parent compounds[93].

The addition/elimination mechanism (equation 11) for formation of phenoxyl radicals by •OH reaction with phenols is now documented for a vast number of substituted phenols[47,86,88,94–98], catechols[97,99–102], resorcinol[94], hydroquinones[92,97] and hydroxylated heterocyclics[103–107]. The addition/elimination mechanism is also operative in •OH reactions with anilines to yield the nitrogen-centered anilinyl radicals (equation 12)[93,97,108–111].

$$\text{ArNH}_2 + \text{•OH} \longrightarrow (\text{HOArNH}_2)\text{•} \longrightarrow \text{ArNH•} + \text{H}_2\text{O} \qquad (12)$$

An essential part of the driving force of the elimination step[86] is the recovery of the aromatic resonance energy in going from the cyclohexadienyl to the benzene system.

The OH adduct of 4-nitrophenol or 4-nitrophenolate is the only phenolic OH adduct which does not observably undergo water or OH$^-$ elimination (equation 13)[112]. In this case the heterolytic elimination step is slowed down to <1 s^{-1} by the pronounced withdrawal of electron density by the NO$_2$ group.

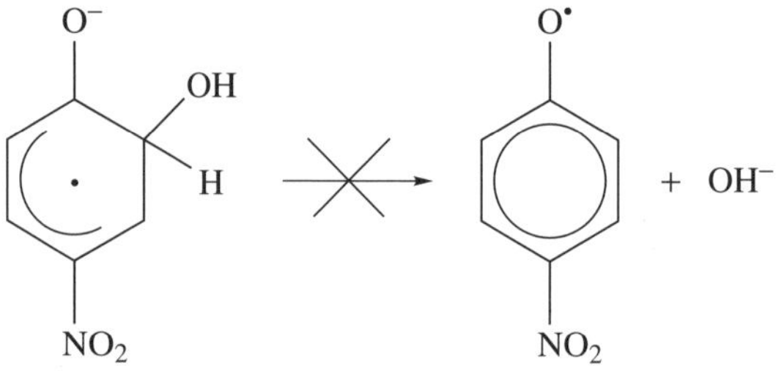

E. Formation of Phenoxyl Radicals by Oxidative Replacement of Substituents

The •OH radical may add to a substituted benzene at the *ipso* position, i.e. at the carbon carrying the substituent X (equation 14).

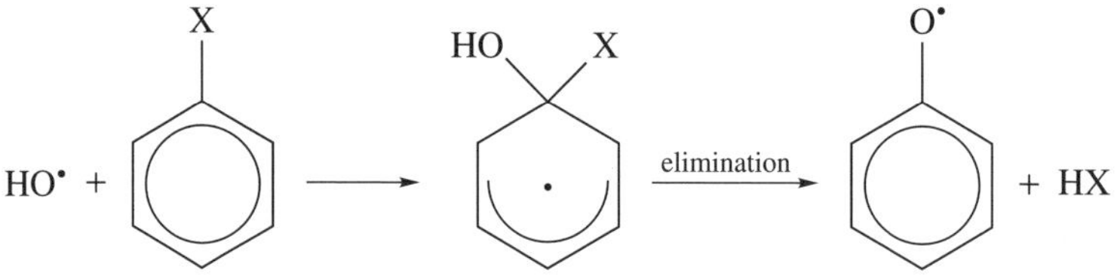

If X is a good leaving group, the resulting OH adduct undergoes elimination of HX to yield a phenoxyl radical. Reaction 14 has been demonstrated to occur for X = halogen, OH, NH_2, NO_2 and alkoxyl, benzyloxyl and phenoxyl substituents. The driving force for elimination of HX from the *ipso* adduct is probably the reconstitution of the aromatic system in going from the OH adduct to the phenoxyl radical. There may also be a contribution from the heat of formation of HX.

With monosubstituted benzenes, the tendency of OH to undergo *ipso* addition seems to be small (<10%). However, if the *ipso* position is activated by a second substituent like CH_3O, HO or O^-, *ipso* addition may contribute up to 25% to the overall reactivity[94]. Due to the small size of the •OH radical, steric effects (i.e. the size of X) do not seem to be of much importance in determining the probability of •OH attack at the *ipso* position. Oxidative replacement of halogen has been observed also with pentafluoro-, pentachloro-, pentabromo- and 2,4,6-triiodo-phenol, where it occurs in parallel to electron transfer[113].

A mechanism analogous to that of equation 14 has been proposed to account for the observation by ESR of the predominant production of phenoxyl radical on reaction of •OH with fluorobenzene at pH 1.8[88], where the fluorobenzene radical cation is assumed to be an intermediate (equation 15).

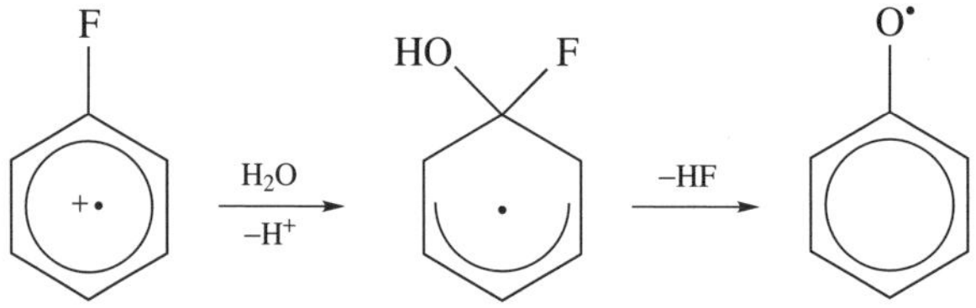

This idea is based on the assumption that hydration of the radical cation occurs predominantly at the *ipso* position. This is reasonable on the basis of estimates[114] of the charge distribution in fluorobenzene radical cation. The mechanism in equation 15 is, however, not likely for oxidative dehalogenation of halophenols[115] at pH 4.5 since the halophenol radical cations would deprotonate[40,41] before they had a chance to react with water. With substituted phenols, therefore, the observed[88] semiquinones are probably produced via equation 14.

A mechanism analogous to equation 14 has been observed with respect to ·OH reactions with halouracils[116-118] and nitrouracils[119], nitro- and bromofurans[120] and chloroethylenes[121], e.g. in the case of uracil (equation 16).

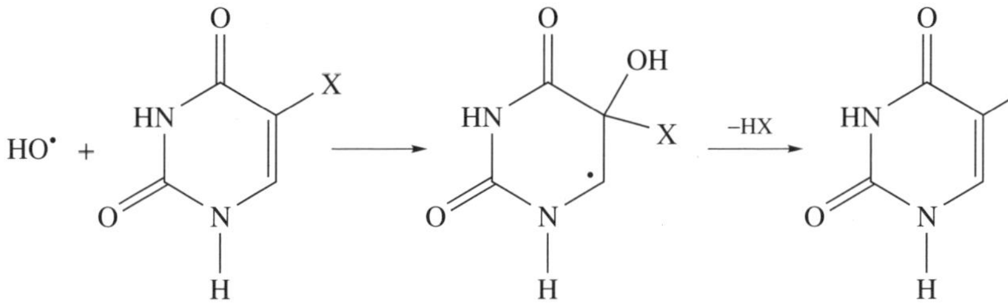

$$HO^{\bullet} + \text{[uracil]} \longrightarrow \text{[intermediate]} \xrightarrow{-HX} \text{[uracil phenoxyl radical]} \quad (16)$$

F. Formation of Phenoxyl Radicals by Intramolecular Electron Transfer

It has been shown by pulse radiolysis that in peptides and enzymes containing both tryptophan (Trp) and tyrosine (Tyr) the radical produced by oxidation of tryptophan can efficiently oxidize tyrosine to yield the tyrosine phenoxyl radical TyrO· (equation 17)[122-125].

$$Trp^{\bullet}\text{—TyrOH} \longrightarrow TrpH\text{—TyrO}^{\bullet} \quad (17)$$

The rate constants of this intramolecular process, which constitutes a transfer of charge and unpaired spin, decrease with increasing distance between tryptophan and tyrosine in an inverse-square distance relationship[124]. Deprotonation of the OH group of tyrosine and protonation of the indolyl radical enhance the electron transfer rates[124]. This is due to the pH-dependent changes in the redox potentials of tyrosine and tryptophan. From the low activation energy (0.22 eV) for the electron transfer in Trp-Tyr it was concluded that an electron-tunneling mechanism is operative[124]. The electron transfer reaction 17 was also observed in enzymes[14,123,125]. Rate constants vary from 10^2 s^{-1} (in lysozyme) to 2×10^4 s^{-1} (in trypsin). For β-lactoglobulin, the activation energy is ca 0.5 eV. On this basis charge conduction along the polypeptide chain and any mechanism involving temperature-labile hydrogen bonds was excluded[123], and electron tunneling was proposed[125]. Moreover, the rate of electron transfer in peptides was found to depend also on the microenvironment[126]. Electron transfer from tyrosine to methionine radicals also has been observed in peptides[126-128]. The possibility of electron transfer means that the initial site of damage produced by reaction of a free radical with an enzyme is not necessarily the site responsible for a consequent loss of activity. More recently, oxidation of tyrosine by the tryptophan radical or radical cation was studied in a series of synthetic peptides using a varying number of proline residues as spacers[129-132]. The mechanism was suggested to involve both through-bond and through-space electron transfer, depending on distance and orientation.

G. Formation of Phenoxyl Radicals from Phenols and Hydroxyphenols by Reaction with O_2 and/or $O_2^{\bullet-}$ (Autoxidation)

The autoxidation of hydroquinone is accompanied by the formation of the *p*-semiquinone radical; this was shown as early as 1938 by Michaelis and coworkers[133-135]. Since then numerous additional examples of this type of reaction have been described, relating not only to hydroquinones but also to catechols, resorcinols, pyrogallols, naphthols and substituted phenols. Most of this material has been reviewed[6,18-20,136-141], including that relating to synthetic application of phenol oxidation and to phenoxyl radicals involved in the biosynthesis of natural products[18,138,139,142].

The autoxidation of phenols is slow in neutral and, especially, in acid solution but becomes very noticeable in alkaline solutions. This base catalysis of phenol oxidation is of course due to the conversion of the neutral phenols to the phenolate ions, which are more easily oxidized by the oxidant Ox (equation 18) than their conjugate acids.

$$ArO^- + Ox \longrightarrow ArO^{\bullet} + Ox^{\bullet-} \tag{18}$$

At present, the exact nature of the oxidant Ox is not yet clear. $O_2^{\bullet-}$ or its conjugate acid, $HO_2^{\bullet}$, have been suggested as candidates by several groups[6,101,124,133-148]. When one looks, however, at the rate constants collected from these sources for reduction of $O_2^{\bullet-}$ by some phenols, mostly catechols, a reasonable correlation between the structure of the electron donor and the rate constant cannot be discerned. For example, if a simple electron transfer mechanism was involved, substitution of the catechol molecule with electron-withdrawing substituents like CHO, $COCH_3$, $COCH_2NHCH_3$ and SO_3^- should decrease and not (as is experimentally observed) increase the rate constant for its oxidation. Furthermore, reported rate constants for reaction of $O_2^{\bullet-}$ with 1,2-dihydroxybenzene-3,5-disulfonic acid (Tiron), a compound proposed[149,150] as a specific $O_2^{\bullet-}$ scavenger, vary between 1×10^{7}[101] and 5×10^8 M^{-1} s^{-1}[144]. The nature of the radical produced from Tiron is not agreed upon either[101,144].

It has been shown[151,152] that, in dimethylformamide solutions, oxidation of Trolox anion or of di-*tert*-butylcatechol monoanion by one-electron transfer to $O_2^{\bullet-}$ is thermodynamically not possible. Therefore, the authors suggested that the experimentally observed oxidation of the substrates occurs by electron transfer to molecular oxygen as the primary oxidant (equation 19), followed by further reactions (equations 20 and 21) that yield the experimentally observed H_2O_2.

$$ArO^- + O_2 \rightleftharpoons O_2^{\bullet-} + ArO^{\bullet} \tag{19}$$

$$2\,ArO^- + O_2 + 2H^+ \rightleftharpoons H_2O_2 + 2\,ArO^{\bullet} \tag{20}$$

$$2\,O_2^{\bullet-} + 2\,H^+ \longrightarrow H_2O_2 + O_2 \tag{21}$$

For reaction 19, taking hydroquinone (H_2Q) as an example, and using the standard electrode potentials (in H_2O, for unit concentration) for the $O_2/O_2^{\bullet-}$ (-0.16 V)[153,154], $Q^{\bullet-}/Q^{2-}$ (0.023 V)[154] and $Q^{\bullet-}/QH_2$ (0.459 V)[154], we calculate $\Delta E = -0.18$ V at pH 13.5 (where hydroquinone exists as the dianion) and $\Delta E = -0.62$ V at pH 7. Thus, the reaction is endothermic. For catechol and Trolox, reaction 19 is even more endothermic. However, since reaction 21 is very rapid in the presence of H^+, equilibrium 19 is pulled to the right. An alternative mechanism is the oxidation of phenols by O_2 via a hydrogen-atom transfer process (equation 22) which also is endothermic[24,60].

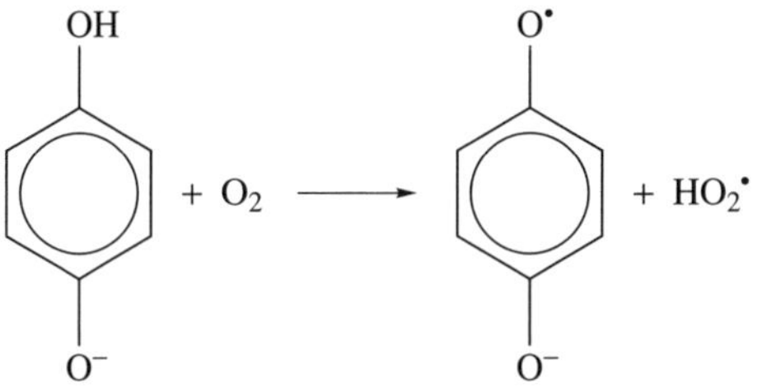

$$\text{HO-C}_6\text{H}_4\text{-O}^- + \text{O}_2 \longrightarrow {}^{\bullet}\text{O-C}_6\text{H}_4\text{-O}^- + \text{HO}_2^{\bullet} \tag{22}$$

(HQ⁻) (Q˙⁻)

Even if reactions 19 and 22 are very slow, the oxidation of hydroquinone by O_2 may still be rapid, due to autocatalysis by the quinone formed as reaction product or present as impurity. The reactions suggested to occur are shown in equations 23–27.

$$HQ^- + O_2 \longrightarrow Q^{\bullet -} + HO_2^{\bullet} \tag{23}$$

$$HO_2^{\bullet} \rightleftharpoons H^+ + O_2^{\bullet -} \tag{24}$$

$$Q^{\bullet -} + O_2 \underset{k_{-25}}{\overset{k_{25}}{\rightleftharpoons}} Q + O_2^{\bullet -} \tag{25}$$

$$Q + H_2Q \underset{k_{-26}}{\overset{k_{26}}{\rightleftharpoons}} 2\,Q^{\bullet -} + 2\,H^+ \tag{26}$$

$$H^+ + HO_2^{\bullet} + Q^{\bullet -} \longrightarrow H_2O_2 + Q \tag{27}$$

The Q formed in equation 25 will be consumed by reaction with H_2Q (equation 26) giving rise to $Q^{\bullet -}$ which (in a reversible reaction) is oxidized by O_2 to recover Q (equation 25). Reaction 27 presents an additional path to Q in which also H_2O_2 is produced.

Reversible electron transfer between semiquinone anions and O_2 (equation 25) has been established by the method of pulse radiolysis[154,155]. The rate constants k_{25} and k_{-25} and the equilibrium constants $K = k_{25}/k_{-25}$ are known for many different quinones Q[155,156]. The k_{25} values are typically in the range $10^4 - 10^9$ M^{-1} s^{-1} and the k_{-25} values are ca 1–100, i.e. for many quinones (e.g. p-benzoquinone) the equilibrium 25 is in favor of $Q^{\bullet -}$ and O_2. However, due to the 'cross' reaction 27, whereby $Q^{\bullet -}$ and $HO_2^{\bullet}/O_2^{\bullet -}$ are removed from reaction 25, O_2 is consumed and ends up oxidizing H_2Q to give H_2O_2. In reaction 26, for which the forward rate constant k_{26} is 2.6×10^8 M^{-1} s^{-1}[136,137], two $Q^{\bullet -}$ are produced for every $Q^{\bullet -}$ consumed in reaction 25. The $Q^{\bullet -}$ can re-enter the reaction cycle at equation 25, thus propagating a chain reaction initiated by traces of $Q^{\bullet -}$. $Q^{\bullet -}$ does not necessarily have to be produced via reaction 23; it could also be generated by reduction of Q by reducing impurities or, more likely, by QH_2 (equation 26). The reaction sequence 23–27 explains a large part if not all of the earlier data[20,140,157,158] on quinol oxidations, such as hydrogen peroxide formation in the air oxidation of phenols[159], the accelerating effect of quinones[140,157,158] and the inhibiting effect of superoxide dismutase (SOD) on, e.g., the autoxidation of catecholamines[101,146,160–163], of pyrogallol[164], of 6,7-dihydroxytryptamine[162], of reduced flavins[165] and of tetrahydropteridines[166]. A very similar, but more detailed mechanism for oxidation of hydroquinones by O_2 has recently been proposed[167].

III. PROPERTIES OF PHENOXYL RADICALS

A. Electron Spin Resonance Spectra of Phenoxyl Radicals

Electron spin resonance spectra of phenoxyl radicals were first recorded with the persistent 2,4,6-tri-substituted phenoxyl, produced by PbO_2 oxidation of the corresponding phenol[28,168]. Autoxidation of 3,4-dihydroxyphenylalanine allowed the observation of ESR spectra of some long-lived secondary radicals[169], while enzymatic oxidation of pyrogallol and other compounds by peroxidase in a flow system enabled the observation of some phenoxyl-type radicals[170]. The first recording of detailed ESR spectra of transient phenoxyl radicals was carried out by Stone and Waters[36], who oxidized phenol and several substituted phenols with ceric ions in a rapid-mix system. Similarly, the reaction of phenol with OH radicals, from the $Ti^{3+}-H_2O_2$ system, afforded a resolved ESR spectrum of the transient phenoxyl radical[84]. Since then, many experiments have been carried out in which phenoxyl radicals were produced *in situ* in the ESR cavity, by chemical[37–39,42,44,88,171–178], photochemical[115,179–185] or radiolytic[47,94,97,186,187] reactions. In numerous cases, the primary phenoxyl radicals were observed along with secondary radicals of the longer-lived semiquinone type, which were produced from hydroxylated products[34,37,39,97,115,171–174,179].

Phenoxyl and semiquinone radicals are important intermediates in numerous biological systems and ESR spectroscopy has been used to detect and identify them in such systems. Studies were carried out on enzymatic reduction of quinone derivatives and enzymatic oxidation of hydroquinone and phenol derivatives. This topic has been reviewed before[14,188–190].

1. Phenoxyl and monosubstituted phenoxyl radicals

The proton hyperfine splitting constants (hfs) for the unsubstituted phenoxyl radical are (in millitesla, mT) as indicated below at the corresponding positions with g = 2.00461[97].

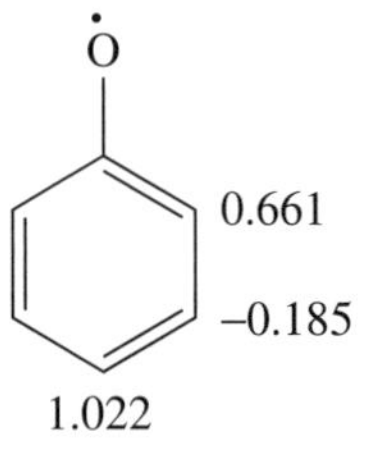

These values indicate that the spin density is mostly at the *ortho* and *para* carbon atoms (*ca* 27% and *ca* 43%, respectively) with a negative spin density at the *meta* position (*ca* −8%) and only *ca* 25% remaining on the oxygen atom. In other words, all the mesomeric structures shown below are of somewhat similar importance.

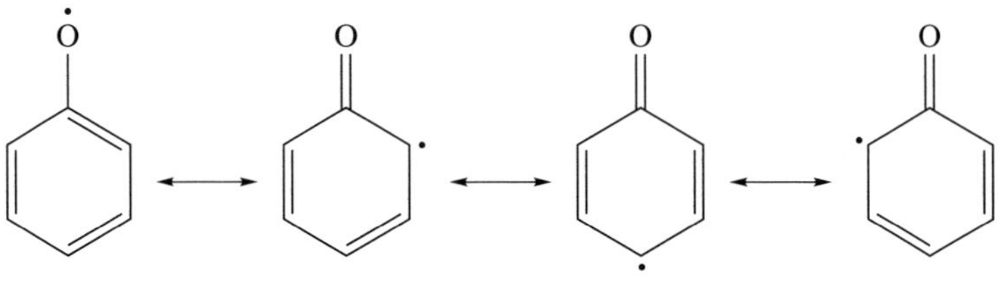

The assignment of the proton hyperfine constants to the particular positions of phenoxyl was supported by the ESR parameters for substituted phenoxyls, for example the carboxylates[97].

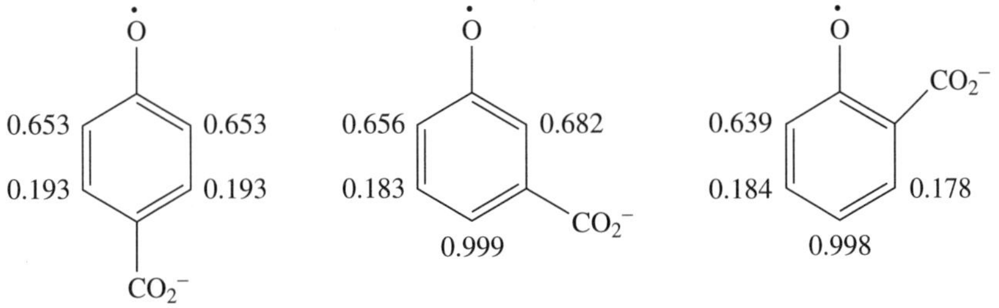

p-carboxylate phenoxyl: 0.653 (2,6), 0.193 (3,5), $g = 2.00477$

m-carboxylate phenoxyl: 0.656, 0.682, 0.183, 0.999, CO_2^-, $g = 2.00472$

o-carboxylate phenoxyl: 0.639, CO_2^-, 0.184, 0.178, 0.998, $g = 2.00476$

Many substituents exert only a mild effect on the proton hyperfine splittings[38]; strong influence is observed in the case of O^-, OH, OR and NH_2. It has been suggested that the effect of substituents on the proton hfs's follows the same order as the electron donating or withdrawing effects of these substituents[38]. However, correlation between a^H and Hammett's σ substituent constants fails because the electron distribution in the radical is different from that in the molecule[191]. Table 3 summarizes the proton hfs constants for a selected set of substituted phenoxyl radicals. The values are taken from Dixon and coworkers[38], who demonstrated qualitative correlations among the various ESR parameters. These correlations allowed assignment of coupling constants to the particular protons even where differences among them were small. Table 3 lists only the ring proton hfs's

TABLE 3. Proton hyperfine splitting constants (in mT) for selected monosubstituted phenoxyl radicals

Substituent	a_2^H	a_3^H	a_4^H	a_5^H	a_6^H
H	0.66	−0.18	1.02	−0.18	0.66
p-NO_2	0.70	−0.24		−0.24	0.70
p-$COCH_3$	0.675	−0.21		−0.21	0.675
p-CH_3	0.61	−0.14		−0.14	0.61
p-Cl	0.64	−0.19		−0.19	0.64
p-OCH_3	0.49	0.00		0.00	0.49
p-NH_2	0.40	0.05		0.05	0.40
p-O^-	0.237	0.237		0.237	0.237
o-NO_2		−0.12	1.025	−0.24	0.725
o-$COCH_3$		−0.15	1.025	−0.20	0.70
o-CH_3		−0.20	0.97	−0.15	0.60
o-Cl		−0.20	0.98	−0.16	0.60
o-OCH_3		−0.19	0.85	0.00	0.43
o-NH_2		−0.09	0.662	0.15	0.26
o-O^-		0.075	0.375	0.375	0.075
m-NO_2	0.735		0.98	−0.21	0.675
m-$COCH_3$	0.71		0.99	−0.19	0.65
m-CH_3	0.59		1.05	−0.19	0.71
m-Cl	0.62		1.05	−0.21	0.75
m-OCH_3	0.35		1.14	−0.23	0.90
m-NH_2	0.31		1.09	−0.20	0.86
m-O^-	−0.07		1.12	−0.28	1.12

for one consistent set of substituted phenoxyl radicals studied under identical conditions. The hfs's for the substituent protons and for other nuclei are omitted and can be found in the original work[38]. Furthermore, a multitude of substituted phenoxyl radicals, studied by ESR under varying conditions, are given in comprehensive compilations[192,193].

Phenoxyl radicals substituted with O⁻, i.e. semiquinone radical anions, should be viewed as a special case because of the equivalence of the two oxygens. Even in the *meta* isomer the two oxygens appear to be equivalent in the ESR spectra (but see below for further details). The ESR parameters are[48,97,186]:

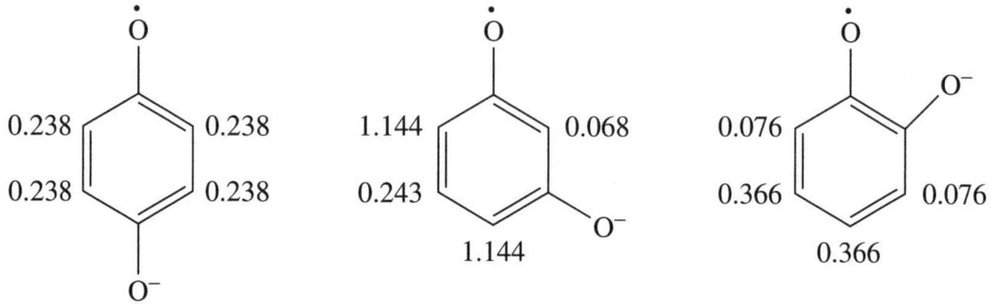

$g = 2.00455$ $g = 2.00383$ $g = 2.00455$

Quantitatively, however, the *meta* isomer is different in that the spin density on its oxygen atoms is at least 3 times smaller that the 60–65% spin density found on the oxygens of *o*- and *p*-semiquinones[37]. Protonation of the O⁻ destroys this equivalence so that the hydroxyphenoxyl radicals shown below[38,39,186,194,195]:

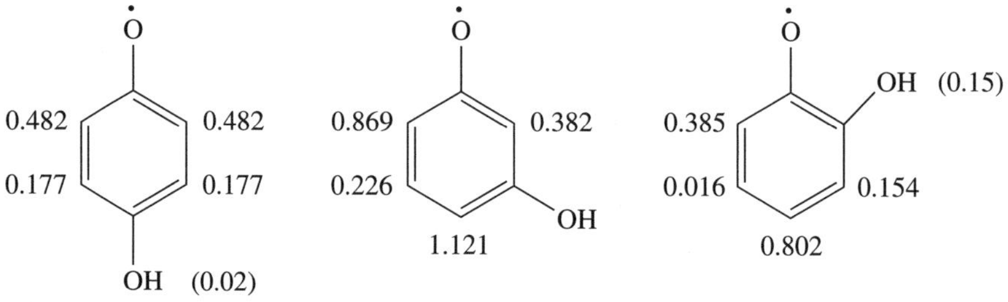

become similar to the methoxyphenoxyl radicals[47].

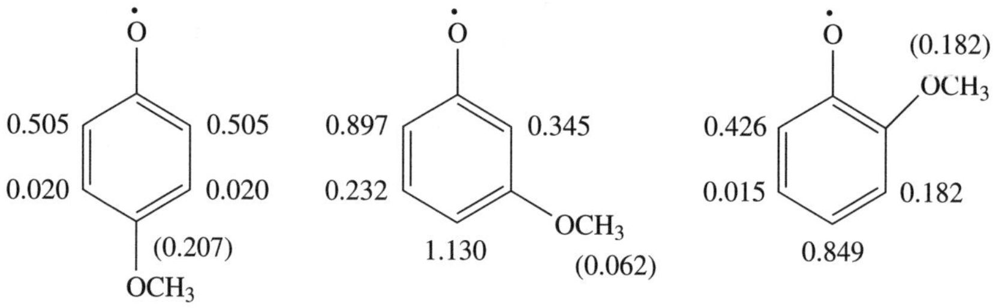

$g = 2.00449$ $g = 2.00431$ $g = 2.00439$

Phenoxyl radicals substituted with OH, NH₂ or OR have intermediate properties between those of the semiquinones and of the simple phenoxyl, due to the contribution of mesomeric

16. Transient phenoxyl radicals: Formation and properties in aqueous solutions   1123

structures with a negative charge on the phenoxyl oxygen and a positive charge on the substituent heteroatom.

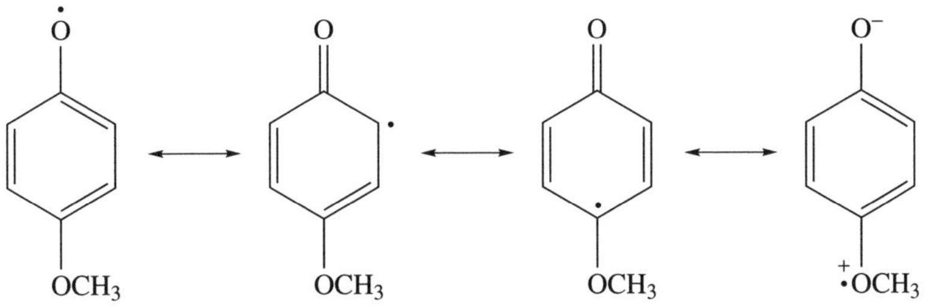

This effect decreases the total spin density on the ring and increases that on the oxygens, as compared with unsubstituted phenoxyl.

The hfs's for the aminophenoxyl radicals[97,196] resemble those of the semiquinones more closely than do the methoxyphenoxyl:

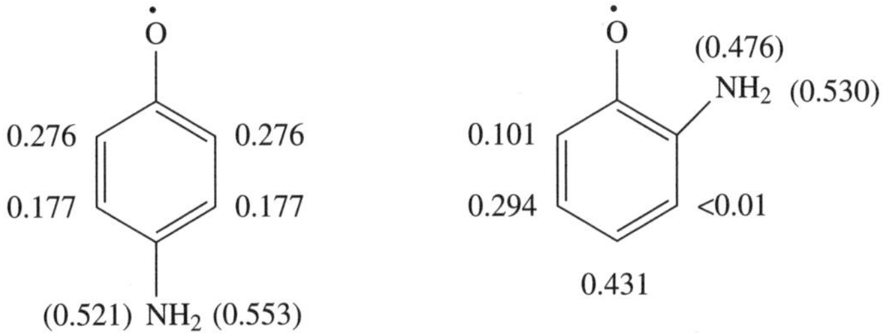

$g = 2.00377$ $g = 2.00372$

It should be noted that the g factors are considerably lower than those observed with the semiquinones or the methoxyphenoxyl radicals. This indicates a considerable transfer of spin density to the nitrogen atom[97]. (For further considerations and theoretical calculations on the p-aminophenoxyl radical, see References 197 and 198).

The hfs constants for the nitrogen atoms (a^N) in aminophenoxyl radicals[30,38,97,196] are slightly lower than the proton hfs (a^H) in the same position, while a^N for nitrophenoxyl is considerably lower[38,42]. The halogen hfs's of halophenoxyl radicals were determined in several cases[38,42,180,199−201]. It was noticed that replacement of H by F or Cl had little effect on the spin density distribution and that the ratio of hfs's $a^H : a^F : a^{Cl}$ at the same position on the ring was approximately 5 : 15 : 1 in all cases.

The hfs constants for ^{13}C were determined for several stable semiquinones[202−206] and sterically hindered persistent phenoxyls[207−215]. In several persistent phenoxyl radicals, the hfs's for ^{13}C were found to be in the range of 1.1−1.3 mT for the *para* carbon, 0.8−1.0 mT for the *meta* and *ipso* and 0.2−0.5 mT for the *ortho* carbon, with slight variations depending on the nature of the substituents[207,209,211,216]. The hfs's were used to derive the various Q values that affect them ($Q^C_{C-C'}$, $Q^C_{C-H'}$, $Q^C_{C-O'}$, Q^C_{O-C}) and to confirm the validity of calculated spin densities. Although the values reported vary considerably[211,216], it is clear that the unpaired spin is distributed mainly at the *ortho* and *para* carbons and the oxygen, with 20−30% at each site.

The ^{13}C hfs's for m-benzosemiquinone[206] are in the same range as those for the phenoxyl radical. In contrast, the o- and p-semiquinones exhibit much lower ^{13}C hfs's. It is interesting to note that while the proton hfs constants of p-benzosemiquinone were not sensitive to solvent composition, the ^{13}C hfs was extremely sensitive to water content[202]. This was explained[217] by the formation of radical–solvent complexes and their effect on spin density.

The spin density on the phenoxyl oxygen was also confirmed by measurements of ^{17}O hfs[208,209,213,216,218–225]. In fact, the observation of the ^{17}O hfs ($a = 1.023$ mT for the 2,4,6-tri-t-butylphenoxyl radical[208] and $a = 0.97$ mT for the 2,4,6-triphenylphenoxyl radical[218]) was the first direct experimental evidence for an appreciable spin density on the phenoxyl oxygen. McLachlan SCF calculations estimated the spin density on the oxygen at 26% which, combined with the experimental hfs's, leads to $Q^O \approx 3.8$ mT[216]. Experiments with ^{17}O-enriched semiquinones[219–225] also indicated $a^O \approx 0.9$–1.0 mT for several benzosemiquinones (and slightly lower a^O for naphtho- and anthra-semiquinones) which again lead to an estimated $Q^O_{OC} \sim 4.0$ mT (the spin density on the oxygen is the main contributor to a^O)[223]. Variations in a^O for p-benzosemiquinone between 0.95 and 0.87 mT with varying water mole fraction from 0 to 1 were rationalized[219] on the basis of the π-electron spin densities calculated from the known proton and ^{13}C hfs's.

2. Phenoxyl radicals with extended π-systems

ESR spectra were reported for various phenoxyl-type radicals with extended π-systems, e.g. biphenyl[34,226,227], tetraphenyl[216], naphthalene[39,227–235], anthracene[229,236–239] and phenyl systems conjugated with aliphatic double bonds[240,241] or triple bonds[242]. In all these cases, the ESR parameters indicate spin density distribution over the extended π-system with presumably lower spin density on the phenoxyl oxygen. For example, the hfs constants (in mT) for α- and β-naphthoxyl are[233,235]:

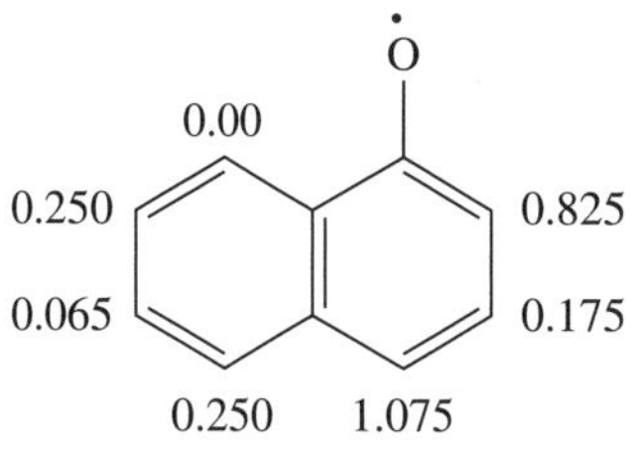

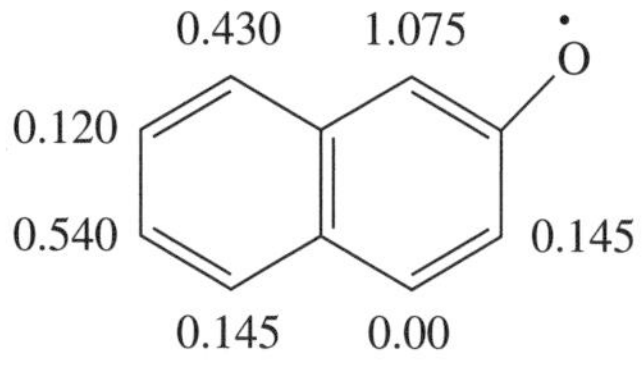

and for anthroxyl[238]:

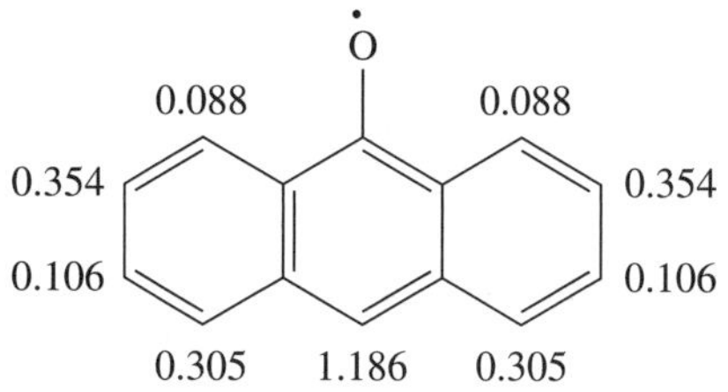

In general, these follow the same pattern as phenoxyl, with *ortho* and *para* positions bearing high spin density and *meta* having little or negative spin density. In these polycyclic phenoxyls the general trend of alternating 'high' and 'low' spin densities appears to hold although β-naphthoxyl is quite different from the α-isomer[233,235]. ESR spectra of naphthosemiquinones and anthrasemiquinones[229,239,243,244] also have been reported.

The radical derived from α-tocopherol (Vitamin E) may be mentioned in this category, although it is basically a p-alkoxyphenoxyl. The hfs's for H or CH$_3$ *ortho* to the phenoxyl oxygen are in the range of 0.5–0.6 mT[245,246], similar to those in *p*-methoxyphenoxyl[47], with a similar *g* factor of 2.0046. The effect of the hetero ring is to make the two positions *ortho* to O• inequivalent, with 20% higher spin density on the *ortho* carbon near the hetero ring than the one on the opposite side[245]. The length of the side chain had little effect on the ESR parameters.

3. Kinetic ESR measurements

ESR experiments were used to measure the kinetics of several types of reactions, those that can be monitored only by ESR, such as proton exchange or electron exchange reactions of radicals, and some that can be measured by other techniques as well, e.g. decay kinetics. Although most decay kinetics of phenoxyl radicals were followed by pulse radiolysis or flash photolysis by monitoring optical absorption, kinetics for some long-lived radicals were frequently monitored by ESR. For example, the second-order decay rates of 4-alkyl-2,6-di-*t*-butylphenoxyl radicals were measured to be 2200, 500 and 2 M^{-1} s^{-1} for the 4-methyl, 4-ethyl and 4-isopropyl derivatives, respectively, in benzene solutions at room temperature[247]. Cross-disproportionation between different phenoxyl radicals[248] and the reaction of persistent phenoxyl with oxygen, peroxyl radical and hydroperoxide[249] were also followed by ESR.

Intermolecular and intramolecular proton exchange reactions in hydroxyphenoxyl radicals were studied by several authors[195,250–255]. In aqueous solutions[250–252], the proton transfer was considered to take place between the radical and the solvent, and the ESR line broadening effects were analyzed in terms of equilibria such as that shown in equation 28 yielding $k(R^• + H^+ = RH^+) \sim 10^9$ M^{-1} s^{-1}. On the other hand, experiments in aprotic solvents indicated intramolecular proton transfer via an internal hydrogen bridge[255] (equation 29) with $k \sim 10^3$ to 10^7 s^{-1} between −100 and +22 °C. A later study[195] of the *o*-semiquinone radical in different media has shown that the effects of both inter- and intramolecular proton transfer processes on the ESR spectra have to be taken into account in such a system. While the intramolecular process is always present, intermolecular hydrogen bonding may predominate in protic solvents. For this reaction the OH proton hfs of *o*-hydroxyphenoxyl was observed in aprotic solvents[255] but not in water[250–252]. In the case of pyrogallol radical the OH proton hfs was observed even in water[250], apparently because of stronger intramolecular hydrogen bonding due to the presence of two OH groups *ortho* to the O• site.

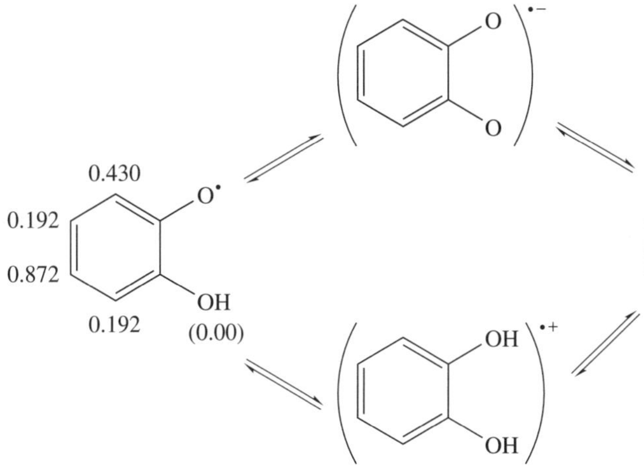

(28)

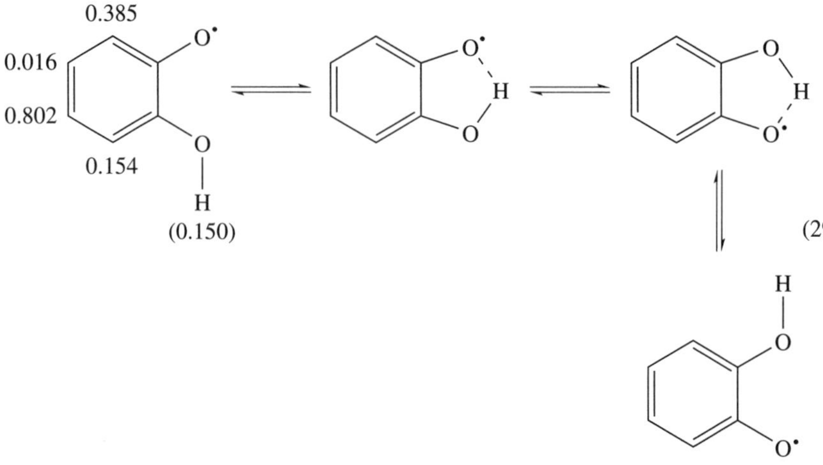

$$\text{(29)}$$

Intermolecular proton exchange between 3,6-di-*t*-butyl-2-hydroxyphenoxyl radical and a variety of organic acids and bases was studied under various conditions[256–259]. Line broadening was analyzed in terms of a mechanism shown in equation 30 for

$$\text{(30)}$$

amine as the base, which involves a proton transfer to the amine and oscillation of the hydrogen bonds between the nitrogen and either of the two oxygens. In the absence of hydrogen bonding compounds, intramolecular hydrogen migration takes place[259–261] as discussed above.

Alternating line-width effects were found in the ESR spectra of some protonated phenoxyl radicals and were interpreted in terms of jumps between geometrical isomers involving the direction of the $O^{\bullet+}$-H group in relation to the ring[262].

Another type of line broadening is caused by electron exchange reactions between the radical and its parent compound, for example between phenoxyl radical and phenolate ion[263] (equation 31, where the labels 1 and 2 mark individual molecules).

$$Ph^1O^{\bullet} + Ph^2O^- \rightleftharpoons Ph^1O^- + Ph^2O^{\bullet} \qquad (31)$$

Such self-exchange reactions cannot be followed by optical spectroscopy since no chemical change is involved. However, ESR spectroscopy provides a unique capability to do so because the spin states of the protons in the radical effectively label a particular radical. The reaction leads to broadening of the ESR lines, which is detectable at rates of transfer $\geqslant 10^6$ s^{-1}. From the dependence of line width on phenolate concentration it was calculated that the rate constant for this self-exchange reaction is 1.9×10^8 M^{-1} s^{-1}[263]. Similarly, the rate constants for the self-exchange reactions between semiquinone radical anions and their parent quinones were determined[264]. They were found to be in the range of $0.5–2.0 \times 10^8$ M^{-1} s^{-1} for benzoquinone and its dimethyl and tetramethyl derivatives. These values were correlated[264] with the rates of electron transfer between semiquinones and different quinones, according to the Marcus theory.

4. Comparison with isoelectronic radicals

The phenoxyl radical can be compared with the isoelectronic anilinyl and benzyl radicals:

Ȯ (phenoxyl) 0.661 (ortho), 0.185 (meta), 1.022 (para)	(0.795) ṄH (1.294) 0.618 (ortho), 0.201 (meta), 0.822 (para)	ĊH$_2$ (1.634) 0.513 (ortho), 0.177 (meta), 0.617 (para)
$g = 2.00461$	$g = 2.00331$	$g = 2.00260$

It is clear from the a_{ortho}^H and a_{para}^H that the spin density on the ring decreases in the order phenoxyl > anilinyl > benzyl and is presumably accompanied by a corresponding increase in spin density on the formal radical site. This trend was also confirmed by theoretical calculations that compare the three radicals[265–267]. Apart from being isoelectronic, the anilinyl radical resembles the phenoxyl in being oxidizing while the benzyl tends to be reducing.

B. Optical Spectra of Phenoxyl Radicals

Optical absorption spectra of transient phenoxyl radicals have been studied by the flash photolysis or pulse radiolysis techniques and for some stable phenoxyl radicals it was possible to record their spectra in a spectrophotometer. Flash photolysis was instrumental in carrying out the first spectral observations of transient phenoxyl radicals under various conditions[268–272]. Pulse radiolysis, however, gave more accurate extinction coefficients owing to the more precise determination of the radiolytic yields of phenoxyl radicals, as compared with the photochemical quantum yields. Pulse radiolysis was also used to obtain very detailed spectra of certain model phenoxyl radicals[263,273] as shown, e.g., in Figure 1.

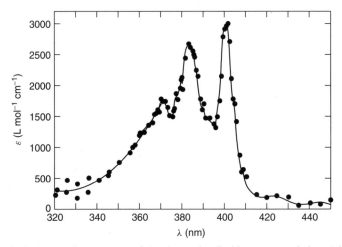

FIGURE 1. Optical absorption spectrum of the phenoxyl radical in aqueous solution. Adapted from Reference 273

TABLE 4. Absorption maxima (λ_{max}, in nm) and molar absorption coefficients (ε, in M^{-1} cm^{-1}) of monosubstituted phenoxyl radicals in aqueous solutions

Substituent	λ_{max}	ε	λ_{max}	ε	λ_{max}	ε	Reference[a]
H	402	3000	385	2100	290	~4000	58, 263 (86, 272)
o-CH$_3$	395	2430	380	1800			58 (57, 91, 272, 274)
m-CH$_3$	414	2700	395	2100			58 (57, 91, 272, 274)
p-CH$_3$	407	3550	390	2300			58 (57, 86, 91, 272, 274)
m-F	407	3100	390	2500			187
p-F	390	2920					57
o-Cl	393	1800	376	1620			275
m-Cl	417	2220	400	1950			275
p-Cl	418	5100	400	4100			275
	417	5000	400	4000			276
o-Br	402	2450	383	1950			275
m-Br	426	2000	407	1500			275
p-Br	430	5500	417	4100			275
	428	5400	412	4100			276
p-I	420	2800					277
o-OCH$_3$	383	2340			280	8770	47
m-OCH$_3$	430	2580			270	4760	47
p-OCH$_3$	420	6360	400		290	13310	47
	417	7030					57
p-NH$_2$	444	6100					197
p-N(CH$_3$)$_2$	490						76
m-OH	428	3000	408	2400			187
p-OH	410–415	4400	399				155, 278 (92, 269)
o-O$^-$	300						92
m-O$^-$	447	ca 2600	ca 425	ca 2400			76, 187
p-O$^-$	430	6100	ca 404	ca 5000	316	40000	155 (32, 92, 278)
	430	6900			310	18000	279
o-C$_6$H$_5$	500				360		272
p-C$_6$H$_5$	545				365		272
	560				350		280
p-CH=CHCO$_2^-$	595	18000	545	15000			281

[a]References in parentheses report similar data to those listed in the Table.

The main absorption maxima and extinction coefficients for a series of monosubstituted phenoxyl radicals in aqueous solutions are summarized in Table 4. Most phenoxyl radicals exhibit a relatively intense absorption (ε ca 2000–6000 M^{-1} cm^{-1}) in the region of 380–450 nm. An additional very intense peak around 300 nm was recorded for some of the phenoxyl radicals. The exceptions are the o-benzosemiquinone anions, which have little absorption above the 300 nm peak, and phenoxyl radicals derived from polycyclic or highly conjugated compounds, which absorb at higher wavelengths (see Tables 5 and 6).

Substituents exert pronounced effects on the absorption spectra. In general, *meta*- and *para*-substituted phenoxyl radicals exhibit absorption maxima at higher wavelengths than the *ortho* analogues. On the other hand, the *para*-substituted phenoxyl radicals have extinction coefficients considerably higher than those of either the *ortho* or the *meta* analogues. These general trends hold for the methyl-, bromo-, methoxy- and hydroxyphenoxyl radicals, and for the semiquinones.

TABLE 5. Absorption maxima (λ_{max}, in nm) and molar absorption coefficients (ε, in M^{-1} cm^{-1}) of selected phenoxyl radicals produced by oxidation of phenols

Phenol	λ_{max}	ε	Reference
4-Aminophenol (protonated on O)	437	7300	197
(neutral)	444	6100	197
(deprotonated at NH$_2$)	474	7500	197
N-Acetyl-4-aminophenol (neutral)	445, 330	6000, 17000	282
(deprotonated at NH)	520, 370	6500, 18000	282
4-Aminoresorcinol	430	3800	283
2,6-Dimethylphenol	375, 390	2950, 3150	284
3,4-Dimethylphenol	400, 415	2900, 3300	284
2,4,5-Trichlorophenol	430	3600	285
Pentachlorophenol	440	2400	286
2,4-Dibromophenol	420	3700	286
Pentabromophenol	470, 330	3200, 3900	286
Eugenol	390, 300		287
Isoeugenol	530, 350		287
4-t-Butylcatechol (acid form)	390, 290	1850, 7700	102
4-t-Butylcatechol (anion form)	313, 350sh	12200, 2400	102
1,2,4-Trihydroxybenzene (neutral)	400	4500	288
(monoanion)	430	5200	288
(dianion)	425		288
1,3,5-Trihydroxybenzene (neutral)	495		289
(monoanion)	550		289
(dianion)	640		289
2,5-Dihydroxybenzoate ion	432, 408	7400, 6600	290
2,4-Dihydroxybenzoate ion	460	3300	290
Tetrafluorohydroquinone	430, 404	6850, 5000	291
Gallic acid	337	3500	76
	340, 400		292
2,4-Dihydroxyacetophenone	460		293
2,4,6-Trihydroxyacetophenone	515		294
Tyrosine	407	3200	58
	395	2400	(56, 95, 122, 295–299)
	260, 405	6000, 2600	300
3-Iodotyrosine	275, 405	8400, 3300	301
3,5-Diiodotyrosine	350, 410	4300, 1700	302
3,4-Dihydroxyphenylalanine	310	5800	76
5-Hydroxydopamine	315	4800	76
6-Hydroxydopamine	440, 420, 345	3500, 2700, 7600	76
2,5-Dihydroxyphenylacetic acid	430, 405, 310	6000, 5200, 11000	76
Norepinephrine	310	5400	76
3-Hydroxykynurenine	460, 690	1300, 700	303
7-Hydroxycoumarin	575, 525	2700, 1750	76
6,7-Dihydroxycoumarin	530	2400	76
α-Tocopherol	425		246, 304

(*continued overleaf*)

TABLE 5. (continued)

Phenol	λ_{max}	ε	Reference
Trolox C	435, 320	6700, 6500	305–308
Catechin	315	5800	76
	315	10000	309
Quercetin	520	17000	76
	525	18000	309
Ellagic acid	525	6400	76
Kaempferol	545	25500	309
Silybin	370, 395		293, 294
5-Hydroxyindole	470, 390	4000, 3800	76
5-Hydroxytryptophan	480, 390, ca 270	3600, 3700, ca 12000	76
Isobarbituric acid	360	4400	76
4,4'-Biphenol	385, 620	2200, 900	310
4,4'-Biphenolate ion	445, 730	2900, 1700	310
4,4'-Thiodiphenol	700–760, 500, 335	ca 3000, 7100, 15600	311
1-Naphthol	400, 530		280
2-Naphthol	360, 490		280
2,3-Dihydroxynaphthalene	366	8500	309
1,4-Dihydroxy-9,10-anthraquinone	540, 390	7000, 3000	312
1,5-Dihydroxy-9,10-anthraquinone	472	10500	312
1,8-Dihydroxy-9,10-anthraquinone	482	8300	312
Quinalizarin	720	17000	76

The absorption bands of phenoxyl and semiquinone radicals are ascribed to $\pi-\pi^*$ transitions and some of the splittings are thought to be due to C—O stretching vibrations in the excited state[32,318]. Photoexcitation of certain phenoxyl radicals at lower wavelengths (<300 nm) often yields a reactive quartet state which can abstract a hydrogen atom from aliphatic solvents[319]. Photolysis at higher wavelengths does not yield such reactive species. Certain semiquinone radicals were also found to be stable toward photolysis at $\lambda > 330$ nm[320].

Absorption spectra of phenoxyl radicals derived from biologically important molecules were recorded in numerous cases. The tyrosyl radical was studied by many investigators[56,58,86,95,122,295–299,321] and its spectrum was used to detect tyrosine oxidation in a protein[322] and to follow intramolecular electron transfer from tyrosine to the tryptophan radical in dipeptides and polypeptides[122–125]. A number of catecholamines, such as adrenaline and dopa, were also studied by kinetic spectrophotometric pulse radiolysis[76,99,101,143,147,323]. The absorption spectra of most of these substituted o-semiquinone anion radicals[76,99,101,102,143,147,323] were similar to those of the unsubstituted radical. The phenoxyl radicals derived from oxidation of Vitamin E[324] and a simpler analogue, Trolox C[76] were found in pulse radiolysis experiments to have an absorption maximum at 430 nm, similar to that of the p-methoxyphenoxyl radical. Absorption spectra of semiquinone radicals derived from reduction of riboflavin[325,326] and FAD[327,328] were also reported. The spectra of many of the compounds mentioned above were used in pulse radiolysis electron transfer experiments aimed at determinations of reduction potentials[76,326,328].

The ultraviolet and visible spectra of persistent phenoxyl radicals have been reviewed[19]. The spectrum of tri-t-butylphenoxyl was also recorded in the infrared region[329].

16. Transient phenoxyl radicals: Formation and properties in aqueous solutions

TABLE 6. Absorption maxima (λ_{max}, in nm) and molar absorption coefficients (ε, in M^{-1} cm^{-1}) of selected semiquinone radicals produced by reduction of quinones

Quinone[a]	λ_{max}	ε	λ_{max}	ε	Reference
2-Methyl-1,4-benzoquinone	430	6200	405	4500	155
	431	6800	315	16000	279
2-t-Butyl-1,4-benzoquinone	432	6500			313
	431	7000	319	16000	279
2,3-Dimethyl-1,4-benzoquinone	430	6700	415	5100	155
	431	6800	319	13700	279
2,5-Dimethyl-1,4-benzoquinone	435	6800	415	5000	155
	440	6800			278
	431	7600	319	15500	279
2,6-Dimethyl-1,4-benzoquinone	430	6100	405	4900	155
	431	7000	319	14400	279
2,3,5-Trimethyl-1,4-benzoquinone	435	6700	410	4300	155
	431	6700	319	12900	279
Tetramethyl-1,4-benzoquinone (duroquinone)	440	7600	420	4700	155
	445	7100			278
2,6-Dimethoxy-1,4-benzoquinone	431	5200	329	13300	279
2,3,5-Trimethoxy-1,4-benzoquinone	435	6000	315	12200	279
1,4-Naphthoquinone	390	12500	370	7100	155
	390	13000			278
2-Methyl-1,4-naphthoquinone	390	12500	370	9500	155
	395	12000			278
2,3-Dimethyl-1,4-naphthoquinone	400	11000	380	7300	155
Ubiquinone	445	8600	425	5300	155
Vitamin K	400	10200	380	9900	155
2-Hydroxy-1,4-naphthoquinone	390	6300			278
1,2-Naphthoquinone	265	4000			278
9,10-Anthraquinone (AQ$^{\bullet-}$)	480	7300	395	7800	278
(AQ$^{\bullet-}$)	490	5200	385	6700	314
(AQH$^{\bullet}$)			390	8900	314
9,10-Anthraquinone-1-sulfonate	500	8000	400	8000	278
9,10-Anthraquinone-2-sulfonate	505	7600	405	8000	315 (316)
1-Hydroxy-AQ (monoanion)	445	7900	390	8100	317
(neutral)	520	1100	390	10500	317
2-Hydroxy-AQ (dianion)	460	5100	405	6500	314
(monoanion)	450	4400	400	6600	314
(neutral)			380	6600	314
2,6-Dihydroxy-AQ (dianion)	500	5700	420	14700	314
(monoanion)	450	4500	410	9000	314
(neutral)			390	11000	314
1,4-Dihydroxy-AQ (monoanion)	475	13700	388	5800	314
(neutral)			410	11600	314
1,5-Dihydroxy-AQ (monoanion)	440	13000	390	8000	314
(neutral)			410	12400	314
1,8-Dihydroxy-AQ (monoanion)	450	14700	380	8500	314
(neutral)			400	15800	314

[a] AQ = 9,10-Anthraquinone.

Resonance Raman spectra have been recorded for transient phenoxyl radicals by pulse radiolysis. From the spectra recorded for various substituted and deuteriated radicals, it was possible to analyze the peaks in terms of the C—O and C—C stretching modes and C—C—C bending modes and to draw conclusions about the structures of the radicals. For example, the frequency assigned to the C—O bond in benzosemiquinone was found to be intermediate between those for the C=O in benzoquinone and the C—O in hydroquinone, which led to the conclusion that the bond order in semiquinone is about 1.5. Raman spectroscopy also yielded information on the excited states of the radicals by examining which frequencies are resonance-enhanced[320,330–332]. Furthermore, time-resolved experiments with Raman spectroscopy allowed kinetic measurements on a specific intermediate unmasked by changes in other species[288].

Comparison of the Raman bands of p-benzosemiquinone anion[330,332] with those of the monoprotonated species (p-hydroxyphenoxyl)[333] indicated certain features in the spectrum of the p-hydroxyphenoxyl that are close to those of the unsubstituted phenoxyl, but the general pattern suggested a stronger similarity with the semiquinone. The diprotonated species (hydroquinone radical cation)[333] was more similar in structure to the semiquinone anion. From the Raman spectrum of the p-aminophenoxyl radical[196,331,334] it was concluded that this radical also is very similar in structure to the p-benzosemiquinone anion rather than to a substituted phenoxyl radical. This was confirmed by ESR parameters and MO considerations[197,198]. Furthermore, the Raman spectrum recorded by pulse radiolysis at pH < 2 indicated that the radical is protonated on the oxygen (and not on the nitrogen) to form the p-aminophenol radical cation[335]. The pK_a for this process was determined to be 2.2. In strongly alkaline solutions, both the absorption spectrum and the Raman spectrum change considerably[197]. From the effect of pH it was concluded that the p-aminophenoxyl radical deprotonates with $pK_a = 14.5$.

A study of the Raman spectra of other p-substituted (CH$_3$, F, Cl, Br, OCH$_3$) phenoxyl radicals indicated a progression from the phenoxyl to the semiquinone character as the substituent becomes a stronger electron donor[336].

The m-benzosemiquinone radical anion exhibits a Raman spectrum[187] that has a CO stretching frequency similar to that of phenoxyl radical[337,338] and another band at a much lower frequency that is ascribed to a second CO stretching. This suggested that the two CO groups are not equivalent and that the m-benzosemiquinone anion is more similar to a 3-hydroxyphenoxyl. ESR spectra of the m-semiquinone anion, however, indicate complete symmetry, probably due to rapid spin interchange.

Comparison of p-benzosemiquinone[330,332] with the tetrafluoro[291] derivative led to the conclusion that fluorination induces an increase in the quinonoidic character of the radical.

While transient phenoxyl radicals for the above resonance Raman measurements were produced by radiolysis, other investigators used photolysis to produce phenoxyl radicals for Raman studies. Such studies were carried out with several tocopherols in various organic solvents and in micellar solutions and phospholipid bilayers[339]. From the solvent effect on the Raman frequencies and the spectra observed in sodium dodecyl sulfate micelles it was concluded that the chromanoxyl group of tocopherol was located in a highly polar environment. However, the spectra in neutral and positively charged micelles and in the membranes suggested that the chromanoxyl group is in an environment of intermediate polarity.

C. Acid–Base Equilibria of Phenoxyl Radicals

The acid–base equilibria of phenoxyl radicals may involve (a) protonation on the phenoxyl oxygen (equation 32), which is important only in strongly acidic solutions,

$$\text{PhOH}^{\bullet+} \rightleftharpoons \text{PhO}^{\bullet} + \text{H}^+ \qquad (32)$$

and (b) dissociation of substituents on the ring of the phenoxyl radical, such as OH and CO_2H (equation 33), which take place under mildly acidic or alkaline conditions.

$$HOArO^{\bullet} \rightleftharpoons {}^-OArO^{\bullet} + H^+ \quad (33)$$

The method most commonly applied to determine pK_a values of radicals is based on the difference in the absorption spectra of the acid and basic forms of the radicals. By monitoring the absorbance at a certain wavelength, where the difference between the two species is large, as a function of pH, one obtains the typical sigmoidal curve with an inflection point at $pH = pK_a$. It is necessary, however, to ascertain that the spectral change is due only to the acid–base equilibrium and that the yield of the radicals in the pulse radiolysis does not change with pH. This technique was applied to the determination of most of the pK_a values for semiquinones. Other pulse radiolytic methods, involving changes in conductance or in reaction rates, were rarely used with phenoxyl radicals.

The ESR technique can be applied to the determination of accurate pK_a values if the acid and basic forms of the radical undergo rapid exchange so that the ESR parameters at any pH are the weighted average of those of the two forms. This method is not dependent on the overall yield of the radicals and is not sensitive to chemical complications as is the optical method. The ESR method has been applied to measure the pK_a for protonation of phenoxyl radicals in strongly acidic solutions[40–42]. The main results of these measurements are summarized in Table 7.

It is clear from Table 7 that most of the phenoxyl radicals protonate on the oxygen to form phenol radical cations with pK_a values about -1 to -2, i.e. $\geqslant 10$ units lower than the pK_a values of the parent phenols. Because of the strong acidities involved and the choice of the appropriate acidity functions, the pK_a values are not as accurate as those measured under milder conditions (pH 2–12). There is no simple correlation between the pK_a values and the σ substituent constants. This is not surprising, since the σ constant reflects the electron distribution in the molecule while the pK_a value depends on the electron distribution in the radical, which is different from that in the parent molecule. There appears to be some correlation between the effect of substituents on the pK_a values and their effect on the spin density distribution in the radical, but not all the substituents

TABLE 7. The pK_a values for the protonation of substituted phenoxyl radicals[a]

Substituent	pK_a		
	p	m	o
H	−2.00	−2.00	−2.00
$COCH_3$	−1.86	−1.81	−2.40
NO_2	−1.79	−1.78	−1.98
CH_3	−1.60	−1.85	−1.99
F	−1.59	−1.95	−1.69
CF_3	−1.46	−1.53	−1.56
OCH_3	−1.41	−2.21	−1.63
OH	−1.30	−2.22	−1.62
Cl	−1.30	−1.75	−1.27
NH_2	+2.2[b]		

[a] From References 40–42, except where noted.
[b] From Reference 335, determined by optical measurements. The pK_a for deprotonation of the NH_2 group was determined to be 14.5[197]. The pK_a for deprotonation of the NH group in the N-acetyl-p-aminophenoxyl radical was found to be lower, 11.1[282,340].

TABLE 8. Dissociation constants of semiquinone radicals

Radical	pK_a	Reference
1,2-Benzosemiquinone	5.0	94
4-t-Butyl-1,2-benzosemiquinone	5.2	102
1,3-Benzosemiquinone	7.1	94
	6.4	186
4-Carboxy-1,3-benzosemiquinone	7.9	290
5-Hydroxy-1,3-benzosemiquinone	6.5; 8.6	289
4-Amino-1,3-benzosemiquinone	3.4, 6.4	283
1,4-Benzosemiquinone	4.0	92
	4.1	278, 347
2-Carboxy-1,4-benzosemiquinone	6.5	290
2-Methyl-1,4-benzosemiquinone	4.5	155
2-t-Butyl-1,4-benzosemiquinone	4.3	313
2,3-Dimethyl-1,4-benzosemiquinone	4.7	155
2,5-Dimethyl-1,4-benzosemiquinone	4.6	278, 348
2,6-Dimethyl-1,4-benzosemiquinone	4.8	155
2,3,5-Trimethyl-1,4-benzosemiquinone	5.0	155
Durosemiquinone	5.1	278, 348
	4.9	155
2-Hydroxy-1,4-benzosemiquinone	4.8; 8.9	288
1,2-Naphthosemiquinone	4.8	278
1,4-Naphthosemiquinone	4.1	278, 348
2-Methyl-1,4-naphthosemiquinone	4.4	155
	4.5	278
	4.7	348
2,3-Dimethyl-1,4-naphthosemiquinone	4.3	155
Vitamin K semiquinone	5.5	155
2-Hydroxy-1,4-naphthosemiquinone	4.7	278
Ubisemiquinone	5.9	155
4,4'-Biphenol semiquinone	7.5	310
9,10-Anthrasemiquinone	5.3	278
	4.4	314
9,10-Anthrasemiquinone-1-sulfonate	5.4	316
9,10-Anthrasemiquinone-2-sulfonate	3.2	316
9,10-Anthrasemiquinone-2,6-disulfonate	3.2	348
1-Hydroxy-9,10-anthrasemiquinone	4.6; >14	317
2-Hydroxy-9,10-anthrasemiquinone	4.7; 10.7	314
2,6-Dihydroxy-9,10-anthrasemiquinone	5.4; 8.7	314
1,4-Dihydroxy-9,10-anthrasemiquinone	3.3; >14	314
1,5-Dihydroxy-9,10-anthrasemiquinone	3.7; >14	314
1,8-Dihydroxy-9,10-anthrasemiquinone	4.0; >14	314
Lumiflavin radical	8.4	349
Riboflavin radical	8.3	325
FMN radical	8.5	349
FAD radical	8.8	349

give a good correlation. Further discussion on the determination of these pK_a values is found in the original papers[40-42].

Optical absorption spectra have not been utilized to measure these pK_a values, although the radical cations of several phenols have been observed by pulse radiolysis[341-343] and by laser flash photolysis[341,344] in organic solvents. The spectra were found to be different than those of phenoxyl radicals, and in the presence of water they underwent very rapid deprotonation to form the corresponding phenoxyl radicals.

The dissociation constants for semiquinone radicals, measured by spectrophotometric pulse radiolysis, were reviewed before[345,346] and are summarized in Table 8. Protonation of semiquinone anion radicals takes place in most cases with pK_a 4–5 with several higher and lower values. The simple benzosemiquinones show a strong effect of the relative positions of the oxygens on the pK_a, i.e. 4 for 1,4-benzosemiquinone, 5 for the 1,2-isomer and ca 7 for the 1,3-isomer. The relatively high pK_a for m-benzosemiquinone is clearly related to the lower spin density on the oxygens of this radical.

This effect is also manifested in electron transfer reactions of the m-semiquinone (see below). The pK_a of ca 7 is somewhat lower than that for the parent resorcinol (9.8) owing to electron withdrawing from OH by the radical. The o- and p-benzosemiquinones have higher spin densities on the oxygens than the *meta* isomer and therefore exhibit lower pK_a values. The o- and p-benzosemiquinones are expected to have similar charge densities on the oxygens and therefore the slightly higher pK_a observed for the *ortho* isomer must be the result of an intramolecular hydrogen bridge between the two oxygens. The same effect is also exhibited by the 1,2- and 1,4-naphthosemiquinones, which have pK_a values of 4.8 and 4.1, respectively. A strong effect of internal hydrogen bonding on the pK_a is also evident in a series of dihydroxyanthrasemiquinones[314,350]. In all semiquinones, substitution by the electron-donating methyl groups increases the pK_a values of the radicals.

The pK_a of a carboxyl group on the phenoxyl radical o-O$^{\bullet}$C$_6$H$_4$CO$_2$H was estimated[97] to be ca 3, i.e. similar to that of the carboxyl group in the parent phenol o-HOC$_6$H$_4$CO$_2$H. It is expected that the pK_a for an amino group on phenoxyl radicals should be considerably lower than that on phenol (pK_a 4–5). This appears to be the case, since ESR spectra of the aminophenoxyl radicals indicated the absence of NH$_3^+$ even in strongly acidic solutions[40].

By comparison with phenoxyl radicals, the isoelectronic anilinyl radicals protonate much more readily; pK_a values in the range 4–7 have been reported for the equilibria between various anilinyl radicals and their corresponding aniline radical cations[108,109,351].

D. Reactions of Phenoxyl Radicals

Most phenoxyl radicals are short-lived intermediates, which react with each other and with other radicals relatively rapidly. Steric hindrance may lower the rates of such reactions to an extent that certain phenoxyl radicals are completely persistent. Some phenoxyl-type radicals are stabilized by thermodynamic factors and may be long-lived or completely stable under certain conditions, such as the semiquinone radicals in anaerobic alkaline solutions.

Phenoxyl radicals react with each other mainly by coupling (or dimerization). Second-order decay of transient phenoxyl radicals takes place with rate constants of the order of 10^9 M^{-1} s^{-1} [272,296,338,352,353] and leads to formation of dimeric products. Various dimers are formed by combination at the various radical sites. Since the unpaired spin is delocalized on the oxygen and on the *ortho* and *para* carbons, dimers result from combination of O with C and of C with C (equation 34). Dimers containing O—O bonds are less stable and generally were not detected.

$$C_6H_5O^{\bullet} + C_6H_5O^{\bullet} \longrightarrow HOC_6H_4C_6H_4OH\ (80\%) + C_6H_5OC_6H_4OH\ (10\%) \quad (34)$$

For example, the products of decay of phenoxyl radical in aqueous solutions are 80% C—C dimerization products, 10% C—O dimerization products and 10% were not identified, possibly including some peroxide products[354]. The major group of products includes 2,2'-, 2,4'- and 4,4'-dihydroxybiphenyl with ratios of 0.7 : 1.7 : 1.0. The second group includes both 2- and 4-phenoxyphenol. The relative abundance of the various products does not correspond to the relative spin populations at the oxygen and the various carbon atoms and suggests that additional factors influence the product distribution[354]. An explanation for these findings may be provided by a suggested dimerization mechanism, which involves a diketo intermediate dimer in equilibrium with the starting radicals[355]. The same mechanism has been invoked to explain the variations in activation energies and pre-exponential factors determined from the temperature dependence of the rate of decay of various substituted phenoxyl radicals[353].

Oxidative coupling of phenols is an important process in biological systems. For example, lignin is formed by coupling of the phenoxyl radicals derived from coniferyl alcohol. The first step, i.e. the dimerization, was shown to take place via radical–radical combination[356,357], although addition of the phenoxyl radical to another phenol molecule has been suggested to occur under certain conditions. Another example is the oxidative polymerization of 3,4-dihydroxyphenylalanine (dopa) to form melanin[358,359]. In this case the mechanism was suggested to involve oxidation of this phenol to an *ortho*-quinone, which undergoes cyclization and further oxidation before forming the polymeric materials.

Phenoxyl radicals react rapidly with $O_2^{\bullet-}$ radicals (Table 9)[360–362]. The reaction has been suggested to proceed via two parallel mechanisms[361]: addition of the $O_2^{\bullet-}$ to the *ortho* or *para* positions of phenoxyl, followed by rearrangement and possibly ring opening, and electron transfer from $O_2^{\bullet-}$ to phenoxyl to form O_2 and phenolate ion. The contribution of the latter reaction depends on the reduction potential of the phenoxyl

TABLE 9. Rate constants for selected reactions of phenoxyl radicals

Phenoxyl radical	Other reactant	Conditions	k (M^{-1} s^{-1})	Reference
PhO$^{\bullet}$	PhO$^{\bullet}$	pH 11	1.3×10^9	338
		pH 1	1.2×10^9	352
4-MeC$_6$H$_4$O$^{\bullet}$	4-MeC$_6$H$_4$O$^{\bullet}$	N$_3^-$, KOH	1.0×10^9	353
(other subst.)	(other subst.)		(0.5 to 1.5×10^9)	
PhO$^{\bullet}$	O$_2^{\bullet-}$	N$_3^-$, HCO$_2^-$, O$_2$	2×10^9	361
substituted PhO$^{\bullet}$	O$_2^{\bullet-}$		0.2 to 3.0×10^9	362
TyrO$^{\bullet}$ (Tyrosine)	O$_2^{\bullet-}$		1.5×10^9	360
related radicals	O$_2^{\bullet-}$		$1-2 \times 10^9$	362
PhO$^{\bullet}$	O$_2$		$<10^5$	364
(subst. PhO$^{\bullet}$)				
2-$^-$O-C$_6$H$_4$O$^{\bullet}$	O$_2$		$10^6 - 10^7$	364
(various subst.)				
PhO$^{\bullet}$	4-BrC$_6$H$_4$OH	pH 11.5	2.0×10^8	263
PhO$^{\bullet}$	ascorbate	pH 11	6.9×10^8	365
2-FC$_6$H$_4$O$^{\bullet}$	ascorbate	pH 11	9.5×10^8	365
3-FC$_6$H$_4$O$^{\bullet}$	ascorbate	pH 11	9.7×10^8	365
4-FC$_6$H$_4$O$^{\bullet}$	ascorbate	pH 11	4.6×10^8	365
2-ClC$_6$H$_4$O$^{\bullet}$	ascorbate	pH 11	1.1×10^9	365
3-ClC$_6$H$_4$O$^{\bullet}$	ascorbate	pH 11	1.3×10^9	365
4-ClC$_6$H$_4$O$^{\bullet}$	ascorbate	pH 11	7.3×10^8	365
2-BrC$_6$H$_4$O$^{\bullet}$	ascorbate	pH 11	7.7×10^8	365
3-BrC$_6$H$_4$O$^{\bullet}$	ascorbate	pH 11	8.9×10^8	365
4-BrC$_6$H$_4$O$^{\bullet}$	ascorbate	pH 11	8.3×10^8	365
4-IC$_6$H$_4$O$^{\bullet}$	ascorbate	pH 11	1.1×10^9	365
4-NCC$_6$H$_4$O$^{\bullet}$	ascorbate	pH 11	2.0×10^9	365
4-(CO$_2^-$)C$_6$H$_4$O$^{\bullet}$	ascorbate	pH 11	4.6×10^8	365
3-HOC$_6$H$_4$O$^{\bullet}$	ascorbate	pH 11	1.1×10^8	365
4-H$_2$NC$_6$H$_4$O$^{\bullet}$	ascorbate	pH 11	5.1×10^7	365
3,5-Cl$_2$C$_6$H$_3$O$^{\bullet}$	ascorbate	pH 11	1.6×10^9	78

TABLE 9. (continued)

Phenoxyl radical	Other reactant	Conditions	k (M^{-1} s^{-1})	Reference
2,4,5-Cl$_3$C$_6$H$_2$O$^\bullet$	ascorbate	pH 11	1.1×10^9	78
C$_6$Cl$_5$O$^\bullet$	ascorbate	pH 11	1.4×10^9	78
C$_6$F$_5$O$^\bullet$	ascorbate	pH 11	1.3×10^9	78
TyrO$^\bullet$ (Tyrosine)	Trolox C	pH 7	3.8×10^8	366
3,5-I$_2$-TyrO$^\bullet$	ascorbate	pH 7.4	3×10^9	367
PhO$^\bullet$	hydroquinone	pH 11.6	2.2×10^9	75
3-$^-$O-C$_6$H$_4$O$^\bullet$	catechol	pH 13.5	7.5×10^8	75
4-MeOC$_6$H$_4$O$^\bullet$	catechol	pH 13.5	8.6×10^8	75
PhO$^\bullet$	resorcinol	pH 13.5	1.7×10^9	75
PhO$^\bullet$	TMPD^a	pH 13.5	3.8×10^9	75
PhO$^\bullet$	Trolox C	N$_3^-$, pH 7	4.1×10^8	308
4-MeC$_6$H$_4$O$^\bullet$	Trolox C	N$_3^-$, pH 7	9.5×10^7	308
3-MeC$_6$H$_4$O$^\bullet$	Trolox C	N$_3^-$, pH 7	2.8×10^8	308
2-MeC$_6$H$_4$O$^\bullet$	Trolox C	N$_3^-$, pH 7	$<10^5$	308
TyrO$^\bullet$ (Tyrosine)	Trolox C	N$_3^-$, pH 7	3.2×10^8	308
TxO$^\bullet$ (Trolox C)	ascorbate	Br$^-$, pH 7	8.3×10^6	308
PhO$^\bullet$	ABTS^b		3.8×10^9	70
TxO$^\bullet$ (Trolox C)	glutathione anion	N$_3^-$, pH 8	1.8×10^6	368
2-ClC$_6$H$_4$O$^\bullet$	SO$_3^{2-}$	pH 11	4.7×10^7	78
3-ClC$_6$H$_4$O$^\bullet$	SO$_3^{2-}$	pH 11	1.1×10^8	78
4-ClC$_6$H$_4$O$^\bullet$	SO$_3^{2-}$	pH 11	1.0×10^7	78
4-BrC$_6$H$_4$O$^\bullet$	SO$_3^{2-}$	pH 11	1.2×10^7	78
2,3-Cl$_2$C$_6$H$_3$O$^\bullet$	SO$_3^{2-}$	pH 11	1.2×10^8	78
3,5-Cl$_2$C$_6$H$_3$O$^\bullet$	SO$_3^{2-}$	pH 11	3.4×10^8	78
2,6-Cl$_2$C$_6$H$_3$O$^\bullet$	SO$_3^{2-}$	pH 11	1.0×10^8	78
2,4,5-Cl$_3$C$_6$H$_2$O$^\bullet$	SO$_3^{2-}$	pH 11	1.4×10^8	78
C$_6$Cl$_5$O$^\bullet$	SO$_3^{2-}$	pH 11	1.5×10^8	78
C$_6$F$_5$O$^\bullet$	SO$_3^{2-}$	pH 11	4.4×10^8	78

^{a}TMPD = N,N,N',N'-tetramethyl-p-phenylenediamine.
^{b}ABTS = 2,2'-azinobis(3-ethylbenzothiazoline-6-sulfonate ion).

radical. More recently, however, it was argued that the direct electron transfer route is of minor importance and that the reaction proceeds predominantly via addition to the ring[362]. Restitution of the phenol is then suggested to occur in a subsequent step, by elimination of O$_2$ from unstable hydroperoxides. In an earlier study[363] the rate constant for reaction of the Trolox C phenoxyl radical with O$_2^{\bullet-}$ was determined to be 4.5×10^8 M^{-1} s^{-1} and the reaction was suggested to proceed mainly via electron transfer, i.e. the superoxide radical can repair vitamin E radicals. Phenoxyl radicals do not react with O$_2$[364], but semiquinone radicals may transfer an electron to O$_2$, depending on their reduction potential relative to that of O$_2$.

Phenoxyl radicals can oxidize various compounds by electron transfer. These reactions depend on the reduction potential of the phenoxyl radical and the other reactant and may appear as equilibrium reactions or may proceed predominantly in one direction. Examples of the latter group of reactions are shown in Table 9 and examples of equilibrium reactions are in Table 10.

It is seen in Table 9 that many phenoxyl radicals oxidize ascorbate (vitamin C, Asc$^-$) and Trolox C (a water-soluble analogue of vitamin E, TxOH) with rate constants of the order of 10^8 to 10^9 M^{-1} s^{-1}. The reactions take place by electron transfer. Taking into account the associated proton transfer equilibria, the reactions at pH 7 can be written as equations 35 and 36.

$$ArO^\bullet + Asc^- + H^+ \longrightarrow ArOH + Asc^\bullet \qquad (35)$$

$$ArO^\bullet + TxOH \longrightarrow ArOH + TxO^\bullet \qquad (36)$$

Such reactions make vitamins C and E better antioxidants than many other phenols. Moreover, the Trolox C radical was found to oxidize ascorbate with a rate constant close to 10^7 M^{-1} s^{-1}[308]. This leads to a synergistic antioxidant effect in the presence of both vitamins (equation 37).

$$TxO^\bullet + Asc^- + H^+ \longrightarrow TxOH + Asc^\bullet \qquad (37)$$

Rate constants for electron transfer equilibrium reactions of phenoxyl radicals (Table 10) have been determined in conjunction with measurements of reduction potentials of phenoxyl radicals. Since most phenoxyl radicals in aqueous solutions are relatively short-lived, it was not possible to determine their reduction potentials by cyclic voltammetry. Therefore, it was necessary to utilize the pulse radiolysis technique to determine the reduction potentials from equilibrium constants, using a reference compound with which a phenoxyl radical can establish equilibrium conditions. Equilibrium concentrations were determined at short times, after the electron transfer equilibrium was achieved but before any significant decay of the radicals took place. The equilibrium constants were determined either from the concentrations at equilibrium, derived from absorbance, or from the rate constants for the forward and reverse reactions, derived from the rate of approach to equilibrium. Further details were given before[24,75,76].

Most of the rate constants in Table 10 were measured at pH $\geq$ 11, where most phenols are dissociated into the phenolate ions. The reason is that electron transfer from phenolate ions takes place much more rapidly than from neutral phenols. Since it is imperative to establish equilibrium conditions before the radicals engage in subsequent decay reactions,

TABLE 10. Rate constants for selected electron transfer equilibrium reactions involving phenoxyl radicals (PhO$^\bullet$ + Ref$^-$ $\rightleftharpoons$ PhO$^-$ + Ref$^\bullet$)^a

PhO$^-$	Ref$^-$	Conditions	k_f (M^{-1} s^{-1})	k_r (M^{-1} s^{-1})	Reference
phenol	phenol	pH 11.5	1.9×10^8	1.9×10^8	263
catechol	hydroquinone	pH 13.5	2.0×10^6	8.5×10^5	75
3,4-dihydroxybenzoate ion	hydroquinone	pH 13.5	6.0×10^6	1.2×10^5	75
2,3-dihydroxybenzoate ion	hydroquinone	pH 13.5	4.2×10^5	9×10^3	75
hydroquinone	p-phenylene-diamine	pH 13.5	2.5×10^5	1.4×10^8	75
resorcinol	2,3-dihydroxy-benzoate ion	pH 13.5	4.7×10^7	5.2×10^4	75
resorcinol	3,4-dihydroxy-benzoate ion	pH 13.5	2.5×10^8	1.9×10^5	75
resorcinol	TMPD	pH 13.5	1.7×10^9	3.6×10^5	75
4-MeOC$_6$H$_4$O$^-$	TMPD	pH 13.5	2.2×10^9	3.7×10^5	75
DMAP	hydroquinone	pH 13.5	1×10^8	3×10^5	76
DMAP	catechol	pH 13.5	3×10^7	2×10^5	76
DMAP	resorcinol	pH 13.5	2×10^4	7×10^7	76

TABLE 10. (continued)

PhO⁻	Ref⁻	Conditions	k_f (M⁻¹ s⁻¹)	k_r (M⁻¹ s⁻¹)	Reference
DMAP	TMPD	pH 13.5	1×10^7	5×10^8	76
Trolox C	catechol	pH 13.5	7×10^7	3×10^5	76
resorcinol	5-hydroxytryptophan	pH 13.5	4×10^8	5.5×10^5	76
4-MeOC₆H₄O⁻	5-hydroxytryptophan	pH 13.5	9.6×10^8	5×10^5	76
DMAP	2-t-butylhydroquinone	pH 13.5	2.9×10^8	2.7×10^5	313
4-CH₃CONHC₆H₄O⁻	resorcinol	pH 12.4	1.7×10^7	2.3×10^6	282
PhO⁻	ClO₂⁻	pH 12	1.3×10^5	3.5×10^7	24
4-MeC₆H₄O⁻	ClO₂⁻	pH 12	2×10^4	2.4×10^8	24
4-FC₆H₄O⁻	ClO₂⁻	pH 12	7×10^4	5.1×10^7	24
4-ClC₆H₄O⁻	ClO₂⁻	pH 12	9×10^4	2.5×10^7	24
4-BrC₆H₄O⁻	ClO₂⁻	pH 12	1.8×10^5	1.7×10^7	24
4-IC₆H₄O⁻	ClO₂⁻	pH 12	2.8×10^5	3.5×10^7	24
4-CH₃COC₆H₄O⁻	ClO₂⁻	pH 12	1.2×10^7	1.4×10^6	24
PhO⁻	4-MeC₆H₄O⁻	pH 12	1.3×10^9	2×10^7	24
PhO⁻	tyrosine	pH 12	4.9×10^8	2.8×10^7	24
PhO⁻	4-IC₆H₄O⁻	pH 12	1.6×10^8	6×10^8	24
4-IC₆H₄O⁻	4-FC₆H₄O⁻	pH 12	1.4×10^9	9×10^7	24
4-IC₆H₄O⁻	4-CO₂⁻C₆H₄O⁻	pH 12	7×10^7	2.2×10^9	24
4-MeC₆H₄O⁻	4-MeOC₆H₄O⁻	pH 12	1.4×10^9	5.5×10^6	24
4-MeOC₆H₄OH	tryptophan	pH 7.5	1×10^6	5.5×10^6	369
4-MeOC₆H₄O⁻	tryptophan	pH 13	1×10^5	3.6×10^7	369
4-MeOC₆H₄OH	tryptamine	pH 7.5	9×10^5	4.8×10^6	369
4-MeOC₆H₄O⁻	tryptamine	pH 13	1×10^5	4.2×10^7	369
tyrosine	tryptophan	pH 13	2.4×10^6	1×10^5	369
3,4-Me₂C₆H₃O⁻	Fe(CN)₆⁴⁻	Br⁻, pH 13.5	6.5×10^5	2.7×10^4	370
3,5-Me₂C₆H₃O⁻	4-MeOC₆H₄O⁻	Br⁻, pH 13.5	8.0×10^8	1.8×10^6	370
3,5-Me₂C₆H₃O⁻	ferrocenedicarboxylate ion	pH 8	7.0×10^8	2×10^6	370
3,4,5-Me₃C₆H₂O⁻	promethazine	pH 3	1×10^6	2.1×10^7	370
PhO⁻	PhS⁻	pH 11–13.5	1.5×10^9	2.9×10^7	371
PhO⁻	4-BrC₆H₄S⁻	pH 11–13.5	1.7×10^9	7.4×10^7	371
4-MeC₆H₄O⁻	4-MeC₆H₄S⁻	pH 11–13.5	6.4×10^8	1.1×10^8	371
4-MeOC₆H₄O⁻	4-MeOC₆H₄S⁻	pH 11–13.5	1.6×10^8	4.6×10^8	371
4-H₂NC₆H₄O⁻	4-H₂NC₆H₄S⁻	pH 11–13.5	3.5×10^6	2.8×10^8	371
4-H₂NC₆H₄O⁻	4-⁻OC₆H₄S⁻	pH 11–13.5	2.9×10^7	3.2×10^6	371
4-H₂NC₆H₄O⁻	TMPD	pH 11–13.5	3.4×10^7	1.3×10^8	371

[a] TMPD = N,N,N',N'-tetramethyl-p-phenylenediamine; DMAP = p-(N,N-dimethylamino)phenol.

it was necessary to carry out the experiments at high pH to achieve rapid equilibrium. The values of the rate constants vary over six orders of magnitude, the upper range being the diffusion-controlled limit and the lower range determined by competing radical–radical reactions. Radical–radical reactions may be minimized by the use of a low dose per pulse, i.e. low radical concentration, but this is restricted by the detection limit of the pulse radiolysis setup. The combined result is that the lower limit of the measured rate constants is of the order of 10^4 M⁻¹ s⁻¹.

The electron transfer rate constants are expected to increase with the driving force of the reaction, i.e. with the difference in reduction potentials of the two radicals involved in the process, according to the Marcus theory. An approximate correlation has been demonstrated[76] for a wide group of reactions of this type.

In principle, phenoxyl radicals can react with other molecules also by a hydrogen-abstraction mechanism. The net result of such reactions may be equivalent to that of the electron transfer processes discussed above. It is likely that in aqueous solutions such reactions are much slower than the electron transfer reactions, as indicated by the fact that most reactions between phenoxyl radicals and other phenols are much slower with the neutral phenols than with the phenolate ions. It is possible that even reactions with neutral phenols in aqueous solutions involve an electron transfer mechanism. On the other hand, reactions in organic solvents may well take place by hydrogen abstraction, as discussed before[5,372–374]. These reactions take place with much lower rate constants than the electron transfer reactions; the most rapid hydrogen abstraction by a phenoxyl radical is probably five orders of magnitude slower than the diffusion-controlled limit and most of them are orders of magnitude slower than that.

E. Reduction Potentials of Phenoxyl Radicals

The reduction potentials of phenoxyl radicals have been determined by pulse radiolysis as discussed above and are summarized in Table 11. Reduction potentials estimated from cyclic voltammetric measurements of irreversible peak potentials, taking into account the decay of the phenoxyl radicals[375], are considered to be less accurate and are not included in Table 11. The primary reference for pulse radiolysis measurements of reduction potentials in the lower range[75,76] was p-benzosemiquinone, whose potential was determined from classical measurements[154]. The primary reference for most monosubstituted phenoxyl radicals[24] was ClO_2, since both the ClO_2 radical and the ClO_2^- anion are stable and the potential $E(ClO_2/ClO_2^-)$ was determined very accurately by electrochemical measurements. Other inorganic radicals were sometimes used as reference; their potentials have been discussed before[376]. In certain cases, measurements using several reference compounds have been conducted to confirm the reduction potential. The values summarized in Table 11 are given at the pH of the measurement. They generally are for the PhO$^\bullet$/PhO$^-$ pair. These values are independent of pH as long as no proton transfer accompanies the electron transfer.

The reduction potentials of p-substituted phenoxyl radicals were found to correlate very well with the Hammett σ^+ substituent constants[24]. Electron-donating substituents stabilize the phenoxyl radical, i.e. lower the reduction potential. A p-OH group has a strong stabilizing effect, but a p-O$^-$ group has a much stronger effect since the two oxygens become equivalent. Thus the reduction potentials of o- and p-benzosemiquinones are about 0.7 V lower than that of phenoxyl. The reduction potential of the m-benzosemiquinone also is considerably (0.4 V) lower than that of phenoxyl, although the effect of the m-O$^-$ group is only about half the effect of the o- and p-O$^-$ groups. The structure of the m-semiquinone has been discussed above. Trolox C is essentially a p-methoxyphenol, but the presence of the additional alkyl substituents results in a reduction potential for the Trolox C phenoxyl radical that is 0.35 V lower than that of the p-methoxyphenoxyl radical. Although this difference in potential may explain why Trolox C (and vitamin E) is a much better antioxidant than p-methoxyphenol, other structural differences also determine the antioxidant efficiency[184].

The dependence on substituent of the reduction potential and other properties of p-substituted phenoxyl radicals has been compared with the properties of the analogous phenylthiyl radicals. From this comparison it is evident that the electronic interaction

TABLE 11. Reduction potentials of phenoxyl radicals (PhO$^\bullet$ + Ref$^-$ ⇌ PhO$^-$ + Ref$^\bullet$)^a

PhOH	pH	Ref$^-$	E(Ref$^\bullet$/Ref$^-$)	E(PhO$^\bullet$/PhO$^-$)	Reference
Hydroquinone	0			1.041	154
	7			0.459	154
	11			0.057	154
	13.5			0.023	154
2-t-Butyl-1,4-hydroquinone	7			0.489	313
Catechol	13.5	hydroquinone	0.023	0.043	76
	11	hydroquinone	0.057	0.139	76
	7			0.53	76
	0			1.06	76
1,2,4-Trihydroxybenzene	13.5	catechol	0.057	−0.110	76
Methoxyhydroquinone	13.5	catechol	0.057	−0.085	76
Ethyl gallate	13.5	hydroquinone	0.023	−0.054	76
Durohydroquinone	13.5	3,4-dihydroxy-benzoate ion	0.119	−0.054	76
	7			0.36	154
2,5-Dihydroxyphenylacetate ion	13.5	catechol	0.057	−0.050	76
3,4-Dihydroxyphenylacetate ion	13.5	DMAP	0.174	0.021	76
Quercetin	13.5	hydroquinone	0.023	−0.037	76
Pyrogallol	13.5	hydroquinone	0.023	−0.009	76
Ascorbate	13.5	catechol	0.057	0.015	76
	7			0.30	76
3,4-Dihydroxyphenylalanine	13.5	hydroquinone	0.023	0.014	76
	13.5	DMAP	0.174	0.022	76
3-Hydroxytyramine	13.5	DMAP	0.174	0.018	76
2,5-Dihydroxybenzoate ion	13.5	catechol	0.057	0.033	76
5-Hydroxydopamine	13.5	DMAP	0.174	0.042	76
Norepinephrine	13.5	DMAP	0.174	0.044	76
Epicatechin	13.5	DMAP	0.174	0.048	76
Catechin	13.5	DMAP	0.174	0.079	76
Quinalizarin	13.5	hydroquinone	0.023	0.073	76
3,4-Dihydroxycinnamate ion	13.5	DMAP	0.174	0.084	76
2,5-Dihydroxyacetophenone	13.5	catechol	0.057	0.118	76
2,3-Dihydroxybenzoate ion	13.5	hydroquinone	0.023	0.118	76
3,4-Dihydroxybenzoate ion	13.5	hydroquinone	0.023	0.119	76
2,4,5-Trihydroxypyrimidine	13.5	TMPD	0.265	0.132	76
p-(N-Methylamino)phenol	13.5	catechol	0.057	0.146	76
		hydroquinone	0.023	0.156	76
		TMPD	0.265	0.146	76
DMAP	13.5	hydroquinone	0.023	0.174	76
	13.5	catechol	0.057	0.174	76
	13.5	2-t-butyl-hydroquinone	−0.08	0.10	313
N-Acetyl-4-aminophenol	12.4	resorcinol	0.386	0.44	282
Adrenalone	13.5	DMAP	0.174	0.18	76
Trolox C	13.5	DMAP	0.174	0.192	76
	13.5	catechol	0.057	0.185	76
	7			0.48	76
Ellagic acid	13.5	hydroquinone	0.023	0.187	76

(*continued overleaf*)

TABLE 11. (continued)

PhOH	pH	Ref⁻	$E(\text{Ref}^\bullet/\text{Ref}^-)$	$E(\text{PhO}^\bullet/\text{PhO}^-)$	Reference
5-Hydroxytryptophan	13.5	DMAP	0.174	0.208	76
5-Hydroxyindole	13.5	DMAP	0.174	0.216	76
	13.5	TMPD	0.265	0.197	76
p-Aminophenol	13.5	DMAP	0.174	0.217	76
	11–13	TMPD	0.265	0.24	371
7-Hydroxycoumarin	13.5	DMAP	0.174	0.315	76
Resorcinol	13.5	DMAP	0.174	0.392	76
	13.5	5-hydroxytryptophan	0.208	0.379	76
	7			0.81	76
Phenol	13	ClO_2^-	0.936	0.80	377
	11–12	ClO_2^-	0.936	0.79	24
	7			0.97	24
4-Methoxyphenol	11–12	4-methylphenol	0.68	0.54	24
4-Methylphenol	11–12	ClO_2^-	0.936	0.68	24
4-Fluorophenol	11–12	ClO_2^-	0.936	0.76	24
4-Chlorophenol	11–12	ClO_2^-	0.936	0.80	24
4-Bromophenol	11–12	ClO_2^-	0.936	0.82	24
4-Iodophenol	11–12	ClO_2^-	0.936	0.82	24
4-Hydroxybenzoate ion	11–12	ClO_2^-	0.936	0.90	24
4-Acetylphenol	11–12	ClO_2^-	0.936	1.00	24
4-Cyanophenol	11–12	$2I^-/I_2^-$	1.06	1.12	24
4-Nitrophenol	11–12	$2(SCN)^-/(SCN)_2^-$	1.331	1.25	24
		1-methylindole	1.23	1.22	24
3-Methylphenol	11–12	ClO_2^-	0.936	0.80	378
2-Methylphenol	11–12	ClO_2^-	0.936	0.76	378
3,5-Dimethoxyphenol	13.5	4-methoxyphenol	0.54	0.71	370
3,4-Dimethoxyphenol	13.5	$Fe(CN)_6^{4-}$	0.36	0.50	370
3,4,5-Trimethoxyphenol	3	promethazine	0.98	0.90	370
	13.5			0.55	370
2,6-Dimethoxyphenol	13.5	TMPD	0.27	0.42	370
4,4′-Thiodiphenol	7	$2I^-/I_2^-$	1.03	0.98	311
4,4′-Biphenol	11.7	$Fe(CN)_6^{4-}$	0.46	0.46	310
4-Hydroxybiphenyl	10.8	$2I^-/I_2^-$	1.03	0.93	280
1-Naphthol	13	3-methylphenol	0.73	0.59	280
3-Methylphenol	13	phenol	0.79	0.73	280
2-Naphthol	13	phenol	0.79	0.69	280
2-Pyridol	13	4-cyanophenol	1.12	1.18	280
3-Pyridol	11	4-cyanophenol	1.12	1.06	280
4-Pyridol	13	4-cyanophenol	1.12	1.24	280
	12.5	N_3^-	1.33	1.26	280
Tyrosine	11–12	phenol	0.80	0.71	24
	11.3	phenol	0.80	0.74	379
Tyrosine[b]	11	various		0.76	380
Tyrosine	7	various		0.93	381
3-Iodotyrosine	11.5	various		0.73	301
	7.4			0.82	
3,5-Diiodotyrosine	7	cysteine	0.92	0.78	302
Silybin	11.3	4-methoxyphenol	0.54	0.58	293

[a] DMAP = p-(N,N-dimethylamino)phenol; TMPD = N,N,N′,N′-tetramethylene-p-phenylenediamine.
[b] Reference 380 contains data for several substituted tyrosines.

16. Transient phenoxyl radicals: Formation and properties in aqueous solutions

TABLE 12. Reduction potentials of phenoxyl radicals and O—H bond dissociation energies of phenols

PhOH	E^0 (V vs. NHE)	pK_a	BDE (kJ mol^{-1})	Reference
Phenol	0.97	10.0	369	24
4-CH$_3$	0.87	10.3	360	24
4-OCH$_3$	0.72	10.1	346	24
4-F	0.93	9.9	366	24
4-Cl	0.94	9.4	367	24
4-Br	0.96	9.4	369	24
4-I	0.96	9.3	368	24
4-CO$_2^-$	1.04	9.4	376	24
4-COCH$_3$	1.06	8.0	378	24
4-CN	1.17	7.9	389	24
4-NO$_2$	1.23	7.1	394	24
4-O$^-$	0.46	11.4	303	24
4-OH	0.46	9.9	336	24
4-NH$_2$	0.42	10.4	316	24
4-N(CH$_3$)$_2$	0.36	10.1	310	24
4-C$_6$H$_5$	1.08	9.6	380	280
3-CH$_3$	0.91	10.0	364	280
1-Naphthol	0.73	9.3	346	280
2-Naphthol	0.84	9.6	358	280
2-Pyridol	1.45	11.6	416	280
3-Pyridol	1.16	8.7	388	280
4-Pyridol	1.49	11.1	420	280

between the sulfur atom and the aromatic ring is much less than that which occurs with the oxygen atom[371]. An analogous comparison can be made for p-substituted anilinyl radicals[382].

The reduction potential changes with pH if either the radical or the molecule undergoes protonation or deprotonation upon pH change. For example, for dihydroxy compounds, where the two OH groups have dissociation constants K_1 and K_2, and the phenoxyl radical has a dissociation constant K_r for the second OH group, the potential at any pH, E_i, is related to the potential at pH 0, E_0, according to equation 38.

$$E_i = E_0 + 0.059 \log \frac{K_1 K_2 + K_1[H^+] + [H^+]^2}{K_r + [H^+]} \quad (38)$$

Using such equations and known pK_a values, the pH dependencies of the reduction potentials of phenoxyl radicals have been calculated for a number of cases.

From the reduction potentials at pH 0 and estimated values for the free energies of solvation of phenol and phenoxyl in water, gas-phase O—H bond dissociation energies have been calculated. The values derived from such calculations are given in Table 12. They are comparable to values determined by other methods which are discussed in Chapter 3.

IV. REFERENCES

1. G. Scott, *Atmospheric Oxidation and Antioxidants*, Elsevier, Amsterdam, 1965.
2. K. U. Ingold, *Adv. Chem. Ser.*, **75**, 296 (1968).
3. L. R. Mahoney, *Angew. Chem., Int. Ed. Engl.*, **8**, 547 (1969).

4. L. Reich and S. S. Stivala, *Autoxidation of Hydrocarbons and Polyolefins*, Dekker, New York, 1969.
5. J. A. Howard, *Adv. Free-Radical Chem.*, **4**, 49 (1972).
6. E. B. Burlakova, *Russ. Chem. Rev.*, **44**, 871 (1975).
7. G. W. Burton and K. U. Ingold, *J. Am. Chem. Soc.*, **103**, 6472 (1981).
8. G. W. Burton and K. U. Ingold, *Acc. Chem. Res.*, **19**, 194 (1986).
9. S. V. Jovanovic, Y. Hara, S. Steenken and M. G. Simic, *J. Am. Chem. Soc.*, **117**, 9881 (1995).
10. S. V. Jovanovic, C. W. Boone, S. Steenken, M. Trinoga and R. B. Kaskey, *J. Am. Chem. Soc.*, **123**, 3064 (2001).
11. H. Erdtman, *Ind. Eng. Chem.*, **49**, 1386 (1957).
12. K. Freudenberg, *Science*, **148**, 595 (1965).
13. V. L. Singleton, *Am. J. Enol. Vitic.*, **38**, 69 (1987).
14. J. A. Stubbe and W. A. van der Donk, *Chem. Rev.*, **98**, 705 (1998).
15. M. L. Gilchrist, J. A. Ball, D. W. Randall and R. D. Britt, *Proc. Natl. Acad. Sci. USA*, **92**, 9545 (1995).
16. C. Tommos, X.-S. Tang, K. Warncke, C. W. Hoganson, S. Styring, J. McCracken, B. A. Diner and G. T. Babcock, *J. Am. Chem. Soc.*, **117**, 10325 (1995).
17. C. W. Hoganson, M. Sahlin, B.-M. Sjöberg and G. T. Babcock, *J. Am. Chem. Soc.*, **118**, 4672 (1996).
18. H. Musso, *Angew. Chem., Int. Ed. Engl.*, **2**, 723 (1963).
19. I. V. Khudyakov and V. A. Kuzmin, *Russ. Chem. Rev.*, **44**, 801 (1975).
20. L. S. Vartanyan, *Russ. Chem. Rev.*, **44**, 859 (1975).
21. D. M. Golden and S. W. Benson, *Chem. Rev.*, **69**, 125 (1969).
22. F. R. Cruickshank and S. W. Benson, *J. Am. Chem. Soc.*, **91**, 1289 (1969).
23. S. W. Benson, *J. Chem. Educ.*, **42**, 502 (1965).
24. J. Lind, X. Shen, T. E. Eriksen and G. Merenyi, *J. Am. Chem. Soc.*, **112**, 479 (1990).
25. J. Lind, T. Reitberger, T. E. Eriksen and G. Merényi, *J. Phys. Chem.*, **97**, 11278 (1993).
26. P. K. Das, M. V. Encinas, S. Steenken and J. C. Scaiano, *J. Am. Chem. Soc.*, **103**, 4162 (1981).
27. E. Mueller, K. Ley, K. Scheffler and R. Mayer, *Chem. Ber.*, **91**, 2682 (1958).
28. J. K. Becconsall, S. Clogh and G. Scott, *Trans. Faraday Soc.*, **56**, 459 (1960).
29. K. Scheffler and H. B. Stegmann, *Z. Phys. Chem.*, **44**, 353 (1965).
30. H. B. Stegmann and K. Scheffler, *Z. Naturforsch., B*, **19**, 537 (1964).
31. K. Mukai, H. Nishinguchi, K. Ishizu, Y. Deguchi and H. Takaki, *Bull. Chem. Soc. Jpn.*, **40**, 2731 (1967).
32. S. Fukuzumi, Y. Ono and T. Keii, *Bull. Chem. Soc. Jpn.*, **46**, 3353 (1973).
33. W. G. B. Huysmans and W. A. Waters, *J. Chem. Soc., B*, 1047 (1966).
34. W. G. B. Huysmans and W. A. Waters, *J. Chem. Soc., B*, 1163 (1967).
35. E. McNelis, *J. Org. Chem.*, **31**, 1255 (1966).
36. T. J. Stone and W. A. Waters, *Proc. Chem. Soc.*, 253 (1962).
37. T. J. Stone and W. A. Waters, *J. Chem. Soc.*, 4302 (1964).
38. W. T. Dixon, M. Moghimi and D. Murphy, *J. Chem. Soc., Faraday Trans. 2*, **70**, 1713 (1974).
39. W. T. Dixon and D. Murphy, *J. Chem. Soc., Perkin Trans. 2*, 850 (1975).
40. W. T. Dixon and D. Murphy, *J. Chem. Soc., Faraday Trans. 2*, **72**, 1221 (1976).
41. W. T. Dixon and D. Murphy, *J. Chem. Soc., Faraday Trans. 2*, **74**, 432 (1978).
42. D. M. Holton and D. Murphy, *J. Chem. Soc., Faraday Trans. 2*, **75**, 1637 (1979).
43. W. T. Dixon, P. M. Kok and D. Murphy, *J. Chem. Soc., Faraday Trans. 2*, **74**, 1528 (1978).
44. W. T. Dixon and D. Murphy, *J. Chem. Soc., Faraday Trans. 2*, **73**, 1475 (1977).
45. W. T. Dixon, P. T. Kok and D. Murphy, *J. Chem. Soc., Faraday Trans. 2*, **73**, 709 (1977).
46. K.-D. Asmus, M. Bonifacic, P. Toffel, P. O'Neill, D. Schulte-Frohlinde and S. Steenken, *J. Chem. Soc., Faraday Trans. 1*, **74**, 1820 (1978).
47. P. O'Neill and S. Steenken, *Ber. Bunsenges. Phys. Chem.*, **81**, 550 (1977).
48. S. Steenken, unpublished results.
49. G. A. Abakumov and V. A. Muraev, *Dokl. Chem.*, **217**, 566 (1974).
50. V. A. Muraev, G. A. Abakumov and G. A. Razuvaev, *Dokl. Chem.*, **217**, 555 (1974).
51. J. D. Rush, J. E. Cyr, Z. W. Zhao and B. H. J. Bielski, *Free Radical Res.*, **22**, 349 (1995).
52. K. Hasegawa and P. Neta, *J. Phys. Chem.*, **82**, 854 (1978).

53. K. Kemsley, J. S. Moore, G. O. Phillips and A. Sosnowski, *Acta Vitaminol. Enzymol.*, **28**, 263 (1974).
54. R. L. Willson, *Biochem. Soc. Trans.*, **2**, 1082 (1974).
55. P. Neta, P. Maruthamuthu, P. M. Carton and R. W. Fessenden, *J. Phys. Chem.*, **82**, 1875 (1978).
56. E. J. Land and W. A. Pruetz, *Int. J. Radiat. Biol.*, **36**, 75 (1979).
57. Z. B. Alfassi and R. H. Schuler, *J. Phys. Chem.*, **89**, 3359 (1985).
58. K. M. Bansal and R. W. Fessenden, *Radiat. Res.*, **67**, 1 (1976).
59. P. Neta and R. H. Schuler, *J. Am. Chem. Soc.*, **97**, 912 (1975).
60. W. Tsang, in *Energetics of Organic Free Radicals* (Eds. J. A. M. Simoes, A. Greenberg and J. F. Liebman), Blackie Academic & Professional, New York, 1996, p. 22.
61. K. U. Ingold, *Spec. Publ.-Chem. Soc.*, **24**, 285 (1971).
62. J. A. Howard, in *Free Radicals* (Ed. J. K. Kochi), Vol. 2, Wiley, New York, 1973, p. 3.
63. J. A. Howard and E. Furimsky, *Can. J. Chem.*, **51**, 3738 (1973).
64. J. A. Howard and K. U. Ingold, *Can. J. Chem.*, **41**, 1744 (1963).
65. L. R. Mahoney and M. A. DaRooge, *J. Am. Chem. Soc.*, **97**, 4722 (1975).
66. J. E. Packer, R. L. Willson, D. Bahnemann and K.-D. Asmus, *J. Chem. Soc., Perkin Trans. 2*, 296 (1980).
67. J. E. Packer, T. F. Slater and R. L. Willson, *Life Sci.*, **23**, 2617 (1978).
68. D. Brault, *Environ. Health Perspect.*, **64**, 53 (1985).
69. P. Neta, R. E. Huie, S. Mosseri, L. V. Shastri, J. P. Mittal, P. Maruthamuthu and S. Steenken, *J. Phys. Chem.*, **93**, 4099 (1989).
70. P. Neta, R. E. Huie, P. Maruthamuthu and S. Steenken, *J. Phys. Chem.*, **93**, 7654 (1989).
71. Z. B. Alfassi, R. E. Huie and P. Neta, *J. Phys. Chem.*, **97**, 7253 (1993).
72. Z. B. Alfassi, R. E. Huie, M. Kumar and P. Neta, *J. Phys. Chem.*, **96**, 767 (1992).
73. S. Steenken, *J. Phys. Chem.*, **83**, 595 (1979).
74. D. M. Camaioni, H. F. Walter, J. E. Jordan and D. W. Pratt, *J. Am. Chem. Soc.*, **95**, 7978 (1973).
75. S. Steenken and P. Neta, *J. Phys. Chem.*, **83**, 1134 (1979).
76. S. Steenken and P. Neta, *J. Phys. Chem.*, **86**, 3661 (1982).
77. J. S. Moore, G. O. Phillips and A. Sosnowski, *Int. J. Radiat. Biol.*, **31**, 603 (1977).
78. P. Neta, unpublished results.
79. G. V. Buxton, C. L. Greenstock, W. P. Helman and A. B. Ross, *J. Phys. Chem. Ref. Data*, **17**, 513 (1988).
80. P. Neta, R. E. Huie and A. B. Ross, *J. Phys. Chem. Ref. Data*, **17**, 1027 (1988).
81. K. U. Ingold, *Can. J. Chem.*, **41**, 2807 (1963).
82. G. Stein and J. Weiss, *J. Chem. Soc.*, 3265 (1951).
83. L. M. Dorfman, I. A. Taub and R. E. Bühler, *J. Chem. Phys.*, **36**, 3051 (1962).
84. W. T. Dixon and R. O. C. Norman, *Proc. Chem. Soc.*, 97 (1963).
85. W. T. Dixon and R. O. C. Norman, *Nature*, **196**, 891 (1962).
86. E. J. Land and M. Ebert, *Trans. Faraday Soc.*, **63**, 1181 (1967).
87. G. E. Adams, B. D. Michael and E. J. Land, *Nature*, **211**, 293 (1966).
88. C. R. E. Jeffcoate and R. O. C. Norman, *J. Chem. Soc., B*, 48 (1968).
89. N. V. Raghavan and S. Steenken, *J. Am. Chem. Soc.*, **102**, 3495 (1980).
90. E. Mvula, M. N. Schuchmann and C. von Sonntag, *J. Chem. Soc., Perkin Trans. 2*, 264 (2001).
91. M. Roder, L. Wojnarovits, G. Foldiak, S. S. Emmi, G. Beggiato and M. D'Angelantonio, *Radiat. Phys. Chem.*, **54**, 475 (1999).
92. G. E. Adams and B. D. Michael, *Trans. Faraday Soc.*, **63**, 1171 (1967).
93. J. Holcman and K. Sehested, *Nukleonika*, **24**, 887 (1979).
94. S. Steenken and P. O'Neill, *J. Phys. Chem.*, **81**, 505 (1977).
95. J. Chrysochoos, *Radiat. Res.*, **33**, 465 (1968).
96. J. Chrysochoos, *J. Phys. Chem.*, **73**, 4188 (1969).
97. P. Neta and R. W. Fessenden, *J. Phys. Chem.*, **78**, 523 (1974).
98. S. Shiga, T. Kishimoto and E. Tomita, *J. Phys. Chem.*, **77**, 330 (1973).
99. M. Gohn, N. Getoff and E. Bjergbakke, *Int. J. Radiat. Phys. Chem.*, **8**, 533 (1976).
100. M. Gohn and N. Getoff, *J. Chem. Soc., Faraday Trans. 1*, **73**, 1207 (1977).
101. W. Bors, M. Saran and C. Michel, *Biochim. Biophys. Acta*, **582**, 537 (1979).

102. H. W. Richter, *J. Phys. Chem.*, **83**, 1123 (1979).
103. S. Steenken and P. O'Neill, *J. Phys. Chem.*, **83**, 2407 (1979).
104. S. Fujita and S. Steenken, *J. Am. Chem. Soc.*, **103**, 2540 (1981).
105. D. K. Hazra and S. Steenken, *J. Am. Chem. Soc.*, **105**, 4380 (1983).
106. A. J. S. C. Vieira and S. Steenken, *J. Phys. Chem.*, **91**, 4138 (1987).
107. H. M. Novais and S. Steenken, *J. Phys. Chem.*, **91**, 426 (1987).
108. H. Christensen, *Int. J. Radiat. Phys. Chem.*, **4**, 311 (1972).
109. L. Qin, G. N. R. Tripathi and R. H. Schuler, *Z. Naturforsch., A*, **40**, 1026 (1985).
110. G. N. R. Tripathi, *J. Am. Chem. Soc.*, **120**, 4161 (1998).
111. G. N. R. Tripathi and Q. Sun, *J. Phys. Chem., A*, **103**, 9055 (1999).
112. P. O'Neill, S. Steenken, H. v. d. Linde and D. Schulte-Frohlinde, *Radiat. Phys. Chem.*, **12**, 13 (1978).
113. X. Fang, H.-P. Schuchmann and C. von Sonntag, *J. Chem. Soc., Perkin Trans. 2*, 1391 (2000).
114. H. Lemmetyinen, J. Konijnenberg, J. Cornelisse and C. Varma, *J. Photochem.*, **30**, 315 (1985).
115. M. Cocivera, M. Tomkiewicz and A. Groen, *J. Am. Chem. Soc.*, **93**, 7102 (1971).
116. O. Volkert, W. Bors and D. Schulte-Frohlinde, *Z. Naturforsch., B*, **22**, 480 (1967).
117. K. M. Bansal, L. K. Patterson and R. H. Schuler, *J. Phys. Chem.*, **76**, 2386 (1972).
118. P. Neta, *J. Phys. Chem.*, **76**, 2399 (1972).
119. P. Neta and C. L. Greenstock, *Radiat. Res.*, **54**, 35 (1973).
120. C. L. Greenstock, I. Dunlop and P. Neta, *J. Phys. Chem.*, **77**, 1187 (1973).
121. R. Köster and K.-D. Asmus, *Z. Naturforsch., B*, **26**, 1108 (1971).
122. W. A. Pruetz and E. J. Land, *Int. J. Radiat. Biol.*, **36**, 513 (1979).
123. W. A. Pruetz, J. Butler, E. J. Land and A. J. Swallow, *Biochem. Biophys. Res. Commun.*, **96**, 408 (1980).
124. W. A. Pruetz, E. J. Land and R. W. Sloper, *J. Chem. Soc., Faraday Trans. 1*, **77**, 281 (1981).
125. J. Butler, E. J. Land, W. A. Pruetz and A. J. Swallow, *Biochim. Biophys. Acta*, **705**, 150 (1982).
126. W. A. Pruetz, F. Siebert, J. Butler, E. J. Land, A. Menez and T. Montenay-Garestier, *Biochim. Biophys. Acta*, **705**, 139 (1982).
127. W. A. Pruetz, J. Butler and E. J. Land, *Int. J. Radiat. Biol.*, **47**, 149 (1985).
128. W. A. Pruetz, J. Butler, E. J. Land and A. J. Swallow, *Free Radical Res. Commun.*, **2**, 69 (1986).
129. M. Faraggi, M. R. DeFelippis and M. H. Klapper, *J. Am. Chem. Soc.*, **111**, 5141 (1989).
130. K. Bobrowski, J. Holcman, J. Poznanski, M. Ciurak and K. L. Wierzchowski, *J. Phys. Chem.*, **96**, 10036 (1992).
131. A. K. Mishra, R. Chandrasekar, M. Faraggi and M. H. Klapper, *J. Am. Chem. Soc.*, **116**, 1414 (1994).
132. K. Bobrowski, J. Poznanski, J. Holcman and K. L. Wierzchowski, *J. Phys. Chem., B*, **103**, 10316 (1999).
133. L. Michaelis, M. P. Schubert, R. K. Reber, J. A. Kuck and S. Granick, *J. Am. Chem. Soc.*, **60**, 1678 (1938).
134. L. Michaelis, M. P. Schubert and S. Granick, *J. Am. Chem. Soc.*, **61**, 1981 (1939).
135. L. Michaelis and S. Granick, *J. Am. Chem. Soc.*, **70**, 624 (1948).
136. M. Eigen and P. Matthies, *Chem. Ber.*, **94**, 3309 (1961).
137. H. Diebler, M. Eigen and P. Z. Matthies, *Z. Elektrochem.*, **65**, 634 (1961).
138. K. Freudenberg, *Fortschr. Chem. Org. Naturstoffe*, **20**, 41 (1962).
139. K. Freudenberg and H. Geiger, *Chem. Ber.*, **96**, 1265 (1963).
140. L. M. Strigun, L. S. Vartanyan and N. M. Emanuel, *Russ. Chem. Rev.*, **37**, 421 (1968).
141. V. D. Pokhodenko, V. A. Khizhnyi and V. A. Bidzilya, *Russ. Chem. Rev.*, **37**, 435 (1968).
142. I. B. Afanasev and N. I. Polozova, *J. Org. Chem. USSR*, **14**, 947 (1978).
143. W. Bors, M. Saran, C. Michel, E. Lengfelder, C. Fuchs and R. Spoettl, *Int. J. Radiat. Biol.*, **28**, 353 (1975).
144. C. L. Greenstock and R. W. Miller, *Biochim. Biophys. Acta*, **396**, 11 (1975).
145. C. L. Greenstock and G. W. Ruddock, *Int. J. Radiat. Phys. Chem.*, **8**, 367 (1976).
146. W. Bors, C. Michel, M. Saran and E. Lengfelder, *Z. Naturforsch., C*, **33**, 891 (1978).
147. W. Bors, M. Saran and C. Michel, *J. Phys. Chem.*, **83**, 3084 (1979).
148. A. D. Nadezhdin and H. B. Dunford, *Can. J. Chem.*, **57**, 3017 (1979).
149. R. W. Miller, *Can. J. Biochem.*, **48**, 935 (1970).

16. Transient phenoxyl radicals: Formation and properties in aqueous solutions

150. R. W. Miller and F. D. H. MacDowall, *Biochim. Biophys. Acta*, **396**, 11 (1975).
151. E. J. Nanni, M. D. Stallings and D. T. Sawyer, *J. Am. Chem. Soc.*, **102**, 4481 (1980).
152. D. T. Sawyer and J. S. Valentine, *Acc. Chem. Res.*, **14**, 393 (1981).
153. P. M. Wood, *FEBS Lett.*, **44**, 22 (1974).
154. Y. A. Ilan, G. Czapski and D. Meisel, *Biochim. Biophys. Acta*, **430**, 209 (1976).
155. K. B. Patel and R. L. Willson, *J. Chem. Soc., Faraday Trans. 1*, **69**, 814 (1973).
156. D. Meisel and G. Czapski, *J. Phys. Chem.*, **79**, 1503 (1975).
157. V. V. Ershov, A. A. Volodkin and G. N. Bogdanov, *Russ. Chem. Rev.*, **32**, 75 (1963).
158. V. V. Ershov, A. A. Volodkin, A. I. Prokof'ev and S. P. Solodovnikov, *Russ. Chem. Rev.*, **42**, 740 (1973).
159. C. Walling, *Free Radicals in Solution*, Wiley, New York, 1957.
160. H. P. Misra and I. Fridovich, *J. Biol. Chem.*, **247**, 3170 (1972).
161. R. E. Heikkala and G. Cohen, *Science*, **181**, 456 (1973).
162. G. Cohen and R. E. Heikkala, *J. Biol. Chem.*, **249**, 2447 (1974).
163. W. Bors, C. Michel, M. Saran and E. Lengfelder, *Biochim. Biophys. Acta*, **540**, 162 (1978).
164. S. Marklund and G. Marklund, *Eur. J. Biochem.*, **47**, 469 (1974).
165. V. Massey, G. Palmer and D. Ballou, in *Oxidases and Related Redox Systems*, Eds. T. King, H. S. Mason and M. Morrison, University Park Press, Baltimore, 1973, p. 25.
166. M. Nishikimi, *Arch. Biochem. Biophys.*, **166**, 273 (1975).
167. V. Roginsky and T. Barsukova, *J. Chem. Soc., Perkin Trans. 2*, 1575 (2000).
168. E. Müller, K. Ley, K. Scheffler and R. Mayer, *Chem. Ber.*, **91**, 2682 (1958).
169. J. E. Wertz, D. C. Reitz and F. Dravnieks, in *Free Radicals in Biological Systems* (Eds. M. S. Blois, Jr., H. W. Brown, R. M. Lemmon, R. O. Lindblom and M. Weissbluth), Academic Press, New York, 1961, p. 183.
170. L. H. Piette, I. Yamazaki and H. S. Mason, in *Free Radicals in Biological Systems* (Eds. M. S. Blois, Jr., H. W. Brown, R. M. Lemmon, R. O. Lindblom and M. Weissbluth), Academic Press, New York, 1961, p. 195.
171. T. J. Stone and W. A. Waters, *J. Chem. Soc.*, 408 (1961).
172. T. J. Stone and W. A. Waters, *J. Chem. Soc.*, 1488 (1965).
173. P. Ashworth and W. T. Dixon, *Chem. Commun.*, 1150 (1971).
174. P. Ashworth and W. T. Dixon, *J. Chem. Soc., Perkin Trans. 2*, 2264 (1972).
175. P. D. Josephy, T. E. Eling and R. P. Mason, *Mol. Pharmacol.*, **23**, 461 (1982).
176. P. R. West, L. S. Harmann, P. D. Josephy and R. P. Mason, *Biochem. Pharmacol.*, **33**, 2933 (1984).
177. K. Scheffler, H. B. Stegmann, C. Ulrich and P. Schuler, *Z. Naturforsch., A*, **39**, 789 (1984).
178. R. C. Sealy, L. Harman, P. R. West and R. P. Mason, *J. Am. Chem. Soc.*, **107**, 3401 (1985).
179. M. Cocivera, M. Tomkiewicz and A. Groen, *J. Am. Chem. Soc.*, **94**, 6598 (1972).
180. F. Graf, K. Loth and H.-H. Guenthard, *Helv. Chim. Acta*, **60**, 710 (1977).
181. A. J. Elliot and J. K. S. Wan, *J. Phys. Chem.*, **82**, 444 (1978).
182. C. C. Felix and R. C. Sealy, *Photochem. Photobiol.*, **34**, 423 (1981).
183. C. C. Felix and R. C. Sealy, *J. Am. Chem. Soc.*, **103**, 2831 (1981).
184. G. W. Burton, T. Doba, E. Gabe, L. Hughes, L. Lee, L. Prasad and K. U. Ingold, *J. Am. Chem. Soc.*, **107**, 7053 (1985).
185. M. T. Craw, M. C. Depew and J. K. S. Wan, *Can. J. Chem.*, **64**, 1414 (1986).
186. C. Jinot, K. P. Madden and R. H. Schuler, *J. Phys. Chem.*, **90**, 4979 (1986).
187. G. N. R. Tripathi, D. M. Chipman, C. A. Miderski, H. F. Davis, R. W. Fessenden and R. H. Schuler, *J. Phys. Chem.*, **90**, 3968 (1986).
188. R. C. Sealy, C. C. Felix, J. S. Hyde and H. M. Schwartz, in *Free Radicals in Biology* (Ed. W. A. Pryor), Vol. 4, Academic Press, New York, 1980, p. 209.
189. R. P. Mason, in *Free Radicals in Biology* (Ed. W. A. Pryor), Vol. 5, Academic Press, New York, 1982, p. 161.
190. P. Reichard, *Science*, **260**, 1773 (1993).
191. W. Espersen and R. W. Kreilick, *J. Phys. Chem.*, **73**, 3370 (1969).
192. W. Uber and H. B. Stegmann, in *Landolt-Boernstein Numerical Data and Functional Relationships in Science and Technology*, Vol. 9, *Magnetic Properties of Free Radicals*, Part C2 (Eds. H. Fischer and K.-H. Hellwege), Chap. 8, Springer-Verlag, Berlin, 1979, p. 29.

193. D. Klotz, G. Deuschle and H. B. Stegmann, in *Landolt-Boernstein Numerical Data and Functional Relationships in Science and Technology*, Vol. 17, subvolume e, *Magnetic Properties of Free Radicals* (Ed. H. Fischer), Chap. 8, Springer-Verlag, Berlin, 1988, p. 35.
194. J. B. Pedersen, C. E. M. Hansen, H. Parbo and L. T. Muus, *J. Chem. Phys.*, **63**, 2398 (1975).
195. K. Loth, F. Graf and H. H. Guenthard, *Chem. Phys. Lett.*, **45**, 191 (1977).
196. G. N. R. Tripathi and R. H. Schuler, *J. Phys. Chem.*, **88**, 1706 (1984).
197. G. N. R. Tripathi, *J. Phys. Chem., A*, **102**, 2388 (1998).
198. D. M. Chipman, *J. Phys. Chem., A*, **103**, 11181 (1999).
199. B. T. Allen and W. Vanneste, *Nature (London)*, **204**, 991 (1964).
200. P. V. Schatsnev, G. M. Zhidomirov and N. D. Chuvylkin, *J. Struct. Chem.*, **10**, 885 (1969).
201. N. M. Bazhin, P. V. Schastnev, N. D. Chuvylkin, V. M. Berdinkov, V. K. Fedorov, A. P. Merkulov and V. D. Shteingarts, *J. Struct. Chem.*, **11**, 935 (1970).
202. E. W. Stone and A. H. Maki, *J. Chem. Phys.*, **36**, 1944 (1962).
203. M. R. Das and G. K. Fraenkel, *J. Chem. Phys.*, **42**, 1350 (1965).
204. T. G. Edwards and R. Grinter, *Mol. Phys.*, **15**, 367 (1968).
205. M. Brustolon, L. Pasimeni and C. Corvaja, *J. Chem. Soc., Faraday Trans. 2*, **71**, 193 (1975).
206. K. P. Madden, H. J. McManus and R. H. Schuler, *J. Phys. Chem.*, **86**, 2926 (1982).
207. A. Rieker, K. Scheffler and E. Mueller, *Justus Liebigs Ann. Chem.*, **670**, 23 (1963).
208. A. Rieker and K. Scheffler, *Tetrahedron Lett.*, 1337 (1965).
209. A. Rieker, *Z. Naturforsch., B*, **21**, 647 (1966).
210. S. P. Solodovnikov, A. I. Prokof'ev, G. N. Bogdanov, G. A. Nikiforov and V. V. Ershov, *Theor. Exp. Chem.*, **3**, 215 (1967).
211. L. Gilbert and R. W. Kreilick, *J. Chem. Phys.*, **48**, 3377 (1968).
212. T. E. Gough and G. A. Taylor, *Can. J. Chem.*, **47**, 3717 (1969).
213. P. D. Sullivan, J. R. Bolton and J. W. E. Geiger, *J. Am. Chem. Soc.*, **92**, 4176 (1970).
214. A. B. Jaffe and R. W. Kreilick, *Chem. Phys. Lett.*, **32**, 572 (1975).
215. W. Lubitz, W. Broser, B. Kirste, H. Kurreck and K. Schubert, *Z. Naturforsch., A*, **33**, 1072 (1978).
216. K. Dimroth, A. Berndt, F. Bar, A. Schweig and R. Volland, *Angew. Chem., Int. Ed. Engl.*, **6**, 34 (1967).
217. J. Gendell, J. H. Freed and G. K. Fraenkel, *J. Chem. Phys.*, **37**, 2832 (1962).
218. K. Dimroth, F. Bar and A. Berndt, *Angew. Chem., Int. Ed. Engl.*, **4**, 240 (1965).
219. W. M. Gulick and D. H. Geske, *J. Am. Chem. Soc.*, **88**, 4119 (1966).
220. B. L. Silver, Z. Luz and C. Eden, *J. Chem. Phys.*, **44**, 4258 (1966).
221. M. Broze, Z. Luz and B. L. Silver, *J. Chem. Phys.*, **46**, 4891 (1967).
222. M. Broze and Z. Luz, *J. Phys. Chem.*, **71**, 3690 (1967).
223. M. Broze and Z. Luz, *J. Chem. Phys.*, **51**, 738 (1969).
224. R. Poupko, B. L. Silver and M. Rubinstein, *J. Am. Chem. Soc.*, **92**, 4512 (1970).
225. E. Melamud and B. L. Silver, *J. Phys. Chem.*, **77**, 1896 (1973).
226. F. R. Hewgill, T. J. Stone and W. A. Waters, *J. Chem. Soc.*, 408 (1964).
227. T. J. Stone and W. A. Waters, *J. Chem. Soc.*, 213 (1964).
228. G. Vincow and G. K. Fraenkel, *J. Chem. Phys.*, **34**, 1333 (1961).
229. S. K. Wong, W. Sytynk and J. K. S. Wan, *Can. J. Chem.*, **50**, 3052 (1972).
230. S. K. Wong and J. K. S. Wan, *J. Am. Chem. Soc.*, **94**, 7197 (1972).
231. H. Yoshida, K. Hayashi and T. Warashina, *Bull. Chem. Soc. Jpn.*, **45**, 3515 (1972).
232. P. B. Ayscough and R. C. Sealy, *J. Chem. Soc., Perkin Trans. 2*, 543 (1973).
233. W. T. Dixon, W. E. J. Foster and D. Murphy, *J. Chem. Soc., Perkin Trans. 2*, 2124 (1973).
234. H. Yoshida, J. Sohma and T. Warashina, *Bull. Chem. Soc. Jpn.*, **47**, 1396 (1974).
235. D. Murphy, *J. Chem. Res. (S)*, 321 (1980).
236. I. Agranat, M. Rabinovitz, H. R. Falle, G. R. Luckhurst and J. N. Ockwell, *J. Chem. Soc., B*, 294 (1970).
237. L. S. Singer, I. C. Lewis, T. Richerzhagen and G. Vincow, *J. Phys. Chem.*, **75**, 290 (1971).
238. P. Devolder and P. Goudmand, *C.R. Acad. Sci. Paris*, **279**, 1001 (1974).
239. B. B. Adeleke, K. Y. Choo and J. K. S. Wan, *J. Chem. Phys.*, **62**, 3822 (1975).
240. W. T. Dixon, M. Moghini and D. Murphy, *J. Chem. Soc., Perkin Trans. 2*, 1189 (1975).
241. G. J. Smith and I. J. Miller, *Aust. J. Chem.*, **28**, 193 (1975).
242. S. Hauff, P. Krauss and A. Rieker, *Chem. Ber.*, **105**, 1446 (1972).
243. D. Schulte-Frohlinde and C. von Sonntag, *Z. Phys. Chem. (Frankfurtam Main)*, **44**, 314 (1965).

16. Transient phenoxyl radicals: Formation and properties in aqueous solutions

244. P. I. Baugh, G. O. Phillips and J. I. C. Arthur, *J. Phys. Chem.*, **70**, 3061 (1966).
245. W. Boguth and H. Niemann, *Biochim. Biophys. Acta*, **248**, 121 (1971).
246. D. Jore, C. Ferradini, K. P. Madden and L. K. Patterson, *Free Radical Biol. Med.*, **11**, 349 (1991).
247. R. D. Parnell and K. E. Russell, *J. Chem. Soc., Perkin Trans. 2*, 161 (1974).
248. W. Adam and W. T. Chiu, *J. Am. Chem. Soc.*, **93**, 3687 (1971).
249. A. P. Griva and E. T. Denisov, *Int. J. Chem. Kinet.*, **5**, 869 (1973).
250. A. Carrington and I. C. P. Smith, *Mol. Phys.*, **8**, 101 (1964).
251. I. C. P. Smith and A. Carrington, *Mol. Phys.*, **12**, 439 (1967).
252. W. T. Dixon and D. Murphy, *J. Chem. Soc., Faraday Trans. 2*, **72**, 135 (1976).
253. K. Loth, M. Andrist, F. Graf and H. H. Guenthard, *Chem. Phys. Lett.*, **29**, 163 (1974).
254. K. Loth, F. Graf and H. H. Guenthard, *J. Photochem.*, **5**, 170 (1976).
255. K. Loth, F. Graf and H. H. Guenthard, *Chem. Phys.*, **13**, 95 (1976).
256. A. I. Prokof'ev, N. N. Bubnov, S. P. Solodovnikov, I. S. Belostotskaya, V. V. Ershov and M. I. Kabachnik, *Bull. Acad. Sci. USSR, Div. Chem. Sci.*, **23**, 2379 (1974).
257. A. I. Prokof'ev, N. N. Bubnov, S. P. Solodovnikov and M. I. Kabachnik, *Bull. Acad. Sci. USSR, Div. Chem. Sci.*, **23**, 2131 (1974).
258. A. S. Masalimov, A. I. Prokof'ev, N. N. Bubnov, S. P. Solodovnikov and M. I. Kabachnik, *Bull. Acad. Sci. USSR, Div. Chem. Sci.*, **25**, 181 (1976).
259. A. I. Prokof'ev, A. S. Masalimov, N. N. Bubnov, S. P. Solodovnikov and M. I. Kabachnik, *Bull. Acad. Sci. USSR, Div. Chem. Sci.*, **25**, 291 (1976).
260. A. I. Prokof'ev, N. N. Bubnov, S. P. Solodovnikov, I. S. Belostotskaya and V. V. Ershov, *Dokl. Chem.*, **210**, 402 (1973).
261. A. I. Prokof'ev, N. N. Bubnov, S. P. Solodovnikov and M. I. Kabachnik, *Tetrahedron Lett.*, 2479 (1973).
262. D. G. Ondercin, P. D. Sullivan, E. van der Drift, W. J. van den Hoek, B. Rousseeuw and J. Smidt, *J. Magn. Res.*, **23**, 39 (1976).
263. R. H. Schuler, P. Neta, H. Zemel and R. W. Fessenden, *J. Am. Chem. Soc.*, **98**, 3825 (1976).
264. D. Meisel and R. W. Fessenden, *J. Am. Chem. Soc.*, **98**, 7505 (1976).
265. A. Hinchliffe, *J. Mol. Struct.*, **43**, 273 (1978).
266. N. M. Atherton, E. J. Land and G. Porter, *Trans. Faraday Soc.*, **59**, 818 (1963).
267. C. Adamo, R. Subra, A. DiMatteo and V. Barone, *J. Chem. Phys.*, **109**, 10244 (1998).
268. G. Porter and F. J. Wright, *Trans. Faraday Soc.*, **51**, 1469 (1955).
269. L. I. Grossweiner and W. A. Mulac, *Radiat. Res.*, **10**, 515 (1959).
270. L. I. Grossweiner and E. F. Zwicker, *J. Chem. Phys.*, **32**, 305 (1960).
271. E. J. Land, G. Porter and E. Strachan, *Trans. Faraday Soc.*, **57**, 1885 (1961).
272. E. J. Land and G. Porter, *Trans. Faraday Soc.*, **59**, 2016 (1963).
273. R. H. Schuler and G. K. Buzzard, *Int. J. Radiat. Phys. Chem.*, **8**, 563 (1976).
274. G. Dobson and L. I. Grossweiner, *Trans. Faraday Soc.*, **61**, 708 (1965).
275. L. Wojnarovits, A. Kovacs and G. Foldiak, *Radiat. Phys. Chem.*, **50**, 377 (1997).
276. A. Kovacs, K. Gonter, G. Foldiak, I. Gyorgy and L. Wojnarovits, *ACH-Models Chem.*, **134**, 453 (1997).
277. I. V. Khudyakov, V. A. Kuzmin and N. M. Emanuel, *Dokl. Phys. Chem., Proc. Acad. Sci. USSR*, **217**, 747 (1974).
278. P. S. Rao and E. Hayon, *J. Phys. Chem.*, **77**, 2274 (1973).
279. V. A. Roginsky, L. M. Pisarenko, W. Bors, C. Michel and M. Saran, *J. Chem. Soc., Faraday Trans.*, **94**, 1835 (1998).
280. T. N. Das and P. Neta, *J. Phys. Chem., A*, **102**, 7081 (1998).
281. K. Bobrowski, *J. Chem. Soc., Faraday Trans. 1*, **80**, 1377 (1984).
282. R. H. Bisby and N. Tabassum, *Biochem. Pharmacol.*, **37**, 2731 (1988).
283. R. H. Bisby, S. A. Johnson and A. W. Parker, *J. Phys. Chem., B*, **104**, 5832 (2000).
284. R. Terzian, N. Serpone and M. A. Fox, *J. Photochem. Photobiol. A*, **90**, 125 (1995).
285. R. B. Draper, M. A. Fox, E. Pelizzetti and N. Serpone, *J. Phys. Chem.*, **93**, 1938 (1989).
286. R. Terzian, N. Serpone, R. B. Draper, M. A. Fox and E. Pelizzetti, *Langmuir*, **7**, 3081 (1991).
287. S. N. Guha and K. I. Priyadarsini, *Int. J. Chem. Kinet.*, **32**, 17 (2000).
288. L. Qin, G. N. R. Tripathi and R. H. Schuler, *J. Phys. Chem.*, **91**, 1905 (1987).
289. D. Wang, I. Gyoergy, K. Hildenbrand and C. von Sonntag, *J. Chem. Soc., Perkin Trans. 2*, 45 (1994).

290. Q. Sun and R. H. Schuler, *J. Phys. Chem.*, **91**, 4591 (1987).
291. G. N. R. Tripathi and R. H. Schuler, *J. Phys. Chem.*, **87**, 3101 (1983).
292. P. Dwibedy, G. R. Dey, D. B. Naik, K. Kishore and P. N. Moorthy, *Phys. Chem. Chem. Phys.*, **1**, 1915 (1999).
293. I. Gyorgy, S. Antus and G. Foldiak, *Radiat. Phys. Chem.*, **39**, 81 (1992).
294. I. Gyorgy, S. Antus, A. Blazovics and G. Foldiak, *Int. J. Radiat. Biol.*, **61**, 603 (1992).
295. R. W. Sloper and E. J. Land, *Photochem. Photobiol.*, **32**, 687 (1980).
296. J. Feitelson and E. Hayon, *J. Phys. Chem.*, **77**, 10 (1973).
297. D. V. Bent and E. Hayon, *J. Am. Chem. Soc.*, **97**, 2599 (1975).
298. M. P. Pileni, *Chem. Phys. Lett.*, **54**, 363 (1978).
299. G. E. Adams, J. E. Aldrich, R. H. Bisby, R. B. Cundall, J. L. Redpath and R. L. Willson, *Radiat. Res.*, **49**, 278 (1972).
300. S. Solar, W. Solar and N. Getoff, *J. Phys. Chem.*, **88**, 2091 (1984).
301. T. N. Das, *J. Phys. Chem., A*, **102**, 426 (1998).
302. T. N. Das and K. I. Priyadarsini, *J. Phys. Chem.*, **98**, 5272 (1994).
303. S. J. Atherton, J. Dillon and E. R. Gaillard, *Biochim. Biophys. Acta*, **1158**, 75 (1993).
304. D. Jore, L. K. Patterson and C. Ferradini, *J. Free Radical Biol. Med.*, **2**, 405 (1986).
305. M. J. Thomas and B. H. J. Bielski, *J. Am. Chem. Soc.*, **111**, 3315 (1989).
306. R. H. Bisby, S. Ahmed, R. B. Cundall and E. W. Thomas, *Free Radical Res. Commun.*, **1**, 251 (1986).
307. D. E. Cabelli and B. H. J. Bielski, *J. Free Radical Biol. Med.*, **2**, 71 (1986).
308. M. J. Davies, L. G. Forni and R. L. Willson, *Biochem. J.*, **255**, 513 (1988).
309. M. Erben-Russ, W. Bors and M. Saran, *Int. J. Radiat. Biol.*, **52**, 393 (1987).
310. T. N. Das, *J. Phys. Chem., A*, **105**, 5954 (2001).
311. H. Mohan and J. P. Mittal, *Int. J. Chem. Kinet.*, **31**, 603 (1999).
312. H. Pal, D. K. Palit, T. Mukherjee and J. P. Mittal, *J. Chem. Soc., Faraday Trans.*, **88**, 681 (1992).
313. J. K. Dohrmann and B. Bergmann, *J. Phys. Chem.*, **99**, 1218 (1995).
314. H. Pal, T. Mukherjee and J. P. Mittal, *J. Chem. Soc., Faraday Trans.*, **90**, 711 (1994).
315. D. Meisel and P. Neta, *J. Am. Chem. Soc.*, **97**, 5198 (1975).
316. B. E. Hulme, E. J. Land and G. O. Phillips, *J. Chem. Soc., Faraday Trans. 1*, **68**, 1992 (1972).
317. H. Pal, T. Mukherjee and J. P. Mittal, *Radiat. Phys. Chem.*, **44**, 603 (1994).
318. E. J. Land, *Prog. React. Kinet.*, **3**, 369 (1965).
319. V. A. Kuzmin, I. V. Khudyakov, A. S. Tatikolov, A. I. Prokof'ev and N. M. Emanuel, *Dokl. Phys. Chem., Proc. Acad. Sci. USSR*, **227**, 383 (1976).
320. G. N. R. Tripathi, *J. Chem. Phys.*, **74**, 6044 (1981).
321. J. Feitelson, E. Hayon and A. Treinin, *J. Am. Chem. Soc.*, **95**, 1025 (1973).
322. G. E. Adams, J. L. Redpath, R. H. Bisby and R. B. Cundall, *J. Chem. Soc., Faraday Trans. 1*, **69**, 1608 (1973).
323. W. Bors, M. Saran and C. Michel, *J. Phys. Chem.*, **83**, 2447 (1979).
324. J. E. Packer, T. F. Slater and R. L. Willson, *Nature (London)*, **278**, 737 (1979).
325. E. J. Land and A. J. Swallow, *Biochem.*, **8**, 2117 (1969).
326. D. Meisel and P. Neta, *J. Phys. Chem.*, **79**, 2459 (1975).
327. B. Holmstroem, *Photochem. Photobiol.*, **3**, 97 (1964).
328. R. F. Anderson, *Ber. Bunsenges. Phys. Chem.*, **80**, 969 (1976).
329. E. Mueller and K. Ley, *Chem. Ber.*, **87**, 922 (1954).
330. G. N. R. Tripathi and R. H. Schuler, *J. Chem. Phys.*, **76**, 2139 (1982).
331. G. N. R. Tripathi and R. H. Schuler, *J. Chem. Phys.*, **76**, 4289 (1982).
332. R. H. Schuler, G. N. R. Tripathi, M. F. Prebenda and D. M. Chipman, *J. Phys. Chem.*, **87**, 5357 (1983).
333. G. N. R. Tripathi and R. H. Schuler, *J. Phys. Chem.*, **91**, 5881 (1987).
334. G. N. R. Tripathi and R. H. Schuler, *J. Chem. Soc., Faraday Trans.*, **89**, 4177 (1993).
335. Q. Sun, G. N. R. Tripathi and R. H. Schuler, *J. Phys. Chem.*, **94**, 6273 (1990).
336. G. N. R. Tripathi and R. H. Schuler, *J. Phys. Chem.*, **92**, 5129 (1988).
337. G. N. R. Tripathi and R. H. Schuler, *Chem. Phys. Lett.*, **98**, 594 (1983).
338. G. N. R. Tripathi and R. H. Schuler, *J. Chem. Phys.*, **81**, 113 (1984).
339. A. W. Parker and R. H. Bisby, *J. Chem. Soc., Faraday Trans.*, **89**, 2873 (1993).
340. R. H. Bisby, R. B. Cundall and N. Tabassum, *Life Chem. Rep.*, **3**, 29 (1985).

16. Transient phenoxyl radicals: Formation and properties in aqueous solutions   1151

341. O. Brede, H. Orthner, V. Zubarev and R. Hermann, *J. Phys. Chem.*, **100**, 7097 (1996).
342. M. R. Ganapathi, R. Hermann, S. Naumov and O. Brede, *Phys. Chem. Chem. Phys.*, **2**, 4947 (2000).
343. O. Brede, M. R. Ganapathi, S. Naumov, W. Naumann and R. Hermann, *J. Phys. Chem., A*, **105**, 3757 (2001).
344. T. A. Gadosy, D. Shukla and L. J. Johnston, *J. Phys. Chem., A*, **103**, 8834 (1999).
345. E. Hayon and M. Simic, *Acc. Chem. Res.*, **7**, 114 (1974).
346. J. K. Dohrmann, in *Landolt-Boernstein Numerical Data and Functional Relationships in Science and Technology*, Vol. 13, subvolume e, *Magnetic Properties of Free Radicals* (Ed. H. Fischer), Chap. 9, Springer-Verlag, Berlin, 1985, p. 5.
347. R. L. Willson, *Trans. Faraday Soc.*, **67**, 3020 (1971).
348. R. L. Willson, *Chem. Commun.*, 1249 (1971).
349. S. P. Vaish and G. Tollin, *Bioenergetics*, **2**, 61 (1971).
350. H. Pal, D. K. Palit, T. Mukherjee and J. P. Mittal, *J. Chem. Soc., Faraday Trans.*, **87**, 1109 (1991).
351. P. S. Rao and E. Hayon, *J. Phys. Chem.*, **79**, 1063 (1975).
352. G. N. R. Tripathi and R. H. Schuler, *Chem. Phys. Lett.*, **88**, 253 (1982).
353. Z. B. Alfassi and L. C. T. Shoute, *Int. J. Chem. Kinet.*, **25**, 79 (1993).
354. M. Ye and R. H. Schuler, *J. Phys. Chem.*, **93**, 1898 (1989).
355. L. R. Mahoney and S. A. Weiner, *J. Am. Chem. Soc.*, **94**, 585 (1972).
356. P. Hapiot, J. Pinson, P. Neta, C. Francesch, F. Mhamdi, C. Rolando and S. Schneider, *Phytochemistry*, **36**, 1013 (1994).
357. P. Hapiot, J. Pinson, C. Francesch, F. Mhamdi, C. Rolando and P. Neta, *J. Phys. Chem.*, **98**, 2641 (1994).
358. M. R. Chedekel, E. J. Land, A. Thompson and T. G. Truscott, *J. Chem. Soc., Chem. Commun.*, 1170 (1984).
359. A. Thompson, E. J. Land, M. R. Chedekel, K. V. Subbarao and T. G. Truscott, *Biochim. Biophys. Acta*, **843**, 49 (1985).
360. F. Jin, J. Leitich and C. von Sonntag, *J. Chem. Soc., Perkin Trans. 2*, 1583 (1993).
361. M. Jonsson, J. Lind, T. Reitberger, T. E. Eriksen and G. Merenyi, *J. Phys. Chem.*, **97**, 8229 (1993).
362. N. d'Alessandro, G. Bianchi, X. Fang, F. Jin, H.-P. Schuchmann and C. von Sonntag, *J. Chem. Soc., Perkin Trans. 2*, 1862 (2000).
363. E. Cadenas, G. Merenyi and J. Lind, *FEBS Lett.*, **253**, 235 (1989).
364. B. Kalyanaraman, W. Korytowski, B. Pilas, T. Sarna, E. J. Land and T. G. Truscott, *Arch. Biochem. Biophys.*, **266**, 277 (1988).
365. R. H. Schuler, *Radiat. Res.*, **69**, 417 (1977).
366. R. H. Bisby, S. Ahmed and R. B. Cundall, *Biochem. Biophys. Res. Commun.*, **119**, 245 (1984).
367. T. N. Das, *Int. J. Radiat. Biol.*, **70**, 7 (1996).
368. W. Bors, C. Michel and K. Stettmaier, in *Antioxidation* (Eds. L. Parker and R. Cadenas), Hippokrates, Stuttgart, 1994, p. 35.
369. S. V. Jovanovic, A. Harriman and M. G. Simic, *J. Phys. Chem.*, **90**, 1935 (1986).
370. S. V. Jovanovic, M. Tosic and M. G. Simic, *J. Phys. Chem.*, **95**, 10824 (1991).
371. D. A. Armstrong, Q. Sun and R. H. Schuler, *J. Phys. Chem.*, **100**, 9892 (1996).
372. J. A. Howard and J. C. Scaiano, in *Landolt-Boernstein. Numerical Data and Functional Relationships in Science and Technology. New Series, Group II; Atomic and Molecular Physics* (Eds. K.-H. Hellwege and O. Madelung), Vol. 13, Part d (Ed. H. Fischer), Springer-Verlag, Berlin, 1984.
373. M. Foti, K. U. Ingold and J. Lusztyk, *J. Am. Chem. Soc.*, **116**, 9440 (1994).
374. J. A. Howard, in *Landolt-Boernstein. Numerical Data and Functional Relationships in Science and Technology. New Series, Group II; Molecules and Radicals*, Vol. 18, Subvol. D1 (Ed. H. Fischer), Chap. 8, Springer-Verlag, Berlin, 1997, p. 231.
375. C. Li and M. Z. Hoffman, *J. Phys. Chem., B*, **103**, 6653 (1999).
376. D. M. Stanbury, *Adv. Inorg. Chem.*, **33**, 69 (1989).
377. G. Merenyi, J. Lind and X. Shen, *J. Phys. Chem.*, **92**, 134 (1988).
378. M. Roder, G. Foldiak and L. Wojnarovits, *Radiat. Phys. Chem.*, **55**, 515 (1999).
379. T. N. Das, R. E. Huie and P. Neta, *J. Phys. Chem., A*, **103**, 3581 (1999).

380. M. R. DeFelippis, C. P. Murthy, F. Broitman, D. Weinraub, M. Faraggi and M. H. Klapper, *J. Phys. Chem.*, **95**, 3416 (1991).
381. M. Faraggi and M. H. Klapper, *J. Chim. Phys.*, **90**, 711 (1993).
382. M. Jonsson, J. Lind, T. E. Eriksen and G. Merenyi, *J. Am. Chem. Soc.*, **116**, 1423 (1994).

CHAPTER **17**

Oxidation of phenols

SHOSUKE YAMAMURA

Department of Chemistry, Faculty of Science and Technology, Keio University, Hiyoshi, Yokohama 223-8522, Japan
Fax: 81-45-566-1697; e-mail: yamamura@chem.keio.ac.jp

I. INTRODUCTION	1154
II. CATALYTIC OXIDATION	1154
A. Electrochemical Oxidation	1154
1. Radical coupling reaction	1156
2. Cationic reaction	1165
B. Oxidation with Dioxygen, Hydrogen Peroxide and Alkyl Hydroperoxide Catalyzed by Metal–Base Complexes	1192
1. Dioxygen–metal complexes	1192
2. Hydrogen peroxide–and *tert*-butyl hydroperoxide–metal complexes	1207
C. Enzymatic Oxidation	1216
III. OXIDATION WITH NONMETAL COMPOUNDS	1224
A. Oxidation with Hypervalent Iodobenzenes	1224
1. Reaction and reaction mechanism	1224
2. Applications in natural products synthesis	1234
B. Oxidation with 2,3-Dichloro-5,6-dicyano-*p*-benzoquinone	1247
C. Oxidation with Dioxirane, Di-*tert*-butyl Peroxide and Di-*tert*-butyl Peroxyoxalate	1255
D. Oxidation with Fremy's Salt, Ammonium Nitrate–Acetic Anhydride, Dioxygen-*tert*-butoxide, Sodium Periodate, Chlorous Acid and Other Oxidants	1259
IV. OXIDATION WITH METAL COMPOUNDS	1273
A. Oxidation with Vanadium, Chromium and Molybdenum Compounds	1273
B. Oxidation with Manganese and Rhenium Compounds	1278
C. Oxidation with Iron, Ruthenium, Cobalt, Nickel and Rhodium Compounds	1287
D. Oxidation with Copper and Silver Compounds	1305

The Chemistry of Phenols. Edited by Z. Rappoport
© 2003 John Wiley & Sons, Ltd ISBN: 0-471-49737-1

E. Oxidation with Thallium, Lead and Bismuth Compounds 1311
F. Oxidation with Other Metal Compounds 1330
V. REFERENCES 1337

I. INTRODUCTION

Birch reduction of aromatic ethers is well known to afford alicyclic compounds such as cyclohexadienes and cyclohexenones, from which a number of natural products have been synthesized. Oxidation of phenols also affords alicyclic cyclohexadienones and masked quinones in addition to C—C and/or C—O coupled products. All of them are regarded as promising synthetic intermediates for a variety of bioactive compounds including natural products. However, in contrast to Birch reduction, systematic reviews on phenolic oxidation have not hitherto appeared from the viewpoint of synthetic organic chemistry, particularly natural products synthesis. In the case of phenolic oxidation, difficulties involving radical polymerization should be overcome. This chapter demonstrates that phenolic oxidation is satisfactorily used as a key step for the synthesis of bioactive compounds and their building blocks.

On electrolysis, a symmetric tetra-substituted phenol **1** will undergo electrochemical 1e oxidation followed by deprotonation resulting in the formation of the corresponding radical **2**. This is also generated by oxidation of **1** with metal or nonmetal compounds as the oxidant. The resulting unstable species undergoes radical coupling reactions to give dimers **3**, **4**, **5**, **6** and/or **7** or is further oxidized to generate a cation **8** which is attacked by a variety of nucleophiles to afford the 2e oxidation products **9** and/or **10**. The cation is also generated directly from the parent phenol through the intermediate **11** (Scheme 1). Herein, the formation of a radical or cation species depends on the choice of oxidants, oxidation conditions and other factors, particularly the substituents attached to the aromatic ring. In the case of thallium trinitrate (M = Tl, X = NO_3, $n = 3$) mediated oxidation, the corresponding cation **8** is generated via such an aryloxy–metal intermediate as **11**, while thallium trifluoroacetate-mediated oxidation often provides radical-coupled dimers.

II. CATALYTIC OXIDATION

From the viewpoint of organic synthesis, catalytic oxidation of phenols with high stereo- and regioselectivities and high yields is more favorable than other reactions using stoichiometric amounts of oxidants, because it is advantageous to obtain only the desired products without formation of byproducts originating from the oxidants used. In the 21st century, more efficient and ideal catalytic systems must be created.

A. Electrochemical Oxidation

Electroorganic chemistry is one of the most useful tools in organic synthesis. In principle, electroorganic reactions take place on the surface of electrodes (anode and cathode). At the anode one-electron transfer occurs from the substrate to the electrode to generate a radical cation. In the case of asymmetric tetra-substituted phenols such as **12**, the resulting radical cation **13** is further deprotonated to the corresponding radical **14**, which undergoes radical coupling reaction to afford dimerization products or is further oxidized to generate a cation **15**, as shown in Scheme 2. Here, the radical coupled dimers are expected to be produced selectively when the oxidation potential for the first step (E_1) is lower than that for the second step (E_2). In contrast, if the oxidation potential E_1 is higher than or comparable to E_2, 2e oxidation products will be formed in competition with the radical

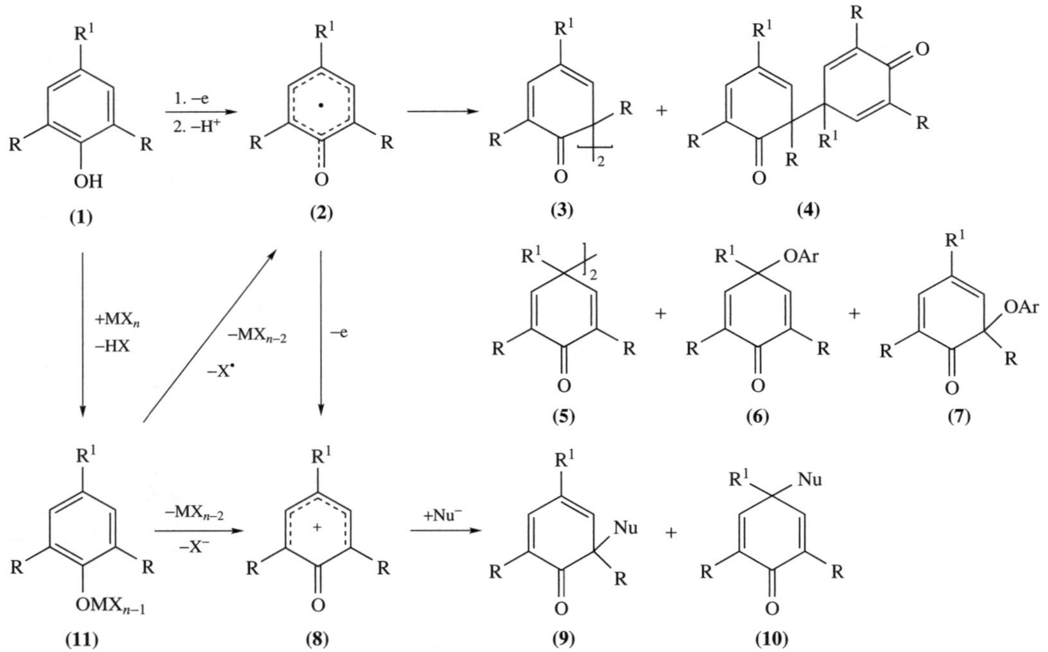

SCHEME 1. Phenolic oxidation process

Ar = 2,6-R$_2$-4-R^{1}C$_6$H$_2$
Nu$^-$ = MeO$^-$ (MeOH), HO$^-$ (H$_2$O), CN$^-$ and others
M = Tl, Pb, Mn and others
X = NO$_3$, AcO, CF$_3$COO and others

coupling reaction. In this case, if the dimers are required to be synthesized, they will be selectively obtained starting from the corresponding phenoxy anion **16** which has a lower oxidation potential (E_3).

As demonstrated in Scheme 2, it is not easy to obtain the desired product in a regio- and stereoselective manner, because several unstable species (Ar•, Ar•+, Ar+) are electro- generated and each one of them shows different reactivity and can react with a nucleophile or dimerize at three or four reactive centers shown by the arrows in the two structures in brackets. Therefore, it is quite important to find the optimum conditions by changing the oxidation potential, the electrode, the supporting electrolyte and other parameters. Partic- ularly, the product selectivity is dependent on the substituents attached to the aromatic ring and the solvents used.

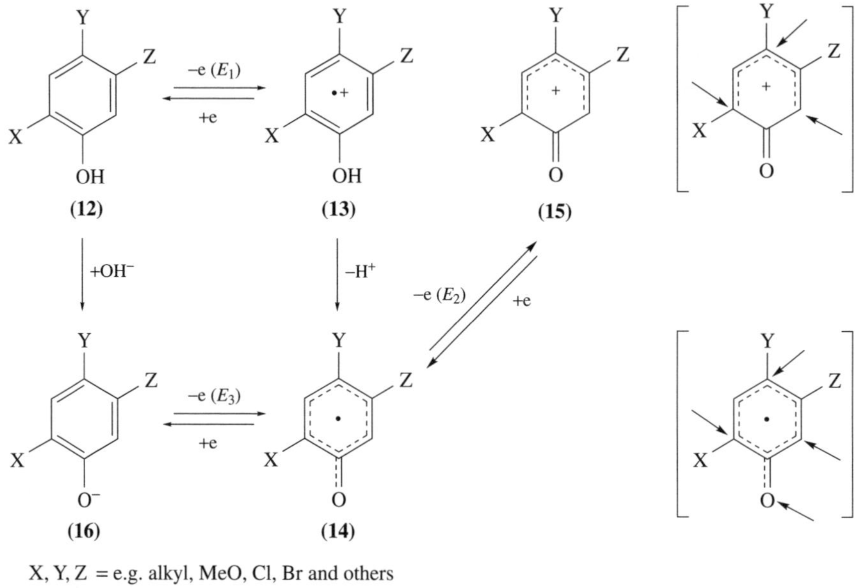

X, Y, Z = e.g. alkyl, MeO, Cl, Br and others

SCHEME 2. Oxidative generation of reactive species

Generally, direct electrolysis is carried out at a controlled potential (CPE) or constant current (CCE) using both undivided and divided cells. In contrast, an indirect method using a mediator is effective for substrates with higher oxidation potentials beyond the achievable region.

Recently, a number of invaluable books on electroorganic chemistry have been published[1–7]. Some of them discuss all aspects of the experimental arrangements, e.g. cells, electrodes, supporting electrolytes, solvents and other parameters, and there are many examples including a variety of both anodic and cathodic reactions followed by chemical reactions[8,9].

1. Radical coupling reaction

Generally, the electrogenerated radical species undergoes dimerization in competition with further 1e oxidation leading to the corresponding cation. On anodic oxidation of

17. Oxidation of phenols

phenol, electropolymerization is well known to take place resulting in the formation of a passivating film on the electrode surface[10,11]. Therefore, both *p*-benzoquinone (**17**) and 4,4′-diphenoquinone (**18**) have been produced as minor products in 20 and 10% yields, respectively, as shown in Scheme 3[10]. The latter is formed through biphenol **19**, a radical coupled dimer. The **17/18** ratio could be varied widely; e.g. electrolysis at more anodic potential provided increased percentage of **17**. Anodic oxidation of 2,6-dimethylphenol also leads to rapid formation of a linear polymer chain, but when phenols bearing a bulky alkyl substituent are used, the resulting radicals are expected to be stable. In fact, the detection of radical formation from 2,6-di(*sec*-butyl)phenol (**20**), based on multiple internal reflection Fourier transform infrared spectroscopy (MIRFTIRS), confirms the radical mechanism during the anodic oxidation of **20** leading to the corresponding 4,4′-diphenoquinone **21** through **22** (Scheme 3)[12]. In these cases, it is difficult to obtain biphenols such as **19**, **22** and **25**.

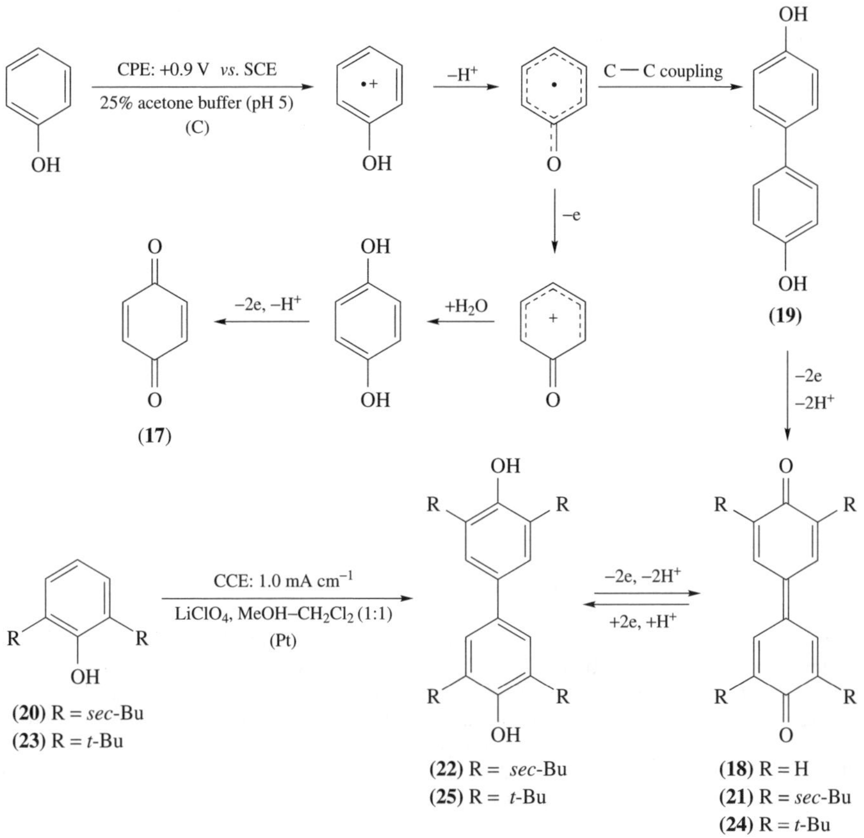

SCHEME 3. Anodic oxidation of simple phenols

On constant current electrolysis (1.0 mA cm^{-2}; 2.5 F mol^{-1}) in MeOH–CH$_2$Cl$_2$ using a divided cell, 2,6-di(*tert*-butyl)phenol (**23**) was converted to 4,4′-diphenoquinone **24** in 84.7% yield. A subsequent electroreduction was performed just by changing the current direction to afford biphenol **25** in 92.5% yield (Scheme 3)[13]. This example is one of the most characteristic features in electroorganic chemistry. Radical coupling reactions of a variety of phenols have been shown in a number of books[1–9]. Some of these couplings were applied to biomimetic synthesis of natural products in view of the oxidative phenol coupling reactions in nature[14]. Duplication will be avoided in this chapter.

From the biogenetic point of view, lignans and neolignans are produced by oxidative phenol couplings between two C6–C3 units. They have a variety of carbon skeletons[15] as well as remarkable bioactivities[16]. Several reviews on lignans and neolignans syntheses have appeared[17].

Lunarine (**26**), one of the typical neolignans, is biosynthesized by the *ortho–para* radical coupling between two molecules of *p*-hydroxycinnamic acid. In this connection, oxidative coupling reactions of 4-substituted phenols have been extensively studied using thallium trifluoroacetate (TTFA), potassium ferricyanide ($K_3[Fe(CN)_6]$) and other reagents. *p*-Cresol (**27**) was also electrolyzed at a controlled potential (+0.25 V *vs.* SCE) in a basic medium to afford Pummerer's ketone **28** in 74% yield[18]. The suggested mechanism is given in Scheme 4.

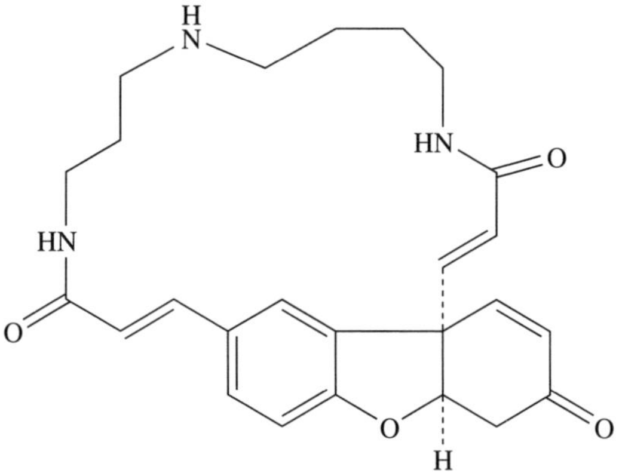

(**26**)

Eugenol is one of the most simple C6–C3 units. As expected from the CV data of eugenol (**29**)[19], it underwent constant current electrolysis (1.5 mA cm^{-2}) in MeOH to afford three 2e oxidation products **30**, **31** and **32** in 1.6, 68 and 4.6% yields, respectively, together with small amounts of dehydrodieugenol (**33**), a radical coupled dimer (7.4%) (Scheme 5). Here, **32** must be produced by the Diels–Alder reaction of the major product **31**[20,21]. As the oxidation potential (500 mV) at the second step is lower than that at the first step (780 mV), the resulting radical will be oxidized easily to the corresponding cation in competition with the radical coupling reaction. In contrast, anodic oxidation of **29** in 1M NaOH–MeOH provided **33** in almost quantitative yield. Electrochemical study on eugenol has also been carried out by Barba and coworkers[22].

trans-Isoeugenol (**34**), having a lower oxidation potential than eugenol (**29**), was electrolyzed at a controlled potential (+800 mV *vs.* SCE) in MeOH to afford four dimers **35**, **36**, **37** and **38** in 6, 5, 29 and 18% yields, respectively (Scheme 6). Herein, the initially generated *p*-quinone methide radical will be dimerized by C–C or C–O coupling. Anodic oxidation of *cis*-isoeugenol provided similar results[23,24].

In the case of sinapic acid (**39**), the CV data indicate that the oxidation potential at the initial step is almost comparable to that at the second step. On controlled potential

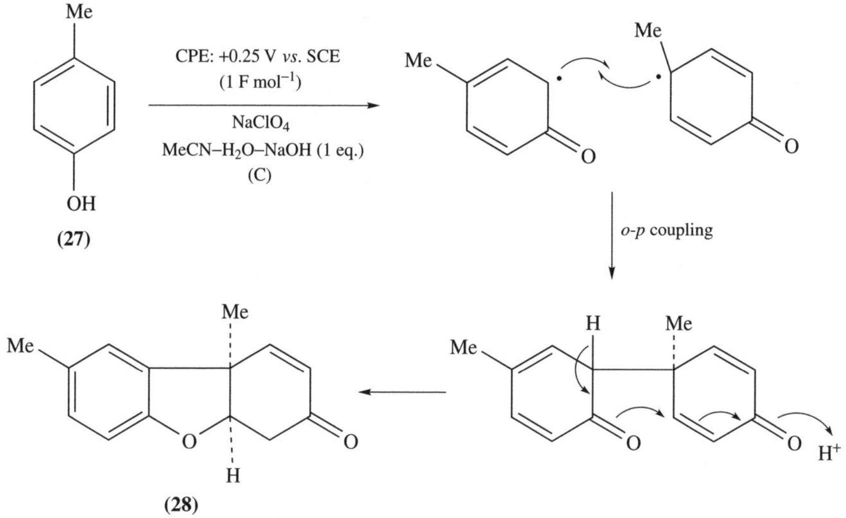

SCHEME 4. Anodic oxidation of *p*-cresol

electrolysis (+840 mV *vs.* SCE) in high concentration (10 mM), **39** was converted to dilactone **40** and an isoasatone-type compound **41** in 60 and 9% yields, respectively, while a lower concentration (1 mM) provided **41** as a sole product (Scheme 7)[25].

3,4-Methylenedioxy-6-propenylphenol (**42**) underwent 1e oxidation followed by radical coupling resulting in the formation of dimeric *o*-quinone methide **43**, which was further converted to carpanone (**44**) and seven-membered ether **45** in 11 and 44% yields, respectively, as shown in Scheme 8. The former is produced by an intramolecular [4 + 2] cycloaddition[24,26]. Carpanone has also been synthesized using oxidants such as palladium chloride[27] and molecular oxygen in the presence of Co(II)salen[28].

From the biogenetic point of view, a series of electrochemical studies on coryalline and related tetrahydroisoquinolines have been performed by Bobbitt and coworkes[29]. On controlled electrolysis (+40 mV *vs.* SCE) in excess base using a divided cell, the racemic tetrahydroquinoline **46** was easily oxidized to generate the corresponding radical, which underwent stereoselective dimerization to afford in 68.9% yield one of three possible isomers (**47**, **48** and **49**) (Scheme 9)[29]. In the case of the *S*-enantiomer **46**, only one of two rotational isomers, **47**, was produced. In contrast, chemical oxidation of racemic **46** using $K_3[Fe(CN)_6]$ provided three dimers (**47**–**49**). From these results, it is evident that electrochemical reaction takes place on or very close to the surface of the electrode, which plays an important role in product selectivity. Of three different kinds of electrodes (graphite felt, platinum and carbon paste anodes) the best result was obtained using the graphite felt anode and the other two electrodes provided very low yields of the carbon–carbon coupled dimer. Benzylisoquinoline alkaloids synthesis has been performed using a variety of oxidants rather than an electrochemical method, as described later.

During the last thirty-five years, a variety of biologically active substances bearing a novel carbon skeleton have been found in marine organisms. Of them, a number of highly brominated diphenyl ethers with antibacterial and antitumor activities were isolated from *Dysidea herbaceae* and *Ptychodera flava laysanica*. These metabolites are regarded as a self-defensive substance. In order to synthesize these metabolites, electrochemical oxidation of bromophenols has been carried out[30]. Some typical examples are shown here.

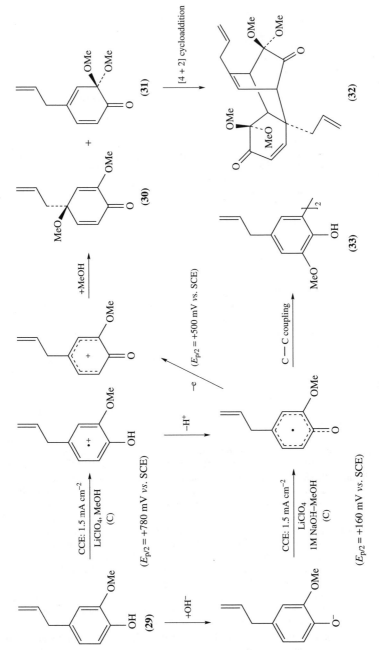

SCHEME 5. Anodic oxidation of eugenol

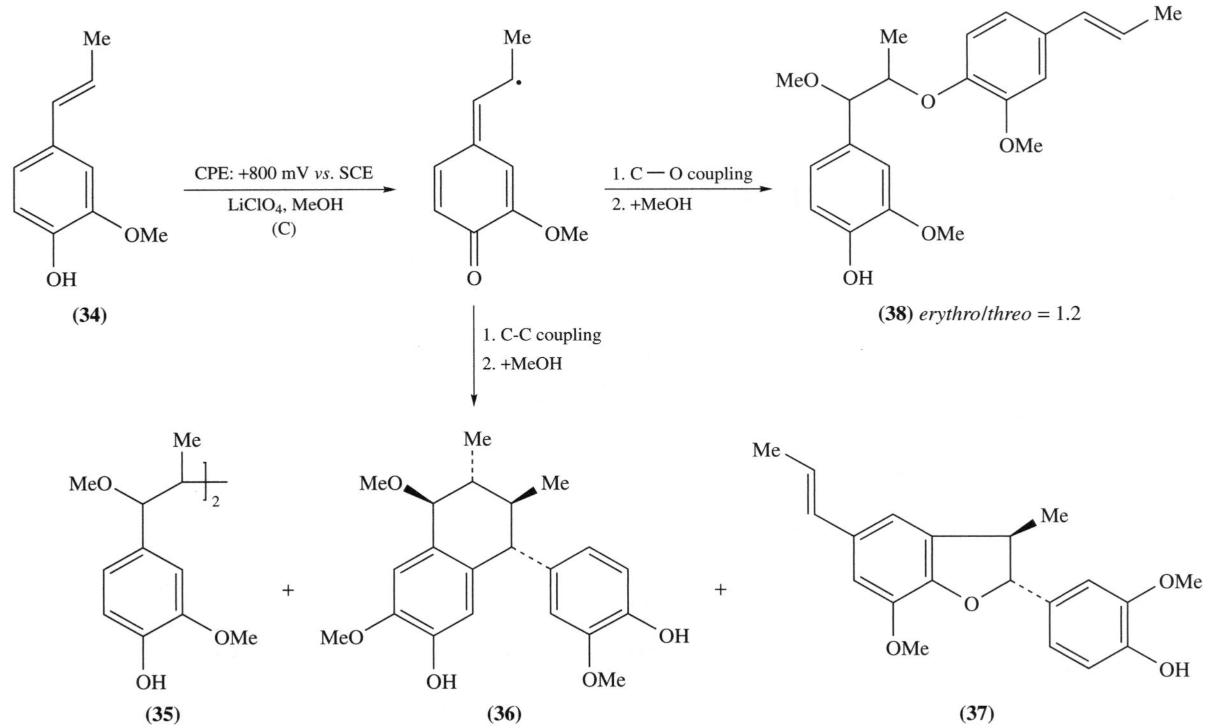

SCHEME 6. Anodic oxidation of *trans*-isoeugenol

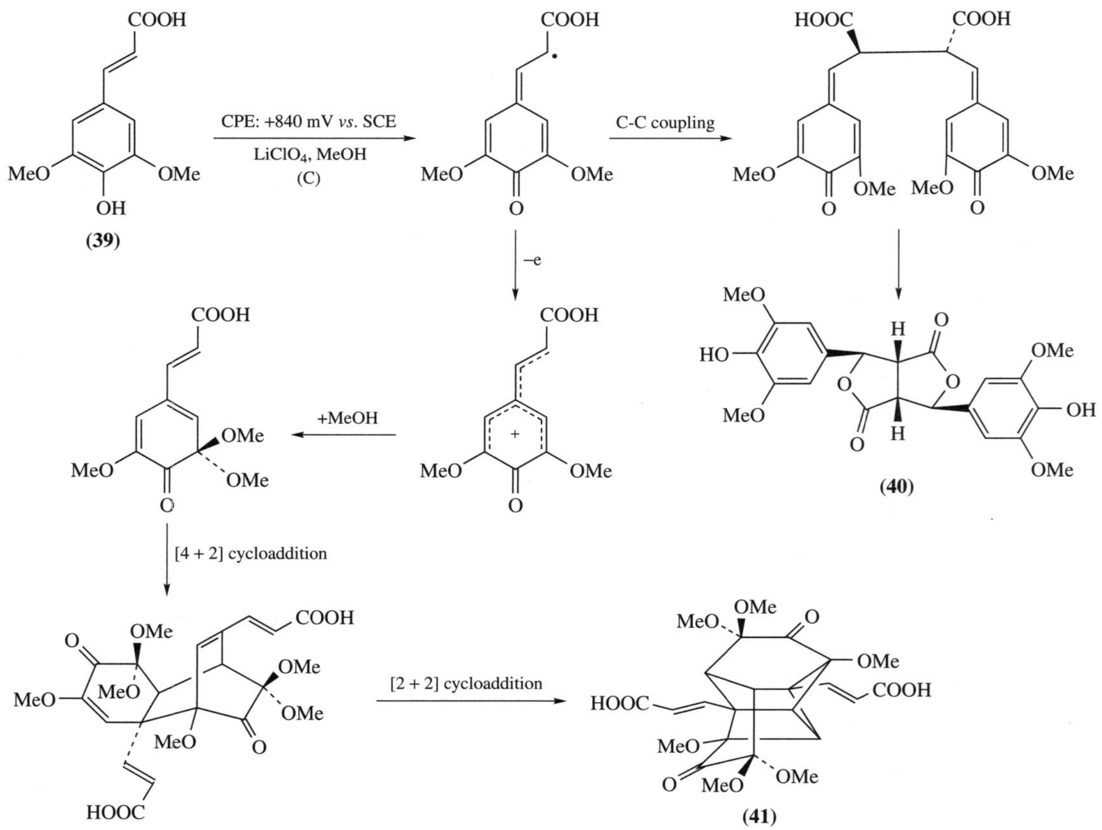

SCHEME 7. Anodic oxidation of sinapic acid

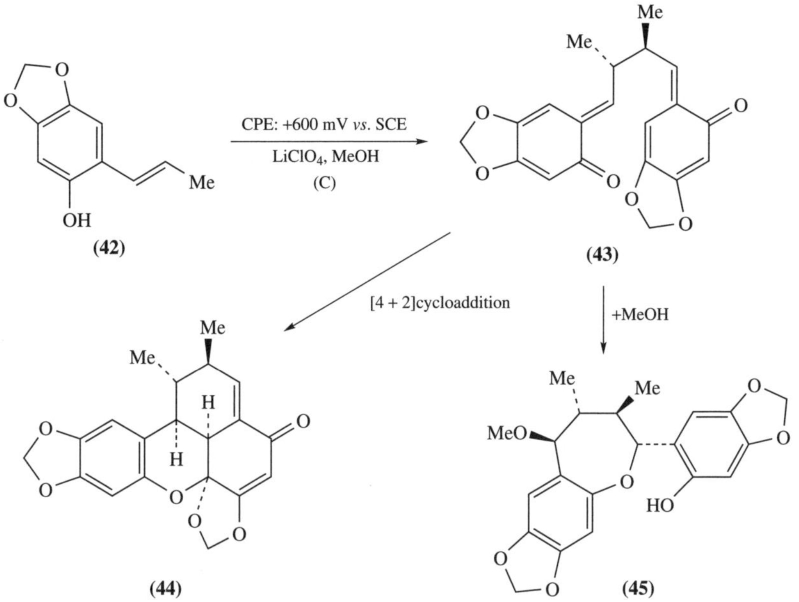

SCHEME 8. Electrolysis of 3,4-methylenedioxy-6-propenylphenol

On controlled electrolysis (+880 mV vs. SCE; 2 F mol^{-1}) in MeOH, 2,6-dibromo-4-methoxyphenol (**50**) underwent 2e oxidation, followed by nucleophilic capture with MeOH to afford 2,6-dibromo-4,4-dimethoxy-2,5-cyclohexadien-1-one (**51**) in quantitative yield. **50** was also electrolyzed at a less positive potential (+440 mV; ca 1 F mol^{-1}) in MeOH containing AcOH–AcONH$_4$ to give two dienones (**52** and **53**) in 32 and 55% yields, respectively, as shown in Scheme 10. Herein, these products must be formed by C–O and C–C couplings with bromine substitution, respectively. Therefore, the selective formation of 2e oxidation products or radical coupling dimers depends on the choice of the solvent.

The highly brominated diphenyl ether **54**, isolated from the marine organism *P. flava laysanica*, was synthesized starting from 2,3,5-tribromo-4-methoxyphenol (**55**). Substrate **55** was electrolyzed at +610 mV vs. SCE (1 F mol^{-1}) in 1:1 MeOH–CHCl$_3$ containing AcOH–AcONH$_4$ and then submitted to zinc reduction leading to two dimers **56** and **57** in 26 and 43% yields, respectively. The former was demethylated with boron tribromide to give rise to the natural **54**.

From the viewpoint of biological activity as well as a novel peptide framework, isodityrosine-class natural products (piperazinomycin, OF 4949-III, K-13 vancomycin), sharing diaryl ethers, are quite attractive[31,32]. Basic isodityrosine (**58**) itself, contributing cross-linked properties of glycoprotein of plant cell wall, has been synthesized by four groups. Three of them employed Ullman reactions of tyrosine derivatives and/or appropriate precursors[33]. Fry adopted phenol-oxidation methodology[34] using potassium ferricyanide to afford isodityrosine (**58**) and dityrosine (**59**), a component of native structural proteins, in 1.8 and 3.4% yields, respectively. Under these conditions, the electrochemical methodology provided the best results in efficiency and simplicity.

The 3,5-dibromotyrosine derivative **60**, easily prepared from tyrosine, was submitted to anodic oxidation (5 mA; +1038–1228 mV vs. SCE) in MeOH, followed by zinc reduction to afford in 45% overall yield the corresponding diaryl ether **61**, which was quantitatively

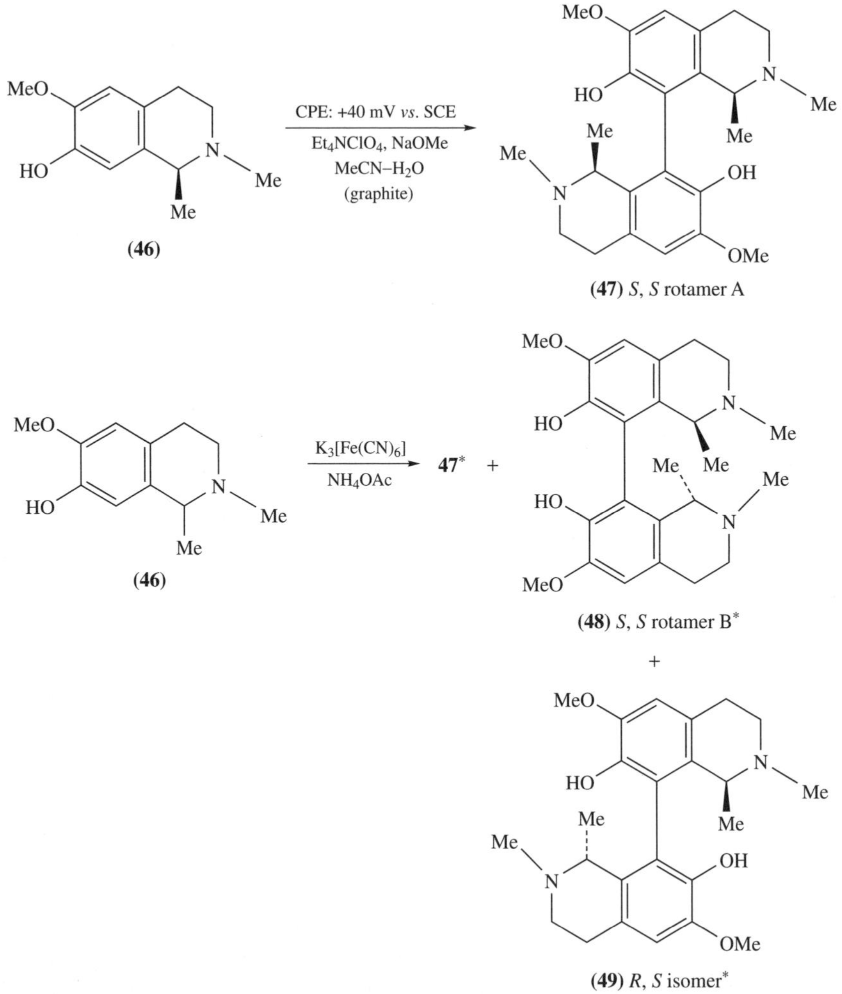

(46)

(47) *S*, *S* rotamer A

(46)

(48) *S*, *S* rotamer B*

(49) *R*, *S* isomer*

* One of the two antipodes in a racemic form is shown here.

SCHEME 9. Electrolysis of 1,2-dimethyl-7-hydroxy-6-methoxy-1,2,3,4-tetrahydroisoquinoline

converted to isodityrosine (**58**) in 2 steps (1. Catalytic hydrogenation, 2. hydrolysis), as shown in Scheme 11[35]. Almost the same result was also obtained in the case of a 3,5-dichlorotyrosine derivative. In contrast, electrolysis of a 3,5-diiodotyrosine derivative **62**, followed by zinc reduction, provided the corresponding diaryl **63** as a sole product (28%), which was converted to dityrosine (**59**) in 2 steps. Furthermore, both isodiphenylglycine and diphenylglycine have also been synthesized based on electrochemical methodology.

As mentioned above, bromine substituents promote the diaryl ether formation, while iodine substitutions prefer to produce diaryls. The *ab initio* calculations[36] indicate that the O-radicals are stable in bromo derivatives, in contrary to the C-radicals in the iodo derivative, although solvent effects were not taken into consideration. Accordingly, the

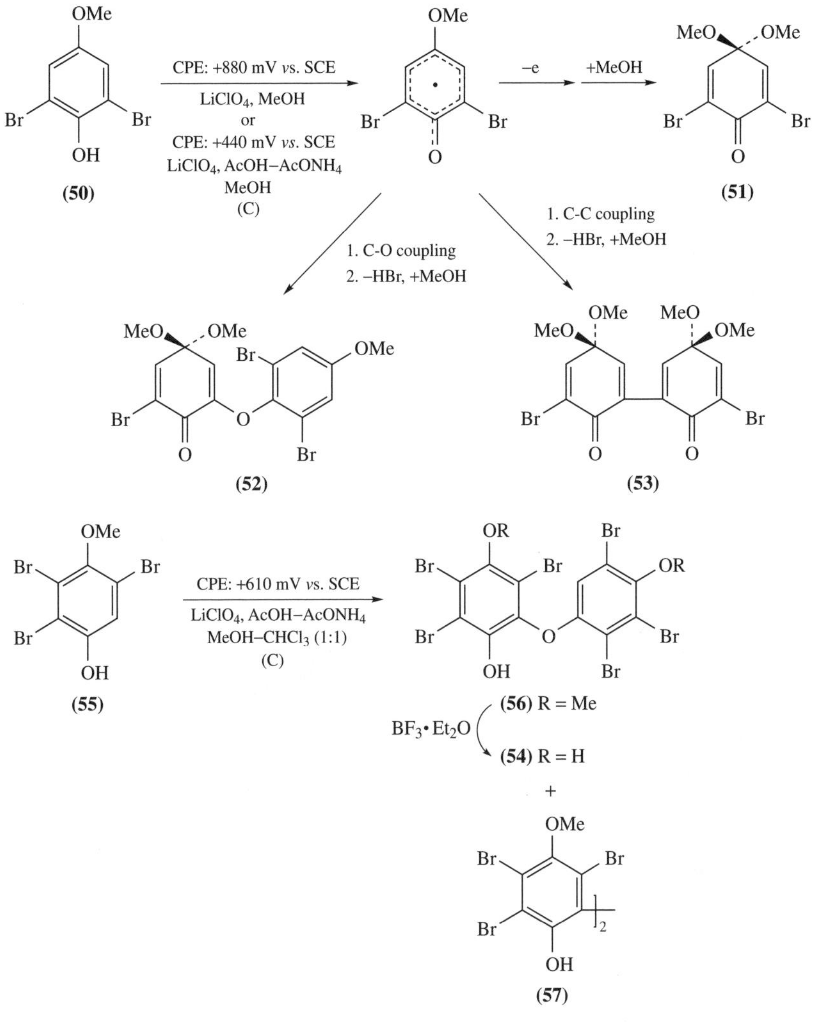

SCHEME 10. Anodic oxidation of bromophenols

C-radicals leading to the C—C coupled dimers were easier to form from diiodophenols than from dibromophenols.

2. Cationic reaction

The resulting phenoxonium ion **15**, cited in Scheme 2, is attacked by a variety of nucleophiles to yield three cyclohexadienones (**64**, **65** and **66**), as shown in Scheme 12, where X, X^1, Y and Z are suitable functional groups such as hydrogen atom, alkyl, aryl, alkoxyl and/or hydroxyl group. Usually, these three compounds are competitively formed depending upon the substituents and their locations on the benzene ring. In the

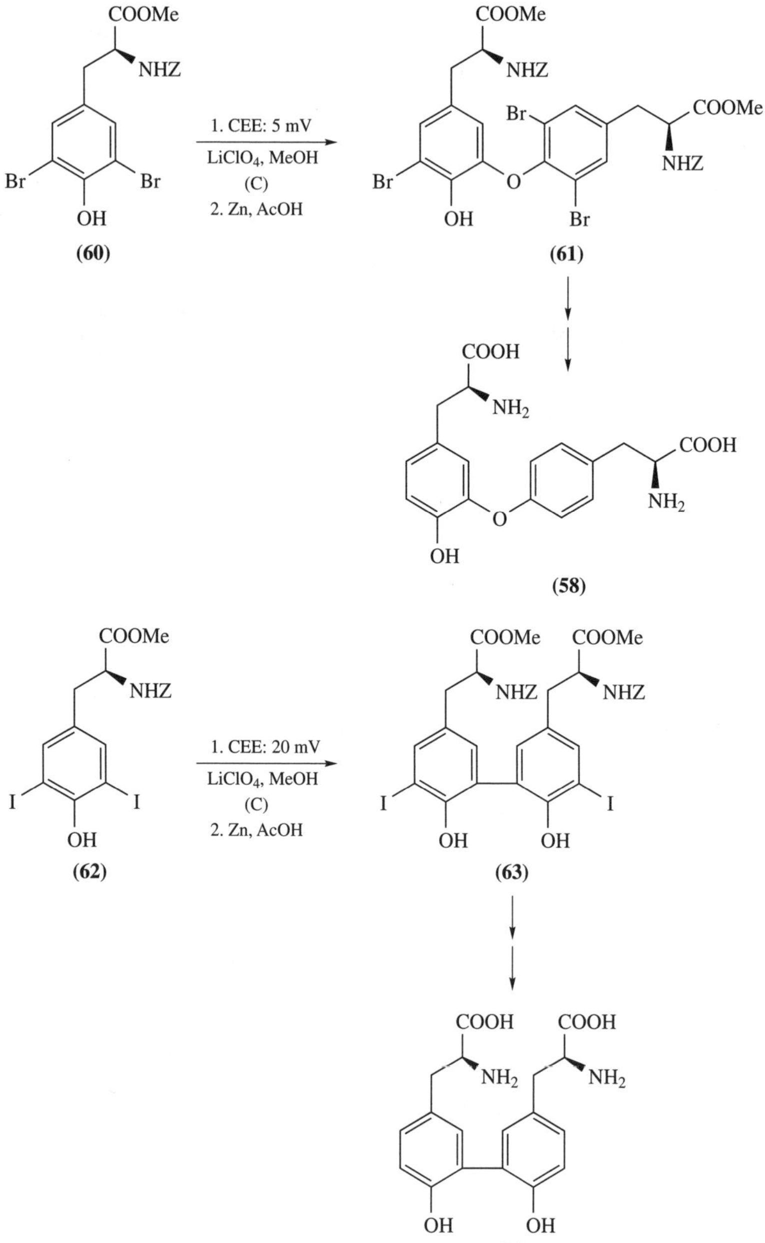

Z = COOCH$_2$Ph

SCHEME 11. Synthesis of isodityrosine and dityrosine

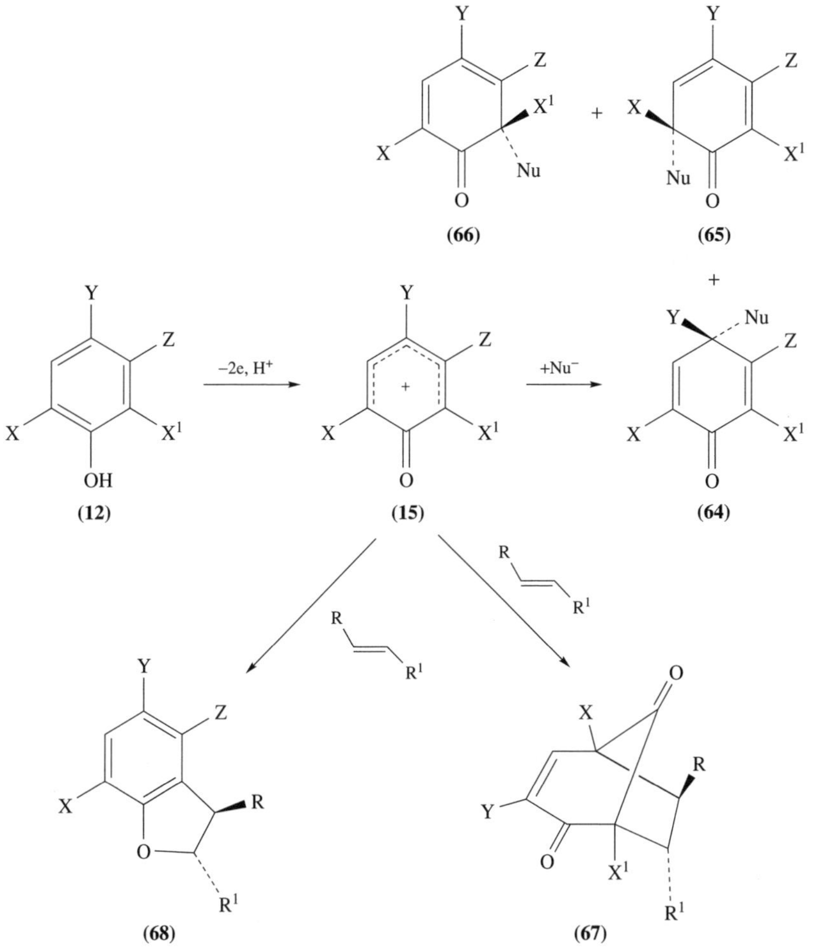

Nu⁻ = MeO⁻ (MeOH), AcO⁻ (AcOH), CN⁻ and others

SCHEME 12. Reactivities of phenoxonium ion

presence of a suitable olefin, a cationic [5 + 2] and formal [3 + 2] cycloaddition will take place to yield the corresponding bicyclo[3.2.1]octenone (**67**) and dihydrobenzofuran (**68**), respectively. Compounds **64**–**68** are promising synthetic intermediates for a variety of natural products.

On controlled current electrolysis (200 mA), 2,6-di(*tert*-butyl)-*p*-cresol (**69**) underwent nucleophilic hydroxylation, methoxylation or acetoxylation depending on the solvent system used (1M H_2O, MeOH or 0.2 M NaOAc–AcOH in MeCN) to afford the corresponding cyclohexa-2,5-dienones **70**, **71** and **72** in 86, 88 and 91% yields, respectively[37]. In the case of 2,4,6-tri(*tert*-butyl)phenol (**73**), 2,6-di(*tert*-butyl)-*p*-benzoquinone (**74**) was produced in 96% yield through cyclohexa-2,5-dienone **75**[38,39], as shown in Scheme 13.

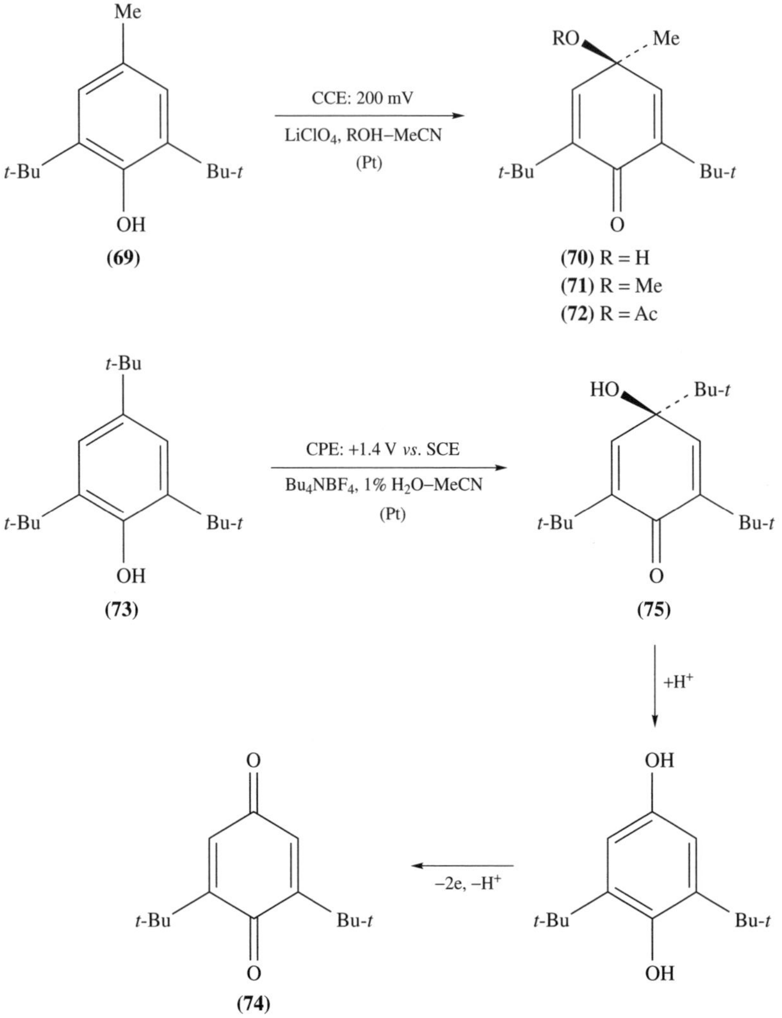

SCHEME 13. Anodic oxidation of 4-substituted 2,6-di(*tert*-butyl)phenol

Anodic halogenation also takes place; e.g. the substrate **69** was electrolyzed at constant current (200 mA) in CH_2Cl_2–pyridine to afford 2,6-di(*tert*-butyl)-4-chloro-4-methylcyclohexa-2,5-dienone (**76**) in 56% yield (Scheme 14)[37]. From the viewpoint of biological activity, fluorinated arenes are quite important because they are used as medicines, agrochemicals and building blocks for the synthesis of such products. Anodic fluorination of phenol was performed at constant current (5 mA cm^{-2}) using $Et_3N\cdot 3HF$ to afford 4,4-difluorocyclohexa-2,5-dienone (**77**) in 25% yield, as shown in Scheme 14. Herein, $Et_3N\cdot 3HF$ is used as a supporting electrolyte as well as a source of fluorine and has good electrical conductivity[40]. The resulting dienone is a useful intermediate in the synthesis of substituted fluorophenols. For example, catalytic hydrogenation of **77** afforded

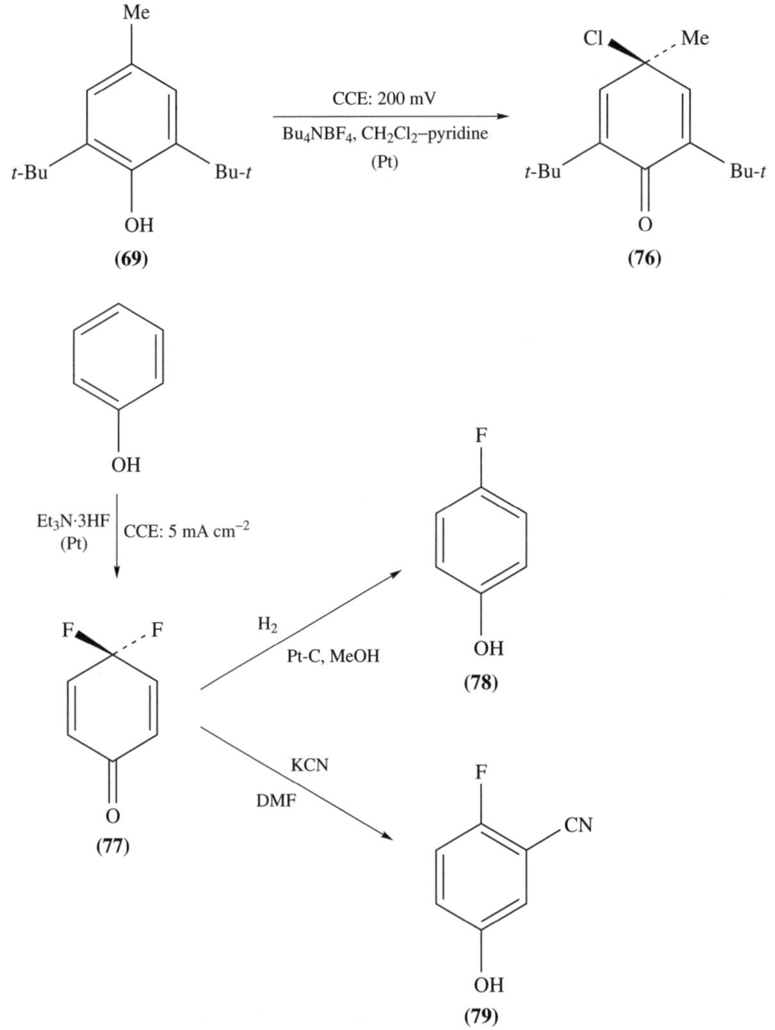

SCHEME 14. Anodic halogenation of phenols

4-fluorophenol (**78**) in 90% yield. **77** also underwent Michael addition by KCN in DMF to give 6-fluoro-3-hydroxybenzonitrile (**79**) in almost quantitative yield.

Similarly, anodic oxidation of 2,4,6-tri(*tert*-butyl)phenol (**73**) in MeCN containing *n*-propylamine provided the corresponding cyclohexa-2,5-dienone **80** (47% yield)[41]. Furthermore, electrochemical oxidation of **73** in MeCN–pyridine (1:1) yielded two pyridinium salts **81** and **82** in 44 and 23% yields, respectively (Scheme 15). Here, pyridine works as a nucleophile[42]. Anodic amination of phenols has been also studied[43,44].

From the viewpoint of synthetic organic chemistry, one of the most characteristic properties in electroorganic chemistry is a direct anodic alkoxylation introducing oxygen functionalities into aromatic rings, resulting in the formation of alkoxy-substituted aromatic

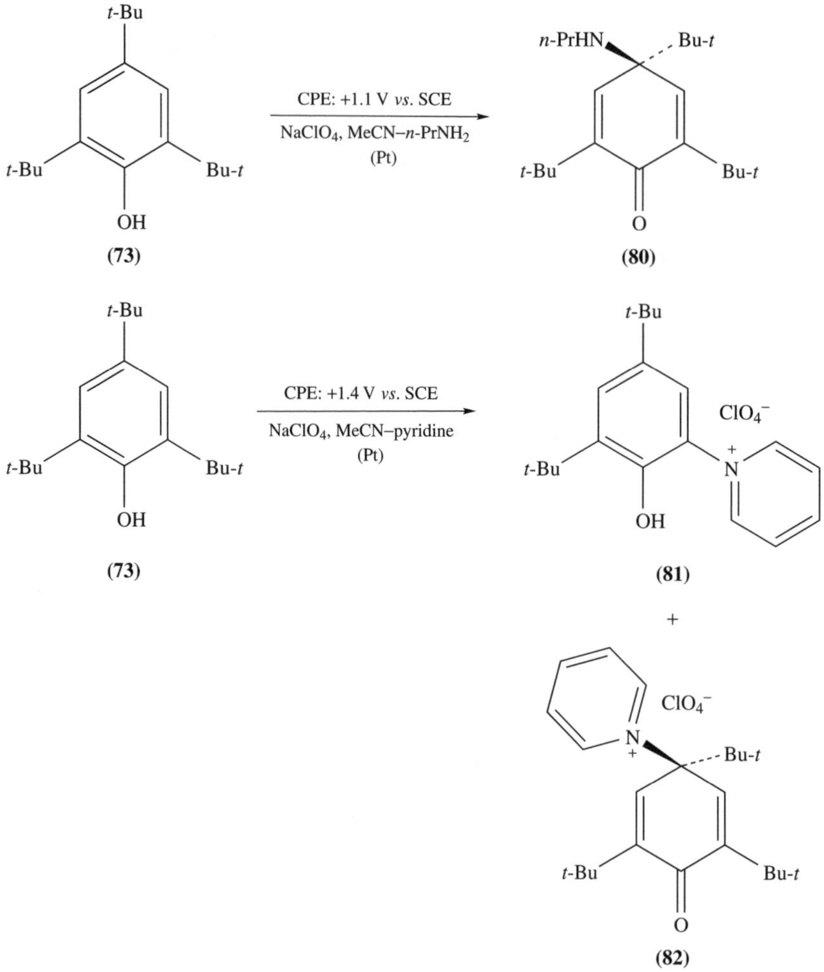

SCHEME 15. Anodic amination of 2,4,6-tri(*tert*-butyl)phenols

compounds, quinones, quinone mono- and bis-ketals. An excellent review on preparations, reactions and mechanistic considerations of both quinone mono- and bis-ketals has appeared[45] and only some recent examples are shown here.

Anthracycline antibiotics represented by daunomycin and adriamycin are well known as anticancer agents and their reaction mechanisms with DNA have been extensively studied. However, an approach to understand the mechanism of drug action based on organic synthesis is still open.

The easily available protected phenol ether **83** was subjected to anodic oxidation (+1.3 V *vs.* SCE) in MeOH containing NaOAc and LiClO$_4$ as a supporting electrolyte to afford in 53% yield quinone monoketal **84**, which reacted with 5-fluoro-3-cyanophthalide (**85**) in the presence of LDA to give anthraquinone **86**. This quinone was converted straightforwardly to the target molecule **87** (Scheme 16)[46]. Compound **87** and its analogs showed inhibitory activities against P388 cell line (IC$_{50}$: 0.2–0.4 μM).

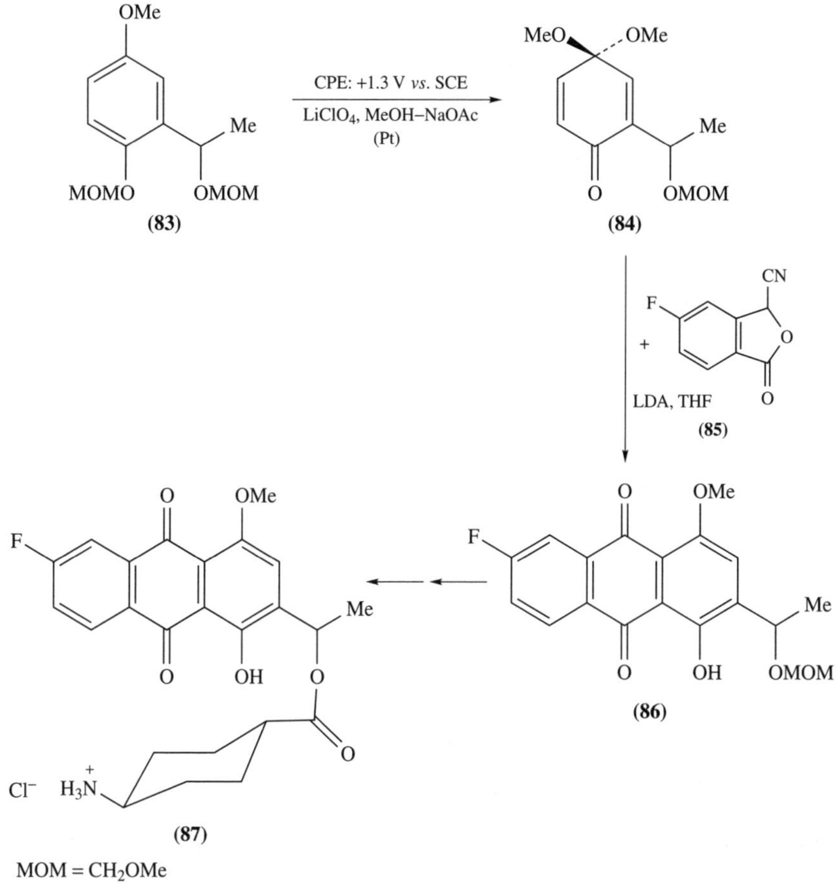

SCHEME 16. Electrochemical synthesis of a quinone monoketal

Quinone imine ketals have also been recognized to be quite useful for heterocycle synthesis. In a series of quinone mono- and bis-ketal chemistry, Swenton and coworkers carried out anodic oxidation of trifluoroacetamido-substituted *p*-methoxyphenols[47]. For example, the readily available *p*-methoxyphenol derivative **88** underwent constant current electrolysis (60 mA) in 2% LiClO$_4$ in methanol, followed by hydrolysis with 5% aqueous KOH to afford quinone imine ketal **89** in 82% overall yield, through quinone monoketal **90** (Scheme 17). Furthermore, acid treatment of **89** with TsOH provided 5-methoxyindole (**91**).

Several neolignans, found in *Piper futokazura* Sieb. et Zucc., are quite interesting because of their antifeedant activity against insects. 4-Substituted 2-allyl-5-methoxyphenol **92** was submitted to constant current electrolysis (10 mA; +900–1090 *vs.* SCE) to afford isodihydrofutoquinol A (**93**) in 51% yield, together with dienone **94** (15%) and a spiro

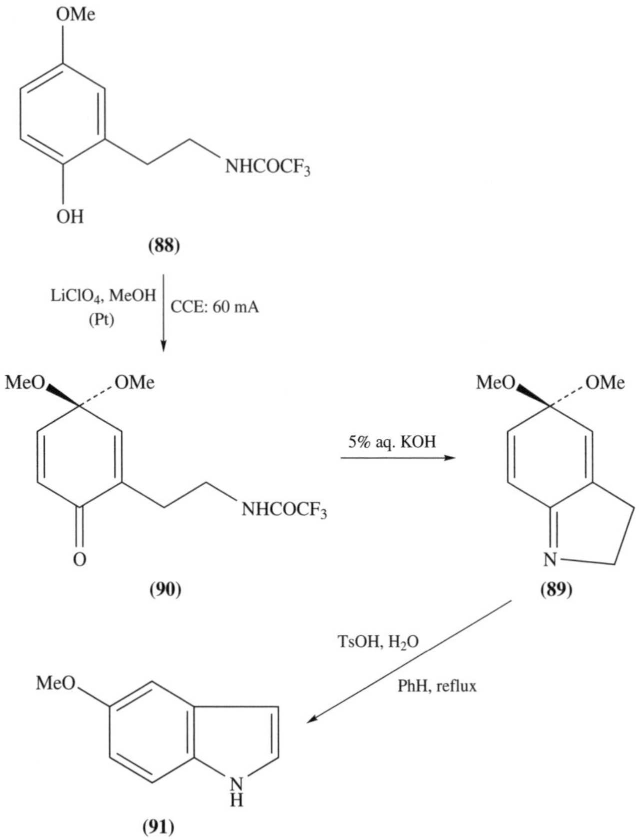

SCHEME 17. Electrochemical formation of a quinone imine ketal

compound **95** (2.3%)[48]. DDQ oxidation of **93** yielded selectively futoquinol (*trans*-**96**), which underwent photochemical reaction in hexane to give rise to isofutoquinol A and B (**97** and *cis*-**96**) in 67 and 16% yields, respectively (Scheme 18).

Intramolecular nucleophilic substitution of electrogenerated phenoxonium ions has been investigated[8,25]. In connection with naturally occurring bromo compounds, methyl 3,5-dibromo-4-hydroxyphenyl pyruvate oxime (**98**) was subjected to anodic oxidation (+1.3 V *vs*. SCE; 2.1 F mol^{-1}) in MeOH to afford spiro-isoxazole **99** in almost quantitative yield[49]. Methyl 3-bromo-4-hydroxyphenyl pyruvate oxime (**100**) was also electrolyzed under similar conditions to give three compounds **101**, **102** and **103** in 34, 14 and 17% yields, respectively. The latter two products are formed by C—O and C—C radical couplings, respectively, as shown in Scheme 19[50].

Tyrosine spirolactones are not only promising synthetic intermediates for bioactive natural products such as alkaloids and antibiotics but also synthons useful in peptide chemistry. For example, N-protected tyrosine derivatives **104** and **105**, prepared from 2,6-di(*tert*-butyl)-4-chloromethylphenol, were electrolyzed at a controlled potential (+1.3–1.4 V *vs*. Ag/Ag$^+$) in MeCN to give spirolactones **106** and **107** (64 and 85%, respectively)[51]. These spirolactones are used for peptide synthesis, as shown in Scheme 19.

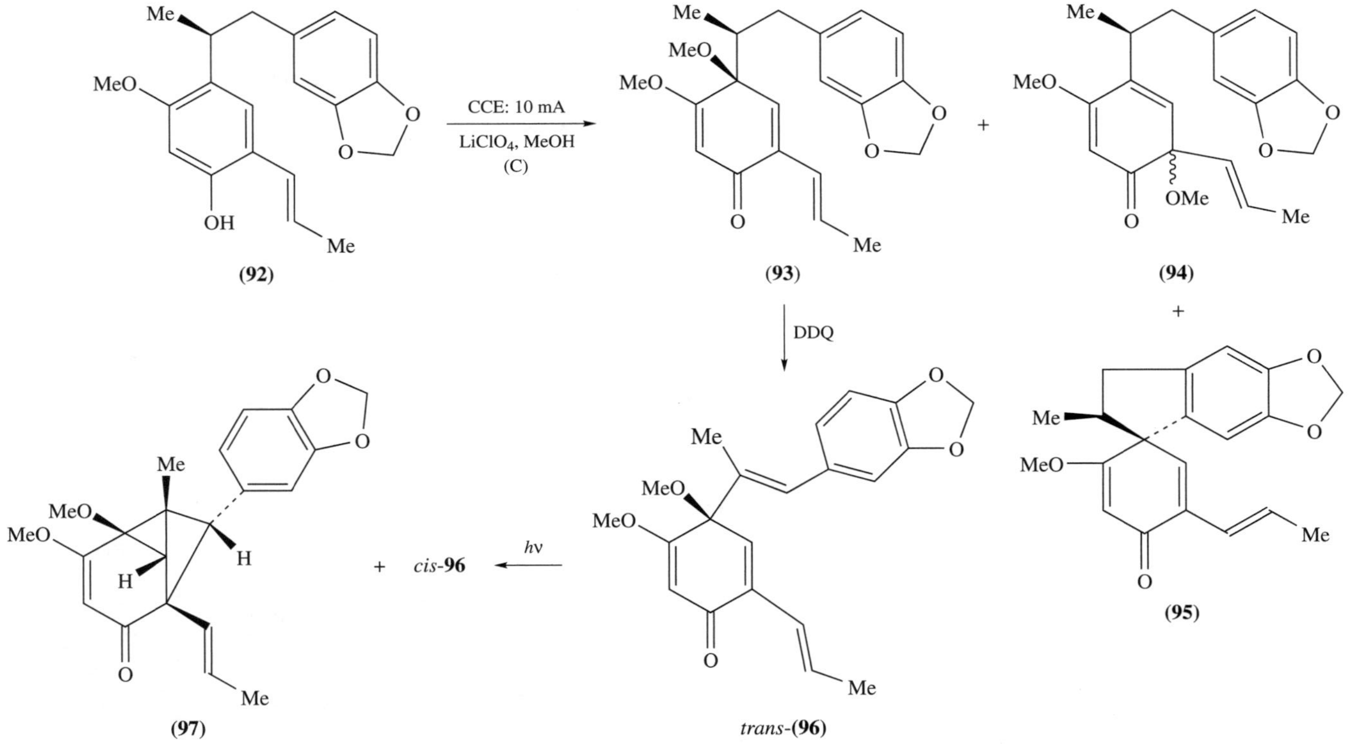

SCHEME 18. Isodihydrofutoquinol A, isofutoquinol A and related neolignans

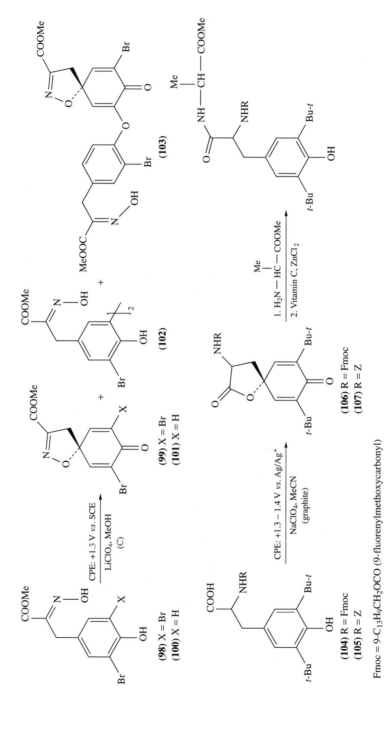

SCHEME 19. Anodic oxidation of tyrosine derivatives

17. Oxidation of phenols

Instead of nucleophiles such as H_2O, MeOH, RNH_2 and halide ions, both the aryl group and olefinic double bond will react with an electrogenerated phenoxonium ion to give carbon–carbon coupled products. In particular, electrooxidative coupling reactions of α,ω-diarylalkanes leading to cyclic diaryl ethers have been known to take place in a radical or cationic manner depending on the oxidation potential, the nature and location of substituents, the solvent systems and other factors, as cited in many books[1–9,14,52]. Electrochemical carbon–carbon bond formations will be described here.

On constant current electrolysis (0.27 mA cm^{-2}; +180–600 mV vs. SCE) in Ac_2O containing ethyl vinyl ether, 4,5-dimethoxy-2-methylphenol (**108**) was converted to two cyclohexa-2,4-dienones **109** and **110** and cyclohexa-2,5-dienone **111**, in 29, 18 and 8% yields, respectively[53] (Scheme 20). The product **110** is formed by nucleophilic substitution at the C-6 position followed by acetal formation with EtOH molecule generated initially from the ethyl vinyl ether while the C-4 position is attacked by ethyl vinyl ether to yield **111**.

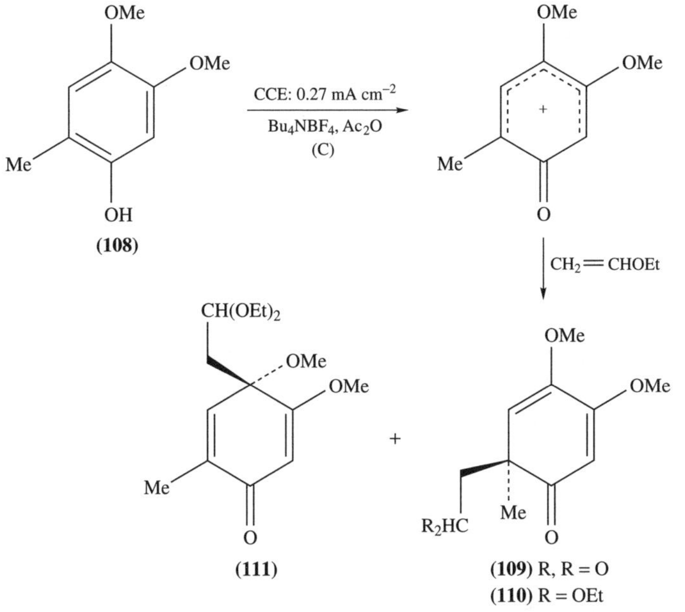

SCHEME 20. Anodic oxidation of 4,5-dimethoxy-2-methylphenol in the presence of ethyl vinyl ether

Intramolecular carbon–carbon bond formation of phenols bearing an olefinic side chain at the C-2 position is effected by using an electrochemical method. Anodic oxidation of 4-(2-alkenylphenyl)phenols (**112a–112c**) in 4:1 MeCN–MeOH provided spirocyclic cyclohexa-2,5-dienones (**113a–113c**) in 85, 70 and 16% yields, respectively, in competition with MeOH addition to the C-4 position leading to 4-methoxycyclohexa-2,5-dienones (**114a–114c**) (Scheme 21)[54]. Only in the case of **112c** was the corresponding dienone **114c** produced in 19% yield. These observations and other results suggest that the remarkable differences between **112a** and **112c** are due to the buttressing effect of an *o*-alkyl group. In this connection, compound **115** was electrolyzed at constant current in 4:1 MeCN–MeOH to afford a 1:1 mixture of dienones (**116** and **117**) in almost quantitative yield and no cyclization product was detected.

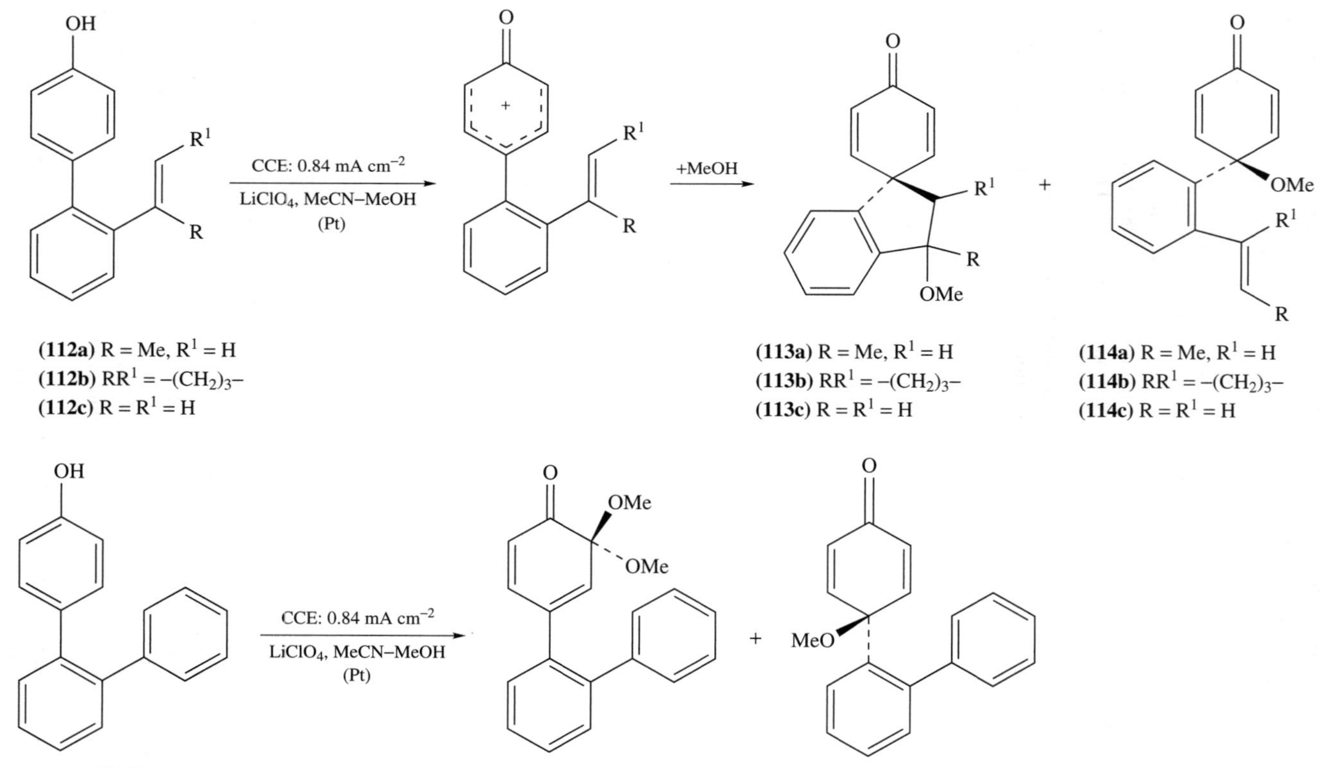

(112a) R = Me, R^1 = H
(112b) RR^1 = $-(CH_2)_3-$
(112c) R = R^1 = H

(113a) R = Me, R^1 = H
(113b) RR^1 = $-(CH_2)_3-$
(113c) R = R^1 = H

(114a) R = Me, R^1 = H
(114b) RR^1 = $-(CH_2)_3-$
(114c) R = R^1 = H

(115) (116) (117)

SCHEME 21. Anodic oxidation of 4-(2-alkenylphenyl)phenols

17. Oxidation of phenols

Two sesquiterpenes, γ- and δ-acoradiene (**118** and **119**), were synthesized efficiently using an electrochemical method as a key step. The readily available 4-substituted phenol **120** was submitted to constant current electrolysis in 2:1 MeOH–THF to afford three spiro compounds (**121**, **122** and **123**) in 43% yield (relative ratio: **121/122/123** = 1:2:1). All of them were readily converted to both **118** and **119**. However, the use of only THF as a solvent provided the corresponding dimer **124** in 80% yield[55] (Scheme 22).

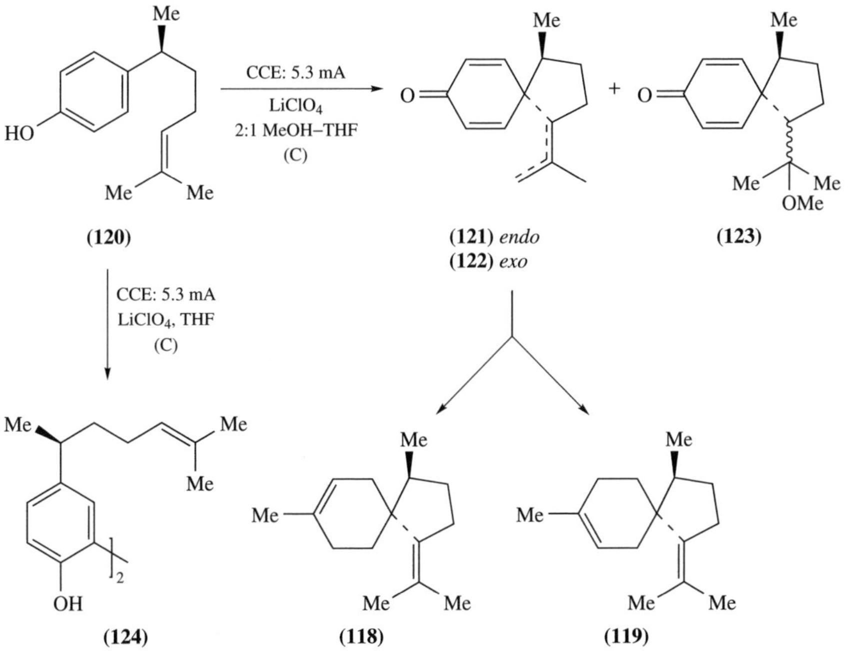

SCHEME 22. Electrochemical synthesis of γ- and δ-acoradiene

Similarly, the 4-substituted anisole **125** underwent constant current electrolysis (11.2 mA, 2F mol^{-1}) in 20% MeOH–CH$_2$Cl$_2$ leading to a spiro compound **126** in 51% yield[56], as shown in Scheme 23.

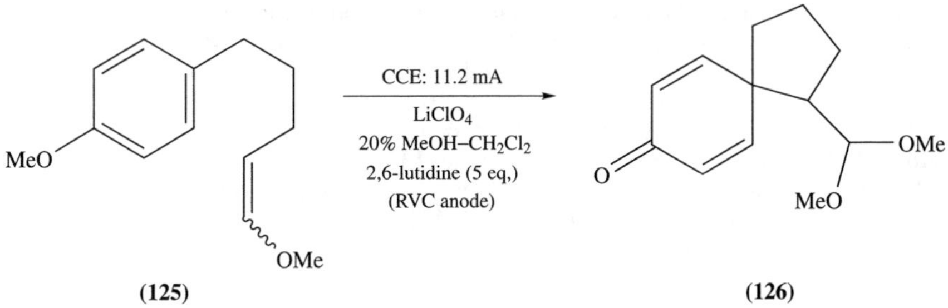

SCHEME 23. Anodic oxidation of an anisole derivative

Of physiologically active substances isolated from marine sources, the pyrroloiminoquinone alkaloids family exhibits antitumor activities derived from the unique highly-fused structure. The first synthesis of discorhabdin C (**127**) was performed by means of an electrochemical method as a key step[57]. The key substrate **128**, efficiently prepared starting from 4,4-dimethoxy-5-nitrobenzaldehyde, was submitted to constant current electrolysis (3 mA; +1.2–1.8 V vs. SCE) in anhydrous MeCN to give rise to discorhabdin C in 24% yield, together with a minor compound **129** (6%) (Scheme 24). After a while, discohabdin C was also synthesized by using PhI(OCOCF$_3$)$_2$-promoted oxidation as a key step[58].

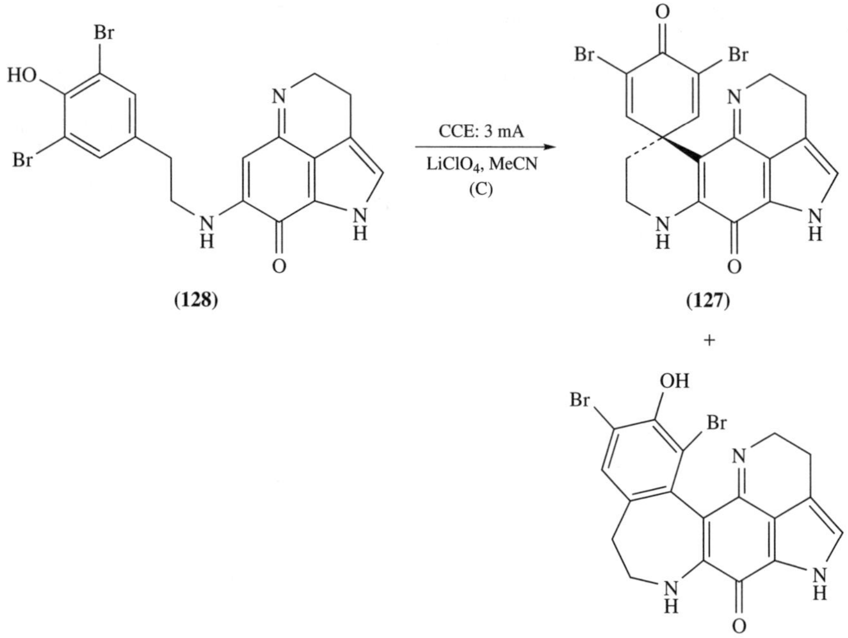

SCHEME 24. Electrochemical synthesis of discorhabdin C

As already shown in Scheme 12, nucleophilic substitution takes place at *ortho*-positions leading to cyclohexa-2,4-dienones such as **65** and **66** in competition with *para*-substitution. In a synthetic study of neolignans isolated from *Heterotropa takaoi* M. and related plants, electrochemical oxidation of 4-allyl-2,6-dimethoxyphenol (**130**) was carried out at constant current (0.31 mA cm^{-2}, +620–660 mV vs. SCE) in MeOH containing LiClO$_4$ to afford in 36% yield the desired cyclohexa-2,4-dienone **131**, which was readily converted to asatone (**132**), isoasatone (**133**), heterotropanone (**134**), heterotropatrione (**135**) and related neolignans, as shown in Scheme 25[21,59]. **134** was synthesized by Diels–Alder reaction of 5-allyl-1,2,3-trimethoxybenzene (ATMB) with **131** regenerated from asatone (**132**) by a retro-Diels–Alder reaction.

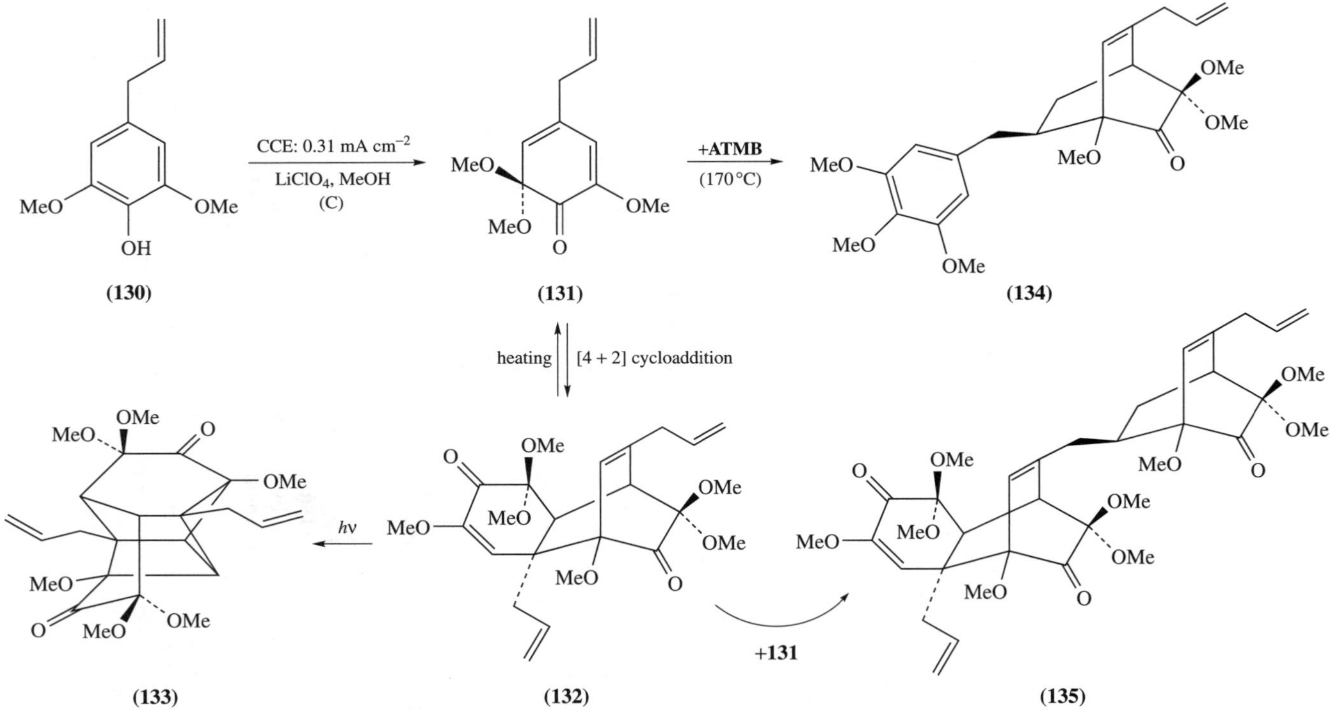

SCHEME 25. Electrochemical synthesis of asatone and related neolignans

Silydianin (**136**), found in the fruits of *Silybum marianum* G., shows an antihepatotoxic activity and has a unique 9-oxaisotwistane skeleton. Generally, 9-oxaisotwist-8-en-2-ones have been synthesized from the corresponding phenols by means of the Wessely oxidation method using lead tetraacetate[60]. However, this method is not applicable to acid-sensitive phenols bearing a methoxymethyl ether group.

On controlled electrolysis (950 mV *vs*. SCE, 2 F mol^{-1}) in 2:1 MeOH–THF, the phenol **137** prepared from 3,4-dihydroxybenzaldehyde underwent 2e oxidation resulting in the formation of 9-oxaisotwist-8-en-2-one **138**, in 82% yield, which was smoothly converted to deoxysilydianin methyl ether **139** (Scheme 26)[61].

On constant current electrolysis (80 mA) in 8:1 MeCN–AcOH including LiClO$_4$ as a supporting electrolyte, formal [3 + 2] cycloaddition took place between *p*-methoxyphenol (**140a**) and *trans*-1,2-dimethoxy-4-propenylbenzene (**141**) in equimolar amounts to afford dihydrobenzofuran **142a** in 61% yield. The use of the *cis*-olefin also provided **142a** in 50% yield, indicating that these reactions proceed in a stepwise manner (Scheme 27)[62]. In the case of 3,4-dimethoxyphenol (**140b**), equimolar amounts of **140b** and **141** gave only a 14% yield of the adduct **142b**. However, the yield of the reaction could be increased to 61% when a 3-fold excess of **141** was used. In particular, the 4-methoxy group is important for obtaining good yields of the cycloaddition products; neither phenol nor *m*-methoxyphenol gave isolatable amounts of product.

Similarly, anodic oxidation of *p*-methoxyphenol (**140a**) was carried out in the presence of the substituted propenylbenzene **143** using teflon-coated electrode to afford the corresponding dihydrobenzofuran **144** in 80% yield[63] (Scheme 27). The hydrophobic coating on the electrode protected the highly reactive intermediate from the solvent and enhanced the reaction with **143**.

p-Methoxyphenol (**140a**) underwent constant current electrolysis (0.27 mA cm^{-2}; +400–800 mV *vs*. SCE) in Ac$_2$O containing dihydrofuran or tetrahydropyran to afford the corresponding dihydrobenzofurans **145a** and **145b** in 11 and 33% yields, respectively. Anodic oxidation of 4,5-dimethoxy-2-methylphenol (**108**) in the presence of furan yielded a 1:1 adduct **146** (30%), as shown in Scheme 28. Herein, the resulting phenoxonium ion must be attacked by furan[53].

From the viewpoints of biogenesis and biological activity, the neolignans found in *Aniba* and *Magnolia* species are quite attractive. A pioneering work in this field was carried out by Büchi and Mak, who could successfully synthesize both guianin and futoenone in short reaction sequences[64]. Electrochemical methodology is used efficiently for syntheses of aniba and magnolia neolignans.

In a series of synthetic studies on these neolignans, 2-allyl-4,5-dimethoxyphenol (**147**) was submitted to constant current or controlled potential electrolysis in 90% aq. MeCN, MeOH and 2:1 MeOH–AcOH containing excess *trans*-isosafrole to afford 2-allyl-5-methoxy-*p*-benzoquinone (**148**), 2-allyl-4,4,5-trimethoxycyclohexa-2,5-dienone (**149**) and one of the Aniba neolignans (**150**) in 83, 87 and 81% yields, respectively. Here, the neolignan **150** is formed selectively by a cationic [5 + 2] cycloaddition. The use of *cis*-safrole instead of the *trans*-isomer provided an *exo*-addition product **151** and futoenone **152** in 25 and 15% yields, respectively[62,65], as shown in Scheme 29. **152** must be produced from the initially formed *endo*-addition product **153**.

4,5-Dimethoxy-2-methylphenol (**108**) which was electrolyzed in 2:3 Ac$_2$O–AcOH including *trans*-1,2-dimethoxy-4-propenylbenzene (**140**) underwent cationic [5 + 2] cycloaddition affording in 80% yield the corresponding bicyclo[3.2.1]oct-3-en-2,8-dione **154**, which was readily converted to helminthosporal (**155**), a toxic sesquiterpene (Scheme 30)[66].

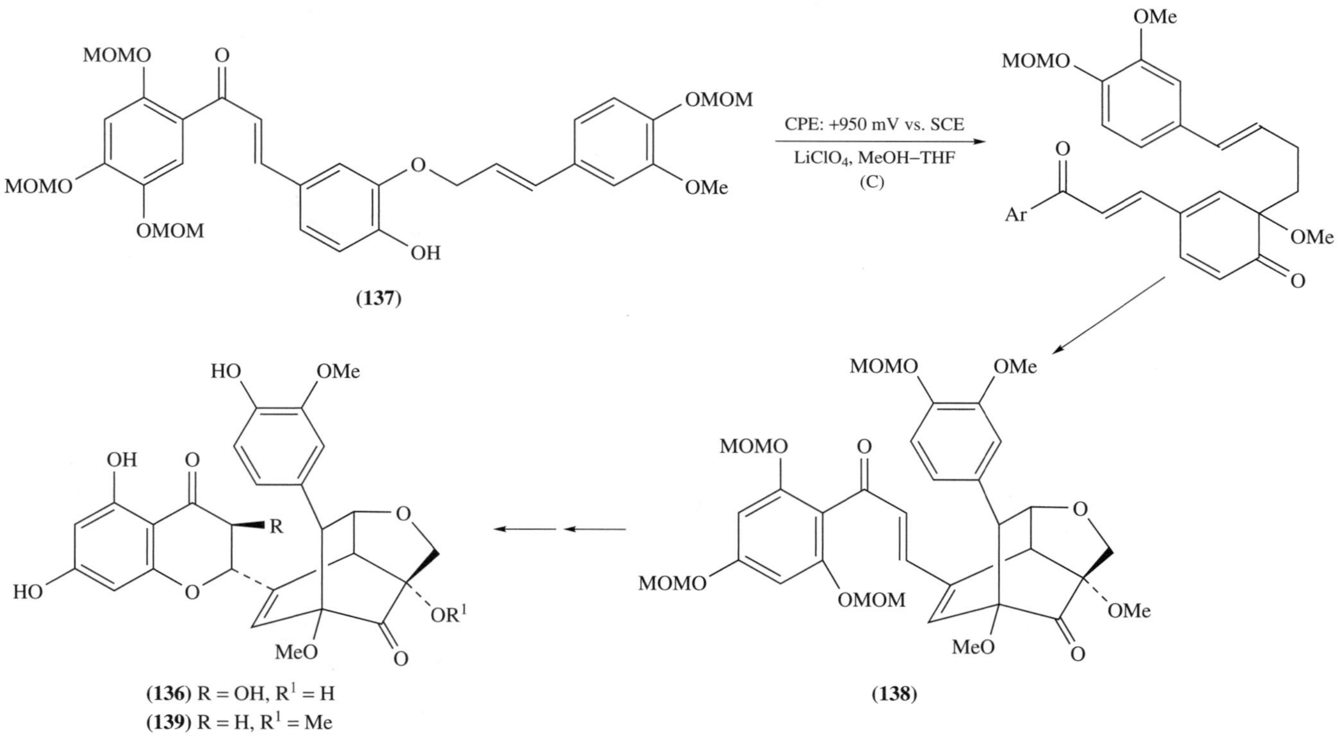

(136) R = OH, R¹ = H
(139) R = H, R¹ = Me

MOM = MeOCH$_2$
Ar = 2,4,5-tri(MeOCH$_2$)C$_6$H$_2$

SCHEME 26. Electrochemical synthesis of deoxysilydianin methyl ether

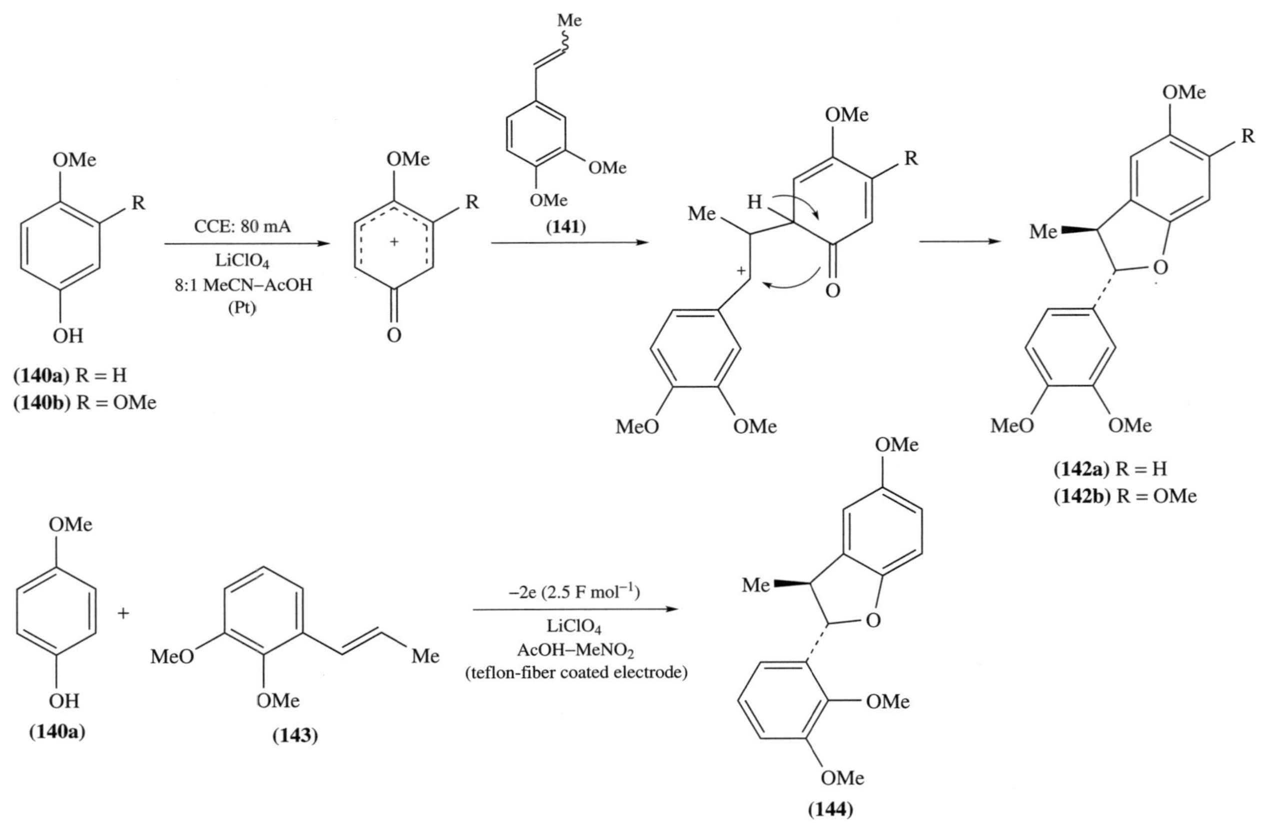

SCHEME 27. Anodic oxidation of *p*-methoxyphenols in the presence of methoxy-substituted propenylbenzenes

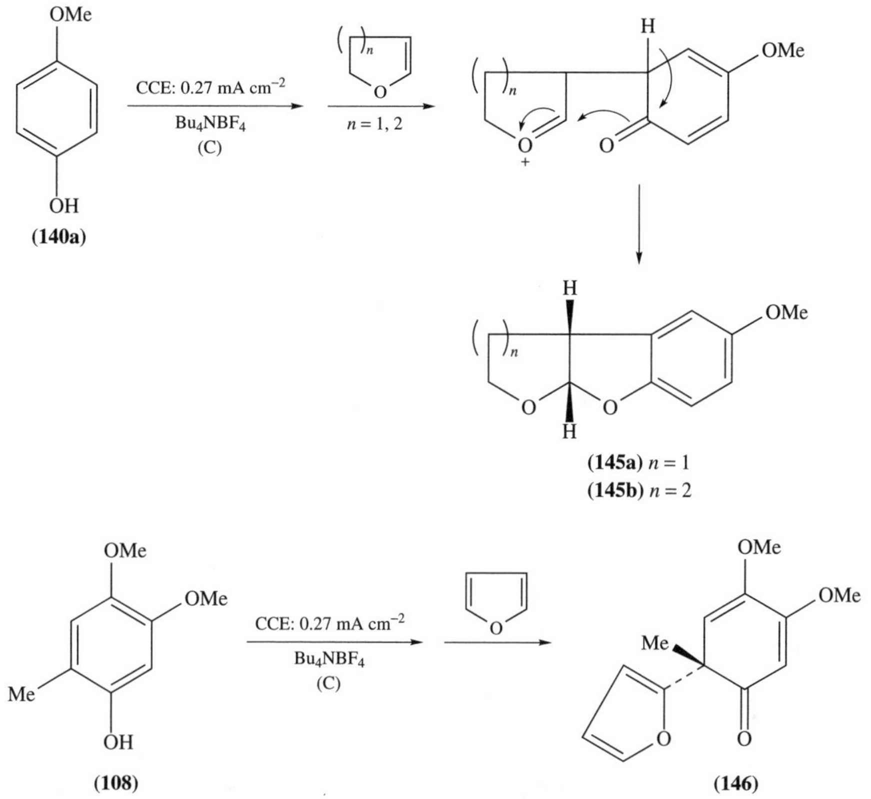

SCHEME 28. Anodic oxidation of p-methoxyphenols in the presence of cyclic enol ethers and furan

3,4-Dimethoxyphenols such as **156** bearing a double bond at the side chain undergo anodic intramolecular cycloaddition resulting in the formation of three possible compounds **157**, **158** and **159** (Scheme 31). These compounds are promising synthetic intermediates for a variety of sesquiterpenes.

Silphinene (**160**), a constituent of the roots of *Silphium perfoliatum*, has attracted a considerable attention of synthetic chemists. Electrochemical methodology is used for its synthesis. On constant current electrolysis (1.18 mA; +750–1200 mV vs. SCE) in Ac$_2$O, the phenol **161**, readily prepared from 3,4-dimethoxyphenol, underwent intramolecular [5 + 2] cycloaddition to give in 59% yield the desired tricyclic compound **162**, which was successfully converted into silphinene through a bicyclic compound **163** (Scheme 32)[62].

Pentalenene (**164**) has the same carbon skeleton as that of silphinene (**160**). However, the methyl substituents are located in different positions. Anodic oxidation of the phenol **165** in 3:2 MeOH–AcOH was carried out at constant current (71.8 mA; +540–1500 mV vs. SCE) to afford two tricyclic epimers **166** and **167** in 64 and 16% yields, respectively (Scheme 32). The major one was further converted to the target molecule (**164**) through an intermediate **168**[67].

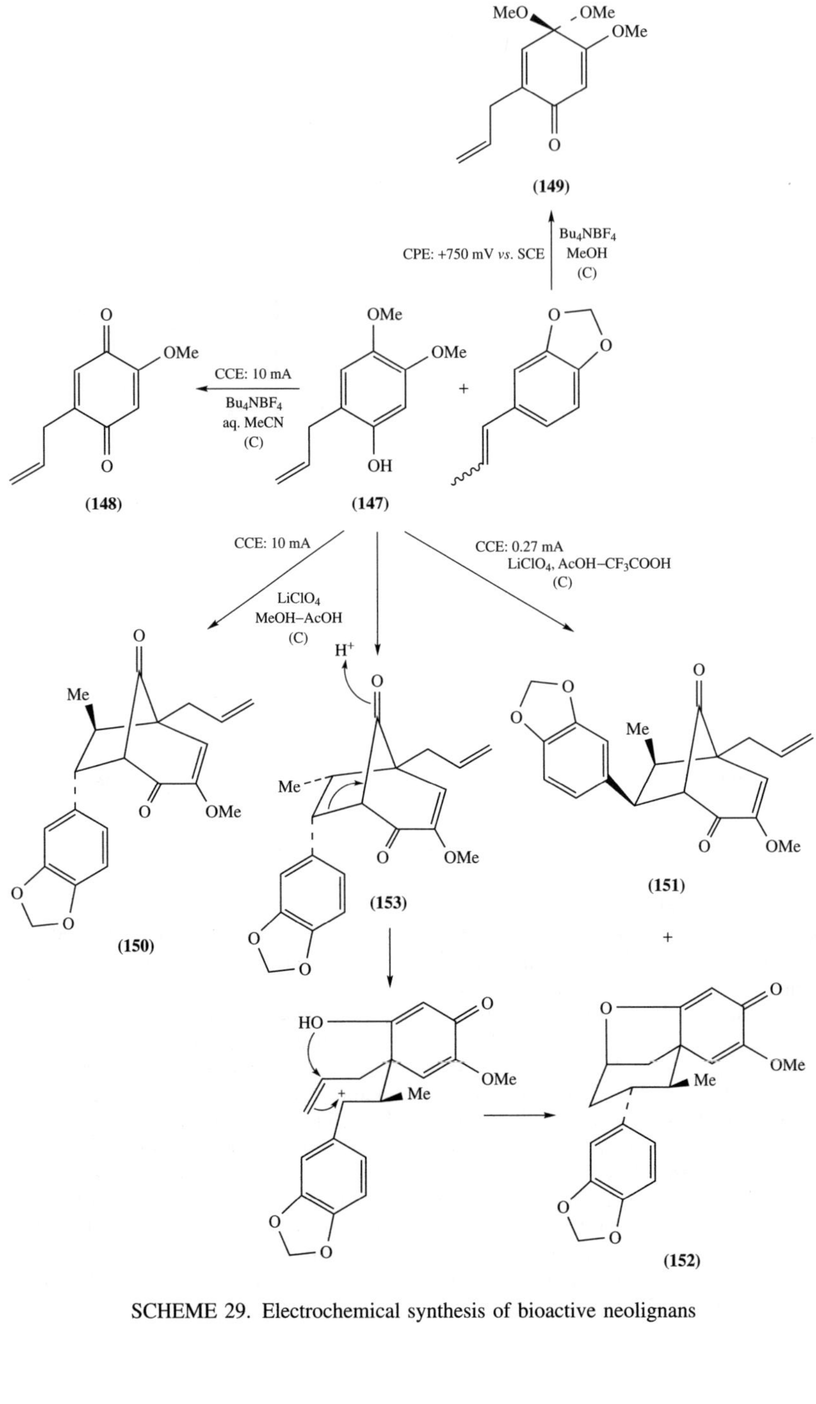

SCHEME 29. Electrochemical synthesis of bioactive neolignans

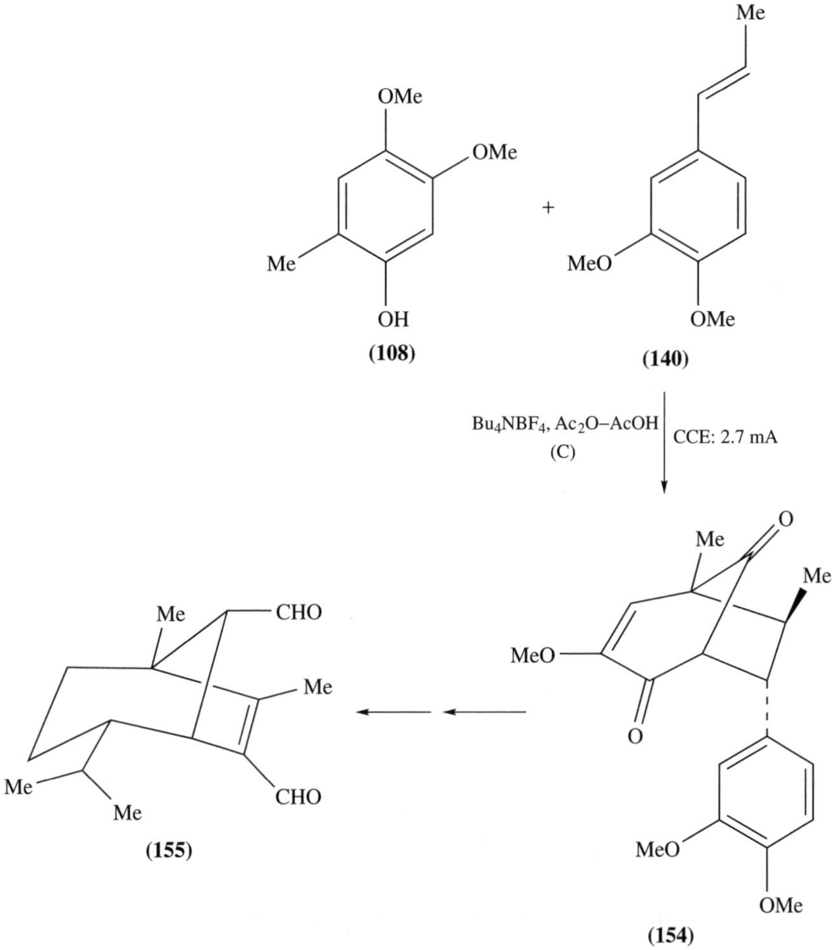

SCHEME 30. Total synthesis of helminthosporal

Acourtia isocedrene (**169**) is one of the highly oxygenated isocedrenes first isolated from *Acourtia Nana*. Retrosynthetic pathways are shown in Scheme 33, where the resulting cycloaddition product (**170**) from penta-substituted phenol **171** has the same carbon skeleton as that of **169**, while addition of one carbon unit to **172** is needed in the case of a tetra-substituted phenol **173**.

On constant current electrolysis (9.4 mA; 2 F mol^{-1}), **171** underwent intramolecular cationic [5 + 2] cycloaddition to afford the β-isomer **170** as a sole product (34%). In the case of the tetra-substituted phenol **173**, it was converted to a mixture of two stereoisomers (**172a** and **172b**) in 70% yield (relative ratio: $\alpha/\beta = 3/1$), as shown in Scheme 34[68]. Both of them were converted successfully to the target molecule **169**[69].

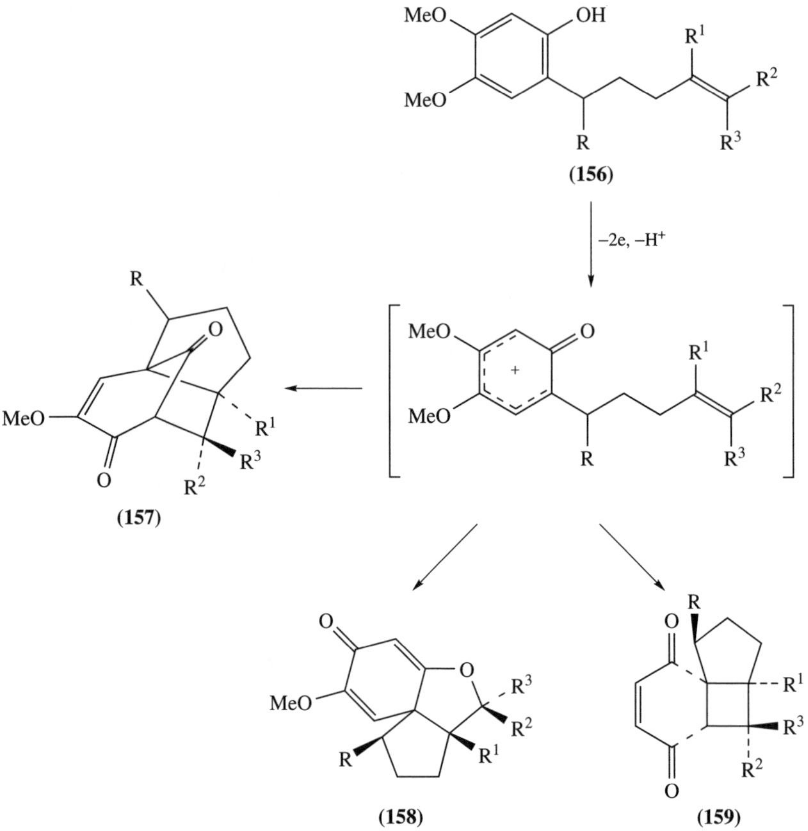

SCHEME 31. Anodic oxidation of 6-substituted 3,4-dimethoxyphenols

R, R^1, R^2, R^3 = H, Me, CH_2OAc, COOMe and others

8,14-Cedranoxide (**174**), a constituent of *Juiperus foetidissima* W., has been synthesized efficiently starting from 3,4-dimethoxyphenol through 6-acetoxymethyl-2,6-dimethyl-9-methoxytricyclo[5.3.1.0^{1,5}]undec-9-en-8,11-dione (**175**), which can be prepared by means of an electrochemical method.

The phenol **176**, prepared from 3,4-dimethoxyphenol, was submitted to constant current electrolysis (2.5 mA; +900–1200 mV vs. SCE) to afford two tricyclic stereoisomers **175** and **177**. Their yields and relative ratio varied with the solvent systems (3:2 and 5:1 Ac_2O–AcOH and Ac_2O). Acetic anhydride as the solvent provided the best result (**175**: 64%; **177**: 16%). The former was converted into 8,14-cedranoxide (Scheme 35)[62,70].

2-*epi*-Cedrene-isoprenologue (**178**), first isolated from *Eremophila georgei* D, constitutes a new class of diterpenes bearing a tricyclic cedrane-type skeleton in their molecule, whose synthesis is shown in Scheme 36. The key intermediate (**179**) has been synthesized electrochemically from the corresponding phenol **180**; electrolysis of **180** in Ac_2O provided a mixture of two tricyclic stereoisomers (**179a** and **179b**) in 68% yield ($\alpha/\beta = 5/2$). Both of them were further converted into the target molecule[71].

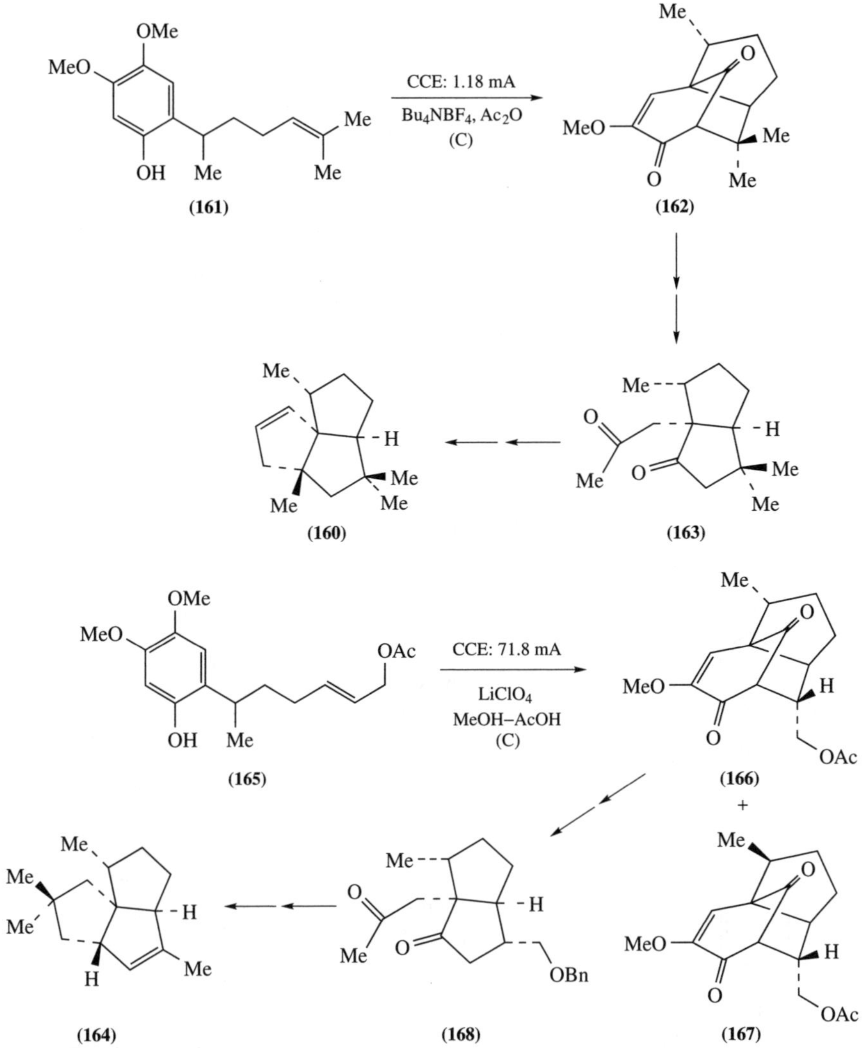

SCHEME 32. Total synthesis of silphinene and pentalenene

From the viewpoint of natural products synthesis, retro-aldol condensation of the electrosynthesized tricyclic compounds **181** and **182** provided the selective formation of *trans*-hydroazulene **183** and triquinane **184** in good yields, respectively[72] (Scheme 37). Herein, the selective attack of a methoxy anion to the β-diketone is due to the stereochemistry of the aryl group introduced to the C6-position.

In the case of 3,4-dimethoxyphenol (**185**) bearing an α,β-unsaturated CO system, one-pot synthesis of the corresponding tricyclic compound (**186**) was performed in *ca* 80% yield by a combination of electro- and photochemical reactions, as shown in Scheme 38[73]. Here, intramolecular cationic [5 + 2] cycloaddition does not take place, because of the

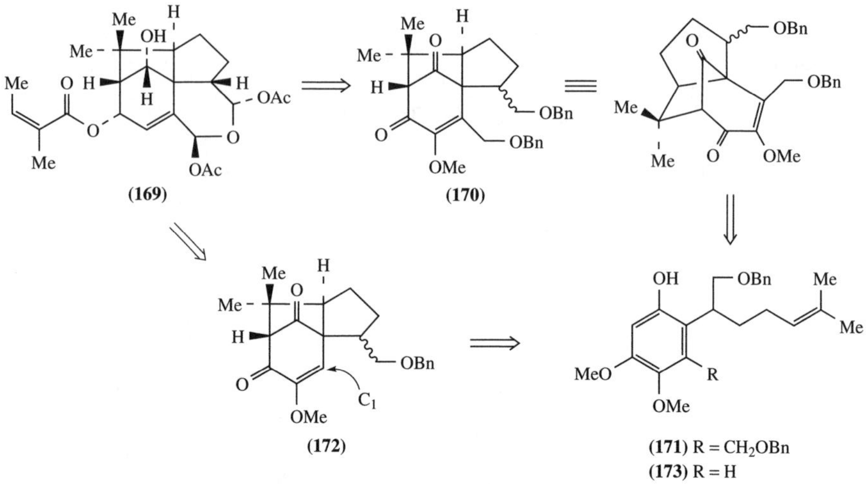

SCHEME 33. Retro-synthetic pathways toward acourtia isocedrene

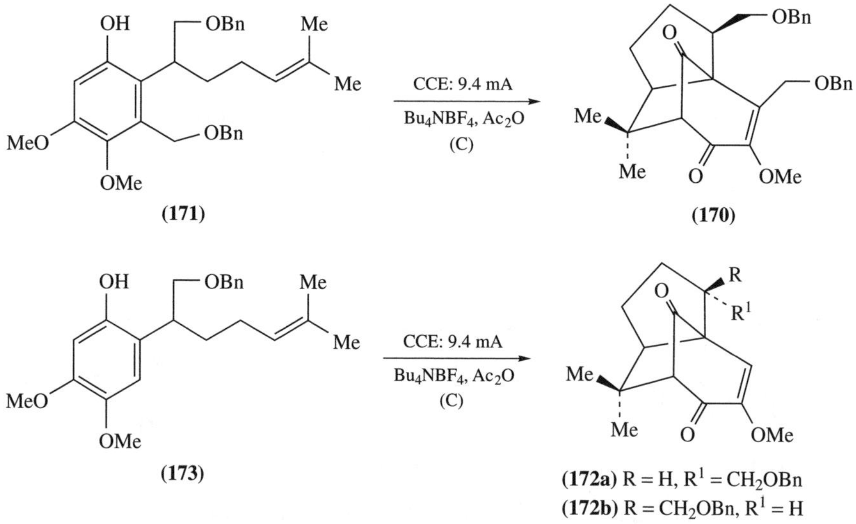

SCHEME 34. Anodic oxidation of penta- and tetra-substituted phenols

electron-deficient double bond. Compound **186** was further converted into angular and linear triquinanes such as **187** and **188**.

One-pot synthesis of isoitalicene (**189**) was also accomplished by similar procedures. The phenol **161**, cited in Scheme 32, was subjected to constant current electrolysis (0.9 mA; 510–1200 mV *vs.* SCE) in 5:1:3 EtOAc–*i*-PrOH–H_2O under irradiation to afford in 80% yield the desired tricyclic compound (**190**), which was readily converted to isoitalicene (Scheme 38)[74].

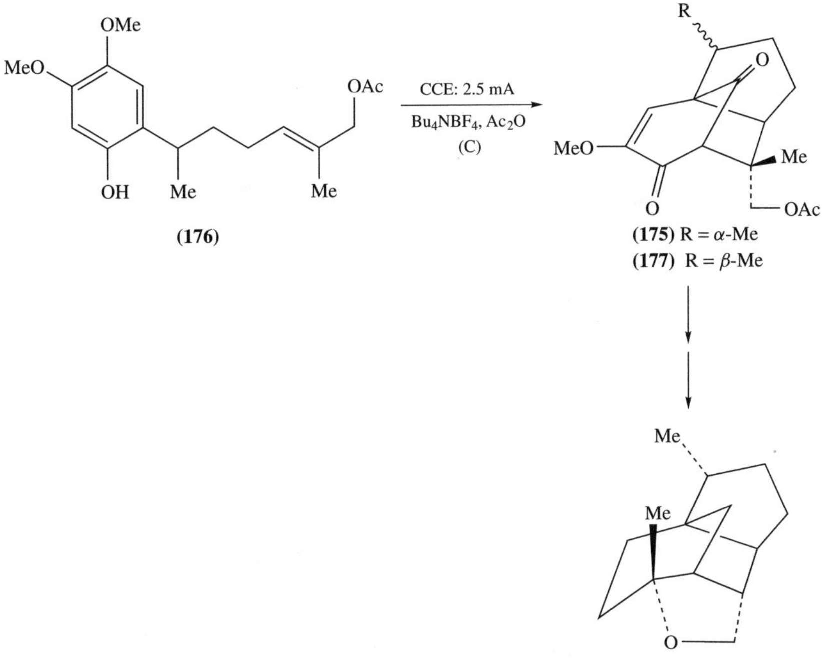

17. Oxidation of phenols

SCHEME 35. Total synthesis of 8,14-cedranoxide

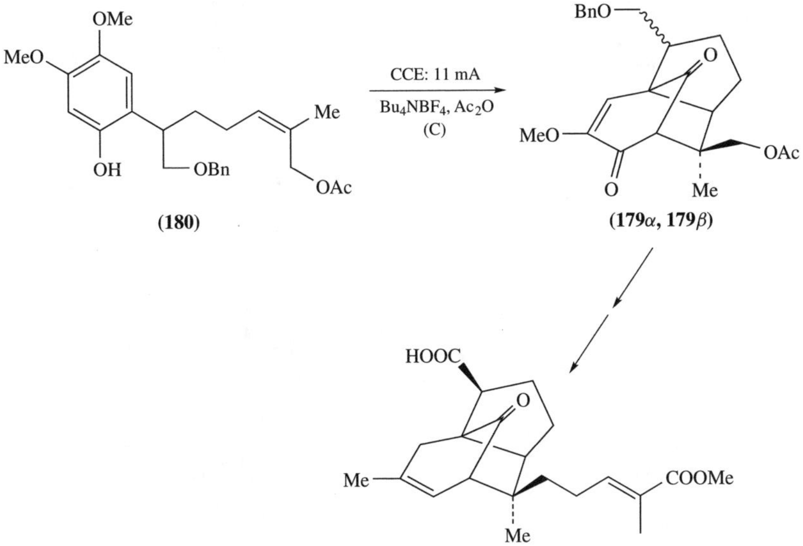

SCHEME 36. Total synthesis of 2-*epi*-cedrene-isoprenologue

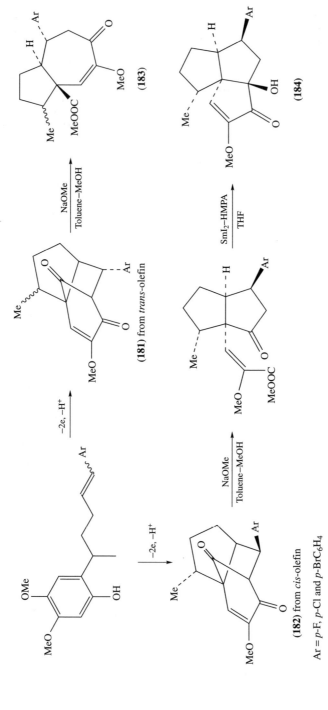

SCHEME 37. Synthesis of hydroazulene and triquinane derivatives

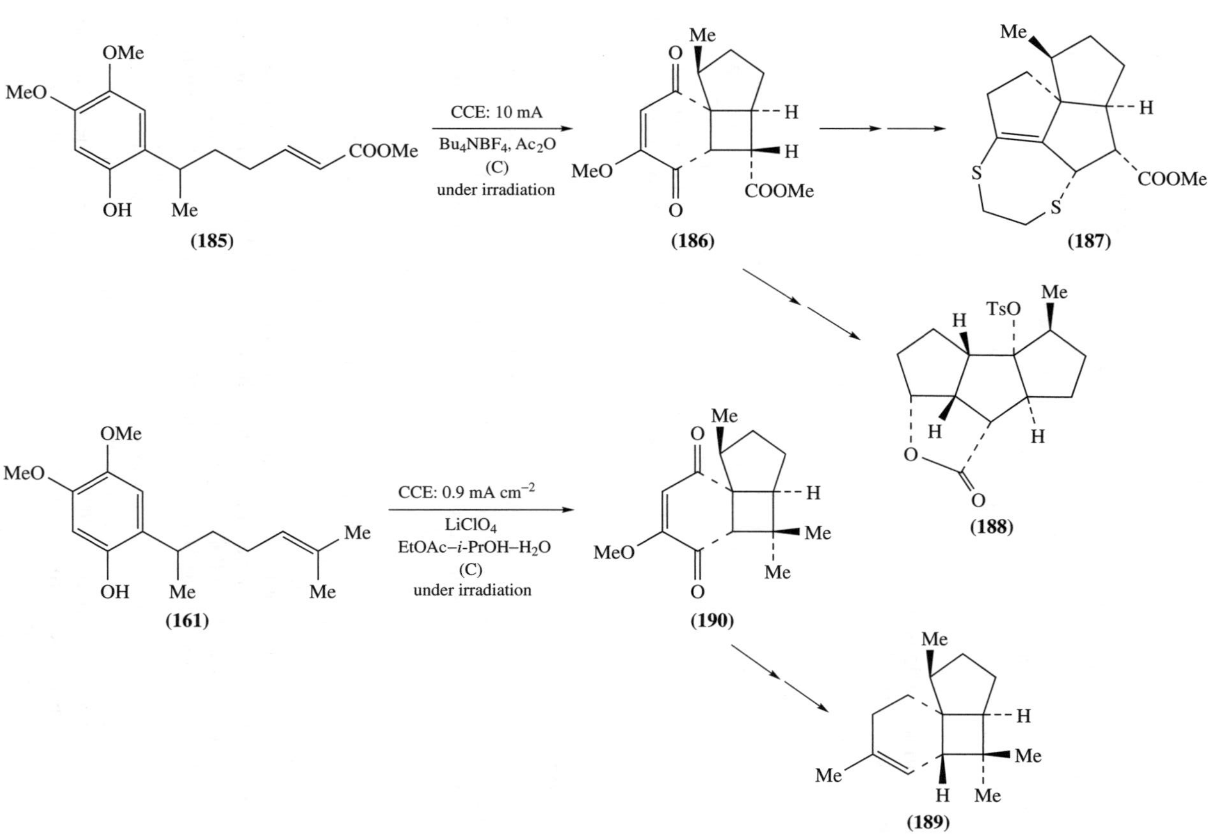

SCHEME 38. Synthesis of triquinanes and isoitalicene

Finally, in the case of such a phenol as grandinol (**191**), an indirect method using a mediator is more favorable than a direct one. Euglobals isolated from *Eucalyptus* sp. show potent inhibitory activity against Epstein–Barr virus activation. From the biogenetic point of view, these euglobals are composed of grandinol and monoterpene moieties.

The phenol **191** was submitted to controlled potential electrolysis (0.45 V *vs.* SCE) in nitromethane containing 0.2 equivalent DDQ in the presence of α-phellandrene (**192**) to afford the corresponding equilibrium mixture of *o*-quinone methide **193a** and **193b**. The redox cycle of DDQ was constructed on the teflon-fiber coated electrode. A Diels–Alder reaction between **193a** or **193b** and **192** afforded euglobal T1 (**194**) and euglobal IIc (**195**) in 51 and 28% yields, respectively (Scheme 39). In the case of pinene (**196**), both euglobal G-3 (**197**) and G-4 (**198**) were produced in 89% yield (G-3/C-4 = 1) in the reaction with **191**. Of a variety of solvents examined, nitromethane was the most effective[75].

B. Oxidation with Dioxygen, Hydrogen Peroxide and Alkyl Hydroperoxide Catalyzed by Metal–Base Complexes

From the viewpoints of organic synthesis including industrial process and understanding the reaction mechanism of a variety of metalloenzymes, selective oxidations of phenols catalyzed by transition metal complexes capable of activating oxygen have long been studied[76,77], so that many efforts have been made to prepare more efficient metal complexes by a combination of metals and new ligands. In parallel, the oxygenation mechanism of phenols has also been examined by using simple phenols such as 2,4,6-tri-, 2,4-di- and 2,6-di(*tert*-butyl)phenol, because of both easy detection of products and simplification of the reaction pathways. A number of invaluable books on these topics have been published[78].

1. Dioxygen–metal complexes

Related to copper-containing enzymes such as laccase and tyrosinase, recent studies have been conducted on the structural characterization of the reactive species generated from molecular oxygen and copper complexes. A continuous effort has also been directed toward the efficient utilization of such oxygen–copper complexes as oxidants, in industrial processes, which will hopefully replace metal compounds such as chromate, manganate and others.

Phenol oxidation has been well known to be effected with cuprous chloride in the presence of nitrogen-containing compounds such as pyridine, oximes and others under an oxygen atmosphere. Oxidation of phenol was performed by CuCl in MeOH containing pyridine for 60 h to afford *cis,cis*-muconic acid monomethyl ester (**199**) as a sole product (44%)[79], as shown in Scheme 40. It is believed that **199** is formed through the intermediacy of catechol (**200**). In fact, on oxidation with the pyridine methoxy cupric chloride complex PyCu(Cl)OMe, which exists as a dimer, in MeOH and pyridine, catechol was readily converted into **199** in 80–85% yield, thus representing a good nonenzymatic model reaction for pyrocatechase[80]. Copper-promoted phenol oxygenation also provided the corresponding *o*-quinone, probably through catechol[81]. However, in the case of 4-methoxycarbonylphenol (**201**), the corresponding *o*-benzoquinone (**202**) was proved to be formed directly from sodium 4-methoxycarbonylphenolate generated *in situ* from **201**. On exposure of the complex formed from the binuclear Cu(I) complex of N,N,N',N'-tetra-[2-(*N*-methylbenzimidazol-2-yl)ethyl]-*m*-xylenediamine and the sodium phenolate to O_2 in MeCN, 40–50% conversion of **201** to **203** through the *o*-quinone

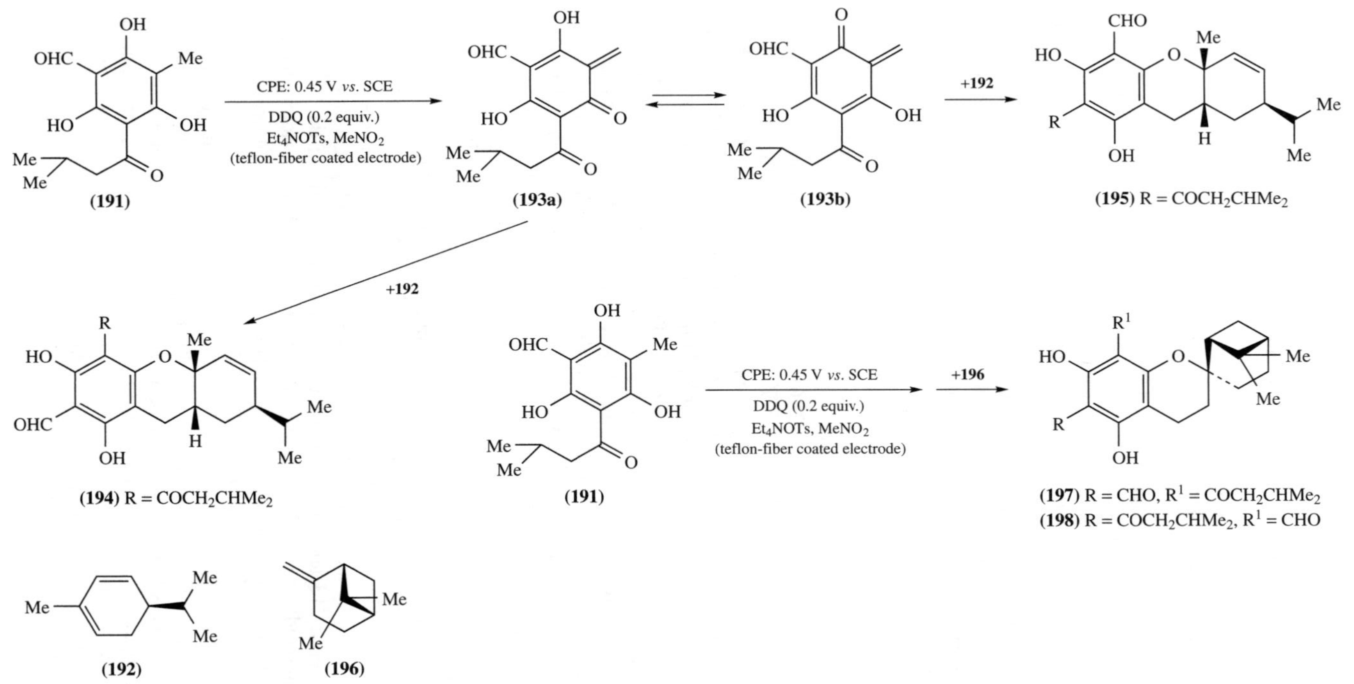

SCHEME 39. Electrochemical synthesis of euglobals

202 was observed (Scheme 40). It should be noted that the *p*-methoxycarbonyl group retards the catechol to *o*-quinone oxidation under these conditions, indicating that 4-methoxycarbonylcatechol is not an intermediate during the oxidation[82]. Furthermore, the yield of **203** could rise to 60% simply by stirring an equimolar mixture of **201** and N,N'-bis[2-(2-pyridyl)ethyl]benzylamine with 1.5 equiv. of copper powder in MeCN under an oxygen atmosphere. In the case of 2,5-dimethylphenol (**204**), a combination of oxidation and Michael addition also provided a 90% yield of the corresponding *o*-quinone **205** (Scheme 40).

2,6-Dimethylphenol (**206**) bearing an electron-donating group shows a different behavior from that of phenol and is known to undergo oxidative C—C and C—O couplings catalyzed by copper–amine complexes to afford mainly 3,3′,5,5′-tetramethyldiphenoquinone (**207**) and a polymer, the linear poly(phenylene ether) (**208**), respectively[83]. Three mechanistic pathways (radical, electrophilic and nucleophilic) were proposed for the oxidative coupling of **206**. Nucleophilic substitutions of the resulting phenoxonium ion from **206** leading to two C—C and C—O dimers were shown to be most plausible routes based on *ab initio* unrestricted Hartree–Fock calculations performed with a 6–31G* basis set on **206** and its deprotonated derivatives. Furthermore, *ab initio* calculations also support the quinone–ketal mechanism for a further C—O coupling oligomerization (Scheme 41). The quinone–ketal **209** may then be converted to a tetramer or split off phenoxy substituents to afford two dimers, a trimer and a monomer. The existence of **209** has been proposed based on several experimental studies[84].

In contrast to 2,6-dimethylphenol (**206**), on Cu–amine complex catalyzed oxidation of 2,6-di(*tert*-butyl)phenol (**23**), only 3,3′,5,5′-tetra(*tert*-butyl)diphenoquinone (**24**) was produced in high yield and no C—O coupled polymer could be detected, because two bulky groups at the *o,o*′-positions presumably prevent the C—O coupling reactions leading to such a polymer as **208**. For example, both stoichiometric and catalytic oxidations of **23** were carried out using a Cu(I)–O$_2$ complex, prepared from the tetra-Shiff base L and Cu(MeCN)$_4$PF$_6$, and Cu(II)-L complex, prepared from CuCl$_2$ and the ligand L, respectively, to afford the corresponding diphenoquinone (**24**). The stoichiometric oxidation reactions are generally first order in the binuclear Cu(I) macrocyclic dioxygen complex and in the substrate. It is evident that 2,6-di(*tert*-butyl)phenol is also catalytically oxidized to **24**, as shown in Scheme 42. Herein, the Cu(II) complex involved in the proposed mechanism must be an effective oxidant[85]. In addition, the most plausible dimerization process is explained by the bridge formation between the phenols and the two Cu(II) centers[84,86].

In connection with iron- and copper-containing metalloenzymes involved in O$_2$ processing, three copper complexes (**210**, **211** and **212**) have been synthesized and the corresponding O$_2$–Cu complexes (**213**, **214** and **215**) are formed reversibly at $-80\,°$C in methylene chloride by addition of O$_2$ (Scheme 43)[87]. Of these O$_2$–Cu complexes, the peroxo group in both **213** and **214** reacts in a manner characteristic of the base/nucleophilic Mn–O$_2$ compounds, while **215** behaves differently and shows a nonbasic/electrophilic reactivity of the peroxodicopper(II) moiety. Thus, 2,4-di(*tert*-butyl)phenol (**216**) acted as a protic acid toward **213** and **214**, but in the presence of **215**-(ClO$_4$)$_2$, 3,3′,5,5′-tetra(*tert*-butyl)-2,2′-dihydroxybiphenyl (**217**) was produced in 93% yield, suggesting a similarity of **215** to the [Cu$_2$–O$_2$] structure in O$_2$ coordinating or activating copper proteins.

Recent extensive studies have been performed on the formation and reactivities of a bis μ-oxodicopper(III) core, [L$_2$Cu(III)$_2$(O$_2$)]$^{2+}$, bearing weakly coordinating anions, at low temperature, where L is one of a variety of peralkylated-diamine or triamine ligands. For example, equimolar quantities of [Cu(I)(PhCN)$_4$](ClO$_4$) and N,N,N',N'-tetramethyl-(1,3)-propanediamine (LTEMPO) reacted rapidly with dioxygen in CH$_2$Cl$_2$ at $-80\,°$C

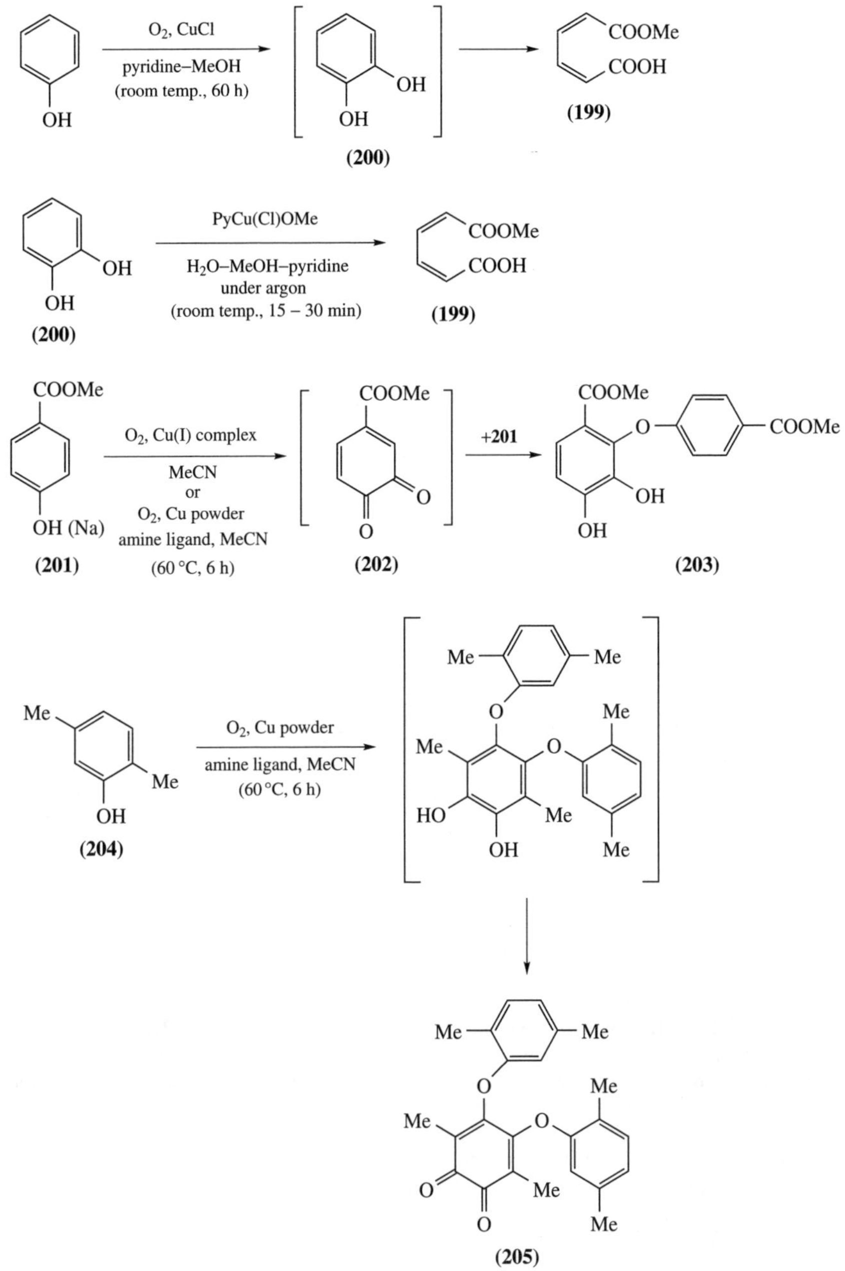

SCHEME 40. Copper-promoted phenol oxygenation

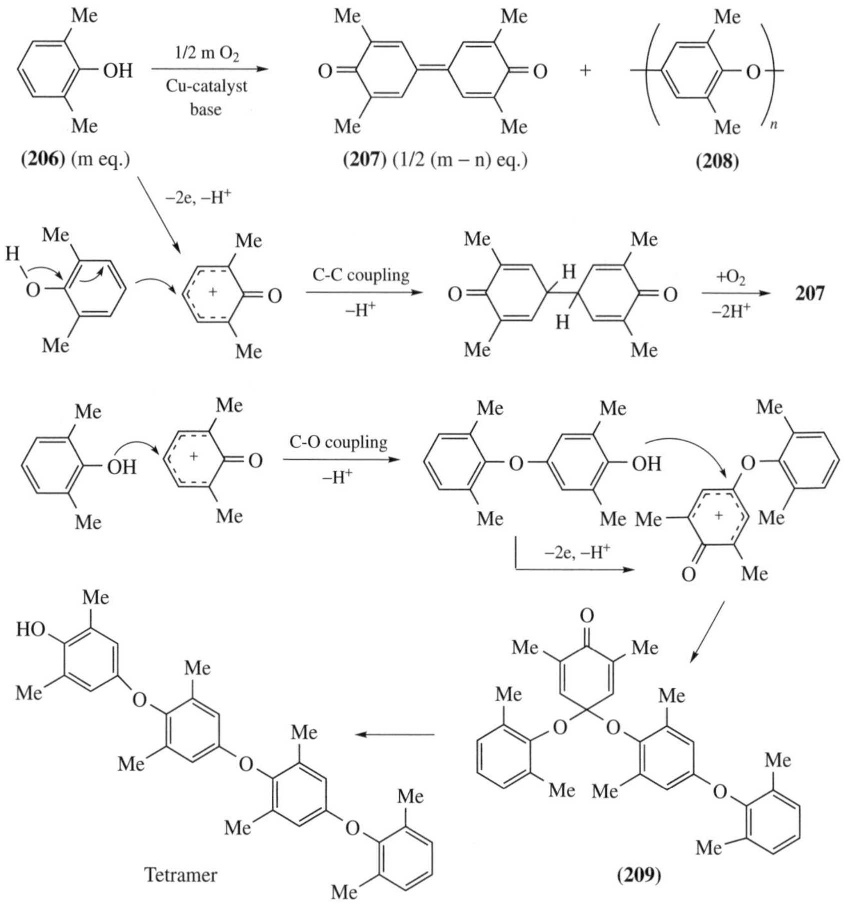

SCHEME 41. Copper-promoted oxidative phenol-couplings of 2,6-dimethylphenol

to generate [(LTEMPO)$_2$Cu(III)$_2$(O$_2$)]$^{2+}$ (ClO$_4$)$_2$, to which structure **218** was proposed (Scheme 44)[88].

Complex **218** could oxidize rapidly and almost quantitatively (>95%) 2,4-di(*tert*-butyl)phenol (**216**) and 3,5-di(*tert*-butyl)catechol (**219**) at $-80\,°$C to the corresponding biphenyl and *o*-benzoquinone (**217** and **220**), respectively.

2,4,6-Trimethylphenol (**221**) was oxidized with dioxygen catalyzed by CuCl$_2$·2H$_2$O to afford 3,5-dimethyl-4-hydroxybenzaldehyde (**222**) and 2,6-dimethyl-*p*-benzoquinone (**223**) in low yields. However, the use of acetone oxime as an additive caused a dramatic change to afford both **222** and **223** in 91.5 and 6.5% yields, respectively[89]. These oxidation products are formed from *p*-quinone methide **225** through 2,6-dimethyl-4-(hexyloxymethyl)phenol (**224**) (Scheme 45).

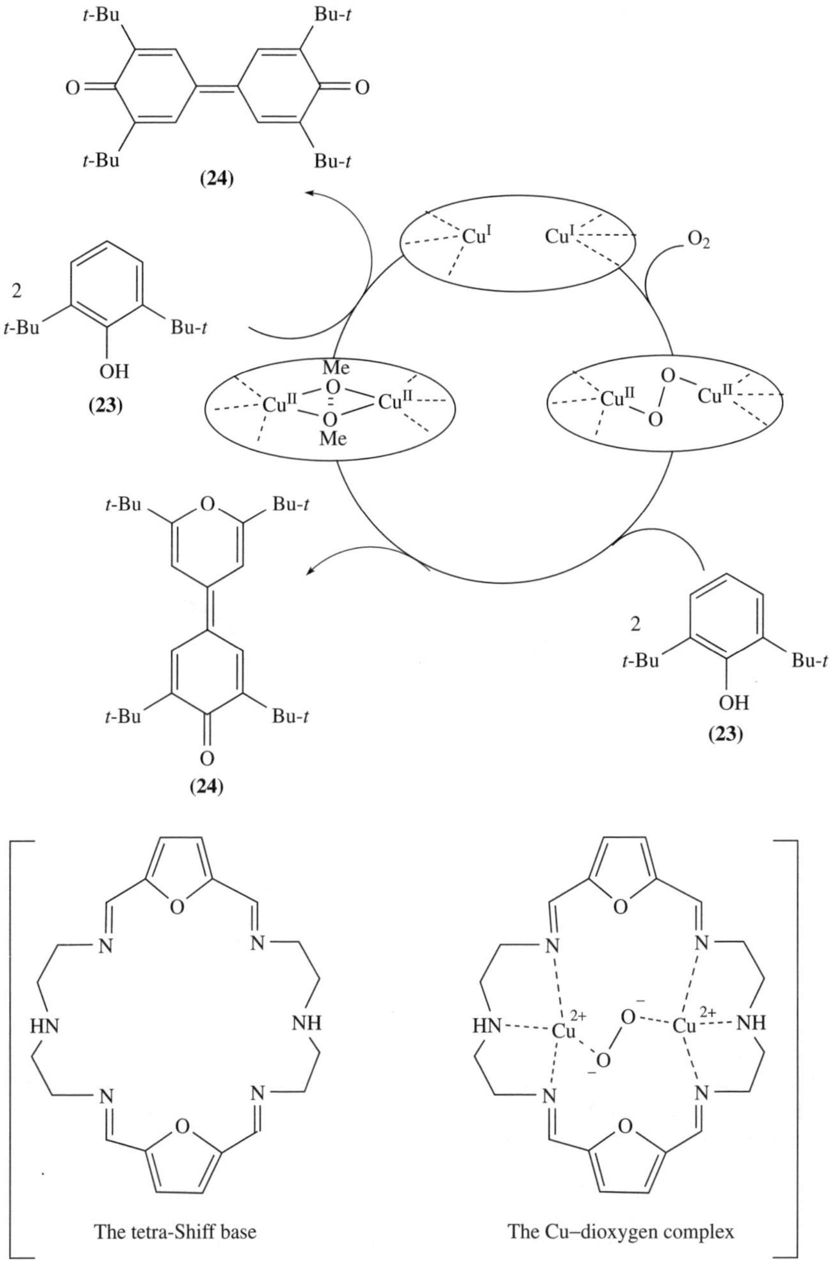

17. Oxidation of phenols

SCHEME 42. Cu–amine catalyzed oxidation of 2,6-di(*tert*-butyl)phenol

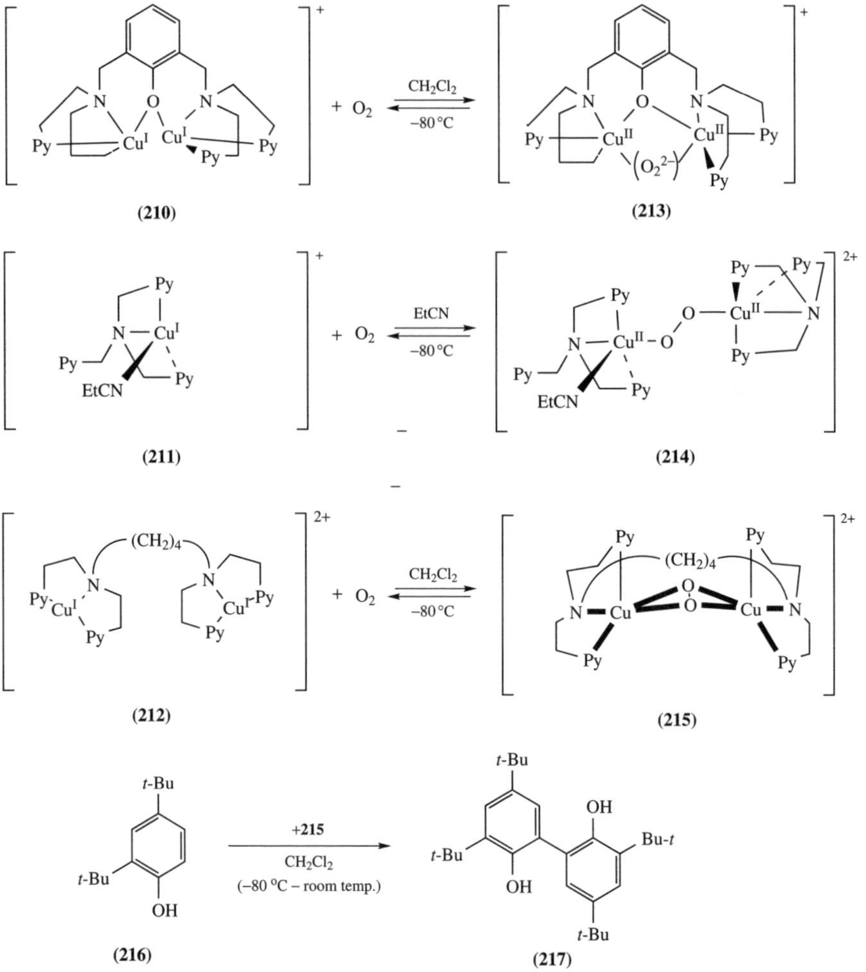

SCHEME 43. Different types of [Cu_2-O_2] complexes and their chemical properties

On treatment with a Cu(I) complex of N,N-bis[2-(N-methylbenzimidazol-2-yl)ethyl]benzylamine in MeCN followed by exposure to O_2, the sodium salt of 4-carbethoxy-2,6-di(*tert*-butyl)phenol (**226**) was selectively oxidized to afford in *ca* 80% yield 4-carbethoxy-3,6-di(*tert*-butyl)-*o*-benzoquinone (**227**), which is probably produced from the Cu(I) peroxide **228** by a 1,2-migration of a *tert*-butyl group (Scheme 45)[90].

From the viewpoints of reaction mechanism and efficiency in organic synthesis, oxidation of phenols with dioxygen catalyzed by cobalt–, manganese– and related metal–amine complexes has been studied[76,77,91]. In particular, much effort has been directed toward constructing new efficient catalysts by a combination of metals with

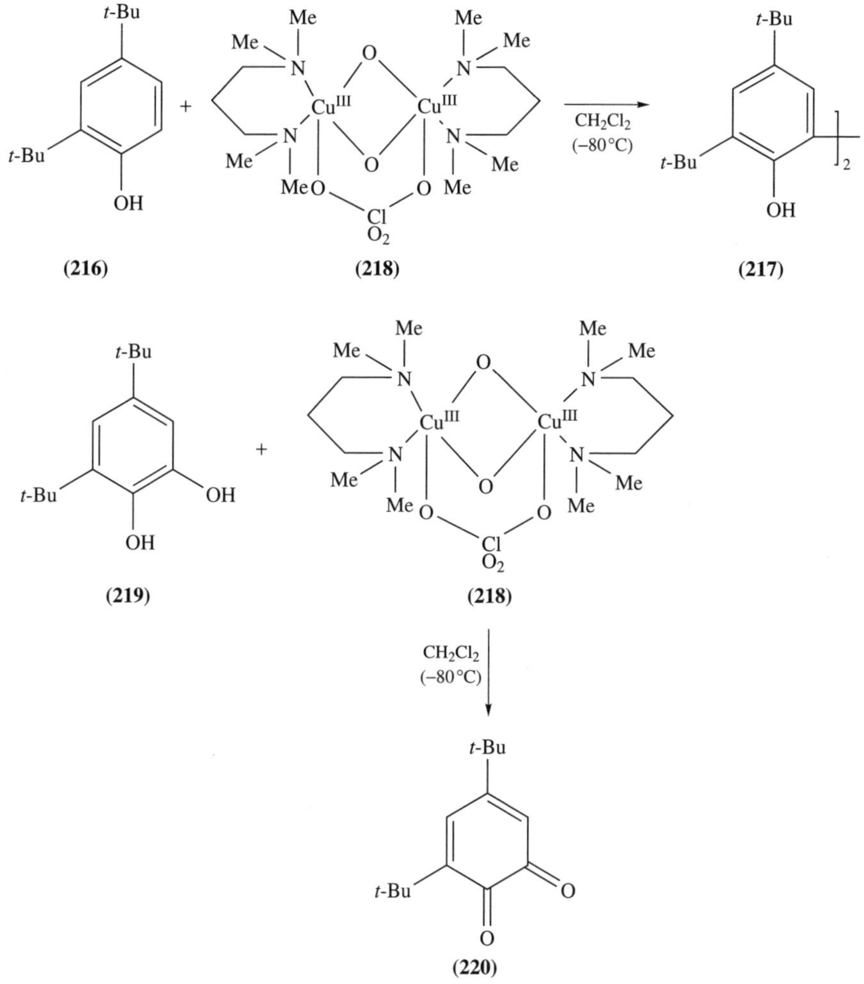

SCHEME 44. Formation and reactivities of $[L_2Cu(III)_2(O_2)](ClO_4)_2$

new ligands. Several metal–ligand complexes (**229–238**) are shown in Chart 1. Of them N,N'-ethylenebis(salicylidene-iminato)cobalt(II) [(salen)Co(II), salcomine, **229**] is the most popular.

Oxidation of 2,6-di(*tert*-butyl)phenol (**23**) provides a useful test for comparing the activity of various catalysts: **23** is oxidized with O_2 catalyzed by metal–amine complexes to give only two products, 2,6-di(*tert*-butyl)-*p*-benzoquinone (**74**) and 3,3′,5,5′-tetra(*tert*-butyl)diphenoquinone (**24**) (Scheme 46). Of the cobalt catalysts **230**, **231**, **232** and **237**, the use of Co(salN-Medpt)[91] in MeCN (room temp., 1 h) provided the most effective results, in which **74** was obtained in 100% yield. The oxidation rate and yield were dependent on

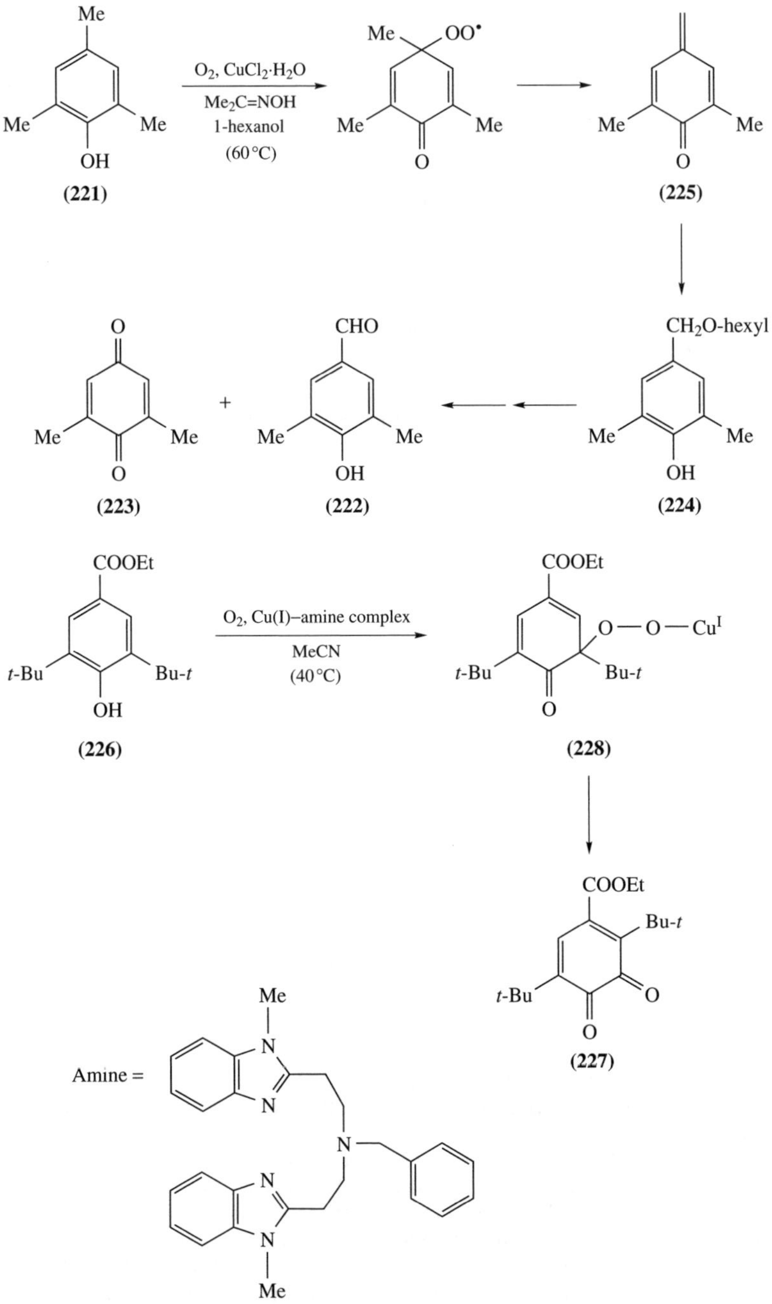

SCHEME 45. Oxidation of 2,4,6-trialkylphenols with O_2 catalyzed by Cu(I) or Cu(II) complexes

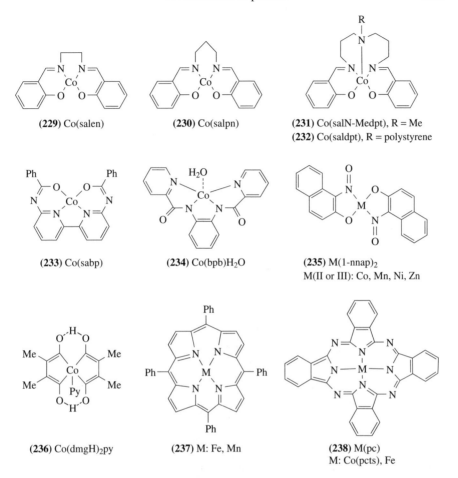

CHART 1. Several metal(II) and (III)–amine complexes

the catalyst and on the solvent. All of these Co–amine complexes catalyze the oxidation of **23** to give **74** as the major product[92].

Manganese porphyrin, Mn^{III}(tpp)Cl (**237**), also catalyzes the oxidation of **23** in the presence of the reducing agent Bu$_4$NBH$_4$ (1 equiv. per mol of phenol) to afford in 90% yield the diphenoquinone (**24**) as a sole product. The role of Bu$_4$NBH$_4$ is to reduce Mn(III) porphyrin to Mn(II) porphyrin, which has an ability to bind O$_2$. Consequently, it is possible to convert selectively **23** to the quinone **74** or the diphenoquinone **24** by suitable choice of a catalyst.

On oxidation with O$_2$ catalyzed by a Co(II) complex of 6,6′-bis(benzoylamino)-2,2′-bipyridine (**233**) in toluene containing an appropriate base such as pyridine (20 °C, 24 h), a quantitative conversion of **23** to **74** was observed. In addition, the durability of this complex as an oxygenation catalyst is much higher than that of **229** [Co(salen)]. Furthermore, the catalytic activity of **233** can be restored by heating it to 200 °C under reduced pressure because of its high thermal stability[93].

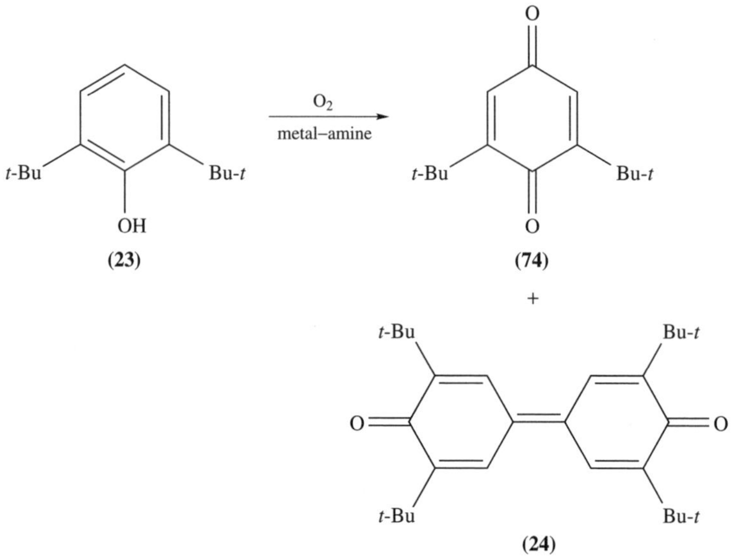

SCHEME 46. Oxidation of 2,6-di(*tert*-butyl)phenol with O_2–Co(II) or Mn(II) complexes

2,6-Di(*tert*-butyl)phenol (**23**) underwent catalytic oxygenation with aqua[N,N'-bis(2'-pyridinecarboxamido)-1,2-benzene]cobalt (II) (**234**) in DMF or DMSO (room temp., 1 h) to afford the corresponding quinone **74** in 100% yield: The metal complex **234** shows high selectivity and ability to work under mild conditions stirring at room temperature under an atmosphere of molecular oxygen[94].

Several phthalocyanines including Mn(II), Co(II), Ni(II) or Cu(II) as the central metal ion were nearly all inactive as the catalysts for the oxidation of **23**, but only the Fe^{II}–PC complex **238** showed a strong catalytic activity; catalytic oxidation of **23** in MeOH was effected with **238** under an oxygen atmosphere (room temp., 18 h) to give an almost quantitative yield of 3,3',4,4'-tetra(*tert*-butyl)diphenoquinone (**24**)[95]. In contrast, the Co^{II}(salen) (**229**) mainly provided the corresponding quinone (**74**). Interestingly, the selective autooxidation of **23** to **24** was accomplished at 35 °C by using Co(II)–phthalocyaninetetrasulfonate [Co(pcts)]$^{4-}$ intercalated into a $Mg_5Al_{2.5}$-layered double hydroxide (LDH)[96]; under homogeneous reaction conditions the complex was deactivated within 25 catalytic turnovers, while the LDH-intercalated catalyst remained fully active even after more than 3200 turnovers. It was also possible to recover the catalyst by filtration and to add more reactants without deactivation of the catalyst.

The proposed reaction mechanism of phenols with O_2 catalyzed by Co–amine complexes is shown in Scheme 47. On oxidation of 2,6-dimethylphenol (**206**) to 2,6-dimethyl-*p*-benzoquinone (**223**), magnetic field effects in the cobalt(II)-catalyzed oxidations were examined by using two different high- and low-spin cobalt(II) complexes. The former complex, Co^{II} bis(3-(salicylideneamino)propyl)methylamine, Co^{II}SMDPT ($S = 3/2$), displays a maximum increase in the initial rate of *ca* 1000 G, while the low-spin cobalt complex, $Co^{II}N,N'$-bis(salicylidene)ethylenediamine, Co^{II}salen ($S = 1/2$), in a 1:10 ratio with pyridine displays a maximum decrease in the initial rate at *ca* 800 G. The difference in the magnetokinetics of both complexes is explained by magnetic field effects on the singlet–triplet (S–T) radical pair and triplet–triplet (T–T) annihilation reactions

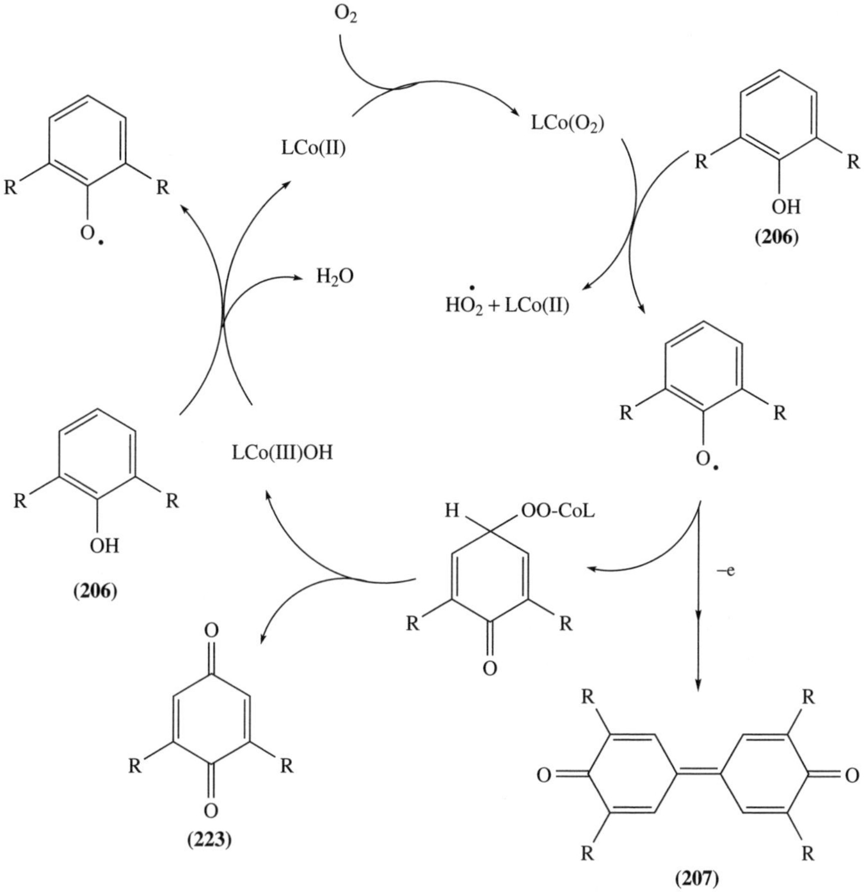

SCHEME 47. Reaction mechanism of phenols with O_2–Co(II) complexes

L = N, N'-bis(3-(salicylideneamino)propyl)methylamine

related to the catalytic regeneration step involving the initial encounter of the diamagnetic $Co^{III}SMDPT(OH)$ and 2,6-dimethylphenol (**206**)[97].

Bis(1-nitroso-2-naphtholato)manganese(II) (**235**) was synthesized by treatment of manganese(II) chloride with sodium 1-nitroso-2-naphthol. Similar reactions of cobalt(II), nickel(II), copper(II) and zinc(II) chlorides with sodium 1-nitroso-2-naphtholate afforded the corresponding bis(1-nitroso-2-naphtholato)metal(II) complexes. Of these complexes, **235** [Mn^{II}(1-nnap)$_2$] was proved to be the most effective catalyst in the oxidation of phenols such as 2,6-di(*tert*-butyl)phenol and 2,6-dimethylphenol under an oxygen atmosphere. Phosphine compounds are essential for this catalytic oxidation of phenols. When a mixture of 2,6-di(*tert*-butyl)phenol (**23**) and a catalytic amount of **235** was stirred in dry CH_2Cl_2 at 23 °C under an oxygen atmosphere (1 atm), the corresponding diphenoquinone **24** was formed in only 5% yield. However, the addition of triphenylphosphine (1 equiv.) as a co-ligand provided the best results with yields of 93% attained after 20 h. The oxygen pressure is also important for product selectivity: Raising the oxygen pressure from 1 to 20 atm provided after 6 h a mixture of **24** and 2,6-di(*tert*-butyl)-*p*-quinone (**74**) in 67 and 29% yields, respectively. The proposed oxidation mechanism of phenols using Mn^{II}(1-nnap)$_2$ is shown in Scheme 48[98].

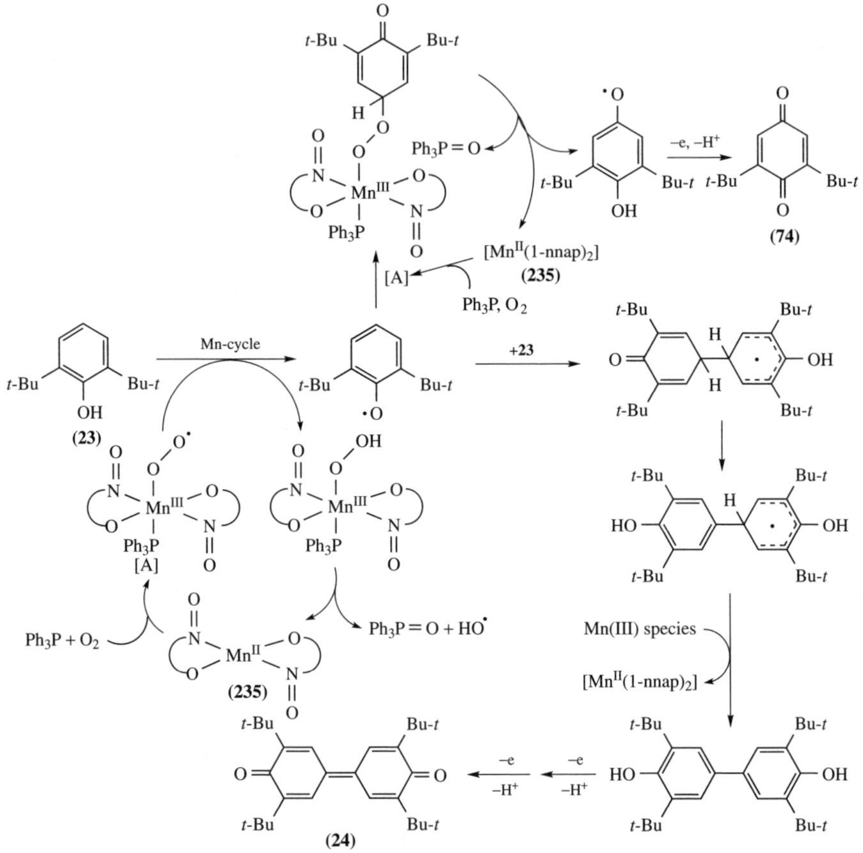

SCHEME 48. [Mn^{II}(1-nnap)$_2$] catalyzed oxidation of 2,6-di(*tert*-butyl)phenol

Heteropolyanions such as $H_5PV_2Mo_{10}O_{40}$ and NPV_6Mo_6/C have also been found to catalyze the highly selective oxidation of dialkylphenols to diphenoquinones[99,100]. Oxidation of 2,6-di(*tert*-butyl)phenol (**23**) was carried out at 25 °C for 4 h in hexane containing the heteropolyanion (0.02 equiv.) under an oxygen atmosphere (1 atm) to afford the corresponding quinone **24** in 96% yield (Scheme 49). In the case of 2,3,5-trimethylphenol (**239**), 2,3,5-trimethyl-*p*-quinone (**240**) was obtained in lower yield (Scheme 49)[99], because of steric hindrance by the two methyl groups at the C3 and C5 positions. The similarity of $H_5PV_2Mo_{10}O_{40}$-catalyzed oxidations with that of $CuCl_2$ oxygenations is noted. However, the former has the significant advantage that the chlorinated side-products are eliminated.

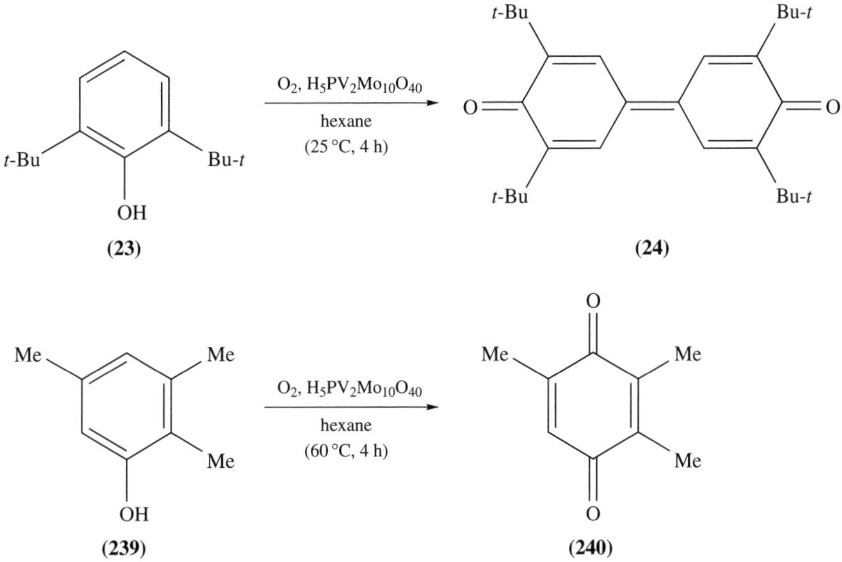

SCHEME 49. $H_5PV_2Mo_{10}O_{40}$ catalyzed oxidation of di- and trialkylphenols

Generally, catalytic oxygenation of trialkyl-substituted phenols such as 2,4,6-(*tert*-butyl)- and 2,4,6-trimethylphenol provides a complex mixture of products, as shown in Scheme 50. Herein, the oxidation products and their distribution vary with the central metal and ligands of the metal complexes, the solvent used, the oxygen pressure and the reaction conditions. Of the five metal complexes [**230**, **231**, **232**, **237** (Co) and **237** (MnCl)], the Mn^{III}(tpp)Cl–Bu_4NBH_4 complex in toluene provided the best results, in which 2,4,6-tri(*tert*-butyl)phenol (**73**) was converted completely to **74**, **243** and **244**, in 51, 36 and 13% yields, respectively[92]. In the case of Co(salN-Medpt) (**231**) in toluene, 30% conversion of **73** took place leading to **74**, **75**, **242**, **243** and **244** in a ratio of 40:7:20:26:7. Oxygenation of **73** with PC-Fe(II) (**238**) as a catalyst yielded selectively **244** in 87% yield, together with small amounts of **74** and **241**[95].

As compared with the well known Co(salen) (**229**), a variety of metal–ligand complexes have been synthesized. Of three complexes [**245**, M = Mn(II), Fe(II) and Co(II)], the manganese complex provided the most efficient conversion of **73** to **74** (48%) together with the oxygenated products (**242**–**244**) (Scheme 50)[101].

When the Co(bpb)H_2O complex **234** was used in MeCN under an oxygen atmosphere (room temp., 4 h), 2,6-di(*tert*-butyl)-4-methylphenol (**69**) was converted into a peroxy-*p*-quinalato–cobalt complex **246**, as a sole product (47%), suggesting that **246** supports the intermediacy of Co(L)-OO• in the reaction (Scheme 51)[91,94,102].

Catalytic oxygenation of **73** with K[Co^{III}(salen)CO_3] in EtOH also yielded **74**, **243** and **244** in 38, 11 and 42% yields, respectively. However, neither K[Co^{III}(salen)(CN)$_2$] nor Na[Co^{III}(salen)(CN)$_2$] gave any amount of the oxygenated products[103].

In connection with lignan chemistry, oxygenation of syringyl alcohol (**247**) with O_2 in the presence of 10% of the 5-coordinate catalyst **231** or **229**–pyridine complex afforded 2,6-dimethoxy-*p*-benzoquinone (**248**) in 71 and 88% yields, respectively (Scheme 52). A peroxy-*p*-quinalato–cobalt complex **249** is a plausible intermediate in the oxidation[104].

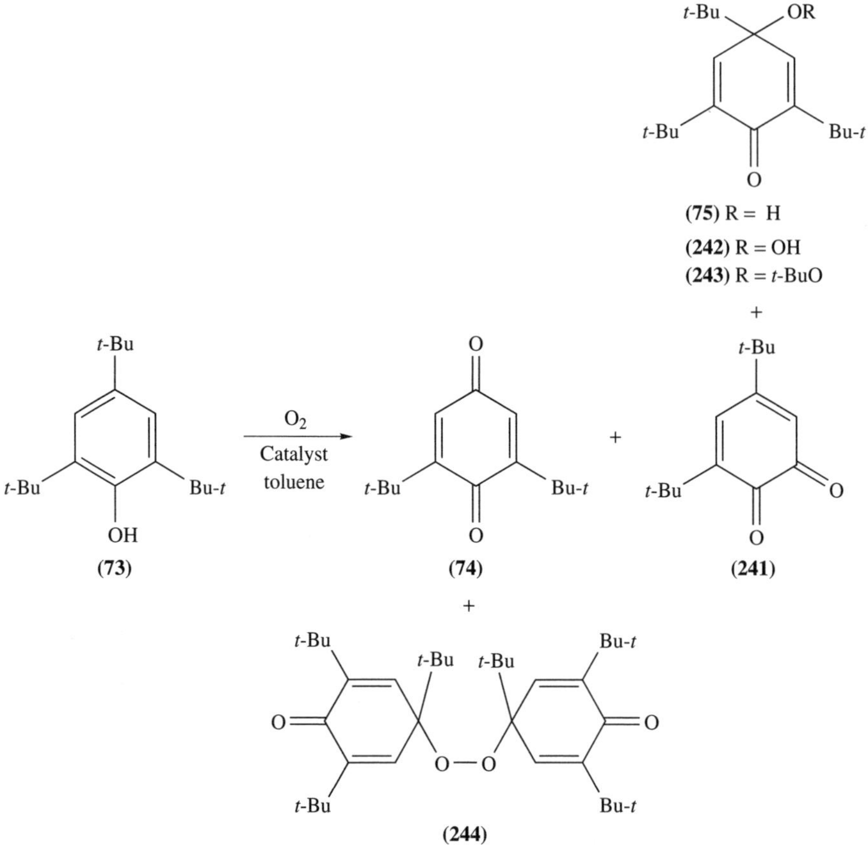

(75) R = H
(242) R = OH
(243) R = t-BuO

Catalyst: Mn(tpp)Cl-BH$_4$, Co(salN-Medpt) and PC-Fe(II)

SCHEME 50. Catalytic oxidation of 2,4,6-tri(*tert*-butyl)phenol

Oxygenation of both 4-alkenyl- and 2-alkenyl-2,6-di(*tert*-butyl)phenols was studied using 1.1 equivalents of Co(salpr) (**231**: R = H) (Scheme 53)[105]. The phenol **250** underwent Co(salpr)-promoted oxygenation in CH$_2$Cl$_2$ (0 °C, 1.0 h) resulting solely in the formation of 3,5-di(*tert*-butyl)-4-hydroxybenzaldehyde (**251**) (88%). In the case of **252**, both **253** and **254** were produced in 28 and 72% yields, respectively. 2-Alkenylphenol **255** was oxidized under similar conditions (0 °C, 3.5 h) to the corresponding benzaldehyde **256** in quantitative yield, while the oxygenation of **257** gave selectively dihydrobenzofuran **258** (78%) together with **259** (5%). Further studies on 4- or 2-alkynylphenols have also been conducted[105].

As already shown in Scheme 8, carpanone (**44**) has been synthesized by an electrochemical method. More efficient synthesis of **44** was effected with O$_2$ catalyzed by metal–Schiff base complexes. Of the four complexes, Co(II)(salpr), Co(II)(salen), Fe(II)(salen) and Mn(II)(salen), Co(II)(salen) provides the best results; its solution with 4,5-methylenedioxy-

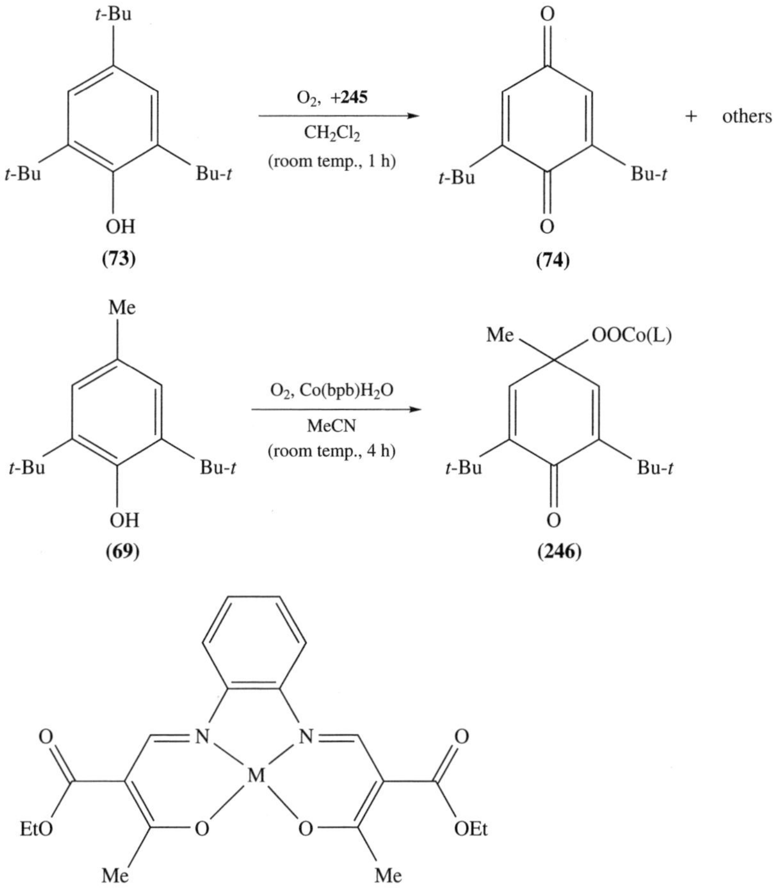

SCHEME 51. Catalytic oxygenation of 2,4,6-trialkylphenols

2-propenylphenol (**42**) in CH$_2$Cl$_2$ was stirred under an oxygen atmosphere at room temperature for 1.5 h to afford carpanone in 94% yield (Scheme 54)[28]. The oxidant PdCl$_2$–NaOAc also provided a 46% yield of carpanone, although the yield is relatively low[27].

2. Hydrogen peroxide– and tert-butyl hydroperoxide–metal complexes

In the previous section, oxygenation of phenols with dioxygen catalyzed by metal complexes was described. From industrial and biological points of view, metal-complex catalyzed oxidation of phenols has also been performed using hydrogen peroxide or *tert*-butyl hydroperoxide instead of dioxygen. Some examples are described briefly in this section.

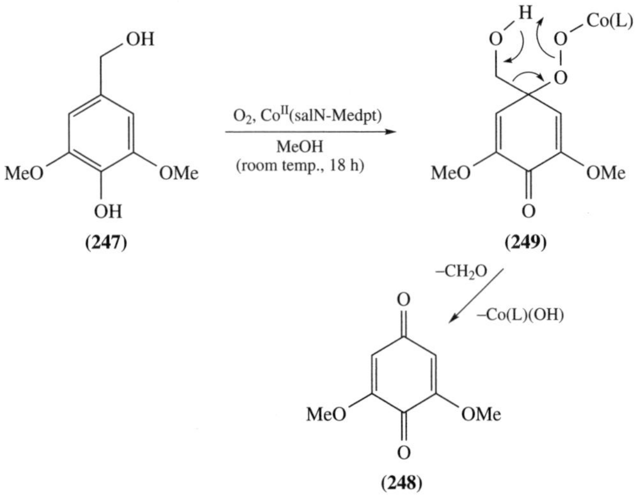

SCHEME 52. Catalytic oxygenation of syringyl alcohol

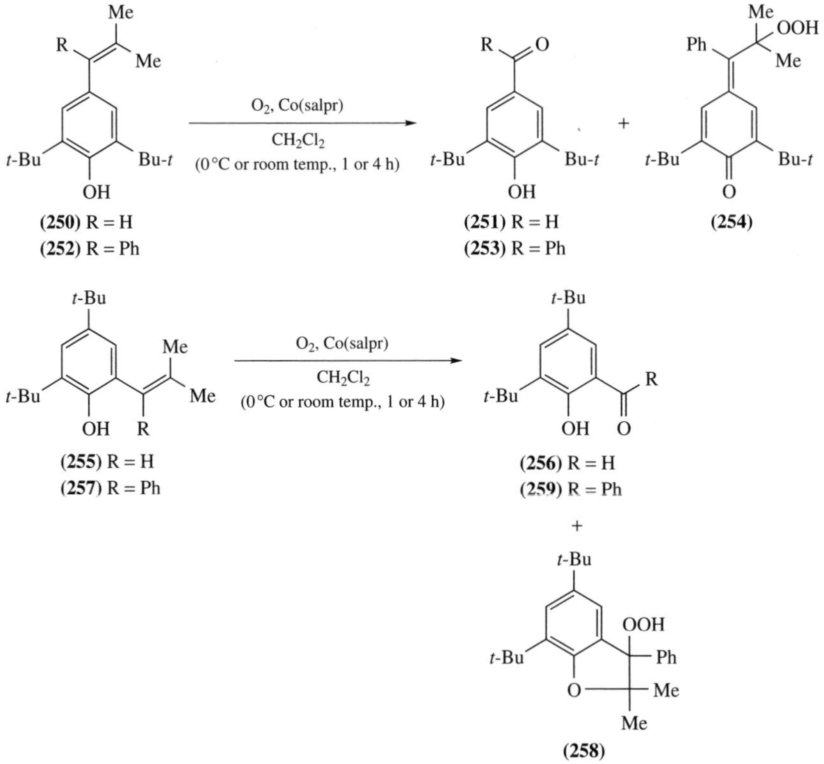

SCHEME 53. Co(salpr)-promoted oxygenation of 4- and 2-alkenyl substituted phenols

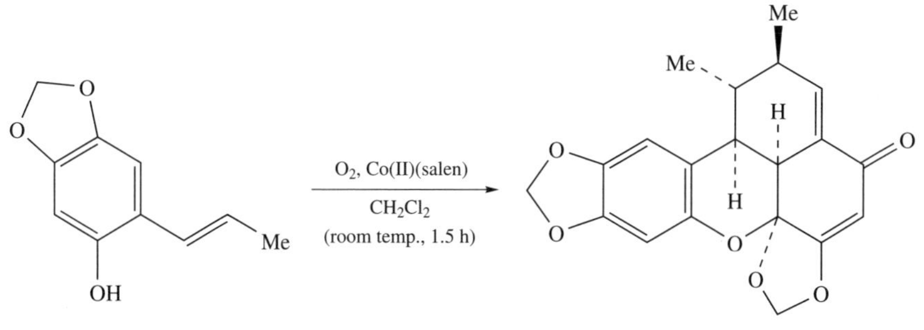

SCHEME 54. Synthesis of carpanone using O_2–Co(II)(salen) complex

Of many copper complexes, one of the interesting Cu(II) complexes is di-μ-hydroxodicopper(II) complex $[Cu_2(OH)_2(hexpy)](X)_2$ [X = ClO_4 or CF_3SO_3; hexpy: 1,2-bis[2-(pyridyl)methyl-6-pyridyl]ethane] (**260**). Its structure has been determined by X-ray crystallographic analysis[106,107]. To a solution of **260** (X = CF_3SO_3) were added 2,4-di(*tert*-butyl)phenol (**216**) and 28% aq. H_2O_2 with vigorous stirring. The reaction was completed in 5 min and afforded the corresponding biphenyl **217** in 86% yield. The suggested reaction mechanism is shown in Scheme 55. In the case of 2,6-dimethylphenol (**206**), almost the same result was obtained.

Co(salen)-catalyzed oxidation of phenols with *tert*-butyl hydroperoxide in CH_2Cl_2 at room temperature provides predominantly *tert*-butylperoxylated products[108]. On catalytic oxidation of 2,6-di(*tert*-butyl)-4-acetylphenol oxime O-methyl ether (**261**), both *o*- and *p*-(*tert*-butylperoxy)quinol ethers (**262** and **263**) were obtained in 8.1 and 87.3% yields, respectively. On the other hand, catalytic oxidation of **264** provided the corresponding *o*- and *p*-quinol ethers (**265** and **266**) in 91.1 and 6.8% yields, respectively (Scheme 56). The remarkable difference of the *o*/*p* ratio in the reactions of **261** and **264** reflects clearly a combination of both steric and electronic factors.

In the case of 4-alkenyl-2,6-di(*tert*-butyl)phenols having three potential reaction sites for attack by *t*-BuOO$^-$, three possible *tert*-butylperoxylated compounds will be produced depending on substituents on the olefinic side chain. Co(salen)-catalyzed oxidation of **267** with *t*-BuOOH provided quinomethane **268** and *p*-(*tert*-butylperoxy)quinol ether **269** (81.5 and 9.5%, respectively). In the case of the substrate **270** bearing a fully substituted olefin, the corresponding *o*-substituted quinol ether **271** was obtained as a sole product (73%) (Scheme 56). Detailed studies on Co(salen)-catalytic oxidation of 2-alkenyl-4,6- and 4-alkynyl-2,6-di(*tert*-butyl)phenols with *t*-BuOOH have also been conducted[108].

Recently, structurally related dimeric Mn complexes with 1,4,7-trimethyl-1,4,7-triazacyclononane ligand (TMATC) were proved to act as potent catalysts for the selective oxidation of alkenes and other substrates. The reaction of $[(TMATC)_2Mn_2^{IV}(\mu\text{-}O)_3](PF_6)_2$ (**272**) with electron-rich phenols such as **273** and **274** in aqueous solution at pH 10.5 was studied (Scheme 57)[109]. The reaction proceeds via a rapid overall one-electron process from the phenolate anion to the Mn^{IV}/Mn^{IV} species **272** to give, initially, a Mn^{III}/Mn^{IV} species and the corresponding phenoxy radical. The Mn^{III}/Mn^{IV} species is ultimately converted into monomeric Mn^{II}. The addition of H_2O_2 accelerates a reoxidation of the manganese species, accompanied by an increase in the rate of the formation of the phenoxy radicals. For example, Trolox (**273**) underwent one-electron oxidation resulting in the formation of the corresponding radical **275**, which was detected by ESR, since it is relatively long-lived. The radical further underwent

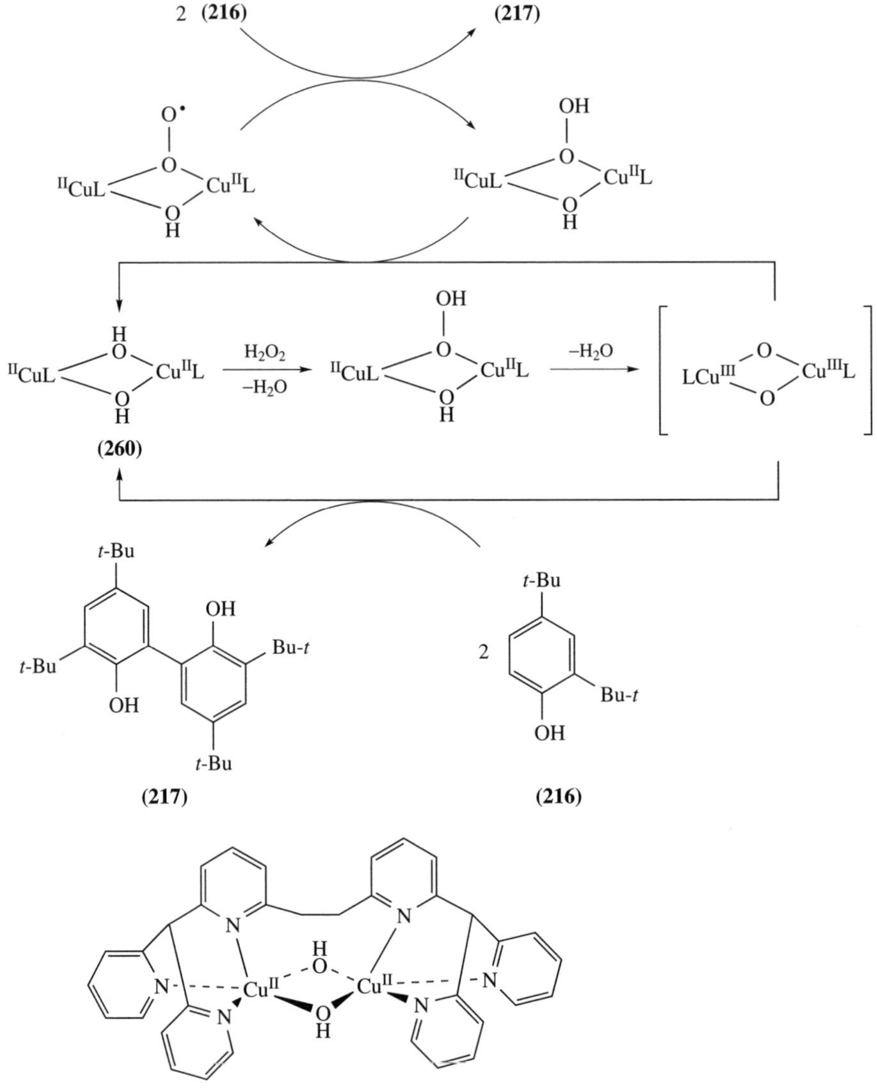

SCHEME 55. Oxidation of 2,4-di(*tert*-butyl)phenol catalyzed by di-μ-hydroxodicopper(II) complex

disproportionation to afford the corresponding quinone (**276**) and the starting Trolox (**273**). In the case of 2,6-dimethoxyphenol (**274**), the resulting phenoxy radical further underwent dimerization, followed by one-electron oxidation to afford mainly 3,3′,5,5′-tetramethoxydiphenoquinone (**277**).

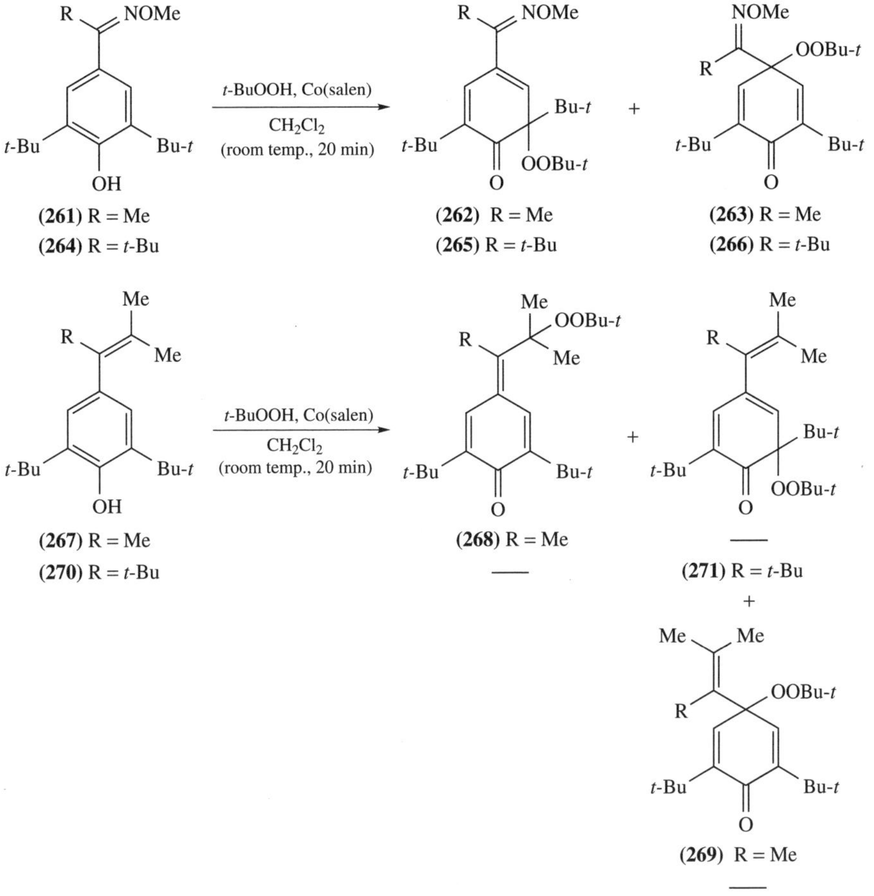

SCHEME 56. Co(salen)-catalyzed oxidation of phenols with *tert*-butyl hydroperoxide

From the viewpoint of the reaction mechanism, the reaction of oxoiron(IV) tetra(2-*N*-methylpyridyl)porphyrin (OFe^{IV}T2MPyP), generated from iron(III) tetra(2-*N*-methylpyridyl)porphyrin and *t*-BuOOH, with phenols has been investigated[110]. Oxidation of Trolox (**273**) with OFe^{IV}T2MPyP generated the ESR observable radical **275**. In addition, kinetic studies on several phenols suggest that the rate-determining step in these oxidations involves hydrogen atom abstraction from the phenol by the oxoiron(IV) species. A plausible mechanism for phenolic oxidation by OFe^{IV}T2MPyP in phosphate buffer (pH 7.7) has been proposed (Scheme 58). Here, the resulting radical species will be further converted into *p*-benzoquinones, dimers, trimers and/or polyphenols, whose distribution depends on the substituents on the phenol ring.

Chlorinated aromatic compounds such as 2,4,6-trichlorophenol (**278**) are well known recalcitrant pollutants because of their slow biodegradation by microorganisms. Hydrogen peroxide oxidation of (U-^{14}C)-**278** catalyzed by FePcS (iron tetrasulfophthalocyanine) was

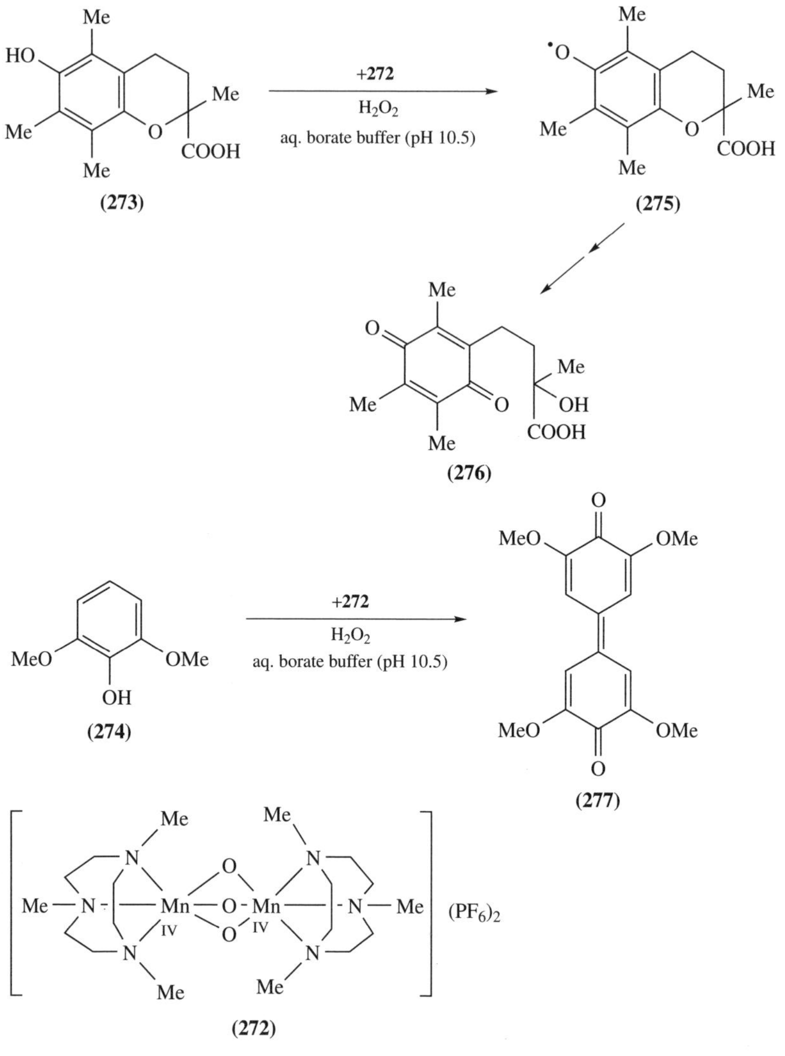

SCHEME 57. One-electron oxidation of phenols by a dimeric Mn(IV/IV) triazacyclononane complex in the presence of H_2O_2

carried out in MeCN–0.5 M phosphate buffer (pH 7.0) to afford mainly chloromaleic acid (**279**) (69%), together with oxidative coupled products (13%), CO_2 (11%) and CO (3%), with 96% recovery of the radioactivity (Scheme 59)[111].

In connection with bioactive quinones such as vitamin E, oxidation of 2,3,5-trimethylphenol (**239**) and related phenols with H_2O_2 or t-BuOOH has been carried out using a variety of metal catalysts. Of the typical catalysts examined ($FeCl_3$, $RuCl_3 \cdot 3H_2O$, $RuCl_2(PPh_3)_3$, CuCl, $CuCl_2$, $CoCl_2$, $RhCl_3 \cdot 3H_2O$, $PdCl_2$, $CeCl_3 \cdot 7H_2O$, $VO(acac)_2$, $MoO_2(acac)_2$ and $P_2O_5 \cdot 24WO_3 \cdot nH_2O$), $RuCl_3 \cdot 3H_2O$ provided the best results, in which **239** was selectively converted into the corresponding quinone **240** in 90%

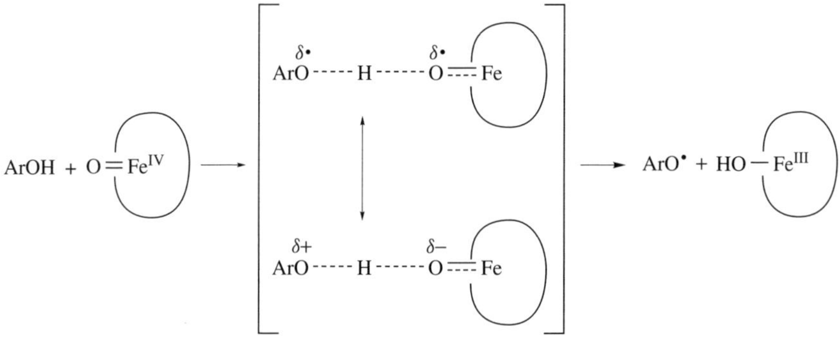

SCHEME 58. A plausible mechanism for phenol oxidation by OFe^{IV}T2MPyP

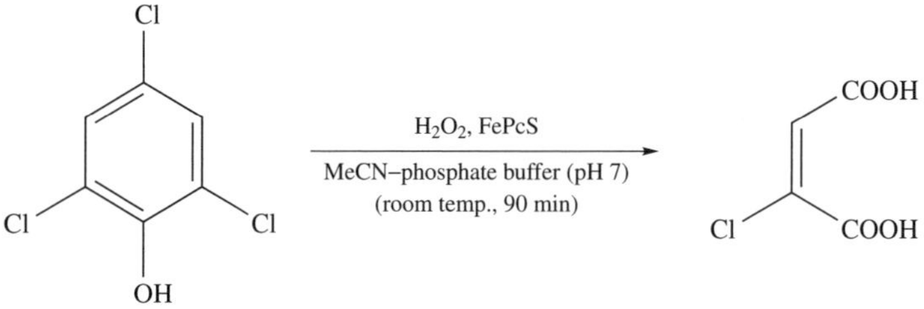

(278) **(279)**

+

oxidative coupling products

+

CO_2, CO

SCHEME 59. Catalytic oxidation of 2,4,6-trichlorophenol with H_2O_2–FePcS

yield[112]. This oxidation system ($RuCl_3{\cdot}3H_2O$–H_2O_2–AcOH) is characteristic of *p*-oxygenation (Scheme 60).

Catalytic oxidation of **239** to the quinone **240** was also effected with H_2O_2 catalyzed by methyltrioxorhenium(VII) ($MeReO_3$) (Scheme 60)[113], where a small amount of hydroxy-substituted quinone **280** was produced in addition to **240** (70%). In this reaction, $MeReO_3$ is stepwise converted by H_2O_2 into the mono- and bis(peroxo)rhenium complex $MeRe(O_2)_2O{\cdot}H_2O$ (**281**). This active oxidant then reacts with the phenol to give the epoxide **282**, which is further converted to the two quinones (**240** and **280**).

Similarly, 2,3,5-trimethylphenol (**239**) was converted into the quinone **240** (*ca* 80%) using $H_3PMo_{12}O_{40}$[114] or titanium substituted aluminophosphate (TiAPO-5) molecular sieves[115]. Efficient oxidation of phenols to the corresponding quinones has also been effected with H_2O_2-V-HMS (vanadium-containing mesoporous molecular sieves)[116].

Oxidation of *p*-substituted phenols with *t*-BuOOH catalyzed by heteropolyacids such as $H_3PMo_{12}O_{40}{\cdot}nH_2O$ (**283**) and $H_4SiW_{12}O_{40}{\cdot}nH_2O$ has been carried out[117]. When 2,6-di(*tert*-butyl)-4-methylphenol (**69**) was stirred with 80% *t*-BuOOH in the presence of **283** in AcOH (30 °C, 3 h), it afforded 2,6-di(*tert*-butyl)-4-(*tert*-butylperoxy)-4-methyl-2,5-cyclohexadienone (**284**) and 2,6-di(*tert*-butyl)-*p*-benzoquinone (**74**) in 62 and 13%

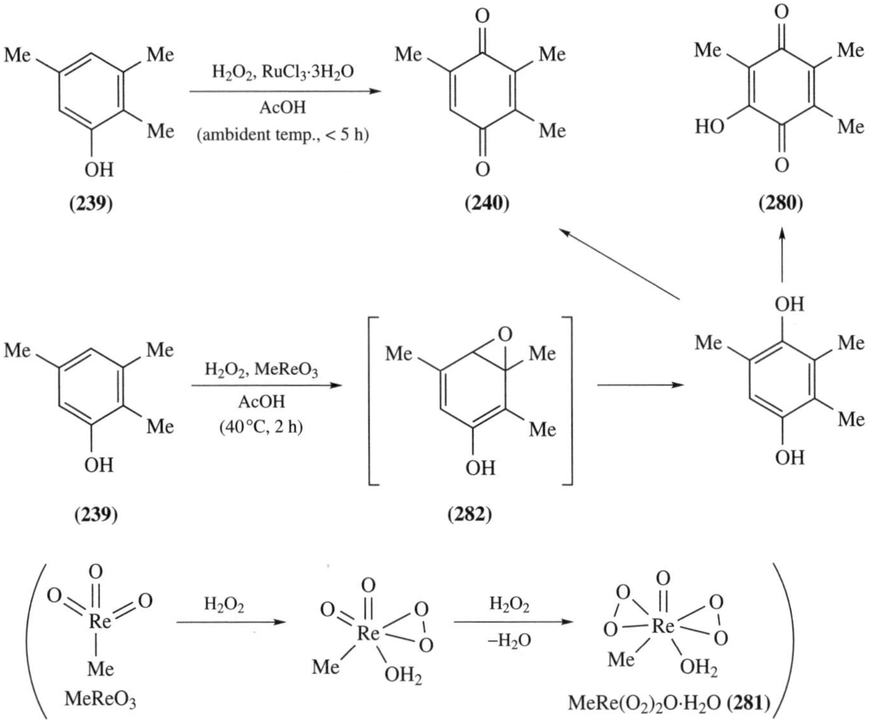

SCHEME 60. Catalytic oxidation of 2,3,5-trimethylphenol with H_2O_2–metal catalysts

yields, respectively. In the cases of 2,6-di(*tert*-butyl)phenols (**285** and **286**) bearing more eliminative substituents (CH_2OAc, MeO) than the methyl group at the *para*-position, **74** was obtained selectively in 65 and 91% yields, respectively (Scheme 61). Clearly, the quinone must be formed easily from **287** as well as from **288**.

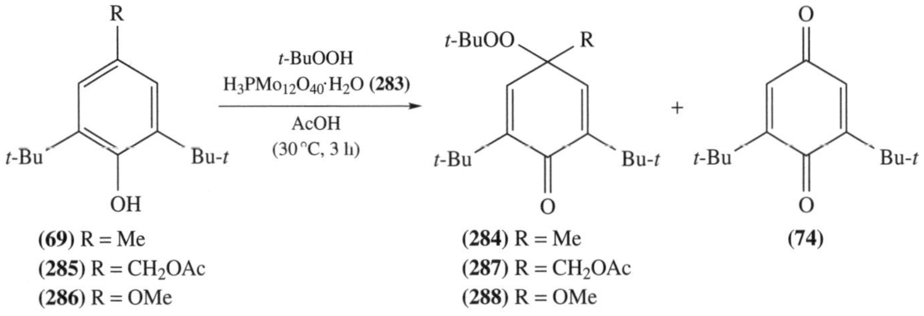

SCHEME 61. Oxidation of 2,4,6-trisubstituted phenols with *t*-BuOOH–heteropolyacid

In the case of *p*-alkyl-substituted phenols, of a variety of *t*-BuOOH–metal complexes investigated, $RuCl_2(PPh_3)_3$ has proved to be the most effective catalyst for the selective formation of 4-alkyl-4-(*tert*-butylperoxy)-2,5-cyclohexadienones[118]. Other ruthenium

17. Oxidation of phenols

catalysts such as $RuCl_3 \cdot nH_2O$ also gave satisfactory results. To a solution of p-cresol (**27**) and $RuCl_2(PPh_3)_3$ in EtOAc was added a solution of t-BuOOH in dry benzene at room temperature over a period of 2 h and stirred for an additional 3 h to afford 4-(*tert*-butylperoxy)-4-methyl-2,5-cyclohexadienone (**289**) in 85% yield. When **289** was treated with $TiCl_4$ in CH_2Cl_2, it afforded 2-methyl-p-benzoquinone (**290**) in 82% yield (Scheme 62). The oxidation which begins with hydrogen abstraction from **27** by the oxoruthenium intermediate derived from the two reagents results in the formation of a phenoxy radical-Ru^{III}(OH) intermediate. A further fast one-electron transfer from the phenoxy radical to the Ru(III) complex gives the corresponding phenoxonium ion and a Ru(II) complex. The former is attacked by t-BuOOH to afford selectively **289**. Estrone (**291**) also underwent t-BuOOH-$RuCl_2(PPh_3)_3$ oxidation leading to a peroxide **292** (89%), whose subsequent reductive acetylation (ZnI_2, $AcOH-Ac_2O$) provided a diacetate **293** in 55% yield.

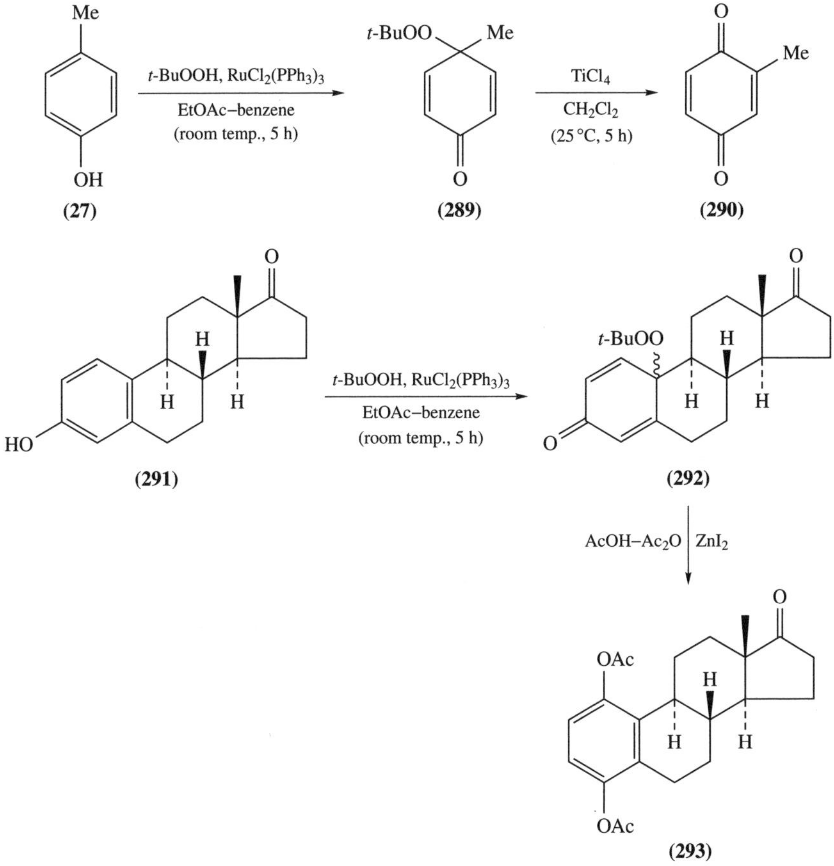

SCHEME 62. Ruthenium catalyzed oxidation of p-substituted phenols with t-BuOOH

C. Enzymatic Oxidation

Generally, the use of enzymes as catalysts in organic synthesis has provided a variety of chiral synthons for bioactive compounds including natural products. A number of invaluable books have hitherto appeared[119]. However, from the viewpoint of organic synthesis, there are only a few examples of enzymatic reactions employed for phenolic oxidation except for biomimetic synthesis of isoquinoline alkaloids[120-122], lignans and neolignans[123], which has been carried out using enzymes such as horseradish peroxidase, potato peelings and rat liver enzyme. Recently, a variety of enzymes have been employed to clarify biosynthetic pathways of these natural products.

(R)-Reticuline (**294**), one of the most fundamental isoquinoline alkaloids, has been known to be converted into morphine alkaloids by intramolecular oxidative phenol-coupling. Thus, **294** was treated with cytochrome P-450 linked microsomal *Papaver* enzyme in a buffer solution (pH 7.5) containing NADPH under aerobic conditions (25 °C, 1 h) to afford in high efficiency and high selectivity salutaridine (**295**), which was chemically converted to thebaine (**296**), morphine (**297**) and related alkaloids. In contrast, the enzyme was not effective in reaction with (S)-reticuline (Scheme 63)[124].

From the biosynthetic point of view, lignans and lignins differ fundamentally in their optical activity, although they are closely related in their chemical structures; the former is optically active, while the latter is inactive. Therefore, lignan biosynthesis must involve an enantioselective process. Thus, ^{2}H-labelled coniferyl alcohol (**298**) underwent H_2O_2 oxidation catalyzed by cell-free extracts of *Forsythia koreana* in potassium phosphate buffer (pH 7.0) containing NADPH, resulting in the enantioselective formation of (−)-secoisolariciresinol (**299**), (−)-lariciresinol (**300**) (88% e.e.) and (−)-pinoresinol (**301**) (91% e.e.). It is noted that both **300** and **301** are unnatural enantiomers, while the *Forsythia koreana* plant produces (+)-**300** and (+)-**301**. The stereoselectivity for the formation of these lignans can be explained, at least in part, by the finding that the enzyme also catalyzed the stereoselective reduction of (+)-lariciresinol, but not of its (−)-enantiomer, to (−)-secoisolariciresinol (**299**), as shown in Scheme 64[125].

When two different phenols having almost the same oxidation potentials are used, both dimerization and cross-coupling reactions may take place. Coniferyl alcohol (**298**) was first oxidized alone with H_2O_2, catalyzed by horseradish peroxidase (HRP) in a 20% buffer solution (pH 3.5) in acetone (room temp., 1 h) to afford three dimers (**301**, **302** and **303**) in 12, 24 and 16% yields, respectively, as shown in Scheme 65, wherein the remaining starting phenol (36%) was further oxidized to oligomers (12%). In the case of a 1:1 mixture of **298** and apocynol (**304**), small amounts of cross-coupled products (**305** and **306**) were obtained in 5–10 and 0–1.5% yields, respectively, in addition to four dimers (**301**, **302**, **303** and **307**). Here, 45% of **304** remained (Scheme 65)[126]. On chemical oxidation of a 1:1 mixture of **298** and **304** with $Mn(OAc)_3$ in AcOH (room temp., 30 min), the yield of **305** increased to 18%.

Biomimetic conversion of ferulic acid derivatives to phenylcoumarans was carried out by using a variety of oxidants, of which the oxidation system (H_2O_2–HRP) gave the best results. However, the enzyme did not effect any stereocontrol. To overcome this difficulty, enantiopure ferulic acid derivatives such a N-ferulyl (S)-alaninate (**308**) were synthesized. The substrate **308** was dissolved in dioxane and phosphate/citric acid buffer (pH 3.5) was added. Aqueous H_2O_2 and HRP were added over 20 min. The mixture was stirred at room temperature for 2.5 h to yield a mixture of two phenylcoumarans **309** and **310** (70%) with a 1:4 ratio (Scheme 66)[127]. In the case of a camphor sultan derivative **311**, a mixture of two phenylcoumarans was also obtained in 40% yield (**312**/**313** = 1 : 9). Furthermore, oxidation of **311** with Ag_2O in CH_2Cl_2 (room temp., 24 h) yielded the same phenylcoumarans (35%) in a 1:12 ratio. The observed enantioselectivity in the oxidation

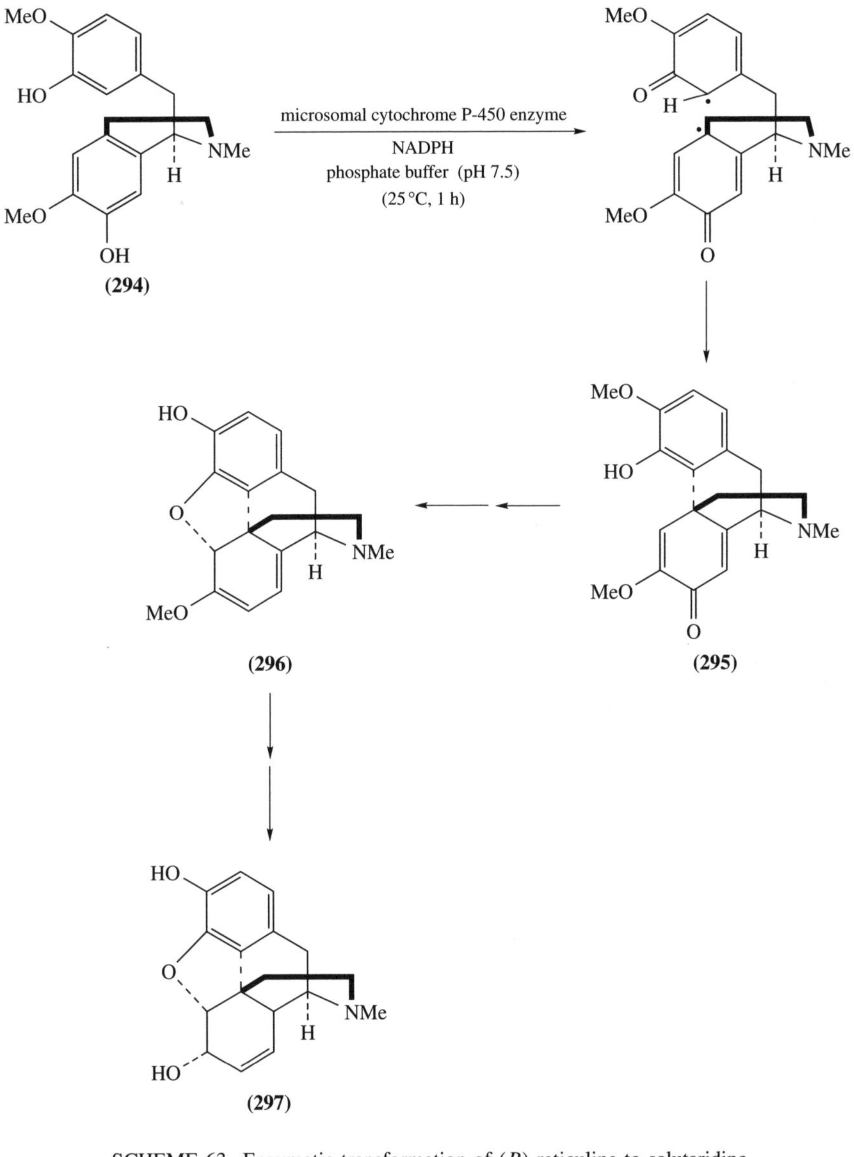

SCHEME 63. Enzymatic transformation of (*R*)-reticuline to salutaridine

step encompasses the range 65–84% and is consistent with the conformational analysis of the quinone methide intermediates at the PM3 level.

Resveratrol (**314**) has been known to undergo hydrogen peroxide oxidation catalyzed by HRP in aqueous acetone to afford dihydrobenzofuran **315** as a main product (41%)[128]. Recently, a variety of neolignans were isolated from the Vitaceaeous plants. Of them, both

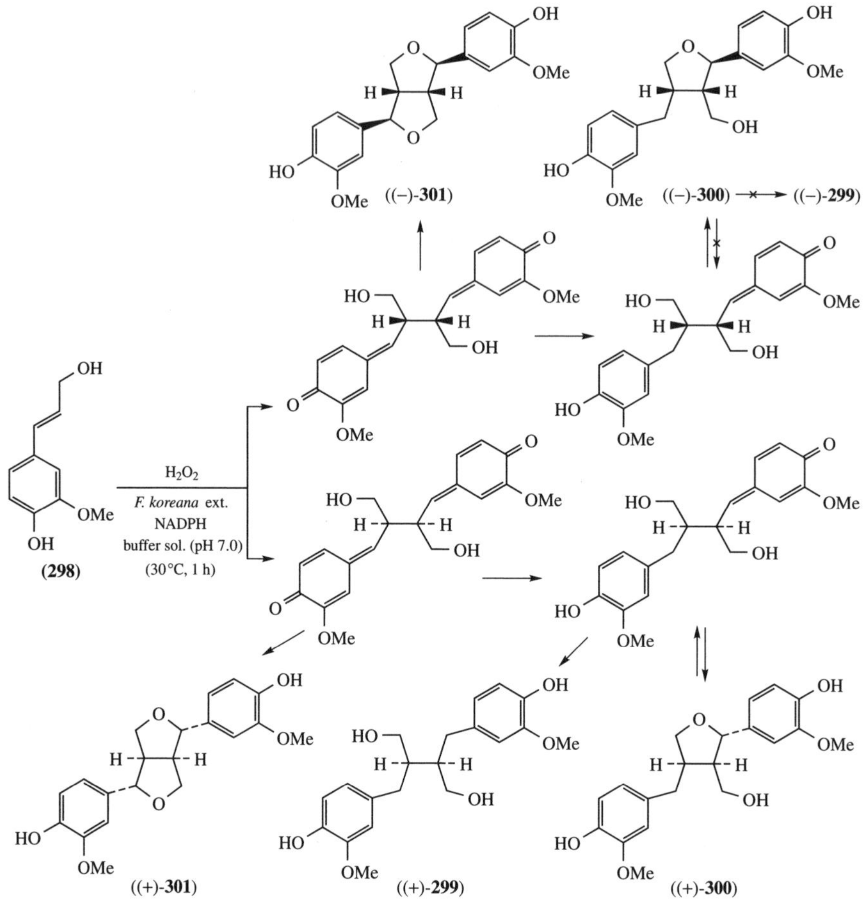

SCHEME 64. Enzymatic oxidation of coniferyl alcohol by *F. koreana* extracts

(−)-vitisin B and (+)-vitisin C (**316** and **317**) were synthesized by enzymatic oxidation of (+)-ε-viniferin (**318**), although the yields (*ca* 5%) were very low (Scheme 67). In this case, a C−C radical coupling reaction takes place[129].

From the viewpoint of organic synthesis, mushroom tyrosinase-mediated oxidation of 2,6-disubstituted phenols (**206** and **274**) was performed only in phosphate buffer (pH 6.8) to afford solely the corresponding 4,4′-diphenoquinones (**207** and **277**) in 96 and 98% yields, respectively. When acetonitrile was used as a co-solvent, biphenols (**319** and **320**) were obtained, though in lower yields (each 20%) than **207** (70%) and **277** (72%), respectively. In the case of 2,6-di(*tert*-butyl)phenol (**23**) bearing more hindered groups than **206**, the corresponding 4,4′-diphenoquinone and biphenol (**24** and **25**) were obtained in rather low yields of 40% and 20%, respectively (Scheme 68)[130]. In contrast, 2,6-dichlorophenol did not undergo enzymatic oxidation at all. These results indicate that the efficiency of enzymatic oxidation depends on steric and electronic effects of the substituents.

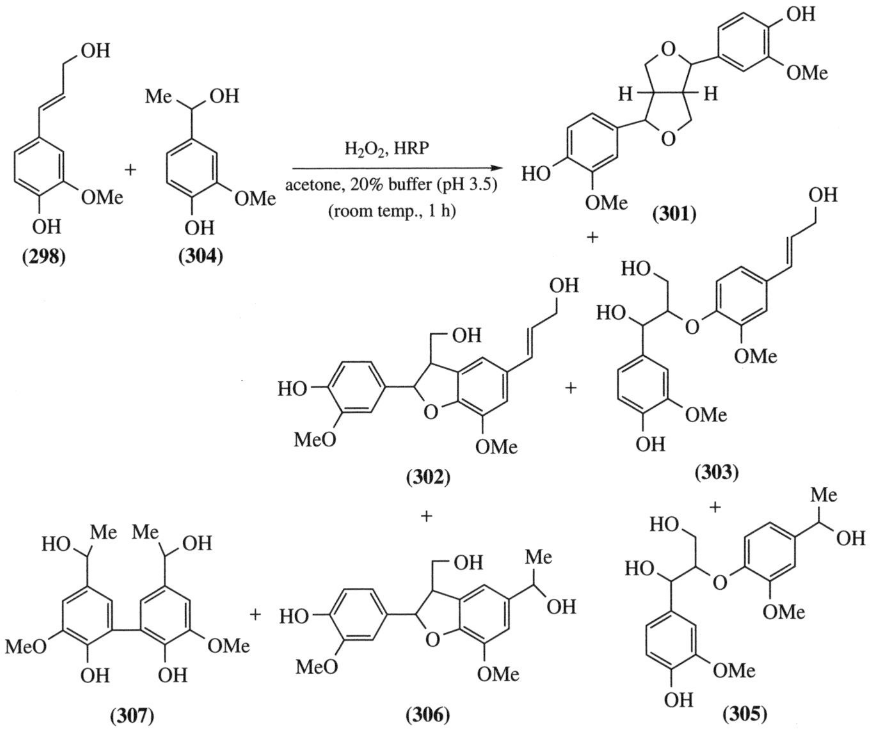

SCHEME 65. Hydrogen peroxide of coniferyl alcohol catalyzed by HRP

Coumestans represented by **321** are an oxygenated class of aromatic natural products, which have phytoalexin and estrogenic activities. From the biogenetic point of view, **321** will be formed from two units, 4-hydroxycoumarin (**322**) and catechol. Thus, the first synthesis of **321** was carried out by an electrochemical method. Catechol was initially oxidized to *o*-quinone, which was attacked by **322** to afford **321** in 95% yield (Scheme 69)[131].

Generally, the regioselective formation of *o*-quinones has been known to be accomplished by using polyphenol oxidase in chloroform and not in water[132], because of rapid inactivation of the enzyme in water. However, catechol underwent mushroom tyrosinase-catalyzed oxidation in phosphate buffer (pH 6.8) containing 4-hydroxycoumarin (**322**) to afford **321** in 96% yield[133], as shown in Scheme 69.

Oxidation of a number of *p*-substituted phenols to the corresponding *o*-benzoquinones was first performed by Kazandjian and Klibanov[132], using mushroom polyphenol oxidase and a quantitative conversion was achieved in CHCl$_3$ as a solvent. Other hydrophobic solvents such as methylene chloride, carbon tetrachloride, benzene, toluene, hexane and butyl acetate can be used, whereas the enzyme is inactive in more hydrophilic solvents such as ether, acetone, ethyl acetate, acetonitrile and other solvents. In addition, an immobilized enzyme on glass powder or beads is more efficient than a free enzyme.

Generally, reactive *o*-quinones are expected to react with a variety of dienophiles to afford the corresponding Diels–Alder products. The diene system of *o*-quinones is rather electron-deficient, so that electron-donating dienophiles such as ethyl vinyl ether must

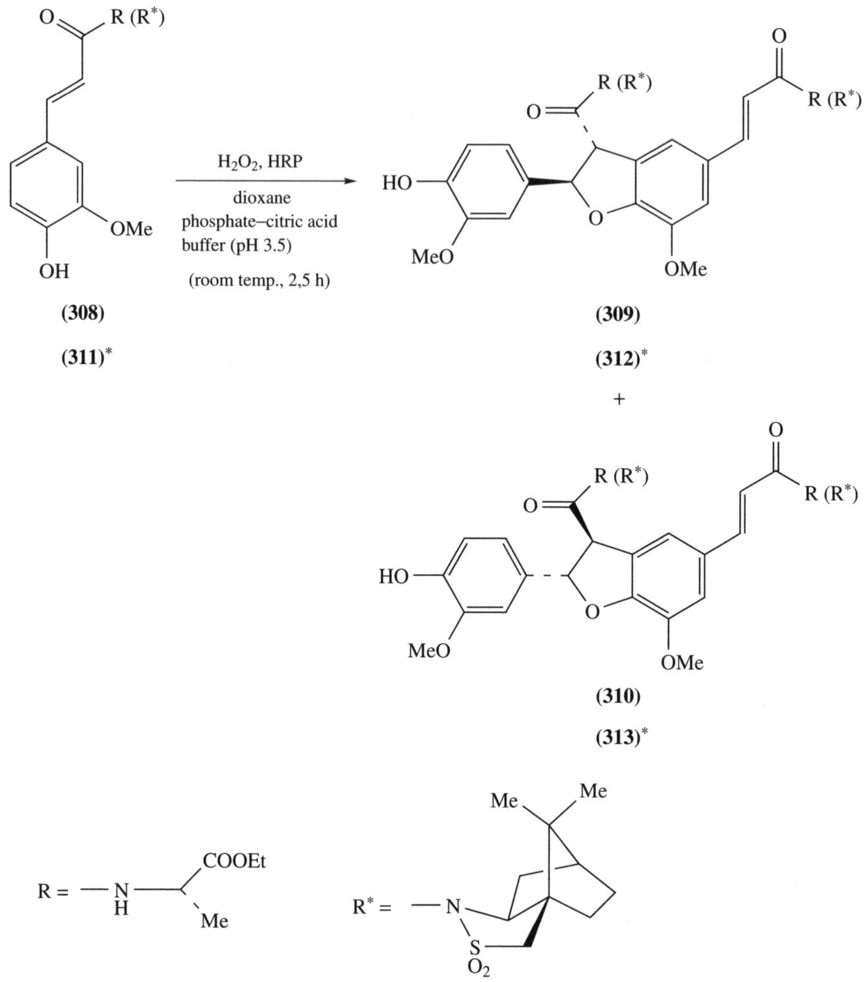

SCHEME 66. Stereoselective oxidation of chiral ferulic acid derivatives

be used. For example, the immobilized tyrosinase and phosphate buffer (0.5 mL, pH 7, 0.05 M) were added to a solution of *p*-cresol (**27**) (1 mM) in a mixed solution of $CHCl_3$ and ethyl vinyl ether [100 mL, 1:1(v/v)] and stirred at room temperature for 2.5 days to afford two isomers (**323** and **324**) (77%) in a 33:1 ratio, as shown in Scheme 70[134], indicating that the combination of enzymatic and nonenzymatic transformations in the three-step reaction cascade provides highly functionalized bicyclo[2.2.2]octenediones in an efficient manner.

4-Substituted 2,6-dimethylphenols (**325**–**327**) underwent enzymatic oxidation with mushroom tyrosinase in 50% MeCN–phosphate buffer (pH 6.8) (room temp., 48–72 h) resulting in the formation of the corresponding optically active compounds (**328**–**330**) in 50–60% yields. It is noted that the intramolecular cyclization of the initially formed quinone methide will take place in the hole of the enzyme or very close to the surface of

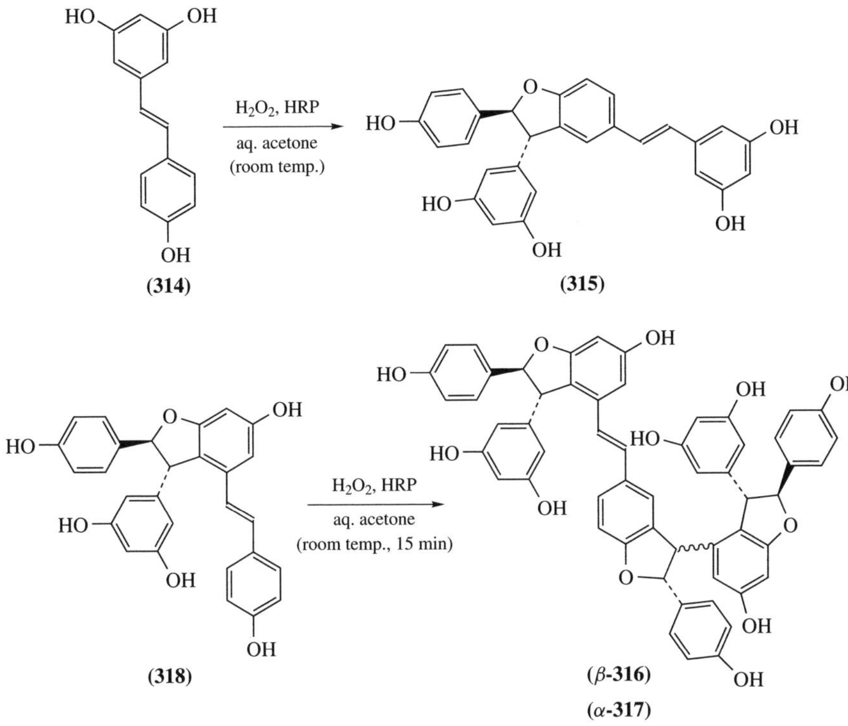

SCHEME 67. Hydrogen peroxide oxidation of stilbenes catalyzed by HRP

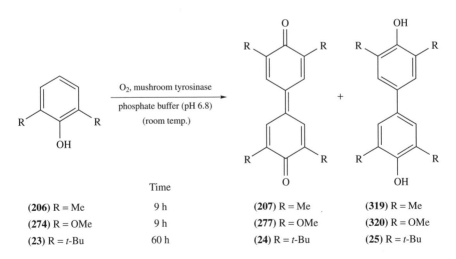

	Time		
(206) R = Me	9 h	(207) R = Me	(319) R = Me
(274) R = OMe	9 h	(277) R = OMe	(320) R = OMe
(23) R = t-Bu	60 h	(24) R = t-Bu	(25) R = t-Bu

SCHEME 68. Mushroom tyrosinase oxidation of 2,6-disubstituted phenols

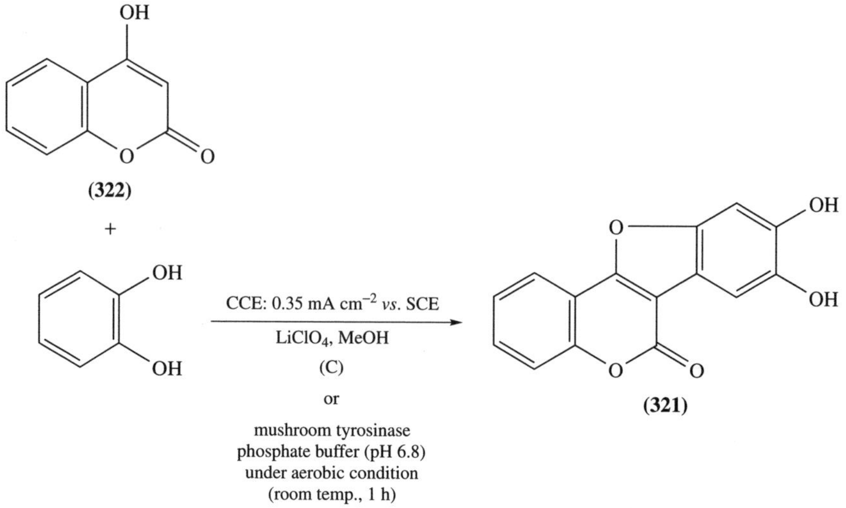

SCHEME 69. Electrochemical and enzymatic synthesis of 11,12-dihydroxycoumestan

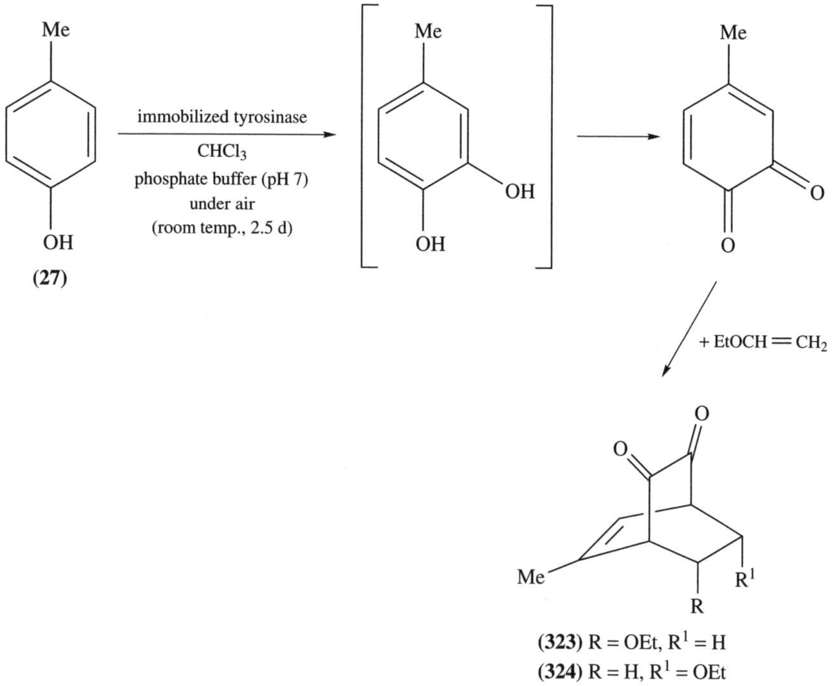

SCHEME 70. Synthesis of bicyclic compounds by a combination of enzymatic and nonenzymatic reactions

the enzyme, leading to the final products, as shown in Scheme 71[135]. The less activated **331** was converted into the cyclization product **332** in only 16% yield. No cyclization product was detected in the case of the nonactivated compound **333**.

Similarly, on mushroom tyrosinase-catalyzed oxidation of both 3,4-dihydroxy- and 4-hydroxybenzyl cyanides (**334** and **335**), the initially formed *o*-quinone (**336**) was converted into the corresponding quinone methide (**337**), which was not isolatable but was spectroscopically detected (Scheme 71)[136].

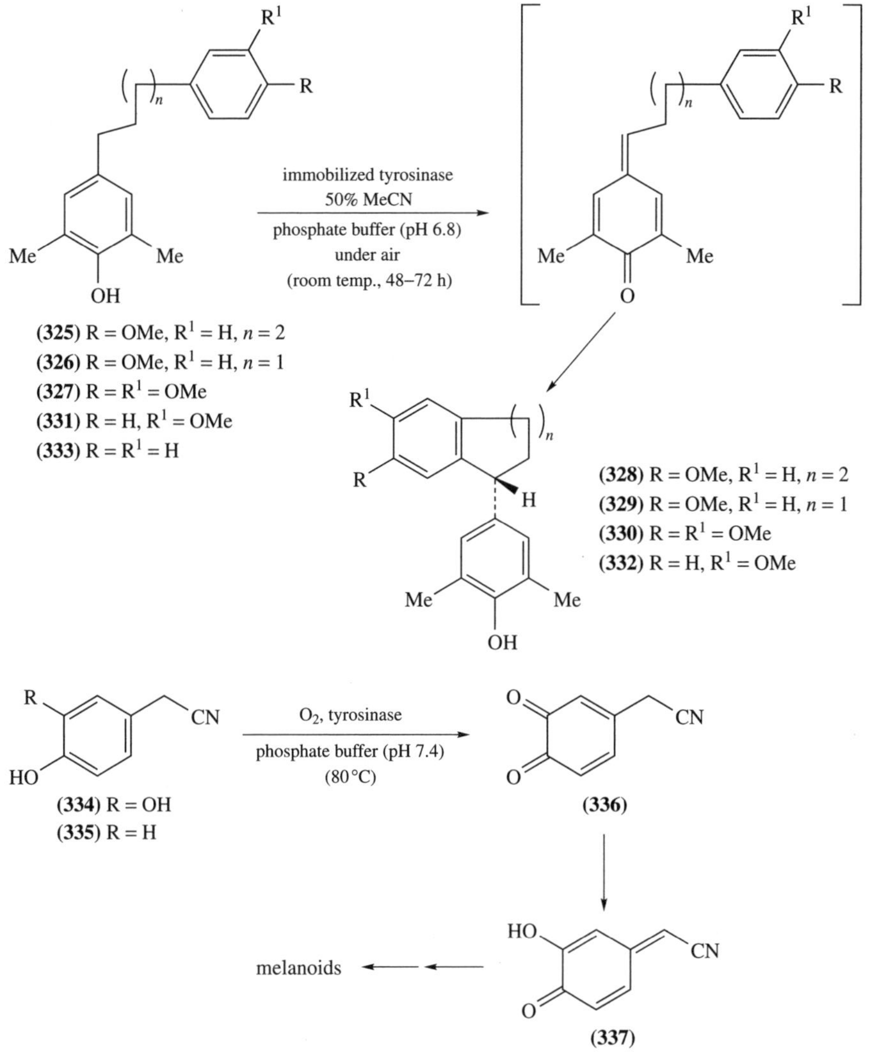

immobilized tyrosinase
50% MeCN
phosphate buffer (pH 6.8)
under air
(room temp., 48–72 h)
(325) R = OMe, R¹ = H, n = 2
(326) R = OMe, R¹ = H, n = 1
(327) R = R¹ = OMe
(331) R = H, R¹ = OMe
(333) R = R¹ = H
(328) R = OMe, R¹ = H, n = 2
(329) R = OMe, R¹ = H, n = 1
(330) R = R¹ = OMe
(332) R = H, R¹ = OMe
O2, tyrosinase
phosphate buffer (pH 7.4)
(80 °C)
(334) R = OH
(335) R = H
(336)
(337)
melanoids

SCHEME 71. Mushroom tyrosinase-catalyzed oxidation of some phenols

Electrochemical oxidation of 3,5-dihalogenated tyrosine derivatives provided the C—O or C—C coupled dimers depending on the halogen substituents. As shown in Scheme 11[35], electrochemical oxidation of both dichloro- and dibromotyrosines provided the corresponding diaryl ethers such as **61**, while the diaryl (**63**) was selectively produced from the 3,5-diiodotyrosine derivative. Quite interestingly, almost the same results have been obtained by enzymatic oxidation[137].

N-Acetyl-3,5-dichlorotyrosine (**338**) underwent hydrogen peroxide oxidation catalyzed by horseradish peroxidase (HRP) in a solution of phosphate buffer (pH 6.0) and MeCN (24 °C, 10 min) resulting in the formation of a mixture of two products (**339** and **340**), which was directly treated with $NaHSO_3$–NaOH to afford the recovered **338** and the corresponding diaryl ether (**341**) in 12 and 76% yields, respectively. Similar oxidation of N-acetyl-3,5-dibromotyrosine (**342**) provided the corresponding diaryl ether (**343**) in 42% yield, which was slightly lower than the 45% yield obtained by the electrochemical method[35]. In contrast, N-acetyl-3,5-diiodotyrosine (**344**) was oxidized by a combination of H_2O_2 and HRP under similar conditions to afford a mixture of two C—C coupled products (**345** and **346**), which was directly reduced with $NaHSO_3$–NaOH to afford the dityrosine derivative (**346**) in 45% overall yield (Scheme 72). Enzymatic oxidation of N-protected D-phenylglycine derivatives has also provided similar results[138].

Oxidative polymerization of phenols has been carried out extensively by using horseradish peroxidase and other enzymes. However, these interesting topics lie far beyond the scope of this chapter.

III. OXIDATION WITH NONMETAL COMPOUNDS

In contrast to the catalytic oxidation of phenols, stoichiometric amounts of oxidants are generally used. Therefore, efficient recycle systems must be deviced. In this section, phenolic oxidations using organic reagents is mainly described. In addition, some well known oxidants such as $NaIO_4$, Fremy's salt and others are briefly described.

A. Oxidation with Hypervalent Iodobenzenes

Hypervalent iodobenzenes have long been known as oxidants and their chemistry is summarized in an early volume of the present series[139]. Some of them are shown in Chart 2. Herein, both (diacetoxyiodo)benzene [$PhI(OAc)_2$] (**347**) and [di(trifluoroacetoxy)iodo]benzene [$PhI(OCOCF_3)_2$] (**348**) are most frequently and widely used for phenolic oxidation. 1-(*tert*-Butylperoxy)-1,2-benziodoxol-3(1*H*)-one (**349**) is also used for phenol oxidation. Many invaluable books and reviews on hypervalent iodobenzene-promoted oxidation of phenols have appeared[140].

1. Reaction and reaction mechanism

2,4-Disubstituted phenols such as **350** undergo $PhI(OAc)_2$-mediated oxidation in the presence of MeOH as a nucleophile resulting in the formation of two possible cyclohexadienones (**351** and **352**) (Scheme 73). The initially formed intermediate **353** is converted to the cyclohexadienones by two plausible routes. In route A, heterolytic dissociation generates a solvated phenoxonium ion **354**, which further reacts with MeOH to afford **351** and/or **352**. In route B, both **351** and **352** are produced by direct attack of MeOH on the intermediate (**353**). In the latter case, the reaction will be strongly influenced by steric factors and a homochiral environment using chiral solvents and chiral oxidants to induce some asymmetric induction, particularly in the formation of **352**.

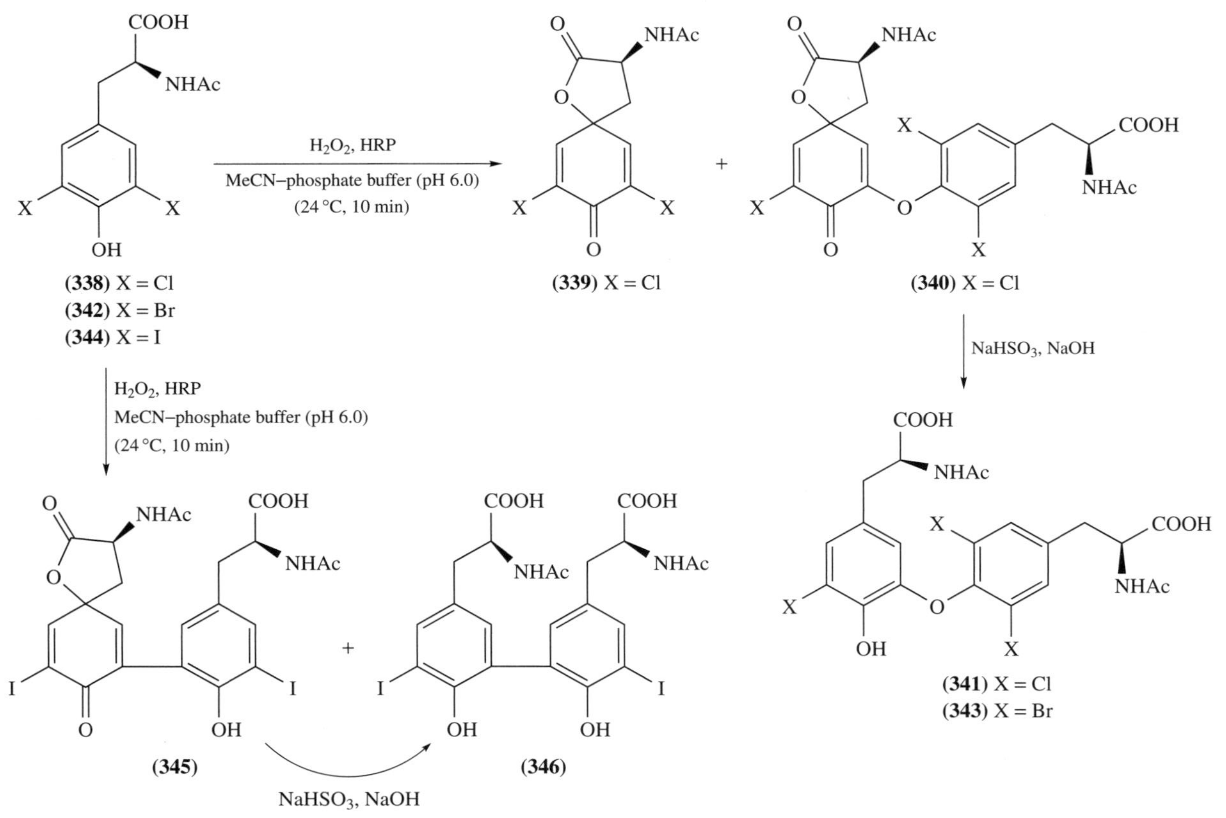

SCHEME 72. HRP-catalyzed oxidation of N-protected 3,5-dihalotyrosine derivatives

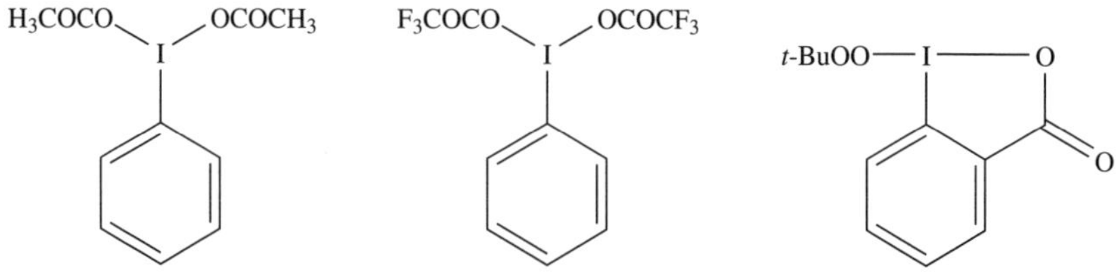

(347) **(348)** **(349)**

CHART 2. Some hypervalent iodobenzenes used for organic synthesis

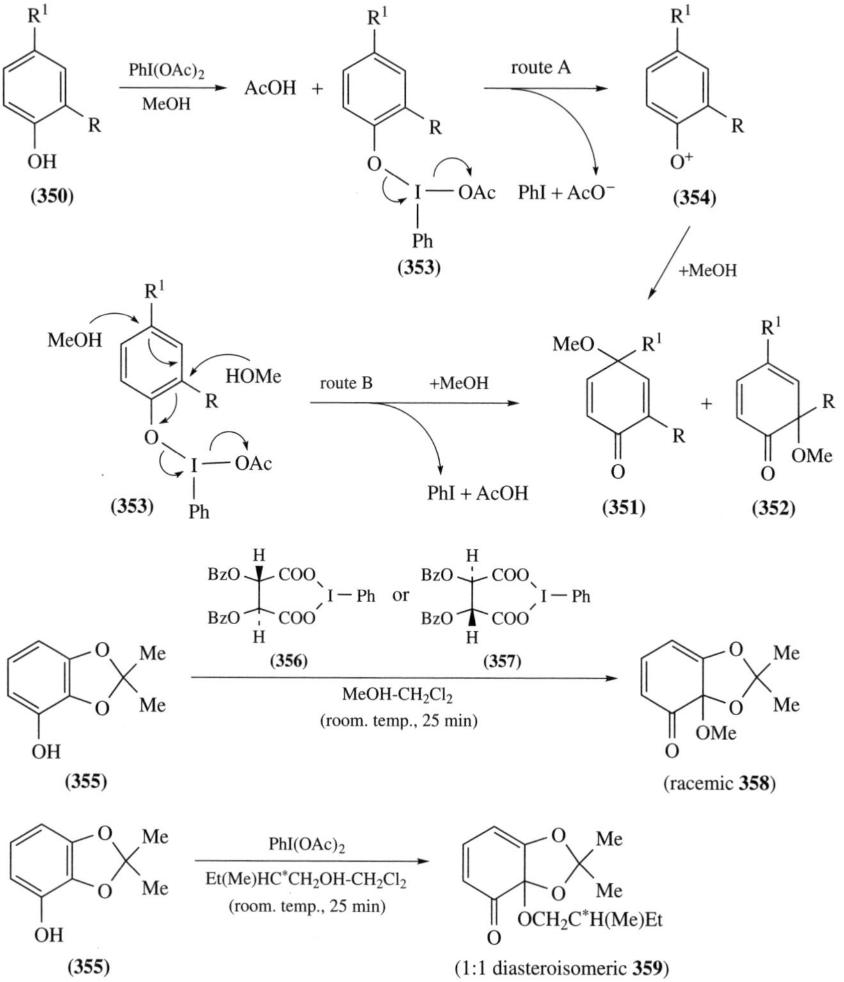

SCHEME 73. Mechanism of phenol oxidation with (diacetoxyiodo)benzene

Thus, oxidation of 2,3-isopropylidenepyrogallol (**355**) was effected with the chiral reagent **356** or **357** in dry CH_2Cl_2 to yield an only racemic mixture (**358**) in each case. In addition, when (S)-$(-)$-2-methylbutan-1-ol was used instead of MeOH as a nucleophile, only a 1:1 diastereomeric mixture (**359**) was obtained (Scheme 73). These experimental results are in good agreement with predictions based on the calculated Mulliken charge distributions and the size of LUMO coefficients for phenoxonium ions **354** in Scheme 73[141]. Although hypervalent iodobenzene-promoted oxidation of phenols would take place via route B depending upon the substituents on the aromatic ring, the solvent systems and other factors, route A must be more favorable. The resulting phenoxonium ions are attacked by a variety of nucleophiles to afford the corresponding 2,5- and/or 2,4-cyclohexadienones. Some examples are shown below.

p-Alkylphenols such as 4-benzyl-, 2,4,6-trimethyl- and 2,4,6-tri(*tert*-butyl)phenol (**360**, **221** and **73**) underwent $PhI(OAc)_2$-promoted oxidation in MeOH at room temperature to afford p-quinol alkyl ethers **361**, **362** and **363** in 65, 72 and 94% yields, respectively[142]. Oxidation of 2,6-dibromo-4-methylphenol (**364**) in MeOH-CH_2Cl_2 was also effected with $PhI(OAc)_2$ to give 2,6-dibromo-4-methoxy-4-methyl-2,5-cyclohexadienone (**365**) in 63% yield[143], while anodic oxidation of **364** at constant current (0.13 mA cm^{-2}; +870–880 mV vs. SCE) provided 2,6-dibromo-4-methoxymethylphenol (**366**) in 52% yield[30]. The remarkable differences in the product selectivity between the chemical and electrochemical reactions must be attributable to the environment surrounding the resulting phenoxonium ion. Oxidation of p-substituted phenols such as **367** and **368** with $PhI(OCOCF_3)_2$ in MeCN produced preferably cyclic compounds **369** and **370** (86 and 59% yields respectively)[144] (Scheme 74).

Oxidative fluorination of the p-substituted phenol **371** was effected with $PhI(OCOCF_3)_2$ and pyridinium polyhydrogen fluoride (PPHF) to afford directly the hydroindolenone (**372**) (35%). In the case of **373**, the corresponding 4-fluorinated cyclohexa-2,5-dienone (**374**) was produced in 43% yield. Cyclization of **374** to **375** was readily effected with Na_2CO_3, as shown in Scheme 75[145]. When MeOH was used instead of PPHF the corresponding methoxy compounds were obtained. The yields are higher than that of the fluoro compounds.

Oxidation of the bicyclic compounds (**376** and **377**) provided the corresponding hydroquinolenones (**378** and **379**) in 39 and 29% yields, respectively (Scheme 75).

p-Substituted phenols (**27**, **221**, **380** and **381**) were effected with $PhI(OCOCF_3)_2$ in aqueous MeCN (0 °C, 5–15 min) to give quinols (**382**–**385**) in moderate and good yields. It is noted that higher yields were obtained when the corresponding tripropylsilyl ethers were used, as shown in Scheme 76. Here, the oxidation required 0.5–2 h to proceed to completion. As compared with p-cresol (**27**) and methyl 4-hydroxyphenyl acetate (**381**), the corresponding silyl ethers were more efficiently converted into the quinols[146].

On $PhI(OAc)_2$-promoted oxidation of phenols bearing an olefinic side chain at the C-4 position, the resulting phenoxonium ion underwent intramolecular cyclization resulting in the formation of the corresponding spiro compounds, although side reaction products were not avoided. Anodic oxidation also provided the same spiro compounds. However, the yields from the $PhI(OAc)_2$-promoted oxidation are quite different from those from electrochemical oxidation[54a,147]. For example, three 2′-isopropenyl-p-arylphenols (**112a**, **386** and **387**) were subjected to $PhI(OAc)_2$-promoted oxidation and to electrochemical oxidation to afford the corresponding spiro dienones **113a**, **388** and/or **389**, respectively. In particular, the phenol **387** gave the spiro dienone **389** in 80% yield via the electrochemical method, but in only 24% yield by the $PhI(OAc)_2$ oxidation. In contrast, **388** was not detected in the electrochemical oxidation (Scheme 77). These differences are due to some environmental factors surrounding the resulting phenoxonium cation. The spiro dienones are regarded as promising synthetic intermediates for natural products synthesis.

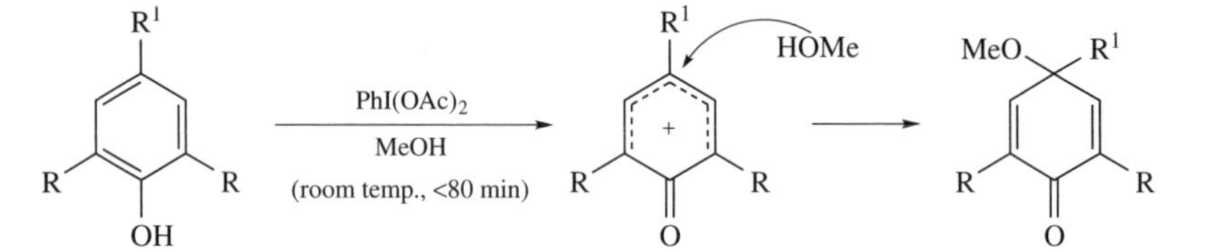

(**360**) R = H, R^1 = CH$_2$Ph
(**221**) R = R^1 = Me
(**73**) R = R^1 = *t*-Bu

(**361**) R = H, R^1 = CH$_2$Ph
(**362**) R = R^1 = Me
(**363**) R = R^1 = *t*-Bu

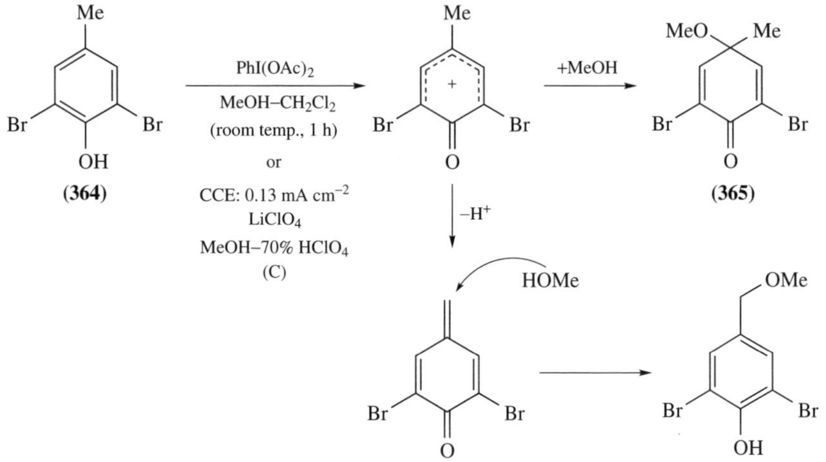

(**364**) (**365**) (**366**)

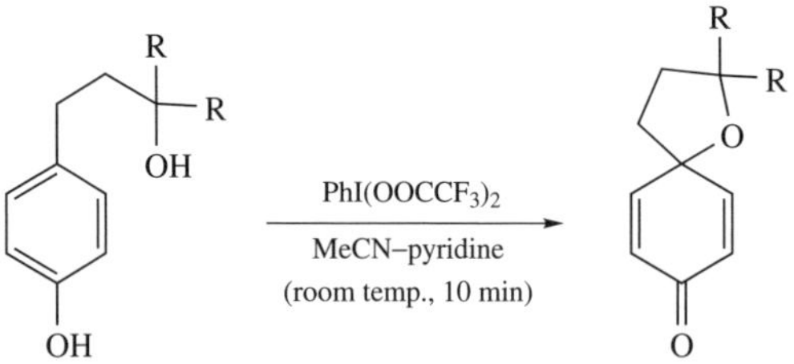

(**367**) RR = O (carbonyl oxygen)
(**368**) R = R = H

(**369**) RR = O (carbonyl oxygen)
(**370**) R = R = H

SCHEME 74. Oxidation of *p*-substituted phenols with (diacyloxyiodo)benzenes

In the case of *p*- or *o*-methoxyphenols, the corresponding quinones and quinone monoketals, which are quite useful in organic synthesis[45], are produced usually in very high yields. *p*-Methoxyphenol (**390**) was oxidized with PhI(OAc)$_2$ in MeOH to afford 4,4-dimethoxycyclohexa-2,5-dienone (**391**) in 99% yield[142]. PhI(OCOCF$_3$)$_2$-mediated oxidation of 2,4,5-trimethoxyphenol (**392**) in MeOH-MeCN yielded the corresponding dienone **393** (86%)[144]. Interestingly, oxidation of 2-benzylphenol (**394**) with 2 equivalents of PhI(OAc)$_2$ in MeOH provided 2-benzyl-4,4-dimethoxycyclohexa-2,5-dienone (**395**) (85%)[142]. When treated with PhI(OCOCF$_3$)$_2$ in MeCN–H$_2$O containing

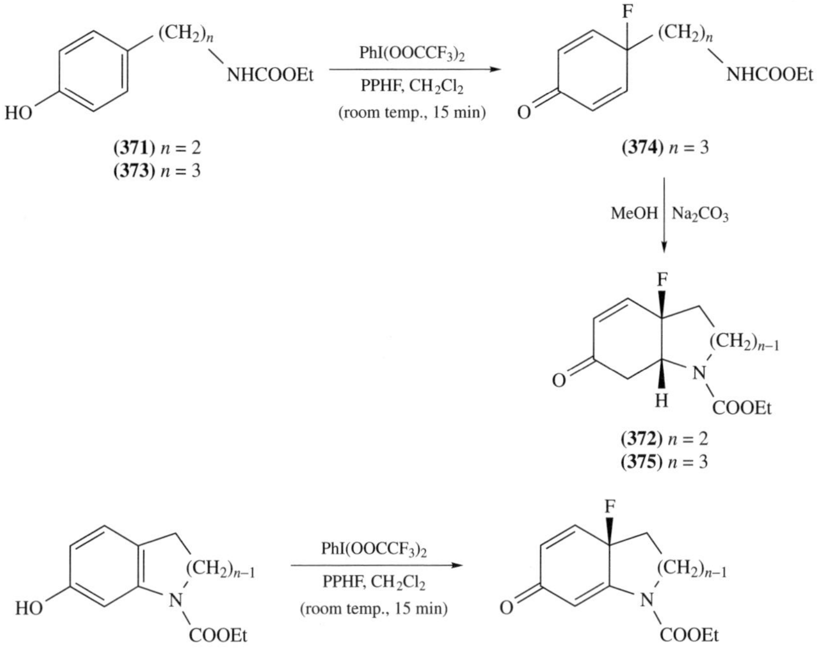

(371) $n = 2$
(373) $n = 3$

(374) $n = 3$

(372) $n = 2$
(375) $n = 3$

(376) $n = 2$
(377) $n = 3$

(378) $n = 2$
(379) $n = 3$

SCHEME 75. Oxidative fluorination of *p*-substituted phenols with bis(trifluoroacetoxy)benzenes and synthesis of hydroindolenones

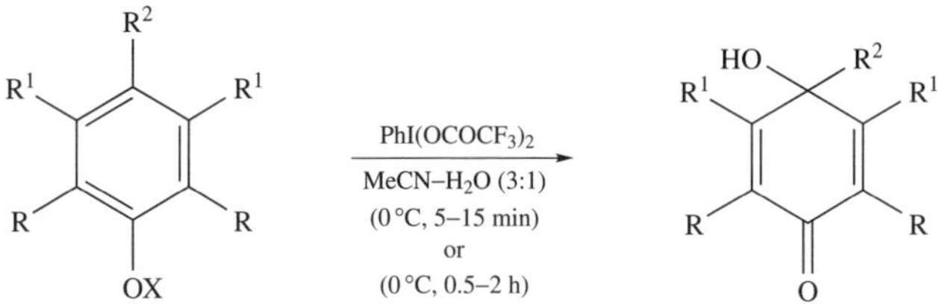

(27) $R = R^1 = H, R^2 = Me$
(221) $R = R^2 = Me, R^1 = H$
(380) $R = H, R^1 = R^2 = Me$
(381) $R = R^1 = H, R^2 = CH_2COOMe$

(382) $R = R^1 = H, R^2 = Me$
(383) $R = R^2 = Me, R^1 = H$
(384) $R = H, R^1 = R^2 = Me$
(385) $R = R^1 = H, R^2 = CH_2COOMe$

Yields (%)	
X = H	$Si(Pr^{-n})_3$
48	73
67	78
63	70
27	59

SCHEME 76. $PhI(OCOCF_3)_2$-promoted oxidation of *p*-substituted phenols and silyl ethers

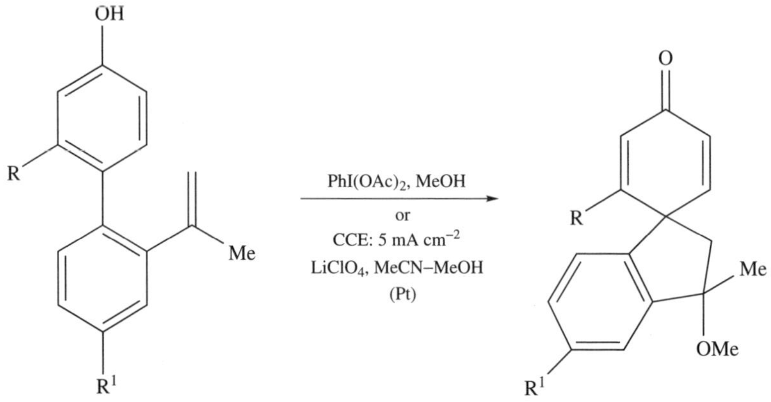

(112a) R = R¹ = H
(386) R = OMe, R¹ = H
(387) R = H, R¹ = OMe

(113a) R = R¹ = H
(388) R = OMe, R¹ = H
(389) R = H, R¹ = OMe

PhI(OAc)$_2$	Anodic oxidation
67	85
74	0
24	80

Yields (%)

SCHEME 77. Formation of spiro dienones by PhI(OAc)$_2$-promoted oxidation and anodic oxidation of 2′-isopropenyl-p-arylphenols

K$_2$CO$_3$, 4-methoxy-2-(tetrahydropyranyloxy)methylphenol (**396**) was converted into the corresponding p-quinone **397** in excellent yield[148] (Scheme 78).

3-Methoxycarbonyl-6-methoxyphenol (**398**) underwent PhI(OAc)$_2$-mediated oxidation in 2:1 MeOH–MeCN resulting in the formation of the corresponding o-quinone monoketal **399**, which was readily dimerized to afford a Diels–Alder product **400** (55%)[149]. A variety of cyclohexa-2,4-dienones such as **399** have been used for natural products synthesis, as shown later. On oxidation with PhI(OAc)$_2$ in CH$_2$Cl$_2$–AcOH (3:1), the phenol **398** was rapidly converted into 6-acetoxy-3-methoxycarbonyl-6-methoxycyclohexan-2,4-dienone (**401**) as a stable product (95%). On silica gel exposure, this compound was cleanly converted into two rearranged products (**402** and **403**) in 45 and 30% yields, respectively (Scheme 79). The former is probably formed by a [3,5] sigmatropic rearrangement. This is incompatible with the Woodward–Hoffmann rule. However, density functional theory calculations indicate that the [3,5] shift leading to **402** is pseudopericyclic, has a remarkably low activation energy of 20.1 kcal mol^{-1} and is favored by 4.5 kcal mol^{-1} over the pericyclic [3,3] shift leading to **403**[149].

In connection with protein tyrosine kinase inhibitors, 4-substituted 2-methoxyphenols such as **404** and **405** were oxidized with PhI(OAc)$_2$ in MeNO$_2$–AcOH (25 °C, 10 min) to afford 2-acetoxy-2-methoxycyclohexa-2,4-dienones (**406** and **407**) in 72 and 61% yields, respectively. On BF$_3$·Et$_2$O treatment, both **406** and **407** were rearranged to phenols **408** and **409**, respectively. When acetonitrile was used as a solvent, PhI(OAc)$_2$-promoted

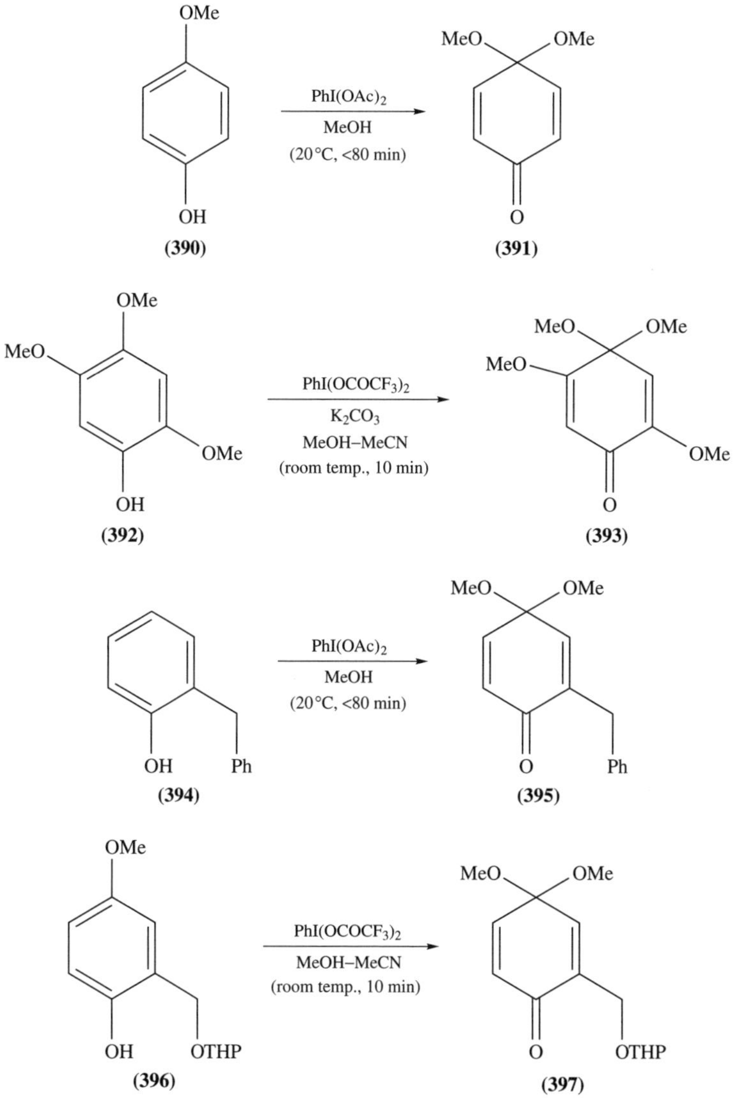

SCHEME 78. Synthesis of *p*-quinones and quinone monoketals

oxidation of **404** provided biphenyl **410** in 52% yield. Substrate **405** was also converted into the corresponding biphenyl **411** (Scheme 80)[150].

4-Hydroxybenzaldehyde (**412**) underwent PhI(OAc)$_2$-mediated oxidation in AcOH under reflux conditions to afford 3-iodo-4-phenoxybenzaldehyde (**413**) in 32% yield. Similarly, methyl 4-hydroxybenzoate (**414**) was also converted into methyl 3-iodo-4-phenoxybenzoate (**415**) (76%). The suggested mechanism involves an initial formation of the zwitterionic intermediate **416** (Scheme 80).

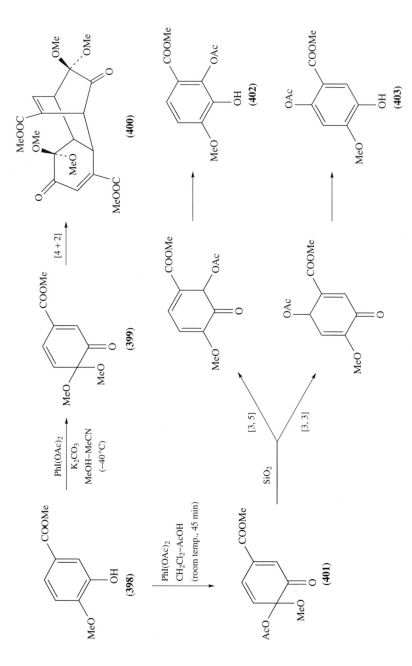

SCHEME 79. Oxidation of 3-methoxycarbonyl-6-methoxyphenol with PhI(OAc)$_2$

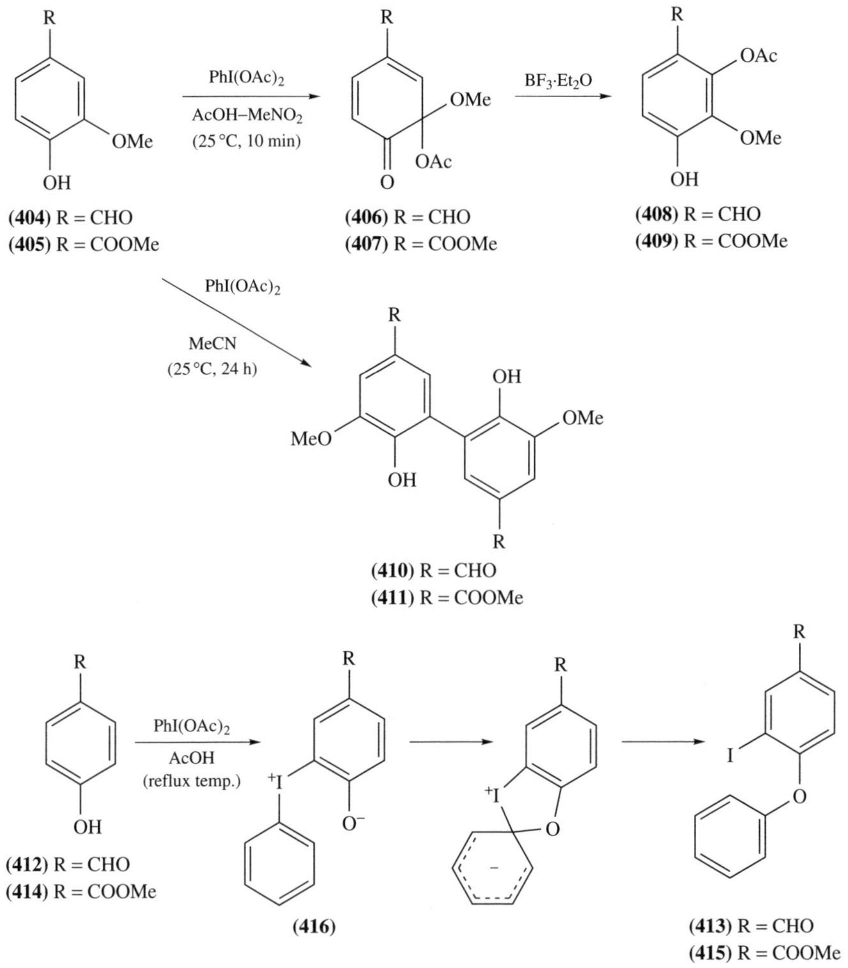

SCHEME 80. Oxidation of *p*-substituted phenols with PhI(OAc)$_2$ in different solvents

Generally, hypervalent (diacyloxyiodo)benzenes such as **347** and **348** react with phenols to generate the corresponding phenoxonium ions, which are attacked by a variety of nucleophiles. In contrast, 1-(*tert*-butylperoxy)-1,2-benziodoxol-3(1*H*)-one (**349**) has been found to undergo homolytic cleavage of the hypervalent *t*-BuOO−I(III) bond at room temperature to generate a *tert*-butylperoxy radical and a *s*-iodanyl radical, which act as an efficient radical oxidant for oxidation of benzylic and allylic C−H bonds[151]. Thus, a variety of *p*-substituted phenols were oxidized with a combination of the peroxyiodane **349** and *t*-BuOOH in EtOAc or benzene to afford the corresponding 4-(*tert*-butylperoxy)-2,5-cyclohexadienones in good yields[152]. *p*-Cresol (**27**) was oxidized with **349** (1.2 equiv.) and *t*-BOOH (6 equiv.) in EtOAc (50 °C, 3.5 h) to afford 4-(*tert*-butylperoxy)-4-methyl-2,5-cyclohexadienone (**289**) (81%) (Scheme 81). The reaction was inhibited by the addition of such a radical scavenger as glavinoxyl. Similarly, both **417** and **418** were converted into the corresponding cyclohexadienones **419** and **420** in 65 and 63% yields, respectively (Scheme 81).

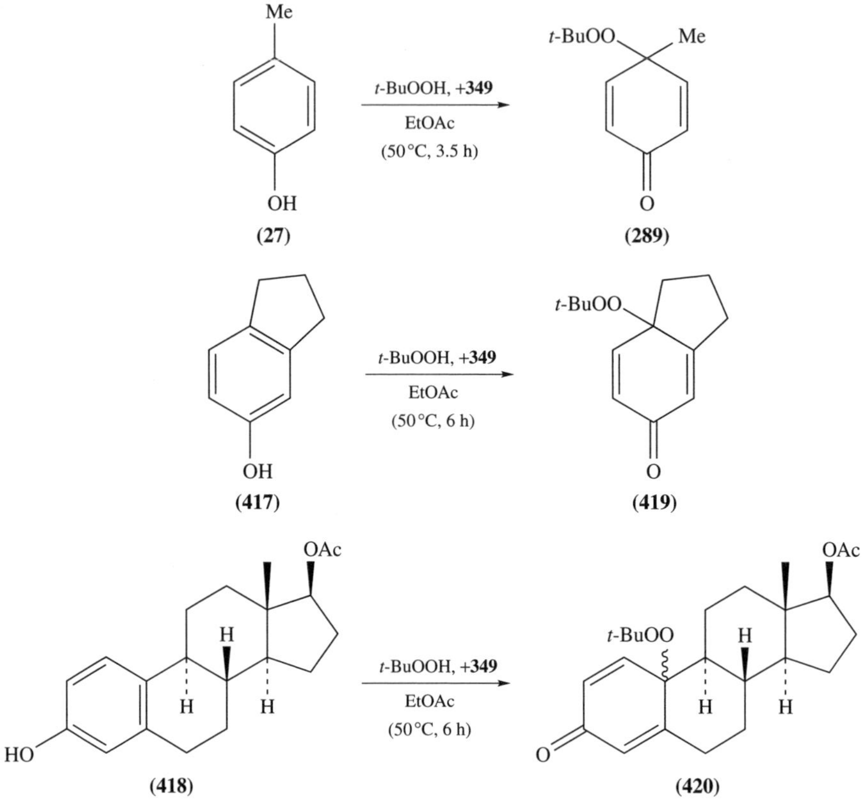

SCHEME 81. Oxidation of *p*-substituted phenols with peroxyiodane **349** and *t*-BuOOH

2. Applications in natural products synthesis

From the viewpoint of organic synthesis, nature provides us with a number of target molecules, which have novel structures and a variety of biological activities. As already shown in Section II.A, electrochemical oxidation of phenols has been applied successfully to natural products synthesis. Hypervalent (diacyloxyiodo)benzenes have also been proved to be effective for natural products synthesis. Generally, oxidation of *o*- and *p*-methoxyphenols in MeOH provides the corresponding *o*- and *p*-quinone monoketals, respectively. They are utilized as promising synthons for natural products and related bioactive compounds, as demonstrated by Swenton[45]. Recently, these quinone monoketals have been utilized for syntheses of terpenoids, neolignans, anthraquinones, alkaloids and related compounds.

4-Methoxycarbonyl-2-methoxyphenol (**421**) underwent PhI(OAc)$_2$-promoted oxidation in MeOH resulting in the formation of 4-carbomethoxy-6,6-dimethoxycyclohexa-2,4-dienone (**422**), which was spontaneously dimerized to **423** in 85% yield[153]. However, in

17. Oxidation of phenols

the presence of an excess (25 equiv.) of dienophiles such as methyl acrylate, the resulting dienone **422** reacted with to give a Diels–Alder product **424** (85%) (Scheme 82).

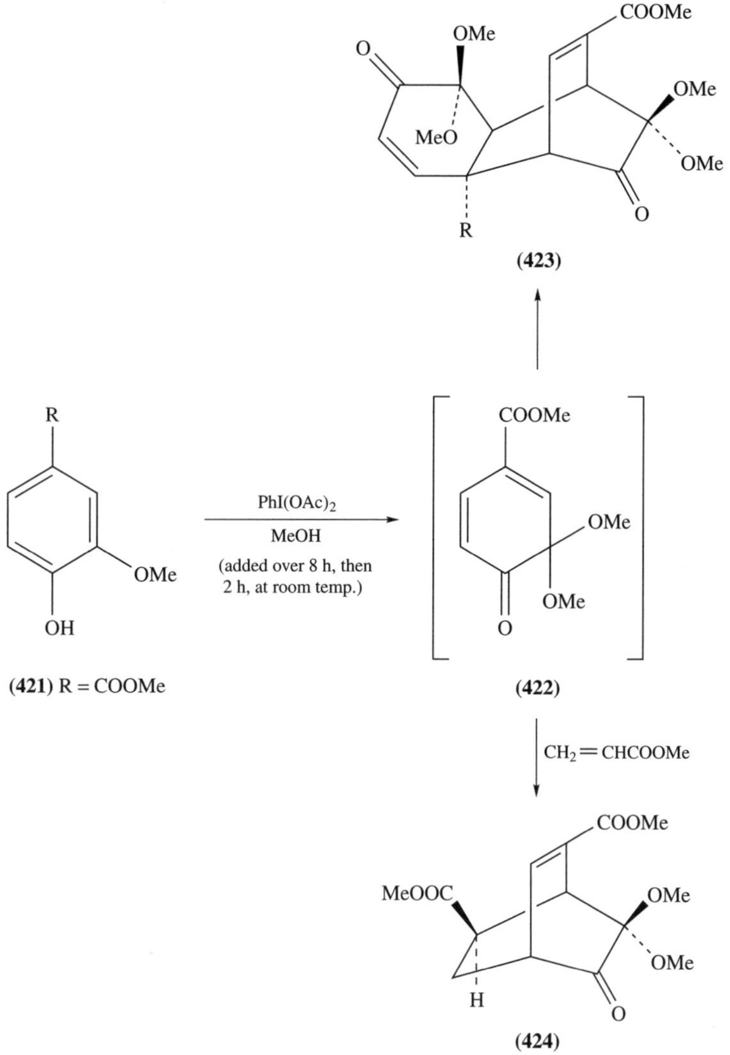

SCHEME 82. Diels–Alder reactions of *o*-quinone monoketals

Generally, on PhI(OAc)$_2$-promoted oxidation of *o*-methoxyphenols in MeOH containing a large excess of electron-rich dienophiles, the resulting *o*-quinone monoketals may undergo an intermolecular Diels–Alder reaction with the dienophiles to afford the corresponding dimers. 4-Methoxycarbonyl-2-methoxyphenol (**421**) was submitted to PhI(OAc)$_2$-promoted oxidation in MeOH containing benzyl vinyl ether (BVE) or dihydrofuran (DHF) at 50 °C to afford the adducts **425** and **426** in 83 and 36%, respectively[154]. Similarly, when dihydrofuran was replaced by 2-methylfuran, the corresponding Diels–Alder product **427** was readily obtained in 85% yield[155] (Scheme 83). 2-Methylfuran acts here as a dienophile, although it is generally utilized as a diene. 2-Methoxyfuran gave similar results[156]. A variety of indoles are also utilized as a dienophile; the initially formed dienone (**422**) reacted with indole (**428**) at 0 °C and then at room temperature to give **429** in 65% yield. In the case of 2-methylindole (**430**), the yield (24%) of the Diels–Alder adduct **431** is low. However, at higher temperature, 3-arylindoles **432** and **433** were obtained in 96 and 86% yields, respectively (Scheme 83)[157]. These 3-arylindoles are proposed to be formed via a Michael addition–aromatization sequence.

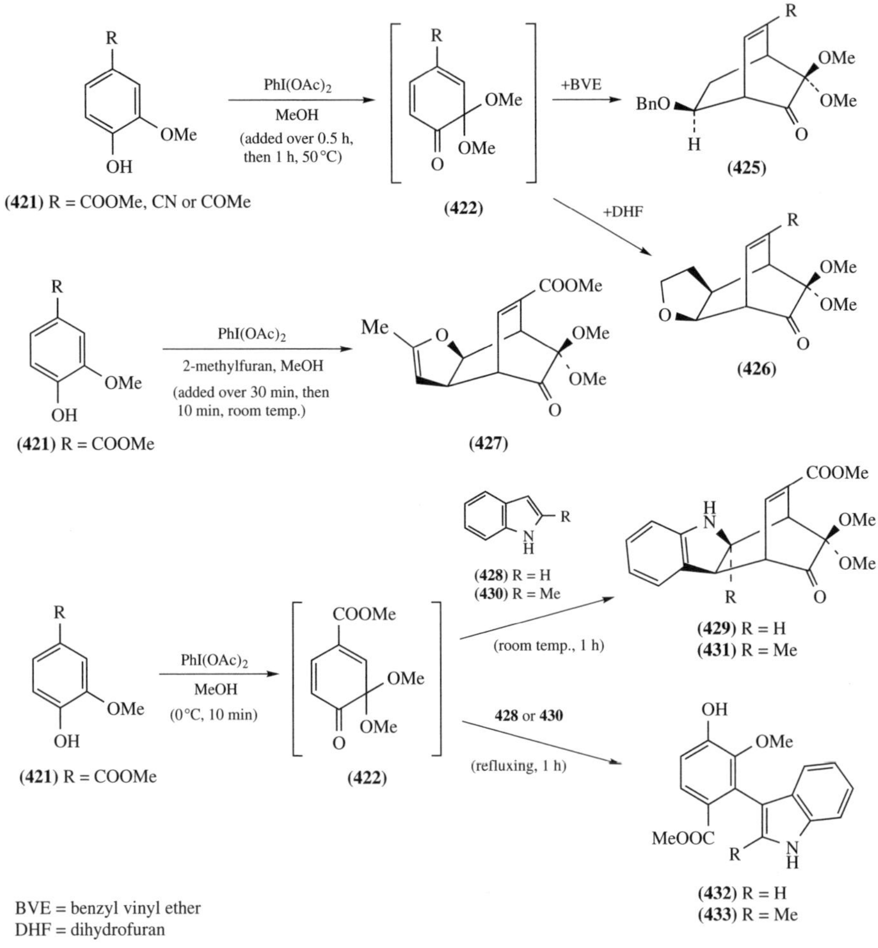

BVE = benzyl vinyl ether
DHF = dihydrofuran

SCHEME 83. Diels–Alder reactions of *o*-quinone monoketals with electron-rich dienophiles

17. Oxidation of phenols

Both linear and angular triquinanes constitute one of the large classes of sesquiterpenoids and have been attracting many synthetic organic chemists, because of their novel structures and biological activities.

Oxidation of 4-methoxycarbonyl-2-methoxyphenol (**421**) with PhI(OAc)$_2$ in MeOH was carried out in the presence of cyclopentadiene (CPD) to afford in 87% yield a Diels–Alder adduct **434**, which was subjected to photochemical reaction followed by Ac$_2$O–BF$_3$·Et$_2$O treatment to give a linear triquinane **435** (47.6% in 2 steps)[158].

Similarly, the combination of PhI(OCOR$_2$)-promoted oxidation and photochemical reaction provided both linear and angular triquinane-type compounds. 2-Methoxyphenol (**436**) bearing an olefinic side chain at the C-3 position was subjected to PhI(OCOCF$_3$)$_2$-mediated oxidation in MeOH, followed by heating in mesitylene at 165 °C to give selectively the corresponding tricyclic compound (**437**) (78%). Compound **437** was successfully converted into the two triquinane-type compounds **438** and **439** (Scheme 84)[159].

(−)-Eremopetasidione (**440**), isolated recently from rhizomes of *Petasites japonicus* MAXIM, has been used in the treatment of tonsillitis, contusion and poisonous snake bites in Chinese medicine. Recently, racemic **440** was synthesized in 9 steps (30% overall yield) starting from 2-methoxy-4-methylphenol (**421**) (Scheme 85)[160]. Oxidation of **421** with PhI(OAc)$_2$ in MeOH in the presence of ethyl vinyl ketone (EVK) afforded in 96% yield a Diels–Alder adduct **441**, which was converted into silyl enolate **442** in 96% yield. This enolate further underwent Cope rearrangement to give regio- and stereoselectively the desired *cis*-decalin **443** (70%). Further conversion of **443** to the target molecule **440** was then accomplished.

On PhI(OAc)$_2$-promoted oxidation in CH$_2$Cl$_2$ containing alkenol (5 equiv.) at room temperature, 2-methoxyphenols such as **421** (R = Me and COOMe) were converted into the corresponding tricyclic compounds (**444** and **445**) in 77 and 75% yields, respectively (Scheme 86)[161] via an intramolecular Diels–Alder reaction of the initially formed cyclohexa-2,4-dienones (**446** and **447**). These tricyclic compounds are recognized as promising synthetic intermediates for synthesis of natural products and related compounds[162].

Pallescensins are a group of furanosesquiterpenoids isolated from the marine sponge *Disidea pallescens*. Of them, pallescensin B (**448**) presents the most complex architecture, with a unique bicyclo[4.2.2]decane system fused to a furan moiety. Thus, pallescensin B was synthesized starting from 2-methoxy-4-methylphenol (**421**). When **421** was submitted to PhI(OAc)$_2$-promoted oxidation followed by an immediate intramolecular Diels–Alder reaction in the presence of 2-methylallyl alcohol, it afforded a tricyclic compound **449** in 58% yield[163]. Grignard reaction on the carbonyl of **449** was effected stereoselectively with vinylmagnesium bromide in the presence of ZnBr$_2$ to afford an 82% yield of **450**, which underwent anionic [1,3]-rearrangement to afford in 80% yield the desired adduct (**451**) bearing the same carbon skeleton as that of pallescensin B. Further conversion of **451** provided pallescensin B (**448**) (Scheme 87)[163].

The plant hormone (+)-abscisic acid (ABA) (**452**) is well known to regulate a wide range of processes in plants, including transpiration through controlling stomatal aperture, responses to environmental stress, inhibition of germination and others. In connection with ABA, chiral synthesis of (+)-8′-demethyl abscisic acid (**453**) was accomplished starting from 2,6-dimethylphenol (**206**), as shown in Scheme 88[164]. 2,6-Dimethylphenol in anhydrous ethylene glycol was oxidized with PhI(OAc)$_2$ in hexane (0 °C–room temp., 2 h) to afford *p*-quinone monoketal **454** in 63% yield. This compound was submitted to yeast reduction to afford in 50% yield (6*R*)-2,6-dimethyl-4,4-ethylenedioxycyclohexa-2-enone (**455**), which was further converted to the target molecule **453**.

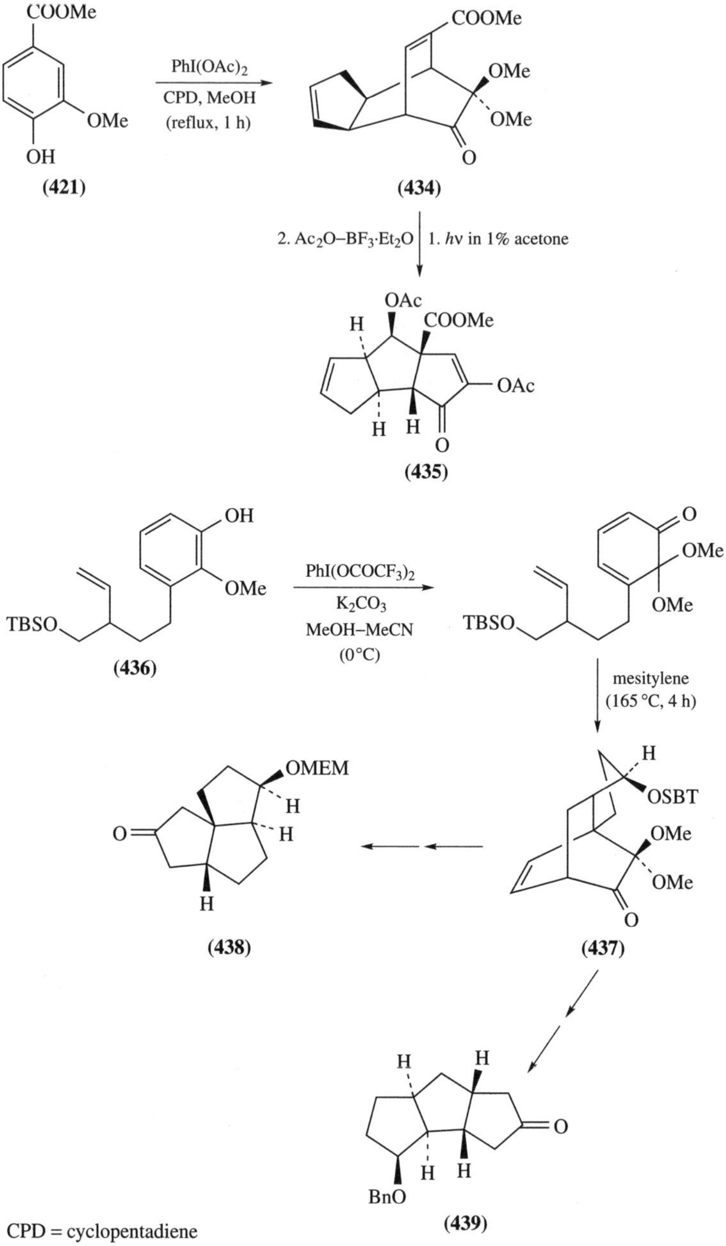

CPD = cyclopentadiene
TBS = *t*-butyldimethylsilyl
MEM = methoxyethoxymethyl

SCHEME 84. Synthesis of triquinane-type compounds

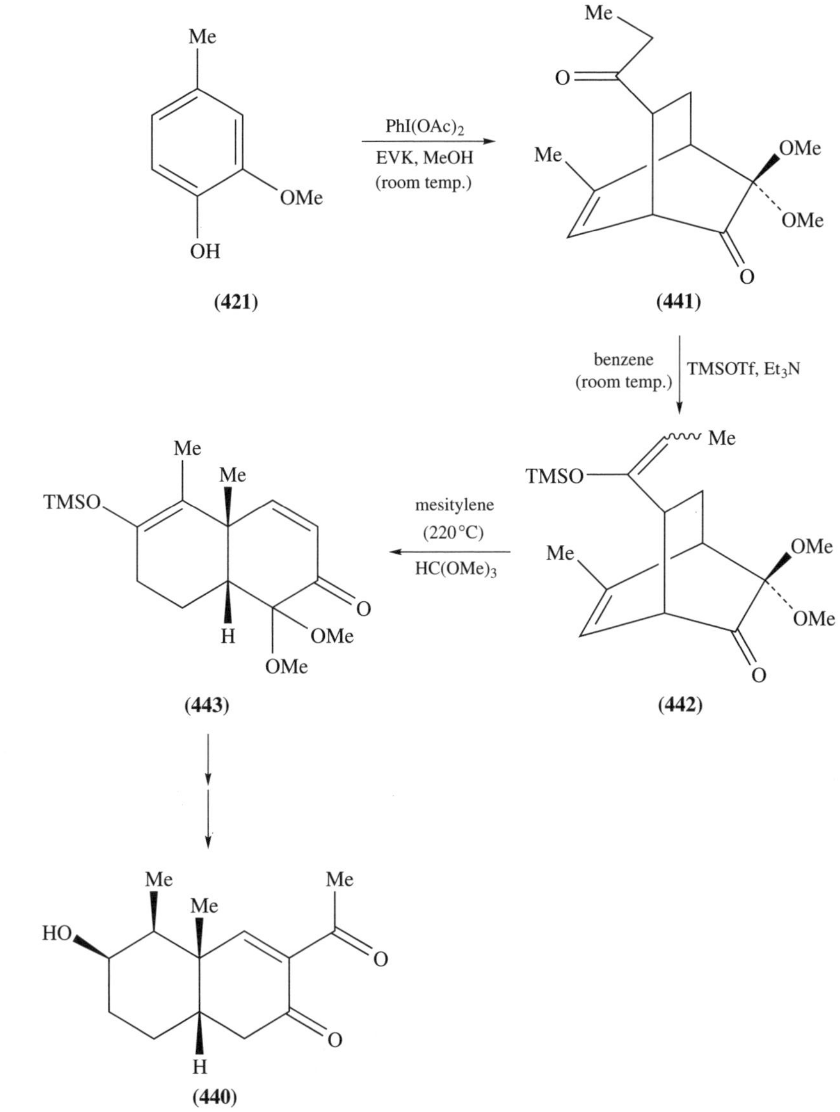

EVK = ethyl vinyl ketone

SCHEME 85. Total synthesis of racemic eremopetasidione

As already shown in Scheme 16, the electrochemically generated *p*-quinone monoacetal (**84**) reacted with 3-cyanophthalide anion **85** to give the anthraquinone **86**. Similarly, $PhI(OAc)_2$-promoted oxidation of 4-substituted phenols in MeOH provides the corresponding cyclohexa-2,5-dienones, which react with the anion of 3-cyanophthalide to yield a variety of anthraquinones[165]. *N*-Acetyltyrosine ethyl ester **456** was subjected to

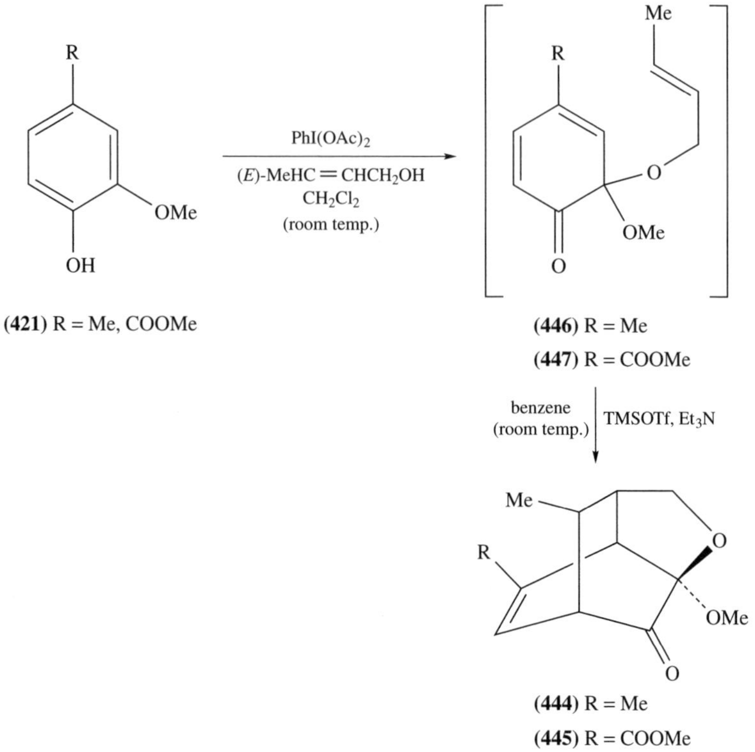

SCHEME 86. PhI(OAc)$_2$-oxidation of 2-methoxyphenols in the presence of an alkenol

PhI(OAc)$_2$-mediated oxidation in MeOH at ambient temperature for 5 min to afford a mixture of crude dienones, which reacted directly with the phthalide anion to give the corresponding anthraquinone **457** in 62.5% overall yield (Scheme 89). In the case of o-quinone monoacetals[165], similar results have been obtained.

2-Methoxycarbonyl-4-methoxyphenol (**458**) was oxidized with PhI(OAc)$_2$ in THF, CF$_3$COOH or CH$_2$Cl$_2$ containing sorbic alcohol (10 equiv.) to afford the corresponding cyclohexa-2,5-dienone (**459**), which immediately underwent an intramolecular Diels–Alder reaction to afford two endo- and exo-adducts **460** and **461**, although the yields (15–27%) were low. In contrast, PhI(OAc)$_2$-promoted oxidation of **462** in MeOH provided the two same adducts in 87% yield (endo/exo = 1.2) (Scheme 90)[166]. In the case of other similar phenols, the endo-adducts were also obtained as either the sole or the predominant product.

Both katsurenone (**463**) and denudatin B (**464**) are representatives of neolignans and show an antifeedant activity. They were synthesized by phenolic oxidation with PhI(OCOCF$_3$)$_2$ in the presence of substituted styrene derivatives. Herein, a 4-allylphenol derivative **465** was oxidized with PhI(OCOCF$_3$)$_2$ in MeCN containing (E)-3,4-dimethoxypropenylbenzene to give in 30% yield the corresponding dihydrobenzofuran **466**, which was further converted to both neolignans (Scheme 91)[167]. Electrochemical methodology has provided similar results (see Scheme 27).

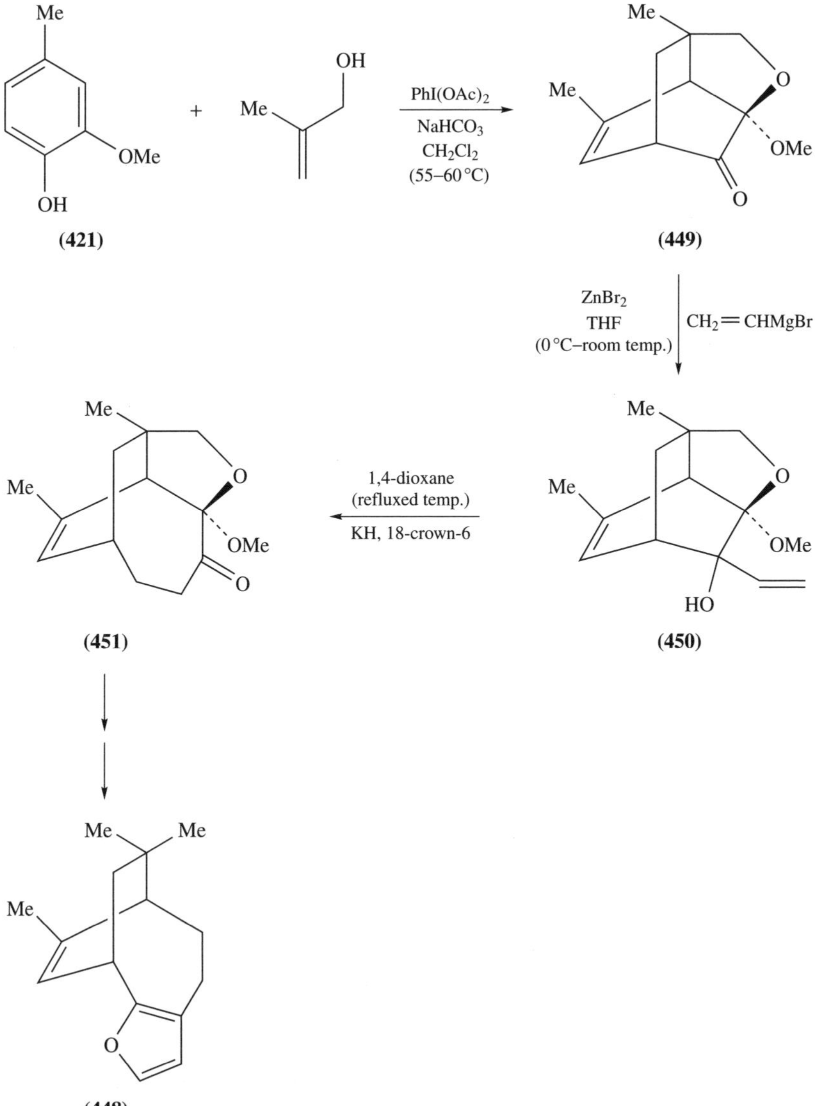

SCHEME 87. Total synthesis of racemic pallescensin B

On 2e oxidation of phenols such as **467**, the resulting phenoxonium ions are expected to undergo intramolecular nucleophilic substitution by the tertiary hydroxyl group, which is in a relatively rigid environment imparted by the presence of the (Z)-alkenyl group, to afford highly functionalized bicyclic compounds. Thus, oxidation of **467** with PhI(OAc)$_2$

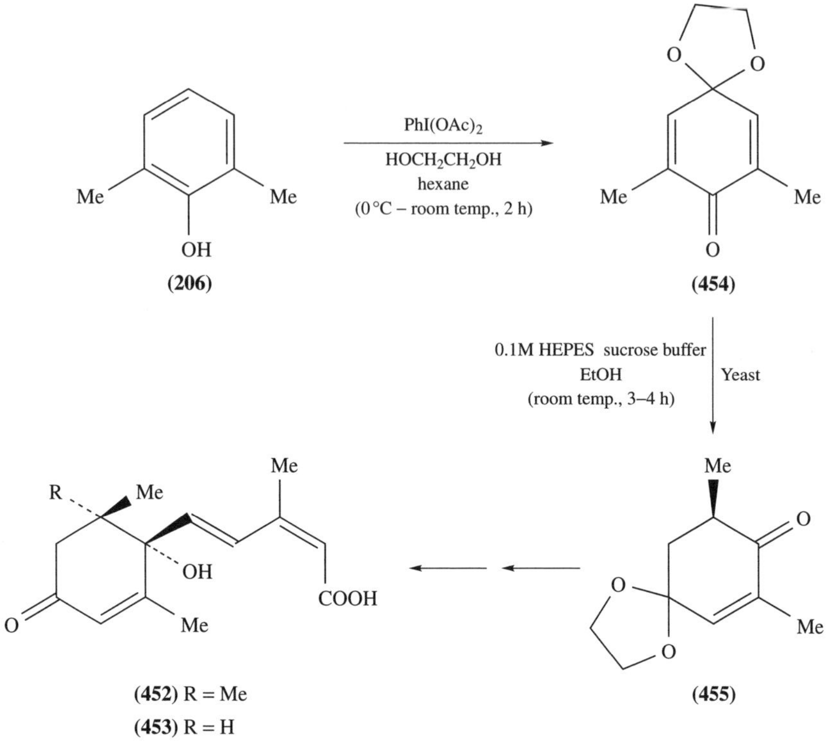

HEPES = *N*-2-hydroxyethylpiperazine-*N*′-2-ethanesulfonic acid

SCHEME 88. Synthesis of (+)-8′-demethyl abscisic acid

in MeOH provided a 4e oxidation product **468** (83%) through 2H-chromene **469**[168]. The desired compound **469** was obtained by DIBAL-H reduction of **468** (Scheme 92).

Similarly, in connection with the cytotoxic meroterpenoid sargaol, the phenol **470** was subjected to $PhI(OAc)_2$-promoted oxidation in MeOH, followed by DIBAL-H reduction to afford in 57% overall yield the target molecule **471** similar to sargaol.

From the biogenetic point of view, both lignan- and neolignan-type compounds are generated from two C6–C3 units. Generally, dibenzocyclooctadiene-type neolignans and their spirodienone precursors are biosynthesized by oxidative phenol coupling in a radical mechanism. In contrast, oxidative C–C couplings of phenols have been effected with hypervalent iodobenzenes in a heterolytic mechanism.

Arctigenin (**472**), prepared from 3,4-dimethoxybenzaldehyde, was submitted to $PhI(OCOCF_3)_2$-promoted oxidation in trifluoroethanol (TFE) (room temp., 24 h) to afford spirodienone **473** and a mixture of two cyclooctadienes (**474** and **475**) in 13 and 14% yields, respectively. When the reaction was repeated in hexafluoroisopropanol for 3.5 h, a 1:1 mixture of stegane **474** and isostegane **475** was obtained in 26% yield (Scheme 93). Acid-catalyzed rearrangement of **473** with $HClO_4$ in $CHCl_3$ provided a quantitative

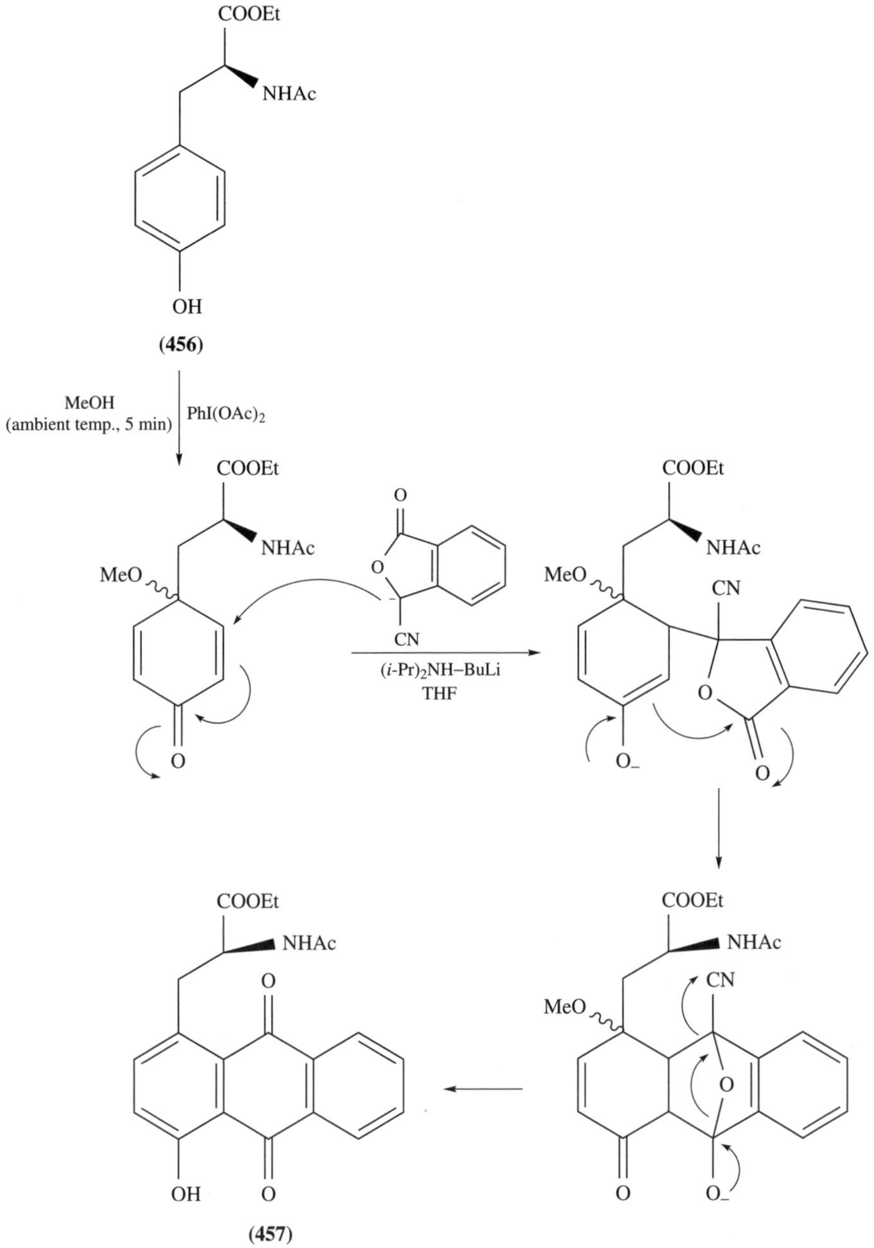

17. Oxidation of phenols

SCHEME 89. Synthesis of an anthraquinone from *N*-acetyltyrosine ethyl ester

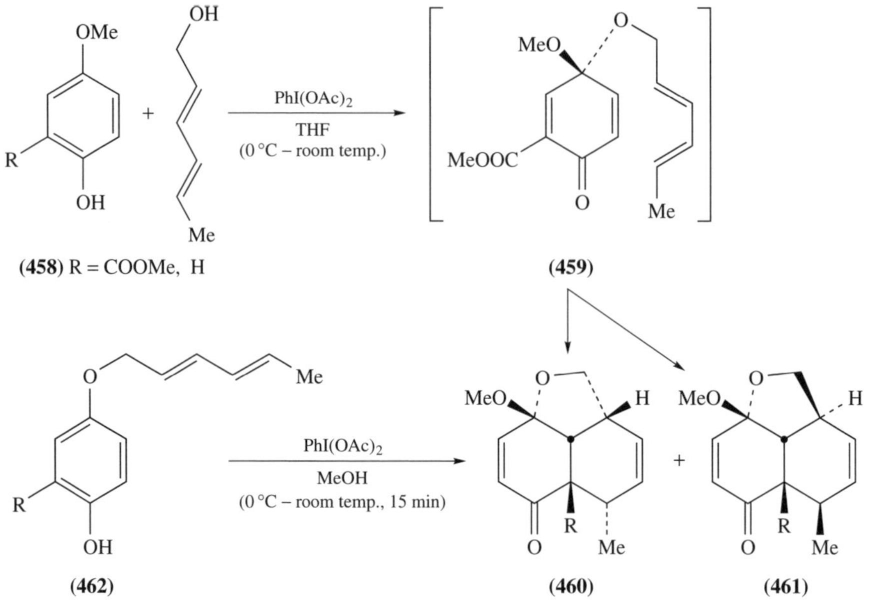

SCHEME 90. Synthesis of highly functionalized cis-decalines

yield of the cyclooctadienes. These reactions provide the first synthesis of spirodienones such as **473**, which are recognized as plausible intermediats in the biosynthesis of dibenzocyclooctadiene-type neolignans.

In the case of prestegane A (**476**), the isostegane derivative **477** was directly produced as a major product (40%) together with three compounds **478**, **479** and **480** (23, 24 and 3%, respectively), as shown in Scheme 93[169].

From the viewpoints of biological activities and structural architectures, a variety of benzylisoquinoline alkaloids have been chosen as synthetic target molecules. These alkaloids are well known to be biosynthesized by oxidative phenol coupling via a radical mechanism. However, White and coworkers demonstrated that hypervalent iodobenzenes are effective oxidants for syntheses of morphinane-type alkaloids such as (−)-codeine[170].

(R)-N-Trifluoroacetyl-6′-bromonorreticuline (**481**), readily prepared from (R)-norreticuline, was oxidized with PhI(OCOCF$_3$)$_2$ in CH$_2$Cl$_2$ at −40 °C to give the desired coupled product **482** in 21% yield. This compound was smoothly converted to (−)-codeine (**483**) (Scheme 94)[170]. Herein, the bromine atom at the C6′-position prevents a *para–para* coupling.

6a-Epipretazettine (**484**) was also synthesized by using PhI(OCOCF$_3$)$_2$-promoted oxidation as a key step[171]. The labile compound **485**, prepared from piperonal and racemic synephrine, was subjected to PhI(OCOCF$_3$)$_2$-promoted oxidation in the presence of propylene oxide (10 equiv.) (−10 °C, 0.5 h) to afford in 13% yield the corresponding *para–para* coupled product **486**, which underwent Zn reduction in 10% 1 M NH$_4$OAc, resulting in the formation of a secondary amine **487**. This amine was cyclized spontaneously to the tetracycle **488** (65%), which was readily converted to 6a-epipretazettine (**484**) (Scheme 94).

Similarly, the norbelladine derivative **489**, prepared from L-tyrosine methyl ester and isovaniline, was oxidized with PhI(OCOCF$_3$)$_2$ in trifluoroethanol (TFE) at −40 °C to afford in 64% yield an intramolecular coupled product **490**. This is known as the key

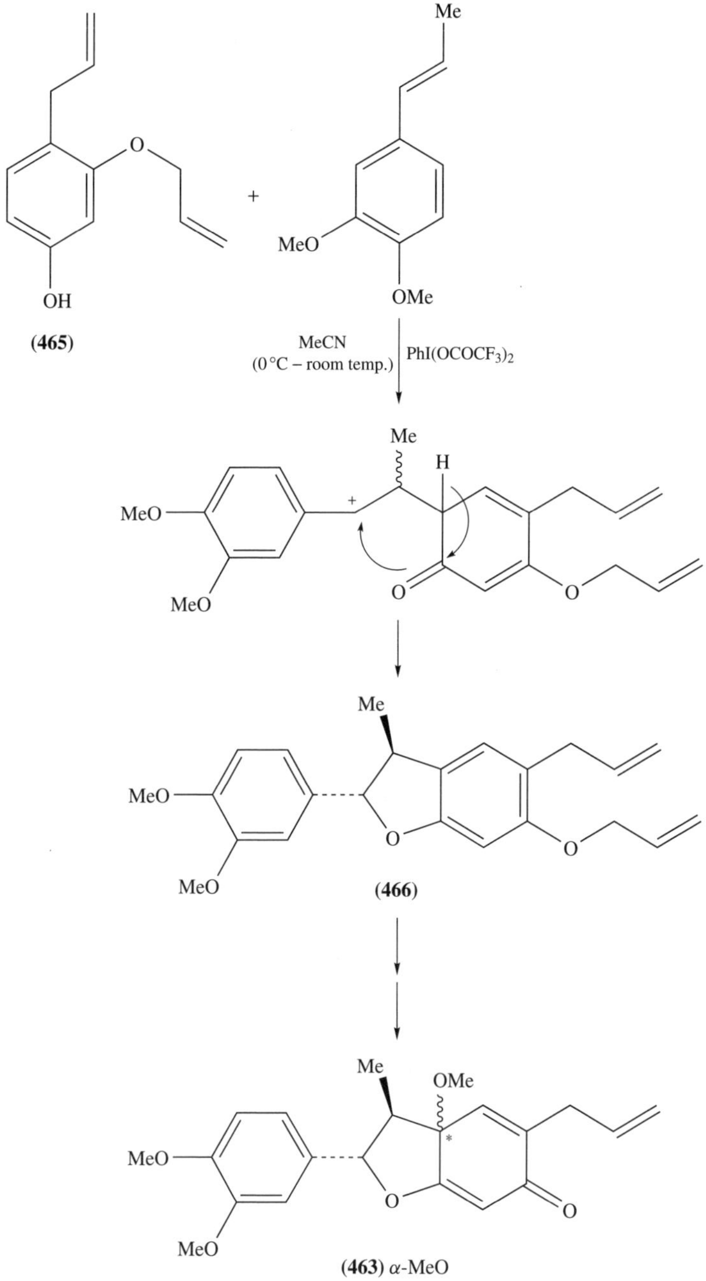

17. Oxidation of phenols

(465)

MeCN
(0 °C – room temp.) | PhI(OCOCF$_3$)$_2$

(466)

(463) α-MeO
(464) β-MeO

SCHEME 91. Synthesis of katsurenone and denudatin B

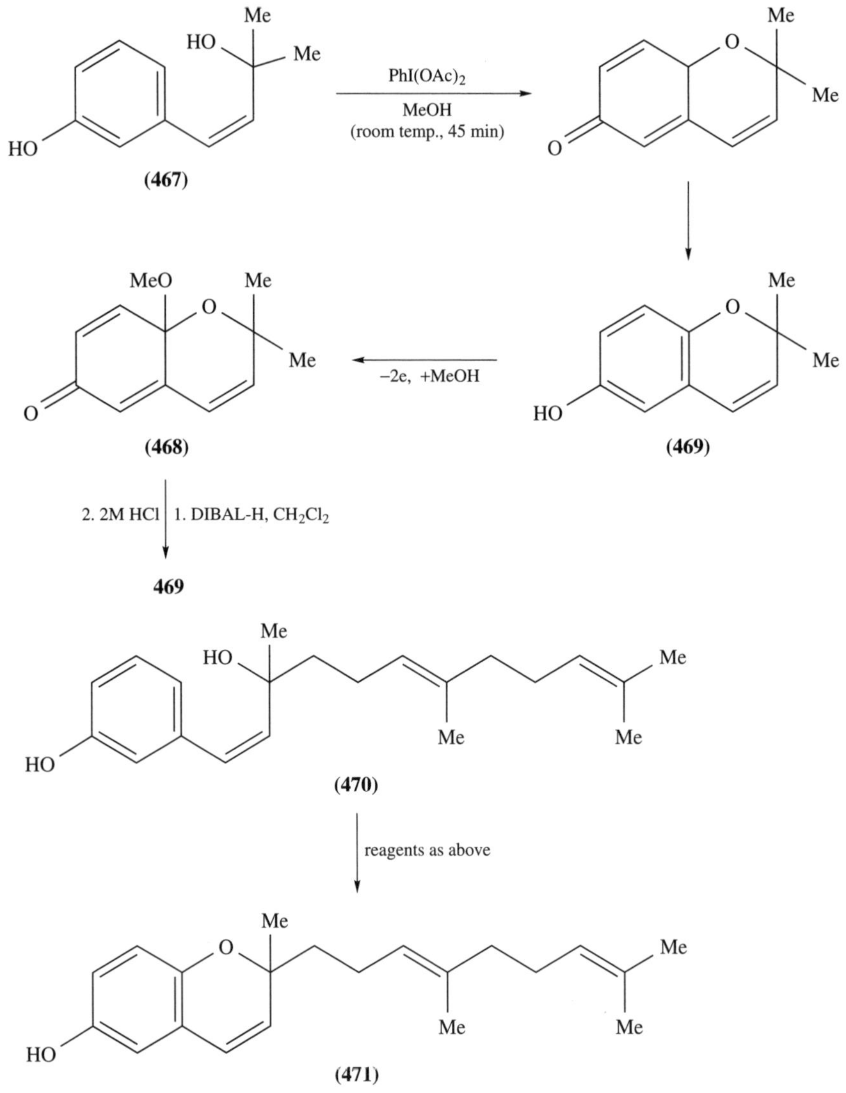

SCHEME 92. Synthesis of 2H-chromenes

synthetic intermediate for (+)-maritidine (**491**)[172] (Scheme 95). Furthermore, effective syntheses of *Amaryllidaceae* alkaloids such as galanthamine, narwedine, lycoramine, norgalanthamine and sanguinine as a racemic form have also been accomplished[173].

As already shown in the case of **489**, the *para–para'* coupled product **490** was selectively obtained. However, when the *p'*-position is protected by an appropriate group, a

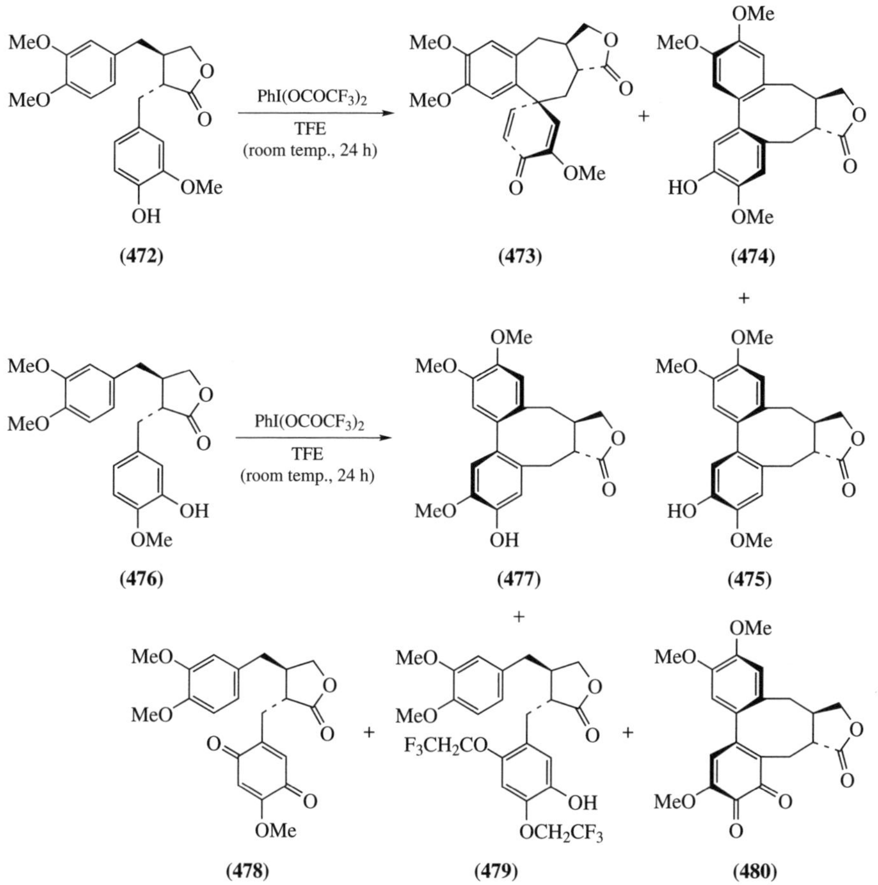

SCHEME 93. Synthesis of dibenzocyclooctadiene-type neolignans and spirodienones

para–ortho' coupling is expected to take place. Thus, the phenol **492**, bearing the TMS group at the p'-position, was treated with PhI(OCOCF$_3$)$_2$ in TFE at $-40\,°$C to afford the para–ortho' coupled product **493** in 36% yield (Scheme 95)[173]. Herein, the two protecting groups are easily removed and the resulting hydroxyl group underwent Michael addition to afford a galanthamine-type compound **494**. From this key intermediate, galanthamine (**495**) and related alkaloids have been synthesized.

B. Oxidation with 2,3-Dichloro-5,6-dicyano-*p*-benzoquinone

A variety of organic compounds such as 2,3-dichloro-5,6-dicyano-*p*-quinone (DD2) and related benzoquinones, *m*-chloroperbenzoic acid, or dioxirane have been utilized as oxidants in organic synthesis. This section will focus on the synthesis of natural products and related compounds using DDQ.

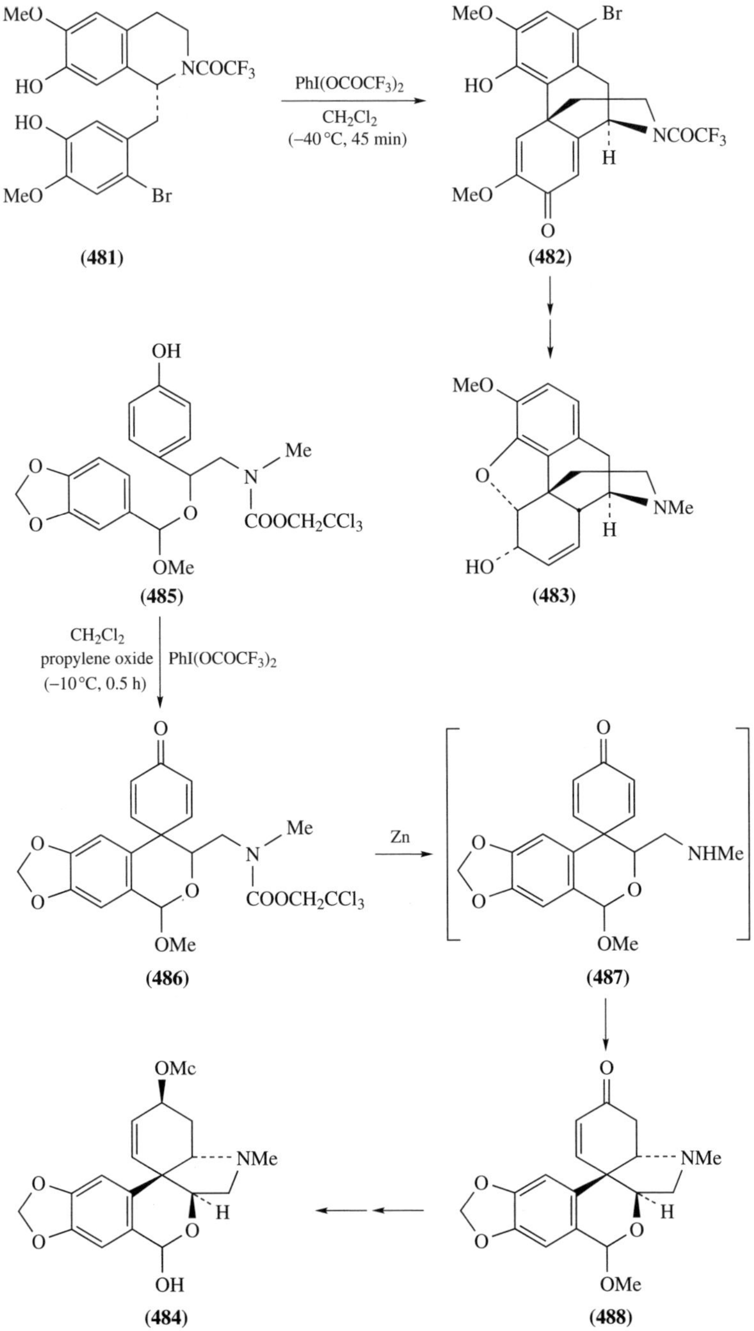

SCHEME 94. Synthesis of (−)-codeine and 6a-epipretazettine

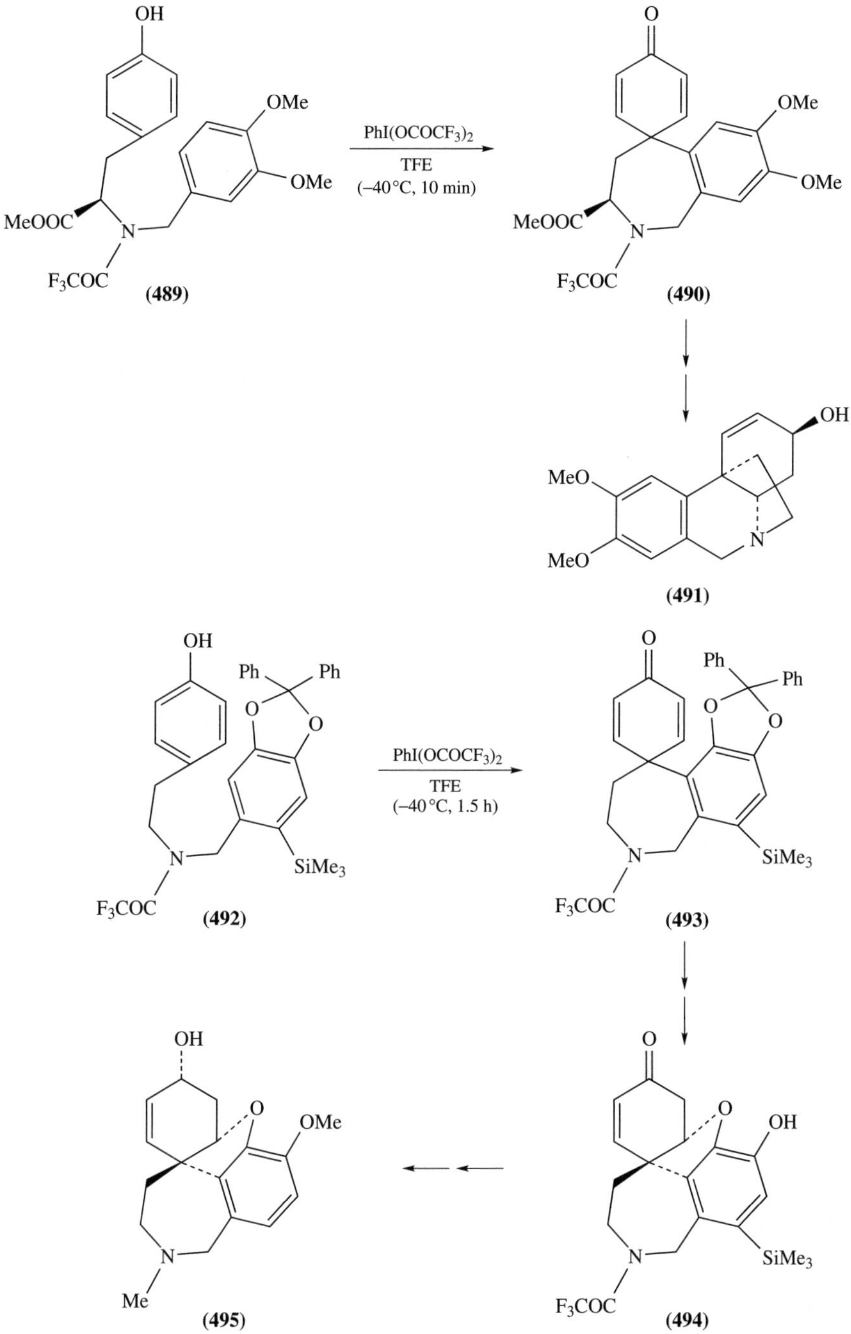

SCHEME 95. Synthesis of (+)-maritine, racemic galanthamine and related alkaloids

Quinone monoketals are usually prepared by oxidation of the corresponding p-alkoxyphenols using a variety of oxidants. Oxidation of a number of p-alkoxyphenols was performed by Büchi and coworkers using DDQ, ferric chloride and thallium(III) nitrate[174]. Of them, oxidation of 2-allyl-4,5-methylenedioxyphenol (**496**) and 2-allyl-3,4-dimethoxyphenol (**497**) in MeOH was effected with DDQ (1 equiv.) in the presence of catalytic amounts of p-nitrophenol (20 °C, 30 min) to afford the corresponding p-quinone monoketals (**498** and **499**) in 88 and 75% yields, respectively. In the case of **497**, neither $FeCl_3$ nor $Tl(NO_3)_3$ could yield **499**. Condensation of monoketal **498** with (E)-isosafrole in MeCN containing 2,4,6-trinitrobenzenesulfonic acid (0 °C, 75 min) afforded bicyclo[3.2.1]octenone **500** and a mixture of diketone **501** and its enol **502** in 27% yields, respectively. Compound **500** was converted easily to guianin **503** by methylation followed by $NaBH_4$ reduction. Burchellin (**504**) was also synthesized from the **501** and **502** mixture, as shown in Scheme 96[64]. When (Z)-isosafrole was used, futoenone (**152**) was obtained. This neolignan was not formed from (E)-isosafrole (see Scheme 29).

The quinone monoketal **499** reacted with (E)-isosafrole in MeCN–MeOH containing 2,4,6-trinitrobenzenesulfonic acid (dry ice temp., 30 min) to afford dihydrobenzofuran **505** and bicyclo[3.2.1]octenone **506** in 42 and 20% yields, respectively. The former was further converted to two neolignans (**507** and **508**) (Scheme 96)[175]. Herein, the dihydrobenzofuran **505** is formed from **509**.

Both megaphone (**510**) and megaphone acetate (**511**), isolated from *Aniba megaphylla* Mez., exhibit inhibitory activity against human KB cells *in vitro*. In a series of Büchi's ingenious studies, these two neolignans were also synthesized based on the same concept used for the synthesis of burchellin (**504**). The benzyl ether (**512**), prepared from 3,4-dimethoxyphenol, was hydrogenated and the resulting phenol **513** was submitted to DDQ oxidation in MeOH (room temp., 15 min) to afford the corresponding quinone monoketal **514**, which was condensed directly with 1,2,3-trimethoxy-5-(1-(Z)-propenyl)benzene (**515**) using stannic chloride in CH_2Cl_2 (−30 °C, 20 min) to give the desired bicyclic compound **516** in 48% overall yield. The compound was further converted to both megaphone and megaphone acetate (Scheme 97)[176].

The tricyclic sesquiterpene gimnomitrol (**517**) was isolated as a major metabolite from the liverwort *Gymnomitrion obtusum* (Lindb.) Pears. An ingenious synthesis of gimnomitrol bearing a novel structure was performed in a short sequence by using a cationic [5 + 2] cycloaddition methodology, as follows[177].

4,5-Dimethoxy-2-methylphenol (**108**) was oxidized with DDQ in MeOH at 0 °C to afford in 63% yield 2-methyl-4,4,5-trimethoxycyclohexa-2,5-dienone (**518**), which reacted with 1,2-dimethylcyclopentene (**519**) in $MeNO_2$–CH_2Cl_2 containing stannic chloride to yield a mixture of two adducts. This mixture was reduced directly with $NaBH_4$ in MeOH to give a separable mixture of **520** and **521** in 10% overall yield. The former was smoothly converted to the target molecule **517** (Scheme 98).

As shown above, Büchi and coworkers accomplished the total synthesis of neolignans and gymnomitrol based on the concept of a cationic [5 + 2] cycloaddition. However, the yields associated with these cycloaddition reactions are not always satisfactory. Recently, this difficulty has been overcome by using trimethylsilyl triflate as an effective reagent. For example, condensation of 2-methyl-4,4,5-trimethoxycyclohexa-2,5-dienone (**518**) with 2,3-dimethyl-2-butene (**522**) was effected with TMSOTf (1.05 equiv.) in 3.0 M $LiClO_4$–EtOAc (−23 °C, 5 min) to afford the corresponding bicyclo[3.2.1]octenone **523** in high yield (84%)[178]. However, when (E)-isosafrole was used as an olefin the initially formed bicyclic compound **524** was readily converted to tetrahydrobenzofuran **525** (89%) under the reaction conditions (Scheme 99).

Phenols bearing an olefinic side chain have been known to be oxidized electrochemically and the resulting cyclohexa-2,5-dienones undergo intramolecular cationic [5 + 2]

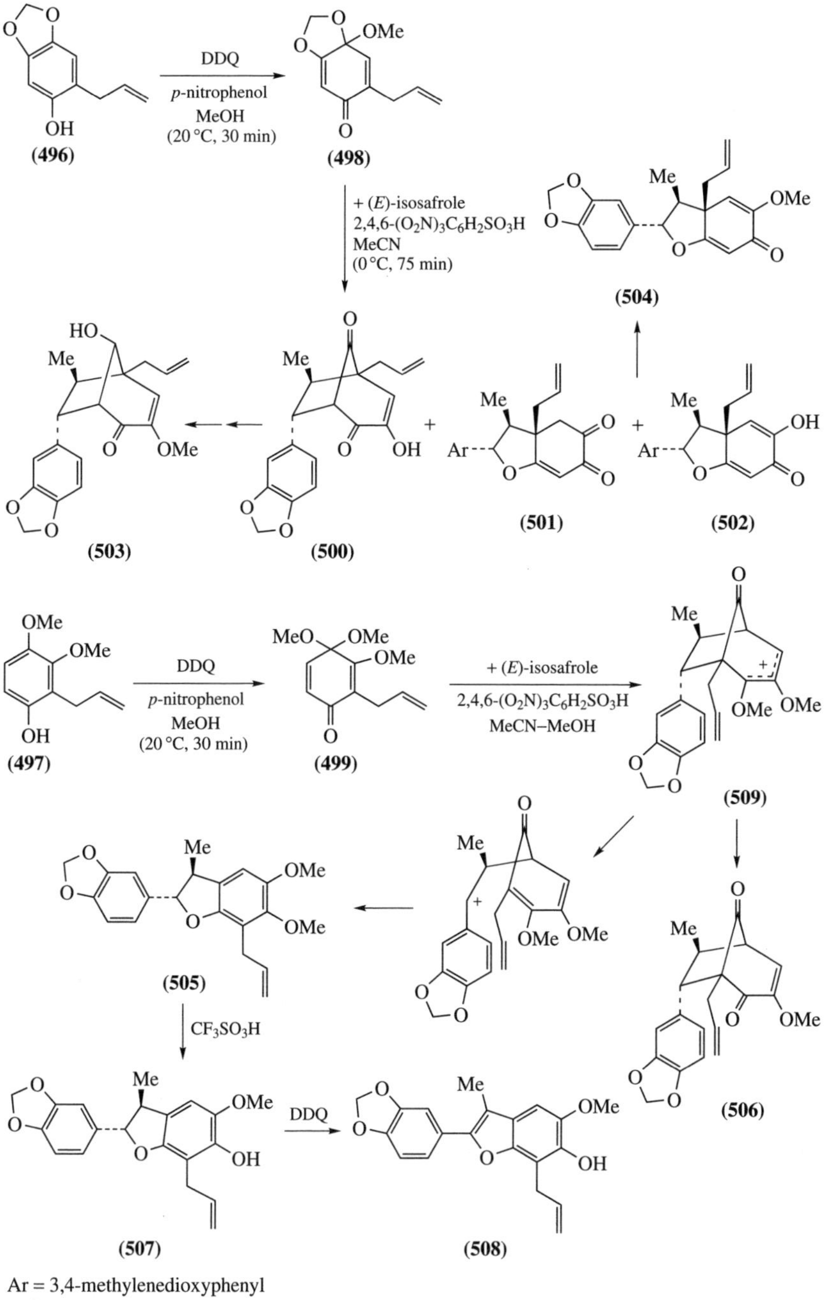

SCHEME 96. Synthesis of guianin, burechellin and related neolignans

Ar = 3,4-methylenedioxyphenyl

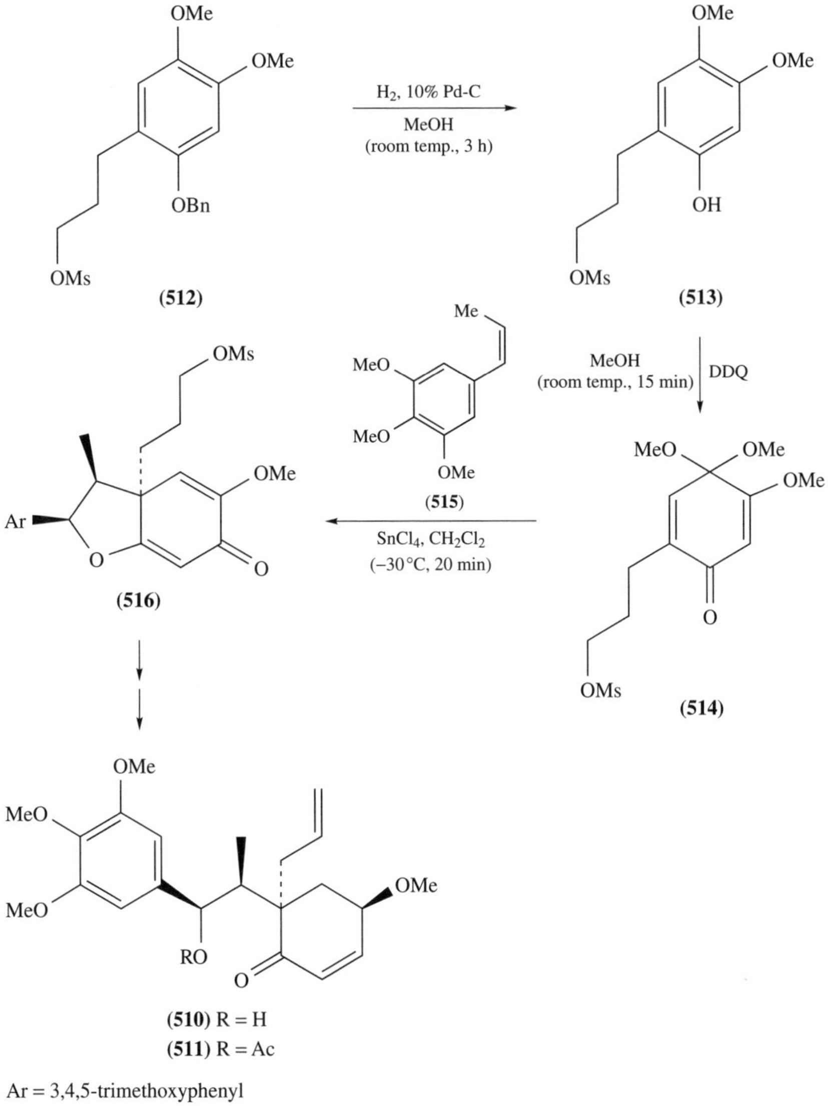

SCHEME 97. Synthesis of megaphone and megaphone acetate

cycloaddition to afford the corresponding tricyclic compounds (see Scheme 31). Thus, a 3,4-dimethoxyphenol derivative **526** was oxidized with DDQ in MeOH at 0 °C to yield cyclohexa-2,5-dienone **527** (>87%). This dienone was treated with TMSOTf in 3.0 M LiClO$_4$–EtOAc at −23 °C to give the desired tricyclic compound **528** (89%), which was further converted to the triquinane isocomene (Scheme 99)[179].

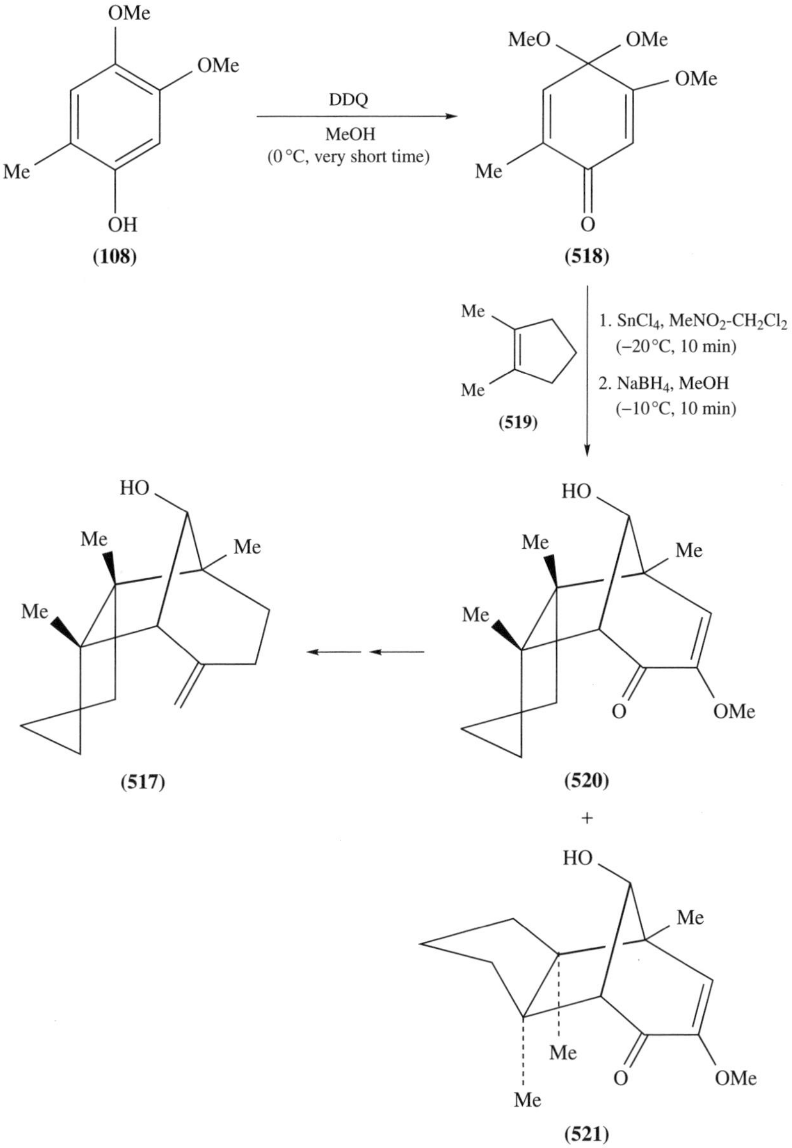

SCHEME 98. Synthesis of gymnomitrol

Generally, oxidations of phenols having an appropriate alkyl group at the *o*- or *p*-position are performed using DDQ and other oxidizing reagents in aprotic solvents such as benzene and ether to afford the corresponding quinone methide, which are useful synthons in organic synthesis.

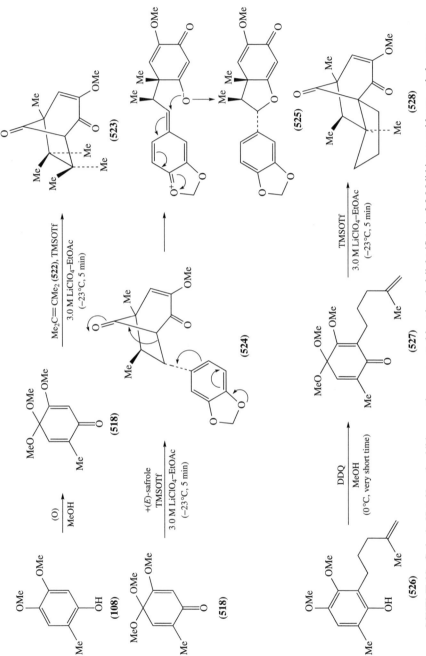

SCHEME 99. Cationic [5 + 2] cycloaddition reactions promoted by trimethylsilyl triflate in 3.0 M lithium perchlorate–ethyl acetate

o-Allylphenols **529** and **530** were oxidized with DDQ (1.1 equiv.) in ether (room temp., 2 h) to give chrom-3-enes **531** and **532** in 85 and 90% yields, respectively (Scheme 100)[180]. Potassium dichromate was also an effective oxidant for the conversion of *o*-allylphenols to chrom-3-enes in good yields.

Oxidation of cinnamylsesamol (**533**) was effected with Ag_2O in ether or by DDQ in acetone to give an almost quantitative yield of the extended *o*-quinomethane **534** as orange-red plates. On heating under reflux in benzene, this compound underwent intramolecular cyclization to afford 6,7-methylenedioxyflav-3-ene (**535**) (77%). Similarly, 2-(4'-methoxybenzyl)-4,5-methylene-dioxyphenol (**536**) was converted to the corresponding quinone methide **537** (50%). A benzene solution of **537** was heated under reflux to yield a dimer **538** (65%) (Scheme 100)[181].

1-(2-Hydroxyphenyl)-2-(4-hydroxyphenyl)ethane (**539**) was oxidized with DDQ (1 equiv.) in benzene (room temp., 24 h) to afford both benzofuran (**540**) and dihydrobenzofuran (**541**) (27 and 22%, respectively). With 2 equivalents of DDQ the yield of the former increased and that of **541** decreased (Scheme 100). The *p*-quinone methide **542** is recognized as a plausible reaction intermediate[182].

A variety of unsymmetrically substituted biaryls exhibiting a variety of bioactivities have been found in nature. Although 4-substituted phenols undergo oxidative coupling to yield symmetrical 3,3'-disubstituted 2,2'-dihydroxybiaryls, the chemoselective direct cross-coupling of different phenols remains an open problem. Thus, the oxidative cross-coupling of *p*-methylphenol (**27**) and *p*-methoxyphenol (**140**) was carried out using different reagents in the presence of $AlCl_3$ (Scheme 101)[183]. The desired selective cross-coupling could be best performed by using DDQ in the presence of 2 equivalents of $AlCl_3$, which probably stabilized the resulting biaryl (**543**) by the formation of aluminum chelates. However, when an alkyl group was introduced at the *ortho*-position of the phenol **27**, cross-coupling was inhibited. In cases of phenols substituted by electron-withdrawing groups such as COMe and CN, no cross-coupling reaction took place.

C. Oxidation with Dioxirane, Di-*tert*-butyl Peroxide and Di-*tert*-butyl Peroxyoxalate

Excellent reviews on dioxirane-mediated oxidations have appeared[184]. One of the most characteristic points is that dioxiranes can be applied to the epoxidation of labile olefins such as enol ethers, enol acrylates, allenes and others. Dioxiranes have also been utilized for phenolic oxidation, but in relatively rare cases. Oxidation of simple phenols and anisoles with dimethyldioxirane (**544**) provided only a complex mixture, so that hindered phenols are more favorable. On treatment with dimethyldioxirane (4 equiv.) in acetone, 2,4-di(*tert*-butyl)phenol (**216**) was oxidized to afford in 79% yield the corresponding *o*-benzoquinone **220**, which reacted with **544** and aq. $NaHSO_3$ to give catechol **545**. Dimethyldioxirane-promoted oxidation of **545** provided again a quantitative yield of **220**. Further oxidation of **220** produced a 52% yield of two epoxides **546** and **547** in a ratio of 1:20. Oxidation of thymol (**548**) was effected with dimethyldioxirane in acetone to afford the four oxidation products **549–552** in 10, 20, 10 and 10% yields, respectively (Scheme 102)[185].

Oxidation of 2,6-di(*tert*-butyl)phenol (**23**), both *ortho*-positions of which are blocked by the bulky *tert*-butyl groups, was effected with 4 equivalents of methyl(trifluoromethyl)dioxirane (**553**) at 0 °C for 1 min to afford three oxygenated products (**554**, **74** and **555**) in 4, 24 and 70% yields, respectively. Dimethyldioxirane-promoted oxidation of **23** required much longer reaction time (48 h) to yield the

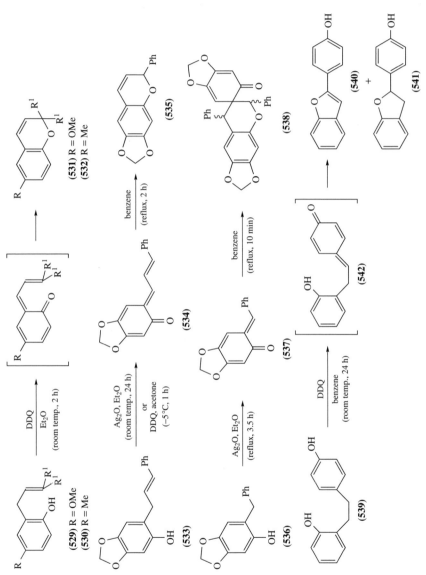

SCHEME 100. Syntheses and reactivities of *o*- and *p*-quinomethanes

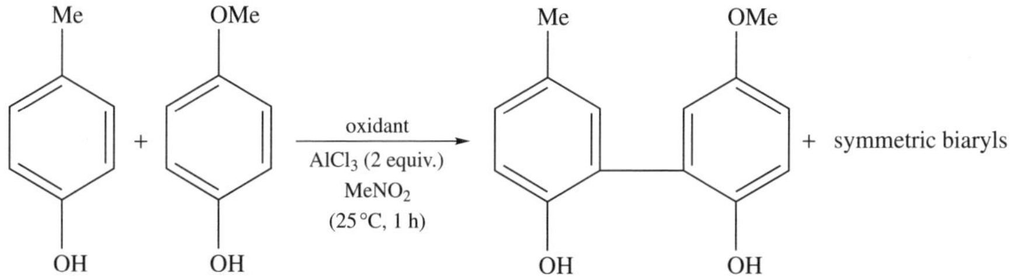

17. Oxidation of phenols

[Scheme showing oxidative cross-coupling reaction of p-methylphenol (27) and p-methoxyphenol (140) with oxidant, AlCl₃ (2 equiv.), MeNO₂ (25 °C, 1 h) to give biaryl product (543) + symmetric biaryls]

oxidant	543 %, yield (% selectivity)
2FeCl$_3$	45 (70)
VOCl$_3$	60 (83)
2CuBr$_2$	15 (84)
DDQ	70 (85)

Selectivity = yield of **543** (%)/reacted **27** (%)

SCHEME 101. Oxidative cross-coupling of *p*-methylphenol and *p*-methoxyphenol

corresponding dehydrodimer **25** (10%) together with **74** and **555** (48 and 34%, respectively), as shown in Scheme 102[186]. Here, **25** must be formed by dimerization of the initially generated radical species. From these results, it is evident that methyl(trifluoromethyl)dioxirane (**553**) is more reactive than dimethyldioxirane (**544**).

In relation to copper-containing enzymes, biomimetic oxidation of catechol (**200**) was performed using dioxirane **553** in acetone-1,1,1-trifluoropropanone (TFP) (−20 °C, 1 h) to produce selectively *cis,cis*-muconic acid (**556**) (88%) (Scheme 102). This result resembles the oxygenation assisted by metal complexes (see Scheme 40).

Organic peroxides such as di-*tert*-butyl peroxide are effective for radical coupling of phenols. 3,5-Dimethylphenol (**557**) having free *ortho*- and *para*-positions was oxidized by heating with di-*tert*-butyl peroxide at 140 °C to afford mainly an *ortho*–*ortho* coupled product **558** (77%) together with an *ortho*–*para* coupled product **559** (16%). In contrast, on oxidation of **557** by di-*tert*-butyl peroxyoxalate at 25 °C, the yield (40%) of the *ortho*–*para* coupled product **559** increased, while the yield (17%) of **558** decreased. In addition, the *para*–*para* coupled product **560** was produced, although the yield (8%) was low (Scheme 103). In the case of phenol, a similar result was also obtained. These results and MINDO-3 calculations with standard parametrization show that stereoelectronic factors can explain the preferred formation of the *ortho*–*ortho* or *ortho*–*para* coupled products[187].

Reactions of substituted bis(3-alkoxybenzoyl) peroxides in neat phenols afford mainly 8-alkoxy-6*H*-dibenzo[*b*, *d*]pyran-6-ones and *ortho*-benzoyloxylation products of the phenol. For example, bis(3,4-dimethoxybenzoyl) peroxide (**561**) in neat *p*-methylphenol was completely decomposed in 1 h at 60 °C with the formation of a dibenzo-α-pyrone derivative **562** (60%) together with an *ortho*–*ortho* coupled product **563** (21%) and benzoate **564** (5%). In contrast, dibenzoyl peroxides having no *meta*-electron-releasing substituents gave mainly *ortho*-benzoyloxyphenols. For example, decomposition of bis(4-methoxybenzoyl)

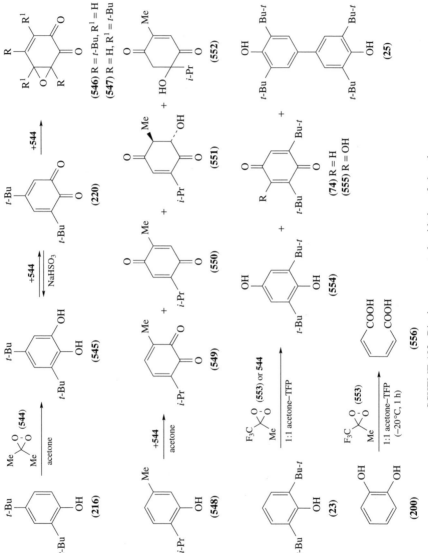

SCHEME 102. Dioxirane-promoted oxidation of phenols

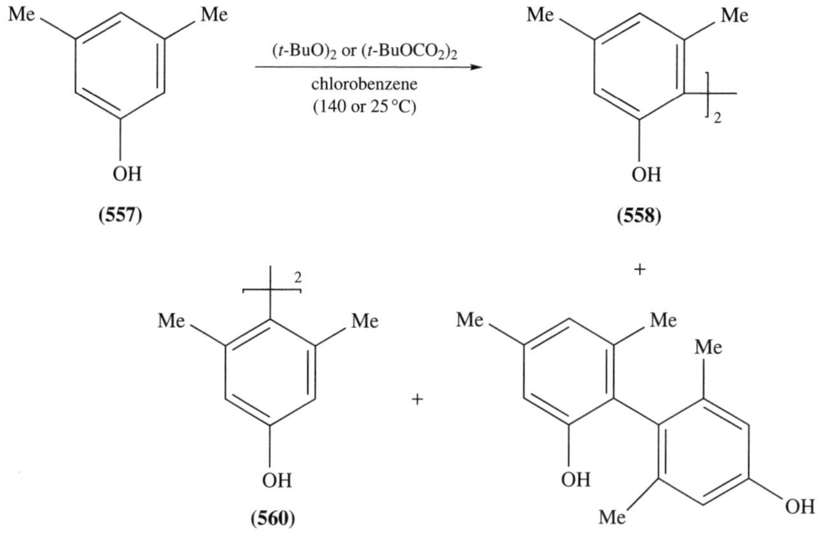

SCHEME 103. Oxidation of phenols with di(*tert*-butyl) peroxide at 140 °C or di(*tert*-butyl) peroxyoxalate at 25 °C

peroxide (**565**) in neat *p*-methylphenol provided the corresponding phenol **566** (71%) together with **563** (5%) and benzoate **567** (20%) (Scheme 104)[188]. Biphenols were always observed when either α-pyrones or *ortho*-benzoyloxylation products were formed as a major product. The yields of the *ortho–ortho* coupled products were found to decrease linearly in all solvents on decreasing the concentration of the phenols. Furthermore, of a variety of solvents used in these reactions, a higher selectivity of the α-pyrone **562** could be reached in nujol. This has higher viscosity than other solvents such as $CHCl_3$, benzene and hexane.

Finally, the formation of biaryls by C—C coupling can take place through two different mechanisms referred to as radical–radical (RRD) and radical–substrate (RSD) dimerization. A mechanism involving 1e oxidation of the phenol by the peroxide and biaryl coupling by preferential addition of the phenol radical cation to the *ortho*-positions to the alkoxy group of the diaroyl peroxide has been suggested.

Many phenols are oxidized efficiently to quinones by acyl *tert*-butyl nitroxides in organic solvents. 2,6-Dimethylphenol (**206**) was oxidized with benzoyl *tert*-butyl nitroxide in CH_2Cl_2 at room temperature to afford 2,6-dimethyl-*p*-benzoquinone (**223**) in 86% yield. Oxidation of 2,4,6-trimethylphenol (**221**) was also effected with the same reagent in ether to afford the *N-tert*-butylbenzohydroxamic acid **568** in 98% yield (Scheme 105)[189].

D. Oxidation with Fremy's Salt, Ammonium Nitrate–Acetic Anhydride, Dioxygen-*tert*-butoxide, Sodium Periodate, Chlorous Acid and Other Oxidants

This section is focused on phenolic oxidation using a variety of oxidants such as Fremy's salt, O_2-NO_2, O_2-*t*-BuOK, chlorous acid, $NaIO_4$ and others. These reagents have long been known, so that some typical examples are presented briefly.

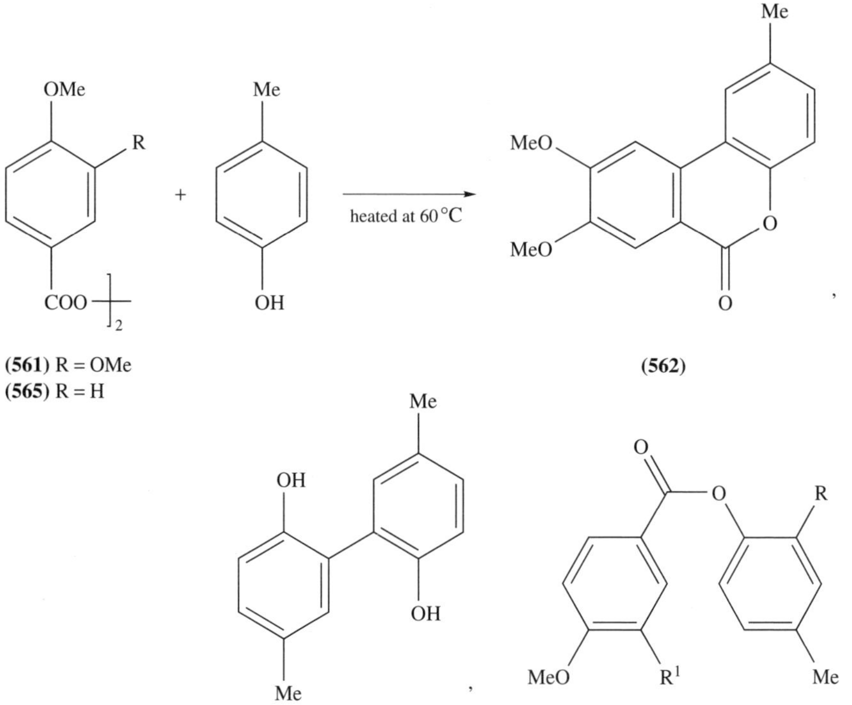

(561) R = OMe
(565) R = H

(562)

(563)

(564) R = H, R[1] = OMe
(566) R = OH, R[1] = H
(567) R = R[1] = H

SCHEME 104. Decomposition of diaroyl peroxides in neat phenols

Generally, Fremy's salt [$^{\bullet}$O-N($SO_3K)_2$] is used for oxidative conversion of phenols (**569**) to the corresponding *o*- and/or *p*-benzoquinones (**570** and **571**) depending on substituents on the aromatic ring. The reaction mechanism of Fremy's salt-mediated oxidation of phenols has been determined as consisting of three steps, as shown in Scheme 106[187,190,191]. The initially formed phenoxy radical undergoes radical coupling with another Fremy's salt at *o*- and/or *p*-positions to yield coupled products **572** and **573**, which release $HN(SO_3K)_2$ with the formation of quinones.

Simple phenols (**574**, **575**, **576** and **247**) were treated with Fremy's salt in acetone and buffer solution to afford the corresponding benzoquinones (**577**, **578**, **579** and **248**) in 53, 60, 58 and 87% yields, respectively. In the cases of both **576** and **247**, the initially formed phenoxy radicals were attacked by the second Fremy's salt at the *ortho*- and *para*-positions, respectively. The resulting radical coupled products were further converted into the corresponding quinones (Scheme 107). From these data, the product selectivity seems to be due to the finely balanced situation of electronic and steric factors[191,192].

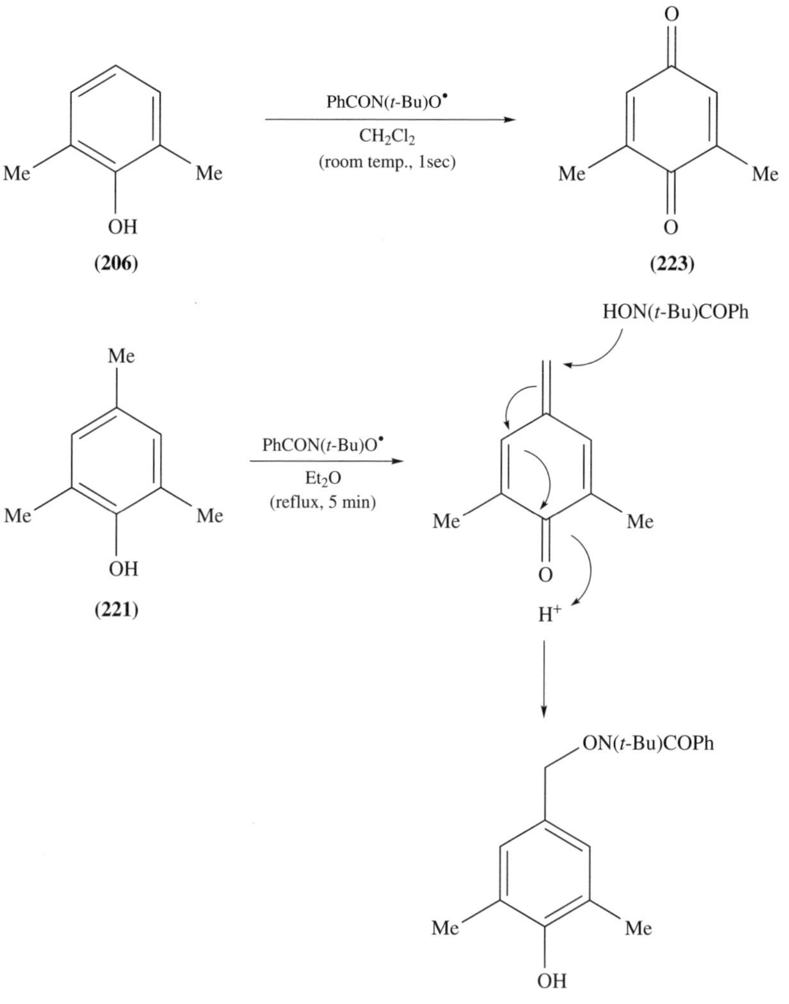

SCHEME 105. Oxidation of phenols with benzoyl *tert*-butyl nitroxide

Of many syntheses of quinone using Fremy's salt, a few examples of natural product syntheses using this reagent will be described herein. Miltirone (**580**), a tricyclic diterpene isolated from the roots of *Salvia miltirrohiza* B, has been synthesized starting from *p*-bromoanisole[193]. The starting phenol was converted into 1,2,3,4-tetrahydro-1,1-dimethyl-6-methoxy-7-isopropylphenanthrene (**581**) through 6-isopropyl-7-methoxy-1-tetralone (**582**). Finally, Fremy's salt-mediated oxidation of **581** provided miltiron in 37% yield (Scheme 108). Scabequinone has been similarly synthesized[194].

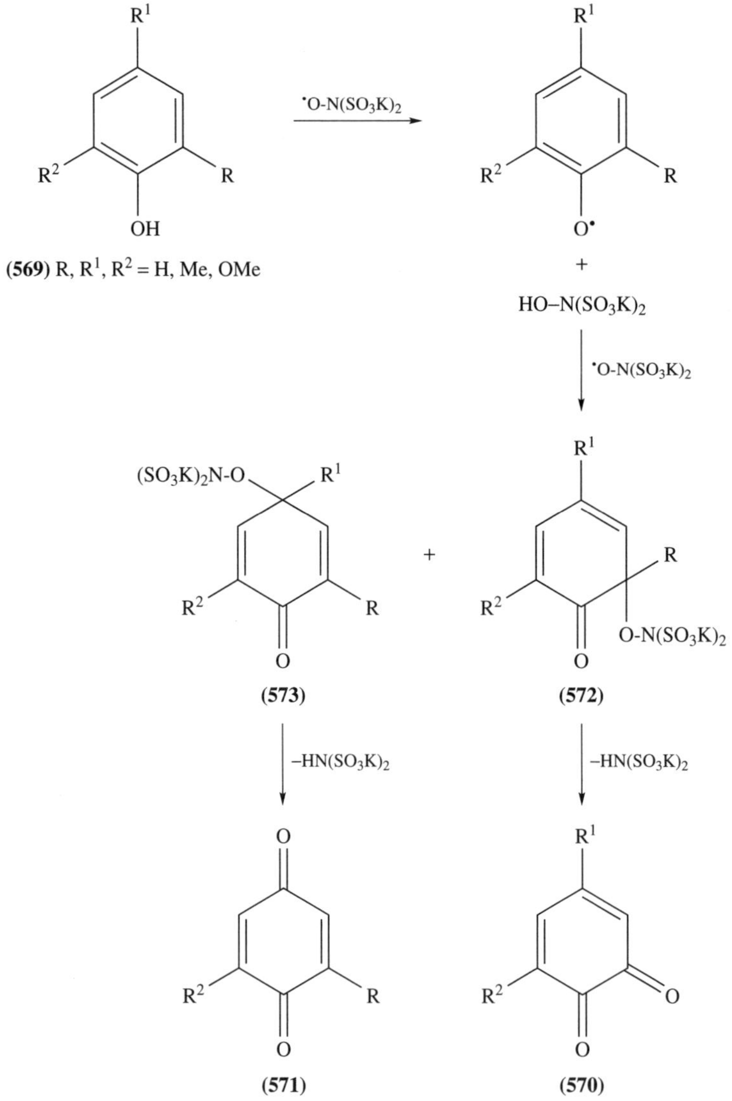

(569) R, R^1, R^2 = H, Me, OMe

SCHEME 106. Reaction mechanism of Fremy's salt-mediated oxidation of phenols

Epoxyquinomicin B (**583**), an antirheumatic agent, was synthesized in 22% overall yield in 8 steps starting from commercially available 3-hydroxy-4-nitrobenzaldehyde (**584**), which was easily converted to the amidophenol **585**. Fremy's salt oxidation of **585** in EtOAc–H_2O was carried out at room temperature overnight to afford selectively in 82% yield the desired *p*-benzoquinone (**586**). This was treated with H_2O_2–$NaHCO_3$ in MeOH to yield the target molecule **583** (Scheme 109)[195].

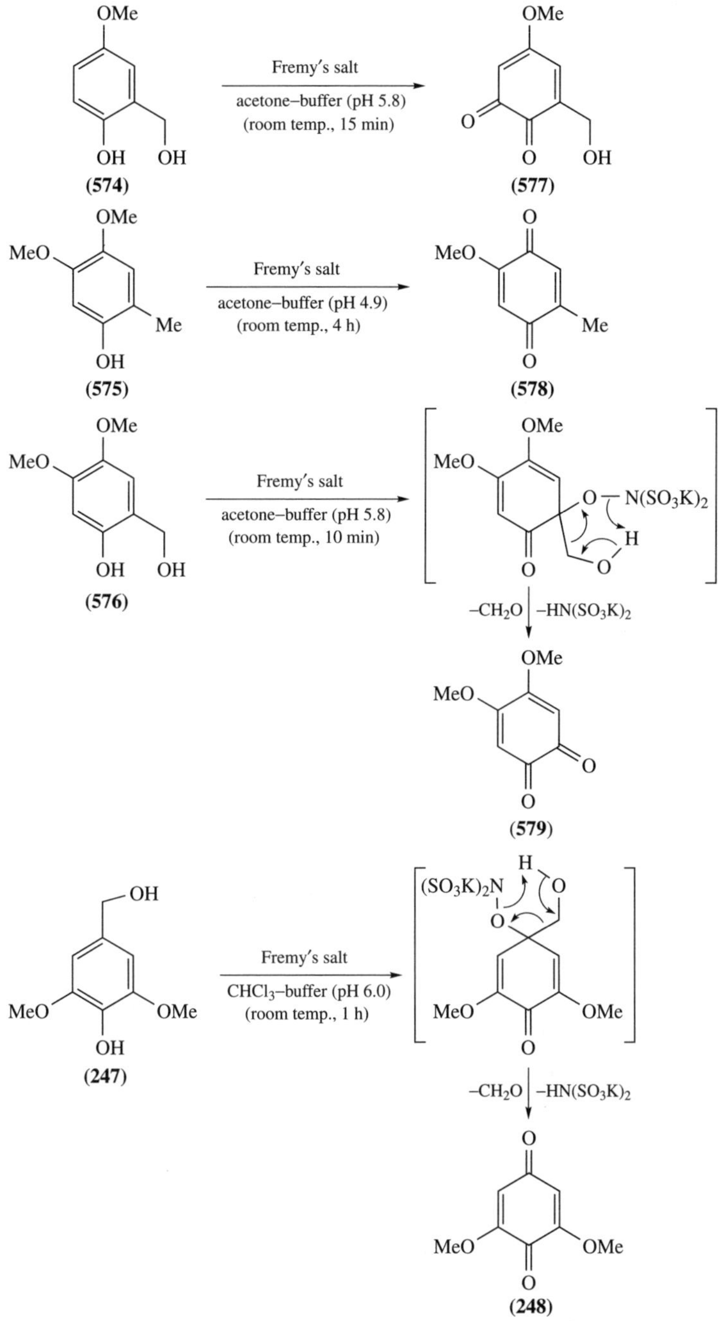

SCHEME 107. Fremy's salt-mediated oxidation of some phenols

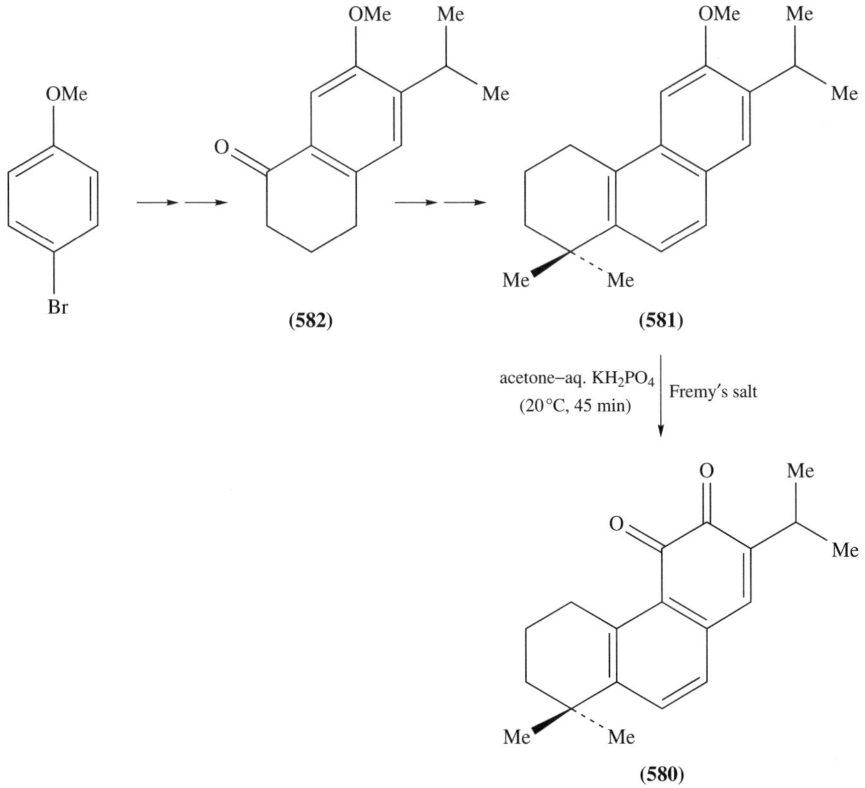

SCHEME 108. Synthesis of miltirone

Makaluvamine C (**587**) is a member of pyrroloiminoquinone alkaloids isolated from marine sponges. These alkaloids show topoisomerase II inhibitory activity and cytotoxic activity against human colon tumor cell line HCT-116. Thus, makaluvamine C was synthesized starting from *p*-anisidine through a dinitro compound **588** in 13 steps (13.1% overall yield). One of the key steps included in this synthesis is the novel use of Fremy's salt. When the protected indole **589**, prepared from **588**, was treated with ceric ammonium nitrate (CAN), only decomposition was observed and attempts to deprotect two functional groups (Boc and OMe) also failed. However, treatment of **589** with TMSCl–NaI in MeCN followed by *in situ* oxidation of the resulting amine with Fremy's salt afforded the target molecule **587** in 73% yield (Scheme 110)[196].

Generally, metal nitrates in trifluoroacetic anhydride have been known to nitrate many aromatic compounds in high yields. Oxidation of substituted phenols with NH_4NO_3–$(CF_3CO)_2O$ affords quinones. 2,6-Dimethylphenol (**206**) was oxidized with NH_4NO_3–$(CF_3CO)_2O$ in AcOH to give 3,3′,5,5′-tetramethyldiphenoquinone (**207**) (45%). Oxidation of 2,6-di(*tert*-butyl)phenol (**23**) with the same reagent in $CHCl_3$ provided the corresponding diphenoquinone **24** in high yield (83%) (Scheme 111)[197].

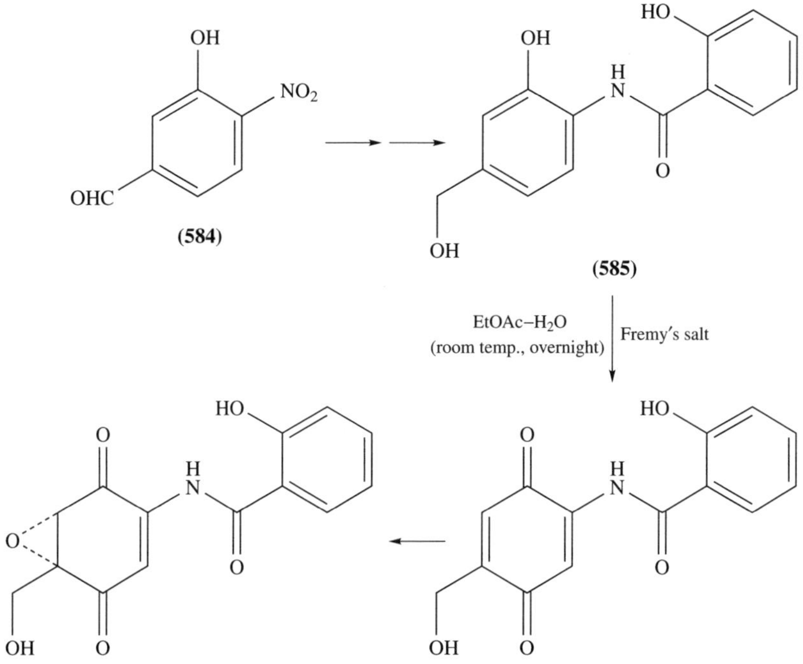

SCHEME 109. Synthesis of racemic epoxyquinomicin

Treatment of pentachlorophenol (**590**) with $NH_4NO_3-(CF_3CO)_2O$ in CH_2Cl_2 resulted in the predominant formation (80%) of tetrachloro-*o*-benzoquinone **591** (Scheme 111).

Related to a model for oxidative conversion of lignin to valuable products, a series of *para*-substituted phenols were oxidized with catalytic amounts of NO_2 under O_2 in MeOH to afford the corresponding benzoquinones in moderate to high yields[198]. Syringyl alcohol (**247**) was treated with a stoichiometric amount of $NaNO_2$ and conc. HNO_3 (a convenient source of NO_2) in MeOH under argon at $-20\,^\circ C$ to afford the *p*-benzoquinone **248** in low yield. However, in the presence of O_2, only catalytic amounts of 20% $NaNO_2$ were sufficient for the conversion of **247** to **248** (80–90%). Similarly, 4-hydroxymethyl-2,6-dimethylphenol (**592**) was converted to 2,6-dimethyl-*p*-benzoquinone (**223**) in quantitative yield (Scheme 112).

Oxygenation of *para*-substituted 2,6-di(*tert*-butyl)phenols with a *tert*-butoxide anion in protic and aprotic solvents has been studied extensively by Nishinaga and coworkers[199]. The dioxygen incorporation depends on the nature of the *para*-substituents and the solvent used. 2,4,6-Tri-(*tert*-butyl)phenol (**73**) was oxygenated in the presence of *t*-BuOK–*t*-BuOH ($0\,^\circ C$, 30 min) to afford mainly the corresponding cyclohexa-2,5-dienone **242** (84%) together with cyclohexa-2,4-dienone **593** (14%). At $30\,^\circ C$ for 10 min, the yield of the former decreased, while **293** was produced in 65% yield. In addition, a new epoxide **594** (8%) was detected. The amount of **242** (3%) further decreased on additional rise in the reaction temperature ($40\,^\circ C$, 10 min), while the amount of **594** (39%) increased

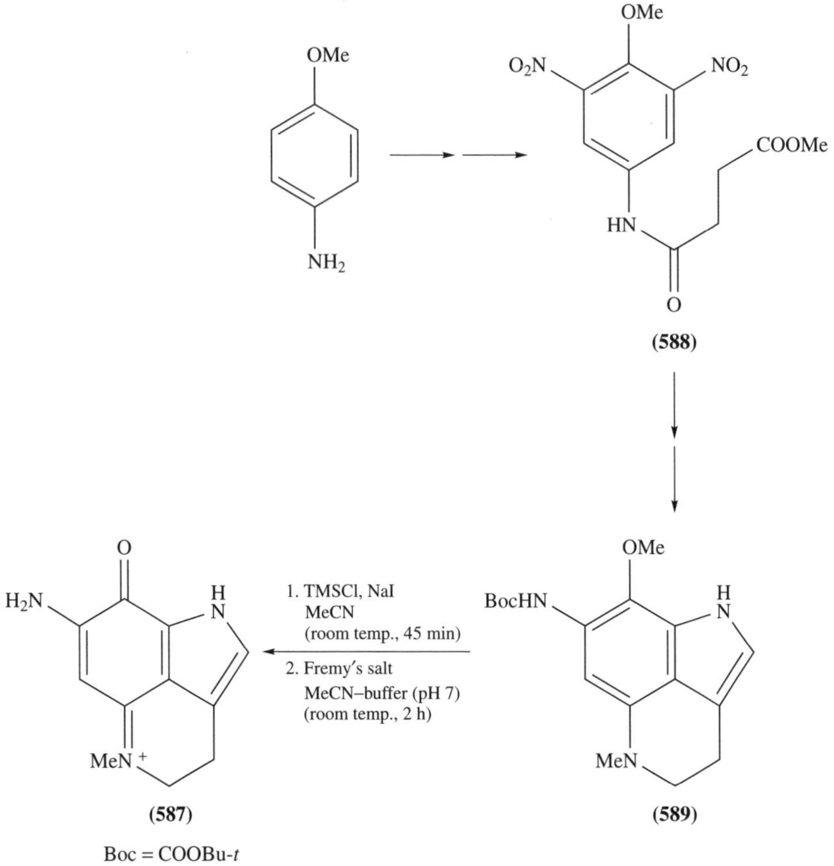

SCHEME 110. Synthesis of makaluvamine C

and **593** was produced in 59% yield (Scheme 113). Evidently, the selective formation of **594** from **242** involves the effective isomerization of **242** to **593**.

Iodine is generally used for iodination of phenols. However, phenolic oxidation has also been effected with iodine in MeOH containing a base such as KOH; 2,4,6-trimethylphenol (**221**) was treated with iodine (1 equiv.) and KOH in MeOH (room temp., 10 min) to afford 2,6-dimethyl-4-(methoxymethyl)phenol (**595**) and 3,5-dimethyl-4-hydroxybenzaldehyde (**222**) in 84 and 5% yields, respectively. The use of 2 equivalents of iodine (room temp., 2 h) provided mainly the aldehyde **222** in 83% yield (Scheme 114)[200].

Cacalol (**596**), a major constituent of *Cacalia delfiniifolia* Sieb. et Zucc., was oxidized with iodine (1 equiv.) and NaOMe in MeOH to give 11-methoxycacalol (**597**) in 62% yield. With 2 equivalents of iodine, a similar reaction provided cacalal (**598**), another sesquiterpene isolated from the same plant, although the yield (0.8%) was very low (Scheme 114).

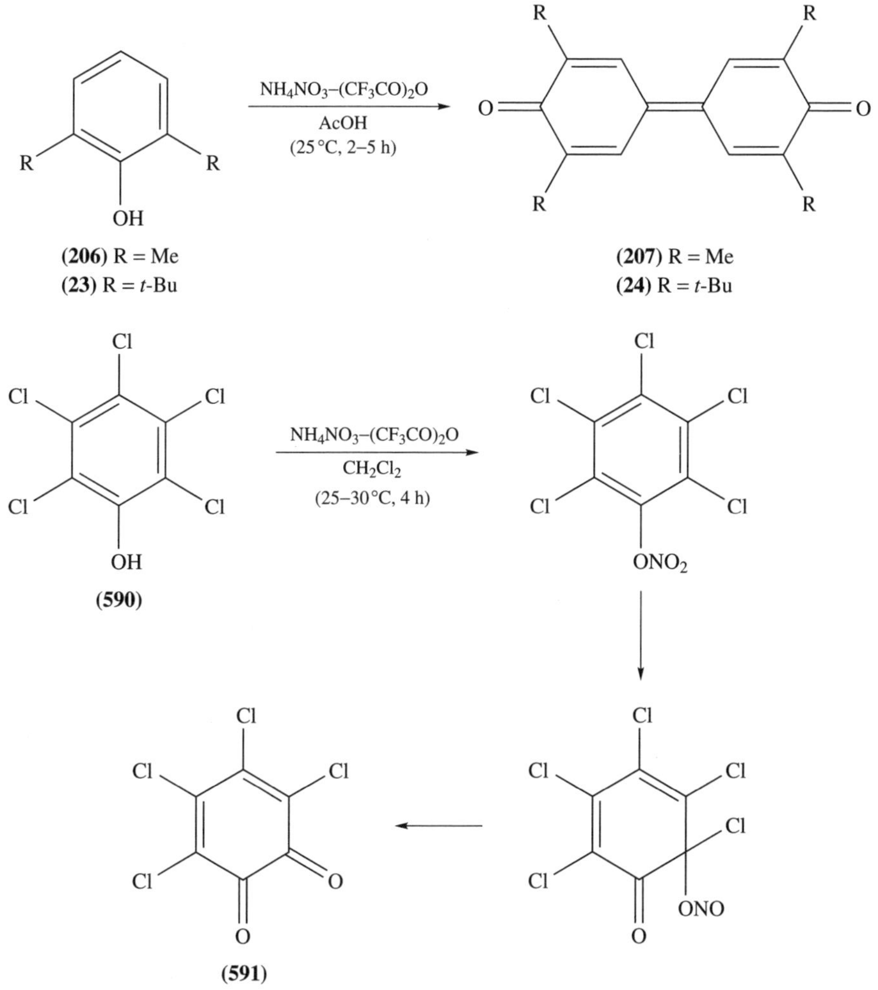

SCHEME 111. Oxidation of phenols with NH_4NO_3–$(CF_3CO)_2O$

Phenolic oxidation is known to be effected with sodium periodate or periodic acid to yield o- and p-quinols, quinones and other products, depending on substituents attached to the aromatic ring. 2,4,6-Trimethylphenol (**221**) was subjected to $NaIO_4$ oxidation in 80% aq. AcOH to afford the four products **383**, **599**, **600** and **601** (10, 17, 28 and 31%, respectively). Of these products, the dimer **601** must be formed from the corresponding cyclohexadienone **602** (Scheme 115)[201].

2,6-Dimethylphenol (**206**) underwent $NaIO_4$ oxidation in 1:1 EtOH–H_2O containing p-benzoquinone resulting mainly in the formation of a 1:1 adduct **603** (68%) together with the o-quinol dimer **604** (6%) (Scheme 115)[202].

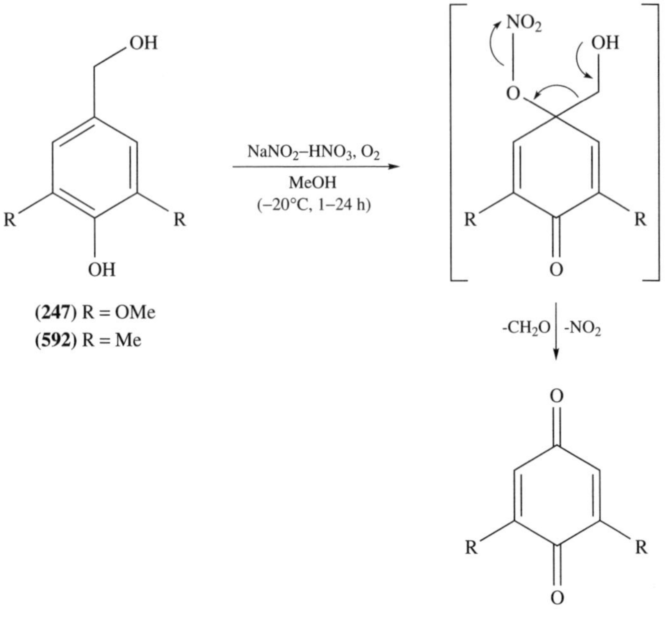

(247) R = OMe
(592) R = Me

(248) R = OMe
(223) R = Me

SCHEME 112. Oxidation of *p*-substituted phenols with NO$_2$ in the presence of O$_2$

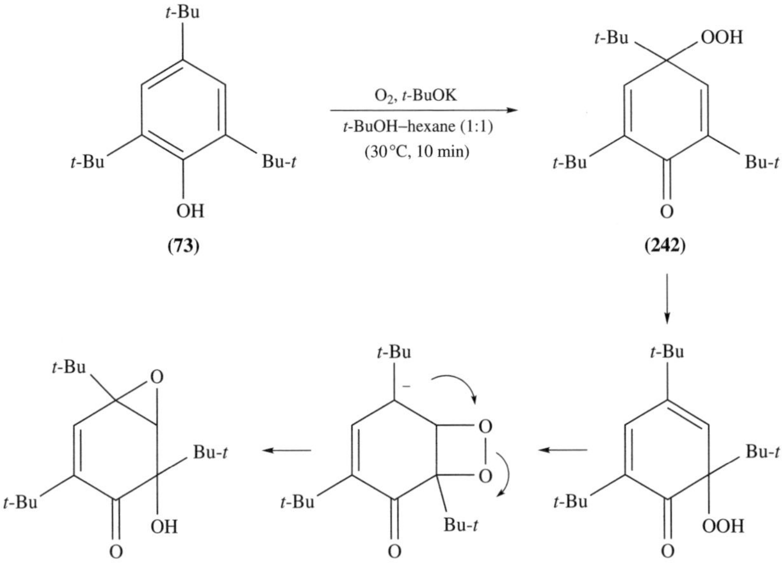

(73)

(242)

(594)

(593)

SCHEME 113. Oxygenation of 2,4,6-tri(*tert*-butyl)phenol with *t*-BuOK–*t*-BuOH

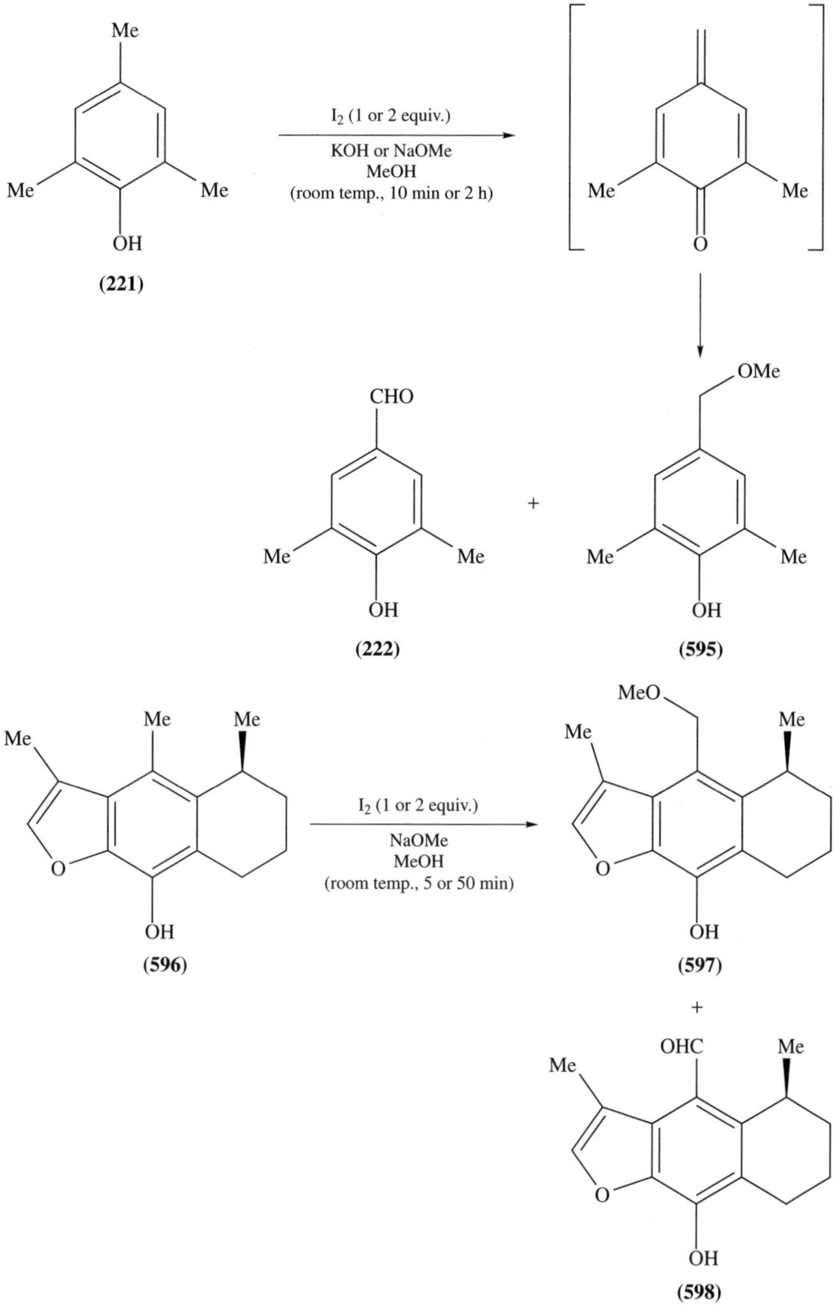

SCHEME 114. Oxidation of phenols with iodine and base in MeOH

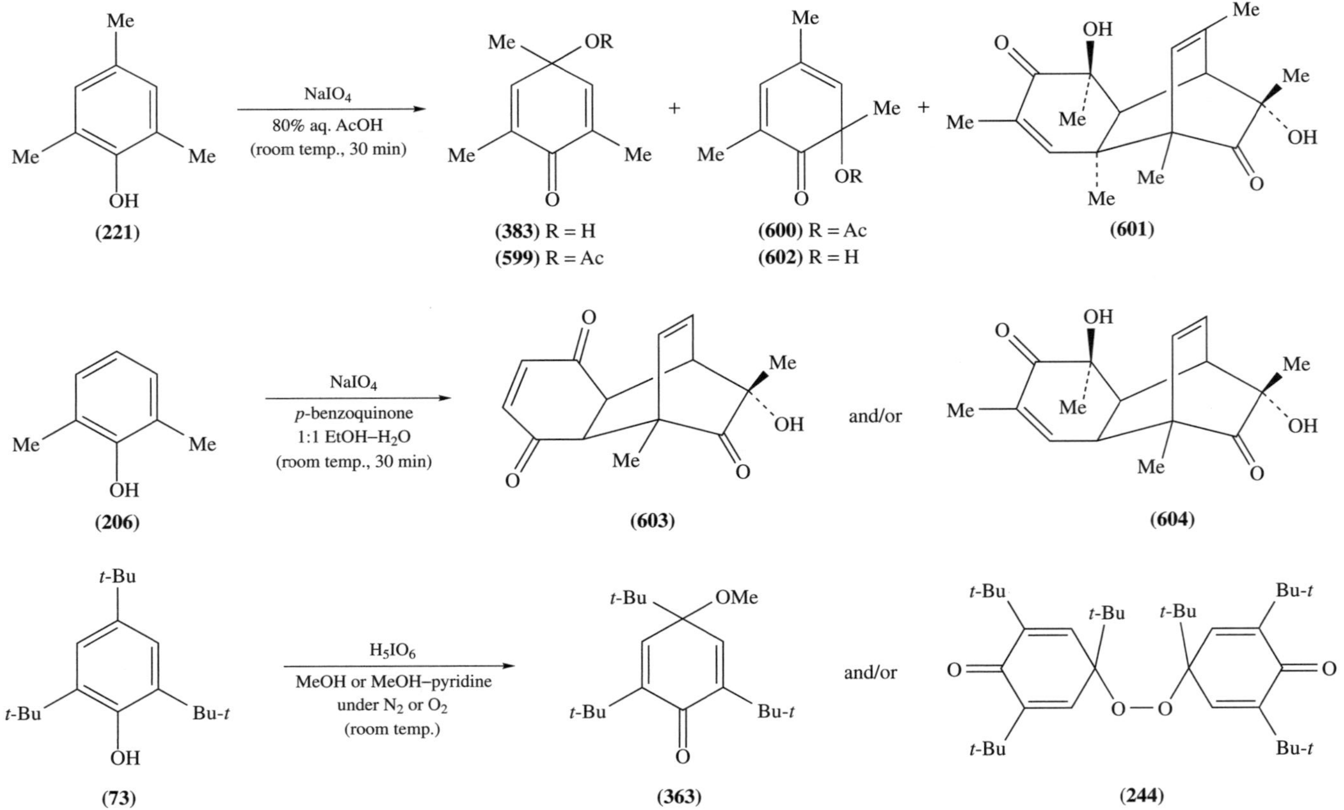

SCHEME 115. Oxidation of phenols with sodium periodate or periodic acid

2,4,6-Tri(*tert*-butyl)phenol (**73**), a sterically hindered phenol, was treated with periodic acid (1 equiv.) in MeOH at room temperature to yield 2,4,6-tri(*tert*-butyl)-4-methoxy-2,5-cyclohexadienone (**363**) (62%), whereas periodic acid oxidation of **73** in 40:1 MeOH–pyridine under an oxygen atmosphere provided selectively the corresponding peroxide **244** (71%) together with small amounts of **363** (5%) (Scheme 115)[203].

Oxidative ring cleavage of catechol is well known to provide *cis,cis*-muconic acid or its monomethyl ester (see Schemes 40 and 102). Similar oxidation of phenols is performed by using chlorous acid. The oxidation products vary with the nature and location of the substituents attached to the aromatic ring. 4-Hydroxy-3-methoxybenzaldehyde (**404**) was treated with $NaClO_2$ in a citrate–phosphate buffer solution (pH 4.0) at 0 °C to afford two isolable products **605** and **606** (22 and 2%, respectively) through a ring-cleaved intermediate **607**. On the other hand, *o*-methoxyphenol (**608**) was oxidized with $NaClO_2$ in aq. H_2SO_4 (pH 0.5) to yield 2-methoxy-*p*-benzoquinone (**609**) (45%) (Scheme 116)[204].

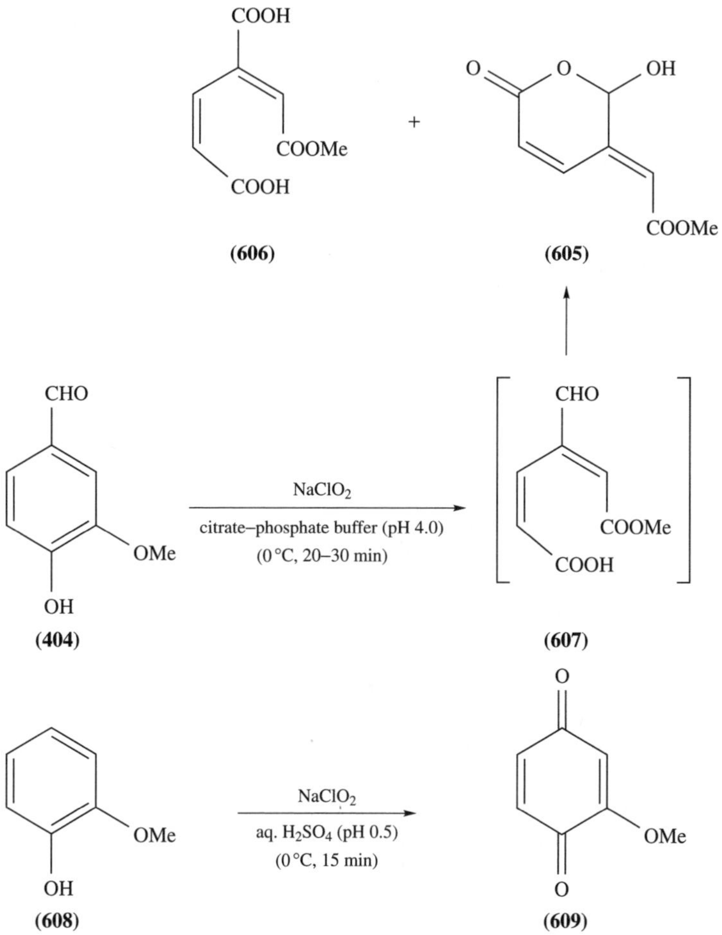

SCHEME 116. Oxidation of phenols with chlorous acid

From the viewpoint of the biological significance of fluorinated steroids, chlorous acid oxidation of (trifluoromethyl)phenols have been examined. In particular, *o*-(trifluoromethyl)phenol (**610**) was oxidized with NaClO$_2$ in 0.3 M H$_2$SO$_4$ (5 °C, 30 min) to afford mainly 5-chloro-4-oxo-5-(trifluoromethyl)cyclopent-2-en-1-ol carboxylic acid (**611**) and 2-chloro-2-(trifluoromethyl)cyclopentene-1,3-dione (**612**) in 60 and 9% yields, respectively. On simple melting or refluxing in MeCN, the former was decarboxylated quantitatively to a diketone **613** (Scheme 117)[205]. Presumably, the acid **611** is produced through the initially formed 3-trifluoromethyl-*o*-benzoquinone **614**, although 2-trifluoromethyl-*p*-benzoquinone **615** is not always ruled out. Compounds **611** and **613**, prepared in good yields, are recognized as valuable intermediates for the synthesis of steroids and other five-membered ring molecules containing a trifluoromethyl group.

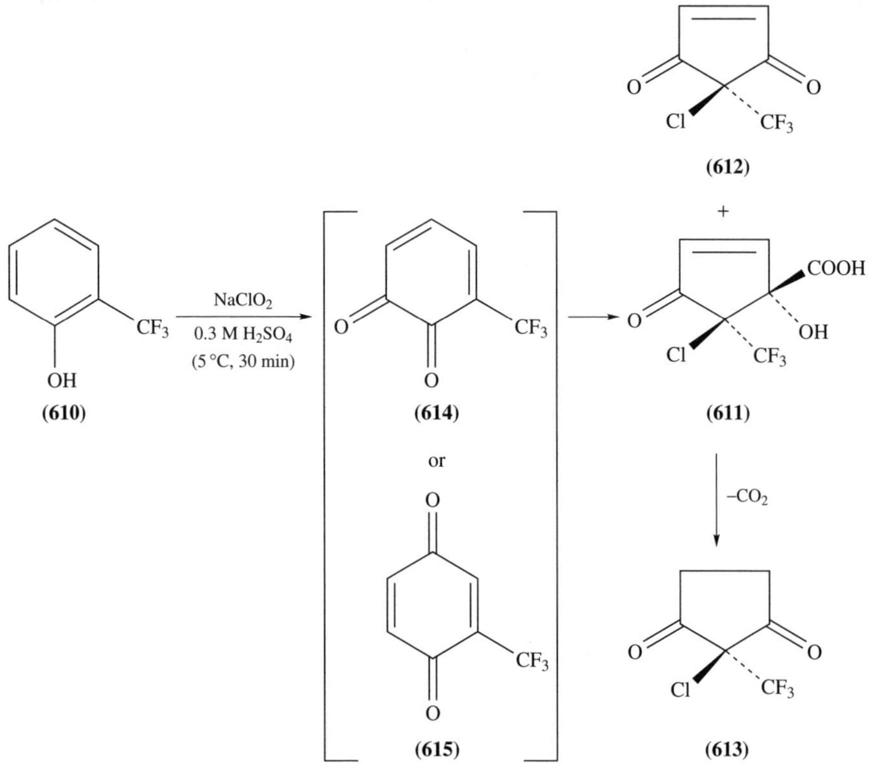

SCHEME 117. Oxidation of 2-trifluoromethylphenol with chlorous acid

Sodium perborate is effective for oxidation of a variety of functional groups. This oxidant is not a mixture of hydrogen peroxide and sodium borate, but its molecular structure has been proved to be represented by **616**. Oxidation of hydroquinone derivatives was effected with sodium perborate in AcOH to afford *p*-benzoquinones in 64–96% yields. However, in the case of phenols bearing no substituent at the C4-position, the corresponding *p*-benzoquinones were obtained, although the yields (42–53%) were relatively low. Both 2,3,5-trimethylhydroquinone (**617**) and 2,3,5-trimethylphenol (**239**) were treated with

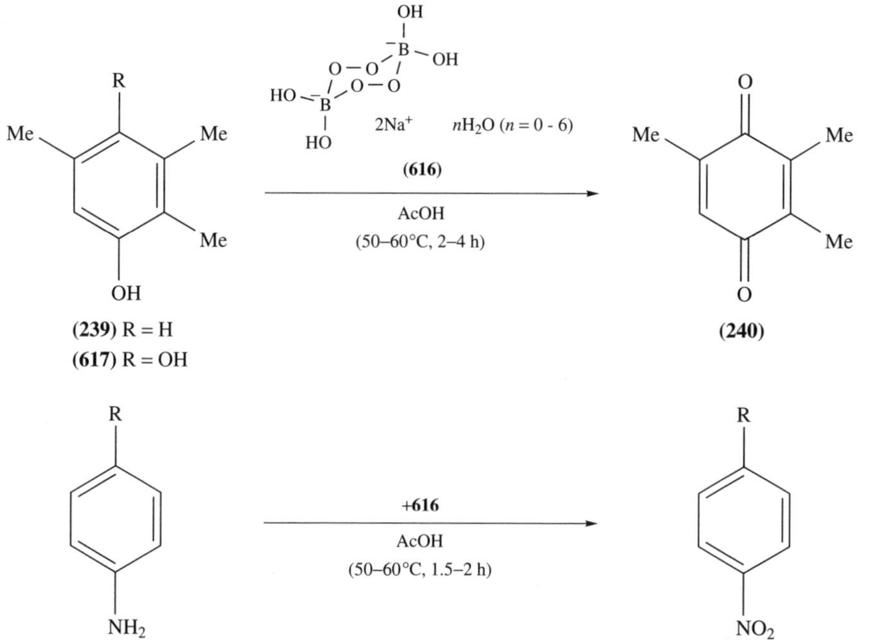

(**239**) R = H
(**617**) R = OH
(**240**)
(**618**) R = OMe
(**619**) R = CN
(**620**) R = OMe
(**621**) R = CN

SCHEME 118. Oxidation of phenols and anilines with sodium perborate

sodium perborate tetrahydrate in AcOH to yield 2,3,5-trimethyl-*p*-benzoquinone (**240**), in 95 and 53% yields, respectively (Scheme 118).

A number of anilines have been oxidized with the same reagent in AcOH (50–60 °C, 1.5–2 h) to afford nitroarenes in 47–92% yields (Scheme 118)[206]. 4-Methoxy- and 4-cyanoanilines (**618** and **619**) underwent sodium perborate oxidation in AcOH to the corresponding nitroarenes **620** and **621** in 70 and 91% yields, respectively.

IV. OXIDATION WITH METAL COMPOUNDS

A variety of metal compounds as an oxidant have long been used for phenolic oxidation. These oxidants are widely used and are very effective for organic synthesis in a laboratory scale. However, they are no longer utilized in an industrial scale, because stoichiometric amounts of these metal compounds are often required. Accordingly, the corresponding oxidation–reduction systems must be constructed for each metal oxidation. In this section, phenolic oxidation using a variety of metal compounds will be described.

A. Oxidation with Vanadium, Chromium and Molybdenum Compounds

Tetra- and pentavalent vanadium compounds such as VCl_4 and $VOCl_3$ are generally used for phenolic oxidation[207]. Phenol was oxidized with VCl_4 in CCl_4 to afford two *ortho–para* and *para–para* coupled biphenyls **622** and **19** in 18 and 34% yields, respectively. Similarly, oxidation of 2,6-dimethylphenol (**206**) with VCl$_4$ provided the corresponding *para–para* coupled biphenyl **319**, while VOCl$_3$-promoted oxidation of **206** afforded mainly 3,3′,5,5′-tetramethyldiphenoquinone **207** (35%) together with small amounts of **319** (6%) (Scheme 119)[208].

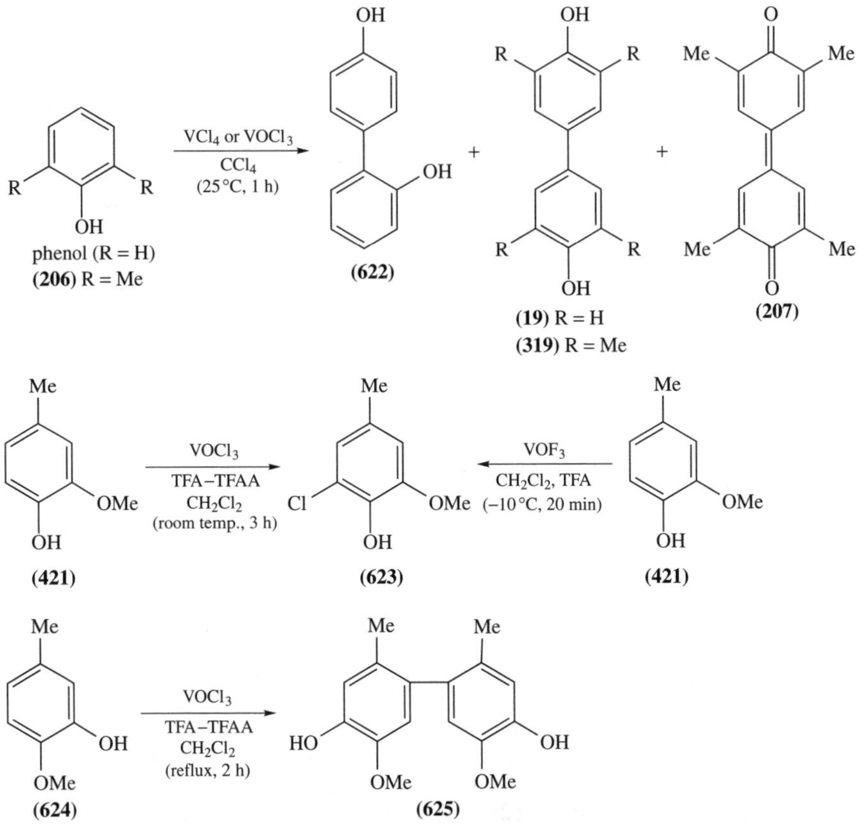

SCHEME 119. Oxidation of simple phenols with vanadium compounds

On oxidation with VOCl$_3$ in CH$_2$Cl$_2$ containing TFA and TFAA (room temp., 3 h), 2-methoxy-4-methylphenol (**421**) was converted into 2-chloro-6-methoxy-4-methylphenol (**623**) in 83% yield. The use of VOF$_3$ as an oxidant also provided **623** under mild conditions ($-10\,°$C, 20 min) and in high yield (91%)[209]. Similarly, there have been another two reports on ring chlorination with VOCl$_3$[210]. Oxidation of 2-methoxy-5-methylphenol (**624**) with VOCl$_3$ afforded the *para–para* coupled biphenyl **625** in 81% yield (Scheme 119). In the case of 2,4,6-tri(*tert*-butyl)phenol, several oxidation products such as *o*- and *p*-benzoquinones, diphenoquinones, major amounts of dealkylated phenols and C–C coupled dimers were produced[211].

Of phenolic oxidations using vanadium compounds, intramolecular oxidative phenol-coupling reactions are quite attractive from the viewpoint of natural products synthesis. A number of benzylisoquinoline alkaloids, lignans and neolignans are well known to be produced, in a key step, by oxidative radical coupling of open phenolic precursors.

17. Oxidation of phenols

In particular, extensive studies on biomimetic syntheses of benzylisoquinoline alkaloids using vanadium compounds were made independently by Kupchan[212,213] and Schwartz[214].

From the viewpoint of biogenetic consideration, intramolecular oxidative coupling reactions of the benzylisoquinolines such as **626** leading to quinonoid oxoaporphines such as **627** were performed by using a variety of metal compounds. Of them, both VOF_3 and $MoOCl_4$ provided good results, as shown in Scheme 120[212].

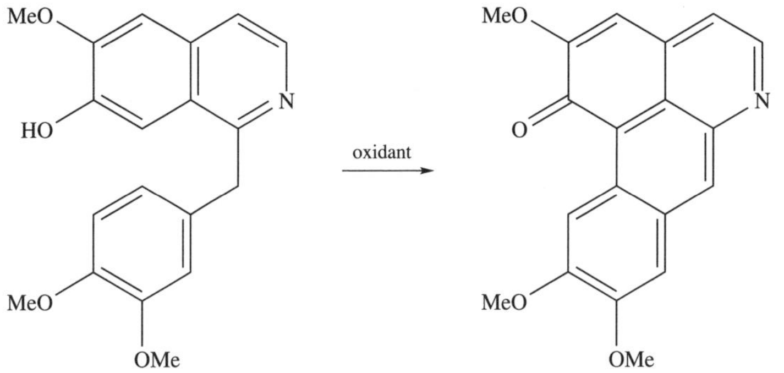

Oxidant	Medium	Yield (%)
MnO_2	CF_3COOH^a	30
CrO_3	aq. H_2SO_4-AcOH^a	25
$Tl(OCOCF_3)_3$	CF_3COOH^b	12
VOF_3	CF_3COOH^a	59
$MoOCl_4$	CF_3COOH-$CHCl_3^b$	62

^{a} 0 °C. ^{b} 25 °C.

SCHEME 120. Oxidative coupling of benzylisoquinoline **626** with metal compounds

Monophenolic benzylisoquinoline **628** was treated with VOF_3 in EtOAc–TFA (CF_3COOH)–TFAA $[(CF_3CO)_2O]$ at $-10\,°C$ to afford mainly trifluoroacetylwilsonirine (**629**) (70%) together with morphinane-type dienone **630** as a minor product (8%) (Scheme 121)[213]. The initially formed intermediate **631** undergoes *ortho–para* and *para–para* couplings to yield **629** and **630**, respectively. In contrast, $VOCl_3$-mediated oxidation of triphenolic hydroxynorreticuline **632** in ether provided predominately the corresponding morphinane-type dienone **633** (64%). Furthermore, this compound was smoothly converted to racemic noroxycodeine (**634**)[214]. Similar oxidative coupling reactions have also been reported[215].

Biomimetic syntheses of dibenzoazonines **635** and **636** were carried out by VOF_3-mediated oxidation of diphenolic benzylisoquinoline **637** and its methyl ether **638**, respectively. The diphenolic compound **637** was submitted to VOF_3-mediated oxidation at $-10\,°C$ to afford in 40% yield the *para–para* coupled spirodienone **639**, which was further converted to dibenzoazonine **635** (Scheme 122)[214]. Similar coupling reactions have also been carried out[214,216,217].

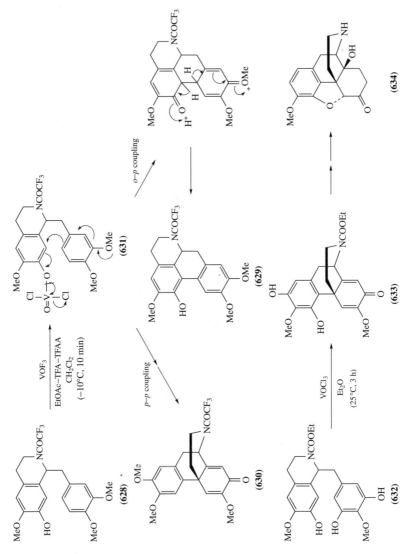

SCHEME 121. Oxidation of phenolic benzylisoquinolines with vanadium compounds

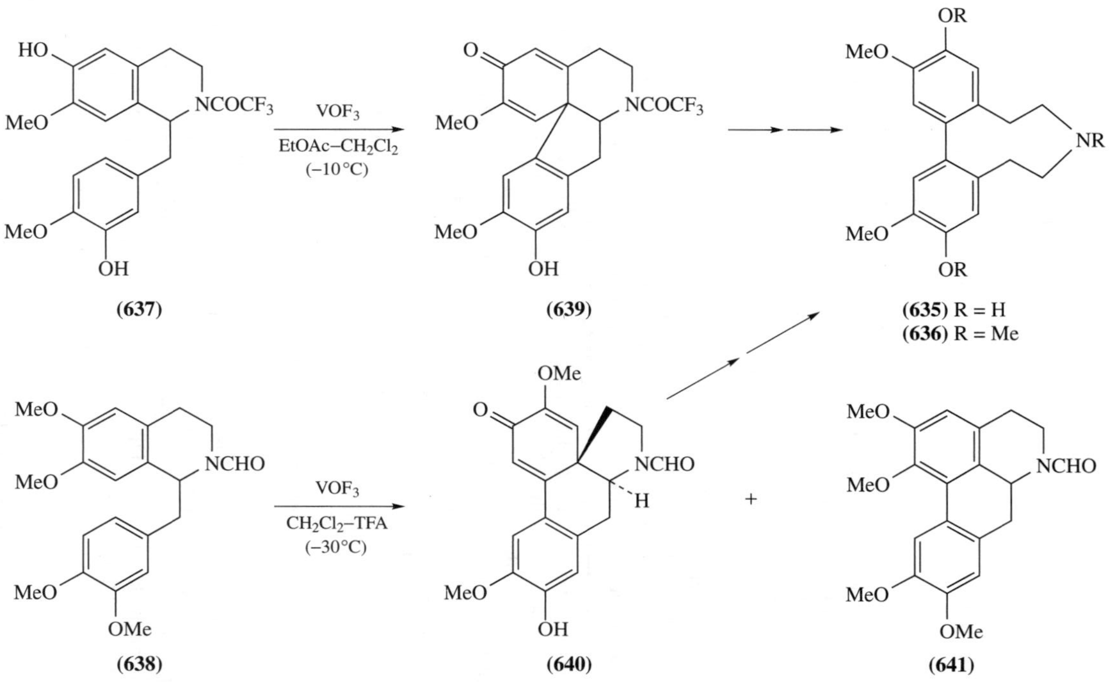

(635) R = H
(636) R = Me

SCHEME 122. Biomimetic synthesis of dibenzoazonines

A nonphenolic compound **638** also underwent VOF$_3$-mediated oxidation in CH$_2$Cl$_2$–TFA at −30 °C to give two coupled products **640** and **641** in 55 and 7%, respectively. The former was converted to an erybidine derivative **636**[218].

Another interesting example is a biomimetic synthesis of maritidine (**491**)[219]. When treated with VOCl$_3$ in Et$_2$O (reflux, 10 h), a diphenolic compound **642** underwent oxidative *para–para* coupling to afford 37% yield of the desired product **643**, which was readily converted to the target molecule (**491**) (Scheme 123). (+)-Maritidine was also synthesized by a similar procedure except for the use of FeCl$_3$ instead of VOCl$_3$, as shown in Scheme 123, wherein the diphenolic compound **644**, prepared from L-tyrosine methyl ester and isovanilline, was oxidized with FeCl$_3$·DMF to give the corresponding coupled product **645** as an optically active form, although the yield (14%) was low. Compound **645** was successfully converted to (+)-maritidine[220]. According to essentially the same synthetic route as shown in Scheme 123, (+)-maritidine (**491**) was also synthesized (see Scheme 95)[172]. In the latter case, the phenolic oxidation was performed using PhI(OCOCF$_3$)$_2$ instead of the FeCl$_3$·DMF complex.

Naturally occurring neolignans, represented by shizandrins and steganes, exhibit a variety of biological activities such as antifeedant against insects and antitumor activities.

Deoxyschizandrin and its related compound (**646** and **647**) have a novel dibenzocyclooctadiene framework, which will be constructed *in vivo* by oxidative phenol-coupling. Thus, the diarylbutane **648**, prepared from ethyl 3,4,5-trimethoxyphenyl ketone, underwent VOF$_3$-mediated oxidation in CH$_2$Cl$_2$–TFAA at −10 °C with intramolecular cyclization resulting in the formation of deoxyschizandrin (**646**) (54%) (Scheme 124)[221].

Recently, a combination of metal oxides such as Re$_2$O$_7$ and fluoro acid media such as CF$_3$COOH was found to be effective for nonphenolic biaryl oxidative couplings[222]. The phenolic diarylbutane **649** was treated with metal oxides in trifluoroacetic acid medium at 20 °C to afford two epi-deoxyschizandrin derivatives **647** and **650** in relatively low yields (24–47%), as shown in Scheme 124. Of three metal oxides (Tl$_2$O$_3$, RuO$_2$–H$_2$O and V$_2$O$_5$), V$_2$O$_5$ provided the best result (47%)[223].

Steganacin (**651**) exihibts highly cytotoxic and antileukemic activities. The first synthesis of steganacin was performed by Kende and Liebeskind[224]. Homopiperonyl alcohol was converted in 3 steps to a diarylbutane derivative **652**, which was subjected to VOF$_3$-mediated oxidation in CH$_2$Cl$_2$–TFAA at 25 °C to yield the desired diarylcyclooctadiene **653** (45%). Further short step manipulations provided the target neolignan **651**. In contrast, oxidation of the diarylbutenolide precursor **654** with VOF$_3$ in CH$_2$Cl$_2$–TFA provided only unnatural isostegane **655** (65–70%) through a plausible spiro intermediate **656**, and no steganacin-type compound was detected (Scheme 125)[225].

Similarly, a variety of metal oxides in CF$_3$COOH or C$_2$F$_5$COOH were effective for oxidation of monophenolic diarylbutenolide **657** to the corresponding isostegane-type dibenzooctadiene **658** (64–18%). Of them, a combination of Tl$_2$O$_3$ and CF$_3$COOH provided the best yield (64%) (Scheme 125)[223].

As already shown in Scheme 120, both chromium and molybdenum compounds were used for phenolic oxidation[212]. In addition, inexpensive chromium reagents such as CrO$_2$Cl$_2$ and Na$_2$Cr$_2$O$_7$·2H$_2$O have been used for the conversion of alkylphenols to the corresponding alkylquinones (30–84%)[226,227]. However, they are scarcely used for phenolic oxidation.

B. Oxidation with Manganese and Rhenium Compounds

Manganese compounds, widely used in organic synthesis, are among the most popular oxidants. Generally, these compounds are recognized as a one-electron oxidant, but in some cases, they act as a two-electron oxidant[226]. On oxidation of 2,6-dimethylphenol (**206**) with excess of MnO$_2$ in benzene (reflux, 2 h), the initially generated phenoxy radical

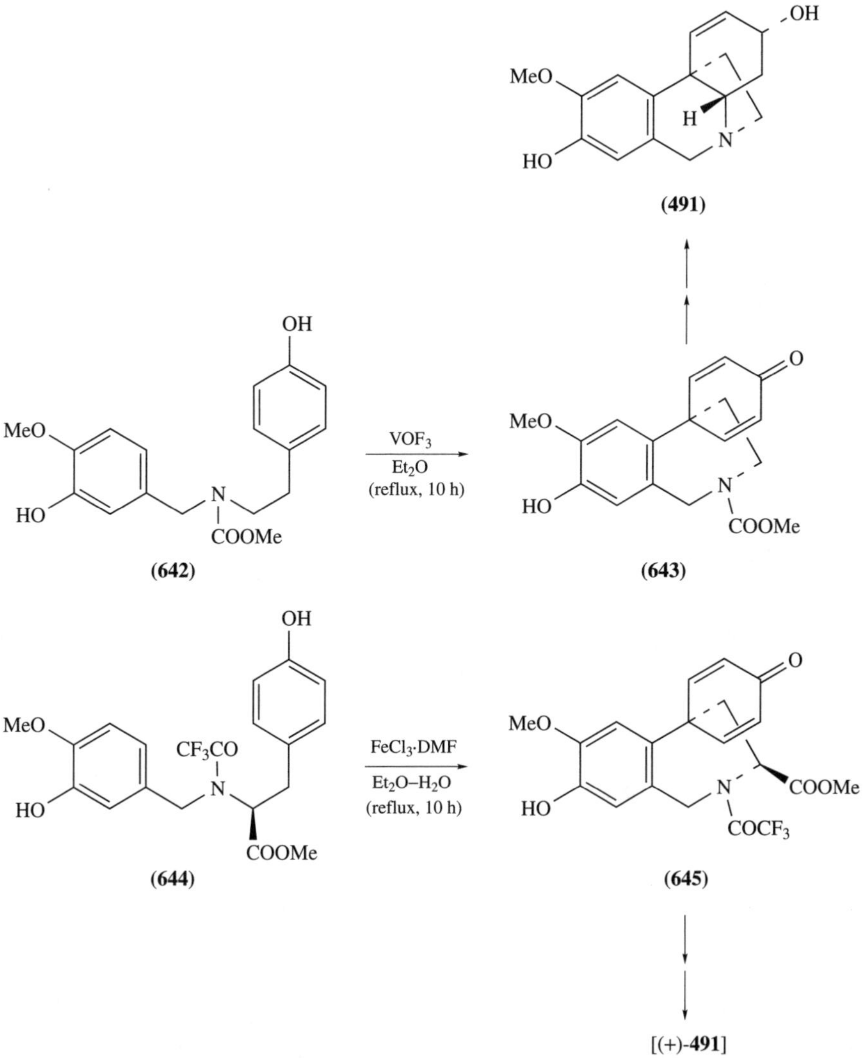

SCHEME 123. Biomimetic synthesis of maritidine

was further polymerized to afford head-to-tail polymers **659** (60–90%) together with small amounts of 3,3′,5,5′-tetramethyldiphenoquinone (**207**). The molecular weight of the polymer varied from 2,000 to 20,000 by the selection of reactant ratios, methods of MnO_2 preparation, the solvent and other factors. The use of a limited amount of MnO_2 (**206**–MnO_2 mole ratio, 20:1) mainly gave 3,3′,5,5′-tetramethylbiphenol (**319**) and a C—O coupled product **660** in 60 and 30% yields, respectively, based on MnO_2 (Scheme 126)[227].

Manganese dioxide and silver oxide are effective for oxidative formation of quinomethanes from the corresponding *p*-alkylphenols. For example, 2,4,5-trialkylphenols such as **661** were submitted to MnO_2-mediated oxidation in benzene at room temperature

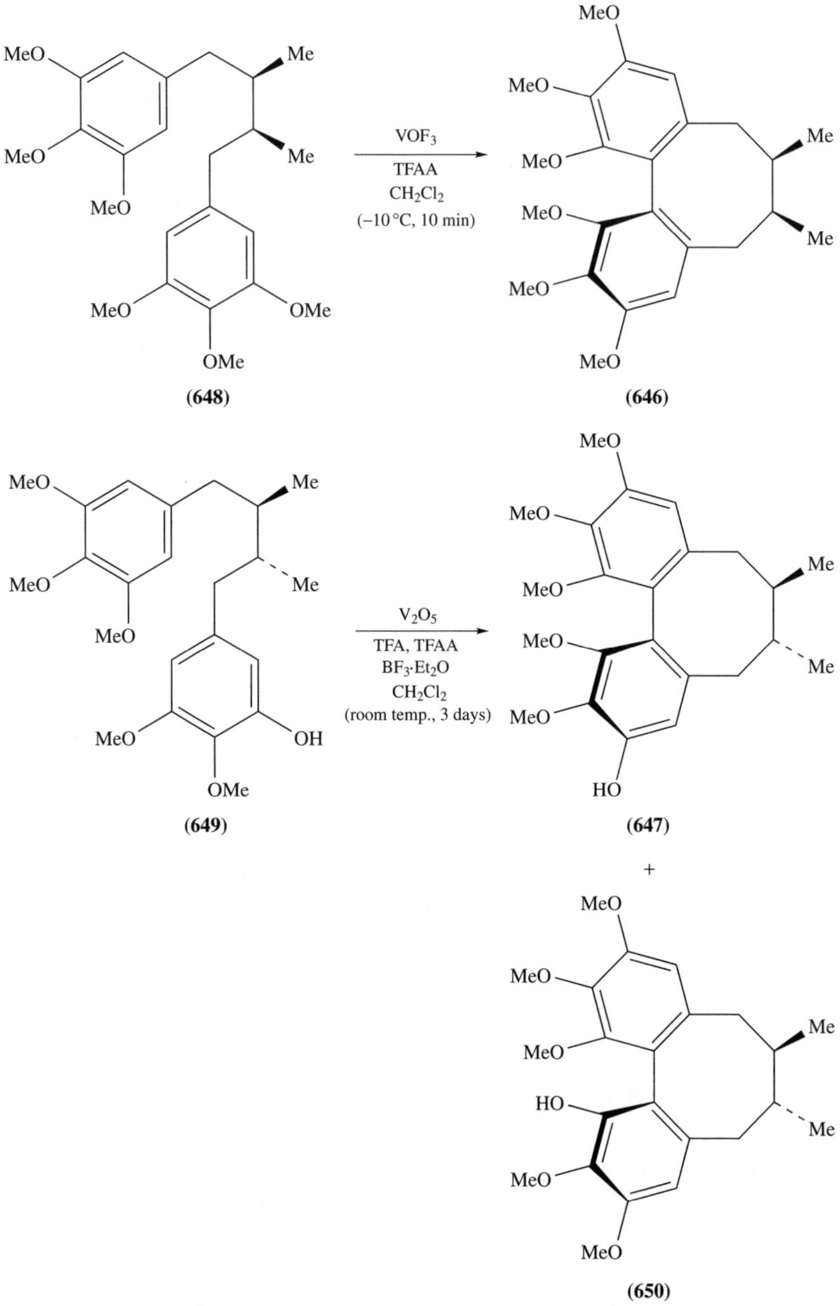

SCHEME 124. Biomimetic synthesis of schizandrin-type compounds

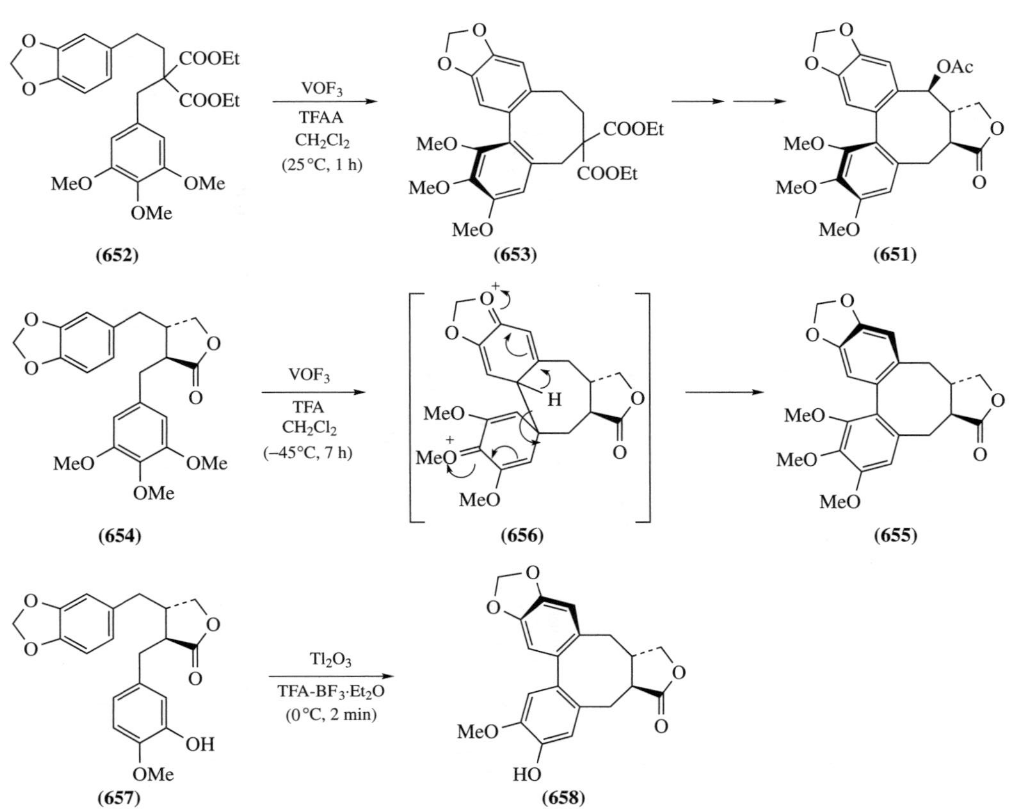

SCHEME 125. Biomimetic synthesis of steganacin and related compounds

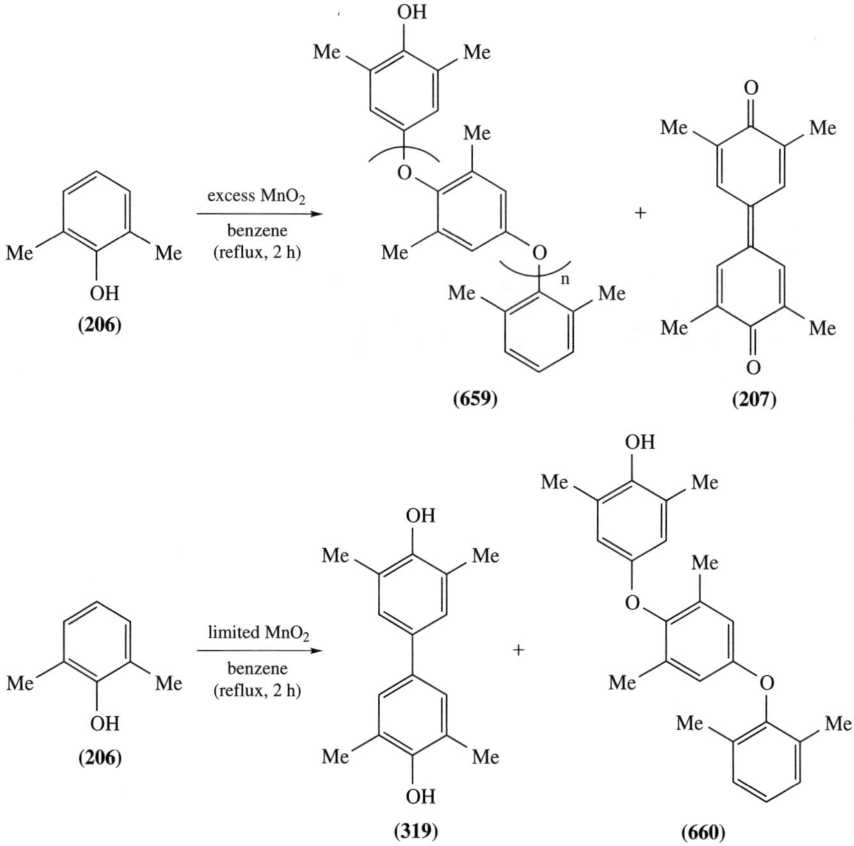

SCHEME 126. MnO$_2$-mediated oxidation of 2,6-dimethylphenol

to afford fuchosones **662** (84–99%) (Scheme 127)[228]. A number of quinone methide have been converted electrochemically to such a peroxide as **663**[229].

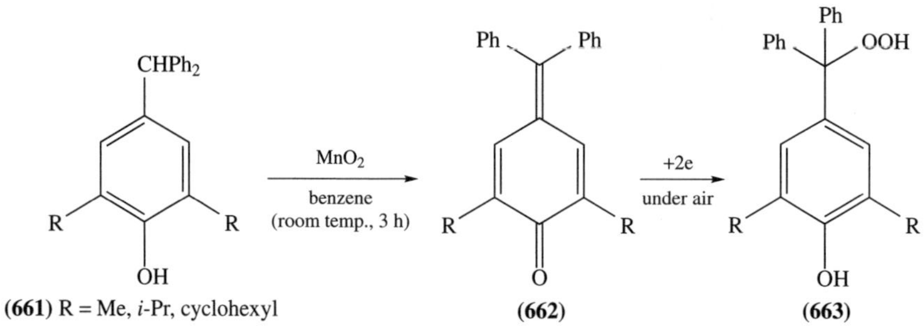

SCHEME 127. MnO$_2$-mediated oxidation of 2,4,6-trialkylphenols to quinomethanes

17. Oxidation of phenols

Manganese dioxide like other oxidants is effective for oxidative conversion of o- and p-hydroquinones into o- and p-benzoquinones, respectively. However, when unstable benzoquinones such as **664** and **665** are produced, the yields are not satisfactory. The synthesis of these quinones could be performed successfully by oxidation of the corresponding hydroquinones **200** and **666** with MnO_2 impregnated with nitric acid in CH_2Cl_2 in 68 and 86% yields, respectively (Scheme 128)[230]. Selection of the solvent used is quite important; generally, methylene chloride is preferred over benzene.

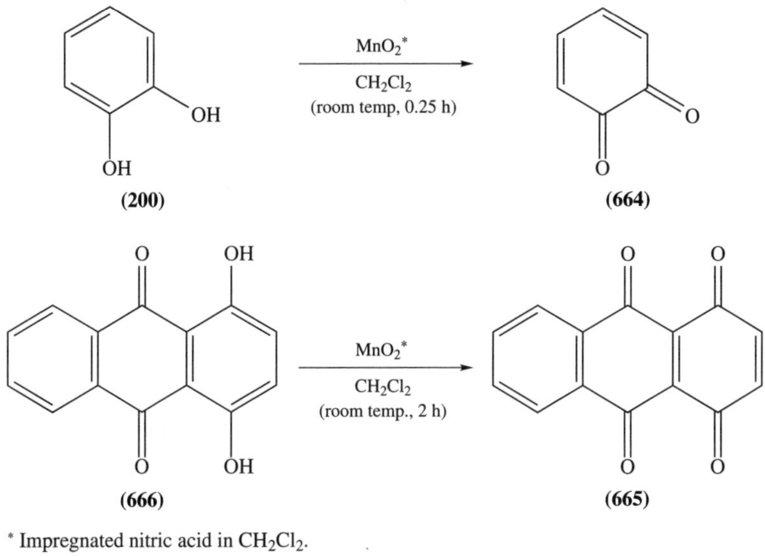

*Impregnated nitric acid in CH_2Cl_2.

SCHEME 128. MnO_2-mediated oxidation of hydroquinones to benzoquinones

Manganese dioxide as an oxidant has been used for biomimetic syntheses of benzylisoquinoline alkaloids and other natural products, but the yields are low[121,122].

Potassium or sodium permanganate under protic and aprotic conditions[231] is well known to be effective for phenolic oxidation leading to quinones and C—C and/or C—O coupled products. Some typical examples demonstrate the utility of barium manganate and methyltributylammonium permanganate in phenolic oxidation.

Barium manganate ($BaMnO_4$), a useful alternative to MnO_2, NiO_2 and Ag_2O as a heterogeneous oxidant, has several advantages over other oxidants such as easy preparation, simple reaction procedure, nontoxicity, and being free from explosion hazards[232].

2,4,6-Tri(tert-butyl)phenol (**73**) was oxidized with $BaMnO_4$ in benzene to afford peroxide **244** in high yield (75–87%). In contrast, 2,6-di(tert-butyl)-4-methylphenol (**69**) was treated with the same reagent in benzene at 60 °C to give 3,3′,5,5′-tetra(tert-butyl)diphenoquinone (**24**) and 2,6-di(tert-butyl)-p-benzoquinone (**74**) (63 and 24%, respectively). Interestingly, $BaMnO_4$-promoted oxidation of 2,4,6-trichlorophenol (**667**) in CH_2Cl_2 provided selectively two diaryl ethers **668** and **669** (81 and 6%, respectively) (Scheme 129).

In studies on a variety of neolignans, oxidative coupling reactions of 4-substituted 2-methoxyphenols were carried out using methyltributylammonium permanganate (MTBAP) in CH_2Cl_2, because the permanganate ion is known to exhibit a lower oxidizing power

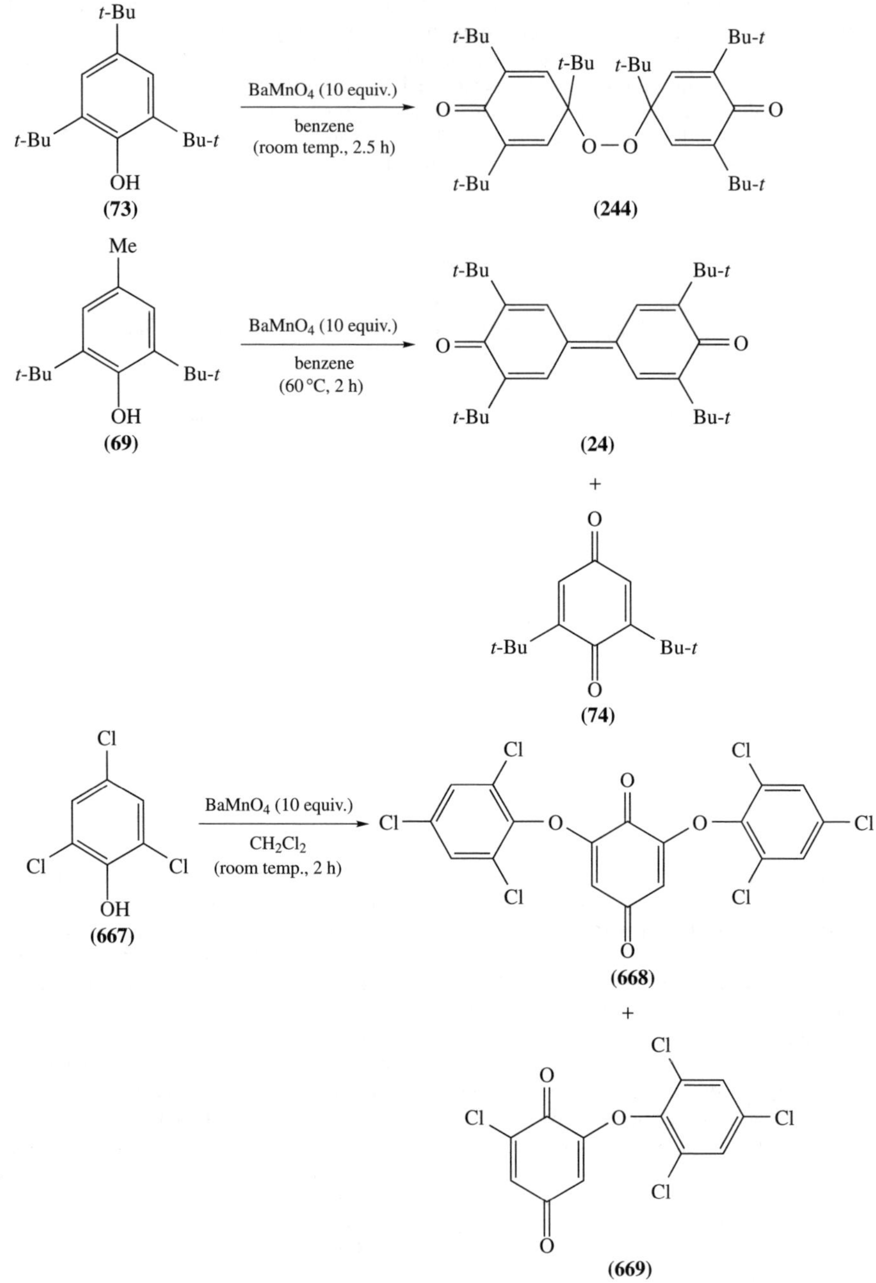

SCHEME 129. Oxidation of some phenols with BaMnO$_4$

in organic solvents than in aqueous solution. This means that the oxidative couplings of phenols bearing easily oxidized functional groups will take place selectively.

Eugenol (**29**) was treated with MTBAP (0.5 equiv.) in CH_2Cl_2 (0–5 °C, 15 min) to afford the corresponding dimer **33** in 52% yield. Another two 4-alkyl-2-methoxyphenols (**670** and **671**) were also oxidized with MTBAP under similar conditions to give dimers **672** and **673** in 52 and 55% yields, respectively (Scheme 130)[233]. In this oxidation, the free phenolic OH group is essential, because the methyl ether of **671** was almost completely recovered on $MeBu_3NMnO_4$-promoted oxidation. In addition, a phenol such as vanillin bearing an electron-attracting group was also resistant to the oxidant.

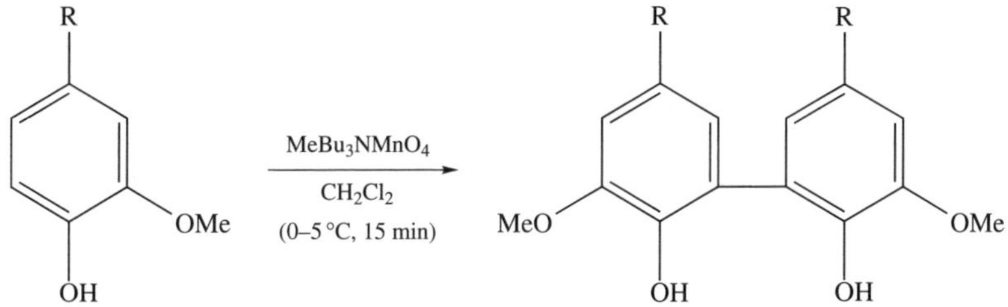

(**29**) R = $CH_2CH=CH_2$
(**670**) R = $CH_2CH\overset{O}{-}CH_2$
(**671**) R = $CH_2CH_2CH_3$

(**33**) R = $CH_2CH=CH_2$
(**672**) R = $CH_2CH\overset{O}{-}CH_2$
(**673**) R = $CH_2CH_2CH_3$

SCHEME 130. Oxidation of 4-substituted 2-methoxyphenols with $MeBu_3NMnO_4$

As manganese(III) compounds are lower in reactivity when compared with other oxidants, higher selectivities in phenolic oxidation can be obtained with these manganese oxidants such as manganese(III)acetate [$Mn(OAc)_3$] and manganese(III)acetylacetonate [$Mn(acac)_3$][234]. Similarly, *trans*-1,2-diaminocyclohexanetetraacetatomanganate(III) [$KMnCyDTA(H_2O)$] is used for phenolic oxidation. Generally, phenolic oxidation with these oxidants initially generates the phenoxy radical[235]. The radical undergoes C—O and/or C—C radical couplings leading to dimers or further oxidation providing hydroquinones, quinones and other compounds[234].

2,6-Disubstituted phenols such as **23** and **274** were oxidized with $Mn(acac)_3$ in AcOH to afford the corresponding biphenols **25** and **320** in high yields (91 and 80%, respectively), while oxidation of both phenols with $Mn(OAc)_3$ in AcOH provided selectively the corresponding diphenoquinones **24** and **277** (98 and 79%, respectively) (Scheme 131).

As already shown in Scheme 101, oxidative cross-coupling reactions of two different phenols using DDQ are quite interesting from the viewpoint of natural products synthesis. A mixture of 2,6-di(*tert*-butyl)phenol (1 equiv.) and 2,6-dimethylphenol (1 equiv.) was also oxidized with $Mn(OAc)_3$ (4 equiv.) to give the desired cross-coupled dimer **674**, although the yield (26–10%) was not satisfactory (Scheme 131)[236].

On treatment with $Mn(acac)_3$ in MeCN, 2-allyl-4-(*tert*-butyl)phenol (**675**) undergoes one-electron oxidation to afford a phenoxy radical. The radical is expected to react with the *ortho*-allyl group to yield a dihydrobenzofuran derivative **676**. However, $Mn(acac)_3$-promoted oxidation of **675** provided a spiro compound **677** (25%) (Scheme 132)[237].

In connection with bioactive neolignans such as schizandrin and steganacin, systematic studies on the oxidative coupling of bis(benzo)cyclooctadiene precursors were carried out

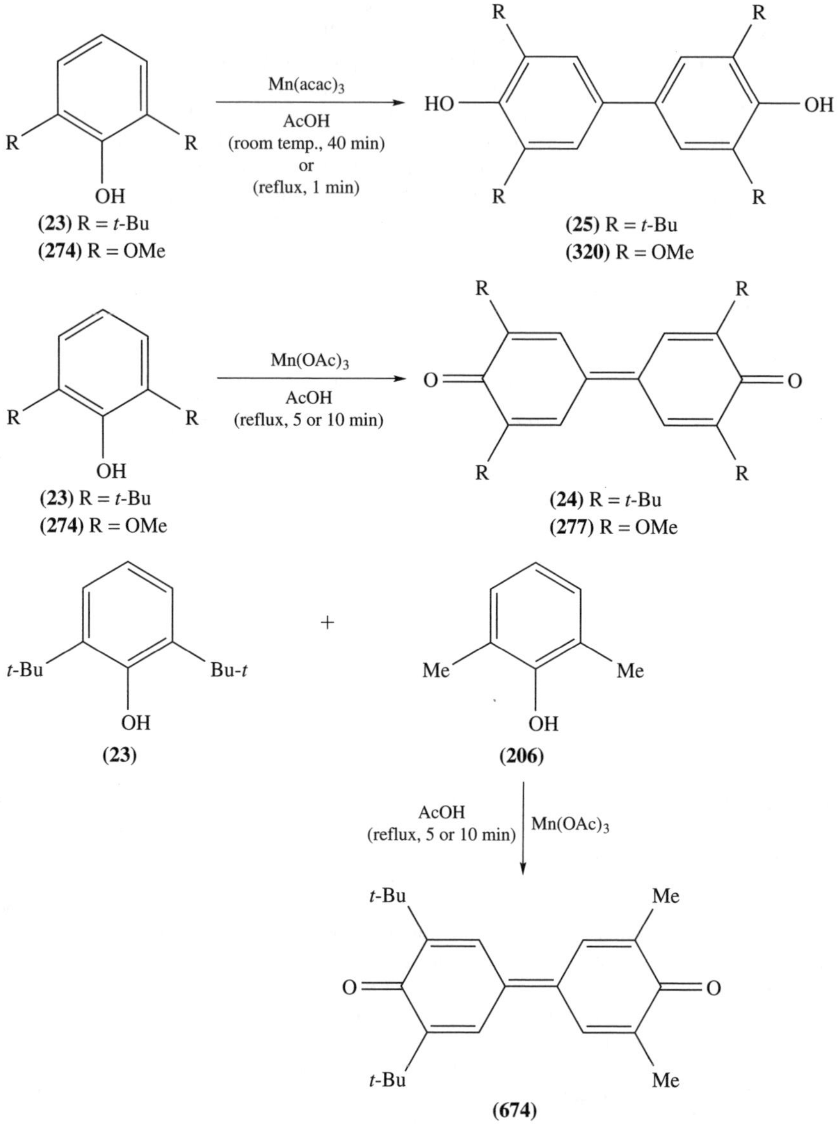

SCHEME 131. Oxidation of 2,6-disubstituted phenols with Mn(III) complexes

using a variety of metal oxides in fluoroacids (CF_3COOH and C_2F_5COOH)[223]. A combination of Re_2O_7 and CF_3COOH was found to be the most effective for the oxidative coupling of prestegane A (**678**) to the corresponding bis(benzo)cyclooctadiene **679**. When V_2O_5, $Cu(OAc)_2 \cdot H_2O$ or $RuO_2 \cdot 2H_2O$ was used as an oxidant, good yields (75–90%) were also obtained. However, $Mn(OAc)_3$ provided a low yield of **679**. In the case of the substrate **657** bearing a methylenedioxy group, Tl_2O_3 was the best oxidant (see Scheme 125),

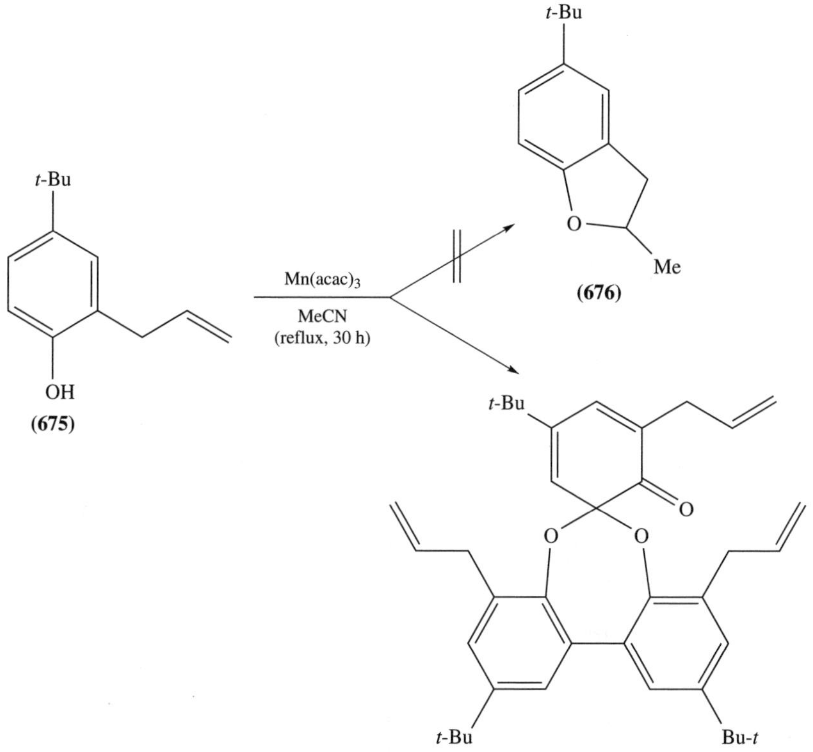

SCHEME 132. Oxidation of 2-allyl-4-(*tert*-butyl)phenol with Mn(acac)$_3$

because of the instability of the methylenedioxy group (Scheme 133). Both CF$_3$COOH and C$_2$F$_5$COOH provide similar results.

C. Oxidation with Iron, Ruthenium, Cobalt, Nickel and Rhodium Compounds

Of a variety of metal compounds described in this section, iron compounds represented by ferric chloride (FeCl$_3$) and potassium ferricyanide [K$_3$Fe(CN)$_6$] have long been used for phenolic oxidation, particularly for biomimetic syntheses of benzylisoquinoline alkaloids and neolignans[120–122].

Generally, on oxidation with Fe(III) compounds, phenol undergoes one-electron oxidation followed by H$^+$ loss, resulting in the formation of the phenoxy radical. The radical undergoes C—C coupling leading to dimers, trimers and polymers, or subsequent oxidation to generate the corresponding phenoxonium ion which is attacked by nucleophiles such as H$_2$O to afford *p*-hydroquinone. Further oxidation provides *p*-benzoquinone (see Scheme 3)[238].

Oxidative aryl–aryl coupling reactions are effected with FeCl$_3$, Fe(ClO$_4$)$_3$, Fe(III) solvates and silica-bound FeCl$_3$. The Fe(III) solvate, [Fe(DMF)$_3$Cl$_2$][FeCl$_4$], is prepared by addition of DMF to a solution of FeCl$_3$ in dry Et$_2$O. On treatment with this oxidant in H$_2$O

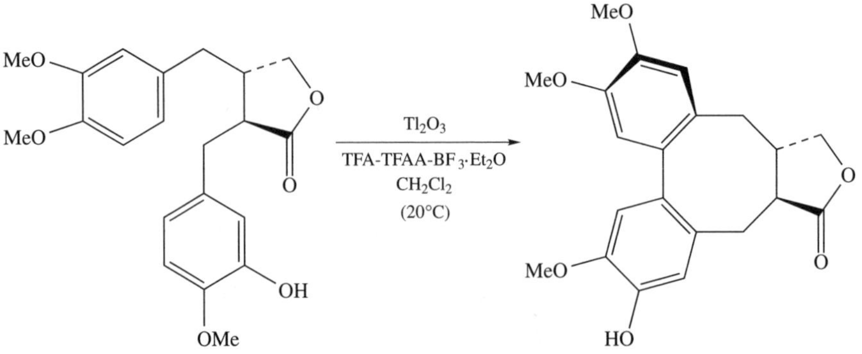

Oxidant	Time	Yield (%)
Tl_2O_3	2 h	56
$Mn(OAc)_3 \cdot 2H_2O$	1 h	34
Re_2O_7	1 h	98
V_2O_5	5 d	75
PrO_2	1 d	58
$RuO_2 \cdot 2H_2O$	15 h	76
TeO_2	40 h	70
$Cu(OAc)_2 \cdot H_2O$	30 min	90
CrO_3	22 h	44
$Fe(OH)(OAc)_2$	24 h	0
Co_3O_4	15 h	72

SCHEME 133. Phenolic oxidation of bisbenzocyclooctadiene presursors with metal oxides

(reflux, 1 h), a 1,3-diarylpropane **680** underwent intramolecular *para–para* coupling reaction to afford the corresponding spiro compound **681** in good yield (67%) (Scheme 134)[239]. Similarly, oxidation of *p*-cresol (**27**) provided Pummer's ketone **28**, although the yield (28%) is low as compared with the electochemical oxidation (see Scheme 4).

Furthermore, oxidative aryl–aryl coupling reactions of phenols and phenol ethers were performed by Tobinaga and coworkers by using tris(2,2′-bipyridyl)iron(III) perchlorate, $Fe(bpy)_3(ClO_4)_3 \cdot 3H_2O$ and some Fe(III) solvates such as $Fe(ClO_4)_3 \cdot 9H_2O$ in MeCN, $Fe(MeCN)_3(ClO_4)_3$, $Fe(ClO_4)_3$ in Ac_2O and $FeCl_3$ in Ac_2O. Solvated $FeCl_3$ in MeCN and Ac_2O have been clarified to be $Fe(MeCN)_6(FeCl_4)_3$ and $Fe(Ac_2O)_3(FeCl_4)_3$, respectively. 1-(4′-Hydroxyphenyl)-3-arylpropane **682** was oxidized with $Fe(byp)_3(ClO_4)_3 \cdot 3H_2O$ in MeCN containing 42% aq. HBF_4 to yield the corresponding *para–para* coupled spiro compound **683** (95%). No reaction took place in the absence of aq. HBF_4. 1-(4′-Methoxyphenyl)-3-arylpropane (**684**) was also oxidized to **683** (56%) with the same reagent in MeCN (Scheme 134)[240].

Similarly, oxidative aryl–aryl coupling reactions of a norbelladine derivative **685** and its methyl ether **686** were carried out using several different Fe(III) solvates to afford the

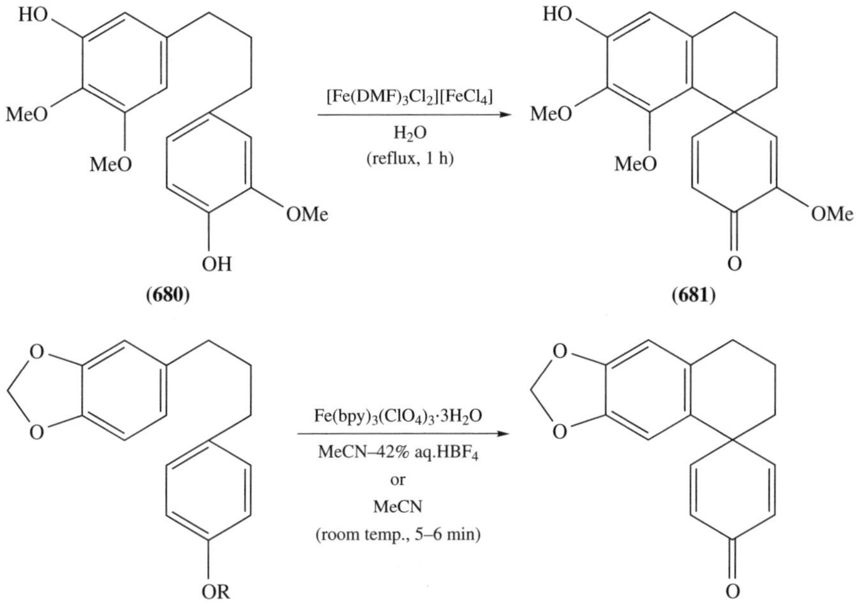

SCHEME 134. Oxidation of 1,3-diarylpropanes with Fe(III) solvates

corresponding spiro compound **687** or its rearranged products **688** and **689** depending on the oxidant. Compounds **688** and **689** may be produced from **687** by the route shown in Scheme 135. On treatment with Na_2CO_3, the spiro compound **687** was further converted easily to crininone (**690**), a precursor of the Amaryllidaceae alkaloid crinine[240].

Silica-bound $FeCl_3$ can act as a one-electron-transfer oxidant, which is very effective for oxidative coupling reactions of aromatic ethers and phenols. 1,2-Diarylethane **691** was oxidized with $FeCl_3$ supported on silica gel in CH_2Cl_2 to give the corresponding *para–para* coupled product **692** in almost quantitative yield (98%). Similar oxidation of 2-methoxy-*p*-hydroquinone (**693**) provided a dibenzofuran **694** (35%) (Scheme 136)[241].

Oxidative coupling reactions of phenols are usually performed by treatment of phenols in solution with more than an equimolar amount of metal salts such as $FeCl_3$. However, the coupling reaction of some phenols with $FeCl_3$ was demonstrated to proceed much faster and more efficiently in the solid state than in solution[242] and the reaction in the solid state is accelerated by irradiation with ultrasound. For example, the irradiation with ultrasound of a mixture of finely powdered *p*-hydroquinone and $[Fe(DMF)_3Cl_2][FeCl_4]$ (2 equiv.) in the solid state (50 °C, 1 h) provided **695** in 64% yield (Scheme 137).

Similarly, oxidation of 2-naphthol (**696**) with $[Fe(DMF)_3Cl_2][FeCl_4]$ (1 equiv.) in the solid state (50 °C, 2 h) gave 2,2′-dihydroxy-1,1′-binaphthol (**697**) in 79% yield (Scheme 137). Interestingly, a mixture of **696** and $FeCl_3 \cdot 6H_2O$ which was kept stationarily at 50 °C for 2 h gave a 95% yield of **697**.

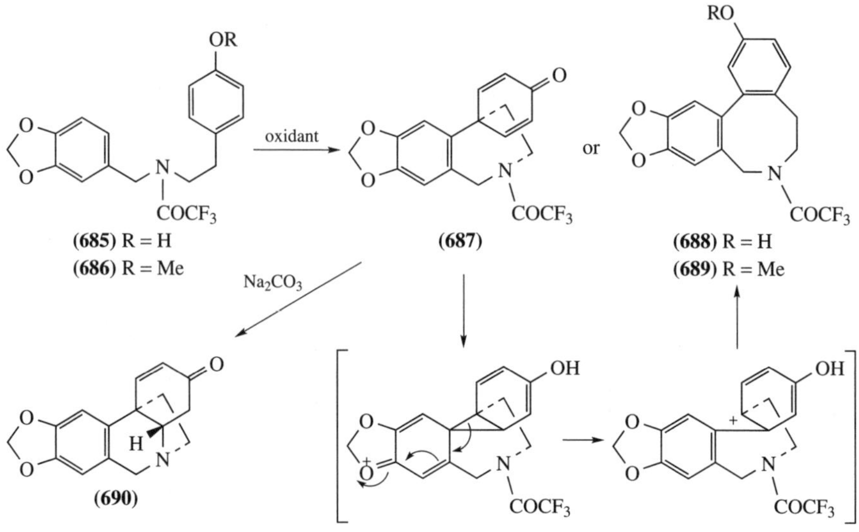

Oxidant	A	B	C	D	E	F
Substrate			Product			
685	687 (50%)	687 (48%)	688 (94%)	688 (58%)		
686		687 (66%)	689 (76%)	689 (96%)	689 (37%)	689 (53%)

A: $Fe(bpy)_3(ClO_4)_3 \cdot 3H_2O$; B: $Fe(ClO_4)_3 \cdot 9H_2O$ in MeCN; C: $Fe(MeCN)_6(ClO_4)_3$; D: $FeCl_3$ in MeCN; E: $Fe(ClO_4)_3$ in Ac_2O; F: $FeCl_3$ in Ac_2O.

SCHEME 135. Oxidation of norbelladine derivatives with Fe(III) solvates

As already described, potassium ferricyanide can act as a one-electron oxidant in phenolic oxidation. From the viewpoint of chemical reactivity, extensive studies on the oxidation of hindered phenols bearing bulky groups such as a *tert*-butyl group have been conducted using $K_3Fe(CN)_6$[243–249] and in rare cases[250] using $H_3Fe(CN)_6$ and $(Bu_4N)_3Fe(CN)_6$. Some interesting examples are shown herein.

On oxidation with $K_3Fe(CN)_6$ in benzene–aq. KOH (room temp., *ca* 10 min), 3,3′-di(*tert*-butyl)-5,5′-ditritylbiphenol (**699**), prepared from 2-*tert*-butyl-4-tritylphenol (**698**), was converted into the corresponding *o*-diphenoquinone **700** (68%). Further thermal isomerization of **700** in isooctane (70 °C, 95 min) provided a quantitative yield of the thermodynamically more stable benzoxete **701** (Scheme 138)[243].

Oxidation of 2-iodo-4,6-di(*tert*-butyl)phenol (**702**) with $K_3Fe(CN)_6$ in aq. KOH (room temp., 20 min) also provided in 82% yield the corresponding benzoxete **703** probably through *o*-diphenoquinone[244]. In contrast, similar oxidation of three 2-halo-4,6-di(*tert*-butyl)phenols (**704, 705** and **706**) mainly afforded a dibenzofuran derivative **707** (81%), a mixture of **708** and **709** (23 and 53%, respectively) and a diaryl ether **710** (63%), respectively (Scheme 138)[245]. Clearly, iodine and bromine substituents promote the diaryl formation, while chlorine and fluorine substituents prefer to produce diaryl ethers. These results seem to be in good agreement with the *ab initio* calculations[36].

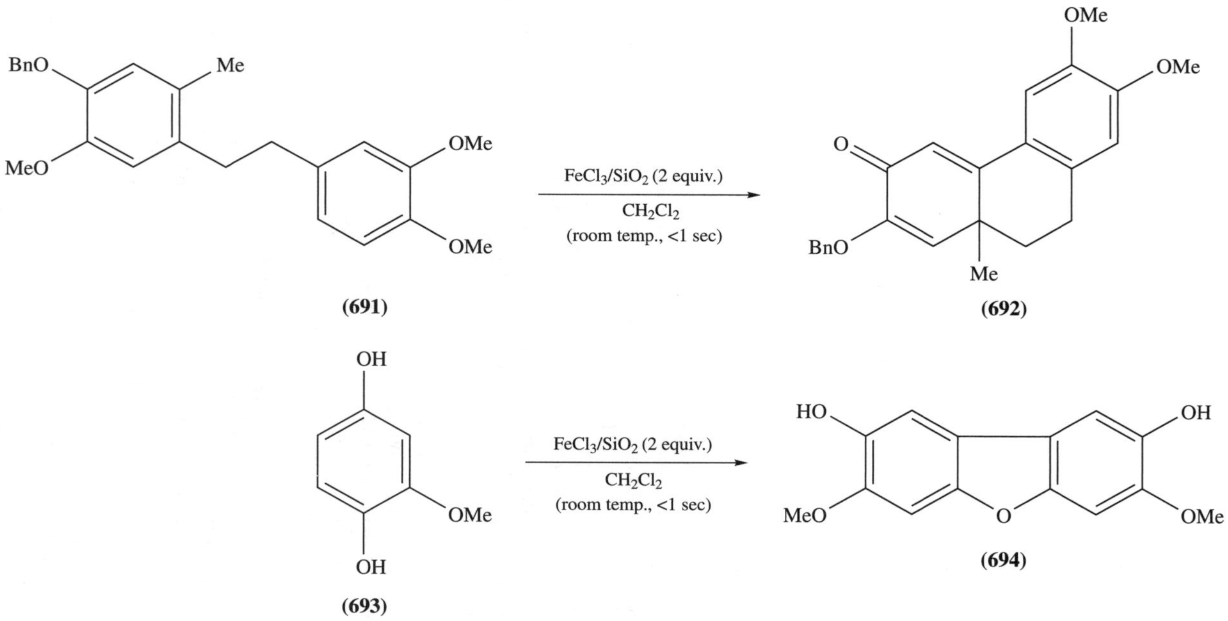

SCHEME 136. Oxidation of phenols with silica-bound ferric chloride

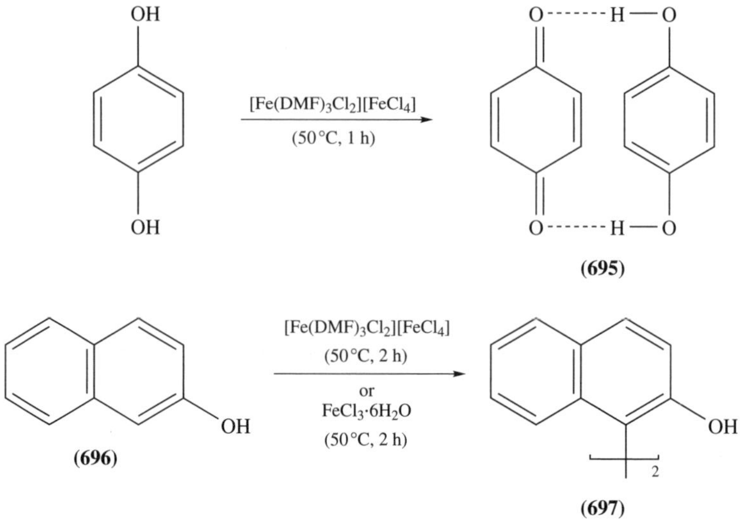

SCHEME 137. Oxidation of phenols with Fe(III) salts in the solid state

From the structural point of view, the acetylenic phenol **711** was oxidized with $K_3Fe(CN)_6$ or PbO_2 in benzene to undergo radical coupling at the β-position leading to the bis-quinobutadiene **712** (40%). Thermal isomerization of **712** further led to the first synthesis of diquinocyclobutene **713** (Scheme 139)[246].

Oxidation of amidine **714** bearing two sterically hindered phenol moieties was performed using $K_3Fe(CN)_6$ in benzene at ambient temperature to afford a dispirocyclohexadienone derivative **715** in 95% yield. During the reaction two stable phenoxy radicals (**716** and **717**) were detected by the ESR spectrum of the reaction mixture, wherein the latter underwent an intramolecular *para–para* coupling leading to **715** (Scheme 140)[247].

In relation to enzyme stereospecificity at the oxidative phenol coupling step, (S)-(+)-2-hydroxy-3,4,-trimethyl-5,6,7,8-tetrahydronaphthalene (**718**) was submitted to $K_3Fe(CN)_6$-promoted oxidation (22 °C, 2 h) to afford the corresponding optically active (S,S)-(+)-*trans*-dinaphthol **719** (62%) in a stereospecific manner (Scheme 141)[251]. This stereoselectivity and other results using racemic **718** (see Scheme 9) indicated that steric interactions play an important role in the stereochemical control during the intermolecular *ortho–ortho* coupling of two molecules **718** leading to the least hindered isomer **719**.

Lunarine (**26**), a constituent of *Lunaria biennis* M., is a member of neolignans (see Scheme 4). Biomimetic synthesis of tetrahydrolunarine (**720**) was performed by $K_3Fe(CN)_6$-promoted oxidation of methyl *p*-hydroxyphenylpropionate (**721**) as a key step. **721** was oxidized with $K_3Fe(CN)_6$ in aq. Na_2CO_3 (0 °C, 3.5 h) to give in 14% yield an *ortho–para* coupled product **722**, which was further converted to the target molecule **720** (Scheme 142)[252].

As already shown in Scheme 101, an oxidative cross-coupling of two different phenols takes place, but the yield is relatively low. In order for satisfactory cross-coupling to occur it is essential that the phenoxy radicals will be generated to a comparable extent from each of the substrates. Based on biogenetic consideration of benzylisoquinoline alkaloids, extensive studies on oxidative cross-coupling of two different phenols have been undertaken

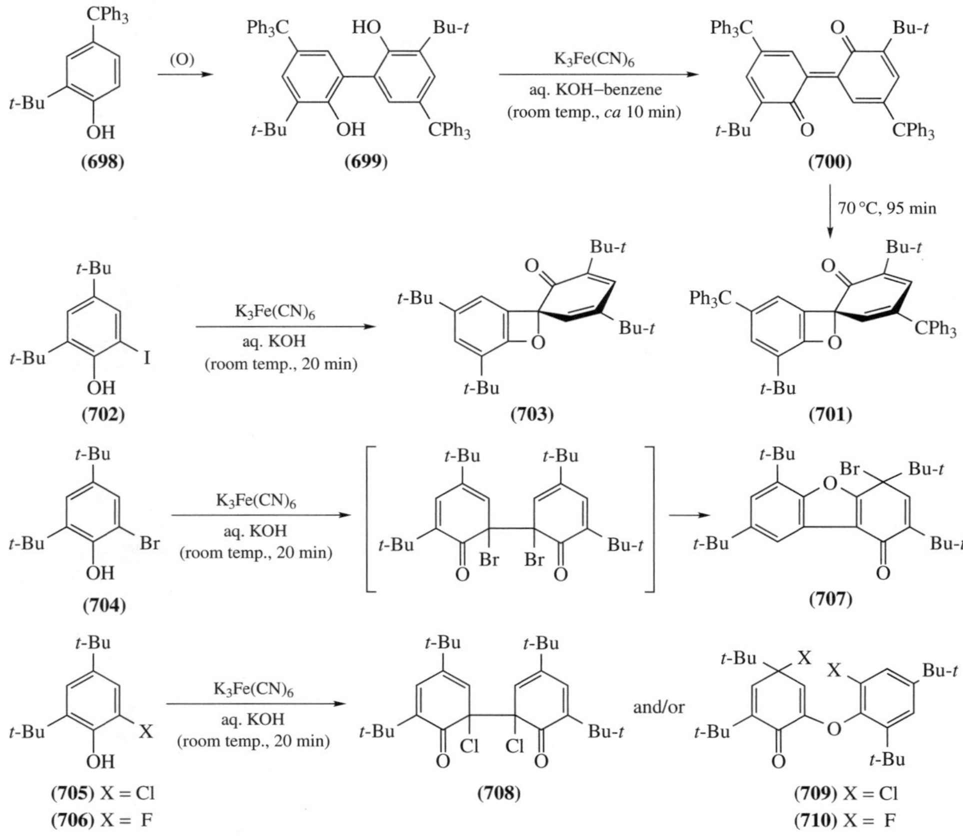

SCHEME 138. Oxidation of some hindered phenols with K$_3$Fe(CN)$_6$

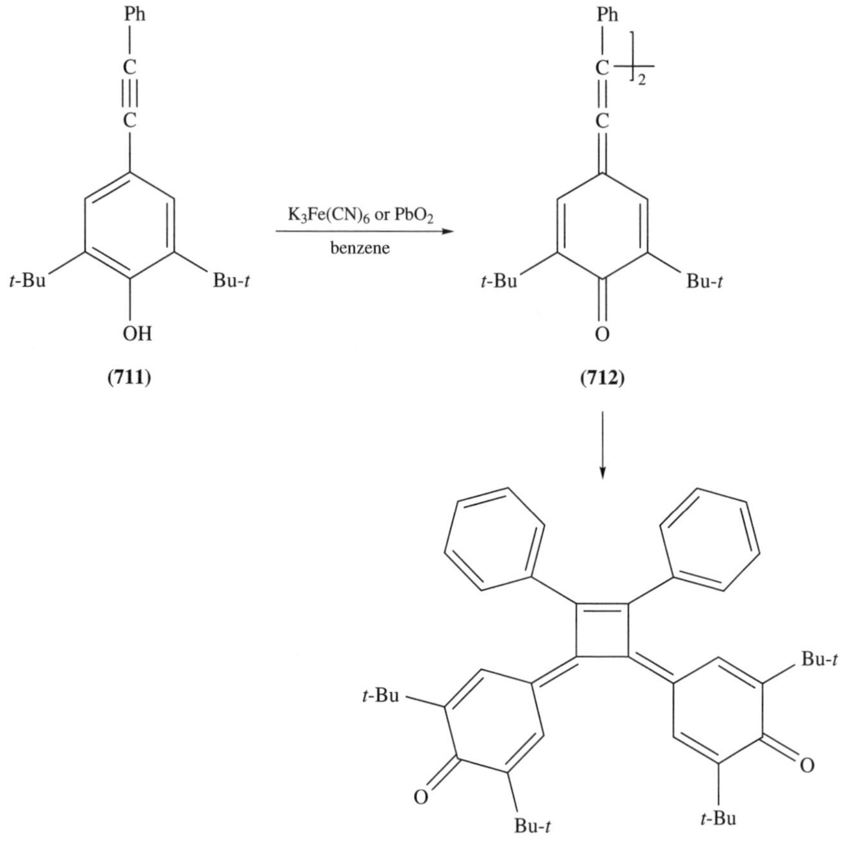

SCHEME 139. Oxidation of 4-acetylenic 2,6-di(*tert*-butyl)phenol with $K_3Fe(CN)_6$ or PbO_2

by Bird and coworkers. A mixture of 2-(*p*-hydroxyphenyl)ethanol (**723**) and 2-naphthol (**696**) was treated with $K_3Fe(CN)_6$ in aq. Na_2CO_3 (0–10 °C, 4 h) to afford the desired *ortho–para* coupled product **725** (15%) through an intermediate **724** (Scheme 143)[253].

From the viewpoint of organic synthesis, oxidation of a variety of *p*-substituted phenols, bearing three- or four-carbon chains terminated by enolic or enolizable groups, at alkaline pH using $K_3Fe(CN)_6$ or K_2IrCl_6 leads to the corresponding spiro cyclization products. For example, phenolic indandione **726** was subjected to $K_3Fe(CN)_6$-mediated oxidation in dilute KOH to afford a spirocyclic compound **727** (88%). When a simple cyclohexa-1,3-dione was used instead of the indandione, oxidative coupling of **728** with alkaline $K_3Fe(CN)_6$ proceeded poorly. In contrast, the use of the more powerful K_2IrCl_6 as an oxidant provided 43% yield of the spirocyclic triketone **729**. In the case of phenolic malononitrile **730**, only K_2IrCl_6 was effective for the oxidative cyclization leading to the corresponding spiro compound **731** (31%), as shown in Scheme 144. Herein, the initially generated enol radical reacts with the phenolate ring or with the phenoxy radical to initiate the cyclization process[254].

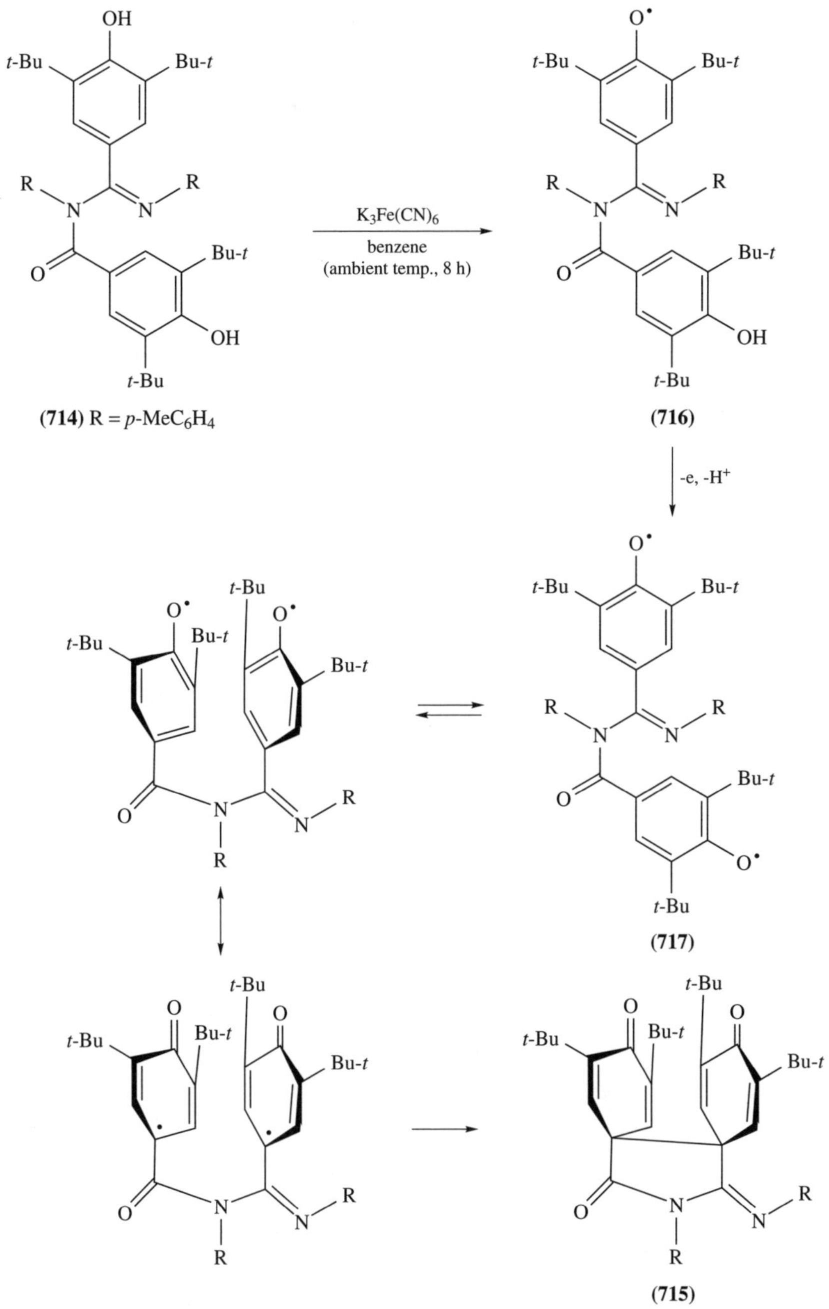

(714) R = p-MeC₆H₄

(716)

(717)

(715)

SCHEME 140. Oxidation of amidine bearing hindered phenol moieties with $K_3Fe(CN)_6$

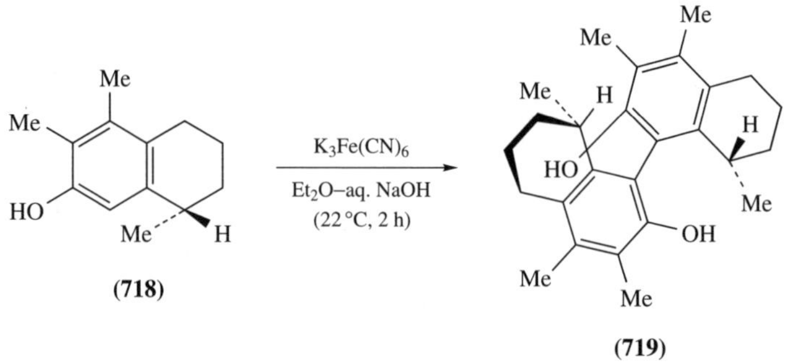

SCHEME 141. Stereoselective oxidative coupling of a chiral tetrahydronaphthol

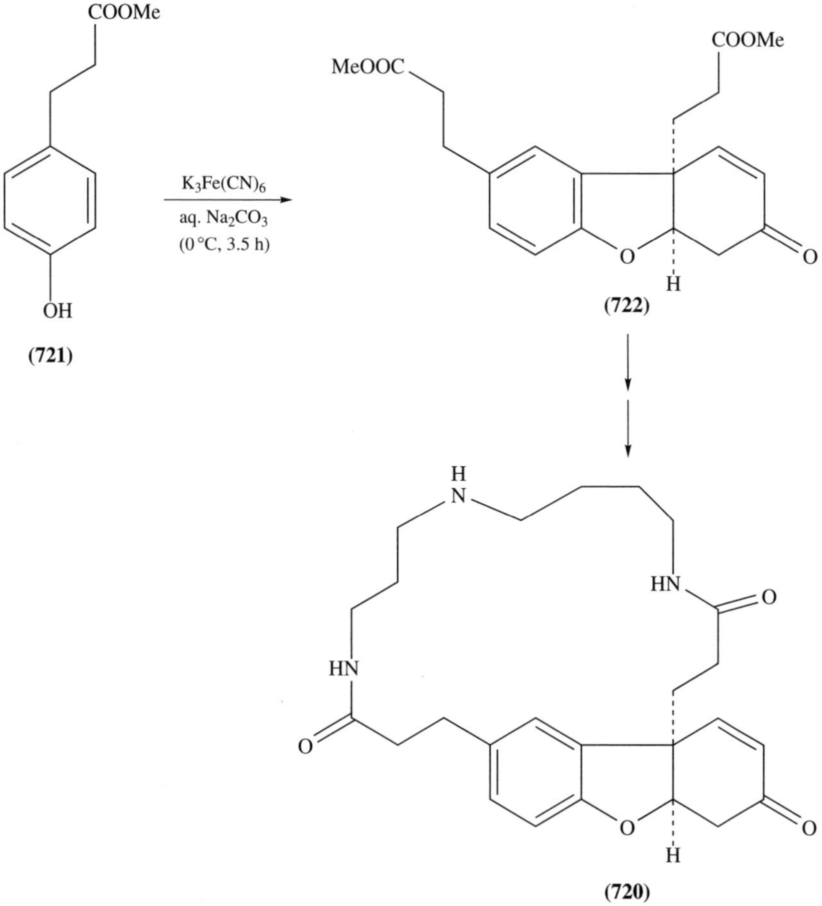

SCHEME 142. Biomimetic synthesis of tetrahydrolunarine

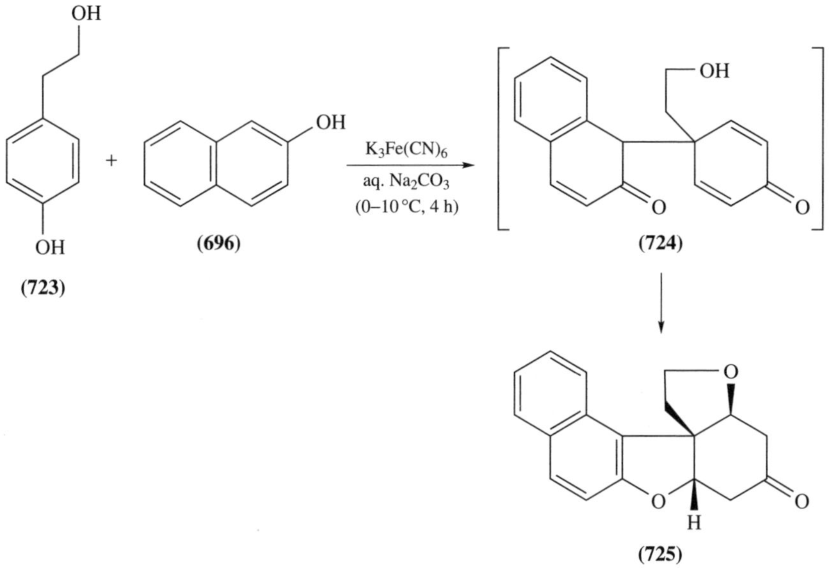

SCHEME 143. Oxidative cross-coupling of 2-naphthol and 2-(p-hydroxyphenyl)ethanol

Biomimetic syntheses of benzylisoquinoline alkaloids have been performed by intramolecular oxidative phenol-coupling reactions using a variety of oxidants. Of them, $K_3Fe(CN)_6$ has long been used for alkaloid syntheses[120-122,255]. The amine **732** bearing two phenol moieties was subjected to $K_3Fe(CN)_6$-mediated oxidation in a mixed solvent of CHCl$_3$ and aq. Na$_2$CO$_3$ to afford erysodienone **733** (35%) through *para–para* coupled dibenzoazonine **734** and then biphenoquinone (Scheme 145)[256]. Addition of benzyltriethylammonium chloride provided an increased yield (44%) of erysodienone[257]. This compound was further converted to dihydroerysodine (**735**)[256]. The $K_3Fe(CN)_6$-mediated oxidation of the corresponding amide **736** yielded a biphenyl derivative **737** (12%) (Scheme 145). In this case, no spirodienone like **733** was detected[257], while similar oxidation of **734** provided the corresponding spirodienone **733** in 80% yield.

A variety of B-homoerythrina alkaloids such as schelhammeridine (**738**) have been found in *Schelhammera* and *Cephalotaxus* plants. When the amide **739** was treated with $K_3Fe(CN)_6$ in a mixed solvent of aq. NaHCO$_3$ and CHCl$_3$, it underwent intramolecular radical coupling to afford the corresponding homoerysodienone **740** (68%)[258]. This compound was readily converted to a dibenzoazonine derivative **741**, which was treated again with the same oxidant to give a schelhammeridine-type compound **742** (61%) (Scheme 146)[259]. Synthetic studies on some interesting alkaloids have also been made using $K_3Fe(CN)_6$[260].

Flavonoides with a variety of biological activities constitute a large group in nature. Related to these natural products, the reaction of a chalcone derivative **743** with 3,5-dimethoxyphenolate (**744**) was effected with $K_3Fe(CN)_6$ in aq. NaOH to afford a diastereomeric mixture of 2-substituted 4,6-dimethoxybenzo[*b*]furan-3(2*H*)-ones (**745**) (26%) and 4'-hydroxy-4,6-dimethoxyaurone (**746**) (23%). A radical species **747** is a plausible intermediate (Scheme 147)[261].

As compared with Fe(III) oxidants, other metal compounds are scarcely used for phenolic oxidation. Both iron and ruthenium are members of the same group of the Periodic

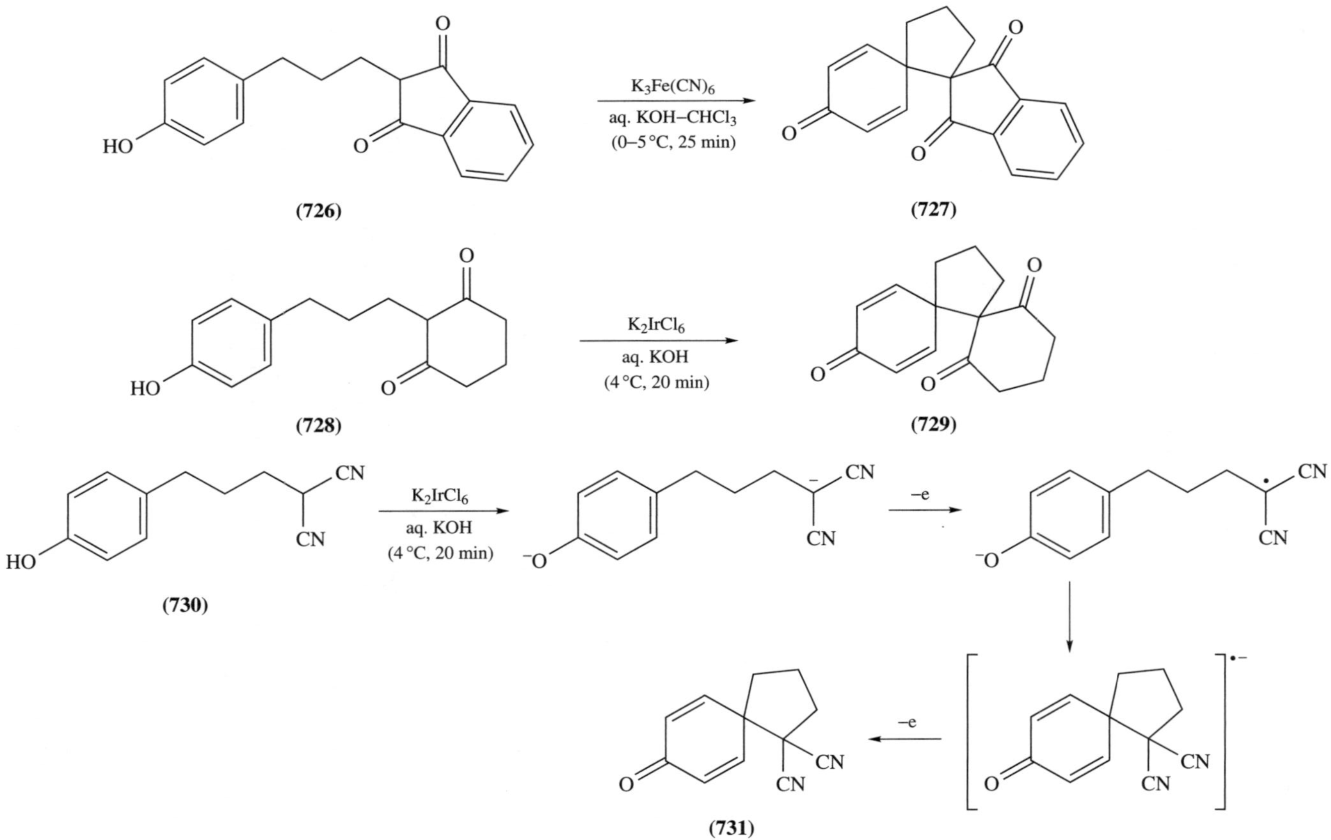

SCHEME 144. Intramolecular radical cyclization of phenolic enolates

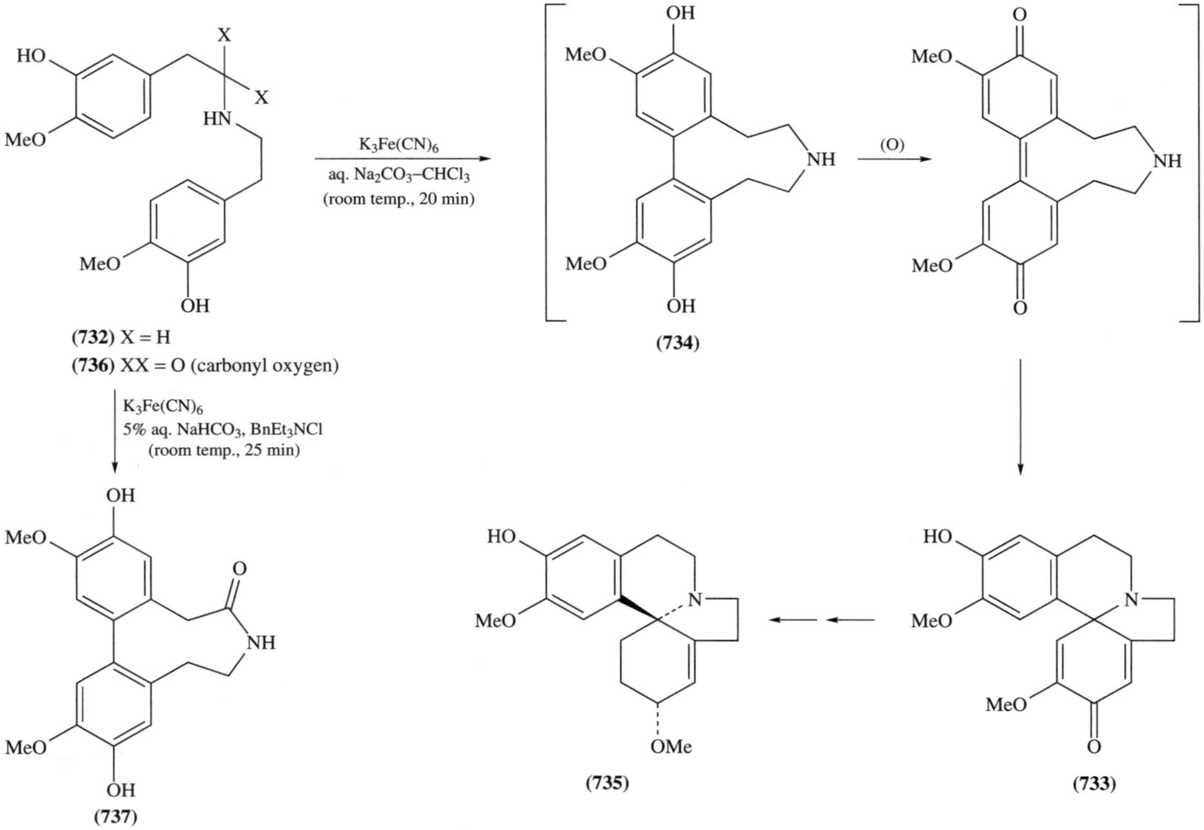

SCHEME 145. Intramolecular oxidative phenol-coupling reactions with $K_3Fe(CN)_6$

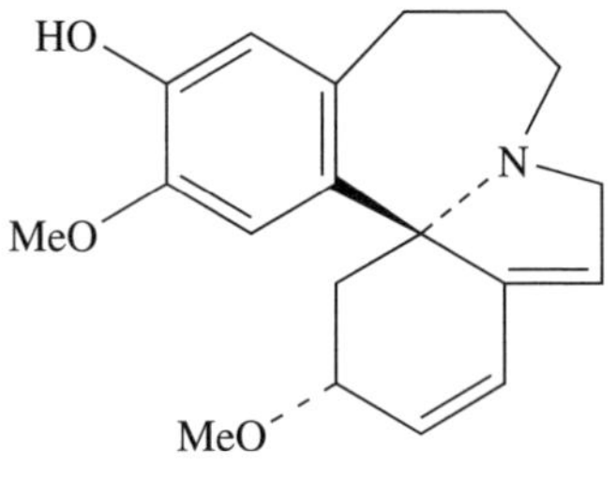
HO
MeO
N
MeO

(738)

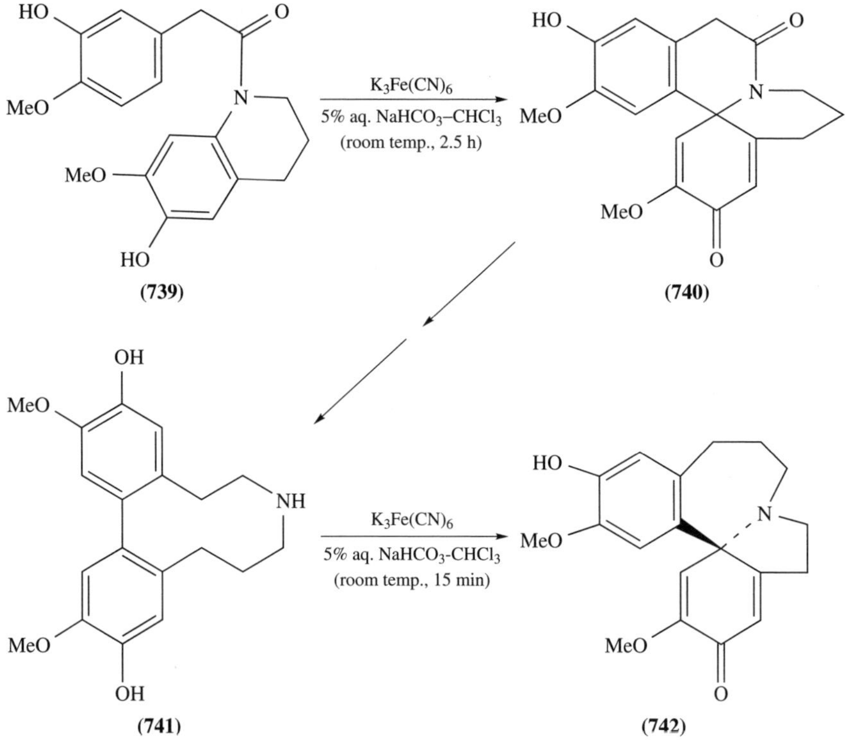
HO
MeO
O
N
MeO
HO
(739)
$K_3Fe(CN)_6$
5% aq. $NaHCO_3$–$CHCl_3$
(room temp., 2.5 h)
HO
O
N
MeO
MeO
O
(740)
OH
MeO
NH
MeO
OH
(741)
$K_3Fe(CN)_6$
5% aq. $NaHCO_3$-$CHCl_3$
(room temp., 15 min)
HO
N
MeO
MeO
O
(742)

SCHEME 146. Biomimetic synthesis of a B-homoerythrina alkaloid

Table. Their reactions have been shown to proceed via radical intermediates[262,263]. Ruthenium dioxide (RuO_2) in fluoro acid medium was effective for intramolecular oxidative phenol coupling reactions leading to the formation of steganacin-type neolignans[264]. In fact, both prestegane A and B (**678** and **748**) were treated with $RuO_2{\cdot}2H_2O$ (1.5 equiv.) in CH_2Cl_2 containing TFA–TFAA and $BF_3{\cdot}Et_2O$ to afford the corresponding isosteganacins (**679** and **749**) in 82 and 80% yields, respectively (Scheme 148).

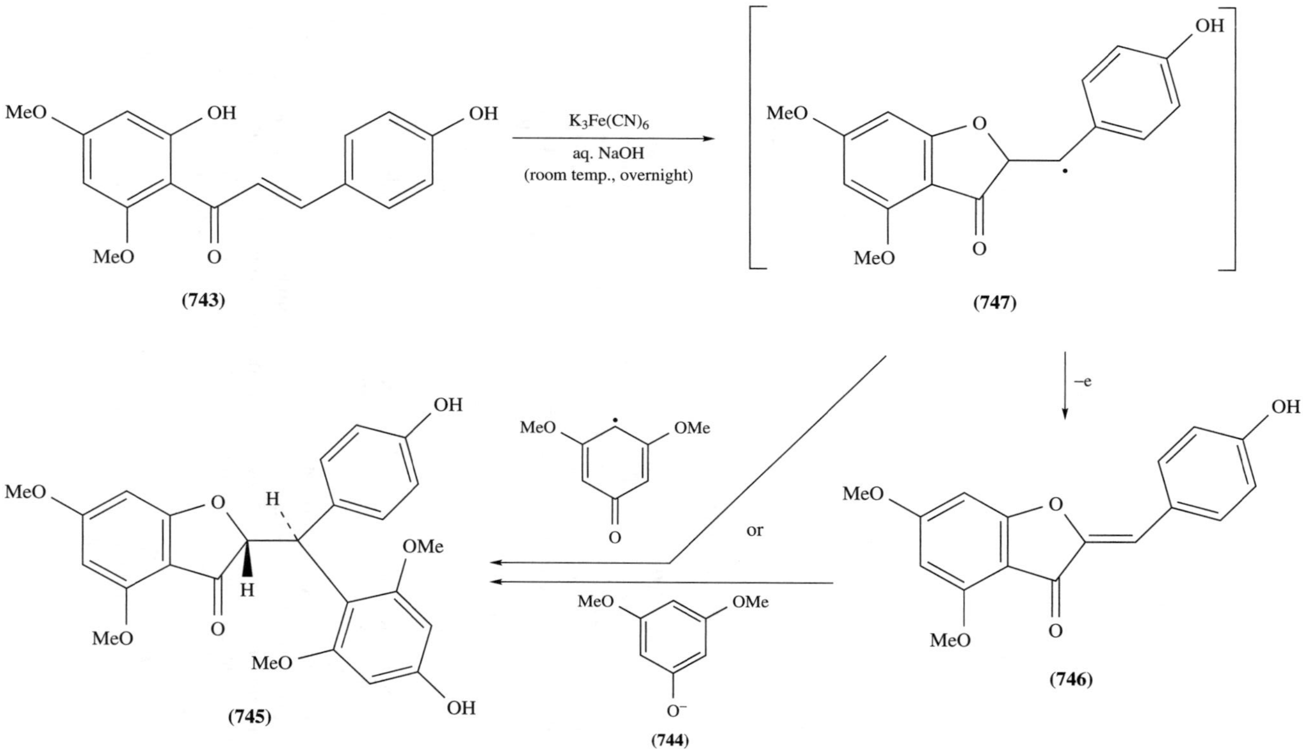

SCHEME 147. Reaction of a chalcone with 3,5-dimethoxyphenol in aq. NaOH–K$_3$Fe(CN)$_6$.

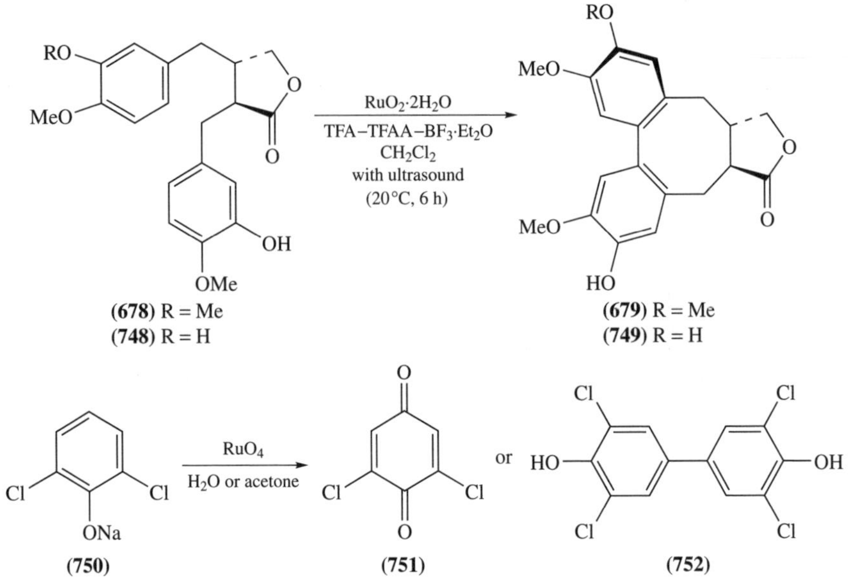

SCHEME 148. Phenolic oxidation with ruthenium dioxide and ruthenium tetroxide

Ruthenium tetroxide (RuO$_4$) is also utilized for phenolic oxidation. Sodium 2,6-dichlorophenoxide (**750**) was oxidized with RuO$_4$ in H$_2$O to afford 2,6-dichloro-*p*-benzoquinone (**751**) (60%), while the use of acetone as a solvent provided the corresponding biphenol **752** as the only isolatable product (20%)[265] (Scheme 148).

Cobalt(III) acetate is an effective reagent for phenolic oxidation. On oxidation with Co(OAc)$_3$ in AcOH, 2,6-disubstituted phenols (**753**: R = Me, *i*-Pr, *t*-Bu, MeO) were converted into the corresponding diphenoquinones (**754**: R = *i*-Pr, *t*-Bu, MeO) in 91–97% yields. In the case of 2,6-dimethylphenol (**753**: R = Me), 2,6-dimethyl-*p*-benzoquinone (**223**) was obtained in 23% yield together with the diphenoquinone (**754**: R = Me) (75%) (Scheme 149)[266].

Oxidation of 2,6-disubstituted 4-methylphenols (**221** and **69**) with Co(OAc)$_3$ in AcOH afforded the corresponding benzyl acetates **755** (49%) and **285** (73%) and benzaldehydes **222** (27%) and **756** (7%), respectively. These results are remarkably different from those of electrochemical and NaIO$_4$-promoted oxidations.

Nickel dioxide (NiO$_2$) is a one-electron oxidant similar to Co(III) acetate. 2,6-Di(*tert*-butyl)phenol underwent NiO$_2$-promoted oxidation in benzene (room temp., 5 h) to afford a quantitative yield of the dibenzoquinone **754** (R = *t*-Bu). In the case of 4-methyl-2,6-di(*tert*-butyl)phenol (**69**), an extended diquinone **757** was produced in 31% yield[267] (Scheme 149).

Oxidation of *p*-cresol (**27**) with NiO$_2$ in benzene was performed under the same conditions as above to afford nearly quantitatively polymeric products together with Pummer's ketone (**28**) (1.7%) and trace amounts of a dimer and a trimer (**758** and **759**) (Scheme 150)[268].

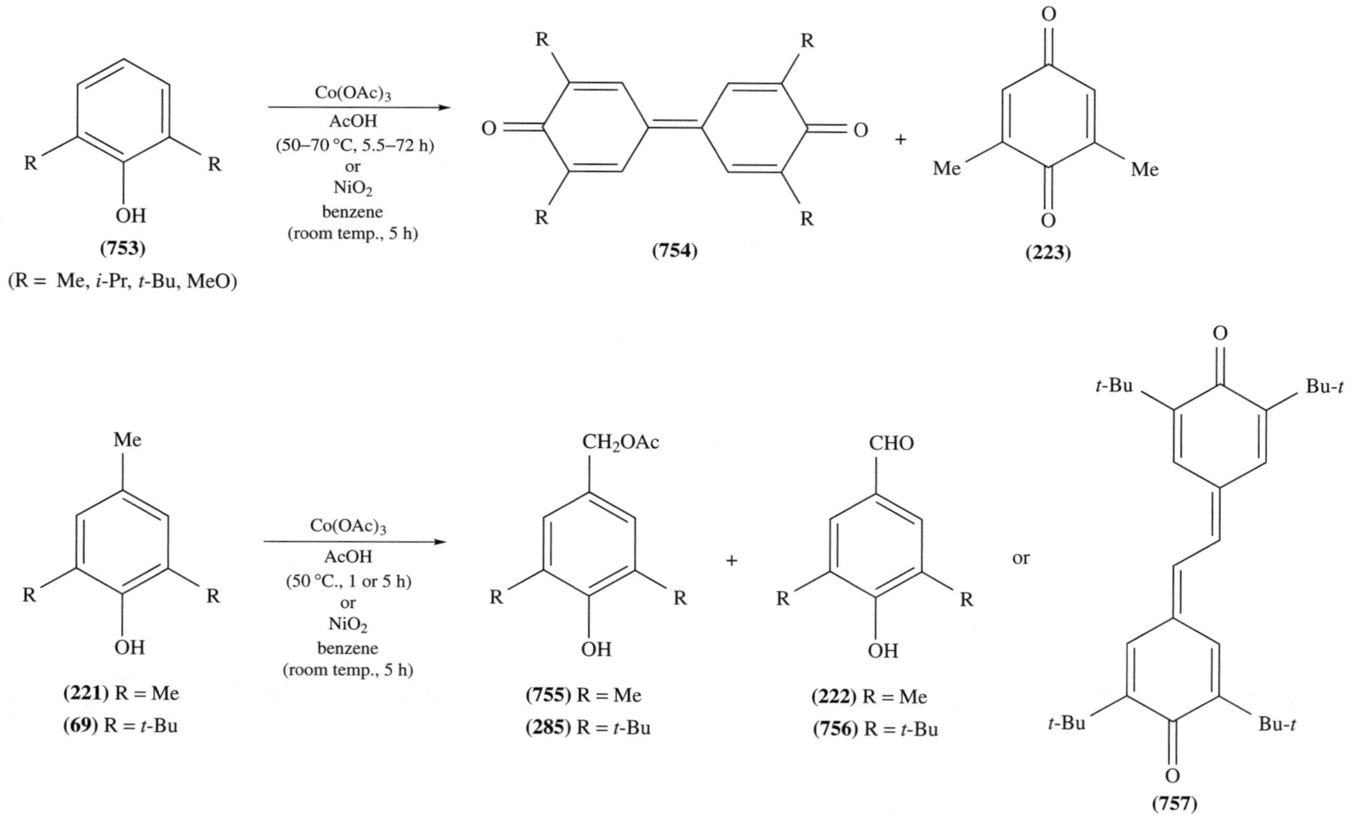

SCHEME 149. Phenolic oxidation with cobalt(III) acetate and nickel dioxide

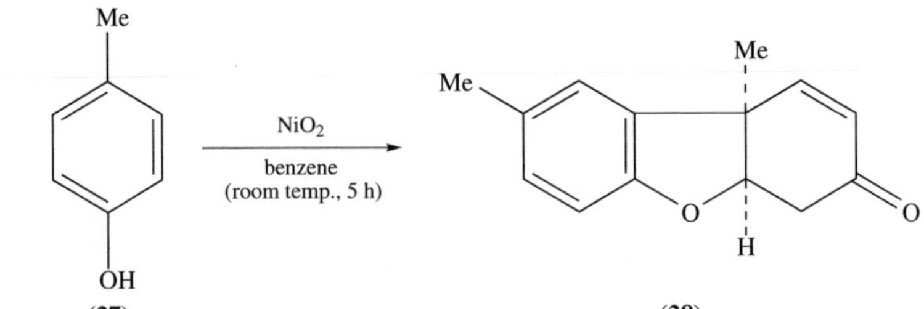

(27) **(28)**

+

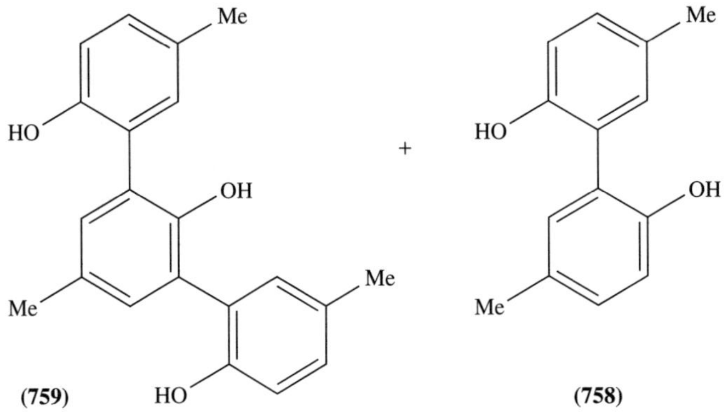

(759) **(758)**

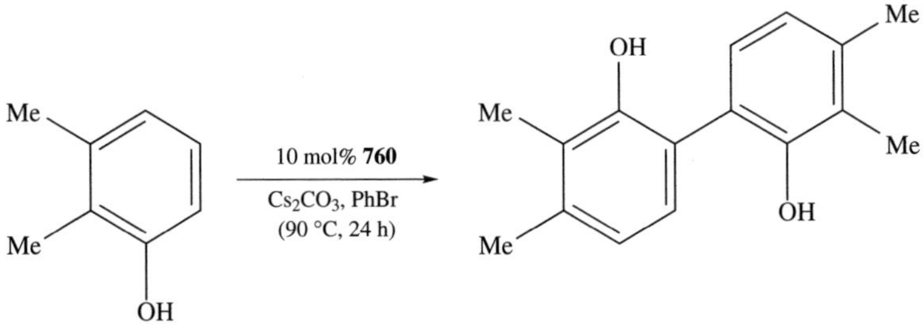

(761) **(762)**

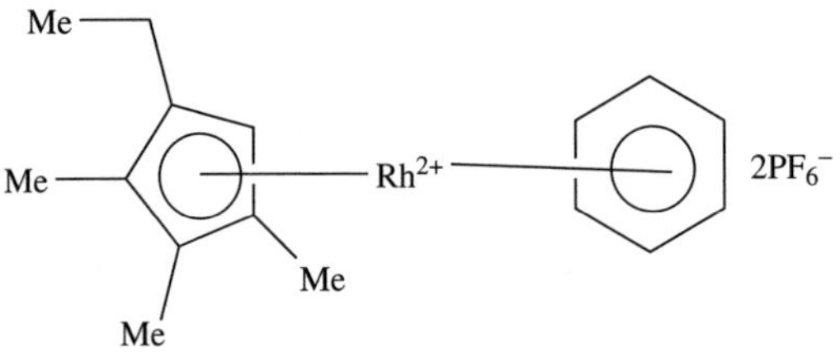

(760)

SCHEME 150. Phenolic oxidation with nickel dioxide or catalytic rhodium complex

17. Oxidation of phenols

Oxidative coupling reaction of *p*-cresol (**27**) was effected with rhodium(III) complex (**760**) and Cs_2CO_3 in bromobenzene (90 °C, 24 h) to give selectively 2,2′-dihydroxy-5,5′-dimethylbiphenyl (**758**) (51–67%). Oxidation of 2,3-dimethylphenol (**761**) also provided the corresponding biphenyl **762** (59%)[269] (Scheme 150).

D. Oxidation with Copper and Silver Compounds

As already described in Section II.B, a combination of cuprous chloride (CuCl) and nitrogen-containing ligands is generally used for phenolic oxidation under oxygen atmosphere. In the absence of these ligands CuCl is also effective for phenolic oxidation. When treated with CuCl in dry MeCN containing *n*-hexanol and $CaSO_4$ under oxygen atmosphere, methyl 2,5-dihydroxybenzoate (**763**) was converted into 3-alkoxylated quinone **764** (88%), as shown in Scheme 151[270]. Herein, the reaction of CuCl with dioxygen

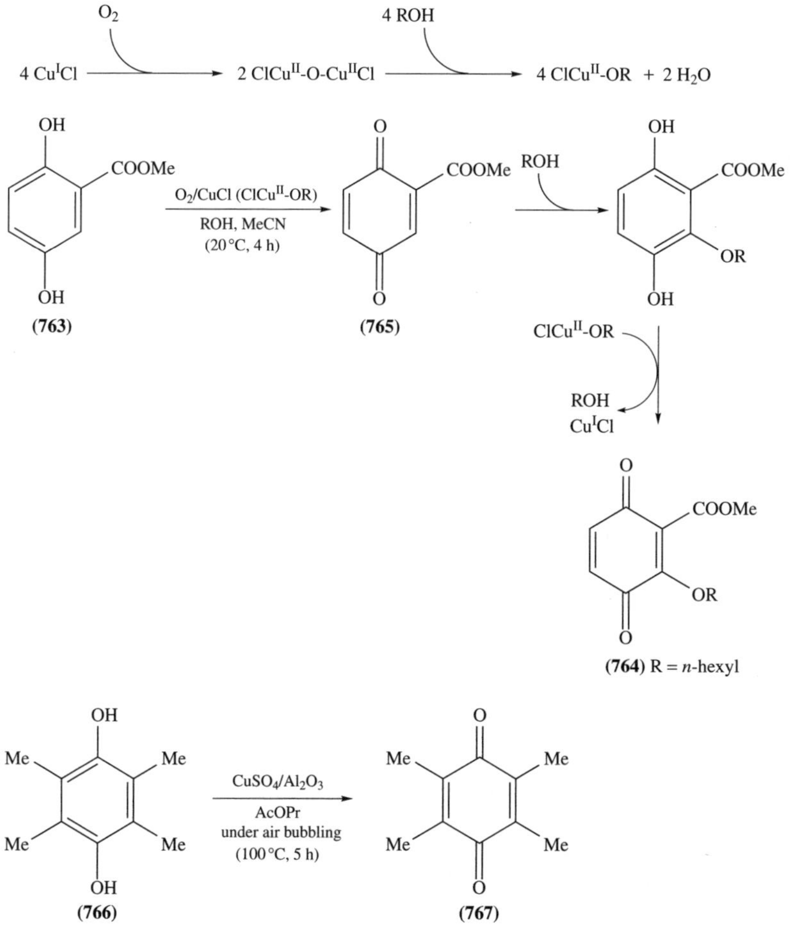

$$4\ Cu^ICl \xrightarrow{O_2} 2\ ClCu^{II}\text{-}O\text{-}Cu^{II}Cl \xrightarrow{4\ ROH} 4\ ClCu^{II}\text{-}OR + 2\ H_2O$$

SCHEME 151. Phenolic oxidation with O_2/CuCl and $CuSO_4/Al_2O_3$

may generate a reactive species $(ClCu^{II})_2O$, which reacts with ROH to yield an oxidant $ClCu^{II}-OR$. This compound oxidizes **763** to the quinone **765**, which undergoes nucleophilic attack by ROH at the C3-position, followed by further oxidation to yield the quinone **764**.

p-Hydroquinones such as 2,3,4,5-tetramethyl-*p*-hydroquinone (**766**) were oxidized efficiently to 1,4-benzoquinones such as **767** (92–98%) under air bubbling with catalytic amounts of supported catalyst $CuSO_4/Al_2O_3$ (Scheme 151). In the case of 2,6-di(*tert*-butyl)phenol, 3,3′,5,5′-tetra(*tert*-butyl)-4,4′-diphenoquinone was produced in 94% yield[271].

In the biosynthesis of benzylisoquinoline alkaloids, a plausible N-oxide intermediate is suggested. Thus, reticuline *N*-oxide (**768**) was treated with CuCl in MeOH in the absence of dioxygen and then with $NaHSO_3$ to afford corytuberine (**769**) (61%), while the use of excess $FeSO_4$ provided corexamine (**770**) and scoulerine (**771**) in 42 and 23% yields, respectively (Scheme 152)[272].

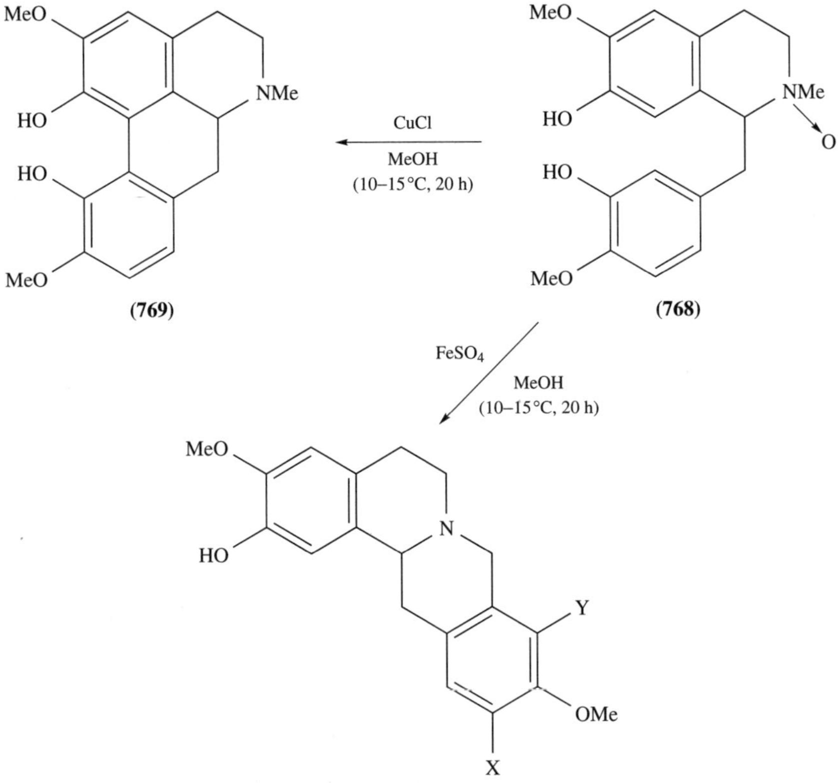

(**770**) X = OH, Y = H
(**771**) X = H, Y = OH

SCHEME 152. Reaction of reticuline N-oxide with CuCl or $FeSO_4$

In a synthetic study on antibiotics such as BE-10988 (**772**), 3-methoxycarbonylindole-4,7-quinone (**773**) was treated with benzhydrylamine and $Cu(OAc)_2$ (1 equiv.) in

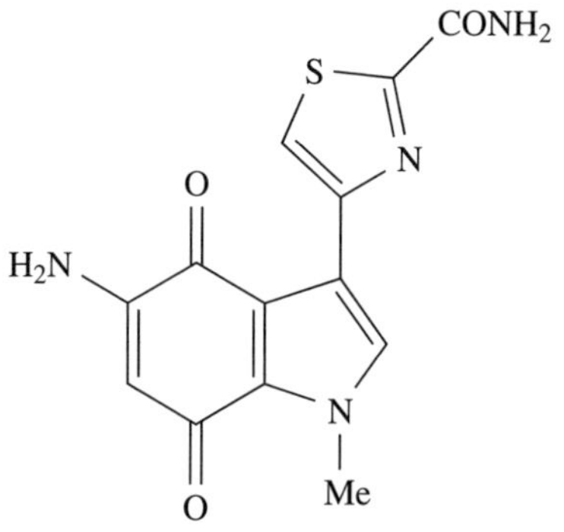

(772)

MeOH–CHCl$_3$ (10 °C, 40 min) to afford two adducts **774** and **775** in 82 and 10% yields, respectively (Scheme 153)[273].

Silver compounds have been used for oxidation of alcohols to aldehydes and ketones. Phenols, *p*-hydroquinones and catechols are also oxidized with Ag$_2$CO$_3$ or Ag$_2$O under mild conditions to afford the corresponding *p*- and *o*-benzoquinones in almost quantitative yields, respectively. Oxidation of 2,6-dimethylphenol (**206**) with Ag$_2$CO$_3$/Celite in benzene (reflux, 30 min) afforded a 98% yield of 3,3′,5,5′-tetramethyldiphenoquinone (**207**). Similar oxidation of 2,4,6-trimethylphenol (**221**) provided the corresponding stilbenequinone (**776**) in 93% yield. Furthermore, chemical transformation of the quinone **776** to a stilbene **777** was readily carried out and *vice versa* (Scheme 154)[274]. An excellent review on silver carbonate on Celite oxidations has appeared[275]. Oxidative phenol-coupling reactions have also been performed using Ag(I)–gelatin complex[276].

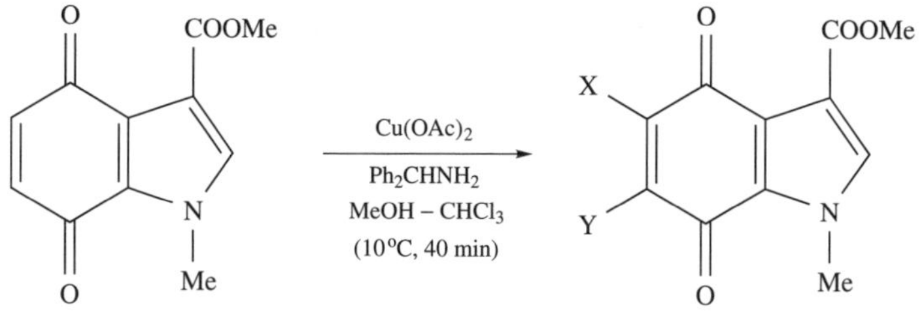

(773) → Cu(OAc)$_2$, Ph$_2$CHNH$_2$, MeOH – CHCl$_3$ (10°C, 40 min)

(774) X = NHCHPh$_2$, Y = H
(775) X = H, Y = NHCHPh$_2$

SCHEME 153. Reaction of indole-4,7-quinone with benzhydrylamine and cupric acetate

As already shown in Schemes 126 and 129, oxidation of phenols with manganese compounds provides diaryl ethers, dimers, trimers and polymers. Similarly, oxidation of phenols with Ag$_2$O also affords diaryl ethers[277,278]. However, one of the most characteristic points is that phenolic oxidation using Ag$_2$O provides a synthetic method for quinomethanes.

4-Allyl-2,6-dimethoxyphenol (**139**) underwent rapid oxidation with Ag$_2$O in benzene or CHCl$_3$ (room temp., 6–9 min) to the extended *p*-quinone methide **778** in quantitative yield (Scheme 155)[279]. This compound is unstable, but isolatable in a pure state.

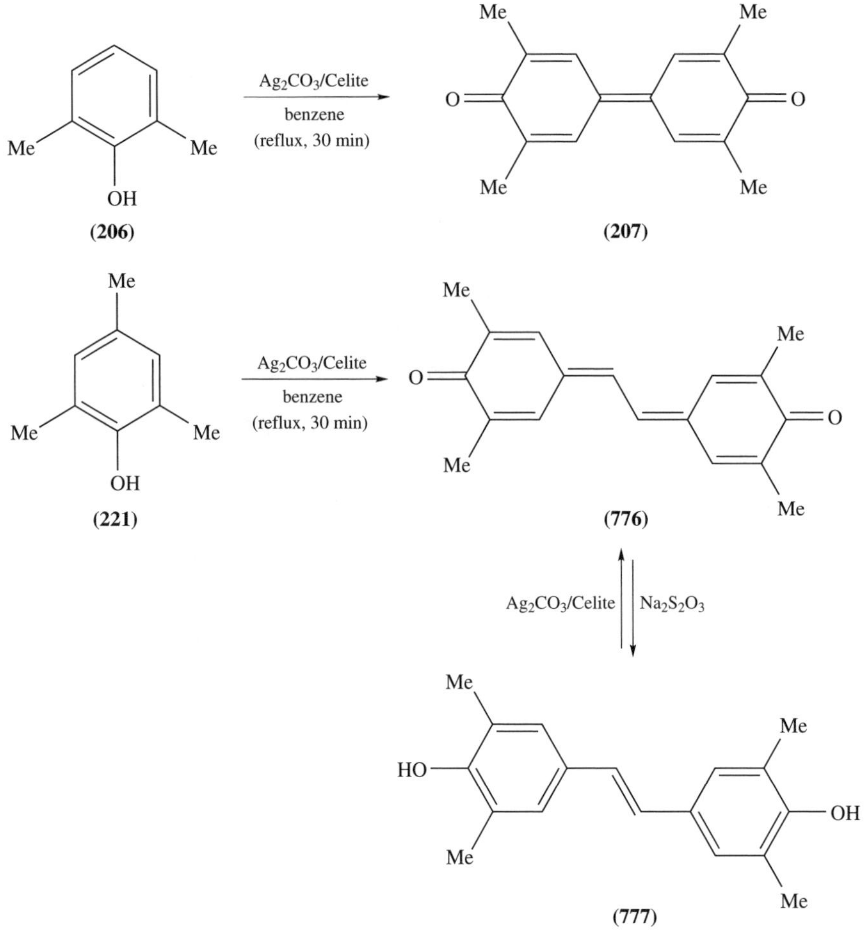

SCHEME 154. Oxidative phenol coupling reactions with silver carbonate/Celite

Subsequent acid-catalyzed methoxylation provided two regioisomers **779** and **780** (47 and 39%, respectively). The quinone methide **778** also underwent nucleophilic acetoxylation with AcOH–AcONa followed by LiAlH$_4$ reduction to afford sinapyl alcohol (**781**) in 80% overall yield.

Highly reactive quinone methide can be utilized as intermediates in organic synthesis. From the viewpoint of biomimetic synthesis, silybin (**782**) bearing a benzodioxane skeleton was synthesized in 44.5% yield, together with isosilybin (**784**) (33.5%), by Ag$_2$O-mediated oxidation of equimolar amounts of 2R,3R-dihydroquercetin (**783**) and coniferyl alcohol (**298**) in benzene–acetone. The p-quinone methide **785** must be generated as a reactive intermediate (Scheme 156)[280].

Biomimetic synthesis of model compounds for dibenzodioxocines occurring in wood lignins was carried out, as follows. On oxidation with Ag$_2$O in CH$_2$Cl$_2$ (room temp., 45 h), the reaction of dehydrodipropylguaiacol (**786**) with coniferyl alcohol (**298**) provided

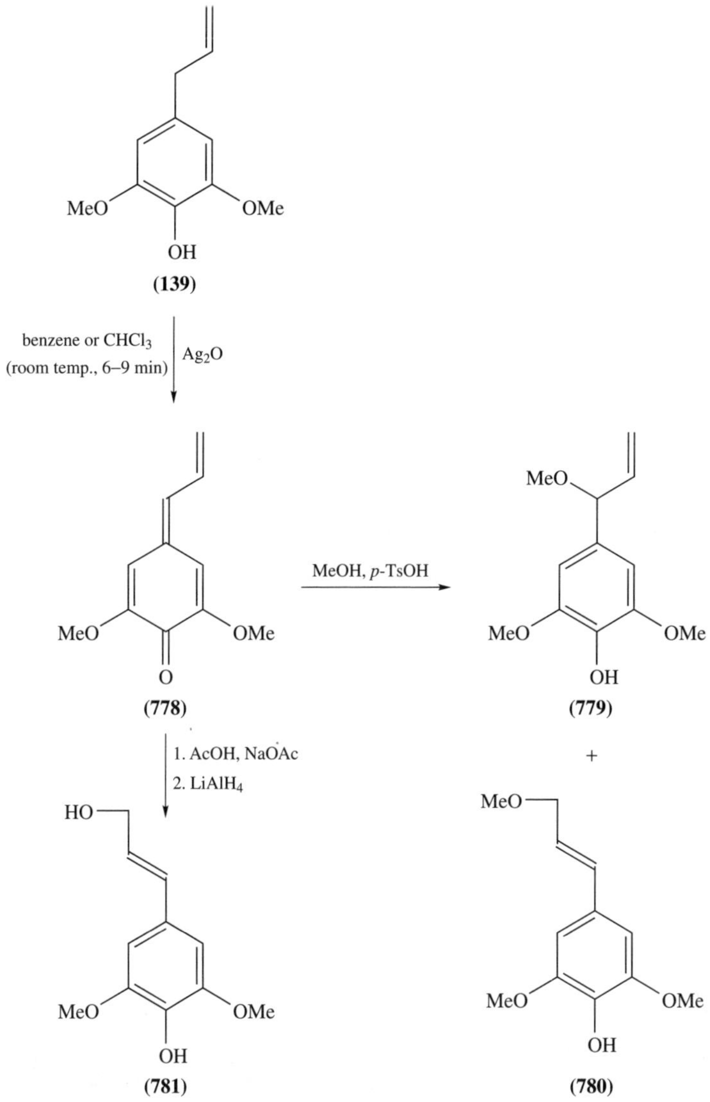

SCHEME 155. Oxidation of 4-allyl-2,6-dimethoxyphenol with silver oxide

two dibenzodioxocine derivatives (**787** and **788**) in 34 and 19% yields, respectively (Scheme 157)[281].

2,4-Di(*tert*-butyl)-6-[(4-methoxyphenyl)methyl]phenol (**789**) sterilizes female housefly and screwworm fly species, because microsomal oxidation of **789** may produce the corresponding reactive *o*-quinone methide (**790**). Thus, the phenol **789** was submitted to Ag$_2$O-promoted oxidation in MeOH–Me$_2$NH (reflux, 1 min) to afford an adduct **791** (60%) through the quinone methide intermediate **790** (Scheme 158)[282].

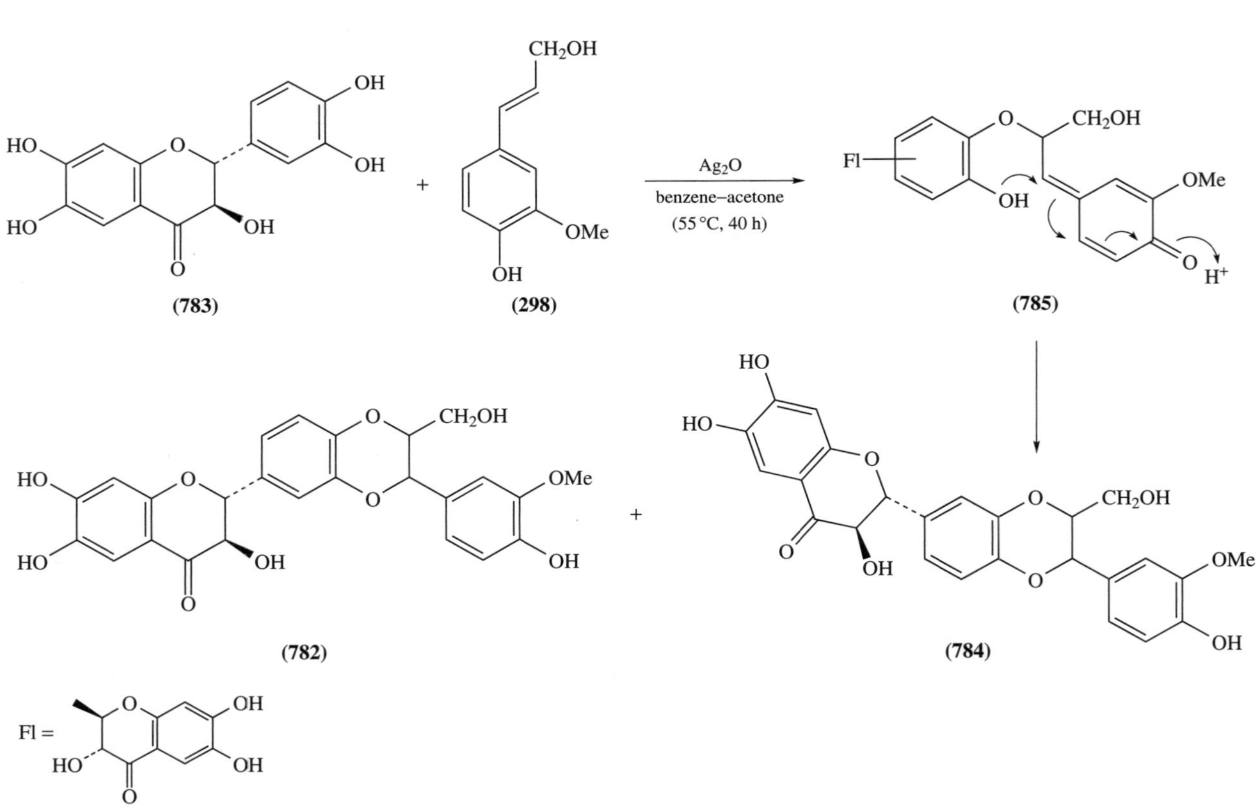

SCHEME 156. Biomimetic syntheses of silybin and isosilybin

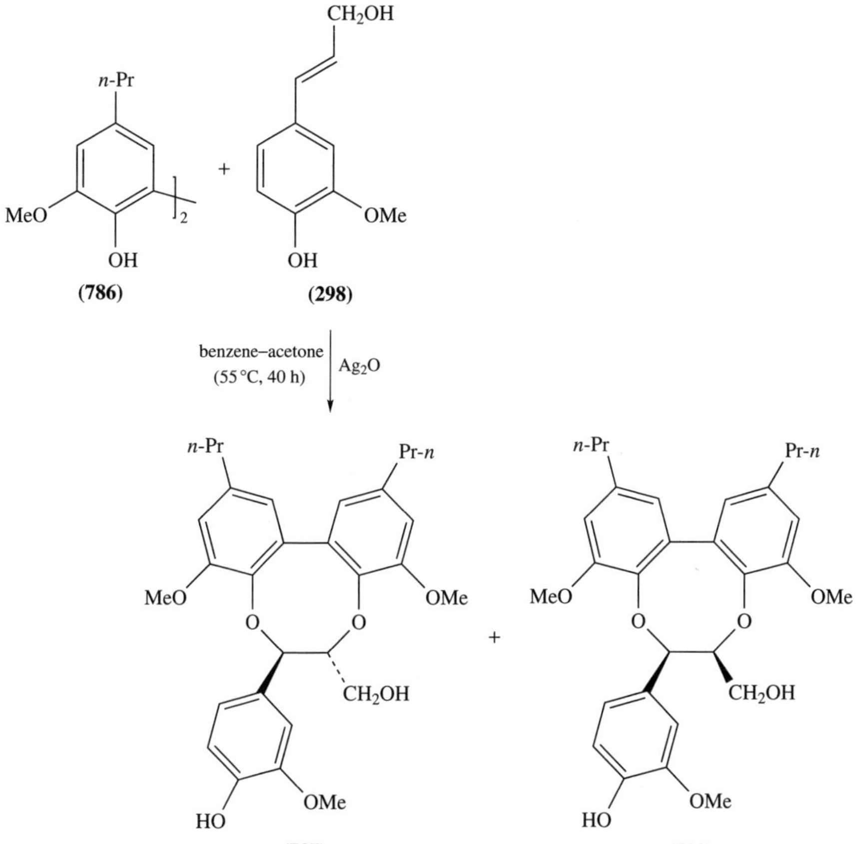

SCHEME 157. Biomimetic syntheses of dibenzodioxocine derivatives

Oxidation of 2-(3′-methyl-2′-butenyl)-4,5-(methylenedioxy)phenol (**792**) was performed using Ag_2O in CH_2Cl_2 to give the bright red o-quinone methide **793**, which on reflux in benzene subsequently underwent intramolecular cyclization leading to 2,2-dimethyl-6,7-(methylenedioxy)-2H-chromene (**794**) in 80% overall yield (Scheme 158)[283]. Carpanone has also been synthesized using Ag_2O as an oxidant instead of $PdCl_2$ and O_2–Co(II)(salen) complex (see Scheme 54).

E. Oxidation with Thallium, Lead and Bismuth Compounds

Of these three metal compounds, thallium compounds such as $Tl(NO_3)_3$ and $Tl(OCOCF_3)_3$ have been widely utilized in organic synthesis. Both Tl^{3+} and Pb^{4+} ions are isoelectronic and the former is a less powerful oxidant than the Pb(IV) ion. Oxidizing reactivities of Tl^{+3} salts vary with the anion associated with the metal, the solvent and other factors. On treatment with $Tl(NO_3)_3$ [TTN], phenols generally undergo two-electron oxidation forming phenoxonium ions which will be attacked by a variety of nucleophiles.

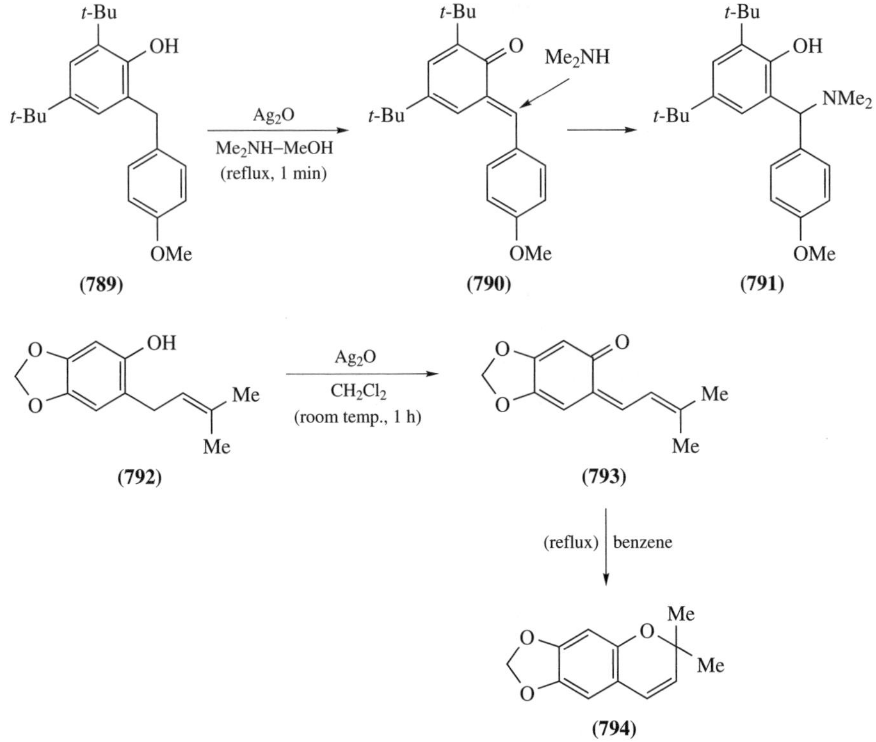

SCHEME 158. Syntheses and reactivities of *o*-quinomethanes

In contrast, phenolic oxidation with Tl(OCOCF$_3$)$_3$ [TTFA] often provides phenoxy radical cations, which undergo C−C or C−O couplings, leading to a variety of natural products. A series of pioneering works on phenolic oxidation using Tl^{3+} salts was carried out by McKillop and Taylor and they have written exellent reviews on the subject[284]. A recent review on the applications of Tl^{3+} salts in organic synthesis has also appeared and covered exhaustively the literature published between 1989 and 1998[285]. Therefore, some typical examples will first be shown in this section and details concerning new synthetic methods of diaryl ethers using TTN in MeOH will be then discussed.

Biomimetic syntheses of benzylisoquinoline alkaloids have been performed by applying oxidative phenol coupling reactions[284]. Morphinane-type alkaloids were synthesized from reticuline or its derivatives using enzyme and other oxidants such as PhI(OCOCF$_3$)$_2$ and VOF$_3$ (see Schemes 63, 94 and 121). Similarly, TTFA was utilized for morphine alkaloids synthesis[286]. However, some differences in the coupling mode are observed between TTFA and VOCl$_3$. For example, a diphenolic compound **795** was treated with TTFA in CH$_2$Cl$_2$ to afford a *para–ortho* coupled product **796** (15%), whereas the use of VOCl$_3$ as an oxidant provided selectively both *para–para* and *ortho–para* coupled products **797** and **798** in 54 and 46% yields, respectively (Scheme 159)[287]. Compound **796** was further converted to cepharamine (**799**). Other interesting benzylisoquinoline alkaloids have also been synthesized using TTFA[284].

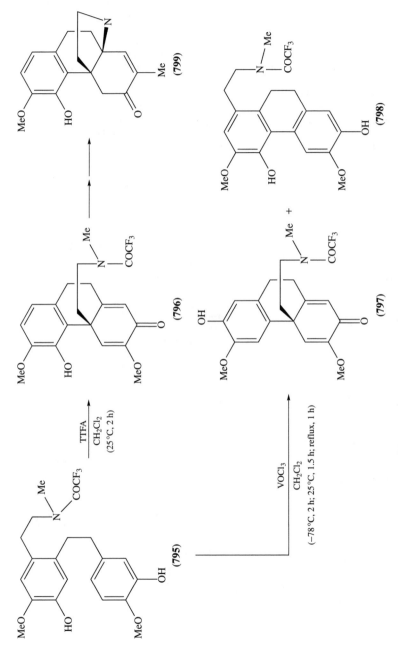

SCHEME 159. Oxidative phenol-coupling reactions with TTFA or VOCl$_3$

Similarly, nonphenolic oxidative coupling reactions with TTFA are effective for the synthesis of benzylisoquinoline alkaloids and neolignans[284]. TTFA-promoted oxidation of 3,4,5-trimethoxycinnamic acid (**800**) was carried out in TFA–CH$_2$Cl$_2$ containing BF$_3$·Et$_2$O at room temperature to afford 2,6-bis(3,4,5-trimethoxyphenyl)-3,7-dioxabicyclo[3.3.0]octane-4,8-dione (**801**) in 54% yield through the intermediate radical cation which presumably undergoes C—C coupling to the dimer **802** (Scheme 160)[288]. 3-(3,4,5-Trimethoxyphenyl)propionic acid (**803**) underwent rapid TTFA-mediated oxidation in TFA containing a catalytic amount of BF$_3$·Et$_2$O to give only the spirocyclohexadienone (**804**). The reaction mechanism is shown in Scheme 160[289].

The diaryl ester **805** was also submitted to TTFA-promoted oxidation in TFA to afford a 13-membered ring lactone **806** (65%) through a conjugated lactone intermediate **807**, as shown in Scheme 160, where on quenching with MeOH instead of H$_2$O the corresponding lactone (**808**) bearing a MeO group was obtained in 53% yield[290].

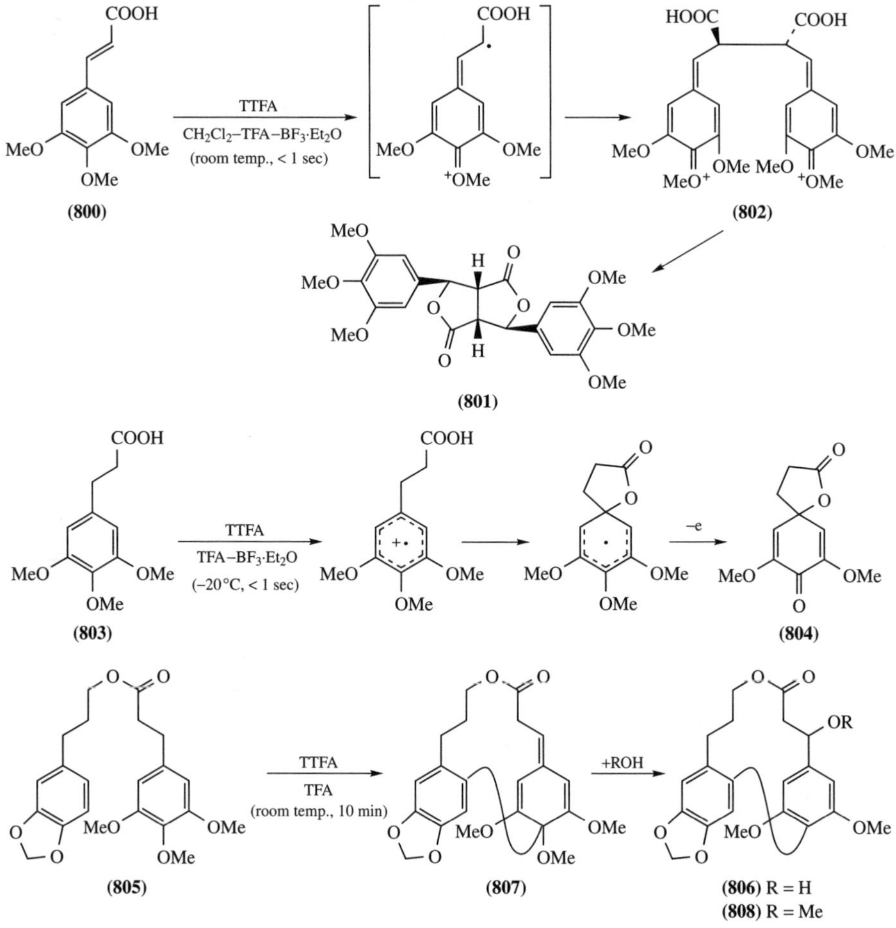

SCHEME 160. Nonphenolic oxidative coupling reactions using TTFA

17. Oxidation of phenols

Aerothionin (**809**), homoaerothionin (**810**) and aerophobin-1 (**811**), metabolites of sponges such as *Aplysia fistularis* and *Veronia thionia*, have a unique spiroisoxazoline framework. These metabolites must be biosynthesized in nature from a common phenylpyruvate oxime intermediate.

On treatment with TTFA in TFA (room temp., 4 h), methyl 2-hydroxyimino-3-(3′,5′-dibromo-2′-hydroxy-4′-methoxyphenyl)propionate (**812**) was converted into the desired spiro compound **813** (27%) together with two compounds (**814** and **815**) in 21 and 3% yields, respectively (Scheme 161)[291]. The compound **813** was converted successfully to the three target molecules.

As compared with TTFA, TTN-mediated oxidation of phenols proceeds by two-electron transfer. On TTN-mediated oxidation in MeOH, 2- and 4-methoxyphenols are converted into quinone monoketals usually in high yields (Scheme 162)[292]. As already demonstrated in Schemes 16, 17, 25 and 78–80, these cyclohexa-2,4- and 2,5-dienones have been utilized for natural products synthesis.

Similarly, TTN-mediated oxidation of 4-alkylphenols in MeOH provided 4-alkyl-4-methoxycyclohexa-2,5-dienones[292] (see Scheme 74). Thallium perchlorate [Tl(ClO$_4$)$_3$] was also applied to the oxidation of phenols **816** and **817** leading to cyclohexadienones **818** and **819**, respectively, each in 80% yield (Scheme 163)[293]. However, when 4-alkylphenols such as **27** and **820** were treated with Tl(ClO$_4$)$_3$–60% HClO$_4$ in CH$_2$Cl$_2$ they gave 2-alkyl-p-benzoquinones **290** and **821** in 70 and 66% yields, respectively (Scheme 163)[294].

Both TTN and TTA [Tl(OAc)$_3$] have been known to act as electrophiles toward olefinic double bonds, enolizable ketones and nitrogen-containing compounds to afford a variety of natural products or their synthons[285]. As already shown in Scheme 162, TTN was applied to phenolic oxidation by Yamamura and Nishiyama[31,32]. In particular, this method which consists of TTN oxidation in MeOH followed by Zn reduction in AcOH is effective for the construction of macrocyclic diaryl ethers from the corresponding open-ring precursors, which are required to possess two o,o'-dihalophenol moieties.

The first successful application of this method was used for the synthesis of bastadin-6 (**822**) having an inhibitory activity against inosine 5′-phosphate dehydrogenase. The appropriately protected substrate **823**, prepared from two tyramines and brominated phenylpyruvate oximes, was submitted to the TTN-mediated oxidation in MeOH to afford macrocyclic dienones **824** and **825**, in 20 and 11% yields, respectively (Scheme 164)[295]. Interestingly, the natural cyclization mode product **824** is preferred to its antipode **825** even in the 28-membered macrocyclic compounds. TTN-promoted oxidation of the benzyl ether (**826**) also provided the corresponding two dienones **827** and **828**[296]. They were further submitted to zinc reduction, followed by hydrogenolysis with Pd-black to give bastadin-6 (**822**) and its antipode **829**, respectively. In contrast, anodic oxidation of **823** in MeOH yielded acyclic cyclohexadienone **830** (10%) as the only isolatable product. Presumably, the TTN-mediated oxidation initially generates the bisphenoxide intermediate (**A**) which enables an intramolecular cyclization by bringing the two phenols to juxtaposition (Scheme 165). In the case of anodic oxidation, two phenols are oxidized independently to the corresponding dienones. Additionally, highly strained cyclization, as in the case of piperazinomycin (**831**)[297], can take place by passing through the intermediate **B** bearing an sp^3 carbon.

Isodityrosine natural products, such as piperazinomycin (**831**), OF 4949-I (**832**), K-13 (**833**), deoxybouvardin (**834**) and vancomycin (**835**), are known to possess interesting biological activities such as antifungus, enzyme-inhibitory, antitumor, antimicrobial and other activities (Chart 3). Structurally, the direction of the diaryl linkage shared by all of these natural products is classified into two types represented by **832** and **833**, respectively.

In order to obtain the desired macrocyclic diaryl ethers, the effect of halogen substituents is utilized to control the directions of intramolecular cyclization reactions mediated by TTN. As shown in Scheme 166, on TTN-mediated oxidation of a compound

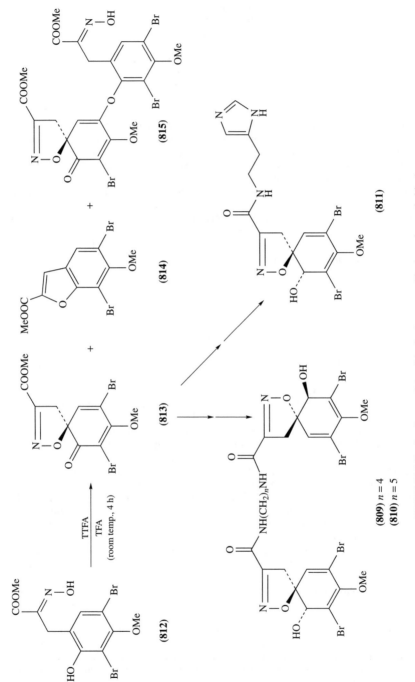

SCHEME 161. Syntheses of aerothionin, homoaerothionin and aerophobin-1

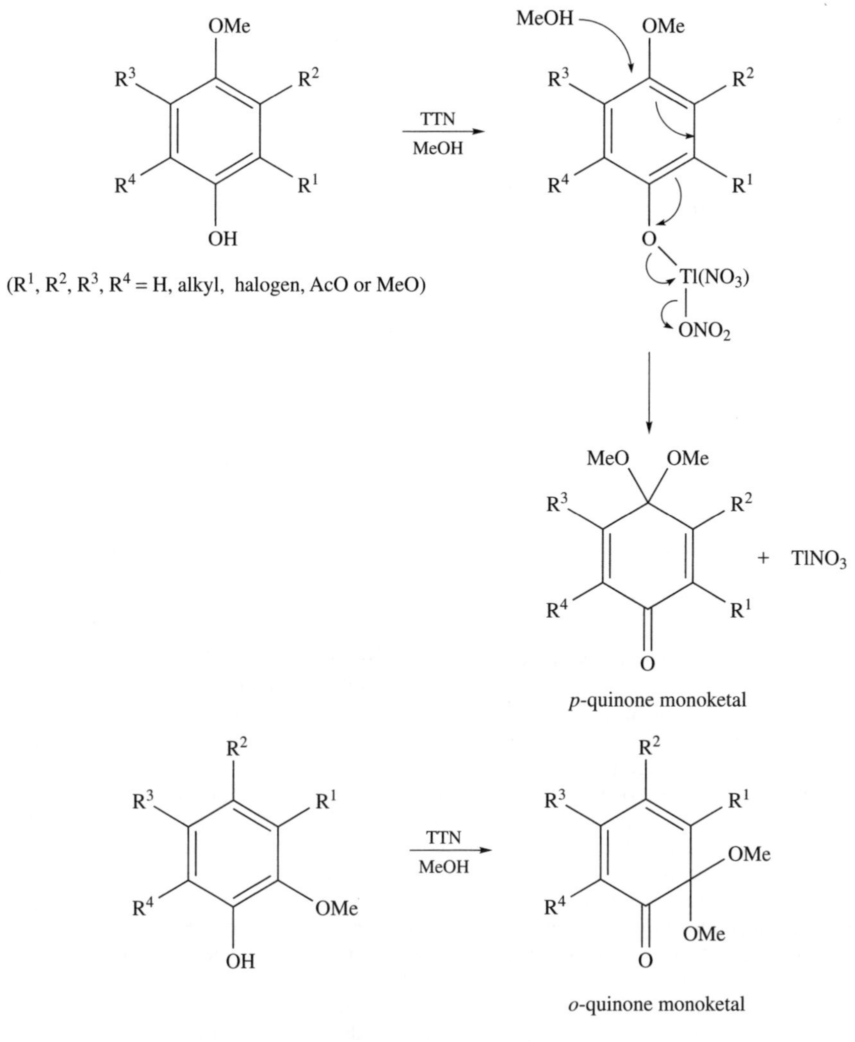

SCHEME 162. TTN oxidation of 2- and 4-methoxyphenols

bearing different halogen couples, the ether linkage is introduced at a halogen atom of weaker electronegativity. For example, TTN-mediated oxidation of compound **C** bearing Cl and I atoms affords macrocyclic diaryl ether **D** with two chlorine and one iodine atoms (Scheme 166).

The tripeptide **836**, prepared easily from L-tyrosine, was treated with TTN (3 equiv.) in MeOH (0 °C, then room temp., overnight) to give the corresponding cyclization product **837** (25%). This was further converted to OF 4949-I (**832**) (Scheme 167)[298].

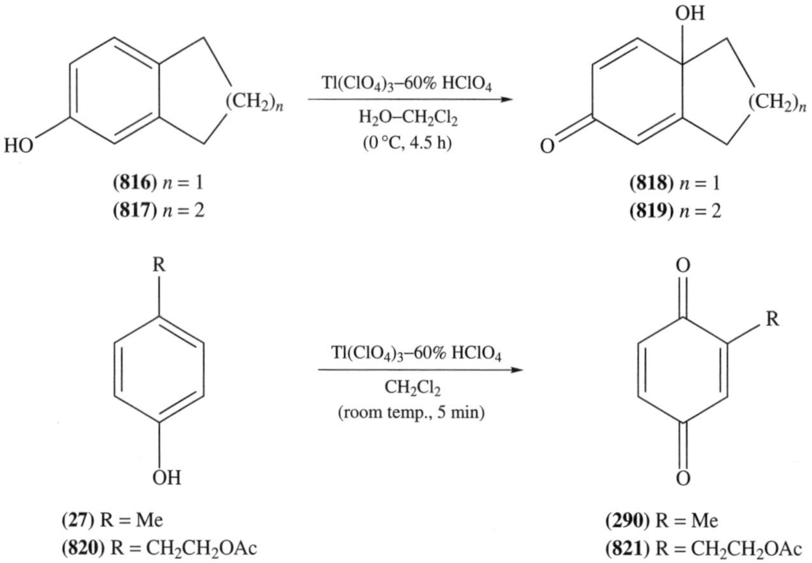

SCHEME 163. Oxidation of 4-alkylphenols with thallium perchlorate

Based on the same protocol as above, piperazinomycin (**831**)[297], K-13 (**833**)[299] and deoxybouvardin (**834**)[300] were synthesized successfully starting from L-tyrosine. Similarly, an excellent synthesis of dichlorovancomycin aglycone (**838**) was accomplished by Evans and coworkers[301] who used TTN·3H$_2$O in MeOH–CH$_2$Cl$_2$ as an oxidation system and CrCl$_2$ as a reducing agent.

Vancomycin (**835**), one of the representative glycopeptide antibiotics, isolated from *Streptomyces orientalis*, is quite attractive from the viewpoints of physiological activity, molecular recognition and natural products synthesis. Recently, total synthesis of vancomycin was accomplished by two groups[302,303]. This antibiotic which is effective for methicillin-resistant *Staphylococcus aureus* (MRSA) is known to inhibit the biosynthesis of bacterial cell wall by high affinity (five hydrogen bondings) to the terminal D-Ala-D-Ala residue of the peptide glycan precursor (Chart 4). Recently, however, serious problems occurred with MRSA, because of the emergence of vancomycin-resistant strains (*Enterococcus faecium* and *E. faecalis*). These strains acquire the resistance by possessing the terminal D-Ala-D-lactate instead of D-Ala-D-Ala. Therefore, the finding of synthetic compounds that are able to bind with high affinity to D-Ala-D-Lactate will provide a powerful strategy for overcoming vancomycin-resistance.

Thus, Ellman and coworkers adopted the TTN-mediated oxidative cyclization strategy to synthesize the macrocyclic diaryl ether as a key compound[304], because of the simple procedures and the ready availability of the amino acid starting materials as compared with other synthetic strategies[305]. The easily available tripeptide **839** was subjected to TTN-mediated oxidation, followed by zinc reduction under similar conditions as reported by Yamamura and coworkers[306] to give a 45–60% overall yield of the desired diaryl ether **840**, from which a number of receptors such as **841** were synthesized (Scheme 168). The oligopeptide **841** exhibited binding to tripeptide N-Ac$_2$-L-Lys-D-Ala-D-Ala that is

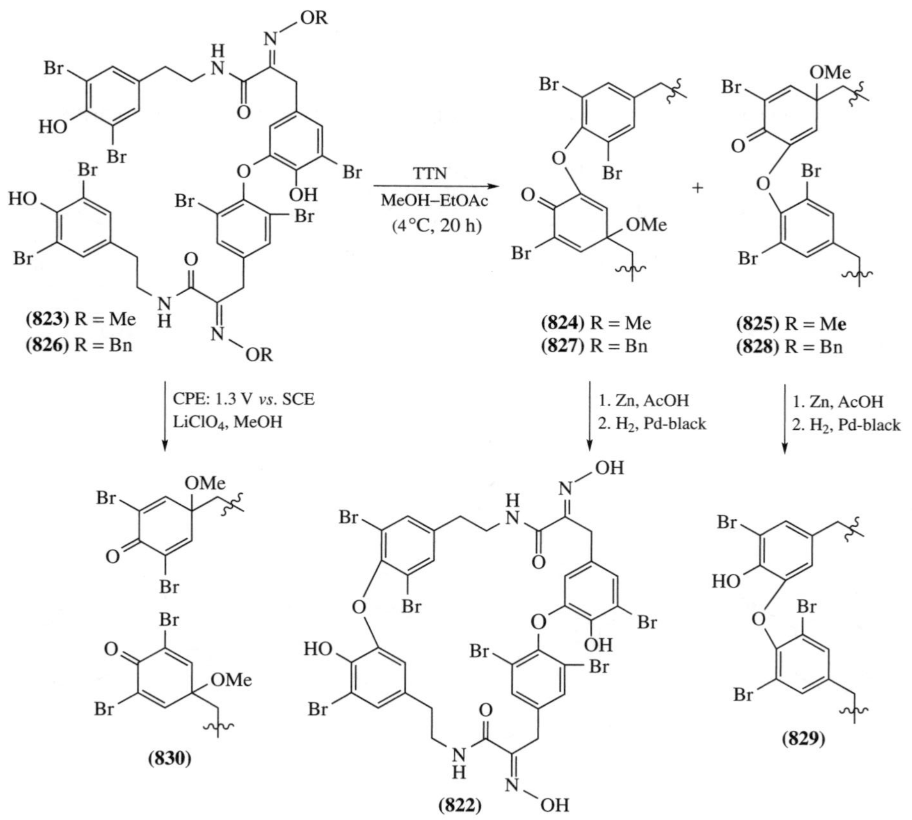

SCHEME 164. Synthesis of bastadin-6 using TTN-mediated oxidation as a key step

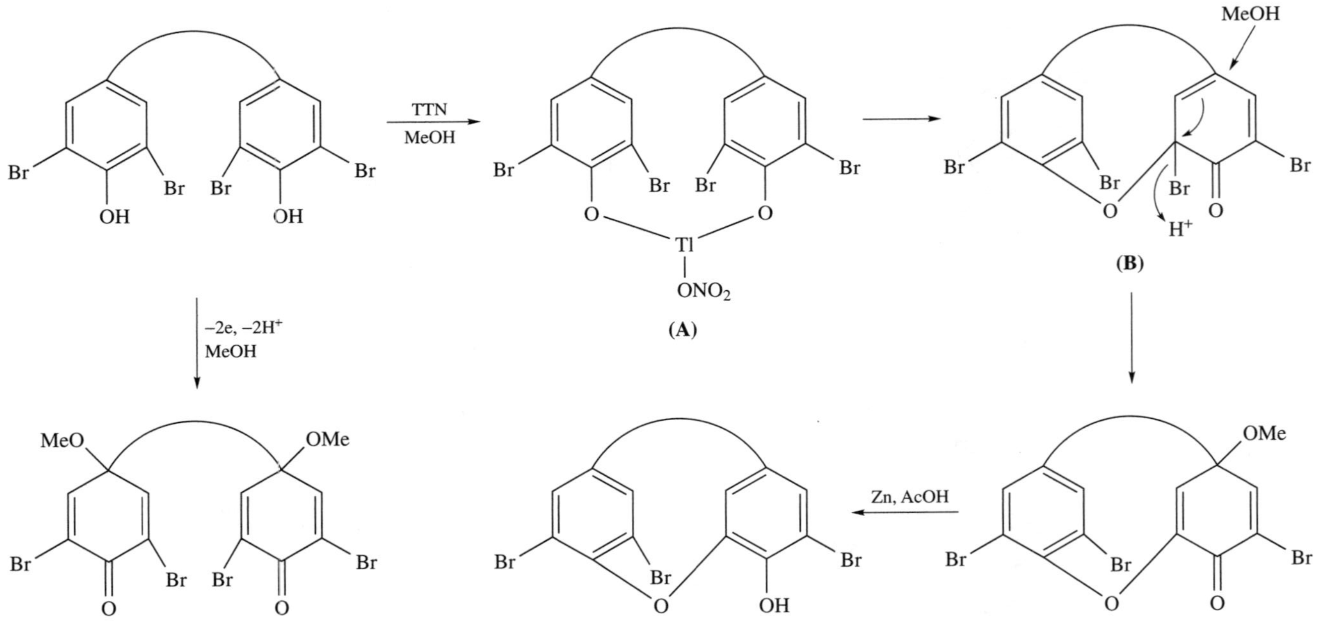

SCHEME 165. Reaction mechanism of TTN-mediated oxidation of *o,o'*-dibromophenols

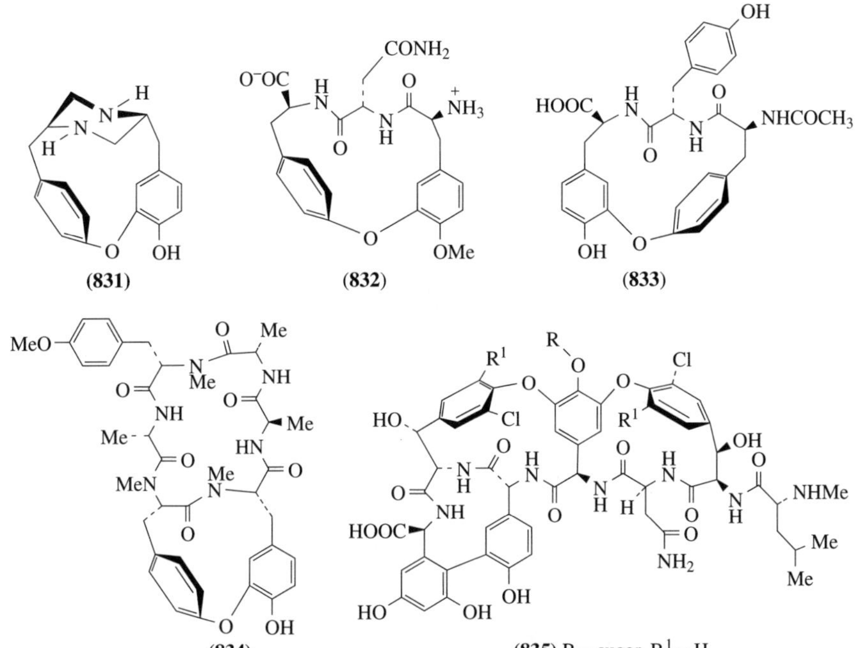

17. Oxidation of phenols

(831) (832) (833)

(834)

(835) R = sugar, R¹ = H
(838) R = H, R¹ = Cl

CHART 3. Selected isodityrosine natural products

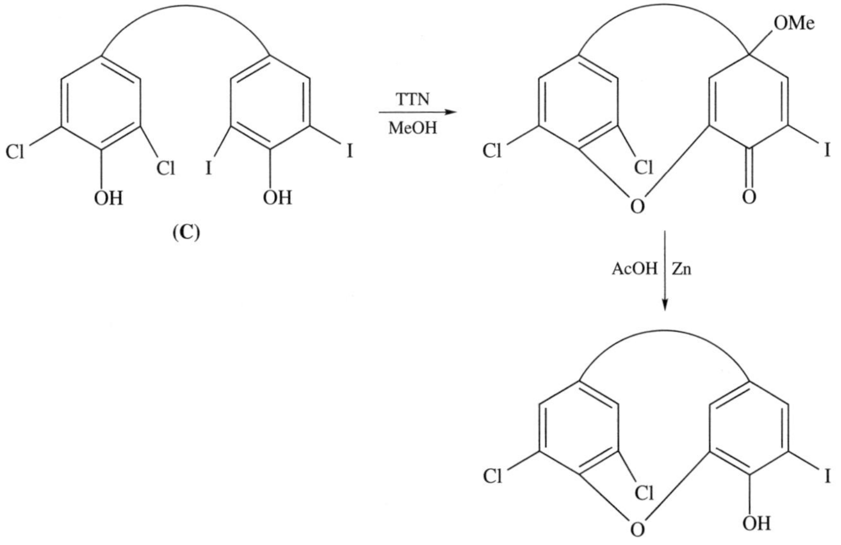

SCHEME 166. Effect of halogen substituents on the directions of intramolecular cyclization

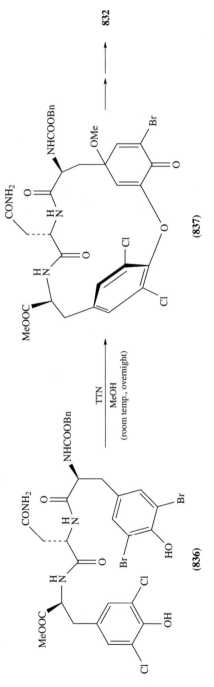

SCHEME 167. Synthesis of OF 4949-I (**832**)

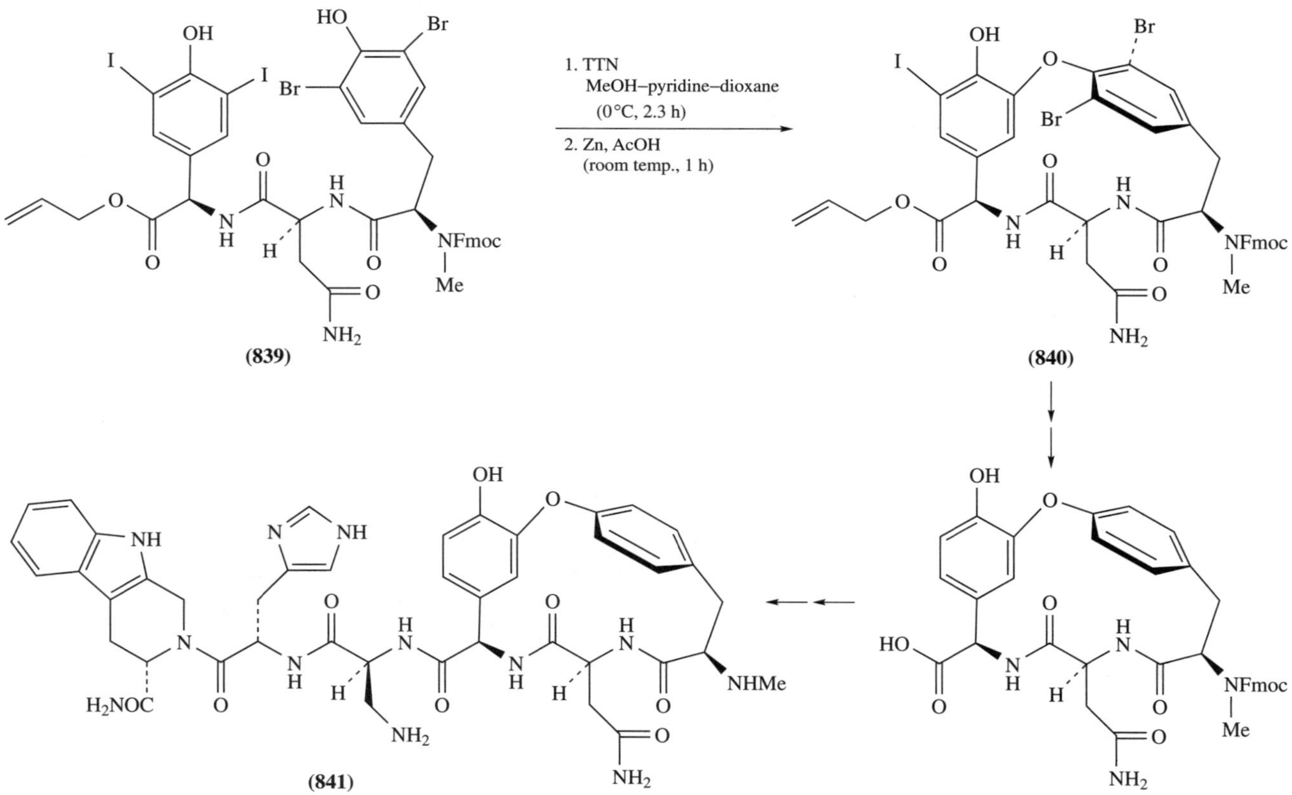

SCHEME 168. Synthesis of receptors binding to N-Ac$_2$-L-Lys-D-Ala-D-Lactate

Fmoc = 9-fluorenylmethoxycarbonyl

only 6-fold weaker than that of vancomycin. More significantly, **841** showed significantly increased binding to N-Ac$_2$-L-Lys-D-Ala-D-Lactate when compared to vancomycin.

In the light of the molecular interaction of vancomycin with the cell wall models (see Chart 4), secoaglucovancomycin (**842**) was synthesized based on TTN-mediated oxidation protocol. The tetrapeptide (**843**), prepared from 3,5-dimethoxyphenylglycine, was treated with TTN (2 equiv.) in THF–MeOH containing CH(OMe)$_3$ to afford the corresponding cyclic diaryl ether **844** (40%) (Scheme 169)[307], wherein after TTN-promoted oxidation the zinc reduction procedure was not needed[308,309]. Compound **844** was further converted to a heptapeptide **845**, which was subjected to TTN-promoted oxidation, followed by Zn reduction to give the bicyclic compound **846** (40%). Further deprotection provided the target molecule **842**, which was employed for binding experiments with N-Ac-D-Ala-D-Ala

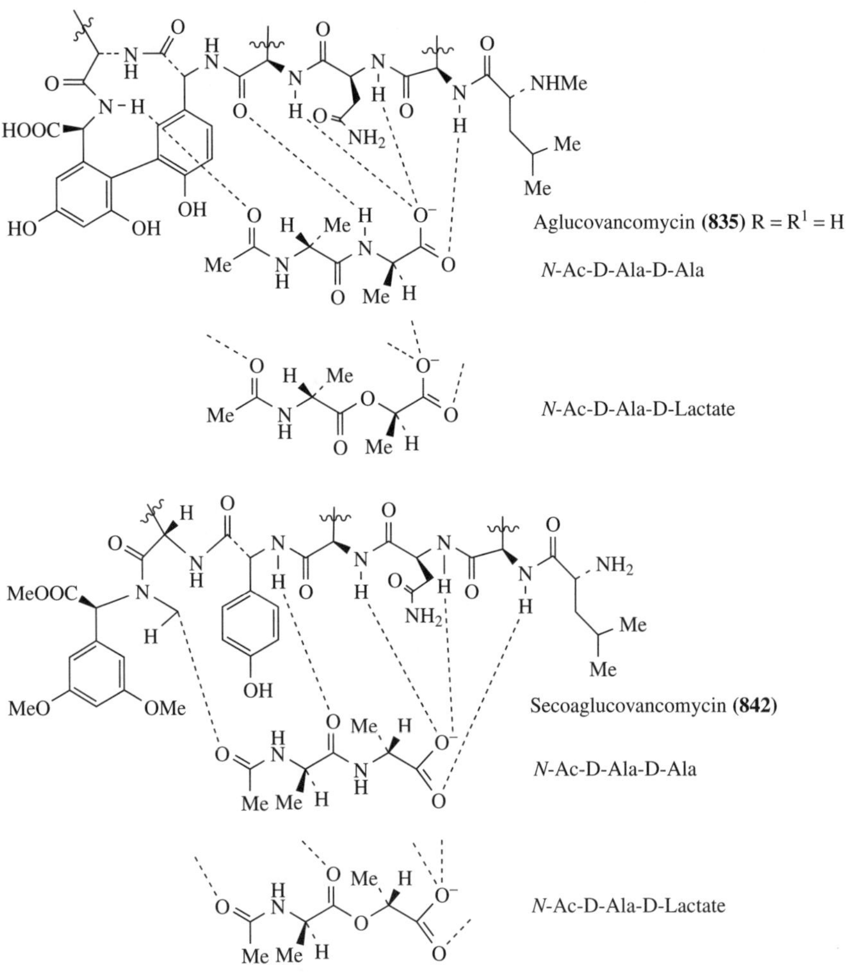

Aglucovancomycin (**835**) R = R^1 = H

N-Ac-D-Ala-D-Ala

N-Ac-D-Ala-D-Lactate

Secoaglucovancomycin (**842**)

N-Ac-D-Ala-D-Ala

N-Ac-D-Ala-D-Lactate

CHART 4. The binding sites and complexation with cell wall models

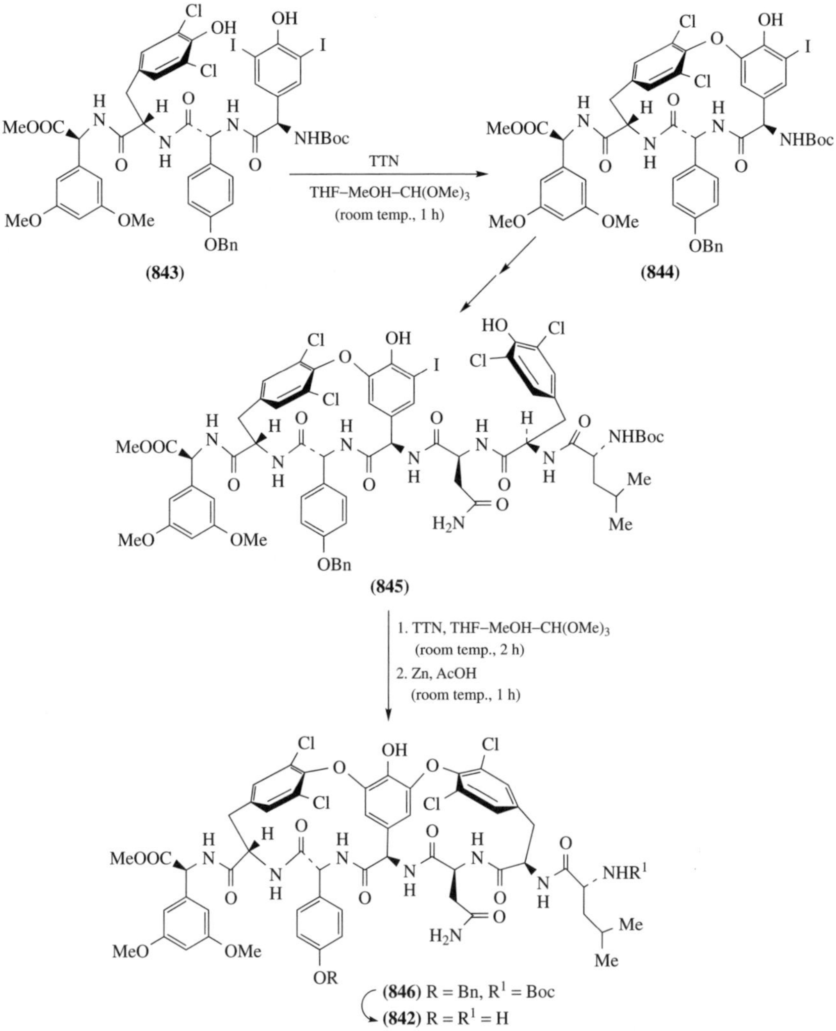

SCHEME 169. Synthesis of secoaglucovancomycin derivatives

as well as with N-Ac-D-Ala-D-Lactate together with MM/MD calculations, indicating that the interaction of **842** with the cell wall models is achieved at the back side of the molecule with five hydrogen bonds (Chart 4)[310].

Lead tetraacetate, Pb(OAc)$_4$, is well known to be effective for phenolic oxidation. From the viewpoint of organic synthesis, it is noted that Wessely oxidation of *ortho*-substituted phenols with Pb(OAc)$_4$ provides a useful method for the synthesis of cyclohexadienones, as shown in Scheme 170[311,312]. Herein, heterolytic cleavage of the initially formed O−Pb bond followed by nucleophilic attack by RCOOH results in the preferential formation of 2-acyloxycyclohexa-2,4-dienones over 4-acyloxycyclohexa-2,5-dienones.

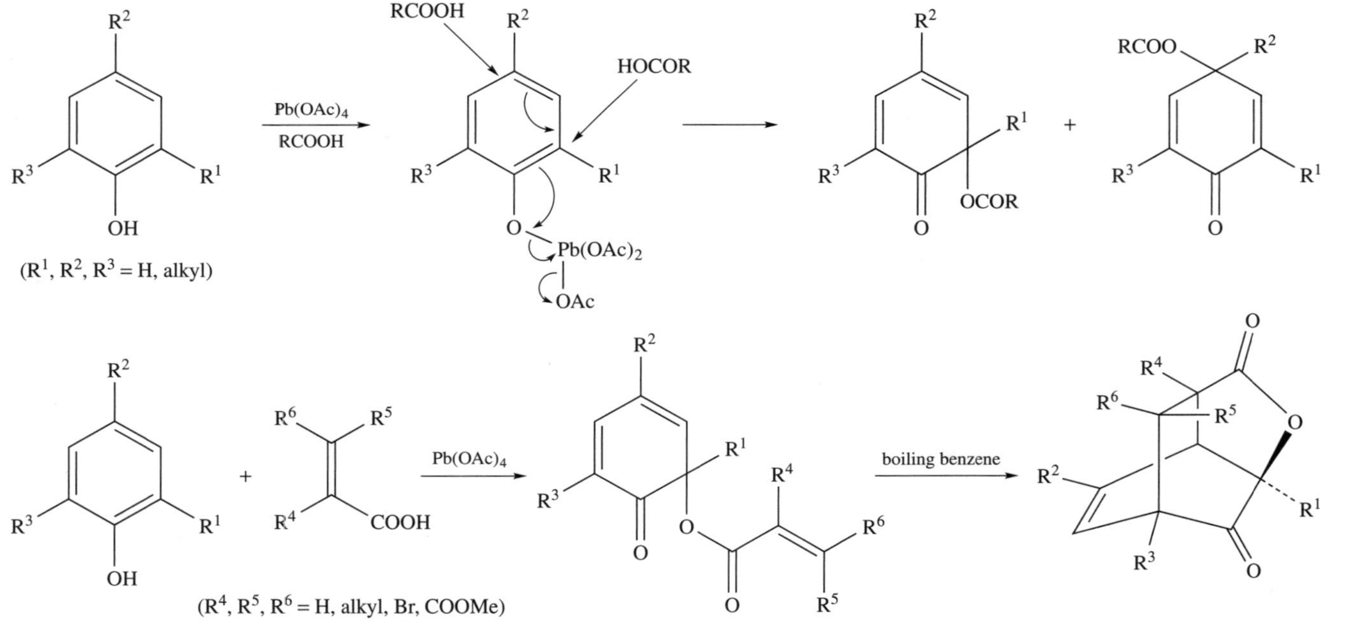

SCHEME 170. Phenolic oxidation with lead tetraacetate

17. Oxidation of phenols

The use of α,β-unsaturated carboxylic acids as a nucleophile provides the corresponding 2-(α,β-unsaturated acyloxy)cyclohexa-2,4-dienones, which on heating undergo an intramolecular Diels–Alder reaction to afford the bicyclo[2.2.2]octenones[60,313]. 2,6-Dimethylphenol (**206**) was oxidized with Pb(OAc)$_4$ in the presence of unsaturated acids **847** and **848** and then heated in boiling benzene to afford the corresponding bicyclo[2.2.2]octenones **849** and **850** in ca 40% overall yields, respectively (Scheme 171). The best yield was obtained with a 4:1 molar ratio of the unsaturated acid and Pb(OAc)$_4$. Further treatment of the latter with aq. NaOH provided the lactonic acid **851**.

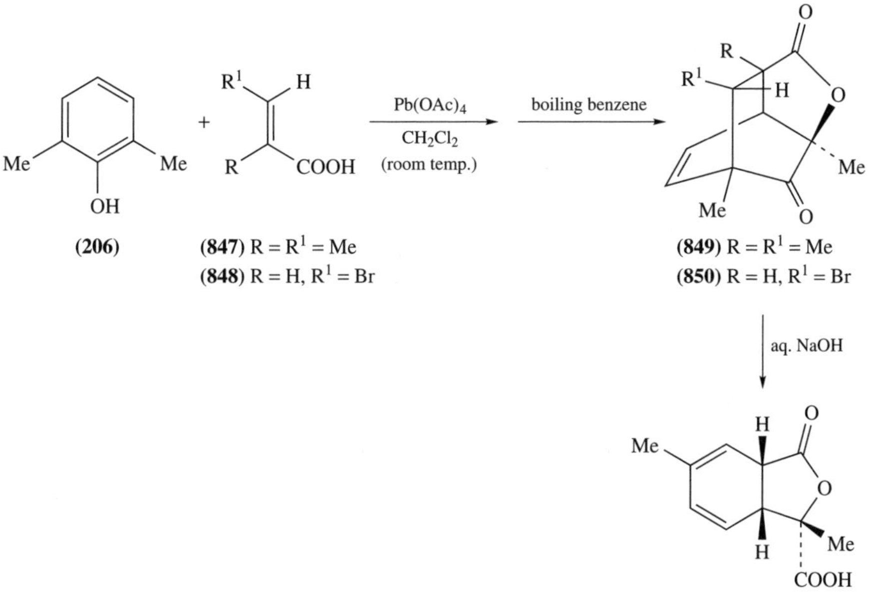

SCHEME 171. Synthesis of bicyclo[2.2.2]octenones via Wessely oxidation of phenols

Wessely oxidation of phenols has been applied successfully to natural product synthesis. Some examples are shown in Scheme 172.

Aeroplysinin-1 (**852**), a metabolite of marine organisms, shows antibiotic activity against *Staphylococcus aureus* and antileukemia activity against L-1210. This metabolite was synthesized by Pb(OAc)$_4$-mediated phenolic oxidation as a key step[314]. Here, 3,5-dibromo-2-hydroxy-4-methoxyphenylacetonitrile (**853**) was oxidized with excess of Pb(OAc)$_4$ in AcOH to give in 35% yield the desired cyclohexa-2,4-dienone **854**, which was converted to the target molecule in 2 steps.

Aspersitin (**855**) is a fungal metabolite of *Aspergillus parasiticus* NRRL 3260. This metabolite was synthesized successfully by Büchi and coworkers[315]. The key compound (**856**), prepared from dimethylphloroglucinol in 5 steps, was treated with Pb(OAc)$_4$ in AcOH to afford the corresponding *o*-quinol acetate **857** in 93% yield. Further treatment of **857** with NH$_4$OH–MeOH provided two 1:1 diastereomers of **855**.

The oxidation of tetrahydroisoquinolines **858** and **859** was carried out using Pb(OAc)$_4$ in CH$_2$Cl$_2$ (room temp., 0.5 h) to afford quantitatively the corresponding cyclohexa-2,4-dienones **860** and **861**, respectively (Scheme 173). The former was further treated with

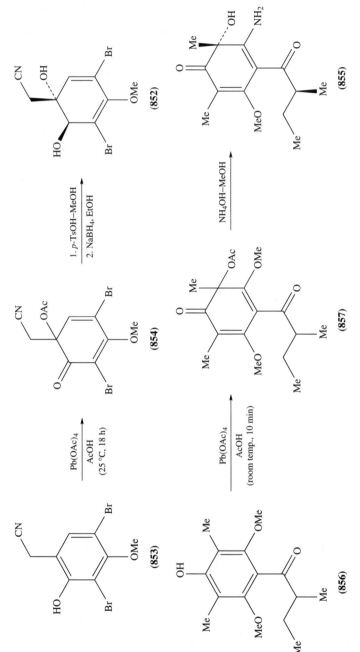

SCHEME 172. Syntheses of aeroplysinin-1 and aspersitin

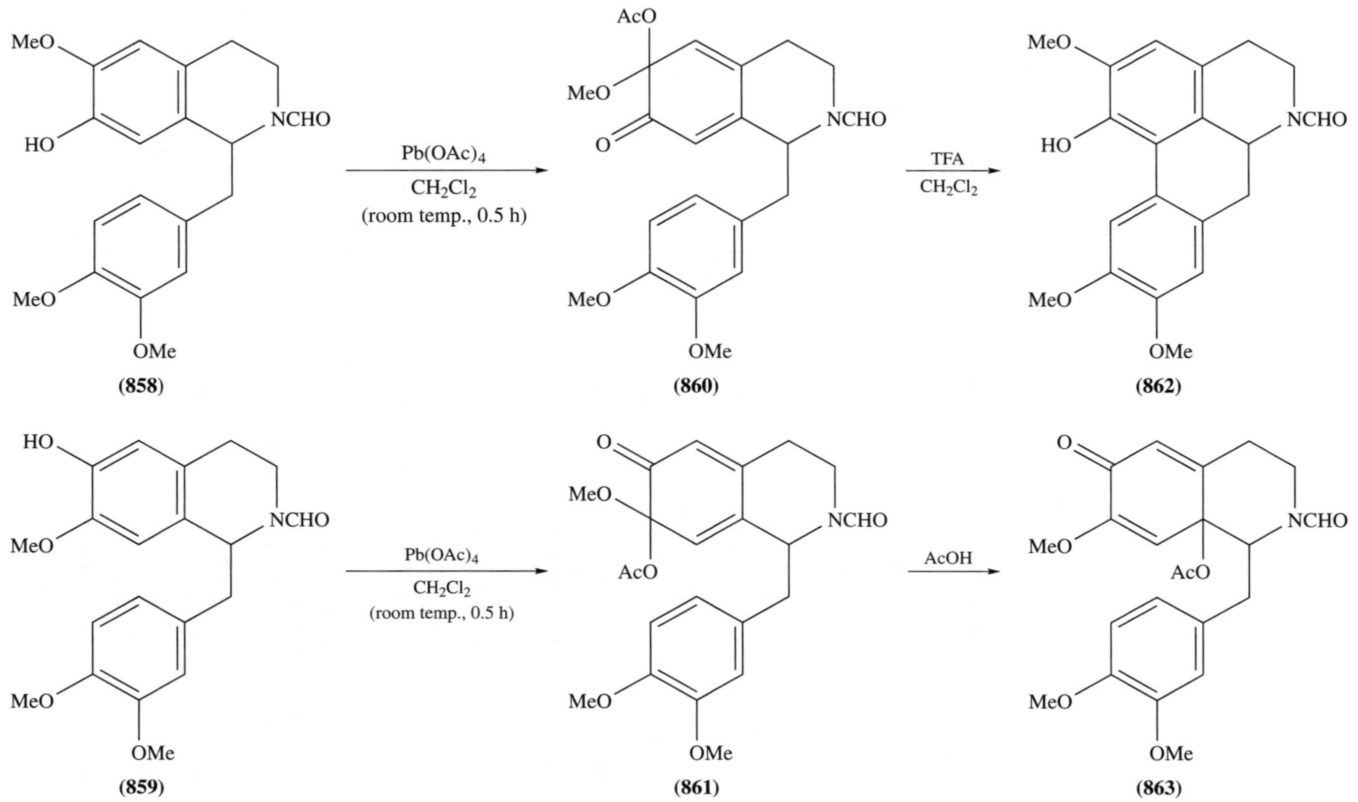

SCHEME 173. Oxidation of tetrahydroisoquinolines with Pb(OAc)$_4$

TFA in CH$_2$Cl$_2$ to give mainly N-formylwilsonirine (**862**) (60%)[316]. On treatment with AcOH at 20–30 °C the dienone **861** was converted into another regioisomer **863** (74%)[317].

Bismuth belongs to the 5B group in the Periodic Table. Bismuth(V) and (III) salts and organobismuth reagents are employed as useful oxidants in organic synthesis[318]. In particular, the bismuth(V) in the form of NaBiO$_3$ is analogous to Pb(OAc)$_4$ in chemical properties, although the oxidizing power of NaBiO$_3$ is relatively weak. NaBiO$_3$ is also effective for phenolic oxidation. Phenols undergo two-electron oxidation with NaBiO$_3$ in AcOH resulting in the formation of quinol acetates. Some typical examples are shown herein.

2,6-Dimethylphenol (**206**) was treated with NaBiO$_3$ in benzene to afford polyphenylene oxide (**659**) and 3,3′,5,5′-tetramethyldiphenoquinone (**207**) in 74 and 12% yields, respectively[319]. This result is similar to that of MnO$_2$ oxidation (see Scheme 126). In contrast, the use of AcOH instead of benzene as a solvent provided the corresponding quinol acetate **864** and **207** in 38 and 15% yields, respectively (Scheme 174)[320]. Oxidation of 2,4,6-tri(*tert*-butyl)phenol (**73**) with NaBiO$_3$ in AcOH afforded the *p*-quinol acetate (**865**) as a major product (62%) and the *o*-quinol acetate (**866**) as a minor product (22%). In contrast, Pb(OAc)$_4$ oxidation of **73** in AcOH provided **866** as a main product (60%)[321] (see Scheme 170). Oxidation of alkoxyphenols and other phenols has also been studied[318,322].

Organobismuth salts such as Ph$_3$BiCl$_2$, Ph$_3$BiCO$_3$ and Ph$_4$BiOTs are also utilized for phenolic oxidation. The reactivity of these oxidants toward hindered phenols under basic conditions was examined by Barton and coworkers (Scheme 175)[323]. 2,6-Di(*tert*-butyl)phenol (**23**) was treated with Ph$_3$BiCl$_2$ in the presence of *N*-(*tert*-butyl)-*N*′,*N*′,*N*″,*N*″-tetramethylguanidine (BTMG) in THF to afford 3,3′,5,5′-tetra(*tert*-butyl)diphenoquinone (**24**) (37%), while the use of Ph$_4$BiOTs provided a 38% yield of 4-phenyl-2,6-di(*tert*-butyl)phenol (**867**) through a plausible bismuth intermediate **868**. In both cases, no reaction took place in the absence of BTMG.

When oxidized with Ph$_3$BiCl$_2$–BTMG in MeOH–THF, 2,6-di(*tert*-butyl)-4-methylphenol (**69**) was converted into 2,6-di(*tert*-butyl)-4-methoxymethylphenol (**869**) (45%). In the case of Ph$_4$BiOTs, phenylation also took place at the *p*- and *o*-positions to give two compounds **870** and **871** in 22 and 20% yields, respectively.

F. Oxidation with Other Metal Compounds

Selenium is a member of the 6B group in the Periodic Table. Selenium reagents as an oxidant were initially shown by Barton and coworkers[324] to be effective for phenolic oxidation[325,326].

Generally, oxidation of phenols with benzeneseleninic anhydride, (PhSeO)$_2$O, provides selectively *o*-benzoquinones, while benzeneseleninic acid (PhSeOOH) can effect phenolic oxidation to afford selectively the corresponding *p*-benzoquinones[326]. In the first case, the oxidant is a moisture-sensitive compound, so that the resulting PhSeOOH can influence the regioselectivity of the oxidation. Barton and coworkers carried out oxidations of 3,5-di(*tert*-butyl)phenol (**872**) with both selenium reagents in a variety of solvents to afford both *o*- and *p*-benzoquinones (**220** and **74**) in a different ratio; THF and benzene are the best solvents for *ortho*-oxidation with (PhSeO)$_2$O (**220**: 73 and 82%, respectively; **74**: 8 and 13%, respectively) and CH$_2$Cl$_2$ is the best solvent for *para*-oxidation with PhSeOOH (**220** and **74**: 6 and 77%, respectively) (Scheme 176). Oxidation of thymol (**548**) with (PhSeO)$_2$O provided selectively tymoquinone (**549**) and the regioisomer

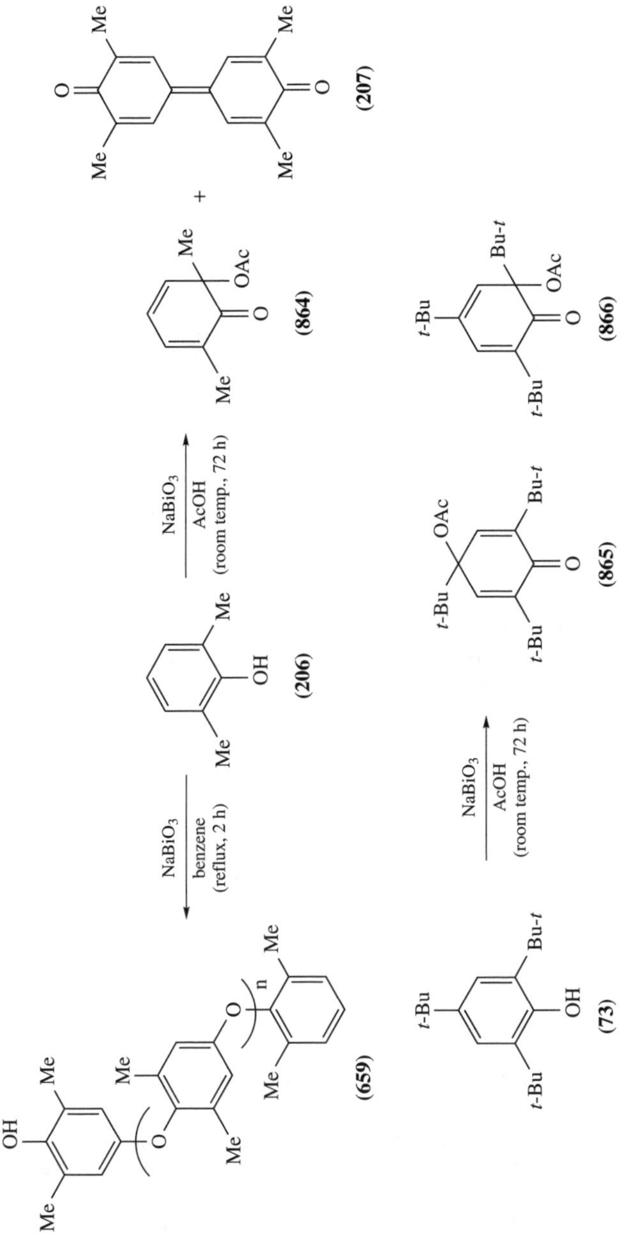

SCHEME 174. Oxidation of phenols with sodium bismuthate

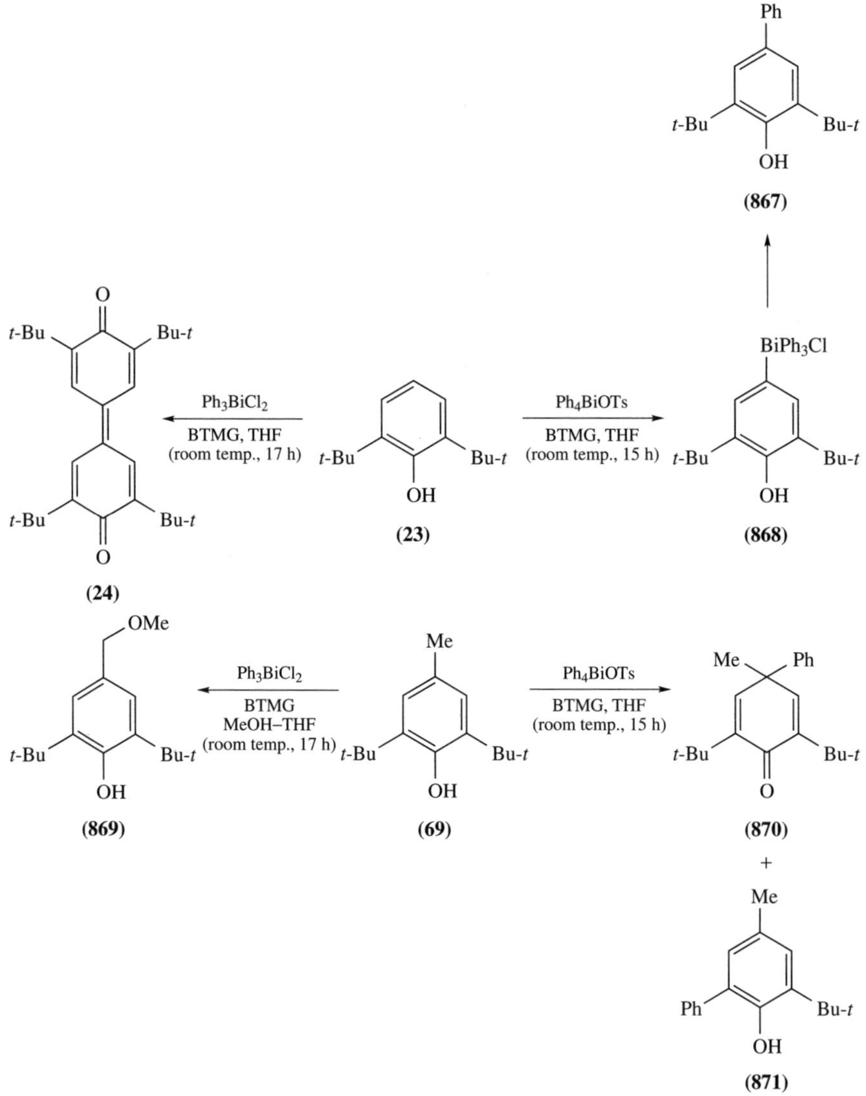

SCHEME 175. Oxidation of hindered phenols with organobismuth salts

(**550**) was mainly produced by using PhSeOOH. On oxidation of 2,6-di(*tert*-butyl)phenol (**23**) with (PhSeO)$_2$O in THF, both 2,6-di(*tert*-butyl)-*p*-benzoquinone (**74**) and 3,3′,5,5′-tetra(*tert*-butyl)diphenoquinone (**24**) were obtained in 11 and 76% yields, respectively. Herein, the radical mechanism of this reaction was supported by ESR experiments. Oxidations of 2,4,6-trimethylphenol and related compounds with (PhSeO)$_2$O have also been studied[325,326].

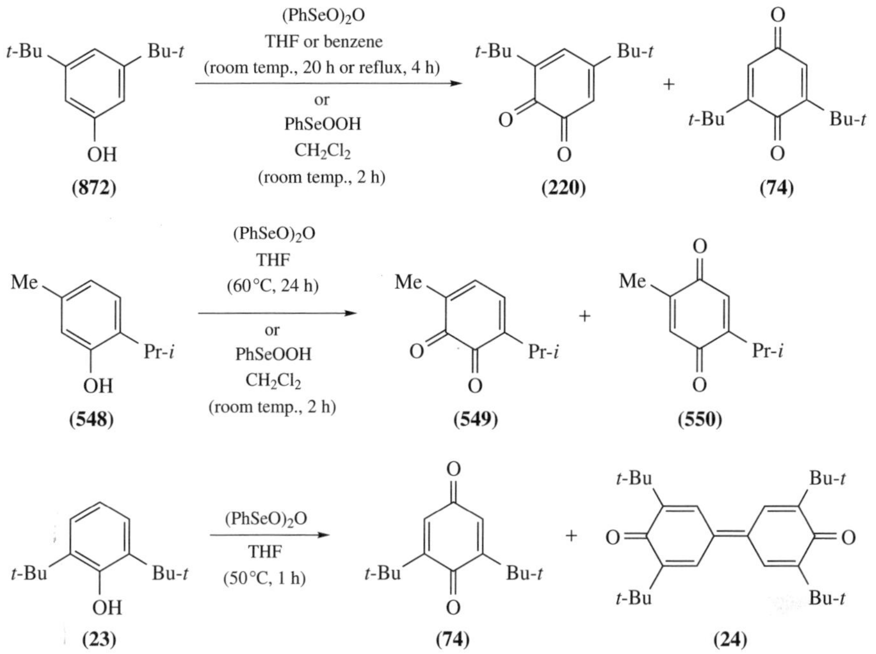

SCHEME 176. Oxidation of phenols with benzeneseleninic anhydride and benzeneseleninic acid

In Scheme 176, the *ortho*-selectivity with (PhSeO)$_2$O is mainly due to the initial formation of aryl benzeneselenates followed by [2,3] sigmatropic rearrangement leading to the corresponding 6-phenylselenoxycyclohexa-2,4-dienones. On oxidation with PhSeOOH, a direct *para*-substitution reaction may take place[326]. In the case of phenol itself, however, another possible mechanism was suggested by Henriksen[327]. Oxidation of phenol with PhSeOOH in CH$_2$Cl$_2$ at 24 °C afforded, beside diphenyl diselenide, *p*-benzoquinone (**17**), 2-(phenylseleno)-*p*-benzoquinone (**873**) and 2,6-bis(phenylseleno)-*p*-benzoquinone (**874**) in the approximate molar ratio 3:4:3 (Scheme 176). The initial addition of diphenyl diselenide to the reaction mixture changed this ratio to 2:5:4 in favor of selenylated products (**873** and **874**). Based on these results together with solvent effects that indicate the participation of an acidic hydrogen atom, ene-reactions may play an important role in both *ortho*-selenylation and *para*-oxidation sequence (Scheme 177).

Cerium is a member of the lanthanides in the Periodic Table and adopts tetra- and tripositive states in its electronic configuration. Among cerium reagents, ceric ammonium nitrate (CAN) is most widely used in organic synthesis. It is well known to convert phenol derivatives to quinones in high yields under mild conditions. An excellent review on cerium(IV) oxidation of organic compounds is available[328], and only a few examples will be described herein.

Oxidation of dihydrobenzofuran **875** with CAN in aq. MeCN afforded the corresponding *p*-benzoquinone **876** in 60% yield (Scheme 178)[329]. Similarly, other substituted *p*-methoxyphenols were converted into the corresponding *p*-benzoquinones in high yields[328]. However, aryl ethers rather than phenols are generally used as the substrates due to the lower reactivity and the easier handling.

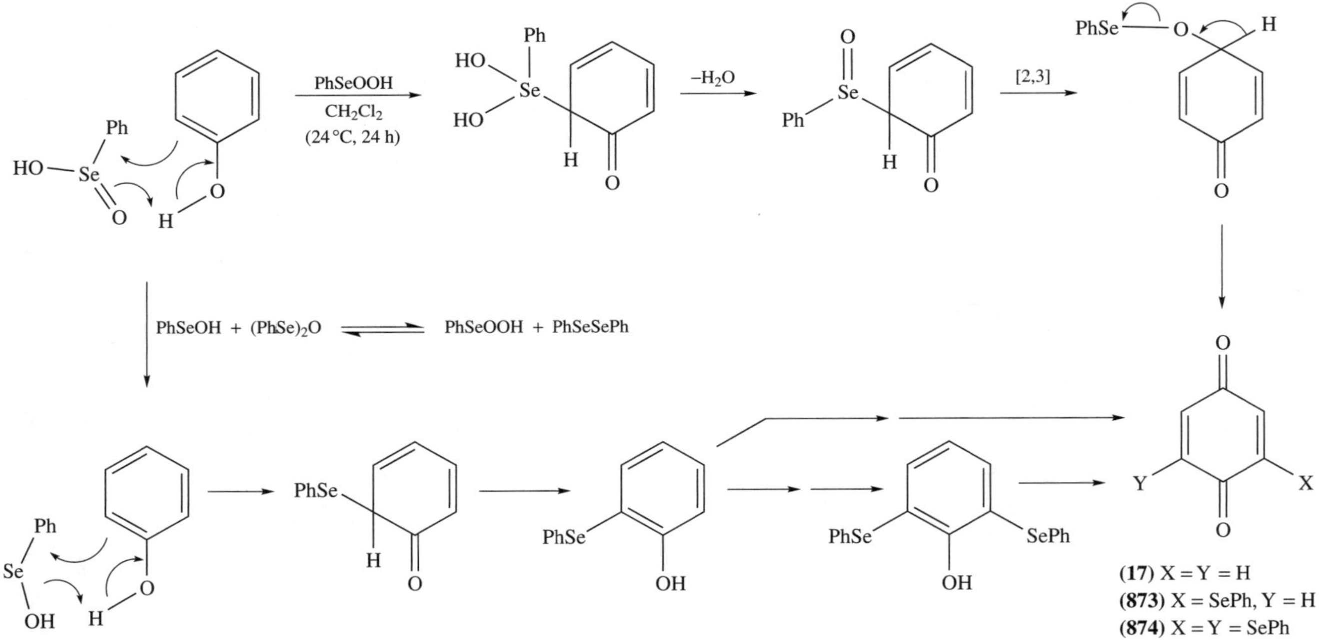

SCHEME 177. Oxidation of phenol with benzeneseleninic anhydride in methylene chloride

(17) X = Y = H
(873) X = SePh, Y = H
(874) X = Y = SePh

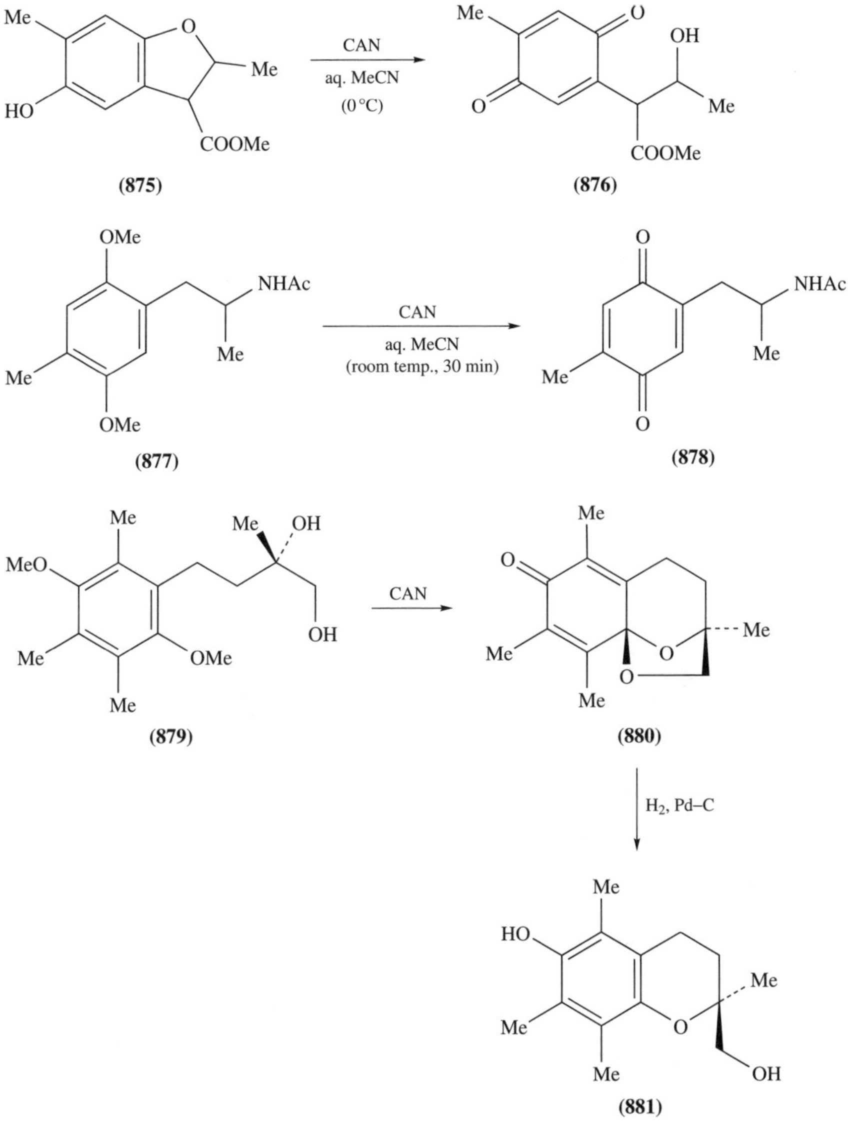

SCHEME 178. Oxidation of phenol derivatives with CAN

Treatment of a 2,5-disubstituted 1,4-dimethoxybenzene **877** with CAN provided a 97% yield of p-benzoquinone **878**[330]. The fully substituted 1,4-dimethoxybenzene derivative **879** was treated with CAN to afford in 64% yield the quinone monoketal **880**. This was submitted to catalytic hydrogenation to give the precursor of α-tocopherol **881** (Scheme 178)[331]. A variety of substituted 1,4-dimethoxybenzenes were also oxidized with CAN to give high yields of p-benzoquinones.

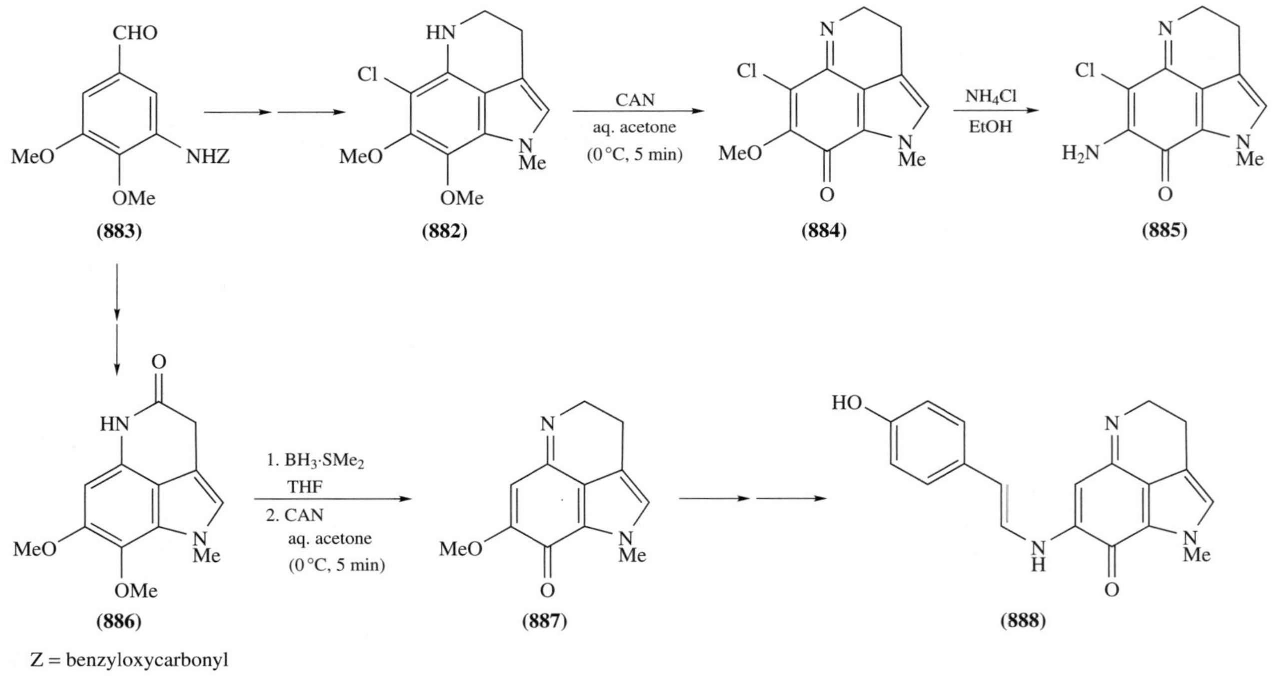

SCHEME 179. Syntheses of isobatzelline C and makaluvamine E

Oxidation of 1-amino-4-methoxybenzenes with CAN is expected to afford *p*-iminoquinones. Thus, batzellines, makaluvamines and discorhabdins, isolated from marine organisms, possess a pyrroloiminoquinone moiety and can be synthesized by CAN-mediated oxidation of the corresponding 4-amino-7-methoxyindole derivative.

The indole derivative **882**, derived from 3-benzyloxycarbonylamino-4,5-dimethoxybenzaldehyde (**883**), was treated with CAN in 70% aq. acetone to afford the desired iminoquinone **884** in 64% yield. Finally, amination of **884** with NH_4Cl provided isobatzelline C (**885**) (Scheme 179)[57]. Similarly, the indole lactam **886** was reduced with $BH_3 \cdot SMe_2$ followed by CAN-mediated oxidation in aq. MeCN to give iminoquinone **887** in 60% overall yield. The key compound **887** was further converted to makaluvamines represented by makaluvamine E (**888**) (Scheme 179)[332].

In this chapter, reagents are classified mainly into three categories: (1) for catalytic oxidation of phenols, (2) for phenolic oxidation with nonmetallic compounds and (3) for phenolic oxidation with metallic compounds. In the 21st century, regardless of metallic or nonmetallic compounds, catalytic oxidation systems with high efficiency must be constructed. If stoichiometric amounts of reagents are employed, efficient oxidation–reduction systems should be invented.

This chapter does not cover all of the literature on phenolic oxidation, but typical examples have been taken up based on the systematization of phenolic oxidation. In addition, phenolic oxidation methodology has been shown to be quite useful for syntheses of natural products and related compounds with a complex structure.

V. REFERENCES

1. M. M. Baizer and H. Lund (Eds.), *Organic Electrochemistry*, Marcel Dekker, New York, 1983.
2. T. Shono, *Electroorganic Chemistry as a New Tool in Organic Synthesis*, Springer-Verlag, New York, 1984.
3. S. Torii, *Electroorganic Synthesis: Methods and Applications, Part 1, Oxidation*, Kodansha, Tokyo, 1985.
4. S. Torii (Ed.), *Recent Advances in Electroorganic Synthesis*, Kodansha, Tokyo, 1987.
5. R. D. Little and N. L. Weinberg (Eds.), *Electroorganic Synthesis: Festschrift for Manuel M. Baizer*, Marcel Dekker, New York, 1991.
6. S. Torii (Ed.), *Novel Trends in Electroorganic Synthesis*, Springer-Verlag, Tokyo, 1998, pp. 97–152.
7. T. Osa (Ed.), *New Challenges in Organic Electrochemistry*, Gordon and Breach Science Publishers, Amsterdam, 1998.
8. N. L. Weinberg, *Technique of Electroorganic Synthesis*, Vol. V, Part 1 of Techniques in Chemical Series (Ed. N. L. Weinberg), Part 3, Wiley, New York, 1974, pp. 410–434.
9. T. Shono, *Electroorganic Synthesis*, Academic Press, New York, 1991.
10. L. Papouchado, R. W. Sandpord, G. Petrie and R. N. Adams, *J. Electroanal. Chem.*, **65**, 275 (1975).
11. M. Gattrell and D. W. Kirk, *J. Electrochem. Soc.*, **140**, 903 (1993).
12. M. C. Pham, F. Adami and P. C. Lacaze, *J. Electrochem. Soc.*, **136**, 677 (1993).
13. S. Torii, A.-L. Dhimane, Y. Araki and T. Inokuchi, *Tetrahedron Lett.*, **30**, 2105 (1989).
14. (a) W. I. Taylor and A. R. Battersby (Eds.), *Oxidative Coupling of Phenols*, Marcel Dekker, New York, 1967.
    (b) A. I. Scott, *Quart. Rev.*, **19**, 1 (1965).
15. R. Gottlieb, *Fortschr. Chem. Org. Naturst.*, **35**, 1 (1978).
16. R. Gottlieb, *New Natural Products and Plant Drugs with Pharmacological, Biological or Therapeutical Activity*, Springer-Verlag, New York, 1977.
17. (a) R. S. Ward, *Chem. Soc. Rev.*, **11**, 75 (1982).
    (b) S. Nishiyama and S. Yamamura, *J. Synth. Org. Chem. Jpn.*, **46**, 192 (1988).

(c) Y. Shizuri, H. Shigemori, K. Suyama, K. Nakamura, M. Okubo and S. Yamamura, in *Studies in Natural Products Chemistry*, Vol 8, Part E (Ed. Att-ur-Rahman), Elsevier Science, Amsterdam, 1991, pp. 159–173.
18. L. L. Miller and R. F. Stewart, *J. Org. Chem.*, **43**, 1580 (1978).
19. M. Iguchi, A. Nishiyama, Y. Terada and S. Yamamura, *Anal. Lett.*, **12**, 1079 (1979).
20. M. Iguchi, A. Nishiyama, Y. Terada and S. Yamamura, *Chem. Lett.*, 451 (1978).
21. A. Nishiyama, H. Eto, Y. Terada, M. Iguchi and S. Yamamura, *Chem. Pharm. Bull.*, **31**, 2820 (1983).
22. I. Barba, R. Chinchilla and C. Gomez, *J. Org. Chem.*, **55**, 3270 (1990).
23. M. Iguchi, A. Nishiyama, M. Hara, Y. Terada and S. Yamamura, *Chem. Lett.*, 1015 (1978).
24. A. Nishiyama, H. Eto, Y. Terada, M. Iguchi and S. Yamamura, *Chem. Pharm. Bull.*, **31**, 2834 (1983).
25. (a) M. Iguchi, Y. Terada and S. Yamamura, *Chem. Lett.*, 1393 (1979).
    (b) A. Nishiyama, H. Eto, Y. Terada, M. Iguchi and S. Yamamura, *Chem. Pharm. Bull.*, **31**, 2845 (1983).
26. M. Iguchi, A. Nishiyama, H. Eto and S. Yamamura, *Chem. Lett.*, 939 (1981).
27. L. Chapman, M. R. Engel, J. P. Springer and J. C. Clardy, *J. Am. Chem. Soc.*, **93**, 6696 (1971).
28. M. Matsumoto and K. Kuroda, *Tetrahedron Lett.*, **22**, 4437 (1981).
29. J. M. Bobbitt, I. Noguchi, H. Yagi and K. H. Weisgraber, *J. Org. Chem.*, **41**, 845 (1976).
30. M. Iguchi, A. Nishiyama, H. Eto, K. Okamoto, S. Yamamura and Y. Kato, *Chem. Pharm. Bull.*, **34**, 4910 (1986).
31. S. Yamamura and S. Nishiyama, in *Studies in Natural Products Chemistry*, Vol. 10 (Ed. Att-ur-Rahman), Elsevier Science, Amsterdam, 1992, pp. 629–669.
32. S. Yamamura and S. Nishiyama, *J. Synth. Org. Chem. Jpn.*, **55**, 1029 (1997).
33. (a) S. Sano, K. Katayama, K. Takesako, T. Nakamura, A. Ohbayashi, Y. Ezure and H. Enomoto, *J. Antibiot.*, **39**, 1685 (1986).
    (b) M. E. Jung, D. Jachiet and J. C. Rohloff, *Tetrahedron Lett.*, **30**, 4211 (1989).
    (c) D. L. Boger and D. Yohann, *J. Org. Chem.*, **55**, 6000 (1990).
34. S. C. Fry, *Methods Enzymol.*, **107**, 388 (1984).
35. S. Nishiyama, M.-H. Kim and S. Yamamura, *Tetrahedron Lett.*, **35**, 8397 (1994).
36. M. Takahashi, H. Konishi, S. Iida, K. Nakamura, S. Yamamura and S. Nishiyama, *Tetrahedron*, **55**, 5295 (1999).
37. A. Ronlán and V. D. Parker, *J. Chem. Soc. (C)*, 3214 (1971).
38. J. A. Richards, P. E. Whitson and D. H. Evans, *J. Electroanal. Chem.*, **63**, 311 (1975).
39. J. A. Richards and D. H. Evans, *J. Electroanal. Chem.*, **81**, 171 (1977).
40. J. H. Meurs, D. W. Sopher and W. Eilenberg, *Angew. Chem., Int. Ed. Engl.*, **28**, 927 (1989).
41. A. Rieker, E.-L. Dreher, H. Geisel and M. H. Khalifa, *Synthesis*, 851 (1978).
42. G. Popp and N. C. Reitz, *J. Org. Chem.*, **37**, 3646 (1972).
43. E.-L. Dreher, J. Bracht, M. El-Mobayed, P. Hütter, W. Winter and A. Rieker, *Chem. Ber.*, **115**, 288 (1982).
44. G. Reischl, M. El-Mobayed, R. Beißwenger, K. Regier, C. Maichle-Mössmer and A. Rieker, *Z. Naturforsch.*, **53b**, 765 (1998).
45. J. S. Swenton, in *The Chemistry of Quinonoid Compounds*, Vol. 2 (Eds. S. Patai and Z. Rappoport), Chap. 15, Wiley, New York, 1988, pp. 899–962.
46. P. Ge and R. A. Russel, *Tetrahedron*, **53**, 17469 (1997).
47. J. S. Swenton, C. Shih, C.-P. Chen and C.-T. Chou, *J. Org. Chem.*, **55**, 2019 (1990).
48. Y. Shizuri, K. Nakamura, S. Yamamura, S. Ohba, H. Yamashita and Y. Saito, *Tetrahedron Lett.*, **27**, 727 (1986).
49. H. Noda, M. Niwa and S. Yamamura, *Tetrahedron Lett.*, **22**, 3247 (1981).
50. S. Nishiyama and S. Yamamura, unpublished results.
51. (a) A. Hutinec, A. Ziogas, M. El-Mobayed and A. Rieker, *J. Chem. Soc., Perkin Trans. 1*, 2201 (1998).
    (b) H. Eickhoff, G. Jung and A. Rieker, *Tetrahedron*, **57**, 353 (2001).
52. K. Yoshida, *Electrooxidation in Organic Chemistry*, Wiley, New York, 1984, pp. 99–156.
53. Y. Shizuri, K. Nakamura and S. Yamamura, *J. Chem. Soc., Chem. Commun.*, 530 (1985).
54. (a) J. S. Swenton, K. Carpenter, Y. Chen, M. L. Kerns and G. W. Morrow, *J. Org. Chem.*, **58**, 3308 (1993).

(b) G. W. Morrow, Y. Chen and J. S. Swenton, *Tetrahedron*, **47**, 655 (1991).
55. S. Yamamura, in *New Challenges in Organic Electrochemistry* (Ed. T. Osa), Chap. 2, Gordon and Breach Science Publishers, Amsterdam, 1998, pp. 166–181.
56. D. G. New, Z. Tesfai and K. D. Moeller, *J. Org. Chem.*, **61**, 1578 (1996).
57. (a) S. Nishiyama, J.-F. Cheng, X.-L. Tao and S. Yamamura, *Tetrahedron Lett.*, **32**, 4151 (1991).
(b) X.-L. Tao, J.-F. Cheng, S. Nishiyama and S. Yamamura, *Tetrahedron*, **50**, 2017 (1994).
58. Y. Kita, H. Tohma, M. Inagaki, K. Hatanaka and Y. Yakura, *J. Am. Chem. Soc.*, **114**, 2175 (1992).
59. S. Yamamura, in Reference 5, pp. 309–315.
60. (a) P. Yates and H. Auksi, *J. Chem. Soc., Chem. Commun.*, 1016 (1976).
(b) T. S. Macus and P. Yates, *Tetrahedron Lett.*, **24**, 147 (1983).
61. Y. Shizuri, H. Shigemori, Y. Okuno and S. Yamamura, *Chem. Lett.*, 2097 (1986).
(b) S. Yamamura, Y. Shizuri, H. Shigemori, Y. Okuno and M. Ohkubo, *Tetrahedron*, **47**, 635 (1991).
62. B. D. Gates, P. Dalidowicz, A. Tebben, S. Wang and J. S. Swenton, *J. Org. Chem.*, **57**, 2135 (1992).
63. (a) K. Chiba, K. Fukuda, S. Kim, Y. Kitano and M. Tada, *J. Org. Chem.*, **64**, 7654 (1999).
(b) K. Chiba, M. Jinno, R. Kuramoto and M. Tada, *Tetrahedron Lett.*, **39**, 5527 (1998).
64. G. H. Büchi and C.-P. Mak, *J. Am. Chem. Soc.*, **99**, 8074 (1977).
65. Y. Shizuri and S. Yamamura, *Tetrahedron Lett.*, **24**, 5001 (1983).
66. Y. Shizuri, K. Suyama and S. Yamamura, *J. Chem. Soc., Chem. Commun.*, 63 (1986).
67. Y. Shizuri, S. Maki, M. Ohkubo and S. Yamamura, *Tetrahedron Lett.*, **31**, 7167 (1990).
68. H. Takakura and S. Yamamura, *Tetrahedron Lett.*, **39**, 3717 (1998).
69. H. Takakura and S. Yamamura, *Tetrahedron Lett.*, **40**, 299 (1999).
70. Y. Shizuri, Y. Okuno, H. Shigemori and S. Yamamura, *Tetrahedron Lett.*, **28**, 6661 (1987).
71. H. Takakura, K. Toyoda and S. Yamamura, *Tetrahedron Lett.*, **37**, 4043 (1996).
72. S. Maki, K. Toyoda, S. Kosemura and S. Yamamura, *Chem. Lett.*, 1059 (1993).
73. S. Maki, K. Toyoda, T. Mori, S. Kosemura and S. Yamamura, *Tetrahedron Lett.*, **35**, 4817 (1994).
74. Y. Harada, S. Maki, H. Niwa, T. Hirano and S. Yamamura, *Synlett*, 1313 (1998).
75. (a) K. Chiba, T. Arakawa and M. Tada, *J. Chem. Soc., Perkin Trans. 1*, 1435 (1996).
(b) K. Chiba, J. Sonoyama and M. Tada, *J. Chem. Soc., Perkin Trans. 1*, 2939 (1998).
76. F. Freeman, in *Organic Syntheses by Oxidation with Metal Compounds* (Eds. W. J. Mijs and C. R. H. I. de Jonge), Plenum Press, New York, 1986, pp. 315–371.
77. R. H. I. de Jong, in *Organic Syntheses by Oxidation with Metal Compounds* (Eds. W. J. Mijs and C. R. H. I. de Jonge), Plenum Press, New York, 1986, pp. 423–443.
78. (a) R. A. Sheldon and J. K. Kochi, *Metal-Catalyzed Oxidations of Organic Compounds*, Academic Press, New York, 1981.
(b) H. Mimoun, in *The Chemistry of Peroxides* (Ed. S. Patai), Wiley, New York, 1983.
(c) A. E. Martell and D. T. Sawyer (Eds.), *Oxygen Complexes and Oxygen Activation by Transition Metals*, Plenum Press, New York, 1988.
(d) D. H. R. Barton, A. E. Martell and D. T. Sawyer, *The Activation of Dioxygen and Homogeneous Catalysis*, Plenum Press, New York, 1993.
(e) S. Fox and K. D. Karlin, in *Active Oxygen in Biochemistry* (Eds. J. S. Valentine, C. S. Foote, A. Greenberg and J. F. Liebman), Blackie Academic and Professional, Glasgow, Scotland, 1995, pp. 188–231.
79. J. Tsuji and H. Takayanagi, *Tetrahedron Lett.*, 1365 (1976).
80. M. M. Rogić, T. R. Demmin and W. B. Hammond, *J. Am. Chem. Soc.*, **98**, 7440 (1976).
81. J. Tsuji and H. Takayanagi, *Tetrahedron*, **34**, 641 (1978).
82. L. M. Sayre and D. V. Nadkarni, *J. Am. Chem. Soc.*, **116**, 3157 (1994).
83. (a) W. J. Mijs, J. H. van Dijk, W. B. G. Huysmans and J. G. Westra, *Tetrahedron*, **25**, 4233 (1969).
(b) K. Maruyama, H. Tsukube and T. Araki, *Chem. Lett.*, 499 (1979).
84. P. J. Baesjou, W. L. Driessen, G. Challa and J. Reedijk, *J. Am. Chem. Soc.*, **119**, 12590 (1997).
85. D. A. Rockcliffe and A. R. Martell, *J. Chem. Soc., Chem. Commun.*, 1758 (1992).
86. N. Kitajima, T. Koda, Y. Iwata and Y. Moro-oka, *J. Am. Chem. Soc.*, **112**, 8833 (1990).

87. P. P. Paul, Z. Tyeklár, R. R. Jacobson and K. D. Karlin, *J. Am. Chem. Soc.*, **113**, 5322 (1991).
88. V. Mahadevan, J. L. DuBois, B. Hedman, K. O. Hodgson and T. D. P. Stack, *J. Am. Chem. Soc.*, **121**, 5583 (1999).
89. M. Shimizu, Y. Watanabe, H. Orita, T. Hayakawa and K. Takehira, *Bull. Chem. Soc. Jpn.*, **66**, 251 (1993).
90. S. Mandal, D. Macikenas, J. D. Protasiewicz and L. M. Sayre, *J. Org. Chem.*, **65**, 4804 (2000).
91. T. Matsuura, *Tetrahedron*, **33**, 2869 (1977).
92. M. Frostin-Rio, D. Pujol, C. Bied-Charreton, M. Perrée and A. Gaudemer, *J. Chem. Soc., Perkin Trans. 1*, 1971 (1984).
93. M. Yamada, K. Araki and S. Shiraishi, *J. Chem. Soc., Perkin Trans. 1*, 2687 (1990).
94. P. A. Ganeshpure, A. Sudalai and S. Satish, *Tetrahedron Lett.*, **30**, 5929 (1989).
95. M. Tada and T. Katsu, *Bull. Chem. Soc. Jpn.*, **45**, 2558 (1972).
96. M. Chibwe and T. J. Pinnavaia, *J. Chem. Soc., Chem. Commun.*, 278 (1993).
97. W. F. Lau and B. M. Pettitt, *J. Am. Chem. Soc.*, **111**, 4111 (1989).
98. H. Nishino, H. Sato, M. Yamashita and K. Kurosawa, *J. Chem. Soc., Perkin Trans. 2*, 1919 (2000).
99. M. Lissel, H. Jansen in de Wal and R. Neumann, *Tetrahedron Lett.*, **33**, 1795 (1992).
100. S. Fujibayashi, K. Nakayama, Y. Nishiyama and Y. Ishii, *Chem. Lett.*, 1345 (1994).
101. J. Knaudt, S. Föster, U. Bartsch, A. Rieker and E.-G. Jäger, *Z. Naturforsch.*, **55b**, 86 (2000).
102. A. Zombeck, R. S. Drago, B. B. Corden and J. H. Gaul, *J. Am. Chem. Soc.*, **103**, 7580 (1981).
103. Y. Aimoto, W. Kanda, S. Meguro, Y. Miyahra, H. Okawa and S. Kido, *Bull. Chem. Soc. Jpn.*, **58**, 646 (1985).
104. J. J. Bozell, B. R. Hames and D. R. Dimmel, *J. Org. Chem.*, **60**, 2398 (1995).
105. A. Nishinaga, H. Iwasaki, T. Shimizu, Y. Toyoda and T. Matsuura, *J. Org. Chem.*, **51**, 2257 (1986).
106. M. Kodera, Y. Tachi, S. Hirota, K. Katayama, H. Shimakoshi, K. Kano, K. Fujisawa, Y. Moro-oka, Y. Naruta and T. Kitagawa, *Chem. Lett.*, 389 (1998).
107. M. Kodera, H. Shimakoshi, Y. Tachi, K. Katayama and K. Kano, *Chem. Lett.*, 441 (1998).
108. K. Maruyama, T. Kusukawa, T. Mashino and A. Nishinaga, *J. Org. Chem.*, **61**, 3342 (1996).
109. B. C. Gilbert, N. W. J. Kamp, J. R. Lindsay and J. Oakes, *J. Chem. Soc., Perkin Trans. 2*, 2161 (1997).
110. N. Colclough and J. R. Linsay Smith, *J. Chem. Soc., Perkin Trans. 2*, 1139 (1994).
111. A. Sorokin, S. De Suzzoni-Dezard, D. Poullain, J.-P. Noël and B. Meunier, *J. Am. Chem. Soc.*, **118**, 7410 (1996).
112. S. Ito, K. Aihara and M. Matsumoto, *Tetrahedron Lett.*, **24**, 5249 (1983).
113. W. Adam, W. A. Herrmann, J. Lin and C. R. Saha-Möller, *J. Org. Chem.*, **59**, 8281 (1994).
114. M. Shimizu, H. Orita, T. Hayakawa and K. Takehira, *Tetrahedron Lett.*, **30**, 471 (1989).
115. R. J. Mahalingam and P. Selvam, *Chem. Lett.*, 455 (1999).
116. J. S. Reddy and A. Sayari, *J. Chem. Soc., Chem. Commun.*, 2231 (1995).
117. M. Shimizu, H. Orita, T. Hayakawa, Y. Watanabe and K. Takehira, *Bull. Chem. Soc. Jpn.*, **64**, 2583 (1991).
118. S. Murahashi, T. Naota, N. Miyaguchi and S. Noda, *J. Am. Chem. Soc.*, **118**, 2509 (1996).
119. (a) W. S. Trahanovsky (Ed.), *Oxidation in Organic Chemistry*, Academic Press, New York, 1973.

    (b) R. Porter and S. Clark (Eds.), *Enzymes in Organic Chemistry*, Pitman Press, London, 1985.

    (c) K. Drauz and H. Waldmann (Eds.), *Enzyme Catalysis in Organic Synthesis*, VCH, Weinheim, 1995.

    (d) T. Loehr (Ed.), *Carrier and Iron Proteins*, VCH Publishers, New York, 1989.

    (e) S. M. Roberts (Ed.), *Biocatalysts for Fine Chemicals Synthesis*, Wiley, New York, 1999.

    (f) R. N. Patel (Ed.), *Stereoselective Biocatalysis*, Marcel Dekker, New York, 2000.
120. D. H. R. Barton, *Chem. Br.*, **3**, 330 (1967).
121. T. Kametani, K. Fukumoto and F. Sato, in *Bioorganic Chemistry* (Ed. E. E. van Tamelen), Vol. II, Chap. 7, Academic Press, New York, 1978, pp. 153–174.
122. (a) T. Kametani and K. Fukumoto, *Synthesis*, 657 (1972).

    (b) P. D. McDonald and G. M. Hamilton, in *Oxidation in Organic Chemistry*, Vol. 5b (Ed. W. S. Trahanovsky), Academic Press, New York, 1973, pp. 97–134.

    (c) S. Tobinaga, *Bioorg. Chem.*, **4**, 110 (1975).

123. D. C. Ayres and J. C. D. Loike, *Lignans: Chemical, Biological and Clinical Properties*, Cambridge University Press, Cambridge, 1990.
124. M. H. Zenk, R. Gerardy and R. Stadler, *J. Chem. Soc., Chem. Commun.*, 1725 (1989).
125. T. Umezawa, H. Kuroda, T. Isohata, T. Higuchi and M. Shimada, *Biosci. Biotech.*, **58**, 230 (1994).
126. (a) K. Syrjänen and G. Brunow, *J. Chem. Soc., Perkin Trans. 1*, 183 (2000).
    (b) K. Syrjänen and G. Brunow, *Tetrahedron*, **57**, 365 (2001).
127. M. Orlandi, B. Rindone, G. Molteni, P. Rummakko and G. Brunow, *Tetrahedron*, **57**, 371 (2001).
128. P. Lanccake and R. J. Pryce, *J. Chem. Soc., Chem. Commun.*, 208 (1977).
129. J. Ito and M. Niwa, *Tetrahedron*, **52**, 9991 (1996).
130. G. Pandey, C. M. Krishna and U. T. Bhalerao, *Tetrahedron Lett.*, **31**, 3771 (1990).
131. Z. Grujic, I. Tabakovic and Z. Bejtovic, *Tetrahedron Lett.*, 4823 (1976).
132. R. Z. Kazandjian and A. M. Klibanov, *J. Am. Chem. Soc.*, **107**, 5448 (1985).
133. (a) U. T. Bhalerao, C. Muralikrishna and G. Pandey, *Synth. Commun.*, **19**, 1303 (1989).
    (b) G. Pandey, C. Muralikrishna and U. T. Bhalerao, *Tetrahedron*, **45**, 6867 (1989).
134. G. H. Müller, A. Lang, D. S. Seithel and H. Waldmann, *Chem. Eur. J.*, **4**, 2513 (1998).
135. U. T. Bhalerao, C. M. Krishna and G. Pandey, *J. Chem. Soc., Chem. Commun.*, 1176 (1992).
136. J. Cooksey, P. J. Garratt, E. J. Land, C. A. Ramsden and P. A. Riley, *Biochem. J.*, **333**, 685 (1998).
137. Z.-W. Guo, G. M. Salamonczyk, K. Han, K. Machida and C. J. Sih, *J. Org. Chem.*, **62**, 6700 (1997).
138. Z.-W. Guo, K. Machida, Y.-A. Ma and C. J. Sih, *Tetrahedron Lett.*, **39**, 5679 (1998).
139. G. F. Koser, in *The Chemistry of Functional Groups, Supplement D. The Chemistry of Halides, Pseudohalides and Azides* (Eds. S. Patai and Z. Rappoport), Wiley, New York, 1983, pp. 721–811.
140. (a) R. M. Moriarty and O. Prakash, *Acc. Chem. Res.*, **19**, 244 (1986).
    (b) R. M. Moriarty and R. K. Vaid, *Synthesis*, 431 (1990).
    (c) A. Varvoglis, *The Organic Chemistry of Polycoordinated Iodine*, VCH Publisher Inc., New York, 1992.
    (d) A. Varvoglis, *Hypervalent Iodine in Organic Synthesis*, Academic Press, New York, 1997.
    (e) A. Varvoglis, *Tetrahedron*, **53**, 1179 (1997).
    (f) A. Pelter and R. Ward, *Tetrahedron*, **57**, 273 (2001).
141. L. Kürti, P. Herczegf, J. Visy, S. Antus and A. Pelter, *J. Chem. Soc., Perkin Trans. 1*, 379 (1999).
142. (a) A. Pelter and S. M. A. Elgendy, *Tetrahedron Lett.*, **29**, 677 (1988).
    (b) A. Pelter and S. M. A. Elgendy, *J. Chem. Soc., Perkin Trans. 1*, 1891 (1993).
143. N. Lewis and P. Wallbank, *Synthesis*, 1103 (1987).
144. Y. Tamura, T. Yakura, J. Haruta and Y. Kita, *J. Org. Chem.*, **52**, 3927 (1987).
145. O. Karam, A. Martin, M. P. Jouannetaud and J. C. Jacquesy, *Tetrahedron Lett.*, **40**, 4183 (1999).
146. A. McKillop, L. McLaren and J. K. Taylor, *J. Chem. Soc., Perkin Trans. 1*, 2047 (1994).
147. A. Callinan, Y. Chen, G. W. Morrow and J. S. Swenton, *Tetrahedron Lett.*, **31**, 4551, (1990).
148. Y. Tamura, T. Yakura, H. Tohma, K. Kikuchi and Y. Kita, *Synthesis*, 126 (1989).
149. S. Quideau, M. A. Looney, L. Pouységu, S. Ham and D. M. Birney, *Tetrahedron Lett.*, **40**, 615 (1999).
150. G. Well, A. Seaton and M. F. G. Stevens, *J. Med. Chem.*, **43**, 1550 (2000).
151. (a) M. Ochiai, T. Ito, H. Takahashi, A. Nakanishi, M. Toyonari, T. Sueda, S. Goto and M. Shiro, *J. Am. Chem. Soc.*, **118**, 7716 (1996).
    (b) M. Ochiai, T. Ito, Y. Masaki and M. Shiro, *J. Am. Chem. Soc.*, **114**, 6269 (1992).
    (c) R. A. Moss and H. Zhang, *J. Am. Chem. Soc.*, **116**, 4471 (1994).
152. M. Ochiai, A. Nakanishi and A. Yamada, *Tetrahedron Lett.*, **38**, 3927 (1997).
153. (a) C.-C. Liao, C.-S. Chu, T.-H. Lee, P. D. Rao, S. Ko, L.-D. Song and H.-C. Shiao, *J. Org. Chem.*, **64**, 4102 (1999).
    (b) C.-H. Lai, Y.-L. Shen and C.-C. Liao, *Synlett*, 1351 (1997).
154. S.-Y. Gao, Y.-L. Lin, P. D. Rao and C.-C. Liao, *Synlett.*, 421 (2000).
155. C.-H. Chen, P. D. Rao and C.-C. Liao, *J. Am. Chem. Soc.*, **120**, 13254 (1998).
156. P. D. Rao, C.-H. Chen and C.-C. Liao, *Chem. Commun.*, 713 (1999).

157. M.-F. Hsieh, P. D. Rao and C.-C. Liao, *Chem. Commun.*, 1441 (1999).
158. D.-S. Hsu, P. D. Rao and C.-C. Liao, *Chem. Commun.*, 1795 (1998).
159. J.-T. Hwang and C.-C. Liao, *Tetrahedron Lett.*, **32**, 6583 (1991).
160. (a) D.-S. Hsu, P.-Y. Hsu and C.-C. Liao, *Org. Lett.*, **3**, 263 (2001).
    (b) P.-Y. Hsu, Y.-C. Lee and C.-C. Liao, *Tetrahedron Lett.*, **39**, 659 (1998).
161. (a) C.-S. Chu, T.-H. Lee, P. D. Rao, L.-D. Song and C.-C. Liao, *J. Org. Chem.*, **64**, 4111 (1999).
    (b) C.-S. Chu, T.-H. Lee and C.-C. Liao, *Synlett*, 635 (1994).
162. (a) T.-H. Lee, P. D. Rao and C.-C. Liao, *Chem. Commun.*, 801 (1999).
    (b) P. D. Rao, C.-H. Chen and C.-C. Liao, *Chem. Commun.*, 155 (1998).
    (c) C.-S. Chu, C.-C. Liao and P. D. Rao, *Chem. Commun.*, 1537 (1996).
    (d) T.-H. Lee and C.-C. Liao, *Terahedron Lett.*, **37**, 6869 (1996).
    (e) C.-C. Liao and C.-P. Wei, *Tetrahedron Lett.*, **32**, 4553 (1991).
163. W.-C. Liu and C.-C. Liao, *Chem. Commun.*, 117 (1999).
164. P. A. Rose, B. Lei, A. C. Shaw, M. K. Walker-Simmons, S. Napper, J. W. Quail and S. R. Abrams, *Can. J. Chem.*, **74**, 1836 (1996).
165. (a) A. S. Mitchell and R. A. Russell, *Tetrahedron*, **53**, 4387 (1997).
    (b) A. S. Mitchell and R. A. Russell, *Tetrahedron Lett.*, **34**, 545 (1993).
166. Y.-F. Tsai, R. Krishna Peddinti and C.-C. Liao, *Chem. Commun.*, 475 (2000).
167. S. Wang, B. D. Gates and J. S. Swenton, *J. Org. Chem.*, **56**, 1979 (1991).
168. A. Pelter, A. Hussain, G. Smith and R. S. Ward, *Tetrahedron*, **53**, 3879 (1997).
169. R. S. Ward, A. Pelter and A. Abd-El-Ghani, *Tetrahedron*, **52**, 1303 (1996).
170. D. White, G. Caravatti, T. B. Kline, E. Edstron, K. C. Rice and A. Brossi, *Tetrahedron*, **39**, 2393 (1983).
171. J. D. White, W. K. M. Chong and K. Thirring, *J. Org. Chem.*, **48**, 2302 (1983).
172. Y. Kita, T. Takada, M. Gyoten, H. Tohma, M. H. Zenk and J. Eichhorn, *J. Org. Chem.*, **61**, 5857 (1996).
173. Y. Kita, M. Arisawa, M. Gyoten, M. Nakajima, R. Hamada, H. Tohma and T. Takada, *J. Org. Chem.*, **63**, 6625 (1998).
174. G. Büchi, P.-S. Chu, A. Hoppmann, C.-P. Mak and A. Pearce, *J. Org. Chem.*, **43**, 3983 (1978).
175. G Büchi and P.-S. Chu, *J. Org. Chem.*, **43**, 3717 (1978).
176. G Büchi and P.-S. Chu, *J. Am. Chem. Soc.*, **103**, 2718 (1981).
177. (a) G. Büchi and P.-S. Chu, *J. Am. Chem. Soc.*, **101**, 6767 (1979).
    (b) G. Büchi and P.-S. Chu, *Tetrahedron*, **37**, 4509 (1981).
178. J. L. Collins, P. A. Grieco and J. K. Walker, *Tetrahedron Lett.*, **38**, 1321 (1997).
179. P. A. Grieco and J. K. Walker, *Tetrahedron*, **53**, 8975 (1997).
180. G. Cardillo, M. Orena, G. Porzi and S. Sandri, *J. Chem. Soc., Chem. Commun.*, 836 (1979).
181. L. Jurd, *Tetrahedron*, **33**, 163 (1977).
182. K. Schofield, R. S. Ward and A. M. Choudhury, *J. Chem. Soc. (C)*, 2834 (1971).
183. G. Sartori, R. Maggi, F. Bigi and M. Grandi, *J. Org. Chem.*, **58**, 7271 (1993).
184. (a) R. Curci, in *Advances in Oxygenated Procedure* (Ed. A. L. Baumstark), Vol. 2, Chap. 1, JAI Press, Greenwich, CT, 1990.
    (b) R. W. Murray, *Chem. Rev.*, **89**, 1187 (1989).
    (c) W. Adam, R. Curci and J. O. Edwards, *Acc. Chem. Res.*, **22**, 205 (1989).
185. K. Crandall, M. Zucco, R. S. Kirsch and D. M. Coppert, *Tetrahedron Lett.*, **32**, 5441 (1991).
186. A. Altamura, C. Fusco, L. D'Accolti, R. Mello, T. Prencipe and R. Curci, *Tetrahedron Lett.*, **32**, 5445 (1991).
187. (a) H.-J, Teuber and W. Rau, *Chem. Ber.*, **86**, 1036 (1953).
    (b) H.-J. Teuber and M. Hasselbach, *Chem. Ber.*, **92**, 674 (1959).
    (c) H.-J. Teuber, *Org. Synth.*, **52**, 88 (1972).
188. S. Auricchio, A. Citterio and R. Sebastiano, *J. Org. Chem.*, **55**, 6312 (1990).
189. S. A. Hussain, T. C. Jenkins and M. J. Perkins, *J. Chem. Soc., Perkin Trans. 1*, 2809 (1979).
190. H. Zimmer, D. C. Lankin and S. W. Horgan, *Chem. Rev.*, **71**, 229 (1971).
191. P. M. Deya, M. Dopico, A. G. Raso, J. Morey and J. M. Saà, *Tetrahedron*, **43**, 3523 (1987).
192. M. Saà, J. Morey and A. Costa, *Tetrahedron Lett.*, **27**, 5125 (1986).
193. D. Nasipuri and A. K. Mitra, *J. Chem. Soc., Perkin Trans. 1*, 285 (1973).
194. K. MacLeod and B. R. Worth, *J. Chem. Soc., Chem. Commun.*, 718 (1973).

195. N. Matsumoto, H. Iinuma, T. Sawa and T. Takeuchi, *Bioorg. Med. Chem. Lett.*, **8**, 2945 (1998).
196. G. A. Kraus and N. Selvakumar, *J. Org. Chem.*, **63**, 9846 (1998).
197. J. V. Crivello, *J. Org. Chem.*, **46**, 3056 (1981).
198. J. J. Bozell and J. O. Hoberg, *Tetrahedron Lett.*, **39**, 2261 (1998).
199. (a) A. Nishinaga, T. Shimizu and T. Matsuura, *J. Org. Chem.*, **44**, 2983 (1979).
    (b) A. Nishinaga, T. Shimizu, T. Fujii and T. Matsuura, *J. Org. Chem.*, **45**, 4997 (1980).
200. K. Omura, *J. Org. Chem.*, **49**, 3046 (1984).
201. E. Adler, K. Holmberg and L.-O. Ryrfors, *Acta Chem. Scand.*, **B 28**, 888 (1974).
202. (a) H. Holmberg, H. Kirudd and G. Westin, *Acta Chem. Scand.*, **B 28**, 913 (1974).
    (b) E. Adler, G. Andersson and E. Edman, *Acta Chem. Scand.*, **B 29**, 909 (1975).
203. H.-D. Becker and K. Gustafsson, *J. Org. Chem.*, **44**, 428 (1979).
204. J. W. Jaroszewski, *J. Org. Chem.*, **47**, 2013 (1982).
205. (a) J.-C. Blazejewski, R. Dorme and C. Wakselman, *J. Chem. Soc., Perkin Trans. 1*, 1861 (1987).
    (b) J.-C. Blazejewski, R. Dorme and C. Wakselman, *Synthesis*, 1120 (1985).
206. A. McKillop and J. A. Tarbin, *Tetrahedron*, **43**, 1753 (1987).
207. F. Freeman, in *Organic Syntheses by Oxidation with Metal Compounds* (Eds. W. J. Mijs and C. R. H. I. de Jonge), Plenum Press, New York, 1986. pp. 8–25.
208. W. L. Carrick, G. L. Karapinka and G. T. Kwiatkowski, *J. Org. Chem.*, **34**, 2388 (1969).
209. J. Quick and R. Ramachandra, *Tetrahedron*, **36**, 1301 (1980).
210. (a) M. A. Schwartz, B. F. Rose, R. A. Halton, S. W. Scott and B. Vishurajjala, *J. Am. Chem. Soc.*, **99**, 2571 (1977).
    (b) D. T. Dalgleish, N. P. Forrest, D. C. Nonhebel and P. L. Pauson, *J. Chem. Soc., Perkin Trans. 1*, 584 (1977).
211. J. F. Harrod and A. Pathak, *Can. J. Chem.*, **58**, 686 (1980).
212. S. M. Kupchan and A. J. Liepa, *J. Am. Chem. Soc.*, **95**, 4062 (1973).
213. S. M. Kupchan, O. P. Dhingra and C.-K. Kim, *J. Org. Chem.*, **41**, 4049 (1976).
214. M. A. Schwartz and M. F. Zoda, *J. Org. Chem.*, **46**, 4623 (1981).
215. T. Kametani, A. Kozuka and K. Fukumoto, *J. Chem. Soc. (C)*, 1021 (1980).
216. M. A. Schwartz, R. A. Holton and S. W. Scott, *J. Am. Chem. Soc.*, **91**, 2800 (1969).
217. J. P. Marino and J. M. Samanen, *Tetrahedron Lett.*, 4553 (1973).
218. S. M. Kupchan, A. J. Liepa, V. Kameswaran and R. F. Bryan, *J. Am. Chem. Soc.*, **95**, 6861 (1973).
219. M. A. Schwartz and R. A. Holton, *J. Am. Chem. Soc.*, **92**, 1090 (1970).
220. S. Yamada, K. Tomioka and K. Koga, *Tetrahedron Lett.*, 57 (1976).
221. J. Liepa and R. E. Suammons, *J. Chem. Soc., Chem. Commun.*, 491 (1973).
222. D. Planchenault, R. Dhal and J.-P. Robin, *Tetrahedron*, **49**, 5823 (1993).
223. D. Planchenault, R. Dhal and J.-P. Robin, *Tetrahedron*, **49**, 1395 (1993).
224. S. Kende and L. S. Liebeskind, *J. Am. Chem. Soc.*, **98**, 267 (1976).
225. R. E. Damon, R. H. Schlessinger and J. F. Blount, *J. Org. Chem.*, **41**, 3772 (1976).
226. J. T. Gupton, J. P. Idoux, G. Baker, C. Colon, D. Crews, C. D. Jurss and R. C. Rampi, *J. Org. Chem.*, **48**, 2933 (1983).
227. F. Freeman, in *Organic Syntheses by Oxidation with Metal Compounds* (Eds. W. J. Mijs and C. R. H. I. de Jonge), Plenum Press, New York, 1986, pp. 81–82.
228. H.-D. Becker, *J. Org. Chem.*, **32**, 2943 (1967).
229. M. O. F. Goulart and J. H. P. Utley, *J. Org. Chem.*, **53**, 2520 (1988).
230. R. Cassis and J. A. Valderrama, *Synth. Commun.*, **13**, 347 (1983).
231. H. Bock and D. Jaculi, *Angew. Chem., Int. Ed. Engl.*, **23**, 305 (1984).
232. R. G. Srivastava and P. S. Venkataramani, *Synth. Commun.*, **22**, 35 (1992).
233. A. Marques, F. Simonelli, A. R. M. Oliveira, G. L. Gohr and P. C. Leal, *Tetrahedron Lett.*, **39**, 943 (1998).
234. W. J. de Klein, in *Organic Syntheses by Oxidation with Metal Compounds* (Eds. W. J. Mijs and C. R. H. I. de Jonge), Plenum Press, New York, 1986, pp. 292–294.
235. R. Stewart and B. L. Poh, *Can. J. Chem.*, **50**, 3437 (1972).
236. H. Nishino, N. Itoh, M. Nagashima and K. Kurosawa, *Bull. Chem. Soc. Jpn.*, **65**, 620 (1992).
237. D. T. Dalgleish, N. P. Forrest, D. C. Nonhebel and P. L. Pauson, *J. Chem. Soc., Perkin Trans. 1*, 584 (1977).

238. W. K. Seok and T. J. Meyer, *J. Am. Chem. Soc.*, **110**, 7358 (1988).
239. S. Tobinaga and E. Kotani, *J. Am. Chem. Soc.*, **94**, 309 (1972).
240. M. Murase, E. Kotani, K. Okazaki and S. Tobinaga, *Chem. Pharm. Bull.*, **34**, 3159 (1986).
241. T. C. Jempty, L. L. Miller and Y. Mazur, *J. Org. Chem.*, **45**, 750 (1980).
242. F. Toda, K. Tanaka and S. Iwata, *J. Org. Chem.*, **54**, 3007 (1989).
243. H.-D. Becker and K. Gustafsson, *Tetrahedron Lett.*, 4883 (1976).
244. E. Müller, R. Mayer, B. Narr, A. Rieker and K. Scheffler, *Justus Liebigs Ann. Chem.*, **645**, 25 (1961).
245. M. Tashiro, H. Yoshiya and G. Fukata, *J. Org. Chem.*, **46**, 3784 (1981).
246. S. Hauff and A. Rieker, *Tetrahedron Lett.*, 1451 (1972).
247. V. I. Minkin, E. P. Ivachnenko, A. I. Shif, L. P. Olekhnovitch, O. E. Kompan, A. I. Yanovskii and Y. T. Struchkov, *J. Chem. Soc., Chem. Commun.*, 990 (1988).
248. E. Manda, *Bull. Chem. Soc. Jpn.*, **46**, 2160 (1973).
249. R. C. Eckert, H. Chang and W. P. Tucker, *J. Org. Chem.*, **39**, 718 (1974).
250. C. Brunow and S. Sumelius, *Acta Chem. Scand.*, **B 29**, 499 (1975).
251. B. Feringa and H. Wynberg, *J. Org. Chem.*, **46**, 2547 (1981).
252. H.-P. Husson, C. Poupat, B. Rodriguez and P. Potier, *Tetrahedron*, **29**, 1405 (1973).
253. (a) J. F. Ajao, C. W. Bird and Y.-P. Chauhan, *Tetrahedron*, **41**, 1367 (1985).
     (b) C. W. Bird and Y.-P. Chauhan, *Tetrahedron Lett.*, 2133 (1978).
254. S. Kende, K. Koch and C. A. Smith, *J. Am. Chem. Soc.*, **110**, 2210 (1988).
255. T. Kametani, *Bioorg. Chem.*, **3**, 430 (1974).
256. (a) A. Mondon and M. Ebrhardt, *Tetrahedron Lett.*, 2557 (1966).
     (b) J. E. Gervay, F. McCapra, T. Money, G. M. Sharma and A. I. Scott, *J. Chem. Soc., Chem. Commun.*, 142 (1966).
257. M. Barrett, D. H. R. Barton, G. Franckowiak, D. Papaioannou and D. A. Widdowson, *J. Chem. Soc., Perkin Trans. 1*, 662 (1979).
258. E. McDonald and A. Suksamrarn, *J. Chem. Soc., Perkin Trans. 1*, 440 (1978).
259. E. McDonald, *J. Chem. Soc., Perkin Trans. 1*, 434 (1978).
260. (a) V. K. Mangla and D. S. Bhakuni, *Tetrahedron*, **36**, 2489 (1980).
     (b) E. McDonald and R. D. Wylie, *Tetrahedron*, **35**, 1415 (1979).
     (c) A. J. Birch, A. H. Jackson, P. V. R. Shannon and W. Stewart, *J. Chem. Soc., Perkin Trans. 1*, 2492 (1975).
261. A.-R. B. Manas and R. A. J. Smith, *J. Chem. Soc., Chem. Commun.*, 216 (1975).
262. L. Courtney, in *Organic Syntheses by Oxidation with Metal Compounds* (Eds. W. J. Mijs and C. R. H. I. de Jonge), Plenum Press, New York, 1986, pp. 448–449.
263. W. K. Seok and T. J. Meyer, *J. Am. Chem. Soc.*, **110**, 7358 (1988).
264. J.-P. Robin and Y. Landais, *J. Org. Chem.*, **53**, 224 (1988).
265. D. C. Ayres and R. Copalan, *J. Chem. Soc., Chem. Comunn.*, 890 (1976).
266. M. Hirano, T. Ishii and T. Morimoto, *Bull. Chem. Soc. Jpn.*, **64**, 1434 (1991).
267. (a) J. Sugita, *Nippon Kagaku Zasshi*, **87**, 607 (1966); *Chem. Abstr.*, **65**, 15522e (1966).
     (b) J. Sugita, *Nippon Kagaku Zasshi*, **87**, 1082 (1966); *Chem. Abstr.*, **66**, 94777w (1967).
268. J. Sugita, *Nippon Kagaku Zasshi*, **87**, 603 (1966); *Chem. Abstr.*, **65**, 15522c (1966).
269. Q. Zheng, Y. Yang and A. R. Martin, *Tetrahedron Lett.*, **34**, 2235 (1993).
270. P. Capdevielle and M. Maumy, *Tetrahedron*, **57**, 379 (2001).
271. T. Sakamoto, H. Yonehara and C. Pac, *J. Org. Chem.*, **62**, 3194 (1997).
272. T. Kametani and M. Ihara, *Heterocycles*, **12**, 893 (1979).
273. D. Edstrom and Z. Jones, *Tetrahedron Lett.*, **36**, 7039 (1995).
274. V. Balogh, M. Fetizon and M. Golfier, *Angew. Chem., Int. Ed. Engl.*, **8**, 444 (1969) and *J. Org. Chem.*, **36**, 1339 (1971).
275. M. Fetizon, M. Golfier, P. Mourgues and J.-M. Louis, in *Organic Syntheses by Oxidation with Metal Compounds* (Eds. W. J. Mijs and C. R. H. I. de Jonge), Plenum Press, New York, 1986, pp. 545–549.
276. T. Pal and A. Pal, *J. Indian Chem. Soc.*, **66**, 236 (1989).
277. D. A. Bolon, *J. Org. Chem.*, **38**, 1741 (1973).
278. F. West and H. W. Moore, *J. Org. Chem.*, **49**, 2809 (1984).
279. A. Zanarotti, *Tetrahedron Lett.*, **23**, 3815 (1982) and *J. Org. Chem.*, **50**, 941 (1985).
280. L. Merlini, A. Zanarotti, A. Pelter, M. P. Rochefort and R. Hänsel, *J. Chem. Soc., Perkin Trans. 1*, 775 (1980).

281. P. Karhunen, P. Rummakko, A. Pajunen and G. Brunow, *J. Chem. Soc., Perkin Trans. 1*, 2303 (1996).
282. L. Jurd, *J. Heterocycl. Chem.*, **21**, 81 (1984).
283. R. Iyer and G. K. Trivedi, *Bull. Chem. Soc. Jpn.*, **65**, 1662 (1992).
284. (a) A. McKillop and E. C. Taylor, in *Comprehensive Organometallic Chem.* (Ed. G. Wilkinson), Vol. 7, Pergamon Press, New York, 1982, p. 465.
    (b) A. McKillop and E. C. Taylor, in *Organic Syntheses by Oxidation with Metal Compounds* (Eds. W. J. Mijs and C. R. H. I. de Jonge), Plenum Press, New York, 1986, pp. 695–740.
285. C. Ferraz, L. F. Silva, Jr. and T. de O. Vieira, *Synthesis*, 2001 (1999).
286. A. Schwartz and I. S. Mami, *J. Am. Chem. Soc.*, **97**, 1239 (1975).
287. A. Schwartz and R. A. Wallace, *Tetrahedron Lett.*, 3257 (1979).
288. E. C. Taylor, J. G. Andrade, G. H. J. Rall, K. Steliou, G. E. Jagdmann and A. McKillop, *J. Org. Chem.*, **46**, 3078 (1981).
289. E. C. Taylor, J. G. Andrade, G. J. H. Rall, I. J. Turchi, K. Steliou, G. E. Jagdmann, Jr. and A. McKillop, *J. Am. Chem. Soc.*, **103**, 6856 (1981).
290. S. Nishiyama and S. Yamamura, *Chem. Lett.*, 1511 (1981).
291. S. Nishiyama and S. Yamamura, *Bull. Chem. Soc. Jpn.*, **58**, 3453 (1985).
292. A. McKillop, D. H. Perry, M. Edwards, S. Antus, L. Farkas, M. Nógrádi and E. C. Taylor, *J. Org. Chem.*, **41**, 282 (1976).
293. Y. Yamada, K. Hosaka, H. Sanjoh and M. Suzuki, *J. Chem. Soc., Chem. Commun.*, 661 (1974).
294. Y. Yamada and K. Hosaka, *Synthesis*, 53 (1977).
295. (a) S. Nishiyama and S. Yamamura, *Tetrahedron Lett.*, **23**, 3247 (1982).
    (b) S. Nishiyama, T. Suzuki and S. Yamamura, *Tetrahedron Lett.*, **23**, 3699 (1982).
296. S. Nishiyama, T. Suzuki and S. Yamamura, *Chem. Lett.*, 1851 (1982).
297. S. Nishiyama, K. Nakamura, Y. Suzuki and S. Yamamura, *Tetrahedron Lett.*, **27**, 4481 (1986).
298. S. Nishiyama, Y. Suzuki and S. Yamamura, *Tetrahedron Lett.*, **29**, 559 (1988).
299. S. Nishiyama, Y. Suzuki and S. Yamamura, *Tetrahedron Lett.*, **30**, 379 (1989).
300. (a) T. Inaba, I. Umezawa, M. Yuasa, T. Inoue, S. Mihashi, H. Itokawa and K. Ogura, *J. Org. Chem.*, **52**, 2957 (1987).
    (b) T. Inoue, T. Inaba, M. Yuasa, H. Itokawa, K. Ogura, K. Komatsu, H. Hara and O. Hoshino, *Chem. Pharm. Bull.*, **43**, 1325 (1995).
301. (a) D. A. Evans, C. J. Dinsmore and A. M. Ratz, *Tetrahedron Lett.*, **38**, 3189 (1997).
    (b) D. A. Evans, J. C. Barrow, P. S. Watson, A. M. Ratz, C. J. Dismore, D. A. Evrard, K. M. DeVries, J. A. Ellman, S. D. Rychnovsky and J. Lacour, *J. Am. Chem. Soc.*, **119**, 3419 (1997).
302. D. A. Evans, M. R. Wood, B. W. Trotter, T. I. Richardson, J. C. Barrow and J. L. Katz, *Angew. Chem., Int. Ed.*, **37**, 2700 (1998).
303. K. C. Nicolaou, S. Natarajan, H. Li, N. F. Jain, R. Hughes, M. E. Solomon, J. M. Ramanjulu, C. N. C. Boody and M. Takayanagi, *Angew. Chem., Int. Ed.*, **37**, 2708 (1998).
304. R. Xu, G. Greiveldinger, L. E. Marenus, A. Cooper and J. A. Ellman, *J. Am. Chem. Soc.*, **121**, 4898 (1999).
305. J. S. Sawyer, *Tetrahedron*, **56**, 5045 (2000).
306. Y. Suzuki, S. Nishiyama and S. Yamamura, *Tetrahedron Lett.*, **31**, 4053 (1990).
307. K. Nakamura, S. Nishiyama and S. Yamamura, *Tetrahedron Lett.*, **36**, 8621 (1995).
308. Y. Suzuki, S. Nishiyama and S. Yamamura, *Tetrahedron Lett.*, **30**, 6043 (1989).
309. H. Konishi, T. Okuno, S. Nishiyama, S. Yamamura, K. Koyasu and Y. Terada, *Tetrahedron Lett.*, **37**, 8791 (1996).
310. (a) S. Yamamura and S. Nishiyama, *J. Synth. Org. Chem., Jpn.*, **55**, 1029 (1997).
    (b) K. Nakamura, S. Nishiyama and S. Yamamura, *Tetrahedron Lett.*, **37**, 191 (1996).
311. (a) F. Wessely, J. Kotlan and W. Metlesics, *Monatsh. Chem.*, **85**, 69 (1954).
    (b) F. Wessely and F. Sinwel, *Monatsh. Chem.*, **84**, 291 (1952).
    (c) F. Wessely, J. Kotlan and W. Metlesics, *Monatsh. Chem.*, **81**, 1055 (1950).
312. M. L. Mihailovic, Z. Cekovic and L. Loreng, in *Organic Syntheses by Oxidation with Metal Compounds* (Eds. W. J. Mijs and C. R. H. I. de Jonge), Plenum, New York, 1986, pp. 781–783.
313. D. J. Bichan and P. Yates, *Can. J. Chem.*, **53**, 2054 (1975).
314. R. J. Andersen and D. J. Faulkner, *J. Am. Chem. Soc.*, **97**, 936 (1975).

315. G. Büchi, M. A. Francisco and P. Michel, *Tetrahedron Lett.*, **24**, 2531 (1983).
316. O. Hoshino, M. Suzuki and H. Ogasawara, *Heterocycles*, **52**, 751 (2000).
317. O. Hoshino, M. Suzuki and H. Ogasawara, *Tetrahedron*, **57**, 265 (2001).
318. P. Kitchin, in *Organic Syntheses by Oxidation with Metal Compounds* (Eds. W. J. Mijs and C. R. H. I. de Jonge), Plenum Press, New York, 1986, pp. 820–824.
319. E. Kon and E. McNelis, *J. Org. Chem.*, **40**, 1515 (1975).
320. E. Kon and E. McNelis, *J. Org. Chem.*, **41**, 1646 (1976).
321. M. J. Harrison and R. O. C. Norman, *J. Chem. Soc. (C)*, 728 (1970).
322. F. R. Hewgill, B. R. Kennedy and D. Kilpin, *J. Chem. Soc.*, 2994 (1965).
323. D. H. R. Barton, J.-P. Finet, C. Giannotti and F. Halley, *Tetrahedron*, **44**, 4483 (1988).
324. S. V. Ley, in *Organoselenium Chemistry* (Ed. D. Liotta), Wiley, New York, 1987, pp. 163–206.
325. (a) D. H. R. Barton, A. G. Brewster, S. V. Ley and M. N. Rosenfeld, *J. Chem. Soc., Chem. Commun.*, 985 (1976).
(b) D. H. R. Barton, S. V. Ley, P. D. Magnus and M. N. Rosenfeld, *J. Chem. Soc., Perkin Trans. 1*, 567 (1977).
(c) D. H. R. Barton, A. G. Brewster, S. V. Ley, C. M. Read and M. N. Rosenfeld, *J. Chem. Soc., Perkin Trans. 1*, 1473 (1981).
326. D. H. R. Barton, J.-P. Finet and M. Thomas, *Tetrahedron*, **44**, 6397 (1988).
327. L. Henriksen, *Tetrahedron Lett.*, **35**, 7057 (1994).
328. T.-L. Ho, in *Organic Syntheses by Oxidation with Metal Compounds* (Eds. W. J. Mijs and C. R. H. I. de Jonge), Plenum Press, New York, 1986, pp. 569–631.
329. K. Maruyama and T. Kozuka, *Chem. Lett.*, 341 (1980).
330. P. Jacob, III, P. S. Callery, A. T. Shulgin and N. Castagnoli, Jr., *J. Org. Chem.*, **41**, 3627 (1976).
331. Y. Sakito and G. Suzukamo, *Tetrahedron Lett.*, **23**, 4953 (1982).
332. T. Izawa, S. Nishiyama and S. Yamamura, *Tetrahedron*, **50**, 13593 (1994).

CHAPTER 18

Environmental effects of substituted phenols

VICTOR GLEZER

National Public Health Laboratory, Ministry of Health, 69 Ben Zvi Rd., Tel Aviv, Israel
Fax: 972-3-6826996; e-mail: victor.glezer@phlta.health.gov.il

I. INTRODUCTION	1347
II. PRODUCTION AND USE	1348
III. ANALYSIS OF PHENOLS, THEIR CONCENTRATION AND SPECIATION IN THE NATURAL ENVIRONMENT	1351
A. Introduction	1351
B. Solubility and pK_a	1352
C. Analysis—Sample Preparation and Methods of Determination	1352
D. Phenols in the Environment	1355
IV. TOXICITY AND HEALTH EFFECTS	1357
A. Introduction	1357
B. Drinking Water Regulations. Taste and Odor	1358
C. Toxicity	1359
D. 'Nontoxic' Biological Activity of Phenols	1361
V. DETOXIFICATION AND DEGRADATION	1361
A. Chemical and Physicochemical Degradation	1361
B. Biodegradation and Biomethylation	1362
VI. REFERENCES	1363

I. INTRODUCTION

Phenols are highly important, well-known and widely used compounds in different fields of the chemical industry. This group of compounds found its application in the manufacture of plastics and plasticizers, explosives, drugs, colors and detergents[1,2]. Different substituted phenols are included among herbicides, insecticides, algaecides, bactericides, molluscicides, fungicides etc.[3]. Many pharmaceuticals contain phenol fragments displaying different kinds of biological activity[4]. They are also widely used in the petrochemical industry and as wood preservative agents[5–8].

The Chemistry of Phenols. Edited by Z. Rappoport
© 2003 John Wiley & Sons, Ltd ISBN: 0-471-49737-1

Widespread use of phenols, often in large-scale production, leads to their unavoidable appearance in the environment. Large amounts of phenols are generated from lignin degradation in paper production[9]. Nitrophenols are formed photochemically in the atmosphere from vehicle exhausts[10]. Phenols can also be formed by degradation of organophosphorous insecticides and chlorophenoxyacetic acids[11]. High solubility of phenols in water[12] explains their easy migration within different aqueous environments and contamination of groundwater[13]. Their toxicity and unpleasant organoleptic properties (a concentration of a few $\mu g\, l^{-1}$ affects the taste and odor of water and fish) was the reason to classify 11 phenols as 'priority pollutants' by the US Environmental Protection Agency (EPA)[14,15]. The European Union (EU) has also classified several phenols as priority contaminants with a maximum concentration of 0.5 $\mu g\, l^{-1}$ of total phenols in drinking water, demanding that each individual concentration be under 0.1 $\mu g\, l^{-1}$ [16,17]. Appearance of phenols in surface water or groundwater leads to formation of more toxic chlorinated phenols during water disinfection processes. One chlorinated phenol representative, i.e. pentachlorophenol, has been used throughout the world as a wood preservative and general biocide[18]. Its residue is widespread in the environment. In an EPA study pentachlorophenol was found in 80% (!) of human urine specimens[18]. Even though pentachlorophenol is included in the EPA priority pollutant list[15], its pyrolysis and combustion reaction products, i.e. polychlorodibenzofurans and polychlorodibenzodioxins, are considerably more toxic[19]. One of the major surfactant groups is alkyl phenol ethoxylates. The surfactants themselves show very low toxicity, but their degradation products, nonyl- and octylphenols, adsorb readily onto suspended soils and are known to exhibit estrogen-like properties, possibly linked to carcinogenic effects and to a decrease in males' sperm count[20]. The wide-ranging use of phenols, combined with their toxicity and unavoidable discharge of considerable amounts into the environment, has promoted extensive research on phenolic compounds and their fate in the environment.

II. PRODUCTION AND USE

In the US, phenols are ranked in the top 50 major chemicals. In 1995 the total annual production of phenols was estimated at 4–5 billion pounds[21–23]. In Japan the production of phenols in the late nineties was estimated approximately on the same level—1,200,000 tons per year[24]. In 1995, 95% of US phenol production was based on oxidation of cumene, the exception being one company that used toluene oxidation and some companies that distilled phenol from petroleum[23]. Two major uses of phenols in 1995 were the production of Bisphenol-A [4,4′-isopropylidenediphenol] (35%) and the production of phenolic resins (34%). Other uses include production of caprolactam (15%), aniline, (5%), alkylphenols (5%), xylenols (5%) and other miscellaneous compounds (1%)[25]. Bisphenol A is one of the raw materials widely used in the production of epoxy resins[26]. Being inert, strong and adhesive with high insulator properties these polymers found their application in construction, coatings and bonding. In addition, Bisphenol A is used for production of polycarbonate plastics, found in such products as baby food bottles, food cans, dental sealants, food packing and coatings. In the US alone, 1.65 billion pounds of this polymeric compound are produced each year[25], and in Japan its production is estimated nowadays at over 200,000 tons per year[27]. Another group of very widely used compounds is phenols with long aliphatic chains R like octyl or nonyl as shown below. These compounds are important intermediates in the production of polyethoxylate surfactants, which are compounds consisting of alkyl chains attached to a phenol ring and combined with a variable number of ethylene oxides. In 1994 their production in EC countries reached 110,000 tons[28], mainly for industrial, agricultural and household uses[29,30]. Moreover, the annual production in all developed countries has been estimated at 0.35 Mton[31].

These compounds yield by their biodegradation the more toxic 4-nonylphenol[32–34]. Owing to their poor ultimate biodegradability and the possible environmental hazard of their metabolites, alkylphenol ethoxylates have been replaced in household applications, mainly by alcohol ethoxylates. However, for industrial applications, this replacement has not been carried out yet due to the excellent performance of alkylphenol ethoxylates and their low production costs[35].

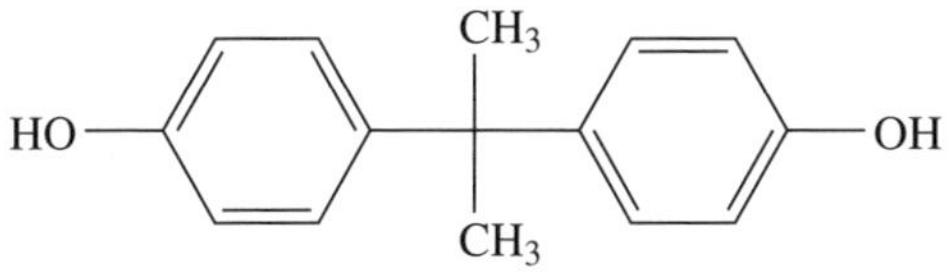

Bisphenol A

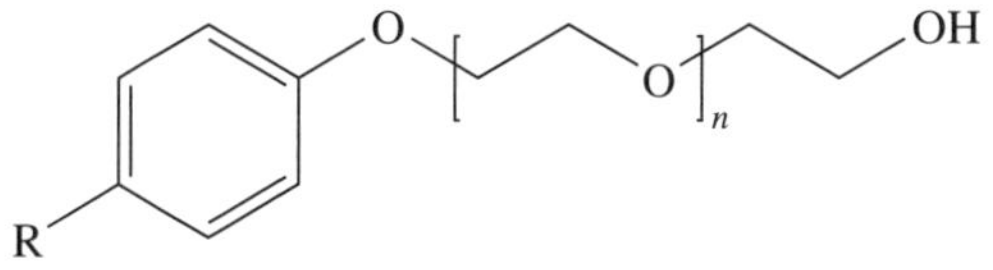

Formula of polyethoxylate surfactants, R = octyl or nonyl

3-*tert*-Butyl-4-methoxyphenol, 2,6-bis(1,1-dimethyl)-4-methylphenol and some other sterically hindered phenols with methyl and *tert*-butyl substituents are generally used as antioxidants in the food industry, primarily in foods with fats, planned for long storage periods, like pastries, cakes, biscuits, frozen meat, frozen fruits, potato chips etc. Alkylphenol compositions are also used in the manufacture of food packing materials such as waxed paper, paperboard and polyethylene. Members of this group of alkylphenols have synergistic effects on antioxidant activity, and influence each other's behavior when more than one is used in the same system[36]. 2,6-Bis(1,1-dimethyl)-4-methylphenol is widely used as an additive in lubricants, turbine and insulating oils, natural and synthetic rubbers, paints, plastics and elastomers. It protects these materials from oxidation by atmospheric oxygen during service and storage conditions[36].

Phenol fragments are an integral part of drugs like analgetic, antipyretic (for example, acetaminophen, better known as paracetamol) and anti-inflammatory (Rowasa, Salsalate) agents, bronchodilators (Albuterol), semisynthetic antibiotics (Amoxyl) and for treatment of Parkinson's disease (levodopa, carbidopa)[4].

Besides alkyl-substituted phenols, other very widely used phenol derivatives are halogenated phenols. Chlorinated phenols, the most common in this group, are manufactured by chlorination of phenol. Likewise, the higher chlorinated phenols are produced by chlorination of less chlorinated phenols at high temperature[37]. Nineteen different chlorinated phenols are commercially available. Both *o*- and *p*-dichlorophenols are used as intermediates in dyestuffs, as preservatives and in the manufacture of disinfectants. The monochlorophenols have been used as antiseptics[38], although in this role they have mostly been replaced by other chemicals[37]. 4-Chlorophenol has been used as a disinfectant for homes, hospitals and farms[37] and as an antiseptic for root canal treatment[39]. 2,4-Dichlorophenol has been used for mothproofing and as a miticide, while the higher phenols have been used as germicides, algaecides and fungicides[37]. 2,4-Dichlorophenol and 2,4,5-trichlorophenol are also used in the large-scale industrial synthesis of the herbicides 2,4-D

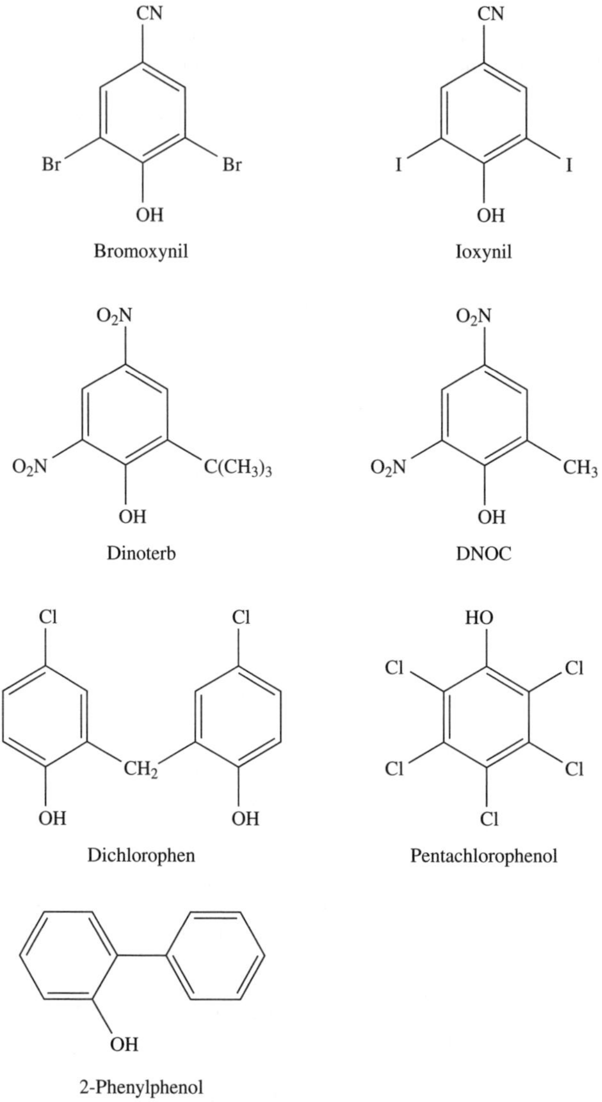

FIGURE 1. Examples of pesticides derived from phenols

and 2,4,5-T (Silvex), respectively[12]. The BASF Corp. in Texas is the largest manufacturer of chlorophenols in the USA with 100,000–900,000 pounds on site[40]. At the top of the list of large-scale produced chlorophenols is pentachlorophenol, used for the preservation of timber against fungal rots and wood-boring insects, and as a general herbicide or general disinfectant, e.g. for trays in mushroom houses[3,41]. One of the main formulations of pentachlorophenol is creosote oil, which also includes polycyclic hydrocarbons and heterocyclic compounds. In the US alone the production of this oil has reached 800 million liters per year[42]. Manufactured pentachlorophenol also contains 4% of tetrachlorophenol and 0.1% of trichlorophenol[43]. 2,4,6-Trichlorophenol and tetrachlorophenols have also been used directly as wood preservatives[38]. North America and Scandinavia are the main regions of the world where chlorinated phenols have been used as wood preservatives. However, the use of these compounds has been banned in Sweden since 1978, and production was banned in Finland in 1984[43]. Some examples of different types of pesticides, based on phenol structure, are presented in Figure 1. These pesticides include mainly dihalophenols, dinitrophenols and diphenol derivatives[3]. DNOC (2-methyl-4,6-dinitrophenol) is used as a herbicide, insecticide, ascaricide and fungicide, dichlorophen is used as an algaecide, bactericide and fungicide, pentachlorophenol is used as an insecticide, fungicide and herbicide, 2-phenylphenol is used as a fungicide and other phenols as herbicides.

Tetrabromobisphenol A, a brominated analog of Bisphenol A, is an important nonflammable additive in the production of synthetic resins, polycarbonates and plastics, used in the manufacture of computer and electronic housings, laminated electronic circuit boards, carpets, upholstery and many other consumer goods[44,45]. Tetrabromobisphenol A is used as a flame retardant to a much larger extent than its chlorinated analog tetrachlorobisphenol A[46,47].

3-Fluoromethyl-4-nitrophenol can be used as an example of a large-scale local distribution of phenols. From 1958 this compound was employed to control the sea lamprey (*Petromyzon marinus*) in four of the North American Great Lakes (Superior, Michigan, Huron and Ontario) by using approximately 50,000 kg per year, such that by 1988 more than 1 million kg had been applied[48]. This compound has also been introduced in order to control tadpole infestations in warm water ornamental fishponds[49].

A large amount of phenols is released in wastewater and can be lost to waste streams. A rapid increase in the distribution and abundance of plastic debris in the ocean around the world was reported, and the adverse influence of plastic's phenol residues has been of great interest[50-52]. Polluted water disinfection, enzymatic oxidation of chlorinated phenols, decomposition of alkylphenol polyethoxylates and combustion of phenols can lead to the formation of highly toxic compounds. High adsorption of phenols on sludge and sediments requires that their distribution in these systems also be followed. All of these facts have promoted extensive research on phenolic compounds and their fate in the environment.

III. ANALYSIS OF PHENOLS, THEIR CONCENTRATION AND SPECIATION IN THE NATURAL ENVIRONMENT

A. Introduction

Wastewaters from plastic and polymer production, fossil fuel refining, pharmaceutical and pesticide factories are the main sources of phenol pollution. Phenols discharged into municipal sewers or rivers can be transported over great distances because of their stability and water solubility. Nonchlorinated phenols are found in aquatic environments as biodegradation products of humic substances, lignins and tannins, or as derivatives of plastics, dye industries and pulp processing. Phenolic resins are utilized as binding materials in semiconductor industry products such as chipboards, paints and insulating materials. Phenolic compounds react rapidly with hypochloric acid by electrophilic attack on phenoxide anions forming the corresponding chlorophenols. Chlorophenols can be generated from phenols by chlorination of drinking water, or formed from different industrial activities (chemicals, conservation agents etc.) or degradation of other pollutants like pesticides etc. Being toxic and only partly biodegradable, phenols nevertheless were found in water[53,54], baby food bottles[55], plastic wastes[56] and living organisms like fishes, humans etc.[57,58]. Polychlorinated phenols can be further transformed into more toxic dimers, such as polychlorinated dibenzofurans and dibenzodioxins, by oxidative processes such as enzymatic reactions[59–61]. Therefore, the detection, identification and quantitation of phenol compounds in water and their subsequent monitoring is of great importance for the control and protection of the environment and for emission control.

B. Solubility and pK_a

Distribution of hazardous materials depends not only on the amount produced and its leakage to the environment, but also on its solubility in water. High concentrations of phenols in water are possible only in the case of highly soluble derivatives. The solubility of phenols depends mainly on the amount and nature of their substituents. For example, the solubility of unsubstituted phenol in water is 77.9 g l^{-1}, 2,4-dichlorophenol solubility is 9.7 g l^{-1}, that of 2,4,6-trichlorophenol is 0.8 g l^{-1} and pentachlorophenol solubility is 14 mg l^{-1} [12]. However, these data are presented for molecular (acidic or unionized) forms of phenolic pollutants and are dramatically different in the case of the ionized form.

Solubility is also a function of the pK_a. The pK_a values are: phenol 9.98, 2,4,6-trichlorophenol 6.15, tetrachlorophenol 5.16 and pentachlorophenol 4.75 [12]. As a rule, the solubility of the anionic form is much higher than that of the molecular form. For example, the solubility of the herbicide Ioxynil (pK_a = 3.96) in water is 50 mg l^{-1} and that of its potassium salt solubility in water is 107 g l^{-1} [13]. The pK_a value of pentachlorophenol is 4.75, and its solubility in water is 14 mg l^{-1} [12], whereas the solubility of the commercially produced sodium pentachlorophenoxide is 330 g l^{-1} [13]. This shows that pentachlorophenol is very soluble in nonacidic wastewater, and its leakage from factories can be very dangerous for the environment.

C. Analysis – Sample Preparation and Methods of Determination

One of the phenol determination methods described in 'Standard Methods', the so-called phenol index number, includes all, water stream distillable, phenolic compounds, which are detected photometrically after derivatization with 4-aminoantipyrine and extraction with chloroform[62]. Here, only the total amount of phenols is measured. It is impossible to distinguish between individual phenols or to estimate the probable toxicity of the analyzed water sample. This method is important only for preliminary information about possible phenol pollution and to determine if further tests are necessary.

GC and LC provide a unique tool for the analysis of complicated aquatic environments, which contain many different classes of organic compounds. The main problems encountered during analysis are (1) separation of complicated and, as a rule, also undesirable matrix components of the investigated samples, (2) achievement of low detection limits and (3) identification of unknown pollutants. The first two goals can be achieved by proper sample preparation, including concentration of phenols from large volumes of water samples to small volumes of organic solvents or water-organic mixtures, followed by matrix removal, elution of retained phenols with a minimum amount of organic solvents, maximizing the compatibility of the solvent with the analytical system and selectivity in the concentration and elution steps[63]. The analytical method, which can provide the last goal, identification of the unknown pollutants, is MS or MS/MS, coupled with a suitable separation technique. For routine analysis, other detectors like ECD and FID in GC or UV in HPLC are widely used. Although high performance liquid chromatography methods are frequently used for the analysis of phenols[64–69], gas chromatography is often preferred. However, liquid chromatography is necessary in some cases, such as for humic substances occurring in environmental samples[70], to overcome the matrix influence.

Generally, preconcentration of pollutants from water samples and sample preparation steps are accomplished by extraction techniques based on enrichment of liquid phase (liquid/liquid extraction) or solid phase (solid/liquid extraction)[71–73]. Historically, liquid/liquid extraction (LLE) was used exclusively to enrich phenols from water samples. LLE is still used as a preconcentration step[74–78]. However, there is an increasing tendency to replace LLE by solid phase extraction (SPE) and solid phase microextraction (SPME). Among the reasons for replacing LLE are foam formation, the large volume of organic solvents needed, the length of the analysis time and difficulties in the automation of LLE procedures. On the other hand, SPE requires incomparable smaller amounts of solvents (SPME requires no solvent at all) and can be easily automated[79]. Finally, SPE and SPME are cheaper in comparison with LLE.

Recently, the extraction of phenols has been performed by SPE using adsorbing materials, mainly of reverse phase[80–82], anion exchange[83,84] and graphitized carbon black (GCB)[85–87]. GCB, known also as Carbopack B or Carbograph 1, was used in the selective extraction of substituted phenols from water[85].

The recovery depends on the pK_a of the phenols. Basic phenols with $pK_a > 8.0$ can be eluted with an organic solution containing methanol; more acidic phenols with $pK_a = 7$ can be eluted with a $CH_2Cl_2-CH_3OH$ mixture (90:10, v/v) containing tetramethylammonium hydroxide. On the other hand, the extraction and recovery of phenols have been found to be independent of the amount of inorganic ions ($I = 0.6$ M)[88]. It is not an easy task to achieve these requirements due to the different behavior of phenols in terms of acidity and polarity. The variability of pK_a values makes a selective isolation, even in the case of 11 phenols from the EPA priority pollutant list with quantitative recoveries, an elusive goal. Some recent reviews of the sample preparation of phenols discuss the application of different sorbents (silica, polymeric, functionalized, carbon based and mixed sorbents), coatings and experimental configurations for SPME to the preconcentration and separation of phenols[63,88,89].

The high polarity of phenols can limit the application of GC to phenol analysis, tending to give broad, tailed peaks and decreasing the lifetime of the chromatographic column[90]. It can be avoided relatively easily using phenol derivatization before or after SPE. Phenol acetylation with acetic anhydride in the presence of carbonate or hydrogen carbonate is one of the most studied and used derivatization methods[90]. This reaction can be performed in aqueous samples before SPE with high efficiency. The acetates obtained are more easily extractable than nonderivatized phenols[91]. Another way to decrease the polarity of phenolic compounds is the formation of ion pairs between phenolate anions and quaternary ammonium salts. Water samples are adjusted to pH 9 and $(C_4H_9)_4NBr$ is added to the sample. Elution is performed with methanol doped with 1% acetic acid to break the ion pairs. The final extract is compatible with HPLC and GC separation techniques[1,92,93]. Both derivatization procedures are applied in water solution before extraction.

Another method is extraction with methylene chloride followed by derivatization with pentafluorobenzyl bromide or diazomethane and subsequent GC/ECD or GC/FID analysis[74,76,94]. The extraction solvent has to be changed before analysis. More than 50 substituted phenols have been derivatized successfully with N-(t-butyldimethylsilyl)-N-methyltrifluoroacetamide by forming the corresponding t-butyldimethylsilyl derivatives. This study includes 21 chlorinated phenols, 13 nitrophenols, 3 aminophenols, 4 alkylphenols, o-phenylphenol, some other substituted phenols including 6 phenolic pesticides and the nonsubstituted phenol. Using SPE with polymeric adsorbents and GC/MS, phenols with very different substituents can be detected in environmental samples with high matrix content at the ppt level[95].

The American Water Work Association (AWWA) and US EPA developed a number of methods for phenol determination[74,76-78,96,97]. EPA Method 528 is dedicated to the determination of phenols in drinking water by solid-phase extraction and GC/MS analysis and is developed for 12 phenols, mainly chlorophenols, nitro- and methyl-substituted phenols[96]. Unfortunately, users have to take into account that the recommended internal standard tetrachlorophenol can also be found in water samples and has to be used with precaution or, better, substituted with another compound. The same problem applies in the case of the recommended surrogate 2,4,6-tribromophenol, which cannot be used in the analysis of water in areas with high bromine ion content. (Some examples of tribromophenol formation by humic or fulvic acid chlorination was mentioned by Richardson[98-100].)

Practically the same list of phenols can be determined by analysis of municipal and industrial wastewater detailed in EPA Method 604 (GC/FID or GC/ECD) and 625 (GC/MS)[76,77]. Method 8041 for determination of phenols in wastewater, presented by the EPA Office of Waste Water, describes the determination of ca 40 phenols specifying extraction and cleanup conditions, derivatization with diazomethane or pentafluorobenzyl bromide and analytical determination by GC/FID, GC/ECD or GC/MS[78].

Matrix effects also play an important role in phenol analysis. Surface and river water, containing fulvic and humic acids at a few mg l^{-1}, give brown extracts after concentration over C18 and polymeric sorbents. In HPLC with UV or electrochemical detector, these extracts give huge peaks at the beginning of the chromatogram that hamper quantitation of less retained particles like phenol and 2-chlorophenol[101-104]. The SPE procedure for phenol extraction can be widely used in different monitoring programs, analyzing a huge amount of samples.

Sample preconcentration in the field provides ample opportunity to transport and store in the lab SPE cartridges instead of large-volume water samples. It saves a lot of space and minimizes the risk of degradation. Analysis of phenols concentrated on C18 disks immediately after loading and after 28 days storage at 3 °C yields the same results[105]. Analog stability studies in the case of pentachlorophenol demonstrate a negligible decrease in phenol peaks after 7 weeks of storage, independent of the moisture level in the environment, while only a 20% decrease of signal was observed at room temperature during the same time[106]. Another concentration method applied mainly to the identification of semivolatile disinfection by-products (DBPs) is resin extraction. This method is used to concentrate large quantities of treated water (40–50 L), which is necessary to detect trace levels of by-products[98].

Besides GC and LC, capillary electrophoresis (CE) has been proposed as a separation technique in the environmental trace analysis of phenols[17,104,107-109]. Sub-ppb levels of phenols can be analyzed in drinking water with GC-MS/MS. Pollutants can be detected from a 10-ml water sample by extraction of preliminary acetylated chlorophenols or by preconcentration of a 1 L sample using a graphite cartridge for solid extraction[110]. Appropriate selection of parent ions and fragmentation conditions ensures high sensitivity and clean product ion spectra, allowing identification of small amounts[110]. Application of liquid chromatography with thermospray MS in the single ion monitoring mode allows the identification of phenols in complex samples, avoiding interference of humic compounds usually present in river water[87].

Relatively simple electrochemical and amperometrical detectors have also been used in combination with reverse-phase LC separation for analysis of environmental water samples[70,111]. Scrupulous studies of phenols' electrochemical oxidation simplified this problem[112].

Determination of phenols in other matrices like food samples must also be mentioned. Capillary liquid chromatography was evaluated as an alternative to conventional HPLC to analyze complex phenolics and polyphenols in apple juice[113]. Determination of polyphenols is of very high importance because of their biological properties, like anti-inflammatory, anti-histaminic and anti-tumor activities, free-radical scavenging and protection against cardiovascular diseases[114–116].

D. Phenols in the Environment

Phenols are released in wastewater and can be lost to waste streams. This explains the many reports on determination of phenols in the environment. Phenols are included among drinking water disinfection by-products (DBPs)[98]. Table 1 lists specific disinfection by-products identified from the interaction of humic material with different kinds of disinfectants like chlorine, ozone, chlorine dioxide, chloramine and combinations thereof. Chlorine dioxide, alone or in combination with free chlorine, and chloramine do not produce any phenolic DBPs. However, after disinfection with free chlorine, different chlorophenols, mainly formed by the reaction of chlorine with phenols present as pollutants in raw water, were detected. High concentration of Br^- in raw water leads to formation of brominated phenol analogs instead of the chlorinated phenols[98–100]. Much smaller amounts of phenols were produced by raw water treatment with ozone, ozone in combination with chlorine or chloramine. DBPs of a phenolic nature are more toxic in the case of free chlorine treatment than other water treatment technologies. While the presence of phenols leads to formation of different chloro- or bromo-phenol derivatives which are generally more toxic than the starting compound, the chlorine dioxide treatment leads to the total disappearance of phenols from surface water[117]. Here, phenols which are not *para*-substituted are oxidized mainly to quinones or chloroquinones. *Para*-substituted phenols undergo oxidative ring cleavage with formation of organic acids, such as oxalic, maleic or fumaric, and carbon dioxide. This generalization has some exceptions; for example, the oxidation of 2,4-dichlorophenol by chlorine dioxide leads to formation of 2,6-dichloro-1,4-benzoquinone.

There are many different studies of organic pollutants in environments like rivers, lakes and seas. Nontarget GC/MS screening of the river Elbe and its tributaries Mulde, Saale, Weisse Elster, Schwarze Elster, Havel was used in 1992–94[118]. 4-*tert*-Butylphenol and different chlorinated phenols were detected in samples from the Elbe and the Mulde[118]. Organic pollutants have been studied in the Ter river and its system of reservoirs supplying water to Barcelona (Spain). During the sampling period 1986–1993, trichlorophenol in the 0.06–0.1 ppb range was found frequently. In more than 75% of the samples, polyethoxylated alkylphenols were found in concentrations of 5–450 ppb[119]. Transformation and biodegradation of alkylphenol polyethoxylates, present in detergents as nonionic surfactants, lead to formation of free alkylphenols. The nature, origin and trend of phenolic

TABLE 1. Phenolic DBPs observed by different water treatment technologies[98]

Disinfectant	By-products
Chlorine	2-Chlorophenol, 3-bromophenol, 2,4-dichlorophenol, bromochlorophenol, 2,4,6-trichlorophenol, 2,4,6-tribromophenol, pentachlorophenol, dichlorodihydroxyphenol, dibromodihydroxyphenol
Chlorine dioxide	—
Chloramine	
Ozone	Methylphenol, 4-methoxy-*tert*-butylphenol
Ozone + chlorine	Trichlorophenol
Ozone + chloramine	2,6-Di-*tert*-butyl-4-nitrophenol
Chlorine dioxide + chlorine	—

compounds were studied in the river Po (Italy) whose water is used as a source of drinking water. The sampling was carried out during a 3-year period (1994–1996) at 15-day intervals. The detected alkylphenols may be divided into two main groups: antioxidants and surfactants[36]. In the antioxidant group 3-*tert*-butyl-4-methoxyphenol and its isomer 1,1-dimethylethyl-4-methoxyphenol were found in all analyzed samples. The concentration of these two compounds and related phenols did not exceed 45 ppb. The presence of these compounds in surface water is undesirable, but presumably not dangerous, as these synthetic antioxidants have been used in the food industry and have been shown to possess low toxicity in animal and human testing[36]. In the same study the low concentrations of chloro- and nitrophenols in the river Po were explained by their high degradation or by their reduced emission by industrial wastes[36]. Chlorinated phenols were detected between different organic micropollutants in lowland rivers[120].

Very interesting results were obtained from studies of urban storm water quality, based on point source discharges to receiving water bodies. The relative pollutant load contribution of nonpoint source discharges has gradually increased. Urban storm water runoff is one such nonpoint discharge. Traditionally, studies in the storm water field have focused on the quantity of water produced and on methods for its safe handling. Only recently has the quality and contamination level of storm water become a major concern[121]. This interest arose from the understanding that the drinking water quality depends not only on the treatment technology, but also on the available raw water quality. This study is very important for Canada with large surface lakes. Phenol in storm water was found in concentrations of $3-10$ $\mu g\,l^{-1}$. The surface water objective of the province of Alberta is 5 $\mu g\,l^{-1}$ for phenol and a limitation of 1 $\mu g\,l^{-1}$ appears in Canadian aquatic guidelines for total phenols[121]. 2-Chlorophenol in storm water was detected at a concentration of 2 $\mu g\,l^{-1}$, and pentachlorophenol at $1-115$ $\mu g\,l^{-1}$ while the Canadian guidelines limit them to 7 and 0.5 $\mu g\,l^{-1}$, respectively. Di- and trichlorophenols were not detected in storm water. On the basis of the results presented, and taking into account that pentachlorophenol was found in 15% of the samples and may contain as impurities dioxin and furan derivatives, making it a more dangerous contaminant, it must be concluded that pentachlorophenol is a possible problem compound in Canadian storm water[121].

Investigations in the South Italian Seas (Tyrrhenian Sea, Ionian Sea and Straits of Messina) were carried out in order to evaluate the anthropogenic inputs of some organic pollutants including phenols[122]. Only the total phenols concentration was detected and was almost always higher than the threshold value of 3 ppb. A maximum concentration of 67 ppb was found in the Tyrrhenian Sea, 12 ppb in the Straits and 8 ppb in the Jonian Sea[122].

Nonylphenols were detected in the water and in sediments from the German Bight of the North Sea. Its concentration in seawater varied from 0.7 to 4.4 $ng\,l^{-1}$, while in the Elbe estuary 33 $ng\,l^{-1}$ was found[123]. Different, independent studies of nonylphenol distribution in Japan allowed one not only to identify marine pollution in the Sea of Japan, but also to compare pollution in the deep-sea area (the so-called semi-enclosed 'small ocean') and in rivers and bays. Nonylphenols were found in the Sea of Japan in the $2-150$ $pg\,l^{-1}$ range[124]. An analogous study of the distribution of alkylphenols in wastewater effluents, river water and riverine and bay sediments was carried out in the Tokyo metropolitan area[125]. The concentration of nonylphenol in Sumidagawa River was in the range $0.1-1.1$ ppb. This concentration interval is much higher than in the Elba river studies, but less than that observed in some European[126,127] and US[128] rivers. In the secondary effluents the nonylphenol concentration reached $0.1-1.2$ ppb. It is much lower than those reported for a Swiss sewage treatment plant ($2.2-44$ ppb[129]).

Another way to study alkylphenol pollution is detection of these compounds in seafood. Four species of edible mollusks, two cephalopods and two bivalves, were studied in

1998 in the Adriatic Sea (Italy). The highest concentration was found for nonylphenol, 200–300 ng g^{-1} in the case of mussels and clams and 400–700 ng g^{-1} in the case of squids[130]. (In comparison, sediments in Jamayca Bay on the Southwestern shore of Long Island, New York contained nonylphenol ethoxylate metabolites in the range 0.05–30 µg g^{-1} [131].) Octylphenol levels were generally 30 times lower than those of nonylphenol[130]. This is because nonylphenol and the corresponding ethoxylate are far more widely used than octylphenol and its ethoxylate[132]. It is also expected that nonylphenol occurs at higher concentrations in fish tissues than its ethoxylate, as nonylphenol is more hydrophobic[132]. Although most chemical contamination originates in northern Italy, where most of the country's population lives, the highest alkylphenol pollution is found in the central and not the northern part of the sea. This is explained by water circulation. Furthermore, the results indicate that alkylphenols are not isolated around urban areas and can be transferred over long distances[132]. This is consistent with the relative stability of nonylphenol in fresh water when it dissipates in 6–22 days[133]. The observed level of alkylphenol contamination does not appear to be harming the mollusks examined in the study and the risk to humans who eat these mollusks is considered low. However, researchers caution that it is difficult to predict the environmental and human health effects because there are insufficient data on the toxicity of alkylphenolic compounds[130].

Four-week incubation of mussels followed by analysis was used in Finland as a sensitive method for monitoring chlorophenols in watercourses. This method was applied at 40 sites to study the influence of pulp mills and to detect possible chlorophenol leakage. Eight chlorophenol derivatives were found to accumulate quite strongly: tri-, tetra- and pentachlorophenols, di-, tri- and tetrachloroguaiacols, and trichlorosyringol. In the first group are wood preservatives and combustion products, while in the second group compounds formed during the bleaching of pulp[134,135].

In agriculture, phenolic compounds are used as pesticides (Figure 1) and can also form from the degradation of chlorinated phenoxycarboxylic acids and organophosphorous insecticides[11]. The herbicide DNOC sorption in a sandy aquifer (Denmark) has been reported[136].

Bisphenol A is a common raw material used to produce paper, such as thermal paper and carbonless copy paper. Therefore, many paper recycling factories are thought to release Bisphenol A into wastewater. Bisphenol A is easily chlorinated by sodium hypochlorite used as a bleaching agent in the paper industry as well as a disinfecting agent in sewage treatment plants. Thus it is important to investigate the release of chlorinated Bisphenol A into the environment. A Japanese research group analyzed the chlorinated Bisphenol A in the Shizuoka prefecture where 100 paper recycling plants are located[137]. In the final effluents of 8 plants chlorinated Bisphenol A was detected in the range from traces to 2 ppb[137].

IV. TOXICITY AND HEALTH EFFECTS

A. Introduction

The current emphasis on the biological properties of natural or anthropogenic compounds depends on studies of possible health hazards or beneficial effects of agents to whom humans are exposed in everyday life. Phenolic compounds, which are ubiquitous among plants, used as food additives and ingested daily in milligram quantities, are a complicated system from this point of view. On the one hand, phenols induce DNA double-strand breaks, DNA adducts, mutations and chromosome aberrations in a great variety of test systems. On the other hand, they suppress the genotoxic activity of carcinogenic compounds *in vitro* as well as *in vivo* studies[138]. The dual function of dietary phenols also becomes evident from the studies of their carcinogenic or, the opposite, anticarcinogenic potential. Some phenols induce precancerous lesions, papillomas and cancers, or act as cocarcinogens, but there are others which are potent inhibitors of carcinogenesis at the initiation and promotion stages[138]. One example of this latter group is vitamin E (tocopherol), which plays an important role in blood cells and nervous system tissues. It must be concluded that health hazard versus protective activity of phenols contained in dietary mixtures remains an unresolved problem and their multiple, occasionally contradictory functions make it difficult to propose their use as chemopreventative agents[138]. This means that each group of phenols has to be examined separately for its biological and toxicological activity.

B. Drinking Water Regulations. Taste and Odor

Drinking water regulations are periodically updated as more information becomes available. For current information it is recommended to check USEPA, WHO or EC guidelines. Some current examples will be presented here. While drinking water regulations are a matter of change, taste and odor standards generally remain constant, as taste and odor threshold concentrations in water seem to retain more or less constant values and will not be changed in time. Chlorophenols may be formed by chlorination of anthropogenic phenol traces or natural organic matter (fulvic and humic acids) even at low concentrations[139]. Table 2 presents the drinking water regulations of the World Health Organization (WHO) and US EPA in $\mu g\, l^{-1}$ or ppb.

Taste and odor complaints from consumers are an important issue for drinking water suppliers. Taste and odor threshold concentrations in water were determined for 59 drinking water contaminants, including phenols. Their determinations are usually based on the WHO drinking water guidelines[140], US EPA[141,142] or European Standards[143]. The odor and taste description can change with concentration: for example, at 0.09 ppm the odor of 2-chlorophenol was described as 'musty, sweet, floral', but in the 0.5–1 ppm range as 'chemical, medical'. In the case of 2,4,5-trichlorophenol, the description changed on increasing its concentration from 'fruity' to 'antiseptic'[144]. While the odor and taste threshold of 2-chlorophenol is 0.1 ppb, the US EPA guideline for drinking water recommends 40 ppb. This shows that there is no correlation between a compound's taste and odor threshold and its health effects. Furthermore, published results[141,144] show that phenols can produce taste or odor in drinking water at concentrations much lower than health-based regulations. This does not mean that it is acceptable to supply water that has an offensive taste or smell. Possible psychomatic effects, such as headaches, stress or upset stomach, must to be taken into account. Although there is an incomparable variation in the level and quality of taste and odor that consumers would regard as acceptable, such effects cannot be ignored and in particular cases a warning to the public not to drink the water must be issued[144]. In order that the concentration of chlorophenols will be lower than can

TABLE 2. Comparison of threshold taste concentrations with maximum contamination level of drinking water[140–142]

Compound	WHO drinking water standard ($\mu g\, l^{-1}$)	EPA drinking water standard ($\mu g\, l^{-1}$)	EPA taste recommendation ($\mu g\, l^{-1}$)
2-Chlorophenol	10	40	0.1
2,4-Dichlorophenol	40	20	0.3
2,4,6-Trichlorophenol	200	30	1
2,3,4,6-Tetrachlorophenol	—	22	1
Pentachlorophenol	9	22	—

be tasted, EPA recommends taste and odor threshold concentrations for these compounds (Table 2). There is practically no difference between taste and odor thresholds.

Short-term exposure limits of pentachlorophenol are higher than long-term limits, namely not more than 1.0 mg l^{-1} for 1 day and 0.3 mg l^{-1} for 10 days. The EPA also decided that any release of more than 10 pounds of pentachlorophenol to the environment should be reported[142].

C. Toxicity

Wide use of phenol and its derivatives led to studies of its occupational exposure and toxicity. Phenol toxicity in humans is not a big surprise, as this compound is toxic to most microorganisms, which explains its common use as a general disinfectant. This fact complicates treatment of phenol-containing wastewater by conventional biological processes[145]. Phenol genotoxicity was determined using Syrian hamster embryo cells. Phenol induced morphological transformation, gene mutation, chromosomal aberrations, sister chromatid exchanges and unscheduled DNA synthesis[146]. 2,4,6-Trichlorophenol induced mononuclear cell leukemia in male rats and liver tumors in mice[147]. In another study, genotoxicity of this compound was established in V79 Chinese hamster cells[148]. Conversely, 2,4-dichlorophenol did not cause any increase in tumors in rats or mice in the 2-year study. In fact, mononuclear cell leukemia in rats and lymphomas in mice were decreased in these studies[149].

The damaging effect of long-term exposures (6+ months) to pentachlorophenol (PCP) on the immune system was studied in 190 patients. The distribution of PCP levels in blood was: 0–10 µg l^{-1} (69%), 11–20 µg l^{-1} (20%), and >20 µg l^{-1} (11%). The patients had various clinical symptoms and complained of the following: general fatigue (64%), rapid exhaustion (59%), sleeplessness (53%), headache (44%), mucous membrane, throat and noise irritation (39%), frequent common diseases (36%), bronchitis (30%) and nausea (13%)[150]. Analogous symptoms were described in previous studies[151,152]. Blood levels of PCP were associated negatively with total lymphocyte counts and several other blood immune parameters. These data provide clear evidence that immunological abnormalities are associated with high levels of PCP in plasma of individuals with long-term exposure[150]. PCP also induces chromosomal aberrations in mammalian cells *in vitro* and in lymphocytes of exposed persons *in vivo*[153]. Several case-control studies have shown significant associations of polychlorophenols with several types of cancer, with the most consistent findings being non-Hodgkin lymphoma and soft-tissue sarcoma[154]. Occupational exposure to chlorophenols can be a risk factor for nasal and nasopharyngeal cancer[155]. In the studies of polychlorophenols, great importance is attached to a compound's purity as its contaminants can include very toxic and carcinogenic dioxins. One has to be sure that the observed toxicity effect is connected with the main compound and not with the impurity[154]. Polychlorophenols are also known to uncouple oxidative phosphorylation[156], alter the electrical conductivity of membranes and inhibit cellular enzymes, such as ATPase, β-galactosidase etc.[157,158]. The genotoxicity of the rodent carcinogen 2,4,6-trichlorophenol was studied in V79 Chinese hamster cells. This compound did not induce mutation or structural chromosome aberrations; however, it did produce dose-related increases in hyperdiploidy and micronuclei. It appears that it causes chromosome malsegregation as a major mode of genotoxic action[148].

As mentioned above, pentachlorophenol has different solubility at acid and neutral pH. It was shown that this compound toxicity also depends on pH[159]. Studies of wastewater from a Baikalsk pulp and paper mill allowed one to evaluate a 'pure' cellulose bleaching process pollution, as Lake Baikal, where it is located, has no agriculture and only little municipal pollution[160]. Although mutagenic activity was effectively decreased during biological and chemical treatment, even modern wastewater purification systems do not totally abolish potential toxicity and mutagenity of the effluents[160].

Chlorinated phenols can degrade with formation of highly carcinogenic dibenzo-p-dioxin and dibenzofuran derivatives. It can occur by thermolysis, slow combustion, photocatalytically, by photochemical degradation and by photolysis[161–166]. Even in the presence of TiO_2, which in many cases leads to the total degradation of organic compounds, photocatalytic degradation includes formation of polychlorinated dibenzo-p-dioxins and dibenzofurans[166]. It was shown that the level of polychlorinated dibenzo-p-dioxins and polychlorinated dibenzofurans in commercial animal products, raised near incinerators, are elevated compared to products from areas with no such industrial sources. It is related primarily to meat, milk from cows and eggs from chicken[167–170].

Some phenol derivatives can act like hormones (e.g. estrogens) and interact with the human hormonal system. Two main phenol groups must be mentioned here—Bisphenol A and octyl- and nonylphenols. Bisphenol A might be a factor in decreasing sperm count in males and increasing rates of breast cancer in women[171]. It was also shown that increased sensitivity to Bisphenol A during the perinatal period causes an increase in body weight soon after birth and in adulthood and a decrease of plasma luteinizing hormone level in adulthood[172]. Octyl- and nonylphenols are formed during anaerobic biodegradation of the corresponding alkylphenol ethoxylates. These compounds are known to cause proliferation of breast cancer cells by acting as estrogenic mimic[33,34]. They also cause endocrine-disrupting effects and 'feminization' of male species[173,174].

In the context of health effects with an emphasis on cancer, phenols, as an independent class of organic compounds, are generally not genotoxic. This means that they cannot modify genes and therefore are not considered to be a direct cancer risk. Laboratory studies have demonstrated that while not genotoxic, phenols can be co-carcinogens or promoters, increasing the effect of environmental genotoxic carcinogens. This promoting effect is highly dependent on the dosage and chronicity of exposure. Recent studies have demonstrated that some phenols found in fruits and vegetables, as well as synthetic phenolic antioxidants, exert protective effects against cancer, demonstrating antimutagenic, anticarcinogenic properties, and can also antagonize the effect of promoters. However, in a high dose range some of them can cause cancer in animals through mechanisms like cytotoxicity, regenerative cell duplication and hydroxyl radical generation[175]. Generally, the neoplastic effects of phenolic antioxidants can be observed at high dietary levels and occur only after effective biological defense mechanisms are overloaded[176]. Therefore, the public needs to be much more aware of the importance of dosage and exposure time. The role of phenols in the mutagenicity of white grape juice in the Ames mutagenicity test was studied. It was concluded that polyphenol oxidase-catalyzed oxidation of phenolic compounds generates toxic species that are responsible, at least partly, for the mutagenicity of grape juice[177].

3-Chloro-4-(dichloromethyl)-5-hydroxy-2(5H)-furanone, better known as MX, is one of the DBP and was found to be one of the most potent direct-acting mutagens ever tested in Ames tester strain[178]. Although the concentration of MX usually reaches only some ng l^{-1}, it comprises 15–57% of the total mutagenicity of drinking water[179]. It was shown that MX formation proceeds via chlorine interaction on the aromatic or phenolic rings of humic substances with subsequent fragmentation[180,181]. Further studies demonstrated that the main MX precursors could be 4-hydroxybenzaldehyde, vanillin (4-hydroxy-3-methoxybenzaldehyde) and syringaldehyde (4-hydroxy-3,5-dimethoxybenzaldehyde), which were formed by humic and fulvic acid mild oxidation[182–184]. These phenol-fragment-containing compounds are the constituent parts of lignin, a phenolic polymer, which is a major component of woody tissues and thought to be a precursor for humic substances[180]. Another possible precursor of MX can be diphenols, like catechol, resorcinol (o- and m-hydroxyphenols, respectively) and p-hydroxyphenol[182,185]. However, these data were not confirmed by any other study[186]. Unfortunately, in each case the exact mechanism of MX formation from substituted phenols, containing aldehyde or an additional hydroxyl group, under disinfection conditions remains unknown.

D. 'Nontoxic' Biological Activity of Phenols

It was demonstrated that the plant phenolic compound's chlorogenic acid and ellagic acid have protective effects against liver, colon and tongue carcinogenesis[187]. According to some data, onion, lettuce, apples and red wine are important sources of dietary flavanoids, which are probably responsible for the anti-mutagenic activity associated with food and beverages[188]. It was further suggested that smokers ingesting dietary phenols, probably flavonoids, are partly protected against harmful effects of tobacco carcinogens within their bladder mucosal cells[188]. It is in good agreement with data demonstrating that smoking increases plasma vitamin E disappearance[189]. These conclusions require additional studies and, if confirmed, it will allow the use of a new chemoprevention strategy.

V. DETOXIFICATION AND DEGRADATION

A. Chemical and Physicochemical Degradation

Contamination of aquatic bodies by different harmful organic pollutants including phenols has stimulated research activity in the development of various treatment technologies to remove, or better to degrade, these pollutants from water and wastewater. Several chemical processes are carried out for this purpose. Generally, these technologies involve oxidation of organic pollutants with various oxidizing agents like ozone, UV radiation, electrochemical methods, hydrogen peroxide etc. Chlorination is not applicable, as it leads to formation of more toxic chloroorganics.

Electrochemical oxidation of chlorinated phenols on different oxide electrodes (PbO_2, SnO_2, IrO_2) was suggested[190]. The same process can also be realized on carbon black-slurry electrodes[191]. A significant increase in carbon black amount achieves full mineralization of 4-chlorophenol[191]. The opposite process—electroreduction under conditions of electrocatalytic dehydrogenation of pentachlorophenol—leads to formation of cyclohexanol (98%) with 2% of cyclohexanone. This means that electrocatalytic hydrogenolysis can accomplish total dehalogenation and further saturation of the chlorinated phenol[192].

Another method, very effective for *in situ* degradation of organic compounds, is ultrasonic irradiation of aqueous solutions, mainly in combination with photochemistry (UV radiation). Sonolysis of aqueous solutions results in the formation and adiabatic collapse of bubbles, generating local high temperatures and pressures and reactive free radicals in the bubble. Application of this method to phenol degradation has been reported[193–195]. Among the various products of phenol degradation identified were maleic acid, polyhydroxybenzenes and quinones[195]. The presence of dissolved oxygen in aqueous solutions was reported to play a very important role in the generation of highly oxidative hydroxyl free radicals and thus might enhance the decomposition of chlorophenols[196,197]. On the other hand, dissolved nitrogen scavenges the free radicals and inhibits their interaction with chlorophenols[197]. These data are confirmed by identification of the first intermediates of chlorophenol photosonochemical degradation. The first intermediates indicate that OH• radicals are involved in the reaction and form compounds containing second OH substituent like hydroquinone, catechol and resorcinol[198]. Addition of ozone did not affect the sonication process of pentachlorophenol degradation[199]. This unexpected result was explained by the theory that O_3 molecules first dissolve in solution and then diffuse into cavitation bubbles, where they undergo thermolytic decomposition[200]. Chemical oxidation by ozone alone is also used for chlorophenol destruction[201,202]. In this process complete removal of chlorophenols and rupture of aromatic rings were reported, but additional UV irradiation leads to complete degradation producing carbon dioxide[203].

Another system—Fenton's reagent—and its advanced form, the photo-Fenton system, is widely used for organic compound degradation. This system includes generation of hydroxyl radicals via the reaction of Fe^{2+} with H_2O_2 and is an effective degradation system also in the case of phenols[204-207]. Extensive mineralization of pentachlorophenol and its total dechlorination was observed in the photo-Fenton reaction[204]. The use of the Fenton process can be recommended also for detoxification of the wood preservative creosote oil used together with pentachlorophenol. This process effects elimination of the acute toxicity of the treated solution to fathead minnows (*Pimephales promelas*) and reduction of its toxicity to daphnia (*Daphnia pulex*)[204]. Besides the use in the Fenton reaction, H_2O_2 is also applied for treatment of phenolic contaminants in combination with horseradish peroxidase. This enzymatic system also has the ability to transform and detoxify aqueous phenolic solutions[208] and soils[209].

A very effective and cheap system for water purification from organic compounds is photocatalytic oxidation of organic species by UV illuminated titania, involving two simultaneous processes—oxidation of the target compound and reduction of dissolved oxygen. In order to promote photocatalytic oxidation of organic compounds, four components are necessary: a target compound, oxygen, solar irradiation or artificial source of light and photocatalyst, mainly using TiO_2 aqueous suspension[210]. In the case of the photocatalytic degradation of Bisphenol A, an endocrine disruptor, its total degradation was reached in 20 hours without generation of any serious secondary pollution. The transcriptional estrogenic activity in response to human estrogen receptors in a yeast hybrid assay decreased drastically to less than 1% of the initial Bisphenol A activity[211]. The same system was applied also to different chlorinated phenols; 360 min irradiation destructive efficiency was 97%[212]. Short UV exposure time leads to formation of different oxidation intermediates, like 2,3,5,6-tetrachlorophenol, tetrachloro-1,4-hydroquinone and *p*-chloranil in the case of pentachlorophenol[212]. Dichlorophenols can also be formed during the photocatalytic degradation of some organic molecules; for example, 2,4-dichlorophenol was detected in the case of photocatalytic oxidation of phenoxyacetic acids[210]. Removal of phenols from aqueous solutions can also proceed by adsorption using Amberlite XAD-4 resins[213], dual-cation organobentonites[214], activated carbon[215] etc.

B. Biodegradation and Biomethylation

Biodegradation would be an effective pathway for detoxification of phenolic compounds. Molasses (residue after sucrose crystallization in the sugar industry) are used further for alcohol production. Vinasse remains after fermentation, and alcohol production forms a number of phenolic compounds which are degradated through *Aspergillus Terreus* and *Geotrichum Candidum* treatment[216]. *Burkholderia* sp. RASC c2 was used for 2,4-dichlorophenol detoxification in soils[217]. Algae blooms also can lead to the disappearance of phenols, as was demonstrated with green algae *Volvox aureus* blooming[218].

Biological processes in nature can also create the opposite effect—formation of more stable, less degradable compounds. Phenols are biomethylated in the environment to their corresponding anisoles which are more stable and lipophilic. This means that phenol pollution studies must also take into account formation and bioaccumulation of anisoles. Biomethylation of phenols to more bioaccumulating anisoles has not only environmental, but also economic consequences, as chloroanisoles are extremely bad-tasting compounds. For example, in sensory panel studies of water solutions, the concentration limit of detectable odor was lowered 3 to 10 orders of magnitude during anisole formation in the phenols methylation process[219]. Analysis in combination with a taste panel study of fish showed that chlorinated anisoles together with veratroles were the main tainting substances of fish in pulp mill recipient waters[220].

VI. REFERENCES

1. D. Martinez, E. Pocurull, R. M. Marce, F. Borrull and M. Calull, *J. Chromatogr. A*, **734**, 367 (1996).
2. C. Schlett and B. Pfeifer, *Vom Wasser*, **79**, 65 (1992).
3. C. Tomlin (Ed.), *The Pesticides Manuel*, 10th edn., Crop Protection Publ., The Royal Society of Chemistry, Cambridge, 1994.
4. *Physicians' Desk Reference*, 52 edn., Medical Economics Co., N.Y., 1998.
5. H. Kontsas and C. Rosenberg, P. Pfaffli and P. Jappinen, *Analyst*, **120**, 1745 (1995).
6. M. D. Mikesell and S. A. Boyd, *Environ. Sci. Technol.*, **22**, 1411 (1988).
7. S. H. Lee and J. B. Carberry, *Water Environ. Res.*, **64**, 682 (1992).
8. M. M. Laine, J. Ahtiainen, N. Wagman, L. G. Oberg and K. S. Jorgensen, *Environ. Sci. Technol.*, **31**, 3244 (1997).
9. A. B. McKague, *J. Chromatogr.*, **208**, 286 (1981).
10. J. Tremp, P. Mattrel, S. Fingler and W. Giger, *Water, Air Soil Pollut.*, **68**, 113 (1993).
11. S. Lacorte and D. Barcelo, *Environ. Sci. Technol.*, **28**, 1159 (1994).
12. R. J. Watts, *Hazardous Wastes, Sources, Pathways, Receptors*, Wiley, New York, 1996, p. 168.
13. J. A. Smith and J. T. Novak, *Water, Air Soil Pollut.*, **33**, 29 (1987).
14. L. H. Keith and W. A. Telliard, *Environ. Sci. Technol.*, **13**, 416 (1979).
15. L. H. Keth (Ed.), *Compilation of Sampling Analysis Methods*, US EPA, Boca Raton, Fl., 1991, p. 389.
16. EEEC Directive 80/77/CEE 15-7-1990; *Official Journal of the European Communities*, 30-8-1990, European Community, Brussels, 1990.
17. I. Rodriguez, M. I. Turnes, M. H. Bollain, M. C. Mejuto and R. Cela, *J. Chromatogr. A*, **778**, 279 (1997).
18. D. G. Crosby, *Pure Appl. Chem.*, **53**, 1051 (1981).
19. A. S. Narang, K. Swami, R. S. Narang and G. A. Eadon, *Chemosphere*, **22**, 1029 (1991).
20. M. J. Scott and M. N. Jones, *Biochim. Biophys. Acta—Biomembranes*, **235**, 1508 (2000).
21. ATCDR, *Toxicological Profile for Phenol*, Agency for Toxic Substances and Disease Registry, Atlanta, GA, 1998.
22. IARC, Phenol, *IARC Monogr. Eval. Carcinog. Risks Humans*, **71**, Part 2, 749–768 (1999).
23. CMR, Olefin stocks as PE sales surge, *Chem. Mark. Rep.*, **250**, 7 (1996).
24. J.-F. Tremblay, *Chem. Eng. News*, December 13, 26 (1999).
25. B. Hileman, *Chem. Eng. News*, March 24, 37 (1997).
26. C. Thurman, *Phenol*, in *Kirk-Othmer Encyclopedia of Chemical Technology*, 3rd edn, Vol. 117, Wiley, New York, 1982, p. 373.
27. T. Kamiura, Y. Tayima and T. Nakahara, *J. Environ. Chem.*, **7**, 275 (1997).
28. *CECIFIC Survey of nonylphenol and nonylphenol ethoxylate production, use, life cycle emission and occupational exposure*, CECIFIC ad hoc Nonylphenol Risk Assessment Task Force: CESIO APE Task Force, 1996.
29. G. G. Naylor, *Tex. Chem. Color*, **27**, 29 (1995).
30. R. W. Tyl, C. B. Myers, M. C. Marr, D. R. Brine, P. A. Fail, J. C. Seely and J. P. Van Miller, *Regul. Toxicol. Pharmacol.*, **30**(2 Pt 1), 81 (1999).
31. D. M. John and G. F. White, *J. Bacteriol.*, **180**, 4332 (1998).
32. M. Ahel, W. Giger and C. Schaffner, *Water Res.*, **28**, 243 (1994).
33. A. M. H. Soto, H. Justicia, J. W. Wray and C. Sonnenschein, *Environ. Health Perspect.*, **92**, 167 (1991).
34. S. Jobling and J. P. Sumpter, *Aquat. Toxicol.*, **27**, 361 (1993).
35. N. Jonkers, T. P. Knepper and P. de Voogt, *Environ. Sci. Technol.*, **35**, 335 (2001).
36. M. L. Davi and F. Gnudi, *Water Res.*, **33**, 3213 (1999).
37. WHO, Environmental Health Criteria, *Chlorophenols other than pentachlorophenol*, World Health Organization, Geneva, Switzerland, 1989.

38. *Hazardous Substances Data Bank, National Library of Medicine*, National Toxicology Information Program, Bethesda, MD, 1998.
39. B. F. Gurney and D. P. Lantenschlager, *J. Dental Assoc. Safr.*, **37**, 815 (1982).
40. TRI96, *Toxic Release Inventory*, Office of Toxic Substances, US EPA, 1998.
41. R. J. Cremlyn, *Agrochemical Preparation and Mode of Action*, Wiley, Chichester, 1991.
42. J. G. Mueller, P. J. Chapman and P. H. Pritchard, *Environ. Sci. Technol.*, **23**, 1197 (1989).
43. P. Kalliokoski and T. Kauppinen, *Complex chlorinated hydrocarbons:occupational exposure in the sawmill industry*, in *Complex Mixtures and Cancer Risk*, (Eds. H. Vainio, M. Sorsa and A. L. McMichael), IARC Scientific Publication, Lion, N 104, 390, 1990.
44. M. Riess and R. van Eldik, *J. Chromatogr. A*, **827**, 65 (1998).
45. Environmental Health Criteria, N 172, *Tetrabromobisphenol A and Derivatives*, International Program on Chemical Safety, WHO, Geneva, 1995.
46. I. Watanabe, T. Kashimoto and R. Tatsukawa, *Bull. Environ. Contam. Toxicol.*, **31**, 48 (1983).
47. U. Sellstrom and B. Jansson, *Chemosphere*, **31**, 3085 (1995).
48. M. L. Hewitt, K. R. Munkittrick, I. M. Scott, J. H. Carry, K. R. Solomon and M. R. Servos, *Environ. Toxicol. Chem.*, **15**, 894 (1996).
49. P. W. Gabbadon and F. A. Chapman, *Prog. Fish Cult.*, **58**, 23 (1998).
50. J. B. Colton, F. D. Knapp, Jr. and B. R. Bruns, *Science*, **185**, 491 (1974).
51. C. S. Wong, D. R. Green and W. J. Cretney, *Nature*, **247**, 30 (1974).
52. J. Sajiki and J. Jonekubo, *Chemosphere*, **46**, 345 (2002).
53. A. Gonzales-Cosado, N. Navas, M. del Olmo and J. L. Vilchez, *J. Chromatogr. Sci.*, **36**, 565 (1998).
54. C. A. Staples, P. B. Dorn, G. M. Klecka, S. T. O'Block and L. R. Harris, *Chemosphere*, **35**, 2149 (1998).
55. K. A. Mountfort, J. Kelly, S. M. Jellis and L. Castle, *Food Additives Contam.*, **14**, 737 (1997).
56. T. Yamamoto and A. Yasuhara, *Chemosphere*, **38**, 2569 (1998).
57. H. Miyakoda, M. Tabata, S. Onodera and K. Takeda, *J. Health Sci.*, **45**, 318 (1999).
58. N. Olea, R. Pulgar, P. Perez, F. Olea-Serano, A. Rivas, A. Novillo-Fertrell, V. Pedraza, A. M. Soto and C. Sonnenschein, *Environ. Health Perspect.*, **104**, 298 (1996).
59. S. M. Maloney, J. Manem, J. Mallevialle and F. Fiesinger, *Environ. Sci. Technol.*, **20**, 249 (1986).
60. L. G. Oberg, B. Glas, S. E. Swanson, C. Rappe and K. G. Paul, *Arch. Environ. Contam. Toxicol.*, **19**, 930 (1990).
61. J. Dec and J.-M. Bollag, *Environ. Sci. Technol.*, **28**, 484 (1994).
62. '5530C, Chloroform Extraction Method', in *Standard Methods for the Examination of Water and Wastewater* (Eds. A. D. Eaton, L. S. Clesceri and A. E. Greenberg), 19th edn., Am. Water Work Assoc., Washington, 1995.
63. I. Rodriguez, M. P. Llompart and R. Cela, *J. Chromatogr. A*, **885**, 291 (2000).
64. B. Schultz, *J. Chromatogr.*, **269**, 208 (1983).
65. O. Busto, J. C. Olucha and F. Borull, *Chromatographia*, **32**, 566 (1991).
66. L. Schmidt, J. J. Sun, J. S. Fritz, D. F. Hagen, C. G. Markell and E. E. Wisted, *J. Chromatogr.*, **641**, 57 (1993).
67. G. Lamprecht and J. F. K. Hurber, *J. Chromatogr. A*, **667**, 47 (1994).
68. M. T. Galceran and O. Jauregui, *Anal. Chim. Acta*, **304**, 75 (1995).
69. D. Puig and D. Baecello, *Chromatographia*, **40**, 435 (1995).
70. E. Pocurull, G. Sanchez, F. Borull and R. M. Marce, *J. Chromatogr. A*, **696**, 31 (1995).
71. J. Horack and R. E. Majors, *LC-GC Int.*, **6**, 208 (1993).
72. L. A. Berrueta, B. Gallo and F. Vicente, *Chromatographia*, **40**, 474 (1995).
73. M. C. Bruzoniti, C. Sarzanini and E. Mentasti, *J. Chromatogr. A*, **902**, 289 (2000).
74. '6420, Phenols', in *Standard Methods for the Examination of Water and Wastewater* (Eds. A. D. Eaton, L. S. Clesceri and A. E. Greenberg), 19th edn., Am. Water Work Assoc., Washington, 1995.
75. *SW-846 On-line*, Method 3500B, 'Organic Extraction and Sample Preparation', 3000 Series Methods, US EPA Office of Solid Waste, www.epa.gov/epaoswer/hazwaste/test/3xxx.htm, 1996.
76. *EPA Method 604*, Phenols in Federal Register, Part 136, 1984.
77. *EPA Method 625*, Base/Neutrals and Acids in Federal Register, Part 136, 1984.
78. *EPA Method 8041*, Phenols by Gas Chromatography, EPA, Washington, DC, 1996.

79. E. M. Thurman and M. S. Mills, *Solid-Phase Extraction—Principles and Practice*, Wiley, New York, 1998.
80. C. E. Werkhoven-Goewie, U. A. Th. Brinkman and R. W. Frei, *Anal. Chem.*, **53**, 2072 (1981).
81. J. Gawdzik, B. Gawdzik and U. Czerwinska-Bil, *Chromatographia*, **25**, 504 (1988).
82. G. Achilli, G. P. Cellerino, G. Melzi d'Eril and S. Bird, *J. Chromatogr. A*, **697**, 357 (1995).
83. L. Renberg, *Anal. Chem.*, **46**, 459 (1974).
84. C. D. Chriswell, R. C. Chang and J. S. Fritz, *Anal. Chem.*, **47**, 1325 (1975).
85. A. Di Corcia, S. Marchese and S. Samperi, *J. Chromatogr.*, **642**, 163 (1993).
86. A. Di Corcia, S. Marchese and S. Samperi, *J. Chromatogr.*, **642**, 175 (1993).
87. A. Di Corcia, A. Bellini, M. D. Madbouly and S. Marchese, *J. Chromatogr. A*, **733**, 383 (1996).
88. M. C. Bruzzoniti, C. Sarzanini and E. Mentasti, *J. Chromatogr. A*, **902**, 289 (2000).
89. M. Moder, S. Schrader, U. Y. Franck and P. Popp, *Fresenius J. Anal. Chem.*, **357**, 326 (1997).
90. P. Mubmann, K. Levsen and W. Radeck, *Fresenius. Z. Anal. Chem.*, **348**, 654 (1994).
91. H. Lee, L. Weng and A. S. Y. Chau, *J. Assoc. Off. Anal. Chem.*, **67**, 789 (1984).
92. E. Pocurull, M. Callul, R. M. Marce and F. Borrull, *Chromatographia*, **38**, 579 (1994).
93. E. Pocurull, M. Callul, R. M. Marce and F. Borrull, *J. Chromatogr. A*, **719**, 105 (1996).
94. *EPA Methods 8041*, Phenols by Gas Chromatography, 1996.
95. T. Heberer and H. J. Stan, *Anal. Chim. Acta*, **341**, 21 (1997).
96. *EPA Method 528*, Determination of phenols in drinking water by solid phase extraction and capillary column GC/MS, EPA, Cincinnati, 2000.
97. *EPA Method 8270*, Semivolatile organic compounds by gas chromatography/mass spectrometry GC/MS, EPA, Cincinnati, 1998.
98. S. D. Richardson, 'Drinking water disinfection by-products', in *Encyclopedia of Environmental Analysis and Remediation* (Ed. R. A. Meyers), Wiley, New York, 1998, p. 1398.
99. S. D. Richardson, Ch. Rav-Acha and V. Glezer, *Environ. Sci. Technol.*, 2003, in press.
100. S. D. Richardson, A. D. Thruston, Jr., T. V. Caughran, P. H. Chen, T. L. Floyd, K. M. Schenck, B. W. Lykins, Jr., Guang-Ri Sun and G. Majetich, *Environ. Sci. Technol.*, **33**, 3378 (1999).
101. C. Crescenzi, A. Di Corcia, G. Passariello, R. Samperi and M. I. Turnes, *J. Chromatogr. A*, **733**, 41 (1996).
102. N. Masque, R. M. Marce and F. Borrull, *Chromatographia*, **48**, 231 (1998).
103. O. Jauregui and M. T. Calceran, *Anal. Chim. Acta*, **340**, 191 (1997).
104. D. Martinez, E. Pocurull, R. M. Marce, F. Borull and M. Callul, *Chromatographia*, **43**, 619 (1996).
105. J. Frebortova and V. Tatarkovicova, *Analyst*, **119**, 1519 (1994).
106. I. Liska and J. Bilikova, *J. Chromatogr. A*, **795**, 61 (1998).
107. D. Martinez, F. Borrull and M. Calull, *Trends Anal. Chem.*, **18**, 282 (1999).
108. S. Morales and R. Cela, *J. Chromatogr. A*, **896**, 95 (2000).
109. G. Li and D. C. Locke, *J. Chromatogr. B*, **669**, 93 (1995).
110. I. Turnes, I. Rodrigez, C. M. Garcia and R. Cela, *J. Chromatogr. A*, **743**, 283 (1996).
111. V. Piangerelli, F. Nerini and S. Cavalli, *Ann. Chim. (Rome)*, **87**, 571 (1997).
112. J. Stradins and B. Hasanli, *J. Electroanal. Chem.*, **353**, 57 (1993).
113. D. B. Gomis, N. F. Palomino and J. J. Mangas Alonso, *Anal. Chim. Acta*, **426**, 111 (2001).
114. D. Treutter and W. Fencht, *J. Hortic. Sci.*, **65**, 511 (1990).
115. J. A. Vinson and B. Hontz, *J. Agric. Food Chem.*, **43**, 401 (1995).
116. J. A. Vinson, Y. A. Dabbagh, M. M. Serry and J. Jang, *J. Agric. Food Chem.*, **43**, 2800 (1995).
117. Ch. Rav-Acha, 'Transformation of aqueous pollutants by chlorine dioxide: reactions, mechanisms and products', in *Handbook of Environmental Chemistry*, Vol. 5, Part C, *Quality and Treatment of Drinking Water* (Ed. J. Hrubec), Springer-Verlag, Berlin, Heidelberg, 1998, p. 62.
118. S. Franke, S. Hildebrandt, J. Schwarzbauer, M. Link and W. Franke, *Fresenius Z. Anal. Chem.*, **353**, 39 (1995).
119. I. Espadaler, J. Caxixach, J. Om, F. Ventura, M. Cortina, F. Paune and J. Rivera, *Water Res.*, **31**, 1996 (1997).
120. S. J. Stangroom, C. D. Collins and J. N. Lester, *Environ. Technol.*, **19**, 643 (1998).
121. D. K. Makepeace, D. W. Smith and S. J. Stainly, *Crit. Rev. Environ. Sci. Technol.*, **25**, 93 (1995).

122. F. Decembrini, F. Azzaro and E. Crisafi, *Water Sci. Technol.*, **32**, 9–10 231 (1995).
123. K. Bester, N. Theobald and H. Fr. Schroeder, *Chemosphere*, **45**, 817 (2001).
124. N. Kannan, N. Yamashita, G. Petrick and J. C. Dunker, *Environ. Sci. Technol.*, **32**, 1747 (1998).
125. T. Isobe, H. Nishiyama, A. Nakashima and H. Takada, *Environ. Sci. Technol.*, **35**, 1041 (2001).
126. M. Ahel, W. Giger and C. Schaffner, *Water Res.*, **28**, 1143 (1994).
127. M. A. Blackburn and M. J. Waldock, *Water Res.*, **29**, 1623 (1995).
128. C. G. Naylor, J. P. Mieure, W. J. Adams, J. A. Weeks, F. J. Castaldi, L. D. Ogle and R. R. Romano, *J. Am. Oil Chem. Soc.*, **69**, 695 (1992).
129. M. Ahel, W. Giger and M. Koch, *Water Res.*, **28**, 1131 (1994).
130. F. Ferrara, F. Fabietti, M. Delise, A. Piccoli Boca and E. Funari, *Environ. Sci. Technol.*, **35**, 3109 (2001).
131. P. L. Ferguson, C. B. Iden and B. J. Brownawell, *Environ. Sci. Technol.*, **35**, 2428 (2001).
132. B. E. Erickson, *Environ. Sci. Technol.*, **35**, 356A (2001).
133. L. J. Heinis, M. L. Knuth, K. Liber, B. R. Sheedy, R. L. Tunnel and G. T. Ankley, *Environ. Toxicol. Chem.*, **18**, 363 (1999).
134. S. Herve, P. Heinonen, R. Paukku, M. Knuutila, J. Koistinen and J. Paasivirta, *Chemosphere*, **17**, 1945 (1985).
135. J. Paasivirta, *Chemical Ecotoxicology*, Lewis Publ., Chelsea, 1991, p. 90.
136. M. M. Broholm, N. Tuxen, K. Rugge and P. L. Bjerg, *Environ. Sci. Technol.*, **35**, 4789 (2001).
137. H. Fukazawa, K. Hoshino, T. Shiozawa, H. Matsushita and Y. Terao, *Chemosphere*, **44**, 973 (2001).
138. H. F. Stich, *Mutat. Res.*, **259**, 307 (1991).
139. R. A. Larson and A. L. Rockwell, *Environ. Sci. Technol.*, **13**, 325 (1979).
140. *World Health Organization (WHO), Guidelines for Drinking Water Quality*, 2nd edn., Vol. 1—Recommendations, Geneva, 1993.
141. *Agency for toxic substances and disease registry*, Atlanta, GA, US Department of Health and Human Services, *Public Health Service. Public Health Statement for Chlorophenols*, www.atsdr.cdc.gov/toxprofiles/phs107.html, updated June 22, 2001.
142. *Agency for toxic substances and disease registry*, Atlanta, GA, US Department of Health and Human Services, *Public Health Service. Public Health Statement for Pentachlorophenols*, www.atsdr.cdc.gov/toxprofiles/phs51.html, updated June 22, 2001.
143. *The Determination of Taste and Odour in Potable Water*, Standing Committee of Analysts, HMSO, London, 1994.
144. W. F. Young, H. Horth, R. Crane, T. Ogden and M. Arnott, *Water Res.*, **30**, 331 (1996).
145. H. H. P. Fang and O.-C. Chan, *Water Res.*, **31**, 2229 (1997).
146. T. Tsutsui, N. Hayashi, H. Maizumi, J. Huff and J. C. Barrett, *Mutat. Res.*, **373**, 113 (1997).
147. *Bioassays of 2,4,6-Trichlorophenol for Possible Carcinogenity*, TR-155, National Cancer Institute, Bethesda, MD, 1979.
148. K. Jansson and V. Jansson, *Mutat. Res.*, **280**, 175 (1992).
149. *Toxicology and Carcinogenesis Studies of 2,4-Dichlorophenol in F344/N Rats and B6C3F1 Mice (Feed Studies)*, TP 353 National Toxicology Program, Research Triangle Park, NC, 1989.
150. V. Daniel, W. Huber, K. Bauer, C. Suesal, J. Mytilineos, A. Melk, Ch. Conradt and G. Opelz, *Arch. Environ. Health*, **56**, 77 (2001).
151. C. Colosio, M. Maroni, W. Barcelini, P. Meroni, D. Alcini, A. Colombi, D. Cavallo and V. Foa, *Arch. Environ. Health*, **48**, 81 (1993).
152. P. Ohnsorge, *Laryngorhynootologie*, **70**, 556 (1991).
153. J. P. Seiler, *Mutat. Res.*, **257**, 27 (1991).
154. J. Huff, *Environ. Health Perspect.*, **109**, 209 (2001).
155. C. Mirabelli, J. A. Hoppin, P. E. Tolbert, R. F. Herrick, D. R. Gnepp and E. A. Brann, *Am. J. Ind. Med.*, **37**, 532 (2000).
156. V. P. Kozak, G. V. Simsiman, G. Chester, D. Stensby and J. Harkin, *Reviews of the Environmental Effects of Pollutants: XI. Chlorophenols*, Report 600/1-79-012, US EPA, Washington, DC, 1979.
157. D. Desaiah, Effect of pentachlorophenol on the ATPases in rat tissues, in *Pentachlorophenol, Chemistry, Pharmacology, and Environmental Toxicology* (Ed. K. R. Rao), Plenum Press, New York, 1978, p. 277.

158. M. Stockdale and M. J. Selwyn, *Eur. J. Biochem.*, **21**, 565 (1971).
159. M. Rutgers, S. Brommel, A. M. Breure, J. G. Andel and W. A. Duetz, *Environ. Toxicol. Chem.*, **17**, 792 (1998).
160. P. Lindstrom-Seppa, S. Huuskonen, S. Kotelevtsev, P. Mikkelson, T. Raanen, L. Stepanova and O. Hanninen, *Mar. Environ. Res.*, **46**, 273 (1998).
161. E. Voncina and T. Solmajer, *Chemosphere*, **46**, 1279 (2002).
162. L. L. Lamparski, R. H. Stehl and R. L. Johnson, *Environ. Sci. Technol.*, **14**, 196 (1980).
163. R. Louw and S. I. Amonkhai, *Chemosphere*, **46**, 1273 (2002).
164. S. Vollmuth, A. Zajc and R. Niessner, *Environ. Sci. Technol.*, **28**, 1145 (1994).
165. D. G. Crosby and A. S. Wong, *Chemosphere*, **5**, 327 (1976).
166. H. Muto, K. Saitoh and H. Funayama, *Chemosphere*, **45**, 129 (2001).
167. P. S. Price, S. H. Su, J. R. Harrington and R. E. Keenan, *Risk Anal.*, **16**, 263 (1996).
168. L. Ramos, E. Ejarrat, L. M. Hernandez, L. Alonso, J. Rivera and J. Gonzalez, *Chemosphere*, **35**, 2167 (1997).
169. F. Schuler, P. Schmid and C. Schlatter, *Chemosphere*, **34**, 711 (1997).
170. M. E. Harnly, M. X. Petreas, J. Flattery and L. R. Goldman, *Environ. Sci. Technol.*, **34**, 1143 (2000).
171. M. I. H. Helaleh, Y. Takabayashi, S. Fujii and T. Korenaga, *Anal. Chim. Acta*, **428**, 227 (2001).
172. B. S. Rubin, M. K. Murray, D. A. Damassa, J. C. King and A. M. Soto, *Environ. Health Perspect.*, **109**, 675 (2001).
173. B. E. Erickson, *Environ. Sci. Technol.*, **36**, 10A (2002).
174. M. Hesselsoe, D. Jensen, K. Skals, T. Olensen, P. Moldrup, P. Roslev, G. K. Mortensen and K. Henriksen, *Environ. Sci. Technol.*, **35**, 3695 (2001).
175. J. H. Weisburger, *ACS Symp. Ser.*, **507**, 35 (1992).
176. F. Iversen, *Cancer Lett.*, **93**, 49 (1995).
177. A. Patrineli, M. N. Clifford and C. Ioannides, *Food Chem. Toxicol.*, **34**, 869 (1996).
178. J. Hemming, B. Holmbom, M. Reunanen and L. Kronberg, *Chemosphere*, **15**, 549 (1986).
179. L. Kronberg and T. Vertianen, *Mutat. Res.*, **206**, 177 (1988).
180. A. D. L. Norwood, R. F. Christman and P. G. Hatcher, *Environ. Sci. Technol.*, **21**, 791 (1987).
181. D. A. Rechow, P. C. Singer and R. L. Malcolm, *Environ. Sci. Technol.*, **24**, 1655 (1990).
182. V. Langvik, O. Hormi, L. Tikkanen and B. Holmbom, *Chemosphere*, **22**, 547 (1991).
183. V. Langvik and O. Hormi, *Chemosphere*, **28**, 1111 (1994).
184. X. Xu, H. Zou and J. Zhang, *Water Res.*, **31**, 1021 (1997).
185. B. Holmbom, L. Kronberg, P. Bachlund, V. Langvic, J. Hemming, M. Reunanen, A. Smeds and L. Tikannen. 'Formation and properties of of 3-chloro-4-(dichloromethyl)-5-hydroxy-2(5H)-furanone, a potent mutagen in chlorinated waters', in *Water Chlorination: Chemistry, Environmental Impact, and Health Effects* (Eds. R. L. Jolley, L. W. Condie, J. D. Johnson, S. Katz, R. A. Minear, J. S. Mattice and V. A. Jacobs), Vol. 6, Lewis Publ., Chelsea, 1990, p. 125.
186. C. Yang, Z. Chen, H. Zou, J. Lu and J. Zhang, *Water Res.*, **34**, 4313 (2000).
187. T. Tanaka, N. Yoshimi, S. Sugie and H. Mori, *ACS Symp. Ser.*, **507**, 326 (1992).
188. A. C. Malaveille, A. Hautefeuille, B. Pignatelli, G. Talaska, P. Vineis and H. Bartsch, *Carcinogenesis*, **17**, 2193 (1996).
189. A. C. Malaveille, A. Hautefeuille, B. Pignatelli, G. Talaska, P. Vineis and H. Bartsch, *Mutat. Res.—Fundamentals and Molecular Mechanisms of Mutagenesis*, **402**, 219 (1998).
190. J. D. Rodgers, W. Jedral and N. I. Bunce, *Environ. Sci. Technol.*, **33**, 1453 (1999).
191. P. Dabo, A. Cyr, F. Laplante, F. Jean, H. Menard and J. Lessard, *Environ. Sci. Technol.*, **34**, 1265 (2000).
192. J.-L. Boudenne and O. Cerclier, *Water Res.*, **33**, 494 (1999).
193. Y. Ku, K. Chen and K. Lee, *Water Res.*, **31**, 929 (1997).
194. C. Petrier, M. Lamy, A. Francony, A. Benahcene and B. David, *J. Phys. Chem.*, **98**, 10514 (1994).
195. N. R. Serpone, P. Terzian, P. Colarusso, C. Minerco, E. Pelizzetti and H. Hidaka, *Res. Chem. Intermed.*, **18**, 183 (1992).
196. N. Serpone, P. Terzian, H. Hidaka and E. Pelizzetti, *J. Phys. Chem.*, **98**, 2634 (1994).
197. J. M. Wu, H. S. Huang and C. D. Livengood, *Environ. Prog.*, **11**, 195 (1992).
198. C. Wu, X. Liu, D. Wei, J. Fan and L. Wang, *Water Res.*, **36**, 3927 (2001).

199. L. K. Weaver, N. Malmstadt and M. R. Hoffmann, *Environ. Sci. Technol.*, **34**, 1280 (2000).
200. L. K. Weaver and M. R. Hoffmann, *Environ. Sci. Technol.*, **32**, 3941 (1998).
201. C. H. Kuo and C. H. Charlen, *J. Hazard. Mater.*, **41**, 31 (1995).
202. F. J. Benitez, J. Beltran-Heredia, J. L. Acero and F. J. Rubio, *J. Hazard. Mater.*, **79**, 271 (2000).
203. C. Cooper and R. Burch, *Water Res.*, **33**, 3695 (1999).
204. M. A. Engwall, J. J. Pignatello and D. Grasso, *Water Res.*, **33**, 1151 (1999).
205. B. G. Kwon, D. S. Lee, N. Kang and J. Yoon, *Water Res.*, **33**, 3695 (1999).
206. M. Fukushima and K. Tatsumi, *Environ. Sci. Technol.*, **35**, 1771 (2001).
207. J. Bertran, J. Torregrosa, J. R. Dominguez and J. A. Peres, *Chemosphere*, **45**, 85 (2001).
208. G. Zhang and J. A. Nicell, *Water Res.*, **34**, 1629 (2000).
209. A. Bhandari and F. Xu, *Environ. Sci. Technol.*, **35**, 3163 (2001).
210. A. Modestov, V. Glezer, L. Marjasin and O. Lev, *J. Phys. Chem. B*, **101**, 4623 (1997).
211. Y. Ohko, I. Ando, C. Niwa, T. Tatsuma, T. Yamamura, T. Nakashima, Y. Kuboto and A. Fujishima, *Environ. Sci. Technol.*, **35**, 2365 (2001).
212. W. F. Jardim, S. G. Morales and M. M. K. Takiyama, *Water Res.*, **31**, 1728 (1997).
213. Y. Ku and K.-C. Lee, *J. Hazard. Mater.*, **80**, 59 (2000).
214. L. Zhu, B. Chen and X. Shen, *Environ. Sci. Technol.*, **34**, 468 (2000).
215. A. R. Khan, T. A. Al-Bahri and A. Al-Hadad, *Water Res.*, **31**, 2102 (1997).
216. I. G. Garcia, J. L. B. Venceslada, P. R. J. Pena and E. R. Gomez, *Water Res.*, **31**, 2005 (1997).
217. Y. Beaton, L. J. Shaw, L. A. Glover, A. A. Meharg and K. Killham, *Environ. Sci. Technol.*, **33**, 4086 (1999).
218. M. I. Gladyshev, N. N. Sushchik, G. S. Kalachova and L. A. Shur, *Water Res.*, **32**, 2769 (1998).
219. J. Paasivirta, *Chemical Ecotoxicology*, Lewis Publ., Chelsea, 1991, p. 93.
220. J. Paasivirta, P. Klein, M. Knuutila, J. Knuutinen, M. Lahtipera, R. Paukku, A. Veijanen, L. Welling, M. Vuorinen and P. J. Vuorinen, *Chemosphere*, **15**, 1429 (1987).

CHAPTER 19

Calixarenes

VOLKER BÖHMER

Johannes Gutenberg-Universität, Fachbereich Chemie und Pharmazie, Abteilung Lehramt Chemie, Duesbergweg 10–14, D-55099 Mainz, Germany
Fax: +49 6131 3925419; e-mail: vboehmer@mail.uni-mainz.de

I. INTRODUCTION AND DEFINITIONS	1370
II. SYNTHESES OF CALIXARENES	1371
A. One-pot Procedures	1371
1. Standard compounds (calixphenols)	1371
2. Modification of the phenolic compound	1372
3. Larger calixarenes	1372
B. Stepwise Syntheses	1373
C. Fragment Condensations	1374
D. Calixarene-like Macrocycles with Other Bridges	1376
1. Homocalixarenes	1376
2. Homooxa- and homoazacalixarenes	1378
3. Thiacalixarenes	1380
E. Syntheses of Resorcarenes (Calixresorcinols)	1381
III. CONFORMATIONAL PROPERTIES	1384
A. Calixphenols	1384
B. Calix[4]resorcinols	1385
IV. REACTIONS OF THE HYDROXY GROUPS IN CALIXPHENOLS	1387
A. Complete Conversions	1387
B. Conformational Isomers	1389
C. Partial Conversions	1392
1. Monoethers and -esters	1392
2. Di- and triethers and -esters of calix[4]arenes	1393
3. Partial ethers (or esters) of larger calixarenes	1394
D. Reactions with Di- and Multifunctional Reagents	1395
1. Calix[4]arenes	1395
2. Calix[5]arenes[194]	1396
3. Calix[6]arenes[196]	1397
4. Calix[8]arenes[101]	1400
E. Replacement of the Hydroxy Groups	1401

The Chemistry of Phenols. Edited by Z. Rappoport
© 2003 John Wiley & Sons, Ltd ISBN: 0-471-49737-1

V. MODIFICATION OF CALIXPHENOLS ON THE WIDE RIM	1402
A. Complete Substitution	1402
B. Selectivity Transfer from the Narrow to the Wide Rim	1402
C. Modification of Substituents	1403
VI. FURTHER REACTIONS OF CALIXPHENOLS	1407
A. Rearrangements	1407
B. Modification of the Methylene Bridges	1408
C. Oxidation of the Aromatic Systems	1409
1. Calixquinones	1409
2. Spirodienones	1410
D. Reduction of the Aromatic Systems[299]	1411
E. Metallation of the π-Electron System	1413
F. Special Reactions with Thiacalixarenes	1415
VII. CHEMICAL MODIFICATION OF CALIXRESORCINOLS	1416
A. Reactions of the Hydroxy Groups	1417
B. Electrophilic Substitutions	1419
C. Reactions of the Substituents at the Bridges	1421
D. Cavitands	1422
VIII. MOLECULES CONSISTING OF SEVERAL CALIXARENE STRUCTURES	1425
A. Double Calixarenes	1425
B. Carcerands	1427
C. Further Combinations	1428
D. Multicalixarenes	1432
IX. CONCLUSIONS AND OUTLOOK	1435
X. ACKNOWLEDGEMENT	1435
XI. REFERENCES AND NOTES	1435

I. INTRODUCTION AND DEFINITIONS

The name 'calix[n]arenes' was coined by C. D. Gutsche originally to describe cyclic oligomers built up by (4-substituted) phenolic units linked in 2- and 6-position via methylene bridges (**I**)[1]. It is deduced from the calix or cup-like conformation assumed especially by the tetra- and pentamer, which resembles an ancient Greek vase, known as 'calix crater', while 'arene' refers to the aromatic units, the number of which is indicated by [n]. All hydroxy groups in the general formula **I** are found in *endo*-position at the 'narrow rim'[2] of the macrocycle.

The basic skeleton of these compounds, calixarenes in the original (narrow) sense, is that of a $[1_n]$metacyclophane and it is appropriate to include into the family of calixarenes also those cyclic oligomers, in which the phenolic hydroxy groups are situated in *exo*-position at the 'wide rim'[2]. Here especially cyclic compounds **II** (nearly exclusively tetramers) built up by resorcinol units (eventually substituted in the 2-position) are important and will be called 'resorcarenes'[3,4] within this article. Alternatively 'calixresorcinols' is used for **II**, in distinction to 'calixphenols' for **I**.

The present chapter will concentrate on these two types, including more or less close modifications of their structure, while cyclic oligomers derived from pyrocatechol which are $[1_n]$orthocyclophanes[5] and various other cyclooligomers for which meanwhile the prefix 'calix' is used[6] will be excluded. The main emphasis will be also on the synthesis and the chemical modification of calixarenes and some basic properties, such as their conformational behaviour. Their host properties towards cations, anions or neutral guests

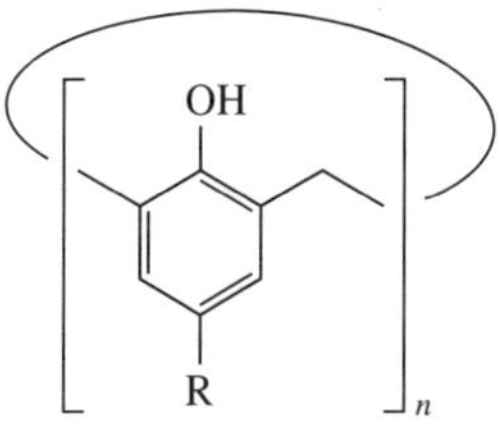

(I) 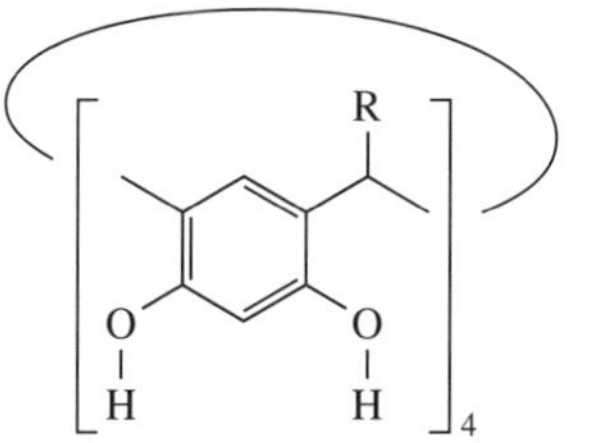   (II)

and various applications in sensors, separation processes, mono- and multilayers are not treated in detail.

II. SYNTHESES OF CALIXARENES

A. One-pot Procedures

1. Standard compounds (calixphenols)

The rapid development of calixarene chemistry in the 1980s followed by an explosion-like development in the 1990s is due to the ease by which larger amounts of t-butyl calixarenes are available on a laboratory scale by alkali-catalysed condensation of p-t-butylphenol **1** with formaldehyde (Scheme 1).

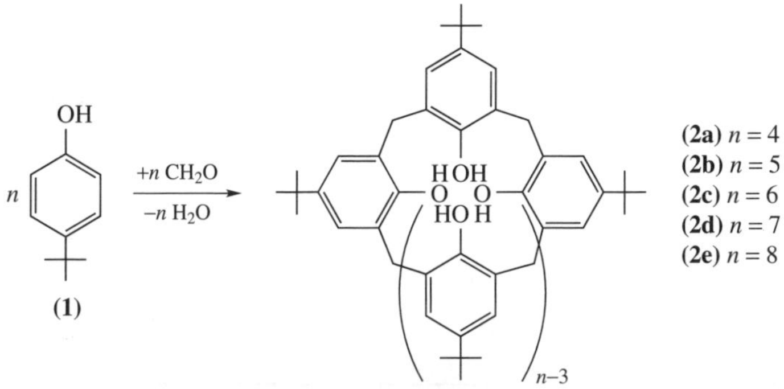

(**2a**) $n = 4$
(**2b**) $n = 5$
(**2c**) $n = 6$
(**2d**) $n = 7$
(**2e**) $n = 8$

SCHEME 1. One-pot condensation of t-butylphenol **1** to form calix[n]arenes **2**. The p-unsubstituted calixarenes (see Section V.A.) will be characterized by 2_H

Especially well elaborated procedures exist for the three 'major' calixarenes **2a**, **2c** and **2e**:

(i) Pre-condensation of **1** with aqueous HCHO using a 0.045 molar amount of NaOH followed by 2 h reflux in diphenyl ether produces about 50% of **2a**[7].

(ii) Heating of **1** and formalin with a 0.34 molar amount of KOH followed by 4 h reflux in xylene yields 83–88% of **2c**[8].

(iii) Refluxing a solution of **1** with paraformaldehyde and a 0.030 molar amount of NaOH in xylene produces **2e** in 62–65% yield[9].

These yields (obtained without dilution techniques) are remarkable, if not unique, considering the fact that, for instance, 16 covalent links are newly formed in the synthesis of **2e**. Even the less favourably formed cyclic penta- and heptamers are available now in multigram quantities in yields of 15–20% for **2b**[10] and 11–17% (LiOH as base) for **2d**[11].

Although it is generally accepted that the *t*-butylcalix[8]arene is the kinetically and the -calix[4]arene the thermodynamically controlled product while the formation of the -calix[6]arene seems to be due to a template effect, not much is really known about the mechanism of calixarene formation. The hypothesis that **2a** is formed from **2e** by an intramolecular step ('molecular mitosis') could not be confirmed by isotopic labelling, which proves a more or less statistical fragmentation and recombination[12].

2. Modification of the phenolic compound

Various other *p*-alkylphenols have been studied in one-pot procedures. However, the yields of calixarenes are generally lower than with **1**, and individual compounds often can be isolated only by chromatography. Table 1 gives a survey. As a rule of thumb it may be concluded that calixarene formation is favoured for those alkylphenols, where a tertiary carbon is attached to the *p*-position[13]. Calixarenes *p*-substituted by electron-withdrawing residues have not been obtained by one-pot syntheses starting with the single phenol, while *p*-benzyloxy- and *p*-phenylphenol were used with some success.

TABLE 1. Yields of the main calix[n]arenes ($n = 4$ to 8) available by one-pot syntheses

R \ n	4	5	6	7	8
Me			74^{14}	22^{15}	
Et				24^{15}	
i-Pr	10^{16}		26^{16}		no yield reported[16]
t-Bu	49^{7}	$15-20^{10}$	$83-88^{8}$	$11-17^{11}$	$62-65^{7}$
t-Pent	$6-7^{17}$		30^{17}		$37-41^{17}$
t-Oct	31^{18}		30^{19}, 16^{18}		
n-Alkyl			10^{20}		10^{20}
1-Adamantyl					71^{21}
Benzyl	60^{22}	$15-20^{23}$	16^{24}	33^{24}, $15-20^{23}$	12^{24}, 30^{22}
Phenyl	10^{25}	15^{25}	10^{26}, 11^{25}		7^{26}, 38^{25}

3. Larger calixarenes

Although larger oligomers than calix[8]arenes have been isolated from one-pot reactions under alkaline conditions[23,27,28] the method of choice to obtain these higher oligomers seems to be an acid-catalysed (*p*-toluenesulphonic acid) condensation of **1** with *s*-trioxane in chloroform[29], where the total yield of calixarenes is nearly quantitative. Procedures to prepare a single, individual compound are not available in this case, but all *t*-butylcalix[n]arenes up to $n = 20$ have been isolated by chromatographic techniques[27,30].

B. Stepwise Syntheses

Calixarenes prepared by one-pot procedures necessarily consist of the same 'repeating unit' (usually a phenol). The stepwise synthesis outlined in Scheme 2 allows one in principle to build up calix[n]arenes with n different p-substituted phenolic units. After protection of one *ortho*-position, usually by bromination, a sequence of hydroxymethylation and condensation steps furnishes a linear oligomer which, after deprotection, is

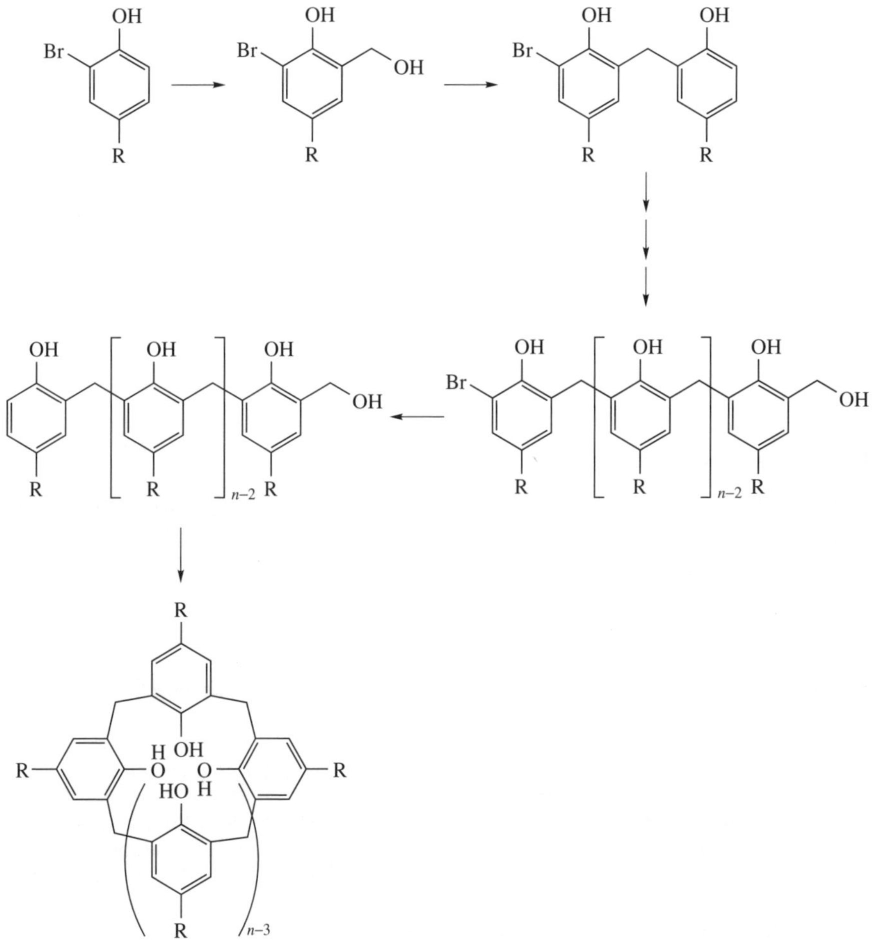

SCHEME 2. Synthesis of calixarenes by the stepwise strategy. In principle, each phenolic unit can have a different substituent (mainly alkyl groups) in the p-position

cyclized under high dilution conditions. Thus, $2n + 2$ steps are necessary to obtain a calix[n]arene. Today this strategy is mainly of historical interest, but the single steps may serve also to built up precursors useful for more convergent strategies.

The last step, for instance, the cyclization of a linear precursor under dilution conditions, was used recently for the synthesis of calix[4–6]arenes **3a–c** with a single carbonyl bridge[31].

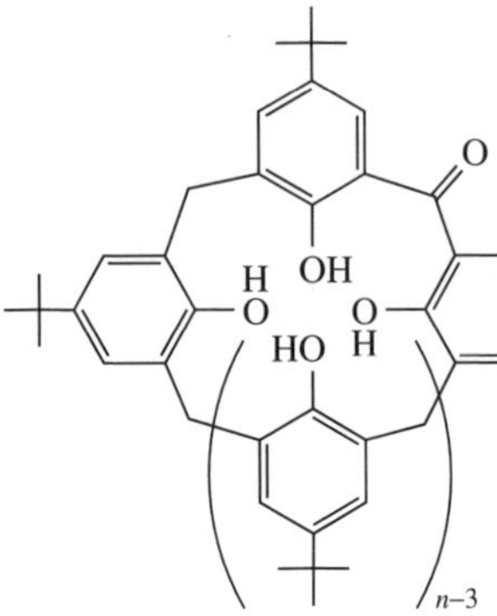

(**3a**) $n = 4$
(**3b**) $n = 5$
(**3c**) $n = 6$

C. Fragment Condensations

A calix[n]arene may be obtained by condensation of two independently synthesized fragments as generally illustrated in Scheme 3. Various calix[4]arenes have been obtained by $3 + 1$ or $2 + 2$ approaches[32] using TiCl$_4$ as catalyst in yields up to 25–30%[33]. For larger calixarenes, however, these conditions are hampered by side reactions (e.g. the cleavage of existing methylene bridges) although some calix[5]-[34] or -[6]arenes[35] have been prepared. A simple 'heat induced' condensation using bishydroxymethylated compounds seems advantageous for the synthesis of calix[5]arenes either by $3 + 2$[36] or $4 + 1$[37] approaches. Its suitability for the synthesis of larger oligomers has still to be checked. Only one example is known for the synthesis of a calix[8]arene in 9% yield by $7 + 1$ condensation of a linear heptamer with a bishydroxymethylated phenol[38], but various calix[4]- (**4**)[39] and calix[5]arenes (**5**)[34,37] with substituted bridges (–CHX–) have been prepared by fragment condensations, following $2 + 2$ or $3 + 2$ strategies.

The condensation of bisbromomethylated phenols with p-bridged diphenols **6** (TiCl$_4$, dioxane, 100 °C) leads to calix[4]arenes **7**, in which two opposite p-positions are connected by an aliphatic chain (Scheme 4)[40]. The yield reaches 30–35% and double calix[4]arenes have been observed as side product in some cases[41].

A regular incorporation of the phenolic units into calix[4]arenes has been observed, if 2- or 6-hydroxymethyl derivatives of 3,4-disubstituted phenols (including cyclic compounds like β-naphthol) are condensed under these conditions (Scheme 5)[42,43]. The resulting calix[4]arenes **8** assume an inherently chiral, C_4-symmetrical *cone* conformation which can be fixed by O-alkylation (see below). In an analogous manner a 2-hydroxymethyl phenol substituted at the 4-position with a porphyrin moiety has been converted to the corresponding calix[4]arene **9** in 60% yield by treatment with NaOH in refluxing diphenyl ether[44].

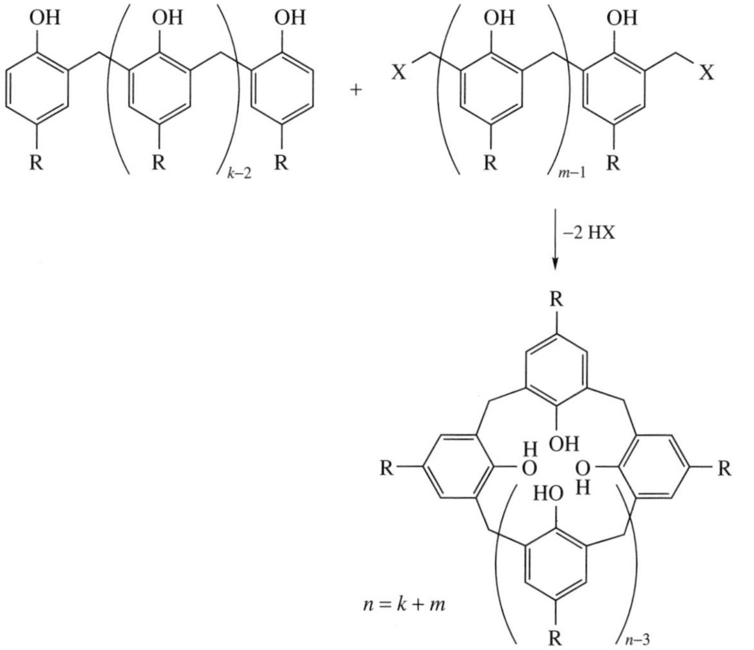

SCHEME 3. Synthesis of calix[n]arenes by fragment condensation '$k + m$'

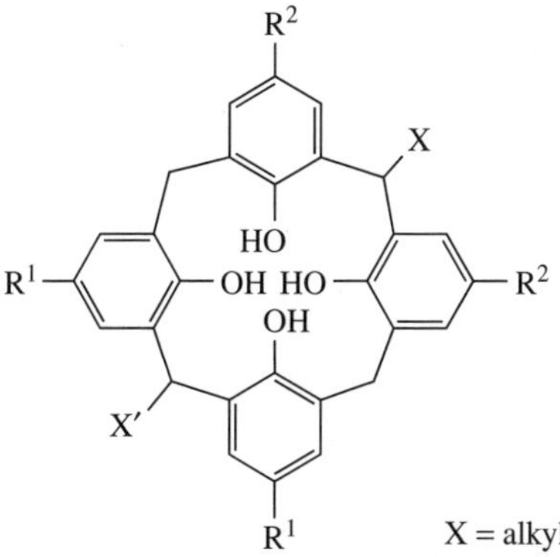

$n = k + m$

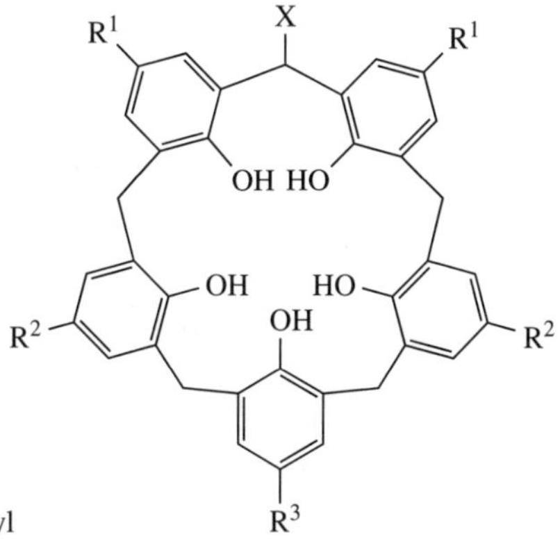

X = alkyl, aryl
X' = H, alkyl, aryl
R^1–R^3 = alkyl; R^3 = NO_2

(4)

(5)

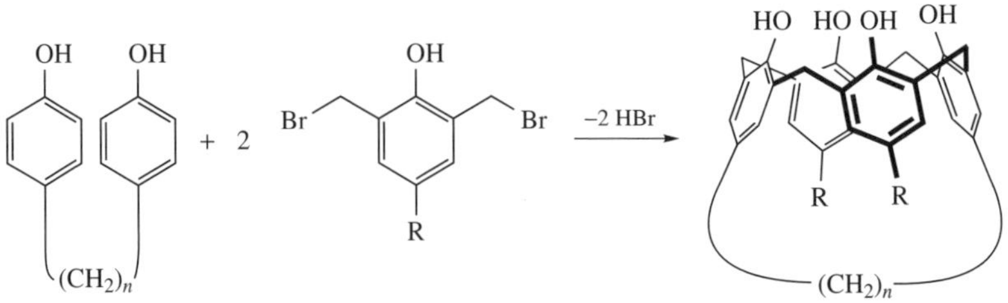

SCHEME 4. Synthesis of *p*-bridged calix[4]arenes by a '2 + 2 × 1' approach

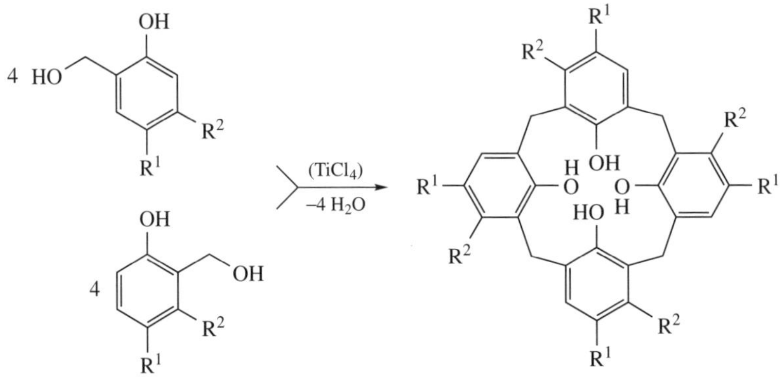

$n > 4$, R = alkyl, Ph, Cl

	R^1	R^2
(a)	CH_3	CH_3
(b)	$CH(CH_3)_2$	CH_3
(c)	Cl	CH_3
(d)	Br	CH_3
(e)	$-(CH_2)_3-$	
(f)	$-(CH_2)_4-$	
(g)	$-CH=CH-CH=CH-$	

SCHEME 5. Synthesis of C_4-symmetrical calix[4]arenes by '1 + 1 + 1 + 1' ('4 × 1') condensation

D. Calixarene-like Macrocycles with Other Bridges

1. Homocalixarenes

Calixarene-like macrocycles such as **10**, **11** and **12** have been also prepared by condensation of the respective bisphenols with formaldehyde under alkaline conditions (Scheme 6).

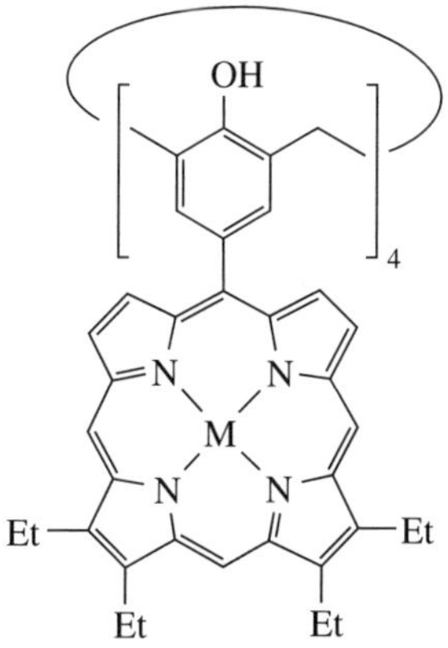

(9)

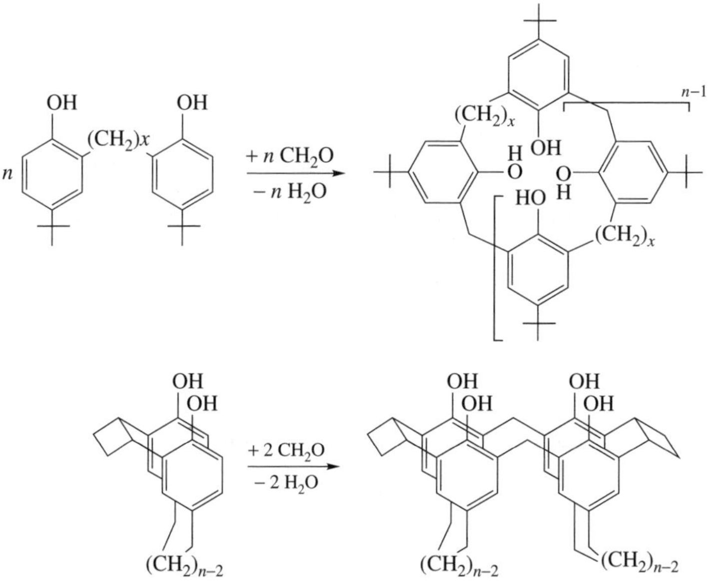

(a) $n = 2$
(b) $n = 3$
(c) $n = 4$

(10) $x = 0$
(11) $x = 2$
(12) $x = 3$

(13) **(14)**

SCHEME 6. One-pot synthesis of calixarene analogues by condensation of bisphenols with formaldehyde

A template effect is concluded from the observation that CsOH favours the formation of the larger macrocycle **11b** in thc case of $x = 2$ while the smaller oligomer **11a** is predominant with NaOH as catalyst[45]. For $x = 0$ a cyclic dimer is not formed, but again the trimer **10b** is formed with NaOH and the tetramer **10c** with CsOH as base[46].

The rigid cyclobutano-bridged bisphenols **13** furnish only the calix[4]arene-like compounds **14** (Scheme 6). The yield is highest for $n = 5$ with LiOH (89%) and for $n = 6$ with CsOH (78%) while the cyclization fails completely for $n = 4$. Not only does this suggest a template effect by the alkali cation, it also clearly demonstrates that further factors (such as rigidity) can be important[47].

'All-homo' calix[4]arenes **16** ($[2_n]$metacyclophanes) were obtained by Müller–Röscheisen cyclization of bisbromomethylated anisole (which furnished a chromatographically separable mixture of the methyl ethers **15** [$n = 5$–8]) and subsequent demethylation[48].

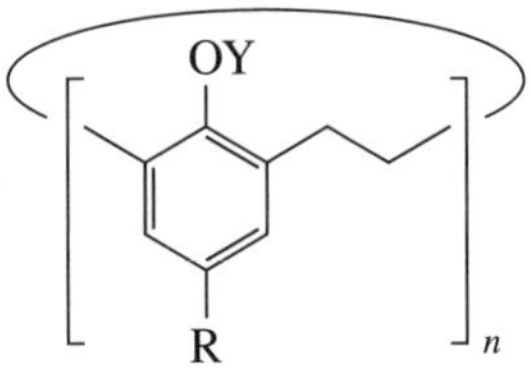

(**15**) Y = CH$_3$
(**16**) Y = H
$n = 3$–8

The acid-catalysed condensation of bisphenols with formaldehyde has been used also to prepare various calix[4]arenes with substituted bridges (**4**, X = X′ [49]) as well as *exo*-calix[4]arenes **17**[50]. In both cases and in the synthesis of *exo–endo* calix[4]arenes **18**[51] linear tetramers have been also cyclized by condensation with (para)formaldehyde[37].

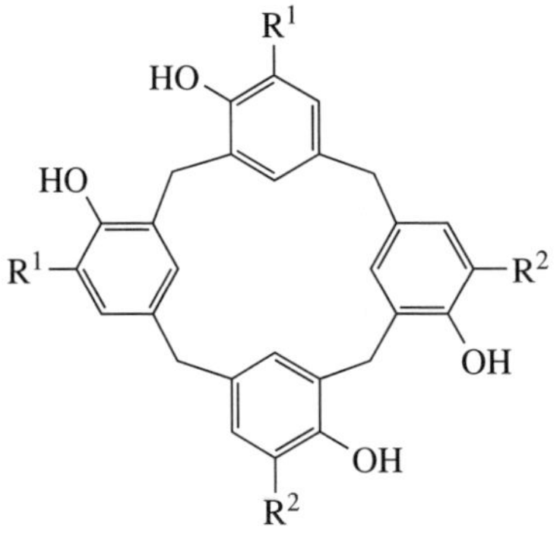

(**17**) 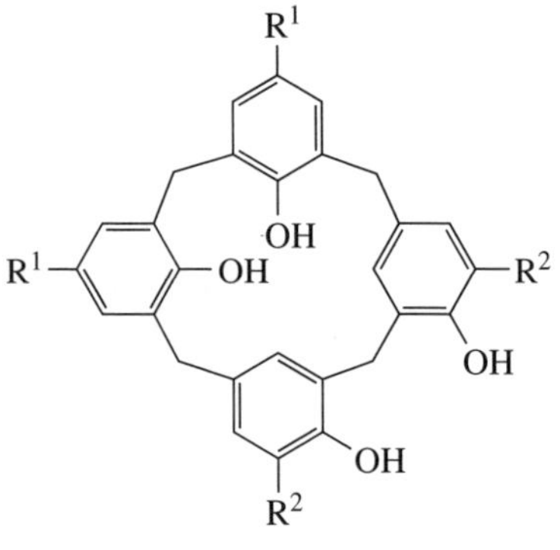   (**18**)

R^1, R^2 = H, alkyl

2. Homooxa- and homoazacalixarenes

As a side product of the one-pot synthesis of calixarenes, compound **23** was isolated already in early studies. Here four t-butylphenol units were linked by three $-CH_2-$bridges and one $-CH_2-O-CH_2-$bridge[52] which explains the name 'bishomooxacalix[4]arene'. Various other macrocyclic compounds are known now, in which the $-CH_2-$bridges of calixarenes are (completely or partly) replaced by $-CH_2-O-CH_2-$ (homooxacalixarenes) or $-CH_2-NR-CH_2-$ (homoazacalixarenes)[53]. The longer (and flexible) bridges allow also the formation of cyclic trimers, e.g. the hexahomotrioxacalix[3]arenes **20**. They are usually prepared in yields up to 30% by thermal dehydration of bishydroxymethylated phenols **19** ($n = 1$) in apolar solvents such as xylene (Scheme 7)[54]. Alternatively,

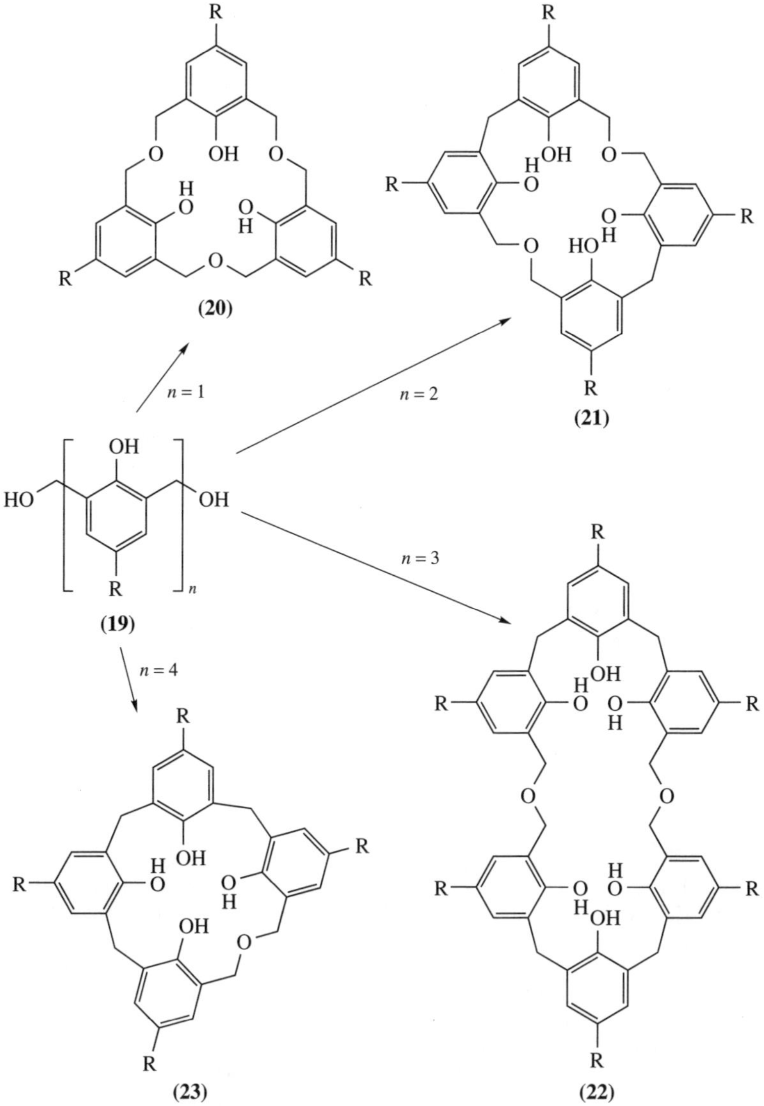

SCHEME 7. Homooxacalixarenes by thermal dehydration of bishydroxymethylated phenols or oligomers (in most cases R = *t*-Bu)

an acid-catalysed cyclization has been proposed, which however requires high dilution conditions[55]. A stepwise procedure using protective groups allowed the synthesis of trimers with three different *p*-substituents[56]. The thermal dehydration was also possible for the bishydroxymethylated dimers to tetramers[54a,57]. The main products (**21, 22, 23**) are indicated in Scheme 7. Various side products containing $-CH_2-$ instead of $-CH_2-O-CH_2-$ bridges are also formed.

Homoazacalixarenes have been prepared in a similar manner reacting bis-hydroxymethylated (or bis-chloromethylated) phenols (or oligomers) with amines[58], a strategy that has been used also to synthesize *N,N*-bridged homoazacalixarenes **24a** and **24b**[59]. The formation of cyclic Schiff bases followed by reduction is a possibility to obtain macrocycles with bridges containing secondary amino groups ($-CH_2NHCH_2-$), such as **25** (Scheme 8)[60].

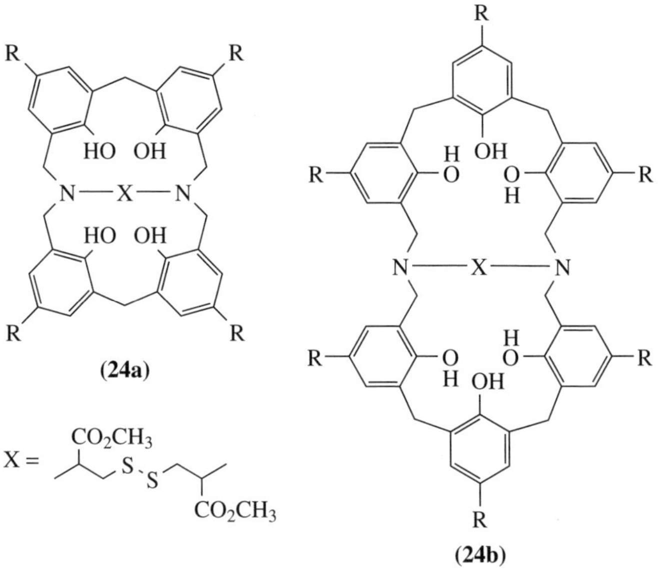

3. Thiacalixarenes

A new and rapidly developing field was opened by the one-step synthesis of tetrathia *t*-butylcalix[4]arene **26** by reaction of **1** with sulphur at 230 °C in tetraethylene glycol dimethyl ether catalysed by NaOH[61], which makes larger quantities of this interesting material available in yields up to 54%. The corresponding *p-t*-octyl compound was obtained under similar conditions (250 °C) in 14%[61b]. The reaction obviously furnishes (almost exclusively) the cyclic tetramer, and only traces of the corresponding hexathia *t*-butylcalix[6]arene have been isolated[62].

Interestingly, also in this case *t*-butylcalix[4]arenes in which one to four methylene bridges have been replaced by sulphur were prepared initially by stepwise procedures[63]. Very recently the cyclization of a sulphur-bridged dimer of **1** (S_8, NaOH, Ph_2O, 130–230 °C) was reported to yield **26** in 83%, while the thiacalix[6]- and -[8]arenes were obtained in 5% and 4%, respectively[64].

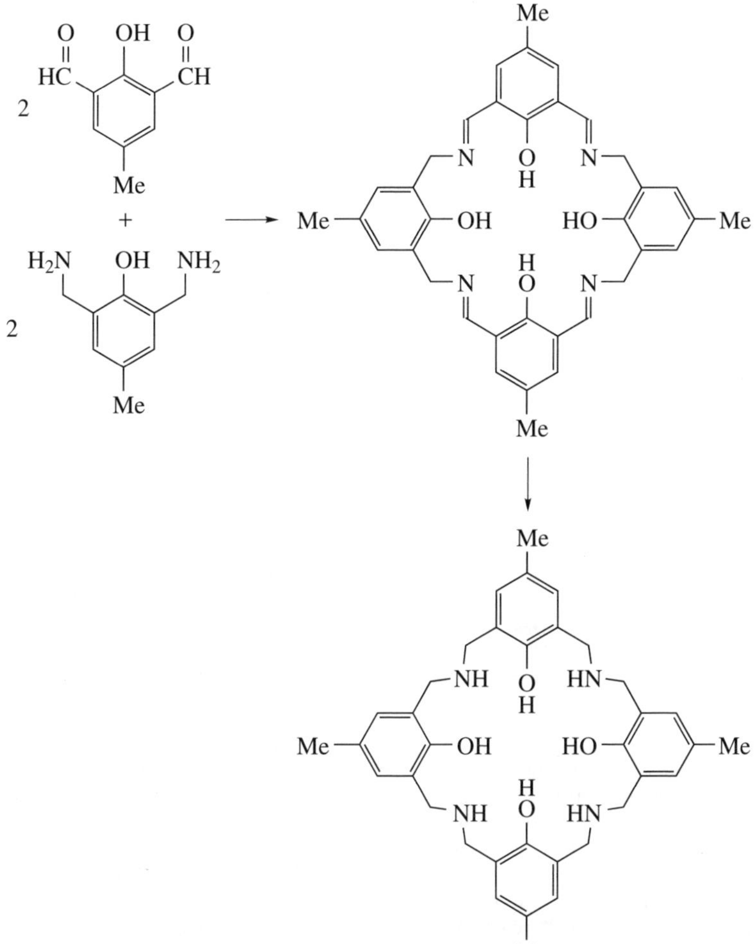

SCHEME 8. Homoazacalix[4]arenes via Schiff base formation

E. Syntheses of Resorcarenes (Calixresorcinols)

The acid-catalysed condensation of resorcinol with different aldehydes than formaldehyde leads (often in high yield) to cyclic tetramers of the general formula **II**. Under typical conditions the reactants are kept for several hours at 80 °C in aqueous ethanol using HCl as catalyst[65]. A solvent-free synthesis has been recently described[66]. Four diastereomeric products are possible in this case which differ in the relative configuration of the −CHR−bridges. For their distinction the following convention is most appropriate: If the macrocycle is considered 'planar' the residues R are found at one or the other side of this plane. If now one of these residues is taken as reference (r) the position of the other residues may be *cis* (*c*) or *trans* (*t*). This situation is illustrated in Figure 1.

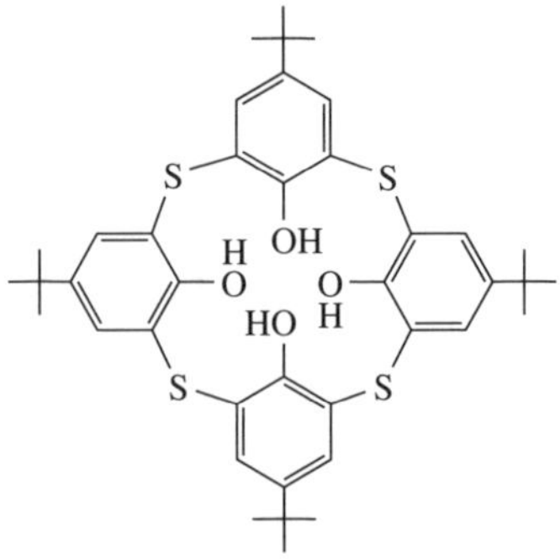

(26)

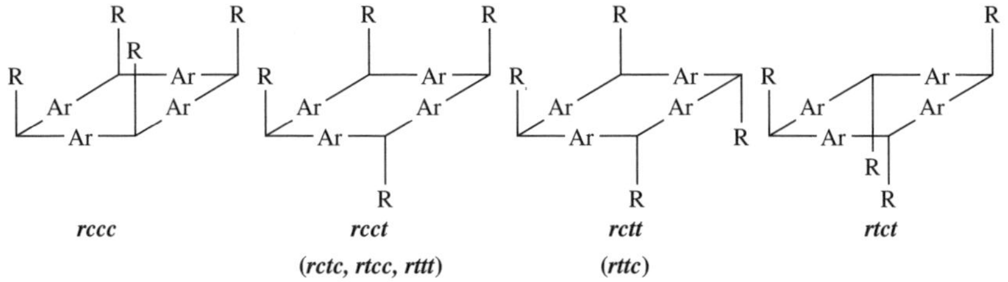

rccc

rcct
(rctc, rtcc, rttt)

rctt
(rttc)

rtct

FIGURE 1. Schematic representation of the different stereoisomers of resorcarenes

Usually only the *rccc* and *rctt* isomers are formed, and often conditions were found under which the *rccc* (or 'all-*cis*') is the only product, e.g. after prolonged heating. The formation of the *rcct* isomer was less frequently observed and the *rtct* isomer has not yet been isolated. The reaction is possible with a large variety of aldehydes (or synthetic equivalents) as shown by the examples given in Figure 2, although the optimal conditions have to be elaborated for each case. 2-Methylresorcinol[67c], other 2-alkyl derivatives and pyrogallol[68] react in a similar fashion. Formaldehyde (or its equivalents like 1,3,5-trioxane or diethoxymethane) can be also used in this case and methylene-bridged macrocyclic products with five or six resorcinol units were obtained in addition to the tetramer[69].

Cyclic products were not obtained with electron-withdrawing substituents like NO_2 or COOH in the 2-position[70]. However, cyclic tetramers of the resorcarene type (**27**) were recently prepared in yields up to 70% from 2,6-dihydroxypyridine and aliphatic or aromatic aldehydes by HCl-catalysed condensation in glycol monoisopropyl ether[71]. Attempts to prepare resorcarene-like macrocycles with 2,7-dihydroxynaphthalene units failed. A 3,5-connected trimer was isolated in 23% yield from the condensation of 3-hydroxymethyl-2,7-dimethoxy-1,8-dipropylnaphthalene[72]. Monoethers of resorcinol (3-alkoxyphenols) react with aldehydes in the presence of Lewis acids to form C_4-symmetrical compounds **28** (as confirmed by X-ray analysis for one example) in high yield (80%)[73].

The fragment condensation of alkylidene-linked dimers with another aldehyde led to resorcarenes with two residues R in alternating order[74]; however, mixtures of the *rccc*, *rctt* and *rcct* isomers which had to be separated chromatographically were obtained under all conditions.

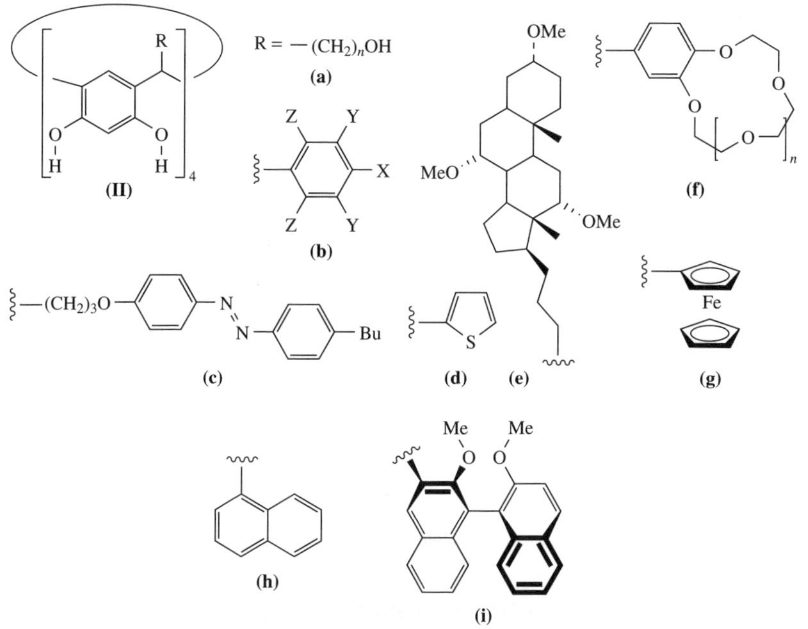

FIGURE 2. Selection of calix[4]resorcinols **II** obtained (in most cases as *rccc* isomer) by condensation with various aldehydes or their synthetic equivalents[67]

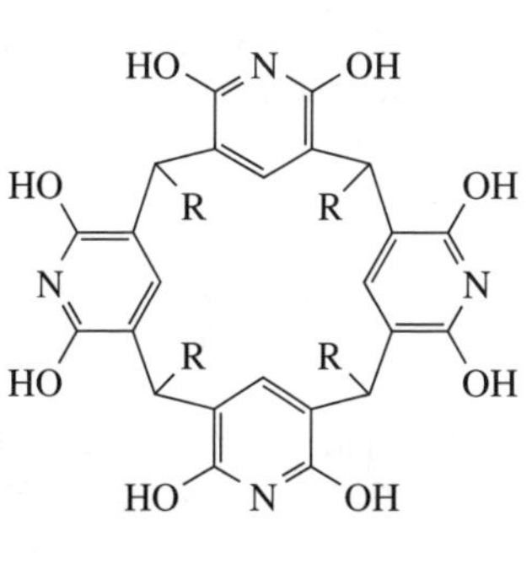

A different access to the resorcarene skeleton was found in the Lewis acid ($BF_3 \cdot Et_2O$) catalysed tetramerization of 2,4-dimethoxycinnamic acid esters or amides (Scheme 9). 2,6-Dimethoxy derivatives rearrange during the reaction and may be also used. A mixture of stereoisomers of **29** is usually obtained in yields of 65–80%, the composition of which depends on R and on the reaction conditions[75].

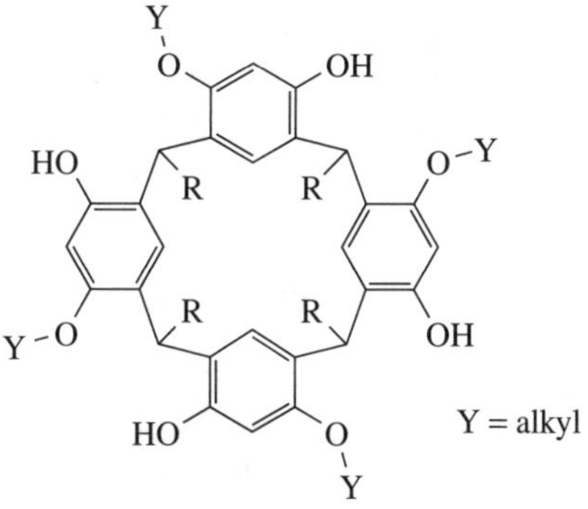

The 'parent' methylene-bridged resorcarene (**II**, R = H) was recently obtained by treatment of 2,4-bis(allyloxy)benzyl alcohol with $Sc(OTf)_3$ in acetonitrile, followed by deallylation by ammonium formate and $PdCl_2(PPh_3)_2$[76].

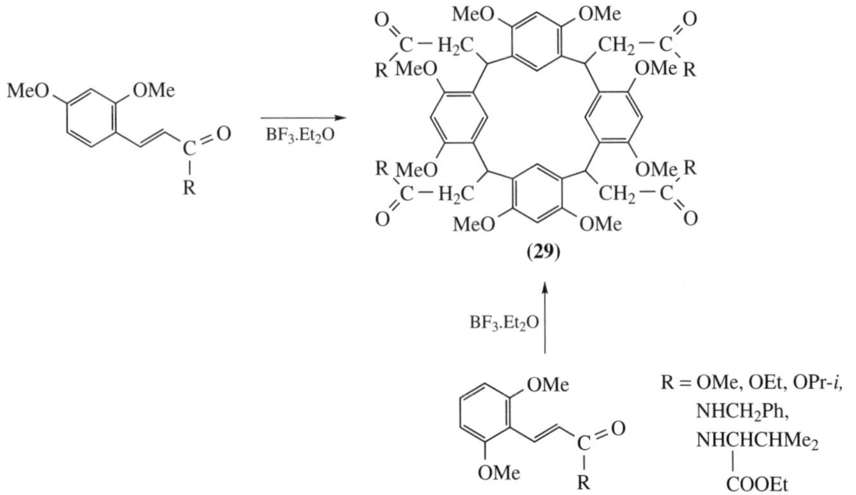

SCHEME 9. Resorcarenes **29** formed by cyclotetramerization of dimethoxycinnamic acid derivatives

III. CONFORMATIONAL PROPERTIES

A. Calixphenols

Characteristic of calix[*n*]arenes and their derivatives is the conformational diversity, which may cause difficulties in the synthesis of narrow rim derivatives (see Section IV), but also offers many additional chances to fine-tune the desired properties. Four basic conformations may be distinguished for a calix[4]arene [differing by the relative orientation of the (*endo*) OH and the *p*-positions] for which Gutsche introduced the names 'cone', 'partial cone', '1,2-alternate' and '1,3-alternate' (Figure 3).

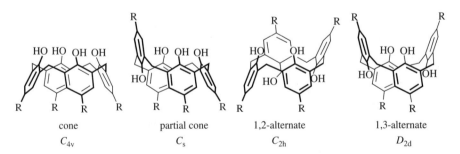

FIGURE 3. The four basic conformations of a calix[4]arene with their symmetry classes

The parent calix[4]arenes are (exclusively) found in the so-called *cone* conformation[77], where all the OH groups point in one direction which is therefore stabilized by an intramolecular array of hydrogen bonds. From variable temperature NMR studies (showing at low temperature a pair of doublets and at high temperature a singlet for the protons of the methylene bridges) the energy barrier. $\Delta G^{\neq}$ for the interconversion between two

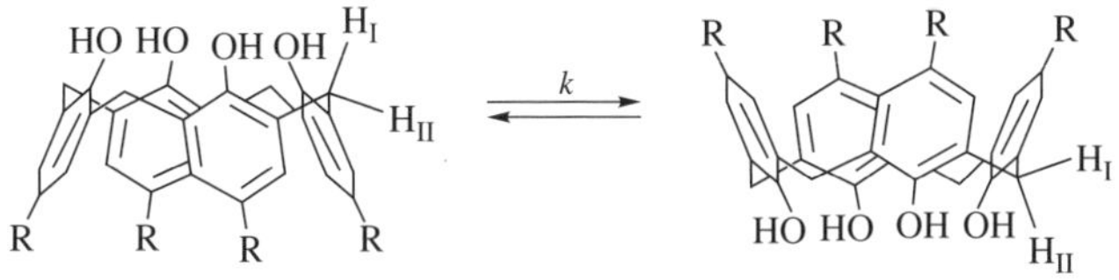

SCHEME 10. *Cone*-to-*cone* ring inversion of calix[4]arenes

identical *cone* conformations (Scheme 10) can be determined. Values for $\Delta G^{\neq}$ range from 14.6 to 15.7 kcal mol^{-1} for calix[4]arenes with different *p*-substituents in CDCl$_3$. They are lower (11.8–12.4 kcal mol^{-1}) in the hydrogen bond breaking [D$_5$]pyridine. The NMR-spectroscopic pattern (a singlet or a pair of doublets for the methylene protons) is identical for calix[5]- and suprisingly also for calix[8]arenes[78], where even analogous energy barriers are found in CDCl$_3$ ($\Delta G^{\neq} = 15.2-15.7$ kcal mol^{-1}) which, however, drop drastically in [D$_5$]pyridine ($\Delta G^{\neq} < 9$ kcal mol^{-1}).

Three pairs of doublets are found for calix[6]arenes at low temperature, indicating a conformation with three different methylene bridges[79], while an asymmetric conformation with seven different methylene bridges is found for calix[7]arenes. For the larger calix[*n*]arenes the $\Delta G^{\neq}$ values determined from the coalescence temperature of the methylene proton signals show slight maxima for $n = 12$, 16 and 20[27].

The conformational mobility of calixarenes can be restricted by bridging of phenolic units (see Section IV) or, for the smaller members of the family, by the introduction of *O*-alkyl or *O*-acyl groups, which are too large to pass the annulus of the macrocycle (see Section IV). Methoxy groups are small enough to pass and tetramethoxy calix[4]arenes have a similar flexibility as the tetrahydroxy compounds. However, due to the absence of hydrogen bonding, the *partial cone* conformer is the most stable and the three other conformers are also found[80], e.g. 85.6% *parco*, 6.1% 1,2-*alt*, 5.5% *cone*, 2.8% 1,3-*alt* for the tetramethyl ether of **2a** in CDCl$_3$ at 243 K[81]. It was even possible to determine the rate constants and the activation parameters for the single interconversion steps[82]. Scheme 11 gives a survey.

Although hydroxy as well as methoxy groups can pass the annulus of a calix[4]arene, all partially *O*-methylated derivatives are found only in the *cone* conformation in the crystalline state as well as in solution up to the highest available temperatures (*ca* 120 °C). No coalescence of the signals of the methylene protons was observed, thus precluding a determination of $\Delta G^{\neq}$ by variable temperature NMR. For inherently chiral calixarenes the *cone*-to-*cone* inversion (which occurs with a similar barrier) means enantiomerization. In fact, the mono-, 1,3-di- and trimethyl ether of **8a** could be resolved by chromatography on Chiralpak AD or Chiralcel OD as chiral stationary phase, and the kinetics of the racemization could be followed as a function of the temperature[83]. This led to the energy barriers collected in Table 2, which are distinctly lower than those calculated with the CHARMM force field[84], while the values calculated with MM3 are in better agreement[85].

B. Calix[4]resorcinols

The preferred conformation of resorcarenes is different for the various diastereomers. The *rccc* isomer assumes a *cone* (*crown*) conformation with an axial orientation of the residues R, which can be distorted to a so-called boat conformation where two opposite rings are more or less parallel, while the remaining two are bent away from the cavity becoming nearly coplanar. The *cone* conformation with an all-equatorial orientation of R was never observed. The *rctt* isomer is always found in a *chair* conformation, in which

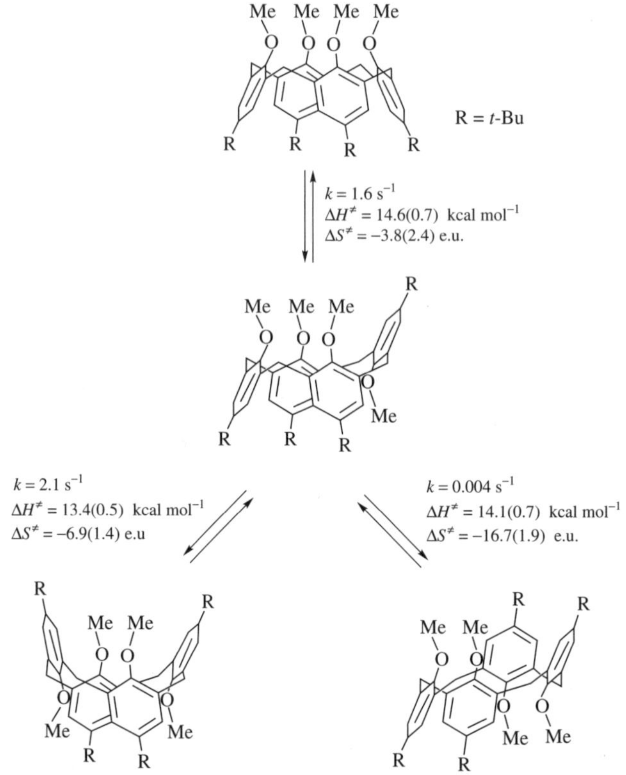

SCHEME 11. Conformational interconversion of the tetramethyl ether of *t*-butylcalix[4]arene **2a**. Rate constants are reported for the conversion of the *partial cone* into the other conformations.

TABLE 2. Energy barriers (in kcal mol^{-1}) for the *cone*-to-*cone* ring inversion of partially *O*-methylated calix[4]arenes

	Experimental values for ethers of **8a**		Calculated values MM3(92)[85]		CHARMM[84]
	$\Delta H^{\neq}$	$\Delta G^{\neq}$	ethers of **8a**	ethers of **2a**	
mono-Me	20.7	24.3	28.8	24.4	35.1
1,2-di-Me			25.3	23.1	32.2
1,3-di-Me	15.6	23.3	27.3	20.3	30.3
tri-Me	15.7	22.7	23.3	20.4	27.0

two opposite rings are nearly coplanar, while the other two are nearly parallel pointing up and downwards. The *rcct* isomer assumes a *diamond* conformation (analogous to the *1,2-alternate* conformation of calix[4]phenols).

The reason for these conformations is the tendency of R to avoid the neighbourhood of the OH groups[86]. This tendency was also found in calix[4/5]phenols with *endo* and *exo* hydroxy groups and one (or two) −CHR−bridges[51,87]. However, although these conformations (*chair, diamond*) are strongly preferred, they are not fixed, which follows for the *rctt* and *rcct* isomers from the fact that cavitands (see Section VII.D) with a (now fixed) *cone* conformation are available, where then two or one of the residues R are/is in equatorial position. Unfortunately the two aspects, (a) the relative configuration at the −CHR−bridges and (b) the conformation adopted by a given diastereomer, are often confused in the literature, probably due to the fact that under drastic conditions (e.g. 140 °C, aqueous solution containing bipyridine) an isomerization can take place[88].

Methylene-bridged calix[4]resorcinols (for synthetic reasons derived from 2-alkylresorcinols) assume a *cone* conformation at lower temperatures, and $\Delta G^{\neq} = 12.0$ kcal mol^{-1} was found at 298 °C for the *cone* → *cone* interconversion[89]. This lower value in comparison to those for calixphenols like **2a** reflects the fact that no substituents have to pass the annulus during the ring inversion. In addition, the intramolecular hydrogen bonds between *exo*-OH groups are distinctly weaker than those of the cyclic array of *endo*-OH groups in **2a** as shown by NMR ($\delta = 6.30$ vs. 10.2) and IR ($\nu = 3420$ vs. 3140 cm^{-1}).

IV. REACTIONS OF THE HYDROXY GROUPS IN CALIXPHENOLS

A. Complete Conversions

The exhaustive *O*-alkylation or *O*-acylation of calix[*n*]arenes is usually not difficult and has been achieved for all ring sizes and a multitude of residues Y (attached to the oxygen) and R (attached to the *p*-position); see general formula **Ia**.

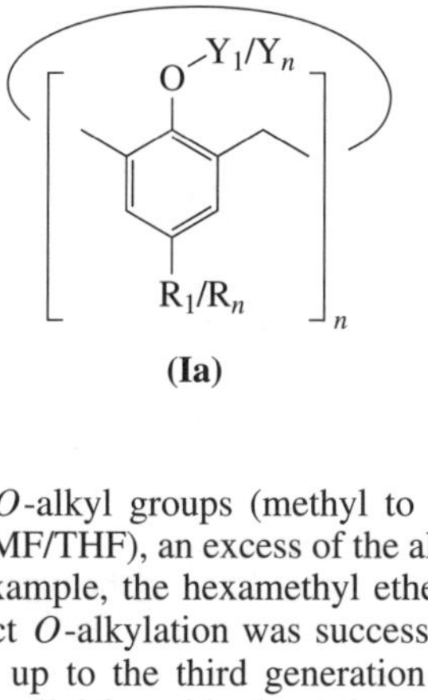

(Ia)

The introduction of simple *O*-alkyl groups (methyl to octadecyl) usually requires a strong base (typically NaH in DMF/THF), an excess of the alkylating agent and sometimes elevated temperatures. As an example, the hexamethyl ether of **2c** was obtained in 99% yield under sonification[90]. Direct *O*-alkylation was successful also with bulky 'dendritic wedges' leading to dendrimers up to the third generation (**30a**) in about 20% yield[91]. More reactive reagents such as allyl bromide, benzyl or picolyl chloride (or bromide) or bromoacetates can be introduced, using carbonates (e.g. K$_2$CO$_3$) as base in refluxing acetone or acetonitrile. The hexa-2'-pyridylmethyl ether, Y = CH$_2$C$_5$H$_4$N-2, for instance, was prepared in 81% in DMF at 70 °C with K$_2$CO$_3$ as base[92]. Phase transfer conditions have also been applied[93]. Compounds with Y = CH$_2$COR (for R = *O*-alk often called tetra- to octaester, although the link to the calixarene is an ether link) have been extensively studied as ligands for spherical cations, especially for alkali and alkaline earth metals and for f-elements [94].

Very recently, the first examples of aryl ethers derived from **2a** were reported. While S$_N$Ar-type reactions with various fluorobenzenes only led to partial etherification, the tetra-*p*-nitrophenyl ether was obtained with K$_2$CO$_3$/CuO in refluxing pyridine (46% *partial cone*, 16% *1,2-alternate*, see below)[95]. Reaction of **2a** or **2$_H$a** with 2-bromopyridine or 2-bromo-4-methylquinoline in refluxing diphenyl ether in the presence of CsCO$_3$ gave the tetraaryl ether in the *1,3-alternate* conformation[96].

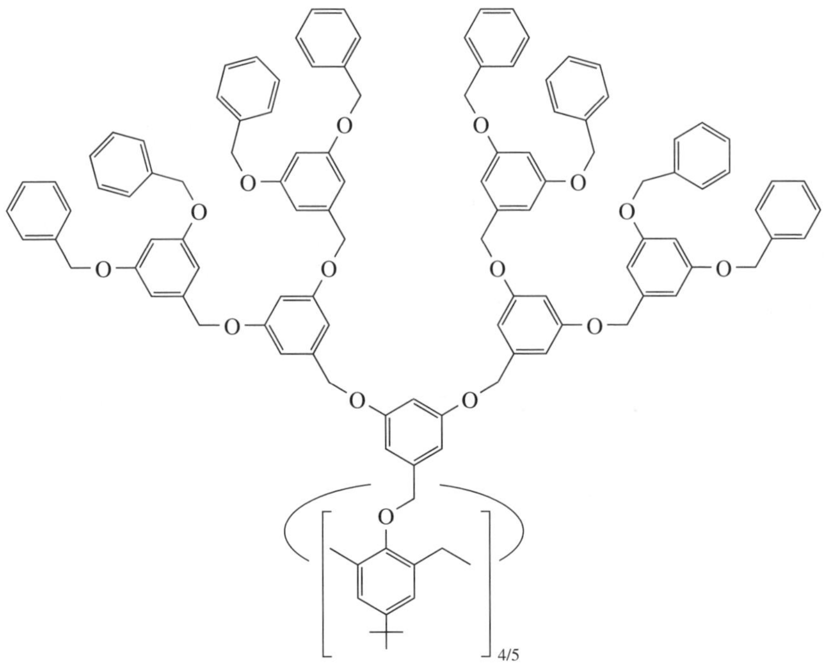

(30a)

Exhaustive O-acylation was less frequently studied. The octaphosphate (Y = PO(OEt)$_2$)[97], the octamesylate (Y = SO$_2$Me[98]), the octaesters with Y = C(O)CH = CHC$_6$H$_3$(OMe)$_2$-2,4[99] or Y = C(O)CHBrCH$_3$[100] may be taken as recent examples for calix[8]arenes[101]. The synthetic conditions described there may be used also in similar cases and with other calixarenes. The octa α-bromopropionates have been used as initiators for the synthesis of star polymers (atomic transfer polymerization), and the hydrolysis of the ester links was used to analyse the branches[100].

Functional groups attached via ether links to the narrow rim can be further modified. Especially, ester groups were used to introduce a multitude of further residues via ester or amide links. The aminosugar dendrimer **30b** is a spectacular recent example, showing high affinity to carbohydrate binding proteins[102], while the cholesteryl hexaester **31** was only prepared to inhibit the rotation of the oxygen functions through the annulus[103]. Reduction of the ester groups followed by tosylation and substitution by various nucleophiles (=Nuc) is another possibility, as shown below.

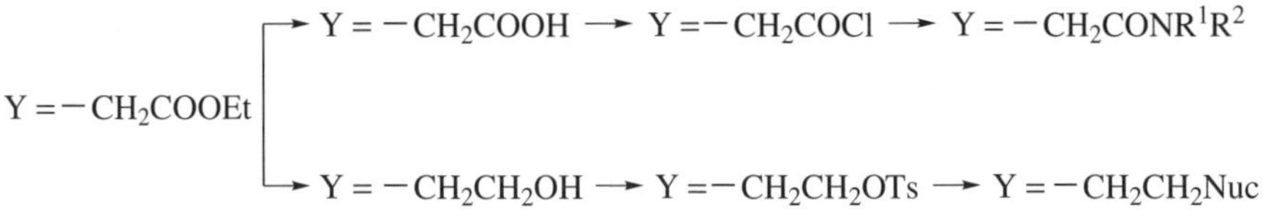

Y = −CH$_2$COOEt
→ Y = −CH$_2$COOH → Y = −CH$_2$COCl → Y = −CH$_2$CONR^{1}R^2
→ Y = −CH$_2$CH$_2$OH → Y = −CH$_2$CH$_2$OTs → Y = −CH$_2$CH$_2$Nuc

Aminoethyl ether groups were obtained by using azide as a nucleophile and subsequent reduction. Longer aminoalkyl ethers were prepared by O-alkylation with ω-bromonitriles or N-(ω-bromoalkyl)phthalimides followed by reduction or hydrazinolysis, respectively.

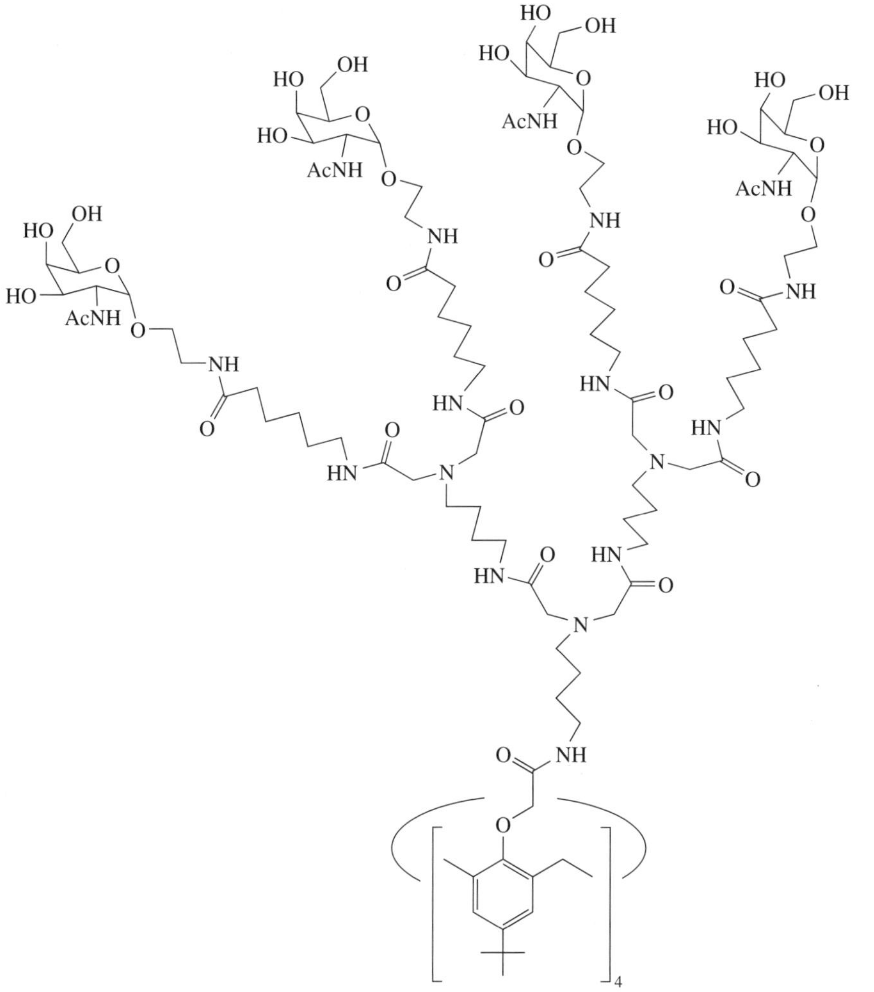

(30b)

Subsequent acylation of the amino functions is another general route to attach various functional groups, as demonstrated by the (thio)urea derivatives **32** (anion receptors)[104] or the CMPO derivatives **33** (ligands for lanthanides and actinides)[105].

B. Conformational Isomers

As already mentioned, the conformational interconversion of calix[4]arenes requires the oxygen function at the narrow rim to pass the annulus and consequently can be hindered by *O*-alkyl or *O*-acyl groups of sufficient size. Thus, the molecule can be fixed in one of the four basic conformations by exhaustive *O*-alkylation or *O*-acylation if the residues introduced are of sufficient size. The same is true for *p*-substituted calix[5]arenes, while a slow rotation of the wide rim through the annulus ($\Delta G^{\neq} = 17.9$–18.8 kcal mol^{-1}) has been observed for derivatives of the *p*-unsubstituted calix[5]arene[106]. For calix[4]arenes, it can be definitely stated that no rotation through the annulus occurs for ether groups equal to or larger than propyl and ester groups larger than acetyl, while the threshold size for calix[5]arenes is slightly larger than *n*-butyl or *n*-butanoyl[107]. A pentaether/ester with Y = CH$_2$COOCH$_2$CH$_3$ is obviously conformationally fixed, while a pentaether with Y = CH$_2$CH$_2$OCH$_2$CH$_3$ exists as a slowly interconverting ($\Delta G^{\neq} = 17.8$ kcal mol^{-1}) mixture of *partial cone/cone* (19:1)[108]. For the larger calixarenes, a passage of the wide rim becomes more and more an alternative, depending of course on the ring size and the size and shape of the *p*-substituents[109]. Finally, a conformational fixation is only possible by bridging (see below).

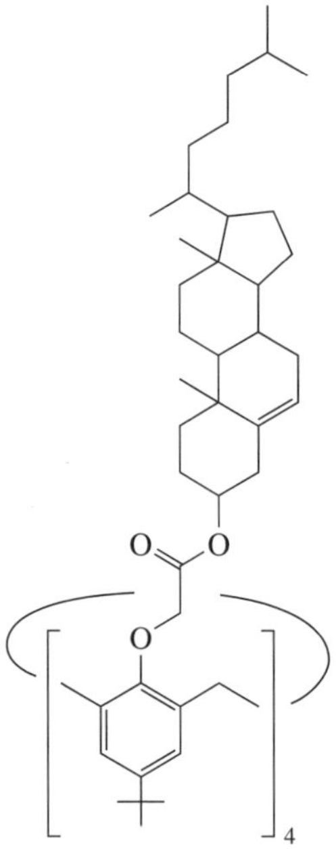

(31)

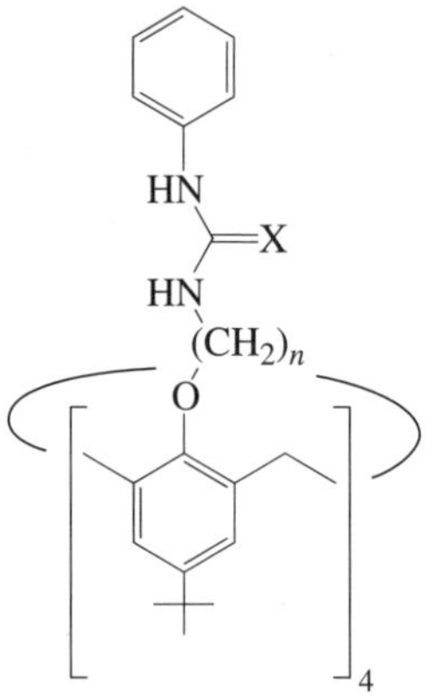

(32) X = O, S

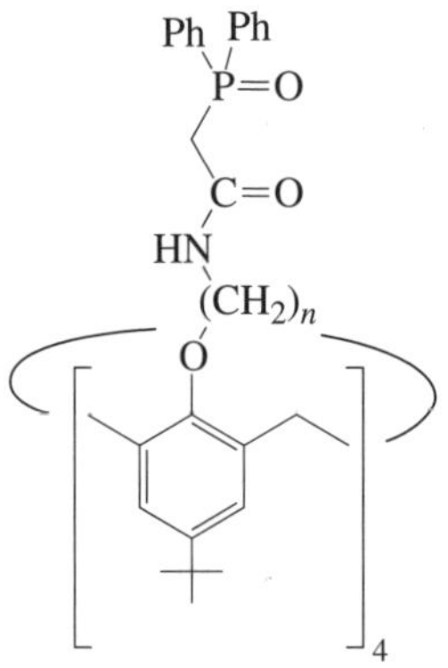

(33)

The existence of stable conformational isomers (atropisomers) may be a complication, since mixtures can be formed during a derivatization but it offers (in principle) additional synthetic possibilities which have already been widely used in the calix[4]arene series, where the four basic conformers are easily distinguished by NMR (Table 3).

TABLE 3. ^{1}H NMR signals in conformational isomers (atropisomers) of calix[4]arene tetraethers **Ia** with Y = CH$_2$R' (R' is a residue which shows no coupling with the adjacent CH$_2$ group) and R = C(CH$_3$)

Conformation Proton	cone C_{4v}	partial cone C_s	1,2-alternate C_{2h}	1,3-alternate D_{2d}
Ar-H	1 s	2 s, 2 d^a	2 d^a	1 s
O-CH$_2$-R'	1 s	2 s, 2 d^b	2 d^b	1 s
Ar-CH$_2$-Ar	2 d^b	4 d^{b,c}	1 s, 2 d^b (2:1:1)	1 s
C(CH$_3$)$_3$	1 s	3 s (1:2:1)	1 s	1 s

^{a} *meta*-coupling.
^{b} Geminal coupling.
^{c} One pair of doublets shows a small difference in chemical shift.

The stereochemical result of a per-*O*-alkylation depends on
(i) the residue Y to be attached,
(ii) the calix[4]arene vis-a-vis its *p*-substituents R and
(iii) the alkylation conditions (solvent, base, temperature).
Template effects by metal cations (due to the base applied) have been especially used to direct the reaction towards a certain conformer.

Usually, the *cone* isomer is formed in the presence of Na$^+$ cations. Na$_2$CO$_3$ in acetone or acetonitrile is often used for reactive alkylating agents such as bromo- or chloroacetates, -acetamides or -ketones (XCH$_2$COR, X = Cl, Br), while NaH in DMF or THF/DMF (sometimes MeCN) is the standard for alkyl bromides, iodides or tosylates[93,110–112]. Derivatives with four bulky residues such as Y = P(C$_6$H$_5$)$_2$[113] or Y = CH(CO$_2$Et)$_2$[114] have been obtained in the *cone* conformation.

Larger alkali cations (K$^+$, Cs$^+$) favour the formation of *partial cone* and *1,3-alternate* isomers, although a general set of conditions is not available for the *O*-alkylation. For example, replacement of Na$_2$CO$_3$ by Cs$_2$CO$_3$ leads to the quantitative formation of the *partial cone* isomer instead of the *cone* isomer in the alkylation of **2a** with ethyl bromoacetate (acetone/reflux)[115] while the effect is less pronounced for the *p*-unsubstituted calix[4]arene **2$_H$a**. Alkylation of **2a** with propyl bromide (KOBu-*t*/benzene/reflux) leads to a 1:1 mixture of *partial cone* and *1,3-alternate*[112]. Use of KOSiMe$_3$ or K$_2$CO$_3$ gives predominantly the tetrabenzyl ether in the *partial cone* conformation[116].

Usually, the *1,2-alternate* isomer is the most difficult to obtain[117,118]. In the case of propyl ethers of **2a** it was synthesized from the *syn*-1,3-dibenzyl ether by alkylation with propyl iodide (NaH/THF/reflux; 67% of the *partial cone* isomer), cleavage of the benzyl ether groups (Me$_3$SiCl/NaI/CHCl$_3$) and alkylation of the thus obtained *anti*-1,3-dipropyl ether with propyl bromide (Cs$_2$CO$_3$/DMF/70 °C)[119].

Partially *O*-alkylated compounds (see below) may also be used as starting material for the synthesis of tetraethers with different ether groups. Thus, tetraethers have been obtained by exhaustive *O*-alkylation of mono-[120], di-[93,121] and tri-ethers[120c,122]. In such cases the sequence of *O*-alkylation steps can also be important for the stereochemical result. Reaction of **2$_H$a** with alkenyl bromides (K$_2$CO$_3$/acetone/reflux) led to the *syn*-1,3-diethers, which were subsequently *O*-alkylated with methyl bromoacetate (NaH/THF) to yield a mixture of *cone* and *1,3-alternate* isomers. The alternative *O*-alkylation sequence (ethyl bromoacetate/K$_2$CO$_3$/acetone, reflux followed by alkenyl bromide/NaH/THF) gave the tetraethers in the *partial cone* conformation with *anti*-oriented alkenyl ether groups[123]. Reasonable to high yields of tetraethers in the *1,3-alternate* conformation were recently obtained from the 1,3-diethoxyethyl ether of **2$_H$a** with chloroethoxyethyl tosylate (Cs_2CO_3/acetone/reflux, 57%)[124], from its 1,3-dibromo analogue ($R^1 = R^3 = Br$) with ethoxyethyl tosylate (Cs_2CO_3/DMF/80 °C, 82%)[125] and from the 1,3-dibutyl ether with methyl bromoacetate (KH/DMF/RT, 88%)[126]. Different ether groups in combination with the *partial cone* conformation may lead also to inherently chiral compounds[122b,127].

Less is known about tetraesters of calix[4]arenes[128,129]. Tetrabutanoates in *partial cone, 1,2-* and *1,3-alternate* conformation have been obtained from **2a**[130], while all four isomers are known for the tetraacetates[131]. The tetratosylate (or *p*-bromophenylsulphonate) was used as protective group while modifying substituents in the *p*-position, and derivatives in the *cone*[132,133] and *1,3-alternate* conformation[134] have been obtained in excellent yield, while the tetratriflate was recently prepared in low yield (11%) as *cone* isomer[135]. Metal ion control has also been used to direct the stereochemical outcome of the diacetylation of 1,3-diethers towards the *cone* (Na^+) and *partial cone* (Tl^+) conformations[136].

Although for calix[5]arenes the number of potential conformational isomers is the same as for calix[4]arenes, special procedures to obtain pentaethers or -esters in conformations different from *cone* have not yet been reported. For calix[6]arenes conformational fixing is only achieved by bridging reactions.

C. Partial Conversions

Calixarenes, in which only some of the hydroxy functions are *O*-alkylated or *O*-acylated, are important derivatives by themselves and interesting intermediates for the construction of more sophisticated compounds. Selected ^{1}H NMR spectroscopic properties of partial ethers of **2a** have been summarized in Table 4. The distinction of regioisomers and conformational isomers becomes more and more complicated for the higher macrocycles.

1. Monoethers and -esters

Mono-*O*-alkylation of calix[4]arenes has been achieved using weak bases (K_2CO_3 in acetonitrile, or CsF in DMF)[137], and monoethers of calix[5]arenes were synthesized under similar conditions[107,138]. 1.1 Equivalents of K_2CO_3 in acetone have been used also to prepare the monomethyl (1.1 mol MeI, 70 °C, 2 bar) or monobenzyl ether (1.1 mol BnCl, reflux) of calix[6]arene **2c**[90] in yields of *ca* 80%. Excess of MeI (15.5 mol), KH (1.9 equivalents) in THF (RT, sonification) is an alternative in the former case. The mono-*O*-alkylated **2e** with a covalently attached C_{60} moiety may be mentioned as an interesting example among the calix[8]arenes[139].

Various other combinations of bases (NaH, KH, $Ba(OH)_2$) have been proposed for calix[4]arenes among which sodium methoxide in methanol (70–80% yield)[122a] and

TABLE 4. ^{1}H NMR signals in partially (*syn*) *O*-alkylated (Y = CH_2R') calix[4]arenes derived from **2a** (R′ is a residue which shows no coupling with the adjacent CH_2 group)

Proton \ Type of ether	Mono C_s	1,2-Di C_s	1,3-Di C_{2v}	Tri C_s
Ar-H	2 s, 2 d (1:1:1:1)	4 d (1:1:1:1)	2 s	2 s, 2 d (1:1:1:1)
O-CH_2-R′	1 s	2 d	1 s	1 s, 2 d (1:1:1)
Ar-CH_2-Ar	4 d	6 d	2 d	4 d
$C(CH_3)_3$	3 s (1:2:1)	2 s (1:1)	2 s	3 s (1:2:1)

bis(butyltin)oxide in boiling toluene[140] are recent examples. The controlled cleavage of 1,3-diethers (see below) or tetraethers by trimethylsilyl iodide (1 or 3 mol) has been also described[141]. O-Alkylation of mono- or triesters (see below) and subsequent hydrolysis of the ester group(s) offers another rational access[142-144]. Reaction of **2a** with tris(dimethylamino)methylsilane or trichloromethylsilane has been used to protect three OH functions. Methylation of the remaining fourth OH group (BuLi/CF$_3$C(O)OMe) and cleavage of the silyl triether gave the monomethyl ether in 83% overall yield[145].

Monoesters have been obtained by direct acylation[135,146,147], from 1,3-diesters by reaction with imidazole[147] or from mono- and 1,3-dibenzyl ethers by acylation and subsequent hydrogenolysis of the benzyl protective groups[148].

2. Di- and triethers and -esters of calix[4]arenes

The functionalization of two hydroxy groups in calix[4]arenes may lead to two regioisomers (1,2 or A,B vs. 1,3 or A,C[149]) and for sufficiently large residues to two conformational isomers (*syn/anti*) for each case, while three conformational isomers exist for three residues (*syn/syn, syn/anti* and *anti/syn*)[150]. It must be emphasized, that in such partial ethers/esters the mutual orientation of the residues Y should not be mixed with the conformation. A *syn*-diether still can assume the *cone* (usually the most stable conformation), the *partial cone*, and one of the *alternate* conformations!

The *syn*-isomers of 1,3-derivatives are easily synthesized in high yields under a variety of conditions (often Na$_2$CO$_3$ or K$_2$CO$_3$ in refluxing acetone or acetonitrile). This comprises diethers[151] and diesters[147,152] with identical residues as well as those with different residues which are obtained in two steps via the respective mono-derivatives[122a,b,153,154]. Compounds with an ether and an ester group in the 1,3-position have been prepared either by acylation of a monoether[155] or by O-alkylation of a monoester[144].

While all known examples of 1,3-diethers with an *anti*-orientation were obtained via protection/deprotection strategies[156], the acylation with excess benzoyl chloride using NaH as base was reported to give the *anti*-isomer in boiling toluene, while the *syn*-isomer is formed in THF at 0 °C[152b].

The ease with which 1,3-derivatives are formed can be rationalized: From two possible monoanions (proximal/distal) of the monosubstituted intermediate the distal anion is formed, since a phenolate group at ring 3 is stabilized by two intramolecular hydrogen bonds. Steric reasons are also in favour of the 1,3-derivative while the 1,2-product would be statistically favoured. Consequently, 1,2-diethers are formed when an excess of a strong base (e.g. NaH in DMF/THF) is applied, while the extent of the O-alkylation is controlled by the amount of alkylating agent (usually 2.2 mol). The dianion in which for electrostatic reasons two opposite hydroxy groups are deprotonated, or even the trianion, are likely intermediates. Although the selectivity is usually less pronounced than with 1,3-derivatives, various *syn*-1,2-diethers have been prepared by direct dialkylation with yields up to 90% in special cases[110,122b,157].

An alternative route to 1,2-diethers is the selective cleavage of neighbouring *syn*-ether groups in tetraethers by TiBr$_4$[158] or Me$_3$SiI[141], where a tetraether in the *partial cone* conformation yields the inherently chiral 1,2-*anti*-diether[159]. Very recently an easy access to 1,2-diethers was found, using a protection of two adjacent oxygens by capping with a disiloxane bridge[160]. While the higher stability of the distal monoanion of a monoether explains the preferred formation of 1,3-derivatives, the monoanion of a 1,3-derivative itself should be less stable than the monoanion of the corresponding 1,2-derivative, which is again stabilized by an intramolecular hydrogen bond. Therefore, 1,3-derivatives should be rearranged into 1,2-derivatives under basic conditions, provided a reaction pathway is available. In fact, this rearrangement has been observed for 1,3-diphosphates[161] and

1,3-ether/phosphates[162,163] as well as for the 1,3-benzyl/(3,5-dinitrobenzoate)[164] during the synthesis of further O-alkylation products (Scheme 12)[165]. At least for the phosphates an *intramolecular* migration of the phosphoryl residue seems reasonable.

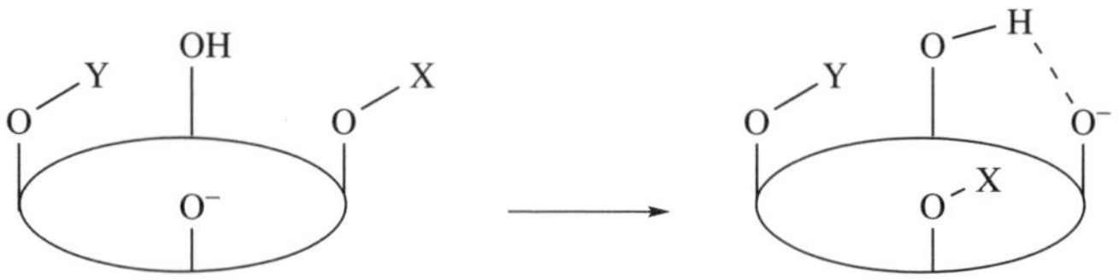

X = PO(OR)$_2$, COC$_6$H$_3$(NO$_2$)$_2$-3,5; Y = X, alk

SCHEME 12. Rearrangement of 1,3- into 1,2-diester derivatives under conditions where the monoanion is formed. No rearrangement takes place when the dianion is formed by an excess of a strong base[166]. The calix[4]arene skeleton is symbolized by a circle

Direct tri-O-alkylation of calix[4]arenes has been reported to lead to the *syn/syn* isomer using bases such as BaO, BaO/Ba(OH)$_2$ or CaH$_2$ in DMF[111,167]. The tribenzoate of the *p*-unsubstituted calix[4]arene, one of the first examples of selectively derivatized calix[4]arenes, was obtained as the *anti–syn* isomer (benzoyl chloride/pyridine)[142,168], while the tribenzoate of **2a** obtained in toluene with *N*-methylimidazole as base (70% yield) was described as the *syn–syn* isomer, assuming a *partial cone* conformation with an inverted phenol ring[169]. Both tris(3,5-dinitrobenzoates) of **2a** were obtained by acylation with 3,5-dinitrobenzoyl chloride/1-methylimidazole. While 95% of the *syn–syn* isomer was formed in acetonitrile, 70% of the *anti–syn* isomer was obtained in chloroform[147].

The dependence on the reaction conditions was shown also for the *syn*-1,3-diallyl ether of **2$_H$a**, which in CH$_3$CN gives *syn–syn*-diether/ester with PhCOCl/NaH, while reaction with PhCOCl in the presence of pyridine led to the *anti–syn*-diether/ester[170]. The formation of an *anti–syn* 1,3-diether/ester derivative was also achieved by barium(II) ion assisted monodeacylation of a 1,3-crown-5 diacetate in the *partial cone* conformation[171].

Various *syn–syn*-triethers, among them inherently chiral derivatives, were prepared starting with mono-, 1,2- or 1,3-diethers[122b,155,172]. Allyl, benzyl or benzoyl groups have been used as protective groups in triether synthesis.

3. Partial ethers (or esters) of larger calixarenes

Efficient procedures for the regioselective functionalization of calix[5]arenes are scarce. Only recently was the 1,2,4-triester obtained in 49% by reaction of **2b** with camphorsulphonyl chloride[173]. (For crown ethers see Section IV.D.)

This contrasts with calix[6]arenes, where for instance all 12 methyl ethers[174] and 10 of the possible 2′-pyridylmethyl ethers[92] of **2c** are known. The 1,3,5-trimethyl ether of **2c**, available in 72% yield (3 K$_2$CO$_3$, 4 MeI, acetone, 70 °C, 2 bar)[90] is an important starting material for further O-alkylation products (e.g. the 2,4,6-tri-*N*-methylimidazolylmethyl derivative **34**[175], or the triurea derivative **35**[176]) and for selectivity transfer to the wide rim (see below). Its formation is favoured by reasons discussed above for di-O-alkylation products of calix[4]arenes, but also by an obviously favourable C_{3v}-symmetric conformation, in which the anisole moieties are inclined with their methoxy groups pointing to the centre. The 1,2-dimethyl ether was also prepared in 81% yield (3.1 KH, 20 Me$_2$SO$_4$, THF,

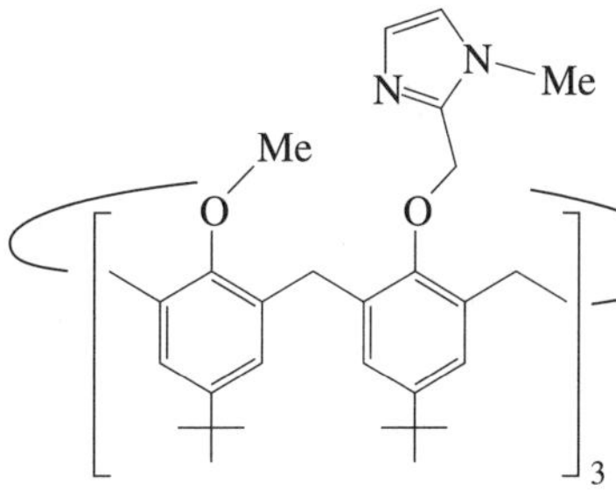

(34)

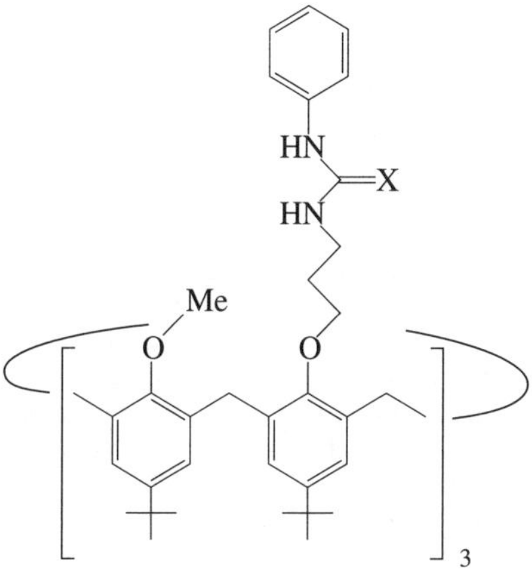

(35) X=O, S

sonification), while the pentamethyl ether, available in 15% yield by direct methylation[90], can be obtained in 76% yield by monobenzylation, subsequent exhaustive methylation and cleavage of the benzyl ether group[177]. Among the ethers available directly in reasonable yields (>50%) by partial O-alkylation are 1,4-diethers (Y = α-picolyl), 1,2,3-triethers (Y = α-picolyl) and 1,2,4,5-tetraethers (Y = α-picolyl, Y = $CH_2CH_2OCH_2CH_2OCH_3$) of **2c**, but the reaction conditions reported do not allow any generalization.

The *partial O*-alkylation of larger calixarenes becomes more and more difficult due to the increasing number of possible products (16, 28 and 46 for calix[7 to 9]arenes). Nevertheless, various 1,3,5,7-tetraethers were obtained in yields up to 50% by alkylation of **2e** in the presence of weak bases such as K_2CO_3 or CsF[178] and 19 out of the 28 possible methyl ethers have been isolated and identified[178]. The benzoylation of **2e** could be optimized to furnish heptabenzoates (with a variety of p-substituents in the ester group) in yields of 40–80%[99], which can be used as 'protected calix[8]arenes' for the synthesis of monosubstituted derivatives (for an example see Section VIII.A). Very recently, the first examples for the regioselective O-substitution of **2d** were reported[179].

D. Reactions with Di- and Multifunctional Reagents

As polyhydroxy compounds, calix[n]arenes can be reacted with multifunctional electrophilic reagents not only inter- but also intramolecularily. Some di- and multicalixarenes as examples of intermolecular reactions are reported in Section VIII. Intramolecular (i.e. bridging) reactions have been used to rigidify the calixarene skeleton, to protect hydroxy groups and to create ligands for metal cations, which in many cases show an unprecedented selectivity.

1. Calix[4]arenes

O-Alkylation of calix[4]arenes with ditosylates of oligoethylene glycols leads to 1,3-dihydroxy calix[4]crowns **36a**, the first examples being described as early as 1983[180]. Alternatively, the 1,3-dimethyl ether of a calix[4]arene can be reacted with the ditosylate, followed by selective removal of the methoxy groups with trimethylsilyl iodide in $CHCl_3$, a reaction sequence obviously accompanied by less by-products[121a]. Compounds **36a** can be modified by further O-alkylation and derivatives fixed in the *cone* or *partial cone* conformation have been synthesized in this way[121,181], while the derivatives fixed in the *1,3-alternate* conformation (**37a**)[182] are usually prepared by reaction of the 1,3-diether with the oligoethylene glycol ditosylate. Further modifications, involving the oxidation of the phenolic units to quinone units, OH-depleted compounds, additional functionalities attached via ether residues in the *p*-positions or in the crown ether bridge, including their rigidification by benzocrown structures or the incorporation of conformationally switchable elements such as azobenzene structures, can be found in Figure 4. 1,3-Crown ethers with a dinaphthol segment incorporated into the crown part (**36b**) show chiral discrimination of guests[183], which can be developed into a visual distinction of enantiomers by colour in chromogenic derivatives[184]. The barium complex of **36a** ($n = 5$) shows catalytic activity in transacylase reactions and has been studied as a simple transacylase mimic[185].

1,2-Crown ethers **38** of calix[4]arenes have been also obtained by direct O-alkylation with ditosylates[118a,186]. Their mono- and 3,4-diether derivatives (*partial cone*) are inherently chiral[172c,181b].

Various examples are known for all possible bis-crown ether derivatives from calix[4]arenes: 1,2;3,4-bis-crowns in the *cone* (**39**) and *1,2-alternate* (**40**) and 1,3;2,4-bis-crowns in the *1,3-alternate* conformation (**41**). Figure 5 gives a survey. Especially, the latter series is well developed, comprising examples with identical and different ether loops, including structures describable as *calixcryptands* (**41g**). Compounds **41a** ($n = 6$) were studied as ligands for the removal of caesium from nuclear wastes in analogy to their mono-crown counterparts in the *1,3-alternate* conformation (**37**)[187]. The 1,2;3,4-bis-crown-3 (**39a**, $n = 3$), on the other hand, is an important building block, since its calix[4]arene skeleton is fixed in a nearly perfect C_{4v}-symmetrical *cone* conformation[188].

Among various other 1,3-O-bridged derivatives *calix[4]spherands* **36f** should be mentioned, which form kinetically very stable complexes with alkali cations (among which Rb^+ is especially interesting for diagnostic purposes)[189]. 1,3-Bridged compounds have also been derived from 1,3-diethers bearing acid or amino functions in the ether residues.

Tri- and tetra-functional reagents have also been used to bridge or cap calix[4]arenes[190]. An especially interesting example is the reaction with $WOCl_4$ or WCl_6[191], which involves all four phenolic oxygens[192]. Two enantiomers of **42a** are formed in the case of C_4-symmetric calix[4]arenes **8a**, which were converted by reaction with a chiral enantiomerically pure diol into diastereomeric alkoxides **42b** (the formulas show only one enantiomer/diastereomer!). After chromatographic separation and cleavage of the diol, the pure enantiomers were obtained which, due to the capping by tungsten, possess an open, 'permanent' cavity[193]. (For double calixarenes obtained with tetrafunctional reagents, see Section VIII.)

2. Calix[5]arenes[194]

Reaction of **2b** with oligoethylene glycol ditosylates leads predominantly to 1,3-crown ethers while 1,2-crowns are isolated only as side product in some cases[195]. In contrast to the calix[4]arene derivatives (having C_s and C_{2v} symmetry) both types are C_s symmetric, but can be distinguished on the basis of their OH signals, one of which is strongly low-field shifted in the case of the 1,2-crowns, due to intramolecular hydrogen bonding. 1,2-Crown ethers, which are inherently chiral, are obtained as the main product in reasonable yield, starting with the mono-α-picolyl ether of **2b**. Chiral derivatives were obtained by selective O-alkylation or O-acylation of 1,3-crown ethers in the 4/5 position.

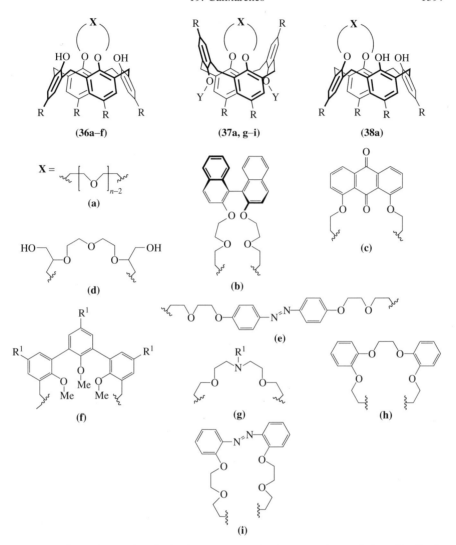

FIGURE 4. Selected examples of calix[4]-crowns and related compounds; usually R = H, *t*-Bu. In addition to the *1,3-alternate* derivatives **37** further di- (*cone* or *partial cone*) and mono- (*syn* or *anti*) *O*-alkyl derivatives are known from **36**, and analogously from **38a**

3. Calix[6]arenes[196]

Various examples of the three possible monobridged derivatives of calix[6]arenes (**2c** or **2$_H$c**) are known, including 1,3- and 1,4-crown ethers[197]. Usually, the 1,4-bridged compounds are most easily formed, while the 1,2-bridging normally requires short bridges. An exception is the bis(chloroacetate) of ethylene glycols (n = 4–6), which gives 1,2-bridged derivatives in yields up to 30%, presumably due to a template effect by K$^+$ ions

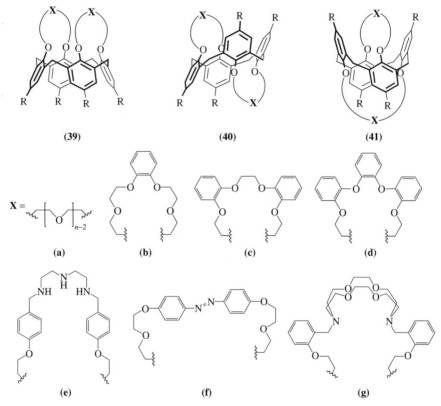

FIGURE 5. Selected examples of the possible calix[4]-bis-crowns **39–41**, usually R = H, *t*-Bu. Compounds of type **41** with identical and different bridges X are known

present[198]. Yields up to 35% (1,2-bis(chloroacetylamino)ethane[199]) or 40% (triethylene glycol ditosylate[197a]) were obtained for 1,3-bridged derivatives.

Examples of 1,4-bridged compounds **43a–g** are shown in Figure 6, which illustrates the diversity of compounds available.

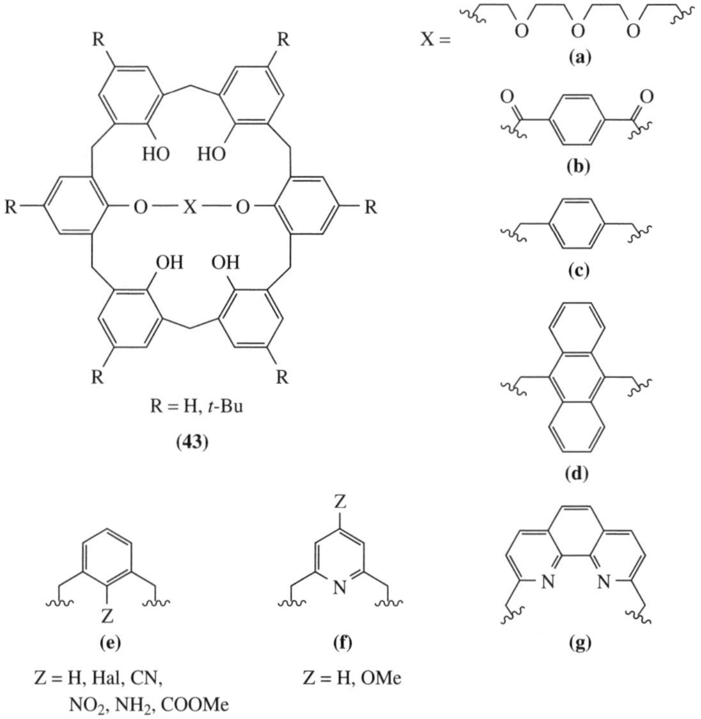

FIGURE 6. Selected examples of 1,4-bridged calix[6]arenes

Flexible bridges (e.g. crown ether **43a**[197b]) are possible as well as rigid ones; ester bridges (e.g. the terephthalate in **43b**[200]) as well as ether bridges and *p*-bis(bromomethylated) aromatic compounds (e.g. **43c,d**) work nearly as well as *m*-bis(bromomethyl)benzenes/pyridines (e.g. **43e,f**). Interesting conformational differences exist for the tetramethyl ethers derived from **43c** and **43d**, the former existing as a 'self-threaded rotaxane'[200b]. Compounds **43e,f** are interesting, since the 2'-position of the *m*-xylylene bridge is sterically protected by the calix[6]arene ring, eventually after further *O*-alkylation of the remaining OH groups. Thus, normally unstable groups Z, such as sulphenic ($-SOH$), selenenic ($-SeOH$) and seleninic ($-SeO_2H$) acid functions, can be stabilized[201]. Compounds **43f** are examples of 'concave bases' with basicities higher by three orders of magnitude in comparison to open-chain analogues[202]. The Cu(I) complexes of **43g** showed surprising selectivities in comparison to other concave phenanthrolines when used as catalyst in the cyclopropanation of alkenes[203].

Eight regioisomers are possible for doubly bridged calix[6]arenes (Figure 7) from which examples (mainly crown and benzocrown ethers) for five types have been realized (1,4;2,3-, 1,2;3,5-, 1,2;4,5- and the 1,3;2,5- and 1,4;2,5-derivatives with 'crossed' bridges). The situation may be even more complicated since two stable conformational isomers (all-up and uuuddd; u = up, d = down) have been obtained from the 1,4-diallyl ether by the introduction of diethylene glycol bridges[204].

Capping of **2c** may be achieved by trifunctional reagents, but the best results are obtained using its 1,3,5-trimethyl ether as starting material[205]. As a recent example the

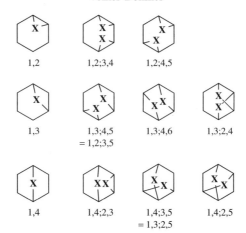

FIGURE 7. Schematic representation of possible regioisomers for singly and doubly bridged calix[6]arenes

ether–ester compound **44** with a triethanolamine derived cap should be mentioned. Its structure was proved by X-ray analysis as the first example of this type. The molecule adopts a *cone* conformation with the anisole units bent outwards[206]. Yields up to 90% were obtained for the capping with 1,3,5-tris(bromomethyl)benzene[207] and 1,4 diethers could be capped analogously by alkylation with 1,2,4,5-tetrakis(bromomethyl)benzene[208].

(44)

4. Calix[8]arenes[101]

Although the situation is even more complicated with calix[8]arenes, crown ether derivatives with 1,2- 1,3-, 1,4- and 1,5- bridging have been obtained in good to excellent yields (88% and 78% for 1,5-crown-2 and 1,5-crown-3)[209]. 1,5-Bridged derivatives have been obtained with *o*- and *p*-bis(bromomethyl)benzene[210], and 1,4-bridged derivatives with *m*- and *p*-bis(bromomethyl)benzene and with 2,7-bis(bromomethyl)naphthalene[210a]. This shows that the calix[8]arene skeleton is flexible enough to adopt various rigid spacers. The level of sophistication may be characterized by the introduction of an acridone-based bridge in **45**[211]. Several biscrowns have been obtained by direct *O*-alkylation[212] among which the D_{2d}-symmetric 1,5;3,7-bis-crown-3 was confirmed by an X-ray structure determination[213]. Bridging via phosphoryl groups may lead to a triphosphate involving all eight phenolic oxygens[214].

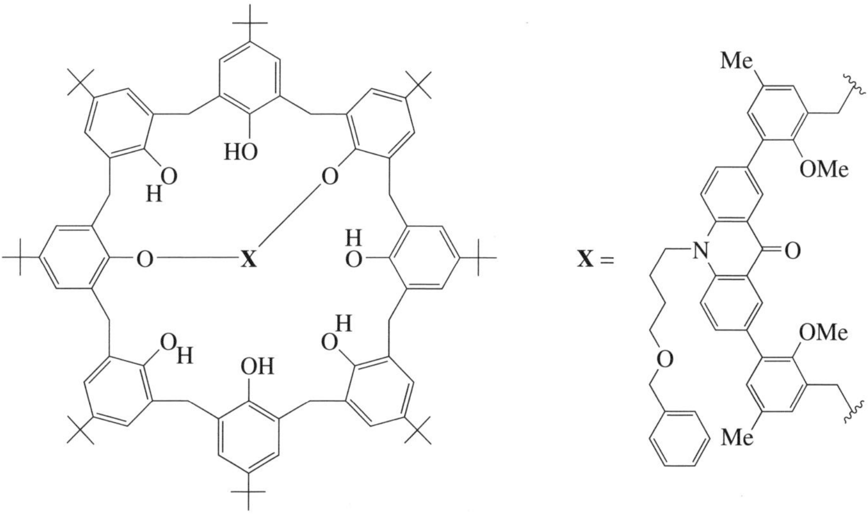

(45)

E. Replacement of the Hydroxy Groups

Various attempts have been made to eliminate the *endo*-OH groups or to replace them by NH$_2$ or SH groups[215]. The complete reductive cleavage of phosphate groups was possible, for instance, for calix[4]-, calix[6]- and calix[8]arenes[216], but the resulting 'OH-depleted' calixarenes could not be substituted again at the narrow rim. A *partial* elimination of OH groups was also reported for calix[4]arenes[217]. Reaction of the diphosphate of **2a** with liquid ammonia led to the introduction of only two amino groups[218] (for the synthesis of amino derivatives of thiacalixarenes, compare Section VI. F).

The Newman–Kwart rearrangement of thiocarbamates has been used to replace all OH groups by SH groups to give **46** in a reaction sequence outlined in Scheme 13[219]. The reaction seems to be sensitive with respect to the correct temperature (310–320 °C being recommended[219b]) and partially rearranged products were also isolated[219a], leading to calix[4]arenes with OH and SH groups.

The 1,3-dimercapto derivative could be obtained also in a rational way, via the bis(dimethylthiocarbamate) in which the remaining OH groups were protected against the rearrangement (360 °C, 20–30 min) by methylation[219a,220]. Mercapto derivatives were also prepared from thiacalix[4]arenes (see Section VI. F), while larger calixarenes with SH instead of OH groups are not yet known.

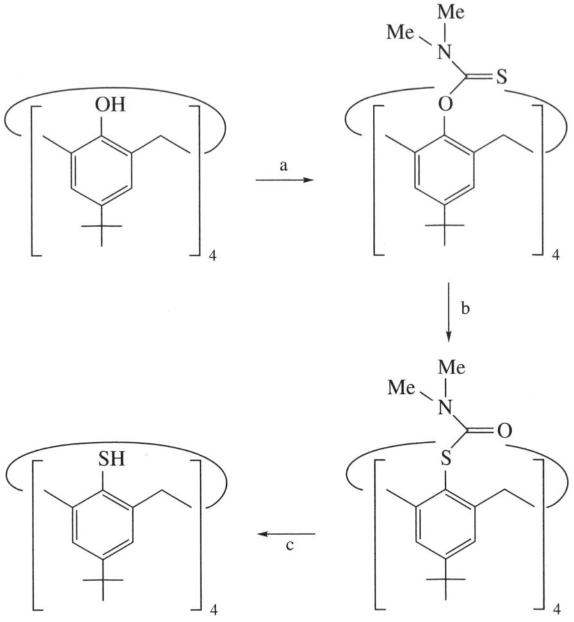

SCHEME 13. Synthesis of mercaptocalix[4]arenes, conditions: (a) Me$_2$NC(S)Cl, NaH, DMF, 25 °C, or K$_2$CO$_3$, acetone, reflux. (b) Heat 310–320 °C. (c) LiAlH$_4$, THF. The tetramercaptocalix[4]arene **46** assumes the *1,3-alternate* conformation. Partial conversions were possible in steps (a) and (b)

V. MODIFICATION OF CALIXPHENOLS ON THE WIDE RIM

A. Complete Substitution

The fact that the one-pot synthesis of calixarenes works best with *p-t*-butylphenol[221] was beneficial for the whole area, since the *t*-butyl groups are easily removed by trans-butylation with AlCl$_3$ in toluene (as solvent and acceptor), converting calixarenes **2a–e** into the *p*-unsubstituted calixarenes **2$_H$a–e**. Thus, the *p*-positions at the wide rim are available for virtually all kinds of electrophilic substitution reactions which are possible with phenols[1c,222]. Sulphonation, nitration, bromination (or iodination), bromomethylation, aminomethylation, formylation, acylation and coupling with diazonium salts[223,224] are examples.

The introduction of nitro-[225] and sulphonic acid groups[226] has been achieved also in excellent yields by *ipso*-substitution of **2a**, while *ipso*-acylation gave the tetraacetyl derivative in only 42% yield[224c,227]. Calixarenes consisting of hydroquinone units have been exhaustively perbrominated to yield compounds **47** with two bromine atoms per phenolic unit[228].

B. Selectivity Transfer from the Narrow to the Wide Rim

Partial *(ipso)* substitutions at the wide rim of calixarenes are known, and in some cases are of preparative importance. The *ipso*-nitration of calix[4]arene tetraethers, for instance,

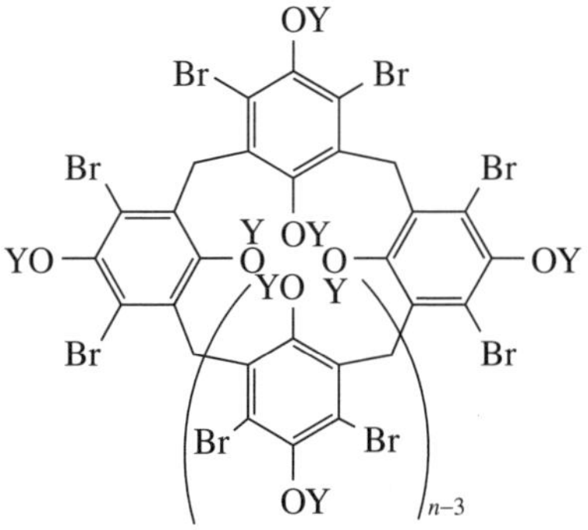

(**47**) $n = 4, 8, Y = Me, n\text{-}C_5H_{11}$

gives mononitro derivatives in yields up to 75%[229]. These are versatile starting materials for further derivatives.

Each carefully designed selective reaction at the wide rim, however, uses the difference in reactivity between phenol and phenol ether or ester units. The selectivity available in O-alkylation or O-acylation reactions can thus be transferred to the wide rim. Scheme 14 gives a schematic survey of reactions, all of which were realised with calix[4]arenes. The principle can be applied also to the larger calixarenes where, however, less examples have been realised.

In the following, selected examples for these reactions are reported. Reaction a) has been discussed in some detail already in Section IV. C; examples for reaction d) are found in Section VI. A. Selective transbutylation (reaction b)) was achieved for various mono-, di- and triethers or esters of calix[4]arene, leading to tri-, di- and mono-t-butyl calix[4]arene derivatives, from which the O-alkyl and especially the O-acyl residues can be cleaved again if desired, or necessary. A single t-butyl group was also eliminated from the tetramethyl ether of **2b**[194], the pentamethyl ether of **2c**[177] and the hepta(p-bromobenzoate) of **2e**[230]. Various other partially debutylated derivatives of **2c** have been prepared analogously[177,231].

Examples of the selective substitution of phenol units in partially O-alkylated (or O-acylated) calixarenes (reaction c)) are the bromination[125,232], iodination[233], nitration[152c,155,177,194,234], formylation[235], chloromethylation[236], alkylation[237] and coupling with diazonium salts[238].

Selective *ipso*-substitutions (reaction e)) are less frequently reported. The *ipso*-nitration of 1,3-diethers of **2a** gave not only the desired product **48a**, but by *ipso*-attack at the methylene bridges also the 6-nitrocyclohexa-2,4-dienone derivative **48b**[239]. The exhaustive substitution of the (remaining) phenol (reaction f))[240] or phenol ether units (reaction g))[241] usually causes no problems. The bromination (NBS, methyl ethyl ketone, RT) as well as the nitration of p-mono- and p-1,3-bis(acetamido) calix[4]arene tetraethers, however, occurs in the m-position (*ortho* to the acetamido groups), which is preferred even over the p-substitution of free phenol ether units[242]. Inherently chiral derivatives were obtained in this way.

C. Modification of Substituents

Substituents introduced at the wide rim by electrophilic substitution can be replaced or modified by further reactions. The following section only contains some typical examples,

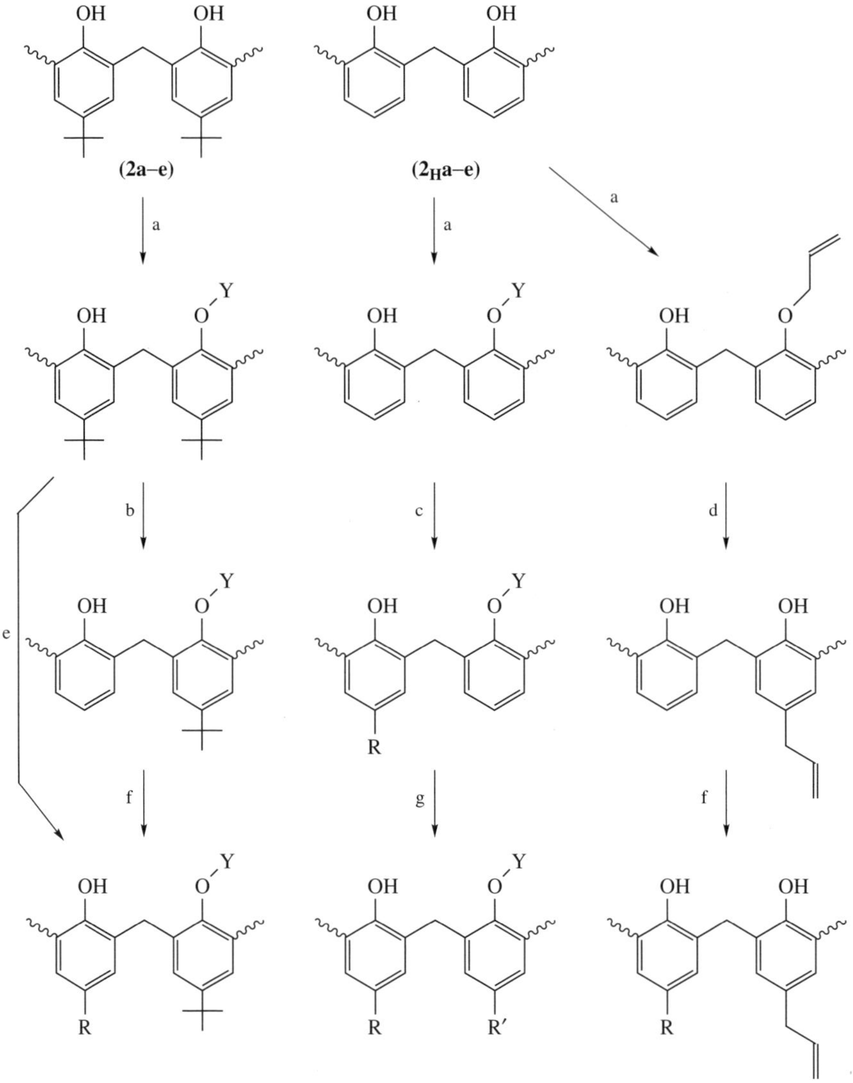

SCHEME 14. Selectivity transfer from the narrow to the wide rim, presented schematically for two units of a calix[n]arene

most of them realized for calix[4]arenes[222]. A complete survey is entirely beyond the scope of this chapter.

Chloromethyl groups[243] are an obvious starting point for the introduction of further functions, e.g. via the Arbuzov reaction[244]. Bidentate N-donor groups were introduced by nucleophilic substitution with suitable diamines[245]. Especially interesting is the intramolecular bridging of adjacent phenolic units by reaction with bis-nucleophiles. Thus, derivatives with C_2 symmetry have been obtained from chloromethylated calix[4]arenes[246], while a

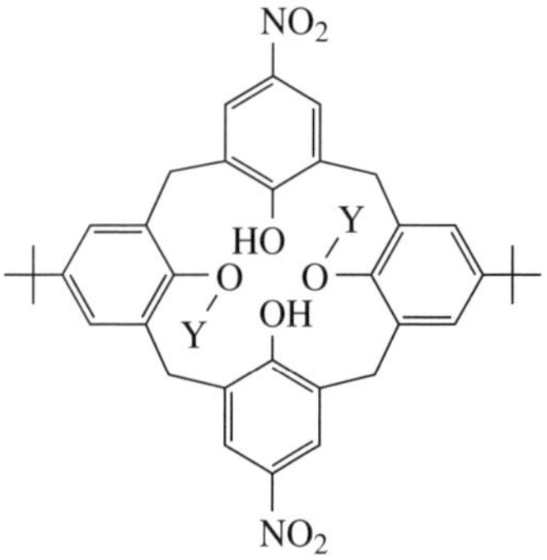

(48a)

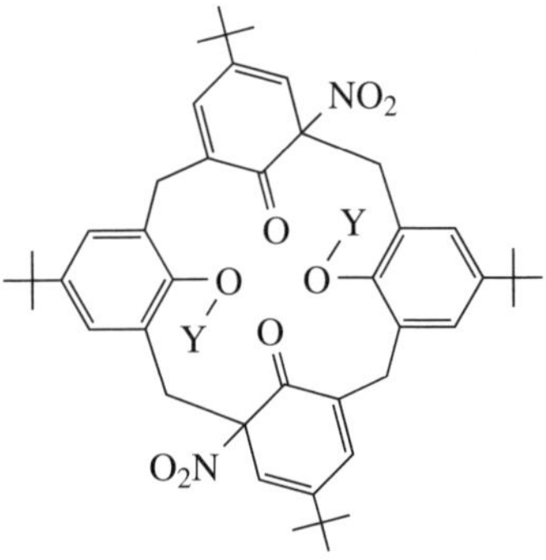

(48b)

Y = C_4H_9, C_5H_{11}

cavity large enough to include C_{60} was constructed on the basis of a calix[6]arene[247]. Trans-cavity bridging of a 1,3-bis(chloromethylated)tetramethyl ether led to the first wide rim crown ethers of calix[4]arenes[248].

Aldehyde functions have been used as such to synthesize Schiff bases[249,250], various stilbene derivatives[224b,251] by Wittig–Horner-type olefination or cinnamic acids or by Knoevenagel condensation with malonic acid[252]. They can also be reduced to CH_2OH[232a,253] or oxidized to COOH[235,254]. As examples for derivatives obtained via alcohol functions, the calixsugars **49a** can be mentioned[255], while the cyclopeptide derivatives **49b** may serve as an example for the attachment via acid functions[256].

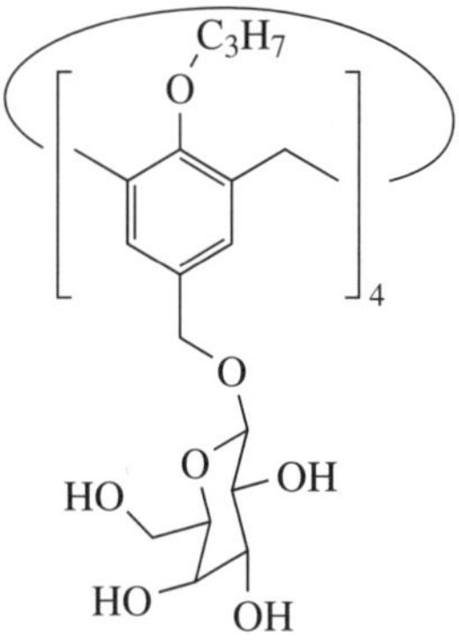

(49a)

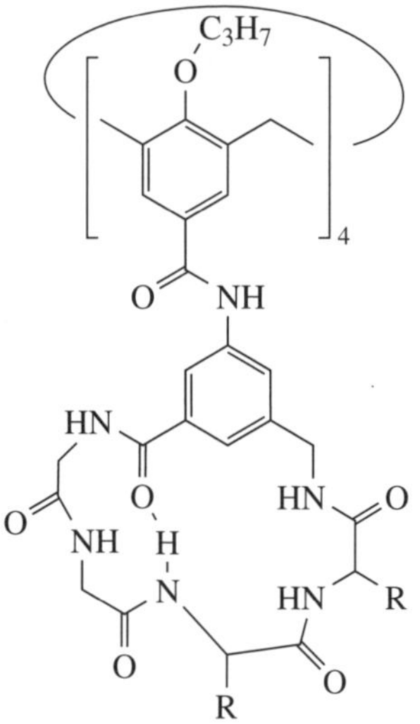

(49b)

R = SCH_2COOH

Aminomethyl groups, introduced by a Mannich reaction with secondary amines, can be quaternized and substituted by various nucleophiles under alkaline conditions, presumably via the intermediate formation of quinone methide units ('quinone methide route'[243]). Cyanomethyl derivatives, easily available in this way, can be hydrolysed and two or four opposite CH$_2$COOH groups can be converted to cyclic mono- or bisanhydride structures. Their ring opening is possible with various nucleophiles[134] and represents an elegant way for the selective introduction of functionalities to the wide rim (Scheme 15). Bisanhydrides fixed in the *1,3-alternate* conformation give chiral (C_2-symmetric) derivatives.

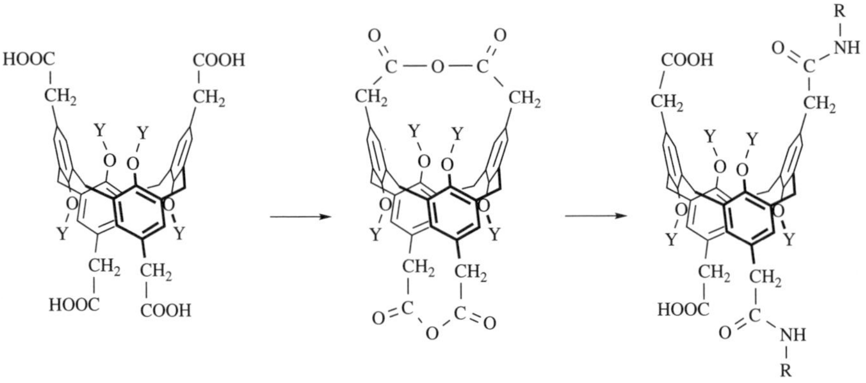

SCHEME 15. Synthesis of C_2-symmetric derivatives fixed in the *1,3-alternate* conformation by ring opening of cyclic anhydrides

Nitro groups can be reduced by catalytic hydrogenation[257], by hydrazine[254a,258] or by Sn(II)[259] and the resulting aminocalixarenes serve as starting materials for the attachment of various residues via acylation[260,261] or Schiff-base formation[257]. Boc-protection has been used in the calix[4]arene series[262] for the introduction of different acyl groups.

Copper-mediated coupling reactions of *p*-iodocalixarenes with phthalimide followed by hydrazinolysis should be mentioned as an alternative and independent strategy to obtain *p*-aminocalixarenes[263]. The carbazole-substituted derivatives **50** (Figure 8) were obtained similarly by Ullman coupling[264]. CMPO derivatives (**51c**), urea compounds (**51b**), available also via the isocyanates (**51a**), may be mentioned additionally. Mono- (**52**) and diimides[265] with acidic functions pointing towards the cavity, and the calix[6]arene-based acetylcholine esterase mimic (**53**)[266] are more sophisticated examples.

Complete lithiation of tetrabromo tetraalkoxycalix[4]arenes can be achieved with an excess of BuLi (THF, −78 °C)[26,267] while controlled amounts allow the mono- or 1,3-dilithiation[267b,268]. Subsequent quenching with electrophiles has been used to introduce various *p*-substituents. 1,3-Di- and tetraboronic acids available in this way could be oxidized (H$_2$O$_2$/OH$^-$) to *p*-hydroxy derivatives[267a,268,269], or underwent Suzuki coupling with iodoarenes[232b]. The alternative way, the coupling of *p*-bromo- or *p*-iodocalixarene ethers with various boronic acids has been also used to synthesize *p*-arylcalixarenes[36b,270]. C−C couplings were also achieved by Heck[271], Negishi or Stille reactions[222]. The oligophenylenevinylene derivative **54** may be cited as a very recent example[272].

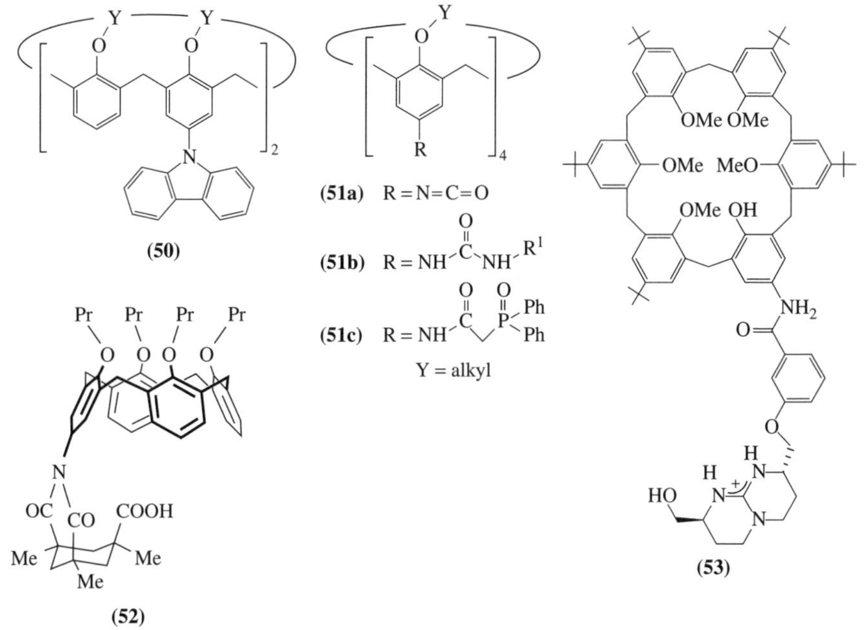

FIGURE 8. Examples of compounds (formally) derived from p-aminocalixarenes

VI. FURTHER REACTIONS OF CALIXPHENOLS

A. Rearrangements

A reaction involving the narrow and the wide rim of a calixarene is the Claisen rearrangement of allyl ethers. In the pioneering times of calixarene chemistry it was regarded as one of the most favourable ways to introduce functionalities onto the *wide rim* via subsequent modification of the *p*-allyl groups[243]. Due to its strict intramolecular course it was appropriate also for a selective *p*-substitution (see Scheme 14, reaction d)), and the first calix[4]arene, monosubstituted at the *wide rim*, was obtained by Claisen rearrangement of the monoallyl ether obtained from the tribenzoate of $2_H a^{142}$. Meanwhile, all variants between mono- and tetraallyl derivatives have been synthesized from calix[4]arenes[143,170,273]. A 1,4-*p*-allyl calix[6]arene was prepared from the corresponding diallyl ether[274].

A recent improvement involves the rearrangement in the presence of bis(trimethylsilyl)-urea to protect intermediately the phenolic hydroxy groups formed during the rearrangement. Thus, the *p*-allylcalix[4]arene was obtained in 99% yield and the larger *p*-allylcalix[4]arenes also became available[275]. Multiple tandem rearrangements have also been used to convert *O*-linked double calixarenes into *p*-linked double calixarenes (see Section VIII. A).

The Fries rearrangement of various calix[4]arene esters was also described[276], including the synthesis of inherently chiral derivatives[277], but has by far not attained the importance of the Claisen rearrangement.

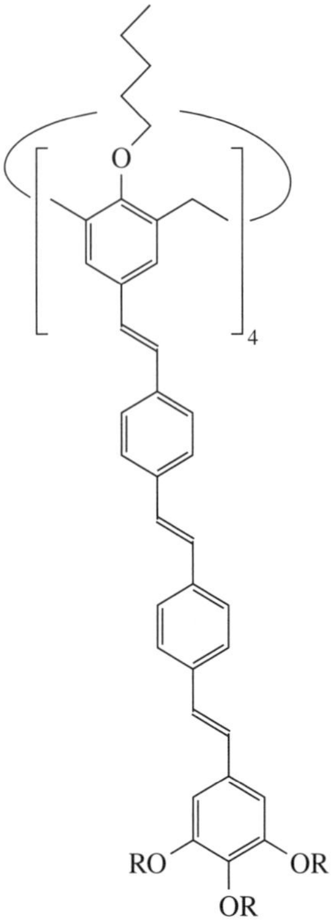

(54)

R = C$_{12}$H$_{25}$

B. Modification of the Methylene Bridges

Although more difficult to address, the methylene bridges are potentially also available for chemical modification[278]. Early attempts at their oxidation to ketone bridges (in the tetraacetate of **2a**) and their subsequent reduction to hydroxy groups met with no response[279]. More recently, the bromination of the tetramethyl ether of **2a** was reported to yield a single stereoisomer with four CHBr bridges in *rccc* configuration[280]. Lithiation of this tetramethyl ether (BuLi) and subsequent reaction with electrophiles such as alkyl halides or carbon dioxide gave derivatives, selectively substituted at one of the methylene bridges in yields up to 75%[281]. The homologous anionic *o*-Fries rearrangement (LDA/THF) was studied with 1,3-biscarbamates in the *cone*, *partial cone* and *1,3-alternate* conformation (Scheme 16)[282], and reaction conditions have been found for the latter, under which certain products are formed regio- and stereoselectively. Reactions of the methylene bridges have not yet been reported for larger calixarenes ($n > 4$), but the spirodienone route (see below) recently described for the stereoselective functionalization of two distal methylene groups in **2a**[283] might well be extended to the larger members.

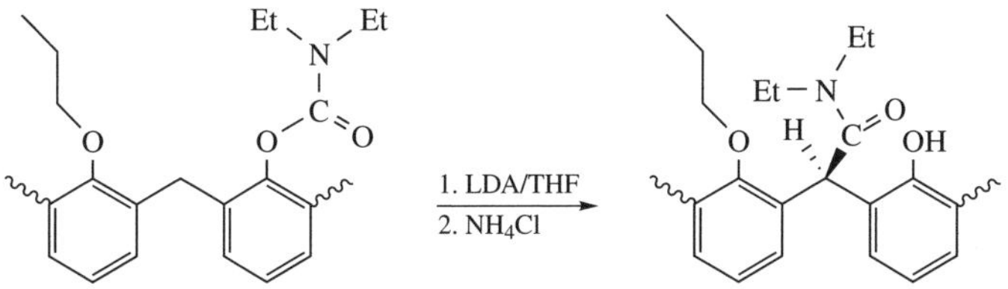

SCHEME 16. *o*-Fries rearrangement of calix[4]arene carbamates

C. Oxidation of the Aromatic Systems

1. Calixquinones

Aromatic systems may be oxidized to quinones and this has been done also with calixarenes[284]. Calix[4]quinone **55a** was first prepared from the *p*-unsubstituted calix[4]arene via azo coupling, reduction to the *p*-amino derivative ($Na_2S_2O_4$) and oxidation with $K_2CrO_4/FeCl_3$[223]. However, the method of choice seems to be the direct oxidation of the phenolic units. Thus, the calixquinones **55b,c** were synthesized by oxidation of the corresponding *p*-unsubstituted calixarene with ClO_2, while with $Tl(OC(O)CF_3)_3$ even the direct oxidation of **2a** to **55a** was possible[285]. A selective oxidation of phenol units beside phenol ether or phenol ester units is also possible with $Tl(OC(O)CF_3)_3$. (Occasionally also Cl_2O, $Tl(NO_3)_3 \cdot 3H_2O$ or $NaBO_3 \cdot 4H_2O$ have been used[284].) Various mono-, di- and triquinones of calix[4]arenes have been obtained in this way, including crown ether derivatives[286], as well as the di-[285] and triquinones[287] (**56a,b**) derived from calix[6]arene.

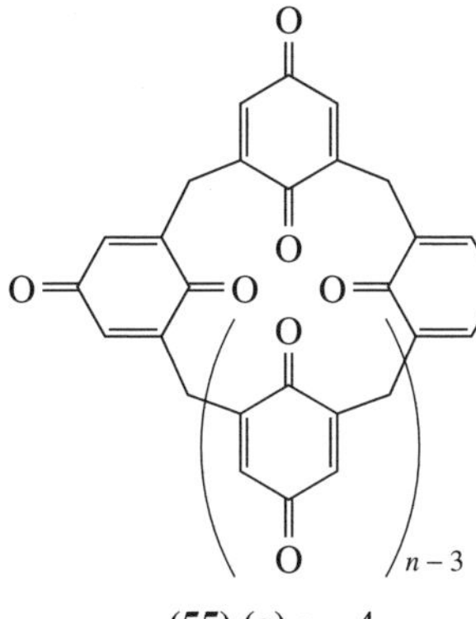

(**55**) (a) *n* = 4
(b) *n* = 5
(c) *n* = 6

Calixquinones can be easily reduced to the corresponding calixhydroquinones (Zn/HCl or $Na_2S_2O_4$)[285,288]. The calix[8]hydroquinone **57b**, however, was prepared from the octabenzyl ether **57a** obtained by one-pot condensation in a mixture with the analogous calix[6]- and -[7]arene. Oxidation of **57b** to the respective octaquinone was not reported, but the *endo*-ether **57c** was obtained by exhaustive *O*-propylation prior to the cleavage of the benzyl ether groups[289]. Inherently chiral derivatives of a calix[4]arene monoquinone have been obtained by 1,4-addition of various nucleophiles[290] to the quinoid system.

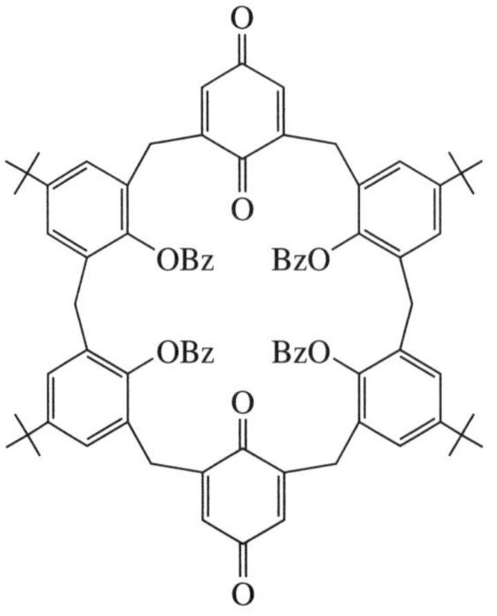

(56a)

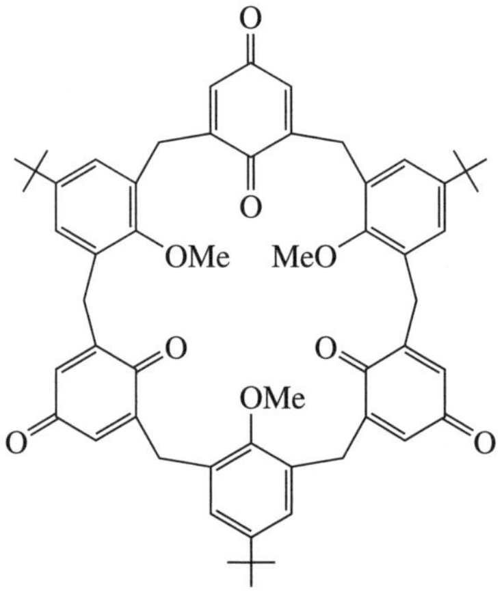

(56b)

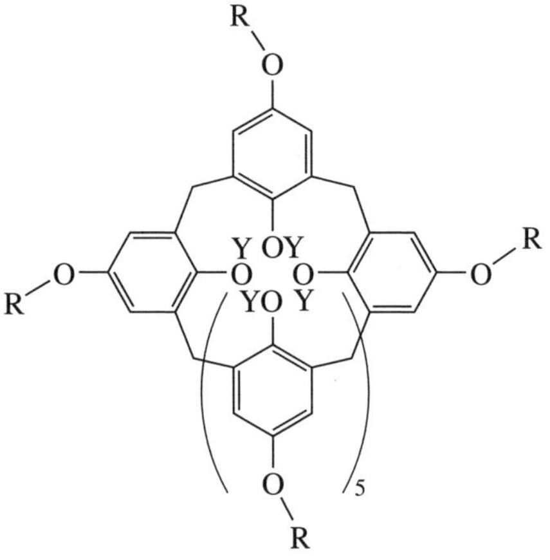

(57) (a) Y = H, R = CH$_2$C$_6$H$_5$
 (b) Y = H, R = H
 (c) Y = Pr, R = H

2. Spirodienones

Mild oxidation of **2a–c** leads to spirodienone derivatives. From **2a**, for instance, the three isomers **58a,b,c** are obtained with phenyltriethylammonium tribromide, which differ in the arrangement of the carbonyl and ether groups and/or the configurations (*R* or *S*) of the spiro centres[291]. The equilibrium mixture of the three interconverting compounds (toluene, 80 °C) contains the ratio 65/10/25 for **58a/58b/58c**, but **58a** can be obtained regio- and stereoselectively in 95% yield by oxidation with I$_2$/PEG200/25%KOH/CHCl$_3$[292]. Various other calix[4]arenes have been oxidized to bisspirodienones, among them β-naphthol derived calixnaphthols (OH groups in *endo*

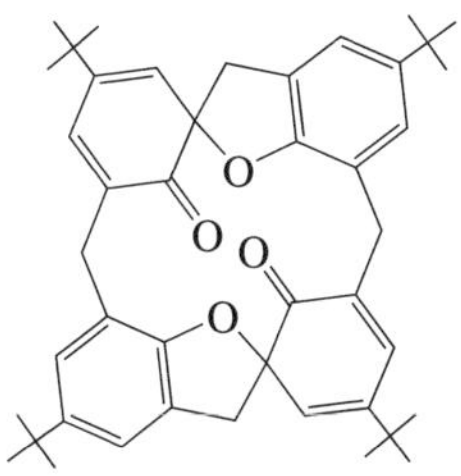

(58a) RS
(58b) RR/SS

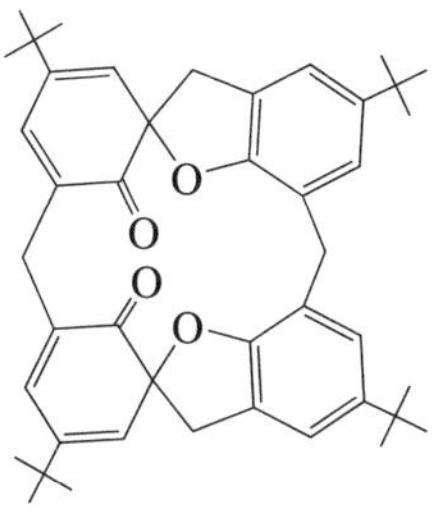

(58c) RS

position)[42c]. Oxidation of **2b** with $K_3Fe(CN)_6$/base gave a bisspirodienone with alternating arrangement of the carbonyl and ether groups[293] and various trisspirodienones have been obtained from **2c**[293,294] and from the spherand-type calixarene **10b**[295]. The formulae **59** and **60** show the most stable isomer. In all cases (**2a–c, 10b**) monospirodienones have been prepared using an equimolar amount of the oxidation reagent and a weaker base[295,296] and some of these compounds have been found to be useful intermediates for the preparation of aminocalixarenes and monodehydroxylated calixarenes[296c,297]. The replacement of two distal OH groups of **2a** by methyl groups was achieved by the reaction sequence shown in Scheme 17[298].

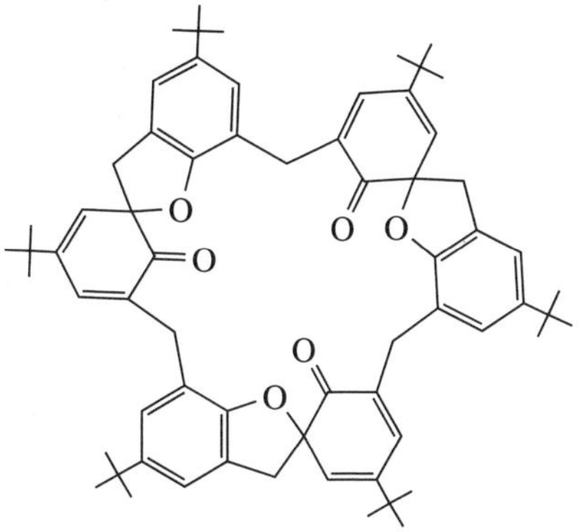

(59) *RRS/SSR*

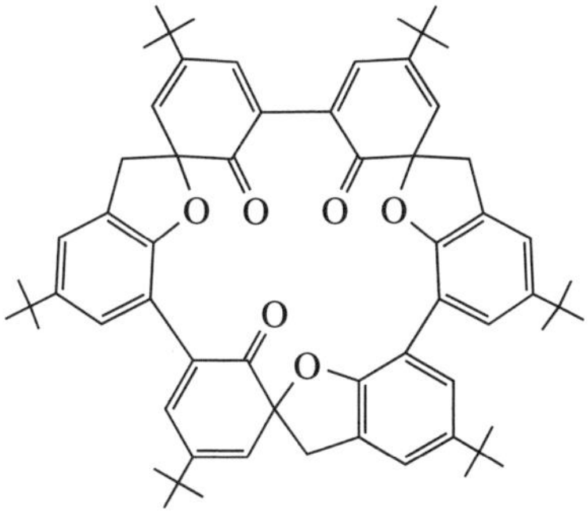

(60) *RRR/SSS*

D. Reduction of the Aromatic Systems[299]

Hydrogenation of the aromatic rings in calixarenes was studied only recently. This may be due to the fact that relatively drastic conditions are required and that numerous new stereocentres are created by this reaction. Therefore, all studies have been carried out with the unsubstituted calix[4]arene **2$_H$a**, where 'only' 3 new stereocentres per ring (12 per molecule) are formed during a complete reduction to the tetracyclohexanol derivative (compare Scheme 18 below).

The outcome of a complete hydrogenation depends on the reaction conditions. The perhydroxanthene derivatives **61a** and **61b**, most probably dehydration products of a calix[4]cyclohexanol, were obtained with Raney-Ni (1450 psi, i-PrOH, 240 °C)[300] and

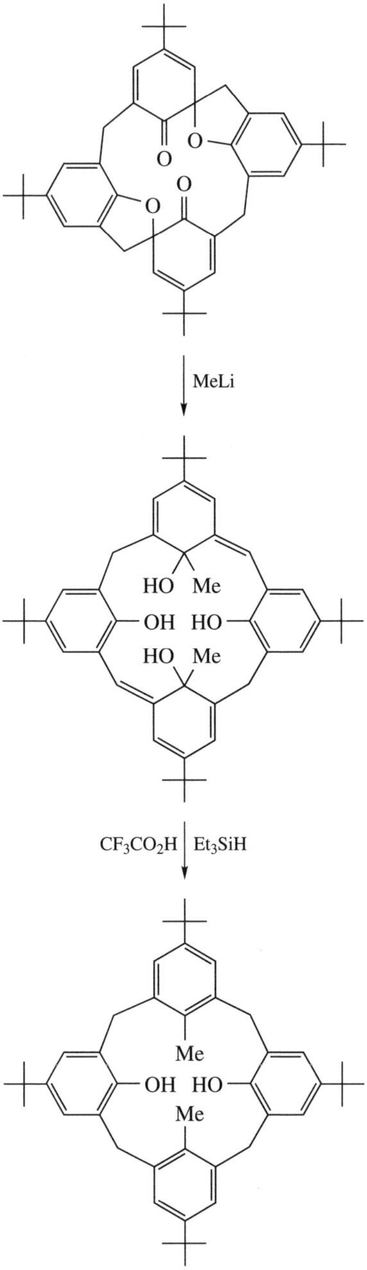

SCHEME 17. Replacement of *endo*-OH groups by methyl groups

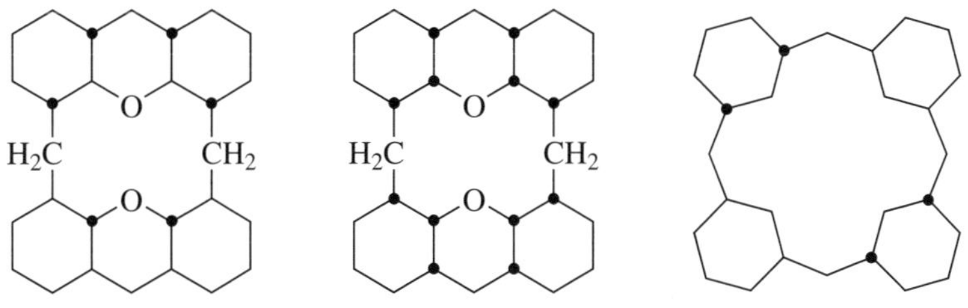

(61a)   (61b)   (62)

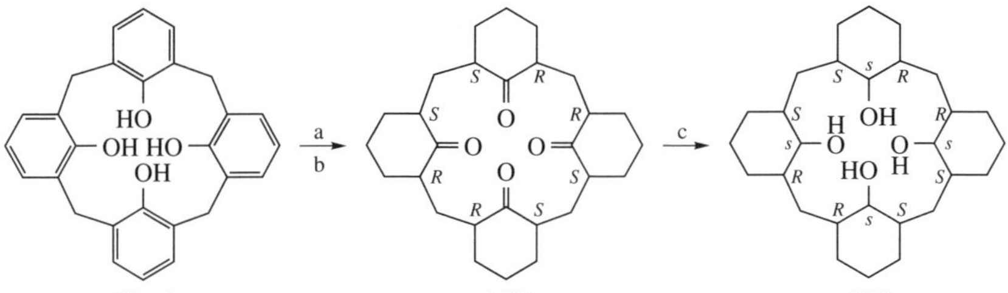

(2_Ha)   (63)   (64)

SCHEME 18. Stereoselective formation of calix[4]cyclohexanone **63** and calix[4]cyclohexanol **64**. (a) RhCl$_3$·3H$_2$O/Aliquat336/H$_2$O/CH$_2$Cl$_2$, 200 psi H$_2$, 90 °C. (b) NaOEt/HOEt; (c) NaBH$_4$. The indicated configuration of **64** was confirmed by X-ray analysis

Pd/C (120 °C)[301], respectively. Using Pd/C under more drastic conditions (250 °C, 600 psi H$_2$) gave the hydrocarbon **62**[301]. A single stereoisomer of the calix[4]cyclohexanone **63** was obtained as outlined in Scheme 18. Obviously, the acidity of the α-hydrogen atoms enabled an epimerization of the initially formed mixture to the most stable isomer upon treatment with base. **63** could be reduced stereospecifically to the calix[4]cyclohexanol **64** which assumes a *cone*-like conformation, and which is substantially more rigid ($\Delta G^{\neq} = 22.1$ kcal mol^{-1} for the *cone*-to-*cone* inversion) than the parent calix[4]phenol[302].

The face selectivity of the hydrogenation was established using the conformationally fixed tetrapropyl ether of **2$_H$a** in the *cone*, *partial cone* and *1,3-alternate* conformation as starting material. An individual product was formed exclusively in each case (Scheme 19) for which X-ray and NMR data indicated that the hydrogenation proceeds in an all-*exo* fashion[303]. Thus, the stereoisomer obtained can be determined by the conformational isomer used as starting material. Using RhCl$_3$/Aliquat336 at room temperature, a single phenolic ring of **2$_H$a** could also be stereospecifically hydrogenated to a cyclohexanone (RS)[304] and, with NaBH$_4$, further to a cyclohexanol (86% *RsS*, 2.2% *RrS*)[305]. Products with one or two opposite cyclohexanol rings were also produced by hydrogenation with Pd/C (100 °C) when the reaction was stopped before completion[301].

E. Metallation of the π-Electron System

Tetrapropyl ethers of **2$_H$a**, fixed in various conformations, have been converted into Cr(CO)$_3$ complexes[306], among them chiral mono derivatives of the *1,2-alternate* and

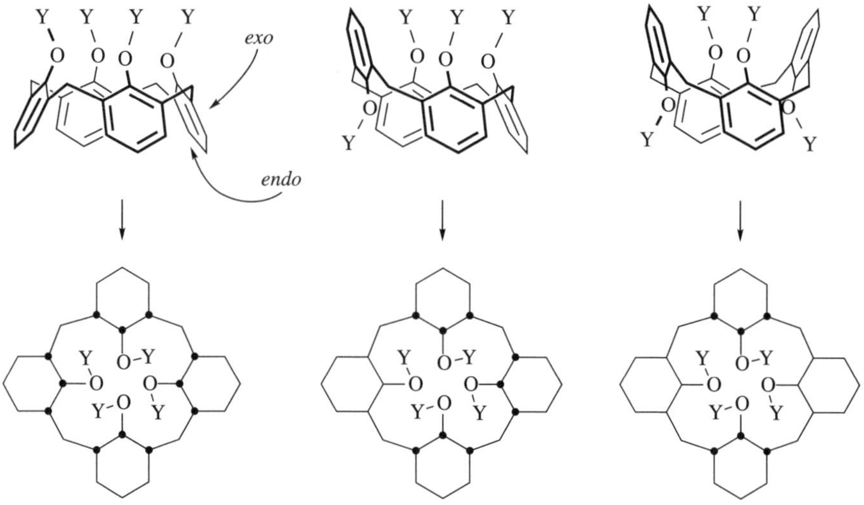

(65a) **(65b)** **(65c)**

$Y = C_3H_7$

SCHEME 19. Face selectivity of the hydrogenation of calix[4]arene tetraethers

the *partial cone* conformation. Complete or partial π-metallation of the free $\mathbf{2_Ha}$ was achieved by reaction with chlorine-bridged dimers in the presence of silver salts; $[\{Ru(\eta^6\text{-}4\text{-}MeC_6H_4Pr\text{-}i)Cl(\mu\text{-}Cl)\}_2]$, for instance, leads to **66** in 80% yield[307]. In a similar way compound **67** was obtained from **2b**[308]. These positively charged calixarene derivatives are able to include anions in their cavity, as shown *inter alia* by several X-ray structures. Simultaneous coordination of two cyclooctadienyl rhodium fragments to adjacent oxygens and to the π-system of **2a** or $\mathbf{2_Ha}$ was recently reported[309].

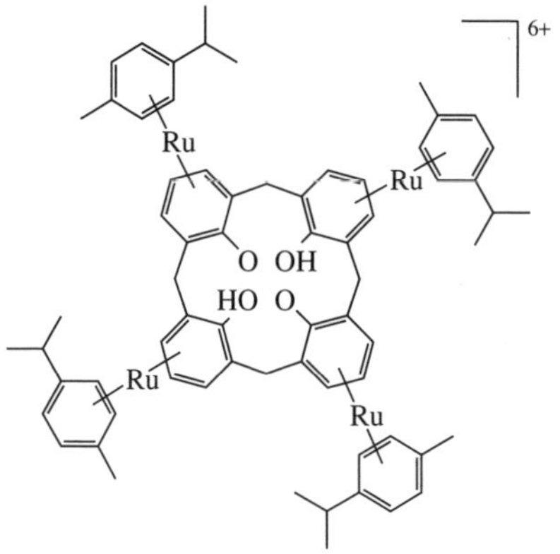

(66)

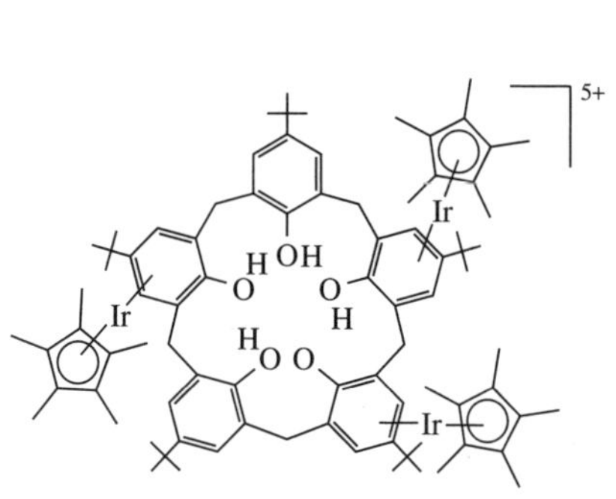

(67)

F. Special Reactions with Thiacalixarenes

Thiacalix[4]arenes can undergo, in principle, all reactions described for calixarenes[310], including, for instance, the de- or transbutylation or the Newman–Kwart rearrangement[311]. However, some qualitative differences were found.

The conformational outcome of O-alkylation reactions is slightly different, with a tendency towards the *1,3-alternate* conformation[312,313], but tetraethers in the *cone* (or *partial cone*) conformation have also been formed[314]. In contrast to calix[4]arenes, the O-propyl group is not large enough to fix a conformation and the tetrapropyl ether in the *1,3-alternate* conformation originally formed by alkylation with PrI/K_2CO_3 in acetone or acetonitrile in 67% (together with <25% of the *partial cone*)[315] is converted into 58% *partial cone*, 31% *cone*, 7% *1,3-* and 4% *1,2-alternate* when heated to 120 °C in $CDCl_2CDCl_2$ for several months[316]. A different chemical behaviour (Scheme 20) was also observed for the 1,3-diacids **68a** and **68b** (or their acid chlorides). They formed the bislactones **69a** and **69b** in the *cone* (up to 69%) and *1,2-alternate* conformation (up to 10%, passage of the O-alkyl residue through the annulus!)[317]. Analogous products were never observed for the methylene-bridged analogues.

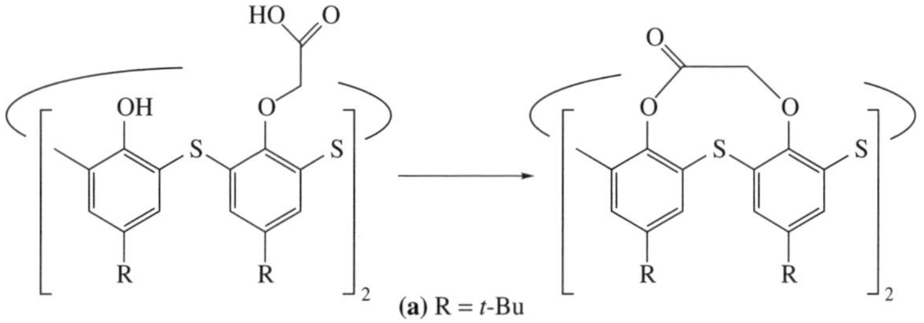

SCHEME 20. Formation of dilactones from thiacalix[4]arenes

In addition to reactions known from C-bridged calixarenes, the sulphur bridges may be oxidized to sulphinyl and sulphonyl bridges[318]. Both reactions have been realized with various tetraethers and recently with the free thiacalixarenes[319]. The complete oxidation with H_2O_2 gives the tetrasulphone usually in high yields (>80%) while the controlled formation of SO bridges (e.g. by $NaBO_3$) is additionally complicated by the potential formation of diastereomers, differing by the relative configuration at the bridges[320]. Two of them could be selectively obtained, starting from a tetraether in the *cone*[321] or *1,3-alternate*[322] conformation.

All possible sulphinyl derivatives (mono-, two di-, tri- and tetra-) of **70** were recently described[323]. Although the S=O group gives rise to additional stereoisomers, only one isomer was isolated in each case which was interpreted in connection with the X-ray structure of the tetrabenzyl ether[321] by the assumption that the oxidation leads to the equatorial disposition of the S=O group pointing away from the O-alkyl group.

Both sulphonyl and sulphenyl bridges enable the nucleophilic substitution of methoxy groups in a chelation-assisted nucleophilic aromatic substitution. Reaction with lithium benzylamide ($PhCH_2NHLi$) in THF, followed by dehydrogenation (NBS-BPO) and hydrolysis led to calix[4]arene analogues **71** (Scheme 21) in which aniline units are linked via SO_2 (**71a**), SO (**71b**) or S-bridges (**71c**, by reduction of SO). This may well be the breakthrough to another interesting class of macrocycles. Interestingly these

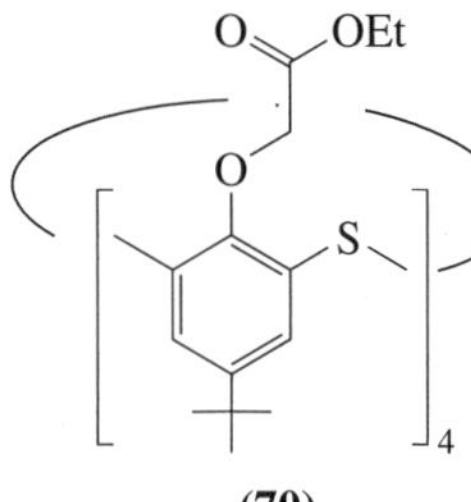

(70)

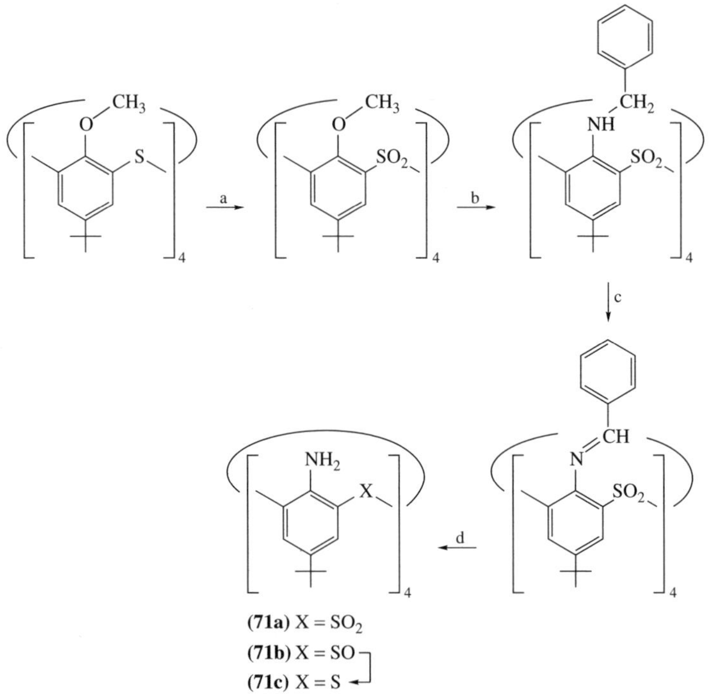

(71a) X = SO_2
(71b) X = SO
(71c) X = S

SCHEME 21. Preparation of thiacalix[4]anilines: (a) H_2O_2, $CHCl_3/CF_3COOH$, reflux; 86%. (b) $PhCH_2NHLi$, THF, rt; 70%. (c) NBS, benzoyl peroxide, benzene, reflux; 90%. (d) conc. HCl, $CHCl_3$, reflux; 78%. The reaction sequence is also possible with the sulphoxides (leading to **71b**) which can be finally reduced to compounds with sulphide bridges (**71c**)

thiacalixanilines **71c** assume the *1,3-alternate* conformation in contrast to the parent thiacalixphenol **25**[324].

VII. CHEMICAL MODIFICATION OF CALIXRESORCINOLS

There are three obvious places in a calixresorcinol where a chemical reaction may occur: The phenolic hydroxy groups may be esterified or etherified, the 2-positions may be substituted by mild electrophiles and functional groups introduced with the aldehyde residue R may be modified.

A. Reactions of the Hydroxy Groups

The phenolic hydroxy groups of resorcarenes can be completely acylated and various octaesters of *rccc*, *rctt* and *rtct* isomers have been prepared[4] (see Figure 9), initially partly to elucidate the structure of the parent compounds[325]. Various derivatives are given in Figure 9. Recent examples (**72**) comprise octaphosphates[326], octaphosphinites[327], octasulphonates[328] and octatrimethylsilyl derivatives[329], respectively. Complete *O*-alkylations of *rccc*-resorcarenes **II** result in octaethers **73a** (Y = Me to Bu)[330]. All twelve OH groups of calix[4]pyrogallol have also been esterified or etherified[331].

The attachment of eight 3,5-dihydroxybenzyl ether groups to the *rccc*-resorcarene led to a first-generation dendrimer **73b**[332]. Second-generation dendrimers of the same type were prepared, starting with the mixture of *rccc*- and *rctt*-isomers obtained with *p*-hydroxy- and 3,5-dihydroxy-benzaldehyde[333].

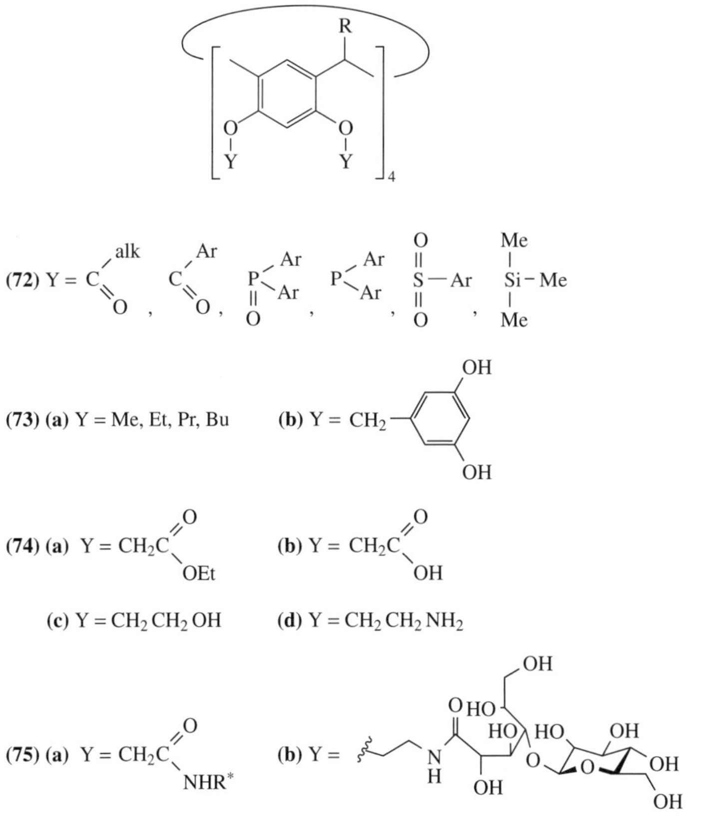

FIGURE 9. Examples of octa-*O*-acyl and -*O*-alkyl derivatives of calixresorcinols

1418   Volker Böhmer

Alkylation of **II** with excess of ethyl bromoacetate led to octaesters **74a**, which were hydrolysed to the corresponding octaacids **74b**[334]. Reduction of **74a** with LiAlH$_4$ gave the octol **74c** which was converted to the octaphthalimide by Mitsunobu reaction (phthalimide/diethyl azodicarboxylate/PPh$_3$) and finally by hydrazinolysis of the phthalimido groups to the corresponding octaamine **74d**[335]. Compounds **74** are versatile starting materials for further derivatization. For instance, aminolysis of **74a** with chiral amines and aminoalcohols resulted in chiral octaamide derivatives **75a**[336]. Reaction of **74d** with a lactonolactone gave a water-soluble resorcarene-sugar cluster **75b**[335].

Regioselective[337] O-acylations (or O-alkylations) of resorcarenes are rare. Examples of several derivates are given in Figure 10. Tetraesters **76**, in which the four hydroxy groups of two opposite resorcinol units in *rccc*-isomers are acylated, are the only examples of a more general character (see Figure 9). Initially, a chiral C_4-symmetric arrangement of the phosphoryl groups was postulated[338,339], but later the C_{2v}-symmetric structure of **76a** (which can be converted to **76b**) was unambiguously proved by NMR spectroscopy and single-crystal X-ray analysis[340].

Selective acylation was also possible with four equivalents of an arylsulphonyl chloride or an aroyl chloride furnishing **76c**[341] and **76d**[342], in yields up to about 50%, while the regioselective acylation fails with aliphatic acid chlorides. Reaction with benzyloxycarbonyl chloride (Et$_3$N, MeCN, RT), however, allowed the *partial* protection of four hydroxy groups to yield **76e**. Compounds **76c–76e** (interesting as building blocks for various self-assembled structures[343]) may be used for further derivatizations.

The subsequent acylation or alkylation of the remaining hydroxy groups in **76c** and **76d** resulted in C_{2v}-symmetric derivatives containing two types of functional groups at the wide rim of the resorcarene, among which the tetra-crown ethers obtained with benzo-15-crown-5-sulphonyl chloride should be mentioned[344,341].

Exhaustive O-acylation of **76e** followed by mild removal of the benzyloxycarbonyl groups (H$_2$, Pd/C, dioxane) gave tetraacylated derivatives including the Boc-protected compound **76f** which are not available by direct acylation of the parent resorcarene **II**. The tetraacid **76g** was obtained in a similar way by O-alkylation of **76d** with ethyl bromoacetate and subsequent hydrolysis[342].

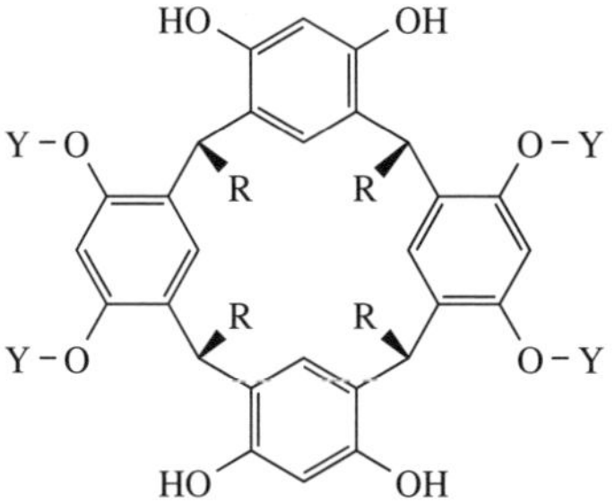

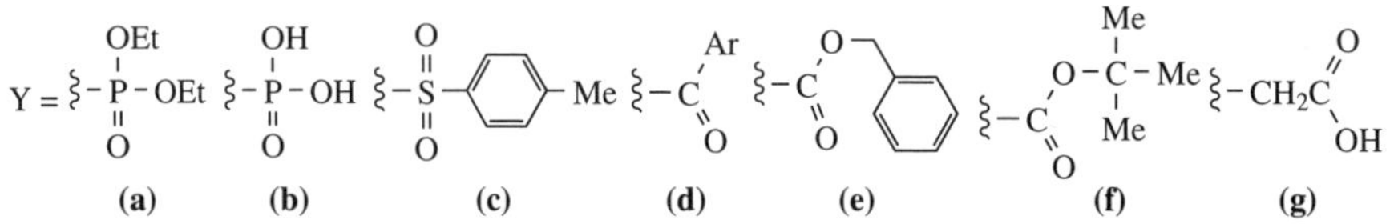

FIGURE 10. Examples of tetra-O-acyl and -O-alkyl derivatives of calixresorcinols

One example of a mono-O-alkylation has been described so far. The reaction of **II** with p-methylbenzyl bromide in a 1:1 molar ratio resulted in the chiral resorcarene monoether **77a**, which was exhaustively acylated to give resorcarene **77b** containing one alkoxy and seven acetoxy groups[345]. Recently, the chiral resorcarene **77d** was obtained in the form of the pure enantiomers by monoacylation of **II** with camphorsulphonyl chloride, separation of the crude mixture by HPLC (11% for each diastereomer **77c**), exhaustive O-methylation of the remaining hydroxy groups and alkaline hydrolysis[346].

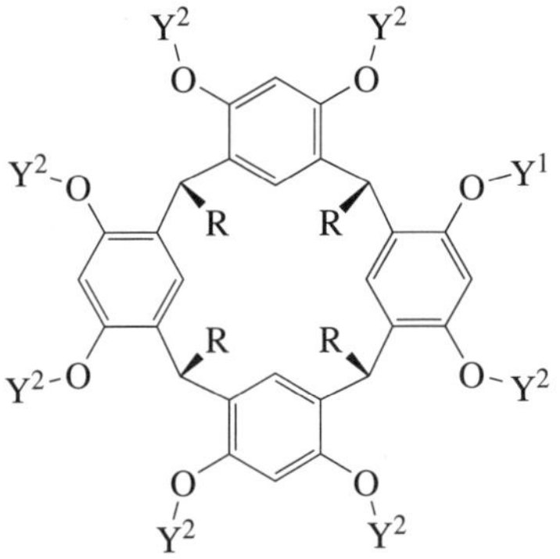

	Y^1	Y^2
(a)	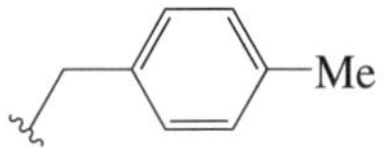 –CH₂–C₆H₄–Me	H
(b)	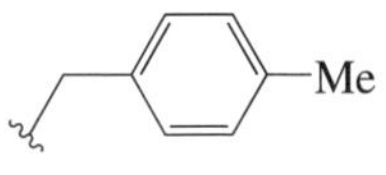 –CH₂–C₆H₄–Me	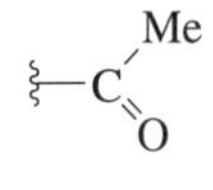 –C(Me)(=O)
(c)	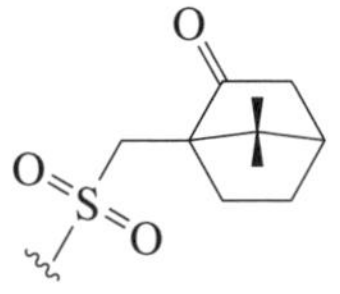 camphorsulphonyl	H
(d)	H	Me

(77)

B. Electrophilic Substitutions

The 2-position of the resorcinol rings may undergo substitution by mild electrophiles, such as bromination, coupling with diazonium salts and Mannich-type reactions, while more drastic reactions such as nitration or sulphonation failed.

The exhaustive coupling of **II** with diazonium salts[347] should also make tetraamino resorcarenes available by reduction of the azo groups, while the tetrabromo resorcarenes[348] are important starting materials for the synthesis of carcerands (see Section VIII. B). The reaction of **II** with NBS in molar ratios from 1:1 to 1:3 resulted in a mixture of all possible partially brominated resorcarenes[349], in which the yield of the distal-dibromo derivative was much higher than statistically predicted. Subsequent thiomethylation (CH₂O/RSH in AcOH) resulted in C_{2v}-symmetric derivatives containing two different functional groups (Br, CH₂SR) at the wide rim of the resorcarene, including distally bridged compounds by reaction with dithiols[350].

Aminomethylation of **II** with secondary amines and formaldehyde readily gives the corresponding tetraamines[351], which exist in apolar solvents in a chiral C_4-symmetric conformation with left- or right-handed orientation of the pendant hydrogen-bonded amino groups[352]. Trisubstituted products have been obtained with bulky amines[353]. Various functional groups including chiral and cation binding functions[354] could be easily attached in this way. Water-soluble derivatives containing four sulphonatomethyl groups were also reported[355].

The aminomethylation of **II** with primary amines leads in an entirely regioselective reaction to chiral C_4-symmetric tetrabenzoxazine derivatives **78**[356–358] as shown by several crystal structures. The subsequent cleavage of the benzoxazine rings (HCl, BuOH, 100 °C) readily gives the corresponding secondary amines as hydrochlorides. If the

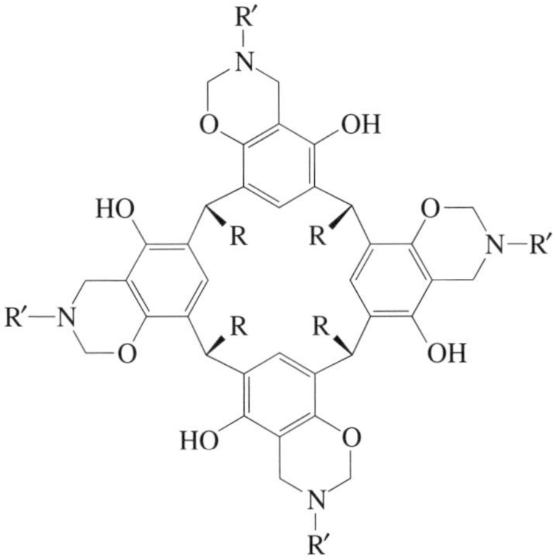

(78)

R' = alkyl, aryl

aminomethylation is carried out with chiral amines (e.g. α-phenylethylamine or its *p*-substituted analogues), only one of the two possible diastereomeric tetrabenzoxazines **78** is formed in high yield[359]. This was proved in two cases by X-ray analysis. Recent studies show that this high diastereoselectivity is due to the preferred crystallization rather than to the preferred formation of a single epimer[360], which is in agreement with the acid-catalysed epimerization in solution already observed earlier.

The diastereomerically pure tetrabenzoxazine derivatives **78a** were used as starting material to synthesize other inherently chiral derivatives. Methylation[361] of the hydroxy groups of the chiral tetrabenzoxazines with dimethyl sulphate or methyl triflate at −78 °C using BuLi as base led to the tetramethylated derivative **79** as a single diastereomer, for which an epimerization is no longer possible. Further chemical modifications furnished various tertiary (e.g. **80a** and **80c**) or secondary (**80b**) amines or benzoxazines directly as single enantiomers, which remain chiral (**80b** and **80c**) also after cleavage of the chiral auxiliary group (Scheme 22)[362].

The reaction of resorcarenes **II** with suitable diamines and CH_2O under high dilution conditions leads to 1,2;3,4-bis-bridged tetrabenzoxazines[363], or in the case of ethylene diamine to a head-to-head connected bis-resorcarene[364]. If 2-aminoalcohols are used in the Mannich reaction with resorcarenes, either benzoxazines, oxazine or oxazolidine rings can be formed. In the case of aminoethanol, predominantly the benzoxazine **78** (R' = CH_2CH_2OH) was detected in solution, while oxazolidines were predominantly obtained with 2-alkylaminoethanols[365].

As with calixphenols **I**, the different reactivity of *O*-acylated and unsubstituted resorcinol rings in the C_{2v}-symmetric derivatives **76** may be used for selective electrophilic substitutions. Distally disubstituted derivatives were obtained, for instance, by bromination or aminomethylation with secondary amines[341,342]. The Mannich reaction of tetrasulphonates **76c** and tetrabenzoates **76d** with primary amines led to C_2-symmetric bis-benzoxazine derivatives in a regioselective manner[366]. Various *trans*-cavity bridged compounds were obtained with primary diamines of different length[366], including enantiomerically pure, distally bridged resorcarenes when a chiral secondary diamine was used[367]. Removal of the Boc-protection in products obtained by aminomethylation

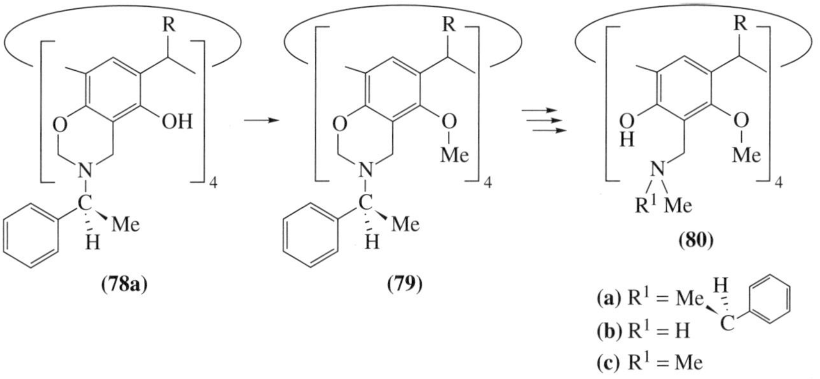

SCHEME 22. Synthesis of C_4-symmetric resorcarenes via Mannich condensation with a chiral primary amine as an auxiliary

of **76f** with secondary amines gave 1,3-diaminomethylated derivatives, which cannot be prepared directly by *partial* aminomethylation[342].

C. Reactions of the Substituents at the Bridges

The acid-catalysed condensation of resorcinol, 2-methylresorcinol and pyrogallol with aldehydes (or their synthetic equivalents) containing hydroxy-, alkoxy-, aryldiazo-, sulphonyl- and $B(OH)_2$ groups, halogens and double bonds introduces additional functional groups[67] which can be further modified.

The tetra-boronic acid **81a** was used, for example, to extend the residues R by a phenyl or biphenyl unit[368]. Acylation of resorcarene **81b** containing four pendant double bonds followed by *anti*-Markovnikov addition of $C_{10}H_{21}SH$[369] resulted in octaacylated resorcarene derivatives containing four thioether fragments at the narrow rim. The smooth cleavage of acyl groups gave the free octol **81c**[369]. Photochemical addition of AcSH to the double bonds of **81b** analogously led to resorcarene **81d** footed with four SH groups[370]. These compounds and their derivatives (Figure 11) were used to form self-assembled monolayers on gold surfaces[371].

The selective benzylation (benzyl bromide, K_2CO_3, NaI) of the phenolic hydroxy groups in **81e** led to the tetrahydroxy derivative **82a**. The subsequent mesylation of the aliphatic hydroxy groups with methanesulphonyl chloride and reaction with NaN_3 resulted in the tetraazide **82b**. Catalytic hydrogenation (Raney-Ni) and reaction with $(Boc)_2O$ led to the *N*-protected amine **82c** from which the benzyl groups could be cleaved to give the Boc-protected **81f**[372]. Alternatively, this compound could be prepared by Mitsunobu reaction of **82a** with EtOC(O)C(O)NHBoc, DEAD (diethyl azodicarboxylate) and PPh_3 in CH_2Cl_2 followed by saponification/decarboxylation ($LiOH/THF-H_2O$)[372]. A series of amphiphilic resorcarenes with azobenzene residues was prepared starting with **81g** which (after etherification of the phenolic hydroxy groups) was converted via the tetraiodide into tetraethers with *p*-hydroxyazobenzenes[373]. The analogue of **81e** containing four methyl groups at the 2-positions of the resorcinol rings was used to synthesize various cavitands footed with hydroxy, acetoxy and dihydroxyphosphoryl groups[67i] (see Section VII. D).

Partial epoxidation of the octapivaloate of **81b** gave the monoepoxide. Hydrogenation of the remaining double bonds (H_2, Pd/C), followed by acid-catalysed hydrolysis of the epoxide ring, oxidative cleavage of the resulting diol and reduction, finally led (after

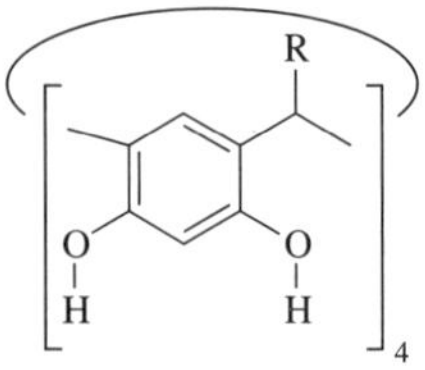

(81)

R =
(a) p-$C_6H_4B(OH)_2$
(b) $(CH_2)_8CH{=}CH_2$
(c) $(CH_2)_{10}SC_{10}H_{21}$
(d) $(CH_2)_{10}SH$
(e) $(CH_2)_3OH$
(f) $(CH_2)_3NH_2$
(g) $(CH_2)_3Cl$

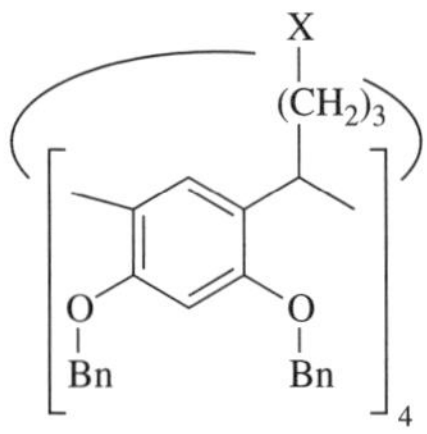

(82)

X =
(a) OH
(b) N_3
(c) NHBoc

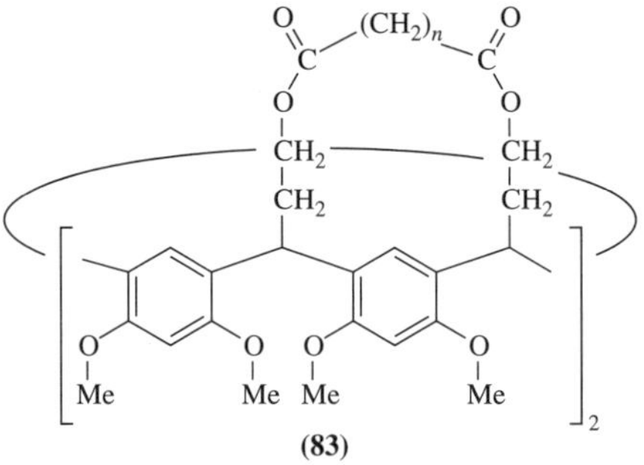

(83)
$n = 3, 4, 6$

FIGURE 11. Resorcarene derivatives **81**–**83** by chemical modification of the substituents at the bridges, introduced by the aldehyde

removal of the protective ester groups) to a resorcarene with a **single** residue R ending in a hydroxy group ($R = (CH_2)_9OH$)[374].

Bridging between residues R was also achieved. The tetraol obtained by reduction of the *rccc* tetraesters **29a** with $LiAlH_4$ in THF[375] was reacted under normal or high dilution conditions with two equivalents of glutaroyl, adipoyl or pimeloyl dichloride (Et_3N, CH_2Cl_2) to give the doubly spanned resorcarenes **83** in moderate yields[75e,375].

D. Cavitands

Probably the most interesting chemical modification of calixresorcinols consists in the intramolecular connection of the adjacent hydroxy functions in neighbouring resorcinol units via suitable bridges. This leads to rigidified, bowl-shaped molecules (general formula **III**) with an enforced cavity ready to include suitable guest molecules, for which D. Cram has coined the named '*cavitands*'[376]. Examples are given in Figure 12.

The most frequently used bridge is a methylene bridge, easily introduced by reaction with CH_2BrCl in yields up to 65% for the cavitands **84**[377]. While originally only the all-*cis* isomers were used as starting material, leading exclusively to cavitands with an all-axial orientation of the residues R, *rctt*-isomers derived from 2-methylresorcinol were recently converted in a similar way (6–8 equivalents CH_2BrCl, K_2CO_3, DMF, 60 °C, 10 h) into cavitands with two axial and two equatorial residues R[378,379]. For *rccc*-isomers, obviously

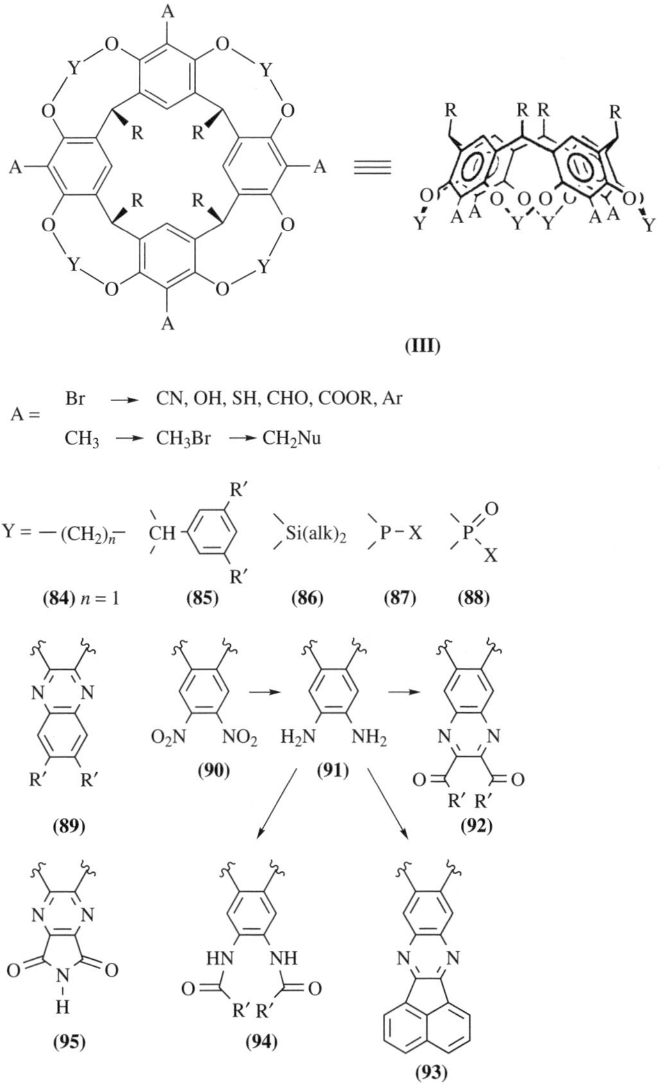

FIGURE 12. General formula of resorcarene-derived cavitands. Virtually all residues R discussed above (see Figures 2 and 11) are possible. The indicated modification of the substituent A was mainly done for Y = CH$_2$. Examples for various bridges Y and their chemical modification are also shown

the introduction of the last bridge is most difficult, making compounds with only three bridges available in which the two remaining OH groups can be used for the introduction of a different bridge. For *rctt*-isomers this is not possible.

The methyl groups in the 2-position of **84** (A = Me) can be easily brominated with NBS[380] and the resulting bromomethyl groups can be further substituted by a variety of nucleophiles[381]. Partial substitution of the bromomethyl groups by potassium phthalimide was also reported[382]. Reduction of the remaining CH_2Br groups ($NaBH_4$) and deprotection by hydrazine led to partially aminomethylated cavitands[383] as starting materials for the introduction of various further functionalities.

Additional functionalities may also be introduced starting with **84** (A = Br), for instance four CN groups[384], or via bromo-lithium exchange and subsequent reaction with appropriate reagents[385] four OH[386], SH, CHO or COOR groups. Suzuki coupling with phenylboronic acids leads to cavitands with a deeper cavity[387] which may be further modified via functional groups introduced in this way by the phenyl residue.

Deepened cavities can be also obtained by bridging with variously substituted benzal bromides[388]. Although six diastereomers are possible with bridges of this type (CHR), a single isomer was isolated in all cases (**85**) with yields as high as 56%. Functional groups introduced by the bridging benzal bromide may be used for further reactions[389]. The cavitand **85a** obtained with 3,5-dibromobenzal bromide (**85**, R' = Br) was further extended by a third row of aromatic residues. Reaction with resorcinol (pyridine, K_2CO_3, CuO) led to **85b** in an outstanding yield of 88% (nearly 97% efficiency for each bridge, more than 98% for each of the 8 covalent links[390]; see Scheme 23 below).

Bridging by a single atom was also achieved by silicon (**86**) and various phosphorus functionalities. Reaction with phenyldichlorophosphine in the presence of pyridine furnished the phosphonito cavitand **87** (X = Ar), a quadridentate ligand for Cu(I), Ag(I), Au(I) and Pt(II)[327,391], while bridging with dichloroarylphosphonate gave the phosphonate cavitands **88** (X = Ar)[392–394]. All six possible diastereomers, having different orientations of the P=O group, were isolated in the latter case, while in the former only the isomer with outward-oriented phenyl groups was formed. Reaction with phosphorous di- and triamides gave the cavitands **87** (X = OEt) and **87** (X = NMe_2, NEt_2), and the latter were converted to the corresponding amidothiophosphates. Bridging with PCl_3[395] and chloromethyldichlorophosphonate[396] led to cavitands **87** (X = Cl) and **88** (X = CH_2Cl) which, due to the presence of four reactive chloro atoms, are starting materials for various further derivatives.

While cavitands with $(CH_2)_2$ and $(CH_2)_3$ bridges, less rigid than the single atom bridged compounds, have not gained much interest, cavitands with o-phenylene bridges represent an important class of cavitands with 'deepened cavities'. Originally they were obtained by reaction with 2,3-dichloroquinoxaline or its 6,7-disubstituted analogues (**89**)[397] while an alternative strategy was recently based on the octanitrocavitand **90** available in yields up to 80% by reaction with 1,2-difluoro-4,5-dinitrobenzene[398]. Reduction of the nitro groups and condensation of the resulting phenylene diamine **91** derivatives with 1,2-diketones gave cavitands **92**, **93** with cavities large enough in the latter case to accommodate C_{60}. Acylation of the eight amino groups in **91**, on the other hand, led to cavitands **94** in which the vase conformation[399] is stabilized by a seam of intramolecular C=O···H−N hydrogen bonds (self-folding cavitands, see Scheme 23b[400]). Reaction of calix[4]resorcinols with 5,6-dichloropyrazin-2,3-dicarboxylic acid imide led to **95**[401], an extended cavitand forming dimers with a large cylindrical cavity, which are held together by intermolecular hydrogen bonds.

Cavitands consisting of more than four resorcinol units have been recently described. Acid-catalysed condensation of 2-methylresorcinol with diethoxymethane in ethanol leads to a mixture of methylene-bridged calixresorcinols with different ring size, which was isolated after 30 min. at 60 °C (higher temperature and longer reaction times favour the calix[4]resorcinols as the thermodynamically controlled products). Subsequent reaction with bromochloromethane (DMA, K_2CO_3, 60 °C) furnished a mixture of cavitands with different ring size ($n = 4$–7) which could be separated by column chromatography (3.6%, 3.6%, 13.9% and 1.1% yields) making these [n]cavitands available in gram quantities[402]. X-ray structures and NMR data revealed a symmetric *cone* conformation (C_{4v}, C_{5v}) for

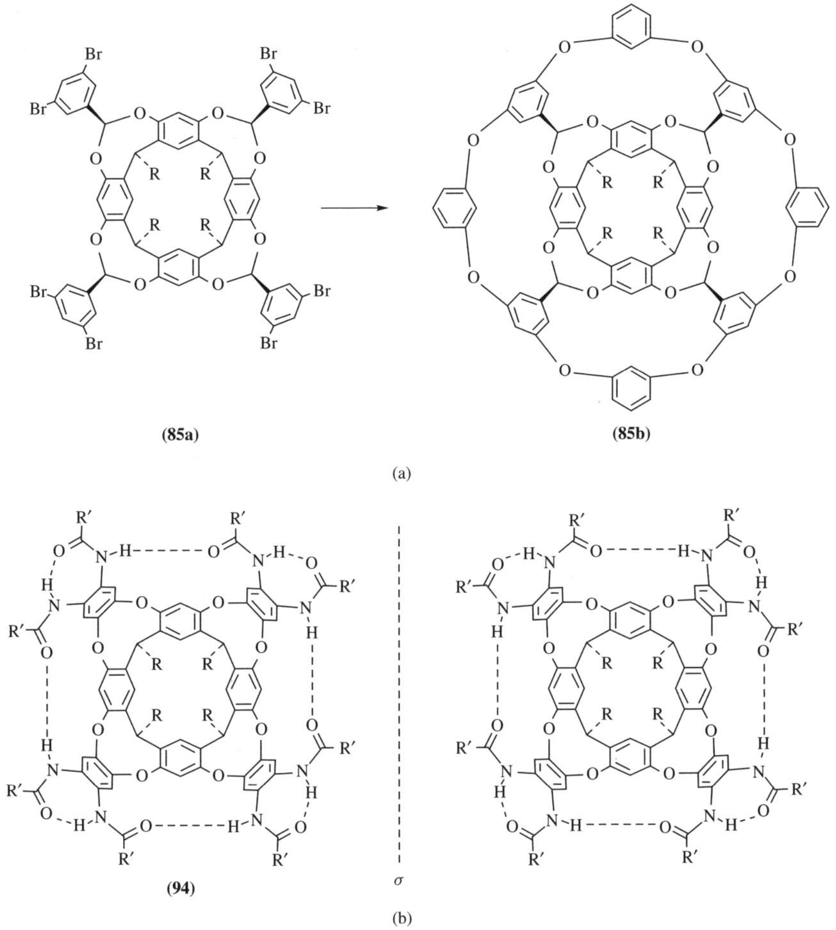

SCHEME 23. (a) Extension of the cavity of **85a** by covalent bridging to **85b**. (b) Stabilization of the deepened cavity by intramolecular hydrogen bonding in **94**. Two enantiomers can be observed due to the directionality of the hydrogen-bonded system

$n = 4, 5$, while pinched conformations (C_{2v} for $n = 6$, C_s for $n = 7$) are assumed by the larger oligomers.

Cavitands **84** have been used as rigid skeleton to attach four peptide chains as substituent A[403]. In these *caviteins* the four α-helical peptides are significantly stabilized by their proximity[404]. Sugar residues (glucose, maltose, maltotriose) have also been attached to cavitands with A = SH[405].

VIII. MOLECULES CONSISTING OF SEVERAL CALIXARENE STRUCTURES

A. Double Calixarenes

The functional groups and reactions discussed above have been used to synthesize various molecules consisting of two or more covalently linked calixarene structures[406]. Figure 13 gives a schematic representation of the main types that have been realized,

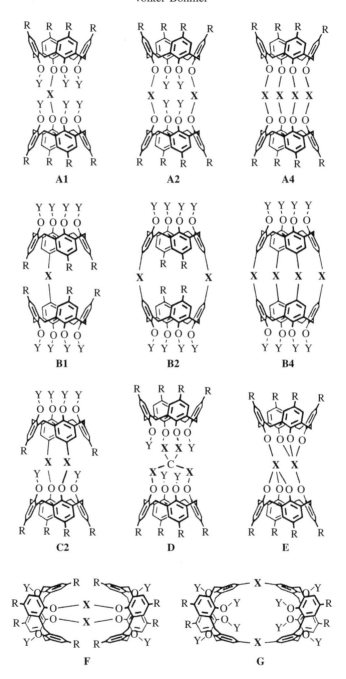

FIGURE 13. Main types of double calix[4]arenes

illustrating simultaneously the versatility of the calixarene skeleton to build up larger structures. Many examples exist in which two calix[4]arenes are connected via one or two bridges between their narrow rims (**A1**, **A2**)[407,408], their wide rims (**B1**, **B2**)[409,410] or between narrow and wide rims (**C2**). In the latter case not only was a covalent connection between 'prefabricated' calix[4]arenes used[411], but also a formation of the second calix[4]arene by 2 + 1 + 1 condensation on the narrow rim[412].

A highly interesting reaction is the tandem Claisen rearrangement, by which double calix[4/6]arenes singly bridged at the narrow rim (type **A1**) can be converted into the corresponding double calixarenes singly bridged at the wide rim (type **B1**). The reaction works even with doubly bridged calix[4/6]arenes as illustrated in Scheme 24[275].

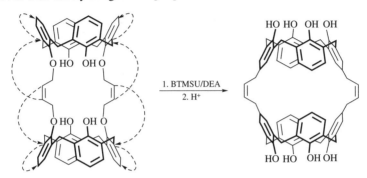

SCHEME 24. Tandem Claisen rearrangement of doubly bridged calix[4]arenes. The intramolecular pathway is indicated by dashed arrows. BTMSU = bistrimethylsilylurea

Oxidation of the *t*-butylphenol units in the corresponding molecules of type **A2** led to tetraquinones of double calix[4]arenes recently described as redox-active ionophors (Cs and Rb selective)[413]. Bis calix[4/5/6/8]arenes connected *directly* via one *p*-position[230,414], a connection easily available by oxidation of the respective tri- to heptaesters, may be seen as a special case of type **B1**. Triply bridged double calix[6]arenes (analogous to type **B2**) have been also prepared recently[415].

Among molecules of type **A4**, double calix[4]arenes with four ethylene bridges (X = CH_2CH_2) (*calix[4]tubes*) should be mentioned as ligands with pronounced selectivity for potassium[416,417]. They are available in surprisingly good yields (about 50% for the connection of the two calix[4]arenes[418]), while a connection via four bridges at the wide rim (type **B4** as analogue to carcerands) is less satisfactory[419], probably due to the flexibility of calix[4]arenes in comparison with cavitands.

Two calix[4]arenes may be connected also by a spiro-linker derived from pentaerythritol (**D** with X = $-(CH_2)_2O(CH_2)_2OCH_2-)$[420] or by two tetravalent atoms (**E**, X = Si, Ti). In the latter case, centro-symmetric molecules with two open cavities pointing in opposite directions (*koilands*) are obtained ($SiCl_4$, NaH, THF, 52%)[421] and may form one-dimensional networks in the crystalline state (*koilates*) when a suitable connector (e.g. hexadiyne) is included in their cavities[422].

Finally, double calix[4]arenes should be mentioned in which two calix[4]arenes in the *1,3-alternate* conformation are connected by two bridges between the narrow (**F**)[423] or the wide rim (**G**)[424].

B. Carcerands

'*Carcerands* are closed-surface, globe-shaped molecules with enforced hollow interiors large enough to incarcerate simple organic compounds (better: molecules), inorganic ions, or both. *Carceplexes* are carcerands whose interiors are occupied by prisoner molecules or ions that cannot escape their molecular cells without breaking covalent bonds between atoms that block their escape'[385].

The first examples of carceplexes have been synthesized by covalent connection of two cavitand molecules **III** (especially **79**) suitably functionalized in the 2-position of the resorcinol rings (A = OH, CH_2SH, CH_2Br etc.). Figure 14 gives a survey on some of the bridges realized (compounds **96–105**)[425], illustrating the diversity of molecules thus available, including water-soluble compounds (**102**)[425] and those with chiral bridges Z (e.g. **103**, **104**)[426]. While a symmetrical carceplex results if a bifunctional bridging reagent is reacted with a cavitand, desymmetrized carceplexes are available by reaction between two cavitands which are complementarily functionalized; e.g. **99** was obtained by reaction of cavitands with A = CH_2SH and A = OCH_2CH_2I[427]. This desymmetrization is more pronounced for molecules composed of a calix[4]resorcinol derived cavitand and a calix[4]phenol (see below).

In general, the formation of the carcerand is not possible in the absence of molecules which can be included, or in other words, the carceplex and not the carcerand is formed. This templating effect was investigated in detail for the smallest bridges ($4 \times OCH_2O$ between the two bowls with Y = CH_2) where template ratios between 1 (for *N*-methylpyrrolidinone) and 1,000,000 (for pyrazine as the best guest) were established by a series of overlapping competition experiments involving two guests with similar template ratios[431]. The differences in template ratios are less pronounced for larger bridges within (e.g. Y = CH_2CH_2) and between the bowls (e.g. Z = $(CH_2)_4$)[432].

Hemicarcerands/hemicarceplexes are similar to carcerands/carceplexes, but distinguished by the possibility of the guest molecule to escape under drastic conditions, e.g. heating under high vacuum. Since the 'empty' hemicarcerand cannot collapse, it will uptake any suitable molecule/atom which is offered and this opens the way to include guests which cannot be present (in sufficient concentration) during the closing reaction. The first examples of Cram consisted of cavitands connected by three instead of four bridges, but hemicarcerands with four longer (flexible) bridges have also been prepared (e.g. **96** ($n = 4$) or **105**). In principle, the 'distinction' between a carceplex and a hemicarceplex depends on portal size, guest size and shape, solvent, temperature, and even the number of guests[433]. Hemicarceplexes have been used to study the reactivity of single molecules included in their internal cavity, and isolated in this way from the bulk surrounding medium[434].

Cavitands functionalized at the 2-position of the resorcinol rings have been transformed into lantern-shaped derivatives **106**, e.g. by bridging the four *m*-hydroxybenzyl ether groups (A = $OCH_2C_6H_4OH$) with a suitable tetrakis(bromomethylated) terphenyl derivative. They may exist in two isomeric forms with the functional group X pointing away from the cavity or into the cavity[435]. In the latter orientation, it is strongly shielded from the environment with drastic consequences for its reactivity. Photolysis of the β-ketosulphide, for instance, yielded an enol which was stable at room temperature in $CDCl_3$ in the presence of TFA over days[436] (Scheme 25).

C. Further Combinations

The connection of a calix[4]resorcinol-based cavitand and a calix[4]phenol to a carcerand via four bridges at the wide rim was already mentioned. Compounds **107** were prepared by 'ring closure' of a precursor, in which the cavitand and the calix[4]arene are connected by two adjacent bridges[263d,437]. A similar compound with O-CH_2CH_2-O-bridges was obtained by reaction of a *p*-substituted (O-CH_2CH_2-I) calix[4]arene with the tetrahydroxy cavitand **84** (A = OH)[438]. Two different orientations can often

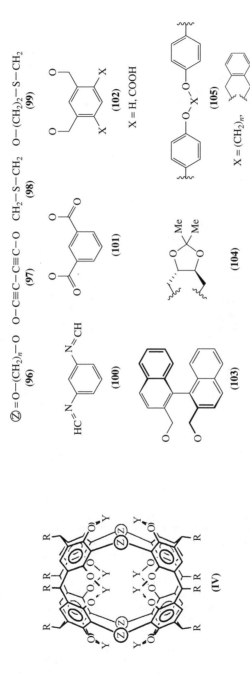

FIGURE 14. General formula **IV** of carcerands derived from two molecules of **III**[428], and survey on some examples illustrating the variety of different bridges Z. Compounds connected by only three bridges Z have been also synthesized[429], as well as compounds with non-identical bridges[426b,430]

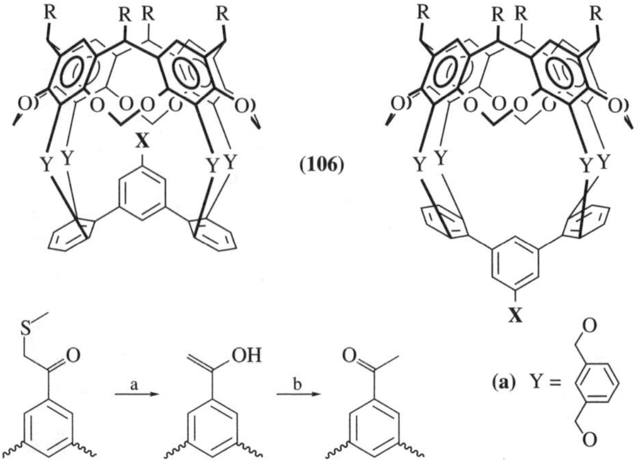

SCHEME 25. Cavitand-based lantern-shaped molecules provide a shielded environment for functional groups X, similar to the shielding of a guest in a (hemi)carceplex. Conditions for the reaction of $X = C(O)CH_2SCH_3$: (a) irradiation, (b) TFA in $CDCl_3$, 3 days

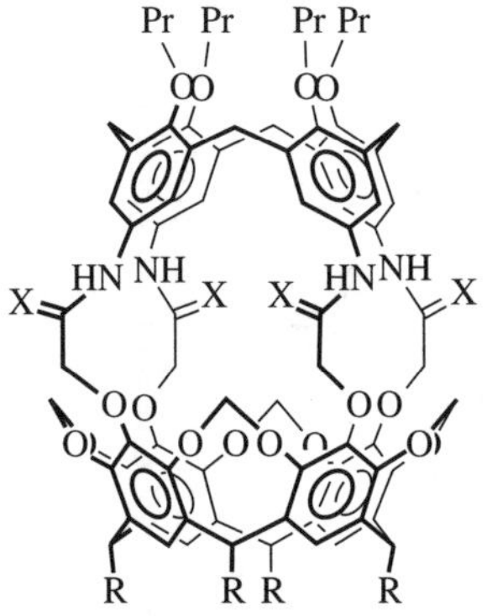

be distinguished for the included guest (omitted in the formula) in such carcerands with different poles ('*carceroisomerism*'). An interesting molecule is the 'head to tail' combination of a calix[4]arene and calix[8]arene in **108**, although the present example shows a collapsed cavity[439].

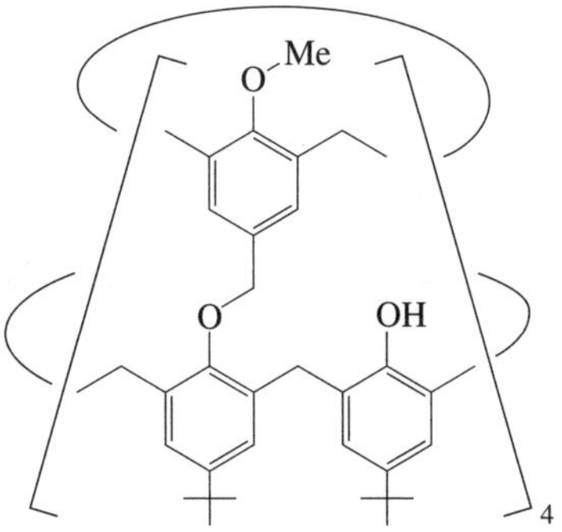

Two calixarenes have also been connected in various ways to porphyrins. **109** may be taken as an example, in which the porphyrin plane separates two 'chambers' confined by the calix[4]arenes[440]. The two self-folding cavitands (compare **94**) in the C-shaped isomer of **110** (formed in mixture with the S-shaped isomer) can simultaneously include two different guests[441].

An appealing combination of a calixarene with the corresponding calixpyrrole (a *heterocalixarene*[6]) was realized with compounds **111**, which were synthesized by condensation of the 2-oxopropyl ethers with pyrrole in 32% and 10% yield, respectively[442].

(109)

(110)

$R = C_{11}H_{23}$

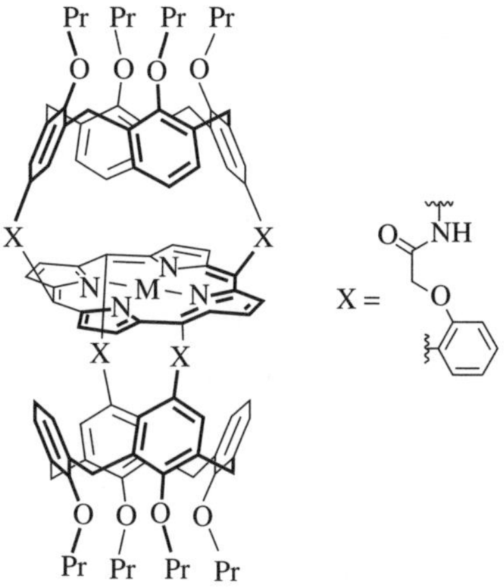

There is also the possibility of constructing molecules consisting of several calixarene systems sharing one or two phenolic units. Two opposite phenols are shared (as 1,3,5-substituted branching points) in *bicyclocalix[4]arenes* **112**, which are prepared

(111)

(112)
R = *t*-Bu, *c*-Hx
Y = H, Me, Bn

from *p*-bridged calix[4]arenes **7**, transforming the $CH_2CH_2COCH_2CH_2$ bridge into a methylene-linked *p*-nitrophenol unit by reaction with nitromalondialdehyde[443]. Two adjacent phenol units are shared in *annelated* calix[4]arenes **113**, available by direct condensation (in analogy to the 2 + 2 strategy) of *exo*-calix[4]arenes with bisbromomethylated dimers[50b,444].

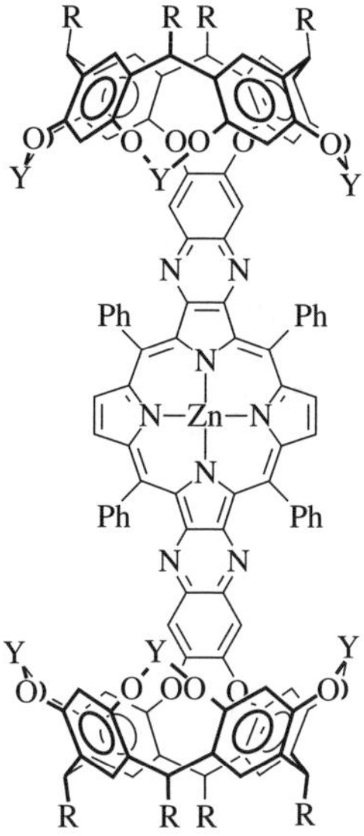

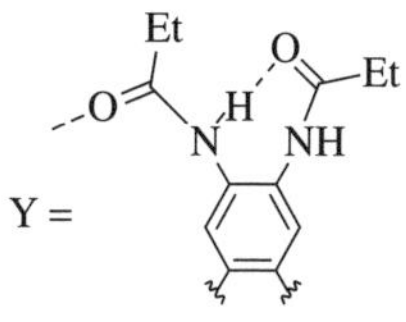

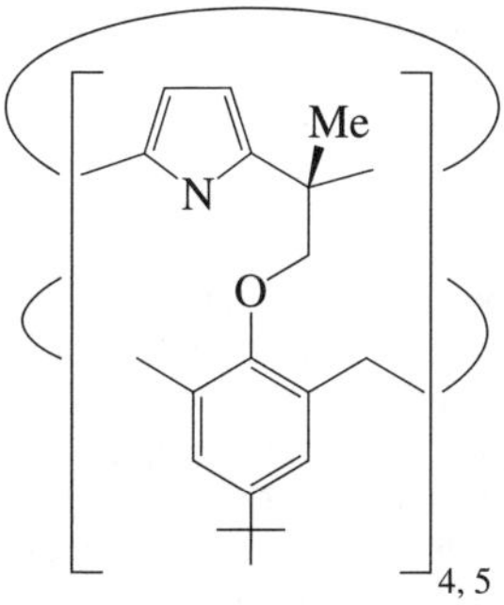

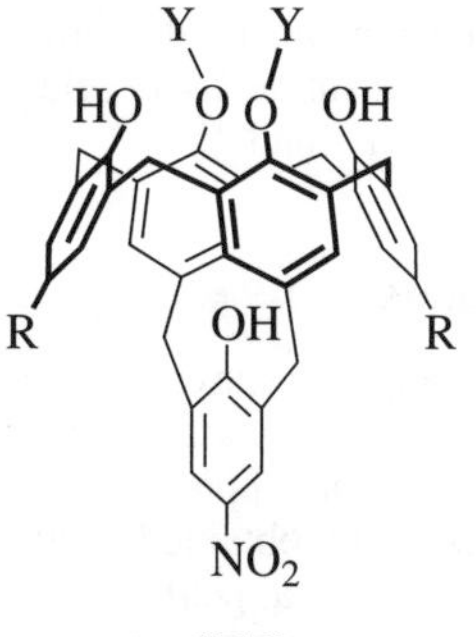

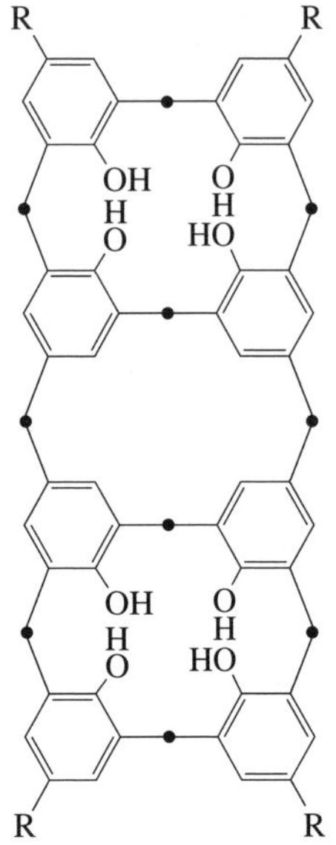

(113)

R = Me, *t*-Bu, Ph • = CH$_2$

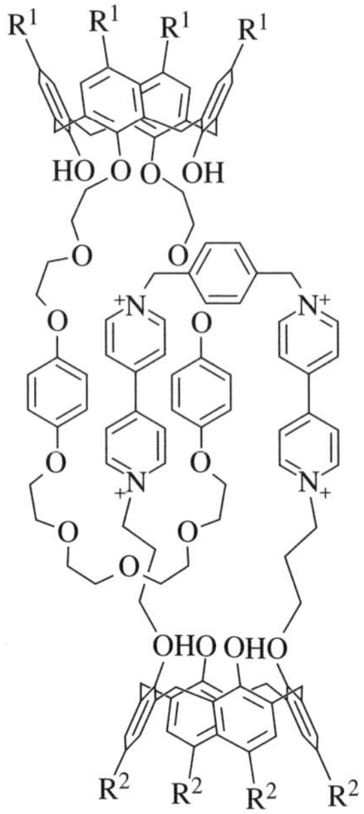

(114)

R^1, R^2 = H, *t*-Bu

Recently, two calix[4]arenes were combined to a catenane via loops connecting the oxygen functions in 1,3-position (**114**)[445] and calixarenes as stoppers were used in a rotaxane formed by a cyclic amide with a dumbbell of type **B1** in Figure 13[446]. Connection of two cavitands via a single C=O bridge between the 2-positions led to a C_2-symmetric, propeller-shaped biscavitand[447].

D. Multicalixarenes

The principles outlined above have also been used to construct even larger molecules consisting of several calixarene substructures. Linear oligomers were constructed by connection between opposite positions at the narrow[122c] or wide rim[448], and cyclic compounds were obtained analogously up to an octamer (analogous to **A2** in Figure 13)[122c,449,450] and a pentamer (analogous to **B2**)[448a]. In most cases a mixture of oligomers is formed and single species have to be isolated chromatographically. The quantitative formation of the trimer **115** by metathesis reaction of the diallyl calix[4]arene is a remarkable exception[448c,d], and also the trimer **116** (compare with **E**) was formed in 60–69% yield[421b]. A trimer consisting of three calix[4]phenols in the *1,3-alternate* conformation, linked analogous to **F**, should also be mentioned[424b]. Various branched molecules are known in which calix[4]arenes are linked by a single bridge from the narrow[122c,d] or wide rim[409a,451] to a central molecule, which can be a calixarene again[229,452,453].

Two bridges between adjacent phenolic units were used in the construction of molecules built up by two calix[4]phenols and one cavitand derived from calix[4]resorcinols[454] or vice versa by two cavitands connected by one calix[4]phenol[455]. Examples for

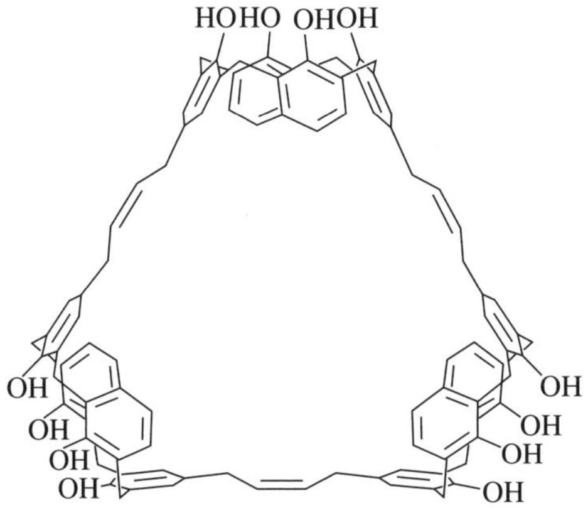

(115)

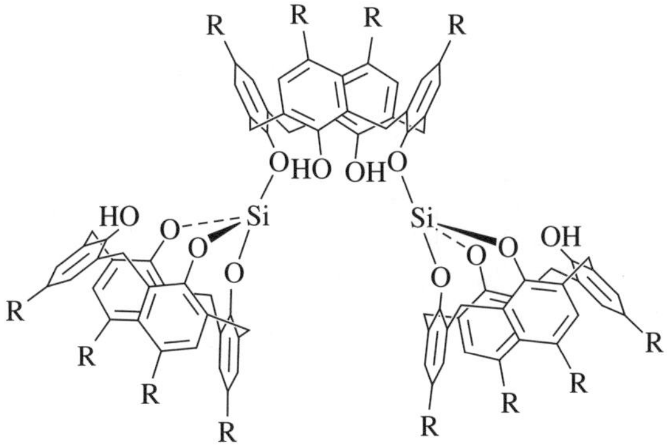

(116)

R = *t*-Bu

the corresponding macrocyclic 2 + 2 combination (*holand*) were also realized[456], but disappointingly these remarkable molecules exhibit no inclusion properties, most probably due to their rigidity.

Cyclization of a tetrahydroxy cavitand (**84**, A = OH,) protected at two opposite rings by benzyl ether groups (**117**) with bromochloromethane under dilute, basic conditions led to a cyclic trimer (**118**) and tetramer (**119**) after hydrogenolysis[457] (Scheme 26). The tetramer can again react with CH_2BrCl in the presence of pyrazine, a good guest and template, to form a bis-carceplex (**120**) with two adjacent cells, each containing one pyrazine[457,458]. A similar bis-capsule (intramolecularly hydrogen bonded via O−H···O⁻ bridges) can be formed by deprotonation with DBU (1,3-diazabicyclo[5.4.0]undecene-7) as a bulky base. Different guests could be observed in the two chambers in this case[459]. A giant carceplex, permanently including 3 DMF molecules, was prepared from the trimer **118** by capping reaction with tris(bromomethylated)mesitylene[460], and the construction of even larger carceplexes seems likely in the future.

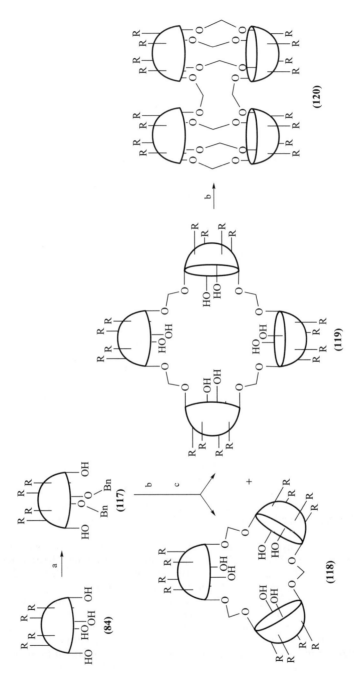

SCHEME 26. Formation of oligo-cavitands and bis-carceplexes (R = CH$_2$CH$_2$C$_6$H$_5$, guest pyrazine omitted): (a) Bn-Br, 15%; (b) CH$_2$BrCl; (c) H$_2$/cat

IX. CONCLUSIONS AND OUTLOOK

This last example demonstrates already one of the future lines. As the chemistry of calixarenes and resorcarenes is more and more understood, these fascinating molecules lend themselves as building blocks for the construction of increasingly larger and more and more sophisticated structures. It is hoped that this fascination could be transferred at least in part to the reader.

A complete survey on calixarenes was far beyond the scope of this chapter. We therefore concentrated mainly on the chemistry, namely the synthesis and the (basic) chemical modification of calixarenes, which form the basis for all further studies. Further interesting aspects, such as inherent chirality of calixarenes[461], catalytic or biomimetic effects[462], larger structures formed via self-assembly in solution[463] or in mono-[464] and multilayers[371] could be stressed only shortly. Important properties such as complexation of cations[94,465], anions[466] and of neutral guests[467], including fullerenes[468], could not be treated at all. The same is true for applications arising from these properties in such different areas as sensor techniques[469], chromatographic separations[470] or treatment of nuclear wastes[187]. In all these cases, the reader is referred to special reviews.

X. ACKNOWLEDGEMENT

I am very grateful to Prof. Silvio E. Biali (Jerusalem) for critical comments and many valuable suggestions.

XI. REFERENCES AND NOTES

1. For general reviews on calixarenes see:
   (a) C. D. Gutsche *Calixarenes Revisited* in *Monographs in Supramolecular Chemistry* (Ed. J. F. Stoddart), Royal Society of Chemistry, Cambridge, 1998.
   (b) Z. Asfari, V. Böhmer, J. Harrowfield and J. Vicens (Eds.), *Calixarenes 2001*, Kluwer, Dordrecht, 2001.
   (c) V. Böhmer, *Angew. Chem.*, **107**, 785 (1995); *Angew. Chem., Int. Ed. Engl.*, **34**, 713 (1995).
2. The original description as 'upper' and 'lower rim' sticks too much to the vision of a calix and is impeding especially for the larger, more flexible members and for molecules built up by several calixarene subunits.
3. No general consent exists on the trivial name of these compounds. This short form 'resorcarenes' has the advantage of comprising the same number of syllables as 'calixarenes'. It would be reasonable to use 'calixarenes' as the general name for the whole class of such [1$_n$]metacyclophanes, and to distinguish, where necessary, between *calixphenols*, *calixresorcinols*, *calixnaphthols*, *calixpyrogallols* etc.
4. For a special review on resorcarenes see: P. Timmerman, W. Verboom and D. N. Reinhoudt, *Tetrahedron*, **52**, 2663 (1996).
5. Most important are 'cyclotriveratrylenes' and 'cryptophanes' derived from them; see: A. Collet, J.-P. Dutasta, B. Lozach and J. Canceill, *Top. Curr. Chem.*, **165**, 103 (1993).
6. Names such as *calixpyrroles*, *calixphyrins*, *calixfurans*, *calixthiophenes*, *calixpyridines*, *calixindoles*, *calixbenzofurans*, *calixureas* have been created in analogy to calixarenes; cationic macrocycles by *N*-benzylation of pyridine or pyrimidine and various cyclic oligomers with S, NR, SiR$_2$ or Pt(II)en bridges could be mentioned. For a short review see: M. Vysotsky, M. Saadioui and V. Böhmer, *Heterocalixarenes*, in Reference 1b, p. 250. One of the most recent examples might be the *calix[4]azulene*: D. A. Colby and T. D. Lash, *J. Org. Chem.*, **67**, 1031 (2002).
7. C. D. Gutsche and M. Iqbal, *Org. Synth.*, **68**, 234 (1990).
8. C. D. Gutsche, B. Dhawan, M. Leonis and D. Stewart, *Org. Synth.*, **68**, 238 (1990).
9. J. H. Munch and C. D. Gutsche, *Org. Synth.*, **68**, 243 (1990).
10. (a) D. R. Stewart and C. D. Gutsche, *Org. Prep. Proced. Int.*, **25**, 137 (1993).
    (b) K. Iwamoto, K. Araki and S. Shinkai, *Bull. Chem. Soc. Jpn.*, **67**, 1499 (1994).

11. F. Vocanson, R. Lamartine, P. Lanteri, R. Longeray and J. Y. Gauvrit, *New J. Chem.*, **19**, 825 (1995).
12. C. D. Gutsche, D. E. Johnston, Jr. and D. R. Stewart, *J. Org. Chem.*, **64**, 3747 (1999).
13. Calix[8,7,6]arenes were obtained from bisphenol A (monoprotected by *t*-butyldimethylsilyl chloride) in 3:2:1 ratio and 63% overall yield: K. H. Ahn, S.-G. Kim and J. S. Un, *Bull. Korean Chem. Soc.*, **21**, 813 (2000).
14. Y. Seki, Y. Morishige, N. Wamme, Y. Ohnishi and S. Kishida, *Appl. Phys. Lett.*, **62**, 3375 (1993).
15. Z. Asfari and J. Vicens, *Makromol. Chem. Rapid Commun.*, **10**, 181 (1989).
16. B. Dhawan, S.-I. Chen and C. D. Gutsche, *Makromol. Chem.*, **188**, 921 (1987).
17. S. R. Izatt, R. T. Hawkins, J. J. Christensen and R. M. Izatt, *J. Am. Chem. Soc.*, **107**, 63 (1985).
18. K. Ohto, M. Yano, K. Inoue, T. Yamamoto, M. Goto, F. Nakashio, S. Shinkai and T. Nagasaki, *Anal. Sci.*, **11**, 893 (1995).
19. F. Vocanson, M. Perrin and R. Lamartine, *J. Incl. Phenom.*, **39**, 127 (2001).
20. Z. Asfari and J. Vicens, *Tetrahedron Lett.*, **29**, 2659 (1988).
21. E. Lubitov, E. A. Shokova and V. V. Kovalev, *Synlett*, 647 (1993).
22. M. Makha and C. L. Raston, *Chem. Commun.*, 2470 (2001).
23. J. L. Atwood, M. J. Hardie, C. L. Raston and C. A. Sandoval, *Org. Lett.*, **1**, 1523 (1999).
24. (a) B. Souley, Z. Asfari and J. Vicens, *Polish J. Chem.*, **66**, 959 (1992).
    (b) For a slightly modified procedure see: J. L. Atwood, L. J. Barbour, P. J. Nichols, C. L. Raston and C. A. Sandoval, *Chem. Eur. J.*, **5**, 990 (1999).
25. M. Makha and C. L. Raston, *Tetrahedron Lett.*, **42**, 6215 (2001).
26. C. D. Gutsche and P. F. Pagoria, *J. Org. Chem.*, **50**, 5795 (1985).
27. D. R. Stewart and C. D. Gutsche, *J. Am. Chem. Soc.*, **121**, 4136 (1999).
28. I. Dumazet, J.-B. Regnouf-de-Vains and R. Lamartine, *Synth. Commun.*, **27**, 2547 (1999).
29. C. D. Gutsche, J. S. Rogers, D. R. Stewart and K.-A. See, *Pure Appl. Chem.*, **62**, 485 (1990).
30. For the crystal structure of *t*-butylcalix[16]arene see: C. Bavoux, R. Baudry, I. Dumazet-Bonnamour, R. Lamartine and M. Perrin, *J. Incl. Phenom.*, **40**, 221 (2001).
31. (a) Y. Ohba, K. Irie, F. Zhang and T. Sone, *Bull. Chem. Soc. Jpn.*, **66**, 828 (1993) (describes the preparation of the linear oligomers).
    (b) K. Ito, S. Izawa, T. Ohba, Y. Ohba and T. Sone, *Tetrahedron Lett.*, **37**, 5959 (1996) (describes the cyclization but with sparse experimental details).
32. The numbers characterize the phenolic units in the two different fragments.
33. (a) V. Böhmer, L. Merkel and U. Kunz, *J. Chem. Soc., Chem. Commun.*, 896 (1987).
    (b) L. Zetta, A. Wolff, W. Vogt, K.-L. Platt and V. Böhmer, *Tetrahedron*, **47**, 1911 (1991).
    (c) M. Backes, V. Böhmer, G. Ferguson, C. Grüttner, C. Schmidt, W. Vogt and K. Ziat, *J. Chem. Soc., Perkin Trans. 2*, 1193 (1997).
34. S. E. Biali, V. Böhmer, I. Columbus, G. Ferguson, C. Grüttner, F. Grynszpan, E. F. Paulus and I. Thondorf, *J. Chem. Soc., Perkin Trans. 2*, 2261 (1998).
35. J. de Mendoza, P. M. Nieto, P. Prados and C. Sanchez, *Tetrahedron*, **46**, 671 (1990).
36. (a) K. No and K. M. Kwan, *Synthesis*, 1293 (1996).
    (b) T. Haino, K. Matsumura, T. Harano, K. Yamada, Y. Saijyo and Y. Fukazawa, *Tetrahedron*, **54**, 12185 (1998).
    (c) C. Schmidt, M. Kumar, W. Vogt and V. Böhmer, *Tetrahedron*, **55**, 7819 (1999).
37. M. Bergamaschi, F. Bigi, M. Lanfranchi, R. Maggi, A. Pastorio, M. A. Pellinghelli, F. Peri, C. Porta and G. Sartori, *Tetrahedron*, **53**, 13037 (1997).
38. (a) H. Tsue, M. Ohmori and K.-I. Hirao, *J. Org. Chem.*, **63**, 4866 (1998).
    (b) For side products due to *ipso*-substitution see: H. Tsue, K. Enyo and K. Hirao, *Org. Lett.*, **2**, 3071 (2000) and *Helv. Chim. Acta*, **84**, 849 (2001).
39. S. E. Biali, V. Böhmer, S. Cohen, G. Ferguson, C. Grüttner, F. Grynszpan, E. F. Paulus, I. Thondorf and W. Vogt, *J. Am. Chem. Soc.*, **118**, 12938 (1996).
40. H. Goldmann, W. Vogt, E. Paulus and V. Böhmer, *J. Am. Chem. Soc.*, **110**, 6811 (1988).
41. M. Tabatabai and V. Böhmer, unpublished results.
42. (a) G. D. Andreetti, V. Böhmer, J. G. Jordon, M. Tabatabai, F. Ugozzoli, W. Vogt and A. Wolff, *J. Org. Chem.*, **58**, 4023 (1993).
    (b) D. K. Fu, B. Xu and T. M. Swager, *J. Org. Chem.*, **61**, 802 (1996).

19. Calixarenes

(c) P. E. Georghiou, M. Ashram, H. J. Clase and J. N. Bridson, *J. Org. Chem.*, **63**, 1819 (1998).
43. 2-Hydroxymethyl-4-benzoylresorcinol (2,4-dihydroxy-3-(hydroxymethyl)benzophenone) gave a calix[5]arene: M. Tabatabai, W. Vogt, V. Böhmer, G. Ferguson and E. F. Paulus, *Supramol. Chem.*, **4**, 147 (1994).
44. R. G. Khoury, L. Jaquinod, K. Aoyagi, M. M. Olmstead, A. J. Fisher and K. N. Smith, *Angew. Chem.*, **109**, 2604 (1997); *Angew. Chem., Int. Ed. Engl.*, **36**, 2497 (1997).
45. (a) T. Yamato, Y. Saruwatari, S. Nagayama, K. Maeda and M. Tashiro, *J. Chem. Soc., Chem. Commun.*, 861 (1992).
    (b) T. Yamato, Y. Saruwatari, L. K. Doamekpor, K.-I. Hasegawa and M. Koike, *Chem. Ber.*, **126**, 2501 (1993).
46. (a) T. Yamato, K.-I. Hasegawa, Y. Saruwatari and L. K. Doamekpor, *Chem. Ber.*, **126**, 1435 (1993).
    (b) T. Yamato, M. Yasumatsu, Y. Saruwatari and L. K. Doamekpor, *J. Incl. Phenom.*, **19**, 315 (1994).
    (c) T. Yamato, F. Zhang and M. Yasumatsu, *J. Chem. Res. Synop.*, 466 (1997).
47. (a) Y. Okada, F. Ishii, Y. Kasai and J. Nishimura, *Chem. Lett.*, 755 (1992).
    (b) Y. Okada, Y. Kasai, F. Ishii and J. Nishimura, *J. Chem. Soc., Chem. Commun.*, 976 (1993).
    (c) Y. Okada, Y. Kasai and J. Nishimura, *Synlett*, 85 (1995).
48. (a) F. Vögtle, J. Schmitz and M. Nieger, *Chem. Ber.*, **125**, 2523 (1992).
    (b) J. Schmitz, F. Vögtle, M. Nieger, K. Gloe, H. Stephan, O. Heitzsch, H.-J. Buschmann, W. Hasse and K. Cammann, *Chem. Ber.*, **126**, 2483 (1993).
49. (a) G. Sartori, R. Maggi, F. Bigi, A. Arduini, A. Pastorio and C. Porta, *J. Chem. Soc., Perkin Trans. 1*, 1657 (1994).
    (b) G. Sartori, F. Bigi, C. Porta, R. Maggi and R. Mora, *Tetrahedron Lett.*, **36**, 2311 (1995).
    (c) For the base-catalysed condensation see: G. Sartori, F. Bigi, C. Porta, R. Maggi and F. Peri, *Tetrahedron Lett.*, **36**, 8323 (1995).
50. (a) V. Böhmer, R. Dörrenbächer, M. Frings, M. Heydenreich, D. de Paoli, W. Vogt, G. Ferguson and I. Thondorf, *J. Org. Chem.*, **61**, 549 (1996).
    (b) T. N. Sorrell and H. Yuan, *J. Org. Chem.*, **62**, 1899 (1997).
51. S. E. Biali, V. Böhmer, J. Brenn, M. Frings, I. Thondorf, W. Vogt and J. Wöhnert, *J. Org. Chem.*, **62**, 8350 (1997).
52. (a) C. D. Gutsche, B. Dhawan, K. H. No and R. Muthukrishnan, *J. Am. Chem. Soc.*, **103**, 3782 (1981).
    (b) For an improved procedure see: C. Bavoux, F. Vocanson, M. Perrin and R. Lamartine, *J. Incl. Phenom.*, **22**, 119 (1995).
53. For an actual review see: B. Masci, *Homooxa- and Homoaza-Calixarenes* in Reference 1b, p. 235.
54. (a) B. Dhawan and C. D. Gutsche, *J. Org. Chem.*, **48**, 1536 (1983).
    (b) K. Araki, N. Hashimoto, H. Otsuka and S. Shinkai, *J. Org. Chem.*, **58**, 5958 (1993).
    (c) B. Masci, *Tetrahedron*, **51**, 5459 (1995).
    (d) A. Ikeda, Y. Suzuki, M. Yoshimura and S. Shinkai, *Tetrahedron*, **54**, 2497 (1998).
    (e) Z. Zhong, A. Ikeda and S. Shinkai, *J. Am. Chem. Soc.*, **121**, 11906 (1999).
55. P. D. Hampton, Z. Bencze, W. Tong and C. E. Daitch, *J. Org. Chem.*, **59**, 4838 (1994).
56. K. Tsubaki, T. Otsubo, K. Tanaka, K. Fuji and T. Kinoshita, *J. Org. Chem.*, **63**, 3260 (1998).
57. B. Masci, *J. Org. Chem.*, **66**, 1497 (2001) and *Tetrahedron*, **57**, 2841 (2001).
58. (a) H. Takemura, K. Yoshimura, I. U. Khan, T. Shinmyozu and T. Inazu, *Tetrahedron Lett.*, **33**, 5775 (1992).
    (b) I. U. Khan, H. Takemura, M. Suenaga, T. Shinmyozu and T. Inazu, *J. Org. Chem.*, **58**, 3158 (1993).
    (c) P. Thuéry, M. Nierlich, J. Vicens and H. Takemura, *J. Chem. Soc., Dalton Trans.*, 279 (2000).
    (d) H. Takemura, K. Yoshimura and T. Inazu, *Tetrahedron Lett.*, **40**, 6431 (1999).
    (e) H. Takemura, T. Shinmyozu, H. Miura and I. U. Khan, *J. Incl. Phenom.*, **19**, 189 (1994).
    (f) P. D. Hampton, W. Tong, S. Wu and E. N. Duesler, *J. Chem. Soc., Perkin Trans. 2*, 1127 (1996).
    (g) K. Ito, T. Ohta, Y. Ohba and T. Sone, *J. Heterocycl. Chem.*, **37**, 79 (2000).

59. K. Ito, Y. Ohba and T. Sone, *Chem. Lett.*, 783 (1996).
60. (a) M. Bell, A. J. Edwards, B. F. Hoskins, E. H. Kachab and R. Robson, *J. Am. Chem. Soc.*, **111**, 3603 (1989).
    (b) M. J. Grannas, B. F. Hoskins and R. Robson, *J. Chem. Soc., Chem. Commun.*, 1644 (1990).
    (c) M. J. Grannas, B. F. Hoskins and R. Robson, *Inorg. Chem.*, **33**, 1071 (1994).
61. (a) H. Kumagai, M. Hasegawa, S. Miyanari, Y. Sugawa, Y. Sato, T. Hori, S. Ueda, H. Kamiyama and S. Miyano, *Tetrahedron Lett.*, **38**, 3971 (1997).
    (b) H. Iki, C. Kabuto, T. Fukushima, H. Kumagai, H. Takeya, S. Miyanari, T. Miyashi and S. Miyano, *Tetrahedron*, **56**, 1437 (2000).
62. N. Iki, N. Morohashi, T. Suzuki, S. Ogawa, M. Aono, C. Kabuto, H. Kumagai, H. Takeya, S. Miyanari and S. Miyano, *Tetrahedron Lett.*, **41**, 2587 (2000).
63. T. Sone, Y. Ohba, K. Moriya, H. Kumada and K. Ito, *Tetrahedron*, **53**, 10689 (1997).
64. N. Kon, N. Iki and S. Miyano, *Tetrahedron Lett.*, **43**, 2231 (2002).
65. Lewis acids have also been used as catalyst:
    (a) O. I. Pieroni, N. M. Rodriguez, B. M. Vuano and M. C. Cabaleiro, *J. Chem. Res.*, 188 (1994).
    (b) A. G. M. Barrett, D. C. Braddock, J. P. Henschke and E. R. Walker, *J. Chem. Soc., Perkin Trans. 1*, 873 (1999).
66. B. A. Roberts, G. W. V. Cave, C. L. Raston and J. L. Scott, *Green Chem.*, **3**, 280 (2001); see also Reference 67(o).
67. (a) P. D. Beer and E. L. Tite, *Tetrahedron Lett.*, **29**, 2349 (1988).
    (b) P. D. Beer, E. L. Tite and A. Ibbotson, *J. Chem. Soc., Chem. Commun.*, 1874 (1989).
    (c) L. M. Tunstad, J. A. Tucker, E. Dalcanale, J. Weiser, J. A. Bryant, J. C. Sherman, R. C. Helgeson, C. B. Knobler and D. J. Cram, *J. Org. Chem.*, **54**, 1305 (1989).
    (d) R. J. M. Egberink, P. L. H. M. Cobben, W. Verboom, S. Harkema and D. N. Reinhoudt, *J. Incl. Phenom.*, **12**, 151 (1992).
    (e) K. Kobayashi, Y. Asakawa, Y. Kato and Y. Aoyama, *J. Am. Chem. Soc.*, **114**, 10307 (1992).
    (f) P. T. Lewis, C. J. Davis, M. C. Saraiva, W. D. Treleaven, T. D. McCarley and R. M. Strongin, *J. Org. Chem.*, **62**, 6110 (1997).
    (g) A. Shivanyuk, V. Böhmer and E. F. Paulus, *Gazz. Chim. Ital.*, **127**, 741 (1997).
    (h) M. Ueda, N. Fukushima, K. Kudo and K. Ichimura, *J. Mater. Chem.*, **7**, 641 (1997).
    (i) A. R. Mezo and J. C. Sherman, *J. Org. Chem.*, **63**, 6824 (1998).
    (j) F. Davis, A. J. Lucke, K. A. Smith and C. J. M. Stirling, *Langmuir*, **14**, 4180 (1998).
    (k) P. Sakhaii, I. Neda, M. Freytag, H. Thönnessen, P. G. Jones and R. Schmutzler, *Z. Anorg. Allg. Chem.*, **626**, 1246 (2000).
    (l) N. Yoshino, A. Satake and Y. Kokube, *Angew. Chem.*, **113**, 471 (2001); *Angew. Chem., Int. Ed.*, **40**, 457 (2001).
    (m) A. J. Wright, S. E. Matthes, W. B. Fischer and P. D. Beer, *Chem. Eur. J.*, **7**, 3474 (2001).
    (n) O. Rusin and V. Kral, *Tetrahedron Lett.*, **42**, 4235 (2001).
    (o) For a recent synthesis, 'using benign procedures effectively embracing the principles of green chemistry', see: G. W. V. Cave, M. J. Hardie, B. A. Roberts and C. L. Raston, *Eur. J. Org. Chem.*, 3227 (2001).
68. G. Cometti, E. Dalcanale, A. Du vosel and A.-M. Levelut, *Liq. Crystals*, **11**, 93 (1992).
69. (a) H. Konishi, K. Ohata, O. Morikawa and K. Kobayashi, *J. Chem. Soc., Chem. Commun.*, 309 (1995).
    (b) H. Konishi, T. Nakamura, K. Ohata, K. Kobayashi and O. Morikawa, *Tetrahedron Lett.*, **37**, 7383 (1996).
70. 2-Butyrylresorcinol reacts with paraformaldehyde in the presence of KOBu-t to give 58% of the cyclic tetramer: H. Konishi and Y. Iwasaki, *Synlett*, 612 (1995).
71. T. Gerkensmeier, J. Mattay and C. Näther, *Chem. Eur. J.*, **7**, 465 (2001).
72. (a) B. J. Shorthill and T. E. Glass, *Org. Lett.*, **3**, 577 (2001).
    (b) For a 3,6-bridged tetramer see: B. J. Shorthill, R. G. Granucci, D. R. Powell and T. E. Glass, *J. Org. Chem.*, **67**, 904 (2002).
73. M. J. McIldowie, M. Mocerino, B. W. Skelton and A. H. White, *Org. Lett.*, **2**, 3869 (2000).
74. G. Rumboldt, V. Böhmer, B. Botta and E. F. Paulus, *J. Org. Chem.*, **63**, 9618 (1998).

75. (a) B. Botta, P. Iacomacci, M. C. Di Giovanni, G. Delle Monache, E. Gacs-Baitz, M. Botta, A. Tafi, F. Corelli and D. Misiti, *J. Org. Chem.*, **57**, 3259 (1992).
    (b) B. Botta, M. C. Di Giovanni, G. Delle Monache, M. C. De Rosa, E. Gacs-Baitz, M. Botta, F. Corelli, A. Tafi, A. Santini, E. Benedetti, C. Pedone and D. Misiti, *J. Org. Chem.*, **59**, 1532 (1994).
    (c) B. Botta, G. Delle Monache, M. C. De Rosa, A. Carbonetti, E. Gacs-Baitz, M. Botta, F. Corelli and D. Misiti, *J. Org. Chem.*, **60**, 3657 (1995).
    (d) B. Botta, G. Delle Monache, P. Salvatore, F. Gasparrini, C. Villani, M. Botta, F. Corelli, A. Tafi, E. Gacs-Baitz, A. Santini, C. F. Carvalho and D. Misiti, *J. Org. Chem.*, **62**, 932 (1997).
    (e) B. Botta, G. Delle Monache, M. C. De Rosa, C. Seri, E. Benedetti, R. Iacovino, M. Botta, F. Corelli, V. Masignani, A. Tafi, E. Gacs-Baitz, A. Santini and D. Misiti, *J. Org. Chem.*, **62**, 1788 (1997).
76. H. Konishi, H. Sakakibara, K. Kobayashi and O. Morikawa, *J. Chem. Soc., Perkin Trans. 1*, 2583 (1999).
77. This cup-shaped conformation was the reason for the name 'calixarene'.
78. A regular (or still time-averaged) *cone* conformation is discussed for calix[5]arenes while calix[8]arenes most probably adopt a 'pleated loop' conformation with a regular 'up and down' arrangement of the phenolic units.
79. A 'winged' or 'pinched' conformation has been suggested:
    (a) M. A. Molins, P. M. Nieto, C. Sanchez, P. Prados, J. de Mendoza and M. Pons, *J. Org. Chem.*, **57**, 6924 (1992).
    (b) W. P. van Hoorn, F. C. J. M. van Veggel and D. N. Reinhoudt, *J. Org. Chem.*, **61**, 7180 (1996).
80. This is in contrast to the hydroxy compound, where *no* intermediate of the *cone*-to-*cone* interconversion has yet been detected experimentally.
81. T. Harada, J. M. Rudzinski and S. Shinkai, *J. Chem. Soc., Perkin Trans. 2*, 2109 (1992).
82. J. Blixt and C. Detellier, *J. Am. Chem. Soc.*, **116**, 11957 (1994).
83. T. Kusano, M. Tabatabai, Y. Okamoto and V. Böhmer, *J. Am. Chem. Soc.*, **121**, 3789 (1999).
84. W. P. van Hoorn, M. G. H. Morshuis, F. C. M. van Veggel and D. N. Reinhoudt, *J. Phys. Chem. A*, **102**, 1130 (1998).
85. I. Thondorf and J. Brenn, unpublished results.
86. (a) L. Abis, E. Dalcanale, A. Du vosel and S. Spera, *J. Org. Chem.*, **53**, 5475 (1988).
    (b) L. Abis, E. Dalcanale, A. Du vosel and S. Spera, *J. Chem. Soc.*, 2075 (1990).
87. S. E. Biali, V. Böhmer, I. Columbus, G. Ferguson, C. Grüttner, F. Grynszpan, E. F. Paulus and I. Thondorf, *J. Chem. Soc., Perkin Trans. 2*, 2777 (1998).
88. B.-Q. Ma, Y. Zhang and P. Coppens, *Cryst. Growth & Design*, **1**, 271 (2001) and **2**, 7 (2002).
89. H. Konishi and O. Morikawa, *J. Chem. Soc., Chem. Commun.*, 34 (1993).
90. R. G. Janssen, W. Verboom, D. N. Reinhoudt, A. Casnati, M. Freriks, A. Pochini, F. Ugozzoli, R. Ungaro, P. M. Nieto, M. Carramolino, F. Cuevas, P. Prados and J. de Mendoza, *Synthesis*, 380 (1993).
91. G. Ferguson, J. F. Gallagher, M. A. McKervey and E. Madigan, *J. Chem. Soc., Perkin Trans. 1*, 599 (1996).
92. P. Neri and S. Pappalardo, *J. Org. Chem.*, **58**, 1048 (1993).
93. See for example: I. Bitter, A. Grun, B. Agai and L. Toke, *Tetrahedron*, **51**, 7835 (1995).
94. F. Arnaud-Neu, M. A. McKervey and M. J. Schwing-Weill, *Metal-Ion Complexation by Narrow Rim Carbonyl Derivatives*, in Reference 1b, p. 385.
95. S. Chowdhury and P. E. Georghiou, *J. Org. Chem.*, **66**, 6257 (2001).
96. X. Zeng, L. Weng and Z.-Z. Zhang, *Chem. Lett.*, 550 (2001).
97. J. M. Harrowfield, M. Mocerino, B. J. Peachey, B. W. Skelton and A. H. White, *J. Chem. Soc., Dalton Trans.*, 1687 (1996).
98. Z. Csok, G. Szalontai, G. Czira and L. Kollar, *Supramol. Chem.*, **10**, 69 (1998).
99. G. M. L. Consoli, F. Cunsolo, M. Piattelli and P. Neri, *J. Org. Chem.*, **61**, 2195 (1996).
100. (a) S. Angot, K. S. Murthy, D. Taton and Y. Gnanou, *Macromolecules*, **31**, 7218 (1998).
     (b) J. Ueda, M. Kamigaito and M. Sawamoto, *Macromolecules*, **31**, 6762 (1998).
101. For a complete survey on *O*-alkyl and *O*-acyl derivatives of calix[8]arenes see: P. Neri, G. M. L. Consoli, F. Cunsolo, C. Geraci and M. Piattelli, *Chemistry of Larger Calix[n]arenes (n = 7,8,9)*, in Reference 1b, p. 89.

102. R. Roy and J. M. Kim, *Angew. Chem.*, **111**, 380 (1999); *Angew. Chem., Int. Ed.*, **38**, 369 (1999).
103. H. Otsuka, K. Araki and S. Shinkai, *Chem. Express*, **8**, 479 (1993).
104. J. Scheerder, M. Fochi, J. F. J. Engbersen and D. N. Reinhoudt, *J. Org. Chem.*, **59**, 3683 (1994).
105. S. Barboso, A. Garcia Carrera, S. E. Matthews, F. Arnaud-Neu, V. Böhmer, J.-F. Dozol, H. Rouquette and M.-J. Schwing-Weill, *J. Chem. Soc., Perkin Trans. 2*, 719 (1999).
106. G. Ferguson, A. Notti, S. Pappalardo, M. F. Parisi and A. L. Spek, *Tetrahedron Lett.*, **39**, 1965 (1998).
107. D. R. Stewart, M. Krawiec, R. P. Kashyap, W. H. Watson and C. D. Gutsche, *J. Am. Chem. Soc.*, **117**, 586 (1995).
108. A. Notti, S. Pappalardo and M. F. Parisi, private communication; see also Reference 187.
109. The cholesterol ester **31** shows a coalescence of the signals of the methylene protons at higher temperatures, suggesting a passage of the *p-t*-butyl groups through the annulus.
110. L. C. Groenen, B. H. M. Ruel, A. Casnati, P. Timmerman, W. Verboom, S. Harkema, A. Pochini, R. Ungaro and D. N. Reinhoudt, *Tetrahedron Lett.*, **32**, 2675 (1991).
111. K. Iwamoto and S. Shinkai, *J. Org. Chem.*, **57**, 7066 (1992).
112. See e.g.: E. Kelderman, L. Derhaeg, G. J. T. Heesink, W. Verboom, J. F. J. Engbersen, N. F. Van Hulst, A. Persoons and D. N. Reinhoudt, *Angew. Chem.*, **104**, 1075 (1992); *Angew. Chem., Int. Ed. Engl.*, **31**, 1107 (1992).
113. C. Floriani, D. Jacoby, A. Chiesi-Villa and C. Guastini, *Angew. Chem.*, **101**, 1430 (1989); *Angew. Chem., Int. Ed. Engl.*, **28**, 1376 (1989).
114. L. Baklouti, R. Abidi, Z. Asfari, J. Harrowfield, R. Rokbani and J. Vicens, *Tetrahedron Lett.*, **39**, 5363 (1998).
115. For the stereochemical outcome of the reaction of **2a** with ethyl bromoacetate under different reaction conditions see:
    (a) K. Iwamoto, K. Fujimoto, T. Matsuda and S. Shinkai, *Tetrahedron Lett.*, **31**, 7169 (1990) and corrigendum, **32**, 830 (1991).
    (b) Reference 111.
116. C. D. Gutsche and P. A. Reddy, *J. Org. Chem.*, **32**, 4783 (1991).
117. K. Iwamoto, K. Araki and S. Shinkai, *J. Chem. Soc., Chem. Commun.*, 1611 (1991).
118. For bis(crown ethers) in the *1,2-alternate* conformation see:
    (a) A. Arduini, L. Domiano, A. Pochini, A. Secchi, R. Ungaro, F. Ugozzoli, O. Struck, W. Verboom and D. N. Reinhoudt, *Tetrahedron*, **53**, 3767 (1997).
    (b) G. Ferguson, A. J. Lough, A. Notti, S. Pappalardo, M. F. Parisi and A. Petringa, *J. Org. Chem.*, **63**, 9703 (1998).
119. For all possible isomeric tetrapropyl ethers of **2a** or *p*-nitrocalix[4]arene see: P. J. A. Kenis, O. F. J. Noordman, S. Houbrechts, G. J. van Hummel, S. Harkema, F. C. J. M. van Veggel, K. Clays, J. F. J. Engbersen, A. Peersoons, N. F. van Hulst and D. N. Reinhoudt, *J. Am. Chem. Soc.*, **120**, 7875 (1998) and references cited there.
120. For recent examples see:
    (a) C. B. Dieleman, D. Matt and P. G. Jones, *J. Organomet. Chem.*, **545–546**, 461 (1997).
    (b) G. Ulrich, R. Ziessel, I. Manet, M. Guardigli, N. Sabbatini, F. Fraternali and G. Wipff, *Chem. Eur. J.*, **3**, 1815 (1997).
    (c) S. E. Matthews, M. Saadioui, V. Böhmer, S. Barboso, F. Arnaud-Neu, M.-J. Schwing-Weill, A. G. Carrera and J.-F. Dozol, *J. Prakt. Chem.*, **341**, 264 (1999).
121. See for instance:
    (a) A. Casnati, A. Pochini, R. Ungaro, F. Ugozzoli, F. Arnaud, S. Fanni, M.-J. Schwing, R. J. M. Egbrink, F. de Jong and D. N. Reinhoudt, *J. Am. Chem. Soc.*, **117**, 2767 (1995).
    (b) G. Arena, A. Casnati, A. Contino, L. Mirone, D. Sciotto and R. Ungaro, *Chem. Commun.*, 2277 (1991).
    (c) A. Casnati, C. Fischer, M. Guardigli, A. Isernia, I. Manet, N. Sabbatini and R. Ungaro, *J. Chem. Soc., Perkin Trans. 2*, 395 (1991).
122. (a) C.-M. Shu, W.-S. Chung, S.-H. Wu, Z.-C. Ho and L.-G. Lin, *J. Org. Chem.*, **64**, 2673 (1999).
    (b) G. Ferguson, J. F. Gallagher, L. Giunta, P. Neri, S. Pappalardo and M. Parisi, *J. Org. Chem.*, **59**, 42 (1994).
    (c) P. Lhotak and S. Shinkai, *Tetrahedron*, **51**, 7681 (1995).

(d) R. K. Castellano and J. Rebek, Jr., *J. Am. Chem. Soc.*, **120**, 3657 (1998).
123. M. Pitarch, J. K. Browne and M. A. McKervey, *Tetrahedron*, **53**, 10503 (1997) and **53**, 16195 (1997).
124. A. Ikeda, T. Tsudera and S. Shinkai, *J. Org. Chem.*, **62**, 3568 (1997).
125. (a) J.-A. Pérez-Adelmar, H. Abraham, C. Sánchez, K. Rissanen, P. Prados and J. de Mendoza, *Angew. Chem.*, **108**, 1088 (1996); *Angew. Chem., Int. Ed. Engl.*, **35**, 1009 (1996).
(b) For an earlier reference see: W. Verboom, S. Datta, Z. Asfari, S. Harkema and D. N. Reinhoudt, *J. Org. Chem.*, **57**, 5394 (1992).
126. V. S. Talanov and R. A. Bartsch, *J. Chem. Soc., Perkin Trans. 1*, 1957 (1999).
127. S. Pappalardo, S. Caccamese and L. Giunta, *Tetrahedron Lett.*, **32**, 7747 (1991).
128. For an early conformational study see: M. Iqbal, T. Mangiafico and C. D. Gutsche, *Tetrahedron*, **43**, 4917 (1987).
129. See for instance:
(a) S. K. Sharma and C. D. Gutsche, *Synthesis*, 813 (1994).
(b) K. No and H. J. Koo, *Bull. Korean Chem. Soc.*, **15**, 483 (1994).
130. K. No, Y. J. Park, K. H. Kim and J. M. Shin, *Bull. Korean Chem. Soc.*, **17**, 447 (1996).
131. For all possible isomeric tetra-acetates see: S. Akabori, H. Sannohe, Y. Habata, Y. Mukoyama and T. Ishi, *Chem. Commun.*, 1467 (1996).
132. C. D. Gutsche, J. A. Levine and P. K. Sujeeth, *J. Org. Chem.*, **50**, 5802 (1985).
133. D. Xie and C. D. Gutsche, *J. Org. Chem.*, **62**, 2280 (1997).
134. S. K. Sharma and C. D. Gutsche, *J. Org. Chem.*, **64**, 3507 (1999).
135. S. Chowdhury, J. N. Bridson and P. E. Georghiou, *J. Org. Chem.*, **65**, 3299 (2000).
136. A. Casnati, A. Pochini, R. Ungaro, R. Cacciapaglia and L. Mandolini, *J. Chem. Soc., Perkin Trans. 1*, 2052 (1991).
137. L. C. Groenen, B. H. M. Ruel, A. Casnati, W. Verboom, A. Pochini, R. Ungaro and D. N. Reinhoudt, *Tetrahedron*, **47**, 8379 (1991).
138. S. Pappalardo and G. Ferguson, *J. Org. Chem.*, **61**, 2407 (1996).
139. M. Takeshita, T. Suzuki and S. Shinkai, *J. Chem. Soc., Chem. Commun.*, 2587 (1994).
140. F. Santoyo-González, A. Torres-Pinedo and A. Sanchéz-Ortega, *J. Org. Chem.*, **65**, 4409 (2000).
141. A. Casnati, A. Arduini, E. Ghidini, A. Pochini and R. Ungaro, *Tetrahedron*, **47**, 2221 (1991).
142. C. D. Gutsche and L. G. Lin, *Tetrahedron*, **42**, 1633 (1986).
143. E. M. Georgiev, J. T. Mague and D. M. Roundhill, *Supramol. Chem.*, **2**, 53 (1993).
144. K. C. Nam, J. M. Kim and D. S. Kim, *Bull. Korean Chem. Soc.*, **16**, 186 (1995).
145. (a) S. Shang, D. V. Khasnis, J. M. Burton, C. J. Santini, M. Fan, A. C. Small and M. Lattman, *Organometallics*, **13**, 5157 (1994).
(b) The analogous compound in which two adjacent oxygens are bridged by $Si(CH_3)_2$ could not be *O*-alkylated: M. Fan, H. Zhang and M. Lattman, *Organometallics*, **15**, 5216 (1996).
146. K. B. Ray, R. H. Weatherhead, N. Pirinccioglu and A. Williams, *J. Chem. Soc., Perkin Trans. 2*, 83 (1994).
147. K. A. See, F. R. Fronczek, W. H. Watson, R. P. Kashyap and C. D. Gutsche, *J. Org. Chem.*, **56**, 7256 (1991).
148. K. C. Nam, S. W. Ko and J. M. Kim, *Bull. Korean Chem. Soc.*, **19**, 345 (1998).
149. It has been suggested to distinguish the single phenolic units by letters A, B, C, ... instead of 1, 2, 3, ... since numbers are also used for the single carbon atoms of the metacyclophane skeleton.
150. The same is true for calix[5]arenes where, however, two regioisomers for trifunctionalized compounds exist: 1,2,3 or A,B,C and 1,2,4 or A,B,D.
151. For the first examples see: J. D. van Loon, A. Arduini, L. Coppi, W. Verboom, A. Pochini, R. Ungaro, S. Harkema and D. N. Reinhoudt, *J. Org. Chem.*, **55**, 5639 (1990).
152. (a) C. D. Gutsche and K. A. See, *J. Org. Chem.*, **57**, 4527 (1992).
(b) C. Shu, W. Liu, M. Ku, F. Tang, M. Yeh and L. Lin, *J. Org. Chem.*, **59**, 3730 (1994).
(c) For a dicarbonate see: P. D. Beer, M. G. B. Drew, M. Hesek, M. Shade and F. Szemes, *Chem. Commun.*, 2161 (1996).
153. G. Ferguson, J. F. Gallagher, A. J. Lough, A. Notti, S. Pappalardo and M. F. Parisi, *J. Org. Chem.*, **64**, 5876 (1999).
154. K. C. Nam, J. M. Kim, S. K. Kook and S. J. Lee, *Bull. Korean Chem. Soc.*, **17**, 499 (1996).
155. K. C. Nam, J. M. Kim and Y. J. Park, *Bull. Korean Chem. Soc.*, **19**, 770 (1998).

156. E.g. L. C. Groenen, J. D. Van Loon, W. Verboom, S. Harkema, A. Casnati, R. Ungaro, A. Pochini, F. Ugozzoli and D. N. Reinhoudt, *J. Am. Chem. Soc.*, **113**, 2385 (1991).
157. (a) F. Bottino, L. Giunta and S. Pappalardo, *J. Org. Chem.*, **54**, 5407 (1989).
    (b) For a chiral example derived from a calix[4]arene of the ABAB type ($R^1 = R^3$, $R^2 = R^4$) see: V. Böhmer, D. Kraft and M. Tabatabai, *J. Incl. Phenom.*, **19**, 17 (1994).
158. A. Arduini, A. Casnati, L. Dodi, A. Pochini and R. Ungaro, *J. Chem. Soc., Chem. Commun.*, 1597 (1990).
159. For a rational synthesis of 1,2-diethers as *syn* or *anti* isomers using benzyl ether groups as protective groups see: K. Iwamoto, H. Shimizu, K. Araki and S. Shinkai, *J. Am. Chem. Soc.*, **115**, 3997 (1993).
160. F. Narumi, N. Morohashi, N. Matsumura, N. Iki, H. Kameyama and S. Miyano, *Tetrahedron Lett.*, **43**, 621 (2002).
161. L. N. Markovsky, M. O. Vysotsky, V. V. Pirozhenko, V. I. Kalchenko, J. Lipkowski and Y. A. Simonov, *Chem. Commun.*, 69 (1996).
162. M. O. Vysotsky, M. O. Tairov, V. V. Pirozhenko and V. I. Kalchenko, *Tetrahedron Lett.*, **39**, 6057 (1998).
163. See also the formation of the 1,2-derivative by *O*-methylation of the mono(3,5-dinitrobenzoate) of $2_H a$: Y. J. Park, J. M. Shin, K. C. Nam, J. M. Kim and S.-K. Kook, *Bull. Korean Chem. Soc.*, **17**, 643 (1996).
164. J. M. Kim and K. C. Nam, *Bull. Korean Chem. Soc.*, **18**, 1327 (1997).
165. It was mentioned already earlier by Gutsche that the 1,2-bis(3,5-dinitrobenzoate) of **2a** can be obtained from the 1,3-isomer in the presence of imidazole; see Reference 147.
166. O. Lukin, M. O. Vysotsky and V. I. Kalchenko, *J. Phys. Org. Chem.*, **14**, 468 (2001).
167. (a) K. Iwamoto, K. Araki and S. Shinkai, *J. Org. Chem.*, **56**, 4955 (1991).
    (b) W. S. Oh, T. D. Chung, J. Kim, H.-S. Kim, H. Kim, D. Hwang, K. Kim, S. G. Rha, J.-I. Choe and S.-K. Chang, *Supramol. Chem.*, **9**, 221 (1998).
    (c) X. Chen, M. Ji, D. R. Fisher and C. M. Wai, *Synlett*, 1784 (1999).
168. Monobenzoylation of 1,3-diethers under similar conditions leads also to the *anti–syn* isomers; see Reference 122a.
169. S. Berthalon, J.-B. Regnouf-de-Vains and R. Lamartine, *Synth. Commun.*, **26**, 3103 (1996).
170. (a) Z.-C. Ho, M.-C. Ku, C.-M. Shu and L.-G. Lin, *Tetrahedron*, **52**, 13189 (1996).
    (b) An *anti–syn* orientation is also claimed for several trisulphonates; see Reference 98.
171. R. Cacciapaglia, A. Casnati, L. Mandolini, S. Schiavone and R. Ungaro, *J. Chem. Soc., Perkin Trans. 2*, 369 (1993).
172. (a) T. Jin and K. Monde, *Chem. Commun.*, 1357 (1998).
    (b) For a chiral triether of the ABBH type see: K. Iwamoto, A. Yanagi, T. Arimura, T. Matsuda and S. Shinkai, *Chem. Lett.*, 1901 (1990).
    (c) For chiral 1,2-crown ethers see: F. Arnaud-Neu, S. Caccamese, S. Fuangswasdi, S. Pappalardo, M. F. Parisi, A. Petringa and G. Principato, *J. Org. Chem.*, **62**, 8041 (1997).
173. L. Motta-Viola, J.-B. Regnouf-de-Vains and M. Perrin, *Tetrahedron Lett.*, **41**, 7023 (2000).
174. H. Otsuka, K. Araki and S. Shinkai, *J. Org. Chem.*, **59**, 1542 (1994).
175. (a) O. Sénèque, M.-N. Rager, M. Giorgi and O. Reinaud, *J. Am. Chem. Soc.*, **122**, 6183 (2000).
    (b) For an analogous tris(α-picolyl) ether derivative see: Y. Rondelez, M.-N. Rager, A. Duprat and O. Reinaud, *J. Am. Chem. Soc.*, **124**, 1334 (2002).
176. J. Scheerder, J. F. G. Engbersen, A. Casnati, R. Ungaro and D. N. Reinhoudt, *J. Org. Chem.*, **60**, 6448 (1995).
177. J. de Mendoza, M. Carramolino, F. Cuevas, P. M. Nieto, P. Prados, D. N. Reinhoudt, W. Verboom, R. Ungaro and A. Casnati, *Synthesis*, 47 (1994).
178. (a) The tetramethyl ether has been exhaustively alkylated or acylated: G. M. L. Consoli, F. Cunsolo and P. Neri, *Gazz. Chim. Ital.*, **126**, 791 (1996).
    (b) F. Cunsolo, G. M. L. Consoli, M. Piattelli and P. Neri, *Tetrahedron Lett.*, **36**, 3751 (1995).
    (c) F. Cunsolo, G. M. L. Consoli, M. Piattelli and P. Neri, *J. Org. Chem.*, **63**, 6852 (1998).
179. M. Martino, L. Gregoli, C. Gaeta and P. Neri, *Org. Lett.*, **4**, 1531 (2002).
180. C. Alfieri, E. Dradi, A. Pochini, R. Ungaro and G. D. Andreetti, *J. Chem. Soc., Chem. Commun.*, 1075 (1983).
181. (a) C. Bocchi, M. Careri, A. Casnati and G. Mori, *Anal. Chem.*, **67**, 4234 (1995).

(b) F. Arnaud-Neu, G. Ferguson, S. Fuangswasdi, A. Notti, S. Pappalardo, M. F. Parisi and A. Petringa, *J. Org. Chem.*, **63**, 7770 (1998).
(c) For recent examples with pendant dithiophene units see: M. Giannetto, G. Mori, A. Notti, S. Pappalardo and M. F. Parisi, *Chem. Eur. J.*, **7**, 3354 (2001).

182. In these series ligands with high selectivity are found.
(a) For potassium (crown-5) see: A. Casnati, A. Pochini, R. Ungaro, C. Bocchi, F. Ugozzoli, R. J. M. Egberink, H. Struijk, R. Lugtenberg, F. de Jong and D. N. Reinhoudt, *Chem. Eur. J.*, **2**, 436 (1996).
(b) For caesium (crown-6) see: A. Casnati, A. Pochini, R. Ungaro, F. Ugozzoli, F. Arnaud, S. Fanni, M.-J. Schwing, R. J. M. Egberink, F. de Jong and D. N. Reinhoudt, *J. Am. Chem. Soc.*, **117**, 2767 (1995).

183. T. Kim, H. Ihm and K. Paek, *Bull. Korean Chem. Soc.*, **18**, 681 (1997).
184. Y. Kubo, S. Maeda, S. Tokita and M. Kubo, *Nature*, **382**, 522 (1996).
185. L. Baldini, C. Bracchini, R. Cacciapaglia, A. Casnati, L. Mandolini and R. Ungaro, *Chem. Eur. J.*, **6**, 1322 (2000).
186. (a) A. Arduini, A. Casnati, M. Fabbi, P. Minari, A. Pochini, A. R. Sicuri and R. Ungaro, *Supramol. Chem.*, **1**, 235 (1993).
(b) H. Yamamoto, T. Sakaki and S. Shinkai, *Chem. Lett.*, 469 (1994).
(c) The first examples were obtained from the 1,2-dimethyl ether; see Reference 158.
187. F. Arnaud-Neu, M.-J. Schwing-Weill and J.-F. Dozol, *Calixarenes for Nuclear Waste Treatment*, in Reference 1b, p. 642.
188. A. Arduini, W. M. McGregor, D. Paganuzzi, A. Pochini, A. Secchi, F. Ugozzoli and R. Ungaro, *J. Chem. Soc., Perkin Trans. 2*, 839 (1996).
189. W. I. Iwema Bakker, M. Haas, C. Khoo-Beattie, R. Ostaszewski, S. M. Franken, H. J. den Hertog, Jr., W. Verboom, D. de Zeeuw, S. Harkema and D. N. Reinhoudt, *J. Am. Chem. Soc.*, **116**, 123 (1994) and references cited there.
190. For a review on calixarene phosphates see: J. Gloede, *Phosphorous, Sulfur, Silicon*, **127**, 97 (1997).
191. (a) B. Xu and T. M. Swager, *J. Am. Chem. Soc.*, **115**, 1159 (1993).
(b) T. M. Swager and B. Xu, *J. Incl. Phenom.*, **19**, 389 (1994).
192. For further similar metalla derivatives and their chemistry see: C. Floriani and R. Floriani-Moro, *Metal Reactivity on Oxo Surfaces Modeled by Calix[4]arenes*, in Reference 1b, p. 536.
193. B. Xu, P. J. Carroll and T. M. Swager, *Angew. Chem.*, **108**, 2238 (1996); *Angew. Chem., Int. Ed. Engl.*, **35**, 2094 (1996).
194. A. Notti, M. F. Parisi and S. Pappalardo, *Selectively Modified Calix[5]arenes*, in Reference 1b, p. 54.
195. (a) D. Kraft, R. Arnecke, V. Böhmer and W. Vogt, *Tetrahedron*, **49**, 6019 (1993).
(b) F. Arnaud-Neu, R. Arnecke, V. Böhmer, S. Fanni, J. L. M. Gordon, M.-J. Schwing-Weill and W. Vogt, *J. Chem. Soc., Perkin Trans. 2*, 1855 (1996).
(c) S. Pappalardo and M. F. Parisi, *J. Org. Chem.*, **61**, 8724 (1996).
(d) R. Arnecke, V. Böhmer, G. Ferguson and S. Pappalardo, *Tetrahedron Lett.* **37**, 1497 (1996).
(e) S. Caccamese, A. Notti, S. Pappalardo, M. F. Parisi and G. Principato, *J. Incl. Phenom.*, **36**, 67 (2000).
196. U. Lüning, F. Löffler and J. Eggert, *Selective Modifications of Calix[6]arenes*, in Reference 1b, p. 71.
197. (a) J. Li, Y. Chen and X. Lu, *Tetrahedron*, **55**, 10365 (1999).
(b) A. Casnati, P. Jacopozzi, A. Pochini, F. Ugozzoli, R. Cacciapaglia, L. Mandolini and R. Ungaro, *Tetrahedron*, **51**, 591 (1995).
(c) Y. Chen, F. Yang and X. Lu, *Tetrahedron Lett.*, **41**, 1571 (2000).
198. Y. Chen and F. Yang, *Chem. Lett.*, 484 (2000).
199. Y.-K. Chen and Y.-Y. Chen, *Org. Lett.*, **2**, 743 (2000).
200. (a) S. Kanamathareddy and C. D. Gutsche, *J. Org. Chem.*, **57**, 3160 (1992).
(b) S. Kanamathareddy and C. D. Gutsche, *J. Am. Chem. Soc.*, **115**, 6572 (1993). Aliphatic diacid dichlorides have been successfully used for 2,5-bridging of 1,4-di-p-tolyl ethers.
201. (a) T. Saiki, K. Goto, N. Tokitoh and R. Okazaki, *J. Org. Chem.*, **61**, 2924 (1996).
(b) K. Goto and R. Okazaki, *Liebigs Ann./Recueil*, 2393 (1997).

(c) T. Saiki, K. Goto and R. Okazaki, *Angew. Chem.*, **109**, 2320 (1997); *Angew. Chem., Int. Ed. Engl.*, **36**, 2223 (1997).
202. H. Ross and U. Lüning, *Liebigs Ann.*, 1367 (1996).
203. F. Löffler, M. Hagen and U. Lüning, *Synlett*, 1826 (1999).
204. M. T. Blanda, D. B. Farmer, J. S. Brodbelt and B. J. Goolsby, *J. Am. Chem. Soc.*, **122**, 1486 (2000).
205. For doubly capped phosphates and similar compounds see Reference 190.
206. Y. Zhang, H. Yuan, Z. Huang, J. Zhou, Y. Kawanashi, J. Schatz and G. Maas, *Tetrahedron*, **57**, 4161 (2001).
207. (a) H. Otsuka, K. Araki, H. Matsumoto, T. Harada and S. Shinkai, *J. Org. Chem.*, **60**, 4862 (1995).
(b) H. Otsuka, Y. Suzuki, A. Ikeda, K. Araki and S. Shinkai, *Tetrahedron*, **54**, 423 (1998).
208. (a) Y. Chen and Y. Chen, *Tetrahedron Lett.*, **41**, 9079 (2000).
(b) K. C. Nam, S. W. Ko, S. O. Kang, S. H. Lee and D. S. Kim, *J. Incl. Phenom.*, **40**, 285 (2001).
209. (a) C. Geraci, M. Piattelli and P. Neri, *Tetrahedron Lett.*, **37**, 3899 (1996).
(b) C. Geraci, M. Piattelli, G. Chessari and P. Neri, *J. Org. Chem.*, **65**, 5143 (2000).
(c) C. Geraci, M. Piattelli and P. Neri, unpublished results.
210. (a) A. Ikeda, K. Akao, T. Harada and S. Shinkai, *Tetrahedron Lett.*, **37**, 1621 (1996).
(b) F. Cunsolo, M. Piattelli and P. Neri, *J. Chem. Soc., Chem. Commun.*, 1917 (1994).
211. Y. S. Tsantrizos, W. Chew, L. D. Colebrook and F. Sauriol, *Tetrahedron Lett.*, **38**, 5411 (1997).
212. (a) C. Geraci, M. Piattelli and P. Neri, *Tetrahedron Lett.*, **36**, 5429 (1995).
(b) C. Geraci, G. Chessari, M. Piattelli and P. Neri, *Chem. Commun.*, 921 (1997).
(c) C. Geraci, G. L. M. Consoli, M. Piattelli and P. Neri, submitted.
(d) C. Geraci, M. Piattelli and P. Neri, *Tetrahedron Lett.*, **37**, 7627 (1996).
213. C. Geraci, A. Bottino, M. Piattelli, E. Gavuzzo and P. Neri, *J. Chem. Soc., Perkin Trans. 2*, 185 (2000).
214. J. Gloede, S. Ozegowski, D. Matt and A. De Cian, *Tetrahedron Lett.*, **42**, 9139 (2001); analogous reactions were carried out with the smaller calixarenes.
215. For a review of hydroxy group replacement in calixarenes see: S. E. Biali, *Isr. J. Chem.*, **37**, 131 (1997).
216. (a) Z. Goren and S. E. Biali, *J. Chem. Soc., Perkin Trans. 1*, 1484 (1990).
(b) J.-B. Regnouf-de-Vains, S. Pellet-Rostaing and R. Lamartine, *Tetrahedron Lett.*, **35**, 8147 (1994).
217. See for instance: K. Araki, H. Murakami, F. Ohseto and S. Shinkai, *Chem. Lett.*, 539 (1992).
218. F. Ohseto, H. Murakami, K. Araki and S. Shinkai, *Tetrahedron Lett.*, **33**, 1217 (1992).
219. (a) C. G. Gibbs and C. D. Gutsche, *J. Am. Chem. Soc.*, **115**, 5338 (1993).
(b) X. Delaigue, J. M. Harrowfield, M. W. Hosseini, A. De Cian, J. Fischer and N. Kyritsakas, *J. Chem. Soc., Chem. Commun.*, 1579 (1994).
220. (a) Y. Ting, W. Verboom, L. G. Groenen, J.-D. van Loon and D. N. Reinhoudt, *J. Chem. Soc., Chem. Commun.*, 1432 (1990).
(b) X. Delaigue, M. W. Hosseini, A. De Cian, N. Kyritsakas and J. Fischer, *J. Chem. Soc., Chem. Commun.*, 609 (1995).
(c) C. G. Gibbs, P. K. Sujeeth, J. S. Rogers, G. G. Stanley, M. Krawiec, W. H. Watson and C. D. Gutsche, *J. Org. Chem.*, **60**, 8394 (1995).
221. Perhaps this is also due to the fact that consequently less effort has been invested in the other phenols.
222. I. Thondorf, A. Shivanyuk and V. Böhmer, *Chemical Modification of Calix[4]arenes and Resorcarenes*, in Reference 1b, p. 26. See also Reference 101, pp. 194, 196.
223. Y. Morita, T. Agawa, E. Nomura and H. Taniguchi, *J. Org. Chem.*, **57**, 3658 (1992).
224. For further examples of tetraazo calix[4]arenes see:
(a) S. Shinkai, K. Araki, J. Shibata, D. Tsugawa and O. Manabe, *J. Chem. Soc., Perkin Trans. 1*, 3333 (1990).
(b) E. Kelderman, L. Derhaeg, W. Verboom, J. F. J. Engbersen, S. Harkema, A. Persoons and D. N. Reinhoudt, *Supramol. Chem.*, **2**, 183 (1993).
(c) Examples for calix[8]arenes: B. Yao, J. Bassus and R. Lamartine, *An. Quim., Int. Ed.*, **94**, 65 (1998).

(d) S. Bouoit-Montésinos, F. Vocanson, J. Bassus and R. Lamartine, *Synth. Commun.*, **30**, 911 (2000).

(e) S. Bouoit, J. Bassus and R. Lamartine, *An. Quim., Int. Ed.*, **94**, 342 (1998).

225. P.-S. Wang, R.-S. Lin and H.-X. Zong, *Synth. Commun.*, **29**, 2225 (1999).
226. J. L. Atwood and S. G. Bott, in: *Calixarenes—A Versatile Class of Macrocyclic Compounds* (Eds. J. Vicens and V. Böhmer), Kluwer, Dordrecht, 1990, pp. 199–210.
227. See also S. Kumar, H. M. Chawla and R. Varadarajan, *Tetrahedron Lett.*, **43**, 2495 (2002).
228. (a) M. Mascal, R. T. Naven and R. Warmuth, *Tetrahedron Lett.*, **36**, 9361 (1995).

(b) M. Mascal, R. Warmuth, R. T. Naven, R. A. Edwards, M. B. Hursthouse and D. E. Hibbs, *J. Chem. Soc., Perkin Trans. 1*, 3435 (1999).
229. O. Mogck, P. Parzuchowski, M. Nissinen, V. Böhmer, G. Rokicki and K. Rissanen, *Tetrahedron*, **54**, 10053 (1998).
230. A. Bottino, F. Cunsolo, M. Piattelli, D. Garozzo and P. Neri, *J. Org. Chem.*, **64**, 8018 (1999).
231. (a) M. Takeshita, S. Nishio and S. Shinkai, *J. Org. Chem.*, **59**, 4032 (1994).

(b) S. Kanamathareddy and C. D. Gutsche, *J. Org. Chem.*, **61**, 2511 (1996).
232. (a) A. Casnati, M. Fochi, P. Minari, A. Pochini, M. Reggiani and R. Ungaro, *Gazz. Chim. Ital.*, **126**, 99 (1996).

(b) S. Shimizu, S. Shirakawa, Y. Sasaki and C. Hirai, *Angew. Chem.*, **112**, 1313 (2000); *Angew. Chem., Int. Ed.*, **39**, 1256 (2000).
233. (a) B. Klenke and W. Friedrichsen, *J. Chem. Soc., Perkin Trans. 1*, 3377 (1998) and references cited therein.

(b) A. Gunji and K. Takahashi, *Synth. Commun.*, **28**, 3933 (1998).
234. (a) P. D. Beer and M. Shade, *Chem. Commun.*, 2377 (1997).

(b) K. C. Nam and D. S. Kim, *Bull. Korean Chem. Soc.*, **15**, 284 (1994).

(c) P. D. Beer and J. B. Cooper, *Chem. Commun.*, 129 (1998).

(d) S. K. Sharma and C. D. Gutsche, *J. Org. Chem.*, **61**, 2564 (1996).

(e) For a first report see: J.-D. van Loon, A. Arduini, W. Verboom, R. Ungaro, G. J. van Hummel, S. Harkema and D. N. Reinhoudt, *Tetrahedron Lett.*, **30**, 2681 (1989).
235. A. Arduini, M. Fabbi, M. Mantovani, L. Mirone, A. Pochini, A. Secchi and R. Ungaro, *J. Org. Chem.*, **60**, 1454 (1995).
236. Z.-T. Huang, G.-Q. Wang, L.-M. Yang and Y.-X. Lou, *Synth. Commun.*, **25**, 1109 (1995).
237. As a recent example see the reaction with 7-methoxy-1,3,5-cycloheptatriene: M. Orda-Zgadzaj, V. Wendel, M. Fehlinger, B. Ziemer and W. Abraham, *Eur. J. Org. Chem.*, 1549 (2001).
238. See for instance: N. J. van der Veen, R. J. M. Egberink, J. F. J. Engbersen, F. J. C. M. van Veggel and D. N. Reinhoudt, *Chem. Commun.*, 681 (1999).
239. O. Mogck, V. Böhmer, G. Ferguson and W. Vogt, *J. Chem. Soc., Perkin Trans. 1*, 1711 (1996).
240. (a) E.g. azo coupling of calix[5]arene: M. Yanase, T. Haino and Y. Fukazawa, *Tetrahedron Lett.*, **40**, 2781 (1999).

(b) See also References 233a, 234e and 448b.
241. See for instance: F. Pinkhassik, V. Sidorov and I. Stibor, *J. Org. Chem.*, **63**, 9644 (1998).
242. W. Verboom, P. J. Bodewes, G. van Essen, P. Timmerman, G. J. van Hummel, S. Harkema and D. N. Reinhoudt, *Tetrahedron*, **51**, 499 (1995).
243. For an early summary of the 'chloromethylation route', the 'Claisen rearrangement route' and the 'quinonemethide route' see C. D. Gutsche, in *Calixarenes: A Versatile Class of Macrocyclic Compounds* (Eds. J. Vicens and V. Böhmer), Kluwer, Dordrecht, 1990, p. 3.
244. M. Almi, A. Arduini, A. Casnati, A. Pochini and R. Ungaro, *Tetrahedron*, **45**, 2177 (1989).
245. D. J. E. Spencer, B. J. Johnson, B. J. Johnson and W. B. Tolman, *Org. Lett.*, **4**, 1391 (2002).
246. (a) A. Ikeda and S. Shinkai, *J. Chem. Soc., Perkin Trans. 1*, 2671 (1993).

(b) A. Ikeda, M. Yoshimura, P. Lhotak and S. Shinkai, *J. Chem. Soc., Perkin Trans. 1*, 1945 (1996).
247. K. Araki, K. Akao, A. Ikeda, T. Suzuki and S. Shinkai, *Tetrahedron Lett.*, **37**, 73 (1996).
248. J.-D. van Loon, L. D. Groenen, S. S. Wijmenga, W. Verboom and D. N. Reinhoudt, *J. Am. Chem. Soc.*, **113**, 2378 (1991).
249. (a) P. Molenveld, J. F. J. Engbersen and D. N. Reinhoudt, *Eur. J. Org. Chem.*, 3269 (1999).

(b) P. Molenveld, W. M. G. Stikvoort, H. Kooijman, A. L. Spek, J. F. J. Engbersen and D. N. Reinhoudt, *J. Org. Chem.*, **64**, 3896 (1999).

250. For the attachment of C_{60} via 1,3-dipolar addition of an azomethine ylide derived from monoformyl calix[4 or 5]arene see: J. Wang and C. D. Gutsche, *J. Org. Chem.*, **65**, 6273 (2000).
251. (a) M. Larsen, F. C. Krebs, M. Jorgensen and N. Harrit, *J. Org. Chem.*, **63**, 4420 (1998).
    (b) J.-B. Regnouf-de-Vains and R. Lamartine, *Tetrahedron Lett.*, **37**, 6311 (1996).
252. P. Lhotak, R. Nakamura and S. Shinkai, *Supramol. Chem.*, **8**, 333 (1997).
253. O. Struck, J. P. M. van Duynhoven, W. Verboom, S. Harkema and D. N. Reinhoudt, *Chem. Commun.*, 1517 (1996).
254. (a) R. H. Vreekamp, W. Verboom and D. N. Reinhoudt, *J. Org. Chem.*, **61**, 4282 (1996).
    (b) O. Struck, W. Verboom, W. J. J. Smeets, A. L. Spek and D. N. Reinhoudt, *J. Chem. Soc., Perkin Trans. 2*, 223 (1997).
255. A. Dondoni, A. Marra, M.-C. Scherrmann, A. Casnati, F. Sansone and R. Ungaro, *Chem. Eur. J.*, **3**, 1774 (1997).
256. (a) Y. Hamuro, M. C. Calama, H. S. Park and A. D. Hamilton, *Angew. Chem.*, **109**, 2797 (1997); *Angew. Chem., Int. Ed. Engl.*, **36**, 2680 (1997).
    (b) H. S. Park, Q. Lin and A. D. Hamilton, *J. Am. Chem. Soc.*, **121**, 8 (1999).
257. R. A. Jakobi, V. Böhmer, C. Grüttner, D. Kraft and W. Vogt, *New. J. Chem.*, **20**, 493 (1996); see also Reference 260.
258. (a) J. D. Van Loon, J. F. Heida, W. Verboom and D. N. Reinhoudt, *Recl. Trav. Chim. Pays-Bas*, **111**, 353 (1992).
    (b) S. K. Sharma and C. D. Gutsche, *J. Org. Chem.*, **64**, 998 (1999).
259. D. M. Rudkevich, W. Verboom and D. N. Reinhoudt, *J. Org. Chem.*, **59**, 3683 (1994).
260. For a recent example see: R. E. Brewster and S. B. Shuker, *J. Am. Chem. Soc.*, **124**, 7902 (2001).
261. For formamides and their conversion to isocyanide groups see: J. Gagnon, M. Drouin and P. D. Harvey, *Inorg. Chem.*, **40**, 6052 (2001).
262. (a) M. Saadioui, A. Shivanyuk, V. Böhmer and W. Vogt, *J. Org. Chem.*, **64**, 3774 (1999).
    (b) L. J. Prins, K. A. Jolliffe, R. Hulst, P. Timmerman and D. N. Reinhoudt, *J. Am. Chem. Soc.*, **122**, 3617 (2000).
263. (a) M. S. Brody, C. A. Schalley, D. M. Rudkevich and J. Rebek, Jr., *Angew. Chem.*, **111**, 1738 (1999); *Angew. Chem., Int. Ed.*, **38**, 1640 (1999).
    (b) P. Timmerman, H. Boerrigter, W. Verboom and D. N. Reinhoudt, *Recl. Trav. Chim. Pays-Bas*, **114**, 103 (1995).
    (c) K. D. Shimizu and J. Rebek, Jr., *Proc. Natl. Acad. Sci. USA*, **92**, 12403 (1995).
    (d) A. M. A. van Wageningen, P. Timmerman, J. P. M. van Duynhoven, W. Verboom, F. C. J. M. van Veggel and D. N. Reinhoudt, *Chem. Eur. J.*, **3**, 639 (1997).
264. M. Larsen, F. C. Krebs, N. Harrit and M. Jorgensen, *J. Chem. Soc., Perkin Trans. 2*, 1749 (1999).
265. (a) H. J. Cho, J. Y. Kim and S. K. Chang, *Chem. Lett.*, 493 (1999).
    (b) J. Y. Kim, Y. H. Kim, J.-I. Choe and S. K. Chang, *Bull. Korean Chem. Soc.*, **22**, 635 (2001).
266. J. O. Magrans, A. R. Ortiz, M. A. Molins, P. H. P. Lebouille, J. Sanchez-Quesada, P. Prados, M. Pons, F. Gago and J. de Mendoza, *Angew. Chem.*, **108**, 1816 (1996); *Angew. Chem., Int. Ed.*, **35**, 1712 (1996).
267. (a) H. Ihm and K. Paek, *Bull. Korean Chem. Soc.*, **16**, 71 (1995).
    (b) M. Larsen and M. Jorgensen, *J. Org. Chem.*, **61**, 6651 (1996).
268. K.-S. Paek, H.-J. Kim and S.-K. Chang, *Supramol. Chem.*, **5**, 83 (1995).
269. H. Ihm, H. Kim and K. Paek, *J. Chem. Soc., Perkin Trans. 1*, 1997 (1997).
270. (a) T. Haino, T. Harano, K. Matsumura and Y. Fukazawa, *Tetrahedron Lett.*, **36**, 5793 (1995).
    (b) T. Haino, K. Nitta, Y. Saijo, K. Matzumura, M. Hirakata and Y. Fukazawa, *Tetrahedron Lett.*, **40**, 6301 (1999).
271. N. Kuhnert and A. Le-Gresley, *J. Chem. Soc., Perkin Trans. 1*, 3393 (2001).
272. T. Gu, P. Ceroni, G. Marceroni, N. Armaroli and J.-F. Nierengarten, *J. Org. Chem.*, **66**, 6432 (2001).
273. (a) K. C. Nam and T. H. Yoon, *Bull. Korean Chem. Soc.*, **14**, 169 (1993).
    (b) C.-M. Shu, W.-L. Lin, G. H. Lee, S.-M. Peng and W.-S. Chung, *J. Chin. Chem. Soc.*, **47**, 173 (2000).
274. K. C. Nam and K. S. Park, *Bull. Korean Chem. Soc.*, **16**, 153 (1995).

275. J. Wang and C. D. Gutsche, *J. Am. Chem. Soc.*, **120**, 12226 (1998).
276. For a recent study see: H. M. Chawla and Meena, *Indian J. Chem., Sect. B: Org. Chem. Incl. Med. Chem.*, **37B**, 28 (1998).
277. K. No and J. E. Kim, *Bull. Korean Chem. Soc.*, **16**, 1122 (1995).
278. For the synthesis of calixarenes with one or two CHR bridges by condensation see Sections II.C and II.D.
279. (a) A. Ninagawa, K. Cho and H. Matsuda, *Makromol. Chem.*, **186**, 1379 (1985).
    (b) G. Görmar, K. Seiffarth, M. Schultz, J. Zimmermann and G. Fläming, *Makromol. Chem.*, **191**, 81 (1990).
280. B. Klenke, C. Näther and W. Friedrichsen, *Tetrahedron Lett.*, **39**, 8967 (1998).
281. P. A. Scully, T. M. Hamilton and J. L. Bennet, *Org. Lett.*, **3**, 2741 (2001).
282. O. Middel, Z. Greff, N. J. Taylor, W. Verboom, D. N. Reinhoudt and V. Snieckus, *J. Org. Chem.*, **65**, 667 (2000).
283. K. Agbaria and S. E. Biali, *J. Am. Chem. Soc.*, **123**, 12495 (2001).
284. For a review see: S. E. Biali, *Oxidation and Reduction of Aromatic Rings*, in Reference 1b, p. 266.
285. P. A. Reddy, R. P. Kashyap, W. H. Watson and C. D. Gutsche, *Isr. J. Chem.*, **32**, 89 (1992).
286. (a) C.-F. Chen, Q.-Y. Zheng and Z.-T. Huang, *Synthesis*, 69 (1999).
    (b) P. D. Beer, P. A. Gale, Z. Chen, M. G. B. Drew, J. A. Heath, M. I. Ogden and H. R. Powell, *Inorg. Chem.*, **36**, 5880 (1997).
287. A. Casnati, L. Domiano, A. Pochini, R. Ungaro, M. Carramolino, J. O. Magrans, P. M. Nieto, J. Lopez-Prados, P. Prados, J. de Mendoza, R. G. Janssen, W. Verboom and D. N. Reinhoudt, *Tetrahedron*, **51**, 12699 (1995).
288. Baeyer–Villiger oxidation of *p*-acetyl or *p*-formyl derivatives is an alternative; see Reference 228 and A. Arduini, L. Mirone, D. Paganuzzi, A. Pinalli, A. Pochini, A. Secchi and R. Ungaro, *Tetrahedron*, **52**, 6011 (1996).
289. (a) T. Nakayama and M. Ueda, *J. Mater. Chem.*, **9**, 697 (1999).
    (b) P. C. Leverd, V. Huc, S. Palacin and M. Nierlich, *J. Incl. Phenom.*, **36**, 259 (2000).
290. P. A. Reddy and C. D. Gutsche, *J. Org. Chem.*, **58**, 3245 (1993).
291. (a) A. M. Litwak and S. E. Biali, *J. Org. Chem.*, **57**, 1943 (1992).
    (b) A. M. Litwak, F. Grynszpan, O. Aleksiuk, S. Cohen and S. E. Biali, *J. Org. Chem.*, **58**, 393 (1993).
292. W.-G. Wang, W.-C. Zhang and Z.-T. Huang, *J. Chem. Res. (S)*, 462 (1998).
293. F. Grynszpan and S. E. Biali, *J. Org. Chem.*, **61**, 9512 (1996).
294. F. Grynszpan and S. E. Biali, *J. Chem. Soc., Chem. Commun.*, 2545 (1994).
295. K. Agbaria, O. Aleksiuk, S. E. Biali, V. Böhmer, M. Frings and I. Thondorf, *J. Org. Chem.*, **66**, 2891 (2001).
296. (a) O. Aleksiuk, F. Grynszpan and S. E. Biali, *J. Chem. Soc., Chem. Commun.*, 11 (1993).
    (b) F. Grynszpan, O. Aleksiuk and S. E. Biali, *J. Org. Chem.*, **59**, 2070 (1994).
    (c) O. Aleksiuk, S. Cohen and S. E. Biali, *J. Am. Chem. Soc.*, **117**, 9645 (1995).
297. O. Aleksiuk, F. Grynszpan and S. E. Biali, *J. Org. Chem.*, **58**, 1994 (1993).
298. J. M. Van Gelder, J. Brenn, I. Thondorf and S. E. Biali, *J. Org. Chem.*, **62**, 3511 (1997).
299. As an additional reaction of the aromatic system the ring expansion with dichlorocarbene may be mentioned: T. Hatsui, H. Ushijima, A. Mori and H. Takeshita, *Tetrahedron Lett.*, **42**, 6855 (2001).
300. F. Grynszpan and S. E. Biali, *Chem. Commun.*, 195 (1996).
301. I. Columbus and S. E. Biali, *J. Am. Chem. Soc.*, **120**, 3060 (1998).
302. I. Columbus, M. Haj-Zaroubi and S. E. Biali, *J. Am. Chem. Soc.*, **120**, 11806 (1998).
303. I. Columbus, M. Haj-Zaroubi, J. S. Siegel and S. E. Biali, *J. Org. Chem.*, **63**, 9148 (1998).
304. A. Bilyk, J. M. Harrowfield, B. W. Skelton and A. H. White, *An. Quim.*, **93**, 363 (1997).
305. A. Bilyk, J. M. Harrowfield, B. W. Skelton and A. H. White, *J. Chem. Soc., Dalton Trans.*, 4251 (1997).
306. H. Iki, T. Kikuchi and S. Shinkai, *J. Chem. Soc., Perkin Trans. 1*, 205 (1993).
307. (a) J. W. Steed, R. K. Juneja, R. S. Burkhalter and J. L. Atwood, *J. Chem. Soc., Chem. Commun.*, 2205 (1994).
    (b) J. W. Steed, R. K. Juneja and J. L. Atwood, *Angew. Chem.*, **106**, 2571 (1994); *Angew. Chem., Int. Ed. Engl.*, **33**, 2456 (1994).

308. M. Staffilani, K. S. B. Hancock, J. W. Steed, K. T. Holman, J. L. Atwood, R. K. Juneja and R. S. Burkhalter, *J. Am. Chem. Soc.*, **119**, 6324 (1997).
309. Y. Ishii, K. Onaka, H. Hirakawa and K. Shiramizu, *Chem. Commun.*, 1150 (2002).
310. M. W. Hosseini, *Thia-, Mercapto-, and Thiamercapto-Calix[4]arenes*, in Reference 1b, p. 110.
311. P. Rao, M. W. Hosseini, A. De Cian and J. Fischer, *Chem. Commun.*, 2169 (1999).
312. For recent examples see: R. Lamartine, C. Bavoux, F. Vocanson, A. Martin, G. Senlis and M. Perrin, *Tetrahedron Lett.*, **42**, 1021 (2001).
313. For bis(crown ethers) in the 1,3-*alternate* conformation see: V. Lamare, J.-F. Dozol, P. Thuéry, M. Nierlich, Z. Asfari and J. Vicens, *J. Chem. Soc., Perkin Trans. 2*, 1920 (2001).
314. (a) N. Iki, F. Narumi, T. Fujimoto, N. Morohashi and S. Miyano, *J. Chem. Soc., Perkin Trans. 2*, 2745 (1998).
    (b) H. Akdas, G. Mislin, E. Graf, M. W. Hosseini, A. De Cian and J. Fischer, *Tetrahedron Lett.*, **40**, 2113 (1999).
315. P. Lhoták, M. Himl, S. Pakhomova and I. Stibor, *Tetrahedron Lett.*, **39**, 8915 (1998).
316. J. Lang, J. Vlach, H. Dvorakova, P. Lhoták, M. Himl, R. Hrabal and I. Stibor *J. Chem. Soc., Perkin Trans. 2*, 576 (2001).
317. P. Lhoták, M. Dudic, I. Stibor, H. Petrickova, J. Sykora and J. Hodacova, *Chem. Commun.*, 731 (2001).
318. (a) This oxidation may be the reason why attempts of nitration or *ipso*-nitration were not successful; see however: C. Desroches, S. Parola, F. Vocanson, M. Perrin, R. Lamartine, J.-M. Létoffé and J. Bouix, *New J. Chem.*, **26**, 651 (2002).
    (b) *p*-Amino derivatives were obtained also via the coupling with diazonium salts and subsequent reduction: P. Lhoták, J. Moravek and I. Stibor, *Tetrahedron Lett.*, **43**, 3665 (2002).
319. N. Morohashi, N. Iki, A. Sugawara and S. Miyano, *Tetrahedron*, **57**, 5557 (2001).
320. Compare the diastereomeric resorcarenes. Unfortunately, a different description is often used (e.g. Reference 319), in which the relative configuration is not related to a reference group but to the preceding S−O group while going around the macrocycle. The following relation exists: $rccc \equiv ccc$, $rcct \equiv cct$, $rctt \equiv ctc$, $rtct \equiv ttt$.
321. N. Morohashi, N. Iki, C. Kabuto and S. Miyano, *Tetrahedron Lett.*, **41**, 2933 (2000).
322. G. Mislin, E. Graf, M. W. Hosseini and A. De Cian, *Tetrahedron Lett.*, **40**, 1129 (1999).
323. P. Lhoták, *Tetrahedron*, **57**, 4775 (2001).
324. H. Katagiri, N. Iki, T. Hattori, C. Kabuto and S. Miyano, *J. Am. Chem. Soc.*, **123**, 779 (2001).
325. For an early example see: A. G. S. Högberg, *J. Am. Chem. Soc.*, **102**, 6046 (1980) and *J. Org. Chem.*, **45**, 4498 (1980).
326. V. I. Kalchenko, D. M. Rudkevich, A. N. Shivanyuk, V. V. Pirozhenko, I. F. Tsymbal and L. N. Markovsky, *Zh. Obshch. Khim.*, **64**, 731 (1994); *Engl. Transl.: Russ. J. Gen. Chem.*, **64**, 663 (1994).
327. W. Hu, J. P. Rourke, J. J. Vital and R. J. Puddephatt, *Inorg. Chem.*, **34**, 323 (1995).
328. V. I. Kalchenko, A. V. Solov'yov, N. R. Gladun, A. N. Shivanyuk, L. I. Atamas', V. V. Pirozhenko, L. N. Markowsky, J. Lipkowski and Y. A. Simonov, *Supramol. Chem.*, **8**, 269 (1997).
329. (a) I. Neda, T. Siedentop, A. Vollbrecht, H. Thönnessen, P. G. Jones and R. Schmutzler, *Z. Naturforsch., B, Chem. Sci.*, **53**, 841 (1998).
    (b) A. Vollbrecht, I. Neda and R. Schmutzler, *Phosphorus, Sulfur, Silicon*, **107**, 173 (1995).
330. (a) M. Urbaniak and W. Iwanek, *Tetrahedron*, **55**, 14459 (1999).
    (b) G. Mann, L. Hennig, F. Weinelt, K. Müller, R. Meusinger, G. Zahn and T. Lippmann, *Supramol. Chem.*, **3**, 101 (1994).
    (c) S. Pellet-Rostaing, J.-B. Regnouf-de-Vains and R. Lamartine, *Tetrahedron Lett.*, **36**, 5745 (1995).
331. See for example: G. Cometti, E. Dalcanale, A. Du vosel and A.-M. Levelut, *J. Chem. Soc., Chem. Commun.*, 163 (1990).
332. O. Haba, K. Haga, M. Ueda, O. Morikawa and H. Konishi, *Chem. Mat.*, **11**, 427 (1999).
333. Y. Yamakawa, M. Ueda, R. Nagahata, T. Takeuchi and M. Asai, *J. Chem. Soc., Perkin Trans. 1*, 4135 (1998).
334. J. R. Fransen and P. J. Dutton, *Can. J. Chem.*, **73**, 2217 (1995).
335. T. Fujimoto, C. Shimizu. O. Hayashida and Y. Aoyama, *J. Am. Chem. Soc.*, **119**, 6676 (1997).
336. W. Iwanek, *Tetrahedron Asymm.*, **9**, 3171 (1998).

337. A partial transformation of functional groups should be called *selective* only, if the yield is (significantly) higher than the statistically expected yield.
338. L. N. Markovsky, V. I. Kalchenko, D. M. Rudkevich and A. N. Shivanyuk, *Mendeleev Commun.*, 106 (1992) and *Phosphorus, Sulfur, Silicon*, **75**, 59 (1993).
339. The reaction of the octamethyl silyl ether of resorcarene **2** with four equivalents of PF_2Cl afforded chiral C_4-symmetric tetrakis-difluorophosphites; compare Reference 329b.
340. J. Lipkowski, O. I. Kalchenko, J. Slowikowska, V. I. Kalchenko, O. V. Lukin, L. N. Markovsky and R. Nowakowsky, *J. Phys. Org. Chem.*, **11**, 426 (1998).
341. O. Lukin, A. Shivanyuk, V. V. Pirozhenko, I. F. Tsymbal and V. I. Kalchenko, *J. Org. Chem.*, **63**, 9510 (1998).
342. A. Shivanyuk, E. F. Paulus, V. Böhmer and W. Vogt, *J. Org. Chem.*, **63**, 6448 (1998).
343. See for instance:
(a) A. Shivanyuk, E. F. Paulus and V. Böhmer, *Angew. Chem.*, **111**, 3091 (1999); *Angew. Chem., Int. Ed.*, **38**, 2906 (1999).
(b) A. Shivanyuk, E. F. Paulus, K. Rissanen, E. Kolehmainen and V. Böhmer, *Chem. Eur. J.*, **7**, 1944 (2001).
(c) A. Shivanyuk, *Chem. Commun.*, 1472 (2001).
344. A. N. Shivanyuk, V. I. Kalchenko, V. V. Pirozhenko and L. N. Markovsky, *Zh. Obshch. Khim.*, **64**, 1558 (1994); *Engl. Transl.: Russ. J. Gen. Chem.*, **64**, 1401 (1994).
345. H. Konishi, T. Tamura, H. Ohkubo, K. Kobayashi and O. Morikawa, *Chem. Lett.*, 685 (1996).
346. C. Agena, C. Wolff and J. Mattay, *Eur. J. Org. Chem.*, 2977 (2001).
347. (a) O. Manabe, K. Asakura, T. Nishi and S. Shinkai, *Chem. Lett.*, 1219 (1990).
(b) O. Omar, A. K. Ray, F. Davis and A. K. Hassan, *Supramol. Sci.*, **4**, 417 (1997).
348. D. J. Cram, S. Karbach, H. Kim, C. B. Knobler, E. F. Maverick, J. L. Ericson and R. C. Helgeson, *J. Am. Chem. Soc.*, **110**, 2229 (1998).
349. H. Konishi, H. Nakamaru, H. Nakatani, T. Ueyama, K. Kobayashi and O. Morikawa, *Chem. Lett.*, 185 (1997).
350. (a) O. Morikawa, K. Nakanishi, M. Miyashiro, K. Kobayashi and H. Konishi, *Synthesis*, 233 (2000).
(b) For a complete tetrathiomethylation of **II** see: H. Konishi, H. Yamaguchi, M. Miyashiro, K. Kobayashi and O. Morikawa, *Tetrahedron Lett.*, **37**, 8547 (1996).
351. U. Schneider and H.-J. Schneider, *Chem. Ber.*, **127**, 2455 (1994).
352. D. A. Leigh, P. Linnane, R. G. Pitchard and G. Jackson, *J. Chem. Soc., Chem. Commun.*, 389 (1994).
353. M. Luostarinen, A. Shivanyuk and K. Rissanen, *Org. Lett.*, **3**, 4141 (2001).
354. P. Linnane and S. Shinkai, *Tetrahedron Lett.*, **36**, 3865 (1995).
355. E. K. Kazakova, N. A. Makarova, A. U. Ziganshina, L. A. Muslinkina, A. A. Muslinkin and W. D. Habicher, *Tetrahedron Lett.*, **41**, 10111 (2000).
356. (a) Y. Matsushita and T. Matsui, *Tetrahedron Lett.*, **34**, 7433 (1993).
(b) R. Arnecke, V. Böhmer, E. F. Paulus and W. Vogt, *J. Am. Chem. Soc.*, **117**, 3286 (1995).
357. A C_4-symmetric derivative was obtained also during the attempt to synthesize a cavitand-derived tetra-acetic acid: H.-J. Choi, D. Buhring, M. L. C. Quan, C. B. Knobler and D. J. Cram, *J. Chem. Soc., Chem. Commun.*, 1733 (1992).
358. K. Airola, V. Böhmer, E. F. Paulus, K. Rissanen, C. Schmidt, I. Thondorf and W. Vogt, *Tetrahedron*, **53**, 10709 (1997).
359. (a) W. Iwanek and J. Mattay, *Liebigs Ann.*, 1463 (1995).
(b) M. T. El Gihani, H. Heaney and A. M. Z. Slawin, *Tetrahedron Lett.*, **36**, 4905 (1995).
(c) R. Arnecke, V. Böhmer, S. Friebe, S. Gebauer, G. J. Kraus and I. Thondorf, *Tetrahedron Lett.*, **36**, 6221 (1995).
360. C. Schmidt, E. F. Paulus, V. Böhmer and W. Vogt, *New J. Chem.*, **25**, 374 (2001).
361. The *O*-acetylation was also reported (see References 359a,c) but could not be reproduced. Cleavage of the benzoxazines and *N*-acetylation occurred instead: C. Schmidt, E. F. Paulus, V. Böhmer and W. Vogt, *New J. Chem.*, **24**, 123 (2000).
362. P. C. Bulman Page, H. Heaney and E. P. Sampler, *J. Am. Chem. Soc.*, **121**, 6751 (1999).
363. C. Schmidt, K. Airola, V. Böhmer, W. Vogt and K. Rissanen, *Tetrahedron*, **53**, 17691 (1997).
364. C. Schmidt, I. Thondorf, E. Kolehmainen, V. Böhmer, W. Vogt and K. Rissanen, *Tetrahedron Lett.*, **39**, 8833 (1998).
365. (a) W. Iwanek, C. Wolf and J. Mattay, *Tetrahedron Lett.*, **36**, 8969 (1995).

(b) C. Schmidt, T. Straub, D. Falabu, E. F. Paulus, E. Wegelius, E. Kolehmainen, V. Böhmer, K. Rissanen and W. Vogt, *Eur. J. Org. Chem.*, 3937 (2000).
366. A. Shivanyuk, C. Schmidt, V. Böhmer, E. F. Paulus, O. V. Lukin and W. Vogt, *J. Am. Chem. Soc.*, **120**, 4319 (1998).
367. G. Arnott, P. C. Bulman Page, H. Heaney, R. Hunter and E. P. Sampler, *Synlett*, 412 (2001).
368. P. T. Lewis and R. M. Strongin, *J. Org. Chem.*, **63**, 6065 (1998).
369. E. U. Thoden van Velzen, J. F. J. Engbersen, P. J. de Lange, J. W. G Mahy and D. N. Reinhoudt, *J. Am. Chem. Soc.*, **117**, 6853 (1995).
370. F. Davis and C. J. M. Stirling, *Langmuir*, **12**, 5365 (1996).
371. A. Lucke, C. J. M. Stirling and V. Böhmer, *Mono- and Multilayers*, in Reference 1b, p. 612.
372. T. Haino, D. M. Rudkevich, A. Shivanyuk, K. Rissanen and J. Rebek, Jr., *Chem. Eur. J.*, **6**, 3797 (2000).
373. M. Fujimaki, S. Kawahara, Y. Matsuzawa, E. Kurita, Y. Hayashi and K. Ichimura, *Langmuir*, **14**, 4495 (1998).
374. S. Shoichi, D. M. Rudkevich and J. Rebek, Jr., *Org. Lett.*, **1**, 1241 (1999).
375. For its conversion to the tetrabromide with PPh_3 and CBr_4 see: B. Botta, G. Delle Monache, P. Ricciardi, G. Zappia, C. Seri, E. Gacs-Baitz, P. Csokasi and D. Misiti, *Eur. J. Org. Chem.*, 841 (2000).
376. D. J. Cram and J. M. Cram, in *Container Molecules and Their Guests*, Monographs in Supramolecular Chemistry (Ed. J. F. Stoddart), Chap. 5, Royal Society of Chemistry, Cambridge, 1994.
377. For an improved procedure using higher temperature in sealed tubes see: E. Román, C. Peinador, S. Mendoza and A. E. Kaifer, *J. Org. Chem.*, **64**, 2577 (1999).
378. O. Middel, W. Verboom, R. Hulst, H. Kooijman, A. L. Spek and D. N. Reinhoudt, *J. Org. Chem.*, **63**, 8259 (1998).
379. For early examples of cavitands derived from *rcct*- and *rctt*-resorcarenes see Reference 86b.
380. (a) T. N. Sorrell and F. C. Pigge, *J. Org. Chem.*, **58**, 784 (1993).
(b) H. Boerrigter, W. Verboom and D. N. Reinhoudt, *J. Org. Chem.*, **62**, 7148 (1997).
381. For very recent examples see: K. Paek, J. Yoon and Y. Suh, *J. Chem. Soc., Perkin Trans. 2*, 916 (2001).
382. (a) H. Boerrigter, W. Verboom, G. J. van Hummel, S. Harkema and D. N. Reinhoudt, *Tetrahedron Lett.*, **37**, 5167 (1996).
(b) H. Boerrigter, W. Verboom and D. N. Reinhoudt, *Liebigs Ann./Recueil*, 2247 (1997).
383. I. Higler, H. Boerrigter, W. Verboom, H. Kooijman, A. L. Spek and D. N. Reinhoudt, *Eur. J. Org. Chem.*, 1597 (1998).
384. P. Jacopozzi and E. Dalcanale, *Angew. Chem.*, **109**, 665 (1997); *Angew. Chem., Int. Ed. Engl.*, **36**, 613 (1997).
385. D. J. Cram and J. M. Cram, in *Container Molecules and Their Guests*, Monographs in Supramolecular Chemistry (Ed.: J. F. Stoddart), Chap. 7, Royal Society of Chemistry, Cambridge, 1994.
386. Mono-, di- and trihydroxy derivatives may be obtained also by incomplete conversion: T. A. Robbins and D. J. Cram, *J. Chem. Soc., Chem. Commun.*, 1515 (1995).
387. S. Ma, D. M. Rudkevich and J. Rebek, Jr., *J. Am. Chem. Soc.*, **120**, 4977 (1998).
388. (a) H. Xi, C. L. D. Gibb, E. D. Stevens and B. C. Gibb, *Chem. Commun.*, 1743 (1998).
(b) J. O. Green, J.-H. Baird and B. C. Gibb, *Org. Lett.*, **2**, 3845 (2000).
389. (a) H. Xi, C. L. D. Gibb and B. C. Gibb, *J. Org. Chem.*, **64**, 9286 (1999).
(b) For cavitands obtained by bridging with 4-(α,α'-dibromomethyl)pyridine and their self-assembly to coordination cages see: L. Pirondini, F. Bertolini, B. Cantadori, F. Ugozzoli, C. Massera and E. Dalcanale, *Proc. Natl. Acad. Sci. USA*, **99**, 4911 (2002).
390. (a) C. L. D. Gibb, E. D. Stevens and B. C. Gibb, *J. Am. Chem. Soc.*, **123**, 5849 (2001).
(b) C. L. D. Gibb and B. C. Gibb, *Proc. Natl. Acad. Sci. USA*, **99**, 4857 (2002).
391. W. Xu, J. J. Vittal and R. J. Puddephatt, *J. Am. Chem. Soc.*, **117**, 8362 (1995).
392. T. Lippmann, H. Wilde, E. Dalcanale, L. Mavilla, G. Mann, U. Heyer and S. Spera, *J. Org. Chem.*, **60**, 235 (1995).
393. P. Jacopozzi, E. Dalcanale, S. Spera, L. A. J. Chrisstoffels, D. N. Reinhoudt, T. Lippmann and G. Mann, *J. Chem. Soc., Perkin Trans. 2*, 671 (1998).
394. P. Delangle and J.-P. Dutasta, *Tetrahedron Lett.*, **36**, 9325 (1995).

395. A. Vollbrecht, I. Neda, H. Thönnessen, P. G. Jones, R. K. Harris, L. A. Crowe and R. Schmutzler, *Chem. Ber./Recueil*, **130**, 1715 (1997).
396. A. I. Konovalov, V. S. Reznik, M. A. Pudovik, E. K. Kazakova, A. R. Burilov, I. L. Nikolaeva, N. A. Makarova, G. R. Davletschina, L. V. Ermolaeva, R. D. Galimov and A. R. Mustafina, *Phosphorus, Sulfur, Silicon*, **123**, 277 (1997).
397. (a) J. R. Moran, J. L. Ericson, E. Dalcanale, J. A. Bryant, C. B. Knobler and D. J. Cram, *J. Am. Chem. Soc.*, **113**, 5707 (1991).
    (b) P. Soncini, S. Bonsignore, E. Dalcanale and F. Ugozzoli, *J. Org. Chem.*, **57**, 4608 (1992).
398. F. C. Tucci, D. M. Rudkevich and J. Rebek, Jr., *J. Org. Chem.*, **64**, 4555 (1999).
399. Cavitands **89** flutter between a C_{2v}-symmetric kite conformation (favoured at low temperature) and a C_{4v}-symmetric vase conformation (favoured at higher temperatures) in which the walls are upright oriented; see Reference 397a.
400. (a) D. M. Rudkevich, G. Hilmersson and J. Rebek, Jr., *J. Am. Chem. Soc.*, **120**, 12216 (1998).
    (b) A. Shivanyuk, K. Rissanen, S. K. Korner, D. M. Rudkevich and J. Rebek, Jr., *Helv. Chim. Acta*, **83**, 1778 (2000).
    (c) For a recent example containing a Zn-phenanthroline fragment as additional binding site see: U. Lücking, J. Chen, D. M. Rudkevich and J. Rebek, Jr., *J. Am. Chem. Soc.*, **123**, 9929 (2001).
401. T. Heinz, D. M. Rudkevich and J. Rebek, Jr., *Nature*, **394**, 764 (1998).
402. C. Naumann, E. Román, C. Peinador, T. Ren, B. O. Patrick, A. E. Kaifer and J. C. Sherman, *Chem. Eur. J.*, **7**, 1637 (2001).
403. (a) B. C. Gibb, A. R. Mezo and J. C. Sherman, *Tetrahedron Lett.*, **36**, 7587 (1995).
    (b) B. C. Gibb, A. R. Mezo, A. S. Causton, J. R. Fraser, F. C. S. Tsai and J. C. Sherman, *Tetrahedron*, **51**, 8719 (1995).
404. A. R. Mezo and J. C. Sherman, *J. Am. Chem. Soc.*, **121**, 8983 (1999).
405. O. Hayashida, K. Nishiyama, Y. Matsuda and Y. Aoyama, *Chem. Lett.*, 13 (1998).
406. M. Saadioui and V. Böhmer, *Double and Multi-Calixarenes*, in Reference 1b, p. 130.
407. Examples for **A1**:
    (a) M. P. Oude Wolbers, F. C. J. M. van Veggel, R. H. J. W. Hofstraat, F. A. J. Geurts, J. van Hummel, S. Harkema and D. N. Reinhoudt, *Liebigs Ann./Recueil*, 2587 (1997).
    (b) P. Lhoták, M. Kawaguchi, A. Ikeda and S. Shinkai, *Tetrahedron*, **52**, 12399 (1996).
    (c) see also Reference 122d
408. Examples for **A2**:
    (a) H. Ross and U. Lüning, *Angew. Chem.*, **107**, 2723 (1995); *Angew. Chem., Int. Ed. Engl.*, **34**, 2555 (1995).
    (b) F. Ohseto and S. Shinkai, *J. Chem. Soc., Perkin Trans. 2*, 1103 (1995).
    (c) G. Ulrich, P. Turek and R. Ziessel, *Tetrahedron Lett.*, **37**, 8755 (1996).
    (d) A. Dondoni, X. Hu, A. Marra and H. D. Banks, *Tetrahedron Lett.*, **42**, 3295 (2001).
409. Examples for **B1**:
    (a) A. Arduini, A. Pochini and A. Secchi, *Eur. J. Org. Chem.*, 2325 (2000).
    (b) V. Kovalev, E. Shokova and Y. Luzikov, *Synthesis*, 1003 (1998).
    (c) See also References 229, 234a,c.
410. Examples for **B2**:
    (a) A. Arduini, S. Fanni, G. Manfredi, A. Pochini, R. Ungaro, A. R. Sicuri and F. Ugozzoli, *J. Org. Chem.*, **60**, 1448 (1995).
    (b) O. Struck, L. A. J. Christoffels, R. J. W. Lugtenberg, W. Verboom, G. J. van Hummel, S. Harkema and D. N. Reinhoudt, *J. Org. Chem.*, **62**, 2487 (1997).
    (c) G. T. Hwang and B. H. Kim, *Tetrahedron. Lett.*, **41**, 10055 (2000).
411. P. D. Beer, P. A. Gale and D. Hesek, *Tetrahedron. Lett.*, **36**, 767 (1995).
412. (a) W. Wasikiewicz, G. Rokicki, J. Kielkiewicz and V. Böhmer, *Angew. Chem.*, **106**, 230 (1994); *Angew. Chem., Int. Ed. Engl.*, **33**, 214 (1994).
    (b) W. Wasikiewicz, G. Rokicki, J. Kielkiewicz, E. F. Paulus and V. Böhmer, *Monatsh. Chem.*, **128**, 863 (1997).
413. P. R. A. Webber, G. Z. Chen, M. G. B. Drew and P. D. Beer, *Angew. Chem.*, **113**, 2325 (2001); *Angew. Chem., Int. Ed.*, **40**, 2265 (2001).
414. (a) P. Neri, A. Bottino, F. Cunsolo, M. Piattelli and E. Gavuzzo, *Angew. Chem.*, **110**, 175 (1998); *Angew. Chem., Int. Ed.*, **37**, 166 (1998).
    (b) J. Wang, S. G. Bodige, W. H. Watson and C. D. Gutsche, *J. Org. Chem.*, **65**, 8260 (2000).

415. A. Arduini, R. Ferdani, A. Pochini and A. Secchi, *Tetrahedron*, **56**, 8573 (2000).
416. (a) P. Schmitt, P. D. Beer, M. G. B. Drew and P. D. Sheen, *Angew. Chem.*, **109**, 1926 (1997); *Angew. Chem., Int. Ed. Engl.*, **36**, 1840 (1997).
 (b) S. E. Matthews, P. Schmitt, V. Felix, M. G. B. Drew and P. D. Beer, *J. Am. Chem. Soc.*, **124**, 1341 (2002).
417. For calixtubes derived from thiacalix[4]arenes see: S. E. Matthews, V. Felix, M. G. B. Drew and P. D. Beer, *New J. Chem.*, **25**, 1355 (2001).
418. For a similar combination of calixarenes with different ring size see: M. Makha, P. J. Nichols, M. J. Hardie and C. L. Raston, *J. Chem. Soc., Perkin Trans. 1*, 354 (2002).
419. (a) K. A. Araki, K. Sisido, K. Hisaichi and S. Shinkai, *Tetrahedron Lett.*, **34**, 8297 (1993).
 (b) M. T. Blanda and K. E. Griswold, *J. Org. Chem.*, **59**, 4313 (1994).
420. (a) J.-S. Li, Y.-Y. Chen and X.-R. Lu, *Eur. J. Org. Chem.*, 485 (2000).
 (b) J.-S. Li, Z. Zhong, Y.-Y. Chen and X.-R. Lu,. *Tetrahedron Lett.*, **39**, 6507 (1998).
421. (a) X. Delaigue, M. W. Hosseini, E. Leize, S. Kieffer and A. van Dorsselaer, *Tetrahedron Lett.*, **34**, 7561 (1993).
 (b) X. Delaigue, M. W. Hosseini, R. Graff, J.-P. Kintzinger and J. Raya, *Tetrahedron Lett.*, **35**, 1711 (1994).
 (c) J. Martz, E. Graf, M. W. Hosseini, A. De Cian and J. Fischer, *J. Mater Chem.*, **8**, 2331 (1998).
 (d) M. W. Hosseini and A. De Cian, *Chem. Commun.*, 727 (1998).
422. (a) F. Hajek, E. Graf and M. W. Hosseini, *Tetrahedron Lett.*, **37**, 1401 (1996).
 (b) F. Hajek, E. Graf, M. W. Hosseini, A. De Cian and J. Fischer, *Angew. Chem.*, **109**, 1830 (1997); *Angew. Chem., Int. Ed. Engl.*, **36**, 1760 (1997).
423. Examples for **F**:
 (a) Z. Asfari, R. Abidi, F. Arnaud and J. Vicens, *J. Incl. Phenom.*, **13**, 163 (1992).
 (b) J. S. Kim, O. J. Shon, J. W. Ko, M. H. Cho, I. Y. Yu and J. Vicens, *J. Org. Chem.*, **65**, 2386 (2000).
424. Examples for **G**:
 (a) S. Kanamathareddy and C. D. Gutsche, *J. Org. Chem.*, **60**, 6070 (1995).
 (b) A. Ikeda and S. Shinkai, *J. Chem. Soc., Chem. Commun.*, 2375 (1994).
 (c) See also Reference 125a.
425. (a) C. von dem Busche-Hünnefeld, D. Bühring, C. B. Knobler and D. J. Cram, *J. Chem. Soc., Chem. Commun.*, 1085 (1995).
 (b) J. Yoon and D. J. Cram, *Chem. Commun.*, 497 (1997).
 (c) E. L. Piatnitsky, R. A. Flowers II and K. Deshayes, *Chem. Eur. J.*, **6**, 999 (2000).
426. (a) B. S. Park, C. B. Knobler, C. N. Eid, Jr., R. Warmuth and D. J. Cram, *Chem. Commun.*, 55 (1998).
 (b) J. Yoon and D. J. Cram, *J. Am. Chem. Soc.*, **119**, 11796 (1997).
427. K. Paek, H. Ihm, S. Yun and H. C. Lee, *Tetrahedron Lett.*, **40**, 8905 (1999).
428. A [5]carceplex derived from two [5]cavitands (see Reference 402) linked via five disulphide bridges was recently described: C. Naumann, S. Place and J. C. Sherman, *J. Am. Chem. Soc.*, **124**, 16 (2002).
429. S. K. Kurdistani, R. C. Helgeson and D. J. Cram, *J. Am. Chem. Soc.*, **117**, 1659 (1995).
430. (a) J. Yoon, C. Sheu, K. N. Houk, C. B. Knobler and D. J. Cram, *J. Org. Chem.*, **61**, 9323 (1996).
 (b) J. Yoon, C. B. Knobler, E. F. Maverick and D. J. Cram, *Chem. Commun.*, 1303 (1997).
431. (a) R. G. Chapman, N. Chopra, E. D. Cochien and J. C. Sherman, *J. Am. Chem. Soc.*, **116**, 369 (1994).
 (b) G. Chapman and J. C. Sherman, *J. Org. Chem.*, **63**, 4103 (1998).
432. D. A. Makeiff, D. J. Pope and J. C. Sherman, *J. Am. Chem. Soc.*, **122**, 1337 (2000).
433. C. Naumann and J. C. Sherman, *Carcerands*, in Reference 1b, p. 199.
434. (a) For *Recent Highlights in Hemicarcerand Chemistry* see: R. Warmuth and J. Yoon, *Acc. Chem. Res.*, **34**, 95 (2001).
 (b) For an actual example see: R. Warmuth, *J. Am. Chem. Soc.*, **123**, 6955 (2001).
435. (a) S. Watanabe, K. Goto, T. Kawashima and R. Okazaki, *Tetrahedron Lett.*, **36**, 7677 (1995).
 (b) S. Watanabe, K. Goto, T. Kawashima and R. Okazaki, *J. Am. Chem. Soc.*, **119**, 3195 (1997).
436. S. Watanabe, K. Goto, T. Kawashima and R. Okazaki, *Tetrahedron Lett.*, **40**, 3569 (1999).

437. (a) A. M. A. van Wageningen, J. P. M. van Duynhoven, W. Verboom and D. N. Reinhoudt, *J. Chem. Soc., Chem. Commun.*, 1941 (1995).
(b) P. Timmerman, W. Verboom, F. C. J. M. van Veggel, J. P. M. van Duynhoven and D. N. Reinhoudt, *Angew. Chem., Int. Ed. Engl.*, **33**, 2345 (1994).
438. H. Ihm and K. Paek, *Bull. Korean Chem. Soc.*, **20**, 757 (1999).
439. A. Arduini, A. Pochini, A. Secchi and R. Ungaro, *J. Chem. Soc., Chem. Commun.*, 879 (1995).
440. D. Rudkevich, W. Verboom and D. N. Reinhoudt, *J. Org. Chem.*, **60**, 6585 (1995).
441. (a) S. D. Starnes, D. M. Rudkevich and J. Rebek, Jr., *J. Am. Chem. Soc.*, **123**, 4659 (2001). For similar bis-cavitands with a rigid connection see:
(b) F. C. Tucci, A. R. Renslo, D. M. Rudkevich and J. Rebek, Jr., *Angew. Chem.*, **112**, 1118 (2000); *Angew. Chem., Int. Ed.*, **39**, 1076 (2000).
(c) U. Lücking, F. C. Tucci, D. M. Rudkevich and J. Rebek, Jr., *J. Am. Chem. Soc.*, **122**, 8880 (2000).
442. (a) P. A. Gale, J. L. Sessler, V. Lynch and P. I. Sansom, *Tetrahedron Lett.*, **37**, 7881 (1996).
(b) P. A. Gale, J. W. Genge, V. Král, M. A. McKervey, J. L. Sessler and A. Walker, *Tetrahedron Lett.*, **38**, 8443 (1997).
443. (a) B. Berger, V. Böhmer, E. Paulus, A. Rodriguez and W. Vogt, *Angew. Chem.*, **104**, 89 (1992); *Angew. Chem., Int. Ed. Engl.*, **31**, 96 (1992).
(b) For crown ether derivatives see: W. Wasikiewicz, M. Slaski, G. Rokicki, V. Böhmer, C. Schmidt and E. F. Paulus, *New J. Chem.*, **25**, 581 (2001).
444. V. Böhmer, R. Dörrenbächer, W. Vogt and L. Zetta, *Tetrahedron Lett.*, **33**, 769 (1992).
445. Z.-T. Li, X.-L. Zhang, X.-D. Lian, Y.-H. Yu, Y. Xia, C.-X. Zhao, Z. Chen, Z.-P. Lin and H. Chen, *J. Org. Chem.*, **65**, 5136 (2000).
446. C. Fischer, M. Nieger, O. Mogck, V. Böhmer, R. Ungaro and F. Vögtle, *Eur. J. Org. Chem.*, 155 (1998).
447. E. S. Barrett, J. L. Irwin, P. Turner and M. S. Sherburn, *Org. Lett.*, **4**, 1455 (2002).
448. (a) Y. S. Zheng and Z. T. Huang, *Tetrahedron Lett.*, **39**, 5811 (1998).
(b) A. Dondoni, C. Ghiglione, A. Marra and M. Scoponi, *J. Org. Chem.*, **63**, 9535 (1998).
(c) M. A. McKervey and M. Pitarch, *Chem. Commun.*, 1689 (1996).
(d) M. Pitarch, V. McKee, M. Nieuwenhuyzen and M. A. McKervey, *J. Org. Chem.*, **63**, 946 (1998).
449. (a) For early examples see: P. D. Beer, A. D. Keefe, A. M. Z. Slawin and D. J. Williams, *J. Chem. Soc., Dalton Trans.*, 3675 (1990).
(b) D. Kraft, J. D. Van Loon, M. Owens, W. Verboom, W. Vogt, M. A. McKervey, V. Böhmer and D. N. Reinhoudt, *Tetrahedron Lett.*, **31**, 4941 (1990).
450. P. Schmitt, P. D. Beer, M. G. B. Drew and P. D. Sheen, *Tetrahedron Lett.*, **39**, 6383 (1998).
451. J. Budka, M. Dudic, P. Lhotak and I. Stibor, *Tetrahedron*, **55**, 12647 (1999).
452. J.-M. Liu, Y.-S. Zheng, Q.-Y. Zheng, J. Xie, M.-X. Wang and Z.-T. Huang, *Tetrahedron*, **58**, 3729 (2002).
453. For dendritic structures see: F. Szemes, M. G. B. Drew and P. D. Beer, *Chem. Commun.*, 1228 (2002).
454. I. Higler, P. Timmerman, W. Verboom and D. N. Reinhoudt, *J. Org. Chem.*, **61**, 5920 (1996).
455. I. Higler, W. Verboom, F. C. J. M. van Veggel, F. de Jong and D. N. Reinhoudt, *Liebigs Ann./Recueil*, 1577 (1997).
456. (a) P. Timmerman, K. G. A. Nierop, E. A. Brinks, W. Verboom, F. C. J. M. van Veggel, W. P. van Hoorn and D. N. Reinhoudt, *Chem. Eur. J.*, **1**, 132 (1995).
(b) A. M. A. van Wageningen, E. Snip, W. Verboom, D. N. Reinhoudt and H. Boerrigter, *Liebigs Ann./Recueil*, 2235 (1997).
457. N. Chopra and J. C. Sherman, *Angew. Chem.*, **109**, 1828 (1997); *Angew. Chem., Int. Ed. Engl.*, **36**, 1727 (1997).
458. See also: J. Yoon and D. J. Cram, *Chem. Commun.*, 2065 (1997).
459. N. Chopra, C. Naumann and J. C. Sherman, *Angew. Chem.*, **112**, 200 (2000); *Angew. Chem., Int. Ed.*, **39**, 194 (2000).
460. N. Chopra and J. C. Sherman, *Angew. Chem.*, **111**, 2109 (1999); *Angew. Chem., Int. Ed.*, **38**, 1955 (1999).
461. For a recent review see: M. Vysotsky, C. Schmidt and V. Böhmer, *Chirality in Calixarenes and Calixarene Assemblies*, in *Adv. Supramol. Chem.* (Ed. G. W. Gokel), Vol. 7, p. 139 (2000).

462. (a) S. Steyer, C. Jeunesse, D. Armspach, D. Matt and J. Harrowfield, *Coordination Chemistry and Catalysis*, in Reference 1b, p. 513.
    (b) F. Sansone, M. Segura and R. Ungaro, *Calixarenes in Bioorganic and Biomimetic Chemistry*, in Reference 1b, p. 496.
463. D. M. Rudkevich, *Self-Assembly in Solution*, in Reference 1b, p. 155.
464. For early examples see:
    (a) Y. Nakamoto, G. Kallinowski, V. Böhmer and W. Vogt, *Langmuir*, **5**, 1116 (1989).
    (b) Y. Ishikawa, T. Kunitake, T. Matsuda, T. Otsuka and S. Shinkai, *J. Chem. Soc., Chem. Commun.*, 736 (1989).
465. (a) D. M. Roundhill and J. Y. Shen, *Phase Transfer Extraction of Heavy Metals*, in Reference 1b, p. 407.
    (b) P. Thuéry, M. Nierlich, J. Harrowfield and M. Ogden, *Phenoxide Complexes of f-Elements*, in Reference 1b, p. 561.
466. S. E. Matthews and P. D. Beer, *Calixarene-based Anion Receptors*, in Reference 1b, p. 421.
467. A. Arduini, A. Pochini, A. Secchi and F. Ugozzoli, *Recognition of Neutral Molecules*, in Reference 1b, p. 457.
468. Z.-L. Zhong, A. Ikeda and S. Shinkai, *Complexation of Fullerenes*, in Reference 1b, p. 476.
469. (a) F. Cadogan, K. Nolan and D. Diamond, *Sensor Applications*, in Reference 1b, p. 627.
    (b) N. Sabbatini, M. Guardigli, I. Manet and R. Ziessel, *Luminescent Probes*, in Reference 1b, p. 583.
    (c) R. Ludwig, *Turning Ionophores into Chromo- and Fluoro-Ionophores*, in Reference 1b, p. 598.
470. R. Milbradt and V. Böhmer, *Calixarenes as Stationary Phases*, in Reference 1b, p. 663.

CHAPTER **20**

Polymers based on phenols

D. H. SOLOMON, G. G. QIAO and M. J. CAULFIELD

Polymer Science Group, Department of Chemical Engineering,
The University of Melbourne, Victoria 3010, Australia
Phone: +61 038344 8200;   Fax: +61 038344 4153;
e-mail: davids@unimelb.edu.au

I. INTRODUCTION	1456
A. The Chemistry	1456
II. PHENOL–CARBONYL RESINS	1457
A. Phenol–Formaldehyde Resins	1457
B. Functionality in Phenol–Formaldehyde Resins	1457
C. Novolac Resin Synthesis	1460
1. Statistical Novolac resins	1461
2. High *ortho–ortho* linked Novolac resins	1461
3. Structurally uniform 'pure' Novolac structures	1462
D. Model Compounds of Novolac Resins	1463
1. *Ortho*-linked pure compounds	1463
2. Other structural isomers	1465
3. *Para*-linked compounds	1465
E. The Structure of Novolac Resins	1469
F. Reaction of Novolacs with HMTA	1470
1. Hexamethylenetetramine (HMTA)	1470
2. Chemistry of crosslinking reactions involving HMTA	1470
3. Reactions of model compounds with HMTA	1471
G. Synthesis of Resole Resins	1475
H. Model Compounds of Resoles	1476
1. *Ortho*-hydroxyl model studies	1476
2. *Ortho*-quinone methide	1478
3. Ether exchange reactions	1483
4. PF ratio effects	1484
I. Phenol–Ketone Novolacs	1487
III. CARBON DERIVED FROM PHENOLIC RESINS	1488
A. From Novolacs	1488
B. From Resoles	1488

The Chemistry of Phenols.   Edited by Z. Rappoport
© 2003 John Wiley & Sons, Ltd   ISBN: 0-471-49737-1

	1. Early stage of carbonization	1489
	2. Carbon from resole resins	1490
IV. OTHER POLYMERS WITH PHENOLIC COMPONENTS		1491
	A. Poly(phenylene oxides)	1491
	B. Epoxy Resins	1492
	C. Polyimides	1493
	1. Historical perspectives	1494
	2. Linear polyimides	1494
	3. Bis(maleimides)	1495
	4. Poly(N-(substituted phenyl)maleimides)	1495
	5. Synthesis and polymerization of N-(substituted phenyl)maleimide	1498
V. REFERENCES		1502

I. INTRODUCTION

Polymers derived from phenols find use in a vast array of industries and the property/cost relationships mean that various phenol-based polymers find applications in areas as diverse and challenging as commodity polymers through to the much more technically demanding fields of specialty polymers with high added value and stringent property requirements. Examples are phenolic photoresists, carbonless copying paper and polymers used as a source of carbon. The latter is a relatively new area of particular importance to revolutionary changes in the steel and aluminium industries where phenolic resins are used as a replacement for coal tar pitch as precursors for carbon because of environmental and property advantages. Phenolic resins are also used as binders for composite materials in producing, for example, electrolytic cells and refractory castings.

There is a vast literature on phenol polymerization and this chapter will only review specific points of interests and recent progress in this field.

In the polymer industry phenolic compounds are important stabilizers; they are used to prevent the free-radical induced polymerization of monomers (e.g. methyl methacrylate, styrene) during transit and as stabilizers for polymer systems where radical induced decomposition and decay mechanisms operate. Examples of the latter include polyolefins, such as polyethylenes or polypropylenes, and polyvinyl chlorides. However, these uses of phenols and phenolic polymers are not considered in this chapter (refer to Chapter 12).

A. The Chemistry

From a chemical point of view the reactions used to prepare polymers from phenol involve either (a) electrophilic substitution at free positions *ortho* and *para* to the hydroxy substituent, or (b) dehydrogenation (oxidation) of the phenol by removing the phenolic hydrogen and an aromatic proton.

Thus much of the challenge to the scientist comes from ways in which to control the polymer-forming reaction by limiting the number of points of attack in, for example, the phenol molecule. In polymer terms this means that the functionality of the phenol is modified by the use of substituents to block reactive positions or by controlling the stoichiometry. For convenience, these aspects are discussed in detail under appropriate sections below but we point out here that the approaches have general application.

II. PHENOL–CARBONYL RESINS

Carbonyl compounds react with phenols at positions *ortho* or *para* to the hydroxy substituents as shown in the generalized Scheme 1. By far the most important aldehyde is formaldehyde, and acetone is the ketone most studied.

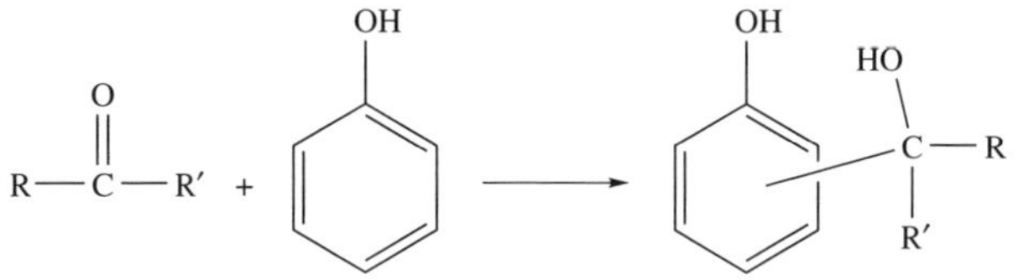

SCHEME 1. Reaction of carbonyl compounds with phenols

A. Phenol–Formaldehyde Resins

Historically, the reaction of phenol with formaldehyde was of vital importance to the polymer industry, being one of the first totally synthetic commercial polymer resin systems developed. In 1907, Leo H. Baekeland commercialized, under the tradename 'Bakelite', a range of cured phenol–formaldehyde resins[1], which were useful in producing heat-resistant molded products[1,2]. Since this early work, phenol–formaldehyde resins have been used in many applications, including refractory compounds, adhesives, thermal insulation and electrical industries[3,4].

Some of the factors identified in determining the final properties of these resins are the phenol–formaldehyde ratio, pH, temperature and the type of catalyst (acid or alkaline) used in the preparation of the resin[5]. The phenol–formaldehyde ratio (P/F) (or formaldehyde to phenol ratio, F/P) is a most important factor as it leads to two different classes of synthetic polymers, namely Novolacs and resoles. The first class of resins, Novolacs, is produced by the reaction of phenol with formaldehyde with a P/F >1 usually under acidic conditions (Scheme 2a). Resoles are produced by the reaction of phenol and formaldehyde with a P/F <1 usually under basic conditions (Scheme 2b).

Novolacs are thermoplastic polymers that require an 'additive' to enable further curing and the formation of insoluble and infusible products.

Often, the additive is a formaldehyde source such as hexamethylenetetramine (HMTA). On the other hand, a resole is capable of forming a network structure by the application of heat.

In effect, the two resin classes result from the deliberate selection of the reaction conditions to control the functionality of the system. However, since there is some confusion and contradiction in the literature regarding functionality in these systems, we will attempt to clarify this issue here.

B. Functionality in Phenol–Formaldehyde Resins

It is important to define the terms used in describing functionality and to clearly distinguish between the *actual* and *potential* functionality and to show the relationship between stoichiometry and functionality. Functionality can be defined as the number of other molecules that a compound can react with. This definition of functionality also means that within step-growth polymerizations the actual functionality is dependent on stoichiometry. The phenol–formaldehyde reaction is a typical step-growth reaction in

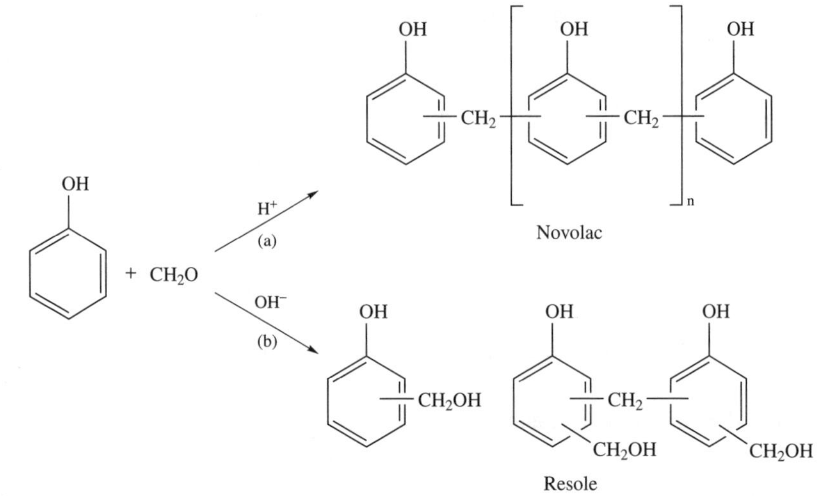

SCHEME 2. Formation of phenol–formaldehyde resins: (a) Novolac resin, (b) resole resin

which the reactants are not present in the required stoichiometric amounts for complete reaction of all functional groups and hence the actual and potential functionalities need to be considered.

Consider phenol that has two *ortho* and one *para* position available for reaction (with either formaldehyde or the methylol group of the reaction product of phenol with formaldehyde). Clearly, phenol has a *potential* functionality of three and similarly formaldehyde has a *potential* functionality of two (Scheme 3).

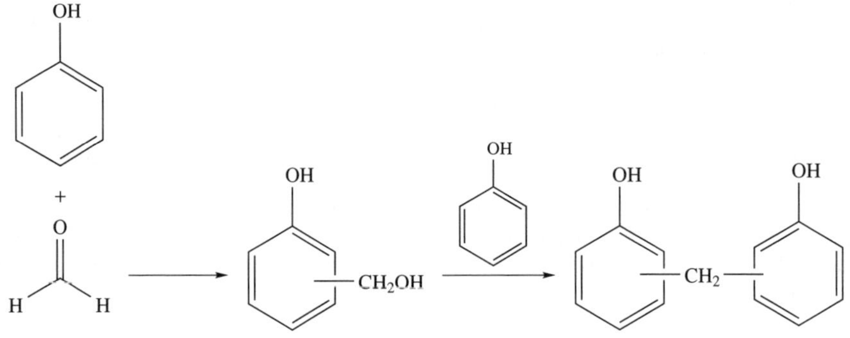

SCHEME 3. Functionality of phenol and formaldehyde

For phenol and formaldehyde to achieve their full potential functionality they require the appropriate stoichiometry. In the above equation, phenol cannot react at three centres as there is insufficient formaldehyde. In fact, the *actual* functionality (f_{actual}) is only one in the equation shown. That of formaldehyde is two.

If we now extend this argument to a Novolac resin, we can clearly conclude that in these systems formaldehyde achieves its full potential functionality of two, which is when the potential and actual functionality are the same. On the other hand, phenol on average achieves a functionality of <2; within chain phenol residues have a functionality of two and the two end groups a functionality of one. In other words, the *potential* functionality of three for phenols is never achieved in a Novolac resin and the actual functionality is <2. We note that some scientists use 2.31 as the functionality of phenol in modelling calculations on Novolacs[6,7], even though it is acknowledged that this value has 'no reliable scientific foundation'[8].

In the commercial synthesis of Novolac resin there exists the strong possibility that some chain branching will occur (Figure 1). Thus the actual functionality of individual phenols will vary depending on its position in the network.

Thus a fully branched phenol residue ($f_{actual} = 3$) is counter-balanced by both linking phenol residues ($f_{actual} = 2$) and chain ends ($f_{actual} = 1$). Hence, the functionality of the phenols in Figure 1 averages out to 1.6. If we now consider the calculated value of $f_{actual} = 2.31$, it is clear that a highly crosslinked structure is required. That is, it is necessary to have extensive crosslinking to minimize phenol end groups ($f_{actual} = 1$). Consequently, to approach a phenol functionality of 2.31, within the established molecular weight ranges for Novolacs (i.e. less than 1000)[9–14], a structure as depicted in Figure 2 is required. Such a structure is extremely unlikely and, in any case, would not be expected to be soluble.

Therefore, the actual functionality of phenol in a Novolac must be less than 2. The figure often quoted of 2.31 has no chemical or physical meaning in terms of the structure of a phenol–formaldehyde resin. An actual functionality above 2 can only eventuate when the P/F ratio is greater than 1, that is, when gelation can occur.

Hence, by controlling the stoichiometry we can control the functionality and the molecular weight of the Novolac. The closer the P/F ratio approaches one, the higher the molecular weight[15].

In resoles, the actual functionality of the formaldehyde is controlled to be less than 2 and the actual functionality of the phenol may be 3 if sufficient formaldehyde is used or slightly less than 3 (Figure 3). In other words, every formaldehyde molecule that only

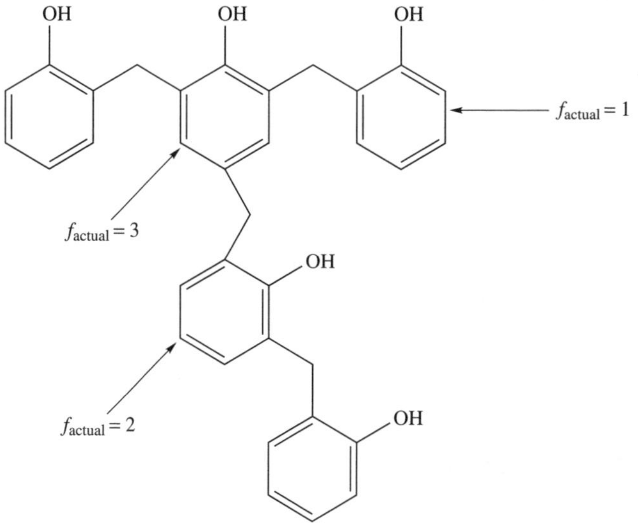

FIGURE 1. A model Novolac structure where f_{actual}(phenol) = 1.6

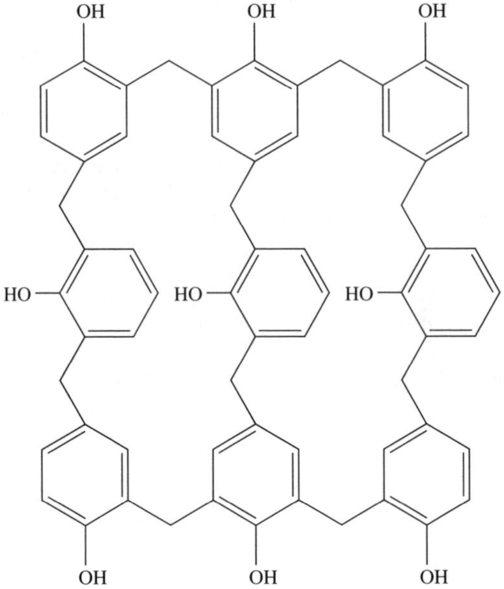

FIGURE 2. A Novolac-type structure required to give a phenol functionality of $f_{actual} = 2.22$

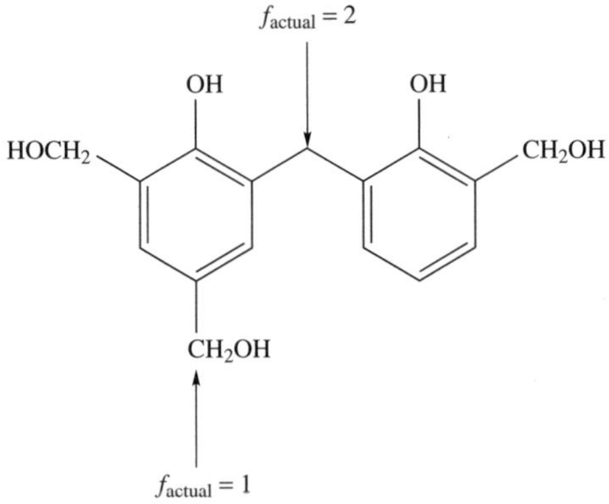

FIGURE 3. A model resole structure which has $f_{actual}(\text{phenol}) = 2.5$ and $f_{actual}(\text{formaldehyde}) = 1.25$

reacts to the methylol stage has an actual functionality of 1 and contributes to the average figure of less than 2.

C. Novolac Resin Synthesis

Novolac resins are generally prepared by the acid-catalysed reaction of phenol and aqueous formaldehyde under reflux. Although strong acids such as sulphuric and hydrochloric acid can be used, the weaker oxalic and phosphoric acids give a less exothermic and more controllable reaction. When the formaldehyde has been consumed, the volatile materials, including water, methanol and phenol are removed by vacuum distillation at temperatures up to 150 °C.

1. Statistical Novolac resins

Under acidic conditions, hydroxymethylation and methylene bridge formation occur preferably at the *para* position. ^{13}C NMR studies have shown that *para–para* bridges are the first to be formed followed by the *ortho–para*, and finally the *ortho–ortho* linkages[5,16]. A study by Natesan and Yeddanapalli showed that *para*-hydroxymethylated phenol condenses more readily than its *ortho*-counterpart at 80 °C and pH of *ca* 1[17]. This observation was supported by Kopf and Wagner and rationalized by proposing that the *ortho*-hydroxymethylphenol may be stabilized by internal hydrogen bonding[18]. Furthermore, both the *ortho*- and *para*-substituted phenols preferentially condensed with the *para* positions of either phenol or dimers.

Thus, in general, Novolac resins typically consist of 8 to 10 phenol units linked via methylene bridges ($-CH_2-$) either *ortho* or *para* to the hydroxy group. A statistical Novolac resin is illustrated in Figure 4.

2. High ortho–ortho linked Novolac resins

The structure of the Novolac resin can be manipulated by adjusting the pH and by the addition of divalent metal salts of Ca, Mg, Zn, Cd, Pb, Cu, Co and Ni as catalysts.

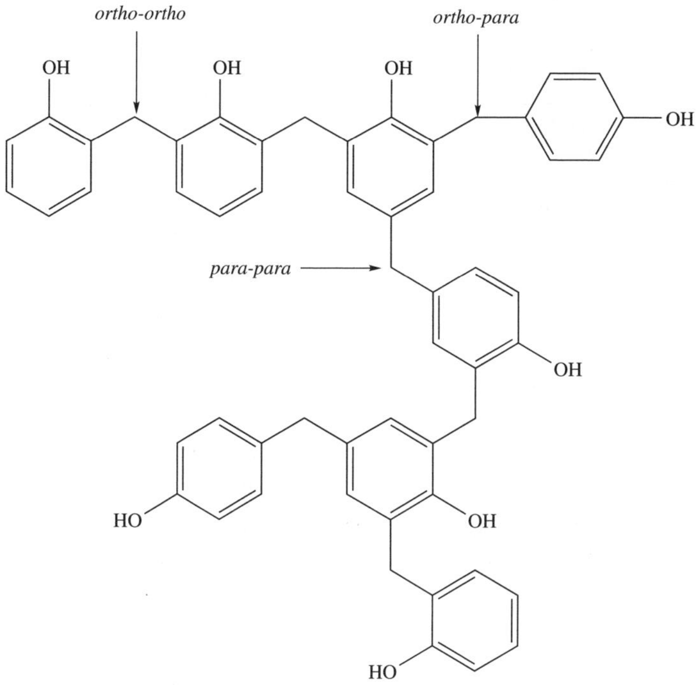

FIGURE 4. A statistical Novolac resin

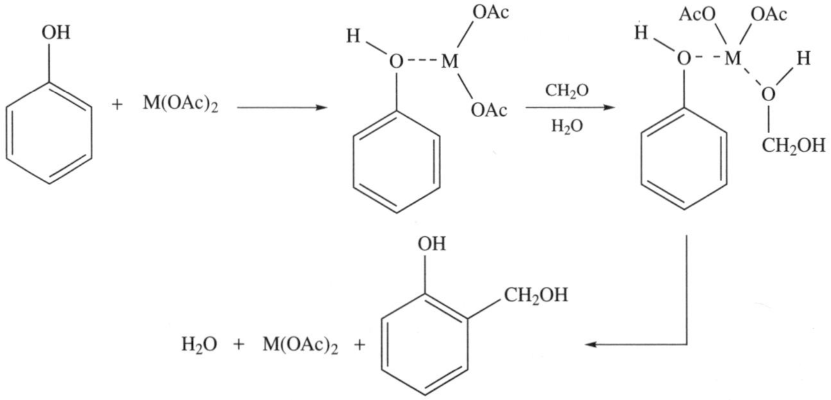

SCHEME 4. Proposed interaction between phenol, metal acetate (M(OAc)$_2$) and formaldehyde

These are commonly termed 'high *ortho*' Novolacs. Zinc acetate is the most commonly used catalyst. The initial reaction is proposed to occur through chelation of phenol and formaldehyde through the metal acetate (Scheme 4)[19–22].

The chelated intermediate is then transformed into *ortho*-methylolphenol, as evidenced by both ^{1}H NMR and gel permeation chromatography (GPC) studies[23].

A series of exclusively *ortho*–*ortho* linked low molecular weight Novolac resins derived from phenol and formaldehyde[24,25], acetaldehyde[26,27] and isobutyraldehyde[28] has also been synthesized.

An interesting property of *ortho*-linked Novolacs is their high acidity, referred to as 'hyperacidity'. Sprengling[29] proposed that the strongly acidic proportion of linear *ortho*-linked di-, tri- and tetra-nuclear oligomers (in comparison to similar isomers) was accounted for by stabilization of the mono-anion via strong intramolecular hydrogen bonding (Scheme 5).

Many workers have since investigated this phenomenon in higher oligomers[25,27,30]. Higher acidity values are observed with increasing chain length as found for linear oligomers with a terminal *p*-nitrophenol unit, which may stabilize the mono-anion. Bulky *ortho* substituents at the other end of the molecule resulted in even higher acidity. An even greater increase in acidity was found when the *p*-nitrophenol was positioned along the chain interior.

3. Structurally uniform 'pure' Novolac structures

Novolac resins may contain a highly complex mixture of homologous compounds. Over 10,000 isomers are possible for linear Novolac containing 10 phenolic nuclei, while branched and cyclic variations further increase this number[31,32]. Possibly of greater importance is the molecular conformations and entanglement encountered by these isomers as a result of *intra*- and *inter*-molecular hydrogen bonding[33]. Investigations of the reactivity of *ortho*-linked oligomers towards formaldehyde under acidic conditions showed that as the chain length is increased, the shielding and deactivating effects of intramolecular hydrogen bonding decreased the reactivity[34]. There is recent evidence[25,35,36] to suggest that in solution at least, a large proportion of the *ortho*-linked oligomers adopt pseudo-cyclic conformations. The molecular freedom of liquid resins may be somewhat restricted by such hydrogen bonding, thereby hindering their ability to adopt conformations favourable for chain extension or crosslinking[31].

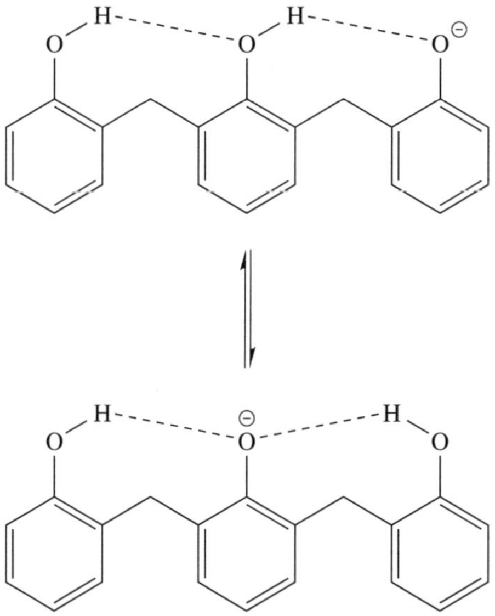

SCHEME 5. Intramolecular hydrogen bonding in oligomers

The commercial importance of phenol–formaldehyde resins has resulted in extensive studies of these systems, with the aim of identifying the reaction mechanisms and intermediates that occur during subsequent polymerization reactions. However, the complexity of Novolac-type systems has made a detailed understanding of the subsequent chemical processes and their relationship to the physical properties of the final polymerized product difficult. Thus, it is necessary to simplify the system in order to more readily unravel this complexity. Model compounds are frequently used to understand complicated chemical systems and their application to phenol–formaldehyde systems has been well documented[18,37,38].

D. Model Compounds of Novolac Resins

Recently, pure compounds which have molecular weights of the same order as commercial Novolacs have been prepared[9,39–40] and used to calibrate GPC systems and to study the chemical reaction, for example, with HMTA.

1. Ortho-linked pure compounds

The synthetic scheme used for the preparation of the pure compounds is based on the reported ion-assisted *ortho*-specific phenol–formaldehyde reaction developed by Casiraghi and coworkers[26]. Thus a series of *ortho*-linked pure phenolic compounds, e.g. **2**, can be synthesized which contain the maximum number of *para*-reactive sites and a small number of *ortho*-sites from **1** (Scheme 6)[41].

Manipulation of the reaction conditions can result in the preparation of a series of pure *ortho*-linked homologues, like those shown below.

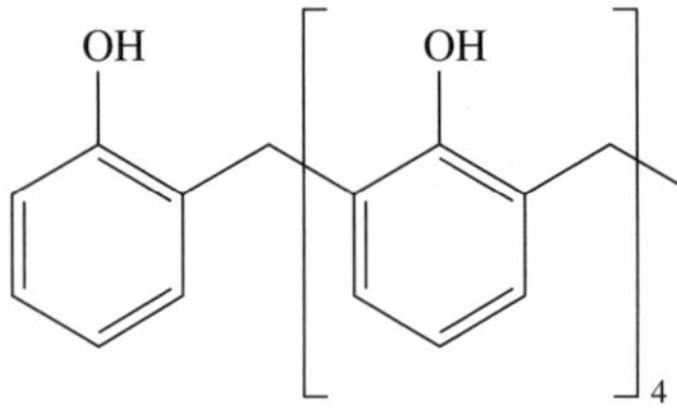

o,o'-Hexamer

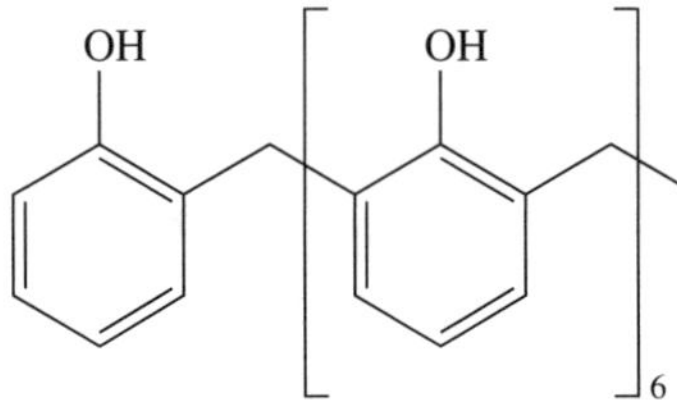

o,o'-Octamer

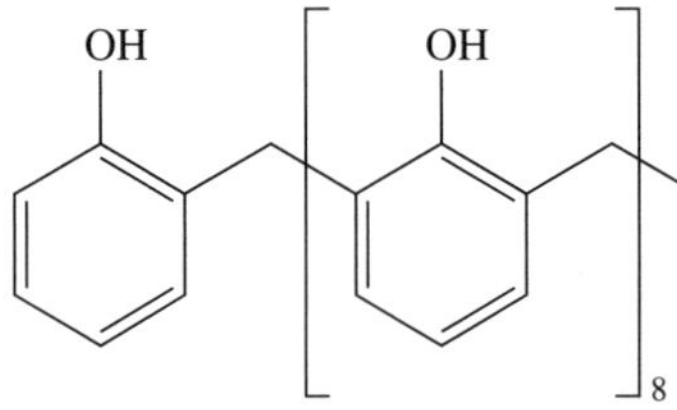

o,o'-Decamer

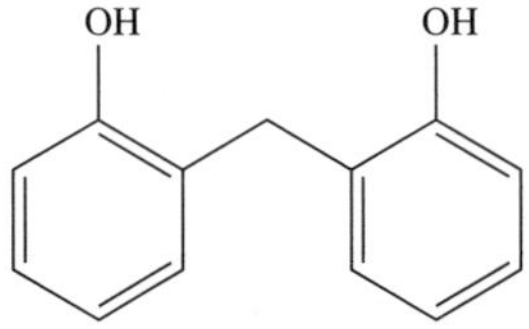

(1)

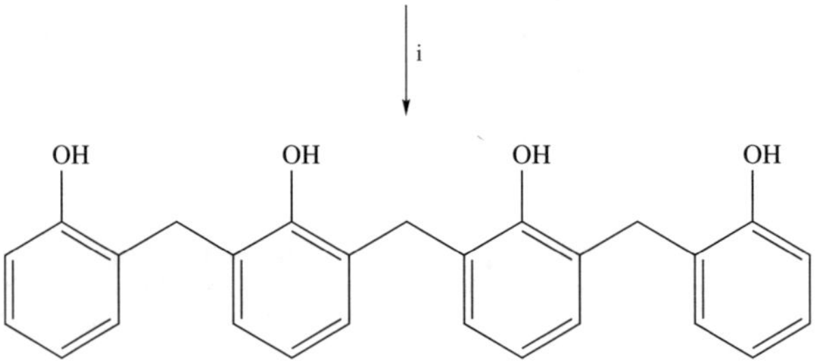

↓ i

(2)

SCHEME 6. *Reagents and conditions*: i. EtMgBr (1 equiv), Et$_2$O, 25 °C, 30 min, then benzene, 25 °C to 80 °C, paraformaldehyde (0.5 equiv), 20 h

2. Other structural Isomers

The synthetic methodology can also be extended to generate a series of compounds **6** that only contain *para*-reactive sites. Thus *ortho*-cresol **3** was directly coupled with **4** to give dimer **5** (Scheme 7). Theoretically, conversion of **3** to **5** requires 0.5 equivalents of formaldehyde. When the amount of formaldehyde is increased, trace amounts of an aldehydic product are formed. This is confirmed by the characteristic ^{1}H NMR signals of the hydroxyl and aldehydic protons. This strongly suggests that excess formaldehyde hinders the coupling of two phenolic units due to the complexing nature of formaldehyde with the metal phenoxide intermediate[42].

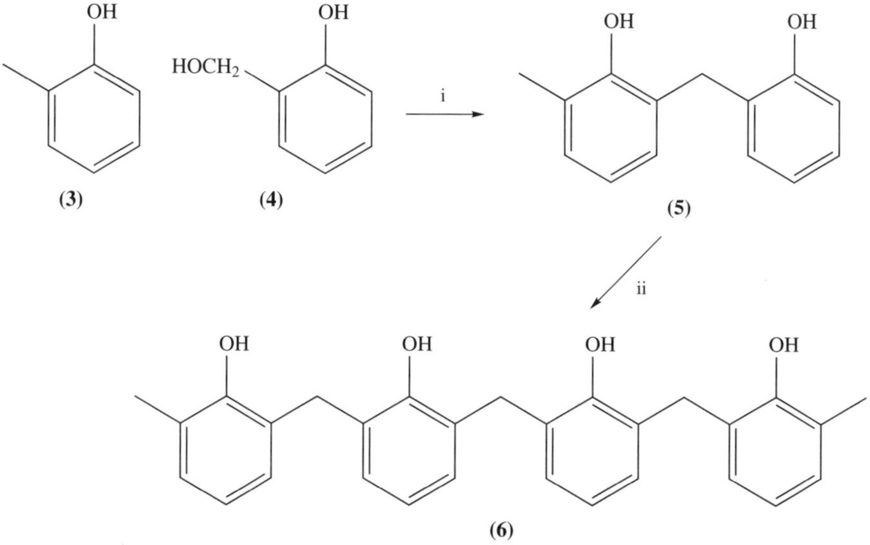

SCHEME 7. *Reagents and conditions*: i. EtMgBr (3 equiv), Et$_2$O, 25 °C, 30 min, then benzene, 25 °C to 80 °C; ii. EtMgBr (2 equiv), Et$_2$O, 25 °C, 30 min, then benzene, 25 °C to 80 °C, paraformaldehyde (0.5 equiv), 20 h

3. Para-linked compounds

The synthesis of an analogous series of model compounds containing the maximum number of free *ortho* positions is more complex, requiring protection/deprotection methodology to control the regioselectivity of the coupling reaction. This type of methodology has been applied to prepare a series of structurally controlled model compounds from **7** via **8** and **9** (Scheme 8).

Compound **9** is a key intermediate in the synthesis of both linear and branched model octamers containing only free *ortho* positions. The tetramer **9** can be conveniently deprotected using tetrabutylammonium fluoride (TBAF)[43] to afford the first model compound, tetramer **10**.

The synthesis of the branched system was carried out in two steps from tetramer **9**, where coupling of the magnesium bromide salt of bis(silylated) tetramer **9**, using the

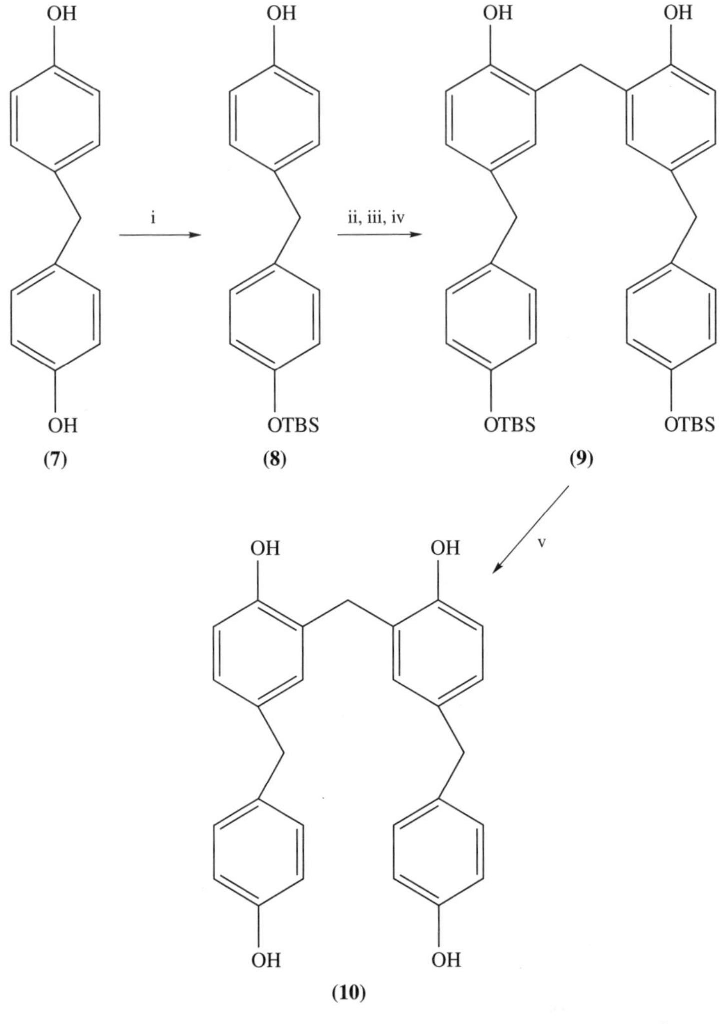

SCHEME 8. *Reagents and conditions*: i. TBSCl (1.2 equiv), imidazole, DMF, 25 °C, 5 h; ii. Mg (1 equiv), EtBr (1 equiv), Et$_2$O, 25 °C, 30 min; iii. **6**, Et$_2$O, 25 °C, 30 min, then benzene, 25 °C to 80 °C; iv. paraformaldehyde (0.5 equiv), 80 °C, 20 h; v. TBAF, THF, 0 °C, 30 min

standard conditions, afforded the carbon skeleton **11** required for the branched octamer, which can be deprotected to afford the free phenolic compound **12** (Scheme 9).

Preparation of the analogous linear system is more complex, since coupling at the alternate *ortho* position requires an additional protection step followed by a deprotection to unmask the *para* terminal end of **9**. The desired tetramer **13** was generated by treating **9** with the more robust silyl protecting group *tert*-butyldiphenylsilyl chloride[44] (TBDP-SCl). Selective deprotection of the fully protected tetramer **13** is achieved by employing hydrogen fluoridepyridine[45] or boron trifluoride etherate[46] at 0 °C to afford **14**. Selective

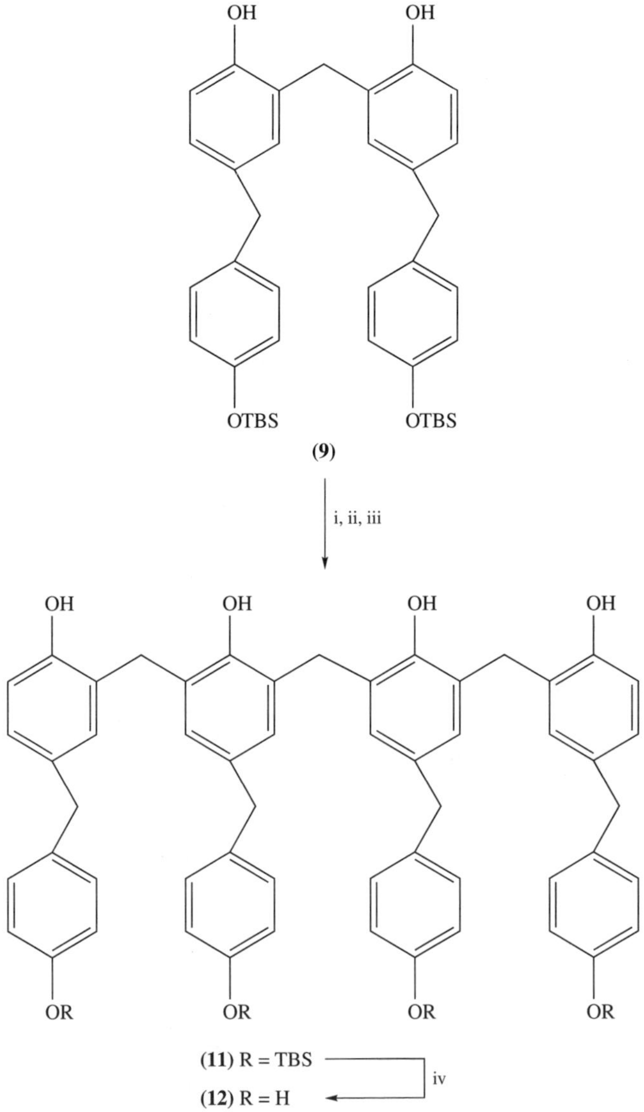

SCHEME 9. *Reagents and conditions*: i. Mg (2 equiv), EtBr (2 equiv), Et$_2$O, 25 °C, 30 min; ii. **9**, Et$_2$O, 25 °C, 30 min, then benzene 25 °C to 80 °C; iii. Paraformaldehyde (0.5 equiv), 80 °C, 20 h; iv. TBAF, THF, 0 °C, 10 min

coupling through the *para*-cresol terminus affords octamer **15** and full deprotection with TBAF gives the linear octamer **16** (Scheme 10).

By applying this methodology a series of *para*-linked structural isomers (**17–20**) has also been synthesized via coupling of the '*ortho–para*' dimer[42].

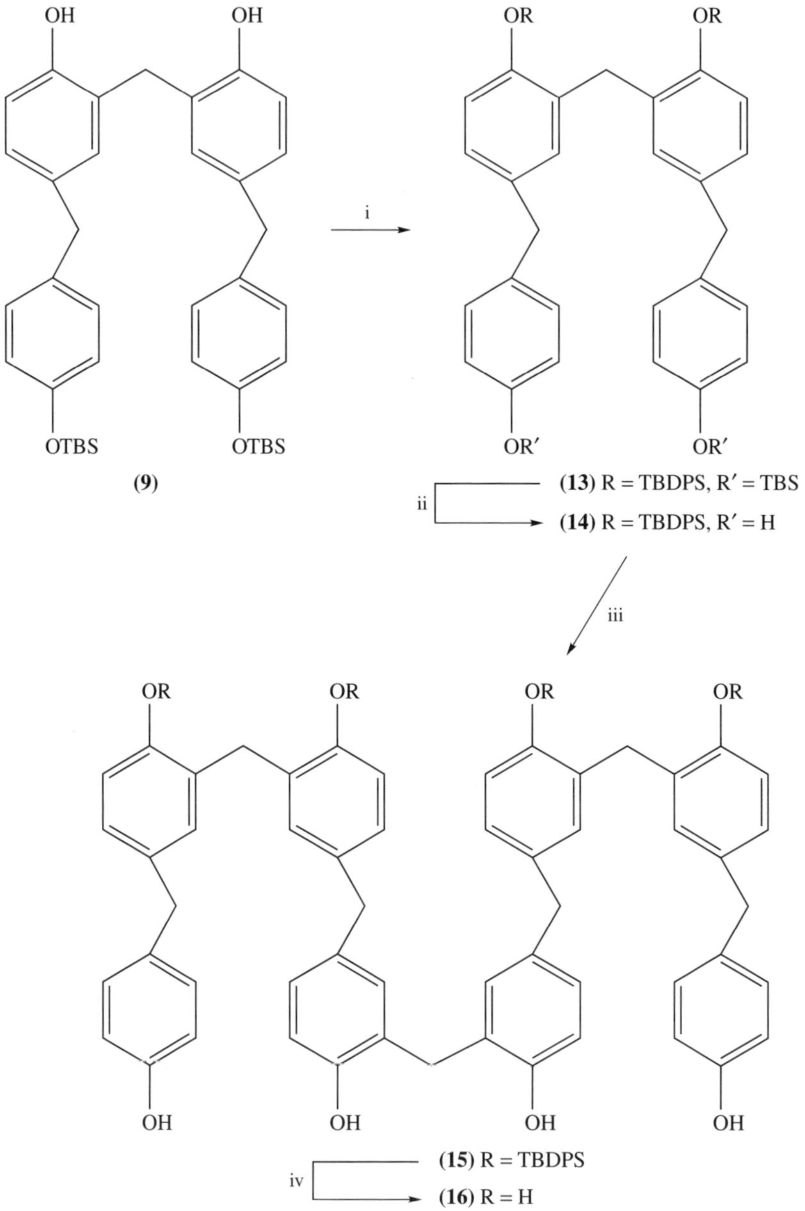

SCHEME 10. *Reagents and conditions*: i. TBDPSCl (4.5 equiv), imidazole (4.5 equiv), DMF, 60 °C, 10 h; ii. HF–pyridine, pyridine, THF, 0 °C to 25 °C, 4.5 h or BF$_3$•Et$_2$O, CHCl$_3$, 0 °C to 25 °C, 3 h; iii. EtMgBr (2 equiv), Et$_2$O, 25 °C, 30 min, then benzene, 25 °C to 80 °C, paraformaldehyde (0.5 equiv) 20 h; iv. TBAF, THF, 25 °C, 6 h

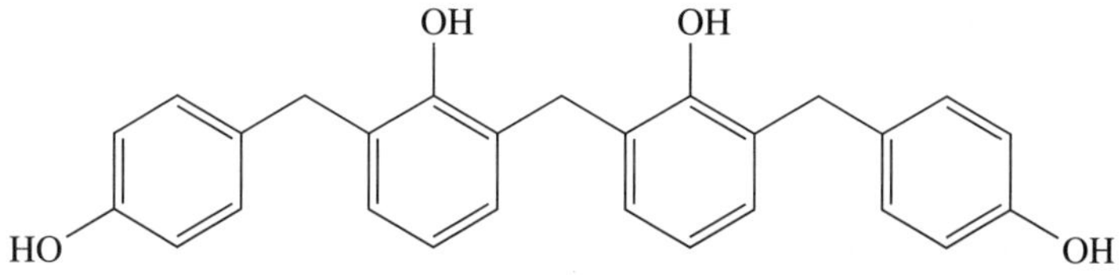

(17), 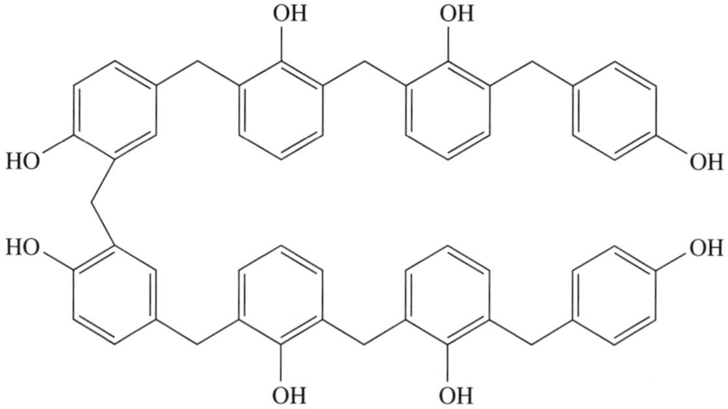 **(18)**, 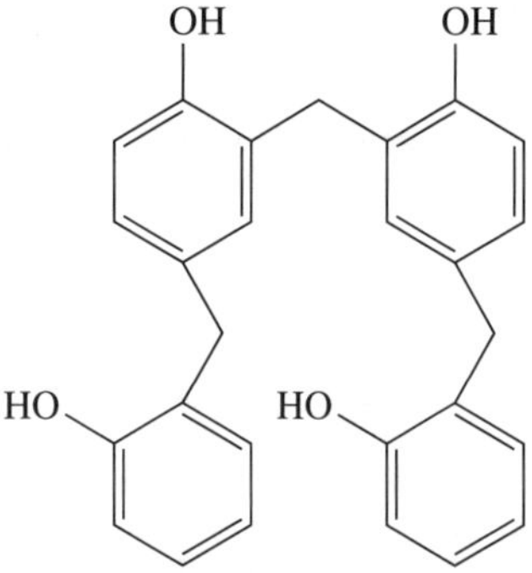 **(19)**, 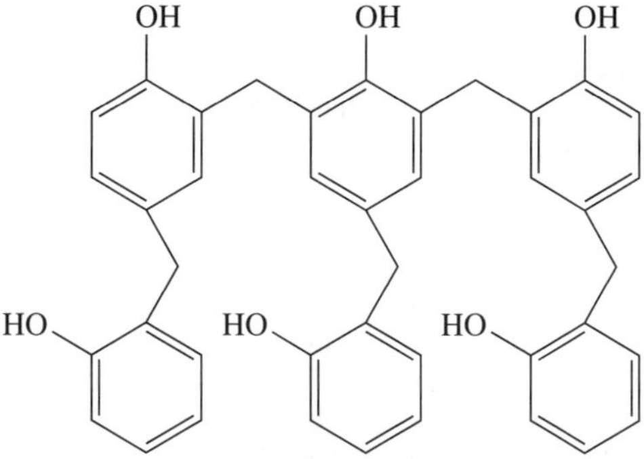 **(20)**

E. The Structure of Novolac Resins

In theory, Novolacs can have *'ortho–ortho'*, *'ortho–para'* and *'para–para'* methylene bridges. A consequence of the different linking structures is that they dictate the free *ortho* or *para* position and, of course, the structure of the Novolac resin.

Similarly, in the preparation of low molecular weight analogue, bis-phenol F, preferential reaction to form *'para–para'* links is achieved using acid catalysis. Bis-phenol F is an important intermediate in the synthesis of epoxy resins (see Section IV.B).

F. Reaction of Novolacs with HMTA

Novolacs are thermoplastic polymers that require the addition of a formaldehyde source to enable further curing and the formation of insoluble and infusible products.

1. Hexamethylenetetramine (HMTA)

The most commonly used crosslinking agent is hexamethylenetetramine (HMTA), which is produced by the reaction of formaldehyde and ammonia, as detailed in Scheme 11[47−50].

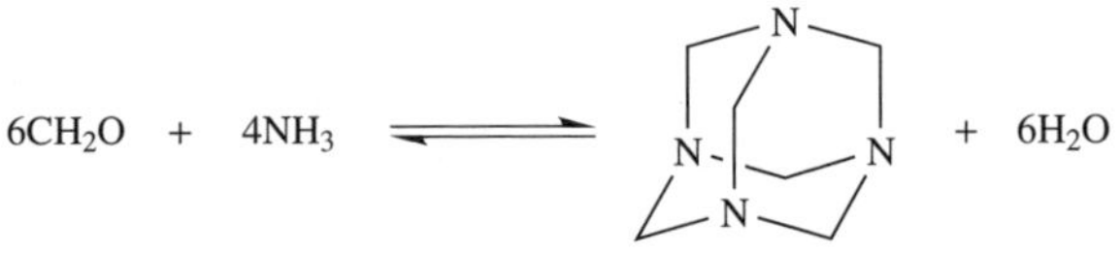

$$6CH_2O + 4NH_3 \rightleftharpoons \text{HMTA} + 6H_2O$$

SCHEME 11. Formation of hexamethylenetetramine (HMTA)

HMTA is very soluble in water, with 87.4 g dissolving in 100 g of water at 20 °C. However, it is less soluble in alcohols such as ethanol or methanol. HMTA readily sublimes at 150 °C[5], while decomposition occurs at elevated temperatures, generally above 250 °C. The thermal decomposition of HMTA occurs via cleavage of N−C bonds to yield methylamines as initial products[51]. Its use as a reagent in organic synthesis has been reviewed[52] as has its derivatives, preparation and properties[53]. In aqueous acid solutions, HMTA will only hydrolyse after several hours at reflux[47] despite having a pK_a value of 4.89 at 25 °C[54]. The hydrogen bonding characteristics of HMTA with water, $CHCl_3$ and $CHBr_3$ have been reported[52]. Hydrogen bonded adducts with phenols[55,56] and salts with acids are also known[57].

^{13}C-labelled or ^{15}N-labelled derivatives of HMTA have been synthesized[58] together with ^{13}C- and ^{15}N-labelled HMTA[59].

2. Chemistry of crosslinking reactions involving HMTA

The reaction between Novolac resins and HMTA forms methylene linkages between the phenolic rings, resulting in an insoluble, infusible polymeric network. The advantages of using HMTA over other formaldehyde sources (e.g. paraformaldehyde or trioxane) include the absence of large amounts of gaseous products (such as formaldehyde or water) and the reduction of the temperature at which crosslinking occurs. Although the properties of the final resin are readily determined, relating these properties to the chemistry of the phenol–formaldehyde resin is more difficult and extremely challenging.

Various reports in the literature suggest that the reaction of a Novolac resin and HMTA proceeds faster at a free *para* position[3,60−63] and hence a resin with predominantly '*ortho–ortho*' linkages is desirable. Several mechanisms for the initial stages of curing of Novolac resins with HMTA have been proposed. Early studies[64,65] suggested that the initial curing is a homogenous acid-catalysed reaction involving a trace amount of water in the Novolac to hydrolyse HMTA to α-amino alcohol. The acidic phenolic units would generate carberium ions from these α-amino alcohols, which then react with phenolic units to form benzylamines. This postulate is supported by the fact that the reaction rate increases with decreasing pH for both the Novolac resins and model systems[5,18,66] and increasing phenol[5,67] or water[66,68−70] content. Other claims suggested that excess water could decrease the reaction rate[71]. Katovic and Stefanic proposed an intermolecular hydrogen-bonding mechanism between Novolac and HMTA[72]. The nitrogens of HMTA can hydrogen bond to the phenolic hydroxyl protons in Novolac chains that are originally self-associated through hydrogen bonding. As the temperature increases, two consecutive and temperature-dependent steps occur in the Novolac reaction with HMTA: (i) the hydrogen of the phenolic hydroxyl transfers to the HMTA nitrogen with the formation of ionized species; (ii) a hydrogen shifts from the *ortho* position of the ring to the oxygen anion. Then the nucleophilic ring carbon anion attacks a methylene group of HMTA and forms an initial methylene bridge at the *ortho* position of the phenolic rings. Once the breakdown of the HMTA molecule has begun, further reactions occur via either protonation of the tertiary amine or all amines, and gives rise to derivatives of Novolac.

The reactions between Novolac resins and HMTA have been examined using a variety of techniques. Recent studies have used NMR and have described the reactions between *ortho*- and *para*-phenolic reactive sites of Novolac resins and HMTA. Thus a combination of ^{13}C and ^{15}N high-resolution solution and solid-state NMR studies has been used to trace the changes in chemical structures through the curing process. As discussed earlier, before curing, the Novolac and HMTA are hydrogen bonded through the phenolic hydroxy group and the nitrogen of HMTA. As the curing temperature initially increases to 90–120 °C, the curing reactions start, and the initial intermediates formed are various substituted benzoxazine and benzylamine-type molecules. Triazine diamine and ether-type structures are also formed during the initial curing stage. The reaction mechanisms involved are complicated and a number of different mechanisms may occur concurrently. A further increase in temperature causes decomposition and reactions of these initial intermediates to produce methylene linkages between phenolic rings for chain extension and crosslinking, together with amide-, imide- and imine-type intermediates by side-reactions such as oxidation and dehydrogenation, whereby NH_3 is liberated from the resins. Methyl-substituted products are also formed in the decomposition/reaction. A small amount of formaldehyde, liberated from the decomposition of the ether intermediates to produce methylene linkages, could also play a role in side-reactions. At high temperatures, various benzoxazine, benzylamine and imine intermediates can be oxidized by air to form numerous amide and imide structures. The aldehyde groups and perhaps even carboxyl groups also form in the oxidation. The various proposed reaction intermediates are outlined in Scheme 12[59,73].

3. Reactions of model compounds with HMTA

Hatfield and Maciel[58] identified 15 possible intermediates involved in the curing of Novolac with HMTA. Recently, the Solomon Group[74–79] conducted a major study investigating the mechanism of the reaction between Novolac and HMTA. Model compounds, the possible reaction intermediates, were produced and their subsequent thermal reactions were investigated. Knop and coworkers[73] reviewed the work recently and this chapter will briefly describe the major findings from these studies.

The study initially established that benzoxazines and benzyl amines are the key intermediates in the curing of the Novolac with HMTA. Benzoxazine **22** was formed by reaction of HMTA with 2,4-xylenol **21** while benzyl amines **24** and **25** are formed with 2,6-xylenol **23** (Scheme 13)[74].

The study found[74] that the reaction pathways for the *ortho* and *para* sites are different and that distinctly different mechanisms apply. In the case of the free *ortho* site, the reaction is dependent on the breakdown of the hydrogen bonding complex to form benzoxazine, in contrast to the free *para* position where the reaction is governed by the breakdown of the amine intermediates. These results strongly suggested that interaction between phenolic entities is primarily controlled by hydrogen bonding, especially when

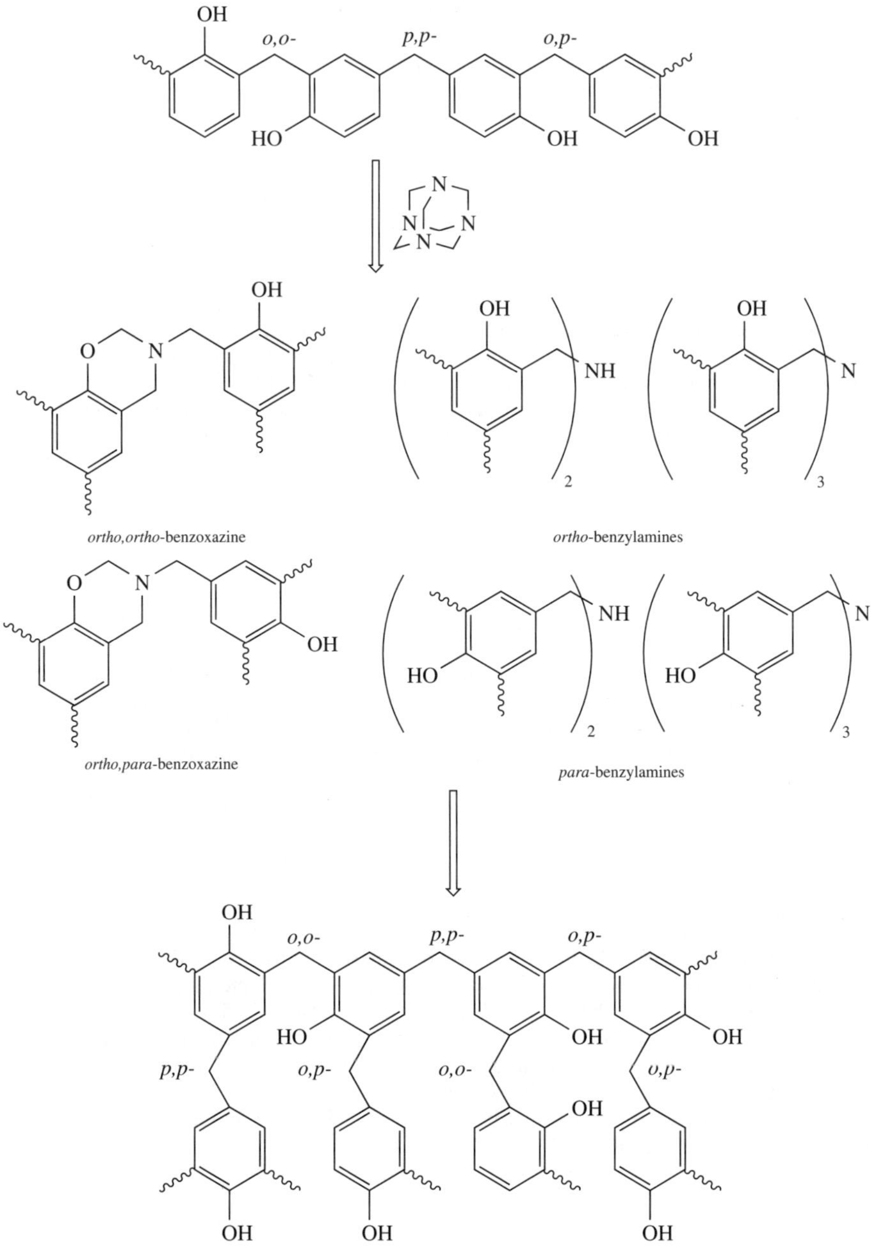

SCHEME 12. Proposed reaction between Novolac and HMTA and involved reaction intermediates

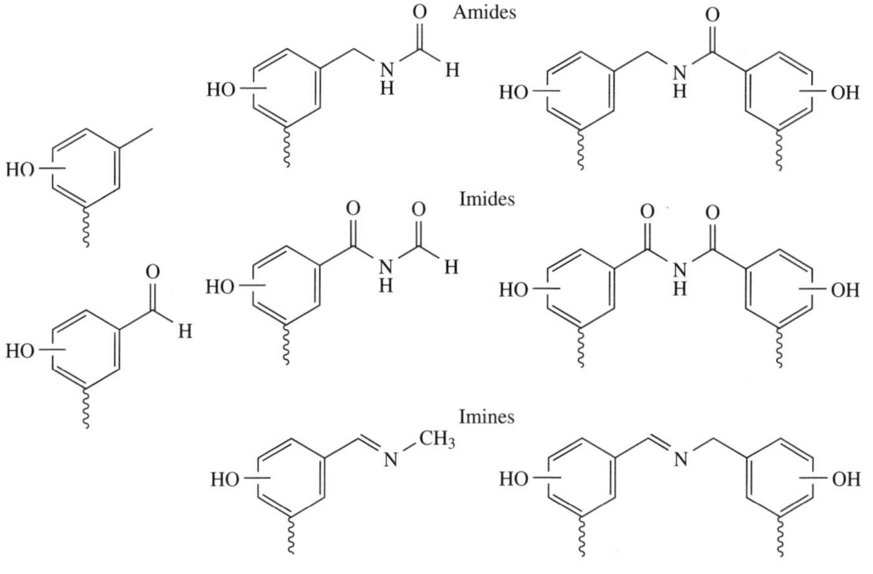

SCHEME 12. (*continued*)

a vacant *ortho* site was present. Although this interaction is relatively strong in the 2,4-xylenol individual case, when subjected to competing conditions such as for a Novolac resin, with 2,6-xylenol present, the reaction is strongly directed to the *ortho* position. An acid–base type relationship must be considered in the case of the 2,6-xylenol–HMTA reaction and, if significant concentrations of 2,4-xylenol are present in the mixture, the HMTA tended to preferentially hydrogen bond to the 2,4-xylenol. Therefore, in a mixed system containing both vacant *ortho* and *para* positions, the *ortho* site was found to preferentially react and the *para* positions take part in the secondary reactions. So only at low concentrations of 2,4-xylenol does the *para* position of the 2,6-xylenol begin to react with HMTA.

The study then investigates the decomposition of these intermediates or their reaction with model compounds.

a. Reactions of benzoxazine with itself and with model phenols. A model benzoxazine, **22**, was heated under carefully controlled conditions, and the structural changes were studied by ^{13}C and ^{15}N NMR spectroscopy[75]. The benzoxazine structure is relatively stable, and detectable decomposition only occurred at about 155 °C with the formation of methylene linkages between phenol rings. Various nitrogen-containing structures, such as amines, amides and imines, together with an alcohol etc. were also formed as side products. At 240 °C, the dominant product is methylene diphenol (**26**). The benzoxazine was then heated in the presence of 2,4-xylenol (**21**) or 2,6-xylenol (**23**) and the formation of the products followed by NMR spectroscopy[76]. The study provided direct evidence of the formation of methylene linkages between phenol rings from the reaction of **22** with **21** or **23**. The reaction pathways of the two systems were found to be different. The benzoxazine can react with **21** at low temperature (even at 90 °C), but with **23**, reaction only occurred

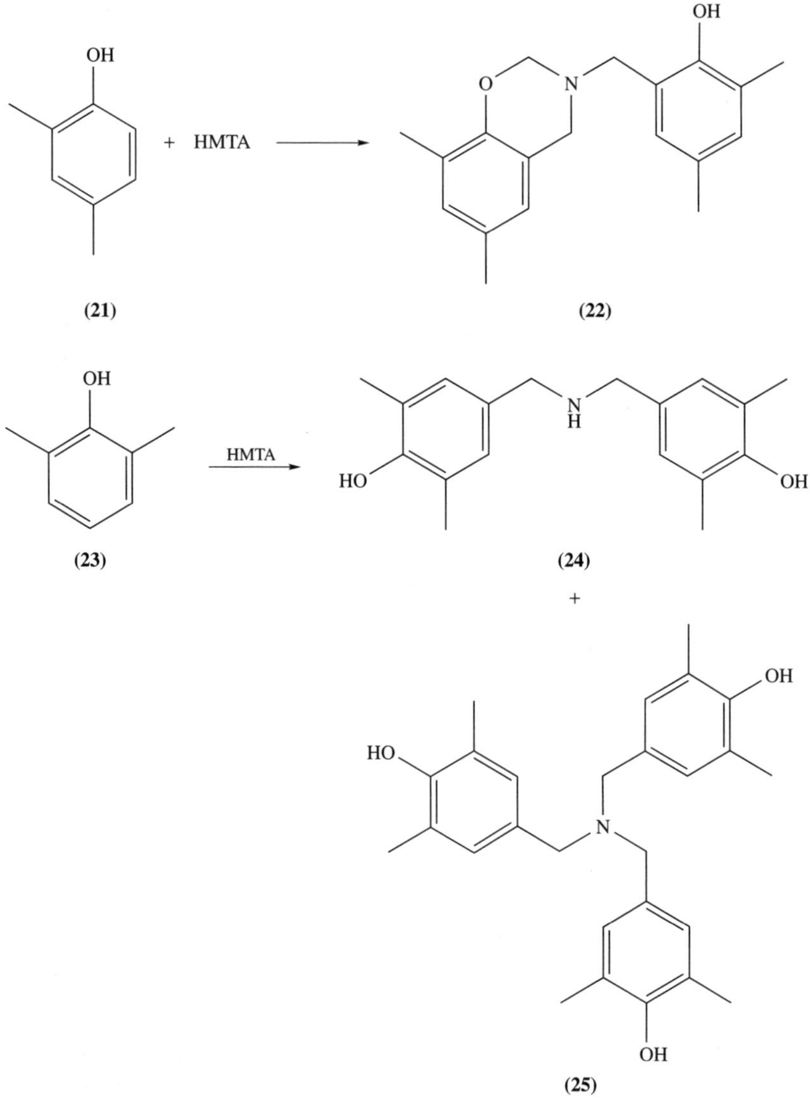

SCHEME 13. Reaction of xylenol **21** and **23** with HMTA

above 135 °C. In addition, **21** can react with **22** directly to form *ortho–ortho* dimer, while **23** reacts with the decomposition species of **22** to form *ortho–ortho*, *ortho–para* and *para–para* dimers.

b. Reactions of para-hydroxybenzylamine with itself and with model phenols. Benzyl amines **24** and **25** were heated and the decomposition products were monitored by NMR spectroscopy[77]. The thermal decomposition resulted in the formation of *para–para* methylene linkages between phenolic rings. Only minor side products formed after heating to 205 °C. The bis(amine) **24** could form a methylene linkage via direct decomposition while the tris(amine) **25** broke down to bis(amine) **24** at about 90–120 °C. Side reaction resulted in the formation of various products during the process, but most of these were converted to the methylene linkage after heating to higher temperatures. When amines **24** and **25** were reacted in the presence of 2,4-xylenol (**21**) or 2,6-xylenol (**23**), the results[78] indicate that **24/25** reacted with **21** to produce *para–para, ortho–para* and *ortho–ortho* methylene linkages between phenolic rings. Heating **24/25** with **23** only produced *para–para* methylene linkages and the reaction occurred at a relatively lower temperature compared to the self-decomposition of **24/25**. Numerous side-products were produced during the process, but most of these reacted further to form methylene linkage. Similarly, when **25** was heated with **21** or **23**[79], both the decomposition of **25** and the reaction between them lead to methylene-bridged phenolic structures.

G. Synthesis of Resole Resins

Resoles are synthesized from a phenol to formaldehyde mole ratio less than one. They will harden (cure) on heating and in this respect contrast with Novolacs, which require an additional crosslinking agent for curing to occur.

Resoles are typically generated in aqueous solution under base-catalysed conditions. Early work focused on the rate of reaction, either by the disappearance of phenol and formaldehyde[80] or by the appearance of hydroxymethyl phenols[81–83]. It was shown that the rate of reaction between phenol and formaldehyde is a function of pH[80], suggesting that the overall reaction proceeds with the generation of a phenolic anion, followed by the addition of formaldehyde[1,6], generating a complex mixture of different hydroxymethyl phenol compounds—the resole resin (Scheme 14).

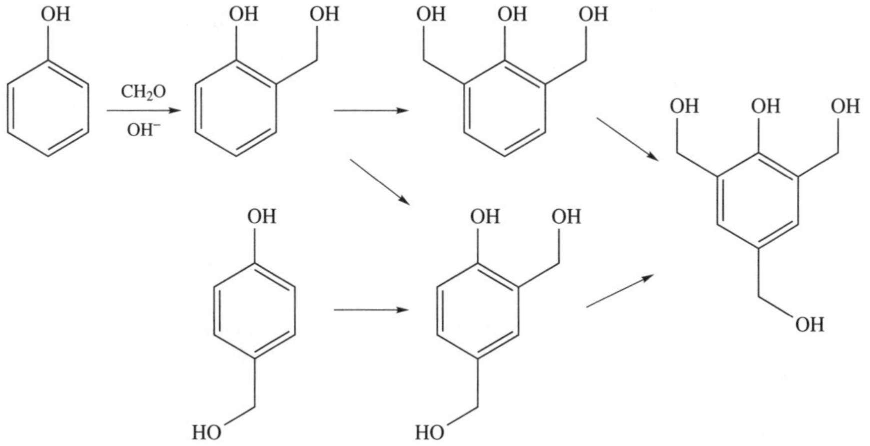

SCHEME 14. Reaction of phenol with formaldehyde under basic reaction conditions

Addition of formaldehyde can occur at three sites; the two sites *ortho* to the phenolic OH and one site *para* to the OH. Once hydroxymethyl compounds are available, there is the potential for reactions that generate dimers, trimers and higher units. These units can further condense to form *ortho–ortho, ortho–para* or *para–para* methylene linkages.

Research has been aimed at understanding the mechanism of these linking reactions. This includes the reactivity of the *ortho* and *para* sites, possible intermediates involved in these linking reactions and behaviour of these higher units to further crosslinking. Attempts have been made to link the properties of the cured resin or carbon derived from these resins to the initial resin formulation and structure. As the crosslinking in a resole is very complicated, various model compounds have been used to investigate the chemistry.

H. Model Compounds of Resoles

Although the overall reaction mechanism is generally understood, the vast commercial importance of phenol–formaldehyde resins has seen numerous studies aimed at a more detailed understanding of the chemistry involved and the structures formed. In these studies extensive use has been made of model compounds, that is, compounds in which the reaction pathways are restricted, and these studies will be considered in this section.

1. Ortho-hydroxyl model studies

Cured Resole resins are hard and insoluble, which makes it difficult to study the reaction by conventional analytic techniques. By using model compounds which have two of the three reactive sites on the aromatic ring blocked, the products of the reaction become relative simple to separate, analyse and characterize. Solomon and coworkers[84] used the model compounds 2,4-dimethylphenol (**21**), 2,6-dimethylphenol (**23**) and 2-hydroxymethyl-4,6-dimethylphenol (**27**), which contain some of the functional groups found in resole resins, to gain an insight into the curing process for *ortho*-hydroxymethyl groups.

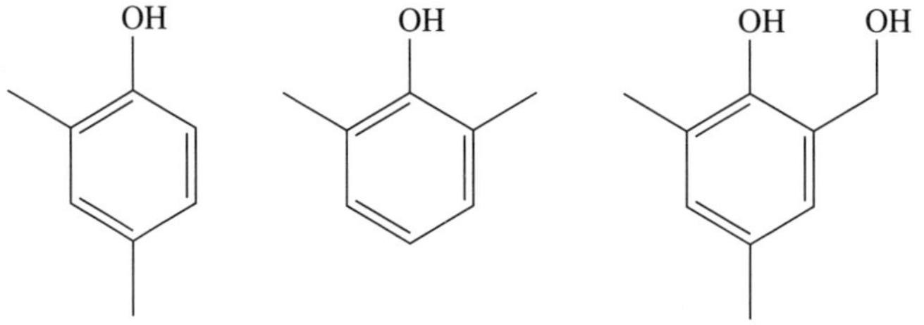

(**21**)　　(**23**)　　(**27**)

a. Reaction of **27**. The self-reaction of 2-hydroxymethyl-4,6-dimethylphenol (**27**) at 120 °C produced bis(2-hydroxy-3,5-dimethylbenzyl) methane (**26**) and bis(2-hydroxy-3,5-dimethylbenzyl) ether (**28**) as major products (Scheme 15), with the ether being produced much faster than the methylene compound.

In contrast to the-self reaction of **27**, the reaction of **27** with one and two molar equivalents of 2,4-dimethylphenol (**21**) gave three products (Scheme 16): the ether (**28**), the methylene compound (**26**) and a phenoxy compound (**29**).

The initial rates of ether formation in the case of a 1:1 mixture of **21** and **27** and the self-reaction of **21** are approximately the same. As more **21** is added, the effect of dilution becomes apparent and the rate of ether formation falls.

The methylene compound **26** forms much faster in the presence of 2,4-dimethylphenol (**21**) than by self-reaction. However, the relative rate of methylene formation does not

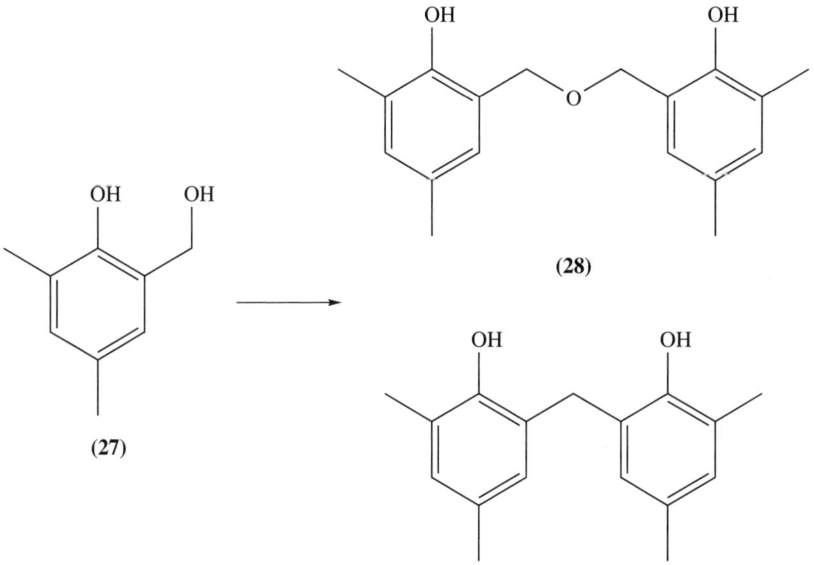

(28)

(27)

(26)

SCHEME 15. Products from the self-reaction of **27**

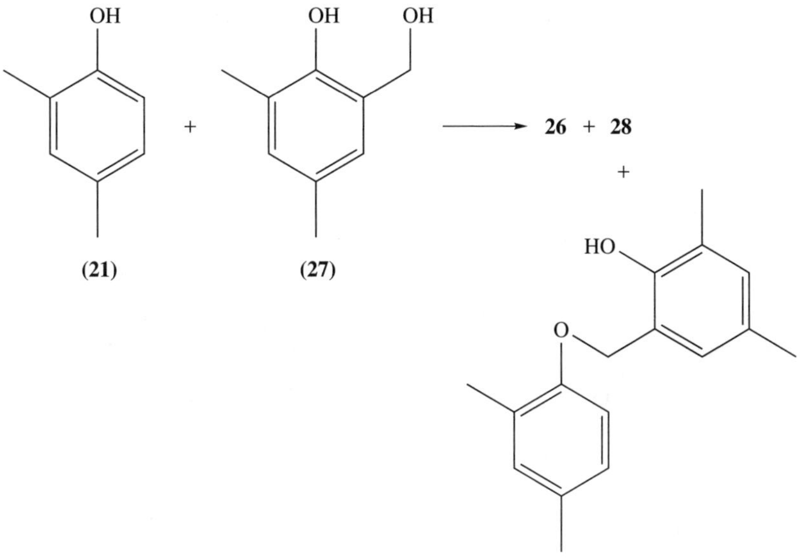

(21) **(27)**

(29)

SCHEME 16. Products from the reaction of **27** with 2,4-dimethylphenol (**21**) change between one and two equivalents of **21**; the time taken to convert a given fraction of **27**, and hence reach a given yield, is independent of the amount of **21** present.

The behaviour of 2-hydroxymethyl-4,6-dimethylphenol (**27**) in the presence of 2,6-dimethylphenol (**23**) was virtually indistinguishable from the self-reaction of 2-hydroxymethyl-4,6-dimethylphenol; thus the rates of formation of ether and methylene compounds are similar. No significant quantities of *ortho–para* linked methylene compound were generated over the timescale studied. A small quantity of the phenoxy derivative **30** was isolated.

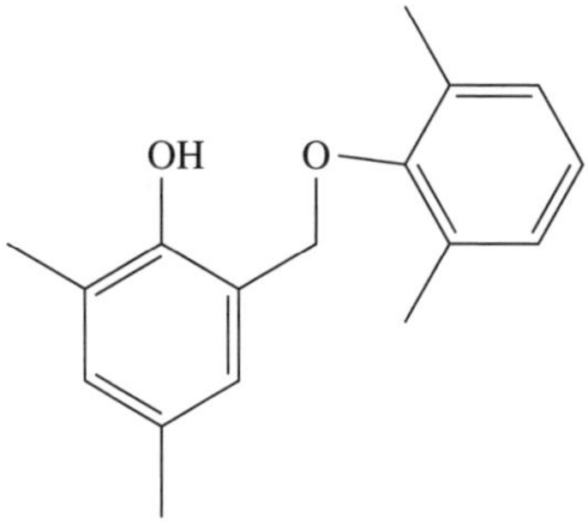

(**30**)

Compound **27** was then heated with **21** and **23** in a 1:1:1 molar ratio, and similar trends were observed. Methylene formation in the 1:1:1 mixture initially (<150 minutes) followed the curve of 1:1 and 1:2, but then dropped away. Ether formation fell midway between 1:1 and 1:2 in the 1:1:1 mixture. The limiting ether yield tended towards an asymptote in the following order: self-reaction of **27** (80%) > 1:1 of **21** and **27** (60%) > 1:1:1 of **21**, **23** and **27** (50%) > 1:2 of **21** and **27** (35%).

2. Ortho-quinone methide

Quinone methide has been previously suggested in resole formation[3]. Solomon, and Wentrup and coworkers[85] have recently observed and isolated quinone methide **31** at low temperature by flash vacuum pyrolysis[86,87] (FVP) of **27**, which was sublimed at *ca* 45 °C in high vacuum (*ca* 4×10^{-6} mbar). The vapour of sublimed **27** was mixed with argon as a carrier gas and passed through a pyrolysis tube. The pyrolysate was immediately condensed on a KBr, BaF$_2$ or CsI target as an argon matrix at 7–12 K and IR spectroscopy of the matrix was conducted at that temperature. At a pyrolysis temperature of 500 °C, 4,6-dimethyl-*o*-quinone methide (**31**) was observed in the matrix together with unchanged **27** (Scheme 17). Above 650 °C, no starting material **27** survived, and **31** and the eliminated H$_2$O were trapped on the target. The IR spectrum of **31** was generated by pyrolysis of **27** at 600 °C and **31** was isolated in an Ar matrix (Figure 5). The main bands of **31** were shown at 1668, 1642/1637 and 1569 cm^{-1}, due to C=O and C=CH$_2$ stretching.

FVP was then performed on a preparative scale, whereby the thermolysate was isolated in a U-tube at 77 K. The use of a U-tube[87], rather than a cold finger, avoids regeneration of the starting material by reaction of quinone methide **31** with eliminated H$_2$O. At a pyrolysis temperature of 800 °C, all starting material had reacted, and a mixture of trimer **32** and tetramer **33** of the quinone methide **31** was isolated from the U-tube in a nearly 1:1 molar ratio and in 98% absolute yield (Scheme 18).

The same trimer and tetramer were also observed by IR spectroscopy in a warm-up experiment of quinone methide **31** isolated neat at 7.6 K on a KBr target. In this

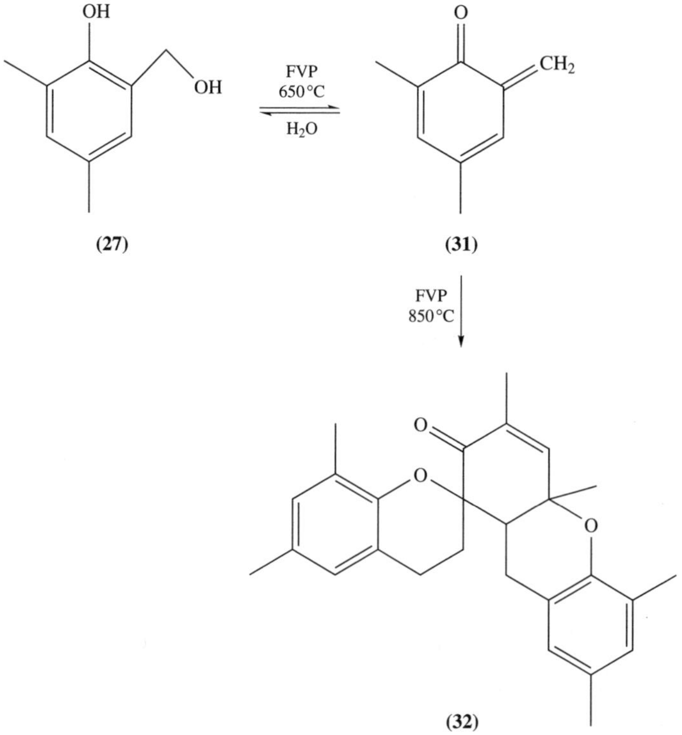

20. Polymers based on phenols

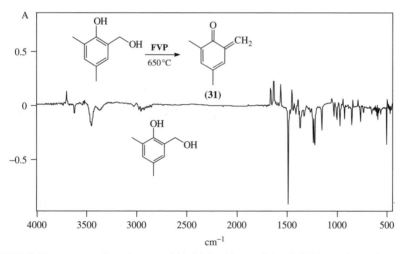

SCHEME 17. Formation of quinone methide **31** by FVP

FIGURE 5. IR spectrum of *o*-quinone methide **31** (positive peaks) at 7.6 K in an Ar matrix, generated by FVP of **27** (negative peaks, arising from a subtraction of the spectrum of **27** from the FVP spectrum) at 650 °C

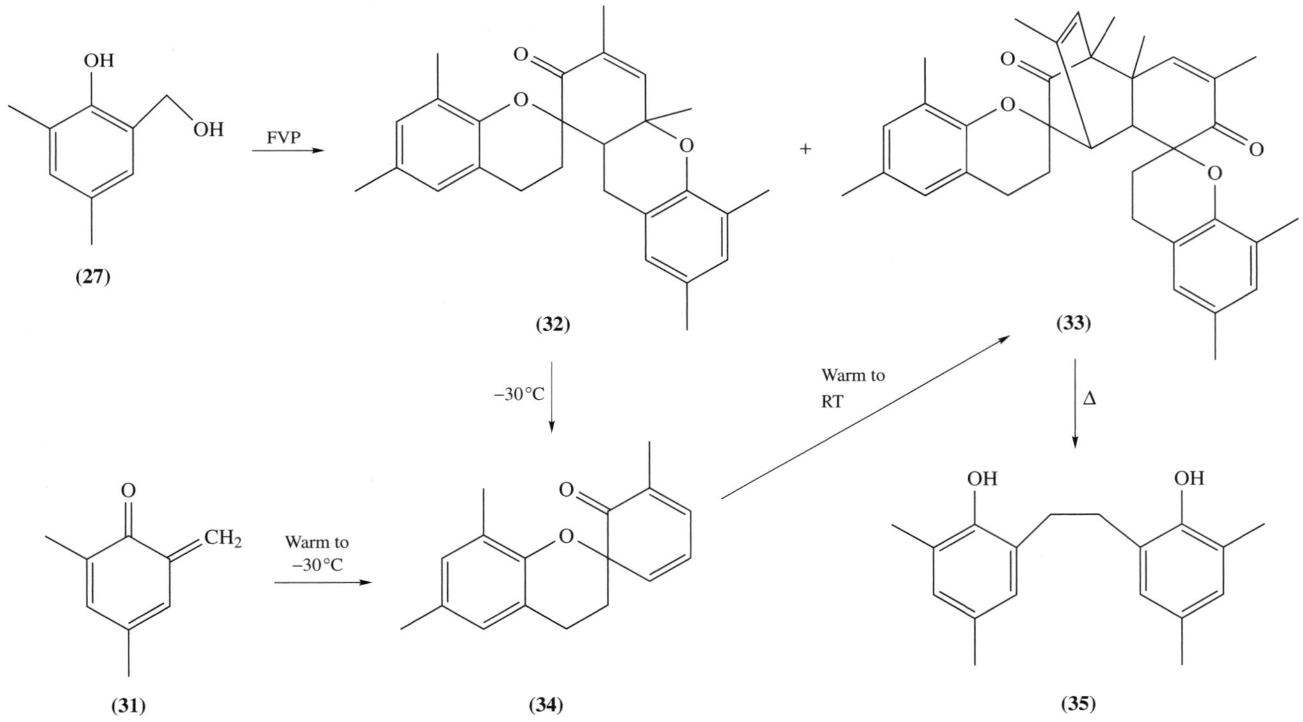

SCHEME 18. Formation of dimer **34**, trimer **32**, tetramer **33** and substituted ethane **35**

experiment, the FVP was conducted at 700 °C without Ar as a carrier gas. A very slow sublimation rate of the precursor **27** was used, since this permits complete conversion of **27** as demonstrated by the IR spectrum. After deposition, the target was slowly warmed to room temperature. IR spectroscopy revealed that the quinone methide **31** had disappeared and precursor **27** had regenerated on the target (reaction starting above −90 °C). Because the water eliminated from **27** was co-condensed on the target, the regeneration of **27** was simply due to reaction of **31** with H_2O. Other major peaks appeared at 1732 and 1696 cm^{-1} and correspond to the main bands of trimer **32** (1695 cm^{-1}) and tetramer **33** (1695 and 1731 cm^{-1}). The material on the target was dissolved in chloroform and examined by GC-MS. Three peaks were observed. Peak 1, containing 94.6% of the mixture according to integration, is the precursor **27**. Peak 2 corresponded to *ca* 3% of the mixture and is due to trimer **32**. The third peak represented *ca* 2.5% of the mixture and had a mass of 270 a.m.u. This is different from the tetramer **33** observed by IR spectroscopy at room temperature in the warm-up experiment. This compound was confirmed as bis(2-hydroxy-3,5-dimethylphenyl)ethane (**35**) (Scheme 18). A quantitative yield of compound **35** was obtained when trimer **32** was thermolysed in the presence of water.

The mass spectrum of trimer **32** indicates that its molecular ion can easily fragment into dimer and monomer. This suggests that **32** is a better precursor of quinone methide **31**, as no by-products would be formed, thus allowing an investigation of the behavior of quinone methide without by-product interference.

Pyrolysis of **32** was carried out at 850 °C. The trimer was sublimed at *ca* 105 °C with argon as a carrier gas. Under these conditions, pure quinone methide **31** was matrix isolated on a 7.6 K KBr target as evidenced by the IR spectrum. In a similar experiment, the pyrolysis of trimer **32** was carried out without argon. When the neat quinone methide **31** was warmed above −92 °C, new IR bands appeared, and the absorptions due to the quinone methide decreased. These newly formed bands became much stronger when the target was warmed further to −65 °C, and they are attributed to the formation of dimer **34** and trimer **32** by comparison with the IR spectrum of the trimer obtained in the preparative FVP work. Moreover, a low temperature NMR experiment revealed the existence of dimer **34**. After the target was warmed to room temperature, additional bands due to the tetramer appeared. It was readily concluded that dimerization would be the first step of reaction of quinone methide **31**.

a. First-order behaviour. The observed behaviour by which both the ether **28** and methylene compound **26** were formed strongly suggests that in both cases the reaction mechanism includes a first-order rate-limiting step. In a first-order step, the time taken for a given fraction of the starting material to react is independent of the starting material concentration. The rate of formation of **26** when **21** and **27** were reacted together was unaffected by doubling the ratio of **21** to **27**. When the concentration of **27** was halved by addition of **23**, the rate of methylene formation and ether formation was unchanged from the self-reaction case.

The presence of a first-order rate-limiting step suggests that the active species is *ortho*-quinone methide (**31**), formed by the intramolecular loss of water from **27**. As described in the previous section, it is the proximity of the hydroxymethyl group to the phenolic OH which allows water loss to occur intramolecularly, and this step would be first-order in **27**.

Compound **23** was found to have minimal reactivity towards **27** in the melt reaction, suggesting that *ortho*-quinone methide does not react with the available *para* position. This was unexpected since *para* preference is observed in phenolic resins[3]. Therefore, this behaviour was further investigated with the quinone methide trimer **32**.

The trimer **32** was heated separately in glass ampules at 150 °C by itself, and with phenol, 2-methylphenol, 2,4-dimethylphenol (**21**) and 2,6-dimethylphenol (**23**) in a 3:1

phenol trimer molar ratio. Under self-reaction conditions a partial retro-Diels–Alder reaction occurred, giving one equivalent each of the *ortho*-quinone methide (**31**) and the bis(2-hydroxy-3,5-dimethylphenyl) ethane (**35**) (Scheme 19).

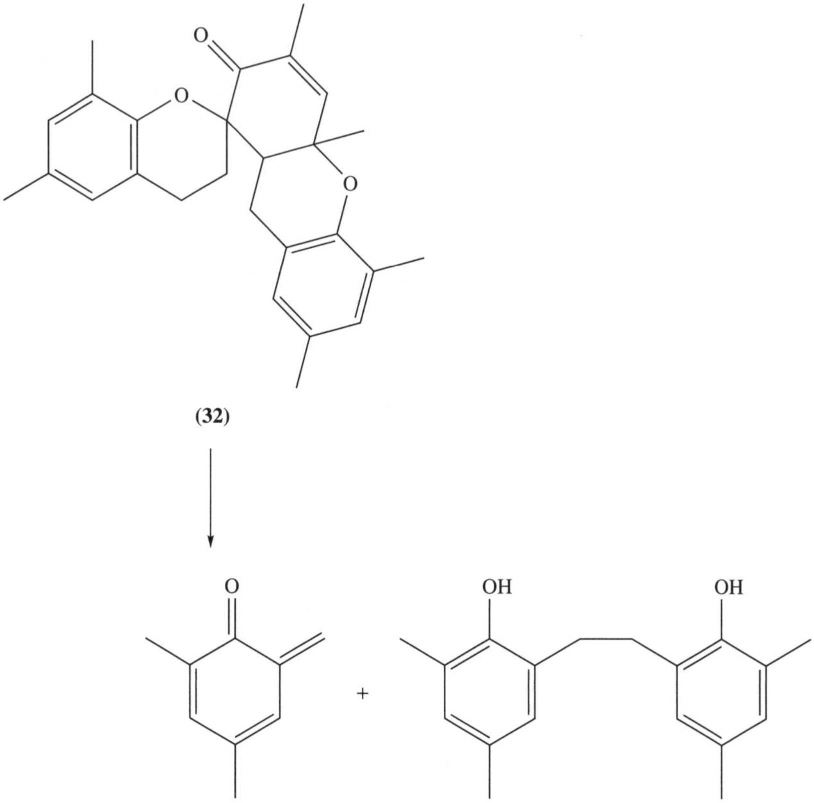

(**32**)

(**31**) + (**35**)

SCHEME 19. The partial retro-Diels–Alder reaction of **32** at 150 °C

The generation of **31** was deduced from the formation of methylene-bridged phenol derivative **36** (Scheme 20). The isolation and direct observation of **31** at cryogenic temperatures has been described previously.

With phenol or 2-methylphenol, the *ortho*-quinone methide (**31**) was found to react entirely at free *ortho* sites. The exclusive *ortho* attack was demonstrated by the distinctive ^{13}C signal of an *ortho–ortho* methylene bridge at 30 ppm[84] with no signal observed at 35 ppm, where an *ortho–para* methylene bridge would appear[88]. The ^1H NMR of the crude reaction mixtures shows the 1:1 molar ratio of methylene bridge to ethylene bridge. With 2,4-dimethylphenol (**21**) the reaction was found to proceed much faster than with 2,6-dimethylphenol (**23**); the ^1H NMR show much greater loss of trimer **32**, and corresponding formation of products, in the presence of **21** compared with **23**.

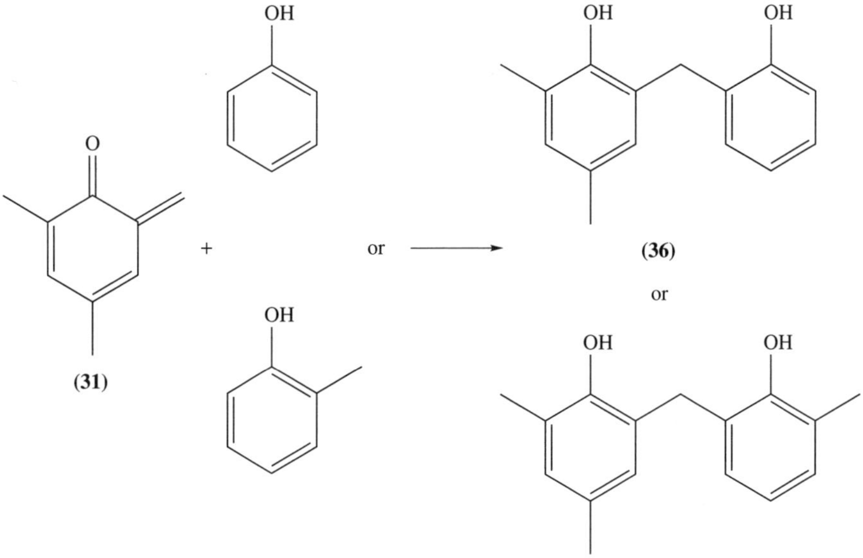

SCHEME 20. The exclusive reaction of **31** at the free *ortho* site of phenol and 2-methylphenol

From these results we would predict that a high *ortho*-bridged resin would be formed when conditions favour the production of *ortho*-quinone methide. This would require a resin which contains predominately *ortho*-hydroxymethyl substituents, and condensation at high temperature, preferably in solvents which encourage dehydration of the *ortho*-hydroxymethyl functionality. The conditions which have been demonstrated to generate a high *ortho*-phenol formaldehyde resin are high condensation temperatures in solvents which generate an azeotrope with water[3]. The catalysts used have been shown to promote *ortho* addition of formaldehyde; subsequent involvement in the condensation reaction has not been demonstrated.

Higuchi and coworkers[89] also demonstrated a genuine first-order kinetics in the condensation reaction of 2-(hydroxymethyl)phenol under basic catalysis. They used LC-MS to monitor the reactant and reaction products. Three main products, dimer **39**, trimer **40** and tetramer **41** (Scheme 21), were observed. By measuring the disappearance of the reactant 2-(hydroxymethyl)phenol, the first-order kinetics is confirmed.

3. Ether exchange reactions

The phenoxy-linked compounds **29** and **30** were isolated from both the reaction of **27** with **21** and **23**. Such phenoxy compounds were not observed in the reaction of *ortho*-quinone methide **31** with any of the phenol and methylphenols. When the ether **28** was heated in the presence of D_2O, it decomposed slowly to **27**. When mixed with **21**, the phenoxy compound **29** was rapidly generated, along with **27** and **26**. From this, it was concluded that the phenolic OH undergoes an ether exchange (Scheme 22), and it is the subsequent reaction of **27** with **21** which generates **26**.

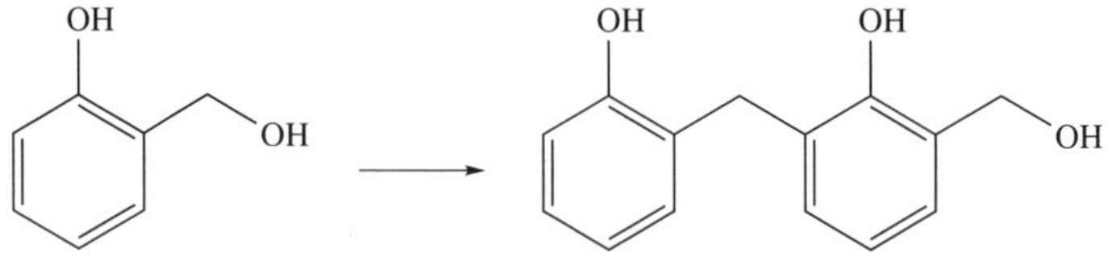

SCHEME 21. Base-catalysed self-condensation of 2-(hydroxymethyl)phenol

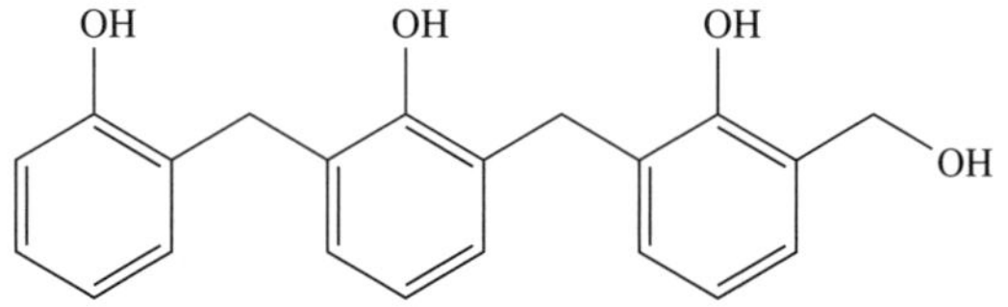

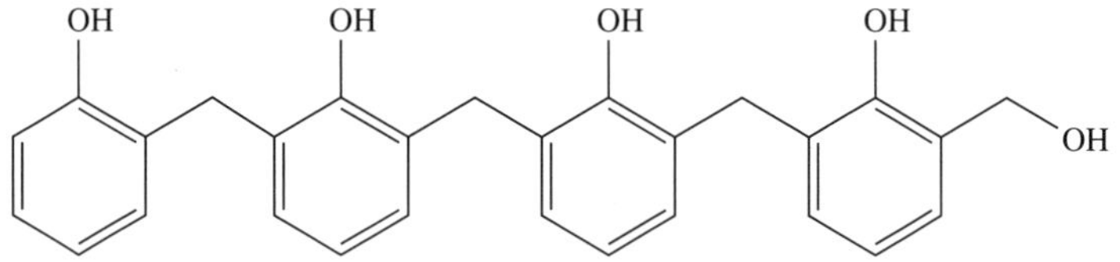

4. PF ratio effects

Many different structures have been identified within cured resole resins[3]. The most common crosslink is the methylene bridge, though ethers can also be present in significant amounts[3,6]. Phenoxy bridge[90], and carbonyl and methyl groups[90-93], have also been identified within the cured structure.

Model studies have shown that the ether can react with unsubstituted phenol to generate a phenoxy bridge[94]. Reaction of the ether bridges has also been suggested as the source of the observed carbonyl and methyl functionalities observed in a cured resole[90-93]. An alternative proposal[91] involves the oxidation of hydroxymethyl groups, and scission of methylene bridges, respectively. Carbonyl and methyl groups can be regarded as broken crosslinks, and may well affect the final properties of the carbonized resin by reducing the degree of crosslinking and by decomposing more readily at high temperatures. If the structure of the cured resin is to be controlled, it is important to understand the processes which result in these types of groups and their subsequent behaviour at higher temperatures. Solomon and coworkers[94] compared the curing and carbonization behaviour of two resole-type resins with a molar formaldehyde to phenol (FP) ratio of 1.2 and 1.8. They focused on the formation of carbonyl and methyl groups during curing, and the differences between the two materials during the heating.

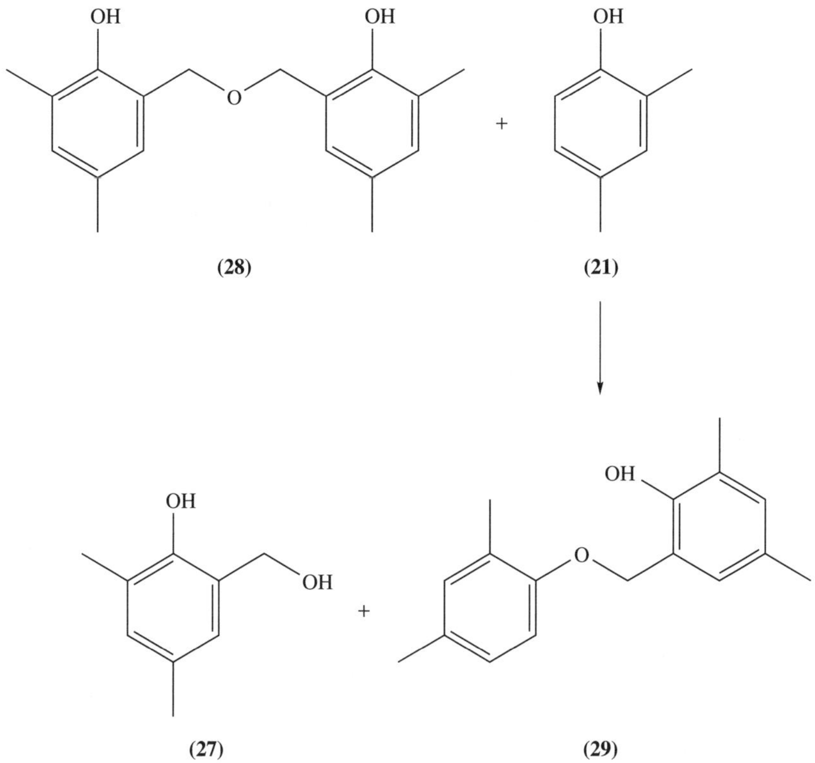

SCHEME 22. Ether exchange between **28** and **21** generates **27** and the phenoxy derivative **29**. **27** will either self-react, regenerating **28**, or react with **21** to give the methylene **26**

There are significant differences between the ^{13}C CP-MAS spectra of the cured resins with a formaldehyde/phenol (FP) ratio = 1.2 and FP = 1.8. There is no peak at 110 ppm in the FP = 1.8 resin, indicating complete substitution at the *ortho* position[95,96], unlike in the FP = 1.2 resin. The cured FP = 1.8 resin contains methyl (10 ppm), phenoxy (150 ppm) and carbonyl groups (190 ppm)[3,91] while the cured FP = 1.2 resin does not. Interrupted decoupling[101,102] identifies the carbonyl in this material as an aldehyde.

The differences between FP = 1.2 and 1.8 materials do not support the interpretation that the carbonyl was generated by oxidation of a hydroxymethyl and the methyl generated by scission of a methylene bridge at typical curing temperatures (up to 200 °C). If the carbonyl and methyl groups were derived from hydroxymethyl and methylene bridges respectively, then the carbonyl and methyl intensities should change in proportion with the FP ratio. But they are undetectable in the FP = 1.2 material, while they are readily apparent in the FP = 1.8 material. It is also unlikely that hydroxymethyl groups will survive to sufficiently high temperatures to be oxidized. These are quite labile groups, readily generating quinone methides which subsequently react to give ether or methylene bridges[84]. Clearly, this mechanism cannot be the primary source of carbonyl and methyl groups.

The compounds with the linkage of *ortho–ortho* (**42**), *para–para* (**43**) and *ortho–para* (**44**) were heated to different temperatures. The CP-MAS spectra obtained from these materials showed that the carbonyl group is detectable after 4 hours at 160 °C and always appears first, and this observation is clearly not consistent with the previously proposed ether fragmentation mechanism, which requires simultaneous formation of the carbonyl and methyl functional groups. The CP-MAS spectra showed the material with all three possible ether orientations and crosslinking with bis(3,5-dihydroxymethyl-4-hydroxyphenyl)methane (THBF) **45** is an example. At the hardened stage and after heating at 160 °C for 4 hours the formation of carbonyl groups (190 ppm) without the corresponding generation of methyl groups can be seen.

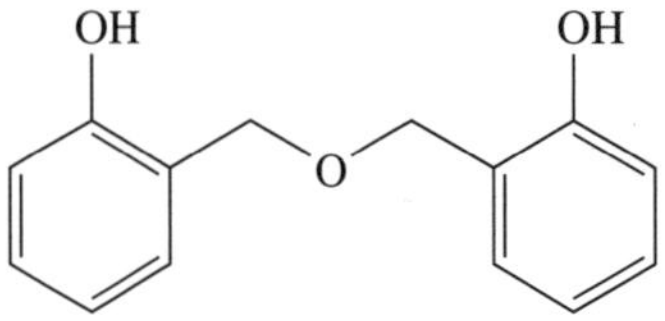

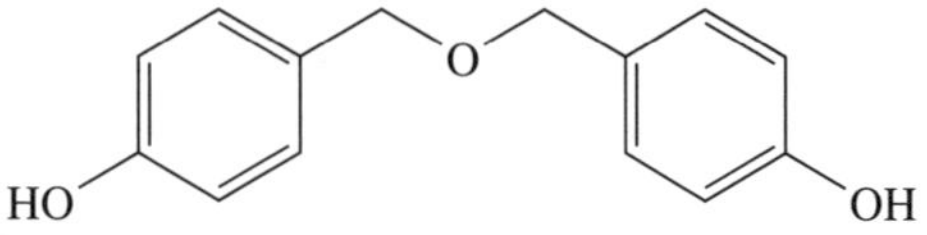

The different behaviour of the FP = 1.2 and FP = 1.8 material shows that the formation of carbonyl and methyl groups is in some way dependent on the formaldehyde content of the starting resin. Hemiformal groups have been found in resole resins, and it was considered possible that they might be contributing to the formation of either or both the methyl and carbonyl groups, since the hemiformal content would be expected to increase with a higher initial formaldehyde content. However, the materials with large quantities of hemiformal groups were not observed to produce either methyl or carbonyl groups at a lower temperature than the THBF **45** cured materials.

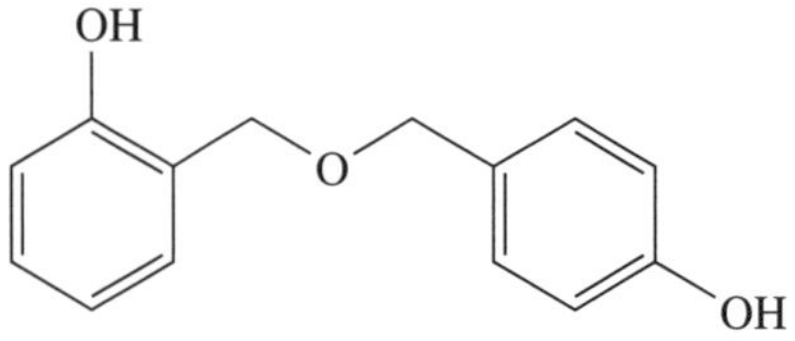

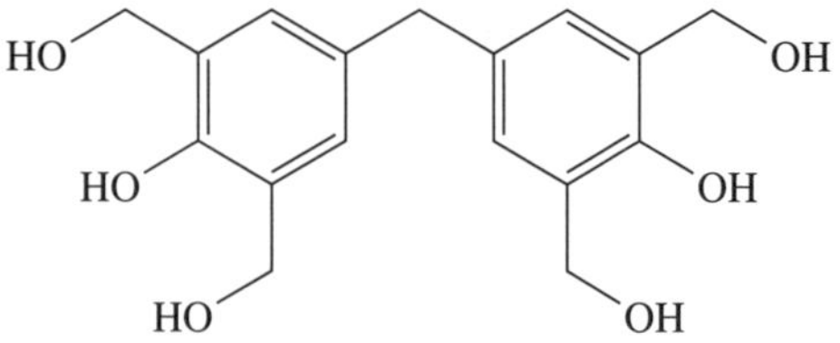

Other points of difference between the FP = 1.2 and FP = 1.8 resins were considered. Phenoxy bridges, shown to be a product of ether exchange at a bridging ether[87], are evident in the FP = 1.8 resin, and not in the FP = 1.2 resin, and it was concluded that the FP = 1.8 material contained significantly more ether bridges than the FP = 1.2 material at some stage of the curing process. Hydroxymethylphenols react faster with formaldehyde than unsubstituted phenols[80,82,83,99–102], and hence hydroxymethyl groups are not distributed evenly on all phenol rings in the resin; instead, a mixture of heavily hydroxymethyl-substituted and much less substituted phenols is created. The formation of dibenzyl ether bridges requires hydroxymethyl groups on separate phenols. The distribution of hydroxymethyl groups makes this situation less likely in the FP = 1.2 resin, compared with the FP = 1.8 resin, than a simple consideration of the formaldehyde to phenol ratio would indicate.

If it is possible for quinone methides to be reduced to methyl phenols, we would predict this to be more strongly favoured, from a kinetic perspective, when the quinone methide is generated in a highly crosslinked material with few or no free aromatic sites available for reaction; that is, when the quinone methide is unable to gain access to all but the nearest phenol rings and the reactive sites on those rings are already occupied. This extreme was tested by hardening THBF **45** itself. As predicted, methyl groups (5 ppm) are observable after 4 hours at 180 °C in this material, but still unobservable in the formaldehyde and THBF **45** hardened resins at these conditions, and higher temperatures are required before methyl groups are observed. The overall proposed mechanism is summarized in Scheme 23.

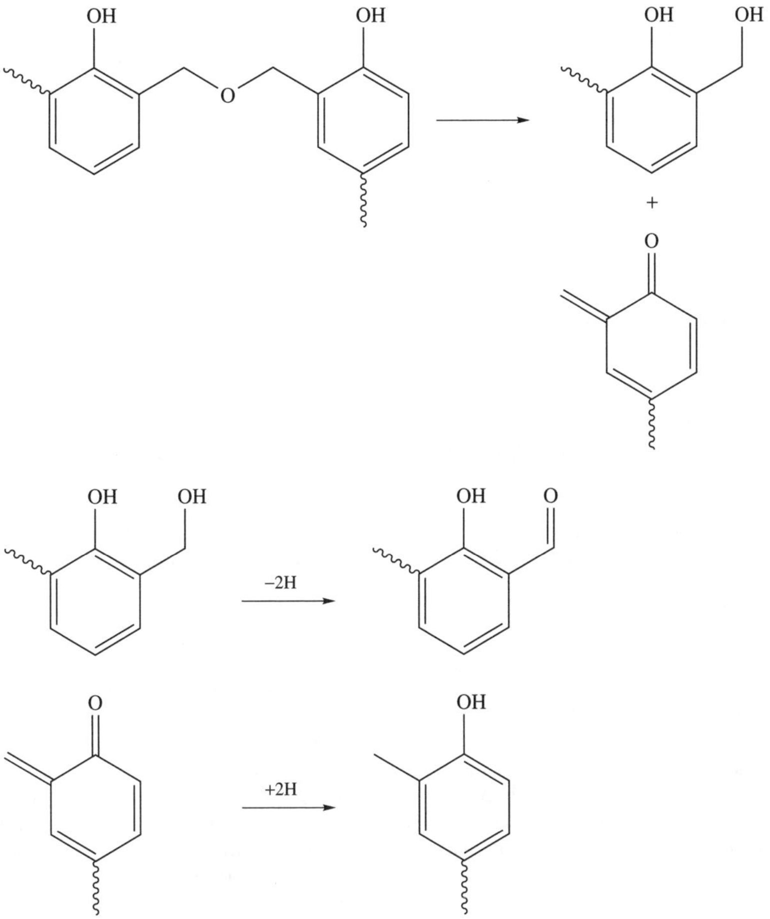

SCHEME 23. Proposed mechanism of ether scission, generating a hydroxymethylphenol and quinone methide, which subsequently react to give a carbonyl and methyl group, respectively

I. Phenol–Ketone Novolacs

Resins derived from ketones are not nearly as common as those prepared from aldehydes. However, an important industrial dimer is Bis-phenol A, made by the controlled condensation of acetone and phenol.

Bis-phenol A is an important intermediate in the manufacture of epoxy resins (see Section IV.B).

III. CARBON DERIVED FROM PHENOLIC RESINS

Phenolic resins have been used commercially as starting materials to produce glassy carbons with high carbon yields. The carbonization reactions of phenolic resins and the properties of the carbon materials derived from the resins have been investigated a great deal for several decades. The resins have also been applied as binding materials in carbon composites, reduction composites and refractories in the aluminium and steel industries. An understanding of the relationship between the structures of the starting polymer resins, the carbonization chemistry and the properties of the carbon materials obtained after pyrolysis is fundamental to the application and modification of the carbon materials. However, few studies have addressed the chemical processes that occur from curing through to subsequent carbonization of phenolic resins. The following section describes some recent publications addressing this area.

A. From Novolacs

Recently, Zhang and Solomon[100] reported on the chemistry of reacting Novolac/furfuryl alcohol (FA) resins with HMTA. A highly crosslinked homogeneous network that incorporates both Novolac and furan entities is formed after curing the mixture to 205 °C. Minor amounts of nitrogen-containing structures are generated in the process. The pyrolysis of Novolac and FA resins proceed by different reaction pathways; therefore, it was of interest to study the carbonization process of the homogeneous mixture of Novolac/FA resins. The chemical structure, especially the nitrogen structure in the carbon products obtained, is another interesting issue to be examined. They further reported the study on the carbonization reactions of HMTA-cured Novolac/FA resins. High-resolution, solid-state NMR techniques were used to follow the changes of chemical structure during the pyrolysis up to 800 °C.

Two different Novolac resins in two Novolac/HMTA/FA formulations were studied with one being a high *ortho*-linked resin and the other a conventional resin. Carbonization reactions of Novolac/HMTA/FA resins mainly occur at a temperature range of 300–600 °C, and aliphatic species disappear above 800 °C. About 2–3% nitrogen still remains in the carbon materials obtained after baking to 800 °C. The pyrolysis process can be influenced by the chemical structure of the starting Novolac resins (*ca* the ratio of *ortho/para* reactive sites) and the FA content in the mixed systems. Where a Novolac resin contains a high ratio of *para*-unsubstituted phenolic positions as reactive sites, the system undergoes a relatively fast reaction and the carbonization occurs at relatively lower temperatures, because the *para* sites are more reactive in both the curing and initial pyrolysis processes. A high FA content slows down the carbonization process, causing the intensities of aliphatic carbons to decrease more slowly and reactions occur at relatively high temperatures. Original Novolac structures and FA content in the systems also vary the nitrogen structures during the carbonization process and the structure distribution in the carbon materials obtained at 800 °C.

B. From Resoles

Carbonized phenolic resins are usually highly microporous, with the amount of open micropores passing through a maximum at a carbonization temperature of 700 °C to

800 °C and then falling at higher carbonization temperatures[104–106]. Attempts to use this property to make molecular sieves[106] have shown some success, but information on how micropores form and develop in carbonizing resins is limited.

Carbon precursors can be broadly divided into two different classes, graphitizable and non-graphitizable. Graphitizable materials develop a graphitic structure on heating to temperatures approaching 3000 °C. Phenol–formaldehyde resins are precursors for non-graphitic carbon[107], which remains highly disordered even on heat treatment to 3000 °C.

Graphitizable materials pass through a liquid crystalline (LC) phase while carbonizing[108], and it is probable that this ordered fluid phase is necessary for graphitization to occur[114]. At its onset, parallel aromatic sheets start to form and grow. As the temperature is increased, this short-range order extends to larger and larger scales, the distance between aromatic sheets starts to approach that of graphite and ripples in the aromatic sheets are smoothed out[109]. Phenol-derived resins are not observed to pass through this liquid crystalline phase[107]. The extended, rigid network formed on the curing of a phenol–formaldehyde resin presumably works against the formation of a fluid phase during carbonization. So it is something of a puzzle that, although phenol-based resins are not graphitizable, 3,5-dimethylphenol (3,5-DMP) resins are reported to be graphitizable[110–112].

Solomon and coworkers have recently reported[113,114] investigations into the carbonization behaviour of a range of resins derived from phenol, *para*-alkylphenols and 3,5-dimethylphenol with particular emphasis on the micropore structure of these carbonized materials. It was anticipated that through comparison of the carbonization behaviours of non-graphitizable (normal resole-type phenols) and graphitizable (3,5-DMP resoles) resins, the mechanism of the carbonization process from phenol could be further understood.

1. Early stage of carbonization

Three novel model compounds, bis(2-hydroxy-4,6-dimethylphenyl)methane (**46**), (2-hydroxy-4,6-dimethylphenyl-4′-hydroxy-2′,6′-dimethylphenyl)methane (**47**) and bis(4-hydroxy-2,6-dimethylphenyl)methane (**48**), were synthesized from 3,5-dimethylphenol. These were used to show that a resole-type resin formed from 3,5-dimethylphenol had a highly condensed, predominately linear structure, linked by *ortho–ortho* and *ortho–para* methylene bridges. This is quite unlike the behaviour of phenol-derived resoles.

It was found that **46** would form 1,3,6,8-tetramethylxanthene **49** on heating in dilute solution (Scheme 24), and would crosslink if heated by itself. ^{13}C CP-MAS solid-state NMR showed that the crosslinked material had a new carbon resonance at 30 ppm, and this was shown to be due to a CH_2 group. Neither of these reactions occurred when bis(2-hydroxyphenyl)methane was used. Solid-state NMR was used to show that the resole resin from 3,5-dimethylphenol also had a CH_2 peak at 30 ppm when heated to 300 °C.

It was concluded that the formation of a xanthene is a key step in the graphitization of 3,5-dimethylphenol resins. Xanthene formation is an efficient way of removing heteroatoms. This step would not be possible if the 3,5-dimethylphenol resin was not significantly *ortho–ortho* linked. However, the *ortho–ortho* methylene orientation, though essential, is not the only influencing factor. The methyl groups in the 3- and 5-positions also influence the xanthene formation process, as this reaction was not detectable in bis(2-hydroxyphenyl)methane under comparable conditions.

The formation of the xanthene also offers an explanation for the reports[111,115,116] that contact with air must be avoided during the resin's synthesis and curing, if a graphitizable material is to be obtained. Oxidation sensitivity is entirely consistent with a xanthene compound being the key intermediate, since they can be readily oxidized to form a xanthene-9-one[117].

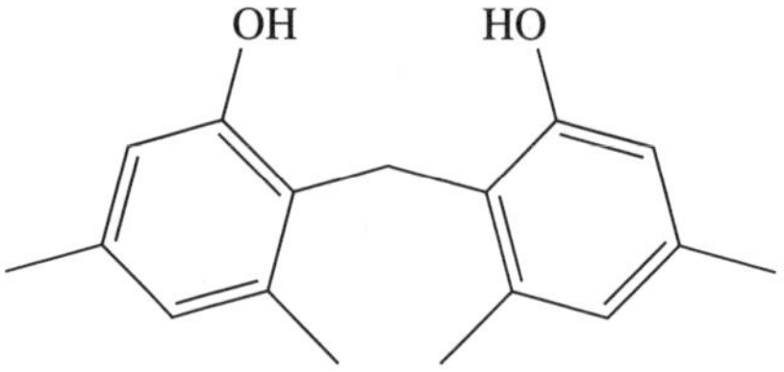

SCHEME 24. Formation of xanthene **49** from **46**

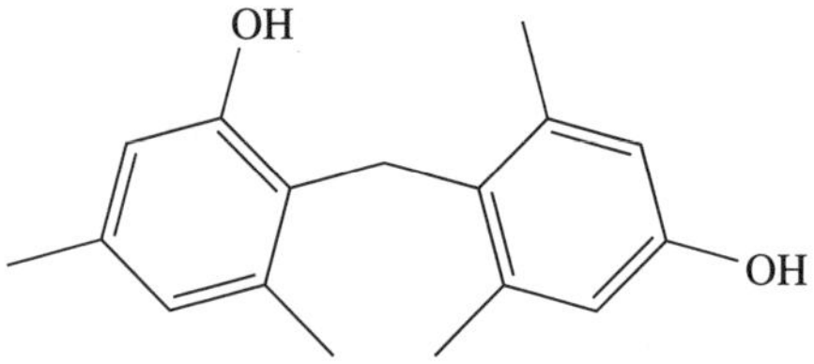

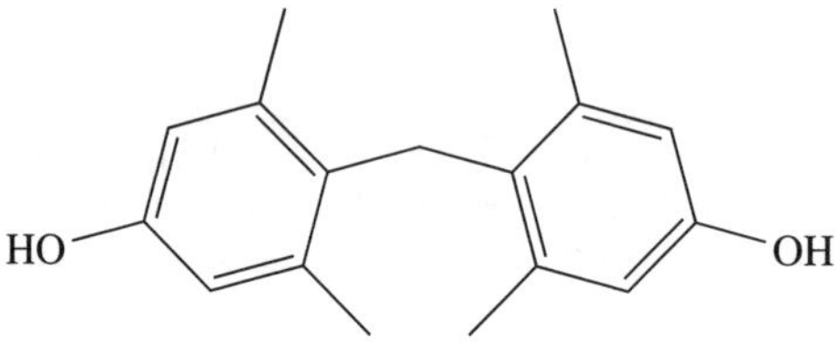

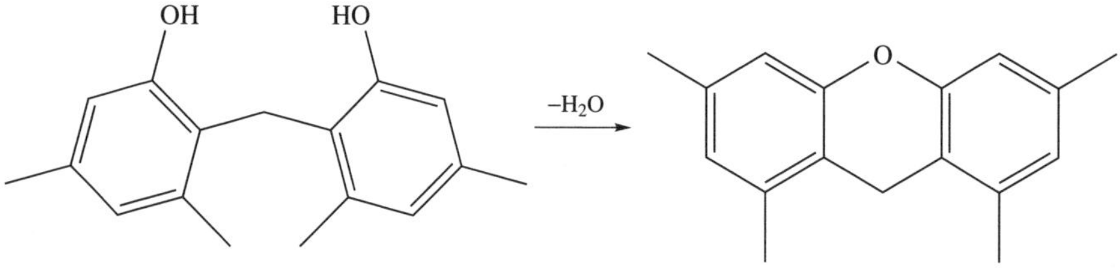

2. Carbon from resole resins

The rewards for being able to understand and control the process of carbonization to give a particular pore structure are potentially enormous, with applications which include catalysis, carbon-in-pulp metal adsorption and separation processing, molecular sieves and bioethical applications.

Investigations into the carbonization behaviour of a range of resins derived from phenol, *para*-alkylphenols and 3,5-dimethylphenol with particular emphasis on the micropore structure of these carbonized materials has been carried out by Solomon and coworkers[113,114].

Carbonized materials based on *para*-alkylphenols (**50–53**) had an unusually high degree of microporosity when compared with conventional phenol–formaldehyde resins. It was possible to generate high surface area materials from conventional phenol–formaldehyde resins by grinding the cured resin prior to carbonization. Carbonization of four phenol–formaldehyde powders containing a narrow particle size distribution showed that surface area increased rapidly as the resin particle size fell. The effect is extremely pronounced, and has not been previously reported.

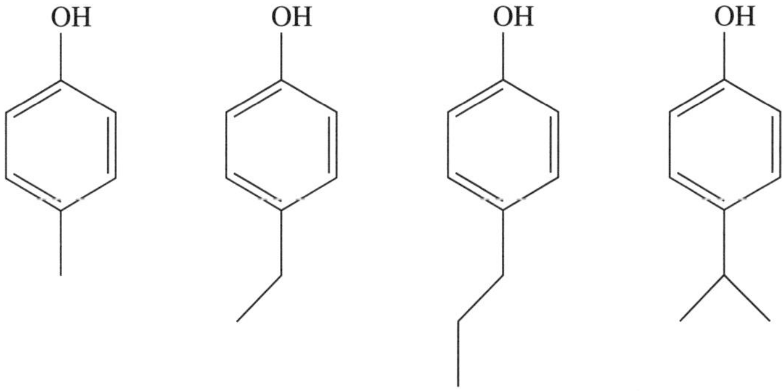

OH   OH   OH   OH

(50) (51) (52) (53)

It is not clear at this stage if the increase in surface area on grinding is due to the same mechanism responsible for the high surface area of the *para*-alkylphenol material. The *para*-alkyl material is extremely brittle and crumbly in texture, and it seems possible that the reduced functionality of the material has created a system of small domains, which on carbonization behave like finely ground phenol–formaldehyde resin.

A resin based on 3,5-dimethylphenol, reported[110,111,116] to pass through a liquid crystal phase and to be graphitizable, was synthesized and carbonized. It was found that the differences between ground and unground samples were much less pronounced with this resin.

A range of behaviours and surface areas is obtainable from carbonized phenolic resins. This extreme variability means that care should be taken when comparing carbonized materials. However, it also potentially gives these materials the ability to be used in a wide range of applications where a well-defined pore size is important, including their use as molecular sieves, catalyst supports and as model carbons for investigating adsorption processes.

IV. OTHER POLYMERS WITH PHENOLIC COMPONENTS

Here we briefly mention some of the commercially important polymers with phenolic components.

A. Poly(phenylene oxides)

Oxidative coupling of phenols was first reported by Hay and coworkers in 1959[118] and has since been developed to produce commercially useful polymers. In these reactions the parent compound, phenol, has a potential functionality of four, that is the two *ortho* and the one *para* position of the aromatic ring and the phenolic group. Not surprisingly, the commercially useful polymers are made from substituted phenols in which the potential functionality is reduced to two. Of these phenols 2,6-dimethylphenol or *ortho*-xylenol has been developed to a commercial polymer, poly(2,6-dimethyl-1,4-phenylene oxide) (**54**). The General Electric Company sells this as a blend with polystyrene under the trade name Noryl.

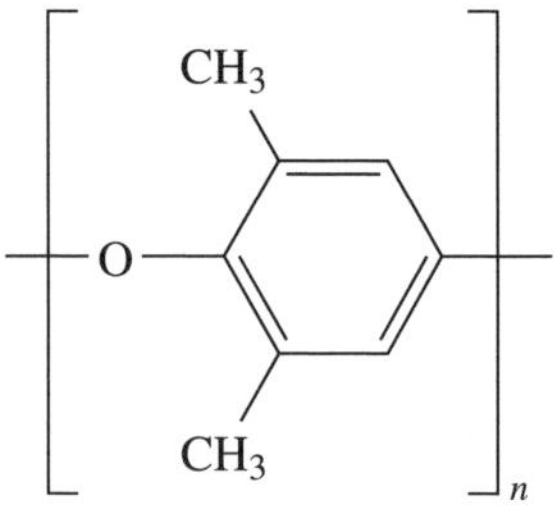

(54)

The oxidative coupling uses a copper-catalysed system and a base, usually an aliphatic or heterocyclic amine, and oxygen as the oxidizing agent. In broad terms, free-radical processes are involved to explain the polymerization pathway which involves formation of the phenoxide radical, and coupling of two radicals through the attack by an oxygen-centred radical at the *para* position of another phenolic molecule (Scheme 25).

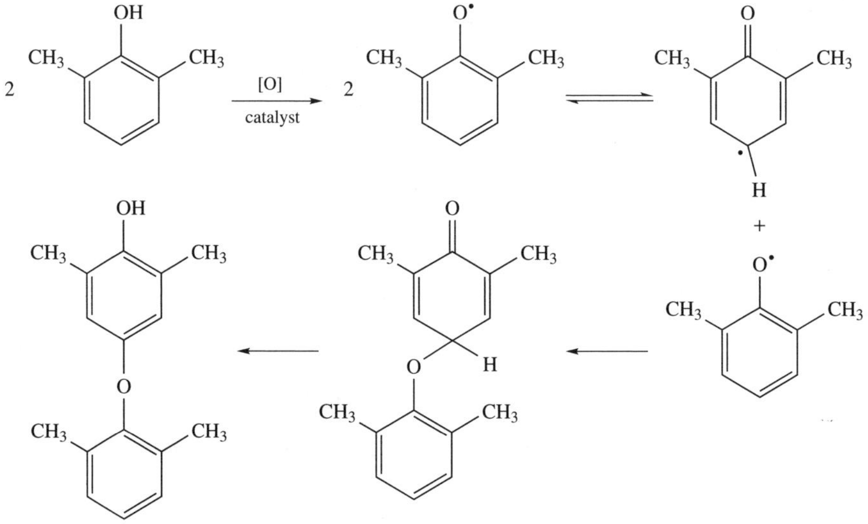

SCHEME 25. Polymerization pathway of phenols under oxidation conditions

B. Epoxy Resins

Polymers that contain an epoxide group include various carbon skeletons, but by far the most important group commercially are formed from reaction of bisphenol A or bisphenol F with epichlorohydrin. By manipulation of the mole ratio of reactants and of the reaction conditions, a range of polymers is formed in which the value of n in formulas A–C varies from 0 up to about 12. Formulas A–C are idealized formulas and it is believed that variations from this occur. However, discussion of such matters is beyond the scope of this chapter and the reader is referred elsewhere[119-121].

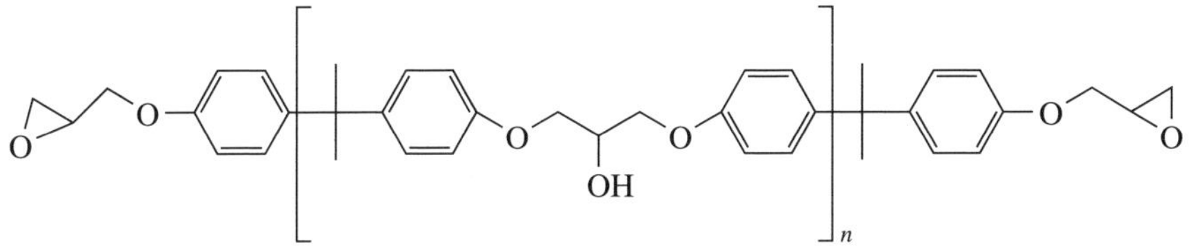

FORMULA A. Bisphenol A diglycidyl ethers ($n = 0, 1, 2, 3, \ldots$)

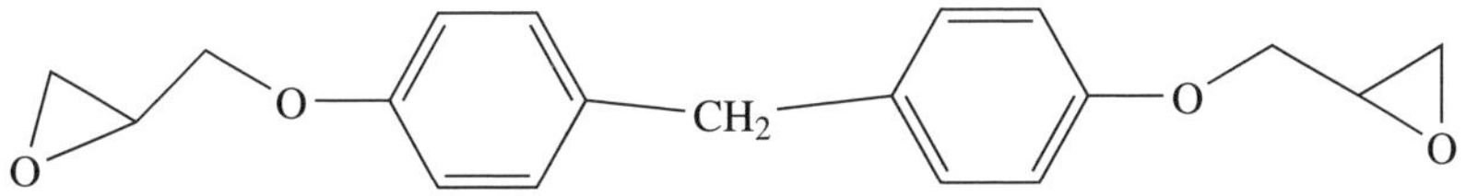

FORMULA B. Bisphenol F diglycidyl ethers

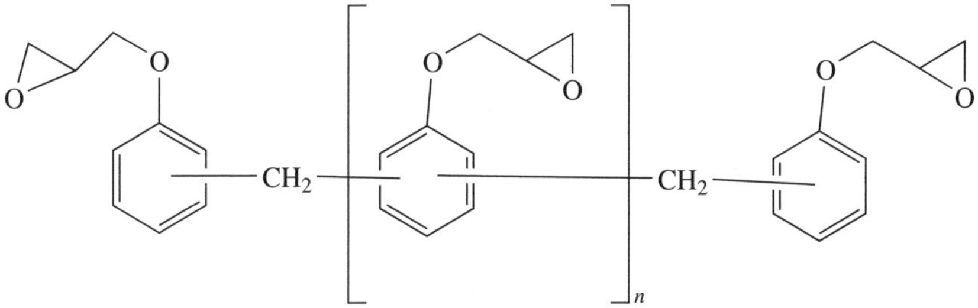

FORMULA C. Epoxy Novolac

Alternatively, a Novolac resin can be used in place of the bisphenol A or F and this gives rise to an epoxy resin with a higher functionality in terms of epoxide groups per molecules, such as formula D. Thus, whilst the bisphenol resins have a maximum of two epoxides per molecule (theoretical maximum, the actual value is slightly less), the Novolac can have up to about 10, the average chain length of commercial Novolacs.

C. Polyimides

The development of high-performance polymeric materials has been at the forefront of scientific endeavours due to the demands of the modern electronic and aerospace industries. Many classes of high-performance polymers have been reported, including poly(aryleneether)s, poly(phenylenesulphide)s, polymaleimides, polybenzimidazoles and polyimides. Polyimides combine good physical properties (i.e. durability, toughness etc.) with excellent thermo-oxidative stability, and as a consequence the markets and applications for polyimides have been expanding at an ever increasing rate. The number of literature references pertaining to polyimides has grown to more than 20,000 since their discovery[122,123]. Further, the number of patents arising from basic research in polyimides has begun to outnumber the journal references, indicating the potential commercial significance of much of the work.

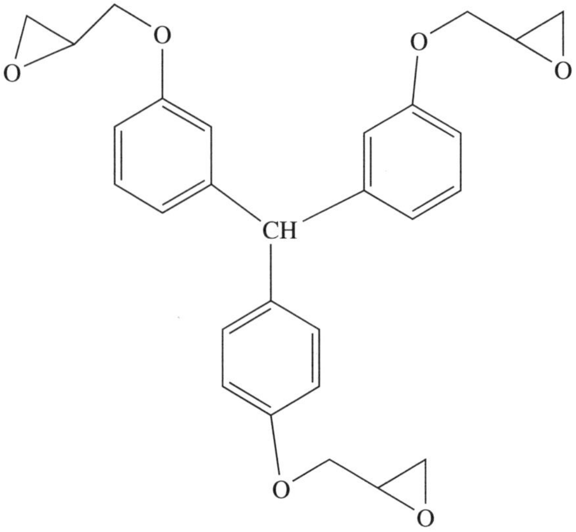

FORMULA D. Tri(hydroxyphenylene)methane triglycidyl ether

1. Historical perspectives

Historically, the origins of polyimides can be dated back to 1908, when Bogert and Renshaw[124] reported that 4-aminophthalic anhydride **55** evolved water upon heating with the possible formation of a 'polymolecular imide' **56** (Scheme 26).

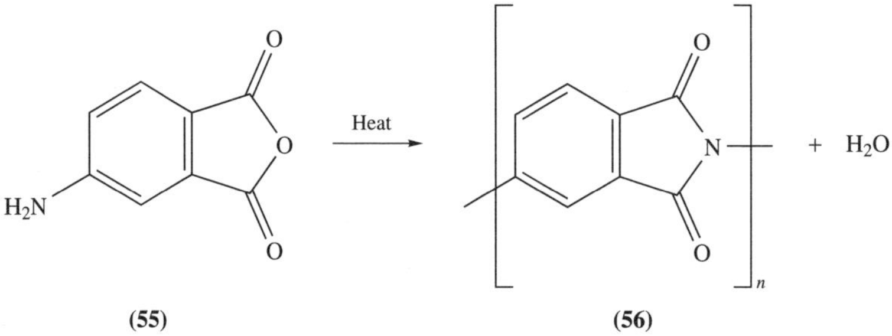

SCHEME 26. First reported reaction to form a polyimide

Improvements to the solubility and processability of the polyimides since then have resulted in a plethora of new discoveries and applications[125–128]. The growth in research and new developments has mirrored the needs of the electronics and aerospace industries for high-performance materials[129].

2. Linear polyimides

Linear polyimides were one of the first types of polyimides to be used industrially. They are generally synthesized via a two-step scheme; one such example is the reaction of the diamine with a dianhydride, in a 1:1 mole ratio, to form a polyamic acid **57**. Following a cyclodehydration reaction, the linear polyimide (**58**) and water is produced, as outlined in Scheme 27[130].

Polyimides of this type are extremely thermally stable, often with glass transition temperatures (T_g) above 300 °C[123,131] and good graphitizability[132-134]. However, their brittleness and insolubility cause severe fabrication problems. Modification of the physical characteristics of the polyimides (i.e. molecular weight, fracture toughness, T_g etc.) can be achieved by altering the nature of the polymeric backbone.

A single material does not generally meet the requirements on high-performance polymers in modern applications. This problem can often be addressed by using a blend of two or more components whereby the desirable physical properties of both components can be expressed in a composite material[5]. The general insolubility and infusability of linear polyimides do not often allow effective blending with other components. It has been found that by blending the polyimide at the polyamic acid stage, prior to the cyclodehydration reaction, some of the fabrication problems can be overcome[135-137]. However, a satisfactory solution is yet to be developed for all the problems associated with linear polyimides[138].

3. Bis(maleimides)

Bis(maleimides) (BMI) are an important class of polyimides that are characterized by excellent thermal, electrical and mechanical properties[139,140]. Bis(maleimides) are low molecular weight oligomers, generally containing terminal reactive groups. A general structure of a bis(maleimide) is shown in Scheme 28.

The maleimide group can undergo a variety of chemical reactions, including polymerizations induced by free radicals[141-143] or anions[144]. Nucleophiles such as primary and secondary amines[145,146], as well as thiophenoxides[147,148], can react via a classical Michael-type addition mechanism[122,149]. The maleimide group can also act as a very reactive dienophile and is thus used in a variety of Diels–Alder reactions[150-153]. By varying the nature of the linkages between the maleimide rings, the physical properties of the bis(maleimide) can be altered.

Bis(maleimides) have been useful in many applications, including the electronic industries, as materials for printed circuit boards and insulators, and in the aerospace industries in matrix resins for structural composites[138,154]. Alternative polyimide systems have been investigated which are formed via the free-radical polymerization of the maleimide ring.

4. Poly(N-(substituted phenyl)maleimides)

The thermal and oxidative stability of polyimides is thought to be related to the combination of both the five-membered cyclic imide ring and the nature of the aromatic ring directly connected to the nitrogen (Figure 6).

Poly(N-substituted maleimides) are formed by the free-radical chain polymerization of the corresponding maleimide monomer[155-161]. Unlike bis(maleimides), this class of polyimide contains only one reactive maleimide ring, producing a polymeric material characterized by high thermo-oxidative stability together with good physical properties (i.e. solubility, tensile strength, flexibility etc.)[162,163]. By altering the nature of the aromatic ring, the chemistry of the polyimide can be manipulated to suit the desired application. Recent studies have focussed on polyimides with phenolic ring substituents, as shown in Figure 7.

Matsumoto and coworkers[164-169] have described the use of poly(N-(hydroxyphenyl)-maleimides) in blends with phenol–formaldehyde resins which, following crosslinking

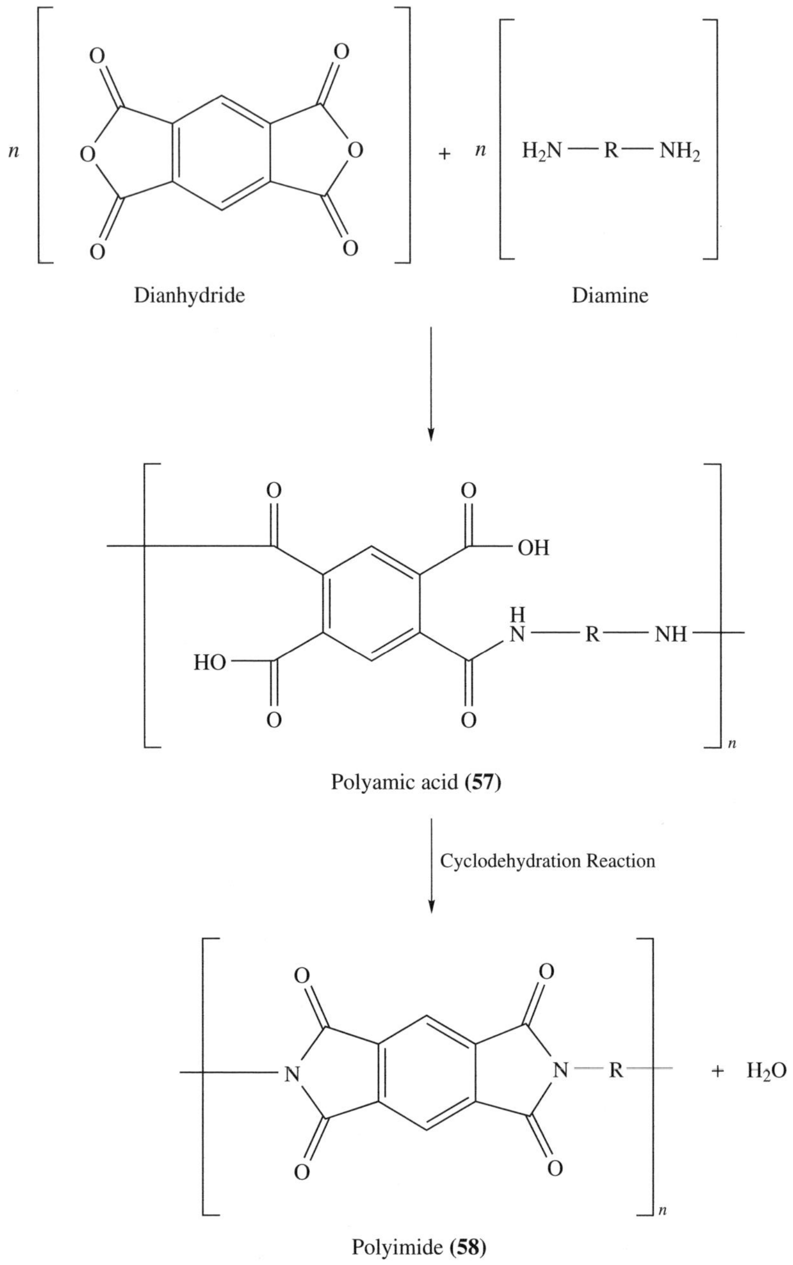

SCHEME 27. Synthesis of linear polyimides, where R = alkene, aromatic moiety etc.

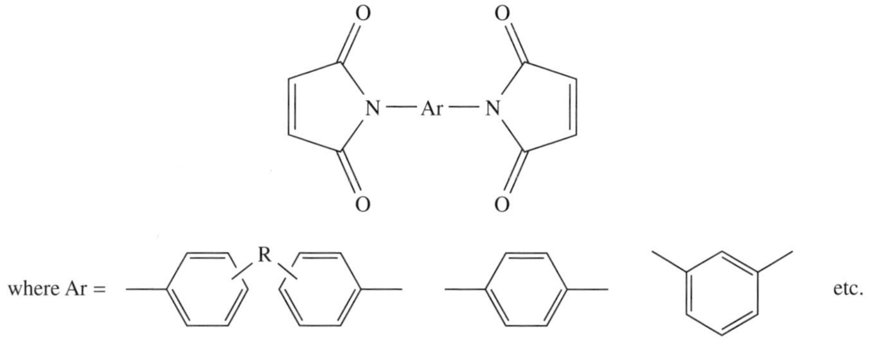

20. Polymers based on phenols

SCHEME 28. A general structure for bis(maleimides), where R = O, S, aromatic rings etc.

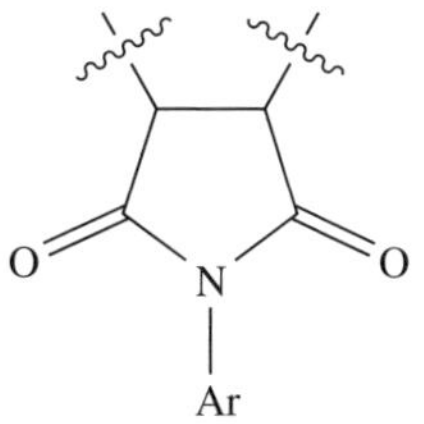

FIGURE 6. Generalized polyimide structure, where Ar = aromatic ring

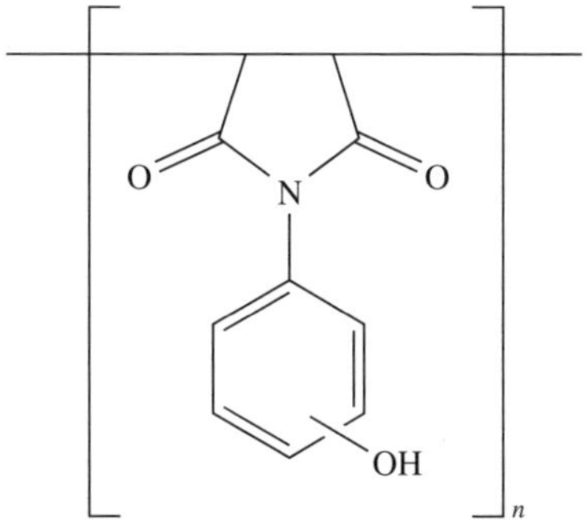

FIGURE 7. Generalized poly(N-(hydroxyphenyl)maleimide) structure

with hexamethylenetetramine (HMTA), produce thermally stable composite materials. Other workers[170,171] have incorporated poly(N-(hydroxyphenyl)maleimides) into the production of photoresists for integrated circuit (IC) technologies.

Copolymerization of the N-(hydroxyphenyl)maleimides with other monomers can produce polyimides with characteristics different from the homopolymer of either monomer. This has been utilized by Chiang and Lu[172–174], who claimed that useful physical characteristics are exhibited by the copolymer of N-(hydroxyphenyl)maleimide with p-trimethylsilylstyrene (TMMS). Polyimide residue of the copolymer contributes excellent thermal stability, while the TMMS component facilitates in the fabrication of the photoresist.

5. Synthesis and polymerization of N-(substituted phenyl)maleimide

The free-radical chain polymerization of maleimides and the N-substituted derivatives has been extensively studied[156,159,160,175,176,184] and both homo- and copolymerization reactions occur readily with a variety of N-substituents and comonomers[142,143,152]. Their reactivity is a consequence of the electron-withdrawing nature of the two adjacent carbonyl groups, which creates a very electron-deficient double bond.

Matsumoto and coworkers[164,169] incorporated poly(N-(hydroxyphenyl)maleimides) into composite materials with phenol–formaldehyde resins. Poly(N-(4-hydroxyphenyl)maleimide) has been shown to form miscible blends with phenolic resins and, after crosslinking, produces composites with good thermal and chemical stability. The hardening or crosslinking agent most commonly used is hexamethylenetetramine (HMTA), to form an insoluble and infusible three-dimensional polymeric network (Scheme 29).

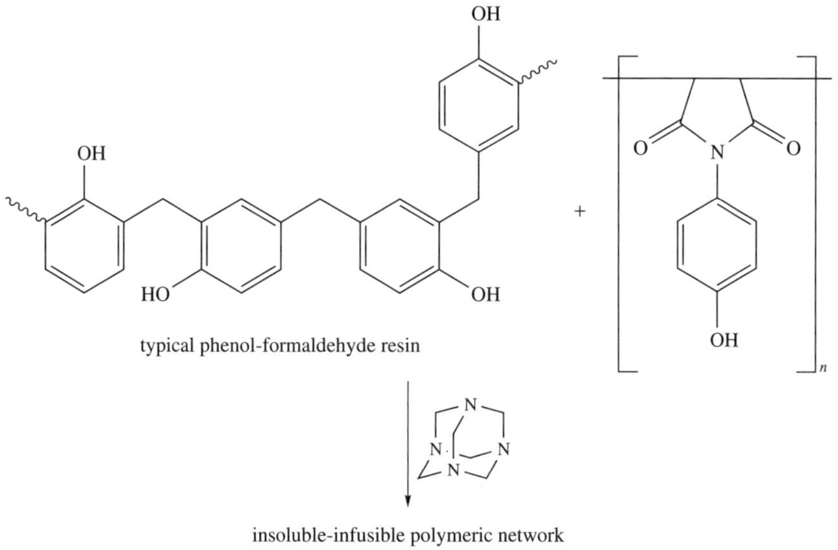

SCHEME 29. Formation of phenol–formaldehyde/polyimide composite materials

The controlled free-radical chain polymerization to form poly(N-(hydroxyphenyl)maleimides) is poorly understood. The choice of solvent for the polymerizations is limited due to the poor solubility of both the monomeric and polymeric materials. Consequently, polar solvents that are often undesirable in free-radical polymerizations are employed (e.g. DMF). The free-radical chain polymerization of N-(hydroxyphenyl)maleimide monomers gives polymers in relatively poor yields with low molecular weights[170,171], which has been attributed to the free phenolic group and chain transfer to the solvent. Masking the phenolic functionality with an acetoxy group gives marginally higher molecular weights, but the effects of the solvent were still controlling the polymerizations.

Protection using a tetrahydropyranyl (THP) protecting substituent gives a similar polymerization pattern (Scheme 30).

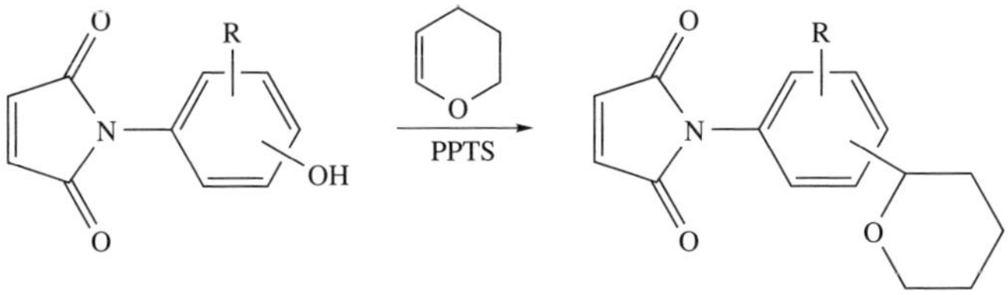

SCHEME 30. Synthesis of *N*-(THP-oxyphenyl)maleimides

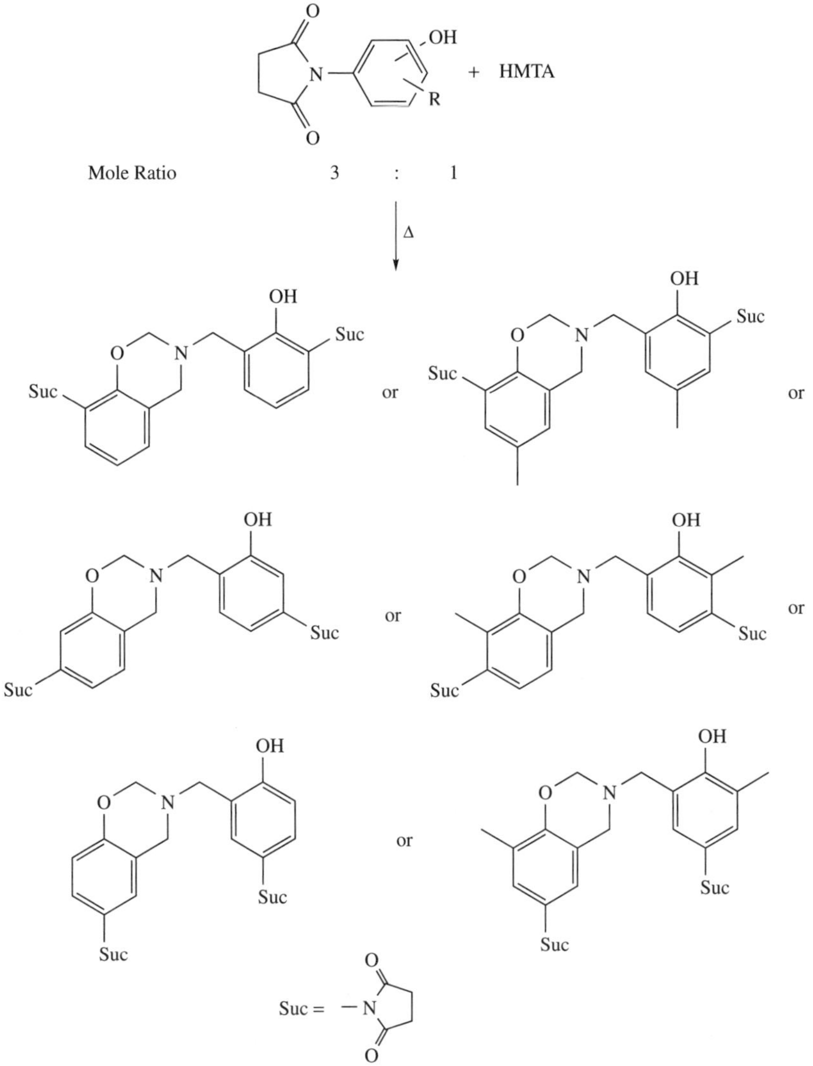

| Mole Ratio | 3 | : | 1 |

Suc = —N(succinimide)

SCHEME 31. Reaction between generalized *N*-(hydroxyphenyl)succinimides and HMTA to form substituted benzoxazines

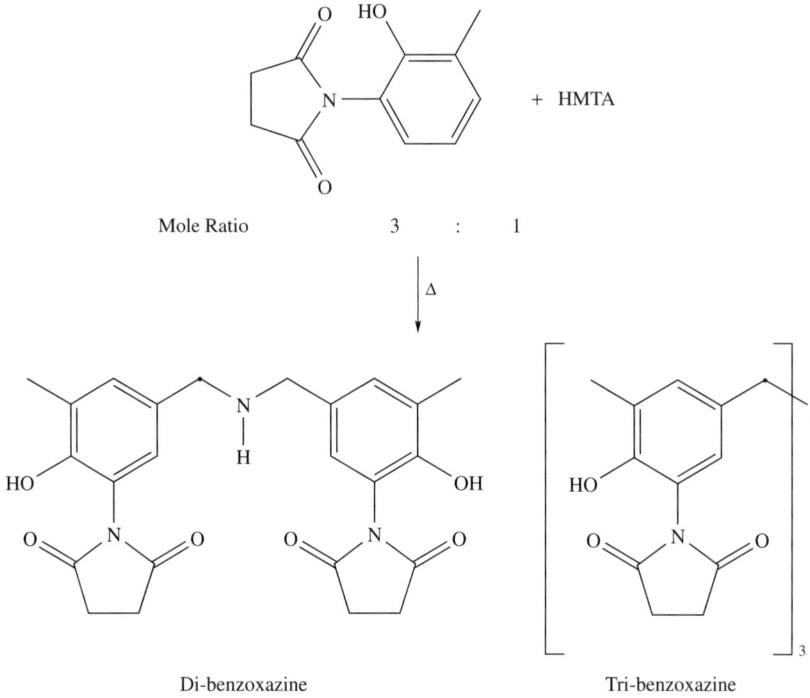

SCHEME 32. Formation of di- and tri-benzylamines

The THP protected monomers have increased solubility in non-polar solvents, such as benzene, and when polymerized in this solvent they give significantly higher molecular weight polymers[177].

The reactivity of the maleimide monomer was dependent on the substitution pattern of the phenyl ring, with the substituents in the *ortho* position tending to lower the molecular weight of the polymer formed. The THP substituent is readily removed either chemically or thermally to yield poly(*N*-(hydroxyphenyl)maleimides). All polymers exhibited excellent thermal stability and showed no evidence of degradation below 360 °C. Reaction occurs between the phenolic ring of the polyimide and HMTA, to form benzoxazine-type derivatives. These reactions have been studied comprehensively using the monomeric model systems, *N*-(hydroxyphenyl)succinimides (Figure 8)[178,179].

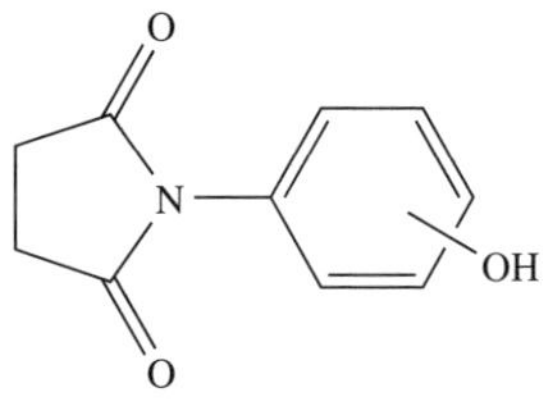

FIGURE 8. Generalized *N*-(hydroxyphenyl)succinimides

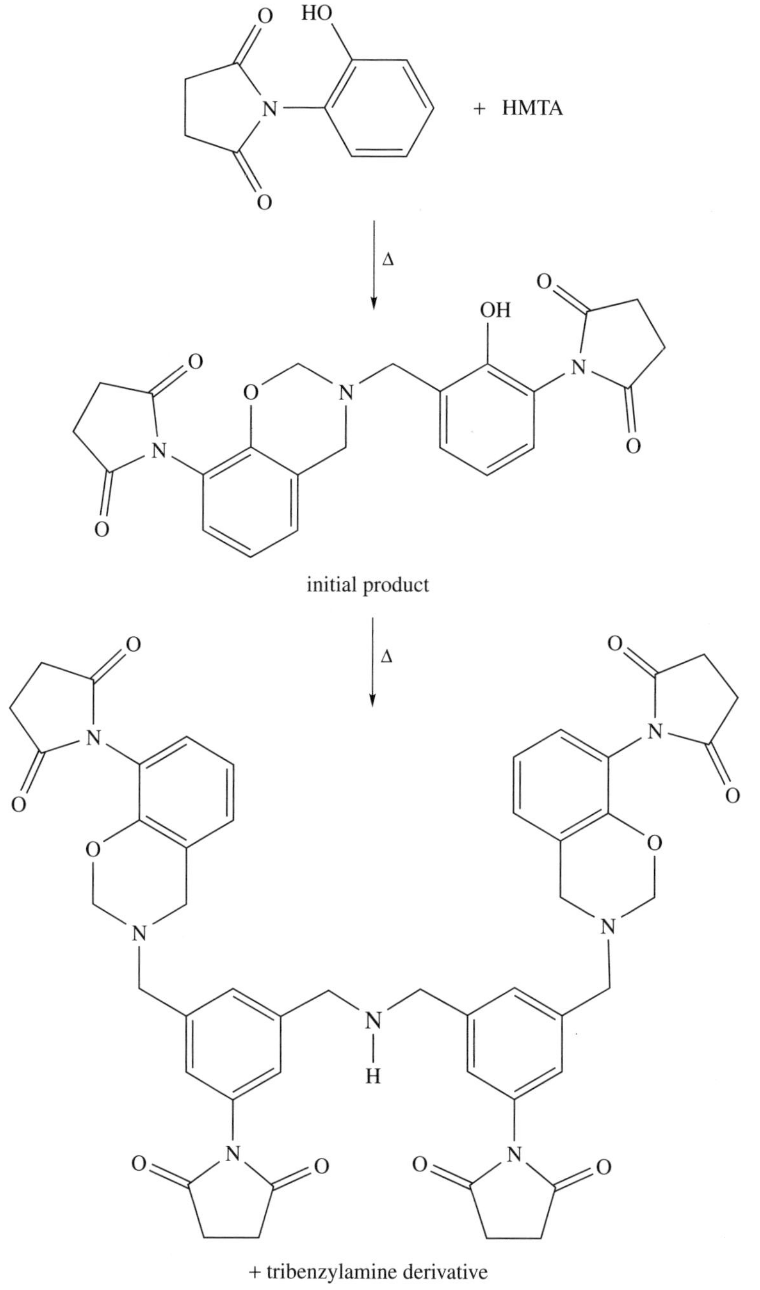

20. Polymers based on phenols

SCHEME 33. Reaction of benzoxazine succinimide derivative with a vacant *p*-position

The mechanistic pathway taken during the reaction of the N-(hydroxyphenyl)succinimides and HMTA is dependent on the substitution of the phenolic ring. With compounds containing a free *ortho* position, the initial intermediates are benzoxazine-type species (Scheme 31).

If there is only a free *para* position, benzylamines are the initial intermediates, with both di- and tri-benzylamines observed (Scheme 32).

Compounds which contain both a vacant *ortho* and *para* position react initially at the *ortho* position to form benzoxazine-type intermediates until most of the *ortho* sites have been consumed; then reaction at the *para* position occurs to form the benzylamine-type products (Scheme 33).

The five-membered succinimide ring does not change the reactive intermediates observed, although it has a marked effect on the rate of their formation. N-(Hydroxyphenyl)succinimides, which have the succinimide ring *ortho* disposed relative to the hydroxyl substituent and which have a free *ortho* position, react up to 7 times faster than the corresponding phenolic compound without a succinimide ring. This increase in reactivity is thought to stem from the effects of the intramolecular hydrogen bonding and its possible consequences on the intermolecular bonding. Intramolecular hydrogen bonding would be most pronounced in the models that contain the succinimide ring *ortho* disposed relative to the hydroxyl group, which is reflected in the increased relative reactivity of those models towards HMTA compared to the models with the succinimide ring *meta* and *para* disposed.

V. REFERENCES

1. L. H. Baekeland, U.S. Patent 942, 699 (1907); U.S. Patent 949, 671 (1907).
2. N. J. L. Megson, *Chem. Ind. (London)*, **20**, 632 (1968).
3. A. Knop and A. Pilato, *Phenolic Resins: Chemistry, Applications and Performance—Future Directions*, Springer-Verlag, Berlin, 1985.
4. A. Gardziella, J. Suren and M. Belsue, *Interceram*, **41**, 461 (1992).
5. W. A. Keutgen, *Encyclopedia of Polymer Science and Technology*, Vol. 10, Interscience, New York, 1969, pp. 1–73.
6. M. F. Drumm and J. R. LeBlanc, in *Kinetics and Mechanisms of Polymerization: Step Growth Polymerizations* (Ed. D. H. Solomon), Marcel Dekker, New York, 1972.
7. J. Borrajo, M. I. Aranguren and R. J. J. Williams, *Polymer*, **23**, 263 (1982).
8. M. I. Aranguren, J. Borrajo and R. J. J. Williams, *Ind. Eng. Chem. Prod. Res. Dev.*, **23**, 370 (1984).
9. T. R. Dargaville, F. N. Guerzoni, M. G. Looney, D. A. Shipp, D. H. Solomon and X. Zhang, *J. Polym. Sci., Polym. Chem.*, **35**, 1399 (1997).
10. S. Mori and A. Yamakawa, *J. Liq. Chromatogr.*, **3**, 329 (1980).
11. S. Mori, *Anal. Chem.*, **53**, 1813 (1981).
12. P. Paniez and A. Weill, *Microelec. Eng.*, **4**, 57 (1986).
13. E. L. Winkler and J. A. Parker, *Macromol. Sci. Rev., Macromol. Chem.*, **C5**, 245 (1971).
14. S. Podzimek and L. Hroch, *J. Appl. Polym. Sci.*, **47**, 2005 (1993).
15. D. A. Shipp and D. H. Solomon, *Polymer*, **38**, 4229 (1997).
16. R. A. Pethrick and B. Thomson, *Br. Polym. J.*, **18**, 380 (1986).
17. R. Natesan and L. M. Yeddanapalli, *Indian J. Chem.*, **11**, 1007 (1973).
18. P. W. Kopf and E. R. Wagner, *J. Polym. Sci., Polym. Chem. Ed.*, **11**, 939 (1973).
19. K. Kamide and Y. Miyakawa, *Makromol. Chem.*, **179**, 359 (1978).
20. D. A. Fraser, R. W. Hall and A. L. J. Raum, *J. Appl. Chem.*, **7**, 676 (1957).
21. D. A. Fraser, R. W. Hall, P. A. Jenkins and A. L. J. Raum, *J. Appl. Chem.*, **7**, 689 (1957).
22. J. C. Woodbrey, H. P. Higginbottom and H. M. Culbertson, *J. Polym. Sci., Part A*, **3**, 1079 (1965).
23. Z. Laszlo-Hedwig, M. Szegstay and F. Tudos, *Angew. Makromol. Chem.*, **57**, 241 (1996).
24. G. Casiraghi, M. Cornia, G. Sartori, G. Casnati, V. Bocchi and G. D. Andreetti, *Makromol. Chem.*, **183**, 2611 (1982).

25. T. Yamagishi, M. Enoki, M. Inui, H. Furukawa, Y. Nakamoto and S.-I. Ishida, *J. Polym. Sci., Polym. Chem.*, **31**, 675 (1993).
26. G. Casiraghi, M. Cornia, G. Balduzzi and G. Casnati, *Ind. Eng. Chem. Prod. Res. Dev.*, **23**, 377 (1984).
27. L. Zetta, A. DeMarco, M. Cornia and G. Casiraghi, *Macromolecules*, **21**, 1170 (1988).
28. G. Casiraghi, M. Corniz, G. Ricci, G. Casnati, G. D. Andreetti and L. Zetta, *Macromolecules*, **17**, 19 (1984).
29. G. R. Sprengling, *J. Am. Chem. Soc.*, **76**, 1190 (1954).
30. V. Böhmer, E. Schade, C. Antes, J. Pachta, W. Vogt and H. Kämmerer, *Makromol. Chem.*, **184**, 3261 (1983).
31. N. J. L. Megson, *Phenolic Resin Chemistry*, Butterworths Scientific Publications, London, 1958.
32. N. J. L. Megson, *Chem.-Ztg.*, **96**, 15 (1972).
33. H. L. Bender, *Industrie Plastiques Mod.*, **5**, 40 (1953).
34. M. Imoto, *Makromol. Chem.*, **113**, 117 (1968).
35. F. L. Tobiason, in Weyerhaeuser Science Symposium, Vol. 2, *Phenolic Resins—Chemistry and Applications*, Weyerhaeuser Company, Washington, 1981, pp. 201–212.
36. F. L. Tobiason, K. Houglum, A. Shanafelt and V. Böhmer, *Polym. Prepr., Am. Chem. Soc., Div. Polym. Chem.*, **24**, 181 (1983).
37. A. Zinke and S. Purcher, *Monatsh Chem.*, **79**, 26 (1948).
38. A. Zinke and G. Ziegler, *Ber. Dtsch. Chem. Ges.*, **77**, 271 (1944).
39. M. G. Looney and D. H. Solomon, *Aust. J. Chem.*, **48**, 323 (1995).
40. A. S. C. Lim, D. H. Solomon and X. Zhang, *J. Polym. Sci., Polym. Chem.*, **37**, 1347 (1999).
41. P. J. de Bruyn, A. S. C. Lim, M. G. Looney and D. H. Solomon, *Tetrahedron Lett.*, **35**, 4627 (1994).
42. P. J. de Bruyn, L. M. Foo, A. S. C. Lim, M. G. Looney and D. H. Solomon, *Tetrahedron*, **53**, 13915 (1997).
43. E. J. Corey and A. Venkateswarlu, *J. Am. Chem. Soc.*, **94**, 6190 (1972).
44. S. Hanessian and P. Lavallee, *Can. J. Chem.*, **53**, 2975 (1975).
45. D. A. Evans, S. W. Kaldor, T. K. Jones, J. Clardy and T. J. Stout, *J. Am. Chem. Soc.*, **112**, 7001 (1990).
46. D. R. Kelly and S. M. Roberts, *Synth. Commun.*, **9**, 295 (1979).
47. A. T. Nielson, D. W. Moore, M. D. Ogan and R. L. Atkins, *J. Org. Chem.*, **44**, 1678 (1979).
48. F. Meissner, US Patent 2, 762, 799 (1956).
49. S. Bulusu, J. Autera and T. Axenrod, *J. Labelled Compd. Radiopharm.*, **18**, 707 (1979).
50. N. Blazevic, D. Kolbah, B. Belin, V. Sunjic and F. Kajfez, *Synthesis*, 161 (1979).
51. E. A. Gusev, S. V. Dalidovich and L. I. Krasovskaya, *Thermochim. Acta*, **93**, 21 (1985).
52. N. Blazevic, D. Kolbar, B. Belin, V. Sunjic and F. Kajfez, *Synthesis*, 161 (1979).
53. A. I. Kuznetsov and N. S. Zefirov, *Usp. Khim.*, **58**, 1815 (1989); *Chem. Abstr.*, **113**, 406188 (1990).
54. A. P. Cooney, M. R. Crampton and P. Golding, *J. Chem. Soc., Perkin Trans. 2*, 835 (1986).
55. E. W. Orrell and R. Burns, *Plast. Polym.*, **36**, 469 (1968).
56. C. S. Tse, T. S. Wong and T. C. W. Mak, *J. Appl. Crystallogr.*, **10**, 68 (1977).
57. R. E. Wasylishen and A. Pettitt, *Can. J. Chem.*, **55**, 2564 (1977).
58. G. R. Hatfield and G. E. Maciel, *Macromolecules*, **20**, 608 (1987).
59. X. Zhang, M. G. Looney, D. H. Solomon and A. K. Whittaker, *Polymer*, **38**, 5835 (1997).
60. A. Knop, V. Böhmer and L. A. Pilato, in *Comprehensive Polymer Science, Vol. 5, General Characteristics of Step Polymerization* (Eds. G. Allen and J. Bevington), Chap. 35, Pergamon Press, Oxford, 1989, pp. 611–647.
61. J. K. Gillham, J. H. Mackey, G. Lester and L. E. Walker, *Polym. Prepr. Am. Chem. Soc., Div. Polym. Chem.*, **22**, 131 (1981).
62. K. Roczniak, T. Biernacka and M. Sharzynski, *J. Appl. Polym. Sci.*, **28**, 531 (1983).
63. D. L. Fraser, R. W. Hall and A. L. J. Raum, *J. Appl. Chem.*, **8**, 478 (1958).
64. K. Hultzsch, *Ber. Dstch Chem. Ges.*, **82**, 62 (1949).
65. J. C. Duff and V. I. Furness, *J. Chem. Soc.*, 1512 (1951).
66. I. V. Kamenskii, J. N. Kuznetsov and A. P. Moiseenko, *Vysokomol. Soedin., Ser. A*, **18**, 1787 (1976); *Chem. Abstr.*, **85**, 560919 (1976).

67. N. Takamitsu, T. Okamura and M. Nishimura, *Nippon Kagaku Kaishi*, 2196 (1973); *Chem. Abstr.*, **80**, 60503 (1974).
68. S. Tonogai, Y. Sakaguchi and S. Seto, *J. Appl. Polym. Sci.*, **22**, 3225 (1978).
69. S. Tonogai, K. Hasegawa and H. Konodo, *Polym. Eng. Sci.*, **20**, 1132 (1980).
70. N. I. Basov, Y. V. Kazankov, A. I. Leonov, V. A. Lynbartovich and V. A. Mironov, *Izv. Vssh. Ucheb. Zaved., Khim. Khim. Teknol.*, **12**, 1578 (1969); *Chem. Abstr.*, **72**, 122279 (1970).
71. X. Wang, B. Riedl, A. W. Christiansen and R. L. Geimer, *Polymer*, **26**, 5685 (1994).
72. Z. Katovic and M. Stefanic, *Ind. Eng. Chem. Prod. Res. Dev.*, **24**, 179 (1985).
73. A. Gardziella, L. A. Pilato and A. Knop, *Phenolic Resins: Chemistry, Applications, Standardization, Safety and Ecology*, Chap. 2, Springer-Verlag, Berlin, 2000, pp. 66–72.
74. T. R. Dargaville, P. J. deBruyn, A. S. C. Lim, M. G. Looney, A. C. Potter, D. H. Solomon and X. Zhang, *J. Polym. Sci., Polym. Chem.*, **35**, 1389 (1997).
75. X. Zhang, A. C. Potter and D. H. Solomon, *Polymer*, **39**, 399 (1998).
76. X. Zhang and D. H. Solomon, *Polymer*, **39**, 405 (1998).
77. X. Zhang, A. C. Potter and D. H. Solomon, *Polymer*, **39**, 1957 (1998).
78. X. Zhang, A. C. Potter and D. H. Solomon, *Polymer*, **39**, 1967 (1998).
79. X. Zhang and D. H. Solomon, *Polymer*, **39**, 6153 (1998).
80. J. I. de Jong and J. de Jong, *Recl. Trav. Chim., Pays-Bas*, **72**, 497 (1953).
81. J. H. Freeman and C. W. Lewis, *J. Am. Chem. Soc.*, **76**, 2080 (1954).
82. H. G. Peer, *Recl. Trav. Chim., Pays-Bas*, **78**, 631 (1959).
83. H. G. Peer, *Recl. Trav. Chim., Pays-Bas*, **78**, 851 (1959).
84. K. Lenghaus, G. G. Qiao and D. H. Solomon, *Polymer*, **41**, 1973 (2000).
85. G. G. Qiao, K. Lenghaus, D. H. Solomon, A. Reisinger, I. Bythway and C. Wentrup, *J. Org. Chem.*, **63**, 9806 (1998).
86. C. Wentrup, R. Blanch, H. Briehl and G. Gross, *J. Am. Chem. Soc.*, **110**, 1874 (1988).
87. G. G. Qiao, M. W. Wong and C. Wentrup, *J. Org. Chem.*, **61**, 8125 (1996).
88. R. A. Pentrick and B. Thomas, *Br. Polym. J.*, **18**, 171 (1986); D. D. Werstler, *Polymer*, **41**, 750 (1986); E. A. Fitzgerald, S. P. Tadros, R. F. Almeida, G. A. Sienko, K. Honda and T. J. Sarubbi, *Appl. Polym. Sci.*, **45**, 363 (1992).
89. M. Higuchi, T. Urakawa and M. Morita, *Polymer*, **42**, 4563 (2000).
90. G. E. Maciel, I.-S. Chuang and L. Gollob, *Macromolecules*, **17**, 1081 (1984).
91. C. A. Fyfe, M. S. McKinnon, A. Rudin and W. J. Tchir, *Macromolecules*, **16**, 1216 (1983).
92. H. Hultzsch, *Ber. Dtsch. Chem. Ges.*, **74**, 898 (1941).
93. A. Zinke, *J. Appl. Chem. 1*, **1**, 257 (1951).
94. K. Lenghaus, G. G. Qiao and D. H. Solomon, *Polymer*, **42**, 3355 (2001).
95. T. Fisher, P. Chao, C. G. Uptoon and A. J. Day, *Magn. Reson. Chem.*, **33**, 717 (1995).
96. D. D. Werstler, *Polymer*, **27**, 750 (1986).
97. S. J. Opella, M. H. Frey and T. A. Cross, *J. Am. Chem. Soc.*, **101**, 5856 (1979).
98. S. J. Opella and M. H. Frey, *J. Am. Chem. Soc.*, **101**, 5854 (1979).
99. M.-F. Grenier-Loustalot, S. Larroque, P. Grenier and D. Bedel, *Polymer*, **37**, 955 (1996).
100. M.-F. Grenier-Loustalot, S. Larroque, J.-P. Leca and D. Bedel, *Polymer*, **35**, 3046 (1994).
101. M.-F. Grenier-Loustalot, S. Larroque and D. Bedel, *Polymer*, **37**, 939 (1996).
102. J. H. Freeman and C. W. Lewis, *J. Am. Chem. Soc.*, **76**, 2080 (1954).
103. X. Zhang and D. H. Solomon, *Chem. Mater.*, **10**, 1833 (1998).
104. P. H. R. B. Lemon, *Br. Ceram. Trans. J.*, **84**, 53 (1985).
105. B. Rand and B. McEnaney, *Br. Ceram. Trans. J.*, **84**, 157 (1985).
106. T. A. Centeno and A. B. Fuertes, *J. Membrane Sci.*, **160**, 201 (1999).
107. B. McEnaney and B. Rand, *Br. Ceram. Trans. J.*, **84**, 193 (1985).
108. H. Honda, H. Kimura, Y. Sanada, S. Sugawara and T. Furuta, *Carbon*, **8**, 181 (1970).
109. M. Shiraishi, G. Terriere and A. Oberlin, *J. Mater. Sci.*, **13**, 702 (1978).
110. K. Kobayashi, S. Sugawara, S. Toyoda and H. Honda, *Carbon*, **6**, 359 (1968).
111. Y. Yamashita and K. Ouchi, *Carbon*, **17**, 365 (1979).
112. Y. Yamashita and K. Ouchi, *Carbon*, **20**, 41 (1982).
113. K. Lenghaus, G. G. Qiao and D. H. Solomon, *Polymer*, **42**, 7523 (2001).
114. K. Lenghaus, G. G. Qiao, D. H. Solomon, C. Gomez, F. Rodriguez-Reinoso and A. Sepulveda-Escribano, *Carbon*, **40**, 743 (2002).
115. M. Shiraishi, G. Terriere and A. Oberlin, *J. Mat. Sci.*, **13**, 702 (1978).

116. S. Yamada and M. Inagaki, *Carbon*, **9**, 691 (1971).
117. P. Hanson, in *Adv. Heterocycl. Chem.*, (Eds. A. R. Katritzky and A. J. Boulton), **27**, (1980).
118. A. S. Hay, H. S. Blanchard, G. F. Endres and J. W. Eustance, *J. Am. Chem. Soc.*, **81**, 6335 (1959).
119. S. Paul, in *Comprehensive Polymer Science*, Vol. 6, (Eds. G. C. Eastmond, A. Ledwith, S. Russo and P. Sigwalt), Pergamon Press, Oxford, 1989, p. 179.
120. K. Hodd, in *Comprehensive Polymer Science*, Vol 6, Pergamon Press, Oxford, 1989, p. 668.
121. D. H. Solomon, *The Chemistry of Organic Film Formers*, Wiley, New York, 1967, p. 189.
122. H. D. Stenzenberger, *Adv. Polym. Sci.*, **117**, 165 (1994).
123. W. Volksen, *Adv. Polym. Sci.*, **117**, 111 (1994).
124. T. M. Bogert and R. R. Renshaw, *J. Am. Chem. Soc.*, **30**, 1135 (1908).
125. W. M. Edwards, US Patent 3, 179, 614 (1965).
126. A. L. Endrey, Canadian Patent, 659, 328 (1962).
127. A. L. Endrey, US Patent 3, 179, 633 (1965).
128. C. E. Sroog, A. L. Endrey, S. V. Abramo, C. E. Berr, W. M. Edwards and K. L. Oliver, *J. Polym. Sci.*, **A3**, 1373 (1985).
129. D. Wilson, H. D. Stenzenberger and P. M. Hergenrother, *Polyimides*, Blackie and Son, Glasgow, 1990.
130. M.-F. Grenier-Loustalot, F. Joubert and P. Grenier, *J. Polym. Sci., Polym. Chem. Ed.*, **29**, 1649 (1991).
131. P. M. Hergenrother, *Polyimides and other High-Temperature Polymers*, Elsevier, Amsterdam, 1991, p. 1.
132. H. Hatori, Y. Yamada and M. Shiraishi, *Carbon*, **31**, 1307 (1993).
133. H. Hatori, Y. Yamada, M. Shiraishi, M. Yoshihara and T. Kimura, *Carbon*, **34**, 201 (1996).
134. M. Inagaki, Y. Hishiyama and Y. Kaburagi, *Carbon*, **32**, 637 (1994).
135. S. Rojstaczer, M. Ree, D. Y. Yoon and W. Volksen, *J. Polym. Sci., Polym. Phys. Ed.*, **30**, 133 (1992).
136. M. Ree, D. Y. Yoon and W. Volksen, *J. Polym. Sci., Polym. Phys. Ed.*, **29**, 1203 (1992).
137. K. Kurita, Y. Suzuki, T. Enari, S. Ishii and S.-I. Nishimura, *Macromolecules*, **28**, 1801 (1995).
138. R. Torrecillas, A. Baudry, J. Dufay and B. Mortaigne, *Polym. Degrad. Stab.*, **54**, 267 (1996).
139. R. Chandra and L. Rajabi, *J. Macromol. Chem., Phys.*, **C37**, 61 (1997).
140. M.-F. Grenier-Loustalot, F. Gouarderes, F. Joubert and P. Grenier, *Polymer*, **34**, 3848 (1993).
141. J. A. Mikroyannidis and A. P. Melissaris, *Br. Polym. J.*, **23**, 309 (1991).
142. I. M. Brown and T. C. Sandreczki, *Macromolecules*, **23**, 4918 (1990).
143. I. M. Brown and T. C. Sandreczki, *Macromolecules*, **23**, 94 (1990).
144. S. D. Pastor and E. T. Hessell, *J. Org. Chem.*, **53**, 5776 (1988).
145. M. N. Khan, *J. Chem. Soc., Perkin Trans. 2*, 891 (1985).
146. M. N. Khan, *J. Chem. Soc., Perkin Trans. 2*, 819 (1987).
147. S. D. Pastor, E. T. Hessell, P. A. Odorisio and J. D. Spivack, *J. Heterocycl. Chem.*, **22**, 1195 (1985).
148. S. Edge, A. Charlton, K. S. Varma, A. E. Underhill, T. K. Hansen, J. Becher, P. Kathirgamanathan and E. Khorsravi, *J. Polym. Sci., Polym. Chem. Ed.*, **30**, 2773 (1992).
149. A. Renner, I. Forgo, W. Hofmann and K. Ramsteiner, *Helv. Chim. Acta*, **61**, 1443 (1978).
150. D. Bryce-Smith, A. Gilbert, I. S. McColl and P. Yianni, *J. Chem. Soc., Chem. Commun.*, 951 (1984).
151. D. Bryce-Smith, A. Gilbert, I. S. McColl, M. G. B. Drew and P. Yianni, *J. Chem. Soc., Perkin Trans. 1*, 1147 (1987).
152. K. G. Olson and G. B. Butler, *Macromolecules*, **17**, 2480 (1984).
153. C. D. Diakoumakos and J. A. Mikroyannidis, *J. Polym. Sci., Polym. Chem. Ed.*, **30**, 2559 (1992).
154. T. Takekoshi, *Advances in Polymer Science*, Springer-Verlag, Berlin, Vol. 94, 1994, p. 1.
155. H. Cheng, G. Zhao and D. Yan, *J. Polym. Sci., Polym. Chem.*, **30**, 2181 (1992).
156. A. Matsumoto, Y. Oki and T. Otsu, *Eur. Polym. J.*, **29**, 1225 (1993).
157. G. S. Prementine and S. A. Jones, *Macromolecules*, **22**, 770 (1989).
158. T. Sato, K. Takahashi, H. Tanaka and T. Ota, *Macromolecules*, **24**, 2330 (1991).
159. T. Sato, K. Masaki, K. Kondo, M. Seno and H. Tanaka, *Polym. Bull.*, **35**, 345 (1995).
160. T. Ostu, A. Matsumoto, T. Kubota and S. Mori, *Polym. Bull.*, **23**, 43 (1990).

161. T. Oishi, K. Kagawa and M. Fujimoto, *J. Polym. Sci., Polym. Chem. Ed.*, **33**, 1341 (1995).
162. A. Matsumoto, T. Kubota and T. Ostu, *Macromolecules*, **23**, 4508 (1990).
163. T. Ostu, A. Matsumoto and T. Kubota, *Polym. Int.*, **25**, 179 (1991).
164. A. Matsumoto, K. Hasegawa, A. Fukuda and K. Otsuki, *J. Appl. Polym. Sci.*, **43**, 365 (1991).
165. A. Matsumoto, K. Hasegawa, A. Fukuda and K. Otsuki, *J. Appl. Polym. Sci.*, **44**, 205 (1992).
166. A. Matsumoto, K. Hasegawa, A. Fukuda and K. Otsuki, *J. Appl. Polym. Sci.*, **44**, 1547 (1992).
167. A. Matsumoto, K. Hasegawa and A. Fukuda, *Polym. Int.*, **30**, 65 (1993).
168. A. Matsumoto, K. Hasegawa and A. Fukuda, *Polym. Int.*, **28**, 173 (1992).
169. A. Matsumoto, K. Hasegawa and A. Fukuda, *Polym. Int.*, **31**, 275 (1993).
170. R. S. Turner, R. A. Arcus, C. G. Houle and W. R. Schleigh, *Polym. Eng. Sci.*, **26**, 1096 (1986).
171. W.-Y. Chiang and J.-Y. Lu, *J. Appl. Polym. Sci.*, **50**, 1007 (1993).
172. W.-Y. Chiang and J.-Y. Lu, *J. Polym. Sci., Polym. Chem. Ed.*, **29**, 399 (1991).
173. W.-Y. Chiang and J.-Y. Lu, *Eur. Polym. J.*, **29**, 837 (1993).
174. W.-Y. Chiang and J.-Y. Lu, *Makromol. Chem.*, **205**, 75 (1993).
175. S. Iwatsuki, M. Kubo, M. Wakita, Y. Matsui and H. Kanoh, *Macromolecules*, **24**, 5009 (1991).
176. T. Oishi, H. Yamasaki and M. Fujimoto, *Polym. J.*, **23**, 795 (1991).
177. M. J. Caulfield and D. H. Solomon, *Polymer*, **40**, 1251 (1999).
178. M. J. Caulfield, M. G. Looney, R. A. Pittard and D. H. Solomon, *Polymer*, **39**, 6541 (1998).
179. M. J. Caulfield and D. H. Solomon, *Polymer*, **40**, 3041 (1999).

Author Index

This author index is designed to enable the reader to locate an author's name and work with the aid of the reference numbers appearing in the text. The page numbers are printed in normal type in ascending numerical order, followed by the reference numbers in parentheses. The numbers in *italics* refer to the pages on which the references are actually listed.

Aakeröy, C. B. 530(17), *597*
Abakumov, G. A. 1109(49, 50), *1144*
Abate, G. 984(383), *1012*
Abboud, J.-L. 503(87–89), *526*, 887(219), *907*
Abboud, J.-L.M. 237(61), *256*, 261(11), *326*, 540(39), 564(170), 575(187), *597, 600*
Abd-El-Ghani, A. 1244(169), *1342*
Abdel-Rahman, A. A.-H. 675(113), *707*
Abdi-Oskoui, S. H. 774(188), *835*
Abdul Malek, S. N. 740(67), *832*
Abe, H. 107(409), 143(409, 574–576, 578), 144(574–576), 145(574), 147(574, 578), *188, 192*
Abe, S. 992(430), *1013*
Abell, C. 459(269), *487*
Abeygunawardana, C. 367(180), *390*
Abeygunwardana, C. 352(94), *388*
Abidi, R. 1391(114), 1427(423a), *1440, 1452*
Abildgaard, J. 336, 343–345, 350, 355(39), 358(116), 366(39, 166), *387–389*
Abis, L. 1386(86a, 86b), 1422(379), *1439, 1450*
Abkowics, A. J. 384(291), *393*
Abkowicz-Bieńko, A. J. 20, 22, 23, 35–38(112), 82(288), 144(608), *182, 186, 193*, 235(52), *255*, 372, 375(236), 384(292), *391, 393*
Abolin, A. H. 432(186), *484*
Abraham, H. 1392, 1403(125a), 1427(424c), *1441, 1452*
Abraham, M. H. 503(89), *526*, 533(22), 534(29), 535(33, 34), 536(33–37), 537(33), 539, 540(37), 544(29), 545(53, 54), 557(33, 145), 560, 564(145), 588(33, 37, 257),
590(145), 591(37), *597–599, 602*, 876–878(186), 879(196), 882(186, 196), 887(219), *906, 907*
Abraham, R. H. 68(261, 262), 70(261), *185*
Abraham, W. 1403(237), *1445*
Abramczyk, H. 170, 171(759), 177(759, 769), *197*
Abramo, S. V. 1494(128), *1505*
Abramova, G. V. 230(27), *254*
Abrams, S. R. 1237(164), *1342*
Abu-Amsha, R. 913(30), *1003*
Abuja, P. M. 851(64), 853(64, 87), 854, 857, 858(64), *903*
Abuzar, S. 412(76), *480*
Acero, J. L. 1080(254–256), 1085(345), *1093, 1095*, 1362(202), *1368*
Achar, B. N. 990(418), *1012*
Achiba, Y. 143(582), *193*
Achilli, G. 953(216, 263), 957(216), 964(263), *1008, 1009*, 1353(82), *1365*
Acker, S. A. B. E. van 871, 873, 898(169), *905*
Acree, W. E. Jr. 227(5, 10), 230(5), *254*
Acuña, A. U. 551(99), *599*, 730(45), *831*, 1025(70b), *1088*
Acuña, J. A. 982(366), *1011*
Acworth, I. N. 953, 960(227), *1008*
Aczel, T. 265(26), *326*
Adachi, K. 687(225), *709*
Adachi, M. 632(86), *658*, 686(214), *709*
Adam, W. 740(61), *832*, 1080(259a, 259c), *1093*, 1125(248), *1149*, 1213(113), 1255(184c), *1340, 1342*
Adami, F. 1157(12), *1337*

The Chemistry of Phenols. Edited by Z. Rappoport
© 2003 John Wiley & Sons, Ltd ISBN: 0-471-49737-1

Adamo, C. 4, 100(2), 132, 133, 136(518), *179, 191*, 368, 372, 373(200), *390*, 1127(267), *1149*
Adamowicz, L. 370, 382(211), *391*, 730(46), *831*
Adams, C. M. 450(239g), *486*
Adams, D. J. 617(33), *656*
Adams, G. E. 1113(87), 1115, 1128(92), 1129(299), 1130(299, 322), 1134(92), *1145, 1150*
Adams, R. N. 1157(10), *1337*
Adams, W. J. 1356(128), *1366*
Adeleke, B. B. 1124(239), *1148*
Adityachaudhury, N. 409(59), *480*
Adkins, K. L. 951, 953(191), *1007*
Adler, E. 1267(201, 202b), *1343*
Adou, E. 322(248), *331*
Aerts, J. 93, 103(363), *187*
Afanasev, I. B. 1118(142), *1146*
Afeefy, H. Y. 224(3), *254*
Agaev, A. A. 612(20), *656*
Agai, B. 1387, 1391(93), *1439*
Agarwal, S. K. 830(347), *838*
Agarwal, U. P. 174(764), *197*
Agavelyan, E. S. 301(130), *328*
Agawa, T. 1402, 1409(223), *1444*
Agbaria, K. 1408(283), *1447*
Agena, C. 320(241), 322(241, 252), *331*
Agenda, C. 1419(346), *1450*
Agmon, N. 493(48–50, 58, 59, 61, 62), 497(58, 59), 498(59), 502(82), 504, 505(49), *526*
Agranat, I. 1124(236), *1148*
Agrawal, O. 988(411), *1012*
Agüí, M. L. 982(362), *1011*
Aguiar, S. 631(72), *657*
Aguilera, M. P. 913(6), *1003*
Ahel, M. 1349(32), 1356(126, 129), *1363, 1366*
Ahlbrecht, H. 744(74), *832*
Ahluwalia, V. K. 677(126), *707*
Ahmed, F. 970(291), *1009*
Ahmed, R. 740(66), *832*
Ahmed, S. 1086(371), *1096*, 1130(306), 1137(366), *1150, 1151*
Ahn, B.-H. 1082(289), *1094*
Ahn, K. 559(184), 575(185), 580(185, 204), *600, 601*
Ahn, K. H. 1372(13), *1436*
Ahtiainen, J. 1347(8), *1363*
Aihara, K. 1213(112), *1340*
Aikens, J. 855(106, 110), 901(110), *904*
Aimoto, Y. 1205(103), *1340*
Airola, K. 1419(358), 1420(363), *1450*
Aizawa, M. 309(173), 310(174), *329*
Aizawa, T. 827(340), *838*
Ajao, J. F. 1294(253a), *1344*
Akabori, S. 1392(131), *1441*

Akai, S. 447(222b–e, 222h, 222j), 448(222b, 225, 226), *485*
Akao, K. 1400, 1401(210a), 1405(247), *1444, 1445*
Akasu, M. 807(275), *837*
Akdas, H. 1415(314b), *1448*
Akimoto, T. 414(90h), *481*
Akita, N. 414(90b), *481*
Akita, S. 323(256), *331*
Akiyama, M. 463(276d), *487*
Aksnes, D. W. 338(13), *386*
Alagona, G. 551, 553(103), *599*
Alam, T. M. 338, 380(15), *386*
Alarcon, S. H. 726(33), *831*
Alarcon, S. H. H. 359(129), *389*
Al-Bahri, T. A. 1362(215), *1368*
Albers, P. 1084(327), *1095*
Alberti, B. N. 411(70a), *480*
Albertini, R. 851(64), 853(64, 87), 854, 857, 858(64), *903*
Albigaard, J. 551(98), *599*
Albini, A. 419(133), *482*, 1081(262), *1093*
Albrecht, G. 144(607), *193*, 364(157), *389*, 593, 595(274), *602*
Alcini, D. 1359(151), *1366*
Alder, A. C. 212(25), *221*
Aldini, G. 953, 966(269), *1009*
Aldoshin, S. M. 1025(69), *1088*
Aldrich, H. S. 139(531), *191*
Aldrich, J. E. 1129, 1130(299), *1150*
Alegria, A. E. 241(79), *256*
Al-Ekabi, H. 1086(366), *1096*
Aleksiuk, O. 1410(291b), 1411(295, 296a–c, 297), *1447*
Alemdar, I. 972(305), *1010*
Alewood, P. F. 774(190), *835*
Alexandratos, S. 261(11), *326*
Alfano, J. C. 20, 24, 106(118), *182*
Alfassi, Z. B. 845(26, 27), 880(26, 201), *902, 906*, 1099(9), 1101(37), *1105, 1106*, 1110(57), 1111(71, 72), 1128(57), 1135, 1136(353), *1145, 1151*
Alfieri, C. 1395(180), *1442*
Al-Hadad, A. 1362(215), *1368*
Al-Hassan, G. 1065(194), *1091*
Alif, A. 1073(220), *1092*
Alimpiev, S. S. 1000(468), *1013*
Al-Jalal, N. A. 1060(183–185), *1091*
Aljosina, L. A. 236(54), *255*
Alkorta, I. 129(479), *190*
Al-Laham, M. A. 4, 100(2), *179*, 368, 372, 373(200), *390*
Allain, R. 322(249), *331*
Allan, E. A. 53(233, 234), *185*
Allen, B. T. 1123(199), *1148*
Allen, C. F. H. 462(273b), *487*
Allen, F. H. 535, 549(32), 550(71), 551(32, 134), 552(134), *597–599*

Allen, I. 397(2a), *478*
Allerhand, A. 170, 171(745), *196*, 589(260), 602
Allott, P. H. 238(68, 69), 239(68), *256*
Almedia, L. M. 853, 856, 857(88), *904*
Almeida, R. F. 1482(88), *1504*
Almi, M. 1404(244), *1445*
Alonso, L. 1360(168), *1367*
Alonso, M. C. 947, 953(171), *1006*
Alpincourt, P. 867, 880, 897(151), *905*
Al-Rashid, W. A. L. K. 551(108), *599*
Al-Rawi, J. M. A. 359(127), *389*
Al-Tamimi, S. A. 995(443), *1013*
Altamura, A. 1257(186), *1342*
Altman, L. J. 349(58), *387*
Alvarez, B. 637(110), *658*
Alvarez, R. J. 145(632), *194*
Alvarez Rodríguez, L. 988(409, 410), *1012*
Alves, R. 901(286), *908*
Alvila, L. 997(445, 446), *1013*
Alwarthan, A. A. 995(443), *1013*
Aly, F. A. 995(443), *1013*
Aly, M. M. 1071(201–203), *1091, 1092*
Amabilino, D. B. 761(116, 117), *833*
Amakura, Y. 953, 957(219), *1008*
Amaral, M. T. 961(245), *1008*
Amat, A. M. 1079(244), *1093*
Amat-Guerri, F. 551(99), *599*, 730(45), *831*, 1025(70b), *1088*
Amborziak, K. 345(48), *387*
Amiot, M. J. 953, 958(225), *1008*
Ammar, M. M. 230(24), *254*
Ammon, H. L. 216(32), *221*
Amonkhai, S. I. 1360(163), *1367*
Amoraati, R. 45, 140(201), *184*
Amouri, H. 748(84–86), *832, 833*
An, J. 762(127), *834*
An, X. W. 227(12), *254*
Anand, R. 109(415), *189*
Anastasiou, D. 739(60), *832*
Anbar, M. 1099(9), *1105*
Ancel, J.-E. 677(132), *707*
Andel, J. G. 1359(159), *1367*
Andersen, K. B. 368(199), 379(273), 380(273, 274, 276), 381(199, 276), 382(199), 383(276, 286), *390, 392, 393*, 721(24), *831*
Andersen, O. K. 69(263), *185*
Andersen, R. J. 1327(314), *1345*
Andersh, B. 650(160), *659*
Anderson, A. A. 627(63), *657*
Anderson, G. B. 314(205–207), *330*
Anderson, J. C. 476(322), *488*
Anderson, R. F. 1130(328), *1150*
Anderson, S. C. 110–112(426), *189*
Anderson, S. L. 129(477), *190*
Andersson, G. 1267(202b), *1343*
Andersson, T. 970(289), *1009*
Ando, I. 1362(211), *1368*

Ando, T. 662(6), 702(332), *705, 712*
Ando, W. 407(51), *479*
Andrade, J. G. 1314(288, 289), *1345*
Andrade, P. 961(245), 973, 981(313), *1008, 1010*
Andrade, P. B. 913(18, 20), 958(20), *1003*
Andreetti, G. D. 1374(42a), 1395(180), *1436, 1442, 1462(24, 28), 1502, 1503*
Andres, J. L. 4, 100(2), *179*, 368, 372, 373(200), *390*
Andrews, L. 129, 131(504), *191*
Andriamiadanarivo, R. 453(244), *486*
Andringa, H. 677(138), *707*
Andrist, M. 1125(253), *1149*
Androulaki, E. 1086(361), *1096*
Andy, P. 631(75), *657*
Anelli, P.-L. 761(117), *833*
Anex, D. S. 137(522), *191*
Angel, A. J. 730(47), *832*
Angelino, S. 953, 962(247), *1008*
Angeloni, L. 383(281), *392*
Angermann, R. 305(147), *329*
Angiolini, L. 684(186), *709*
Angioni, E. 851, 854(74), *903*
Agnes, L. 978(348), *1011*
Angot, S. 1388(100a), *1439*
Angren, H. 179(775), *197*
Ángyán, G. 69(269), *185*
Anjaneyulu, A. S. R. 631(84), *658*
Ankley, G. T. 1357(133), *1366*
Anlauf, S. 300(113), *328*
Annapurna, P. 447(222i), *485*
Annis, D. A. 668(53), *706*
Annoura, H. 447(222g), *485*
Anoshin, A. N. 383(280), *392*
Antes, C. 1462(30), *1503*
Antolovich, M. 942(136), 999(466), *1006, 1013*
Antona, D. 549(67), 550(68), *598*
Antonelli, A. 303(142), *329*
Antunes, F. 843(283), 851(60, 70), 852(60), 887(60, 220), 888(60), 901(283, 286, 287), *903, 907, 908*
Antus, S. 1122(194), 1129, 1130, 1142(293), *1150*, 1227(141), 1315(292), *1341, 1345*
Antypas, W. G. 336(6), *386*
Anulewicz, R. 93(362), *187*
Anvia, F. 311(186), 312(193), *329, 330*, 586, 587(245), *602*
Aoki, H. 1099(11–13), *1105*
Aoki, K. 322(243), *331*
Aoki, T. 683(174), *708*
Aoki, Y. 694(263), *710*
Aono, M. 1380(62), *1438*
Aoyagi, K. 1374(44), *1437*
Aoyama, T. 699(306), *711*
Aoyama, Y. 1383(67e), 1418(335), 1421(67e), *1438, 1448*

Aparicio, A. 953, 961(238), *1008*
Apeloig, Y. 115(462), *190*
Appelqvist, L.-A. 846, 847(41), *902*
Aqiao, M. 702(335, 336), *712*
Arabei, S. M. 384(287, 288), *393*
Arai, H. 1100(27), *1106*
Arai, T. 695(270, 280), *710, 711*, 1028(93, 95), *1089*
Arakawa, T. 1192(75a), *1339*
Araki, K. 634(95), *658*, 1201(93), *1340*, 1372(10b), 1379(54b), 1388(103), 1391(117), 1393(159), 1394(167a, 174), 1400(207a, 207b), 1401(217, 218), 1402(224a), 1405(247), *1435, 1437, 1440, 1442, 1444, 1445*
Araki, K. A. 1427(419a), *1452*
Araki, M. 674(108), *707*
Araki, T. 1194(83b), *1339*
Araki, Y. 129, 179(491), *190*, 881(202), *906*, 1158(13), *1337*
Araldi, G. 622(51, 52), *657*, 683(179), *708*
Aramaki, Y. 855(117), *904*
Aramendia, M. A. 972(304), *1010*
Arana, G. 984(384), *1012*
Aranda, F. J. 887(223, 224), *907*
Aranguren, M. I. 1459(7, 8), *1502*
Aranyos, A. 671(80), *706*
Aratani, T. 697(287, 288), *711*
Archelas, A. 412(74a), *480*
Archer, M. D. 358(112), *388*
Arcus, R. A. 1497, 1498(170), *1506*
Ardenne, M. von 310(175), *329*
Arduini, A. 1378(49a), 1391(118a), 1393(141, 151, 158), 1396(118a, 186a, 186c, 188), 1403(234e, 235, 240b), 1404(244), 1405(235), 1409(288), 1427(409a, 410a, 415), 1430(439), 1432(409a), 1435(467), *1437, 1440–1443, 1445, 1447, 1451–1454*
Areias, F. 913(18), *1003*
Arena, G. 1391, 1396(121b), *1440*
Arends, I. 139(533), *191*
Aresta, M. 686(212), *709*
Argyropoulos, D. S. 998(457), 999(464), *1013*
Arias, J. J. 987(406), *1012*
Arienti, A. 608, 612(10), *656*, 680(155), 683(170), *708*
Arikawa, T. 309(172), *329*
Arimoto, H. 937, 938(124), *1005*
Arimura, T. 634(95), *658*, 1020(34), *1087*, 1394(172b), *1442*
Aripov, T. F. 204(10), *220*
Arisawa, M. 680(157–159), 681(161), 682(166), *708*, 1246, 1247(173), *1342*
Arite, K. 762(126), *834*
Arkema, S. 1424(382a), *1450*
Arkhipov, Yu. M. 1081(263, 264), *1093*
Arkhipova, M. B. 1081(263, 264), *1093*
Armaroli, N. 1406(272), *1446*

Armspach, D. 1435(462a), *1454*
Armstrong, D. A. 1139, 1142, 1143(371), *1151*
Arnason, J. T. 913(2), 994(438), *1002, 1013*
Arnaud, F. 1391(121a), 1396(121a, 182b), 1427(423a), *1440, 1443, 1452*
Arnaud, M. 432(185), *484*
Arnaud-Neu, F. 1387(94), 1389(105), 1391(120c), 1394(172c), 1396(172c, 181b, 187, 195b), 1435(94, 187), *1439, 1440, 1442, 1443*
Arnaut, L. G. 492, 493, 503(15), *525*, 1021(37), 1082(276), *1088, 1093*
Arnecke, R. 1396(195a, 195b), 1419(356b), 1420(359c, 361), *1443, 1450*
Arnett, E. M. 586(237), 589, 590(237, 258), 592(258), *602*
Arnold, S. F. 925(50), *1004*
Arnold-Stanton, R. 669(60), *706*
Arnone, A. 721(25), *831*
Arnott, G. 1420(367), *1450*
Arnott, M. 1358(144), *1366*
Arora, A. 774(185, 186), *835*, 895(264), *908*
Arora, K. K. 677(126), *707*
Arora, R. 970(291), *1009*
Arques, A. 1079(244), *1093*
Arredondo, Y. 677(119), *707*
Arrendondo, Y. 607(5), *656*
Arsenyev, V. G. 824(331), *838*
Arsequell, G. 651(172), *660*
Artemchenko, S. S. 725(28), *831*
Artem'ev, Y. M. 1086(378), *1096*
Artem'eva, M. A. 1086(378), *1096*
Arthen-Engeland, T. 1025(84), *1089*
Arthur, J. I. C. 1124(244), *1149*
Arts, I. C. W. 953, 961(242), *1008*
Artynkhin, A. B. 1073(221), *1092*
Artyukhov, V. Ya. 1017(5, 6), *1087*
Artz, J. D. 851, 858, 889(58), *903*
Arvanaghi, M. 643(141), *659*
Asai, M. 953, 966(268), *1009*, 1417(333), *1448*
Asakawa, Y. 1383, 1421(67e), *1438*
Asakura, K. 1419(347a), *1450*
Aschmann, S. M. 1105(63), *1106*
Asfari, Z. 635(101), *658*, 1370(1b), 1372(15, 20, 24a), 1391(114), 1392, 1403(125b), 1415(313), 1427(423a), *1435, 1436, 1440, 1441, 1448, 1452*
Ashby, K. D. 1045(137), *1090*
Asher, S. A. 129, 133(493), *191*
Ashida, Y. 624(56), *657*
Ashram, M. 1374, 1411(42c), *1437*
Ashton, P. R. 761(116–118), *833*
Ashworth, P. 1120(173, 174), *1147*
Asmus, K.-D. 1109(46), 1111(66), 1117(121), *1144–1146*
Asselin, M. 37(178), *183*

Assfeld, X. 867, 880, 897(151), *905*
Astarloa, G. 377(251), *392*
Astles, D. P. 472(313), *488*
Astruc, M. 946, 953(162), *1006*
Atamas', L. I. 1417(328), *1448*
Atherton, N. M. 1127(266), *1149*
Atherton, S. J. 1129(303), *1150*
Atkins, R. L. 1470(47), *1503*
Atkinson, E. 880(200), *906*
Atkinson, I. M. 144(681), *195*
Atkinson, J. 994(438), *1013*
Atkinson, R. 1105(63), *1106*
Atland, H. W. 416(110b), *482*
Atovmyan, L. O. 1025(69), *1088*
Attinà, M. 309(169–171), *329*, 932, 937(93), *1005*
Atwood, J. L. 1372(23, 24b), 1402(226), 1414(307a, 307b, 308), *1436, 1445, 1447, 1448*
Aubert, S. Y. 953, 958(225), *1008*
Audier, H. E. 300(110, 115, 122), *328*
Augugliaro, V. 1083(296, 301), *1094*
Auhuja, S. 1083(299), *1094*
Auksi, H. 1180, 1327(60a), *1339*
Aulbach, M. 461(290), *487*
Aumann, R. 669(59), *706*
Auret, B. J. 412(71a), *480*
Auricchio, S. 1259(188), *1342*
Aurich, J. 1073(223), *1092*
Aust, S. D. 854, 855(101), *904*
Austen, M. A. 69(267), *185*
Autera, J. 1470(49), *1503*
Avdovich, H. W. 380, 381(275), *392*
Avent, A. G. 813(296), *837*
Avery, M. Y. 325(278), *332*
Avila, D. V. 877(189, 192), 879, 881, 883(192), *906*
Axelrod, J. 405(43a), *479*
Axenrod, T. 1470(49), *1503*
Ayala, P. Y. 4, 100(2), *179*, 368, 372, 373(200), *390*
Äyras, P. 349(63), *387*
Ayres, D. C. 1216(123), 1302(265), *1341, 1344*
Ayscough, P. B. 1124(232), *1148*
Ayuzawa, T. 300(125), *328*
Azad, S. M. 691(240), *710*
Azcarate, A. A. 1084(324), *1095*
Azevedo, A. L. 1100(34), *1106*
Azócar, J. A. 471(305), *488*
Azzaro, F. 1356(122), *1366*

Baar, B. L. M. van 296(86), *327*
Baba, H. 47, 143(203, 205, 206), 144(203, 594), 148(203), *184, 193*, 493, 503, 519, 520(34), *525*
Babakhina, G. M. 357(111), *388*
Babcock, G. T. 129(483), *190*, 1108(16, 17), *1144*
Baboul, A. G. 368, 372, 373(200), *390*
Babu, B. S. 675(114), *707*
Baccin, C. 701(318), *712*
Bach, R. D. 417(117), *482*
Bachand, B. 551, 553(82), *598*
Bachler, V. 129, 132, 133, 135(514), *191*
Bachlund, P. 1360(185), *1367*
Bachmann, D. 260(3), 323(3, 253), *325, 331*
Bačič, Z. 149, 161(735), *196*
Baciocchi, C. 947, 955(170), *1006*
Baciocchi, R. 932, 937(93), *1005*
Backes, M. 1374(33c), *1436*
Bacon, G. E. 200(6), *220*, 550(70), *598*
Badalian, I. E. 979(353), *1011*
Badamali, S. K. 616(28, 29), 620(44), *656, 657*
Baddar, F. G. 472(311d), *488*
Badenhoop, J. K. 4(5), *179*
Bader, R. F. W. 31, 35, 68(141), 69(267), 89(309), *183, 185, 186*
Badger, G. M. 419(134a), *482*
Badger, R. M. 20(103), 48, 53(231), 56(231, 241), *182, 185*, 555, 589(139), *599*
Badr, M. Z. A. 1071(201–203), *1091, 1092*
Baecello, D. 1352(69), *1364*
Baeckstrom, P. 1048(155), *1090*
Baekeland, L. H. 1457, 1475(1), *1502*
Baer, S. 314(201), *330*
Baer, T. 110, 112, 121(444), *189*, 265(24), 288(58), *326, 327*
Baesjou, P. J. 1194(84), *1339*
Baeten, A. 89, 90(319), *186*
Baeyens, W. R. G. 951, 953(193), *1007*
Baeyer, A. 718(15), *831*
Bag, N. 735(54), *832*
Bagchi, P. 425(159, 160), *483*
Bagi, F. 683(170), *708*
Bagirova, N. A. 977(342), *1010*
Baglio, D. 1086(357), *1096*
Bagnell, L. 762(127), *834*
Bagno, A. 89(310), *186*
Bagryanskaya, I. Y. 616(32), *656*
Bahnemann, D. 1111(66), *1145*
Bahr, U. 260, 323(3), *325*
Bailey, A. M. H. 851, 887(52), *903*
Bailey, J. E. Jr. 358(112), *388*
Bailey, W. F. 770(152), *834*
Baillet, A. 953(202), 954(199), 955(202), 970(199), *1007*
Baird, J.-H. 1424(388b), *1450*
Baizer, M. M. 1156, 1158, 1175(1), *1337*
Bakalova, R. A. 887(222), *907*
Baker, A. W. 47(223, 227), 48(227), 53(223), 56(223, 227), *184*, 554, 555(136), *599*
Baker, B. R. 400(19), *478*
Baker, G. 1278(226), *1343*

Baker, P. F. 845, 846(38), *902*
Baker, R. 441(206), *484*
Baker, S. 1033(108), *1089*
Baker, W. 408(56, 57), 419(136b), *480, 482*
Baklouti, L. 1391(114), *1440*
Bakry, R. S. 990(423), *1012*
Bala, P. 144(651), *194*
Balaban, A. T. 432(184a, 186), *484*, 823(330), *838*
Balachandran, K. S. 745(79), *832*
Balakrishnan, P. 349(75), *388*
Balani, S. K. 412(71a), *480*
Balasubramanian, K. K. 301(134), *328*, 663(23, 25, 26), 675(114, 115), 680(147), *705, 707, 708*, 763(132), *834*
Balasubramanian, T. 301(134), *328*
Balasubramanian, V. V. 617(34), *656*
Balavoine, G. 460(285), *487*
Bald, E. 460(283a), *487*
Baldini, L. 1396(185), *1443*
Baldioli, M. 998(452, 453), *1013*
Balduzzi, G. 686(213), *709*, 1462, 1463(26), *1503*
Balesdent, J. 303(145), *329*
Balkan, S. 434(192), *484*
Ball, J. A. 1108(15), *1144*
Ball, J. C. 138(528), *191*, 1104(61), *1106*
Ball, J. W. 925, 929(54), *1004*
Ball, R. G. 457(263), *487*
Ballardini, R. 761(118), *833*
Ballester, P. 750(102, 103), *833*
Ballesteros, E. 932, 937(92), *1005*
Ballone, P. 6(27), 93, 105(354), *180, 187*
Ballou, D. 1119(165), *1147*
Ballou, D. P. 411(69), *480*
Balogh, V. 1307(274), *1344*
Baloghova, M. 953, 962(255), *1009*
Baltas, M. 400(18b), *478*
Balthazor, T. M. 449(234b), *486*
Baltussen, E. 937, 938(125), *1005*
Balzani, V. 761(118), *833*
Bamberger, E. 419(122a, 122b), *482*, 801(240), *836*
Bandara, J. 1084(327), *1095*
Banerjee, A. K. 471(305), *488*
Banerjee, D. K. 425(159), *483*
Banerji, A. 1062(187), *1091*
Banik, B. K. 633(92), *658*
Banks, H. D. 1427(408d), *1451*
Banks, J. T. 852, 868(86), 877(191), 878(86, 191), 879(191, 195), 881(191, 203), 882(86, 191, 195), 883(191), *903, 906*
Banks, R. E. 647(147–149), *659*
Banni, S. 851, 854(74), *903*
Banno, H. 460(283b), *487*, 689(234, 235), 693(255), *710*
Bansal, K. M. 1110(58), 1117(117), 1128–1130(58), *1145, 1146*

Banville, J. 441(207), *484*
Bao, J. 457(263), *487*, 693(253), *710*
Bao, M. L. 933(102), 934(108), 937(102, 108), *1005*
Bar, F. 1123(216), 1124(216, 218), *1148*
Barak, T. 493(46, 47), 498(47), 504, 505(91), 506(46, 91), 510(91), 511(47), 516–519, 521(91), *526, 527*
Barakat, A. 236(55), *255*
Baranova, N. V. 725(28), *831*
Baranowska, I. 929(68), 953, 962(251), 984(68), *1004, 1008*
Barańska, H. 170, 171, 177(759), *197*
Baraton, M.-I. 170, 171(758), *197*
Baratta, M. T. 894(262), *908*
Barba, I. 1158(22), *1338*
Barbara, P. F. 1020(35b), 1025(85a), *1087, 1089*
Barbas, C. 953, 956(212), *1007*
Barber, R. B. 424(154b), 444(218), *483, 485*
Barberis, G. 178(771), *197*
Barberousse, V. 609(17), *656*
Barbieri, K. 933(102), 934(108), 937(102, 108), *1005*
Barbieux-Flammang, M. 84(303), 110(448, 449), 115(303), *186, 189, 190*
Barboso, S. 1389(105), 1391(120c), *1440*
Barbour, L. J. 1372(24b), *1436*
Barcelini, W. 1359(151), *1366*
Barceló, D. 296(86), 319(229), *327, 330*, 928(56), 946(158, 159), 947(171), 953(158, 159, 171, 201, 206, 262), 955(201, 206), 964(262), 977(335), *1004, 1006, 1007, 1009, 1010*, 1348, 1357(11), *1363*
Barcelo, G. 663(16), *705*
Barclay, L. R. C. 842(15), 843(283), 845(30), 851(44–60, 62, 70), 852(60), 855(116), 856(50), 857(62), 858(56, 58), 862(49), 863(49, 55), 868, 870(51), 871(159), 872(50), 876(184), 877(116), 878(50), 882(55), 884(44, 45, 207, 208), 885(45–47, 54, 55, 207, 213–215), 886(55), 887(52, 55, 60, 220), 888(60), 889(45, 56, 58), 890(53, 214, 238), 893(256), 899(279), 901(279, 283, 284), *902–908*
Barcley, L. R. C. 867(150, 152), 868, 869, 871(152), 897(150), *905*
Barich, D. H. 171(762), *197*
Barillier, D. 929, 954(62), *1004*
Barison, G. 979(354), *1011*
Barker, B. 1033(110), 1034(111), *1089*
Barker, J. L. 19(101), *182*
Barkhash, V. A. 616(32), *656*
Barlow, I. 949, 953(185), *1007*
Barluenga, J. 651(172), *660*, 770(153), *834*
Barner, R. 460(280a), *487*
Barnes, I. 18(83a), *181*
Barnes, S. 637(110), *658*

Barnett, D. 978(345), *1011*
Barnette, W. E. 687(220), *709*
Barni, B. 1082(287), *1094*
Barone, V. 4, 100(2), 132, 133, 136(518), *179, 191*, 368, 372, 373(200), *390*, 1127(267), *1149*
Barral, L. 234(43), *255*
Barrash-Shiftan, N. 493(63), *526*
Barratt, D. G. 536, 539, 540, 588, 591(37), *597*
Barre, L. 647(153), *659*
Barrett, A. G. M. 428(170), *483*, 1381(65b), *1438*
Barrett, D. G. 782(221), *836*
Barrett, E. S. 1432(447), *1453*
Barrett, J. C. 1359(146), *1366*
Barrett, M. 1297(257), *1344*
Barroso, C. G. 947(175), 953(175, 232), 960(232), *1007, 1008*
Barrow, J. C. 1318(301b, 302), *1345*
Barrows, J. N. 358(112), *388*
Barry, A. O. 1002(480), *1014*
Barry, B. 129(483), *190*
Barsukova, T. 1119(167), *1147*
Barsukova, T. K. 894(261), *908*
Bartak, P. 932(87, 90), 937(87), *1005*
Bartels, O. 1082(277), *1094*
Barth, H.-D. 147(714), *195*
Barthel, N. 617(35), *656*
Bartholini, G. 475(318), *488*
Bartl, F. 144(676), *195*
Bartlett, W. 1044(131), *1090*
Bartmess, J. E. 5(18), 98(376), *180, 188*, 261(7), 265, 268, 269, 272(29a), 310(182, 184, 185), 311, 312(7), *325, 326, 329*, 719(20), *831*
Bartolo, M. M. S. S. F. 234(46), *255*
Barton, D. H. R. 671(81, 82), 681(163), *706, 708*, 1192(78d), 1216, 1287(120), 1297(120, 257), 1330(323, 325a–c, 326), 1332(325a–c, 326), 1333(326), *1339, 1340, 1344, 1346*
Bartsch, H. 1361(188, 189), *1367*
Bartsch, R. A. 724, 725(27), *831*, 1392(126), *1441*
Bartsch, U. 1205(101), *1340*
Bartulewicz, E. 930(82), *1005*
Bartulewicz, J. 930(82), *1005*
Baruah, J. B. 677(123), *707*
Bascal, Z. A. 856, 857(131), *904*
Baser, K. H. C. 944, 953(142), *1006*
Bashford, D. 101(385), *188*
Bashir, S. 260, 325(5), *325*
Baskin, K. 992(425), *1012*
Baskin, K. A. 851(47, 54, 55), 863, 882(55), 885(47, 54, 55, 215), 886, 887(55), *903, 906*
Basov, N. I. 1470(70), *1504*

Basses, M. P. 493(43), *525*
Bassus, J. 609(18), *656*, 1402(224c–e), *1444, 1445*
Bast, A. 871, 873, 898(169), *905*
Bastic, M. B. 1001(475, 476), 1002(478), *1014*
Bates, C. G. 670(77), *706*
Bates, R. B. 677(139), *707*
Batista, V. S. 385(301), *393*
Bats, J. W. 450(236, 237), *486*
Batterham, T. 719(18), *831*
Battersby, A. R. 1158, 1175(14a), *1337*
Bauder, A. 35(162), *183*
Baudry, A. 1495(138), *1505*
Baudry, M. 609(17), *656*
Baudry, R. 1372(30), *1436*
Bauer, K. 913(19), *1003*, 1359(150), *1366*
Bauer, R. 1081(269), *1093*
Bauer, S. H. 56(241), *185*, 555, 589(139), *599*
Bauerle, G. F. Jr. 308(161), *329*
Baugh, P. I. 1124(244), *1149*
Baugh, S. 942(137), *1006*
Baulch, D. L. 1105(64, 65), *1106*
Bauman, J. G. 444(218), *485*
Baumgartner, M. T. 1073(226), *1092*
Baumstark, A. L. 340, 341(29), 349(29, 75), 352(88), *387, 388*, 551(94, 95), *598*
Bautista, F. M. 608(13), 618(36), *656*
Bautz, D. 933, 937(101), *1005*
Bavoux, C. 219(37), *221*, 1372(30), 1378(52b), 1415(312), *1436, 1437, 1448*
Bax, B. M. 454(253), *486*, 735(52), *832*
Baxendale, I. R. 677(124), *707*
Bayerlein, R. 802(249), *836*
Baylocq-Ferrier, D. 953, 955(202), *1007*
Bazhin, N. M. 1123(201), *1148*
Bazon, J. N. 895(266), *908*
Bazyl', O. K. 1017(5, 6), *1087*
Beadles, S. 1084(312), *1094*
Beam, C. F. 730(47), *832*
Beaton, Y. 1362(217), *1368*
Beauchamp, J. L. 291(65), *327*
Beaulieu, F. 782(217), *836*
Beavis, R. C. 323(255), *331*
Becconsall, J. K. 1109, 1120(28), *1144*
Becher, G. 931, 937(83), *1005*
Becher, J. 1495(148), *1505*
Becht, J. 639(128), *659*
Beck, H. 315(215), *330*
Beck, S. M. 129, 133(492), *190*
Becke, A. D. 68(255), 89(314), *185, 186*
Becke, A. L. 71(272), *185*
Becker, A. 662(5), *705*
Becker, C. H. 323(260), *331*
Becker, F. F. 633(92), *658*
Becker, H. 731(49), *832*
Becker, H.-D. 951, 953(191), *1007*, 1016, 1020(1), *1087*, 1271(203), 1282(228), 1290(243), *1343, 1344*

Becker, K. H. 18(83a), *181*
Becker, K. Y. 953, 955(204), *1007*
Becker, P. 200(8), 220, 375(246), *392*, 549(67), 550(68), *598*
Beckering, W. 548(63), *598*
Beckert, C. 913(19), *1003*
Beckman, B. S. 925(50), *1004*
Beckmann, K. 144, 147(596), *193*
Beckmann, O. 701(314), *711*
Bedel, D. 1485(101), 1486(99–101), *1504*
Beenackers, A. A. C. M. 1084(311), *1094*
Beer, P. D. 1383(67a, 67b, 67m), 1393(152c), 1403(152c, 234a, 234c), 1409(286b), 1421(67a, 67b, 67m), 1427(409c, 411, 413, 416a, 416b, 417), 1432(449a, 450, 453), 1435(466), *1438, 1441, 1445, 1447, 1451–1454*
Beeson, M. 551(94), *598*
Beggiato, G. 1115, 1128(91), *1145*
Behr, A. 685(196), *709*
Behrman, E. J. 408(54g, 54h), 409(62b, 63c), *480*, 801(243), *836*
Beilin, L. J. 913(30), *1003*
Beißwenger, R. 1169(44), *1338*
Beja, A. M. 233(39), *255*
Bejtovic, Z. 1219(131), *1341*
Bekkum, H. van 645(144), *659*, 774(179), *835*
Belal, F. 988(412), *1012*
Belal, S. F. 990(423), 995(439), *1012, 1013*
Bèlanger, G. 37(178), *183*
Bélanger, J. M. R. 930(79), 932(97), 937(79), *1004, 1005*
Belhakem, M. 368(195), *390*
Belicchi, M. F. 680(148), 683(176, 184), *708*
Belin, B. 1470(50, 52), *1503*
Bell, M. 1380(60a), *1438*
Bell, R. P. 83(290), *186*, 501, 502, 507(80), *526*
Bell, S. J. 358(112), *388*
Bellini, A. 1353, 1354(87), *1365*
Bellobono, I. R. 1082(287), *1094*
Bellon, L. 540(39), 564(170), *597, 600*
Bellucci, F. 552(135), *599*
Belluš, D. 476(321), *488*
Belokon, Y. 640(132), *659*
Belostotskaya, I. S. 651, 654(167), *659*, 747(82, 83), *832*, 1126(256, 260), *1149*
Belov, M. E. 1000(468), *1013*
Bel'skii, V. E. 740(64, 65), *832*
Bel'skii, V. K. 200(4), *220*
Belskikh, N. V. 1001(475), 1002(478), *1014*
Belsue, M. 1457(4), *1502*
Beltrán, G. 913(6), *1003*
Beltrán, S. 961(235), *1008*
Beltran-Heredia, J. 1080(255, 256), 1085(345), *1093, 1095*, 1362(202), *1368*
Belussi, G. 413(80a), *480*
Belyakov, S. A. 434(193, 194), *484*

Belyakov, V. A. 871, 872, 874(168), *905*
Benahcene, A. 1361(194), *1367*
Benamou, C. 540(39), *597*
Benassi, M. T. 944(143), *1006*
Bencivenni, L. 375(241), 383(279), *392*
Bencze, Z. 1380(55), *1437*
Bendale, P. M. 662(10), *705*
Bender, H. L. 1462(33), *1503*
Benderskii, V. A. 371(223, 228), *391*
Benedetti, E. 1383(75b, 75e), 1422(75e), *1439*
Benedetto, C. 141(540), *192*
Benedict, C. 367(171), *390*
Benedict, H. 364(155), 367(178), *389, 390*
Benet-Buchholz, J. 322(246), *331*
Benezra, S. A. 299(90, 94), 307(153), *327, 329*
Bengtssan, S. 424(154c), *483*
Benham, G. Q. 359(127), *389*
Benitez, F. J. 1080(254–256), 1085(345), *1093, 1095*, 1362(202), *1368*
Bennet, J. L. 1408(281), *1447*
Bennetau, B. 630(69), *657*
Bennett, D. M. 450, 453(239f), *486*
Bennett, J. M. 716(11), *831*
Bennett, J. S. 110(438), *189*
Bennett, P. A. 1084(312), *1094*
Bennett, S. M. W. 691(240), *710*, 782(220), *836*
Benoit, D. M. 148(724–726), *196*
Benoit, F. M. 300(127), *328*
Benoit-Marquie, F. 1080(253), *1093*
Ben Salah, M. 375(246), *392*
Benson, S. W. 5, 137, 138(21), *180*, 1108(21–23), *1144*
Bent, D. V. 110(417), *189*, 1129, 1130(297), *1150*
Bentley, F. F. 36, 45, 50, 58(172), 87(307), *183, 186*, 238(72), *256*, 551(131), 554, 555(137), *599*
Beone, T. 977(334), *1010*
Berberan-Santos, M. N. 887(224), *907*
Berchtold, G. A. 829(342–345), *838*
Berckmans, D. 170(756), *197*
Berden, G. 20, 24, 106(127), 147(698), *182, 195*, 581(215), *601*
Berdinkov, V. M. 1123(201), *1148*
Berenguer, V. 932(96), *1005*
Beresford, N. 925(53), *1004*
Berg, A. 338(17), 350(77), *386, 388*, 511(103), *527*
Berg, D.-J. van den 871, 873, 898(169), *905*
Berg, M. 971(293), *1009*
Bergamaschi, M. 1374, 1378(37), *1436*
Bergel, A. 977(340), *1010*
Berger, B. 1431(443a), *1453*
Berger, S. 338(11), *386*
Bergmann, B. 1131, 1134, 1139, 1141(313), *1150*
Bergmann, E. J. 408(54b), *480*

Bergmann, F. 425(159), *483*
Bergvinson, D. 994(438), *1013*
Berho, F. 138(527), *191*
Berlman, I. 107, 109(411), *188*
Berman, Z. 411(70a), *480*
Bermejo, J. M. 769(148), *834*
Bermudez, G. 143, 147(585), *193*
Bernabei, M. 953, 970(288), *1009*
Bernal, J. L. 953(209, 256), 955(209), 962(256), *1007, 1009*
Bernard, A. M. 764(138, 139), *834*
Bernardinelli, G. 783(222), *836*
Berndt, A. 1123(216), 1124(216, 218), *1148*
Berr, C. E. 1494(128), *1505*
Berrier, C. 405(39a), *479*
Berrueta, L. A. 1353(72), *1364*
Berset, C. 856(125), *904*, 982(369, 371), *1011*
Berstein, M. A. 683(175), *708*
Berthalon, S. 1394(169), *1442*
Berthelon, G. 229, 230(23), *254*
Berthelot, M. 238–240(66), *256*, 531(18, 19), 532(18), 533(22), 534(23), 543(50), 544, 545(51), 547(18, 51), 548, 549(51), 551, 553(87), 554, 555(138), 558(19, 146), 559(138, 154), 568(154), 575(186, 188), 576(19, 146, 189), 580(188, 211), 581(211), 583(223), 586(18, 19, 23, 146, 154, 186, 188, 241–256), 587(18, 19, 23, 146, 186, 188, 241–256), 589(18, 19, 244, 248, 252, 254, 256), 590(138), *597–602*
Berthod, A. 953, 956(210), *1007*
Berthomieu, C. 94, 105(366), 132, 133, 135(516), *187, 191*, 369(205–207), *391*
Bertolasi, V. 338, 342(9), 363(146), *386, 389*, 552(135), *599*
Bertolini, F. 1424(389b), *1450*
Bertran, J. 129, 144, 148, 149, 153(473), *190*, 577, 581(192), *600*, 1362(207), *1368*
Bertrand, M. 290(61), *327*
Bertrand, W. 115(461), *190*
Berzélius 7(36, 37), *180*
Besheesh, M. K. 647(147–149), *659*
Besombes, J. L. 977(343), *1011*
Besseau, F. 531(19), 558(19, 146), 575(188), 576(19, 146), 580(188), 586, 587(19, 146, 188, 249, 252, 255), 589(19, 252), *597, 599, 600, 602*
Besson, T. 680(150), *708*
Bester, K. 1356(123), *1366*
Beswick, C. 233(41), *255*
Beta, T. 913(14), *1003*
Beugelmans, R. 420(141), *483*
Beveridge, D. L. 5(14), *179*
Beveridge, K. A. 1036(119), *1089*
Beyer, J. 675(116), *707*
Beyersdorf-Radeck, B. 978(351), *1011*
Beylen, M. van 585(236), *601*

Beynon, J. H. 263(20), 264(22), 290(61), 301(137), 302(137, 138), *326–328*
Bezuglaya, E. Y. 178(772), *197*
Bhagat, S. D. 941(134), *1006*
Bhakuni, D. S. 807(277), *837*, 1297(260a), *1344*
Bhalerao, U. T. 1218(130), 1219(133a, 133b), 1223(135), *1341*
Bhandari, A. 1362(209), *1368*
Bhandari, J. M. 990(418), *1012*
Bhanu, V. A. 359(128), *389*
Bhasikuttan, A. C. 1021(38), *1088*
Bhaskar, K. R. 856(132), *904*
Bhatacharjee, S. K. 216(32), *221*
Bhatia, K. 1099(16), *1105*
Bhatkhande, D. S. 1084(311), *1094*
Bhatnagar, N. Y. 681(163), *708*
Bhatnagar, R. 1017(13), *1087*
Bhattacharjee, M. 988(414), *1012*
Bhattachatarjee, S. 1081(267), *1093*
Bhattcharyya, K. 1025(71a, 71b), *1088*
Biali, S. E. 341(31), *387*, 1374(34, 39), 1378(51), 1386(51, 87), 1401(215, 216a), 1410(291a, 291b), 1411(293–295, 296a–c, 297, 298, 300), 1413(301–303), *1436, 1437, 1439, 1444, 1447*
Biall, S. E. 1408(283), 1409(284), *1447*
Bialopiotrowicz, T. 1002(482), *1014*
Bianchi, C. L. 1085(339), *1095*
Bianchi, G. 1136(362), *1151*
Bianchi, L. 953, 968(275), *1009*
Bianchi, M. 413(84), *481*
Bichan, D. J. 1327(313), *1345*
Bicke, C. 305(147), *329*
Bidzilya, V. A. 1118(141), *1146*
Bied-Charreton, C. 1201, 1205(92), *1340*
Biekofsky, R. R. 366(167), *390*
Bielski, B. H. J. 845(37), *902*, 1109(51), 1130(305, 307), *1144, 1150*
Bienaymé, H. 677(132), *707*
Bieniek, G. 920(34), *1003*
Bieńko, D. C. 20, 22, 23(112), 35(112, 170), 36–38(112), 47, 58(170), 82(288), *182, 183, 186*, 285(52), *255*, 372, 375(236), 384(291, 292), *391, 393*
Bienwald, F. 699(304), *711*
Bieri, J. H. 212(25), *221*
Biermann, H. W. 300(126), *328*
Biernacka, T. 1470(62), *1503*
Bieske, E. J. 144(681), *195*
Biggers, K. E. 786(229, 230), *836*
Bighelli, A. 365(159), *389*
Bigi, F. 1255(183), *1342*
Bigi, F. 608, 612(10), 616(26), 622(51–53), 623(54), 627(62), *656, 657*, 677(122, 131, 133), 680(153, 155), 683(179, 181–184), 686(206–208, 213), *707–709*, 740(62),

Bigi, F. (*continued*)
832, 1374(37), 1378(37, 49a–c), *1436, 1437*
Bignan, G. 673(94), *706*
Bignoli, F. 1082(287), *1094*
Bigot, A. 420(141), *483*
Bikádi, Z. 179(782), *198*, 369(209), *391*
Bilikova, J. 1354(106), *1365*
Billard, J.-P. 1054(164), *1091*
Billes, F. 20, 22, 23, 35, 36, 38(111), *182*, 368, 370–375(198), *390*
Billion, M. A. 639(125), *659*
Billmers, J. M. 417(117), *482*
Bilton, C. 551, 552(134), *599*
Bilyk, A. 1413(304, 305), *1447*
Binkley, J. S. 368(201), *391*
Binkley, R. W. 314, 315(208), *330*, 1043(130b), *1090*
Biradha, K. 211(21), *221*
Birch, A. J. 13(65), *181*, 426, 427(166b), *483*, 1297(260c), *1344*
Bird, C. W. 1294(253a, 253b), *1344*
Bird, S. 953, 964(263), *1009*, 1353(82), *1365*
Birge, R. 93–96(357), *187*
Birkett, A. M. 953, 969(281), *1009*
Birknes, B. 579(198), *601*
Birkofer, L. 823(324, 325), *838*
Birney, D. M. 807(274), *837*, 1230(149), *1341*
Bisarya, S. C. 651(171), *660*
Bisby, R. H. 844(20), *902*, 1129(282, 283, 299), 1130(299, 306, 322), 1132(339), 1133(282, 340), 1134(283), 1137(366), 1141(282), *1149–1151*
Bist, H. D. 35(153, 156, 159), 36, 37, 66, 67(153), 107(156), *183*
Bitsch, I. 953, 957(218), 960(226), *1008*
Bitter, I. 1387, 1391(93), *1439*
Bittner, D. 18(83a), *181*
Bizarro, M. M. 229(32), *255*
Bjerg, P. L. 1357(136), *1366*
Bjergbakke, E. 1115, 1130(99), *1145*
Bjorsvik, H. R. 607(6), *656*
Blackburn, M. A. 1356(127), *1366*
Blacklock, C. J. 957(220), *1008*
Blackwell, W. 934, 937(109), *1005*
Bladek, J. 953, 965(264), *1009*
Blair, L. K. 310(183), 312(191), *329, 330*
Blaise, P. 371(225, 226), *391*
Blake, A. J. 691(241), *710*
Blake, D. M. 129, 132, 133(498), *191*
Blanch, R. 1478(86), *1504*
Blanchard, H. S. 1491(118), *1505*
Blanchard, M. 405(45), *479*
Blanco, B. 932, 937(94), *1005*
Blanco, J. 1086(372), *1096*
Blanco Gomis, D. 953, 958(223), *1008*
Blanco-Labra, A. 913(2), *1002*

Blanda, M. T. 322(247, 248), *331*, 1399(204), 1427(419b), *1444, 1452*
Blank, A. B. 987(407), *1012*
Blanpain-Avet, P. 1002(482), *1014*
Blasco, T. 316(220), *330*
Blasko, A. 769(147), *834*
Blatt, A. H. 639(127), *659*
Blatt, A. M. 472(308d), *488*
Blatz, P. E. 593, 595, 596(279), *602*
Blazejewski, J.-C. 671(81), 681(163), *706, 708*, 1272(205a, 205b), *1343*
Blazevic, N. 1470(50, 52), *1503*
Blazkova, A. 1082(284), *1094*
Blazovics, A. 1129, 1130(294), *1150*
Bleton, J. 913(5), *1003*
Blixt, J. 1385(82), *1439*
Blomberg, M. R. A. 45(196), *184*
Blount, J. F. 1278(225), *1343*
Blum, W. 291(67), *327*
Boatella, J. 961(243), *1008*
Bobbitt, J. M. 1159(29), *1338*
Bobrowski, K. 1117(130, 132), 1128(281), *1146, 1149*
Bocchi, C. 953, 955(203), *1007*, 1396(181a, 182a), *1442, 1443*
Bocchi, V. 1462(24), *1502*
Bocchinfuso, G. 953, 970(288), *1009*
Bocchini, P. 303(142), *329*, 937, 941(131), *1006*
Bocelli, G. 622(52, 53), 623(54), *657*, 683(179, 182), *708*
Boch, C. W. 82(284), *186*
Bock, C. W. 93(362), *187*, 551, 553(77, 78), *598*, 826(337), *838*
Bock, H. 1283(231), *1343*
Boddy, C. N. C. 670(71), *706*
Bodewes, P. J. 1403(242), *1445*
Bodibo, J. P. 631(77), *657*
Bodige, S. G. 1427(414b), *1451*
Bodot, H. 727(37), *831*
Boehme, W. R. 432(183b), *484*
Boerrigter, H. 1406(263b), 1423(380b), 1424(382a, 382b, 383), 1433(456b), *1446, 1450, 1453*
Boese, R. 219(35), *221*, 322(246), *331*
Boeseken, J. 422(144), *483*
Bogaerts, S. 571(179), *600*
Bogdal, D. 662(8), *705*
Bogdanov, B. 86(306), *186*
Bogdanov, G. N. 745(78), *832*, 1119(157), 1123(210), *1147, 1148*
Boger, D. L. 450(246f), *486*, 670(73), *706*, 1163(33c), *1338*
Bogert, T. M. 1494(124), *1505*
Boguslavskaya, N. I. 1104(58), *1106*
Boguth, W. 1125(245), *1149*
Bohle, D. S. 637(113), *658*
Bohm, J. L. 929, 954(62), *1004*

Böhmer, V. 320(238, 239), *331*, 635(100), *658*, 1370(1b, 1c, 6), 1374(33a–c, 34, 36c, 39–41, 42a, 43), 1378(50a, 51), 1382(74), 1383(67g), 1385(83), 1386(51, 87), 1389(105), 1391(120c), 1393(157b), 1396(195a, 195b), 1402(1c, 222), 1403(229, 239), 1404(222), 1406(222, 257, 262a), 1411(295), 1418(342, 343a, 343b), 1419(356b, 358), 1420(342, 359c, 361, 360, 361, 363, 364, 365b, 366), 1421(67g, 342, 371), 1425(406), 1427(409c, 412a, 412b), 1430(6), 1431(443a, 443b, 444), 1432(229, 446, 449b), 1435(371, 461, 464a, 470), *1435–1440, 1442–1447, 1450, 1451, 1453, 1454*, 1462(30, 36), 1470(60), *1503*
Bohr, F. 867, 880, 897(151), *905*
Bois, C. 549(64), *598*
Bois, J. D. 702(328–331), *712*
Bois-Choussy, M. 420(141), *483*
Boivin, J. L. 421(130), *482*
Bojesen, G. 325(277), *332*
Bökelmann, V. 323(262), *331*
Bol, R. 303(145), *329*
Bolitt, V. 422(147c), *483*
Bollag, J.-M. 1352(61), *1364*
Bollain, H. M. 971(294), *1009*
Bollain, M. H. 1348, 1354(17), *1363*
Bolm, C. 699(302, 304), 701(314–317), 702(337), *711, 712*
Bolon, D. A. 1307(277), *1344*
Bolshov, V. A. 18(85), *181*
Bolte, M. 1017(15–17), 1018(18, 19), 1020(31), 1082(273, 275), 1085(347), 1086(359), *1087, 1093, 1095, 1096*
Bolton, J. L. 874(178), *906*
Bolton, J. R. 1072(211), 1073(211, 213), *1092*, 1123, 1124(213), *1148*
Bolvig, S. 336(5, 39, 40), 340(28), 342(35, 38), 343(39, 40, 42), 344(39, 40), 345(5, 39, 40, 47), 346(35, 47, 51), 347(35, 42, 55), 348(55), 349(55, 60), 350(39), 351(87), 354(5, 40, 47, 51), 355(39, 40, 47, 51), 356(42, 47), 358(117), 361(47, 142), 363(142), 366(39, 40, 51), *386–389*, 551(96, 98), *598, 599*
Bolz, U. 925(51, 52), 936(51), *1004*
Bolzan, R. M. 637(107), *658*
Bonchio, M. 413(83, 84), 414(85), *481*
Bondarenko, L. I. 631(79), *657*
Bondet, V. 856(125), *904*
Bondy, H. F. 419(136b), *482*
Bonet, J.-J. 813(298), *837*
Bonfanti, G. 740(62), *832*
Boni, M. R. 932, 937(93), *1005*
Bonifacic, M. 1109(46), *1144*
Bonn, G. 974(315), *1010*
Bonn, G. K. 945(157), *1006*
Bonnely, S. 982(371), *1011*
Bonnichon, F. 1020(31), 1054(163), 1057(174), *1087, 1091*
Bonrad, K. 211(18), *220*
Bonsignore, S. 1424(397b), *1451*
Bonvicino, G. E. 469(297a, 297b), *488*
Boody, C. N. C. 1318(303), *1345*
Boon, J. J. 293(71), 303(143), *327, 329*
Boone, C. W. 868(156, 157), 870(157), *905*, 1108(10), *1144*
Boquet, G. 583(223), *601*
Borau, V. 972(304), *1010*
Borchardt, R. T. 415(107), *481*
Borchers, F. 110(427), *189*, 299(99), *328*
Borden, W. T. 110, 112, 139(442), *189*
Bordwell, F. G. 5(24), 110(421), 139(24), *180, 189*, 311(190), *330*
Borello, E. 1083(293), 1084(320), *1094, 1095*
Borgarello, E. 1082(281), *1094*
Borges dos Santos, R. M. 98(375), *188*, 229(32), *255*
Borgese, M. 1082(288), *1094*
Borgulya, T. 460(280a), *487*
Bories, G. 316(218), *330*
Borisenko, K. B. 82(284), *186*, 244(86, 87), *256*, 551, 553(76, 77), *598*
Bornaz, C. 230(31), *255*
Boron, A. 662(8), *705*
Borowicz, P. 1025(68), *1088*
Borrajo, J. 1459(7, 8), *1502*
Borredon, E. 663(19), *705*
Borrull, F. 944(147), 945(147, 152, 153, 156), 946(164), 953(152, 164, 209, 254, 256, 261), 955(209), 962(254, 256), 963(261), 971(295), *1006–1009*, 1347(1), 1353(1, 92, 93), 1354(102, 107), *1363, 1365*
Bors, W. 855(109), 856(109, 121), 871, 872, 874(168), 894(261), 895(269), *904, 905*, 908, 1115(101), 1117(116), 1118(101, 143, 146, 147), 1119(101, 146, 163), 1128(279), 1130(101, 143, 147, 309, 323), 1131(279), 1137(368), *1145–1147, 1149–1151*
Borsdorf, R. J. 252(96), *257*
Borsiov, E. V. 349(60), *387*
Bortolini, R. 683(181), *708*
Borull, F. 1352(65), 1353(70), 1354(70, 104), *1364, 1365*
Borzatta, V. 644(143), *659*
Bos, M. E. 450(246b), *486*, 735(53), *832*
Bosca, F. 1079(243), *1092*
Bosch-Reig, F. 942(135), 953, 954(198), *1006, 1007*
Bosco, D. 607(6), *656*
Boshoff, A. 411(70b), *480*
Boskou, A. 998(454), *1013*
Bosner, B. 385(298), *393*
Boswell, D. E. 400(16), *478*
Bott, S. G. 1402(226), *1445*

Botta, B. 1382(74), 1383(75a–e), 1422(75e, 375), *1438, 1439, 1450*
Botta, M. 1383(75a–e), 1422(75e), *1439*
Böttcher, J. H. 997(444), *1013*
Bottino, A. 1401(213), 1403(230), 1427(230, 414a), *1444, 1445, 1451*
Bottino, F. 1393(157a), *1442*
Bottling, R. M. 10(45), *180*
Bouchard, G. 544(52), *598*
Boucher, D. 171(762), *197*
Bouchoule, C. 405(45), *479*
Bouchoux, G. 83(296b), 110(452), 115(461), *186, 190,* 299(95), 300(124), *327, 328*
Bouchy, M. 1080(249), *1093*
Boudenne, J.-L. 1361(192), *1367*
Bougauchi, M. 695(270), *710*
Bouix, J. 1415(318a), *1448*
Boule, P. 1054(164), 1072(210), 1073(217, 220), 1080(251), 1083(300), 1084(319), 1085(344), *1091–1095*
Boule, P. 1057(175–177), 1072(177), *1091*
Boullais, C. 132, 133, 135(516), *191*
Bouma, W. J. 114(460), *190*
Bouoit, S. 1402(224e), *1445*
Bouoit-Montésinos, S. 1402(224d), *1445*
Bourassa-Bataille, H. 47, 53, 56(229), *185*
Bourelle-Wargnier, F. 818(314), *838*
Bourguignon, B. 954, 970(199), *1007*
Bourne, L. C. 968(276), *1009*
Boussac, A. 94, 105(366), 132, 133, 135(516), *187, 191,* 369(205–207), *391*
Boussenadji, R. 953, 956(210), *1007*
Bowden, B. F. 807(279), *837*
Bowen, R. D. 263(19), 300(106), *326, 328*
Bower, J. F. 670(72), *706*
Bowers, A. R. 178(773), *197*
Bowie, J. H. 310(177, 180), 314(209), 315(210), 316(180, 209), *329, 330,* 467(296), *488*
Bowry, V. W. 141, 142(538), *192,* 846(42), 876(185), 892(185, 252, 253), *902, 906, 907*
Boyar, W. C. 15(72), *181*
Boyatzis, S. 1086(361), *1096*
Boyd, D. R. 412(71a, 72), *480,* 721(21), 824(21, 335), 829(342, 344, 345), 830(347), *831, 838*
Boyd, S. A. 450(235), *486,* 1347(6), *1363*
Boyd, S. E. 761(116), *833*
Boyd, T. J. 933, 937(100), *1005*
Boykin, D. W. 339(22), 340(23, 24, 28, 29), 341(29, 30), 349(29, 30, 75, 76), 350(30), 352(88), 356(108), *386–388,* 551(94, 95), *598*
Boykin, W. 551, 553(80), *598*
Boyland, E. 409(62a, 63b), *480,* 801(241, 244), *836*
Bozan, B. 944, 953(142), *1006*

Bozell, J. J. 676(118), *707,* 1205(104), 1265(198), *1340, 1343*
Brabant-Govaerts, H. 583(226), *601*
Bracchini, C. 1396(185), *1443*
Bracey, A. 764(137), *834*
Bracht, J. 1169(43), *1338*
Brack, T. L. 1028(96), *1089*
Braddock, D. C. 1381(65b), *1438*
Bragina, N. V. 953, 957(214), *1007*
Bram, G. 460(285), *487*
Brand, J. C. D. 35(153, 156), 36, 37, 66, 67(153), 107(156), *173*
Brandes, B. D. 698(297), *711*
Brandsma, L. 677(138), *707*
Brandsteterova, E. 1082(271, 272), *1093*
Brandt, O. 493(55), *526*
Brandvold, T. A. 450(246b), 454(253), *486,* 735(52), *832*
Brand-Williams, W. 856(125), *904*
Brann, E. A. 1359(155), *1366*
Brar, J. S. 774(185, 186), *835*
Bräse, S. 670(71), *706*
Braslavsky, S. E. 1085(330), *1095*
Brass, I. J. 559, 574(155), *600*
Brassard, P. 441(207), 444(215a, 215b, 216a, 216b), *484, 485*
Bratan-Mayer, S. 361(145), *389*
Brattan, D. 1072(212), *1092*
Brauer, B. 493(63), *526,* 948(177, 178), *1007*
Bräuer, M. 385(302), *393,* 1025(70a), *1088*
Braughler, J. M. 857(135), *905*
Brault, D. 1111(68), *1145*
Brauman, J. I. 95, 109(369), *188,* 310(183), 312(191), 314(201, 202), *329, 330*
Braun, A. M. 1080(253), 1085(334), *1093, 1095*
Braun, D. 816(308), *838*
Braun, S. 338(11), *386*
Bray, T. L. 460(279), *487*
Braybrook, C. 121(465), *190*
Brede, O. 110(420, 422–424), 114(424), 121(423, 424), *189,* 1017(12), *1087,* 1101(38–40), 1102(40, 45–47), *1106,* 1134(341–343), *1151*
Breder, C. V. 650(163), *659*
Breen, J. J. 493(21, 23), *525*
Bregman-Reisler, H. 1099(9), *1105*
Bremner, P. D. 957(220), *1008*
Bren, V. A. 729(44), *831*
Brenes, M. 930(78), *1004*
Brenn, J. 1378, 1386(51), 1411(298), *1437, 1447*
Brennan, J. A. 400(16), *478*
Brenner, V. 78(282), 129, 144(474), *186, 190,* 578(195), *601*
Breton, J. 369(205–207), *391*
Breuer, S. W. 414(95), 416(112), *481, 482*
Breuker, K. 324(269, 275), 325(276), *331, 332*

Breure, A. M. 1359(159), *1367*
Brewer, E. J. 67(251), *185*
Brewer, W. E. 1025(85b), *1089*
Brewster, A. G. 1330, 1332(325a, 325c), *1346*
Brewster, R. E. 1406(257, 260), *1446*
Brezezinski, B. 551, 594, 595(126), *599*
Brezinski, B. 355(107), *388*
Breznan, M. 1082(284), *1094*
Brezova, V. 1082(284), 1084(328), *1094, 1095*
Bribois, R. G. 450(241c, 246h), 453(241c), *486*
Brice, K. A. 319(230), *330*
Bridle, P. 973(309, 310), *1010*
Bridson, J. N. 1374(42c), 1392, 1393(135), 1411(42c), *1437, 1441*
Briehl, H. 1478(86), *1504*
Brigati, G. 45, 140(198), *184*, 900(280), *908*
Brigaud, T. 428(171), *483*
Brillas, E. 1085(348), *1095*
Brimble, M. A. 623(55), *657*, 776(197, 198), *835*
Brinck, T. 45(193), 105(403), 140(193), *184, 188*
Brine, D. R. 1348(30), *1363*
Bringmann, G. 428(168), *483*
Brinkman, U. A. T. 296(86), *327*
Brinkman, U. A. Th. 1353(80), *1365*
Brinks, E. A. 1433(456a), *1453*
Brinn, I. M. 1026(89), *1089*
Brisson, C. 444(216b), *485*
Britt, R. D. 15(71), *181*, 1108(15), *1144*
Brittain, J. M. 606, 649(3), *656*, 806(267), *837*
Brittain, W. J. 997(448), *1013*
Britton, D. 216(30, 31), *221*
Britz, H. E. 986(393), *1012*
Brodbelt, J. 296(84, 85), 308(160, 161), *327, 329*
Brodbelt, J. S. 309(166), 322(247, 248), *329, 331*, 1399(204), *1444*
Brodersen, R. 335(2), *386*
Brodersen, S. 67(250), *185*
Brodie, B. B. 405(43a), *479*
Brodilova, J. 1080(250), *1093*
Brody, J. G. 925(47), *1004*
Brody, M. S. 322(251), *331*, 1406(263a), *1446*
Broholm, M. M. 1357(136), *1366*
Broitman, F. 1142(380), *1152*
Brommel, S. 1359(159), *1367*
Brønsted, J. N. 491(6, 7), 522(98), *525, 527*
Brooks, V. L. 1104(54), *1106*
Broquier, M. 78(282), *186*
Brose, T. 405(48a), *479*
Broser, W. 1123(215), *1148*
Brossi, A. 1244(170), *1342*
Broster, F. A. 414(95), *481*
Brousmiche, D. W. 1033(108), 1035(114), *1089*
Brown, C. E. 877(189), *906*

Brown, C. T. 1020(34), *1087*
Brown, E. V. 35(158), *183*
Brown, G. R. 761(117), *833*
Brown, H. C. 226(6), *254*, 414(97c), 415(98a, 98b), *481*
Brown, I. M. 1495, 1498(142, 143), *1505*
Brown, J. 121(465), *190*
Brown, J. H. 414(92), *481*
Brown, J. S. 631(81), *658*
Brown, M. G. 148, 160(721), *196*
Brown, N. C. 408(56), *480*
Brown, R. G. 1024(60), 1025(67, 88), 1047(147), 1057(173), 1072(212), *1088–1092*
Brown, S. M. 691(240), *710*, 782(220), *836*
Brownawell, B. J. 1357(131), *1366*
Brownbridge, P. 439(200, 201a, 201b, 202), 443(201a, 201b), 449(234a), *484, 486*
Browne, J. K. 1391(123), *1441*
Broze, M. 1124(221–223), *1148*
Bruce, J. M. 227, 248(15), 252(94), *254, 257*
Brueckner, B. 1080(260), *1093*
Bruhn, P. R. 673(90), *706*
Bruice, T. C. 14(68, 69), *181*, 769(147), *834*
Brun, P. 448(229a, 229b, 230), *485*
Brunow, C. 1290(250), *1344*
Brunow, G. 1216(126a, 126b, 127), 1309(281), *1341, 1345*
Bruns, B. R. 1351(50), *1364*
Brus, L. E. 129, 133(492), *190*, 1025(85a), *1089*
Brusset, H. 200(8), *220*
Brustolon, M. 1123(205), *1148*
Brutschy, B. 145(636), 147(714), 156(636), 179(781), *194, 195, 198*
Bruyn, P. J. de 1463(41), 1465, 1467(42), *1503*
Bruzoniti, M. C. 1353(73), *1364*
Bruzzoniti, M. C. 1353(88), *1365*
Bryan, R. F. 1278(218), *1343*
Bryant, J. A. 1382, 1383, 1421(67c), 1424(397a, 399), *1438, 1451*
Bryce-Smith, D. 1495(150, 151), *1505*
Brycki, B. 145(632, 633), *194*, 364(156), *389*, 593(273), 594(289), 595(273, 289), *602, 603*
Bryl-Sandelewska, T. 1100(29), *1106*
Bryson, A. 498(73), *526*
Bryzki, B. 355(107), *388*
Brzeziński, B. 144(609, 671–673, 675, 676), 145(628–633), 146(642), *193–195*, 353(95, 98, 101), 355(106), 377(259, 263, 264, 267), 386(307), *388, 392, 393*, 593(273, 277), 595(273, 277, 294, 295, 298, 299), 596(294, 295, 298), *602, 603*, 742(68), *832*
Bu, X. R. 685(201), *709*
Buben, I. 874(179), *906*

Bubnov, N. N. 1125(254), 1126(257–261), *1149*
Bucala, R. 855, 868, 901(111), *904*
Buchanan, M. V. 293(70), *327*
Buchberger, W. 974(316), *1010*
Buche, K. 950(188), *1007*
Bucherer, H. T. 402(23), *479*
Buchhold, K. 147(714), *195*
Buchholz, K. D. 932(88, 89), *1005*
Büchi, G. 1250(174–176, 177a, 177b), 1327(315), *1342, 1346*
Büchi, G. H. 1180(64), *1339*
Buchmeiser, M. R. 945(157), *1006*
Buchwald, S. L. 670(74), 671(80), *706*
Buck, E. 670(78), *706*
Buckingham, D. A. 170(751), *196*
Buckman, N. G. 1002(479), *1014*
Buczak, G. 146(638), *194*, 593, 595(281, 282), *602, 603*
Budka, J. 1432(451), *1453*
Budzikiewicz, H. 260(1), 310, 315(176, 178), *325, 329*
Buehler, C. A. 410(65a), *480*
Buerki, S. 1085(334), *1095*
Buettner, G. R. 890(239), 896(270), *907, 908*
Bugge, G. 6, 7, 133(31), *180*
Bühler, R. E. 1113(83), *1145*
Bühring, D. 1419(357), 1428(425a), *1450, 1452*
Bui, H.-H. 871, 898(164), *905*
Buiarelli, F. 942, 943, 953(138), *1006*
Buist, G. J. 503(89), *526*
Buker, H. H. 93(355), *187*
Bullock, A. T. 559, 574(155), *600*
Bullock, W. H. 668(58), *706*
Bulman Page, P. C. 1420(362, 367), *1450*
Bultmann, T. 1025(84), *1089*
Buluku, E. N. L. E. 241(74), *256*
Bulusu, S. 1470(49), *1503*
Bumble, S. 920(32), *1003*
Bumgardner, C. L. 669(61), *706*
Bunce, N. I. 1361(190), *1367*
Bunce, N. J. 419(134b), *482*, 1071(208), *1092*
Buncel, E. 419(127, 128b, 129), *482*, 802(248), *836*
Bunch, J. E. 319(225), *330*, 669(61), *706*, 932, 937(91), *1005*
Bund, O. 953, 962(250), *1008*
Bungh, W. B. 170(739), *196*
Bunnett, J. F. 397, 467(5), *478*
Buntowsky, G. 503(90), *527*
Buraev, V. I. 806(261), *837*
Burant, J. C. 4, 100(2), *179*, 368, 372, 373(200), *390*
Burch, R. 1362(203), *1368*
Burckel, P. 233(40), *255*

Bureiko, S. F. 143(557–560), 145(626, 627), 146(641), 147(643–645), *192–194*, 352(89, 90), *388*
Burel, R. 144(691), *195*
Burestedt, E. 953, 969(280), 977(335), *1009, 1010*
Bürgi, H. B. 147(709), *195*
Bürgi, T. 20, 35(122), 36(122, 175), 78(278), 147(122, 175, 709–711), 148(710, 711), *182, 183, 186, 195*, 375(242, 243), *392*, 577(191), 581(219), *600, 601*
Burie, J. 171(762), *197*
Burilov, A. R. 1424(396), *1451*
Burinsky, D. J. 293(76), 308(163), *327, 329*
Burke, S. D. 416(109b, 115), *482*
Burkert, U. 322(246), *331*
Burkhalter, R. S. 1414(307a, 308), *1447, 1448*
Burlakova, E. B. 891(243), *907*, 1108, 1118(6), *1144*
Burlingame, A. L. 129(485), *190*
Burmistrov, K. S. 725(28), *831*
Burmistrov, S. I. 725(28), *831*
Burns, E. 934, 937(109), *1005*
Burns, R. 1470(55), *1503*
Burr, J. A. 845(39), 852(79), *902, 903*
Burrini, D. 933(102), 934(108), 937(102, 108), *1005*
Burrow, P. D. 315(211), *330*
Burrows, C. J. 702(325), *712*
Burrows, H. D. 1017(15), 1018(18, 19), 1082(276), 1086(359), *1087, 1093, 1096*, 1101(35), *1106*
Bursey, J. T. 280(49), *326*
Bursey, M. M. 280(49), 299(90, 94), 307(153, 154), *326, 327, 329*
Burton, G. W. 15, 129(76), 141(76, 539, 540), 142(76), 143(552), *181, 192*, 840(3), 842(14), 844, 847(18), 851(43, 62), 853(92), 855(18), 857(62), 858(43), 860(14, 18, 141), 861, 862(18), 864(18, 143, 144), 865(144), 884(207), 885(207, 212), 887(217), 890(212), 891(241), 897, 898(18), 900(14), 901(288), *902–908*, 1108(7, 8), 1111(7), 1120, 1140(184), *1144, 1147*
Burton, J. M. 1393(145a), *1441*
Burton, R. D. 325(278), *332*
Burton, S. G. 411(70b), *480*, 980(358), *1011*
Buryan, P. 265, 268, 269(27), *326*
Busch, J. H. 721(23), *831*
Busche-Hünnefeld, C. von dem 1428(425a), *1452*
Buschmann, H.-J. 1378(48b), *1437*
Buser, H. R. 315(216), *330*
Bushkov, A. Y. 1025(69), *1088*
Bussandri, A. 1016(4), *1087*
Bussche, L. von der 681(165), *708*
Busto, O. 1352(65), *1364*

Author Index

But, P. P. H. 212(22), *221*
Butler, G. B. 1495, 1498(152), *1505*
Butler, J. 1100(21), *1105*, 1117(123, 126–128), 1130(123), *1146*
Butters, B. 1086(366), *1096*
Buttery, R. G. 419(134a), *482*
Buttrill, S. E. Jr. 291(68), *327*
Buvari-Barcza, A. 358(117), *388*
Buxton, G. V. 1098(2), *1105*, 1112(79), *1145*
Buzbee, L. R. 398(9b), *478*
Buzzard, G. K. 129, 131(503), *191*, 1127(273), *1149*
Byers, J. R. 400(20), *478*
Bythway, I. 1478(85), *1504*

Cabaleiro, M. C. 1381(65a), *1438*
Cabaleiro-Lago, E. M. 171(762), 174(763), *197*
Caballol, R. 110, 121(428), *189*
Cabani, S. 101(384), *188*
Cabelli, D. E. 1130(307), *1150*
Cabiddu, S. 139(532), *191*
Cablewski, T. 762(127), *834*
Caboot, J. B. 951, 953(191), *1007*
Cabral, B. J. C. 179(777), *198*
Cabredo, S. 934, 937(107), *1005*
Cabrini, L. 852(77), *903*
Cabrol, B. J. C. C. 20, 23(117), *182*
Cacace, F. 307(156), 309(169–171), *329*
Caccamese, S. 1392(127), 1394(172c), 1396(172c, 195d), *1441–1443*
Cacciapaglia, R. 1392(136), 1394(171), 1396(185), 1397, 1399, 1403(197b), *1441–1443*
Caceres, J. 1086(372), *1096*
Cadahía, E. 953, 965(267), *1009*
Cadenas, E. 1136, 1137(363), *1151*
Cadogan, F. 1435(469a), *1454*
Cai, L.-N. 674(104), *707*
Cai, M.-S. 674(104), *707*
Cai, W. 666(48), *705*
Caira, M. R. 550(72), *598*
Cairns, N. 472(313), *488*
Calama, M. C. 1405(256a), *1446*
Calba, P. J. 323(259), *331*
Calceran, M. T. 1354(103), *1365*
Caldwell, G. W. 314(204), *330*
Caldwell, J. 935, 937(114), *1005*
Califano, S. 4(7), *179*
Caliskan, E. 448(228), *485*
Callery, P. S. 1335(330), *1346*
Callinan, A. 1227(147), *1341*
Calloni, M. 953, 966(269), *1009*
Callul, M. 1353(92, 93), 1354(104), *1365*
Calull, M. 945, 953(152), 971(295), *1006*, *1009*, 1347, 1353(1), 1354(107), *1363*, *1365*

Camacho, D. H. 664(35), *705*, 763(134), *834*
Camacho-Frias, E. 945(150), *1006*
Camaioni, D. M. 110, 112(441), *189*, 1111(74), *1145*
Camaioni, O. M. 405(41d), *479*
Camarero, S. 937, 941(131), *1006*
Cambie, R. C. 416(114), *482*
Cameron, D. 308(163), *329*
Cameron, D. W. 444(215c, 216g, 216h), 448(228), *485*
Caminati, W. 551, 553(73), *598*
Cammann, K. 1378(48b), *1437*
Cammi, R. 4, 100(2), *179*, 368, 372, 373(200), *390*
Campagnaro, G. E. 85(304), *186*
Campana, J. E. 293(76), *327*
Campanella, L. 975(321), 977(334), *1010*
Campbell, C. B. 749(99), *833*
Campbell, C. D. 367(170), *390*
Campbell, D. 1103(51–53), *1106*
Campbell, J. M. 170(747), *196*
Campbell, R. M. 412(72), *480*
Campbell, T. W. 845(28), *902*
Campelo, J. M. 608(13), 618(36), *656*, 774(182), *835*
Campíns Falcó, F. 942(135), *1006*
Campíns-Falcó, P. 953, 954(198), *1007*
Campo Dall'Orto, V. 953, 956(211), *1007*
Campostrini, R. 1082(290), 1084(326), *1094*, *1095*
Camps, F. 422(146), *483*, 663(20), *705*
Campsteyn, H. 583(224), *601*
Camy-Payret, C. 150(738), *196*
Cancelli, J. 1370(5), *1435*
Canella, K. A. 677(137), *707*
Canesson, P. 405(45), *479*
Canonica, S. 1017(14), 1085(332), *1087*, *1095*
Cantadori, B. 1424(389b), *1450*
Canters, G. W. 977(344), *1011*
Cantoni, A. 622(53), *657*
Cantoni, O. 895(265), *908*
Canuto, S. 728(40), *831*
Cao, G. 871(162), *905*
Cao, Z. 666(48), *705*
Cap, L. 932(87, 90), 937(87), *1005*
Capdevielle, P. 1305(270), *1344*
Capecchi, T. 462(273a), *487*
Capiello, A. 953, 961(236), *1008*
Capo, M. 750(102, 103), *833*
Caponio, F. 913(7), *1003*
Capponi, M. 38(184), 127(184, 468), *184*, *190*, 716(7), *831*, 1035(116), *1089*
Caprioli, R. M. 264(22), *326*
Carabatos-Nédelec, A. 375(246), *392*
Caralp, F. 138(527), *191*
Caramão, E. B. 933, 937(103), *1005*
Caravatti, G. 1244(170), *1342*
Carberry, J. B. 1347(7), *1363*

Carbonetti, A. 1383(75c), *1439*
Cardellicchio, N. 945(155), 953(155, 246), 962(246), *1006, 1008*
Cárdenas, S. 319(226, 227), *330*
Cardillo, G. 1255(180), *1342*
Cardona, C. 1083(297), 1084(315), *1094, 1095*
Cardoso, D. 631(72), *657*
Cardwell, T. J. 983(375, 377), 984(380), *1011*
Careira, E. M. 702(328–331), *712*
Careri, M. 953(203, 236), 955(203), 961(236), *1007, 1008*, 1396(181a), *1442*
Carini, M. 953, 966(269), *1009*
Carle, R. 953, 958(224), *1008*
Carles, M. 727(37, 38), *831*
Carlon, R. P. 651(172), *660*
Carloni, S. 677(131), 680(153), *707, 708*
Carlsen, L. 339(18), *386*
Carlson, G. L. 36, 45, 50(172), 53(237), 58(172), 87(307), *183, 185, 186*, 233(42), 238(72), *255, 256*, 551(131), 554, 555(137), *599*
Carlson, G. R. 36, 45, 50, 53(171), *183*
Carlson, R. E. 1086(374), *1096*
Carnacini, A. 303(142), *329*
Carney, R. L. 430(181c), *484*
Carpenter, D. J. 20, 23, 139, 140(116), *182*, 897(275), *908*
Carpenter, J. E. 4(5), *179*
Carpenter, K. 1175, 1227(54a), *1338*
Carpenter, T. A. 366(168), *390*, 726(31), *831*
Carr, R. A. E. 428(170), *483*
Carramolino, M. 1387, 1392, 1394(90), 1395(90, 177), 1409(287), *1439, 1442, 1447*
Carrasco, R. 670(76), *706*
Carreira, E. M. 702(339–341), *712*
Carrell, H. L. 375(250), *392*
Carreno, M. C. 650(162), *659*
Carrera, A. G. 1391(120c), *1440*
Carrick, W. L. 1274(208), *1343*
Carrieri, G. 945, 953(155), *1006*
Carrillo, J. R. 662(4), *705*
Carrington, A. 1125(250, 251), *1149*
Carroccio, S. 953, 969, 1000(286), *1009*
Carroll, A. R. 807(281), *837*
Carroll, D. I. 294(79), 312(194), *327, 330*
Carroll, P. J. 1396(193), *1443*
Carrott, M. J. 953, 969(285), *1009*
Carrozzo, P. 953, 970(288), *1009*
Carrupt, P.-A. 544(52), *598*
Carry, J. H. 1351(48), *1364*
Carsten, D. H. 698(296), *711*
Carta, G. 851, 854(74), *903*
Carter, C. 148, 160(721), *196*
Carton, P. M. 1110, 1112(55), *1145*
Cartoni, G. 942, 943, 953(138), *1006*
Carturan, G. 1082(290), 1084(326), *1094, 1095*

Caruso, D. 1000(467), *1013*
Carvalho, C. E. M. 1026(89), *1089*
Carvalho, C. F. 1383(75d), *1439*
Carvalho, R. M. de 929, 981(65), *1004*
Casalnuovo, A. 670(75), *706*
Casanova, J. 365(159), *389*
Case, D. A. 101(385), *188*, 511(94), *527*
Casebier, D. S. 450(241a, 241b, 241e), 453(241e), *486*
Casell, L. 699(307), *711*
Caserio, M. 506(93), *527*
Cases, J. M. 944(144), *1006*
Casiraghi, G. 626(59), *657*, 677(122, 133), 680(148), 683(168, 169, 176–178, 183), 685(200), 686(206, 207, 213), *707–709*, 1462(24, 26–28), 1463(26), *1502, 1503*
Casnati, A. 1387(90), 1391(110, 121a–c), 1392(90, 136, 137), 1393(110, 141, 156, 158), 1394(90, 171, 176), 1395(90, 177), 1396(121a–c, 181a, 182a, 182b, 185, 186a, 186c), 1397, 1399(197b), 1403(197b, 232a), 1404(244), 1405(232a, 255), 1409(287), *1439–1443, 1445–1447*
Casnati, G. 622(51, 52), 626(59), *657*, 677(122), 683(168, 169), 685(200), *707–709*, 1462(24, 26, 28), 1463(26), *1502, 1503*
Cassady, J. M. 403(28), *479*
Cassis, R. 813(301, 303), *837, 838*, 1283(230), *1343*
Castagnoli, N. Jr. 1335(330), *1346*
Castaldi, F. J. 1356(128), *1366*
Castani, G. 677(133), 683(176, 179, 181, 183, 184), 686(206–208, 213), *707–709*
Castellano, R. K. 322(251), *331*, 1391(122d), 1427(407c), 1432(122d), *1441, 1451*
Castello, G. 937(121), *1005*
Castillo, M. 946, 953(159), *1006*
Castillo-Meléndez, J. A. 471(305), *488*
Castle, L. 858, 887(136), *905*, 1352(55), *1364*
Castleman, A. W. 143, 147–149, 168(590), *193*
Castleman, A. W. Jr. 147(708), *195*
Castner, E. W. 1045(137), *1090*
Castola, V. 365(159), *389*
Catalán, J. 101, 102(387, 389), *188*, 371, 372, 375, 377, 378(235), *391*, 551(84, 99, 106, 107), 552(106), 553(84, 106, 107), 554(107), *598, 599*
Cater, S. R. 1071(208), *1092*
Cattrall, R. W. 983(377), 984(380), *1011*
Caughran, T. V. 1354, 1355(100), *1365*
Caulfield, M. J. 1500(177–179), *1506*
Causton, A. S. 1425(403b), *1451*
Cavalli, S. 945(151, 154, 155), 953(151, 154, 155, 246), 962(246), *1006, 1008*, 1354(111), *1365*
Cavallo, D. 1359(151), *1366*

Cave, G. W. V. 1381(66), 1383, 1421(67o), *1438*
Cavicchioli, A. 1082(287), *1094*
Caxixach, J. 1355(119), *1365*
Cazaux, L. 400(18b), *478*
Cecchi, H. M. 944(143), *1006*
Cecchini, G. 946, 953(165), *1006*
Ceković, Z. 1325(312), *1345*
Cela, R. 930(79), 932(94, 97), 937(79, 94), 971(294, 296, 297), *1004, 1005, 1009*, 1348(17), 1352, 1353(63), 1354(17, 108, 110), *1363–1365*
Cela-Torrijos, R. 933(104), *1005*
Celeste, M. 1001(472), *1013*
Cellerino, G. P. 953(216, 263), 957(216), 964(263), *1008, 1009*, 1353(82), *1365*
Centenera, J. I. 953, 969(284), *1009*
Centeno, T. A. 1489(106), *1504*
Ceppan, M. 1082(284), 1084(328), *1094, 1095*
Ceraulo, L. 301(135, 136), *328*
Cercek, B. 1099(15), *1105*
Cerdá, V. 929, 984(67), 1001(472), *1004, 1013*
Cerelier, O. 1361(192), *1367*
Cerioni, G. 341(31), *387*
Ceroni, P. 1406(272), *1446*
Cetina Rosado, R. 551(93), *598*
Ceyer, S. T. 83(291), *186*
Cha, D. D. 450(239d), *486*
Chabanel, M. 559, 568, 586(154), *600*
Chadda, S. K. 1048(157), *1090*
Chadenson, M. 778(211), *836*
Chadha, H. S. 545(54), *598*
Chait, B. T. 323(255), *331*
Chakrabarthy, D. K. 749(100), *833*
Chakrabarty, D. K. 620(42), *657*
Chakraborty, R. 774(182), *835*
Chakravorty, A. 735(54, 55), *832*
Chalais, S. 471(304), *488*, 806(262), *837*
Challa, G. 1194(84), *1339*
Challacombe, M. 4, 100(2), *179*, 368, 372, 373(200), *390*
Challis, B. C. 638(114, 115), *658*
Chalmers, J. M. 368(186), *390*
Chaly, T. 440(203), *484*
Chamberlin, A. R. 814(305), *838*
Chamberlin, S. 450(246b), *486*
Chambers, C. C. 57(242), *185*
Chambers, R. D. 410(64a), 448(229c), *480, 485*
Chaminade, P. 954, 970(199), *1007*
Champagne, J. 854(102), *904*
Chan, A. S. 697(281, 282), *711*
Chan, D. M. T. 671(83), *706*
Chan, H. W.-S. 852(75, 76, 82), *903*
Chan, K. F. 1081(268), *1093*
Chan, K. S. 454(253), *486*, 697(283), *711*, 735(52), *832*
Chan, O.-C. 1359(145), *1366*

Chan, T.-H. 439(200, 201a, 201b, 202), 440(203), 443(201a, 201b), 449(234a), *484, 486*
Chan, T. L. 210(15), *220*
Chan, W. H. 994(436), *1013*
Chandalia, S. B. 650(166), *659*
Chandler, C. J. 1073(219), *1092*
Chandler, K. 621(48, 49), *657*
Chandra, A. K. 83(289), 89(326, 330, 331, 334–340), 90(326, 330, 331, 334–341, 344), *186, 187*
Chandra, R. 1495(139), *1505*
Chandrakumar, N. S. 692(249), *710*
Chandran, S. S. 19(101), *182*
Chandrasakar, R. 419(123c), *482*
Chandrasekan, S. 340, 341, 349(29), 356(108), *387, 388*
Chandrasekar, R. 1117(131), *1146*
Chandrasekaran, S. 551(95), *598*
Chandrasekhar, J. 1047(148), *1090*
Chandrasekharam, M. 668(57), *706*
Chaney, S. T. 737(56), *832*
Chang, C. 89(309), *186*
Chang, C. C. 895(268), *908*
Chang, C.-J. 349(65), 359(121), *387, 389*, 874(174), *905*
Chang, H. 1290(249), *1344*
Chang, H. M. 129, 131(506), *191*
Chang, R. C. 1353(84), *1365*
Chang, S.-K. 1394(167b), 1406(265a, 265b, 268), *1442, 1446*
Chang, S.-M. 1085(337), 1086(370), *1095, 1096*
Chantegel, B. 453(244), *486*
Chantooni, M. K. Jr. 216(30, 31), *221*
Chantranupong, L. 93, 94, 97, 132, 133(358), *187*
Chantrapromma, S. 377(263), *392*
Chao, H. S.-I. 829(343, 344), *838*
Chao, P. 1485(95), *1504*
Chao, Y.-C. 971(299), *1009*
Chapell, B. J. 795(234), *836*
Chapman, D. M. 129(478), 144(478, 688), *190, 195*
Chapman, F. A. 1351(49), *1364*
Chapman, L. 1159, 1207(27), *1338*
Chapman, P. J. 1351(42), *1364*
Chapman, R. G. 1428(431a, 431b), *1452*
Chapoteau, E. 724, 725(27), *831*
Charalambous, J. 290(62), *327*
Chardin, A. 586(238, 247, 250, 251, 253, 254), 587(247, 250, 251, 253, 254), 589(254), 590(238), *602*
Charlen, C. H. 1362(201), *1368*
Charles, M. J. 937(119), *1005*
Charlton, A. 1495(148), *1505*
Charpiot, B. 671(81), 681(163), *706, 708*
Chase, R. L. 857(135), *905*

Chatfield, D. A. 307(154), *329*
Chatrathi, M. P. 971(301), *1010*
Chattopadhyay, N. 1022(51), 1025(87), *1088, 1089*
Chattopadhyay, S. 786(231), *836*
Chau, A. S. Y. 1353(91), *1365*
Chaube, V. D. 645(145), *659*
Chaudhary, A. J. 1079(246), *1093*
Chauhan, Y.-P. 1294(253a, 253b), *1344*
Chavagnac, A. X. 148(724), *196*
Chawla, H. M. 634(97), *658*, 1402(227), 1407(276), *1445, 1447*
Chedekel, M. R. 1136(358, 359), *1151*
Chee, K. K. 946, 953(161), *1006*
Cheeseman, I. R. 368, 372, 373(200), *390*
Cheeseman, J. R. 4, 100(2), *179*
Cheeseman, K. H. 141(540), *192*
Chen, B. 1362(214), *1368*
Chen, C. 378(270), *392*
Chen, C.-F. 1409(286a), *1447*
Chen, C.-H. 1085(337), *1095*, 1236(155, 156), 1237(162a), *1341, 1342*
Chen, C.-P. 1171(47), *1338*
Chen, C.-W. 1085(343), *1095*
Chen, D.-Z. 982(372), *1011*
Chen, G. 697(285), *711*
Chen, G. Z. 1427(413), *1451*
Chen, H. 1432(445), *1453*
Chen, H.-Y. 983(378), *1011*
Chen, I.-C. 972(302), *1010*
Chen, J. 937, 940(129), *1006*, 1424(400c), *1451*
Chen, J.-S. 999(463), *1013*
Chen, K. 1361(193), *1367*
Chen, L. H. 1018(20), *1087*
Chen, P. 855(115), *904*
Chen, P. C. 82(285), *186*, 236(53), *255*
Chen, P. H. 1354, 1355(100), *1365*
Chen, S. C. 82(285), *186*
Chen, S.-I. 1372(16), *1436*
Chen, W. 4, 100(2), *179*, 368, 372, 373(200), *390*
Chen, X. 1394(167c), *1442*
Chen, X.-T. 463(275), *487*
Chen, Y. 806(263), *837*, 986(396), *1012*, 1175(54a, 54b), 1227(54a, 147), *1338, 1339, 1341*, 1397(197a, 197c), 1398(197a, 198), 1400(208a), 1403(197a, 197c), *1443, 1444*
Chen, Y.-J. 684(191, 192), *709*
Chen, Y.-K. 1398(199), *1443*
Chen, Y.-Y. 1398(199), 1427(420a, 420b), *1443, 1452*
Chen, Z. 1361(186), *1367*, 1409(286b), 1432(445), *1447, 1453*
Chen, Z. H. 997(448), *1013*
Chen, Z. L. 973(314), *1010*
Chênevert, R. 477(326c), *489*

Cheng, F.-C. 953, 969(283), *1009*
Cheng, H. 1495(155), *1505*
Cheng, J.-F. 1178, 1337(57a, 57b), *1339*
Cheng, J.-P. 5(24), 110(421), 139(24), *180, 189*
Cheng, P. C. 82(285), *186*
Cheng, S. 1082(286), *1094*
Cheng, Y. 674(106), *707*
Cheong, C. 1086(373), *1096*
Chermette, H. 89, 90(321), *187*
Cherry, W. R. 477(328), *489*
Cheshnovsky, O. 493(19, 20), *525*
Chesnut, D. B. 136, 137(520), *191*
Chessari, G. 1400(209b), 1401(212b), *1444*
Chester, G. 1359(156), *1366*
Cheung, J. 319(223), *330*, 931, 937(84), *1005*
Cheung, L. M. 1081(268), *1093*
Chevalier, Y. 559, 567(156), *600*
Chew, W. 1401(211), *1444*
Cheynier, V. 949(183), *1007*
Cheyran, M. 1000(470), *1013*
Chia, L. S. 975(323), *1010*
Chiang, H.-C. 1055(168), *1091*
Chiang, M. Y.-N. 377(261), *392*
Chiang, P. K. 403(27), *479*
Chiang, W.-Y. 1497(171–174), 1498(171), *1506*
Chiang, Y. 1034(113), *1089*
Chiappe, C. 675(112), *707*
Chiarelli, M. P. 323(261), *331*
Chiba, K. 636(105), *658*, 1180(63a, 63b), 1192(75a, 75b), *1339*
Chibisova, T. A. 776(200), *835*
Chibwe, M. 1202(96), *1340*
Chickos, J. S. 225(4), 227(4, 5), 230(5), *254*
Chi'en, C.-H. 1055(168), *1091*
Chien, C. J. 937(119), *1005*
Chiesi-Villa, A. 1391(113), *1440*
Childs, R. F. 1048(152–154, 156, 157), *1090*
Chilla, C. 947, 953(175), *1007*
Chinchilla, R. 1158(22), *1338*
Chiong, K. G. 447(220), *485*
Chipman, D. 129, 131–133(510), *191*
Chipman, D. M. 1120(187), 1123(198), 1128(187), 1132(187, 198, 332), *1147, 1148, 1150*
Chipot, C. 179(780), *198*
Chirico, R. D. 226, 227(7), *254*
Chitnis, S. R. 763(130), *834*
Chiu, W. T. 1125(248), *1149*
Chmutova, G. A. 744(74), *832*
Cho, H. J. 1406(265a), *1446*
Cho, K. 1408(279a), *1447*
Cho, M. H. 1427(423b), *1452*
Cho, S. H. 663(14), *705*
Choe, J.-I. 1394(167b), 1406(265b), *1442, 1446*
Choi, H.-J. 1419(357), *1450*

Choi, H. S. 149, 158, 161(736), *196*
Choi, S. H. 981(361), *1011*
Choi, Y. H. 454(258), *487*
Chong, W. K. M. 1244(171), *1342*
Choo, K. Y. 1124(239), *1148*
Chopin, M. J. 778(211), *836*
Chopineaux-Courtois, V. 544(52), *598*
Chopra, N. 1428(431a), 1433(457, 459, 460), *1452, 1453*
Chopra, S. K. 941(134), *1006*
Chou, C.-T. 1171(47), *1338*
Chou, P.-T. 1022(50), 1025(85b), 1028(94), *1088, 1089*
Chou, T. C. 447(224), *485*
Choudhury, A. M. 1255(182), *1342*
Choudry, C. G. 1074(231), *1092*
Choudry, G. G. 1073(215), *1092*
Chowdhury, S. 100, 102(381), *188*, 1387(95), 1392, 1393(135), *1439, 1441*
Chrisstoffels, L. A. J. 1424(393), *1450*
Christensen, D. H. 383(282), *393*
Christensen, H. 1115, 1135(108), *1146*
Christensen, J. J. 1372(17), *1436*
Christensen, S. B. 757(108), *833*
Christian, R. 12(60b), *181*
Christiansen, A. W. 1471(71), *1504*
Christie, I. M. 983(379), *1011*
Christman, R. F. 1360(180), *1367*
Christoffels, L. A. J. 1427(410b), *1451*
Christoffersen, J. 35(157), *183*
Christoffersen, M. 336, 345, 354(5), *386*
Christophersen, M. J. 983(375), *1011*
Christov, S. 976(331), *1010*
Chriswell, C. D. 1353(84), *1365*
Chronister, E. L. 110(437), *189*
Chrysochoos, J. 1098(5), *1105*, 1115(95, 96), 1129, 1130(95), *1145*
Chu, C. C. 895(268), *908*
Chu, C.-S. 1234(153a), 1237(161a, 161b, 162c), *1341, 1342*
Chu, P. P. 367(175), *390*
Chu, P.-S. 1250(174–176, 177a, 177b), *1342*
Chuang, I.-S. 1484(90), *1504*
Chuche, J. 818(314), *838*
Chudziak, A. 1100(29), *1106*
Chum, K. 349(70), *387*
Chung, C. 551, 553(79), *598*
Chung, G. 242(82), 244(85), *256*, 551(102, 116), 553(102), 556(116), *599*
Chung, K. L. 925(48), *1004*
Chung, T. D. 1394(167b), *1442*
Chung, W.-S. 1391–1393(122a), 1394(168), 1407(273b), *1440, 1442, 1446*
Churbakov, A. M. 18(90), *181*
Churbakov, M. A. 396(1), *478*
Church, D. F. 886, 887(216), *906*
Chuvylkin, N. D. 1123(200, 201), *1148*
Chyall, L. J. 123(467), *190*

Cichelli, A. 930(77), *1004*
Cillard, J. 892(248, 249), *907*
Cillard, P. 892(248, 249), *907*
Cimerman, Z. 385(298, 300), *393*, 726(34, 35), *831*
Cinar, Z. 1083(294), *1094*
Ciocazanu, I. 230(31), *255*
Cioslowski, J. 4(2), 69(268), 100(2), *179, 185*, 368, 372, 373(200), *390*
Cipollini, R. 309(167), *329*
Cipollone, M. 45(190), *184*, 855(120), *904*
Ciranni, G. 309(169–171), *329*
Cirilli, L. 946, 953(165), *1006*
Cisowski, W. 953, 967(272), *1009*
Citterio, A. 408(54e), *480*, 683(180), 686(210), *708, 709*, 1259(188), *1342*
Ciucu, A. 976(329), *1010*
Ciurak, M. 1117(130), *1146*
Cladera, A. 929, 984(67), 1001(472), *1004, 1013*
Claisen, L. 460(278), *487*
Claramunt, R. 559, 572(157), *600*
Claramunt, R. M. 349(59), *387*
Clardy, J. 454(253), *486*, 735(52), *832*, 1466(45), *1503*
Clardy, J. C. 1159, 1207(27), *1338*
Clark, C. T. 405(43a, 43b), *479*
Clark, J. H. 493, 498(42), *525*, 617(33), *656*
Clark, S. 1216(119b), *1340*
Clark, T. 385(302), *393*
Clark, T. J. 319(225), *330*, 932, 937(91), *1005*
Clarke, M. F. 399(11), *478*
Clary, D. C. 148(721, 724–726), 160(721), *196*
Clase, H. J. 1374, 1411(42c), *1437*
Clauson-Kaas, N. 432(182b–d, 183a), *484*
Claxton, T. A. 137(521), *191*
Clayden, J. 786(232), *836*
Clays, K. 1391(119), *1440*
Clerici, F. 674(102), *707*
Clerici, M. G. 414(90a), *481*
Clesceri, L. S. 1352(62), 1353, 1354(74), *1364*
Clifford, M. N. 1360(177), *1367*
Clifford, S. 4, 100(2), *179*, 368, 372, 373(200), *390*
Clogh, S. 1109, 1120(28), *1144*
Clotman, D. 534(26), *597*
Cobben, P. L. H. M. 1383, 1421(67d), *1438*
Cobos, C. J. 1105(64, 65), *1106*
Coburn, C. A. 450(246d), *486*
Coccioli, F. 942, 943, 953(138), *1006*
Cocco, M. T. 764(138, 139), *834*
Cochien, E. D. 1428(431a), *1452*
Cochran, J. C. 685(198), *709*
Cocivera, M. 1117(115), 1120(115, 179), *1146, 1147*
Cockett, M. C. R. 110, 112, 121(444), 129, 144(478), *189, 190*

Coda, A. 216(27), *221*
Codding, P. W. 233(41), *255*
Codony, R. 961(243), *1008*
Coelho, A. L. 786(231), *836*
Coffey, D. L. 947, 953(173), *1006*
Cogan, D. A. 699(305), *711*
Cohen, A. O. 522(99), *527*
Cohen, G. 874(173), *905*, 1119(161, 162), *1147*
Cohen, S. 1374(39), 1410(291b), 1411(296c), *1436, 1447*
Cohen, T. 404(33), *479*
Cohen, W. D. 422(144), *483*
Colandrea, V. J. 665(41), *705*
Colarusso, P. 1361(195), *1367*
Colby, D. A. 1370, 1430(6), *1435*
Colclough, N. 1211(110), *1340*
Colding, A. 127(469), *190*, 718(17), *831*
Cole, B. M. 702(345, 346), *712*
Cole, E. R. 880(200), *906*
Colebrook, L. D. 1401(211), *1444*
Coleman, C. A. 367(177), *390*, 541(44), *597*
Coleman, R. S. 653(187), *660*
Colivicchi, M. A. 953, 968(275), *1009*
Coll, J. 422(146), *483*, 663(20), *705*
Collet, A. 1370(5), *1435*
Collette, Y. 609(17), *656*
Collins, C. D. 1356(120), *1365*
Collins, C. J. 470(300b), *488*, 806(250), *836*
Collins, J. L. 1250(178), *1342*
Collins, M. 141(540), *192*
Collomb, D. 818(317), *838*
Colombi, A. 1359(151), *1366*
Colombo, L. 953, 966(269), *1009*
Colombo, R. 1000(467), *1013*
Colomina, M. 228, 229(16), 230(25), 232(38), 241(80), *254–256*
Colon, C. 1278(226), *1343*
Colon, G. 1080(248), *1093*
Colonna, F. 69(265, 266), 70(265), 148(266), *185*
Colonna, S. 699(307), *711*
Colosio, C. 1359(151), *1366*
Colson, S. D. 30(134), 129(475), 143(585–587, 591), 147(585–587, 693, 694), *182, 190, 193, 195*
Colthup, N. B. 369, 371(208), *391*
Colton, J. B. 1351(50), *1364*
Coluccia, S. 1083(293), 1084(320), *1094, 1095*
Columbus, I. 1374(34), 1386(87), 1413(301–303), *1436, 1439, 1447*
Colussi, A. J. 5, 137, 138(21), *180*
Comer, F. W. 426(166a), 429(179e), 430(179e, 181e), *483, 484*
Cometti, G. 1382(68), 1417(331), *1438, 1448*
Comor, J. J. 931, 935, 937(85), *1005*
Compton, R. G. 982(368), *1011*
Concepcion, A. B. 689(232), 693(252), *710*

Conde, E. 953, 965(267), *1009*
Conesa, J. C. 1083(296), *1094*
Conforti, M. L. 627(62), *657*
Cong, P. 493(21), *525*
Conlon, D. A. 632(89), *658*
Conole, G. 782(220), *836*
Conradt, Ch. 1359(150), *1366*
Consoli, G. L. M. 1401(212c), *1444*
Consoli, G. M. L. 1388(99, 101), 1395(99, 178a–c), 1402, 1404, 1406(222), *1439, 1442, 1444*
Constantinescu, T. 855(118), *904*
Consula, F. 1388(101), 1402, 1404, 1406(222), *1439, 1444*
Conte, L. 646(146), *659*
Conte, V. 413(83, 84), 414(85), *481*
Conti, S. 1028(96), *1089*
Contini, M. S. 851, 854(74), *903*
Contino, A. 1391, 1396(121b), *1440*
Contreras, M. 851(63), *903*
Contreras, R. H. 366(167), *390*
Cooke, M. 937, 938(123), 953, 965(266), *1005, 1009*
Cooks, R. G. 110(432, 422), *189*, 264(22), 290(61), 308(163–165), 309(165, 166), *326, 327, 329*
Cooksey, J. 1223(136), *1341*
Coolen, S. A. J. 953, 968(277, 278), *1009*
Coombes, R. G. 806(270), *837*
Cooney, A. P. 1470(54), *1503*
Cooper, A. 1318(304), *1345*
Cooper, C. 1362(203), *1368*
Cooper, D. L. 89(312), *186*
Cooper, J. B. 1403(234c), 1427(409c), *1445, 1451*
Cooper, M. E. 994(437), *1013*
Cooper, W. C. 1028(94), *1089*
Cooper, W. J. 1100(33), *1106*
Copalan, R. 1302(265), *1344*
Coppa, F. 413(84), *481*
Coppens, P. 1387(88), *1439*
Coppert, D. M. 1255(185), *1342*
Coppi, L. 1393(151), *1441*
Coppinger, G. M. 845(28), *902*
Corden, B. B. 1205(102), *1340*
Cordes, E. 110, 112, 121(444), 129(476), *189, 190*
Cordova, E. 761(117), *833*
Corelli, F. 1383(75a–e), 1422(75e), *1439*
Corey, E. J. 1465(43), *1503*
Coricchiato, M. 365(159), *389*
Corke, H. 990(417), *1012*
Cormier, M. 892(248), *907*
Cornelisse, J. 1117(114), *1146*
Cornforth, J. W. 402(22), *479*
Cornia, M. 680(148), 683(176–178), *708*, 1462(24, 26, 27), 1463(26), *1502, 1503*
Cornicelli, J. A. 851, 852, 894(67), *903*

Corniz, M. 1462(28), *1503*
Corongui, F. P. 851, 854(74), *903*
Corrie, J. E. T. 1044(132), *1090*
Cortina, M. 1355(119), *1365*
Corvaja, C. 1123(205), *1148*
Cosnier, S. 977(338, 343), *1010, 1011*
Cosse-Barbi, A. 576(190), *600*
Cossi, M. 4, 100(2), *179*, 368, 372, 373(200), *390*
Cossy, J. 770(156), *834*
Costa, A. 1260(192), *1342*
Costa Cabral, B. J. 148(727), *196*
Costain, C. C. 171(762), *197*
Costas, M. 546(57, 58), *598*
Cote, D. 370(214), *391*
Cotgreave, I. A. 865(145), 890(240), *905, 907*
Cottet, F. 93(355), *187*
Coupar, P. I. 580(208), *601*
Courteix, A. 977(340), *1010*
Courtney, L. 1300(262), *1344*
Courty, A. 129, 144(474), *190*, 578(195), *601*
Coustard, J. M. 630(71), 651(174–177), 652(185), 653(186), *657, 660*
Coutinho, A. 887(224), *907*
Coutts, R. T. 412(76), *480*
Couture, A. 784(227), *836*
Couture, R. 953, 957(221), *1008*
Couzi, M. 37, 170(179), *183*
Covarrubias, R. 944(140), *1006*
Cox, G. G. 666(47), *705*
Cox, J. R. 801(246), *836*
Cox, R. A. 1105(64, 65), *1106*
Coxon, D. T. 852(75, 82), *903*
Craig, H. C. 412(72), *480*
Craik, D. J. 1073(219), *1092*
Cram, D. J. 651(169), *660*, 1382, 1383(67c), 1419(348, 357), 1421(67c), 1422(376), 1424(385, 386, 397a, 399), 1428(385, 425a, 425b, 426a, 426b), 1429(426b, 429, 430a, 430b), 1433(458), *1438, 1450–1453*
Cram, J. M. 1422(376), 1424, 1428(385), *1450*
Cramer, C. J. 20, 25(130), 57(242), 114, 129, 135(459), *182, 185, 190*
Cramer, F. 778(210), *836*
Cramers, C. 937, 938(125), *1005*
Crampton, M. R. 1470(54), *1503*
Crandall, K. 1255(185), *1342*
Crane, R. 1358(144), *1366*
Craw, M. T. 1120(185), *1147*
Creaser, C. S. 307(157, 158), 308(158), *329*
Credi, A. 761(118), *833*
Cremlyn, R. J. 1350(41), *1364*
Crepin, C. 144(601), *193*
Crescentini, G. 975(321), *1010*
Crescenzi, C. 1354(101), *1365*
Crespín, M. A. 319(226, 227), *330*, 932, 937(92), 946(160), *1005, 1006*
Cretney, W. J. 1351(51), *1364*

Crews, D. 1278(226), *1343*
Crews, O. P. 400(19), *478*
Crippa, P. R. 1104(57), *1106*
Crisafi, E. 1356(122), *1366*
Crist, D. R. 251, 252(92), *257*, 385(297), *393*
Crivello, J. V. 770(150, 151), *834*, 1264(197), *1343*
Croft, K. D. 913(30), *1003*
Croley, T. R. 320(235), *331*
Crombie, L. 430(181a, 181b), *484*, 807(278), *837*
Crombie, M. L. 807(278), *837*
Crosby, D. G. 420(140), *482*, 1348(18), 1360(165), *1363, 1367*
Cross, C. E. 636(104), 637(109), *658*
Cross, T. A. (97), *1504*
Crowe, E. 901(284), *908*
Crowe, L. A. 1424(395), *1451*
Crozier, A. 957(220), *1008*
Cruickshank, F. R. 1108(22), *1144*
Cruz Vieira, I. da 978(350), *1011*
Csok, Z. 1388(98), 1394, 1407(170b), *1439, 1442*
Csokasi, P. 1422(375), *1450*
Csonka, G. I. 551, 555(113), *599*
Csöregi, E. 976(324), *1010*
Cueto, R. 637(108), *658*
Cuevas, F. 1387, 1392, 1394(90), 1395(90, 177), *1439, 1442*
Cuevas, J. L. 953, 969(284), *1009*
Cui, H. 982(364), *1011*
Cui, Q. 4, 100(2), *179*, 368, 372, 373(200), *390*
Cukier, R. I. 144(690), *195*
Culbertson, H. M. 1462(22), *1502*
Cumming, J. B. 5(17), *180*, 313(200), *330*
Cundall, R. B. 1129(299), 1130(299, 306, 322), 1133(340), 1137(366), *1150, 1151*
Cunningham, A. 691(241, 242), *710*
Cunsolo, F. 1388(99), 1395(99, 178a–c), 1400, 1401(210b), 1403(230), 1427(230, 414a), *1439, 1442, 1444, 1445, 1451*
Cuquerella, M. C. 1079(243), *1092*
Curci, R. 1255(184a, 184c), 1257(186), *1342*
Curti, C. 895(266), *908*
Curtin, D. Y. 206(11), *220*
Curtiss, L. A. 4(4a), *179*
Custer, L. J. 913(24), *1003*
Custodio, R. 632(90), *658*
Cuvelier, M. E. 982(369), *1011*
Cynshi, O. 856, 863, 864(126), *904*
Cyr, A. 1361(191), *1367*
Cyr, J. E. 1109(51), *1144*
Czapski, G. 1083(298), *1094*, 1118(154), 1119(154, 156), 1141(154), *1147*
Czarnik-Matusewicz, B. 38(183), 83(289), 145(616), 170(755), *183, 186, 193, 197,*

Czarnik-Matusewicz, B. (continued)
371, 372(221), *391*, 559, 570, 571(158), 573(181), 575(158), *600*
Czech, B. P. 724, 725(27), *831*
Czerwinska-Bil, U. 1353(81), *1365*
Czira, G. 1388(98), 1394, 1407(170b), *1439, 1442*
Czygan, F. C. 913(19), *1003*

Dabbagh, Y. A. 913, 920(27), *1003*, 1355(116), *1365*
Dabo, P. 1361(191), *1367*
D'Accolti, L. 1257(186), *1342*
Dachà, M. 895(265), *908*
DaCosta, W. F. 928(55), *1004*
Dagley, S. 411(69), *480*
Dahhani, F. 300(115), *328*
Dai, P.-S.E. 616(30), *656*, 763(129), *834*
Dai, S. 1086(380), *1096*
Daifuku, K. 865(149), *905*
Dailey, B. P. 20(106), *182*
Dainty, F. R. 472(309c), *488*
Dainty, R. F. 472(315), *488*, 774(173), *835*
Daire, E. 413(81), *480*
Daishima, S. 295(82), *327*
Daitch, E. 1380(55), *1437*
Dakin, H. D. 419(136a), *482*
Dakin, K. A. 851, 863, 882, 885–887(55), 890(238), *903, 907*
Dalcanale, E. 322(250), *331*, 1382(67c, 68), 1383(67c), 1386(86a, 86b), 1417(331), 1421(67c), 1422(379), 1424(384, 389b, 392, 393, 397a, 399, 397b), *1438, 1439, 1448, 1450, 1451*
Dalenko, A. 419(127), *482*
d'Alessandro, N. 930(77), *1004*, 1136(362), *1151*
Dalgleish, D. T. 1274(210b), 1285(237), *1343*
Dali, H. 748(87, 88), *833*
Dalidovich, S. V. 1470(51), *1503*
Dalidowicz, P. 1180, 1183, 1186(62), *1339*
Dall'Orto, V. C. 976(326), *1010*
Dalprato, C. 683(181), *708*
D'Alva Torres, G. S. F. 561(164), *600*
Daly, J. 405(44), *479*
Daly, J. W. 412(72), *480*
Damassa, D. A. 1360(172), *1367*
Damon, R. E. 1278(225), *1343*
Danforth, R. H. 416(110b), *482*
D'Angelantonio, M. 1115, 1128(91), *1145*
Dangles, O. 871(167), *905*
Danheiser, D. L. 450(246h, 246i), *486*
Danheiser, R. L. 416(109b, 115), 450(239a, 239c–f, 241a–e), 452(239c), 453(239f, 241c, 241e), *482, 486*
Daniel, V. 1359(150), *1366*
Danieli, R. 551, 553(73), *598*

Daniels, A. D. 4, 100(2), *179*, 368, 372, 373(200), *390*
Danielsen, K. 338(13), *386*
Danielsson, J. 93, 95(361), *187*
Danilova, T. 640(132), *659*
Danilowicz, C. 953, 956(211), 976(326), *1007, 1010*
Danishefsky, S. 441(204, 205, 208–210), 447(224), *484, 485*
Danishefsky, S. J. 450(246a), 463(275), *486, 487*, 674(110, 111), *707*
Danjou, J. 925(46), *1004*
Dapprich, S. 4, 100(2), *179*, 368, 372, 373(200), *390*
Darby, M. R. 265, 319(28), *326*
Dargaville, T. R. 1459, 1463(9), 1471(74), *1502, 1504*
Dargel, T. K. 110(451), *190*
Darling, S. D. 415(105a, 105b), *481*
DaRooge, M. A. 45(189), *184*, 889(227), *907*, 1111(65), *1145*
Das, C. G. 299(98), *328*
Das, D. K. 840, 901(10), *902*
Das, K. 1025(71a, 71b), 1045(137–139), *1088, 1090*
Das, K. G. 110(431), *189*
Das, M. R. 1123(203), *1148*
Das, P. K. 1021(42), *1088*, 1109, 1110, 1112(26), *1144*
Das, R. 1017(13), 1021(45–49), 1047(144), *1087, 1088, 1090*
Das, T. N. 45(199), *184*, 1099(18), *1105*, 1128(280), 1129(301, 302), 1130(280, 310), 1134(310), 1137(367), 1142(280, 301, 302, 379), 1143(280), *1149–1151*
Dasgupta, A. 934, 937(109), *1005*
Dasgupta, P. K. 972(306), *1010*
Datta, D. K. 425(160), *483*
Datta, S. 1392, 1403(125b), *1441*
Dauben, H. J. 95(368), *188*
Daum, G. 823(324, 325), *838*
Davi, M. L. 1349, 1356(36), *1363*
David, B. 1361(194), *1367*
David, F. 937, 938(125), *1005*
David-Oudjehani, K. 1085(344), *1095*
Davidson, G. 953, 969(285), *1009*
Davies, M. J. 854(95), *904*, 1130, 1137, 1138(308), *1150*
Davies, M. M. 35(148, 149), 47(202, 221), 48(232), *183–185*
Davies, R. E. 852, 854, 855, 857(83), *903*
Davietschina, G. R. 1424(396), *1451*
Davis, B. 290(59), *327*
Davis, C. J. 1383, 1421(67f), *1438*
Davis, D. W. 102(393), *188*
Davis, F. 1383(67j), 1419(347b), 1421(67j, 370), *1438, 1450*
Davis, F. A. 416(116), 417(116–118), *482*

Davis, H. F. 1120, 1128, 1132(187), *1147*
Davis, J. C. Jr. 336(7), *386*
Davis, M. E. 631(75), *657*
Davis, S. L. 677(124), *707*
Davison, J. 37(180), *183*
Davydov, I. 1080(252), *1093*
Dawson, I. M. 472(311b), *488*
Day, A. I. 745, 746(80), *832*
Day, A. J. 1485(95), *1504*
Day, J. 1025(73), *1088*
Day, M. W. 702(328, 330), *712*
Day, R. 806(265), *837*
De, A. 783(223), *836*, 1081(267), *1093*
De, R. 1019(25), *1087*
Deagro, M. 937, 940(129), *1006*
Deak, G. 621(50), *657*
De Angelis, C. 953, 970(288), *1009*
Deardorff, D. R. 664(29, 30), *705*
Deb, C. 783(223), *836*
Deb, K. K. 336(7), *386*
Debies, T. P. 110, 112(445), *189*
De Blic, A. 400(18b), *478*
Debonder-Lardeux, C. 493, 500(31), *525*
Debrauwer, L. 316(218, 219), *330*
DeBruyn, P. J. 1471(74), *1504*
De Buyck, L. 471(303), *488*
Dec, J. 1352(61), *1364*
Decembrini, F. 1356(122), *1366*
De Cian, A. 1401(214, 219b, 220b), 1415(311, 314b, 322), 1427(421c, 421d, 422b), *1444, 1448, 1452*
Deck, J. 999(465), *1013*
Deck, L. 338, 380(15), *386*
Decker, O. H. W. 450(238), *486*
Decouzon, M. 322, 325(244), *331*
Dedonder-Lardeus, C. 109(414), *189*
Dedov, A. G. 985(389), *1012*
Defaye, D. 83(296b), *186*
DeFelippis, M. R. 1117(129), 1142(380), *1146, 1152*
DeFrees, D. F. 311(189), *330*
DeFrees, D. J. 5(22), 83, 84(297), *180, 186*, 261, 263(13), *326*
Dega-Szafran, S. 580(205), *601*
Dega-Szafran, Z. 143(565, 566), 146(638), *192, 194*, 593, 595(281, 282), *602, 603*
De Giorgi, A. 1082(287), *1094*
DeGourrierec, D. 1025(67), *1088*
Deguchi, Y. 1109(31), *1144*
De Heer, M. I. 45(195), *184*
Deiana, M. 851, 854(74), *903*
DeJong, J. 1475, 1486(80), *1504*
De Koning, C. B. 462(273a), *487*
delaBreteche, M. L. 639(125), *659*
Delaere, D. 127(470), *190*
Delaigue, X. 1401(219b, 220b), 1427(421a, 421b), 1432(421b), *1444, 1452*
Delaire, J. 1073(216), 1085(353), *1092, 1096*

Delaire, S.-M. 1085, 1086(336), *1095*
De La Mare, P. B. D. 606, 649(3), *656*
DelaMater, M. R. 416(113), *482*
Delangle, P. 1424(394), *1450*
Delany, D. 1086(366), *1096*
Del Carlo, M. 976(326), *1010*
Delépine, M. 230(28), *254*
Delfly, M. 472(314), *488*
Delgado, J. 1040(124), 1042(128), *1090*
Delise, M. 1357(130), *1366*
Della, E. W. 350(79), *388*
Della Corte, L. 953, 968(275), *1009*
Delle Monache, G. 1383(75a–e), 1422(75e, 375), *1439, 1450*
Dell'Erba, C. 1073(225), *1092*
Delmas, M. 685(197), *709*
DeLuca, H. F. 840(7), *902*
Demaison, J. 171(762), *197*
Demarchi, A. C. C. O. 949(184), *1007*
DeMarco, A. 1462(27), *1503*
De Mare, G. R. 89(313), *186*
De Maria, P. 301(135, 136), *328*
De Mayo, P. 748(93), *833*
Demerseman, P. 472(314), 474(317), *488*
De Meyer, C. 110(431), *189*
Demidkina, T. V. 748(90), *833*
Deming, L. S. 47, 48(217), *184*
Demmin, T. R. 1192(80), *1339*
Demyanov, P. I. 933, 949(98), *1005*
Demyashkevich, A. B. 492, 493, 501(13), 515(96), *525, 527*, 1020(35d), *1087*
Deng, F. 171(762), *197*
Deng, F. H. 621(48), *657*
Deng, L. 698(295), *711*
Deng, Q. 984(381), *1011*
Deniau, E. 784(227), *836*
De Nino, A. 953, 961(239), *1008*
Denis, A. 664(28), *705*
Denis, G. 685(197), *709*
Denisenko, V. I. 301(130), *328*
Denisov, E. T. 15(77), *181*, 1125(249), *1149*
Denisov, G. S. 143(554, 555, 557), 145(626), 147(643), *192–194*, 207(13), *220*, 364(155), 366(165), 367(178), *389, 390*, 503(90), *527*
Denisova, N. K. 887(222), *907*
Denney, D. B. 801(238), *836*
Denney, D. Z. 801(238), *836*
Dennis, M. R. 782(219, 220), *836*
Denny, R. W. 853(93), *904*
De Palma, D. 47, 143(207), *184*
De Paz, J. L. G. 371, 372, 375, 377, 378(235), *391*, 551(84, 106, 107), 552(106), 553(84, 106, 107), 554(107), *598, 599*
Depew, M. C. 1018(22), *1087*, 1120(185), *1147*
De Pooter, H. 471(303), *488*

De Proft, F. 84(301), 89(301, 319, 324, 331–333, 336), 90(301, 319, 324, 331–333, 336, 341), *186, 187*
Derbyshire, D. H. 404(36), *479*
Derhaeg, L. 1391(112), 1402, 1405(224b), *1440, 1444*
Deril, G. V. M. 953, 957(216), *1008*
Dermer, R. 408(54d), *480*
De Rosa, M. C. 1383(75b, 75c, 75e), 1422(75e), *1439*
De Rossi, R. H. 477(324b, 326b), *489*
Derrick, P. J. 260, 325(5), *325*
Desai, D. H. 724, 725(27), *831*
Desai, P. D. 241(75), *256*
Desai, R. B. 409(61), *480*
Desaiah, D. 1359(157), *1366*
Desarmenien, M. 475(318), *488*
Desbene, P. L. 974(317), *1010*
Descoings, M.-C.M. 350(83), *388*
Descotes, G. 609(17), *656*
Deshayes, C. 453(244), *486*, 818(317), *838*
Deshayes, K. 1428(425c), *1452*
Deshpande, S. S. 1000(470), *1013*
Desiraju, G. R. 219(35), *221*, 530(16), *597*
DesMarteau, D. D. 312(193), *330*, 647(150), *659*, 687(221), *709*
De Sousa, B. 419(135b), *482*
Desroches, C. 1415(318a), *1448*
Dess, D. 673(96), *707*
Dessent, C. E. H. 129(478), 144(478, 685, 687), 147(717), *190, 195*
Dessi, M. A. 851, 854(74), *903*
DeStefano, A. J. 292, 293(69), *327*
De Suzzoni-Dezard, S. 1212(111), *1340*
Desvergnes, L. 397(6), *478*
De Taeye, J. 145, 146(610, 618), *193*, 583(228, 231), 584(234), *601*
Detellier, C. 1385(82), *1439*
Detmer, C. A. 336(8), *386*
Dettman, H. D. 346, 365(52), *387*
DeTuri, V. F. 5(20), *180*
Deuschle, G. 1122(193), *1148*
Deusen, C. 143(577), 144(577, 678), *192, 195*, 581(217), *601*
Devall, L. J. 851, 852, 894(67), *903*
Devlin, J. L. III 263, 296(16), *326*
Devolder, P. 1124(238), *1148*
Devon, T. K. 13(62), *181*
DeVries, K. M. 1318(301b), *1345*
De Wael, K. 145, 146(615), *193*
Dewar, M. J. S. 639(129), *659*
Dey, G. R. 1129(292), *1150*
Deya, P. M. 1260(191), *1342*
Dhal, R. 1278(222, 223), 1286(223), *1343*
Dhawan, B. 1371(8), 1372(8, 16), 1378(52a), 1379, 1380(54a), *1435–1437*
Dhimane, A.-L. 1158(13), *1337*
Dhindsa, K. S. 757(107), *833*

Dhingra, O. P. 1275(213), *1343*
D'Hondt, J. 143(561), *192*
Diakoumakos, C. D. 1495(153), *1505*
Diamond, D. 1435(469a), *1454*
Diao, L. 1033(109, 110), 1034(111), *1089*
Diaz-Mondejar, M. R. 1055(167), *1091*
Dicker, I. B. 651(169), *660*
Dickey, J. B. 400(20), *478*
Di Corcia, A. 946(165), 947(167), 953(165, 167), *1006*, 1353(85–87), 1354(87, 101), *1365*
Dideberg, O. 583(224), *601*
Didier, F. 944(144), *1006*
Diebler, H. 1118, 1119(137), *1146*
Diederich, F. 322(245), *331*
Dieffenbacher, A. 925, 961(45), *1004*
Dieleman, C. B. 1391(120a), *1440*
Diemann, E. 143, 177(553), *192*
Diep, O. H. 856(123), *904*
Dietrich, A. M. 928(55), *1004*
Dietrich, S. W. 53(239), *185*
Dietz, A. G. 404(33), *479*
Dietz, R. 129(482), *190*
Dîez-Barra, E. 662(4), *705*
Di Furia, F. 413(82b, 83, 84), 414(85), *480, 481*
Di Giovanni, M. C. 1383(75a, 75b), *1439*
Dijk, J. H. van 1194(83a), *1339*
Dijkstra, G. 290(60), *327*
DiLabio, G. A. 31(140), 45(140, 192, 198), 110, 113(440), 139(192), 140(140, 192, 198), *183, 184, 189*, 867, 871, 895–898(154), 899(154, 278), 900(280), *905, 908*
Dileo, C. 686(212), *709*
Dill, J. D. 275(45), *326*
Dillon, J. 1129(303), *1150*
Dillow, A. K. 621(48), *657*
Dillow, G. W. 100, 102(381), *188*
Di Matteo, A. 132, 133, 136(518), *191*, 1127(267), *1149*
Dimicoli, B. 578(195), *601*
Dimicoli, I. 109(414), 129(474), 144(474, 683), *189, 190, 195*
Dimitroff, M. 449(234c), *486*
Dimitrova, Y. 581(220), *601*
Dimmel, D. R. 1205(104), *1340*
Dimoticali, D. 1086(361), *1096*
Dimroth, K. 1123(216), 1124(216, 218), *1148*
Dinculescu, A. 823(330), *838*
Diner, B. A. 15(71), *181*, 369(206, 207), *391*, 1108(16), *1144*
Dinh, P. A. 913(25), *1003*
Dinis, T. C. P. 853, 856, 857(88), *904*
Dinjus, E. 6(26), *180*
Dinsmore, C. J. 1318(301a, 301b), *1345*
Diogo, H. P. 148(727), *196*
Diosady, L. L. 990(422), *1012*

Diplock, A. T. 840(6), *902*
Dirlam, N. L. 663(22), *705*
D'Ischia, M. 1079(242), *1092*
Divakar, S. 365(161), *389*
Dix, T. A. 855(106, 110), 901(110), *904*
Dixon, J. P. 545(54), *598*
Dixon, S. L. 925, 929(54), *1004*
Dixon, W. T. 113(457), 129, 135(457, 489), *190*, 1109(38–41, 43–45), 1113(84, 85), 1117(40, 41), 1120(38, 39, 44, 84, 173, 174), 1121(38), 1122(38, 39), 1123(38), 1124(39, 233, 240), 1125(252), 1133(40, 41), 1135(40), *1144, 1145, 1147–1149*
Dizdaroglu 129(486), *190*
Dizdaroglu, M. 1100(22, 23), *1105*
Djafari, S. 147(714), *195*
Djaneye-Boundjou, G. 405(45), *479*
Djerassi, C. 260(1), 299(93), 300(118), *325, 327, 328*
Dmitrienko, S. G. 987(408), 988(413), *1012*
Doamekpor, L. K. 1377(45b, 46a, 46b), *1437*
Doba, T. 844, 847, 855, 860–862, 864(18), 885, 890(212), 897, 898(18), *902, 906*, 1120, 1140(184), *1147*
Dobel, S. 339(18), *386*
Dobrodumov, A. V. 1030(105), *1089*
Dobson, G. 1128(274), *1149*
Dobson, J. F. 70(270, 271), *185*
Doctor, B. P. 403(27), *479*
Doczi, J. 420(138), *482*
Dodd, J. A. 314(201), *330*
Dodd, J. P. 928(55), *1004*
Dodge, J. A. 814(305), *838*
Dodgson, I. 178(771), *197*
Dodi, L. 1393(158), 1396(186c), *1442, 1443*
Doering, W. E. 425(158), *483*
Doherty, R. M. 503(89), *526*
Dohrmann, J. K. 1131, 1134(313), 1135(346), 1139, 1141(313), *1150, 1151*
D'Oliviera, J. C. 1086(368), *1096*
Dolmanova, I. F. 978(347), *1011*
Dom, P. B. 1351(54), *1364*
Domagala, S. 748(92), *833*
Domash, L. 226(6), *254*
Domcke, W. 106, 107(408), *188*
Domenech, X. 1086(367), *1096*
Domenicano, A. 20, 22(109), *182*
Domiano, L. 1391, 1396(118a), 1409(287), *1440, 1447*
Domínguez, E. 953, 969(284), 975(322), 976(330, 332), 977(335), *1009, 1010*
Dominguez, J. R. 1362(207), *1368*
Dominiak, P. 344(44), *387*
Dona-Rodriguez, J. M. 1084(324), *1095*
Donati, I. 19(100), *182*
Dondoni, A. 1403(240b), 1405(255), 1427(408d), 1432(448b), *1445, 1446, 1451, 1453*

Dong, C. 1085(343), *1095*
Dong, Q. 1084(310), *1094*
Dong, S. 976(328), *1010*
Dong-Sheng-Ming 212(22), *221*
Donk, W. A. van der 1108, 1117, 1120(14), *1144*
Donovan, F. W. 426, 427(166b), *483*
Donovan, T. 296(84, 85), 308(160), *327, 329*
Dooly, G. L. 932, 937(86), *1005*
Doong, R.-A. 1085(337), 1086(370), *1095, 1096*
Dopfer, O. 30(136), 83, 84(300), 110, 112, 121(444), 129(476), 143(591), 144(677, 689, 691), 147(712), *183, 186, 189, 190, 193, 195*
Dopico, M. 1260(191), *1342*
Dopp, D. 1043(130a), *1090*
Doppler, E. 823(326), *838*
Doppler, T. 419(135a), *482*
Doretti, L. 979(354), *1011*
Dorfman, L. M. 1098(4), *1105*, 1113(83), *1145*
Döring, C. E. 319(222), *330*
Dorme, R. 1272(205a, 205b), *1343*
Dormer, P. G. 670(78), *706*
Dorofeenko, G. N. 823(327, 330), 824(333), *838*
Dorr, F. 1025(82), *1089*
Dörrenbächer, R. 1378(50a), 1431(444), *1437, 1453*
Dorrestijn, E. 139(533), *191*
Dotsenko, I. N. 631(79), *657*
Doty, B. J. 717, 745(13), *831*
Dötz, K. H. 450(246e), 451(248a, 251a–c), 454(251a–c), 455(265b), 456(259), 457(262, 264), *486, 487*
Doucet, J.-P. 350(83), 364(154), *388, 389*, 576(190), *600*
Douchal, A. 1025(70b), *1088*
Dougherty, D. A. 322(242), *331*
Dougherty, R. C. 293(72), 314(203), *327, 330*
Douglas, J. L. 429, 430(179a, 179d), *483*
Douhal, A. 730(45), *831*, 1047(149), *1090*
Douteau, M. 652(185), *660*
Doutheau, A. 453(244), *486*, 818(317), *838*
Downing, D. F. 405(43b), *479*
Downing, R. S. 460(286), *487*, 763(128), *834*
Doye, S. 670(74), *706*
Dozol, J.-F. 322(249), *331*, 1389(105), 1391(120c), 1396(187), 1415(313), 1435(187), *1440, 1443, 1448*
Dradi, E. 1395(180), *1442*
Dragisich, V. 457(263), *487*
Drago, R. S. 548(62), 590(263), *598, 602*, 1205(102), *1340*
Drake, N. L. 400(18a), *478*
Drakenberg, T. 350(80), *388*, 998(451), *1013*
Dransfeld, A. 95(367), *188*

Draper, R. B. 1100(25, 26), *1105*, 1129(285, 286), *1149*
Draths, K. M. 19(101, 102), *182*
Drauz, K. 1216(119c), *1340*
Dravnieks, F. 1120(169), *1147*
Dreher, E.-L. 1169(41, 43), *1338*
Dreiding, A. S. 403(29), *479*
Dreier, D. 211, 216(20), *221*
Drenth, J. 412(73), *480*
Dressman, S. F. 953, 955(207), *1007*
Drew, M. G. B. 1393, 1403(152c), 1409(286b), 1427(413, 416a, 416b, 417), 1432(450, 453), *1441, 1447, 1451–1453*, 1495(151), *1505*
Drewes, S. E. 766(141), *834*
Driessen, W. L. 1194(84), *1339*
Drift, E. van der 1126(262), *1149*
Drilleau, J. F. 913(10), *1003*
Drinkwater, D. E. 110, 121(434), *189*
Drioli, E. 1082(288), *1094*
Dross, A. 315(215), *330*
Drouin, M. 1406(261), *1446*
Droz, T. 577(191), *600*
Drumm, M. F. 1459, 1475, 1484(6), *1502*
Du, C. M. 545(53), *598*
Du, Y. 272(37), *326*
Duarte, C. L. 1100(34), *1106*
DuBois, J. L. 1196(88), *1340*
Dubovitskii, F. I. 1104(58), *1106*
Dubrelle, A. 171(762), *197*
Duce, P. P. 535(33), 536(33, 36, 37), 537(33), 539, 540(37), 557(33), 588(33, 37), 591(37), *597*, 879, 882(196), *906*
Ducrocq, C. 639(125), *659*
Duddeck, H. 357(109), *388*
Duddek, E. P. 359(132), *389*
Duddek, G. 359(132), *389*
Dudic, M. 1432(451), *1453*
Dudicc, M. 1415(317), *1448*
Dudley, G. B. 450(239d), *486*
Duesler, E. N. 206(11), 220, 1380(58f), *1437*
Duetz, W. A. 1359(159), *1367*
Dufay, J. 1495(138), *1505*
Duff, J. C. 1470(65), *1503*
Dufour, C. 871(167), *905*
Dufresne, C. 683(175), *708*
Duguay, G. 586, 587(241), *602*
Duijn, D. van 86(306), *186*
Dulenko, V. I. 433(187a), *484*
Dulin, D. 845, 846(31), *902*
Dumas 7(38), *180*
Dumazet, I. 1372(28), *1436*
Dumazet-Bonnamour, I. 1372(30), *1436*
Dumont, J. 953(221, 222), 957(221), 958(222), *1008*
Dunbar, R. C. 83(294), *186*
Duncalf, L. J. 776(197, 198), *835*
Duncan, L. A. 857(135), *905*

Duncan, S. M. 460(279), *487*
Dunford, H. B. 1118(148), *1146*
Dunker, J. C. 1356(124), *1366*
Dunlop, I. 1117(120), *1146*
Dunn, M. F. 801(246), *836*
Dunn, M. M. 664(30), *705*
Dunogues, J. 630(69), *657*
Dupeyron, S. 946, 953(162), *1006*
Dupont, J. P. 143(561), *192*
Dupont, L. 583(224), *601*
Duprat, A. 1394(175b), *1442*
Duran, M. 455(266, 267), *487*, 730(46), *831*, 1025(77), *1089*
Duran, N. 974(319), *1010*
Durand, A.-P.Y. 1057(173), 1072(212), *1091, 1092*
Durie, A. 635(101), *658*
Durig, J. R. 35, 36(154), *183*
Durocher, G. 37(177), *183*
Durust, N. 944, 953(142), *1006*
Durust, Y. 944, 953(142), *1006*
Dushenko, G. A. 770(154), *834*
Dushin, R. G. 674(111), *707*
Dutasta, J.-P. 1370(5), 1424(394), *1435, 1450*
Dutta, B. K. 1081(267), *1093*
Dutton, P. J. 1418(334), *1448*
Duve, C. de 840(4), *902*
Du vosel, A. 1382(68), 1386(86a, 86b), 1417(331), 1422(379), *1438, 1439, 1448, 1450*
Duynhoven, J. P. M. van 1405(253), 1406(263d), 1428(263d, 437a, 437b), *1446, 1453*
Dvorakova, H. 1415(316), *1448*
Dwibedy, P. 1129(292), *1150*
Dwivedi, C. P. D. 82(287), *186*
Dworschak, E. 856, 857(122), *904*
Dynak, J. N. 829(344), *838*
Dyszlewski, A. D. 668(58), *706*
Dzidic, I. 312(194), *330*
Dziegiec, J. 748(92), *833*
Dziembowska, T. 144(652, 660), *194*, 342(32, 33), 345(48), 348(56), 353(56, 96, 98, 101), 355(105), 359(122–124, 126), 360(134), 361(122), 371, 382(32), *387–389*, 551(127, 128), *599*, 1024(63, 64), *1088*
Dzimbowska, T. 386(307), *393*

Eadon, G. A. 1348(19), *1363*
Eastman, J. F. 470(300b), *488*, 806(250), *836*
Eaton, A. D. 1352(62), 1353, 1354(74), *1364*
Ebata, T. 107(410), 143(580), 144(598, 686), 147(580, 703, 705–707), 148(705), 149(703, 705), 150, 157, 167, 168(705), *188, 192, 193, 195*, 375(245), 377(255), *392*, 551(85, 130), *598, 599*

Ebby, S. T. N. 727(38), *831*
Ebeler, S. E. 365(160), *389*
Eberson, L. 639(126), *659*
Ebert, K. 551(93), *598*
Ebert, M. 1098(3), *1105*, 1113–1115, 1128, 1130(86), *1145*
Ebine, S. 818(318–320), *838*
Ebisuzaki, Y. 232(36), *255*
Ebrhardt, M. 1297(256a), *1344*
Eckart, K. 315(215), *330*
Ecker, J. R. 698(295), *711*
Eckert, C. A. 621(48, 49), 631(81), *657, 658*
Eckert, R. C. 1290(249), *1344*
Eckert-Maksić, M. 83, 84(298), 97, 101(374), *186, 188*, 261, 263, 296(14, 15), 308(162), *326, 329*
Edamatsu, R. 855, 856(105), *904*
Eden, C. 1124(220), *1148*
Eder, K. 945(157), *1006*
Edge, S. 1495(148), *1505*
Edgecombe, K. E. 68(255), 89(314), *185, 186*
Edgehill, R. U. 944(146), *1006*
Edlund, U. 357(110), *388*
Edman, E. 1267(202b), *1343*
Edstrom, D. 1307(273), *1344*
Edstron, E. 1244(170), *1342*
Edward, J. T. 409(63a), *480*
Edwards, A. J. 1380(60a), *1438*
Edwards, C. D. 851(50, 57), 856, 872(50), 876(184), 878(50), 901(284), *903, 906, 908*
Edwards, J. 1104(55), *1106*
Edwards, J. O. 1255(184c), *1342*
Edwards, M. 983(373), *1011*, 1315(292), *1345*
Edwards, R. A. 1402(228b), 1409(288), *1445, 1447*
Edwards, T. G. 1123(204), *1148*
Edwards, W. 411(70b), *480*
Edwards, W. M. 1494(125, 128), *1505*
Effenberger, F. 472(312), *488*, 631(80), *657*, 686(202), *709*, 773(165), *834*
Egan, W. 348, 349(57), *387*
Egawa, Y. 851(57, 60), 852(60), 876(184), 887, 888(60), 894(260), *903, 906, 908*
Egberink, R. J. M. 635(101), *658*, 1383(67d), 1391(121a), 1396(121a, 182a, 182b), 1403(238), 1421(67d), *1438, 1440, 1443, 1445*
Eggert, J. 1397(196), *1443*
Eggert, L. 1073(222, 223), *1092*
Egler, R. S. 438(199), *484*
Eguchi, H. 650(164), *659*
Egyed, O. 179(782), *198*, 369(209), *391*
Ehrin, H. 324(268), *331*
Eichhorn, J. 1246(172), *1342*
Eichinger, K. 434(192), *484*
Eichinger, P. C. H. 467(296), *488*
Eickhoff, H. 1172(51b), *1338*
Eid, C. N. Jr. 1428(426a), *1452*

Eigen, M. 1118, 1119(136, 137), *1146*
Eikema Hommes, N. R. J.van 95(367), *188*
Eikhman, R. K. 18(89), *181*
Eilenberg, W. 1168(40), *1338*
Eisen, M. S. 619(40), *656*
Eiserich, J. P. 636(104), 637(109), *658*
Eisleb, O. 460(278), *487*
Ejarrat, E. 1360(168), *1367*
Ekaeva, I. 647(153), *659*
Ekiel, I. H. 887(217), *906*
Eksperiandova, L. P. 987(407), *1012*
El-Abbady, A. M. 472(311d), *488*
Elaloui, A. 1084(327), *1095*
Elango, V. 807(275), *837*
El-Ashry, S. M. 988(412), *1012*
El Azmirly, A. 230(24), *254*
El-Bermani, M. F. 551(108), *599*
Elbs, K. 408(54a), *479*
Elchinger, P. C. H. 314(209), 315(210), 316(209), *330*
El-Deeb, M. K. 129(483), *190*
Eldik, R.van 1351(44), *1364*
Elermand, Y. 1024(62), *1088*
El-Gendy, A. E. 995(440), *1013*
Elgendy, S. M. A. 1227, 1228(142a, 142b), *1341*
El-Gharkawy, El-S. R. H. 5(11), *179*
El Ghomari, M. J. 575, 586, 587(186), *600*
El Gihani, M. T. 1420(359b), *1450*
Elguero, J. 110, 123(450), 129(479), *190*, 349(59), *387*, 559, 572(157), 580, 581(212), *600, 601*
El Hadrami, I. 913(4), *1003*
Eliason, R. 776(200), *835*
Eling, T. E. 1120(175), *1147*
Elings, J. A. 460(286), *487*, 763(128), *834*
Elix, J. A. 424(154a), *483*
El-Kerdawy, M. M. 988(412), *1012*
El-Kholy, I. E.-S. 436(197), *484*
Elkobaisi, F. M. 759(110), *833*
Elleingand, E. 129(490), *190*
Ellington, J. C. 884(208), *906*
Elliot, A. J. 1120(181), *1147*
Ellis, D. A. 1084(314), *1095*
Ellman, J. A. 699(305), *711*, 1318(301b, 304), *1345*
Elmali, A. 1024(62), *1088*
Elming, K. 127(469), *190*, 718(17), *831*
Elming, N. 432(182b–d), *484*
El-Mobayed, M. 1169(43, 44), 1172(51a), *1338*
El Sayed, N. 230(24), *254*
Elschnig, G. H. 778(210), *836*
Elseewi, A. A. 955(205), *1007*
El Telbani, E. 776(205), *835*
El Walili, A. F. M. 995(439), *1013*
El Walily, A. F. M. 990(423), *1012*
El Wasseef, D. R. 988(412), *1012*

El Watik, L. 229, 250, 251(18), *254*
Emanuel, N. M. 1118, 1119(140), 1128(277), 1130(319), *1146, 1149, 1150*
Emeline, A. V. 1086(369), *1096*
Emery, S. 141(540), *192*
Emmett, J. C. 549(65), *598*
Emmi, S. S. 1115, 1128(91), *1145*
Emmons, W. D. 424(152a, 152b, 156), *483*
Emnéus, J. 953, 969(280), 974(320), 975(322), 976(324, 329), 977(333, 335, 337), *1009, 1010*
Emori, S. 1018(22), *1087*
Emsley, J. 8, 12(44), *180*, 370, 371, 382(210), *391*
Emslie, N. D. 766(141), *834*
Emura, M. 211(19), *220*
Enari, T. 1495(137), *1505*
Enchev, V. 360(136, 140), 361(140), 385(296), *389, 393*, 1024(65), *1088*
Encinas, M. V. 1021(42), *1088*, 1109, 1110, 1112(26), *1144*
Endo, A. 684(193), *709*
Endo, K. 1080(259d), *1093*
Endres, G. F. 1491(118), *1505*
Endrey, A. L. 1494(126–128), *1505*
Engbersen, J. F. G. 1394(176), *1442*
Engbersen, J. F. J. 1389(104), 1391(112, 119), 1402(224b), 1403(238), 1405(224b, 249a, 249b), 1421(369), *1440, 1444, 1445, 1450*
Engdahl, C. 357(110), *388*
Engel, M. R. 1159, 1207(27), *1338*
Engen, D. V. 674(106), *707*
English, D. S. 1045(136–138), *1090*
Engman, L. 865(145), *905*
Engstrom, M. 179(775), *197*
Engwall, M. A. 1362(204), *1368*
Enkelmann, V. 211(18), *220*
Enoki, M. 1462(25), *1503*
Enomoto, H. 1163(33a), *1338*
Enyo, K. 1374(38b), *1436*
Erben-Russ, M. 1130(309), *1150*
Erdem, A. 978(349), *1011*
Erdmann, K. 998(456), *1013*
Erdtman, H. 1108(11), *1144*
Erhardt, P. W. 233(40), *255*
Erhardt-Zabik, S. 315(212), *330*
Ericksen, T. E. 139(534), *191*
Erickson, B. E. 1357(132), 1360(173), *1366, 1367*
Ericson, J. L. 1419(348), 1424(397a, 399), *1450, 1451*
Eriksen, T. E. 1108(24, 25), 1118(24), 1136(361), 1138–1140, 1142(24), 1143(24, 382), *1144, 1151, 1152*
Erikson, S. K. 13(67), *181*
Eriksson, L. A. 45(196), 132, 133, 136(517), 143(549), *184, 191, 192*
Erim, F. B. 972(305), *1010*

Erin, A. N. 887(225), *907*
Erkasov, R. S. 230(27), *254*
Erker, G. 461(290), *487*, 684(185), *708*
Erlebacher, J. 375(250), *392*
Ermolaeva, L. V. 1424(396), *1451*
Ermolaeva, T. N. 983(374), 984(382), *1011*
Ernst, L. 350(78), *388*
Ershov, V. V. 651, 654(167), *659*, 715, 717, 718, 720, 722, 724, 731(2), 745(78), 747(82, 83), 811(292), *830, 832, 837*, 1119(157, 158), 1123(210), 1126(256, 260), *1147–1149*
Erthal, B. 718(16), *831*
Erusting, N. P. 1025(84), *1089*
Ervin, K. M. 5(20), *180*
Erway, M. 608(16), *656*, 683(171), *708*
Escamilla, J. E. 851(63), *903*
Escarpa, A. 913(9, 15), 953, 957(217), *1003, 1008*
Escobar, L. 851(63), *903*
Escrig-Tena, I. 988(410), *1012*
Espadaler, I. 1355(119), *1365*
Espenson, J. H. 666(49), *705*
Espersen, W. 1121(191), *1147*
Espinado-Pérez, G. 551(93), *598*
Espinos, A. 1040(124), 1042(128), *1090*
Espuna, G. 651(172), *660*
Essen, G. van 1403(242), *1445*
Esser, C. 1105(64), *1106*
Estel, D. 319(222), *330*
Estela, J. M. 929, 984(67), 1001(472), *1004, 1013*
Esterbauer, H. 854(100), *904*
Esteve Romero, J. S. 988(409, 410), *1012*
Estévez, C. M. 551, 553(105), *599*
Eswaramoorthy, M. 414(91), *481*
Etheredge, S. J. 441(210), *484*
Etievant, P. 948, 953(176), *1007*
Etlzhofer, C. 345(50), *387*
Etmetchenko, L. N. 824(333), *838*
Eto, H. 1158(21, 24), 1159(24, 25b, 26, 30), 1172(25b), 1178(21), 1227(30), *1338*
Ettre, L. S. 929(72), *1004*
Etxebarria, N. 984(384), *1012*
Eugster, C. H. 212(21), *221*
Eustance, J. W. 1491(118), *1505*
Evans, D. A. 671(84), *706*, 1318(301a, 301b, 302), *1345*, 1466(45), *1503*
Evans, D. H. 1167(38, 39), *1338*
Evans, F. E. 999(465), *1013*
Evans, J. C. 35(151), *183*
Evans, M. G. 404(35), *479*
Evans, P. A. 664(38), *705*
Everett, J. R. 346(53), *387*
Evrard, D. A. 1318(301b), *1345*
Exarchou, V. 998(454), *1013*
Exner, O. 559(152), *600*

Eyler, J. R. 325(278), *332*
Ezure, Y. 1163(33a), *1338*

Fabbi, M. 1396(186a), 1403, 1405(235), *1443, 1445*
Fabietti, F. 1357(130), *1366*
Facelli, J. C. 366(167), *390*
Facino, R. M. 953, 966(269), *1009*
Factor, A. 466(292), *487*
Fagan, P. J. 670(75), *706*
Fagley, T. F. 251(93), *257*
Fagnou, K. 665(44), *705*
Fahmy, A. M. 1071(202, 203), *1091, 1092*
Fail, J. 549(65), *598*
Fail, P. A. 1348(30), *1363*
Failor, R. 1104(54), *1106*
Fairhurst, R. A. 684(188), *709*
Faix, O. 303(143), 305(147), *329*, 997(444), *1013*
Fajardo, V. 807(276), *837*
Falabella, E. 631(72), *657*
Falabu, D. 1420(365b), *1450*
Falck, J. R. 418(120), *482*
Faleev, N. G. 748(90), *833*
Falk, H. 345(50), *387*, 1045(140, 143), *1090*
Falle, H. R. 1124(236), *1148*
Fallon, L. 1045(135), *1090*
Famiglini, G. 953, 961(236), *1008*
Famini, G. R. 99, 101(378), *188*
Fan, J. 1361(198), *1367*
Fan, M. 1393(145a, 145b), *1441*
Fananas, F. J. 770(153), *834*
Fang, C. 1085(354), *1096*
Fang, C.-Y. 999(463), *1013*
Fang, H. H. P. 1359(145), *1366*
Fang, W.-H. 106, 107(407), 148(728), *188, 196*
Fang, X. 763(136), *834*, 1100(32), *1106*, 1116(113), 1136(362), *1146, 1151*
Fang, X. W. 1086(375), *1096*, 1100(31), *1106*
Fang, Y. 434(193), *484*, 1085(354), *1096*
Fanni, S. 1391(121a), 1396(121a, 182b, 195b), 1427(410a), *1440, 1443, 1451*
Fanta, P. E. 472(311g), *488*
Faraggi, M. 1117(129, 131), 1142(380, 381), *1146, 1152*
Farant, R. D. 366(168), *390*
Fargeix, G. 871(167), *905*
Faria, M. D. G. 561(165), *600*
Farkas, L. 1315(292), *1345*
Farkas, O. 4, 100(2), *179*, 368, 372, 373(200), *390*
Farmer, D. B. 322(247), *331*, 1399(204), *1444*
Farneth, W. E. 314(202), *330*
Farnini, G. R. 57(243), *185*
Farrant, R. D. 359(125), *389*, 726(30, 31), *831*
Fasani, E. 419(133), *482*

Fastabend, U. 284(53), 293(74), *327*
Fateley, W. 551(131), *599*
Fateley, W. G. 36, 45, 50(171, 172), 53(171, 237), 58(172), 87(307), *183, 185, 186*, 233(42), 238(72), *255, 256*, 369, 371(208), *391*, 554, 555(137), *599*
Fatibello, O. 981(359), *1011*
Fatibello-Filho, O. 978(350), *1011*
Fattuoni, C. 139(532), *191*
Faulkner, D. J. 1327(314), *1345*
Fausto, R. 561(164, 165), *600*
Fava, G. G. 680(148), 683(176, 184), *708*
Fawcett, W. R. 353(97), *388*
Fecka, I. 953, 967(272), *1009*
Fedorov, A.Yu. 979(353), *1011*
Fedorov, V. K. 1123(201), *1148*
Fedorowicz, A. 144(651, 669), *194*
Fedotov, M. A. 340, 341(26), *387*
Fehlinger, M. 1403(237), *1445*
Feiring, A. E. 669(62), *706*
Feitelson, J. 1098(6), *1105*, 1129(296), 1130(296, 321), 1135(296), *1150*
Fekete, J. 759(115), *833*
Felix, C. C. 1120(182, 183, 188), *1147*
Felix, V. 1427(416b, 417), *1452*
Felker, P. M. 35(163), 143, 144, 147(164), *183*
Feller, D. 20, 35, 148(121), *182*
Fellin, P. 319(230), *330*
Feltz, P. 475(318), *488*
Fencht, W. 913(1), *1002*, 1355(114), *1365*
Fendler, H. 493(54, 55), *526*
Ferdani, R. 1427(415), *1452*
Ferguson, G. 580(208–210), *601*, 635(100), *658*, 1374(33c, 34, 39, 43), 1378(50a), 1386(87), 1387(91), 1390(106), 1391(118b, 122b), 1392(122b, 138), 1393(122b, 153), 1394(122b), 1396(181b), 1403(239), *1436, 1437, 1439–1441, 1443, 1445*
Ferguson, P. L. 1357(131), *1366*
Feria, L. S. 952, 953, 990(197), *1007*
Feringa, B. 1292(251), *1344*
Feringa, B. L. 359(130), *389*
Fernández, L. A. 984(384), *1012*
Fernández, M. F. 925(48), *1004*
Fernandez-Ibanez, P. 1086(372), *1096*
Fernández Laespada, M. E. 948, 953(180, 181), *1007*
Fernelius, W. C. 778(208), *835*
Ferradini, C. 1125(246), 1129(246, 304), *1149, 1150*
Ferrao, M. L. C. C. H. 238, 239(67), *256*
Ferrara, D. 979(354), *1011*
Ferrara, F. 1357(130), *1366*
Ferraz, C. 1312, 1315(285), *1345*
Ferreira, M. A. 913, 958(20), *1003*
Ferreres, F. 961(245), 973, 981(313), *1008, 1010*

Ferretti, V. 338, 342(9), 363(146), *386, 389*, 552(135), *599*
Ferrier, D. 954, 970(199), *1007*
Ferriere, J. J. 913(4), *1003*
Ferris, F. C. 45(189), *184*
Ferrugia, M. 301(135, 136), *328*
Fessenden, R. W. 1017(11), *1087*, 1098(7), 1099(17), *1105*, 1110(55, 58), 1112(55), 1115(97), 1120(97, 187), 1121–1123(97), 1126(263, 264), 1127(263), 1128(58, 187, 263), 1129, 1130(58), 1132(187), 1135(97), 1136, 1138(263), *1145, 1147, 1149*
Fetizon, M. 1307(274, 275), *1344*
Fettig, A. 353(102), *388*
Feutrill, G. I. 444(215c), *485*
Feyereisen, M. W. 20, 35, 148(121), *182*
Fiamegos, Y. C. 986(398), *1012*
Fiehn, O. 953, 963(259), *1009*
Fields, D. L. 818(316), *838*
Fiesinger, F. 1352(59), *1364*
Fife, W. K. 460(271b), 463(276c), *487*
Figeys, H. 170(756), *197*
Filarovski, A. 551, 556(119), *599*
Filarowski, A. 144(621, 654, 659–661, 665, 666, 668), 145, 146(621), *193, 194*, 355(104, 105), 371, 377, 385(216), 386(303), *388, 391, 393*
Filges, U. 300(111), *328*
Filip, S. V. 662(9), *705*
Filipenko, O. S. 1025(69), *1088*
Filippi, A. 274(41), *326*
Filippini, L. 683(180), 686(210), *708, 709*
Finch, A. 231(33), 234(33, 47), 236(55), 238(68, 69), 239(68), 241(33, 77), *255, 256*
Finck, G. 472(308a), *488*
Finefrock, A. E. 730(47), *832*
Finet, J.-P. 671(82), 681(163), *706, 708*, 1330(323, 326), 1332, 1333(326), *1346*
Fingler, S. 1348(10), *1363*
Finiels, A. 617(35), 645(145), *656, 659*
Fink, M. 441(211), *484*
Finkbeiner, H. 466(292), *487*
Finkel, T. 141(536), *191*
Fiorentini, D. 852(77), *903*
Firooznia, F. 450(241a), *486*
Firouzabadi, H. 638(123), *659*
First, P. R. 990(421), *1012*
Fischer, A. 806(268), *837*
Fischer, C. 1391, 1396(121c), 1432(446), *1440, 1453*
Fischer, E. 953, 969(282), *1009*
Fischer, G. W. 823(330), *838*
Fischer, H. 132, 136(519), *191*
Fischer, I. 492(18), 493, 515(18, 28), *525*
Fischer, J. 413(81), *480*, 1401(219b, 220b), 1415(311, 314b), *1444, 1448*
Fischer, M. 1024(56, 57), 1033(110), 1038(120), *1088, 1089*

Fischer, R. 674(107), *707*
Fischer, W. 953, 962(250), *1008*
Fischer, W. B. 1383, 1421(67m), *1438*
Fishbein, J. C. 419(124a, 124b), *482*
Fisher, A. J. 1374(44), *1437*
Fisher, D. R. 1394(167c), *1442*
Fisher, I. P. 274(40), *326*
Fisher, J. 1427(421c, 422b), *1452*
Fisher, P. 353(102), *388*
Fisher, T. 36, 147(175), *183*, 1485(95), *1504*
Fishman, E. 47, 53(224), *184*
Fitzgerald, E. A. 1482(88), *1504*
Fitznar, H. P. 953, 962(249), *1008*
Flad, J. 69(263), *185*
Fläming, G. 1408(279b), *1447*
Flammang, R. 84(303), 110(431–433, 448–451), 115(303), 121(465, 466), 123(450), *186, 189, 190*, 263(20), 299(98), 301, 302(137), *326, 328*
Flattery, J. 1360(170), *1367*
Flaud, J. M. 150(738), *196*
Flechtner, T. W. 1043(130b), *1090*
Fleming, G. R. 493, 498(45), 503(83, 84), 522(83, 97), 523(97), *525–527*
Fleming, I. 689(231), *709*
Flett, M. St. C. 47(228), *184*
Flood, T. C. 14(70), *181*
Florêncio, H. 290(60), *327*
Florián, J. 5(15), 179(778), *180, 198*
Floriani, C. 1391(113), 1396(192), *1440, 1443*
Floriani-Moro, R. 1396(192), *1443*
Floss, H. G. 359(121), *389*
Flowers, R. A. II 1428(425c), *1452*
Floyd, T. L. 1354, 1355(100), *1365*
Fluegge, A. P. 1026(90), *1089*
Foa, V. 1359(151), *1366*
Foces-Foces, C. 580, 581(212), *601*
Fochi, M. 1389(104), 1403, 1405(232a), *1440, 1445*
Fogarasi, G. 4(6), *179*
Fokin, V. N. 930(81), 937(121), *1004, 1005*
Fokina, I. I. 987(407), *1012*
Foldiak, G. 1100(30), *1106*, 1115(91), 1128(91, 275, 276), 1129, 1130(293, 294), 1142(293, 378), *1145, 1149–1151*
Fomenko, V. V. 616(32), *656*
Fomin, V. V. 993(434), *1013*
Fonesca, R. G. B. 20, 23(117), *182*
Fonseca, R. G. B. 179(777), *198*
Fontana, F. 607(6), *656*
Fontecave, M. 129(490), *190*
Foo, L. M. 1465, 1467(42), *1503*
Foote, C. S. 853(93), *904*
Force, D. A. 15(71), *181*
Ford, A. 691(241), *710*
Forehand, J. B. 932, 937(86), *1005*
Fores, M. 730(46), *831*, 1025(77), *1089*

Foresman, J. B. 4, 100(2), *179*, 368(194, 200), 372, 373(200), *390*
Forest, H. 20(106), *182*
Forgo, I. 1495(149), *1505*
Formosinho, S. J. 492, 493, 503(15), *525*, 1021(37), *1088*
Forni, L. 18(87), *181*
Forni, L. G. 1130, 1137, 1138(308), *1150*
Forrest, B. J. 884(208), *906*
Forrest, N. P. 1274(210b), 1285(237), *1343*
Forsén, S. 348(57), 349(57, 58), 353, 354(100), *387, 388*
Förster, Th. 491, 492(1–4), 493(2, 52), *525, 526*
Foster, D. O. 891(241), *907*
Foster, K. L. 1033(108), *1089*
Foster, M. D. 997(448), *1013*
Föster, S. 1205(101), *1340*
Foster, W. E. J. 1124(233), *1148*
Foti, M. 851, 852(68), 855, 856(112), 871, 872(68), 891(112), 894(262), 899, 901(279), *903, 904, 908*, 992(429), *1013*, 1140(373), *1151*
Foti, S. 301(135, 136), *328*
Foucher, D. 534, 586, 587(23), *597*
Fourmentin, S. 992(433), *1013*
Fournier, F. 316(218–221), *330*
Fowler, J. S. 416(111), *482*
Fox, D. 4, 100(2), *179*
Fox, D. J. 368, 372, 373(200), *390*
Fox, M. A. 1100(25, 26), *1105*, 1129(284–286), *1149*
Fox, S. 1192(78e), *1339*
Fox, T. 896(271), *908*
Fraccaro, C. 646(146), *659*
Fradera, X. 69(267), *185*
Fraenkel, G. K. 1123(203), 1124(217, 228), *1148*
Fraga Palomino, N. 953, 958(223), *1008*
Fraile, S. 985(388), *1012*
France, P. 1080(252), *1093*
Frances, G. W. 338(13), *386*
Francesch, C. 1136(356, 357), *1151*
Franchi, P. 877, 878, 881, 884(193), *906*
Francisco, M. A. 1327(315), *1346*
Franck, H.-G. 19(95), *182*
Franck, U. Y. 1353(89), *1365*
Francke, S. 1355(118), *1365*
Franckowiak, G. 1297(257), *1344*
François, J.-P. 83, 84(298), *186*, 311(187), *329*
Francony, A. 1361(194), *1367*
Frange, B. 540(39), *597*
Frank, P. 1105(64, 65), *1106*
Franke, A. A. 913(24), *1003*
Franke, W. 1355(118), *1365*
Frankel, E. N. 857, 858(133), *904*
Franken, S. M. 1396(189), *1443*
Fransen, J. R. 1418(334), *1448*

Franz, K. J. 219(38), *221*
Fraser, D. A. 1462(20, 21), *1502*
Fraser, D. L. 1470(63), *1503*
Fraser, J. R. 1425(403b), *1451*
Fraser, P. K. 782(220), *836*
Fraser-Monteiro, L. 110, 112, 121(444), *189*, 265(24), *326*
Frascr-Monteiro, M. L. 110, 112, 121(444), *189*, 265(24), *326*
Fráter, G. 463(277), *487*, 753, 764(105), *833*
Fraternali, F. 1391(120b), *1440*
Frebortova, J. 1354(105), *1365*
Freccero, M. 1081(262), *1093*
Freean, P. K. 1074(227–230), *1092*
Freed, J. H. 1124(217), *1148*
Freed, K. F. 93–96(357), *187*
Freedman, J. 663(24), *705*
Freeman, B. A. 636(104), 637(110), *658*
Freeman, F. 1192, 1198(76), 1273(207), 1278, 1279(227), *1339, 1343*
Freeman, J. H. 1475(81), 1485, 1486(102), *1504*
Freeman, W. P. 300(126), *328*
Freese, U. 774(178), *835*
Frei, R. W. 1353(80), *1365*
Freiermuth, B. 229, 250(20), *254*
Freindorf, M. 95, 101(371), *188*
Freiser, B. S. 291(65), *327*
French, A. 313(200), *330*
Frenzel, U. 1065(194), *1091*
Frenzel, W. 950, 953(190), 987(404), *1007, 1012*
Frére, S. 680(150), *708*
Freriks, M. 1387, 1392, 1394, 1395(90), *1439*
Freudenberg, K. 1108(12), 1118(138, 139), *1144, 1146*
Frey, M. H. (97, 98), *1504*
Freytag, M. 1383, 1421(67k), *1438*
Fridovich, I. 1119(160), *1147*
Friebe, S. 1420(359c, 361), *1450*
Friedman, J. P. 405(42), *479*
Friedrichsen, W. 1403(233a, 240b), 1408(280), *1445, 1447*
Fries, K. 472(308a, 308b), *488*
Friess, S. L. 424(157a), *483*
Friestad, H. O. 987(403), *1012*
Frimer, A. A. 471(302), *488*, 806, 814(254), *836*
Frings, M. 1378(50a), 1411(295), *1437, 1447*
Frisch, M. J. 4, 100(2), *179*, 368(194, 200, 201), 372, 373(200), *390, 391*
Fristch, N. 441(205), *484*
Fritsch, J. 145(620), *193*
Fritsch, K. H. 987(402), *1012*
Fritz, B. 953, 963(260), *1009*
Fritz, J. S. 1352(66), 1353(84), *1364, 1365*
Frnkova, P. 932, 937(87), *1005*
Frohlich, H. 546, 547(55), *598*, 999(460), *1013*

Fröhlich, R. 669(59), *706*
Fromentin, E. 630(71), *657*
Fronczek, F. R. 1393(147), 1394(147, 165), *1441, 1442*
Frongs, M. 1378, 1386(51), *1437*
Frost, J. W. 19(101, 102), *182*
Frost, T. M. 450(236, 237), *486*
Frostin-Rio, M. 1201, 1205(92), *1340*
Frumin, H. 1104(54), *1106*
Fry, S. C. 1163(34), *1338*
Fu, D. K. 1374(42b), *1436*
Fu, P. P. 425(162), *483*
Fu, Y. J. 1085(355), *1096*
Fuangswasdi, S. 1394(172c), 1396(172c, 181b), *1442, 1443*
Fuchita, Y. 414(86b), *481*
Fuchs, C. 1118, 1130(143), *1146*
Fuente, C.de la 982(366), *1011*
Fuerst, D. E. 463(276a), *487*
Fuertes, A. B. 1489(106), *1504*
Fuhrmann, B. 979(355), *1011*
Fuji, K. 1380(56), *1437*
Fujibayashi, S. 1204(100), *1340*
Fujibuchi, T. 1025(86), *1089*
Fujii, A. 144(598, 686), 147(706, 707), *193, 195*, 375(245), *392*, 551(85, 130), *598, 599*
Fujii, M. 37, 46, 110, 111(182), 147(692, 703), 149(703), *183, 195*
Fujii, S. 1360(171), *1367*
Fujii, T. 92(347), *187*, 937, 938(124), *1005*, 1265(199b), *1343*
Fujii, Y. 447(222d), *485*
Fujikawa, C. 275(44), *326*
Fujikawa, M. 651(170), *660*
Fujimaki, E. 551(130), *599*
Fujimaki, M. 1421(373), *1450*
Fujimori, Y. 641(136), *659*
Fujimoto, K. 410(68), *480*, 1391(115a), *1440*
Fujimoto, M. 1495(161), 1498(176), *1506*
Fujimoto, S. 1030(104), *1089*
Fujimoto, T. 1415(314a), 1418(335), *1448*
Fujimoto, Y. 680(144), *708*
Fujio, M. 5(19), *180*
Fujioka, A. 674(105), *707*
Fujioka, H. 447(222b, 222g, 222h), 448(222b, 225), *485*
Fujisaki, S. 650(164), *659*
Fujisawa, K. 1209(106), *1340*
Fujishige, M. 300(125), *328*
Fujishima, A. 1084(307), *1094*, 1362(211), *1368*
Fujita, J. 699(303, 306), *711*
Fujita, S. 1115(104), *1146*
Fujita, T. 699(308), *711*
Fujiwara, H. 953, 970(292), *1009*
Fujiwara, Y. 414(86a, 86b), *481*, 1025(78), *1089*
Fukai, T. 338(12), *386*

Fukakawa, H. 1357(137), *1366*
Fukami, S. 687(223), *709*
Fukata, F. 447(222a), *485*
Fukata, G. 1290(245), *1344*
Fukazawa, Y. 1374(36b), 1403(240a), 1406(36b, 270a, 270b), *1436, 1445, 1446*
Fuke, K. 107(412a), 143(412a, 581, 582), 147(412a), *189, 192, 193*
Fukuda, A. 674(100), *707*, 1495(164–169), 1498(164, 169), *1506*
Fukuda, K. 865(148), 866(147), 890(231), 896(147, 148), *905, 907*, 1180(63a), *1339*
Fukuda, T. 697(289, 290), *711*
Fukuhara, T. 648(154), *659*
Fukui, H. 414(87), *481*
Fukumoto, K. 1216(121, 122a), 1275(215), 1283, 1287, 1297(121, 122a), *1340, 1343*
Fukumoto, S. 624(56), *657*
Fukumoto, Y. 891(246), *907*
Fukushima, H. 692(248), *710*
Fukushima, M. 1086(376), *1096*, 1362(206), *1368*
Fukushima, N. 1383, 1421(67h), *1438*
Fukushima, T. 951, 953(192), *1007*, 1380(61b), *1438*
Fukuyama, T. 684(193), *709*
Fukuzumi, S. 1109, 1128, 1130(32), *1144*
Fulcher, R. G. 994(438), *1013*
Fulian, Q. 982(368), *1011*
Fülöp, F. 740(63), *832*
Fülscher, M. 20, 106(119), *182*
Fun, H. K. 377(263), *392*
Funabashi, K. 695(269), *710*
Funari, E. 1357(130), *1366*
Funayama, H. 1360(166), *1367*
Fung, A. P. 405(37b, 38), *479*
Fung, Y. S. 953, 965(265), *1009*
Funke, T. 948(177, 178), *1007*
Furimsky, E. 1111(63), *1145*
Furka, A. 472(310b, 310c), *488*
Furness, V. I. 1470(65), *1503*
Furniss, B. S. 626(61), *657*
Furstoss, R. 412(74a), *480*
Furukawa, M. 143, 147(580), *192*
Furukawa, N. 398(9a), *478*
Furukawa, Y. 1462(25), *1503*
Furuta, H. 1019(28, 29), *1087*
Furuta, T. 1489(108), *1504*
Fusco, C. 1257(186), *1342*
Fuster, F. 47, 48, 56(220), 71, 73(274), 75(275), 78(276), 89(220, 315, 316, 317b), 90(316), *184–186*
Fyfe, C. A. 1484, 1485(91), *1504*
Fygle, K. E. 851, 852(69), *903*
Fysh, R. R. 290(62), *327*

Gabbadon, P. W. 1351(49), *1364*
Gabe, E. 1120, 1140(184), *1147*
Gabe, E. J. 844, 847, 855, 860–862, 864, 897, 898(18), *902*
Gabhe, S. Y. 416(113), *482*
Gabowska, A. 1025(68), *1088*
Gacs-Baitz, E. 1383(75a–e), 1422(75e, 375), *1439, 1450*
Gadonas, R. 1020(33), *1087*
Gadosy, T. A. 38, 39(186), 110, 111, 113, 114(425), 127(186), *184, 189*, 715(6), *830*, 1019(27a), *1087*, 1134(344), *1151*
Gaeta, C. 1395(179), *1442*
Gagnon, J. 1406(261), *1446*
Gagnon, P. E. 421(130), *482*
Gago, F. 1406(266), *1446*
Gahrni, P. 460(280a), *487*
Gaier, H. 441(211), *484*
Gaikowski, M. 319(222), *330*
Gaillard, E. R. 1129(303), *1150*
Gal, J.-F. 322(244), 324(270), 325(244), *331*, 503(89), *526*
Galassi, S. 972(308), *1010*
Galaup, J. P. 384(287, 288), *393*
Galban, J. 934, 937(107), *1005*
Galceran, M. T. 944(148), 946(166), 947(172), 953(148, 166, 172), *1006*, 1352(68), *1364*
Gale, P. A. 1409(286b), 1427(411), 1430(442a, 442b), *1447, 1451, 1453*
Galeano Díaz, T. 983(376), *1011*
Galetti, G. C. 303(142), *329*
Galià, M. 945(153), *1006*
Galic, N. 385(298, 300), *393*, 726(34, 35), *831*
Galich, P. N. 620(46), *657*
Galimberti, L. 607(6), *656*
Galimov, R. D. 1424(396), *1451*
Galindo, F. 1079(244), *1093*
Gallagher, J. F. 1387(91), 1391, 1392(122b), 1393(122b, 153), 1394(122b), *1439–1441*
Gallego, M. 319(226, 227), *330*, 932, 937(92), 946(160), *1005, 1006*
Gallegos-Perez, J. L. 945(150), *1006*
Galletti, G. C. 937, 941(131), *1006*
Galli, G. 1000(467), *1013*
Gallo, B. 1353(72), *1364*
Gallot, S. 985(385, 386), *1012*
Galloy, J. 580(206), *601*
Galluci, J. C. 575(185), 580(185, 204), *600, 601*
Gamache, P. 953, 960(227), *1008*
Gamache, P. H. 953, 957(216), *1008*
Gambaretto, G. P. 646(146), *659*
Gamble, A. A. 299(91), *327*
Gamble, G. B. 444(215c), *485*
Games, D. E. 430(181b), *484*
Gammill, R. B. 441(205, 208), *484*
Gamoh, K. 319(231), *331*
Gan, C.-Y. 444(216g), *485*

Ganapathi, M. R. 110(420), *189*, 1017(12), *1087*, 1101, 1102(40), *1106*, 1134(342, 343), *1151*
Gandhidasan, R. 874(176, 177), *905, 906*
Gandolfi, M. T. 761(118), *833*
Ganeshpure, P. A. 1202, 1205(94), *1340*
Ganguly, A. K. 668(58), *706*
Ganguly, T. 1019(25), *1087*
Gantzel, P. K. 451(248b), *486*
Gao, C. 1084(310), *1094*
Gao, J. 95, 101(371), *188*
Gao, L. 913, 966(8), *1003*
Gao, S.-Y. 1236(154), *1341*
Gao, Y. 685(199), *709*
Garavito, R. M. 11(51, 52), *180*
García, A. 608(13), 618(36), *656*, 930(78), *1004*
Garcia, B. 769(147), *834*
García, C. 971(293), *1009*
Garcia, C. M. 1354(110), *1365*
Garcia, H. 1055(165), *1091*
García, I. 972(304), *1010*
Garcia, I. G. 1362(216), *1368*
Garcia, J. 1086(367), *1096*
Garcia, J. M. 782(216), *836*
Garcia, N. A. 1085(330), *1095*
García, P. 930(78), *1004*
García Alvarez-Coque, M. C. 988(409, 410), *1012*
Garcia-Blanco, S. 211(17), *220*
García-Carmona, F. 979(356), *1011*
Garcia Carrera, A. 1389(105), *1440*
Garcia-Fraga, J. M. 987(406), *1012*
Garcia-Martinez, E. 580(207), *601*
Garcia-Martinez, J. 631(75), *657*
García-Moreno, E. 953, 956(212), *1007*
Garcia-Raso, A. 137(523), *191*
Garcias, X. 750(103), *833*
García Sánchez, F. 952, 953, 990(197), *1007*
García-Sánchez, M. T. 948, 953(179), *1007*
García-Vallejo, M. C. 953, 965(267), *1009*
García-Viguera, C. 973(309, 310), *1010*
Gardner, P. J. 231, 234, 241(33), *255*
Gardziella, A. 1457(4), 1471(73), *1502, 1504*
Garozzo, D. 1403, 1427(230), *1445*
Garratt, P. J. 1223(136), *1341*
Garrido, A. 930(78), *1004*
Garver, L. C. 448(232b), *486*
Gaset, A. 685(197), *709*
Gasic, M. J. 813(297), *837*
Gasparrini, F. 1383(75d), *1439*
Gassman, P. G. 367(179), *390*
Gastilovich, E. A. 383(280), *392*
Gates, B. D. 1180, 1183, 1186(62), 1240(167), *1339, 1342*
Gates, J. W. 462(273b), *487*
Gatilov, Y. V. 616(32), *656*
Gatlin, C. L. 305(148), *329*

Gattrell, M. 1157(11), *1337*
Gaudemer, A. 1201, 1205(92), *1340*
Gaul, J. H. 1205(102), *1340*
Gauvrit, J. Y. 1372(11), *1436*
Gavard, J.-P. 472(314), *488*
Gavelanes, L. S. 234(43), *255*
Gavlick, W. K. 953, 970(287), *1009*
Gavuzzo, E. 1401(213), 1427(414a), *1444, 1451*
Gawdzik, B. 1353(81), *1365*
Gawdzik, J. 1353(81), *1365*
Gawlowski, J. 930(82), *1005*
Gazaryan, I. G. 977(341), 978(347), *1010, 1011*
Gazda, K. 947(169), *1006*
Gazy, A. A. K. 995(439), *1013*
Gbaria, K. 1411(295), *1447*
Gdaniec, M. 207(13), *220*
Ge, C.-S. 684(191, 192), *709*
Ge, P. 1171(46), *1338*
Gebauer, C. R. 724, 725(27), *831*
Gebauer, S. 1420(359c, 361), *1450*
Gebicki, J. 551(88), *598*
Gee, S. K. 450(239a, 239c), 452(239c), *486*
Geerlings, J. D. 1047(150), *1090*
Geerlings, P. 84(301), 89(301, 319, 324, 330–333, 336), 90(301, 319, 324, 330–333, 336, 341), 101–103(390), 170(756), *186–188, 197*
Geetha, B. 1063(192), 1064(193), *1091*
Geetha, N. 763(132), *834*
Gefen, S. 110(435), *189*
Geiger, H. 913(19), *1003*
Geiger, J. W. E. 1123, 1124(213), *1148*
Geiger, W. 551, 553(83), *598*
Geimer, R. L. 1471(71), *1504*
Geisel, H. 1169(41), *1338*
Geller, A. M. 933, 937(103), *1005*
Gelmi, M. L. 674(102), *707*
Gendell, J. 1124(217), *1148*
Genêt, J.-P. 664(28), *705*
Genge, J. W. 1430(442b), *1453*
Gennaro, M. C. 953, 962(247), *1008*
Geno, P. W. 925(47), *1004*
Genov, D. G. 351(87), *388*
Gentric, E. 575(187), *600*
Genuit, W. 303(143), *329*
George, B. E. 1048(156), *1090*
George, M. V. 745(79), *832*
George, P. 826(337), *838*
Georghiou, P. E. 1374(42c), 1387(95), 1392, 1393(135), 1411(42c), *1437, 1439, 1441*
Georgiev, E. M. 1393, 1407(143), *1441*
Geppert, W. D. 129, 144(478), *190*
Geraci, C. 1388(101), 1400(209a–c), 1401(212a–d, 213), 1402, 1404, 1406(222), *1439, 1444*
Geraldes, C. F. G. C. 360(138), *389*

Gerardy, R. 1216(124), *1341*
Gerbaux, P. 84(303), 110(448–450, 452), 115(303), 121(466), 123(450), *186, 189, 190*
Gerber, J. J. 550(72), *598*
Gerbino, T. C. 938(122), *1005*
Gerdes, R. 1082(277), *1094*
Gerecs, A. 773(163), *834*
Geresh, S. 1038(122a), *1090*
Gerez, C. 129(490), *190*
Gerhards, J. 67(254), 110, 113(439), *185, 189*
Gerhards, M. 20(114, 126), 23(114), 24(126), 35, 36(169), 106(114, 126), 107(114, 126, 169), 143(577), 144(577, 596, 678), 147(596, 696, 699, 700), 148(696, 699, 700), 149(699), 160(700), *182, 183, 192, 193, 195*, 375(247), *392*, 581(214, 217, 218), *601*
Gerhardt, Ch. 7(39), *180*
Gerhardt, W. 639(128), *659*
Gerkensmeier, T. 1382(71), *1438*
Gerlach, H. 441(211), *484*
Gerlt, J. A. 367(179), *390*
Gerothanassis, I. P. 998(454), *1013*
Gerrard, D. L. 144(592), *193*
Gerrard, W. 460(280b), *487*
Gerratt, J. 89(312), *186*
Gervay, J. 674(110), *707*
Gervay, J. E. 1297(256b), *1344*
Gescher, A. 913(25), *1003*
Geske, D. H. 1124(219), *1148*
Gesson, J. C. 651(173–175, 178–181), *660*
Gesson, J.-P. 444(216c, 216c, 216e, 216f, 217, 219), *485*
Getoff, N. 110, 129, 131(416), *189*, 1082(282), 1083(295), *1094*, 1098(8), 1100(8, 24), *1105*, 1115(99, 100), 1129(300), 1130(99), *1145, 1150*
Geurts, F. A. J. 1427(407a), *1451*
Gevert, T. 997(447), *1013*
Gfaf, S. 493, 515(27), *525*
Ghaemi, E. 632(87), *658*
Ghaffar, A. 649(157), *659*
Ghaffarzadeh, M. 774(188), *835*
Ghersetti, S. 558(151), *600*
Ghidini, E. 1393(141), *1441*
Ghiglione, C. 1403(240b), 1432(448b), *1445, 1453*
Ghigo, G. 18(83b), *181*
Ghio, C. 551, 553(103), *599*
Ghio, D. 138(526), *191*
Ghosh, A. K. 1025(71b), *1088*
Ghosh, P. 735(54, 55), *832*
Giacomello, P. 309(169–171), *329*
Giacosa, D. 947, 955(170), *1006*
Giandomenico, S. 953, 962(246), *1008*
Giannetto, M. 1396(181c), *1443*
Gianni, P. 101(384), *188*

Giannotti, C. 1073(216), 1085(336, 353), 1086(336), *1092, 1095, 1096*, 1330(323), *1346*
Gianturco, F. 1082(287), *1094*
Giavarini, F. 1000(467), *1013*
Gibb, B. C. 1424(388a, 388b, 389a, 390a, 390b), 1425(403a, 403b), *1450, 1451*
Gibb, C. L. D. 1424(388a, 389a, 390a, 390b), *1450*
Gibbs, C. G. 1401(219a, 220c), *1444*
Gibson, J. L. 472(311a), *488*
Gibson, J. M. 19(101), *182*
Gier, H. 20, 23, 106, 107(114), *182*
Giessen, D. J. 57(242), *185*
Gigante, B. 633(91), *658*
Giger, W. 1348(10), 1349(32), 1356(126, 129), *1363, 1366*
Giguere, R. J. 460(279), *487*
Gil, M. 129, 132, 133(498), *191*
Gil, M. I. 973(309, 313), 981(313), *1010*
Gilbert, A. 1059(180), *1091*, 1495(150, 151), *1505*
Gilbert, A. M. 454(253, 255), 456(255), *486, 487*, 735(52), *832*
Gilbert, B. C. 1209(109), *1340*
Gilbert, J. C. 862, 863(142), *905*
Gilbert, J. R. 299(91), *327*
Gilbert, L. 1123(211), *1148*
Gilchrist, G. 143(552), *192*
Gilchrist, M. L. 1108(15), *1144*
Giles, R. G. F. 766(144), *834*
Gilinsky-Sharon, P. 471(302), *488*, 806, 814(254), *836*
Gill, P. M. W. 4, 100(2), *179*, 368, 372, 373(200), *390*
Gilles, M. K. 95, 98, 99, 131, 132, 139(370), *188*
Gillespie, R. J. 31(143, 144), 68(143, 144, 258, 259), *183, 185*
Gillham, J. K. 1470(61), *1503*
Gilli, G. 338, 342(9), 363(146), 371(231), *386, 389, 391*, 552(135), *599*
Gilli, P. 338, 342(9), 363(146), 371(231), *386, 389, 391*
Gillier-Pandraud, H. 200(8), *220*, 549(67), *598*, 999(462), *1013*
Gilligan, W. H. 591(264), *602*
Gillis, R. G. 314(207), *330*
Gilski, M. 207(13), *220*
Gin, D. L. 994(437), *1013*
Giordano, C. 408(54e), *480*
Giorgi, M. 1394(175a), *1442*
Giovannardi, S. 383(281), *392*
Giovannini, E. 419(135b), *482*
Girardin, M. 944(139), *1006*
Girault, H. H. 544(52), *598*
Giray, M. 229, 230(23), *254*
Girling, R. B. 557(144), *599*

Girotti, A. W. 855(107), *904*
Gist, L. 219(38), *221*
Giunta, L. 1391(122b), 1392(122b, 127), 1393(122b, 157a), 1394(122b), *1440–1442*
Givens, R. S. 1044(131), *1090*
Gladun, N. R. 1417(328), *1448*
Gladyshev, M. I. 1362(218), *1368*
Glas, B. 1352(60), *1364*
Glaser, R. 631(81), *658*
Glass, T. E. 1382(72a, 72b), *1438*
Glasser, R. 11(56), *180*
Gleichmann, M. M. 455(265b), *487*
Gleixner, G. 303(145), *329*
Glendening, E. D. 4(5), *179*
Gless, R. D. 444(218), *485*
Glezer, V. 1354, 1355(99), 1362(210), *1365, 1368*
Glidewell, C. 580(208–210), *601*
Glish, G. L. 110(432), *189*, 310(181), *329*
Gloe, K. 1378(48b), *1437*
Gloede, J. 1396(190), 1399(205), 1401(214), *1443, 1444*
Glombik, H. 826(339), *838*
Glover, L. A. 1362(217), *1368*
Glowiak, T. 144(654, 659, 660, 665), *194*, 355(104), 377(264), *388, 392*, 551, 556(119), *599*
Glückmann, M. 323(264), *331*
Gluika, T. 405(47), *479*
Glusker, J. P. 826(337), *838*
Gnaim, J. M. 649(156), *659*
Gnanou, Y. 1388(100a), *1439*
Gnappi, G. 680(155), *708*
Gnep, S. 616(31), *656*
Gnepp, D. R. 1359(155), *1366*
Gnudi, F. 1349, 1356(36), *1363*
Gobaru, K. 1079(240), *1092*
Goddard, J. D. 129, 132(508), *191*
Goddard, R. 695(272), *710*
Godfrey, I. M. 424(154a), *483*
Godfrey, J. D. Jr. 665(41), *705*
Godinez, L. A. 761(117), *833*
Goetghebeur, M. 953(221, 222), 957(221), 958(222), *1008*
Goethals, G. 350(83), 364(154), *388, 389*
Goethals, M. 145(616), *193*, 559(158), 561(162), 568(176), 570, 571(158), 573(180), 575(158), *600*
Goetzen, T. 627(63), *657*
Goggin, P. 410(64a), *480*
Gogoll, A. 357(110), *388*
Goh, N. K. 975(323), *1010*
Gohn, M. 1115(99, 100), 1130(99), *1145*
Gohr, G. L. 1285(233), *1343*
Gokgunnec, L. 978(349), *1011*
Goldberg, D. M. 935, 937(112), 953, 960(230), *1005, 1008*
Goldberg, D. R. 693(254), *710*

Golden, D. M. 1108, 1118, 1119(20), *1144*
Golding, P. 1470(54), *1503*
Goldman, L. R. 1360(170), *1367*
Goldmann, H. 1374(40, 42a), *1436*
Goldstein, S. 1083(298), *1094*
Golfier, M. 1307(274, 275), *1344*
Gollob, L. 1484(90), *1504*
Golounin, A. V. 631(82), *658*
Golovchenko, L. S. 357(111), *388*
Goluben, N. S. 143(559, 560), *192*
Golubev, N. S. 143(557), 145(627), 146(641), 147(643–645), *192–194*, 352(89, 90), 364(155), 367(178), *388–390*, 503(90), *527*
Golumbic, C. 889(228), *907*
Golyanskaya, O. M. 776(196), *835*
Gomes, T. 913(7), *1003*
Gomez, C. 1158(22), *1338*, 1489, 1490(114), *1504*
Gomez, E. R. 1362(216), *1368*
Gómez-Fernández, J. C. 887(223, 224), *907*
Gomez-Lopez, M. 761(116), *833*
Gomis, D. B. 1354(113), *1365*
Gomperts, R. 4, 100(2), *179*, 368, 372, 373(200), *390*
Gong, Q. Q. 448(229d), *485*
Gong, Y. 608(14, 15), *656*, 683(172, 173), *708*
Gonohe, N. 143, 144(576), *192*
Gonsalves, A. M. D. 1082(276), *1093*
Gonter, K. 1100(30), *1106*, 1128(276), *1149*
Gonzales-Cosado, A. 1351(53), *1364*
Gonzales-Sierra, M. 726(33), *831*
Gonzáles-Tablas, F. 551(99), *599*
Gonzalez, C. 4, 100(2), *179*, 368, 372, 373(200), *390*
Gonzalez, H. 631(75), *657*
Gonzalez, J. 1360(168), *1367*
Gonzalez, J. M. 651(172), *660*
González, M. C. 913(9, 15), 953, 957(217), *1003, 1008*
Gonzalez, T. 1080(255, 256), *1093*
González-SanJosé, M. L. 961(235), *1008*
Gonzalez-Sierra, M. 359(129), *389*
Good, W. D. 237(70), *256*
Goodman, L. 110–112(426), *189*, 400(19), *478*
Goodyear, G. 471(301), *488*, 811(286), *837*
Goolsby, B. J. 322(247, 248), *331*, 1399(204), *1444*
Goon, D. J. W. 419(134b), *482*
Gopinathan, C. 639(131), *659*
Gopinathan, S. 639(131), *659*
Gopinathan, V. 854(95), *904*
Goralski, P. 559, 568, 586(154), *600*
Gorbunov, N. V. 887(225), *907*
Gorczynska, K. 992(426), 998(458), *1012, 1013*
Gordeishuk, S. S. 638(122), *658*

Gordes, E. 144(677), *195*
Gordon, D. M. 450(246a), *486*
Gordon, J. E. 145(624), *193*
Gordon, J. L. M. 1396(195b), *1443*
Gordon, M. 20, 25(110), *182*
Gordon, P. N. 436(195), *484*
Gordon, R. K. 403(27), *479*
Gordy, W. 113, 129, 135(458), *190*
Gorelik, M. V. 383(280), *392*
Goren, Z. 1401(216a), *1444*
Gorfinkel, M. I. 269(34), *326*
Gorie, T. M. 586, 589, 590(237), *602*
Gorler, K. 1065(194), *1091*
Görmar, G. 1408(279b), *1447*
Gorodetsky, L. M. 340(27), *387*
Gorrichon, L. 400(18b), *478*
Gorski, A. 129, 132, 133(498), *191*
Gorton, L. 975(322), 976(324, 329, 330), 977(333, 335, 337), *1010*
Gosciniak, D. J. 417(117), *482*
Gosteminskaya, T. V. 725(28), *831*
Goswami, D. Y. 1084(308, 309), *1094*
Goto, A. 147(692), *195*
Goto, H. 851, 868, 870(51), *903*
Goto, K. 1399(201a–c), 1428(435a, 435b, 436), *1443, 1444, 1452*
Goto, M. 1372(18), *1436*
Goto, R. 419(132), *482*
Goto, S. 1233(151a), *1341*
Goto, T. 953, 961(241), *1008*, 1024(58), *1088*
Gottlieb, R. 1158(15, 16), *1337*
Gottlieb, S. 950(189), *1007*
Gouali, M. 234(44, 45), *255*
Gouarderes, F. 1495(140), *1505*
Goudgaon, N. M. 415(99), *481*
Goudmand, P. 1124(238), *1148*
Gough, T. E. 1123(212), *1148*
Goulart, M. O. F. 1282(229), *1343*
Gourand, F. 647(153), *659*
Goursaud, S. 913(5), *1003*
Goux, C. 664(37), *705*
Gowenlock, B. G. 252(95), *257*
Goycoolea, C. 813(302), *837*
Gozlan, I. 397(7), *478*
Gozler, B. 1070(199), *1091*
Grabner, G. 1054(163), 1057(171, 174), 1082(282), 1083(295), *1091, 1094*
Grabowska, H. 608(8, 9, 12), *656*
Grabowski, A. 1025(83), *1089*
Graf, E. 1415(314b, 322), 1427(421c, 422a, 422b), *1448, 1452*
Graf, F. 1120(180), 1122(195), 1123(180), 1125(195, 253–255), *1147–1149*
Graff, R. 1427, 1432(421b), *1452*
Grafton, A. K. 35, 111, 112, 114(168), *183*
Graham, D. E. 93, 97, 132, 133(359), *187*
Gramstad, T. 219(33), *221*, 558(147), 579(199–201), *600, 601*

Gramstadt, T. 170, 171(746, 757), *196, 197*
Granata, A. 998(457), 999(464), *1013*
Grand, A. 84, 89–91(302), *186*
Grandelaudon, P. 784(227), *836*
Grandi, M. 1255(183), *1342*
Grandjean, A. 371(225), *391*
Grandmaison, J.-L. 444(216a), *485*
Granick, S. 1118(133–135), *1146*
Grannas, M. J. 1380(60b, 60c), *1438*
Granucci, G. 20, 23, 78, 93–96, 101, 106, 107, 109(115), *182*, 500(78), *526*
Granucci, R. G. 1382(72b), *1438*
Graovac, M. 931, 935, 937(85), *1005*
Graselli, J. G. 369, 371(208), *391*
Graß, B. 936, 937(116), *1005*
Grassbauer, M. 319(229, 232), *330, 331*, 953, 955(206), *1007*
Grasserbauer, M. 947, 953(171), *1006*
Grassmann, M. 801, 802(245), *836*
Grasso, D. 1362(204), *1368*
Graton, J. 581(213), 586, 587(255), *601, 602*
Gray, J. I. 854(98), *904*
Gray, K. A. 1085(351, 352), *1095*
Gray, M. 795(234), *836*
Greatorex, D. 1101(35), *1106*
Grebenshchikov, S. Yu. 371(223), *391*
Grech, E. 143(568), 144(568, 609), *192, 193*, 352(91), 359(123), 377(259), *388, 389, 392*, 551, 557(121), 593(276, 277), 594(276), 595(276, 277), *599, 602*
Grechishcheva, N. Yu. 339(19), *386*
Gredel, F. 211, 216(20), *221*
Greef, J. van der 299(101), *328*
Green, D. R. 1351(51), *1364*
Green, F. J. 924, 925, 927(38), *1003*
Green, I. R. 766(143, 144), 776(195), *834, 835*
Green, J. 691(240), *710*
Green, J. B. 934, 937(106), *1005*
Green, J. H. S. 35(152, 163), 36(152), 50(165), *183*
Green, J. O. 1424(388b), *1450*
Green, N. J. 816(311), *838*
Green, R. J. 129(477), *190*
Green, R. S. 174(764), *197*
Green, W. H. 877, 879, 881, 883(192), *906*
Greenberg, A. E. 1352(62), 1353, 1354(74), *1364*
Greenberg, R. S. 830(346), *838*
Greene, R. M. E. 412(71a), *480*
Greene, T. W. 662, 666(1), *704*
Greenstock, C. L. 1098(2), *1105*, 1112(79), 1117(119, 120), 1118(144, 145), *1145, 1146*
Greff, Z. 1408(282), *1447*
Gregoire, G. 109(414), *189*, 493, 500(31), *525*
Gregoli, L. 1395(179), *1442*
Gregolin, C. 890(234), *907*
Gregor, I. K. 315(213), *330*

Gregory, J. K. 148, 160(721), *196*
Gregson, R. M. 580(209, 210), *601*
Greiveldinger, G. 1318(304), *1345*
Grellier, P. L. 533(22), 535–537, 557(33), 588(33, 257), *597, 602*, 876–878(186), 879(196), 882(186, 196), *906*
Grellier, R. A. 503(89), *526*
Grellmann, K. H. 1025(79–81), *1089*
Grenier, p. 1486(99), *1504*
Grenier, P. 1495(130, 140), *1505*
Grenier-Loustalot, M.-F. 1485(101), 1486(99–101), 1495(130, 140), *1504, 1505*
Grenouillat, D. 663(16), *705*
Grenzer, E. M. 451(248b), *486*
Griebenow, N. 695(272), *710*
Grieco, P. A. 1250(178), 1252(179), *1342*
Griffen, K. 178(771), *197*
Griffini, O. 933(102), 934(108), 937(102, 108), *1005*
Griffith, S. S. 719(20), *831*
Griffiths, P. G. 444(216g, 216h), 448(228), *485*
Griffiths, P. R. 368(186), *390*
Griller, D. 845(30), 897(272), *902, 908*
Grimes, S. M. 1079(246), *1093*
Grinter, R. 1123(204), *1148*
Griswold, K. E. 1427(419b), *1452*
Griva, A. P. 1125(249), *1149*
Grodkowski, J. 1101(36), *1106*
Groen, A. 1117(115), 1120(115, 179), *1146, 1147*
Groenen, L. C. 1391(110), 1392(137), 1393(110, 156), *1440–1442*
Groenen, L. D. 1405(248), *1445*
Groenen, L. G. 1401(220a), *1444*
Gröger, H. 695(268), *710*
Groot, M. J.de 871, 873, 898(169), *905*
Groppi, F. 953, 955(203), *1007*
Gross, G. 1478(86), *1504*
Gross, K. C. 101–103(391), *188*
Gross, M. L. 110(429, 430), *189*, 264(21), 299(97), 300(21), 308(159), *326, 327, 329*
Grossman, N. 806(259), *837*
Grossweiner, L. I. 1127(269, 270), 1128(269, 274), *1149*
Grotemeyer, J. 324(273, 274), 325(279), *332*
Groth, C. 502(82), *526*
Grove, D. 93(364), *187*
Grubb, H. M. 268(32), *326*
Gruber, H. J. 1045(140), *1090*
Grujic, Z. 1219(131), *1341*
Grun, A. 1387, 1391(93), *1439*
Grundig, B. 976(327), *1010*
Grundwald-Wyspiańska, M. 143(565), *192*, 593, 595(282), *603*
Gruszecka, E. 419(134d), *482*
Grüttner, C. 1374(33c, 34, 39), 1386(87), 1406(257), *1436, 1439, 1446*

Grützmacher, H. F. 268(33), 270(35, 36), 277(47), 283(52), 285(56), *326, 327*
Gruzdev, I. V. 937(121), *1005*
Grynszpan, F. 1374(34, 39), 1386(87), 1410(291b), 1411(293, 294, 296a, 296b, 297, 300), *1436, 1439, 1447*
Grzechulska, J. 1083(305), *1094*
Gschwend, H. W. 414(94b), *481*
Gu, M. B. 981(361), *1011*
Gu, S. X. 639(124), *659*
Gu, T. 1406(272), *1446*
Gualano, E. 110(448), *189*
Guallar, V. 385(301), *393*
Guang-Ri Sun 1354, 1355(100), *1365*
Guardia, M. de la 987(406), *1012*
Guardigli, M. 1391(120b, 121c), 1396(121c), *1440*
Guardiola, F. 961(243), *1008*
Guardugli, M. 1435(469b), *1454*
Guastini, C. 1391(113), *1440*
Gubelmann, M. 631(77), *657*, 686(205), *709*
Gubernick, S. 416, 417(116), *482*
Guder, H.-J. 674(107), *707*
Gudrinize, E. 347(54), *387*
Guedes, R. C. 148(727), *196*
Guenthard, H.-H. 1120(180), 1122(195), 1123(180), 1125(195, 253–255), *1147–1149*
Guernon, J. M. 653(187), *660*
Guerzoni, F. N. 1459, 1463(9), *1502*
Gugliemetti, R. 770(157), *834*
Guha, D. 144(668), *194*, 1021(45), 1022(51), 1047(144, 145), *1088, 1090*
Guha, S. N. 1129(287), *1149*
Guibe, F. 460(285), *487*
Guiberteau Cabanillas, A. 983(376), *1011*
Guidarelli, A. 895(265), *908*
Guihéneuf, G. 559, 572(157), 575(187), *600*
Guilford, H. 429(179b), 430(179b, 181d), *483, 484*
Guilhaus, M. 315(213), *330*
Guillén, D. A. 947(175), 953(175, 232), 960(232), *1007, 1008*
Guinaudeau, H. 807(276), *837*, 1070(199), *1091*
Guindy, N. M. 1085(334), *1095*
Guisnet, M. 616(31), 630(71), 631(72, 77, 78), 639(130), *656, 657, 659*, 686(205), *709*, 774(176), *835*
Gujadhur, R. K. 670(77), *706*
Güler, M. L. 698(294), *711*
Gulick, W. M. 1124(219), *1148*
Gullotti, M. 699(307), *711*
Gumb, J. 419(136b), *482*
Gunduz, N. 727(39), *831*
Gunion, R. F. 95, 98, 99, 131, 132, 139(370), *188*
Gunji, A. 1403(233b), *1445*

Gunnarson, G. 348(57), 349(57, 58), *387*
Gunnewegh, E. A. 631, 640(74), *657*, 774(179), *835*
Gunthard, Hs. H. 35(162), *183*
Gunther, F. A. 987(403), *1012*
Günther, H. 824(336), *838*
Guo, Z. 472(307), *488*
Guo, Z.-W. 1224(137, 138), *1341*
Gupta, V. K. 988(411, 414), *1012*
Gupton, J. T. 1278(226), *1343*
Gurka, D. 531(20, 21), 586, 593(20), *597*, 853(87), *903*
Gurney, B. F. 1349(39), *1364*
Gusev, E. A. 1470(51), *1503*
Gusso, O. 701(318), *712*
Gustafsson, K. 1271(203), 1290(243), *1343, 1344*
Gustavsson, T. 493(64, 66), 502, 507, 511, 515, 522(66), *526*
Gut, I. 38(184), 127(184, 468), *184, 190*
Gut, I. G. 716(7), *831*, 1035(116), *1089*
Gutierrez, A. R. 403(32), *479*
Gutman, M. 493(53), *526*, 1020(35c), *1087*
Gutmann, R. 472(312), *488*, 686(202), *709*
Gutow, J. H. 37(180), *183*
Gutsche, C. D. 320(237), *331*, 1370(1–6), 1371(7–9), 1372(7, 8, 10a, 12, 16, 26, 27, 29), 1378(52a), 1379, 1380(54a), 1385(27), 1390(107), 1391(116), 1392(107, 128, 129a, 132–134), 1393(142, 147, 152a), 1394(142, 147, 165), 1399(200a, 200b), 1401(219a, 220c), 1403(231b, 234d), 1404(243), 1405(250), 1406(26, 134, 243, 258b), 1407(142, 243, 275), 1409(285, 290), 1417(4), 1427(275, 414b, 424a), 1430(6), *1435–1437, 1440–1447, 1451, 1452*
Gutteridge, J. M. C. 129(480), *190*, 840(1), 851, 855(66), 874, 892(1), *902, 903*
Guyan, C. 1073(217), *1092*
Guyan, P. M. 227, 248(15), 252(94), *254, 257*
Guyon, C. 1057(175–177), 1072(177), *1091*
Guyot, G. 1057(174), *1091*
Guyot, S. 913(10), 949(183), *1003, 1007*
Gyoergy, I. 1129, 1134(289), *1149*
Gyorgy, I. 1100(30), *1106*, 1128(276), 1129, 1130(293, 294), 1142(293), *1149, 1150*
Gyoten, M. 1246(172, 173), 1247(173), *1342*
Gyra, A. K. 542(48), *597*

Ha, H. 668(57), *706*
Ha, T. K. 89(313), 91(345), *186, 187*
Haas, M. 1396(189), *1443*
Haba, O. 1417(332), *1448*
Habata, Y. 1392(131), *1441*
Habich, A. 463(277), *487*
Häbich, D. 773(165), *834*

Habicher, W. D. 1419(355), *1450*
Hachiya, I. 472(309a, 309b), 473(309b), *488*, 629(65, 66), 630(67, 68), 639(65, 67), *657*, 686(203, 204), *709*, 774(168–172), *835*
Hadjoudis, E. 385(299), *393*
Hadži, D. 145(624), *193*, 385(295), *393*, 530(2, 11), *596, 597*
Haeberlein, M. 45(193), 105(403), 140(193), *184, 188*
Haga, K. 1417(332), *1448*
Hage, E. R. E.van der 293(71), *327*
Hage, R. 359(130), *389*
Hagen, D. F. 1352(66), *1364*
Hagen, M. 1399(203), *1444*
Hagenbruch, B. 662(5), *705*, 806(258), *837*
Hagenmaier, H. 925(51, 52), 936(51), *1004*
Haggert, B. E. 721(23), *831*
Haglund, E. A. 749(97), *833*
Hahn, L. C. 984(383), *1012*
Hahn, S. 559(183), *600*
Hahn, Y.-S. P. 807(272, 273), *837*
Haid, A. 371, 377(233), *391*
Haigh, D. 666(46), *705*
Haines, A. H. 405(41a), 408(55), *479, 480*
Haines, S. R. 129(478), 144(478, 685, 687), *190, 195*
Haino, T. 1374(36b), 1403(240a), 1406(36b, 270a, 270b), 1421(372), *1436, 1445, 1446, 1450*
Hajek, F. 1427(422a, 422b), *1452*
Haji, M. 874(171), *905*
Haj-Zaroubi, M. 1413(302, 303), *1447*
Halaburda, P. 995(441), *1013*
Halkyard, C. E. 730(47), *832*
Hall, A. 548(61), *598*
Hall, L. E. 174(765, 766), 177(766), *197*
Hall, R. W. 1462(20, 21), 1470(63), *1502, 1503*
Hall, W. L. 823(328), *838*
Halley, F. 1330(323), *1346*
Halliwell, B. 129(480), *190*, 636(104), 637(109), *658*, 840(1), 853(87), 857, 858(134), 874, 892(1), *902, 903, 905*
Halsam, E. 13(66), *181*
Halton, R. A. 1274(210a), *1343*
Ham, S. 807(274), *837*, 1230(149), *1341*
Hamachi, K. 701(321, 322), *712*
Hamada, A. 855(117), *904*
Hamada, M. 414(90g), *481*
Hamada, N. 811(294), *837*
Hamada, R. 1246, 1247(173), *1342*
Hamada, T. 699(301), 701(322), *711, 712*, 1084(322), *1095*
Hamaguchi, H. 1027(91, 92), *1089*
Hamamoto, K. 419(123a), *482*
Hamanaka, N. 211(19), *220*
Hamashita, Y. 691(238), 694(267), *710*

Hameka, H. F. 93–95, 106, 107, 109, 110(356), *187*
Hames, B. R. 676(118), *707*, 1205(104), *1340*
Hamilton, A. D. 1405(256a, 256b), *1446*
Hamilton, G. A. 405(42), *479*
Hamilton, G. M. 1216, 1283, 1287, 1297(122b), *1340*
Hamilton, J. V. 251(93), *257*
Hamilton, K. 1076(236), *1092*
Hamilton, T. M. 1408(281), *1447*
Hamlin, R. 826(338), *838*
Hammad, M. A. 776(205), *835*
Hammarström, L. 865(145), *905*
Hämmerle, M. 455(265a), *487*
Hammett, L. P. 101(386), *188*, 506(92), *527*
Hammond, W. B. 1192(80), *1339*
Hampton, P. D. 1380(55, 58f), *1437*
Hamuro, Y. 1405(256a), *1446*
Han, K. 1224(137), *1341*
Han, M. 620(45), *657*
Han, S.-K. 874(173), *905*
Han, X. 664(36), *705*
Han, Z.-X. 855(114, 115), *904*
Hanafusa, T. 93(362), *187*, 662(6), *705*
Hanai, T. 954(200), 970(291), *1007, 1009*
Hancock, K. S. B. 1414(308), *1448*
Handrick, G. R. 237(63), *256*
Hanessian, S. 1466(44), *1503*
Hanf, V. 925, 936(51), *1004*
Hanifin, J. W. 405(42), *479*
Hannaford, A. J. 626(61), *657*
Hannan, R. L. 424(154b), *483*
Hanninen, O. 1359, 1360(160), *1367*
Hansch, C. 537, 539, 559(38), *597*
Hänsel, R. 1308(280), *1344*
Hänsel, W. 361(145), *389*
Hansen, C. E. M. 1122(194), *1148*
Hansen, H.-J. 419(135a), 460(280a, 284), 463(277), 466(291), *482, 487*, 753(105), 755(106), 764(105), 823(326), *833, 838*
Hansen, P. E. 335(2), 336(4, 5, 39, 40), 338(16, 17), 339(4, 18, 20), 340(28), 342(4, 35, 36, 38), 343(4, 39, 40, 42), 344(4, 20, 39, 40, 43–45), 345(5, 39, 40, 47), 346(35, 36, 47, 51, 52), 347(35, 42, 55), 348(55, 56), 349(55, 60, 62), 350(39, 62, 77), 351(87), 353(56), 354(5, 40, 47, 51), 355(39, 40, 47, 51, 105), 356(42, 47), 357(4, 111), 358(113, 115–117), 359(122, 124), 360(134), 361(47, 115, 122, 142), 363(142), 365(52), 366(39, 40, 51, 166), *386–389*, 551(96–98, 128), *598, 599*
Hansen, T. K. 1495(148), *1505*
Hanson, J. R. 813(295, 296), *837*
Hanson, P. 699(311), *711*, 1489(117), *1505*
Hanus, V. 691(239), *710*
Hanzlik, R. P. 320(236), *331*
Hapiot, P. 1136(356, 357), *1151*

Haque, M. S. 417(118), *482*
Hara, H. 1318(300b), *1345*
Hara, M. 1158(23), *1338*
Hara, T. 855(117), *904*
Hara, Y. 871(163), 874(170), 898(163), 899(170), *905*, 913(26, 28, 29), *1003*, 1108(9), *1144*
Harada, K. 424(155b), *483*
Harada, T. 697(284), *711*, 1385(81), 1400(207a, 210a), 1401(210a), *1439, 1444*
Harada, Y. 1080(257), *1093*, 1188(74), *1339*
Haraguchi, Y. 774(192, 193), *835*
Harano, T. 1374(36b), 1406(36b, 270a), *1436, 1446*
Harasawa, K. 687(223), *709*
Harashina, H. 1079(241), *1092*
Harayama, T. 770, 773(158), *834*
Hardcastle, I. R. 783(224), *836*
Hardie, M. J. 233(40), *255*, 1372(23), 1427(418), *1436, 1452*
Hardinger, S. A. 470(300c, 300d), *488*, 816(310), *838*
Hardle, M. J. 1383, 1421(67o), *1438*
Hardy, R. A. 469(297a), *488*
Hargis, K. J. 951, 953(191), *1007*
Hargittai, I. 20, 22(109), 31, 68(144), *182, 183*, 238(71, 73), 244(86, 87), *256*, 551(76–78, 111–114, 125), 553(76–78), 555(111–114), *598, 599*
Hargittai, I. J. 82(284), *186*
Hargittai, M. 551(125), *599*
Harino, H. 972(303), *1010*
Haripoglu, A. 1083(294), *1094*
Harkema, S. 1383(67d), 1391(110, 119), 1392(125b), 1393(110, 151, 156), 1396(189), 1402(224b), 1403(125b, 234e, 240b, 242), 1405(224b, 253), 1421(67d), 1427(407a, 410b), *1438, 1440–1446, 1451*
Harkin, J. 1359(156), *1366*
Harman, D. J. 140(535), *191*
Harman, L. 1120(178), *1147*
Harman, W. D. 732(50), 735, 748(51), *832*
Harmann, L. S. 1120(176), *1147*
Harmer, M. A. 644(142), *659*, 774(180, 181), *835*
Harnish, D. 300(120), *328*
Harnly, M. E. 1360(170), *1367*
Harnois, J. 937, 940(130), *1006*
Haroutounian, S. A. 1070(200), *1091*
Harrelson, J. A. Jr. 5, 139(24), *180*
Harriman, A. 1139(369), *1151*
Harrington, J. R. 1360(167), *1367*
Harris, C. M. 428(167), *483*, 524(102), *527*
Harris, H. C. 400(18a), *478*
Harris, L. R. 1351(54), *1364*
Harris, R. K. 367(170, 172, 173), *390*, 1424(395), *1451*
Harris, T. K. 352(92, 93), 367(93), *388*

Harris, T. M. 428(167), 429(174), 430(180a, 180b, 181c), *483, 484*
Harrison, A. G. 274(39), 300(109, 127), *326, 328*
Harrison, D. J. 35(163), 50(165), *183*
Harrison, D. M. 824(334), *838*
Harrison, M. J. 1330(321), *1346*
Harrit, N. 1405(251a), 1406(264), *1446*
Harrity, J. P. A. 454(256, 257), *487*, 702(345), *712*
Harrod, J. F. 1274(211), *1343*
Harrowfield, J. 1370(1b), 1391(114), 1435(462a, 465b), *1435, 1440, 1454*
Harrowfield, J. M. 1388(97), 1401(219b), 1413(304, 305), *1439, 1444, 1447*
Harrowven, D. C. 472(309c, 315), *488*, 774(173), *835*
Harrup, M. K. 19(101), *182*
Hart, D. J. 811(291), *837*
Hart, H. 410(65a, 65b), *480*, 749(97), *833*
Hart, L. S. 472(311a, 311b), *488*, 639(129), *659*
Hart, W. J.van der 299(101), *328*
Harting, N. 283(52), *327*
Hartland, G. V. 35(163), 143, 144, 147(164), *183*
Hartman, T. G. 937, 940(129), *1006*
Hartman, W. W. 399(10), 400(20), *478*
Hartner, F. W. 632(89), *658*
Hartwig, J. F. 671(79), *706*
Haruta, A. M. 813(299), *837*
Haruta, J. 1227, 1228(144), *1341*
Harvey, D. F. 451(248b), *486*
Harvey, P. D. 1406(261), *1446*
Harvey, R. G. 425(162), 426(165), *483*
Harvilchuck, L. 770(151), *834*
Harwood, L. M. 472(313), *488*
Hasan, B. A. 987(406), *1012*
Hasanli, B. 1354(112), *1365*
Hase, A. 998(451), *1013*
Haseba, Y. 614(25), *656*
Hasegawa, K. 1110, 1112(52), *1144*, 1470(69), 1495(164–169), 1498(164, 169), *1504, 1506*
Hasegawa, K.-I. 1377(45a, 46a), *1437*
Hasegawa, M. 1380(61a), *1438*
Hasegawa, R. 412(74b), *480*
Hashem, T. M. 1085(334), *1095*
Hashida, Y. 402(26b), *479*
Hashimoto, N. 1379(54b), *1437*
Hashimoto, Y. 851, 868, 870(51), *903*
Hashmi, A. S. K. 450(236, 237), *486*
Haslinger, E. 144(647), *194*
Hassall, C. H. 424(150), *483*
Hassan, A. K. 1419(347b), *1450*
Hasse, W. 1378(48b), *1437*
Hasselbach, M. 1257, 1260(187b), *1342*
Hassell, C. H. 799(236), *836*

Hässner, R. 252(96), *257*
Hastings, T. G. 874(175), *905*
Hatakenaka, S. 1025(86), *1089*
Hatanaka, K. 1178(58), *1339*
Hatano, H. 954(200), *1007*
Hatano, T. 744(73), *832*
Hatayama, A. 699(300), *711*
Hatcher, P. G. 1360(180), *1367*
Hatfield, G. R. 1470, 1471(58), *1503*
Hatori, H. 1495(132, 133), *1505*
Hatrik, S. 953, 962(255), *1009*
Hatsuda, M. 697(284), *711*
Hatsui, T. 1411(299), *1447*
Hattori, T. 1416(324), *1448*
Hau, S.-Y. 673(90), *706*
Haubrich, J. E. 492(8, 9), 493(8, 9, 47), 498(9, 47), 511(47), 522(8, 9), *525, 526*
Hauck, H. E. 953, 962(250), *1008*
Hauff, S. 1124(242), *1148*, 1290, 1292(246), *1344*
Haulait-Pirson, M. C. 571(179), *600*
Hauptman, E. 670(75), *706*
Hauser, C. R. 425(161), 472(311e), *483, 488*
Hautefeuille, A. 1361(188, 189), *1367*
Hauteville, M. 778(211), *836*
Haverbeke, Y. V. 299(98), *328*
Haw, J. F. 171(762), *197*
Hawkins, G. D. 57(242), *185*
Hawkins, R. T. 1372(17), *1436*
Hawranek, J. P. 377(256), *392*
Hawthorne, M. 424(152a), *483*
Hawthorne, M. F. 415(104), 424(156), *481, 483*
Hay, A. S. 1491(118), *1505*
Hayaishi, O. 840(4), *902*
Hayakawa, T. 1196(89), 1213(114, 117), *1340*
Hayama, E. 265, 272(29b), *326*, 614(25), *656*
Hayashi, A. 680(156), *708*
Hayashi, K. 677(128), *707*, 953, 970(292), *1009*, 1124(231), *1148*
Hayashi, M. 702(334, 338), *712*
Hayashi, N. 1359(146), *1366*
Hayashi, T. 669(66, 67), *706*, 762(125, 126), *833, 834*
Hayashi, Y. 360(137), *389*, 1421(373), *1450*
Hayashida, O. 1418(335), 1425(405), *1448, 1451*
Hayes, R. N. 314(209), 315(210), 316(209), *330*, 467(296), *488*
Hayes, W. 761(116, 117), *833*
Hayman, G. 1105(65), *1106*
Hayon, E. 110(417), *189*, 1098(6), *1105*, 1128(278), 1129(296, 297), 1130(296, 297, 321), 1131, 1134(278), 1135(296, 345, 351), *1149–1151*
Hayvali, Z. 727(39), *831*
Hazra, D. K. 1115(105), *1146*
He, F. 1085(354), *1096*

He, P. 668(56), *706*
He, X. 1045(142), *1090*
He, Y. K. 1100(31), *1106*
He, Y. Z. 1105(62), *1106*
He, Z. 93–96(357), *187*
Head, A. J. 237(70), *256*
Head-Gordon, M. 4, 100(2), *179*, 368, 372, 373(200), *390*
Headley, P. M. 475(318), *488*
Heaney, H. 684(187–189), *709*, 773(166), *834*, 1420(359b, 362, 367), *1450*
Heath, E. A. 232(36), *255*
Heath, J. A. 1409(286b), *1447*
Heath, P. 1059(180), *1091*
Heberer, T. 319(224), *330*, 934, 937(110), *1005*, 1353(95), *1365*
Hedgecock, H. C. 415(103a), *481*
Hedman, B. 1196(88), *1340*
Heer, M. I. de 45(200), *184*, 867(153, 155), 876, 882, 892(155), 893(153), *905*
Heerma, W. 290(60), *327*
Heesink, G. J. T. 1391(112), *1440*
Hegarty, A. F. 89(313, 327), 90(327), 91(346), 179(776), *186, 187, 197*, 651(182), *660*
Hegedus, L. 451, 454(249), *486*
Hehre, W. J. 5(22), 53(237), 83, 84(297), 101(388), *180, 185, 186, 188*, 261(11, 13), 263(13, 16), 296(16), 311(189), *326, 330*
Heida, J. F. 1406(258a), *1446*
Heidekum, A. 644(142), *659*, 774(180, 181), *835*
Heikal, A. 493(21, 23), *525*
Heikkala, R. E. 1119(161, 162), *1147*
Heilser, G. 1081(269), *1093*
Heimlich, F. 936, 937(116), *1005*
Heinis, L. J. 1357(133), *1366*
Heinonen, P. 1357(134), *1366*
Heinrich, A. 1025(80), *1089*
Heinrich, F. 774(178), *835*
Heinrich, N. 300(108), *328*
Heinz, T. 1424(401), *1451*
Heitner, C. 8(42), *180*
Heitzsch, O. 1378(48b), *1437*
Helaleh, M. I. H. 1360(171), *1367*
Helbert, M. 531, 532, 547(18), 576(189), 586, 587(18, 243–245), 589(18, 244), 591, 592(267), *597, 600, 602*
Helgason, A. L. 450(241d), *486*
Helgeson, R. C. 1382, 1383(67c), 1419(348), 1421(67c), 1429(429), *1438, 1450, 1452*
Heller, D. P. 693(254), *710*
Heller, W. 895(269), *908*
Hellrung, B. 127(468), *190*, 229, 250(20), *254*, 716(7), *831*, 1017(14), 1035(116), *1087, 1089*
Helm, R. M. 147(713, 715), *195*, 992(427), *1012*
Helma, W. P. 1098(2), *1105*, 1112(79), *1145*

Helmreich, B. 303(144), *329*
Hemert, M. C.van 1047(150), *1090*
Hemingway, R. W. 744(73), *832*
Hemming, J. 1360(178, 185), *1367*
Henderson, D. E. 851–853(72), *903*
Henderson, G. N. 806(268), *837*
Henderson, S. A. 434(194), *484*
Henderson, S. K. 851–853(72), *903*
Hendricks, S. B. 47(214, 215), *184*
Hendry, D. G. 879(197), *906*
Hennequin, D. 929, 954(62), *1004*
Hennig, D. 1033(107), *1089*
Hennig, L. 1417(330b), *1448*
Henri, G. 143(591), *193*
Henrichs, U. 144(678), *195*
Henriksen, K. 1360(174), *1367*
Henriksen, L. 1333(327), *1346*
Henri-Rousseau, O. 371(225, 226), *391*
Henschke, J. P. 1381(65b), *1438*
Hensel, J. 979(352), *1011*
Henson, B. F. 35(163), 143, 144, 147(164), *183*
Henson, R. M. C. 498(75), *526*
Henze, G. 953, 969(282), *1009*
Hepburn, C. 983(379), *1011*
Hepler, L. G. 558(148), *600*
Herbert, C. G. 290(62), *327*
Herbert, J. A. 204(9), *220*
Herbstein, F. H. 216(29), *221*
Hercules, D. 515(95), *527*
Hercules, D. M. 323(261), *331*, 476(323), *488*
Herczegf, P. 1227(141), *1341*
Hergenrother, P. M. 1494(129), 1495(131), *1505*
Hering, P. 581(217), *601*
Herington, E. F. G. 95, 107, 109(373), *188*
Herlt, T. 999(466), *1013*
Hermann, R. 110(420, 422–424), 114(424), 121(423, 424), *189*, 1017(12), *1087*, 1101(38–40), 1102(40, 45), *1106*, 1134(341–343), *1151*
Hermes-Lima, M. 854(97), *904*
Hernández, C. A. 853(91), *904*
Hernandez, L. M. 1360(168), *1367*
Hernández-Córdoba, M. 953, 960(234), *1008*
Hernandez-Ortega, S. 551(93), *598*
Herrera, J. C. 853(91), *904*
Herrera-Melian, J. A. 1084(324), *1095*
Herrick, R. F. 1359(155), *1366*
Herrington, J. W. 1104(59), *1106*
Herrmann, J.-M. 1080(247), *1093*
Herrmann, W. A. 1213(113), *1340*
Hertog, H. J.den Jr. 1396(189), *1443*
Hertog, M. G. L. 913(23), *1003*
Herve, S. 1357(134), *1366*
Herzig, J. 718(16), *831*
Herzschuh, R. 319(233, 234), *331*

Hesek, D. 1393, 1403(152c), 1427(411), *1441, 1451*
Hess, B. A. 455(265b), *487*
Hesse, D. G. 225, 227(4), *254*
Hessell, E. T. 1495(144, 147), *1505*
Hesselsoe, M. 1360(174), *1367*
Hewawasam, P. 608(16), *656*, 683(171), *708*
Hewel, J. 578, 581(193), *601*
Hewgill, F. R. 1124(226), *1148*, 1330(322), *1346*
Hewitt, M. L. 1351(48), *1364*
Heydenreich, M. 1378(50a), *1437*
Heydkamp, W. 802(249), *836*
Heyer, U. 1424(392), *1450*
Hguang, C. P. 1086(358), *1096*
Hibbert, F. 354(103), 370, 371, 382(210), *388, 391*
Hibbs, D. E. 1402(228b), 1409(288), *1445, 1447*
Hibdon, S. H. 367(177), *390*
Hickinbottom, W. J. 759(110), *833*
Hidaka, H. 1084(318), 1086(369), *1095, 1096*, 1361(195, 196), *1367*
Hidalgo, M. C. 1080(248), *1093*
Hidalgo Hidalgo de Cisneros, J. L. 982(363), *1011*
Hidebrandt, P. 384(289), *393*
Hienerwadel, R. 369(205–207), *391*
Higashita, N. 1034(112), *1089*
Higginbottom, H. P. 1462(22), *1502*
Higgins, J. 93, 94, 132(360), *187*
Higgins, R. J. 638(114, 115), *658*
Higgs, P. I. 416(114), *482*
Highet, R. 719(18), *831*
Higler, I. 1424(383), 1432(454, 455), *1450, 1453*
Higuchi, K. 447(222b, 222h), 448(222b, 225), *485*
Higuchi, M. 1483(89), *1504*
Higuchi, T. 1216(125), *1341*
Hikino, H. 1080(259d), *1093*
Hilal, A. 913(4), *1003*
Hilbert, G. E. 47(215), *184*
Hildebrand, J. P. 699(302), *711*
Hildebrand, P. 129, 132, 133, 135(514), *191*
Hildebrandt, S. 1355(118), *1365*
Hildenbrand, K. 1085(350), *1095*, 1129, 1134(289), *1149*
Hileman, B. 1348(25), *1363*
Hill, J. O. 1002(479), *1014*
Hill, J. R. 342(37), *387*
Hill, S. M. 925(50), *1004*
Hillenkamp, F. 260(3), 323(3, 253–255), 324(268), *325, 331*, 1047(146), *1090*
Hiller, G. 925, 936(51), *1004*
Hiller, W. 550(71), *598*
Hills, D. J. 1104(55), *1106*
Hilmersson, G. 1424(400a), *1451*

Hilmi, A. 971(298), *1009*
Hilton, M. J. 677(127), *707*, 765(140), *834*
Hiltunen, R. 953, 968(273), 973(312), *1009, 1010*
Himi, M. 1415(315, 316), *1448*
Himo, F. 45(196), 179(775), *184, 197*
Hinchliffe, A. 1127(265), *1149*
Hine, J. 101(383), *188*, 559(183, 184), 575(185), 580(185, 204), *600, 601*
Hine, N. 953, 966(268), *1009*
Hiner, S. 441(209), *484*
Hipple, W. G. 732(50), *832*
Hirai, C. 1403, 1406(232b), *1445*
Hirakata, M. 1406(270b), *1446*
Hirakawa, H. 1414(309), *1448*
Hiraki, K. 414(86b), *481*
Hirama, M. 680(156–158), *708*
Hiramatsu, M. 855, 856(105), *904*
Hirano, J. L. 1080(257), *1093*
Hirano, M. 1302(266), *1344*
Hirano, T. 1188(74), *1339*
Hirao, I. 686(211), *709*, 1028(97), *1089*
Hirao, K.-I. 1374(38a, 38b), *1436*
Hirata, J.-I. 413(77), *480*
Hirata, T. 403(31), *479*
Hiratani, K. 20, 44(120), *182*, 551, 556(117), *599*, 766(142), *834*
Hiratsuka, A. 827(340), *838*
Hiratsuka, H. 1029(102), *1089*
Hirokawa, A. 1065(195), *1091*
Hirokawa, S. 107, 143, 147(412b), *189*
Hirosawa, Y. 950, 953(187), *1007*
Hirota, K. 1019(27b, 30), *1087*
Hirota, N. 1021(36), *1088*
Hirota, S. 1209(106), *1340*
Hisaichi, K. 1427(419a), *1452*
Hisatome, K. 636(105), *658*
Hishiyama, Y. 1495(134), *1505*
Hiskia, A. 1086(361), *1096*
Hites, R. A. 310(179), *329*
Ho, D. M. 726(36), *831*
Ho, P. 953, 960(233), *1008*
Ho, P.-S. 664(33), *705*
Ho, T.-L. 1333(328), *1346*
Ho, Z.-C. 1391–1393(122a), 1394(168, 170a), 1407(170a), *1440, 1442*
Hoberg, J. O. 1265(198), *1343*
Hobza, P. 144(691), *195*
Hock, H. 19(93), *182*, 799(235), *836*
Hocking, M. B. 419, 420(137), *482*
Hodacova, J. 1415(317), *1448*
Hodd, K. 1492(120), *1505*
Hodgdon, R. B. 406(50), *479*
Hodgson, D. 17(81), *181*
Hodgson, K. O. 1196(88), *1340*
Hoefnagel, A. J. 631, 640(74), 645(144), *657, 659*, 774(179), *835*
Hoek, J. B. 888(226), *907*

Hoek, W. J.van den 1126(262), *1149*
Hoelderich, W. F. 644(142), *659*, 774(180, 181), *835*
Hoffman, M. A. 451(247), *486*
Hoffman, M. Z. 1082(273–275), *1093*, 1140(375), *1151*
Hoffman, P. 455(265a), *487*
Hoffmann, K.-L. 433(187b), *484*
Hoffmann, M. R. 1361(199, 200), *1368*
Hoffmann, N. 1062(188–190), 1063(191), *1091*
Hofmann, W. 1495(149), *1505*
Hofslokken, N. U. 626(60), *657*
Hofstraat, R. H. J. W. 1427(407a), *1451*
Hoganson, C. W. 1108(16, 17), *1144*
Högberg, A. G. S. 1417(325), *1448*
Hogberg, K. 320(236), *331*
Högberg, T. 424(154c), *483*
Hogenheide, M. P. 93(364), *187*
Hoger, S. 211(18), *220*
Hogg, T. A. 953, 960(233), *1008*
Hogue, C. 12(59), *181*
Hoigne, J. 1084(316), 1085(332), *1095*
Hoke, H. 110, 112, 139(442), *189*
Hökelek, T. 551, 557(122), *599*, 729(43), *831*
Hol, W. G. J. 412(73), *480*
Holbrook, N. J. 141(536), *191*
Holcman, J. 1115(93), 1117(130, 132), *1145, 1146*
Holder, S. J. 1000(469), *1013*
Hollas, J. M. 35(157), *183*
Hollinger, A. B. 67(247–249), *185*
Hollman, P. C. H. 913(23), 953, 961(242), *1003, 1008*
Holly, S. 179(782), *198*, 369(209), *391*
Holman, K. T. 1414(308), *1448*
Holmberg, H. 1267(202a), *1343*
Holmberg, K. 1267(201), *1343*
Holmbom, B. 1360(178, 182, 185), *1367*
Holmes, J. L. 98(376), 139(529), *188, 191*, 261(7), 265, 268, 269, 272(29a), 275, 277(42, 43), 300(120), 302(139), 311, 312(7), *325, 326, 328*
Holmstroem, B. 1130(327), *1150*
Holopainen, T. 997(445, 446), *1013*
Holoubek, J. 933, 937, 953(99), *1005*
Holthausen, M. C. 368(192), *390*
Holtom, G. R. 493(24), *525*
Holton, D. M. 1109, 1120, 1123, 1133(42), *1144*
Holton, R. A. 1275(216), 1278(219), *1343*
Holzscheiter, F. 405(48a), *479*
Homan, M. 928(55), *1004*
Homma, M. 968(274), *1009*
Honda, H. 1489(108, 110), 1491(110), *1504*
Honda, K. 1482(88), *1504*
Honda, T. 447(221a, 221b), *485*
Hong, J. 702(328–331), *712*, 1086(373), *1096*

Hong, R.-K. 664(33), *705*
Hong, S.-S. 1082(289), *1094*
Hong, Y. 685(199), *709*
Hongu, M. 673(98), *707*
Hongyou, R. L. 932, 933(95), *1005*
Honig, B. 5(12), *179*
Honing, M. 296(86), *327*
Hontz, B. 1355(115), *1365*
Hooff, J. H. C.van 941(133), *1006*
Hoom, W. P.van 1433(456a), *1453*
Hoorn, W. P.van 1385(79b, 84), 1386(84), *1439*
Hopf, H. 782(221), *836*
Hopkins, T. L. 305(148), *329*
Hoppin, J. A. 1359(155), *1366*
Horack, J. 1353(71), *1364*
Horak, M. 170(752, 753), *196*
Horák, V. 251, 252(92), *257*, 385(297), *393*
Horgan, S. W. 1260(190), *1342*
Hori, K. 677(136), *707*
Hori, T. 1380(61a), *1438*
Horiguchi, A. 674(100), *707*
Horiuchi, H. 1029(102), *1089*
Hormadaly, J. 542(47), *597*
Hormi, O. 1360(182, 183), *1367*
Hormi, O. E. O. 668(55), *706*
Horne, S. 476(320), *488*, 785(228), *836*
Horner, L. 801(247), *836*
Horning, E. C. 294(79), 312(194), *327, 330*
Hornstra, G. 852(78), *903*
Horsley, D. B. 438(199), *484*
Horswill, E. C. 845(32, 33), 859, 872(138), *902, 905*
Horth, H. 1358(144), *1366*
Horwood, S. 399(13a), *478*
Hosaka, K. 1315(293, 294), *1345*
Hoshikawa, M. 769(149), *834*
Hoshimoto, E. 1021(36), *1088*
Hoshino, K. 1357(137), *1366*
Hoshino, M. 1021(39), *1088*
Hoshino, O. 1065(195), 1066(196), *1091*, 1318(300b), 1330(316, 317), *1345, 1346*
Hoshino, Y. 693(256), *710*
Hoskins, B. F. 1380(60a–c), *1438*
Hosokawa, K. 414(90i), *481*
Hosokawa, T. 669(64, 65), *706*
Hosose, H. 865, 866, 896(146), *905*
Hosoya, N. 699(299–301, 312), 702(326), *711, 712*
Hosoya, T. 680(145), *708*, 776(201), *835*
Hosseini, M. W. 1401(219b, 220b), 1415(310, 311, 314b, 322), 1427(421a–d, 422a, 422b), 1432(421b), *1444, 1448, 1452*
Hossenlopp, I. A. 226, 227(7), *254*
Hostettmann, K. 953, 967(270), *1009*
Hotta, Y. 420(142b), 424(155a), *483*
Houbrechts, S. 1391(119), *1440*
Houglum, K. 1462(36), *1503*

Houjou, H. 20, 44(120), *182*, 551, 556(117), *599*
Houk, K. N. 1429(430a), *1452*
Houle, C. G. 1497, 1498(170), *1506*
House, H. O. 676(117), *707*
Hovari, J. 856, 857(122), *904*
Hoveyda, A. H. 702(345, 346), *712*
Howard, J. A. 142(542), 143(543–545), *192*, 841, 843(11), 844(17, 19), 859(137), 860(19, 139, 140), 872(139), 877(190), 879(198), 884(206), *902, 905, 906*, 1108(5), 1111(62–64), 1140(5, 372, 374), *1144, 1145, 1151*
Howard, J. A. K. 551, 552(134), *599*, 1036(118), *1089*
Howard, J. B. 18(83a), *181*
Howard, P. C. 999(465), *1013*
Howarth, O. W. 830(347), *838*
Howe-Grant, M. 419(125), *482*
Hoyer, H. 234(48), *255*
Hoz, A.de la 680(154), *708*
Hrabal, R. 1415(316), *1448*
Hrdlovič, P. 476(321), *488*
Hristova, M. 636(104), *658*
Hroch, L. 1459(14), *1502*
Hrovat, D. A. 110, 112, 139(442), *189*
Hsiao, G. 851, 853, 857(61), *903*
Hsieh, C.-T. 944(145), *1006*
Hsieh, M.-F. 1236(157), *1342*
Hsieh, Y.-H. 1085(338, 341, 342), *1095*
Hsiue, G. H. 608(7), *656*
Hsu, D.-S. 1237(158, 160a), *1342*
Hsu, P.-Y. 1237(160a, 160b), *1342*
Hsung, R. P. 454(253, 255), 456(255), *486, 487*, 735(52), *832*
Hu, C. 1082(280), 1084(306), *1094*
Hu, H. 786(229, 230), *836*
Hu, L. Q. 312(193), *330*
Hu, Q.-S. 694(264, 266), *710*
Hu, W. 1417, 1424(327), *1448*
Hu, X. 1081(268), *1093*, 1427(408d), *1451*
Hu, Y. 662(7), *705*
Huang, C. C. 82(285), *186*
Huang, C. G. 1036(119), *1089*
Huang, C. H. 620(41), *657*
Huang, C. P. 1085(335), 1086(364), *1095, 1096*
Huang, H.-N. 687(221), *709*
Huang, H. S. 1361(197), *1367*
Huang, K.-S. 1085(341, 342), *1095*
Huang, R. 305(151), *329*
Huang, S.-D. 953, 970(290), *1009*
Huang, T. 684(190), 695(271), *709, 710*
Huang, W.-S. 694(264–266), *710*
Huang, X. 668(56), *706*
Huang, X. X. 1100(27), *1106*
Huang, Y. 323(258), *331*
Huang, Z. 1400(206), *1444*

Author Index 1551

Huang, Z.-T. 634(99), *658*, 1403(236), 1409(286a), 1410(292), 1432(448a, 452), *1445, 1447, 1453*
Huber, E. W. 663(24), *705*
Huber, J. F. K. 953, 962(258), *1009*
Huber, W. 1359(150), *1366*
Huboux, A. H. 450(241b), *486*
Huc, V. 1409(289b), *1447*
Hückel, W. 737(57), *832*
Hudhomme, P. 586, 587(241), *602*
Hudson, E. A. 913(25), *1003*
Hudson, R. A. 47, 143, 170(209), *184*, 592(268), *602*
Huf, F. A. 953, 968(277, 278), *1009*
Huff, J. 1359(146, 154), *1366*
Hug, R. 755(106), *833*
Huggins, C. M. 547(60), *598*
Hughes, D. E. 412(75a), *480*
Hughes, D. L. 663(21), *705*
Hughes, E. D. 801(239), *836*
Hughes, L. 844, 847, 855, 860–862(18), 864(18, 143, 144), 865(144), 887(217), 897, 898(18), *902, 905, 906*, 1120, 1140(184), *1147*
Hughes, P. F. 786(229, 230), *836*
Hughes, R. 1318(303), *1345*
Hugo, V. I. 766(143, 144), 776(195), *834, 835*
Huh, J. 1085(346), *1095*
Huie, R. E. 845(25, 26), 880(25, 26, 201), 881(25), *902, 906*, 1111(69–72), 1112(80), 1137(70), 1142(379), *1145, 1151*
Huisgen, R. 467(295b), *488*, 802(249), *836*
Hulme, B. E. 1131, 1134(316), *1150*
Hulme, J. 983(379), *1011*
Hulst, N. F.van 1391(119), *1440*
Hulst, R. 1406(262b), 1422(378), *1446, 1450*
Hultzsch, K. 1470(64), 1484(92), *1503, 1504*
Hummel, G. J.van 1391(119), 1403(234e, 240b, 242), 1424(382a), 1427(410b), *1440, 1445, 1450, 1451*
Hummel, J.van 1427(407a), *1451*
Hunad, L. 359(131), *389*
Hünig, S. 806(258), *837*, 987(402), *1012*
Hunt, A. L. 412(75a), *480*
Hunt, D. F. 291(66), *327*
Hunt, J. D. 416(111), *482*
Hunter, E. P. 5(10), 83(296a), *179, 186*
Hunter, E. P. L. 261, 265, 310(8), *326*
Hunter, J. A. 1076(236), *1092*
Hunter, R. 1420(367), *1450*
Huong, P. V. 35(161), 37, 170(179), *183*
Huppert, D. 493(44, 48–50, 53, 56–62), 497(58–60), 498(59), 504, 505(49), *525, 526*, 1020(35c), *1087*
Hurber, J. F. K. 1352(67), *1364*
Hurh, B. 412(75c), *480*
Hurley, M. D. 138(528), *191*, 1104(61), *1106*
Hurst, D. R. 730(47), *832*

Hursthouse, M. B. 1402(228b), 1409(288), *1445, 1447*
Husain, S. 806(269), *837*
Husebye, S. 219(33), *221*, 579(199–201), *601*
Hush, N. S. 171(762), *197*
Hussain, A. 1242(168), *1342*
Hussain, S. A. 1259(189), *1342*
Husson, H.-P. 1292(252), *1344*
Huter, O. 450(246f), *486*
Hutinec, A. 1172(51a), *1338*
Hütter, P. 1169(43), *1338*
Huttl, R. 977(336), *1010*
Hutzinger, O. 1074(231), *1092*
Huuskonen, S. 1359, 1360(160), *1367*
Huyskens, P. 47(210), *184*, 595, 596(296), *603*
Huyskens, P. L. 377(260), *392*, 530(8), 534(25), 593(272), 594(283), 595, 596(272), *597, 602, 603*
Huysmans, W. B. G. 1194(83a), *1339*
Huysmans, W. G. B. 1109(33, 34), 1120, 1124(34), *1144*
Huystee, R. B.van 977(342), *1010*
Hwang, D. 1394(167b), *1442*
Hwang, G. T. 1427(410c), *1451*
Hwang, H. I. 937, 940(129), *1006*
Hwang, J.-T. 1237(159), *1342*
Hy, Z. Y. 233(40), *255*
Hyatt, J. A. 206(11), *220*
Hyde, J. S. 1120(188), *1147*
Hynes, J. T. 20, 23, 78, 93–96, 101, 106, 107, 109(115), *182*, 493(64–66), 500(78), 502, 507(66), 511(65, 66), 515(66), 522(66, 100, 101), *526, 527*
Hyötyläinen, T. 970(289), *1009*

Iacomacci, P. 1383(75a), *1439*
Iacovino, R. 1383, 1422(75e), *1439*
Iannucci, B. 375(250), *392*
Ibbotson, A. 1383, 1421(67b), *1438*
Ibragimov, B. T. 204(10), *220*
Ibrahim, A. R. 377(263), *392*
Ibsen, S. N. 346, 354, 355, 366(51), *387*, 551(96), *598*
Ichikawa, H. 92(347), *187*
Ichimura, K. 1383(67h), 1421(67h, 373), *1438, 1450*
Iden, C. B. 1357(131), *1366*
Ido, Y. 323(256), *331*
Idoux, J. P. 1278(226), *1343*
Igaz, S. 937(120), *1005*
Ignat'ev, N. V. 312(193), *330*
Ignatov, O. V. 979(353), *1011*
Ignatova, Z. N. 953, 957(214), *1007*
Ignelzi, S. 942(137), *1006*
Iguchi, H. 855(104), *904*

Iguchi, M. 1027(91, 92), *1089*, 1158(19–21, 23, 24), 1159(24, 25a, 25b, 26, 30), 1172(25a, 25b), 1178(21), 1227(30), *1338*
Ihara, M. 665(40), *705*, 1306(272), *1344*
Ihm, H. 1396(183), 1406(267a, 269), 1428(427, 438), *1443, 1446, 1452, 1453*
Iida, S. 1164, 1290(36), *1338*
Iida, T. 667(50), *705*
Iida, Y. 295(82), *327*
Iimura, M. 409(58), *480*
Iinuma, H. 1262(195), *1343*
Iio, K. 447(222b, 222h, 222j), 448(222b, 225, 226), *485*
Iizuka, M. 291(63), *327*
Ikawa, S. 551, 554(109, 110), 555(109), *599*
Ikeda, A. 320(240), *331*, 1379(54d, 54e), 1392(124), 1400(207b, 210a), 1401(210a), 1404(246a, 246b), 1405(247), 1427(407b, 424b), 1432(424b), 1435(468), *1437, 1441, 1444, 1445, 1451, 1452, 1454*
Ikeda, H. 677(136), *707*
Ikeda, I. 1054(162), *1091*
Ikeda, M. 664(34), *705*
Iki, H. 1380(61b, 62), 1413(306), *1438, 1447*
Iki, N. 1380(64), 1393(160), 1415(314a, 319–321), 1416(324), *1438, 1442, 1448*
Ikonomou, M. G. 314(204), *330*
Ila, H. 420(142a), 434(191), *483, 484*
Ilan, Y. A. 1118, 1119, 1141(154), *1147*
Ilczyszyn, M. 143(571, 572), 144(602, 603), *192, 193*, 335(1), 363(1, 147, 148), 364(147–153), *386, 389*, 593(271), 594(271, 288), 595(271, 288, 293), *602, 603*
Iliev, I. 976(331), *1010*
Ilika, T. I. 1086(378), *1096*
Illera, M. 961(243), *1008*
Il'yasov, A. V. 740(64), *832*
Imada, I. 403(31), *479*
Imagawa, T. 1038(122b), *1090*
Imai, K. 951, 953(192), *1007*
Imakubo, T. 1027(91, 92), *1089*
Imasaka, T. 107, 143, 147(412b), *189*
Imbert, F. E. 616(31), *656*
Imgartinger, H. 211, 216(20), *221*
Imhof, P. 35, 36(169), 78(280, 281), 107(169), 145(622, 623), *183, 186, 193*, 581(221, 222), *601*
Imonigie, J. A. 637(113), *658*
Imoto, H. 688(229), *709*
Imoto, M. 1462(34), *1503*
Inaba, K. 818(320), *838*
Inaba, T. 1318(300a, 300b), *1345*
Inabe, T. 367(171), *390*
Inagaki, M. 1178(58), *1339*, 1489, 1491(116), 1495(134), *1505*
Inauen, A. 578, 581(193), *601*

Inazu, T. 353(99), *388*, 674(105), *707*, 1380(58a, 58b, 58d), *1437*
Inbaraj, J. J. 874(176, 177), *905, 906*
Incorvia Mattina, M. J. 315, 316(217), *330*
Inel, Y. 1084(329), *1095*
Ingallina, P. 414(90a), *481*
Ingemann, S. 86(306), *186*
Ingold, C. K. 801(239), *836*
Ingold, K. U. 15(75, 76), 20, 23(116), 44(188), 45(191, 200), 129(76), 139(116, 191, 531), 140(116), 141(76, 538–540), 142(76, 538, 542), 143(543–545, 552), *181, 182, 184, 191, 192*, 840(3), 841(12), 842(14, 16), 843(283), 844(17–19), 845(29, 30, 32, 33), 846(42), 847(18), 851(43–45, 62), 852(86), 853(92), 855(18, 112), 856(112), 857(62), 858(43), 859(137, 138), 860(14, 18, 19, 139–141), 861, 862(18), 864(18, 143, 144), 865(144), 867(155), 868(86), 872(138, 139), 876(155, 185), 877(188–192), 878(86, 188, 191), 879(191, 192, 194, 195, 198), 881(188, 191, 192, 203), 882(86, 155, 191, 194, 195), 883(188, 191, 192, 194, 205), 884(44, 45, 206, 207), 885(45, 207, 212), 887(205, 217), 889(45), 890(212), 891(112, 241), 892(155, 185, 253), 897(18, 273, 275), 898(18), 900(14, 282), 901(283, 288), *902–908*, 1108(2, 7, 8), 1111(7, 61, 64), 1112(81), 1120(184), 1140(184, 373), *1143–1145, 1147, 1151*
Inokuchi, T. 1158(13), *1337*
Inoue, H. 951, 953(195, 196), *1007*, 1076(234), *1092*
Inoue, K. 865(149), *905*, 1372(18), *1436*
Inoue, M. 463(275), *487*, 844, 891(21), *902*
Inoue, S. 677(134–136), 702(333, 344), *707, 712*
Inoue, T. 702(334, 338), *712*, 1318(300a, 300b), *1345*
Inoue, Y. 945(149), *1006*
Inouye, M. 323(259), *331*, 763(135), *834*
Insole, J. M. 402(25), *479*
Inui, M. 1462(25), *1503*
Inui, S. 669(64), *706*
Inyengar, N. R. 38, 127(185), *184*
Inzebejkin, A.Ju. 145, 146(615), *193*, 371, 377(230), *391*
Ioannides, C. 1360(177), *1367*
Ioffe, D. 397(7), *478*
Ione, K. G. 616(32), *656*
Iqbal, M. 1371, 1372(7), 1392(128), *1435, 1441*
Iqbal, T. 424(153a), *483*
Iranpoor, N. 638(123), *659*
Ireland, F. 492, 493, 501, 507(12), *525*
Ireland, J. 1086(366), *1096*
Irgibaeva, I. S. 371(228), *391*

Irico, A. 322(250), *331*
Irie, K. 1374(31a), *1436*
Irie, R. 699(298–301, 312), 701(321, 322), 702(326), *711, 712*
Irth, H. 977(335), *1010*
Irto, Y. 702(326), *712*
Irwin, J. L. 1432(447), *1453*
Isaks, M. 1022(52), *1088*
Isern-Flecha, I. 308(164, 165), 309(165), *329*
Isernia, A. 1391, 1396(121c), *1440*
Ishi, T. 1392(131), *1441*
Ishida, H. 211(16), *220*, 1084(322), *1095*
Ishida, S.-I. 1462(25), *1503*
Ishida, T. 424(155b), *483*, 866, 896(147), *905*
Ishida, Y. 129, 179(491), *190*, 881(202), *906*
Ishigaki, M. 112(456), *190*
Ishihara, K. 677(129), 691(243, 244), 693(257, 259–261), *707, 710*, 844, 845(24), *902*
Ishihata, K. 414(90c), *481*
Ishii, F. 1378(47a, 47b), *1437*
Ishii, H. 424(155b), *483*, 770(158), 773(158, 159), *834*, 874(174), *905*
Ishii, S. 1495(137), *1505*
Ishii, T. 987(405), *1012*, 1302(266), *1344*
Ishii, Y. 1204(100), *1340*, 1414(309), *1448*
Ishikawa, T. 770(158), 773(158–160), *834*
Ishikawa, Y. 1435(464b), *1454*
Ishitani, H. 695(276–278), *711*, 774(168), *835*
Ishiuchi, S. I. 37, 46, 110, 111(182), *183*
Ishizu, K. 856(129), 865(148), 890(231), 896(148), *904, 905, 907*, 1109(31), *1144*
Ismaev, I. E. 740(64), *832*
Ismail, F. M. D. 677(127), *707*, 765(140), *834*
Isobe, T. 1356(125), *1366*
Isohata, T. 1216(125), *1341*
Isota, Y. 774(194), *835*
Isotani, S. 1104(57), *1106*
Israili, Z. H. 424(157b), *483*
Itayama, Y. 951, 953(194), *1007*
Ito, A. 414(89b), *481*
Ito, H. 460(288), *487*, 624(58), *657*, 691(236, 237), *710*, 769(146), *834*
Ito, J. 1079(240), *1092*, 1218(129), *1341*
Ito, K. 694(262, 263), *710*, 992(430), *1013*, 1066(197), *1091*, 1374(31b), 1380(58g, 59, 63), *1436–1438*
Ito, M. 107(409), 143(409, 574–576, 578–580), 144(574–576, 595), 145(574), 147(574, 578, 580, 692), *188, 192, 193, 195*, 414(87), *481*, 689(230, 233), *709, 710*
Ito, S. 414(89a, 90c), *481*, 1213(112), *1340*
Ito, T. 1233(151a, 151b), *1341*
Ito, Y. 699(298), *711*
Itoh, H. 1028(95), *1089*
Itoh, J. 950, 953(187), *1007*
Itoh, K. 702(342), *712*
Itoh, M. 1025(78), *1089*

Itoh, N. 414(91), *481*, 695(279), *711*, 1285(236), *1343*
Itoh, S. 851, 862, 863(49), *903*
Itoh, T. 688(228), 693(250), *709, 710*, 844(23), *902*
Itoho, K. 1084(321), *1095*
Itokawa, H. 1318(300a, 300b), *1345*
Ivachnenko, E. P. 1290, 1292(247), *1344*
Ivanosvskaia, L. Yu. 269(34), *326*
Ivanov, A. V. 1073(221), *1092*
Ivanov, B. E. 740(64, 65), *832*
Ivanov, V. L. 1073(221–224), *1092*
Ivanova, G. 360(136, 140), 361(140), 385(296), *389, 393*
Iversen, F. 1360(176), *1367*
Ivkova, T. I. 987(407), *1012*
Iwaki, T. 774(193), *835*
Iwaki, Y. 856, 863, 864(126), *904*
Iwamoto, K. 1372(10b), 1391(111, 115a, 115b, 117), 1393(159), 1394(111, 167a, 172b), *1435, 1440, 1442*
Iwamoto, M. 413(77), *480*
Iwanek, W. 1417(330a), 1418(336), 1420(359a, 361, 365a), *1448, 1450*
Iwao, M. 415(106a), *481*
Iwasaki, A. 107(410), 144(598), *188, 193*
Iwasaki, F. 216(26), *221*
Iwasaki, H. 1206(105), *1340*
Iwasaki, Y. 1382(70), *1438*
Iwasawa, N. 448(231), *486*
Iwata, K. 1027(91, 92), *1089*
Iwata, S. 95, 106, 107(372), 148(729, 730), 149, 153, 156(729), 160(729, 730), 170(729), *188, 196*, 1289(242), *1344*
Iwata, Y. 1194(86), *1339*
Iwatsuki, M. 882(204), *906*
Iwatsuki, S. 1498(175), *1506*
Iwema Bakker, W. I. 1396(189), *1443*
Iwuoba, E. I. 977(339), *1010*
Iyengar, N. R. 716(11), *831*
Iyer, R. 1311(283), *1345*
Izatt, R. M. 1372(17), *1436*
Izatt, S. R. 1372(17), *1436*
Izawa, S. 1374(31b), *1436*
Izawa, T. 447(222f), *485*, 1337(332), *1346*
Izod, T. P. J. 307(152), *329*
Izumi, T. 1028(99), *1089*
Izvekov, V. 244(84), *256*, 370(212), 384(293, 294), 385(294), *391, 393*, 551, 553(74, 75), *598*
Izvekov, V. 82(286), *186*

Jabbar, A. 649(157), *659*
Jabconova, A. 1082(271), *1093*
Jablonski, J. 608(8), *656*
Jaccard, G. 340, 341, 352(25), *386*
Jachiet, D. 671(86), *706*, 1163(33b), *1338*

Jackman, L. M. 105(401), *188*
Jackson, A. H. 1297(260c), *1344*
Jackson, G. 1419(352), *1450*
Jackson, L. B. 811(287–289), *837*
Jackson, W. R. 739(60), *832*
Jacniacki, J. 632(88), *658*
Jacob, P. III 1335(330), *1346*
Jacobsen, E. N. 667(51), 668(52, 53), 685(199), 688(226), 698(291, 293, 294, 296, 297), 699(311), 701(320), 702(324), *705, 706, 709, 711, 712*
Jacobson, R. R. 1194(87), *1340*
Jacobsson, U. 1048(155), *1090*
Jacoby, C. 581(217), *601*
Jacoby, Ch. 143, 144(577), 147, 148(697, 701, 702), 149(697), 156, 160(702), *192, 195*
Jacoby, D. 1391(113), *1440*
Jacopozzi, P. 1397, 1399, 1403(197b), 1424(384, 393), *1443, 1450*
Jacquesy, J.-C. 405(39a, 39b), 444(216c–f, 217), *479, 485*, 648(155), 651(155, 173–181, 183), 652(184, 185), 653(186), *659, 660*, 1227(145), *1341*
Jacquesy, R. 651(173, 178–181), 652(184, 185), *660*
Jacquier, J. C. 974(317), *1010*
Jacquot, R. 617(35), *656*
Jaculi, D. 1283(231), *1343*
Jaeger, C. B. 400(18a), *478*
Jaffe, A. B. 1123(214), *1148*
Jaffé, H. H. 53, 56(240), 129, 131(506), *185, 191*, 261(12), *326*, 419(134c), *482*
Jagdmann, G. E. 1314(288, 289), *1345*
Jäger, E.-G. 1205(101), *1340*
Jagielska, A. 929, 977(64), *1004*
Jagodzinska, E. 144(652), *194*, 359(123), *389*
Jain, N. 631(83), *658*
Jain, N. F. 1318(303), *1345*
Jain, S. 807(277), *837*, 1032(106), *1089*
Jakabsen, S. 111(454), *190*
Jakob, L. 1085(334), *1095*
Jakobi, R. A. 1406(257), *1446*
Jakobsen, H. J. 350(78), *388*
Jakobsen, R. J. 67(251), *185*
Jakoubkova, M. 170(752), *196*
Jamali, F. 412(76), *480*
James, A. W. G. 430(181a), *484*
Jammu, K. S. 67(249), *185*
Jamroz, D. J. 170(741), *196*
Janak, K. 931, 937(83), *1005*
Janes, N. 888(226), *907*
Janes, S. M. 129(485), *190*
Janetka, J. W. 673(93), *706*
Jang, J. 913, 920(27), *1003*, 1355(116), *1365*
Jang, J. H. 149, 158, 161(736), *196*
Jankowski, C. K. 322(249), *331*
Jansen, A. 375(247), *392*
Jansen in de Wal, H. 1204(99), *1340*

Janssen, H. G. 937, 938(125), *1005*
Janssen, R. G. 377(251), *392*, 1387, 1392, 1394, 1395(90), 1409(287), *1439, 1447*
Jansson, B. 1351(47), *1364*
Jansson, K. 1359(148), *1366*
Jansson, V. 1359(148), *1366*
Janzen, C. 145(622, 623), *193*, 581(221, 222), *601*
Janzen, Ch. 147, 148(697, 702), 149(697), 156, 160(702), *195*
Janzen, E. G. 864, 865(144), *905*
Jappinen, P. 1347(5), *1363*
Jaquinod, L. 1374(44), *1437*
Jaracz, S. 691(239), *710*
Jardim, W. F. 1362(212), *1368*
Jardine, D. 999(466), *1013*
Jardon, P. 384(287, 288), *393*
Jaroszewski, J. W. 1271(204), *1343*
Jarret, R. M. 367(176), *390*
Jarva, M. 561(163), *600*
Jasiński, T. 594, 595(291), *603*
Jauhiainen, T. P. 236, 248(56), *255*
Jáuregui, O. 944(148), 946(166), 947(172), 953(148, 166, 172), *1006*, 1352(68), 1354(103), *1364, 1365*
Javor, T. 974(316), *1010*
Jawed, J. 177(767), *197*
Jayaram, B. 5(14), *179*
Jayat, F. 631(72, 78), 639(130), *657, 659*, 686(205), *709*, 774(176), *835*
Jean, F. 1361(191), *1367*
Jedral, W. 1361(190), *1367*
Jeffcoate, C. R. E. 1113–1117, 1120(88), *1145*
Jeffrey, G. A. 498, 499, 502(77), *526*, 530(13, 14), *597*
Jeffries, H. E. 937(119), *1005*
Jeger, O. 433(189a), *484*
Jekel, M. 953, 963(259), *1009*
Jelling, M. 854(99), *904*
Jellis, S. M. 1352(55), *1364*
Jemenez, E. 987(406), *1012*
Jemmett, A. E. 399(13a), *478*
Jempty, T. C. 1289(241), *1344*
Jen, J.-F. 953, 969(283), *1009*
Jeng, M.-L. H. 563(169), *600*
Jenkins, P. A. 1462(21), *1502*
Jenkins, T. C. 1259(189), *1342*
Jennings, K. R. 263(18), *326*
Jennings, W. B. 830(347), *838*
Jensen, D. 1360(174), *1367*
Jensen, F. 368(193), *390*
Jensen, H. B. 67(250), *185*
Jensen, J. 607(4), *656*
Jensen, M. W. 382(277), *392*
Jensen, O. J. 93–95, 106, 107, 109, 110(356), *187*
Jeon, J.-H. 664(32), *705*
Jepsen, O. 69(263), *185*

Jereb, M. 655(190), *660*
Jerina, D. 405(44), *479*
Jerina, D. M. 14(69), *181*, 412(72), *480*, 721(21, 22), 824(21), 829(344), *831, 838*
Jerussi, R. A. 466(292), *487*
Jeskey, H. 716(10), *831*
Jeunesse, C. 1435(462a), *1454*
Ji, H. 1077(237), *1092*
Ji, M. 1394(167c), *1442*
Jiang, L. 414(89c), *481*
Jiang, L.-F. 414(90d), *481*
Jiang, Z. H. 999(464), *1013*
Jiao, H. 95(367), *188*
Jiao, K. S. 937, 938(126), *1006*
Jiménez, A. 913(6), 932(96), *1003, 1005*
Jimenez, A. I. 987(406), *1012*
Jiménez, C. 972(304), *1010*
Jimenez, M. C. 744(71, 72), *832*
Jiménez, P. 230(25), 232(38), 237(61), 241(80), *254–256*
Jimenez, R. 511(94), *527*
Jiminez, M. C. 1038(122c), 1039(123), 1040(124–126), 1042(127, 128), 1055(169), 1071(204), *1090–1092*
Jin, F. 1136(360, 362), *1151*
Jin, T. 1394(172a), *1442*
Jin, Z. 377(261), *392*
Jing, H. W. 639(124), *659*
Jinno, M. 1180(63b), *1339*
Jinot, C. 1120, 1122, 1134(186), *1147*
Jintoku, T. 414(86a, 86b), *481*
Jiye, F. 238, 239(67), *256*
Jobling, S. 925(49), *1004*, 1349, 1360(34), *1363*
Jocelyn Paré, J. R. 930(79), 932(97), 937(79), *1004, 1005*
Joerg, E. 953, 961(244), *1008*
Joesten, M. D. 368(182), *390*, 530, 551, 554, 557, 559, 588–590(4), *596*
John, A. 378(271), *392*
John, D. M. 1348(31), *1363*
John, V. T. 365(164), *389*
Johns, R. B. 314(205–207), *330*
Johnsen, P. 367(170), *390*
Johnson, B. 4, 100(2), *179*, 368, 372, 373(200), *390*
Johnson, B. J. 1404(245), *1445*
Johnson, C. R. 129, 133(493), *191*
Johnson, E. 899(277), *908*
Johnson, E. R. 31, 45, 140(140), *183*, 867, 871, 895–899(154), *905*
Johnson, F. 682(167), *708*
Johnson, G. J. 405(40), *479*
Johnson, P. M. 109(415), 143(583), *189, 193*
Johnson, R. A. 405(41c), *479*
Johnson, R. L. 1360(162), *1367*
Johnson, S. A. 1129, 1134(283), *1149*
Johnson, W. J. 1104(59), *1106*

Johnston, D. E.Jr. 1372(12), *1436*
Johnston, L. D. 129, 133(497), *191*
Johnston, L. J. 110, 111, 113, 114(425), *189*, 860(141), *905*, 1019(27a), *1087*, 1134(344), *1151*
Johri, M. H. 290(62), *327*
Jolliffe, K. A. 1406(262b), *1446*
Jolly, R. S. 677(126), *707*
Joly, G. 651(181), *660*
Jona, H. 641(137), *659*, 679(142, 143), *708*
Jonas, V. 1074(230), *1092*
Jonekubo, J. 1351(52), *1364*
Jones, A. 774(190), *835*
Jones, A. D. 636(104), 637(109), *658*
Jones, C. W. 631(75), *657*
Jones, D. A. K. 53(235), *185*
Jones, E. J. 35(174), 47(216, 217), 48(217), *183, 184*
Jones, G. P. 953, 969(281), *1009*
Jones, J. 933, 937(101), *1005*
Jones, M. N. 1348(20), *1363*
Jones, P. G. 1383(67k), 1391(120a), 1417(329a), 1421(67k), 1424(395), *1438, 1440, 1448, 1451*
Jones, R. L. 35(149), *183*
Jones, R. M. 549(65), *598*
Jones, R. O. 6(27), 93, 105(354), *180, 187*
Jones, S. A. 1495(157), *1505*
Jones, T. K. 1466(45), *1503*
Jones, W. A. 647(151), *659*
Jones, Z. 1307(273), *1344*
Jong, F.de 1391(121a), 1396(121a, 182a, 182b), 1432(455), *1440, 1443, 1453*
Jong, J. I.de 1475, 1486(80), *1504*
Jong, R. H. I.de 1192, 1198(77), *1339*
Jonkers, N. 1349(35), *1363*
Jönsson, J.Å 947(168), 953(168, 280), 969(280), *1006, 1009*
Jonsson, M. 45(193), 139(534), 140(193), *184, 191*, 865(145), *905*, 1136(361), 1143(382), *1151, 1152*
Jönsson-Pettersson, G. 976(330), *1010*
Jordan, J. E. 1111(74), *1145*
Jordan, K. D. 20(123), 148, 160(720), *182, 196*, 315(211), *330*
Jordanov, B. 368(195), *390*
Jordon, J. G. 1374(42a), *1436*
Jore, D. 1125(246), 1129(246, 304), *1149, 1150*
Jorgensen, E. C. 53(239), *185*
Jorgensen, K. S. 1347(8), *1363*
Jorgensen, M. 1405(251a), 1406(264, 267b), *1446*
Jørgensen, T. J. D. 325(277), *332*
Joris, L. 531(21), 559(153), 586(153, 237), 589(237, 262), 590(237), 593(153), *597, 600, 602*, 877(187), *906*
Josephsen, J. 358(116), *388*

Josephy, P. D. 990(424), *1012*, 1120(175, 176), *1147*
Joshi, H. C. 1047(148), *1090*
Josien, M. C. 35(161), *183*
Jouannetaud, M.-P. 405(39a, 39b), *479*, 648(155), 651(155, 183), 653(186), *659, 660*, 1227(145), *1341*
Joubert, F. 1495(130, 140), *1505*
Joule, J. A. 748(95), *833*
Jouvet, C. 109(414), 143(588), 144(588, 600), *189, 193*, 493, 500(31), *525*
Jovall, P. A. 887(217), *906*
Jovanovic, S. V. 868(156, 157), 870(157), 871(163, 165), 874(170), 897(274), 898(163), 899(170), *905, 908*, 1019(26), *1087*, 1108(9, 10), 1139(369, 370), 1142(370), *1144, 1151*
Joyce, A. 840(3), *902*
Ju, C.-S. 1082(289), *1094*
Jude, R. J. 200(6), *220*, 550(70), *598*
Judson, C. M. 320(236), *331*
Jujio, M. 92, 99–101(351), *187*
Julia, M. 414(94a), *481*
Juneja, R. K. 1414(307a, 307b, 308), *1447, 1448*
Jung, A. 1044(131), *1090*
Jung, D. M. 365(160), *389*
Jung, G. 1172(51b), *1338*
Jung, K.-H. 778(209), *835*
Jung, M. E. 422(147b), *483*, 668(54), 671(86), 681(165), *706, 708*, 776(207), *835*, 1163(33b), *1338*
Jung, M.-W. 946(163), *1006*
Jung, S. Y. 1086(373), *1096*
Jungbluth, G. 739(59), *832*
Junge, W. 143, 177(553), *192*
Junjappa, H. 420(142a), 434(191), *483, 484*, 668(57), *706*
Jurd, L. 1255(181), 1309(282), *1342, 1345*
Jurs, P. C. 925, 929(54), *1004*
Jursic, B. 663(13), *705*
Jurss, C. D. 1278(226), *1343*
Just, T. 1105(64, 65), *1106*
Justicia, H. 1349, 1360(33), *1363*

Käb, G. 740(61), *832*
Kabalka, G. W. 415(98a, 98b, 99, 100, 103a, 103b), 422(148), *481, 483*
Kabashnik, M. I. 1126(256–259, 261), *1149*
Kaboudin, B. 474(316), *488*, 640(133), *659*, 774(174), *835*
Kaburagi, Y. 1495(134), *1505*
Kabuto, C. 1380(61b, 62), 1415(321), 1416(324), *1438, 1448*
Kabuto, K. 680(157), *708*
Kachab, E. H. 1380(60a), *1438*
Kaczmarek, L. 1025(68), *1088*

Kad, G. L. 774(185, 186), *835*
Kada, T. 913(28), *1003*
Kader, F. 944(139), *1006*
Kadowaki, D. 855, 856(105), *904*
Kaeding, W. W. 19(96–98), 47, 53, 56(223), *182, 184*
Kafari, S. A. 5(11), *179*
Kaftory, M. 216(29), *221*
Kagan, V. E. 887(222, 225), *907*
Kagawa, K. 1495(161), *1506*
Kagawa, S. 413(77), *480*
Kagen, B. S. 687(215), *709*
Kageyama, H. 1028(98), *1089*
Kageyama, Y. 866, 896(147), *905*
Kahlig, H. 339, 366(21), *386*
Kahn, P. H. 770(156), *834*
Kahne, D. 674(106), *707*
Kaifer, A. E. 761(117), *833*, 1422(377), 1424(402), 1429(428), *1450–1452*
Kaisheva, A. 976(331), *1010*
Kajfez, F. 1470(50, 52), *1503*
Kaji, A. 419(132), *482*
Kajigaeshi, S. 651(170), *660*
Kajii, Y. 129, 131(505), *191*, 1017(11), *1087*
Kakinami, T. 651(170), *660*
Kakitani, T. 1057(170), *1091*
Kakiuchi, K. 1048(158, 159), *1090, 1091*
Kalachova, G. S. 1362(218), *1368*
Kalanderopoulos, P. 1022(52, 53), *1088*
Kalchenko, O. I. 1418(340), *1450*
Kalchenko, V. I. 1393(161), 1394(162, 166), 1417(326, 328), 1418(338, 340, 341, 344), 1420(341), *1442, 1448, 1450*
Kaldor, S. W. 1466(45), *1503*
Kalena, G. P. 1062(187), *1091*
Kaleta, W. 619(37), *656*
Kalinin, A. V. 670(72), *706*, 786(231), *836*
Kalinowski, H. O. 338(11), *386*
Kalir, R. 472(311j), *488*
Kallas, J. 1085(331), *1095*
Kallinowski, G. 1435(464a), *1454*
Kalliokoski, P. 1351(43), *1364*
Kalmus, C. E. 476(323), *488*
Kalninsh, K. 1020(33), *1087*
Kalucki, K. 1083(305), *1094*
Kalyan, Yu. B. 763(131), *834*
Kalyanaraman, B. 1136, 1137(364), *1151*
Kalyanaraman, R. 237(62), *256*
Kamada, K. 1021(41), *1088*
Kamal-Eldin, A. 846, 847(41), *902*
Kamat, P. V. 1085(351, 352), *1095*
Kamelova, G. P. 235(51), *255*
Kamenetskaya, I. A. 770(154), *834*
Kamenskii, I. V. 1470(66), *1503*
Kameswaran, V. 1278(218), *1343*
Kametani, T. 1216(121, 122a), 1275(215), 1283, 1287(121, 122a), 1297(121, 122a, 255), 1306(272), *1340, 1343, 1344*

Kameyama, H. 1393(160), *1442*
Kamide, K. 1462(19), *1502*
Kamienski, B. 1024(63), *1088*
Kamieth, M. 322(245, 246), *331*
Kamigaito, M. 1388(100b), *1439*
Kamigata, N. 669(63), *706*
Kaminski, M. 947(169), *1006*
Kamiura, T. 1348(27), *1363*
Kamiya, Y. 853(90), 855, 856(90, 119), 886(233), 887(221), 890(232, 233, 235), *904, 907*
Kamiyama, H. 1380(61a), *1438*
Kamlet, M. J. 503(87–89), *526*, 541(45), 575(187), 591(264, 265), 592(265), *597, 600, 602*, 887(219), *907*
Kämmerer, H. 1462(30), *1503*
Kamp, N. W. J. 1209(109), *1340*
Kamura, A. 855(104), *904*
Kan, T. 684(193), *709*
Kan, Y. 211(19), *220*
Kanai, M. 691(238), 694(267), 695(269), *710*
Kanai, M. M. 695(268), *710*
Kanamathareddy, S. 1399(200a, 200b), 1403(231b), 1427(424a), *1443, 1445, 1452*
Kanda, F. 295(82), *327*
Kanda, W. 1205(103), *1340*
Kandaswami, C. 894(259), *907*
Kane, S. A. 977(339), *1010*
Kaneeda, M. 691(243, 244), *710*
Kang, N. 1362(205), *1368*
Kang, S.-K. 664(32, 33), *705*
Kang, S. O. 1400(208b), *1444*
Kang, X. 1081(265), *1093*
Kania, A. 143(565), *192*
Kanki, T. 1083(304), *1094*
Kannan, N. 1356(124), *1366*
Kano, K. 1209(106, 107), *1340*
Kano, T. 681(164), *708*
Kanoh, H. 1498(175), *1506*
Kapila, S. 955(205), *1007*
Kaplan, J.-P. 475(318), *488*
Kapoor, V. B. 941(134), *1006*
Kappe, T. 342(38), *387*
Kappus, H. 851, 852, 854, 855(65), *903*
Kar, S. 972(306), *1010*
Karadakov, P. B. 89(312), *186*
Karakulski, K. 1083(305), *1094*
Karam, L. R. 1100(22), *1105*
Karam, O. 648(155), 651(155, 183), *659, 660*, 1227(145), *1341*
Karapinka, G. L. 1274(208), *1343*
Karas, M. 260(3), 305(147), 323(3, 253–255, 264), 324(268), *325, 329, 331*
Karaseva, S. Ya. 227(13), *254*
Karavaev, S. A. 986(399), *1012*
Karayannis, M. I. 986(398), *1012*
Karbach, S. 1419(348), *1450*

Karbach, V. 323(263), 324(269), *331*, 493, 515(27), *525*
Karhunen, P. 1309(281), *1345*
Karibe, N. 677(128), *707*
Karlberg, B. 971(293), *1009*
Karlin, K. D. 1192(78e), 1194(87), *1339, 1340*
Karlson, K. E. 1070(200), *1091*
Karmanova, L. P. 241, 251, 252(81), *256*
Karmas, K. 937, 940(129), *1006*
Karmilov, A. Yu. 747(83), *832*
Karodia, N. 766(141), *834*
Karovicova, J. 947(174), *1006*
Karpfen, A. 20, 21, 24, 35, 37(124), *182*, 339, 366(21), 371, 372, 375(234), 377(234, 268), 378, 380(268), 386(305, 306), *386, 391–393*, 551, 553(104), *599*
Karpyuk, A. D. 953, 957(214), *1007*
Karr, C. 414(97a), *481*
Karrer, W. 13(61), *181*
Kartal, I. 729(42), *831*
Karthien, R. 129(482), *190*
Karyakin, N. V. 235(51), *255*
Kasai, Y. 662(12), *705*, 1378(47a–c), *1437*
Kasakara, A. 1028(99), *1089*
Kasende, O. 144(674), 145–147(619), *193, 195*, 569(178), 583(225, 229), *600, 601*
Kasey, R. B. 868, 870(157), *905*
Kasham, S. 593, 595(278), *602*
Kashimoto, T. 1351(46), *1364*
Kashino, S. 211(16), *220*, 811(294), *837*
Kashyap, R. P. 1390, 1392(107), 1393(147), 1394(147, 165), 1409(285), *1440–1442, 1447*
Kaskey, R. B. 1108(10), *1144*
Kasperek, G. J. 14(68, 69), *181*
Kaspzycka-Guttman, T. 367(174), *390*
Kassir, J. M. 818(313), *838*
Kast, B. 913(19), *1003*
Kasumov, V. T. 729(42), *831*
Katagiri, H. 1416(324), *1448*
Kataja, K. 998(451), *1013*
Katayama, K. 1156, 1158(1), 1163(33a), 1175(1), 1209(107), *1338, 1340*
Kathirgamanathan, P. 1495(148), *1505*
Kato, K. 608(14, 15), *656*, 669(66, 67), 683(172, 173), *706, 708*, 845(36, 40), *902*, 1076(234), *1092*
Kato, S. 1079(241), *1092*
Kato, T. 686(211), *709*, 769(149), *834*
Kato, Y. 856, 863, 864(126), *904*, 1159, 1227(30), *1338*, 1383, 1421(67e), *1438*
Katovic, Z. 1471(72), *1504*
Katritzky, A. R. 434(193, 194), *484*
Katrusiak, A. 146(638), *194*, 593, 595(281, 282), *602, 603*
Katsu, T. 1202, 1205(95), *1340*
Katsuki, H. 641(137), *659*
Katsuki, M. 674(99), *707*, 776(202), *835*

Katsuki, T. 694(262, 263), 697(289, 290), 698(292), 699(298–301, 309, 310, 312, 313), 701(319, 321–323), 702(326, 327), *710–712*
Katsumata, S. 844(23), 891(246), *902, 907*
Kattner, G. 953, 962(249), *1008*
Katuki, M. 679(141–143), *708*
Katz, J. L. 671(84), *706*
Katzenellenbogen, J. A. 1070(200), *1091*
Kauber, O. 402(22), *479*
Kauffmann, J. M. 983(376), *1011*
Kaufman, R. J. 673(96), *707*
Kaufmann, K. J. 1020(35c), *1087*
Kaufmann, R. 323(262), *331*
Kauppinen, T. 1351(43), *1364*
Kaur, J. 774(185), *835*
Kaus, S. 129(485), *190*
Kausbisch, N. 14(69), *181*
Kaushal, R. 433(190), *484*
Kawada, K. 687(222, 223), *709*
Kawada, M. 403(31), *479*
Kawaguchi, H. 1081(266), 1084(317), 1085(333), *1093, 1095*
Kawaguchi, K.-I. 448(226), *485*
Kawaguchi, M. 1427(407b), *1451*
Kawahara, F. K. 368(185), *390*
Kawahara, S. 1421(373), *1450*
Kawahara, S. I. 90(344), *187*
Kawakami, A. 886(233), 887(221), 890(233, 235), *907*
Kawakishi, S. 1099(11–14), *1105*
Kawamoto, A. 219(36), *221*
Kawamura, K. 669(63), *706*
Kawanashi, Y. 1400(206), *1444*
Kawanishi, M. 398(9a), *478*
Kawanisi, M. 1038(122b), *1090*
Kawano, Y. 216(26), *221*
Kawase, M. 415(107), *481*
Kawashima, T. 1428(435a, 435b, 436), *1452*
Kawate, T. 662(6), *705*
Kawecki, R. 340(28), *387*
Kaya, K. 107(412a), 143(412a, 581, 582), 147(412a), *189, 192, 193*
Kayami, K. 397(3), *478*
Kaysen, K. L. 845, 846(38), 852(79), *902, 903*
Kazakova, E. K. 1419(355), 1424(396), *1450, 1451*
Kazandjian, R. Z. 1219(132), *1341*
Kazankov, M. V. 730(48), *832*
Kazankov, Y. V. 1470(70), *1504*
Kazareva, R. 976(331), *1010*
Kazlauskas, R. J. 638(121), *658*
KcKinnon, J. T. 17(81), *181*
Keane, M. A. 178(774), *197*
Kearns, D. R. 853(94), *904*
Kebarle, P. 5(17), 83(295), 92(350), 100(381), 101(350), 102(381), *180, 186–188*,
261(10), 313(197–200), 314(199), 324(271, 272), *326, 330–332*
Keck, J. 1026(90), *1089*
Kecki, Z. 366(165), *389*
Keefe, A. D. 1432(449a), *1453*
Keenan, R. E. 1360(167), *1367*
Keii, T. 1109, 1128, 1130(32), *1144*
Keil, T. 145(629, 630, 632, 633), *194*, 593(273), 595(273, 298), 596(298), *602, 603*
Keith, L. H. 1348(14, 15), *1363*
Keith, T. 4, 100(2), *179*, 368, 372, 373(200), *390*
Kelder, G. D.-O.den 871, 873, 898(169), *905*
Kelderman, E. 1391(112), 1402, 1405(224b), *1440, 1444*
Keller, P. 953, 958(224), *1008*
Keller, W. E. 662(2), *704*
Kelley, W. T. 947, 953(173), *1006*
Kelly, D. R. 1466(46), *1503*
Kelly, J. 1352(55), *1364*
Kelly, S. 89, 90(327), *187*
Kelly, T. R. 692(249), *710*
Kelsh, L. P. 735, 748(51), *832*
Kemister, G. 109(413), *189*
Kemp, T. J. 1086(371), *1096*, 1101(35), *1106*
Kemsley, K. 1110, 1112(53), *1145*
Kende, A. S. 807(282), *837*
Kende, S. 1278(224), 1294(254), *1343, 1344*
Kenis, P. J. A. 1391(119), *1440*
Kennard, O. 535, 549(32), 550(71), 551(32), *597, 598*
Kennedy, B. R. 1330(322), *1346*
Kennedy, D. A. 829(345), *838*
Kennedy, J. A. 960(228), *1008*
Kennedy, J. P. 621(50), *657*
Kentosova, I. 1082(272), *1093*
Keogh, J. P. 179(776), *197*, 651(182), *660*
Keough, T. 292, 293(69), *327*
Keresztury, G. 20, 22, 23, 35, 36, 38(111), 82(286), 179(782), *182, 186, 198*, 244(84), *256*, 368(198), 369(209), 370–375(198), 384(293, 294), 385(294), *390, 391, 393*, 551, 553(74, 75), *598*
Kermasha, S. 953(221, 222), 957(221), 958(222), *1008*
Kerns, E. 305(151), *329*
Kerns, M. L. 1175, 1227(54a), *1338*
Kerr, J. A. 1105(64, 65), *1106*
Kerr, M. 786(231), *836*
Kerr, W. J. 454(256, 257), *487*
Kerwin, J. L. 305(148–150), *329*
Kessel, J.van 884(207, 208), 885(207), *906*
Kettämaa, H. I. 123(467), *190*
Keumi, T. 405(37b, 38), *479*
Keutgen, W. A. 1457, 1461, 1470(5), *1502*
Khadilkar, B. M. 662(10), 680(151), *705, 708*, 774(187), *835*

Khaimova, M. 470(299), *488*
Khalaf, K. D. 987(406), *1012*
Khalfin, R. 619(40), *656*
Khalifa, M. H. 1169(41), *1338*
Khamis, E. F. 995(439), *1013*
Khamsl, J. 671(82), *706*
Khan, A. R. 1362(215), *1368*
Khan, I. U. 353(99), *388*, 1380(58a, 58b, 58e), *1437*
Khan, J. A. 885(209), *906*
Khan, M. A. 773(161), *834*
Khan, M. N. 1495(145, 146), *1505*
Khan, S. I. 671(86), *706*
Khandekar, A. C. 680(151), *708*
Khandelwal, D. P. 35(159), *183*
Khang, K. 1044(132), *1090*
Khanna, R. N. 478(329), *489*
Kharitonov, A. S. 18(84), *181*, 413(79), *480*
Kharlanov, V. 1025(67, 88), *1088, 1089*
Kharybin, A. N. 977(333), *1010*
Khasnis, D. V. 1393(145a), *1441*
Khizhnyi, V. A. 1118(141), *1146*
Khlebalkin, V. A. 985(389), *1012*
Kho, R. 1084(323), *1095*
Khodja, A. A. 1084(319), *1095*
Khodzhaev, O. 421(126), *482*
Khokhar, S. 953, 961(242), *1008*
Khoo-Beattie, C. 1396(189), *1443*
Khopde, S. M. 868(158), *905*
Khor, J. A. Y. 890(238), *907*
Khorsravi, E. 1495(148), *1505*
Khoschsorur, G. 920(35), *1003*
Khoury, R. G. 1374(44), *1437*
Khrapova, N. G. 891(243), *907*
Khryaschevsky, A. V. 962(252), *1008*
Khudoyan, G. G. 301(130), *328*
Khudyakov, I. V. 15(77), *181*, 1108, 1118(19), 1128(277), 1130(19, 319), *1144, 1149, 1150*
Khundkar, L. R. 143(584), *193*
Khurana, A. L. 953, 957(215), *1008*
Kido, S. 1205(103), *1340*
Kido, Y. 680(158), *708*
Kidwell, R. L. 415(105a, 105b), *481*
Kiefer, P. M. 522(100, 101), *527*
Kieffer, S. 1427(421a), *1452*
Kielkiewicz, J. 1427(412a, 412b), *1451*
Kiely, J. S. 434(193), *484*
Kienzle, F. 416(110a), *482*
Kierzek, A. 929, 977(64), *1004*
Kifayatullah, M. 806(269), *837*
Kikuchi, K. 1230(148), *1341*
Kikuchi, S. 856(128), *904*
Kikuchi, T. 1413(306), *1447*
Kilényi, S. N. 428(169), *483*
Kiliç, Z. 551, 557(122), *599*, 727(39), 729(43), *831*
Kilinc, E. 978(349), *1011*

Killham, K. 1362(217), *1368*
Kilpin, D. 1330(322), *1346*
Kim, A. 811(291), *837*
Kim, B. H. 1427(410c), *1451*
Kim, C. 671(86), 681(165), *706, 708*
Kim, C.-K. 1275(213), *1343*
Kim, D. 1047(151), *1090*
Kim, D. G. 1086(373), *1096*
Kim, D. S. 635(102), *658*, 1393(144), 1400(208b), 1403(234b), *1441, 1444, 1445*
Kim, D.-Y. 664(33), *705*
Kim, H. 1394(167b), 1406(269), 1419(348), *1442, 1446, 1450*
Kim, H.-J. 1406(268), *1446*
Kim, H.-S. 830(346), *838*, 1394(167b), *1442*
Kim, H. T. 129(477), *190*
Kim, J. 149, 158, 161(736), *196*, 1394(167b), 1407(277), *1442, 1447*
Kim, J. M. 1388(102), 1393(144, 148, 154, 155), 1394(155, 163, 164), 1403(155), *1440–1442*
Kim, J. N. 773(162), *834*
Kim, J. S. 1427(423b), *1452*
Kim, J. Y. 1406(265a, 265b), *1446*
Kim, K. 20(123), 148, 160(720), *182, 196*, 1394(167b), *1442*
Kim, K. H. 1392(130), *1441*
Kim, K. J. 1086(373), *1096*
Kim, K. S. 149, 158(736), 161(736, 737), *196*, 454(258), *487*
Kim, M.-H. 1164, 1224(35), *1338*
Kim, S. 1085(346), *1095*
Kim, S.-G. 1372(13), *1436*
Kim, S. H. 982(367), *1011*
Kim, S. K. 493(22, 23), *525*, 1017(7), *1087*
Kim, T. 1396(183), *1443*
Kim, T. K. 1086(373), *1096*
Kim, Y. 11(54), *180*, 232(36), *255*, 1047(151), *1090*
Kim, Y. H. 1406(265b), *1446*
Kimoto, H. 608(14, 15), *656*, 683(172, 173), *708*
Kimura, A. 614(25), *656*
Kimura, H. 1489(108), *1504*
Kimura, K. 106(406), 110, 112, 121(444), 143(582), *188, 189, 193*, 414(88), *481*
Kimura, T. 1495(133), *1505*
King, D. J. 412(71b), *480*
King, G. S. 93, 103(363), *187*
King, J. 450(246c), *486*
King, J. C. 1360(172), *1367*
King, K. 773(164), *834*
King, M. 843, 901(283), *908*
King, T. P. 400(17), *478*
Kingston, D. G. I. 280(49), *326*
Kingston, E. E. 263(20), 301(137), 302(137, 138), *326, 328*
Kinoshita, S. 951, 953(193, 194), *1007*

Author Index

Kinoshita, T. 954(200), 970(291), *1007, 1009*, 1380(56), *1437*
Kinsel, G. R. 323(257), *331*
Kintzinger, J.-P. 1427, 1432(421b), *1452*
Kinugawa, M. 1048(159), *1091*
Kip, C. J. 422(144), *483*
Kiparisova, E. G. 235(51), *255*
Kiralj, R. 385(300), *393*, 726(34), *831*
Kirby, G. H. 35(157), *183*
Kirby, S. P. 224, 227, 228, 230, 241(2), *254*
Kirchner, J. J. 227(10), *254*
Kirichenko, A. I. 631(79), *657*
Kiricsi, I. 613, 644(23), *656*
Kirihara, M. 447(222c–e), *485*
Kirk, D. W. 1157(11), *1337*
Kirk, M. 637(110), *658*
Kirsch, R. S. 1255(185), *1342*
Kirste, B. 1123(215), *1148*
Kirtaniya, C. L. 409(59), *480*
Kirthivasan, N. 619(39), *656*
Kirudd, H. 1267(202a), *1343*
Kir'yanov, K. V. 235(51), *255*
Kisch, H. 1085(350), *1095*
Kise, M. 398(9a), *478*
Kishi, H. 100, 102(381), *188*, 937, 938(124), *1005*
Kishida, E. 855(104), *904*
Kishida, S. 1372(14), *1436*
Kishimoto, T. 1115(98), *1145*
Kishore, K. 359(128), *389*, 1129(292), *1150*
Kishore, Y. 82(287), *186*
Kita, Y. 447(221a, 221b, 222a–e, 222g, 222h, 222j), 448(222b, 225–227), *485*, 1178(58), 1227, 1228(144), 1230(148), 1246(172, 173), 1247(173), *1339, 1341, 1342*
Kitagaki, S. 447(222b, 222h), 448(222b), *485*
Kitagawa, K. 219(36), *221*
Kitagawa, S. 275(44), *326*
Kitagawa, T. 1209(106), *1340*
Kitai, S. 951, 953(196), *1007*
Kitajima, H. 694(262, 263), *710*
Kitajima, N. 414(87), *481*, 1194(86), *1339*
Kitamura, T. 1038(122b), *1090*
Kitamura, Y. 890(231), *907*
Kitano, F. 414(89b), *481*
Kitano, T. 414(89a, 89c, 90d), *481*
Kitano, Y. 1180(63a), *1339*
Kitao, T. 763(135), *834*
Kitchin, P. 1330(318), *1346*
Kiwi, J. 1084(327), 1086(360), *1095, 1096*
Kiyomori, A. 671(80), *706*
Kjällstrand, J. 303(141), *329*
Klapper, M. H. 1117(129, 131), 1142(380, 381), *1146, 1152*
Klärner, F. G. 322(245, 246), *331*
Klausner, J. F. 1084(308, 309), *1094*
Klecka, G. M. 1351(54), *1364*
Klein, C. 992(431), *1013*
Klein, P. 1363(220), *1368*
Klein, W. J.de 1285(234), *1343*
Kleine, H. P. 673(96), *707*
Kleinermanns, K. 20(114, 126, 127), 23(114), 24(126, 127), 35, 36(169), 78(280, 281), 106(114, 126, 127), 107(114, 126, 169), 143(577), 144(577, 596, 678), 145(622, 623), 147(596, 695–702), 148(695–697, 699–702), 149(697, 699), 156(702), 160(700, 702), *182, 183, 186, 192, 193, 195*, 375(247), *392*, 581(214, 215, 217, 218, 221, 222), *601*
Kleir, D. A. 336(6), *386*
Klementova, S. 1079(245), *1093*
Klemm, L. H. 807(271), *837*
Klenk, H. 631(80), *657*
Klenke, B. 1403(233a, 240b), 1408(280), *1445, 1447*
Kleshchevnikova, V. N. 737(58), *832*
Klessinger, M. 83, 84(298), *186*
Kletskii, M. E. 1024(59), *1088*
Klibanov, A. M. 411(70a), *480*, 1219(132), *1341*
Kline, T. B. 1244(170), *1342*
Klinman, J. P. 129(485), *190*
Kloc, K. 424(152c), *483*
Klockow, D. 936, 937(116), *1005*
Klohr, S. 305(151), *329*
Klopffer, W. 1020(35a), *1087*
Klopman, G. 102(395), *188*
Klotz, D. 1122(193), *1148*
Klotz, D. M. 925(50), *1004*
Knapp, F. D.Jr. 1351(50), *1364*
Knaudt, J. 1205(101), *1340*
Knee, J. L. 143(584), 144(684), *193, 195*
Knepper, T. P. 1349(35), *1363*
Knežević, A. 261, 263, 296(15), 308(162), *326, 329*
Knickmeier, M. 461(290), *487*
Knight, A. E. W. 144(681), *195*
Knight, M. H. 430(181b), *484*
Knights, S. G. 813(295), *837*
Knobler, C. B. 651(169), *660*, 1382, 1383(67c), 1419(348, 357), 1421(67c), 1424(397a, 399), 1428(425a, 426a), 1429(430a, 430b), *1438, 1450–1452*
Knochenmuss, R. 260(4), 323(4, 263, 265, 267), 324(269, 275), *325, 331, 332*, 492(18), 493(18, 20, 24–29), 515(18, 25, 27, 28), *525*
Knop, A. 1457(3), 1470(3, 60), 1471(73), 1481, 1483–1485(3), *1502–1504*
Knuth, M. L. 1357(133), *1366*
Knutsson, M. 947(168), 953(168, 280), 969(280), *1006, 1009*
Knuutila, M. 1357(134), 1363(220), *1366, 1368*
Knuutinen, J. 1363(220), *1368*

Knyazhansky, M. L. 1024(59), 1025(69), *1088*
Ko, F.-N. 851, 853, 857(61), 895(268), *903, 908*
Ko, J. W. 1427(423b), *1452*
Ko, S. 1234(153a), *1341*
Ko, S. W. 1393(148), 1400(208b), *1441, 1444*
Kobayakawa, K. 1084(307), *1094*
Kobayashi, S. 1054(162), *1091*
Kobayashi, H. 420(142b), 424(155a), *483*
Kobayashi, K. 680(159, 160), 681(161), 682(166), *708*, 874(171), *905*, 1027(91, 92), *1089*, 1382(69a, 69b), 1383(67e, 76), 1419(345, 349, 350a, 350b), 1421(67e), *1438, 1439, 1450*, 1489, 1491(110), *1504*
Kobayashi, M. 811(294), *837*
Kobayashi, S. 472(309a, 309b), 473(309b), *488*, 629(65, 66), 630(67, 68), 639(65, 67), *657*, 686(203, 204), 695(276–278), *709, 711*, 774(168–172), *835*
Kobayashi, T. 110, 112(447), *189*, 677(135), *707*
Koblik, A. V. 761, 766(123), 770(155), 806(253), 823(327, 330), 824(331, 332), *833, 834, 836, 838*
Koch, K. 807(282), *837*, 1294(254), *1344*
Koch, M. 1356(129), *1366*
Koch, W. 110(451), *190*, 368(192), *390*
Kochany, J. 1073(214, 215), *1092*
Kochi, J. K. 1192(78a), *1339*
Koch-Pomeranz, U. 460(284), *487*
Kocianova, S. 933, 937, 953(99), *1005*
Kocovsky, P. 677(124, 125), 691(239), *707, 710*
Kocturk, G. 1083(294), *1094*
Koda, T. 1194(86), *1339*
Kodama, T. 856, 863, 864(126), *904*
Kodera, M. 1209(106, 107), *1340*
Koelle, U. 353, 354(100), *388*
Koepp, T. 1080(260), *1093*
Koether, M. 1080(260), *1093*
Koga, K. 1278(220), *1343*
Koh, Y. 814(304), *838*
Kohler, G. 110, 129, 131(416), *189*, 992(431), *1013*, 1057(171), *1091*
Kohlrausch, K. W. F. 35(147), *183*
Kohno, Y. 891(247), *907*
Kohnstam, G. 419(123b), *482*
Koike, M. 1377(45b), *1437*
Koistinen, J. 1357(134), *1366*
Koizumi, K. 970(291), *1009*
Koizumi, M. 608(11), *656*
Kojima, M. 699(303, 306, 308), *711*
Kojima, T. 20, 35(105), *182*
Kojlo, A. 995(441), *1013*
Kojo, S. 855(104), *904*
Kok, P. M. 1109(43), *1144*
Kok, P. T. 1109(45), *1144*
Kokatnur, V. R. 854(99), *904*

Kokot, S. 929, 981(66), *1004*
Koksal, F. 729(42), *831*, 1104(56), *1106*
Kokube, Y. 1383, 1421(67l), *1438*
Kokubo, C. 699(313), *711*
Kokubun, H. 47, 143(206), *184*
Kokubun, T. 913(25), *1003*
Kolbar, D. 1470(50, 52), *1503*
Kolchmainen, E. 1418(343b), 1420(364, 365b), *1450*
Kolev, S. D. 983(377), 984(380), *1011*
Koleva, V. 367(169), *390*, 729(41), *831*
Koll, A. 20, 25(129), 144(621, 648–652, 654–661, 664–670), 145(612, 621, 625), 146(621, 639, 640), 171, 173(612), *182, 193, 194*, 355(104), 358(119, 120), 361(143, 144), 370(215), 371(216, 224, 230), 377(216, 230, 269), 385(216), 386(215, 303, 305, 306), *388, 389, 391–393*, 551(118–120), 556(118, 119), 557(120), *599*
Kollar, L. 1388(98), 1394, 1407(170b), *1439, 1442*
Kollman, P. 92(348), *187*
Kollman, P. A. 53(239), *185*, 896(271), *908*
Kolodziejski, W. 367(174), *390*
Kolonits, P. 759(115), *833*
Kolossváry, I. 551, 555(113), *599*
Koltunov, K. Yu. 744(75, 76), *832*
Komano, N. 773(160), *834*
Komaromi, I. 4, 100(2), *179*, 368, 372, 373(200), *390*
Komasa, A. 593, 595(281, 282), *602, 603*
Komata, M. 950, 953(187), *1007*
Komatsu, K. 1318(300b), *1345*
Komatsu, M. 702(332), *712*
Kometani, T. 641(136), *659*, 680(144), *708*
Kominos, D. 983(373), *1011*
Komissarova, N. D. 747(83), *832*
Komissarova, N. L. 651, 654(167), *659*, 747(82), *832*
Komiya, T. 1099(11–14), *1105*
Komiyama, S. 695(278), *711*
Kompan, O. E. 432(186), *484*, 1290, 1292(247), *1344*
Komura, E. 856, 863, 864(126), *904*
Komuro, E. 882(204), *906*
Kon, E. 1330(319, 320), *1346*
Kon, N. 1380(64), *1438*
Kondo, A. 699(309), *711*
Kondo, H. 447(222g), *485*, 641(136), *659*, 680(144), *708*
Kondo, K. 1495, 1498(159), *1505*
Kondo, S. 447(222f), *485*, 693(259), *710*
Kondo, T. 674(101), *707*
Kondrat, R. W. 300(128, 129), 301(128), *328*
Kondratenko, N. V. 312(193), *330*
Kondratenok, B. M. 937(121), *1005*
Kondrateva, E. I. 978(347), *1011*

Kong, D. 851(46, 52), 885(46, 215), 887(52), *903, 906*
Kong, S. 770(150, 151), *834*
Kongshaug, M. 1099(15), *1105*
Königsbacher, K. 433(189c), *484*
Konijnenberg, J. 1117(114), *1146*
Konishi, H. 275(44), *326*, 1164, 1290(36), 1324(309), *1338, 1345*, 1382(69a, 69b, 70), 1383(76), 1387(89), 1417(332), 1419(345, 349, 350a, 350b), *1438, 1439, 1448, 1450*
Konodo, H. 1470(69), *1504*
Konovalov, A. I. 1424(396), *1451*
Kontsas, H. 1347(5), *1363*
Koo, H. J. 1392(129b), *1441*
Kooijman, H. 1405(249b), 1422(378), *1445, 1450*
Koojiman, H. 1424(383), *1450*
Kook, S.-K. 1393(154), 1394(163), *1441, 1442*
Koopmans, T. A. 30(132), *182*
Kopach, M. E. 732(50), 735, 748(51), *832*
Kopecek, J. 15(73), *181*
Kopeckova, P. 15(73), *181*
Kopecni, M. M. 931, 935, 937(85), *1005*
Kopf, P. W. 1461, 1463, 1470(18), *1502*
Koppel, I. A. 311(186), 312(193), *329, 330*, 589(259), *602*
Koppenol, W. H. 637(111, 112), *658*
Kopteva, T. S. 383(280), *392*
Koptyug, V. A. 744(75, 76), 806(261), *832, 837*
Kopylova, T. N. 1017(8), *1087*
Kopytug, V. A. 269(34), *326*
Korbonits, D. 740(63), *832*
Korchagina, D. V. 616(32), *656*
Korenaga, T. 1360(171), *1367*
Korenman, Ya. I. 930(76, 81), 937(121), 983(374), 984(382), 986(399), 1001(475, 476), 1002(478), *1004, 1005, 1011, 1012, 1014*
Korner, S. K. 1424(400b), *1451*
Körner, W. 925(51, 52), 936(51), *1004*
Korobov, M. S. 1024(66), *1088*
Korshak, V. V. 235(51), *255*
Korth, H.-G. 45(195), 139(533), *184, 191*, 867(153, 155), 876, 882, 892(155), 893(153), *905*
Korvenranta, J. 361(141), *389*
Korytowski, W. 852(84), *903*, 1136, 1137(364), *1151*
Kosciuszko-Panek, B. 726(36), *831*
Kosemura, S. 1187(72, 73), *1339*
Koser, G. F. 1224(139), *1341*
Kossmehl, G. 950(188), *1007*
Köster, R. 415(102), *481*, 1117(121), *1146*
Kosturkiewicz, Z. 580(205), *601*
Kotake, Y. 864, 865(144), *905*
Kotani, E. 1288(239, 240), 1289(240), *1344*
Kotelevtsev, S. 1359, 1360(160), *1367*

Koten, G.van 93(364), *187*
Koten, M. A.van 807(279), *837*
Kotila, S. 669(59), *706*
Kotlan, J. 1325(311a, 311c), *1345*
Koto, S. 674(108), *707*
Kottas, P. 1082(284), *1094*
Kotte, H. 976(327), *1010*
Kotulák, L. 874(180), *906*
Koutek, B. 1048(155), *1090*
Kouwenhoven, H. W. 631, 640(74), *657*, 774(175, 179), *835*
Kovacek, D. 89(311), 97, 101(374), *186, 188*
Kováčik, V. 303, 305(146), *329*
Kovacova, J. 1082(272), *1093*
Kovács, A. 82(286), *186*, 238(71), 244(84), 256, 370(212), 384(293, 294), 385(294), *391, 393*, 551(74, 75, 113, 114), 553(74, 75), 555(113, 114), *598, 599*, 1100(30), *1106*, 1128(275, 276), *1149*
Kovacs, P. 759(115), *833*
Kovalev, V. 1427(409b), *1451*
Kovalev, V. V. 1372(21), *1436*
Kovaleva, T. A. 78(279), *186*
Kowalczyk, J. J. 450(241c, 246h), 453(241c), *486*
Kowalski, C. J. 452(242), *486*
Koyasu, K. 1324(309), *1345*
Kozak, V. P. 1359(156), *1366*
Kozaki, M. 1018(24), *1087*
Kozerski, L. 340(28), *387*
Kozikowski, A. P. 443(213), *485*
Kozishi, K. A. 414(93), *481*
Kozlowska, H. 853, 857(89), *904*
Kozmin, S. A. 450(239g, 239h), *486*
Kozuka, A. 1275(215), *1343*
Kozuka, T. 1333(329), *1346*
Kozukue, E. 913(13), *1003*
Kozukue, N. 913(13), *1003*
Kracke, B. 442(212), *485*
Kraft, D. 1393(157b), 1396(195a), 1406(257), 1432(449b), *1442, 1443, 1446, 1453*
Krajewski, P. 340(28), *387*
Krajnik, P. 1100(28), *1106*
Král, V. 1383, 1421(67n), 1430(442b), *1438, 1453*
Kramer, H. E. A. 1026(90), *1089*
Kramer, K. J. 305(148), *329*
Krämer, R. 144(672), *194*, 593(275), 595(275, 299), *602, 603*
Krantz, A. 551(88), *598*
Krapcho, A. P. 397(8), *478*
Krasauskas, V. 1020(33), *1087*
Krashakov, S. A. 891(243), *907*
Krasovskaya, L. I. 1470(51), *1503*
Kraus, G. A. 1045(136, 138), *1090*, 1264(196), *1343*
Kraus, G. J. 1420(359c, 361), *1450*

Krauss, M. 93–95, 106, 107, 109, 110(356), 112, 129, 132(455), *187, 190*
Krauss, P. 1124(242), *1148*
Krausz, F. 472(311h), *488*
Kravtchenko, D. V. 776(200), *835*
Kravtsov, D. N. 357(111), *388*
Krawiec, M. 1401(220c), *1444*
Krawlec, M. 1390, 1392(107), *1440*
Krebs, F. C. 1405(251a), 1406(264), *1446*
Kreevoy, M. M. 367(178), *390*
Kreider, E. M. 467(295a), *487*
Kreilick, R. W. 1121(191), 1123(211, 214), *1147, 1148*
Krekler, S. 950, 953(190), *1007*
Kremer, T. 93, 95, 104, 105(352), *187*, 311(188), *329*
Kremers, F. 460(278), *487*
Kresge, A. J. 1034(113), *1089*
Kretzschmar, H. J. 953, 963(260), *1009*
Krichna Peddinti, R. 1240(166), *1342*
Krimer, M. Z. 763(131), *834*
Krinsky, N. I. 851, 855(73), *903*
Krishna, C. M. 1218(130), 1223(135), *1341*
Krishna, M. C. 874(177), *906*
Krishnamurthy, R. 811(291), *837*
Krishnamurthy, V. V. 643(141), *659*
Krishnamurti, G. S. R. 973(314), *1010*
Krishnamurty, H. 631(83), *658*
Kristensen, T. 346, 354, 355, 366(51), *387*, 551(96), *598*
Krivdin, L. B. 350(79), *388*
Krivopalow, V. P. 340, 341(26), *387*
Kroebusch, P. L. 83(291), *186*
Krogh-Jespersen, K. 110–112(426), *189*
Krol, E. S. 874(178), *906*
Krol-Starzomska, I. 1024(64), *1088*
Kronberg, L. 1360(178, 179, 185), *1367*
Kropf, H. 799(235), *836*
Kroschwitz, J. I. 419(125), *482*
Krueger, C. A. 702(345, 346), *712*
Krueger, P. J. 234(49), *255*
Krüger, C. 461(290), *487*
Kryachko, E. 143(551), *192*
Kryachko, E. S. 31(145), 47, 48, 56(220), 58(244), 68, 69(145), 83(293, 299), 84–86, 88(299), 89(220, 299, 317a, 317b), 90, 91(299), 144(680), 145(632), 148(722, 723, 731, 732), 149, 158(731), 160(722, 723), 162(723), 171(761), 177(732), *183–186, 194–197*
Krygowski, M. T. 353(97), *388*
Krygowski, T. M. 93(362), *187*
Krylova, A. I. 357(111), *388*
Kryzanowski, E. 1067(198), *1091*
Kržan, A. 251, 252(92), *257*, 385(297), *393*
Krzykawski, J. 724, 725(27), *831*
Ksi, Y. 359(131), *389*
Ku, M. 1393(152b), *1441*

Ku, M.-C. 1394, 1407(170a), *1442*
Ku, Y. 1361(193), 1362(213), *1367, 1368*
Kuban, P. 971(293), *1009*
Kubelka, V. 265, 268, 269(27), *326*
Kubicki, M. 233(41), *255*
Kubinec, R. 1102(44), *1106*
Kubinyi, M. 20, 22, 23, 35, 36, 38(111), *182*, 368, 370–375(198), *390*
Kubo, M. 1396(184), *1443*, 1498(175), *1506*
Kubo, Y. 1396(184), *1443*
Kubota, L. T. 929(65), 974(319), 981(65), *1004, 1010*
Kubota, M. 677(129), *707*
Kubota, T. 501(79), *526*, 1495(160, 162, 163), 1498(160), *1505, 1506*
Kuboto, Y. 1362(211), *1368*
Kuc, T. 377(262), *392*
Kuchel, P. W. 338(16), *386*
Kuchmenko, T. A. 930(76), 983(374), 984(382), 985(390), 986(399), 1001(476), *1004, 1011, 1012, 1014*
Kuck, D. 83(292), *186*, 260(5, 6), 263(6, 17), 265(25), 268(25, 33), 270(25, 35, 36), 273(25), 277(46–48), 284(17, 25, 46, 53, 54), 285(57), 293(17, 25, 46, 74, 75), 296(6), 298(25), 300(111–117), 302(140), 303(25, 54), 325(5), *325–328*
Kuck, J. A. 1118(133), *1146*
Kudin, K. N. 4, 100(2), *179*, 368, 372, 373(200), *390*
Kudo, K. 1383, 1421(67h), *1438*
Kudo, M. 702(344), *712*
Kudryavtsev, V. V. 1030(105), *1089*
Kudryavtseva, L. A. 740(64, 65), *832*
Kuhn, W. 450(246e), *486*
Kuhnert, N. 1406(271), *1446*
Kuivilla, H. G. 414(97b), *481*
Kulagin, V. N. 235(51), *255*
Kulagowski, J. J. 666(47), *705*
Kulbida, A. 377(257), *392*
Kulbida, A. I. 143(555), *192*, 377(258), *392*
Kulkarni, G. U. 216(28), *221*
Kulkarni, S. J. 630(70), 631(73), *657*, 680(152), *708*
Kulmacz, R. J. 15(72), *181*
Kulshrestha, S. K. 1050(161), *1091*
Kumada, H. 1380(63), *1438*
Kumagai, H. 1380(61a, 61b, 62), *1438*
Kumagai, N. 695(273, 274), *710, 711*
Kumamoto, M. 170(740), *196*
Kumanotani, J. 272(37), *326*
Kumar, A. 349(76), *388*, 724, 725(27), *831*
Kumar, I. S. 617(34), *656*
Kumar, M. 845, 880(26), *902*, 1111(72), *1145*, 1374(36c), *1436*
Kumar, R. 774(177), *835*
Kumar, S. 634(97), *658*, 1402(227), *1445*
Kumaradhas, P. 216(28), *221*

Kunai, A. 414(89c, 90c, 90d), *481*
Kündig, E. P. 783(222), *836*
Kunert, M. 6(26), *180*
Kunitake, T. 1435(464b), *1454*
Kunitomo, J. 807(275), *837*
Kuntz, K. W. 702(346), *712*
Kunz, U. 1374(33a), *1436*
Kunze, K. L. 721(23), *831*
Kuo, C. H. 1362(201), *1368*
Kuo, W. J. 608(7), *656*
Kuo, W. S. 1085(349), 1086(363), *1095, 1096*
Kuo, Y. H. 1018(20), *1087*
Kupchan, S. M. 1275(212, 213), 1278(212, 218), *1343*
Küpper, J. 144(679), *195*, 377(253), *392*, 578(194), 581(216), *601*, 992(428), *1012*
Kuramoto, R. 1180(63b), *1339*
Kuran, P. 1102(43, 44), *1106*
Kuranaka, A. 844(22), 856, 865(127), *902, 904*
Kurdistani, S. K. 1429(429), *1452*
Kuribayashi, M. 953, 962(248), *1008*
Kurihara, H. 693(260, 261), *710*
Kurita, E. 1421(373), *1450*
Kurita, K. 1495(137), *1505*
Kurkovskaya, L. N. 349(66), *387*
Kuroda, H. 1216(125), *1341*
Kuroda, K. 1159, 1207(28), *1338*
Kuroda, N. 951, 953(193, 194), *1007*
Kuroda, Y. 414(89a–c, 90d), *481*
Kuromizu, H. 762(126), *834*
Kurosawa, K. 1204(98), 1285(236), *1340, 1343*
Kurreck, H. 1123(215), *1148*
Kürti, L. 1227(141), *1341*
Kuruc, J. 1102(43, 44), *1106*
Kurucz, C. N. 1100(33), *1106*
Kurur, N. D. 634(97), *658*
Kurz, M. E. 405(40), *479*
Kusakabe, K. 695(278), *711*
Kusama, H. 811(293), *837*
Kusano, T. 1385(83), *1439*
Kuskov, V. L. 550(71), *598*
Kusukawa, T. 1209(108), *1340*
Kusunoki, J. 1025(86), *1089*
Kutal, C. 1082(270), *1093*
Kuthan, J. 823(329), *838*
Kutnetsov, S. L. 347(54), *387*
Kutty, T. R. N. 1083(299), *1094*
Kuzmin, M. G. 492, 493, 501(13), 515(96), 525, 527, 1020(35d), 1073(223), *1087, 1092*
Kuzmin, V. A. 1108, 1118(19), 1128(277), 1130(19, 319), *1144, 1149, 1150*
Kuznetsov, A. I. 1470(53), *1503*
Kuznetsov, E. V. 432(186), *484*
Kuznetsov, J. N. 1470(66), *1503*
Kuznetsova, N. 1085(356), *1096*
Kuznetsova, R. T. 1016(3), 1017(8), *1087*

Kwan, K. M. 1374(36a), *1436*
Kwart, H. 773(164), *834*
Kwiatkowski, E. 726(36), *831*
Kwiatkowski, G. T. 1274(208), *1343*
Kwiatkowski, M. 726(36), *831*
Kwiecien, B. 340(28), *387*
Kwo, O. 244(85), *256*
Kwon, B. G. 1362(205), *1368*
Kwon, O. 242(82), *256*, 551(79, 102, 116), 553(79, 102), 556(116), *598, 599*
Kwon, Y. 242(82), 244(85), *256*, 551(79, 102, 116), 553(79, 102), 556(116), *598, 599*
Kynaston, W. 35(163), 50(165), 95, 107, 109(373), *183, 188*
Kyritsakas, N. 1401(219b, 220b), *1444*

Laaksonen, A. 93, 95(361), *187*, 384(290), *393*
Laarhoven, L. J. J. 45(197), 139(531), 179(197), *184, 191*
Laatikainen, R. 349(74), *388*
Labbe, P. 977(343), *1011*
Labib, A. 472(311d), *488*
Labrosse, J.-R. 664(39), *705*
Labudzińska, A. 170, 171, 177(759), *197*, 992(426), *1012*
Lacaze, P. C. 1157(12), *1337*
Lach, F. 462(274), 463(276b), *487*
Lachkar, A. 586, 587(243), *602*
Lacorte, S. 1348, 1357(11), *1363*
Lacour, J. 1318(301b), *1345*
Ladd, J. A. 144(602, 603), *193*, 363, 364(147), *389*
Lafleur, L. E. 933, 937(101), *1005*
Lafont, F. 972(304), *1010*
Lafontaine, P. 937, 940(130), *1006*
Lagier, C. M. 367(172, 173), *390*
Lagouri, V. 998(454), *1013*
Lahmani, F. 78(282), *186*
Lahtipera, M. 1363(220), *1368*
Lai, C.-H. 1234(153b), *1341*
Lai, D. K. W. 20, 45, 140(125), *182*
Lai, J.-Y. 418(120), *482*
Laine, J. 1080(247), *1093*
Laine, M. M. 1347(8), *1363*
Laines, G. L. 1079(239), *1092*
Laing, D. G. 978(345), *1011*
Lal, G. S. 452(242), *486*, 687(216), *709*
Lalande, M. 1002(482), *1014*
Lam, E. 1071(207), *1092*
Lam, F. L. Y. 1081(268), *1093*
Lam, Y. S. 994(436), *1013*
Lamare, V. 1415(313), *1448*
Lamartine, M. 200(1), *220*
Lamartine, R. 609(18), *656*, 1372(11, 19, 28, 30), 1378(52b), 1394(169), 1401(216b), 1402(224c–e), 1405(251b), 1415(312,

318a), 1417(330c), *1436, 1437, 1442, 1444–1446, 1448*
Lambert, C. 104(399), *188*
Lamparski, L. L. 1360(162), *1367*
Lampe, J. W. 786(229, 230), *836*
Lampert, H. 20, 21, 24, 35, 37(124), *182*, 339, 366(21), 371, 372, 375(234), 377(234, 268), 378, 380(268), *386, 391, 392*, 551, 553(104), *599*
Lamprecht, G. 953, 962(258), *1009*, 1352(67), *1364*
Lamuela-Raventós, R. M. 920, 965(31), 990(419), 1001(473), *1003, 1012, 1014*
Lamy, M. 1361(194), *1367*
Lancaster, J. E. 913(11, 12), *1003*
Lanccake, P. 1217(128), *1341*
Land, C. M. 323(257), *331*
Land, E. J. 129, 131(500), *191*, 1098(3), 1100(21), *1105*, 1110(56), 1113(86, 87), 1114, 1115(86), 1117(122–128), 1118(124), 1127(266, 271, 272), 1128(86, 272), 1129(56, 122, 295), 1130(56, 86, 122–125, 295, 318, 325), 1131(316), 1134(316, 325), 1135(272), 1136(358, 359, 364), 1137(364), *1145, 1146, 1149–1151*, 1223(136), *1341*
Landais, Y. 1300(264), *1344*
Landells, R. G. M. 402(26b), *479*
Landi, L. 852(77), *903*
Landis, P. S. 400(16), *478*
Landis, W. 405(44), *479*
Landrieu, P. 227(9), *254*
Landuyt, L. 119(463), *190*
Landy, D. 992(433), *1013*
Landzettel, W. J. 951, 953(191), *1007*
Lane, C. 414, 415(97d), *481*
Lane, D. C. 293(73), *327*
Lanfranchi, M. 1374, 1378(37), *1436*
Lang, A. 1220(134), *1341*
Lang, G. L. 325(278), *332*
Lang, J. 1415(316), *1448*
Lang, K. 1080(250), *1093*
Lang, M. J. 522, 523(97), *527*
Lang, S. 19(93), *182*
Langbein, H. 236, 237(58), *256*
Lange, I. Ya. 145(627), 147(643, 645), *193, 194*
Lange, P. J. de 1421(369), *1450*
Langenaeker, W. 89, 90(324, 331), *187*
Langer, P. 442(212), *485*
Langer, U. 367(171), *390*
Langgård, M. 368(199), 380(274), 381, 382(199), *390, 392*
Langner, R. 145(628), *193*
Langoor, M. H. 368, 369(181), *390*, 551, 556(115), *599*
Langvic, V. 1360(185), *1367*
Langvik, V. 1360(182, 183), *1367*

Lanjun, W. 702(335), *712*
Lankin, D. C. 1260(190), *1342*
Lantenschlager, D. P. 1349(39), *1364*
Lanteri, P. 1372(11), *1436*
Lapa, R. A. S. 986(395, 397), *1012*
Lapachev, V. V. 340, 341(26), *387*
Lapasset, J. 727(38), *831*
Lapcik, L. 1084(328), *1095*
Lapinski, L. 370, 382(211), *391*
Laplante, F. 1361(191), *1367*
Lappert, M. F. 460(280b), *487*
Lapworth, A. 715, 716(5), *830*
Laraba, D. 913(10), *1003*
Laramee, J. A. 110(432), *189*
Larbig, H. 451(248a), *486*
Lardeux-Dedonder, C. 144(600), *193*
Larock, R. C. 664(31, 36), *705*
Larrauri, J. A. 856, 857(124), *904*
Larroque, S. 1485(101), 1486(99–101), *1504*
Larrow, J. F. 685(199), 701(320), *709, 712*
Larsen, M. 1405(251a), 1406(264, 267b), *1446*
Larsen, N. W. 20, 22(107, 108), 24(108), 35(107, 155), 53(155), 85(305), *182, 183, 186*
Larson, R. A. 1358(139), *1366*
Lascombe, J. 35(161), *183*
Lash, T. D. 737(56), *832*, 1370, 1430(6), *1435*
Lasne, M. C. 647(153), *659*
Lassen, P. 339(18), *386*
Laszlo, P. 471(304), *488*, 806(262), *837*
Laszlo-Hedwig, Z. 1462(23), *1502*
Latajka, Z. 20, 22, 23, 35–38(112), 82(288), 144(608), *182, 186, 193*, 372, 375(236), 384(291, 292), *391, 393*
Lathioor, E. C. 1021(43, 44), *1088*
Lattman, M. 1393(145a, 145b), *1441*
Lau, C. K. 683(175), *708*
Lau, W. F. 1203(97), *1340*
Lau, Y. 83(295), *186*, 261(10), *326*
Laue, T. 640(135), *659*
Lauer, M. 651(169), *660*
Laungani, D. 349(58), *387*
Lauransan, J. 575(187), *600*
Laurence, C. 531(18, 19), 532(18), 533(22), 534(23), 541(43), 543(50), 547(18), 551, 553(87), 554, 555(138), 558(19, 146), 559(138), 575(186, 188), 576(19, 146, 189), 580(188, 211), 581(211), 583(223), 586, 587(18, 19, 23, 146, 186, 188, 239–256), 589(18, 19, 244, 248, 252, 254, 256), 590(138), 591(266, 267), 592(267), *597–602*
Laurent, A. 6(33–35), 144, 145(33), *180*
Laureys, C. 583(230), *601*
Lautens, M. 665(44), *705*
Lauterbach, M. 542(48), *597*
Lauterwein, J. 340, 341, 352(25), *386*
Lavallee, P. 1466(44), *1503*

Lavee, S. 953, 961(240), *1008*
Lavigne, J. B. 400(17), *478*
Lavorato, D. J. 110(451), *190*
Lawrence, R. V. 814(307), *838*
Laws, W. R. 106, 148(404), *188*
Lawson, A. J. 638(115), *658*
Lay, J. O. Jr. 308(159), *329*
Laya, M. S. 471(305), *488*
Lazarev, G. G. 1018(23), *1087*
Lazarova, T. I. 422(147b), *483*, 776(207), *835*
Lazaru, G. G. 550(71), *598*
Le, H. T. 84(303), 110(448–450, 452), 115(303), 123(450), *186, 189, 190*
Le, N. T. 89, 90(331), *187*
Le, T. H. D. 230(26), *254*
Leader, H. 403(27), *479*
Leaffer, M. A. 400(19), *478*
Leahy, D. K. 664(38), *705*
Leal, A. I. 1080(254), *1093*
Leal, P. 1039(123), 1040(125), 1055(169), *1090, 1091*
Leal, P. C. 1285(233), *1343*
Leanza, R. 18(87), *181*
Leary, G. J. 8(43), *180*
Leavy, P. J. 745, 746(80), *832*
Lebedev, Ya. S. 550(71), *598*
Lebedev, Y. S. 1104(58), *1106*
LeBlanc, J. R. 1459, 1475, 1484(6), *1502*
Lebouille, P. H. P. 1406(266), *1446*
Le Bras, J. 748(84–86), *832, 833*
Lebrun, S. 784(227), *836*
Leca, J.-P. 1486(100), *1504*
Le Calve, J. 144(682, 683), *195*
LeClaire, J. E. 109(415), *189*
Leclerc, P. 370(214), *391*
Lecomte, L. 663(19), *705*
Lederer, K. 949, 953(186), *1007*
Lee, A. W. M. 994(436), *1013*
Lee, B. L. 920, 949, 953(33), *1003*
Lee, C.-H. 769(147), *834*
Lee, D. S. 1085(346), *1095*, 1362(205), *1368*
Lee, D. W. 946(163), *1006*
Lee, F. L. 844, 847, 855, 860–862, 864, 897, 898(18), *902*
Lee, G. 631(75), *657*
Lee, G.-D. 1082(289), *1094*
Lee, G. H. 1407(273b), *1446*
Lee, H. 1353(91), *1365*
Lee, H. B. 932, 933(95), *1005*
Lee, H. C. 1428(427), *1452*
Lee, H. K. 946, 953(161), *1006*
Lee, J. 493(40, 42, 43), 498(42), *525*, 1017(7), *1087*
Lee, J. B. 799(237), *836*
Lee, J. C. 663(14), *705*
Lee, J. L. 149, 158, 161(736), *196*
Lee, J. Y. 161(737), *196*
Lee, K. 1361(193), *1367*

Lee, K.-C. 1362(213), *1368*
Lee, L. 1120, 1140(184), *1147*
Lee, M. K. T. 555(140), *599*
Lee, N. H. 664(31), *705*
Lee, S. A. 403(32), *479*
Lee, S.-G. 418(120), *482*
Lee, S. H. 1347(7), *1363*, 1400(208b), *1444*
Lee, S. J. 1393(154), *1441*
Lee, S. K. 460(286), *487*
Lee, T.-H. 1234(153a), 1237(161a, 161b, 162a, 162d), *1341, 1342*
Lee, W. 702(339, 340), *712*
Lee, W.-H. 944(141), *1006*
Lee, Y. 663(17), *705*
Lee, Y. C. 1237(160b), *1342*
Lee, Y. F. 1082(286), *1094*
Lee, Y. R. 1044(133), 1045(135), *1090*
Lee, Y. T. 83(291), 137(522), *186, 191*
Leese, T. 456(259), *487*
Lefebvre, O. 428(171), *483*
Legars, P. 663(19), *705*
Legnaro, E. 646(146), *659*
Legoff, D. 559, 568, 586(154), *600*
LeGourrierec, D. 1025(88), 1047(147), *1089, 1090*
Le-Gresley, A. 1406(271), *1446*
Lehman, L. S. 884(208), 885(210), *906*
Lehmann, E. 323(265–267), *331*
Lehn, J.-M. 761(119), *833*
Lehnert, E. K. 678(140), *707*
Lehnig, M. 637(106), *658*
Lehotay, J. 953, 962(255), *1009*
Lei, B. 1237(164), *1342*
Leigh, D. A. 1419(352), *1450*
Leigh, W. J. 1021(43, 44), *1088*
Lein, E. J. 871, 898(164), *905*
Leis, J. R. 638(117), *658*
Leitich, J. 1136(360), *1151*
Leity, J. L. 95(368), *188*
Leize, E. 1427(421a), *1452*
Lemaire, J. 1057(175, 176), 1073(217, 220), *1091, 1092*
Lemal, D. M. 669(60), *706*, 826(338), *838*
Lcmanski, D. 580(205), *601*
Lembach, G. 30(136), *183*
Lemercier, J. N. 637(107, 108), *658*
Lemmetyinen, H. 1073(224), *1092*, 1117(114), *1146*
Lemon, P. H. R. B. 1489(104), *1504*
Lemoult, P. 234(50), 251, 252(90), *255, 257*
Lempers, H. E. B. 460(286), *487*
Lenchitz, C. 236(55), *255*
Lenga, R. E. 924, 925, 927(39, 40), *1003, 1004*
Lengfelder, E. 1118(143, 146), 1119(146, 163), 1130(143), *1146, 1147*

Lenghaus, K. 1476(84), 1478(85), 1482(84), 1484(94), 1485(84), 1489, 1490(113, 114), *1504*
Lenicek, J. 933, 937, 953(99), *1005*
Leo, A. 537, 539, 559(38), *597*
Leo, A. J. 543(50), 545(54), *598*
Leong, K. 724, 725(27), *831*
León-González, M. E. 953, 962(253, 257), *1008, 1009*
Leonis, M. 1371, 1372(8), *1435*
Leonov, A. I. 1470(70), *1504*
Lepage, G. 854(102), *904*
Le Perchec, P. 559, 567(156), *600*
Lepoittevin, J. P. 650(159), *659*
Leppard, D. 1026(90), *1089*
Leprori, L. 101(384), *188*
Le Questel, J.-Y. 575(186), 580, 581(211), 583(223), 586, 587(186, 241, 243–245, 255), 589(244), *600–602*
Lerner, R. A. 806(263), *837*
Le Roux, E. 949(183), *1007*
Leroux, N. 37(181), 145, 146(613), *183, 193*, 371, 372, 377(220), *391*, 566(174), 573(180), *600*
Leroy, E. 236, 237(59), *256*
Les, A. 1025(68), *1088*
Lesclaux, R. 138(527), *191*
Lessard, J. 1361(191), *1367*
Lessinger, L. 546(56), *598*, 998(459), *1013*
Lester, D. J. 681(163), *708*
Lester, G. 1470(61), *1503*
Lester, G. R. 264(22), *326*
Lester, J. N. 1356(120), *1365*
Leszczynska, I. 998(458), *1013*
Létoffé, J.-M. 1415(318a), *1448*
Lettmann, C. 1085(350), *1095*
Letzel, M. C. 320(241), 322(241, 252), *331*
Letzel, T. 319(232), *331*
Leuchs, D. 433(188b), *484*
Leunens, V. 566(175), *600*
Leung, H. W. 300(109), *328*
Leutwyler, S. 20, 35(122), 36(122, 175), 78(278), 147(122, 175, 709–711), 148(710, 711), *182, 183, 186, 195*, 375(242, 243), *392*, 493(19, 20, 27), 515(27), *525*, 577(191), 578(193), 581(193, 219), *600, 601*
Lev, O. 1362(210), *1368*
Levelut, A.-M. 1382(68), 1417(331), *1438, 1448*
Leverd, P. C. 1409(289b), *1447*
Levetsovitou, Z. 942, 943, 953(138), *1006*
Levett, G. 852(76), *903*
Levin, R. D. 98(376), *188*, 261(7), 265, 268, 269, 272(29a), 311, 312(7), *325, 326*
Levine, J. A. 1392(132), *1441*
Levine, R. 778(208), *835*

Levsen, K. 110(427), *189*, 299(99, 100), *328*, 930(75), *1004*, 1353(90), *1365*
Levy, D. 415(101), *481*
Levy, D. H. 20, 24, 106(118), *182*
Levy, J. B. 551(125, 129), *599*
Levy, O. 1038(122a), *1090*
Lewis, C. E. 913(11, 12), *1003*
Lewis, C. W. 1475(81), 1485, 1486(102), *1504*
Lewis, D. A. 874(175), *905*
Lewis, E. S. 402(25), *479*
Lewis, G. E. 402(26b), *479*
Lewis, G. N. 68(256, 257), *185*
Lewis, I. C. 1124(237), *1148*
Lewis, M. D. 424(153b), *483*
Lewis, N. 1086(366), *1096*, 1227(143), *1341*
Lewis, N. J. 416(113), *482*
Lewis, P. T. 1383(67f), 1421(67f, 368), *1438, 1450*
Lewis, R. J. Sr. 924, 925, 927(41), *1004*
Ley, K. 1109(27), 1120(168), 1130(329), *1144, 1147, 1150*
Ley, S. V. 1330(324, 325a–c), 1332(325a–c), *1346*
Lhoste, P. 664(37, 39), *705*
Lhoták, P. 1391(122c), 1404(246b), 1405(252), 1415(315–317, 318b, 323), 1427(407b), 1432(122c, 451), *1440, 1445, 1446, 1448, 1451, 1453*
Li, A. X. 632(85), *658*
Li, C. 968(274), *1009*, 1082(274), *1093*, 1140(375), *1151*
Li, C.-J. 684(190), *709*
Li, G. 763(136), *834*, 1082(282), 1083(295), *1094*, 1354(109), *1365*
Li, G.-Y. 1082(285), *1094*
Li, H. 670(71), *706*, 1318(303), *1345*
Li, H.-D. 1082(285), *1094*
Li, H. X. 620(41), *657*
Li, J. 57(242), *185*, 975(323), *1010*, 1397, 1398, 1403(197a), *1443*
Li, J. P. 424(157b), *483*
Li, J.-S. 1427(420a, 420b), *1452*
Li, K. 19(102), *182*, 930(79), 932(97), 937(79), *1004, 1005*
Li, L. 281(51), *327*
Li, M. 305(151), *329*
Li, N. 95, 101(371), *188*
Li, T. S. 632(85), *658*
Li, W. 990(417), *1012*
Li, X. 1024(61), *1088*
Li, X. W. 620(45), *657*
Li, Y. 1024(61), *1088*
Li, Y.-S. 563(169), *600*
Li, Z. 763(136), *834*
Li, Z.-J. 674(104), *707*
Li, Z.-T. 1432(445), *1453*
Lialulin, A. L. 977(341), *1010*
Lian, X.-D. 1432(445), *1453*

Lianfeng, Z. 1083(304), *1094*
Liang, Y. M. 639(124), *659*
Liao, C.-C. 1234(153a, 153b), 1236(154–157), 1237(158, 159, 160a, 160b, 161a, 161b, 162a–e, 163), 1240(166), *1341, 1342*
Lias, S. G. 5(10), 83(296a), 98(376), *179, 186, 188*, 261(7, 8), 265(8, 29a), 268, 269, 272(29a), 310(8), 311, 312(7), *325, 326*
Liashenko, A. 368, 372, 373(200), *390*
Liashenko, M. A. 4, 100(2), *179*
Liber, K. 1357(133), *1366*
Liberatore, L. 930(77), *1004*
Liberio, R. 686(212), *709*
Libus, W. 558(149), *600*
Lichtig, J. 984(383), *1012*
Liddel, U. 47(213–215), *184*, 555(141), *599*
Liebeskind, L. S. 1278(224), *1343*
Liebler, D. C. 845(38, 39), 846(38), 852(79), *902, 903*
Liebman, J. F. 224(1, 3), 225(1, 4), 227(1, 4, 5), 229(1), 230(5), 233(1), 237(61, 64), 247(89), *254, 256, 257*, 261(7), 265, 268, 269, 272(29a), 311, 312(7), *325, 326*
Liebskind, L. S. 451(247), *486*
Liehr, J. G. 291(67), 301, 302(137), *327, 328*
Lieman, J. F. 98(376), *188*
Liepa, A. J. 1275(212), 1278(212, 218), *1343*
Liepa, J. 1278(221), *1343*
Liepins, E. 347(54), *387*
Lifshitz, C. 110(435, 436), 121(436), *189*, 265(23), *326*
Lightner, D. A. 268(30), *326*
Ligor, T. 953, 968(278), *1009*
Ligtenbarg, A. G. J. 359(130), *389*
Liguori, L. 607(6), *656*
Likhohyot, A. 83(296b), *186*
Lilla, G. 274(41), *326*
Lillo, M. P. 730(45), *831*, 1025(70b), *1088*
Lim, A. S. C. 1463(40, 41), 1465, 1467(42), 1471(74), *1503, 1504*
Lim, C.-J. 1082(289), *1094*
Lim, K.-T. 1082(289), *1094*
Lima, A. W. O. 978(348), *1011*
Lima, E. C. 984(383), *1012*
Lima, J. L. F. C. 986(395, 397), *1012*
Limbach, H.-H. 364(155), 367(171, 178), *389, 390*, 503(90), *527*
Lin, C.-I. M. 818(321), *838*
Lin, C. N. 895(268), *908*
Lin, C. Y. 129, 135, 137(487), *190*
Lin, G.-Q. 776(199), *835*
Lin, J. 1213(113), *1340*
Lin, J. G. 1085(340), *1095*
Lin, K. J. 1100(33), *1106*
Lin, L. 1393(152b), *1441*
Lin, L.-G. 1391, 1392(122a), 1393(122a, 142), 1394(142, 168, 170a), 1407(142, 170a), *1440–1442*
Lin, M. C. 129, 135(487), 137(487, 524), 138(524, 525), *190, 191*
Lin, Q. 493, 515(28), *525*, 1405(256b), *1446*
Lin, R.-S. 634(98), *658*, 1402(225), *1445*
Lin, T. 47, 53(224), *184*
Lin, W.-L. 1407(273b), *1446*
Lin, X.-F. 674(109), *707*
Lin, Y.-L. 1236(154), *1341*
Lin, Z. 929, 985(69), *1004*
Lin, Z.-P. 1432(445), *1453*
Lina, M. 616(26), *656*
Lind, J. 139(534), *191*, 1108(24, 25), 1118(24), 1136(361, 363), 1137(363), 1138–1140(24), 1142(24, 377), 1143(24, 382), *1144, 1151, 1152*
Lindberg, J. J. 236, 248(56), *255*
Lindblom, R. O. 19(96, 98), *182*
Linde, H. v. d. 1115(112), *1146*
Linde, R. G. II 664(30), *705*
Lindeman, S. V. 200(7), *220*
Linden, S. M. 575(185), 580(185, 204), *600, 601*
Lindgren, A. 976(324), 977(337), *1010*
Lindholm, N. 237(61), *256*
Lindley, J. 670(68), *706*
Lindner, R. 143(591), *193*
Lindon, J. C. 359(125), 366(168), *389, 390*, 726(30, 31), *831*
Lindsay, J. R. 1209(109), *1340*
Lindsey, A. S. 716(10), *831*
Lindstrom-Seppa, P. 1359, 1360(160), *1367*
Lineberger, W. C. 95, 98, 99, 131, 132, 139(370), *188*
Linell, R. H. 530, 557(5), *596*
Link, M. 1355(118), *1365*
Linnane, P. 1419(352, 354), *1450*
Linnell, R. H. 368(183), *390*
Linsay Smith, J. R. 1211(110), *1340*
Linstrom, P. J. 261(9), *326*
Lin-Vien, D. 369, 371(208), *391*
Lios, J. L. 357(109), *388*
Liotard, D. A. 57(242), *185*
Liotta, C. L. 621(48, 49), 631(81), *657, 658*
Liou, C. C. 308(160), *329*
Liou, J. 296(85), *327*
Lipczynska-Kochany, E. 1072(211), 1073(211, 213, 214), *1092*
Lipert, R. J. 30(134), 129(475), 143(585–587), 147(585–587, 693, 694), *182, 190, 193, 195*
Lipkowski, J. 1417(328), 1418(340), *1448, 1450*
Lipowski, J. 1393(161), *1442*
Lippmann, T. 1417(330b), 1424(392, 393), *1448, 1450*
Lis, T. 352(91), *388*, 551, 557(121), 580(202), *599, 601*
Lisá, E. 874(179–181), *906*

Liska, I. 1354(106), *1365*
Lissel, M. 663(15), *705*, 1204(99), *1340*
Liti, K. 148, 160(721), *196*
Little, R. D. 1156, 1158, 1175(5), *1337*
Littlejohn, D. 985(391), *1012*
Littler, J. S. 472(311b), *488*
Litvinenko, L. M. 631(79), *657*
Litwak, A. M. 1410(291a, 291b), *1447*
Liu, D. 982(364), *1011*
Liu, G. 4, 100(2), *179*, 368, 372, 373(200), *390*, 699(305), *711*
Liu, H. 450(246g), *486*, 1084(310), 1087(382), *1094, 1096*
Liu, J. 855, 856(105), *904*, 929, 985(69), *1004*, 1100(31), *1106*
Liu, J.-M. 1432(452), *1453*
Liu, M. 1081(265), *1093*
Liu, M. H. 955(205), *1007*
Liu, R. 93, 94(360), 129(509, 510), 131(510), 132(360, 509, 510), 133(509, 510), 137, 138(524), *187, 191*, 982(364), *1011*
Liu, R.-Z. 148(728), *196*
Liu, S. Y. 296(83), *327*, 620(41), *657*
Liu, W. 1393(152b), *1441*
Liu, W.-C. 1237(163), *1342*
Liu, W. Z. 311(190), *330*
Liu, X. 1361(198), *1367*
Liu, X. Y. 620(45), *657*
Liu, Y.-C. 855(114, 115), 892(250), *904, 907*
Liu, Z.-L. 855(114, 115), 892(250), *904, 907*, 1080(258), *1093*
Livengood, C. D. 1361(197), *1367*
Llamas-Saiz, A. L. 580, 581(212), *601*
Llauro, M. F. 559, 567(156), *600*
Llompart, M. 932, 937(94), *1005*
Llompart, M. P. 930(79), 932(97), 937(79), *1004, 1005*, 1352, 1353(63), *1364*
Llompart-Vizoso, M. P. 933(104), *1005*
Lloyd, G. T. 949, 953(185), *1007*
Lluch, J. M. 1047(149), *1090*
Lo Balbo, A. 953, 956(211), *1007*
Lobanov, G. A. 241, 251, 252(81), *256*
Lobbes, J. M. 953, 962(249), *1008*
Lochmann, L. 677(138), *707*
Locke, D. C. 1354(109), *1365*
Locke, S. 901(288), *908*
Locke, S. J. 851(47, 54, 55, 62), 857(62), 863, 882(55), 884(207), 885(47, 54, 55, 207, 213–215), 886, 887(55), 890(214), *903, 906*
Loebach, J. 450, 453(239f), *486*
Loebach, J. L. 450, 453(241e), *486*, 698(293), *711*
Loehlin, J. H. 219(38), *221*
Loehr, T. 1216(119d), *1340*
LoFaro, A. D. 45, 139, 140(192), *184*
Löffler, F. 1397(196), 1399(203), *1443, 1444*
Logani, M. K. 852, 854, 855, 857(83), *903*

Loginov, D. B. 977(341), *1010*
Logue, E. A. 806(256, 257), *837*
Loh, E. 1104(57), *1106*
Loh, W. 957(213), *1007*
Loike, J. C. D. 1216(123), *1341*
Loizou, G. 766(141), *834*
Lokshin, V. 770(157), *834*
Loll, P. J. 11(51, 52), *180*
Lombardi, G. 932, 937(93), *1005*
Lombardi, P. 953, 961(236), *1008*
Lombardo, N. 953, 961(239), *1008*
Lommatzsch, U. 147(714), *195*
Long, Y. H. 953, 965(265), *1009*
Longchambon, F. 550(68), *598*
Longeray, R. 1372(11), *1436*
Longevialle, P. 300(102), *328*
Longfield, T. M. 412(74c), *480*
Looney, M. A. 807(274), *837*, 1230(149), *1341*
Looney, M. G. 1459(9), 1463(9, 39, 41), 1465, 1467(42), 1470(59), 1471(59, 74), 1500(178), *1502–1504, 1506*
Loos, M. 69(269), *185*
Lopes, J. L. C. 895(266), *908*
López, J. 932(96), *1005*
López-Erroz, C. 953, 960(234), *1008*
Lopez-Mardomingo, C. 237(61), *256*
Lopez-Munoz, M. J. 1083(296), *1094*
Lopez-Prados, J. 1409(287), *1447*
Lora, S. 979(354), *1011*
Loreng, L. 1325(312), *1345*
Lorente, P. 503(90), *527*
Lorentzon, J. 20, 106(119), *182*
Lorenzo, R. A. 930(79), 932(97), 937(79), *1004, 1005*
Lorenzo-Ferreira, R. A. 933(104), *1005*
Lork, E. 743(70), *832*
Loshadkin, D. 874, 875, 893(182), *906*
Lossing, F. P. 139(529), *191*, 274(40), *326*
Loth, K. 1120(180), 1122(195), 1123(180), 1125(195, 253–255), *1147–1149*
Lotter, A. P. 550(72), *598*
Lou, Y.-X. 1403(236), *1445*
Louch, J. R. 933, 937(101), *1005*
Loudon, J. D. 470(298), *488*
Lough, A. 697(285), *711*
Lough, A. J. 580(210), *601*, 1391(118b), 1393(153), *1440, 1441*
Louis, J.-M. 1307(275), *1344*
Loupy, A. 662(3, 4), *705*
Louw, R. 40, 127(187), *184*, 716(8), *831*, 1360(163), *1367*
Loveless, C. 676(118), *707*
Lovillo, J. 952, 953, 990(197), *1007*
Lowenstein, J. M. 412(75a), *480*
Lowry, T. 7(40), *180*
Loyd, K. G. 475(318), *488*
Lozach, B. 1370(5), *1435*

Lu, C.-S. 953, 970(290), *1009*
Lu, F. 953, 955(204), *1007*
Lu, G. Q. 944(146), *1006*, 1081(268), *1093*
Lu, J. 1361(186), *1367*
Lu, J.-Y. 1497(171–174), 1498(171), *1506*
Lu, K. 296(83), *327*
Lu, X. 1397(197a, 197c), 1398(197a), 1403(197a, 197c), *1443*
Lu, X.-R. 1427(420a, 420b), *1452*
Lubitov, E. 1372(21), *1436*
Lubitz, W. 1123(215), *1148*
Luboch, E. 724, 725(27), *831*
Lucarini, M. 45(190, 201), 140(201), *184*, 855(120), 877, 878(193), 879, 880(199), 881, 884(193), *904, 906*
Lucariri, M. 139(532), *191*
Lucas, G. B. 424(152b), *483*
Luck, W. A. P. 371(232), *391*, 530(8), *597*
Lucke, A. 1421, 1435(371), *1450*
Lucke, A. J. 1383, 1421(67j), *1438*
Luckhurst, G. R. 1124(236), *1148*
Lücking, U. 1424(400c), 1430(441c), *1451, 1453*
Luçon, M. 544, 545, 547–549(51), 551, 553(87), 558, 576(146), 586, 587(146, 248), 589(248), *598, 599, 602*
Lucy, C. A. 971, 972(300), *1009*
Ludemann, H. C. 1047(146), *1090*
Ludeña, E. V. 31, 68, 69(145), *183*
Ludwig, M. N. 129, 133(493), *191*
Ludwig, R. 1435(469c), *1454*
Lugasi, A. 856, 857(122), *904*
Lugtenberg, R. 1396(182a), *1443*
Lugtenberg, R. J. W. 1427(410b), *1451*
Luhus, D. 493, 515(28), *525*
Lukeman, M. 1035(115), *1089*
Lukes, R. 432(184b), *484*
Lukin, O. 1394(166), 1418, 1420(341), *1442, 1450*
Lukin, O. V. 1418(340), 1420(366), *1450*
Lukyanov, D. B. 631(78), *657*
Lukyanov, S. M. 761, 766(123), 770(155), 806(253), 824(331–333), *833, 834, 836, 838*
Luk'yashina, V. A. 1030(105), *1089*
Lumpkin, H. E. 265(26), *326*
Luna, D. 608(13), 618(36), *656*
Lund, H. 1156, 1158, 1175(1), *1337*
Lund, T. 350(78), *388*
Lundanes, E. 931, 937(83), *1005*
Lundqvist, M. J. 143(549), *192*
Lunelli, B. 877, 878, 881, 884(193), *906*
Lüning, U. 1397(196), 1399(202, 203), 1427(408a), *1443, 1444, 1451*
Lunte, C. E. 973(311), *1010*
Luong, J. H. T. 971(298), *1009*
Luostarinen, M. 1419(353), *1450*
Lupon, P. 1080(259c), *1093*

Luque, F. J. 5(13), *179*, 1025(77), *1089*
Luque de Castro, M. D. 983(377), 984(380), *1011*
Lusa, A. 558(151), *600*
Lusztyk, E. 45(191), 139(191, 531), *184, 191*, 897(273), *908*
Lusztyk, J. 852(86), 855, 856(112), 864, 865(144), 867(155), 868(86), 876(155), 877(188, 189, 191, 192), 878(86, 188, 191), 879(191, 192, 194, 195), 881(188, 191, 192, 203), 882(86, 155, 191, 194, 195), 883(188, 191, 192, 194, 205), 887(205), 891(112), 892(155), *903–906*, 1140(373), *1151*
Lüttke, W. 47, 53(218), *184*, 375(240), *392*
Luttringhaus, A. 466(293), 467(293, 294), *487*
Lutz, B. T. G. 368, 369(181), 377(251), *390, 392*
Lutz, E. S. M. 976(332), *1010*
Lutz, M. 977(335), *1010*
Lutz, R. P. 460(271a), *487*, 761, 763(121), *833*
Luukko, P. 997(445), *1013*
Luz, Z. 340(27), *387*, 1124(220–223), *1148*
Luzhkov, V. B. 129, 131(507), *191*
Luzikov, Y. 1427(409b), *1451*
Lwowski, W. 801, 802(245), *836*
Lyalyulin, A. L. 978(347), *1011*
Lyashkevich, S. Yu. 1073(224), *1092*
Lycka, A. 358(113–117), 361(115), *388*, 725(29), *831*
Lyga, J. W. 426(163), *483*
Lykins, B. W. Jr. 1354, 1355(100), *1365*
Lynbartovich, V. A. 1470(70), *1504*
Lynch, D. C. 816(309), *838*
Lynch, G. C. 57(242), *185*
Lynch, J. E. 632(89), *658*
Lynch, V. 1430(442a), *1453*
Lynn, B. C. Jr. 320(235), *331*
Lysenko, K. 640(132), *659*
Lysenko, Z. 406(49c), *479*

Ma, B.-Q. 1387(88), *1439*
Ma, J. C. 322(242), *331*
Ma, J.-Q. 1082(285), *1094*
Ma, S. 1424(387), *1450*
Ma, Y.-A. 1224(138), *1341*
Ma, Y. S. 1085(340), *1095*
Ma, Y. Z. 503(84), *526*
Maarmann-Moe, K. 579(199–201), *601*
Maartman-Moe, K. 219(33), *221*
Maas, G. 433(187b), *484*, 1400(206), *1444*
Maas, J. H. van der 368, 369(181), 377(251), *390, 392*, 551, 556(115), *599*
Mabic, S. 650(159), *659*
Mabury, S. A. 1084(314), *1095*
Macák, J. 265, 268, 269(27), *326*
Macchioni, A. 998(452, 453), *1013*

Macdonald, T. L. 678(140), *707*
MacDowall, F. D. H. 1118(150), *1147*
MacFaul, P. A. 877, 878, 881, 883(188), *906*
MacFerrin, K. D. 664(29), *705*
MacGillivray, L. R. 207(12), *220*
Macharacek, K. 236(54), *255*
Macheix, J. J. 913(4), *1003*
Machida, K. 11(54), *180*, 232(36), *255*, 1224(137, 138), *1341*
Machida, R. 414(88), *481*
Machlin, L. J. 141(541), *192*, 840(6), *902*
Macias, A. 101, 102(387, 389), *188*
Maciejewska, D. 367(169), *390*, 729(41), *831*
Maciejewska, H. 144(675), 145(631), *194, 195*
Maciejewska-Urjasz, H. 355(106, 107), *388*, 742(68), *832*
Maciel, G. E. 1470, 1471(58), 1484(90), *1503, 1504*
Macielag, M. 470(300d), *488*
Macikenas, D. 1198(90), *1340*
Mackey, J. H. 1470(61), *1503*
MacKinnon, A. 1036(118), *1089*
Macko, T. 949, 953(186), *1007*
MacLeod, K. 1261(194), *1342*
MacNamee, R. W. 35, 36(154), *183*
MacNeil, J. M. 884(207, 208), 885(207, 213, 214), 890(214), *906*
Macśari, I. 238(71), *256*, 370(212), *391*
Macus, T. S. 1180, 1327(60b), *1339*
Macyk, W. 1085(350), *1095*
Madbouly, M. D. 1353, 1354(87), *1365*
Maddams, W. F. 144(592), *193*
Madden, K. P. 1098(7), *1105*, 1120, 1122(186), 1123(206), 1125, 1129(246), 1134(186), *1147–1149*
Maddrix, S. P. 141(540), *192*
Madeira, V. M. C. 853, 856, 857(88), *904*
Madeja, R. 460(280a), *487*
Madigan, E. 1387(91), *1439*
Madrakian, E. 632(87), *658*
Madsen, F. 373, 383, 384(238), *392*
Madsen, R. 675(116), *707*
Madyar, V. R. 774(187), *835*
Maeda, H. 447(222d), *485*
Maeda, K. 1377(45a), *1437*
Maeda, M. 992(430), *1013*
Maeda, S. 1396(184), *1443*
Maekawa, H. 410(68), *480*
Maerker, C. 95(367), *188*
Maes, G. 562(167), 583(228, 232), *600, 601*
Maeta, K. 774(191), *835*
Mafatle, T. 982(365), *1011*
Magee, R. J. 1002(479), *1014*
Maggi, R. 608, 612(10), 616(26), 622(53), 623(54), 627(62), *656, 657*, 677(131), 680(153, 155), 683(170, 182), *707, 708*, 740(62), *832*, 1255(183), *1342*, 1374(37), 1378(37, 49a–c), *1436, 1437*

Maggiani, A. 448(229a, 229b, 230), *485*
Magnes, B.-Z. 493(46, 47, 51), 498(47), 504, 505(91), 506(46, 51, 91), 509(51), 510(91), 511(47), 512–514(51), 516–519(91), 520(51), 521(51, 91), 522, 523(97), *526, 527*
Magnoux, P. 631(77, 78), *657*, 686(205), *709*
Magnus, P. D. 1330, 1332(325b), *1346*
Magoński, J. 594, 595(291), *603*
Magoub, S. A. 1071(201–203), *1091, 1092*
Magrans, J. O. 1406(266), 1409(287), *1446, 1447*
Mague, J. T. 1393, 1407(143), *1441*
Mahadevan, V. 1196(88), *1340*
Mahalaxmi, G. R. 110, 114, 121(424), *189*
Mahalingam, R. J. 1213(115), *1340*
Mahan, B. H. 83(291), *186*
Maheshwary, S. 1047(148), *1090*
Mahindaratne, M. P. D. 612(19), *656*
Mahling, J.-A. 776(203, 204), 778(209), *835*
Mahmud, M. 370(214), *391*
Mahon, H. I. 19(98), *182*
Mahoney, L. R. 15(74), 45(189), *181, 184*, 841, 842(13), 889(13, 227), 891(242), *902, 907*, 1108(3), 1111(3, 65), 1136(355), *1143, 1145, 1151*
Mahy, J. W. G. 1421(369), *1450*
Maichle-Mössmer, C. 1169(44), *1338*
Maidment, M. J. 252(95), *257*
Maier, G. V. 1017(5, 6), *1087*
Maier, J. P. 110, 112(446), *189*
Maier, W. F. 1085(350), *1095*
Maillard, J. P. 150(738), *196*
Maillard, M. 953, 967(270), *1009*
Mainagashev, I. Ya. 340, 341(26), *387*
Maiorino, M. 890(234), *907*
Maithreepala, R. A. 1085(337), 1086(370), *1095, 1096*
Maizumi, H. 1359(146), *1366*
Majerz, I. 144(605, 606), 146(639, 640), 178(770), *193, 194, 197*, 377(264), *392*, 579(196), 580(202, 203), 593(280), 594(284), 595(280, 284), *601–603*
Majetich, G. 460(279), *487*, 1354, 1355(100), *1365*
Majewski, E. 144(660), *194*, 345(48), 353(98), 359(124), *387–389*
Majors, R. E. 1353(71), *1364*
Majumdar, D. J. 1025(71a, 71b), *1088*
Majumder, D. 144(661, 666), *194*
Mak, C.-P. 1180(64), 1250(174), *1339, 1342*
Mak, T. C. 210(15), *220*
Mak, T. C. W. 212(22), *221*, 1470(56), *1503*
Makarova, N. A. 1419(355), 1424(396), *1450, 1451*
Makashiro, K. 413(78a), *480*
Makeiff, D. A. 1428(432), *1452*
Makepeace, D. K. 1356(121), *1365*

Makha, M. 1372(22, 25), 1427(418), *1436, 1452*
Maki, A. H. 1123, 1124(202), *1148*
Maki, S. 1080(257), *1093*, 1183(67), 1187(72, 73), 1188(74), *1339*
Maki, T. 129, 179(491), *190*, 881(202), *906*
Maki, Y. 403(31), *479*, 1019(27b, 30), *1087*
Makosza, M. 405(46a, 46b, 47), 406(49a, 49b), *479*
Makower, A. 977(344), *1011*
Maksić, Z. B. 83, 84(298), 89(311), 97, 101(374), *186, 188*, 261, 263, 296(14, 15), 308(162), *326, 329*
Makuch, B. 947(169), *1006*
Malarski, Z. 143(568), 144(568, 606, 609), *192, 193*, 352(91), 353(96), 377(259), *388, 392*, 551(121, 127), 557(121), 593(276, 277, 280), 594(276), 595(276, 277, 280), *599, 602*
Malato, S. 1073(216), 1085(336, 353), 1086(336, 372), *1092, 1095, 1096*
Malaveille, A. C. 1361(188, 189), *1367*
Malcolm, N. 32(146), *183*
Malcolm, R. L. 1360(181), *1367*
Males, Z. 929(63), 953, 967(63, 271), *1004, 1009*
Malesa, M. 632(88), *658*
Malhotra, R. 606(2), 613(22), *656*
Malick, D. K. 4, 100(2), *179*, 368, 372, 373(200), *390*
Malik, M. A. 740(66), *832*
Malinovich, Y. 110, 121(436), *189*, 265(23), *326*
Malkin, J. 1045(141), *1090*
Malkov, A. V. 677(124, 125), *707*
Mallard, W. G. 98(376), *188*, 261(7, 9), 265, 268, 269, 272(29a), 311, 312(7), *325, 326*, 1105(62), *1106*
Mallavadhani, U. V. 631(84), *658*
Mallevialle, J. 1352(59), *1364*
Malmqvist, P.-A. 20, 106(119), *182*
Malmstadt, N. 1361(199), *1368*
Malmström, J. 865(145), *905*
Malone, J. F. 450(246c), *486*, 829(345), *838*
Maloney, S. M. 1352(59), *1364*
Malova, T. N. 231(34), *255*
Malshe, V. C. 628(64), *657*
Maltby, D. 129(485), *190*
Malterud, K. E. 856(123), *904*
Malysheva, R. D. 811(292), *837*
Mamaev, V. P. 340, 341(26), *387*
Mami, I. S. 1312(286), *1345*
Man, E. H. 472(311e), *488*
Manabe, N. 935(115), *1005*
Manabe, O. 419(123a), *482*, 633(94), 634(95), *658*, 1402(224a), 1419(347a), *1444, 1450*
Manas, A.-R. B. 1297(261), *1344*
Manchado, C. M. 853(91), *904*

Manda, E. 1290(248), *1344*
Mandal, A. 144(661, 666, 668), *194*, 1021(45), 1022(51), 1047(145), *1088, 1090*
Mandal, S. 1198(90), *1340*
Mandal, S. S. 783(223), *836*
Mandix, K. 127(469), *190*, 718(17), *831*
Mandolini, L. 1392(136), 1394(171), 1396(185), 1397, 1399, 1403(197b), *1441–1443*
Manem, J. 1352(59), *1364*
Manet, I. 1391(120b, 121c), 1396(121c), 1435(469b), *1440, 1454*
Manfredi, A. 699(307), *711*
Manfredi, G. 1427(410a), *1451*
Mangas Alonso, J. J. 953, 958(223), *1008*, 1354(113), *1365*
Mangia, A. 953, 955(203), *1007*
Mangiafico, T. 1392(128), *1441*
Mangini, F. 953, 961(236), *1008*
Mangla, V. K. 1297(260a), *1344*
Manickam, G. 695(268), *710*
Manini, P. 953, 955(203), *1007*
Mann, G. 671(79), *706*, 1417(330b), 1424(392, 393), *1448, 1450*
Manocha, A. S. 238(72), *256*, 554, 555(137), *599*
Manolache, N. 451(250), 456(261), *486, 487*
Mansilla-Koblavi, F. 727(37, 38), *831*
Manson, D. 409(62a), *480*, 801(241, 242), *836*
Mantovani, M. 1403, 1405(235), *1445*
Maquestiau, A. 110(432–434), 121(434), *189*, 263(20), 299(98), 301, 302(137), *326, 328*
Marbach, M. 946, 953(162), *1006*
Marbury, G. D. 288(58), *327*
Marcé, R. M. 944(147), 945(147, 152, 153, 156), 946(164), 953(152, 164, 209, 254, 256), 955(209), 962(254, 256), *1006–1009*, 1347(1), 1353(1, 70, 92, 93), 1354(70, 102, 104), *1363–1365*
Marcelocurto, M. J. 633(91), *658*
Marceroni, G. 1406(272), *1446*
March, J. 103(397), *188*, 759, 801(113), *833*
Marchand, A. P. 447(222i), *485*
Marchese, S. 946(165), 947(167), 953(165, 167), *1006*, 1353(85–87), 1354(87), *1365*
Marchesi, S. 686(206), *709*
Marci, G. 1083(301), 1084(326), *1094, 1095*
Marconi, G. 992(431), *1013*
Marcoux, J.-F. 670(74), *706*
Marcus, E. 429(176), *483*
Marcus, R. A. 17, 18(82), *181*, 522(99), *527*
Marcus, Y. 541(46), 542(47), *597*
Mare, P. B. D. de la 806(267), *837*
Marengo, E. 947, 955(170), *1006*
Marenus, L. E. 1318(304), *1345*
Marguet, S. 845(27), *902*
Maria, P.-C. 312(193), 322(244), 324(270), 325(244), *330, 331*, 503(89), *526*

Mariam, Y. H. 93, 94, 97, 132, 133(358), *187*, 383(285), *393*
Marigorta, E. M. de 689(231), *709*
Marín, M. L. 932(96), *1005*, 1079(243), *1092*
Marinas, J. M. 608(13), 618(36), *656*, 774(182), *835*, 972(304), *1010*
Marín-Hernández, J. J. 953, 960(234), *1008*
Marinho, H. S. 901(286), *908*
Marino, J. P. 1275(217), *1343*
Marjanovic, B. 871(165), *905*
Marjasin, I. 1362(210), *1368*
Mark, M. 933, 937(101), *1005*
Markell, C. G. 1352(66), *1364*
Markides, K. E. 953, 955(208), *1007*
Marklund, G. 1119(164), *1147*
Marklund, S. 1119(164), *1147*
Marko-Varga, G. 953, 969(280), 974(320), 975(322), 976(324), 977(333, 335, 337), *1009, 1010*
Markovits, Y. 1038(122a), *1090*
Markovsky, L. N. 1393(161), 1417(326), 1418(338, 340, 344), *1442, 1448, 1450*
Markowsky, L. N. 1417(328), *1448*
Marks, V. 471(302), *488*, 806, 814(254), *836*
Marnet, N. 913(10), *1003*
Maroni, M. 1359(151), *1366*
Maros, L. 937(120), *1005*
Marovatsanga, L. T. 913(14), *1003*
Marques, A. 1285(233), *1343*
Marquez, F. 744(72), *832*, 1038(122c), *1090*
Marquis, E. 535, 548(30), *597*
Marr, M. C. 1348(30), *1363*
Marra, A. 1403(240b), 1405(255), 1427(408d), 1432(448b), *1445, 1446, 1451, 1453*
Marsden, J. E. 68(262), *185*
Marsel, C. J. 397(2b), *478*
Marsh, P. B. 950(189), *1007*
Marshall, J. L. 349(61), *387*
Marshall, P. J. 15(72), *181*
Marta, G. 1083(293), *1094*
Martell, A. E. 1192(78c, 78d), *1339*
Martell, A. R. 1194(85), *1339*
Martin, A. 648, 651(155), *659*, 1227(145), *1341*, 1415(312), *1448*
Martin, A. R. 1305(269), *1344*
Martin, C. 1084(325, 326), *1095*
Martin, C. H. 93–96(357), *187*
Martin, F. 985(387), *1012*
Martin, I. 1084(325), *1095*
Martin, J. B. 405(43b), *479*
Martin, J. M. L. 4(8), 83, 84(298), *179, 186*, 311(187), *329*
Martin, N. H. 551(125), *599*
Martin, R. 338(14a, 14b), *386*, 472(308f, 311h, 311i, 314), 474(317), *488*, 642(138), *659*, 773(167), *835*, 890(237), *907*
Martin, R. L. 4, 100(2), 102(392), *179, 188*, 368, 372, 373(200), *390*

Martínez, A. T. 937, 941(131), *1006*
Martínez, D. 971(295), *1009*, 1347, 1353(1), 1354(104, 107), *1363, 1365*
Martinez, M. L. 1022(50), 1025(85b), 1028(94), *1088, 1089*
Martinez, S. J. III 20, 24, 106(118), *182*
Martinho Simões, J. A. 20, 23(117), 45(194), 98(375), 148(727), 179(777), *182, 184, 188, 196, 198*, 229(32), *255*
Martino, M. 1395(179), *1442*
Martins, A. F. 933, 937(103), *1005*
Martinsen, D. P. 291(68), *327*
Martinson, M. M. 1086(374), *1096*
Martire, D. O. 1085(330), *1095*
Martrenchard, S. 493, 500(31), *525*
Martrenchard-Barra, S. 109(414), *189*
Martsinkovsky, R. M. 145(626), *193*
Martynov, I. Y. 492, 493, 501(13), *525*, 1020(35d), *1087*
Martynova, I. A. 1081(263), *1093*
Martz, J. 1427(421c), *1452*
Maruoka, K. 460(283b), *487*, 688(227–229), 689(230, 232, 234, 235), 693(250–252, 255, 258, 256), *709, 710*
Maruthamuthu, P. 845, 880, 881(25), *902*, 1110(55), 1111(69, 70), 1112(55), 1137(70), *1145*
Maruyama, K. 419(132), 460(281a, 281b), *482, 487*, 1019(28, 29), *1087*, 1194(83b), 1209(108), 1333(329), *1339, 1340, 1346*
Marx, J. N. 807(272, 273), *837*
Marzi, E. 93(355), *187*
Marzocchi, M. 383(281), *392*
Marzocchi, M. P. 383, 384(283), *393*
Masad, A. 493(44), *525*
Masaki, K. 1495, 1498(159), *1505*
Masaki, Y. 1233(151b), *1341*
Masalimov, A. S. 1126(258, 259), *1149*
Masalitinova, T. N. 236, 237(57), *255*
Mascagna, D. 1079(242), *1092*
Mascal, M. 1402(228a, 228b), 1409(288), *1445, 1447*
Masci, B. 1379(53, 54c), 1380(57), *1437*
Mascini, M. 976(326), *1010*
Mashige, F. 953, 962(248), *1008*
Mashino, T. 1209(108), *1340*
Masignani, V. 1383, 1422(75e), *1439*
Masini, J. C. 984(383), *1012*
Masmandis, C. A. 129, 131(506), *191*
Mason, H. S. 1120(170), *1147*
Mason, K. E. 840(2), *902*
Mason, R. P. 1120(175, 176, 178, 189), *1147*
Masqué, N. 945(153, 156), *1006*, 1354(102), *1365*
Massart, D. L. 954, 970(199), *1007*
Massat, A. 576(190), *600*
Massaxcret, M. 664(37), *705*
Massera, C. 1424(389b), *1450*

Massey, V. 411(69), *480*, 1119(165), *1147*
Masucci, J. A. 314(204), *330*
Mataga, N. 501(79), *526*, 1021(41), *1088*
Matarski, Z. 580(202, 203), *601*
Mateo, M. E. 1077(238), *1092*
Mater, I. 179(782), *198*
Matevosyan, R. O. 301(130), *328*
Mathar, W. 315(215), *330*
Mathias, L. J. 607(4), *656*
Mathiasson, L. 947(168), 953(168, 280), 969(280), *1006, 1009*
Mathier, E. 35(162), *183*
Mathies, R. A. 551(90), *598*
Mathivanan, N. 129, 133(497), *191*
Mathy, A. 471(304), *488*, 806(262), *837*
Matos, J. 1080(247), *1093*
Matos, M. A. R. 229(32), 230(21, 30), 233(30), 242, 247, 248(83), *254–256*
Matouskova, J. 1079(245), *1093*
Matras, G. 1084(320), *1095*
Matsubara, Y. 410(68), *480*
Matsuda, H. 1408(279a), *1447*
Matsuda, T. 1391(115a), 1394(172b), 1435(464b), *1440, 1442, 1454*
Matsuda, Y. 1425(405), *1451*
Matsuhasi, S. 1100(27), *1106*
Matsui, K. 855, 857(108), *904*
Matsui, M. 677(128, 130), *707*
Matsui, T. 1419(356a), *1450*
Matsui, Y. 551(133), *599*, 1498(175), *1506*
Matsukami, K. 413(77), *480*
Matsumaru, Y. 129, 179(491), *190*
Matsumoto, A. 1495(156, 160, 162–169), 1498(156, 160, 164, 169), *1505, 1506*
Matsumoto, H. 1400(207a), *1444*
Matsumoto, K. 448(232a), *486*
Matsumoto, M. 420(142b), 424(155a), *483*, 693(261), *710*, 1159, 1207(28), 1213(112), *1338, 1340*
Matsumoto, N. 699(298), *711*, 1262(195), *1343*
Matsumoto, T. 291(63), *327*, 447(223), *485*, 641(137), *659*, 674(99), 679(141–143), 680(145), *707, 708*, 776(201, 202), *835*
Matsumoto, Y. 377(255), *392*
Matsumura, H. 632(86), *658*
Matsumura, K. 90(344), *187*, 1374(36b), 1406(36b, 270a, 270b), *1436, 1446*
Matsumura, N. 1393(160), *1442*
Matsumura, Y. 881(202), *906*
Matsunaga, S. 695(273, 274), *710, 711*
Matsunga, K. 1076(234), *1092*
Matsuo, G. 642(139), *659*, 680(146), *708*
Matsushima, R. 1028(97, 98), 1029(100, 102), 1030(103, 104), *1089*
Matsushita, H. 1357(137), *1366*
Matsushita, T. 1059(178), *1091*
Matsushita, Y. 1419(356a), *1450*

Matsuura, S. 409(58), *480*
Matsuura, T. 1080(261), *1093*, 1198, 1199, 1205(91), 1206(105), 1265(199a, 199b), *1340, 1343*
Matsuyama, A. 47, 143(206), *184*
Matsuzaki, T. 913(26, 29), *1003*
Matsuzawa, Y. 1421(373), *1450*
Matt, D. 1391(120a), 1401(214), 1435(462a), *1440, 1444, 1454*
Mattay, J. 320(241), 322(241, 252), *331*, 1059(181, 182), *1091*, 1382(71), 1419(346), 1420(359a, 361, 365a), *1438, 1450*
Mattersteig, G. 405(48a, 48b), *479*
Mattews, R. W. 498(73), *526*
Matthes, S. E. 1383, 1421(67m), *1438*
Matthews, I. 613(24), *656*
Matthews, R. W. 1082(278), *1094*
Matthews, S. E. 1389(105), 1391(120c), 1427(416b, 417), 1435(466), *1440, 1452, 1454*
Matthias, C. 300(112–114), *328*
Matthies, P. Z. 1118, 1119(136, 137), *1146*
Mattill, H. A. 889(228), *907*
Mattinen, J. 143(559), *192*, 352(89), *388*
Mattrel, P. 1348(10), *1363*
Matunaga, S. 667(50), *705*
Mauer, M. 1080(259a), *1093*
Maumy, M. 1305(270), *1344*
Maunov, S. 110, 121(423), *189*
Maurette, M.-T. 1080(253), *1093*
Maverick, E. F. 1419(348), 1429(430b), *1450, 1452*
Mavilla, L. 1424(392), *1450*
Mavri, J. 144(651), *194*
Mayanta, A. G. 638(122), *658*
Mayer, B. 992(431), *1013*
Mayer, G. Y. 1017(8), *1087*
Mayer, I. 69(269), *185*, 369(209), *391*
Mayer, J. P. 403(28), *479*
Mayer, P. S. 300(123), *328*
Mayer, R. 1109(27), 1120(168), *1144, 1147*, 1290(244), *1344*
Mayer, W. 719(19), *831*
Mayr, E. 345(50), *387*
Mazas, N. 985(386), *1012*
Mazellier, P. 1085(347), *1095*
Mazur, Y. 340(27), 384(289), *387, 393*, 1045(141), *1090*, 1289(241), *1344*
Mazurek, A. P. 367(174), *390*
Mazza, G. 913, 966(8), *1003*
Mazzacani, A. 616(26), *656*
Mazzini, S. 338, 345(10), *386*, 721(26), *831*
Mazzola, E. P. 358(112), *388*
Mazzola, I. 18(87), *181*
Mazzoni, V. 365(159), *389*
Mbiya, K. 450(246f), *486*
McAdoo, D. J. 300(105, 107), *328*

McAllister, K. L. 851, 852, 887, 888(60), *903*
McCallum, K. S. 424(152a, 156), *483*
McCapra, F. 1297(256b), *1344*
McCarley, T. D. 1383, 1421(67f), *1438*
McClellan, A. L. 20, 25(128), 35(173), 47, 53(225), 171(173), *182–184*, 503(86), *526*, 530, 559(3), *596*
McClelland, R. A. 38, 39, 127(186), *184*, 419(124a, 124b), *482*, 715(6), *830*
McClure, J. D. 410(64b), *480*
McColl, I. S. 1495(150, 151), *1505*
McCombie, S. W. 668(58), *706*
McConnell, H. M. 136, 137(520), *191*
McCord, J. D. 953, 960(229), *1008*
McCormick, D. 607(4), *656*
McCracken, J. 1108(16), *1144*
McCurry, P. M. 441(205), *484*
McDonald, E. 1297(258, 259, 260b), *1344*
McDonald, P. D. 1216, 1283, 1287, 1297(122b), *1340*
McEneney, B. 1489(105, 107), *1504*
McEvoy, S. R. 1082(278), *1094*
McEwen, C. N. 291(66), *327*
McGeorge, G. 367(172), *390*
McGhee, E. M. 415(107), *481*
McGibbon, G. A. 110(451), *190*
McGillivray, G. 416(110b, 111), *482*
McGlashen, M. L. 94, 105(365), 129, 133(496), *187, 191*, 997(450), *1013*
McGowen, J. 607(4), *656*
McGregor, W. M. 1396(188), *1443*
McGuckin, M. R. 830(347), *838*
McGuire, M. 293(73), *327*
Mchedlov-Petrosyan, N. O. 737(58), *832*
McIldowie, M. J. 1382(73), *1438*
Mcill, R. A. 503(89), *526*
McIver, R. T. 92(349, 351), 99, 100(351), 101(349, 351), *187*
McIver, R. T. Jr. 5(19, 22), 83(294, 297), 84(297), *180, 186*, 261, 263(13), 310(182, 184, 185), 311(189), 312(192), *326, 329, 330*
McKague, A. B. 1348(9), *1363*
McKay, D. J. 20, 23, 139, 140(116), *182*, 897(275), *908*
McKee, V. 1432(448d), *1453*
McKelvey, J. M. 261(11), *326*
McKervey, M. A. 1387(91, 94), 1391(123), 1430(442b), 1432(448c, 448d, 449b), 1435(94), *1439, 1441, 1453*
McKillop, A. 416(108, 109a, 110a, 110b, 111), 422(149), *481–483*, 1227(146), 1273(206), 1312(284a, 284b), 1314(284a, 284b, 288, 289), 1315(292), *1341, 1343, 1345*
McKinnon, M. S. 1484, 1485(91), *1504*
McLachlan, J. A. 925(50), *1004*
McLafferty, F. J. 237(61), *256*

McLafferty, F. W. 110, 121(434), *189*, 260, 264(2), 299(96), *325, 327*
McLaren, L. 1227(146), *1341*
McLaughlin, K. A. 851(59), *903*
McLean, E. J. 580(210), *601*
McLuckey, S. A. 310(181), *329*
McMahon, T. 83(296b), *186*
McMahon, T. B. 92, 101(350), *187*, 324(271, 272), *331, 332*
McManus, H. J. 1123(206), *1148*
McManus, M. J. 829(345), *838*
McNelis, E. 1109(35), *1144*, 1330(319, 320), *1346*
McPartlin, M. 782(220), *836*
McPherson, G. L. 365(164), *389*
Mebel, A. M. 137(524), 138(524, 525), *191*
Mechan, P. R. 580(209), *601*
Mecik, M. 558(149), *600*
Mecke, R. 35(150), 47, 53(218), *183, 184*, 375(240), *392*
Medard, L. 241(78), *256*
Medic-Saric, M. 929(63), 953, 967(63, 271), *1004, 1009*
Meena 1407(276), *1447*
Meerts, W. L. 20, 24, 106(127), 147(698), *182, 195*, 581(215), *601*
Meetsma, A. 359(130), *389*
Meffert, A. 324(273, 274), 325(279), *332*
Megson, N. J. L. 1457(2), 1462(31, 32), *1502, 1503*
Meguro, S. 1205(103), *1340*
Meharg, A. A. 1362(217), *1368*
Meharg, V. E. 397(2a), *478*
Mehdizadeh, A. 300(116), *328*
Mehler, E. L. 67(254), 110, 113(439), *185, 189*
Mehnert, R. 1102(45), *1106*
Mehra, R. K. 1084(323), *1095*
Mehta, P. G. 470(300d), *488*
Meidar, D. 613(22), *656*
Meier, B. 816(308), *838*
Meier, G. A. 426(163), *483*
Meijer, E. W. 685(195), *709*
Meilland, P. 677(132), *707*
Meisel, D. 1118(154), 1119(154, 156), 1126(264), 1130(326), 1131(315), 1141(154), *1147, 1149, 1150*
Meissner, F. 1470(48), *1503*
Mejanelle, P. 913(5), *1003*
Mejuto, M. C. 971(294), *1009*, 1348, 1354(17), *1363*
Melamud, E. 1124(225), *1148*
Melf, M. 110, 112, 121(444), *189*
Melikova, S. 371(224), 377(269), *391, 392*
Melikova, S. M. 144(649), 145(612, 625), 171, 173(612), *193, 194*, 371, 377(230), 386(305, 306), *391, 393*
Melis, M. R. 851, 854(74), *903*

Melissaris, A. P. 1495(141), *1505*
Melk, A. 1359(150), *1366*
Mello, C. 929, 981(65), *1004*
Mello, R. 1257(186), *1342*
Mellors, A. 852, 856(85), *903*
Melly, S. J. 925(47), *1004*
Melo, T. B. 851, 852(69), *903*
Meltzer, R. I. 420(138), *482*
Melville, M. G. 685(198), *709*
Melzi d'Eril, G. 953, 964(263), *1009*, 1353(82), *1365*
Melzig, M. 1025(82), *1089*
Menard, H. 1361(191), *1367*
Mendes, G. C. 913, 958(20), *1003*
Mendez, F. 101–103(390), *188*
Mendoza, J. de 1374(35), 1385(79a), 1387(90), 1392(90, 125a), 1394(90), 1395(90, 177), 1403(125a), 1406(266), 1409(287), 1427(424c), *1436, 1439, 1441, 1442, 1446, 1447, 1452*
Mendoza, S. 1422(377), *1450*
Menez, A. 1117(126), *1146*
Menezes, M. L. 949(184), *1007*
Meng, J. K. 359(131), *389*
Meng, Q. J. 685(201), *709*
Mennucci, B. 4, 100(2), *179*, 368, 372, 373(200), *390*
Mentasti, E. 1353(73, 88), *1364, 1365*
Menzer, S. 761(118), *833*
Meot-Ner, M. 273(38), 312(195), *326, 330*
Mereiter, K. 350(86), *388*, 551(92), *598*
Merenyi, G. 139(534), *191*, 1108(24, 25), 1118(24), 1136(361, 363), 1137(363), 1138–1140(24), 1142(24, 377), 1143(24, 382), *1144, 1151, 1152*
Merino, I. 451, 454(249), *486*
Merk, C. 695(272), *710*
Merkel, L. 1374(33a), *1436*
Merkulov, A. P. 1123(201), *1148*
Merlini, L. 338, 345(10), *386*, 721(25, 26), *831*, 1308(280), *1344*
Meroni, P. 1359(151), *1366*
Merriman, G. H. 811(291), *837*
Merz, H. 144(674), 145(634, 635), *194, 195*
Messeguer, A. 422(146), *483*, 636(103), *658*
Metcalf, T. A. 141(539), *192*
Metche, M. 944(139), *1006*
Metelista, A. V. 1024(59, 66), 1025(69), *1088*
Metlesics, W. 1325(311a, 311c), *1345*
Metzger, J. 432(185), *484*
Metzger, J. O. 305(147), *329*
Metzler, O. 433(189a, 189b), *484*
Meunier, B. 1212(111), *1340*
Meurs, J. H. 1168(40), *1338*
Meusinger, R. 1417(330b), *1448*
Meutermans, W. D. F. 774(190), *835*
Meyer, A. S. 857, 858(133), *904*
Meyer, C. D. 299(98), *328*
Meyer, K. H. 247, 250(88), *256*
Meyer, T. J. 1287(238), 1300(263), *1344*
Meyerson, S. 268(32), *326*
Meyrant, P. 110(433), *189*, 263(20), 301, 302(137), *326, 328*
Mezheritskii, V. V. 776(196), 778(213), 823(330), *835, 836, 838*
Mezo, A. R. 1383, 1421(67i), 1425(403a, 403b, 404), *1438, 1451*
Mhamdi, F. 1136(356, 357), *1151*
Miah, M. A. J. 786(231), *836*
Mian, O. I. 1084(323), *1095*
Michael, A. C. 953, 955(207), *1007*
Michael, B. D. 1113(87), 1115, 1128, 1134(92), *1145*
Michael, J. P. 462(273a), *487*
Michaelis, L. 1118(133–135), *1146*
Michalak, A. 89, 90(338), *187*
Michalowski, J. 995(441), *1013*
Michalska, D. 20, 22, 23(112), 35(112, 170), 36–38(112), 47(170), 58(170, 245), 67(245), 82(288), *182, 183, 185, 186*, 235(52), *255*, 372, 375(236), 384(291), *391, 393*
Michejda, J. A. 315(211), *330*
Michel, C. 855, 856(109), 895(269), *904, 908*, 1115(101), 1118(101, 143, 146, 147), 1119(101, 146, 163), 1128(279), 1130(101, 143, 147, 323), 1131(279), 1137(368), *1145–1147, 1149–1151*
Michel, P. 1327(315), *1346*
Michl, J. 368, 373(190, 197), 374(190), *390*, 511(103), *527*
Michot, L. J. 944(144), *1006*
Middel, O. 1408(282), 1422(378), *1447, 1450*
Middleton, E. 894(259), *907*
Miderski, C. A. 1120, 1128, 1132(187), *1147*
Midland, M. M. 415(98a, 98b), *481*
Mieure, J. P. 1356(128), *1366*
Migchela, P. 565(172, 173), 566(174, 175), *600*
Migchels, P. 377(265, 266), *392*, 583(232), 584(235), *601*
Mighels, P. 145, 146(613, 614), 147(614), *193*
Miguel, M. D. 1082(276), *1093*
Mihailović, M. L. 1325(312), *1345*
Mihara, J. 701(322), *712*
Mihara, S. 1050(160), *1091*
Mihashi, S. 1318(300a), *1345*
Mihm, J. A. 913(2), *1002*
Mijs, W. J. 1194(83a), *1339*
Mikami 107, 143(409), *188*
Mikami, K. 695(275), *711*
Mikami, M. 90(344), *187*
Mikami, N. 107(410), 112(456), 143(574–576, 578), 144(574–576, 598, 599, 686), 145(574), 147(574, 578, 692, 703–707), 148(704, 705), 149(703–705), 150(704,

705), 157, 167(705), 168(704, 705), *188, 190, 192, 193, 195*, 375(245), 377(255), *392*, 497, 498(70), *526*, 551(85, 130), *598, 599*
Mikenda, W. 20, 21, 24, 35, 37(124), *182*, 339(21), 343(41), 350(84–86), 366(21), 371(227, 234), 372, 375(234), 377(227, 234, 268), 378(268, 272), 379(272), 380(268), 383(284), *386–388, 391–393*, 551(81, 86, 91, 92, 104, 123), 553(81, 86, 104), 557(123), *598, 599*
Mikesell, M. D. 1347(6), *1363*
Miketova, P. 305(151), *329*
Mikhailov, I. E. 770(154), *834*
Miki, M. 890(236), *907*
Mikkelsen, K. V. 111(454), *190*
Mikkelson, P. 1359, 1360(160), *1367*
Mikroyannidis, J. A. 1495(141, 153), *1505*
Milart, P. 93(362), *187*
Milas, N. A. 408(52a, 52b), *479*
Milbradt, R. 1435(470), *1454*
Milby, T. 415(107), *481*
Milczak, T. 632(88), *658*
Mildvan, A. S. 352(92–94), 367(93, 180), *388, 390*
Miles, D. 419(136b), *482*
Miles, D. E. 559(183), *600*
Milián, V. 670(76), *706*
Mill, T. 845, 846(31), *902*
Millam, J. M. 4, 100(2), *179*, 368, 372, 373(200), *390*
Millé, P. 78(282), *186*
Miller, B. 470(300e), *488*, 806(251, 260, 265, 266), *836, 837*
Miller, D. 676(118), *707*, 925(53), *1004*
Miller, D. J. 666(47), *705*
Miller, F. A. 66(246), *185*
Miller, I. J. 1124(241), *1148*
Miller, J. 405(46c), *479*
Miller, J. A. 476(319), *488*, 783(226), *836*
Miller, L. L. 1158(18), 1289(241), *1338, 1344*
Miller, M. W. 12, 13(57), *180*
Miller, N. J. 854(95), 871(161), 894(161, 257), *904, 905, 907*
Miller, R. A. 450(246b), *486*
Miller, R. E. 149, 161(735), *196*
Miller, R. F. 450(241c, 246h), 453(241c), *486*
Miller, R. W. 1118(144, 149, 150), *1146, 1147*
Miller, S. S. 994(438), *1013*
Miller, W. H. 385(301), *393*
Millié, P. 20, 23, 78, 93–96, 101, 106, 107, 109(115), 129, 144(474), *182, 190*, 500(78), *526*, 578(195), *601*
Millie, Ph. 493, 511(65), *526*
Milliet, A. 300(115, 122), *328*
Millov, A. A. 1024(59), *1088*
Mills, M. S. 1353(79), *1365*
Milner, A. 854(95), *904*

Mimoun, H. 413(81), *480*, 1192(78b), *1339*
Minakata, S. 702(332), *712*
Minami, T. 429(172), *483*, 674(100), *707*
Minari, P. 1396(186a), 1403, 1405(232a), *1443, 1445*
Minerco, C. 1361(195), *1367*
Minero, C. 1082(281), 1086(361, 365, 368), *1094, 1096*
Minesinger, R. E. 591(264), *602*
Mineva, T. 84(302), 89(302, 323, 325), 90(302, 323, 325, 342), 91(302, 342), *186, 187*
Mingatto, F. E. 895(266), *908*
Miniati, E. 998(452), *1013*
Minisci, F. 45, 140(201), *184*, 408(54e), *480*, 607(6), *656*
Minkin, V. I. 371(217), *391*, 729(44), 743(69), 770(154, 157), *831, 832, 834*, 1024(66), *1088*, 1290, 1292(247), *1344*
Mino, M. 890(236), *907*
Minoura, N. 766(142), *834*
Mintz, E. A. 685(201), *709*
Mintz, M. 683(175), *708*
Minyaev, R. M. 371(217), *391*
Mioskowski, C. 422(147c), *483*
Mirabelli, C. 1359(155), *1366*
Miranda, L. P. 774(190), *835*
Miranda, M. A. 744(71, 72), *832*, 1038(122c), 1039(123), 1040(124–126), 1042(127–129), 1055(165–167, 169), 1071(204), 1079(243, 244), *1090–1093*
Miranda, M. S. 229(32), 230(21, 30), 233(30), 242, 247, 248(83), *254–256*
Mirgorodskaya, A. B. 740(65), *832*
Mirlohi, S. 928(55), *1004*
Miró, M. 929, 984(67), *1004*
Mirone, L. 1391, 1396(121b), 1403, 1405(235), 1409(288), *1440, 1445, 1447*
Mironov, V. A. 1470(70), *1504*
Mirza, U. A. 293(77), *327*
Mirzoyan, R. G. 301(130), *328*
Misaki, S. 687(217), *709*
Miser, J. R. 404(33), *479*
Misharev, A. D. 301(131), *328*
Mishina, A. V. 983(374), *1011*
Mishra, A. K. 1117(131), *1146*
Mishra, H. 1047(148), *1090*
Mishra, S. P. 1104(55), *1106*
Mishrikey, M. M. 436(197), *484*
Misiti, D. 1383(75a–e), 1422(75e, 375), *1439, 1450*
Mislin, G. 1415(314b, 322), *1448*
Misra, H. P. 1119(160), *1147*
Mista, W. 608(8, 9, 12), *656*
Mitani, M. 291(63), *327*
Mitani, T. 219(36), *221*
Mitchell, A. S. 745, 746(81), *832*, 1239, 1240(165a, 165b), *1342*

Mitchell, E. J. 586(237), 589, 590(237, 258), 592(258), *602*
Mitchell, J. 454(253), *486*, 735(52), *832*
Mitchell, S. C. 935, 937(114), *1005*
Mitchell, W. L. 677(124), *707*
Mitera, J. 265, 268, 269(27), *326*
Mitra, A. 620(42), *657*, 749(100), *833*
Mitra, A. K. 1261(193), *1342*
Mitra, S. 1021(45–49), 1022(51), 1047(144), *1088*, *1090*
Mitra, S. S. 170(744), *196*
Mitsky, J. 559, 586, 593(153), *600*
Mitsuzuka, A. 551(85), *598*
Mittal, J. P. 110, 121(423), *189*, 1021(38), *1088*, 1099(19), 1101(39), *1105*, *1106*, 1111(69), 1130(311, 312), 1131, 1134(314, 317), 1135(314, 350), 1142(311), *1145*, *1150*, *1151*
Miura, H. 353(99), *388*, 1380(58e), *1437*
Miura, M. 664(34), *705*
Miwa, C. 773(160), *834*
Mixon, S. T. 69(268), *185*
Miyafuji, A. 701(323), *712*
Miyagi, Y. 702(342), *712*
Miyaguchi, N. 1214(118), *1340*
Miyahra, Y. 1205(103), *1340*
Miyakawa, K. 1029(100), *1089*
Miyakawa, T. 992(432), *1013*
Miyakawa, Y. 1462(19), *1502*
Miyake, M. 1062(186), *1091*
Miyake, T. 414(90g), *481*
Miyakoda, H. 1352(57), *1364*
Miyakoshi, T. 432(182a, 183c), *484*
Miyamoto, O. 677(136), *707*
Miyamoto, Y. 702(334, 338), *712*
Miyanari, S. 1380(61a, 61b, 62), *1438*
Miyano, S. 1380(61a, 61b, 62, 64), 1393(160), 1415(314a, 319–321), 1416(324), *1438*, *1442*, *1448*
Miyasaka, H. 1021(41), *1088*
Miyashi, T. 1380(61b), *1438*
Miyashiro, M. 1419(350a, 350b), *1450*
Miyata, T. 1100(27), *1106*
Miyazaki, M. 144(686), *195*
Mizukami, F. 414(91), *481*
Mizumo, T. 410(68), *480*
Mizutani, A. 773(160), *834*
Mlochowski, J. 424(152c), *483*
Mlynski, V. V. 1000(468), *1013*
Mó, O. 83(296b), *186*, 322, 325(244), *331*, 580, 581(212), *601*
Moane, S. 973(311), *1010*
Mocerino, M. 1382(73), 1388(97), *1438*, *1439*
Mochalov, S. S. 763(133), *834*
Modena, G. 413(82b, 83, 84), 414(85), *480*, *481*
Möder, M. 319(233, 234), *331*, 1353(89), *1365*
Modestov, A. 1362(210), *1368*

Moehren, G. 888(226), *907*
Moeller, K. D. 1177(56), *1339*
Moga-Gheorghe, S. 230(31), *255*
Mogck, O. 635(100), *658*, 1403(229, 239), 1427(409c), 1432(229, 446), *1445*, *1451*, *1453*
Mogensen, B. B. 339(18), *386*
Moghaddam, F. M. 774(188), *835*
Moghimi, M. 129, 135(489), *190*, 1109, 1120–1123(38), *1144*
Moghini, M. 1124(240), *1148*
Mohan, H. 110, 121(423), *189*, 1101(39), *1106*, 1130, 1142(311), *1150*
Mohan, T. 998(455, 456), *1013*
Mohanta, P. K. 420(142a), *483*
Mohanty, J. 1038(121), *1090*
Mohialdin-Khaffaf, S. N. 647(147, 149), *659*
Mohri, S.-I. 447(222c), *485*
Moiseenko, A. P. 1470(66), *1503*
Moisio, M. R. 668(55), *706*
Moldéus, P. 890(240), *907*
Moldoveanu, S. C. 932, 937(86), *1005*
Moldrup, P. 1360(174), *1367*
Molenveld, P. 1405(249a, 249b), *1445*
Molinari, R. 1082(288), *1094*
Molins, M. A. 1385(79a), 1406(266), *1439*, *1446*
Moll, G. 695(273), *710*
Möller, J. 987(404), *1012*
Möller, K. 997(447), *1013*
Mollica, V. 101(384), *188*
Molnar, R. 613, 644(23), *656*
Molteni, G. 1216(127), *1341*
Monaco, K. L. 671(83), *706*
Monde, K. 1394(172a), *1442*
Mondelli, R. 338, 345(10), *386*, 721(25, 26), *831*
Mondon, A. 1297(256a), *1344*
Mondon, M. 444(216f, 217, 219), *485*
Money, T. 426(166a), 429(179a, 179c–e), 430(179a, 179c–e, 181e), *483*, *484*, 1297(256b), *1344*
Mons, M. 129(474), 144(474, 682, 683), *190*, *195*, 578(195), *601*
Montanari, B. 6(27), *180*
Montaudo, G. 953, 969, 1000(286), *1009*
Monte, A. P. 764(137), *834*
Monte, M. J. S. 234(46), *255*
Montedoro, G. F. 998(452, 453), *1013*
Monteiro, C. 300(110), *328*
Montenay-Garestier, T. 1117(126), *1146*
Montgomery, J. A. Jr. 4, 100(2), *179*, 368, 372, 373(200), *390*
Moody, C. J. 462(274), 463(276b), *487*, 666(45, 47), *705*, 761(122), *833*
Mookerjee, P. K. 101(383), *188*
Moore, A. N. J. 900(282), *908*
Moore, B. S. 663(22), *705*

Moore, D. W. 1470(47), *1503*
Moore, H. W. 450(238, 239b, 239i, 239j, 246g), 452(239b, 243), *486*, 1307(278), *1344*
Moore, J. S. 1110(53), 1112(53, 77), *1145*
Moore, R. H. 219(38), *221*
Moore, R. N. 814(307), *838*
Moorthy, P. N. 1129(292), *1150*
Mootz, D. 200(5), *220*, 550(69), *598*
Mora, R. 1378(49b), *1437*
Morais, V. M. F. 230(21, 30), 233(30), 242, 247, 248(83), *254, 256*
Morales, S. 971(296, 297), *1009*, 1354(108), *1365*
Morales, S. G. 1362(212), *1368*
Morales-Rubio, A. 987(406), *1012*
Moran, J. R. 1424(397a, 399), *1451*
Moran, P. J. S. 632(90), *658*
Moravec, J. 170(752, 753), *196*
Moravek, J. 1415(318b), *1448*
Morawski, A. W. 1083(305), *1094*
Mordzinski, A. 1025(79, 80, 83), *1089*
Moreau, C. 617(35), *656*
Moreau, P. 645(145), *659*
Moreau-Descoings, M. C. 364(154), *389*
Morellet, G. 405(39b), *479*
Moreno, A. 680(154), *708*
Moreno, M. 1047(149), *1090*
Moreno Cordero, B. 948, 953(179–181), *1007*
Moreno-Esparza, R. 546(57), *598*
Moreno-Manas, M. 607(5), *656*, 677(119), *707*
Moretó, M. 663(20), *705*
Moretti, N. 680(155), *708*
Morey, J. 1260(191, 192), *1342*
Morgan, M. A. 551(89), *598*
Mori, A. 702(333, 344), *712*, 855, 856(105), *904*, 1411(299), *1447*
Mori, G. 953, 955(203), *1007*, 1396(181a, 181c), *1442, 1443*
Mori, H. 1361(187), *1367*
Mori, K. 776(206), *835*
Mori, M. 414(89a, 89b), *481*
Mori, S. 1459(10, 11), 1495, 1498(160), *1502, 1505*
Mori, T. 1187(73), *1339*
Moriarty, R. M. 1224(140a, 140b), *1341*
Morikawa, O. 1382(69a, 69b), 1383(76), 1387(89), 1417(332), 1419(345, 349, 350a, 350b), *1438, 1439, 1448, 1450*
Morimoto, H. 347–349(55), *387*, 851, 862, 863(49), 891(245), *903, 907*
Morimoto, T. 1302(266), *1344*
Morimoto, Y. 1018(24), *1087*
Morishige, Y. 1372(14), *1436*
Morishima, N. 674(108), *707*
Morita, K. 1021(41), *1088*
Morita, M. 1483(89), *1504*
Morita, T. 695(280), *711*
Morita, Y. 415(102), *481*, 1402, 1409(223), *1444*
Morivaki, M. 774(168–172), *835*
Moriwaki, M. 472(309a, 309b), 473(309b), *488*, 629(65, 66), 630(67, 68), 639(65, 67), *657*, 686(203, 204), *709*
Moriya, K. 1380(63), *1438*
Morizur, J. P. 300(123), *328*
Morley, J. O. 1076(236), *1092*
Mormann, M. 260(5), 285(57), 293(75), 325(5), *325, 327*
Moro, G. L. 675(112), *707*
Moro, S. 413(84), 414(85), *481*
Morohashi, N. 1380(62), 1393(160), 1415(314a, 319–321), *1438, 1442, 1448*
Morokuma, K. 4(2), 89, 90(340), 100(2), 137, 138(524), *179, 187, 191*, 368, 372, 373(200), *390*
Moroni, M. 419(133), *482*
Moro-oka, Y. 414(87, 90b), *481*, 1194(86), 1209(106), *1339, 1340*
Morozov, V. I. 740(64), *832*
Morris, D. G. 586, 587(247, 248, 250, 251, 254), 589(248, 254), *602*, 1019(26), *1087*
Morris, J. J. 503(89), *526*, 533(22), 535(33), 536(33, 36, 37), 537(33), 539, 540(37), 557(33, 145), 560, 564(145), 588(33, 37, 257), 590(145), 591(37), *597, 599, 602*, 876–878(186), 879(196), 882(186, 196), *906*
Morrison, H. 1039(122d), *1090*
Morrow, G. W. 717, 745(13), *831*, 1175(54a, 54b), 1227(54a, 147), *1338, 1339, 1341*
Morshuis, M. G. H. 1385, 1386(84), *1439*
Morsi, S. E. 230(24), *254*
Mortaigne, B. 1495(138), *1505*
Mortensen, G. K. 1360(174), *1367*
Mortier, W. J. 89, 90(322), *187*
Morton, T. H. 110(437, 438), *189*, 300(103, 104, 107, 119, 121, 123, 126, 128, 129), 301(128), *328*
Moskvin, L. 993(434), *1013*
Moskvin, L. N. 993(434), *1013*
Mosny, K. K. 447(224), *485*
Mosquera, M. 385(302), *393*, 1025(70a, 72, 74–76), *1088, 1089*
Mosrati, R. 929, 954(62), *1004*
Moss, R. A. 1233(151c), *1341*
Mosseri, S. 1101(37), *1106*, 1111(69), *1145*
Moszhuchin, A. V. 993(434), *1013*
Motamed, B. 928(61), 929, 954(62), *1004*
Motherwell, W. B. 671(81), 681(163), *706, 708*
Motomizu, S. 974(318), *1010*
Motoyama, T. 890(236), *907*
Motoyama, Y. 695(275), *711*
Mottadelli, S. 674(102), *707*
Motta-Viola, L. 1394(173), *1442*

Moualla, S. 851, 855(73), *903*
Mouchahoir, E. G. 358(112), *388*
Moulin, C. 322(249), *331*
Mountfort, K. A. 1352(55), *1364*
Mourgues, P. 300(122), *328*, 1307(275), *1344*
Mowat, J. J. 851, 858, 889(58), *903*
Moyano, E. 947, 953(172), *1006*
Moyes, W. 67(252), 110, 112(443), *185, 189*
Moylan, C. R. 314(201), *330*
Mrsulja, B. B. 874(174), *905*
Mu, D. 129(485), *190*
Muathen, H. A. 651(168), *659*
Mubmann, P. 1353(90), *1365*
Mucchino, C. 953, 961(236), *1008*
Muchetti, C. 677(131), 680(153), *707, 708*
Muci, A. R. 698(295), *711*
Mueller, E. 1109(27), 1123(207), 1130(329), *1144, 1148, 1150*
Mueller, H. 147, 148(695), *195*
Mueller, J. G. 1351(42), *1364*
Mueller, R. H. 665(41), *705*
Mueller, T. C. 947, 953(173), *1006*
Mühlermeier, J. 457(262), *487*
Muiño, P. L. 493(26), *525*
Muir, J. G. 953, 969(281), *1009*
Mukai, K. 844(22, 23), 851(49, 50, 57, 60), 852(60), 856(50, 127–129), 862, 863(49), 865(127, 146, 148, 149), 866(146, 147), 872(50), 875(183), 876(184), 878(50), 887, 888(60), 890(231), 891(244–247), 893(183), 894(260), 896(146–148), *902–908*, 1025(86), *1089*, 1109(31), *1144*
Mukaiyama, T. 448(231), 460(283a), *486, 487*
Mukawa, F. 813(300), *837*
Mukerjee, S. K. 748(94), *833*, 1050(161), *1091*
Mukherjee, A. 129, 133(496), *191*, 997(450), *1013*
Mukherjee, B. 409(59), *480*
Mukherjee, S. 144(661, 666, 668), *194*, 1047(144, 145), *1090*
Mukherjee, S. J. 1021(45–49), 1022(51), *1088*
Mukherjee, T. 1130(312), 1131, 1134(314, 317), 1135(314, 350), *1150, 1151*
Mukhtar-ul-Hassan 1079(246), *1093*
Mukoyama, Y. 1392(131), *1441*
Mulac, W. A. 1127, 1128(269), *1149*
Muldagaliev, Kh. Kh. 78(279), *186*
Mulder, P. 45(191, 195, 197, 200), 139(191, 531, 533), 179(197), *184, 191*, 716(8), *831*, 867(153, 155), 876(155), 879(195), 882(155, 195), 892(155), 893(153), 897(272), *905, 906, 908*
Mulkey, D. 997(448), *1013*
Mullen, W. 957(220), *1008*
Mullens, J. 311(187), *329*
Müller, A. 143, 177(553), *192*
Muller, D. P. R. 141(539), *192*

Müller, E. 1120(168), *1147*, 1290(244), *1344*
Müller, G. H. 1220(134), *1341*
Müller, H. 144(678), *195*, 674(103), *707*
Müller, J. 15(76), 129(76, 514), 132, 133, 135(514), 141, 142(76), *181, 191*
Muller, J. F. 323(259), *331*
Müller, J. P. 534(27), *597*
Müller, K. 110, 112, 121(444), *189*, 1417(330b), *1448*
Müller, P. 702(337, 343), *712*
Müller, W. M. 580(206), *601*
Müller-Dethlefs, K. 30(135, 136), 129(476, 478), 143(591), 144(478, 677, 685, 687, 688, 691), 147(712, 717), *182, 183, 190, 193, 195*, 375(244), *392*
Mullers, J. 83, 84(298), *186*
Mulliken, R. S. 30(133), *182*
Munasinghe, V. R. N. 1044(132), *1090*
Munch, J. H. 1371(9), *1435*
Munforte, P. 675(112), *707*
Muniz, K. 699(302), *711*
Munkittrick, K. R. 1351(48), *1364*
Munoz, C. 241(79), *256*
Munoz, G. 854(102), *904*
Muñoz Leyva, J. A. 982(363), *1011*
Munson, B. 294(80, 81), *327*
Muntau, H. 1082(287), *1094*
Mur, C. 953, 955(202), *1007*
Muradyan, L. A. 824(331, 332), *838*
Muraev, V. A. 1109(49, 50), *1144*
Murahashi, S. 1214(118), *1340*
Murahashi, S.-I. 669(64, 65), *706*
Murai, H. 1018(21), *1087*
Murakami, H. 1401(217, 218), *1444*
Murakami, T. 1030(103), *1089*
Muralikrishna, C. 1219(133a, 133b), *1341*
Murase, M. 1288, 1289(240), *1344*
Murase, N. 693(258), *710*
Murata, I. 219(36), *221*
Murayama, K. 322(243), *331*
Murkherjee, A. 94, 105(365), *187*
Murphy, D. 113(457), 129, 135(457, 489), *190*, 1109(38–45), 1117(40, 41), 1120(38, 39, 42, 44), 1121(38), 1122(38, 39), 1123(38, 42), 1124(39, 233, 235, 240), 1125(252), 1133(40–42), 1135(40), *1144, 1148, 1149*
Murray, C. J. 367(177), *390*, 541(44), *597*
Murray, J. S. 543(49), *598*
Murray, M. K. 1360(172), *1367*
Murray, N. G. 419(134b), *482*
Murray, R. W. 1255(184b), *1342*
Murrells, T. 1105(65), *1106*
Murthy, C. P. 1142(380), *1152*
Murthy, D. L. 650(160), *659*
Murthy, K. S. 1388(100a), *1439*
Murthy, M. 415(105b), *481*
Murtinho, D. 1082(276), *1093*

Murty, T. S. S. R. 586(237), 589, 590(237, 258), 592(258), *602*
Murugesan, R. 874(176, 177), *905, 906*
Murugesan, V. 617(34), *656*
Musgrave, W. K. R. 410(64a), *480*
Musin, R. N. 383(285), *393*
Muslinkin, A. A. 1419(355), *1450*
Muslinkina, L. A. 1419(355), *1450*
Mussmann, P. 930(75), *1004*
Musso, H. 1108, 1109, 1118(18), *1144*
Mustafa, A. 472(311f), *488*
Mustafina, A. R. 1424(396), *1451*
Mustroph, H. 252(96), *257*
Muszyński, A. S. 377(256), *392*
Muthukrishnan, R. 1378(52a), *1437*
Muto, H. 681(164), *708*, 1360(166), *1367*
Muus, L. T. 1122(194), *1148*
Muzart, J. 664(28), *705*
Mvula, E. 143, 144(550), *192*, 1114(90), *1145*
Myers, C. B. 1348(30), *1363*
Myles, D. C. 664(29), *705*
Mylonas, A. 1086(361, 362), *1096*
Mylrajan, M. 384(289), *393*
Myshak, E. N. 987(408), 988(413), *1012*
Mytilineos, J. 1359(150), *1366*
Mytilineou, C. 874(173), *905*
Mytych, D. T. 724, 725(27), *831*

Naberfeld, G. 695(272), *710*
Nabeta, T. 414(90j), *481*
Naczk, M. 992(425), *1012*
Nadezhdin, A. D. 1118(148), *1146*
Nadkarni, D. V. 1194(82), *1339*
Nagae, K. 769(149), *834*
Nagahata, R. 1417(333), *1448*
Nagai, N. 460(281a, 281b), *487*
Nagai, S. 972(307), *1010*
Nagakura, S. 47(203, 204), 106(406), 143(203, 204), 144, 148(203), *184, 188*
Nagaoka, S. 1021(36), 1025(86), *1088, 1089*
Nagaoka, S.-I. 844(21–24), 845(24), 856, 865(127), 891(21, 244–246), 894(260), *902, 904, 907, 908*
Nagaraj, P. 990(418), *1012*
Nagaraju, A. 650(165), *659*
Nagasaki, T. 1372(18), *1436*
Nagasawa, T. 412(74b, 75c), *480*
Nagashima, M. 1285(236), *1343*
Nagashima, U. 844(22), 856, 865(127), 891(244, 246), *902, 904, 907*, 1025(86), *1089*
Nagata, W. 683(174), *708*
Nagatomy, H. R. 984(383), *1012*
Nagawa, Y. 20, 44(120), *182*, 551, 556(117), *599*
Nagayama, S. 774(168), *835*, 1377(45a), *1437*
Nagayoshi, M. 687(225), *709*

Nageswer, B. 477(324a), *488*
Nagl, A. 447(222i), *485*
Nagra, D. S. 281(51), *327*
Nagy, G. 662(9), *705*
Nagyrei, A. 145(637), *194*
Nagyrevi, A. 595, 596(292), *603*
Naicker, K. P. 422(143), *483*
Naidu, R. 973(314), *1010*
Naik, D. B. 1129(292), *1150*
Nair, J. 414(91), *481*
Nair, M. G. 895(264), *908*
Nairn, M. R. 623(55), *657*
Nakadai, M. 681(164), *708*
Nakagaki, K. 414(90h), *481*
Nakagawa, K. 448(227), *485*
Nakagawa, Y. 890(240), *907*
Nakahara, T. 1348(27), *1363*
Nakajima, K. 699(306, 308), *711*
Nakajima, M. 1246, 1247(173), *1342*
Nakakura, S. 110, 112(447), *189*
Nakamaru, H. 1419(349), *1450*
Nakamoto, Y. 1435(464a), *1454*, 1462(25), *1503*
Nakamura, H. 651(170), *660*
Nakamura, I. 664(35), *705*, 763(134), *834*
Nakamura, K. 953, 970(292), *1009*, 1158(17c), 1164(36), 1172(48), 1175, 1180(53), 1290(36), 1315(297), 1324(307), 1325(310b), *1338, 1345*
Nakamura, M. S. 984(383), *1012*
Nakamura, R. 1405(252), *1446*
Nakamura, S. 691(244), *710*, 986(394), *1012*
Nakamura, T. 1163(33a), *1338*, 1382(69b), *1438*
Nakamura, Y. 913(28), *1003*
Nakanishi, A. 1233(151a, 152), *1341*
Nakanishi, K. 1419(350a), *1450*
Nakashima, A. 1356(125), *1366*
Nakashima, H. 429(172), *483*
Nakashima, K. 319(231), *331*, 951, 953(193, 194), *1007*
Nakashima, N. 129, 131(505), *191*
Nakashima, T. 1362(211), *1368*
Nakashima, Y. 345(49), *387*
Nakashio, F. 1372(18), *1436*
Nakasuji, K. 219(36), *221*
Nakata, H. 275(44), 299(92), *326, 327*
Nakata, M. 642(139), *659*
Nakatani, H. 1419(349), *1450*
Nakatani, K. 1034(112), *1089*
Nakatani, M. 1079(241), *1092*
Nakatsuji, H. 148, 149, 158(731), *196*
Nakatuji, K. 1018(24), *1087*
Nakayama, K. 699(303), *711*, 1204(100), *1340*
Nakayama, T. 1409(289a), *1447*
Nakazaki, M. 692(248), *710*
Nakazawa, H. 987(405), *1012*
Nakijima, K. 424(155b), *483*

Nalewajski, R. 89, 90(338), *187*
Nam, K. C. 635(102), *658*, 1393(144, 148, 154, 155), 1394(155, 163, 164), 1400(208b), 1403(155, 234b), 1407(273a, 274), *1441, 1442, 1444–1446*
Nam, K. S. 955(205), *1007*
Namba, T. 414(91), *481*
Namiki, M. 1099(11–14), *1105*
Nanayakkara, A. 4, 100(2), *179*, 368, 372, 373(200), *390*
Nangia, A. 219(35), *221*
Nanni, E. J. 1118(151), *1147*
Naota, T. 1214(118), *1340*
Napoli, M. 646(146), *659*
Napolitano, A. 1079(242), *1092*
Napper, S. 1237(164), *1342*
Narang, A. S. 1348(19), *1363*
Narang, R. S. 1348(19), *1363*
Narang, S. C. 606(2), 613(22), *656*
Naranjo Rodríguez, I. 982(363), *1011*
Narasaka, K. 460(283a), *487*, 811(293), *837*
Narayana, C. 422(148), *483*
Narita, N. 407(51), *479*
Narr, B. 1290(244), *1344*
Narumi, F. 1393(160), 1415(314a), *1442, 1448*
Naruta, Y. 460(281a, 281b), *487*, 1209(106), *1340*
Narváez, A. 977(335), *1010*
Nascimento, V. B. 978(348), *1011*
Nashed, N. T. 721(22), *831*
Nasim, K. T. 778(212), *836*
Nasini, G. 338, 345(10), *386*, 721(25, 26), *831*
Nasipuri, D. 1261(193), *1342*
Nastainczyk, W. 129(482), *190*
Natarajan, G. S. 1002(481), *1014*
Natarajan, S. 670(71), *706*, 1318(303), *1345*
Natesan, R. 1461(17), *1502*
Nath, D. N. 1021(47), 1022(51), 1025(71b), *1088*
Näther, C. 1382(71), 1408(280), *1438, 1447*
Nauck, M. 6(26), *180*
Naumann, C. 1424(402), 1428(433), 1429(428), 1433(459), *1451–1453*
Naumann, W. 1017(12), *1087*, 1101, 1102(40), *1106*, 1134(343), *1151*
Naumov, P. 377(263), *392*
Naumov, S. 110(420, 424), 114, 121(424), *189*, 1017(12), *1087*, 1101(39, 40), 1102(40), *1106*, 1134(342, 343), *1151*
Navas, N. 1351(53), *1364*
Navas, N. A. 993(435), *1013*
Navas Díaz, A. 952, 953(197), 981(360), 990(197), *1007, 1011*
Naven, R. T. 1402(228a, 228b), 1409(288), *1445, 1447*
Navio, J. A. 1080(248), *1093*
Nawata, H. 874(171), *905*
Naylor, C. G. 1356(128), *1366*

Naylor, G. G. 1348(29), *1363*
Naylor, R. D. 224, 227, 228, 230, 241(2), *254*
Nayyar, S. 774(185), *835*
Nazarov, G. B. 204(10), *220*
Neda, I. 1383(67k), 1417(329a, 329b), 1418(339), 1421(67k), 1424(395), *1438, 1448, 1450, 1451*
Nedenskov, P. 432(183a), *484*
Nefdt, S. 766(143), *834*
Neffe, S. 953, 965(264), *1009*
Negishi, E. 414(96), *481*
Negri, F. 129, 133(497), *191*
Neish, A. C. 13(64), *181*
Nekrasova, V. V. 985(389), *1012*
Nelander, B. 35(163),(166), *183*
Nemethy, G. 144(593), *193*
Nemoto, H. 665(40), *705*
Nenitzescu, C. D. 432(184a), *484*
Neri, P. 1387(92), 1388(99, 101), 1391–1394(122b), 1395(99, 178a–c, 179), 1400(209a–c, 210b), 1401(210b, 212a–d, 213), 1402(222), 1403(230), 1404, 1406(222), 1427(230, 414a), *1439, 1440, 1442, 1444, 1445, 1451*
Nerini, F. 945, 953(151, 154), *1006*, 1354(111), *1365*
Neshev, N. 89, 90(325), *187*
Nespurek, S. 748(89), *833*
Nesterenko, P. N. 962(252), *1008*
Nesterova, T. N. 227(13), 231(34), *254, 255*
Neta, P. 845(25–27), 880(25, 26, 201), 881(25), *902, 906*, 1098(4), 1099(10, 17, 19), 1101(36, 37, 41, 42), *1105, 1106*, 1110(52, 55, 59), 1111(69–72, 75, 76), 1112(52, 55, 78, 80), 1115(97), 1117(118–120), 1120–1123(97), 1126, 1127(263), 1128(76, 263, 280), 1129(76), 1130(76, 280, 326), 1131(315), 1135(97), 1136(78, 263, 356, 357), 1137(70, 75, 78), 1138(75, 76, 263), 1139(76), 1140(75, 76), 1141(76), 1142(76, 280, 379), 1143(280), *1144–1146, 1149–1151*
Neuert, U. 284(55), *327*
Neuhauser, R. G. 30(137), *183*
Neuman, B. 663(15), *705*
Neumann, B. 978(346, 351), 979(352), *1011*
Neumann, J. M. 132, 133, 135(516), *191*
Neumann, R. 1204(99), *1340*
Neuser, N. J. 147(715), *195*
Neusser, H. J. 30(137), 147(713, 716), *183, 195*, 992(427), *1012*
Neuzil, J. I. 892(254), *907*
Neves, R. 631(77, 78), *657*, 686(205), *709*
Neville, D. 776(197, 198), *835*
Neville, G. A. 380, 381(275), *392*
New, D. G. 1177(56), *1339*
Newman, P. A. 806(267), *837*
Newmark, R. A. 342(37), *387*

Neyen, V. 953, 963(260), *1009*
Neykov, G. A. 1024(65), *1088*
Neykov, G. D. 360(136), *389*
Ng, C. Y. 83(291), *186*
Ng, J. C. F. 170(748, 749), *196*, 557(142, 143), *599*
Ng, L.-K. 937, 940(130), *1006*
Ng, S. W. 377(263), *392*, 740(67), *832*
Nguen, T. 786(231), *836*
N'Guessan, T. Y. 727(37), *831*
Nguyen, A. 226, 227(7), *254*
Nguyen, A.-L. 971(298), *1009*
Nguyen, L. T. 89(331–333, 336), 90(331–333, 336, 341), *187*
Nguyen, M. T. 47, 48, 56(220), 58(244), 83(289, 293, 299), 84(299, 303), 85, 86, 88(299), 89(220, 299, 313, 317a, 317b, 326–340), 90(299, 326–341), 91(299, 345, 346), 110(448–450, 452), 114(460), 115(303), 119(463), 123(450), 127(470–472), 139(530), 143(551), 144(472), 148(732), 171(761), 177(732), *184–187, 189–192, 196, 197*
Nguyen, V. 110(438), *189*
Ni, Y. 929, 981(66), *1004*
Nibbering, N. M. M. 93(355), 110(427, 429, 430), *187, 189*, 264(21), 299(97, 99, 101), 300(21), 302(138), *326–328*
Nicell, J. A. 1362(208), *1368*
Nichel, C. 856(121), *904*
Nichols, D. E. 403(28), *479*, 764(137), *834*
Nichols, P. J. 1372(24b), 1427(418), *1436, 1452*
Nicholson, J. L. 776(195), *835*
Nickelsen, M. G. 1100(33), *1106*
Nicolaisen, F. M. 35, 53(155), 85(305), *183, 186*
Nicolaou, K. C. 670(71), *706*, 1318(303), *1345*
Nicolet, P. 503(89), *526*, 541(43), 586, 587(239), 591(266, 267), 592(267), *597, 602*
Nicoletti, R. 268(30), *326*
Nie, X. 685(199), *709*
Niedzielski, J. 930(82), *1005*
Nieger, M. 1378(48a, 48b), 1432(446), *1437, 1453*
Nielsen, B. B. 212(23), *221*
Nielsen, J. T. 432(182c), *484*
Nielsen, K. W. 358(112), *388*
Nielsen, O. J. 138(528), *191*, 1104(61), *1106*
Nielsen, S. F. 757(108), *833*
Nielson, A. T. 1470(47), *1503*
Nielson, J. B. 342(34), *387*
Niemann, H. 1125(245), *1149*
Nierengarten, J.-F. 1406(272), *1446*
Nierlich, M. 1380(58c), 1409(289b), 1415(313), 1435(465b), *1437, 1447, 1448, 1454*

Nierop, K. G. A. 1433(456a), *1453*
Niessner, R. 319(232), *331*, 1360(164), *1367*
Nieto, P. M. 1374(35), 1385(79a), 1387, 1392, 1394(90), 1395(90, 177), 1409(287), *1436, 1439, 1442, 1447*
Nieuwenhuyzen, M. 1432(448d), *1453*
Niki, E. 851, 852(71), 853(90), 855(90, 113, 119), 856(90, 119, 126, 130), 863, 864(126), 882(204), 886(233), 887(218, 221), 890(232, 233, 235, 236), 892(71, 251), 893(255), 900(281), *903, 904, 906–908*
Nikiforov, G. A. 715, 717, 718, 720, 722, 724, 731(2), *830*, 1123(210), *1148*
Nikiforov, S. M. 1000(468), *1013*
Nikitin, P. I. 1000(469), *1013*
Nikolaeva, I. L. 1424(396), *1451*
Nilsson, M. 414(94c), *481*
Nimlos, M. R. 17(81), *181*
Nimura, N. 954(200), *1007*
Ninagawa, A. 1408(279a), *1447*
Nishi, N. 174, 177(766), *197*
Nishi, T. 851(57, 60), 852(60), 876(184), 887, 888(60), *903, 906*, 1419(347a), *1450*
Nishida, A. 450(239e), *486*, 650(164), *659*
Nishida, S. 1062(186), *1091*
Nishiguchi, I. 410(68), *480*
Nishihata, N. 1029(100), *1089*
Nishikimi, M. 1119(166), *1147*
Nishikori, H. 702(327), *712*
Nishimoto, A. 1048(158), *1090*
Nishimoto, J. 170(740), *196*
Nishimura, J. 662(12), *705*, 1378(47a–c), *1437*
Nishimura, M. 702(332), *712*, 856(128), *904*, 1470(67), *1504*
Nishimura, S.-I. 1495(137), *1505*
Nishimura, T. 265, 272(29b), *326*, 614(25), *656*
Nishimura, Y. 309(172, 173), 310(174), *329*, 447(222f), *485*
Nishinaga, A. 1206(105), 1209(108), 1265(199a, 199b), *1340, 1343*
Nishinguchi, H. 1109(31), *1144*
Nishino, H. 1204(98), 1285(236), *1340, 1343*
Nishino, R. 937(118), *1005*
Nishio, S. 1403(231a), *1445*
Nishio, Y. 1071(209b), *1092*
Nishioka, C. 844, 891(21), *902*
Nishioku, Y. 844, 891(21), *902*
Nishiwaki, K. 871, 874(160), *905*
Nishiyama, A. 1158(19–21, 23, 24), 1159(24, 25b, 26, 30), 1163(31, 32), 1172(25b), 1178(21), 1227(30), 1315(31, 32), *1338*
Nishiyama, H. 1356(125), *1366*
Nishiyama, K. 232(37), *255*, 1425(405), *1451*
Nishiyama, S. 1158(17b), 1164(35, 36), 1172(50), 1178(57a, 57b), 1224(35), 1290(36), 1314(290), 1315(291, 295a,

Nishiyama, S. (*continued*)
  295b, 296, 297), 1317(298), 1318(299, 306), 1324(307–309), 1325(310a, 310b), 1337(57a, 57b, 332), *1337–1339, 1345, 1346*
Nishiyama, T. 143(546, 547), 179(546), *192*, 871, 874(160), *905*
Nishiyama, Y. 1204(100), *1340*
Nishizawa, M. 211(19), *220*
Nissen, J. 338(16), *386*
Nissinen, M. 1403(229), 1427(409c), 1432(229), *1445, 1451*
Nistor, C. 976(329), *1010*
Nito, S. 938(128), *1006*
Nitta, H. 702(344), *712*
Nitta, K. 1406(270b), *1446*
Nitta, M. 414(89b), *481*
Nivorozhkin, L. E. 1024(66), *1088*
Niwa, C. 1362(211), *1368*
Niwa, H. 1080(257), *1093*, 1188(74), *1339*
Niwa, J. 100(379), *188*
Niwa, M. 1079(240), *1092*, 1172(49), 1218(129), *1338, 1341*
Niwa, S.-I. 414(91), *481*
Niwa, T. 953, 969(279), *1009*
No, K. 634(96), *658*, 1374(36a), 1392(129b, 130), 1407(277), *1436, 1441, 1447*
No, K. H. 1378(52a), *1437*
Nobeli, I. 535(31), *597*
Noda, H. 1172(49), *1338*
Noda, K. 699(298, 312), 702(326), *711, 712*
Noda, S. 1214(118), *1340*
Nodes, J. T. 141(540), *192*
Noël, J.-P. 1212(111), *1340*
Nogaj, B. 580(205), *601*
Nogami, H. 695(268), *710*
Nógrádi, M. 1315(292), *1345*
Noguchi, I. 1159(29), *1338*
Noguchi, N. 856, 863, 864(126), 893(255), *904, 907*
Nogudhi, N. 900(281), *908*
Noguera-Ortí, J. F. 953, 961(237), *1008*
Noh, Y. 634(96), *658*
Nolan, K. 1435(469a), *1454*
Nolte, J. 936, 937(116), *1005*
Nolte, M. 461(290), *487*
Nolte, R. J. M. 1000(469), *1013*
Nomura, E. 662(11), *705*, 1402, 1409(223), *1444*
Nomura, M. 664(34), *705*
Nonella, M. 93, 94, 105, 132(353), *187*
Nonhebel, D. C. 1274(210b), 1285(237), *1343*
Nonoshita, K. 688(228), 689(234), *709, 710*
Nonoyama, N. 636(105), *658*
Noodleman, L. 101(385), *188*
Noordman, O. F. J. 1391(119), *1440*
Norberg, J. 953, 969(280), *1009*
Nordén, B. 368(191), *390*

Norell, J. R. 472(311c), *488*, 749(98), 774(98, 189), *833, 835*
Norikane, Y. 1028(93, 95), *1089*
Norin, T. 414(94c), *481*, 1048(155), *1090*
Norman, R. O. C. 405(41b), *479*, 1113(84, 85, 88), 1114–1117(88), 1120(84, 88), *1145*, 1330(321), *1346*
Normura, T. 338(12), *386*
Norris, J. Q. 843(283), 867–869, 871(152), 901(283), *905, 908*
Norris, K. K. 360(135), *389*
Norwood, A. D. L. 1360(180), *1367*
Nosova, G. I. 1030(105), *1089*
Notario, R. 312(193), *330*
Notté, P. P. 413(80c), *480*
Notti, A. 1390(106, 108), 1391(118b), 1393(153), 1396(181b, 181c, 194, 195d), 1403(194), *1440, 1441, 1443*
Nour El-Dien, F. A. 988(415), *1012*
Noury, S. 69, 148(266), *185*
Novais, H. M. 1115(107), *1146*
Novak, A. 145(624), *193*
Novak, I. 89(311), *186*
Novak, J. T. 1348(13), *1363*
Novak, L. 759(115), *833*
Novi, M. 1073(225), *1092*
Novillo-Fertrell, A. 1352(58), *1364*
Nowak, M. J. 370, 382(211), *391*
Nowakowska, M. 1086(377), *1096*
Nowakowsky, R. 1418(340), *1450*
Nowicka-Sceibe, J. 352(91), *388*
Nowicka-Scheibe, J. 551, 557(121), *599*
Nowinska, K. 619(37), *656*
Nowlin, J. G. 294(79), *327*
Noyori, R. 692(247), *710*
Nozaki, F. 608(11), *656*
Nozal, M. J. 953, 962(256), *1009*
Nukatsuka, I. 986(394), *1012*
Nummy, L. J. 827(341), *838*
Nuñez, L. 234(43), *255*
Nurakhmetov, N. N. 230(27), *254*
Nury, P. 702(343), *712*
Nussbaumer, P. 434(192), *484*
Nuutinen, J. M. J. 322(250), *331*
Nwobi, O. 93, 94, 132(360), *187*
Nye, M. J. 7(41), *180*
Nygaard, L. 20, 22, 35(107), *182*
Nyholm, L. 953, 955(208), *1007*
Nyholm, R. S. 68(258), *185*
Nyokong, T. 982(365), *1011*, 1085(356), *1096*
Nyquist, R. A. 53(236), *185*, 368–371(188), *390*

Oae, S. 398(9a), 407(51), 422(131), *478, 479, 482*
Oakes, J. 1209(109), *1340*
Oberg, L. G. 1347(8), 1352(60), *1363, 1364*

Oberhauser, T. 650, 651(161), *659*
Oberlin, A. 1489(109, 115), *1504*
Obermueller, R. A. 1045(140), *1090*
Obi, K. 129, 131(505), *191*
O'Block, S. T. 1351(54), *1364*
Obrecht, D. 444(214), *485*
O'Brien, D. H. 345, 349(46), *387*
Occhiucci, G. 274(41), *326*
Occolowitz, J. L. 268(31), *326*
Ochiai, M. 1233(151a, 151b, 152), *1341*
Ochiucci, G. 307(156), *329*
Ochterski, J. 4, 100(2), *179*, 368, 372, 373(200), *390*
Ockwell, J. N. 1124(236), *1148*
O'Connor, K. J. 702(325), *712*
Oda, M. 1018(24), *1087*
Oda, T. 811(290), *837*
Odaira, Y. 1048(159), *1091*
Odo, M. 1018(21), *1087*
Odorisio, P. A. 1495(147), *1505*
Ogan, M. D. 1470(47), *1503*
Ogasawara, H. 1065(195), 1066(196), *1091*, 1330(316, 317), *1346*
Ogasawara, K. 463(276d), *487*
Ogasawara, M. 669(67), *706*
Ogata, Y. 408(53), 422(145), 472(310a), *479, 483, 488*
Ogawa, K. 211(19), *220*
Ogawa, S. 1380(62), *1438*
Ogden, M. 1435(465b), *1454*
Ogden, M. I. 1409(286b), *1447*
Ogden, T. 1358(144), *1366*
Ogle, L. D. 1356(128), *1366*
Oguni, N. 702(334, 338, 342), *712*
Ogura, K. 1318(300a, 300b), *1345*
Oguri, M. 414(90g), *481*
Oh, W. S. 1394(167b), *1442*
Ohara, K. 844, 891(21), *902*
Ohara, M. 477(326a), *489*
Ohara, S. 1019(27b, 30), *1087*
Ohata, K. 1382(69a, 69b), *1438*
Ohba, S. 699(306), *711*, 855, 856(105), *904*, 1172(48), *1338*
Ohba, Y. 1374(31a, 31b), 1380(58g, 59, 63), *1436–1438*
Ohbayashi, A. 1163(33a), *1338*
Ohguchi, C. 844, 891(21), *902*
Ohira, K. 1066(197), *1091*
Ohko, Y. 1362(211), *1368*
Ohkubo, H. 1419(345), *1450*
Ohkubo, K. 1084(322), *1095*
Ohkubo, M. 1180(61b), 1183(67), *1339*
Ohkubo, N. 774(194), *835*
Ohlemeier, L. A. 953, 970(287), *1009*
Ohmori, M. 1374(38a), *1436*
Ohnishi, T. 414(89c), *481*
Ohnishi, Y. 1372(14), *1436*
Ohno, T. 408(53), *479*, 990(421), *1012*

Ohnsorge, P. 1359(152), *1366*
Ohseto, F. 1401(217, 218), 1427(408b), *1444, 1451*
Ohshima, T. 695(273, 274), *710, 711*
Ohta, C. 699(309), *711*
Ohta, N. 397(3), *478*
Ohtaka, S. 614(25), *656*
Ohto, K. 1372(18), *1436*
Ohyashiki, T. 855, 857(108), *904*
Ohzeki, K. 986(394), *1012*
Oikawa, A. 143, 147(578), *192*
Oishi, T. 1495(161), 1498(176), *1506*
Oka, K. 968(274), *1009*
Oka, W. 894(260), *908*
Okabe, A. 144(599), *193*
Okabe, K. 865(146, 149), 866, 896(146), *905*
Okada, K. 683(174), *708*, 762(126), *834*, 1018(21, 24), *1087*
Okada, M. 953, 957(219), *1008*
Okada, Y. 662(12), *705*, 1378(47a–c), *1437*
Okajima, A. 447(222j), 448(225, 226), *485*
Okamoto, A. 650(164), *659*
Okamoto, H. 219(36), *221*
Okamoto, K. 1159, 1227(30), *1338*
Okamoto, T. 651(170), *660*
Okamoto, Y. 1385(83), *1439*
Okamura, N. 953, 966(268), *1009*
Okamura, T. 1470(67), *1504*
Okaniwa, K. 219(36), *221*
Okauchi, Y. 891(244, 245), *907*
Okawa, H. 1205(103), *1340*
Okazaki, H. 1071(209a, 209b), *1092*
Okazaki, K. 1288, 1289(240), *1344*
Okazaki, R. 1399(201a–c), 1428(435a, 435b, 436), *1443, 1444, 1452*
Oki, Y. 1495, 1498(156), *1505*
Okte, A. N. 1084(329), *1095*
Oktyabrskii, V. P. 143(558), 147(644), *192, 194*
Oku, A. 697(284), *711*
Oku, Y. 773(160), *834*
Okubo, K. 1018(21), *1087*
Okubo, M. 1158(17c), *1338*
Okuda, A. 1100(20), *1105*
Okuda, C. 669(65), *706*
Okumoto, S. 762(125), *833*
Okunaka, R. 447(221a, 221b, 222d, 222e, 222k), *485*
Okuno, T. 1324(309), *1345*
Okuno, Y. 1180(61a, 61b), 1186(70), *1339*
Okutsu, T. 1029(102), *1089*
Okuyama, K. 110, 112, 121(444), *189*
Okuyama, M. 551, 554(109, 110), 555(109), *599*
Olah, G. A. 405(37b, 38), 472(308e), *479, 488*, 606(2), 613(21–23), 643(141), 644(23), 645(21), *656, 659*, 671(85), *706*
Olah, J. A. 613(22), *656*

Old, D. W. 671(80), *706*
Olea, M. 925(48), *1004*
Olea, N. 1352(58), *1364*
Olea-Serrano, F. 1352(58), *1364*
Olea-Serrano, M. F. 925(48), *1004*
Olechnowicz, A. 726(36), *831*
Oleinikova, T. P. 236, 237(57), *255*
Olejnik, J. 377(267), *392*
Olejnik, Z. 352(91), *388*, 551, 557(121), *599*
Olejnok, J. 144(672, 673), *194*
Olekhnovich, L. P. 743(69), 770(154), *832, 834*
Olekhnovitch, L. P. 1290, 1292(247), *1344*
Olensen, T. 1360(174), *1367*
Olesky-Frenzel, J. 987(404), *1012*
Oleszek, W. 953, 958(225), *1008*
Oliva, A. 129, 144, 148, 149, 153(473), *190*, 577, 581(192), *600*
Oliveira, A. P. de 632(90), *658*
Oliveira, A. R. M. 1285(233), *1343*
Olivella, S. 137(523), *191*
Oliver, J. V. 1000(469), *1013*
Oliver, K. L. 1494(128), *1505*
Oliver, M. 358(112), *388*
Oliver, S. 503(89), *526*
Oliveros, E. 1080(253), *1093*
Olivieri, A. C. 349(59), 359(129), 367(172, 173), *387, 389, 390*, 726(33), *831*
Ollis, W. D. 806(264), *837*
Olmo, M. del 993(435), *1013*, 1351(53), *1364*
Olmstead, M. M. 1374(44), *1437*
Olovsson, I. 144(664), *194*
Olsen, C. E. 757(108), *833*
Olskær, K. 373, 383, 384(238), *392*
Olson, K. G. 1495, 1498(152), *1505*
Olson, R. J. 650(160), *659*
Olszanowski, A. 1067(198), *1091*
Olucha, J. C. 1352(65), *1364*
Om, J. 1355(119), *1365*
O'Malley, P. J. 131–133(515), *191*
Omar, O. 1419(347b), *1450*
Omata, K. 680(157), *708*
Omura, A. 650(164), *659*
Omura, K. 650(158), 654(158, 188), 655(189), *659, 660*, 717(14), 748(91), *831, 833*, 1266(200), *1343*
Onaka, K. 1414(309), *1448*
Ondercin, D. G. 1126(262), *1149*
O'Neill, P. 1109(46, 47), 1114(47), 1115(47, 94, 103, 112), 1116(94), 1120(47, 94), 1122, 1125, 1128(47), 1134(94), *1144–1146*
Ong, A. S. H. 840, 871(9), *902*
Ong, C. N. 920, 949, 953(33), *1003*
Ong, H. Y. 920, 949, 953(33), *1003*
Ong, S. K. 178(773), *197*
Onjia, A. E. 1001(474), *1014*
Önnerfjord, P. 975(322), *1010*

Onnis, V. 764(138, 139), *834*
Ono, Y. 413(78a), *480*, 1109, 1128, 1130(32), *1144*
Onodera, S. 1352(57), *1364*
Onomura, O. 129, 179(491), *190*, 881(202), *906*
Onopchenko, A. 749(99), *833*
Onuska, F. I. 930, 937(80), *1004*
Onyiriuka, S. O. 633(93), *658*
Ooshima, M. 1028(99), *1089*
Oosthuizen, F. 766(144), *834*
Opaprakasit, P. 377(252), *392*
Opella, S. J. (97, 98), *1504*
Opelz, G. 1359(150), *1366*
Opovecz, O. 759(115), *833*
Orda-Zgadzaj, M. 1403(237), *1445*
Orena, M. 1255(180), *1342*
Orendt, A. M. 366(167), *390*
Organero, J. A. 1047(149), *1090*
Oribe, M. 855(104), *904*
Orita, H. 1196(89), 1213(114, 117), *1340*
Orlandi, M. 1216(127), *1341*
Orlando, R. 294(80, 81), *327*
Orlov, V. M. 301(131), *328*
Ormad, M. P. 1086(360), *1096*
Ormson, S. M. 1024(60), 1047(147), *1088, 1090*
Ornstein, R. L. 105(402), *188*
Orozco, M. 5(13), *179*, 1025(77), *1089*
Orphanos, D. G. 420(139), *482*
Orrell, E. W. 1470(55), *1503*
Orrell, K. G. 252(95), *257*
Ortega, F. 953, 969(284), 975(322), 976(330), 977(335), *1009, 1010*
Orthner, H. 110(422), *189*, 1101(38), *1106*, 1134(341), *1151*
Orthofer, R. 990(419), 1001(473), *1012, 1014*
Ortiz, A. R. 1406(266), *1446*
Ortiz, J. M. 983(376), *1011*
Ortiz, J. V. 4, 100(2), *179*, 368, 372, 373(200), *390*
Orton, E. 551(89), *598*
Osa, T. 1156, 1158, 1175(7), *1337*
Osaki, K. 11(54), *180*, 232(36), *255*
Osamura, Y. 20, 22, 23, 31, 129, 144, 148, 149(113), *182*
O'Shea, K. E. 1083(297), 1084(315), *1094, 1095*
Oshikata, M. 807(275), *837*
Oshima, R. 272(37), *326*
Oshima, Y. 1079(240), *1092*
Oshuev, A. G. 18(91), *182*
Osmanoglu, S. 1104(56), *1106*
Ostaszewski, R. 1396(189), *1443*
Ostromyslenskii, I. I. 638(118), *658*
Ostu, T. 1495(156, 160, 162, 163), 1498(156, 160), *1505, 1506*
Osuka, A. 1019(28), *1087*

Ota, T. 677(135, 136), *707*, 1495(158), *1505*
Otake, H. 855(117), *904*
Othman, A. H. 740(67), *832*
Otsubo, T. 1380(56), *1437*
Otsuda, K. 414(90j), *481*
Otsuji, Y. 662(11), *705*
Otsuka, H. 1379(54b), 1388(103), 1394(174), 1400(207a, 207b), *1437, 1440, 1442, 1444*
Otsuka, K. 414(90e, 90h), *481*
Otsuka, T. 1435(464b), *1454*
Otsuki, K. 1495(164–166), 1498(164), *1506*
Otsura, K. 414(90i), *481*
Ott, D. E. 987(403), *1012*
Otto, C. A. 447(220), *485*
Otto, M. 985(387), *1012*
Ouchi, K. 1489(111, 112), 1491(111), *1504*
Oude Wolbers, M. P. 1427(407a), *1451*
Oudjehani, K. 1072(210), *1092*
Ougass, Y. 913(4), *1003*
Ouk, S. 663(19), *705*
Ould-Mame, S. M. 1080(249), *1093*
Ouvrard, C. 554, 555, 559(138), 586, 587, 589(256), 590(138), *599, 602*
Ovelleiro, J. L. 1086(360), *1096*
Owen, L. N. 399(11), *478*
Owens, M. 1432(449b), *1453*
Oyama, K. 674(101), *707*
Ozawa, K. 677(134, 135), *707*
Ozden, S. 944, 953(142), *1006*
Ozegowski, S. 1401(214), *1444*
Ozkabak, A. G. 110–112(426), *189*
Ozoemena, K. 1085(356), *1096*
Ozsoz, M. 978(349), *1011*

Paasivirta, J. 1357(134, 135), 1363(219, 220), *1366, 1368*
Pac, C. 1306(271), *1344*
Pachta, J. 1462(30), *1503*
Pachuta, S. J. 308(164), *329*
Packer, J. E. 890(230), *907*, 1111(66, 67), 1130(324), *1145, 1150*
Packer, K. J. 367(170), *390*
Packer, L. 840, 871(9), 887(222), *902, 907*, 913(21), *1003*, 1071(207), *1092*
Padilla, F. N. 403(27), *479*
Padmaja, S. 637(108, 111, 112), *658*
Padwa, A. 449(234c), *486*, 668(58), *706*, 818(312, 313), *838*
Paek, K. 1396(183), 1406(267), 1423(381), 1428(427, 438), *1443, 1446, 1450, 1452, 1453*
Paek, K.-S. 1406(268), *1446*
Paemel, C. V. 1104(54), *1106*
Paeng, K.-J. 946(163), *1006*
Paganga, G. 871(161), 894(161, 257), *905, 907*, 957(220), *1008*
Paganuzzi, D. 1396(188), 1409(288), *1443, 1447*
Page, B. D. 925, 961(44), *1004*
Page, D. 139(531), *191*
Pagoria, P. F. 1372, 1406(26), *1436*
Pai, B. R. 1063(192), 1064(193), *1091*
Painter, P. 377(252), *392*
Paixo, J. A. 233(39), *255*
Pajonk, G.-M. 1084(327), *1095*
Paju, A. I. 589(259), *602*
Pajunen, A. 1309(281), *1345*
Pakarinen, J. M. H. 322(250), *331*
Pakhomova, S. 1415(315), *1448*
Pakkanen, T. A. 143(548), *192*
Pakkanen, T. T. 997(445, 446), *1013*
Pal, A. 1307(276), *1344*
Pal, H. 1038(121), *1090*, 1130(312), 1131, 1134(314, 317), 1135(314, 350), *1150, 1151*
Pal, T. 1307(276), *1344*
Palacin, S. 1409(289b), *1447*
Palanichamy, M. 617(34), *656*
Palit, D. K. 1021(38), *1088*, 1130(312), 1135(350), *1150, 1151*
Palkowitz, J. A. 673(88), *706*
Pallalardo, S. 1392(138), *1441*
Palmacci, E. R. 642(140), *659*
Palmer, G. 15(72), *181*, 1119(165), *1147*
Palmer, M. H. 67(252), 110, 112(443), *185, 189*
Palmer, T. F. 274(40), *326*
Palmisano, L. 1082(288, 290), 1083(293, 301, 302), 1084(320, 325, 326), *1094, 1095*
Palomar, J. 371, 372, 375, 377, 378(235), *391*, 551(84, 106, 107), 552(106), 553(84, 106, 107), 554(107), *598, 599*
Palomino, N. F. 1354(113), *1365*
Palozza, P. 851, 855(73), *903*
Palucki, M. 699(311), *711*
Palui, G. D. 729(44), *831*
Pan, Y. 377(261), *392*
Panda, M. 1047(148), *1090*
Pandarese, F. 216(27), *221*
Pandey, A. K. 631(76), *657*
Pandey, G. 1218(130), 1219(133a, 133b), 1223(135), *1341*
Pandiaraju, S. 697(285), *711*
Pangarkar, V. G. 1084(311), *1094*
Panicker, C. Y. 378(271), *392*
Paniez, P. 1459(12), *1502*
Panov, G. I. 18(84, 85), *181*, 413(79, 80b), *480*
Panster, P. 620(47), *657*
Pantani, F. 933(102), 934(108), 937(102, 108), *1005*
Paoli, D. de 1378(50a), *1437*
Paone, D. V. 450(239g, 239h), *486*
Papaconstantinou, E. 1086(361, 361), *1096*

Papadopoulos, M. 1075(232c), *1092*
Papageorgiou, G. 684(188, 189), *709*
Papaioannou, D. 1297(257), *1344*
Papouchado, L. 1157(10), *1337*
Papoula, M. T. B. 681(163), *708*
Pappalardo, S. 1387(92), 1390(106, 108), 1391(118b, 122b), 1392(122b, 127), 1393(122b, 153, 157a), 1394(122b, 172c), 1396(172c, 181b, 181c, 194, 195c, 195d), 1403(194), *1439–1443*
Parbo, H. 1122(194), *1148*
Pariente, L. P. 1084(324), *1095*
Paris, A. 316(218, 219), *330*
Parisi, M. 1391–1394(122b), *1440*
Parisi, M. F. 1390(106, 108), 1391(118b), 1393(153), 1394(172c), 1396(172c, 181b, 181c, 194, 195c, 195d), 1403(194), *1440–1443*
Park, B. S. 1428(426a), *1452*
Park, C. H. 1044(131), *1090*
Park, D.-C. 664(32), *705*
Park, H. S. 1405(256a, 256b), *1446*
Park, J. 1045(137), *1090*
Park, J. G. 673(92), *706*
Park, J. S. O. 623(55), *657*
Park, K. S. 1407(274), *1446*
Park, S. 973(311), *1010*
Park, Y. C. 1086(373), *1096*
Park, Y. J. 1392(130), 1393(155), 1394(155, 163), 1403(155), *1441, 1442*
Park, Y. S. 170(747–749), *196*, 557(142, 143), *599*
Parkanyi, C. 403(32), *479*
Parker, A. W. 844(20), *902*, 1129(283), 1132(339), 1134(283), *1149, 1150*
Parker, J. A. 1459(13), *1502*
Parker, K. A. 414(93), 424(153a), 450(246d), *481, 483, 486*, 814(304), *838*
Parker, M. G. 925(49), *1004*
Parker, V. D. 1167, 1168(37), *1338*
Parkhurst, R. M. 845(34), *902*
Parmentier, J. 145, 146(610, 615), *193*, 562(166), 583(227), 584(234), 585(236), *600, 601*
Parnell, R. D. 1125(247), *1149*
Parola, S. 1415(318a), *1448*
Parpinello, G. P. 972(308), *1010*
Parr, A. 913(3), *1003*
Parr, R. G. 89, 90(320), *187*
Parr, W. J. E. 349(71), *387*
Parrington, B. D. 1048(152, 153), *1090*
Parsons, A. S. 782(216), *836*
Parsons, G. H. 230(29), *254*
Partington, J. R. 6, 7(32), *180*
Parzuchowski, P. 1403(229), 1427(409c), 1432(229), *1445, 1451*
Paseiro Losada, P. 937, 941(132), *1006*
Pashaev, Z. M. 612(20), *656*

Pasimeni, L. 1123(205), *1148*
Pasqualone, A. 913(7), *1003*
Passariello, G. 1354(101), *1365*
Pastor, S. D. 1495(144, 147), *1505*
Pastorio, A. 680(155), *708*, 740(62), *832*, 1374(37), 1378(37, 49a), *1436, 1437*
Pasutto, F. M. 412(76), *480*
Patai, S. 4(9), *179*
Patchornik, A. 472(311j), *488*
Patel, K. B. 1119, 1128, 1131, 1134(155), *1147*
Patel, R. N. 1216(119f), *1340*
Patelli, R. 1000(467), *1013*
Pathak, A. 1274(211), *1343*
Pathak, V. P. 478(329), *489*
Patoiseau, J. F. 652(184), *660*
Patricia, J. J. 438(198), *484*
Patrick, B. O. 1424(402), 1429(428), *1451, 1452*
Patrick, T. B. 687(215), *709*
Patrineli, A. 1360(177), *1367*
Pattermann, J. A. 444(216g), *485*
Patterson, D. 546(57), *598*
Patterson, I. L. J. 580(209), *601*
Patterson, J. M. 368(184), *390*
Patterson, L. K. 1117(117), 1125(246), 1129(246, 304), *1146, 1149, 1150*
Paukku, R. 1357(134), 1363(220), *1366, 1368*
Paul, G. J. C. 313(198, 199), 314(199), *330*
Paul, I. C. 206(11), *220*
Paul, K. G. 1352(60), *1364*
Paul, P. P. 1194(87), *1340*
Paul, S. 1492(119), *1505*
Paul, S. O. 383(278), *392*
Pauling, L. 20(104), 47(104, 212), *182, 184*, 530(1), *596*
Paulins, J. 347(54), *387*
Paulus, E. F. 1374(34, 39, 40, 43), 1382(74), 1383(67g), 1386(87), 1418(342, 343a, 343b), 1419(356b, 358), 1420(342, 360, 361, 365b, 366), 1421(67g, 342), 1427(412b), 1431(443a, 443b), *1436–1439, 1450, 1451, 1453*
Paune, F. 1355(119), *1365*
Pauson, P. L. 1274(210b), 1285(237), *1343*
Pauwels, P. 110(433), *189*
Pawelka, Z. 352(91), 377(260, 262), *388, 392*, 534(28), 551, 557(121), 563(168), 574(28, 182), 593(28, 272), 595, 596(28, 182, 272), *597, 599, 600, 602*
Pawlak, D. 367(169), *390*, 729(41), *831*
Pawlak, Z. 594, 595(291), *603*
Pawliszyn, J. 932(88, 89), *1005*
Paya-Perez, A. B. 1086(357), *1096*
Payne, C. S. 403(27), *479*
Payne, J. 241(77), *256*
Paz Abuín, S. 937, 941(132), *1006*
Peachey, B. J. 1388(97), *1439*

Pearce, A. 1250(174), *1342*
Pearl, I. A. 462(272), *487*
Pearlstein, R. 983(373), *1011*
Pearson, A. J. 673(90–92, 94), *706*
Pearson, D. E. 650(163), *659*
Pearson, R. G. 101(382), 103(396), *188*
Peart, T. E. 932, 933(95), *1005*
Pecchi, G. 1087(383), *1096*
Pedersen, J. A. 913(17), *1003*
Pedersen, J. B. 1122(194), *1148*
Pedersen, S. U. 111(454), *190*
Pedersen, T. 20, 22, 35(107), *182*
Pederson, K. J. 522(98), *527*
Pedley, J. B. 224, 227, 228, 230, 241(2), *254*
Pedone, C. 1383(75b), *1439*
Pedraza, V. 1352(58), *1364*
Pedrielli, P. 139(532), *191*, 871, 872(166), *905*
Pedulli, G. F. 45(190, 198, 201), 139(532), 140(198, 201), *184, 191*, 855(120), 871, 872(166), 877, 878(193), 879, 880(199), 881, 884(193), 900(280), *904–906, 908*
Peel, J. B. 144(688), *195*
Peer, H. G. 1475, 1486(82, 83), *1504*
Peers, K. E. 852(75, 82), *903*
Peersoons, A. 1391(119), *1440*
Peeters, D. 565(173), 566(175), *600*
Pehle, W. 319(222), *330*
Pei, Z. F. 620(45), *657*
Peinador, C. 1422(377), 1424(402), 1429(428), *1450–1452*
Pelizzetti, E. 1082(281), 1086(361, 365, 368), *1094, 1096*, 1100(25, 26), *1105*, 1129(285, 286), *1149*, 1361(195, 196), *1367*
Pellet-Rostaing, S. 1401(216b), 1417(330c), *1444, 1448*
Pellinghelli, M. A. 1374, 1378(37), *1436*
Pellón, R. F. 670(76), *706*
Pelter, A. 1224(140f), 1227(141, 142a, 142b), 1228(142a, 142b), 1242(168), 1244(169), 1308(280), *1341, 1342, 1344*
Pena, P. R. J. 1362(216), *1368*
Pendery, J. J. 447(220), *485*
Penedo, J. C. 1025(72), *1088*
Peng, C. 101(385), *188*
Peng, C. Y. 4, 100(2), *179*, 368, 372, 373(200), *390*
Peng, L. W. 493(21, 23), *525*
Peng, S.-M. 1407(273b), *1446*
Peng, W. 1002(480), *1014*
Penner, N. A. 962(252), *1008*
Pentrick, R. A. 1482(88), *1504*
Pepe, N. 309(167, 168), *329*
Peperle, W. 234(48), *255*
Perego, C. 413(80a), *480*
Peregudov, A. S. 357(111), *388*
Pereira, M. M. 1082(276), *1093*
Peremans, A. 78(282), *186*
Peres, J. A. 1362(207), *1368*

Pérez, J. J. 450, 452(239c), *486*
Perez, M. A. 769(148), *834*
Perez, P. 1352(58), *1364*
Pérez-Adelmar, J.-A. 1392, 1403(125a), 1427(424c), *1441, 1452*
Pérez-Arribas, L. V. 953, 962(253, 257), *1008, 1009*
Perez-Bendito, D. 929, 987(70), *1004*
Pérez-Bustamante, J. A. 947(175), 953(175, 232), 960(232), *1007, 1008*
Perez-Casas, S. 546(57, 58), *598*
Perez-Garcia, L. 761(118), *833*
Pérez Lamela, C. 937, 941(132), *1006*
Pérez-Lustres, J. L. 385(302), *393*, 1025(70a), *1088*
Pérez-Magariño, S. 961(235), *1008*
Pérez Pavón, J. L. 948, 953(179–181), *1007*
Pergal, M. 432(184b), *484*
Peri, F. 1374(37), 1378(37, 49c), *1436, 1437*
Pericàs, M. A. 422(146), *483*
Peringer, P. 1084(327), *1095*
Perkin, M. J. 858, 887(136), *905*
Perkins, M. J. 1259(189), *1342*
Perl, W. 144(678), *195*
Perlat, M. C. 316(221), *330*
Perminova, I. R. 339(19), *386*
Pernecker, T. 621(50), *657*
Perosa, A. 663(18), *705*
Pérot, G. 631(77), *657*, 686(205), *709*
Perrakyla, M. 143(548), *192*
Perrée, M. 1201, 1205(92), *1340*
Perri, E. 953, 961(239), *1008*
Perrin, C. L. 342(34), *387*
Perrin, M. 200(1), *220*, 549(66), *598*, 1372(19, 30), 1378(52b), 1394(173), 1415(312, 318a), *1436, 1437, 1442, 1448*
Perrin, R. 200(1), *220*, 229, 230(23), *254*
Perry, D. H. 1315(292), *1345*
Persoons, A. 1391(112), 1402, 1405(224b), *1440, 1444*
Persy, G. 127(468), *190*, 716(7), *831*, 1035(116), *1089*
Pertlik, F. 350(85), *388*, 551(91), *598*
Pervez, H. 633(93), *658*
Pesakova, J. 990(416), *1012*
Petasis, N. A. 671(85), *706*
Petch, W. A. 419(123b), *482*
Pete, J.-P. 1062(188–190), 1063(191), *1091*
Peteanu, L. A. 551(90), *598*
Petek, W. 920(35), *1003*
Peter, F. A. 1099(10), *1105*
Peterli, S. 229, 250(20), *254*
Peters, E. M. 1080(259a), *1093*
Peters, K. 1080(259a), *1093*
Peters, K. S. 5(23), *180*
Petersen, A. 293(74), *327*
Peterson, G. A. 456(260), *487*
Peterson, J. R. 438(199), *484*

Petersson, G. 303(141), *329*
Petersson, G. A. 4, 100(2), *179*, 368, 372, 373(200), *390*
Pethrick, R. A. 1461(16), *1502*
Petránek, J. 874(180), *906*
Petreas, M. X. 1360(170), *1367*
Petrich, J. W. 1045(136–139), *1090*
Petrick, G. 1356(124), *1366*
Petrickova, H. 1415(317), *1448*
Petrie, G. 1157(10), *1337*
Petrier, C. 1361(194), *1367*
Petrillo, G. 1073(225), *1092*
Petringa, A. 1391(118b), 1394(172c), 1396(172c, 181b), *1440, 1442, 1443*
Petropoulos, J. C. 459(268), *487*
Petrosyan, V. S. 339(19), *386*
Petrova, M. V. 347(54), *387*
Petruska, J. 106(405), *188*
Pettitt, B. A. 1470(57), *1503*
Pettitt, B. M. 1203(97), *1340*
Pettus, T. R. R. 463(275), *487*
Pews, R. G. 406(49c), *479*
Peyrat-Maillard, M. N. 982(371), *1011*
Pez, G. O. 687(216), *709*
Pfaffli, P. 1347(5), *1363*
Pfaltz, A. 697(286), *711*
Pfartz, A. 688(226), *709*
Pfeifer, B. 1347, 1352(2), *1363*
Pfeiffer, D. 979(357), *1011*
Pfletschinger, J. 783(225), *836*
Phadke, R. 680(149), *708*
Pham, M. C. 1157(12), *1337*
Pham-Tran, N. N. 83–86, 88–91(299), *186*
Philip, D. 378(271), *392*
Phillip, B. 15, 129, 141, 142(76), *181*
Phillips, G. O. 1110(53), 1112(53, 77), 1124(244), 1131, 1134(316), *1145, 1149, 1150*
Phillips, L. A. 493, 498(42), *525*
Phillips, R. S. 748(90), *833*
Philp, D. 761(117), *833*
Phukan, P. 616(27), *656*
Piangerelli, V. 945(151, 154), 953(151, 154, 246), 962(246), *1006, 1008*, 1354(111), *1365*
Piatak, D. M. 813(297), *837*
Piatnitsky, E. L. 1428(425c), *1452*
Piatteli, M. 894(262), *908*
Piattelli, M. 1388(99, 101), 1395(99, 178b, 178c), 1400(209a–c, 210b), 1401(210b, 212a–d, 213), 1402(222), 1403(230), 1404, 1406(222), 1427(230, 414a), *1439, 1442, 1444, 1445, 1451*
Piccoli Boca, A. 1357(130), *1366*
Piccolo, O. 683(180), 686(210), *708, 709*
Pichat, P. 1086(364, 368, 369), *1096*
Pichon, C. 671(82), *706*
Picket, J. E. 1080(259b), *1093*

Pickles, G. M. 416(112), *482*
Picot, D. 11(51), *180*
Picot, M. J. S. 639(130), *659*, 774(176), *835*
Pielichowski, J. 662(8), *705*
Pierini, A. B. 1073(226), *1092*
Pieroni, O. I. 1381(65a), *1438*
Pierretti, A. 375(241), *392*
Pieszko, C. 929(68), 953, 962(251), 984(68), *1004, 1008*
Pietra, F. 822(322, 323), *838*
Pietra, S. 419(133), *482*
Pietrzak, L. 913(2), *1002*
Pietta, P.-G. 894, 901(258), *907*
Piette, L. H. 1120(170), *1147*
Pigge, F. C. 1423(380a), *1450*
Pignatelli, B. 1361(188, 189), *1367*
Pignatello, J. J. 1086(374), *1096*, 1362(204), *1368*
Pignatoro, F. 178(771), *197*
Pihlaja, K. 301(131), *328*, 352(89, 90), *388*
Pihlaya, K. 143(558–560), *192*
Pike, J. E. 402(22), *479*
Pikus, A. L. 778(213), *836*
Pilas, B. 1136, 1137(364), *1151*
Pilato, A. 1457, 1470, 1481, 1483–1485(3), *1502*
Pilato, L. A. 1470(60), 1471(73), *1503, 1504*
Pilcher, G. 227(10, 15), 228, 229(17), 234(43), 241(17, 76), 248(15), 252(94), *254–257*
Pileni, M. P. 1129, 1130(298), *1150*
Pilichowski, J.-F. 1054(164), 1084(319), *1091, 1095*
Pilidis, G. A. 986(398), *1012*
Pilkington, J. W. 807(283, 284), *837*
Pilley, B. A. 745, 746(80), *832*
Pilling, M. J. 1105(64, 65), *1106*
Pil'shchikov, V. A. 231(34), *255*
Pimentel, G. 35, 171(173), *183*
Pimentel, G. C. 47, 53(225), *184*, 503(85, 86), *526*, 530(3), 547(60), 551(89), 559(3), *596, 598*
Pinalli, A. 1409(288), *1447*
Pineiro, M. 1082(276), *1093*
Pines, D. 493(46, 47), 498(47), 503(84), 506(46), 511(47), *526*
Pines, E. 493(45–47, 51, 53, 56–63, 63), 496(67), 497(58–60, 67), 498(45, 47, 59, 67), 503(83, 84), 504, 505(91), 506(46, 51, 91), 509(51), 510(91), 511(47), 512–514(51), 516–519(91), 520(51), 521(51, 91), 522(83, 97), 523(97), *525–527*
Pingarrón, J. M. 953, 956(212), 982(362), *1007, 1011*
Pink, D. 992(425), *1012*
Pink, J. H. 786(232), *836*
Pinkerton, A. A. 233(40), *255*
Pinkhassik, F. 1403(241), *1445*

Pinna, F. 701(318), *712*
Pinnavaia, T. J. 1202(96), *1340*
Pino, G. 493, 500(31), *525*
Pinson, J. 1136(356, 357), *1151*
Pinto, A. V. 1026(89), *1089*
Pinto, I. V. O. S. 986(395, 397), *1012*
Pinto, M. 862, 863(142), *905*
Pinto, M. C. F. R. 1026(89), *1089*
Pinto, R. E. 851(70), 887(220), 901(286, 287), *903, 907, 908*
Piras, P. P. 764(138, 139), *834*
Pires, J. 609(17), *656*
Pirinccioglu, N. 1393(146), *1441*
Pirogov, N. O. 891(243), *907*
Pirola, C. 1085(339), *1095*
Pirondini, L. 1424(389b), *1450*
Pirozhenko, V. V. 1393(161), 1394(162), 1417(326, 328), 1418(341, 344), 1420(341), *1442, 1448, 1450*
Pirrung, M. C. 1044(133, 134), 1045(135), *1090*
Pisarenko, L. M. 1128, 1131(279), *1149*
Piskorz, P. 4, 100(2), *179*, 368, 372, 373(200), *390*
Piskula, M. 894(263), *908*
Pissolatto, T. M. 933, 937(103), *1005*
Pitarch, M. 1391(123), 1432(448c, 448d), *1441, 1453*
Pitchard, R. G. 1419(352), *1450*
Pitchumani, K. 459(270), *487*, 774(183), *835*
Pitt, B. M. 408(54b), *480*
Pittard, R. A. 1500(178), *1506*
Piuzzi, F. 129(474), 144(474, 683), *190, 195*, 578(195), *601*
Piva, C. N. 1019(26), *1087*
Pizzocaro, C. 1082(273, 275), *1093*
Place, S. 1429(428), *1452*
Plaffendorf, W. 472(308b), *488*
Plagens, A. 640(135), *659*
Planas, A. 813(298), *837*
Planchenault, D. 1278(222, 223), 1286(223), *1343*
Platonov, V. A. 227(11), *254*
Platt, J. R. 495(68a, 68b), 508, 512(68b), *526*
Platt, K.-L. 953, 969(282), *1009*, 1374(33b), *1436*
Platteborze, K. 562(166), 568(176), 594(290), 595, 596(290, 296), *600, 603*
Platteborze-Stienlet, K. 595, 596(297), *603*, 997(449), *1013*
Platts, J. A. 545(53), *598*
Platz, J. 138(528), *191*, 1104(61), *1106*
Pleixats, R. 607(5), *656*
Pleizats, R. 677(119), *707*
Pliss, E. 874, 875, 893(182), *906*
Pliva, J. 170(752), *196*
Plugge, J. 933(105), 935, 936(113), *1005*
Plumb, G. W. 895(267), *908*

Pniewska, B. 93(362), *187*
Poblet, J. M. 110, 121(428), *189*
Pochini, A. 683(169), *708*, 1387(90), 1391(110, 118a, 121a), 1392(90, 136, 137), 1393(110, 141, 151, 156, 158), 1394(90), 1395(90, 180), 1396(118a, 121a, 182a, 182b, 186a, 188), 1397, 1399(197b), 1403(197b, 232a, 235), 1404(244), 1405(232a, 235), 1409(287, 288), 1427(409a, 410a, 415), 1430(439), 1432(409a), 1435(467), *1439–1443, 1445, 1447, 1451–1454*
Pocurull, E. 944(147), 945(147, 152, 156), 946(164), 953(152, 164, 209, 254, 256), 955(209), 962(254, 256), *1006–1009*, 1347(1), 1353(1, 70, 92, 93), 1354(70, 104), *1363–1365*
Podesta, J. C. 416(112), *482*
Podosenin, A. 237(61), *256*
Podzimek, S. 1459(14), *1502*
Pogodin, S. S. 763(133), *834*
Pogrebnoi, S. I. 763(131), *834*
Poh, B. L. 1285(235), *1343*
Pokhodenko, V. D. 1118(141), *1146*
Pol, A. J. H. P. van der 941(133), *1006*
Polak, M. L. 95, 98, 99, 131, 132, 139(370), *188*
Polakova, J. 170(752), *196*
Polasek, M. 691(239), *710*, 986(400), *1012*
Poletaeva, I. L. 953, 957(214), *1007*
Politi, M. J. 493(54, 55), *526*
Polizer, P. 82(283), *186*
Pollaud, G. M. 613(24), *656*
Polo-Díez, L. M. 953, 962(253, 257), *1008, 1009*
Polozova, N. I. 1118(142), *1146*
Poluzzi, E. 644(143), *659*
Polyakov, A. V. 432(186), *484*
Pomelli, C. 4, 100(2), *179*, 368, 372, 373(200), *390*
Pomio, A. B. 366(167), *390*
Pommeret, S. 493(64), *526*
Pongor, G. 82(286), *186*, 384(293), *393*, 551, 553(74), *598*
Ponomarev, D. A. 236, 237(57), *255*
Ponomareva, D. A. 301(131), *328*
Ponomareva, R. P. 1076(235), *1092*
Pons, M. 1385(79a), 1406(266), *1439, 1446*
Popa, E. 855(118), *904*
Popall, M. 457(264), *487*
Pope, D. J. 1428(432), *1452*
Popelier, P. L. A. 31, 35(142), 71(273), *183, 185*
Popescu, I. C. 977(338), *1010*
Pople, J. A. 4(2), 20, 25(110), 53(237), 100(2), *179, 182, 185*, 275(45), *326*, 368, 372, 373(200), *390*
Popp, G. 1169(42), *1338*

Popp, P. 1353(89), *1365*
Porschmann, J. 933(105), 935, 936(113), *1005*
Porta, C. 680(155), 683(170), *708*, 740(62), *832*, 1374(37), 1378(37, 49a–c), *1436, 1437*
Portalone, G. 20, 22(109), *182*
Portella, C. 428(171), *483*
Porter, G. 129, 131(499, 500), *191*, 1127(266, 268, 271, 272), 1128, 1135(272), *1149*
Porter, N. A. 852(80, 81), 884(208), 885(81, 209–211), *903, 906*
Porter, Q. N. 314(205–207), *330*
Porter, R. 1216(119b), *1340*
Porthault, M. 953, 956(210), *1007*
Porzi, G. 1255(180), *1342*
Pöschl, U. 319(232), *331*
Posner, G. H. 677(137), *707*
Pospíšil, J. 748(89), *833*, 874(179–181), *906*
Pospisil, P. J. 698(296), *711*
Pospisilova, M. 986(400), *1012*
Postel, M. 413(81), *480*
Potier, P. 1292(252), *1344*
Potrzebowski, M. J. 593, 595(282), *603*
Potter, A. C. 1471(74, 75, 77, 78), 1473(75), 1474(77), 1475(78), *1504*
Potterat, O. 15(78), *181*
Potts, K. T. 448(233), *486*
Pouchan, C. 561(164), *600*
Pouchert, C. J. 368(189), *390*, 924, 925, 927(36, 37), *1003*
Pouilloux, Y. 631(77), *657*
Poullain, D. 1212(111), *1340*
Poulsen, O. K. 338(17), 350(77), *386, 388*
Poupat, C. 1292(252), *1344*
Poupko, R. 1124(224), *1148*
Pouyséu, L. 807(274), *837*, 1230(149), *1341*
Powell, D. H. 325(278), *332*
Powell, D. L. 555(140), *599*
Powell, D. R. 1382(72b), *1438*
Powell, H. R. 1409(286b), *1447*
Powell, T. 1086(366), *1096*
Powlowski, J. B. 411(69), *480*
Poznanski, J. 1117(130, 132), *1146*
Pozzo, J.-L. 770(157), *834*
Pradham, P. 1062(187), *1091*
Prados, P. 1374(35), 1385(79a), 1387(90), 1392(90, 125a), 1394(90), 1395(90, 177), 1403(125a), 1406(266), 1409(287), 1427(424c), *1436, 1439, 1441, 1442, 1446, 1447, 1452*
Prakash, G. K. S. 613, 645(21), *656*, 671(85), *706*
Prakash, O. 1224(140a), *1341*
Praly, J.-P. 609(17), *656*
Pramanic, A. 735(55), *832*
Prasad, A. V. R. 631(84), *658*
Prasad, L. 844, 847, 855, 860–862, 864, 897, 898(18), *902*, 1120, 1140(184), *1147*

Pratt, D. A. 45(192, 198, 200), 110, 113(440), 139(192), 140(192, 198), *184, 189*, 900(280), *908*
Pratt, D. W. 1111(74), *1145*
Prayer, C. 493(64, 65), 511(65), *526*
Prazeres, A. O. 633(91), *658*
Prebenda, M. F. 1132(332), *1150*
Predieri, G. 683(170), 686(208), *708, 709*
Prein, M. 1080(259a), *1093*
Preis, S. 1085(331), *1095*
Prelog, V. 433(189a–c), *484*
Prementine, G. S. 1495(157), *1505*
Prencipe, T. 1257(186), *1342*
Prenzler, P. 942(136), 999(466), *1006, 1013*
Prescott, S. 358(112), *388*
Preston, P. N. 1076(236), *1092*
Preuss, H. 69(263), *185*
Price, K. R. 895(267), *908*
Price, P. S. 1360(167), *1367*
Price, S. L. 535(31), *597*
Price, S. P. 953, 960(229), *1008*
Prieto, M. J. E. 887(224), *907*
Prilipko, L. L. 887(225), *907*
Principato, G. 1394(172c), 1396(172c, 195d), *1442, 1443*
Prins, L. J. 1406(262b), *1446*
Prins, R. 774(175), *835*
Prior, D. V. 503(89), *526*, 533(22), 535(33), 536(33, 37), 537(33), 539, 540(37), 557(33, 145), 560, 564(145), 588(33, 37, 257), 590(145), 591(37), *597, 599, 602*, 876–878(186), 879(196), 882(186, 196), *906*
Prior, R. L. 871(162), *905*
Prisbylla, M. P. 441(209), *484*
Pritchard, P. H. 1351(42), *1364*
Pritzkow, W. 405(48a, 48b), *479*
Priyadarsini, K. I. 868(158), *905*, 1129(287, 302), 1142(302), *1149, 1150*
Procida, G. 930(77), *1004*
Procopio, A. 953, 961(239), *1008*
Procopio, D. R. 877, 879, 881, 883(192), *906*
Prodi, L. 761(118), *833*
Prokeš, I. 252(95), *257*
Prokof'ev, A. I. 747(83), *832*, 1020(32), *1087*, 1119(158), 1123(210), 1126(256–261), 1130(319), *1147–1150*
Prokof'eva, T. I. 747(82, 83), *832*
Prosen, E. J. 226(6), *254*
Pross, A. 67(253), 99, 100, 103(377), 109(413), *185, 188, 189*
Protasiewicz, J. D. 1198(90), *1340*
Proudfoot, J. M. 913(30), *1003*
Proudfoot, K. 141(540), *192*
Prout, K. 549(65), *598*
Pruetz, W. A. 1100(21), *1105*, 1110(56), 1117(122–128), 1118(124), 1129(56, 122), 1130(56, 122–125), *1145, 1146*

Pruskil, I. 457(262), *487*
Pryce, R. J. 1217(128), *1341*
Pryor, W. A. 141(537), *191*, 637(107, 108), *658*, 851, 852(67), 886, 887(216), 894(67), 901(285), *903, 906, 908*
Przeslawska, M. 144(655), *194*
Ptasiewicz-Bak, H. 144(664), *194*
Pu, L. 691(245), 694(264–266), *710*
Puddephatt, R. J. 1417(327), 1424(327, 391), *1448, 1450*
Puddey, I. B. 913(30), *1003*
Pudovik, M. A. 1424(396), *1451*
Puech-Costes, E. 1080(253), *1093*
Puglia, G. 626(59), *657*, 683(168), 685(200), *708, 709*
Puglisi, C. 953, 969, 1000(286), *1009*
Pugzhlys, A. 1020(33), *1087*
Puig, D. 319(229), *330*, 928(56), 946(158, 159), 947(171), 953(158, 159, 171, 201, 206, 262), 955(201, 206), 964(262), 977(335), *1004, 1006, 1007, 1009, 1010*, 1352(69), *1364*
Pujol, B. 453(244), *486*
Pujol, D. 1201, 1205(92), *1340*
Pulay, P. 4(6), 129, 131–133(510), *179, 191*
Pulgar, R. 1352(58), *1364*
Pulgarin, C. 1084(327), *1095*
Pulley, S. R. 451(252), *486*
Pullin, D. 129, 131(504), *191*
Pullman, A. 21, 30, 38(131), *182*
Pullman, B. 21, 30, 38(131), *182*
Punzalan, E. R. 770(152), *834*
Purcher, S. 1463(37), *1503*
Purkayastha, M. L. 668(57), *706*
Purkayastha, P. 1025(87), *1089*
Purrington, S. T. 647(151), *659*, 687(215, 218), *709*
Pyreseva, K. G. 638(122), *658*

Qian, C. 695(271), *710*
Qian, J. 129(477), *190*
Qian, X. 365(158), *389*
Qian, Z.-H. 776(206), *835*
Qiao, G. G. 1476(84), 1478(85, 87), 1482(84), 1484(94), 1485(84), 1486(87), 1489, 1490(113, 114), *1504*
Qin, L. 1115(109), 1129, 1132, 1134(288), 1135(109), *1146, 1149*
Qin, Y. 31, 111–113(138), 132, 133, 136(511, 513), *183, 191*
Quadri, S. M. 170(750), *196*
Quail, J. W. 1237(164), *1342*
Quan, M. L. C. 1419(357), *1450*
Quaranta, E. 686(212), *709*
Quardaoui, A. 1057(172), *1091*
Quattropani, A. 783(222), *836*
Quayle, P. 450(246c), *486*, 783(224), *836*

Quetglas, M. B. 1084(324), *1095*
Quick, J. 1274(209), *1343*
Quideau, S. 807(274), *837*, 1230(149), *1341*
Quint, R. 1082(282), 1083(295), *1094*
Quint, R. M. 1082(282), 1083(295), *1094*, 1100(28), *1106*
Qureshi, T. H. 429, 430(179c), *483*

Raanen, T. 1359, 1360(160), *1367*
Rabalais, J. W. 102(393), 110, 112(445), *188, 189*
Rabani, J. 1083(298), *1094*
Rabinovich, I. B. 235(51), *255*
Rabinovitz, M. 1124(236), *1148*
Rablen, P. R. 87(308), *186*
Rabold, A. 144(653, 671), 145(628), 147(646), *193, 194*, 377(257), 386(304), *392, 393*, 551, 594, 595(126), *599*
Rabuck, A. D. 4, 100(2), *179*, 368, 372, 373(200), *390*
Raczyńska, E. D. 324(270), *331*, 586, 587(239–242, 246, 255), *602*
Radeck, W. 930(75), *1004*, 1353(90), *1365*
Radeke, K. H. 1080(260), *1093*
Radi, R. 637(110), *658*
Radinov, R. 470(299), *488*
Radner, F. 639(126), *659*
Radojevic, S. 782(220), *836*
Radom, L. 53(237), 67(253), 99, 100, 103(377), 109(413), 114(460), *185, 188–190*, 368(202), *391*
Radu, C. 1028(96), *1089*
Rădulescu, S. 855(118), *904*
Radziejewski, P. 144(671), *194*, 595, 596(295), *603*
Radziszewski, J. G. 129, 132, 133(498), *191*, 368(197), 372(237), 373(197, 237), *390, 391*
Rae, C. D. 338(16), *386*
Rae, W. E. 971, 972(300), *1009*
Rafecas, M. 961(243), *1008*
Raffaelli, A. 953, 961(239), *1008*
Rafla, F. K. 436(197), *484*
Ragaini, V. 1085(339), *1095*
Rager, M.-N. 1394(175a, 175b), *1442*
Ragg, E. 338, 345(10), *386*, 721(25, 26), *831*
Ragone, P. 953, 962(246), *1008*
Ragone, P. P. 945, 953(155), *1006*
Rahbek-Nielsen, H. 325(277), *332*
Rahm, A. 454, 456(255), *487*
Rahman, A. R. H. A. 776(205), *835*
Rahman, W. 778(212), *836*
Rai, M. M. 35(159), *183*

Raimondi, M. 89(312), *186*
Rainbird, M. W. 144(681), *195*
Raines, D. 813(295), *837*
Rainina, E. I. 979(353), *1011*
Rainio, J. 997(445, 446), *1013*
Raizou, B. M. 475(318), *488*
Raj, A. 414(91), *481*
Raj, N. K. K. 639(131), *659*
Raj, S. S. S. 377(263), *392*
Rajabi, L. 1495(139), *1505*
Rajagopalan, K. V. 237(62), *256*
Rajakovic, L. 1001(475), *1014*
Rajakovic, L. V. 1001(474, 476), 1002(478), *1014*
Rajan, R. 639(131), *659*
Rajarison, F. 630(69), *657*
Rajoharison, H. G. 432(185), *484*
Rakshys, J. W. 877(187), *906*
Rakshys, W. J. 531(21), *597*
Rall, G. J. H. 1314(288, 289), *1345*
Ramachandra, R. 1274(209), *1343*
Rama Devi, A. 294(78), *327*
Ramamurthy, V. 477(324a), *488*
Ramana, D. V. 301(132–134), *328*, 418(120), *482*
Ramanjulu, J. M. 670(71), *706*, 1318(303), *1345*
Ramano, R. R. 1356(128), *1366*
Rama Rao, A. V. 680(152), *708*
Ramaswamy, A. V. 645(145), *659*
Ramesh, N. G. 447(222j), 448(225), *485*, 680(147), *708*
Ramezanian, M. S. 637(111, 112), *658*
Ramis-Ramos, G. 953, 961(237), 988(409), *1008, 1012*
Ramnäs, O. 303(141), *329*
Ramondo, F. 375(241), 383(279), *392*
Ramos, L. 1360(168), *1367*
Ramos, M. C. 981(360), *1011*
Rampey, M. E. 730(47), *832*
Rampi, R. C. 1278(226), *1343*
Ramsden, C. A. 1223(136), *1341*
Ramshaw, E. H. 949, 953(185), *1007*
Ramsteiner, K. 1495(149), *1505*
Rand, B. 1489(105, 107), *1504*
Randall, D. W. 15(71), *181*, 1108(15), *1144*
Rangelova, D. S. 887(222), *907*
Rannou, J. 559, 568, 586(154), *600*
Rao, A. V. R. 630(70), 631(73), *657*
Rao, B. C. S. 414(97c), *481*
Rao, C. N. R. 216(28), *221*, 547, 548(59), *598*, 856(132), *904*
Rao, D. 316(218, 219), *330*
Rao, M. N. A. 868(158), *905*
Rao, P. 1415(311), *1448*
Rao, P. D. 1234(153a), 1236(154–157), 1237(158, 161a, 162a–c), *1341, 1342*
Rao, P. S. 1128, 1131, 1134(278), 1135(351), *1149, 1151*
Rao, R. 651(171), *660*
Rao, V. B. 441(206), *484*
Rao, V. S. S. 774(184), *835*
Rao, X. 1086(380), *1096*
Rao, Y. V. S. 630(70), 631(73), *657*
Raos, G. 89(312), *186*
Rapoport, H. 400(17), 424(154b), 444(218), *478, 483, 485*
Rappe, C. 315(216), *330*, 1352(60), *1364*
Rappoport, Z. 341(31), *387*
Rappo-Rabusio, M. 559, 567(156), *600*
Rasco, A. G. 1260(191), *1342*
Raspoet, G. 89, 90(327–329), 127(470–472), 144(472), *187, 190*
Rassu, G. 680(148), 683(177), *708*
Rastelli, M. 677(131), *707*
Rastetter, W. H. 424(153b), *483*, 827(341), *838*
Rastogi, P. P. 144(674), *195*
Raston, C. L. 1372(22, 23, 24b, 25), 1381(66), 1383, 1421(67o), 1427(418), *1436, 1438, 1452*
Ratajczak, E. 138(527), *191*
Ratajczak, H. 20, 25(129), 143(553, 567, 571, 572), 144(602–604), 177(553), *182, 192, 193*, 335(1), 363(1, 147, 148), 364(147, 148, 150, 152, 153), 371(229), *386, 389, 391*, 593(271), 594, 595(271, 285, 288), *602, 603*
Ratz, A. M. 1318(301a, 301b), *1345*
Rau, W. 1257, 1260(187a), *1342*
Raulins, N. R. 677(121), *707*, 761(120), *833*
Raum, A. L. J. 1462(20, 21), 1470(63), *1502, 1503*
Rauzy, S. 925(46), *1004*
Rav-Acha, Ch. 1354(99), 1355(99, 117), *1365*
Ravenscroft, P. D. 441(206), *484*
Ravichandran, R. 1071(205), *1092*
Ray, A. 144(593), *193*
Ray, A. K. 1419(347b), *1450*
Ray, B. 619(40), *656*
Ray, D. 493(24), *525*
Ray, J. K. 426(165), *483*
Ray, K. B. 1393(146), *1441*
Raya, J. 1427, 1432(421b), *1452*
Raychaudhuri, S. 748(94), *833*
Rayez, M. T. 138(527), *191*
Razuvaev, G. A. 1109(50), *1144*
Re, S. 20, 22, 23, 31, 129, 144, 148, 149(113), *182*
Read, C. M. 1330, 1332(325c), *1346*
Read, R. W. 807(279, 280), *837*
Ready, J. M. 667(51), 668(52), *705, 706*
Real, F. J. 1080(254), *1093*
Reamer, R. A. 632(89), *658*

Rebek, J. Jr. 322(251), *331*, 1391(122d), 1406(263a, 263c), 1421(372), 1422(374), 1424(387, 398, 401), 1427(407c), 1430(441a–c), 1432(122d), *1441, 1446, 1450, 1451, 1453*
Rebek, J. Jr. 1424(400a–c), *1451*
Reber, R. K. 1118(133), *1146*
Rechow, D. A. 1360(181), *1367*
Reddy, E. P. 1080(252), *1093*
Reddy, G. S. 110(431), *189*, 299(98), *328*
Reddy, J. S. 1213(116), *1340*
Reddy, K. R. 420(142a), *483*
Reddy, N. K. 422(148), *483*
Reddy, P. A. 1391(116), 1409(285, 290), *1440, 1447*
Reddy, S. 983(379), *1011*
Reddy, S. M. 1001(477), *1014*
Redmond, R. W. 1047(146), *1090*
Redpath, J. L. 1129(299), 1130(299, 322), *1150*
Ree, M. 1495(135, 136), *1505*
Reed, A. E. 4(4a, 5), *179*
Reed, J. N. 415(106a), *481*
Reed, M. W. 450, 452(239b), *486*
Reedijk, J. 1194(84), *1339*
Rees, L. 633(93), *658*
Reese, C. B. 476(322), *488*
Reetz, M. T. 695(272), *710*
Reeves, L. W. 53(233, 234), *185*
Reggiani, M. 1403, 1405(232a), *1445*
Regier, K. 1169(44), *1338*
Régimbal, J. M. 275, 277(42), *326*
Regitz, M. 433(187b), *484*
Regnouf-de-Vains, J.-B. 1372(28), 1394(169, 173), 1401(216b), 1405(251b), 1417(330c), *1436, 1442, 1444, 1446, 1448*
Rehman, A.-U. 451(250), 456(261), *486, 487*
Reich, L. 1108(4), *1144*
Reichard, P. 1120(190), *1147*
Reichardt, C. 540(42), 541(43), 542(48), *597*
Reichenbach, C. F. v. 6(30), *180*
Reider, P. J. 670(78), *706*
Reijenga, J. C. 953, 968(277), *1009*
Reimann, B. 147(714), *195*
Reimers, J. R. 171(762), 174(765, 766), 177(766), *197*
Reimschüssel, W. 170, 171(759), 177(759, 769), *197*
Reinaud, O. 1394(175a, 175b), *1442*
Reinholz, E. 662(5), *705*
Reinhoudt, D. N. 377(251), *392*, 635(101), *658*, 782(218), *836*, 1370(4), 1383(67d), 1385(79b, 84), 1386(84), 1387(90), 1388(101), 1391(110, 112, 118a, 119, 121a), 1392(90, 125b, 137), 1393(110, 151, 156), 1394(90, 176), 1395(90, 177), 1396(118a, 121a, 182a, 182b, 189), 1401(220a), 1402(224b), 1403(125b, 234e, 238, 240b, 242), 1405(224b, 248, 249a, 249b, 253, 254a, 254b), 1406(254a, 258a, 259, 262b, 263b, 263d), 1408(282), 1409(287), 1417(4), 1421(67d, 369), 1422(378), 1423(380b), 1424(382a, 382b, 383, 393), 1427(407a, 410b), 1428(263d, 437a, 437b), 1430(440), 1432(449b, 454, 455), 1433(456a, 456b), *1435, 1438–1447, 1450, 1451, 1453*
Reinolds, T. 925(49), *1004*
Reis, A. M. M. V. 234(46), *255*
Reischl, G. 1169(44), *1338*
Reiser, G. 143(591), *193*
Reisinger, A. 1478(85), *1504*
Reitberger, T. 1108(25), 1136(361), *1144, 1151*
Reitz, D. C. 1120(169), *1147*
Reitz, N. C. 1169(42), *1338*
Rela, P. R. 1100(34), *1106*
Rella, R. 979(354), *1011*
Remaud, B. 316(220), *330*
Remorova, A. A. 894(261), *908*
Ren, S. 871, 898(164), *905*
Ren, T. 1424(402), 1429(428), *1451, 1452*
Renberg, L. 1353(83), *1365*
Rendon, E. T. 1084(324), *1095*
Renner, A. 1495(149), *1505*
Renoux, A. 405(39a), *479*
Renoux, B. 444(216c–e), *485*
Renshaw, R. R. 1494(124), *1505*
Rensio, A. R. 1430(441b), *1453*
Rentzepis, D. M. 1020(35b), *1087*
Rentzepis, P. M. 1025(85a), *1089*
Ren-Wang-Jiang 212(22), *221*
Repinskaya, I. B. 744(75, 76), *832*
Replogle, E. S. 4, 100(2), *179*, 368, 372, 373(200), *390*
Reppy, M. A. 994(437), *1013*
Resat, M. S. 1084(329), *1095*
Rettig, W. 502(82), *526*, 1025(67, 88), *1088, 1089*
Reuben, J. 350(82), *388*
Reunanen, M. 1360(178, 185), *1367*
Reverdy, G. 977(343), *1011*
Reviejo, A. J. 982(362), *1011*
Revill, E. 953, 960(231), *1008*
Revilla, I. 961(235), *1008*
Rey, M. 403(29), *479*
Reyes, P. 1087(383), *1096*
Reymond, F. 544(52), *598*
Reymond, J.-L. 806(263), *837*
Reynolds, W. F. 358(112), *388*
Reyntjens-Van Damme, D. 145, 146(617), *193*
Reza, J. 944(140), *1006*
Reznik, V. S. 1424(396), *1451*
Rezzano, I. 953, 956(211), 976(326), *1007, 1010*
Rha, S. G. 1394(167b), *1442*

Rhee, J.-S. 946(163), *1006*
Rhee, K. S. 454(258), *487*
Rheingold, A. L. 693(253), *710*
Rho, H.-S. 664(32), *705*
Rhoads, S. J. 677(121), *707*, 761(120), *833*
Rhodes, M. J. C. 895(267), *908*
Ribe, S. 461(289), *487*
Ribeiro, F. R. 631(77), *657*, 686(205), *709*
Ribeiro da Silva, M. A. V. 227(15), 228(17), 229(17, 32), 230(21), 234(46), 237(60), 238, 239(67), 241(17, 60, 76), 248(15), 252(94), *254–257*
Ribeiro da Silva, M. D. M. C. 227(15), 228, 229(17), 237(60), 241(17, 60, 76), 248(15), 252(94), *254, 256, 257*
Ricci, A. 418(119), *482*
Ricci, G. 1462(28), *1503*
Ricciardi, P. 1422(375), *1450*
Rice, K. C. 1244(170), *1342*
Rice-Evans, C. 854(95, 96), 857, 858(96), 894(257), *904, 907*
Rice-Evans, C. A. 871, 894(161), *905*, 957(220), 968(276), *1008, 1009*
Rich, D. H. 673(93), *706*
Richard, C. 890(237), *907*, 1054(163), 1057(171, 174), 1080(251), 1083(300), 1086(357), *1091, 1093, 1094, 1096*
Richard, H. 982(369), *1011*
Richard, T. J. 424(153b), *483*
Richards, J. A. 1167(38, 39), *1338*
Richards, J. H. 53, 56(238), *185*
Richardson, J. H. 95, 109(369), *188*
Richardson, S. D. 1354, 1355(98–100), *1365*
Richardson, T. I. 1318(302), *1345*
Richardson, W. H. 101(385), *188*
Richard-Viard, M. 144(600), *193*
Richerzhagen, T. 1124(237), *1148*
Riches, A. G. 444(216h), *485*
Richter, C. 1073(216), 1085(336, 353), 1086(336), *1092, 1095, 1096*
Richter, D. T. 737(56), *832*
Richter, H. W. 1115, 1129, 1130, 1134(102), *1146*
Richter, W. J. 291(67), *327*
Rickards, R. W. 13(63), *181*
Ricker, M. K. A. 550(71), *598*
Ridd, J. H. 638(116), *658*
Riddick, J. A. 170(739), *196*
Ridge, D. P. 294(80, 81), *327*
Ridyard, J. N. A. 67(252), 110, 112(443), *185, 189*
Riebel, P. 670(72), *706*
Riedel, K. 978(351), 979(352), *1011*
Riedl, B. 1002(480), *1014*, 1471(71), *1504*
Riehn, C. 145, 156(636), 179(781), *194, 198*
Rieker, A. 814(306), *838*, 1018(23), *1087*, 1123(207–209), 1124(208, 209, 242), *1148*, 1169(41, 43, 44), 1172(51a, 51b),
1205(101), 1290(244, 246), 1292(246), *1338, 1340, 1344*
Riekkola, M. L. 970(289), 973(312), *1009, 1010*
Rienstra-Kiracofe, J. C. 93, 97, 132, 133(359), *187*
Riess, M. 1351(44), *1364*
Rigaud, J. 949(183), *1007*
Rigby, J. H. 1077(238), *1092*
Rigo, A. 890(234), *907*
Riley, J. S. 288(58), *327*
Riley, P. A. 1223(136), *1341*
Rinaldi, D. 57(242), *185*
Rindone, B. 1216(127), *1341*
Ríos, A. 638(117), *658*, 986(392), *1012*
Ríos, M. 171(762), 174(763), *197*
Rios, M. A. 551, 553(105), *599*
Rios-Fernandez, M. A. 1025(74), *1088*
Rios-Rodriguez, M. C. 1025(72, 74–76), *1088, 1089*
Rishton, G. M. 472(306), *488*
Rissanen, K. 1392(125a), 1403(125a, 229), 1418(343b), 1419(353, 358), 1420(363, 364, 365b), 1421(372), 1424(400b), 1427(409c, 424c), 1432(229), *1441, 1445, 1450–1452*
Ritchey, J. G. 990(420), *1012*
Riva, E. 1082(287), *1094*
Rivals, P. 230(28), *254*
Rivas, A. 1352(58), *1364*
Rivera, J. 1355(119), 1360(168), *1365, 1367*
Rives, V. 1084(325, 326), *1095*
Robards, K. 942(136), 953, 961(240), 999(466), *1006, 1008, 1013*
Robb, M. 4, 100(2), *179*
Robb, M. A. 368, 372, 373(200), *390*
Robbat, A. 937, 938(126), *1006*
Robbins, T. A. 1424(386), *1450*
Roberts, B. A. 1381(66), 1383, 1421(67o), *1438*
Roberts, E. L. 1025(73), *1088*
Roberts, J. D. 293(72), 314(203), *327, 330*, 506(93), *527*
Roberts, J. R. 252(95), *257*
Roberts, S. M. 1216(119e), *1340*, 1466(46), *1503*
Robey, R. L. 416(110a), *482*
Robillard, B. 864, 865(144), *905*
Robin, D. 300(110), *328*
Robin, J.-P. 1278(222, 223), 1286(223), 1300(264), *1343, 1344*
Robinson, E. A. 31(143), 47, 48, 53(222), 68(143), *183, 184*
Robinson, G. W. 493(40–43), 498(42), *525*
Robinson, R. 402(22), *479*
Robson, R. 1380(60a–c), *1438*
Roche, A. J. 448(229c), *485*
Rochefort, M. P. 1308(280), *1344*

Rochester, C. H. 230(29), *254*, 530(7), *597*
Rock, M. H. 448(229c), *485*
Rockcliffe, D. A. 1194(85), *1339*
Rockwell, A. L. 1358(139), *1366*
Roczniak, K. 1470(62), *1503*
Roder, M. 1115, 1128(91), 1142(378), *1145, 1151*
Rodés, L. 670(76), *706*
Rodger, A. 368(191), *390*
Rodgers, J. D. 1361(190), *1367*
Rodrigez, I. 1348(17), 1354(17, 110), *1363, 1365*
Rodrigo, R. 476(320), *488*, 785(228), *836*
Rodrigues, J. A. R. 632(90), *658*
Rodrigues, J. A. R. G. O. 234(46), *255*
Rodriguez, A. 1431(443a), *1453*
Rodriguez, B. 1292(252), *1344*
Rodríguez, H. R. 414(94b), *481*
Rodríguez, I. 971(294), *1009*, 1352, 1353(63), *1364*
Rodríguez, J. 551, 553(100, 105), *599*
Rodriguez, J. G. 211(17), *220*
Rodriguez, M. A. 1025(84), *1089*
Rodríguez, M. C. 953, 960(232), *1008*
Rodriguez, M. S. 241(79), *256*
Rodriguez, N. M. 1381(65a), *1438*
Rodriguez-Prieto, F. 1025(70a, 72, 74–76), *1088, 1089*
Rodriguez-Reinoso, F. 1489, 1490(114), *1504*
Rodriguez-Sanchez, L. 638(117), *658*
Rodwell, P. W. 1059(180), *1091*
Roebber, J. L. 129, 131(501), *191*
Roeschenthaler, G.-V. 743(70), *832*
Roessner, F. 774(178), *835*
Rogers, D. W. 237(61, 61), *256*
Rogers, J. S. 1372(29), 1401(220c), *1436, 1444*
Rogers, K. R. 953, 955(204), *1007*
Rogers, L. 515(95), *527*
Rogers, L. B. 498(71), *526*
Rogers, M. L. 370(213), *391*
Roggero, M. A. 947, 955(170), *1006*
Rogić, M. M. 1192(80), *1339*
Roginsky, V. 874, 875, 893(182), *906*, 1119(167), *1147*
Roginsky, V. A. 871, 872, 874(168), 894(261), *905, 908*, 1128, 1131(279), *1149*
Rohloff, J. C. 1163(33b), *1338*
Röhm, D. 383(278), *392*
Rojano, B. 853(91), *904*
Rojstaczer, S. 1495(135), *1505*
Rokbani, R. 1391(114), *1440*
Rokicki, G. 1403(229), 1427(409c, 412a, 412b), 1431(443b), 1432(329), *1445, 1451, 1453*
Roland, B. 93, 103(363), *187*
Roland, J. T. 653(187), *660*
Rolando, C. 1136(356, 357), *1151*

Román, E. 1422(377), 1424(402), 1429(428), *1450–1452*
Romanowski, H. 143(569), *192*
Romero, A. A. 608(13), 618(36), *656*, 774(182), *835*
Romero, M. de L. 101–103(390), *188*
Romesberg, F. E. 511(94), *527*
Rondelez, Y. 1394(175b), *1442*
Ronlán, A. 1167, 1168(37), *1338*
Rony, C. 974(317), *1010*
Rooney, J. R. 633(93), *658*
Rooney, L. W. 913(14), *1003*
Roos, B. O. 20, 106(119), *182*
Rop, J. S. de 365(160), *389*
Röpenack, E. von 913(3), *1003*
Rosales, J. 419(135b), *482*
Rosas-Romero, A. J. 853(91), *904*
Rosatto, S. S. 974(319), *1010*
Rosay, M. 338, 380(15), *386*
Rose, A. 979(357), *1011*
Rose, B. F. 1274(210a), *1343*
Rose, P. A. 1237(164), *1342*
Rose, P. D. 411(70b), *480*
Rosenberg, C. 1347(5), *1363*
Rosenberg, E. 319(232), *331*
Rosenberg, L. S. 498(72b), *526*
Rosenfeld, M. N. 1330, 1332(325a–c), *1346*
Rosi, J. A. 1000(471), *1013*
Roslev, P. 1360(174), *1367*
Rospenk, M. 37(181), 38(183), 143(556, 562, 563, 568), 144(568, 609, 650, 652, 653, 670), *183, 192–194*, 361(143, 144), 371(219–221), 372(220, 221), 377(219, 220, 257–259), 386(304), *389, 391–393*, 583(232), 593(276, 277), 594(276, 287), 595(276, 277), *601–603*, 1024(64), *1088*
Ross, A. B. 1098(2), *1105*, 1112(79, 80), *1145*
Ross, H. 1399(202), 1427(408a), *1444, 1451*
Ross, J. B. A. 106, 148(404), *188*
Rossetti, I. 18(87), *181*
Rossi, A. 1017(17), *1087*
Rossi, I. 57(242), *185*
Rossi, M. 375(250), *392*
Rossi, R. A. 1073(226), *1092*
Rossi, R. H. de 365(162), *389*
Rossignol, S. 551, 553(87), *598*
Rossmy, G. 35(150), 47, 53(218), *183, 184*
Rostkowa, H. 370, 382(211), *391*
Rostkowski, A. 953, 965(264), *1009*
Roth, H. D. 1042(128), *1090*
Roth, W. 20, 23(114), 35, 36(169), 78(280), 106(114), 107(114, 169), 145(622, 623), 147, 148, 156, 160(702), *182, 183, 186, 193, 195*, 581(221, 222), *601*
Rothe, S. 979(352), *1011*
Rothenberg, G. 617(33), *656*
Rothenberg, S. 53(239), 92(348), *185, 187*
Rotkiewicz, P. 929, 977(64), *1004*

Roundhill, D. M. 1393, 1407(143), 1435(465a), *1441, 1454*
Rouquette, H. 1389(105), *1440*
Rourick, R. 305(151), *329*
Rourke, J. P. 1417, 1424(327), *1448*
Roush, W. R. 674(109), *707*
Rousseau, C. 890(237), *907*
Rousseeuw, B. 1126(262), *1149*
Roussel, C. 432(185), *484*, 575(187), *600*
Rousslang, K. W. 106, 148(404), *188*
Roux, M. V. 228, 229(16), 230(25), 232(38), 237(61), 241(80), *254–256*
Rovel, B. 944(139), *1006*
Rowbotham, J. B. 349(67–70), *387*
Rowlands, G. J. 691(246), *710*
Roy, A. 420(142a), *483*
Roy, C. C. 854(102), *904*
Roy, R. 673(97), *707*, 1388(102), *1440*
Roy, R. K. 84, 89, 90(301), *186*
Royer, R. 338, 380(15), *386*
Rozas, I. 129(479), *190*
Rozenberg, M. M. 227(11), *254*
Rozenberg, V. 640(132), *659*
Rozhnov, A. M. 227(13), 231(34), *254, 255*
Rozwadowski, Z. 342(33), 345(48), 348(56), 353(56, 98, 101), 355(105), 359(122, 124, 126), 360(134), 361(122), 386(307), *387–389, 393*, 551(128), *599*, 1024(63, 64), *1088*
Rtishchev, N. I. 1030(105), *1089*
Ruana, J. 953, 963(261), *1009*
Ruano, J. L. G. 650(162), *659*
Rubbo, H. 637(110), *658*
Ruberto, G. 851, 852, 871, 872(68), 894(262), *903, 908*, 992(429), *1013*
Rubin, B. S. 1360(172), *1367*
Rubin, E. 888(226), *907*
Rubinstein, M. 1124(224), *1148*
Rubio, F. J. 1080(254), 1085(345), *1093, 1095*, 1362(202), *1368*
Rubio, M. 551(93), *598*
Rubio, P. 211(17), *220*
Rücker, G. 409(60), *480*
Ruddock, G. W. 1118(145), *1146*
Rudel, R. A. 925(47), *1004*
Rudin, A. 1484, 1485(91), *1504*
Rudkevich, D. 1430(440), *1453*
Rudkevich, D. M. 322(251), *331*, 1406(259, 263a), 1417(326), 1418(338), 1421(372), 1422(374), 1424(387, 398, 400a–c, 401), 1430(441a–c), 1435(463), *1446, 1448, 1450, 1451, 1453, 1454*
Rudolph, J. 695(272), *710*
Rudzinski, J. M. 1385(81), *1439*
Ruedi, P. 212(25), *221*
Ruel, B. H. M. 1391(110), 1392(137), 1393(110), *1440, 1441*
Ruf, H. H. 129(482), *190*

Rugge, K. 1357(136), *1366*
Rühling, I. 739(59), *832*
Rui, X. 929, 987(70), *1004*
Ruiz, M. A. 953, 956(212), *1007*
Ruiz-López, M. F. 867, 880, 897(151), *905*
Rulinda, J. B. 559(159–161), 568(159, 160), *600*
Rumboldt, G. 1382(74), *1438*
Rummakko, P. 1216(127), 1309(281), *1341, 1345*
Rumynskaya, I. G. 143(556), *192*
Runge, F. F. 6(28, 29), *180*
Runov, V. K. 988(413), *1012*
Ruppert, L. 18(83a), *181*
Rush, J. D. 1109(51), *1144*
Rusin, O. 1383, 1421(67n), *1438*
Russel, D. H. 110(429, 430), *189*
Russel, R. A. 1171(46), *1338*
Russell, D. H. 264(21), 299(97), 300(21), 323(258), *326, 327, 331*
Russell, G. A. 879(197), *906*
Russell, I. M. 980(358), *1011*
Russell, K. E. 1125(247), *1149*
Russell, R. A. 745, 746(80, 81), *832*, 1239, 1240(165a, 165b), *1342*
Russo, N. 84(302), 89(302, 323, 325), 90(302, 323, 325, 342), 91(302, 342), *186, 187*, 898(276), *908*
Rutgers, M. 1359(159), *1367*
Rutkowski, K. 144(649, 657), *194*, 377(269), *392*, 551, 557(120), *599*
Rutledge, P. S. 416(114), *482*
Ruzgas, T. 974(320), 976(324), 977(337), *1010*
Ruzicka, L. 433(189b), *484*
Ryabov, M. A. 730(48), *832*
Ryan, D. 942(136), 953, 961(240), 999(466), *1006, 1008, 1013*
Ryan, E. 953, 960(227), *1008*
Ryan, J. J. 429, 430(179b), *483*
Ryan, J. M. 953, 960(231), *1008*
Rybalkin, V. P. 729(44), *831*
Rychnovsky, S. D. 1318(301b), *1345*
Ryo, I. 702(332), *712*
Ryrfors, L.-O. 1267(201), *1343*
Ryskalieva, A. K. 230(27), *254*
Ryu, E. K. 773(162), *834*
Ryu, S. 1017(7), *1087*
Ryuichi, O. 447(222c), *485*
Ryzhkina, I. S. 740(64, 65), *832*

Saà, J. M. 750(102, 103), *833*, 1260(191, 192), *1342*
Saadioui, M. 1370(6), 1391(120c), 1406(262a), 1425(406), 1430(6), *1435, 1440, 1446, 1451*
Saaf, G. W. 466, 467(293), *487*

Saarinen, H. 361(141), *389*
Saasted, O. W. 897(272), *908*
Sabbah, R. 227(12), 229(18), 230(26), 234(44, 45), 241(74), 250, 251(18), *254–256*
Sabbatini, N. 1391(120b, 121c), 1396(121c), 1435(469b), *1440, 1454*
Sadighi, J. P. 671(80), *706*
Sadik, O. 978(345), *1011*
Sadikov, A. S. 204(10), *220*
Saebø, J. 579(200), *601*
Saenger, W. 530(13), *597*
Safar, M. 575, 580, 586, 587(188), *600*
Saha, U. K. 1086(379), 1087(381), *1096*
Saha-Möller, C. R. 1213(113), *1340*
Sahlin, M. 1108(17), *1144*
Sahoo, M. K. 1102(43, 44), *1106*
Saijo, Y. 1406(270b), *1446*
Saijyo, Y. 1374, 1406(36b), *1436*
Saiki, T. 1399(201a, 201c), *1443, 1444*
Saini, R. D. 1038(121), *1090*
Saint-Jalmes, V. P. 414(94a), *481*
Saito, B. 699(310), *711*
Saito, I. 1034(112), 1080(261), *1089, 1093*
Saito, K. 673(98), *707*
Saito, M. 887(221), *907*
Saito, S. 664(35), 681(164), 688(229), 689(231–233, 235), *705, 708–710*,763(134), *834*
Saito, T. 886, 890(233), *907*, 937(118), 938(127), *1005, 1006*
Saito, Y. 699(306), *711*, 1172(48), *1338*
Saitoh, K. 1360(166), *1367*
Sajiki, J. 1351(52), *1364*
Sakaguchi, Y. 1470(68), *1504*
Sakai, S. 90(341), 143(546, 547), 179(546), *187, 192*
Sakaki, T. 1396(186b), *1443*
Sakakibara, H. 1383(76), *1439*
Sakamoto, T. 1306(271), *1344*
Sakano, T. K. 170(739), *196*
Sakhail, P. 1383, 1421(67k), *1438*
Sakito, Y. 1335(331), *1346*
Sakiyama, M. 232(37), *255*
Sako, M. 1019(27b, 30), *1087*
Sakthivel, A. 616(28, 29), 620(43, 44), *656, 657*
Sakuragi, H. 174, 177(766), *197*
Sakurai, H. 275(44), *326*
Sakurai, M. 688(228), *709*
Salakhutdinov, N. F. 616(32), *656*
Salamonczyk, G. M. 1224(137), *1341*
Salcedo-Loaiza, J. 551(93), *598*
Saleh, N. A. I. 726(32), *831*
Saletti, R. 301(135, 136), *328*
Salinas, F. 983(376), *1011*
Salles, C. 948, 953(176), *1007*

Salman, S. R. 349(72, 73), 359(125, 127), 366(168), *387, 389, 390*, 726(30–32), *831*, 929(71), *1004*
Salome, M. 992(433), *1013*
Salunkhe, D. K. 1000(470), *1013*
Salvador, A. 901(286, 287), *908*
Salvador, C. 851(63), *903*
Salvatore, P. 1383(75d), *1439*
Salzbrunn, S. 671(85), *706*
Samajdar, S. 633(92), *658*
Samanen, J. M. 1275(217), *1343*
Samanta, S. S. 783(223), *836*
Samat, A. 770(157), *834*
Sammartino, M. P. 975(321), 977(334), *1010*
Sampa, M. H. O. 1100(34), *1106*
Samperi, F. 953, 969, 1000(286), *1009*
Samperi, R. 946(165), 947(167), 953(165, 167), *1006*, 1354(101), *1365*
Samperi, S. 1353(85, 86), *1365*
Sampler, E. P. 1420(362, 367), *1450*
Samreth, S. 609(17), *656*
Samyn, C. 585(236), *601*
San, N. 1083(294), *1094*
Sanada, Y. 1489(108), *1504*
San Andrés, M. P. 953(238, 253), 961(238), 962(253), *1008*
Sanches Freire, R. 974(319), *1010*
Sánchez, A. N. 477(324b), *489*
Sánchez, C. 1374(35), 1385(79a), 1392, 1403(125a), 1427(424c), *1436, 1439, 1441, 1452*
Sánchez, G. 946, 953(164), *1006*, 1353, 1354(70), *1364*
Sánchez Batanero, P. 982(366), *1011*
Sánchez-Moreno, C. 856, 857(124), *904*
Sanchéz-Ortega, A. 1393(140), *1441*
Sanchez-Quesada, J. 1406(266), *1446*
Sandermann, H. 895(269), *908*
Sanderson, R. T. 89(318), *186*
Sanderson, W. R. 422(149), *483*
Sandorfy, C. 37(177, 178), 47(226, 229, 230), 53(226, 229), 56(229), *183–185*, 530(6), 589(261), *596, 602*
Sandoval, C. A. 1372(23, 24b), *1436*
Sandpord, R. W. 1157(10), *1337*
Sandra, P. 937, 938(125), *1005*
Sandreczki, T. C. 1495, 1498(142, 143), *1505*
Sandri, S. 1255(180), *1342*
Sandström, J. 170, 171(746), *196*, 350(80), *388*
Sandusky, P. O. 129(483), *190*
Sangwan, N. K. 757(107), *833*
Sanhueza, P. 1087(383), *1096*
Sanjoh, H. 1315(293), *1345*
Sannohe, H. 1392(131), *1441*
Sano, N. 1083(304), *1094*
Sano, S. 1163(33a), *1338*
Sanoner, P. 913(10), *1003*

Sansom, P. I. 1430(442a), *1453*
Sansone, F. 1405(255), 1435(462b), *1446, 1454*
Sansoulet, J. 662(3, 4), *705*
Santhosh, K. C. 663(25), *705*
Santini, A. 1383(75b, 75d, 75e), 1422(75e), *1439*
Santini, C. J. 1393(145a), *1441*
Santoro, D. 40, 127(187), *184*
Santos, A. C. 895(266), *908*
Santos, J. G. 813(302), *837*
Santos, L. 1047(149), *1090*
Santos, R. M. B. dos 45(194), *184*
Santoyo-González, F. 1393(140), *1441*
Sanz, D. 349(59), *387*
Sanz, G. 650(162), *659*
Sanz, R. 770(153), *834*
Sanz-Vicente, I. 934, 937(107), *1005*
Sappok-Stang, A. 403(32), *479*
Sapre, A. V. 1021(38), 1038(121), *1088, 1090*
Sarada, K. 1063(192), 1064(193), *1091*
Saraiva, M. C. 1383, 1421(67f), *1438*
Sarakha, M. 1017(15–17), 1018(18, 19), 1020(31), 1085(347), 1086(359), *1087, 1095, 1096*
Saran, M. 855, 856(109), *904*, 1115(101), 1118(101, 143, 146, 147), 1119(101, 146, 163), 1128(279), 1130(101, 143, 147, 309, 323), 1131(279), *1145–1147, 1149, 1150*
Sarasa, P. 110, 121(428), *189*
Sarazin, C. 364(154), *389*
Sargent, M. V. 424(154a), *483*
Sarin, V. N. 35(159), *183*
Saritha, N. 620(43), *657*
Sariyar, G. 1070(199), *1091*
Sarkar, N. 1025(71a, 71b), *1088*
Sarma, J. 650(165), *659*
Sarna, T. 852(84), *903*, 1136, 1137(364), *1151*
Sartori, G. 608, 612(10), 616(26), 622(51–53), 623(54), 626(59), 627(62), *656, 657*, 677(122, 131, 133), 680(153, 155), 683(168–170, 179, 181–184), 685(200), 686(206–208, 213), *707–709*, 740(62), *832*, 1374(37), 1378(37, 49a–c), *1436, 1437*, 1462(24), *1502*
Sartory, G. 1255(183), *1342*
Sarubbi, T. J. 1482(88), *1504*
Saruwatari, Y. 1377(45a, 45b, 46a, 46b), *1437*
Sarzanini, C. 1353(73, 88), *1364, 1365*
Sasage, S. 669(63), *706*
Sasai, H. 695(270, 279, 280), *710, 711*
Sasaki, C. 699(306, 308), *711*
Sasaki, H. 699(300), *711*, 774(192, 193), *835*
Sasaki, K. 414(89a–c, 90c, 90d), *481*
Sasaki, Y. 414(90g), *481*, 1403, 1406(232b), *1445*
Sasakura, K. 686(209), *709*
Sasho, M. 447(221b, 222e), *485*

Sasidharan, M. 774(177), *835*
Sasson, Y. 421(126), *482*
Satake, A. 1383, 1421(67l), *1438*
Sathyamurthy, N. 1047(148), *1090*
Sathyanarayana, D. N. 359(128), *389*
Satish, S. 1202, 1205(94), *1340*
Sato, A. 460(288), *487*, 624(58), *657*, 769(146), *834*
Sato, F. 1216, 1283, 1287, 1297(121), *1340*
Sato, H. 1204(98), *1340*
Sato, J. 460(283b), *487*
Sato, K. 143(582), *193*, 677(134–136), *707*
Sato, M. 818(318–320), *838*
Sato, S. 112(456), *190*, 608(11), *656*, 677(136), *707*
Sato, T. 972(303), *1010*, 1071(209a), *1092*, 1495(158, 159), 1498(159), *1505*
Sato, Y. 811(290), *837*, 1084(307), *1094*, 1380(61a), *1438*
Satoh, J. Y. 813(299), *837*
Satoh, K. 813(299), *837*
Satoh, T. 664(34), *705*, 813(299), *837*
Satyanarayana, C. V. V. 749(100), *833*
Satyanarayana, C. V. V. 620(42), *657*
Sauer, J. 467(295b), *488*
Sauer, R. 928(55), *1004*
Sauleda, R. 1085(348), *1095*
Saunders, J. K. 880(200), *906*
Saunders, M. 367(176), *390*
Saunders, W. H. 424(151), *483*
Saura-Calixto, F. 856, 857(124), *904*
Sauriol, F. 1401(211), *1444*
Saussine, L. 413(81), *480*
Sauter, M. 740(61), *832*
Sauvageau, P. 47, 53, 56(229), *185*
Savage, R. 408(57), *480*
Savard, J. 444(215a, 215b), *485*
Savariar, S. 450(239e), *486*
Savich, V. I. 725(28), *831*
Savin, A. 69(263–266), 70(265), 148(266), *185*
Savolainen, A. 236, 248(56), *255*
Sawa, T. 1262(195), *1343*
Sawada, D. 694(267), *710*
Sawada, H. 291(63), *327*
Sawada, K. 891(246, 247), *907*
Sawada, M. 93(362), *187*
Sawaguchi, M. 648(154), *659*
Sawahata, T. 827(340), *838*
Sawaki, Y. 408(53), 422(145), *479, 483*
Sawamoto, M. 1388(100b), *1439*
Sawamura, T. 147(706), *195*
Sawant, D. P. 639(131), *659*
Sawka-Dobrowolska, W. 580(203), *601*
Sawyer, D. T. 1118(151, 152), *1147*, 1192(78c, 78d), *1339*
Sawyer, J. S. 662(7), 670(70), 673(88), 678(140), *706, 707*, 1318(305), *1345*

Saxo, M. 447(221a, 222a, 222c), 448(227), *485*
Sayari, A. 1213(116), *1340*
Sayer, J. M. 721(22), *831*
Saykally, R. J. 148, 160(721), *196*
Sayre, L. M. 1194(82), 1198(90), *1339, 1340*
Scaglioni, L. 338, 345(10), *386*, 721(25, 26), *831*
Scaiano, J. C. 8(42), *180*, 843, 901(283), *908*, 1019(26), 1021(40, 42), 1040(125), *1087, 1088, 1090*, 1109, 1110, 1112(26), 1140(372), *1144, 1151*
Scaroni, A. 377(252), *392*
Scarpa, M. 890(234), *907*
Scarpf, W. G. 432(183b), *484*
Scavarda, F. 1054(163), *1091*
Schaad, L. J. 368(182), *390*, 530, 551, 554, 557, 559, 588–590(4), *596*
Schade, E. 1462(30), *1503*
Schadelin, J. 13(67), *181*
Schaefer, H. F. 93, 97, 132, 133(359), *187*
Schaefer, H. F. III 368(201), *391*
Schaefer, T. 349(67–71, 73), 350(81), *387, 388*
Schaefer, T. D. 851, 885(47), *903*
Schaeffer, T. 551(132), *599*
Schäfer, J. 323(264), *331*
Schäfer, S. 662(5), *705*
Schaff, P. A. van den 93(364), *187*
Schaffner, C. 1349(32), 1356(126), *1363, 1366*
Schah-Mohammedi, P. 364(155), *389*
Schaldach, B. 285(56), *327*
Schaller, F. 978(351), *1011*
Schalley, C. A. 322(251), *331*, 1406(263a), *1446*
Schamp, N. 471(303), *488*
Schanze, K. S. 1084(308, 309), *1094*
Schatsnev, P. V. 1123(200, 201), *1148*
Schatz, J. 1400(206), *1444*
Schaube, J. H. 721(23), *831*
Scheerder, J. 1389(104), 1394(176), *1440, 1442*
Scheffler, K. 1109(27, 29, 30), 1120(168, 177), 1123(207, 208), 1124(208), *1144, 1147, 1148*, 1290(244), *1344*
Scheiner, A. C. 368(201), *391*
Scheiner, S. 144(597), *193*, 530(9, 10), 551, 553(101), *597, 599*
Scheler, U. 367(172), *390*
Scheller, F. 976(331), 978(351), 979(352), *1010, 1011*
Scheller, F. W. 977(344), 979(357), *1011*
Schelton, B. W. 1388(97), *1439*
Schenck, K. M. 1354, 1355(100), *1365*
Scheppingen, W. van 139(533), *191*
Scheringer, C. 999(461), *1013*
Scherrmann, M.-C. 1405(255), *1446*

Schiavello, M. 1082(288, 290), 1083(293, 301), 1084(320, 325), *1094, 1095*
Schiavone, S. 1394(171), *1442*
Schick, C. P. 110(418, 419), 121(419), *189*, 1017(9, 10), *1087*
Schieber, A. 953, 958(224), *1008*
Schiefke, A. 143, 144(577), *192*, 581(217), *601*
Schiegl, C. 303(144), *329*
Schilf, W. 1024(63), *1088*
Schink, B. 15, 129, 141, 142(76), *181*
Schiraldi, D. 621(49), *657*
Schlag, E. W. 30(135), 143(591), 144(691), 147(718), *182, 193, 195, 196*
Schlatter, C. 1360(169), *1367*
Schlegel, H. B. 4, 100(2), *179*, 368, 372, 373(200), *390*
Schleigh, W. R. 1497, 1498(170), *1506*
Schlemper, E. O. 207(14), 212(24), *220, 221*
Schlessinger, R. H. 1278(225), *1343*
Schlett, C. 1347, 1352(2), *1363*
Schleyer, P. v. R. 93(352), 95(352, 367), 104(352, 398, 399), 105(352), 170, 171(745), *187, 188, 196*, 275(45), 311(188), *326, 329*, 531(21), 555(140), 586(237), 589(237, 260, 262), 590(237), *597, 599, 602*, 877(187), *906*
Schlingloff, G. 701(315–317), *711, 712*
Schlosberg, L. P. 236(55), *255*
Schlosser, M. 93(355), *187*, 759(114), *833*
Schlossler, P. 933, 937(103), *1005*
Schmeling, U. 13(67), *181*
Schmerling, L. 405(37a), *479*
Schmid, E. 753, 764(105), *833*
Schmid, H. 419(135a), 460(280a, 284), 463(277), 466(291), *482, 487*, 753(105), 755(106), 764(105), 823(326), *833, 838*
Schmid, P. 1360(169), *1367*
Schmid, S. 1100(28), *1106*
Schmid, W. 436(196), *484*
Schmideder, H. 144(674), *195*
Schmider, H. L. 71(272), *185*
Schmidt, C. 1374(33c, 36c), 1419(358), 1420(360, 361, 363, 364, 365b, 366), 1431(443b), 1435(461), *1436, 1450, 1453*
Schmidt, D. M. 469(297b), *488*
Schmidt, L. 1352(66), *1364*
Schmidt, M. 144(682), *195*
Schmidt, R. 226, 232(8), *254*
Schmidt, R. R. 673(95), 675(113), *707*, 776(203–205), 778(209), *835*
Schmidt, S. 663(15), *705*
Schmidt, U. 429(177, 178), *483*
Schmiesing, R. 443(213), *485*
Schmitt, A. 662(5), *705*
Schmitt, M. 20, 24, 106(127), 143(577), 144(577, 678, 679), 147(695, 697–702), 148(695, 697, 699–702), 149(697, 699),

Schmitt, M. (*continued*) 156(702), 160(700, 702), *182, 192, 195,* 377(253), *392,* 578(194), 581(214–217), *601,* 992(428), *1012*
Schmitt, P. 1427(416a, 416b), 1432(450), *1452, 1453*
Schmittling, E. A. 673(87, 88), *706*
Schmitz, J. 1378(48a, 48b), *1437*
Schmoltner, A. M. 137(522), *191*
Schmutzler, R. 1383(67k), 1417(329a, 329b), 1418(339), 1421(67k), 1424(395), *1438, 1448, 1450, 1451*
Schnatter, W. F. K. 451(250), 456(261), *486, 487*
Schneider, G. 1082(277), *1094*
Schneider, H.-J. 365(163), *389,* 1419(351), *1450*
Schneider, S. 428(168), *483,* 1025(82), *1089,* 1136(356), *1151*
Schneider, U. 365(163), *389,* 1419(351), *1450*
Schneider, W. F. 138(528), *191,* 1104(61), *1106*
Schnepf, R. 129, 132, 133, 135(514), *191*
Schnering, H. G. v. 69(263), *185*
Schnering, H. G. von 1080(259a), *1093*
Schoch, J.-P. 419(134b), *482*
Schofield, K. 1255(182), *1342*
Scholz, M. 813(301, 303), *837, 838*
Schonberg, A. 472(311f), *488*
Schoth, R.-M. 743(70), *832*
Schott, H. H. 13(67), *181*
Showen, R. L. 367(178), *390*
Schrader, S. 1353(89), *1365*
Schram, K. H. 305(151), *329*
Schreiber, H. D. 47, 48, 53(222), *184*
Schreiber, V. 377(257, 269), *392*
Schreiber, V. M. 143(554–556), 144(648), *192, 194,* 377(258), *392*
Schreuder, H. A. 412(73), *480*
Schröder, D. 731(49), *832*
Schroeder, G. 144(609), 145(630), *193, 194,* 377(259), *392,* 595, 596(298), *603*
Schroeder, H. Fr. 1356(123), *1366*
Schroth, W. 823(330), *838*
Schubert, K. 1123(215), *1148*
Schubert, M. P. 1118(133, 134), *1146*
Schuchmann, H.-P. 1086(375), *1096,* 1100(32), *1106,* 1116(113), 1136(362), *1146, 1151*
Schuchmann, M. N. 143, 144(550), *192,* 1114(90), *1145*
Schuetz, G. J. 1045(140), *1090*
Schuler, F. 1360(169), *1367*
Schuler, P. 1120(177), *1147*
Schuler, R. H. 129(494, 503), 131(503), 133(494), *191,* 1098(7), 1099(16, 17), *1105,* 1110(57, 59), 1115(109), 1117(117), 1120(186, 187), 1122(186), 1123(196, 206), 1126(263), 1127(263, 273), 1128(57, 187, 263), 1129(288, 290, 291), 1132(187, 196, 288, 291, 330–338), 1133(335), 1134(186, 288, 290), 1135(109, 338, 352, 354), 1136(263, 338, 352, 354, 365), 1138(263), 1139, 1142, 1143(371), *1145–1151*
Schuller, W. 925, 936(51), *1004*
Schulman, S. G. 78(277), *186,* 492, 493(11), 498(72a, 72b), *525, 526*
Schulte, K. E. 409(60), *480*
Schulte-Frohlinde, D. 1109(46), 1115(112), 1117(116), 1124(243), *1144, 1146, 1148*
Schultz, A. G. 470(300c, 300d), 472(307), *488,* 816(310, 311), *838*
Schultz, B. 1352(64), *1364*
Schultz, G. 20, 22(109), *182*
Schultz, J. 928(55), *1004*
Schultz, M. 1408(279b), *1447*
Schulz, G. 434(192), *484*
Schulz, M.-F. 229, 250(20), *254*
Schulz-Ekloff, G. 1082(277), *1094*
Schulze-Lefert, P. 913(3), *1003*
Schumm, S. 20(114, 126), 23(114), 24(126), 35, 36(169), 106(114, 126), 107(114, 126, 169), *182, 183*
Schuster, H. 467(294), *487*
Schuster, P. 47, 53(226), *184,* 371, 377(227), *391,* 530(6), *596*
Schutte, C. J. H. 383(278), *392*
Schütz, M. 20, 35(122), 36(122, 175), 78(278), 147(122, 175, 709–711), 148(710, 711), *182, 183, 186, 195,* 375(242, 243), *392*
Schüürmann, G. 5(16), *180*
Schwalm, W. J. 877(190), *906*
Schwartz, A. 1312(286, 287), *1345*
Schwartz, H. M. 1120(188), *1147*
Schwartz, M. A. 472(306), *488,* 1274(210a), 1275(214, 216), 1278(219), *1343*
Schwarz, H. 110(451), *190,* 280(50), 300(108), *327, 328,* 731(49), *832*
Schwarzbauer, J. 1355(118), *1365*
Schwarzinger, S. 345(50), *387*
Schweig, A. 1123, 1124(216), *1148*
Schwing, M.-J. 1391(121a), 1396(121a, 182b), *1440, 1443*
Schwing-Weill, M.-J. 1387(94), 1389(105), 1391(120c), 1396(187, 195b), 1435(94, 187), *1439, 1440, 1443*
Schwochau, M. 429(178), *483*
Sciotto, D. 1391, 1396(121b), *1440*
Sclafani, A. 1082(290), 1083(293, 301, 302), 1084(320, 325, 326), *1094, 1095*
Scoponi, M. 1403(240b), 1432(448b), *1445, 1453*
Scorrano, G. 89(310), *186*
Scott, A. I. 13(62), *181,* 426(166a), 429(179b–e), 430(179b–e, 181d, 181e),

483, 484, 1158, 1175(14b), 1297(256b), *1337, 1344*
Scott, A. P. 368(202), *391*
Scott, G. 1108(1), 1109, 1120(28), *1143, 1144*
Scott, I. M. 1351(48), *1364*
Scott, J. A. 310(184, 185), *329*, 470(298), *488*
Scott, J. L. 1381(66), *1438*
Scott, M. J. 1348(20), *1363*
Scott, R. 47, 143(207–209), 170(209), *184*
Scott, R. M. 592(268), 593, 595(278), *602*
Scott, S. W. 1274(210a), 1275(216), *1343*
Scribner, A. W. 1070(200), *1091*
Scully, P. A. 1408(281), *1447*
Scuseria, G. E. 4, 100(2), *179*, 368, 372, 373(200), *390*
Seabra, R. M. 913(18, 20), 958(20), *1003*
Sealy, R. C. 1120(178, 182, 183, 188), 1124(232), *1147, 1148*
Searls, L. J. 315, 316(217), *330*
Seaton, A. 1231(150), *1341*
Sebastian, R. 349(73), *387*
Sebastiano, R. 1259(188), *1342*
Secchi, A. 1391(118a), 1396(118a, 188), 1403, 1405(235), 1409(288), 1427(409a, 415), 1430(439), 1432(409a), 1435(467), *1440, 1443, 1445, 1447, 1451–1454*
Sedergran, T. C. 665(41), *705*
See, K. A. 1393(147, 152a), 1394(147, 165), *1441, 1442*
Seebach, D. 105(400), *188*
Seeberger, P. H. 642(140), *659*
Seeboth, H. 401(21), *479*
Seely, J. C. 1348(30), *1363*
Sega, M. 1086(365), *1096*
Segal, J. A. 673(89), *706*
Seguin, J.-P. 350(83), *388*
Segura, M. 1435(462b), *1454*
Sehested, J. 138(528), *191*, 1104(61), *1106*
Sehested, K. 1115(93), *1145*
Sehili, T. 1084(319), *1095*
Seiffarth, K. 319(222), *330*, 1408(279b), *1447*
Seiji, Y. 419(123a), *482*
Seiler, J. P. 1359(153), *1366*
Seita, J. 350(80), *388*
Seithel, D. S. 1220(134), *1341*
Seki, S. 232(37), *255*
Seki, Y. 1372(14), *1436*
Sekiguchi, O. 291(64), 300(125), *327, 328*
Sekiguchi, S. 551, 554(110), *599*
Sekihachi, J.-I. 447(222c, 222e), *485*
Sekiya, A. 219(34), *221*
Sekyra, M. 933, 937, 953(99), *1005*
Seleznev, V. N. 1076(235), *1092*
Selinger, B. K. 524(102), *527*
Sellens, R. J. 354(103), *388*
Selles, E. 985(388), *1012*
Selli, E. 1085(339), *1095*
Sellstrom, U. 1351(47), *1364*

Selva, A. 757(109), *833*
Selva, M. 663(18), *705*
Selvaggini, R. 998(452, 453), *1013*
Selvakumar, N. 1264(196), *1343*
Selvam, P. 616(28, 29), 620(43, 44), *656, 657*, 1213(115), *1340*
Selwyn, M. J. 1359(158), *1367*
Selzer, S. 1071(207), *1092*
Semensi, V. 748(87, 88), *833*
Semiat, R. 619(40), *656*
Sen, A. 994(438), *1013*
Sen, C. K. 840, 901(10), *902*
Sen, S. 451(252), *486*
Sénèque, O. 1394(175a), *1442*
Senet, J.-P. 663(16), *705*
Sengupta, D. 89, 90(334), *187*
Senlis, G. 1415(312), *1448*
Sennyey, G. 663(16), *705*
Seno, M. 1495, 1498(159), *1505*
Seok, W. K. 1287(238), 1300(263), *1344*
Sepulveda-Escribano, A. 1489, 1490(114), *1504*
Sequin, J. P. 364(154), *389*
Sera, A. 448(232a), *486*
Serbinova, E. A. 887(222, 225), *907*
Sergeeva, E. 640(132), *659*
Seri, C. 1383(75e), 1422(75e, 375), *1439, 1450*
Serna, M. L. 953(209, 256), 955(209), 962(256), *1007, 1009*
Serpone, N. 1082(281), 1084(313, 318), 1086(369), *1094–1096*, 1100(25, 26), *1105*, 1129(284–286), *1149*
Serpone, N. R. 1361(195, 196), *1367*
Serra, A. C. 233(39), *255*
Serra, F. 1086(367), *1096*
Serrano, A. 728(40), *831*
Serratos, J. A. 913(2), *1002*
Serry, M. M. 913, 920(27), *1003*, 1355(116), *1365*
Servili, M. 998(452, 453), *1013*
Servos, M. R. 1351(48), *1364*
Seshadri, T. R. 748(94), *833*
Sessier, J. L. 1020(34), *1087*
Sessler, J. L. 1430(442a, 442b), *1453*
Sestili, P. 895(265), *908*
Sethna, S. 409(61), *480*, 680(149), *708*
Sethna, S. M. 408(54c), *480*
Seto, S. 1470(68), *1504*
Sevilla, C. L. 1104(59), *1106*
Sevilla, M. D. 1104(54, 59), *1106*
Sevin, A. 71, 73(274), 89(315), *185, 186*
Sexton, K. G. 937(119), *1005*
Seya, K. 1080(259d), *1093*
Seybold, P. G. 101–103(391), *188*
Sgaragli, G. P. 953, 968(275), *1009*
Sha, Y. A. 608(7), *656*
Shabarov, Yu. S. 763(133), *834*
Shackelford, S. A. 627(63), *657*

Shade, M. 1393(152c), 1403(152c, 234a), 1427(409c), *1441, 1445, 1451*
Shagidullina, R. A. 740(65), *832*
Shahidi, F. 840, 871(8), *902*, 992(425), *1012*
Shahrzad, S. 953, 957(218), 960(226), *1008*
Shair, M. D. 447(224), *485*
Shakirov, M. M. 744(75), *832*
Shaler, T. A. 300(121), 323(260), *328, 331*
Shambayati, S. 664(30), *705*
Shamma, M. 807(276), *837*, 1070(199), *1091*
Shanafelt, A. 1462(36), *1503*
Shang, S. 1393(145a), *1441*
Shani, A. 1038(122a), *1090*
Shankar, B. B. 668(58), *706*
Shannon, P. V. R. 1297(260c), *1344*
Shannon, T. W. 274(39), *326*
Shaofeng, L. 227(10), *254*
Shapiro, R. 670(75), *706*
Shapiro, R. H. 299(89), *327*
Shapovalova, E. N. 987(408), *1012*
Sharghi, H. 474(316), *488*, 640(133), *659*
Sharif, I. 647(147, 149), *659*
Sharkey, A. G. Jr. 323(261), *331*
Sharma, G. M. 1297(256b), *1344*
Sharma, H. K. 890(239), *907*
Sharma, M. M. 763(130), *834*
Sharma, S. K. 1392(129a, 134), 1403(234d), 1406(134, 258b), *1441, 1445, 1446*
Sharma, S. N. 82(287), *186*
Sharma, V. 551, 553(82), *598*
Sharp, K. A. 5(12), *179*
Sharpless, K. B. 14(70), *181*, 413(82a), *480*
Sharzynski, M. 1470(62), *1503*
Shastri, L. V. 1111(69), *1145*
Shaw, A. C. 1237(164), *1342*
Shaw, C. D. 68, 70(261), *185*
Shaw, G. S. 1048(154), *1090*
Shaw, L. J. 1362(217), *1368*
Shchepkin, D. 371(224), *391*
Shchepkin, D. E. 377(269), *392*
Shchepkin, D. N. 145(612, 625), 171, 173(612), *193*, 371, 377(230), *391*
Shcherbakova, I. V. 432(186), *484*
Shchutskaya, A. B. 178(772), *197*
She, L. Q. 620(45), *657*
Sheban, G. V. 730(48), *832*
Sheedy, B. R. 1357(133), *1366*
Sheen, P. D. 1427(416a), 1432(450), *1452, 1453*
Sheikheldin, S. Y. 983(377), 984(380), *1011*
Sheilds, G. P. 551, 552(134), *599*
Shekhovtsova, T. N. 977(341, 342), 978(347), *1010, 1011*
Sheldon, R. A. 460(286), *487*, 649(156), *659*, 763(128), *834*, 1192(78a), *1339*
Shemyakin, M. M. 18(89), *181*
Shen, H. R. 15(73), *181*
Shen, J. 983(373), *1011*

Shen, J. Y. 1435(465a), *1454*
Shen, T. 1045(142), *1090*
Shen, W. 1085(354), *1096*
Shen, X. 1108, 1118, 1138–1140(24), 1142(24, 377), 1143(24), *1144, 1151*, 1362(214), *1368*
Shen, Y.-L. 1234(153b), *1341*
Shenderovich, I. G. 366(165), *389*, 503(90), *527*
Sheppard, A. C. 416, 417(116), *482*
Sherburn, M. S. 1432(447), *1453*
Sherman, J. C. 1382(67c), 1383, 1421(67c, 67i), 1424(402), 1425(403a, 403b, 404), 1428(431a, 431b, 432, 433), 1429(428), 1433(457, 459, 460), *1438, 1451–1453*
Sheu, C. 1429(430a), *1452*
Shevade, A. 937, 940(129), *1006*
Shi, C. Y. 913(24), 920, 949, 953(33), *1003*
Shi, G. 666(48), 668(56), *705, 706*
Shi, H. 893(255), *907*
Shi, J. 1084(310), *1094*
Shi, X. 806(265, 266), *837*
Shi, Y. 1035(117), 1036(117, 118), *1089*
Shi, Y. J. 1033(110), 1038(120), *1089*
Shiao, H.-C. 1234(153a), *1341*
Shiao, S. J. 608(7), *656*
Shiau, J.-C. 999(463), *1013*
Shibakami, M. 219(34), *221*
Shibamoto, T. 1050(160), *1091*
Shibasaki, M. 667(50), 691(238), 694(267), 695(268–270, 273, 274, 279, 280), *705, 710, 711*
Shibata, J. 1402(224a), *1444*
Shibata, M. 1066(197), *1091*
Shibato, K. 429(172), *483*
Shieh, T.-L. 359(121), *389*
Shif, A. I. 1290, 1292(247), *1344*
Shiga, S. 1115(98), *1145*
Shigemori, H. 1158(17c), 1180(61a, 61b), 1186(70), *1338, 1339*
Shigorin, D. N. 383(280), *392*
Shih, C. 1171(47), *1338*
Shim, I. 127(469), *190*, 718(17), *831*
Shim, Y. B. 982(367), *1011*
Shima, K. 1022(54, 55), *1088*
Shima, Y. 953, 962(248), *1008*
Shimada, K. 689(231), *709*
Shimada, M. 1216(125), *1341*
Shimakoshi, H. 1209(106, 107), *1340*
Shimamura, T. 1079(240), *1092*
Shimao, I. 422(131), *482*
Shimizu, C. 1418(335), *1448*
Shimizu, H. 699(309), *711*, 1393(159), *1442*
Shimizu, I. 663(17), *705*
Shimizu, K. D. 702(345, 346), *712*, 1406(263c), *1446*
Shimizu, M. 1196(89), 1213(114, 117), *1340*
Shimizu, N. 945(149), *1006*

Shimizu, S. 1403, 1406(232b), *1445*
Shimizu, T. 1206(105), 1265(199a, 199b), *1340, 1343*
Shimizu, Y. 174, 177(766), *197*
Shimoi, K. 913(28), *1003*
Shin, E.-J. 178(774), *197*
Shin, G. C. 454(258), *487*
Shin, H. 673(91), *706*
Shin, J. M. 1392(130), 1394(163), *1441, 1442*
Shin, S. C. 454(258), *487*
Shin, S. Y. 773(162), *834*
Shine, H. J. 419(121, 128a, 134d), 470(300a), 472(308c), *482, 488*
Shinkai, S. 320(240), *331*, 419(123a), *482*, 633(94), 634(95), *658*, 744(77), *832*, 1372(10b, 18), 1379(54b, 54d, 54e), 1385(81), 1388(103), 1391(111, 115a, 115b, 117, 122c), 1392(124, 139), 1393(159), 1394(111, 167a, 172b, 174), 1396(186b), 1400(207a, 207b, 210a), 1401(210a, 217, 218), 1402(224a), 1403(231a), 1404(246a, 246b), 1405(247, 252), 1413(306), 1419(347a, 354), 1427(407b, 408b, 419a, 424b), 1432(122c, 424b), 1435(464b, 468), *1435–1437, 1439–1447, 1450–1452, 1454*
Shinmyozu, T. 353(99), *388*, 1380(58a, 58b, 58e), *1437*
Shinoda, H. 92(347), *187*
Shinozuka, N. 953, 962(248), *1008*
Shinra, K. 360(137), *389*
Shintani, F. 636(105), *658*
Shiota, Y. 18(86), *181*
Shiozawa, M. 689(233), *710*
Shiozawa, T. 1357(137), *1366*
Shipp, D. A. 1459(9, 15), 1463(9), *1502*
Shiraishi, M. 624(56), *657*, 1489(109, 115), 1495(132, 133), *1504, 1505*
Shiraishi, S. 1201, 1205(92), *1340*
Shirakawa, S. 1403, 1406(232b), *1445*
Shiramizu, K. 1414(309), *1448*
Shirasaka, T. 693(250, 256), *710*
Shirley, D. A. 102(392), *188*
Shiro, M. 1233(151a, 151b), *1341*
Shisuka, H. 492, 493(14), *525*
Shitomi, H. 37, 46, 110, 111(182), *183*
Shivanyuk, A. 1383(67g), 1402, 1404(222), 1406(222, 262a), 1418(341, 342, 343a–c), 1419(353), 1420(341, 342, 366), 1421(67g, 342, 372), 1424(400b), *1438, 1444, 1446, 1450, 1451*
Shivanyuk, A. N. 1417(326, 328), 1418(338, 344), *1448, 1450*
Shizuka, H. 493(36–39), 512(38, 39), *525*
Shizuri, Y. 1158(17c), 1172(48), 1175(53), 1180(53, 61a, 61b, 65, 66), 1183(67), 1186(70), *1338, 1339*
Shklover, V. E. 200(7), *220*
Shkrob, I. A. 1103(48), *1106*
Shoichi, S. 1422(374), *1450*
Shoji, H. 414(91), *481*
Shokova, E. 1427(409b), *1451*
Shokova, E. A. 1372(21), *1436*
Shon, O. J. 1427(423b), *1452*
Shono, T. 360(137), *389*, 1156, 1158, 1175(2, 9), *1337*
Shoolery, J. N. 547(60), *598*
Shorshnev, S. V. 776(200), *835*
Shorthill, B. J. 1382(72a, 72b), *1438*
Shoup, T. M. 415(99, 100), *481*
Shoute, L. C. T. 1099(19), 1101(41, 42), *1105, 1106*, 1135, 1136(353), *1151*
Shpigun, O. A. 962(252), 987(408), *1008, 1012*
Shreve, N. R. 397(2b), *478*
Shroeder, G. 593, 595(277), *602*
Shteingarts, V. D. 1123(201), *1148*
Shu, C. 1393(152b), *1441*
Shu, C.-M. 1391–1393(122a), 1394(168, 170a), 1407(170a, 273b), *1440, 1442, 1446*
Shuely, W. J. 503(89), *526*
Shuker, S. B. 1406(257, 260), *1446*
Shukla, D. 110, 111, 113, 114(425), *189*, 492, 493(16), *525*, 1019(27a), *1087*, 1134(344), *1151*
Shuler, R. H. 1103(49), *1106*
Shulgin, A. T. 554, 555(136), *599*, 1335(330), *1346*
Shulz, R. A. 557, 560, 564, 590(145), *599*
Shur, L. A. 1362(218), *1368*
Shurukhina, A. 377(269), *392*
Shurvell, H. F. 170(747–750), *196*, 557(142–144), *599*
Shutie, W. M. 290(62), *327*
Shutz, M. 581(219), *601*
Shyu, S.-F. 378(270), *392*
Siahaan, T. J. 677(139), *707*
Sibi, M. P. 782(215), *836*
Sicuri, A. R. 1396(186a), 1427(410a), *1443, 1451*
Siddiq, M. 649(157), *659*
Sidhu, R. S. 673(96), *707*
Sidorov, V. 1403(241), *1445*
Sie, E.-R. H. B. 666(47), *705*
Siebert, F. 1117(126), *1146*
Siebrand, W. 129, 133(497), 179(779), *191, 198*, 371(222), 377(222, 254), *391, 392*
Sieck, L. W. 312(195), *330*
Siedentop, T. 1417(329a), *1448*
Siegbahn, P. E. M. 45(196), *184*
Siegel, G. G. 145, 146(611), *193*, 569(177), *600*
Siegel, J. S. 1413(303), *1447*
Siegl, R. 949, 953(186), *1007*
Siegrist, J. 948, 953(176), *1007*
Sieler, J. 6(26), *180*

Sienkiewicz, K. 406(49a, 49b), *479*
Sienko, G. A. 1482(88), *1504*
Sierra, M. G. 367(172), *390*
Siglow, K. 30(137), 147(716), *183, 195*
Sih, C. J. 1224(137, 138), *1341*
Silcoff, E. R. 691(237), *710*
Silgoner, I. 319(229), *330*, 947(171), 953(171, 206), 955(206), *1006, 1007*
Silicia, E. 89(323, 325), 90(323, 325, 342), 91(342), *187*
Silva, B. M. 913, 958(20), *1003*
Silva, J. 853(91), *904*
Silva, L. F. Jr. 1312, 1315(285), *1345*
Silva, M. 929, 987(70), *1004*
Silva, M. C. M. 953, 960(233), *1008*
Silva, M. I. F. 360(138), *389*
Silva, M. R. 233(39), *255*
Silveira, C. G. 1100(34), *1106*
Silver, B. L. 1124(220, 221, 224, 225), *1148*
Silver, H. B. 460(280b), *487*
Silver, J. H. 92, 101(349), *187*
Silvers, J. H. 312(192), *330*
Silvestro, G. 236(55), *255*
Silvestro, T. 100(380), *188*
Silvi, B. 47, 48, 56(220), 69(264–266), 70(265), 71, 73(274), 75(275), 78(276), 89(220, 315, 316, 317b), 90(316), 148(266), *184–186*
Simal Lozano, J. 937, 941(132), *1006*
Simard, M. 551, 553(82), *598*
Simchen, G. 783(225), *836*
Simeone, A. M. 953, 955(207), *1007*
Simic, M. 1135(345), *1151*
Simic, M. G. 129(486), *190*, 868(156), 871(163, 165), 874(170), 897(274), 898(163), 899(170), *905, 908*, 1100(22, 23), *1105*, 1108(9), 1139(369, 370), 1142(370), *1144, 1151*
Simkin, B. Ya. 371(217), *391*
Simko, P. 947(174), *1006*
Simmonds, M. S. J. 913(25), *1003*
Simon, J. 671(85), *706*
Simonato, J.-P. 677(132), *707*
Simonelli, F. 1285(233), *1343*
Simonian, A. L. 979(353), *1011*
Simonov, Y. A. 1393(161), *1442*
Simonov, Yu. A. 1417(328), *1448*
Simonyi, M. 179(782), *198*, 369(209), *391*
Simova, E. 470(299), *488*
Simova, S. 470(299), *488*
Simperler, A. 378, 379(272), 383(284), *392, 393*, 551(81, 123), 553(81), 557(123), *598, 599*
Sims, P. 409(62a, 63b), *480*, 801(244), *836*
Simsiman, G. V. 1359(156), *1366*
Simulin, Y. N. 227(11), *254*
Sinder, B. B. 438(198), *484*
Sindo, M. 447(221a), *485*

Sindona, G. 953, 961(239), *1008*
Singdren, J. 170(741), *196*
Singer, L. S. 1124(237), *1148*
Singer, P. C. 1360(181), *1367*
Singer, R. A. 702(339–341), *712*
Singh, A. K. 477(327a, 327b), *489*, 1021(38), *1088*
Singh, A. P. 619(38), 631(76), 645(145), *656, 657, 659*
Singh, G. P. 420(141), *483*
Singh, J. 421(130), *482*
Singh, L. W. 434(191), *484*
Singh, N. N. 412(76), *480*
Singh, P. 232(36), *255*, 421(130), *482*
Singh, R. K. 441(205, 208), *484*
Singh, S. 547, 548(59), *598*, 687(221), *709*, 856(132), *904*
Singleton, V. L. 990(419), 1000(471), 1001(473), *1012–1014*, 1108(13), *1144*
Sinha, S. 1019(25), *1087*
Sinhababu, A. K. 415(107), *481*
Sinke, G. C. 238–240(66), *256*
Sinn, E. 691(240), *710*
Sinou, D. 664(37, 39), *705*
Sinwel, F. 1325(311b), *1345*
Sioli, G. S. 19(100), *182*
Sirén, H. 973(312), *1010*
Sisido, K. 1427(419a), *1452*
Sitkoff, D. 5(12), *179*
Sitkowski, J. 353(101), 359(122, 123), 361(122), *388, 389*
Siukolan, L. M. 336(6), *386*
Siuzdak, G. 322(251), *331*
Sivasubramanian, S. 459(270), *487*
Sjöberg, M. B.-M. 1108(17), *1144*
Sjödin, M. 865(145), *905*
Skaanderup, P. R. 675(116), *707*
Skals, K. 1360(174), *1367*
Škamla, J. 303, 305(146), *329*
Skarzynski, J. 998(458), *1013*
Skattebol, L. 626(60), *657*
Skelton, B. W. 1382(73), 1413(304, 305), *1438, 1447*
Skibsted, L. H. 871, 872(166), *905*
Skibsted, U. 338(16), *386*
Skingle, D. 429(179b), 430(179b, 181d), *483, 484*
Skinner, W. A. 845(34, 35), *902*
Skopec, S. 978(345), *1011*
Skowronek, K. 143(572), *192*, 335, 363(1), *386*
Skowronska-Ptasinska, M. 782(218), *836*
Skupinski, W. 632(88), *658*
Slaski, M. 1431(443b), *1453*
Slater, T. F. 141(540), *192*, 890(230), *907*, 1111(67), 1130(324), *1145, 1150*
Slavinskaya, R. A. 78(279), *186*

Slawin, A. M. Z. 761(117), *833*, 1420(359b), 1432(449a), *1450, 1453*
Slayden, S. W. 224, 225, 227, 229, 233(1), 247(89), *254, 257*, 415(103b), *481*
Slickman, A. M. 851–853(72), *903*
Sloper, R. W. 1117, 1118(124), 1129(295), 1130(124, 295), *1146, 1150*
Slowikowska, J. 1418(340), *1450*
Small, A. C. 1393(145a), *1441*
Smedarchina, Z. 371, 377(222), *391*
Smedarchina, Z. K. 179(779), *198*
Smeds, A. 1360(185), *1367*
Smeets, W. J. J. 1405(254b), *1446*
Smejkal, R. M. 403(27), *479*
Smidt, J. 1126(262), *1149*
Smirniotis, P. G. 1080(252), *1093*
Smirnov, A. V. 1045(138, 139), *1090*
Smirnova, I. V. 178(772), *197*
Smirnow, S. N. 364(155), 367(178), *389, 390*
Smissman, E. E. 424(157b), *483*
Smit, W. A. 763(131), *834*
Smith, A. B. III 428(169), 450(239g, 239h), *483, 486*
Smith, A. E. 234(47), *255*
Smith, A. J. 129(485), *190*
Smith, B. D. 105(401), *188*
Smith, B. J. 114(460), *190*
Smith, C. A. 807(282), *837*, 1294(254), *1344*
Smith, D. A. 530(12), *597*
Smith, D. E. 493, 515(25), *525*
Smith, D. W. 1356(121), *1365*
Smith, G. 1242(168), *1342*
Smith, G. F. 748(95), *833*
Smith, G. J. 1124(241), *1148*
Smith, I. C. P. 887(217), *906*, 1125(250, 251), *1149*
Smith, J. A. 1348(13), *1363*
Smith, J. R. L. 405(41b), *479*
Smith, K. 613(24), *656*
Smith, K. A. 1383, 1421(67j), *1438*
Smith, K. J. 884(208), 885(210), *906*
Smith, K. N. 1374(44), *1437*
Smith, M. 349(68), *387*, 953, 965(266), *1009*
Smith, N. K. 226, 227(7), *254*
Smith, P. W. G. 626(61), *657*
Smith, R. A. J. 1297(261), *1344*
Smith, S. H. 786(229, 230), *836*
Smith, W. E. 1024(66), *1088*
Smith, W. J. III6 73(88), *706*
Smith, W. T. Jr. 368(184), *390*
Smithers, J. L. 994(437), *1013*
Smolarz, H. D. 913(16), *1003*
Smolders, A. 562(167), *600*
Smolskii, G. M. 985(390), *1012*
Smrcina, M. 691(239), *710*
Smulevich, G. 383(281, 283), 384(283), *392, 393*
Smyth, M. R. 973(311), 977(339), *1010*

Snapper, M. L. 702(345, 346), *712*
Snelgrove, D. W. 879, 882(195), *906*
Snieckus, V. 414(97d), 415(97d, 106a, 106b), *481*, 670(72), *706*, 781(214), 782(215–217), 786(231), 795(233, 234), *836*, 1038(120), *1089*, 1408(282), *1447*
Snijman, P. W. 766(143), 776(195), *834, 835*
Snip, E. 1433(456b), *1453*
Snobl, D. 358(114), *388*
Sntorey, J. 1086(366), *1096*
Snyder, C. 414(97c), *481*
Snyder, M. D. 1045(139), *1090*
Sobczyk, L. 20, 25(129), 47(211), 143(567–570), 144(568, 605, 606, 609, 648, 650, 653, 662, 663, 670), *182, 184, 192–194*,352(91), 358(118), 361(143, 144), 377(257–259, 262, 264), 386(304), *388, 389, 392, 393*,551, 557(121), 593(276, 277, 280), 594(276, 284), 595(276, 277, 280, 284), *599, 602, 603*
Sobolev, V. I. 18(84), *181*, 413(79), *480*
Sobolewski, A. L. 106, 107(408), *188*
Sobral, S. 953, 956(211), *1007*
Sobrinho, E. V. 631(72), *657*
Sodupe, M. 577, 581(192), *600*
Sodupe, M. S. 129, 144, 148, 149, 153(473), *190*
Sofic, E. 871(162), *905*
Sohma, J. 1124(234), *1148*
Soicke, H. 1065(194), *1091*
Sokolova, I. V. 1016(3), 1017(5, 6, 8), *1087*
Sokolowski, A. 129, 132, 133, 135(514), *191*
Sol, V. M. 716(8), *831*
Solà, M. 455(266, 267), *487*, 730(46), *831*, 1025(77), *1089*
Solana, G. 1084(326), *1095*
Solar, M. 990(416), *1012*
Solar, S. 1098(8), 1100(8, 24, 28), *1105, 1106*, 1129(300), *1150*
Solar, W. 1129(300), *1150*
Solcá, N. 83, 84(300), 144(689), *186, 195*
Sole, A. 137(523), *191*
Soleas, G. J. 935, 937(112), 953, 960(230), *1005, 1008*
Solgadi, D. 109(414), 143(588), 144(588, 600), *189, 193*
Solgadi, M. D. 493, 500(31), *525*
Soliman, G. 436(197), *484*
Solmajer, T. 1360(161), *1367*
Solntsev, K. M. 492(17), 493(17, 29, 48–50), 504, 505(49), 522(17), *525, 526*
Solodovnikov, S. P. 1119(158), 1123(210), 1126(256–261), *1147–1149*
Solomon, D. H. 1459(9, 15), 1463(9, 39–41), 1465, 1467(42), 1470(59), 1471(59, 74–79), 1473(75, 76), 1474(77), 1475(78, 79), 1476(84), 1478(85), 1482(84), 1484(94), 1485(84), 1488(103), 1489,

Solomon, D. H. (*continued*)
1490(113, 114), 1492(121), 1500(177–179), *1502–1506*
Solomon, K. R. 1351(48), *1364*
Solomon, M. E. 1318(303), *1345*
Solovskaya, N. A. 1030(105), *1089*
Solov'yov, A. V. 1417(328), *1448*
Soloway, A. H. 424(157a), *483*
Soltani, H. 432(185), *484*
Somanathan, R. 806(264), *837*
Somassari, E. S. R. 1100(34), *1106*
Somlai, L. 607(4), *656*
Sommer, J. 613, 645(21), *656*
Sommerdijk, N. A. J. M. 1000(469), *1013*
Sonar, S. M. 477(327a), *489*
Soncini, P. 683(184), *708*, 1424(397b), *1451*
Sone, T. 419(123a), *482*, 633(94), *658*, 1022(54, 55), *1088*, 1374(31a, 31b), 1380(58g, 59, 63), *1436–1438*
Song, L.-D. 1234(153a), 1237(161a), *1341, 1342*
Song, W. 171(762), *197*
Song, W.-L. 987(401), *1012*
Song, Z. J. 670(78), *706*
Sonnenberg, F. M. 460(282), *487*
Sonnenschein, C. 925(48), *1004*, 1349(33), 1352(58), 1360(33), *1363, 1364*
Sonntag, C. von 143, 144(550), *192*, 1086(375), *1096*, 1100(32), *1106*, 1114(90), 1116(113), 1124(243), 1129, 1134(289), 1136(360, 362), *1145, 1146, 1148, 1149, 1151*
Sonoyama, J. 1192(75b), *1339*
Sontag, G. 953, 961(244), *1008*
Sopher, D. W. 1168(40), *1338*
Soria, J. 1083(296), *1094*
Sorokin, A. 1212(111), *1340*
Sorrell, T. N. 1378(50b), 1423(380a), 1431(50b), *1437, 1450*
Sosnovsky, G. 414(92), *481*
Sosnowski, A. 1110(53), 1112(53, 77), *1145*
Soto, A. M. 925(48), *1004*, 1352(58), 1360(172), *1364, 1367*
Soto, A. M. H. 1349, 1360(33), *1363*
Soukhomlinov, B. P. 987(407), *1012*
Souley, B. 1372(24a), *1436*
Soundararajan, N. 665(41), *705*
Souquet, J. M. 949(183), *1007*
Sousa, M. H. F. A. 229(32), *255*
Sousa Lopes, M. S. 170(743), *196*
Southwell-Keely, P. T. 880(200), *906*
Sowmya, S. 675(115), *707*
Sozzi, G. 300(122), *328*
Spacek, W. 1081(269), *1093*
Spadoni, M. 699(307), *711*
Spagnol, M. 617(35), *656*
Spandet-Larsen, J. 368, 373(196), *390*
Spange, S. 542(48), *597*

Spangenberg, D. 145(622, 623), 147, 148(697, 702), 149(697), 156, 160(702), *193, 195*, 581(216, 221, 222), *601*, 992(428), *1012*
Spanget-Larsen, J. 129, 132, 133(498), *191*, 368(199, 204), 371(204), 373(238), 380(274), 381, 382(199), 383(238, 282), 384(238), *390–393*
Sparapani, C. 274(41), 307(155), *326, 329*, 1104(60), *1106*
Spatz, M. 874(174), *905*
Speers, L. 425(158), *483*
Speirs, M. 67(252), 110, 112(443), *185, 189*
Spek, A. L. 93(364), *187*, 1390(106), 1405(249b, 254b), 1422(378), 1424(383), *1440, 1445, 1446, 1450*
Spencer, D. J. E. 1404(245), *1445*
Spencer, J. N. 47, 48, 53(222), *184*
Spengler, B. 323(262), *331*
Spenser, N. 761(117), *833*
Spera, S. 1386(86a, 86b), 1422(379), 1424(392, 393), *1439, 1450*
Speranza, M. 274(41), 307(155, 156), 309(167, 168), *326, 329*, 1104(60), *1106*
Spikes, J. D. 15(73), *181*
Spinelli, D. 301(135, 136), *328*
Spinks, J. W. T. 1098(1), *1105*
Spirina, S. N. 748(90), *833*
Špirko, V. 144(691), *195*
Spiro, T. G. 94, 105(365), 129, 133(496), *187, 191*, 997(450), *1013*
Spivack, J. D. 1495(147), *1505*
Spoettl, R. 1118, 1130(143), *1146*
Spohn, U. 979(355), *1011*
Spoliti, M. 375(241), *392*
Spoor, P. 677(125), *707*
Sprecher, M. 806, 814(254), *836*
Sprechter, M. 471(302), *488*
Spreingling, G. R. 1462(29), *1503*
Springer, J. P. 1159, 1207(27), *1338*
Springs, S. L. 1020(34), *1087*
Sprous, D. 5(13), *179*
Spyroudis, S. P. 1075(232a–c,233), *1092*
Squadrito, G. L. 637(107, 108), *658*
Squire, R. H. 419(134c), *482*
Sraïdi, K. 531, 532(18), 540(40, 41), 541, 542(41), 547(18), 559, 572(157), 575(187), 586, 587, 589(18), *597, 600*
Sridar, V. 774(184), *835*
Srinivas, R. 293(77), 294(78), *327*
Srinivasa, R. 1074(228, 229), *1092*
Srivastava, R. G. 1283(232), *1343*
Sroog, C. E. 1494(128), *1505*
Stach, J. 319(233, 234), *331*
Stack, T. D. P. 1196(88), *1340*
Stadelhofer, J. W. 19(95), *182*
Staden, J. F. van 986(393), *1012*
Stadler, R. 1216(124), *1341*
Staeger, M. A. 450(235), *486*

Staffilani, M. 1414(308), *1448*
Stafford, U. 1085(351, 352), *1095*
Stahl, K. 147–149(699), *195*
Stahl, W. 581(214), *601*
Stainly, S. J. 1356(121), *1365*
Stalikas, C. D. 986(398), *1012*
Stallings, M. D. 1118(151), *1147*
Stan, H. J. 319(224), *330*, 934(110), 936(117), 937(110, 117), *1005*, 1353(95), *1365*
Stanbury, D. M. 1140(376), *1151*
Standen, S. 413(84), *481*
Stanforth, S. P. 681(163), *708*
Stangret, J. 170(741), *196*
Stangroom, S. J. 1356(120), *1365*
Stanimirovic, D. B. 874(174), *905*
Staninets, V. I. 144(680), *195*
Stankovic, J. 813(297), *837*
Stanley, G. G. 1401(220c), *1444*
Stanley, R. J. 143(590), 147(590, 708), 148, 149, 168(590), *193, 195*
Staples, C. A. 1351(54), *1364*
Starczedska, H. 227, 248, 249, 251(14), 252(14, 94), *254, 257*
Starikova, Z. 640(132), *659*
Starkey, L. S. 668(54), *706*
Starnes, S. D. 1430(441a), *1453*
Staudinger, H. 13(67), *181*
Stavber, S. 655(190), *660*, 687(219), *709*
Stavrakis, J. 444(215c), *485*
Steadman, J. 143(589), *193*, 296(87), 298(88), *327*, 493(30), *525*
Steckhan, E. 404(34), *479*
Steed, J. W. 1414(307a, 307b, 308), *1447, 1448*
Steel, J. P. 434(194), *484*
Steele, W. V. 226, 227(7), *254*
Steenken, S. 845(25), 868(156, 157), 870(157), 871(163, 165), 874(170), 880, 881(25), 898(163), 899(170), *902, 905*, 1108(9, 10), 1109(26, 46–48), 1110(26), 1111(69, 70, 73, 75, 76), 1112(26, 73), 1113(48, 89), 1114(47, 89), 1115(47, 48, 94, 103–107, 112), 1116(94), 1120(47, 94), 1122(47, 48), 1125(47), 1128(47, 76), 1129, 1130(76), 1134(94), 1137(70, 75), 1138(75, 76), 1139(76), 1140(75, 76), 1141, 1142(76), *1144–1146*
Steenvorden, R. J. J. M. 325(276), *332*
Stefaniak, L. 353(101), 359(122, 123), 361(122), *388, 389*
Stefanic, M. 1471(72), *1504*
Stefanov, B. B. 4, 100(2), *179*, 368, 372, 373(200), *390*
Stefinovic, M. 677(127), *707*, 765(140), *834*
Steglinska, V. 748(92), *833*
Stegmann, H. B. 1109(29, 30), 1120(177), 1122(192, 193), *1144, 1147, 1148*
Stehl, R. H. 1360(162), *1367*

Stein, G. 1113(82), *1145*
Stein, M. 1026(90), *1089*
Stein, S. E. 224(3), *254*
Steiner, T. 178(770), *197*, 551(124), 579(196, 197), *599, 601*
Steinfelder, K. 310(175), *329*
Steinwender, E. 343(41), 350(84–86), *387, 388*, 551(86, 91, 92), 553(86), *598*
Steirert, J. G. 1086(374), *1096*
Stekhova, S. A. 340, 341(26), *387*
Steliou, K. 1314(288, 289), *1345*
Stellman, J. M. 12(60b), *181*
Stellman, S. D. 12(60b), *181*
Stemmler, E. A. 310(179), *329*
Stensby, D. 1359(156), *1366*
Stenzenberger, H. D. 1493(122), 1494(129), 1495(122), *1505*
Stepanova, L. 1359, 1360(160), *1367*
Stephan, H. 1378(48b), *1437*
Stephan, J. S. 1025(81), *1089*
Stephen, J. f. 429(176), *483*
Stephenson, L. M. 95, 109(369), *188*
Steppan, H. 801(247), *836*
Steren, C. A. 1057(172), *1091*
Sternhell, S. 360(135), *389*
Sternson, L. A. 419(123c), *482*
Stetter, H. 429(175), *483*
Stettmaier, K. 856(121), *904*, 1137(368), *1151*
Stevens, E. D. 1424(388a, 390a), *1450*
Stevens, M. F. G. 1231(150), *1341*
Stewart, D. R. 1371(8), 1372(8, 10a, 12, 27, 29), 1385(27), 1390, 1392(107), *1435, 1436, 1440*
Stewart, J. J. P. 4(3), *179*
Stewart, R. 501, 502(81), *526*, 1285(235), *1343*
Stewart, R. F. 1158(18), *1338*
Stewart, W. 1297(260c), *1344*
Steyer, S. 1435(462a), *1454*
Stibor, I. 1403(241), 1415(315–317, 318b), 1432(451), *1445, 1448, 1453*
Stich, H. F. 1357, 1358(138), *1366*
Sticha, M. 691(239), *710*
Stikvoort, W. M. G. 1405(249b), *1445*
Stillwell, R. N. 294(79), 312(194), *327, 330*
Stipanovitch, R. D. 345, 349(46), *387*
Stirling, C. J. M. 1383(67j), 1421(67j, 370, 371), 1435(371), *1438, 1450*
Stivala, S. S. 1108(4), *1144*
Stockdale, M. 1359(158), *1367*
Stocker, R. 876(185), 892(185, 252, 254), *906, 907*
Stoddart, J. F. 761(116–118), *833*
Stohmann, F. 236, 237(58), *256*
Stohmann, G. 226, 232(8), *254*
Stolz, B. M. 463(276a), *487*
Stone, E. W. 1123, 1124(202), *1148*
Stone, J. A. 275, 277(43), *326*

Stone, L. 891(241), *907*
Stone, T. J. 129, 135(488), *190*, 1109(36, 37), 1120(36, 37, 171, 172), 1124(226, 227), *1144, 1147, 1148*
Storer, J. W. 57(242), *185*
Storey, K. B. 854(97), *904*
Stork, K. C. 735, 748(51), *832*
Storozhok, N. M. 891(243), *907*
Stottmeister, U. 976(327), *1010*
Stout, T. J. 1466(45), *1503*
Stoytchev, Ts. S. 887(225), *907*
St. Pierre, M. J. 1021(43, 44), *1088*
Straccia, A. M. 138(528), *191*, 1104(61), *1106*
Strachan, E. 129, 131(500), *191*, 1127(271), *1149*
Strachan, M. G. 314(205–207), *330*
Stradins, J. 1354(112), *1365*
Strain, M. C. 4, 100(2), *179*, 368, 372, 373(200), *390*
Stranadko, T. N. 962(252), *1008*
Strangeland, L. J. 558(147), *600*
Strasburg, G. M. 895(264), *908*
Strashnikova, N. 493, 506, 509, 512–514, 520, 521(51), *526*
Stratmann, R. E. 4, 100(2), *179*, 368, 372, 373(200), *390*
Straub, T. 1420(365b), *1450*
Strauss, C. R. 762(127), *834*
Streffer, K. 977(344), *1011*
Strehlitz, B. 976(327), *1010*
Strein, T. G. 951, 953(191), *1007*
Streitwiesser, A. Jr. 261(11), *326*
Stremple, P. 935, 937(111), *1005*
Strickland, T. 886, 887(216), *906*
Strigun, L. M. 1118, 1119(140), *1146*
Strobel, F. 294(80), *327*
Strohbusch, F. 361(145), *389*
Strongin, R. M. 1383(67f), 1421(67f, 368), *1438, 1450*
Struchkov, Y. T. 432(186), *484*, 1290, 1292(247), *1344*
Struchkov, Yu. T. 200(7), *220*, 824(331), *838*
Struck, O. 1391, 1396(118a), 1405(253, 254b), 1427(410b), *1440, 1446, 1451*
Struijk, H. 1396(182a), *1443*
Strukul, G. 701(318), *712*
Strupat, K. 305(147), *329*
Stubbe, J. 129(481), *190*
Stubbe, J. A. 1108, 1117, 1120(14), *1144*
Studer, S. L. 1022(50), *1088*
Studer-Martinez, S. L. 730(47), *832*
Studzinskii, O. P. 1076(235), *1092*
Stull, D. R. 238–240(66), *256*
Stymme, B. 558–560(150), *600*
Stymme, H. 558–560(150), *600*
Styring, S. 1108(16), *1144*
Su, S. H. 1360(167), *1367*
Su, Y. 975(321), *1010*

Suammons, R. E. 1278(221), *1343*
Suarna, C. 880(200), *906*
Suau, R. 477(325), *489*
Subbarao, K. V. 1136(359), *1151*
Subba Rao, Y. V. 680(152), *708*
Subotkowski, W. 419(134d), *482*
Subra, R. 132, 133, 136(518), *191*, 1127(267), *1149*
Subrahmanyam, M. 630(70), 631(73), *657*
Subramanian, R. 682(167), *708*
Subramanian, R. S. 663(23, 26), *705*
Subramanian, S. 530(15), *597*, 620(42), *657*, 749(100), *833*
Subramanyam, M. 680(152), *708*
Suckling, C. J. 633(93), *658*
Sudalai, A. 616(27), *656*, 1202, 1205(94), *1340*
Sudha, M. S. 301(132–134), *328*
Sueda, T. 1233(151a), *1341*
Sueishi, Y. 992(432), *1013*
Suenaga, M. 1380(58b), *1437*
Suesal, C. 1359(150), *1366*
Sugasawa, T. 632(86), *658*, 686(209, 214), *709*
Sugawa, Y. 1380(61a), *1438*
Sugawara, A. 1415(319), *1448*
Sugawara, S. 1489(108, 110), 1491(110), *1504*
Sugie, S. 1361(187), *1367*
Sugita, J. 1302(267a, 267b, 268), *1344*
Sugitani, Y. 662(12), *705*
Sugiura, K. 219(36), *221*
Sugumaran, M. 748(87, 88), *833*
Suguna, H. 1063(192), 1064(193), *1091*
Suh, Y. 1423(381), *1450*
Sujatha, E. S. 628(64), *657*
Sujeeth, P. K. 1392(132), 1401(220c), *1441, 1444*
Suksamrarn, A. 1297(258), *1344*
Sukumar, N. 82(283), *186*
Sulck, W. 558(149), *600*
Sullivan, P. D. 1123, 1124(213), 1126(262), *1148, 1149*
Sul'timova, N. B. 1016(3), *1087*
Sumarno, M. 880(200), *906*
Sumelius, S. 1290(250), *1344*
Sumi, S. 662(6), *705*
Summanen, D. 953, 968(273), *1009*
Summanen, J. 973(312), *1010*
Sumpter, J. P. 925(49, 53), *1004*, 1349, 1360(34), *1363*
Sun, G. 925(47), *1004*
Sun, H. 365(158), *389*, 1082(273), *1093*
Sun, J. J. 1352(66), *1364*
Sun, Q. 1115(111), 1129(290), 1132, 1133(335), 1134(290), 1139, 1142, 1143(371), *1146, 1150, 1151*
Sun, Y.-M. 982(372), *1011*
Sund, B. C. 668(55), *706*
Sund, R. B. 856(123), *904*

Sundaram, K. M. S. 948, 953(182), *1007*
Sundaresan, M. 237(62), *256*
Sundhom, E. G. 336(3), *386*
Sundius, T. 20, 22, 23, 35, 36, 38(111), *182*, 368, 370–375(198), *390*
Sunesen, L. 127(469), *190*, 718(17), *831*
Sung, K. S. 312(193), *330*
Sunjic, V. 1470(50, 52), *1503*
Sur, A. 143(583), *193*
Surducan, E. 662(9), *705*
Surducan, V. 662(9), *705*
Suren, J. 1457(4), *1502*
Surpateanu, G. 992(433), *1013*
Sushchik, N. N. 1362(218), *1368*
Suslov, S. G. 985(389), *1012*
Sussman, S. 408(52b), *479*
Sußmuth, W. 925, 936(51), *1004*
Susuki, E. 413(78a), *480*
Suter, H. U. 93, 94, 105, 132(353), *187*
Sutherland, I. O. 806(264), *837*
Sutton, K. H. 913(11, 12), *1003*
Suyama, K. 1158(17c), 1180(66), *1338, 1339*
Suzdalev, K. F. 823(327), *838*
Suzukamo, G. 1335(331), *1346*
Suzuki, H. 636(105), *658*
Suzuki, I. 144(599), *193*
Suzuki, J. 1071(209a, 209b), *1092*
Suzuki, K. 447(223), *485*, 641(137), *659*, 674(99), 679(141–143), 680(145), 699(301), *707, 708, 711*, 776(201, 202), *835*
Suzuki, M. 770(158), 773(158, 159), *834*, 1029(101), *1089*, 1315(293), 1330(316, 317), *1345, 1346*
Suzuki, S. 47, 143(205), 144(594), *184, 193*, 493, 503, 519, 520(34), *525*, 1071(209a, 209b), *1092*
Suzuki, T. 143, 147(580), *192*, 669(63), 695(273, 279), *706, 710, 711*, 913(29), *1003*, 1315(295b, 296, 297), 1317(298), 1318(299), *1345*, 1380(62), 1392(139), 1405(247), *1438, 1441, 1445*
Suzuki, Y. 1318(306), 1324(308), *1345*, 1379(54d), 1400(207b), *1437, 1444*, 1495(137), *1505*
Svobodova, D. 986(400), *1012*
Swager, T. M. 1374(42b), 1396(191a, 191b, 193), *1436, 1443*
Swain, C. J. 441(206), *484*
Swallow, A. J. 1117(123, 125, 128), 1130(123, 125, 325), 1134(325), *1146, 1150*
Swami, K. 1348(19), *1363*
Swanson, E. 451(252), *486*
Swanson, S. E. 1352(60), *1364*
Swarts, F. 237, 238(65), *256*
Swenton, J. S. 1170(45), 1171(47), 1180, 1183, 1186(62), 1227(147), 1228, 1234(45), 1240(167), *1338, 1339, 1341, 1342*
Swenton, S. 1175(54a, 54b), 1227(54a), *1338, 1339*
Swetlitchnyi, V. A. 1017(8), *1087*
Swientoslawski, W. 227, 248, 249(14), 251(14, 91), 252(14, 94), 253(91), *254, 257*
Syage, J. A. 143(589), 147(719), *193, 196*, 296(87), 298(88), *327*, 493(23, 30), *525*
Syamala, M. S. 477(324a), *488*
Sykes, B. D. 346, 365(52), *387*
Sykora, J. 1082(270–272), *1093*, 1415(317), *1448*
Symonds, M. 859(137), *905*
Symons, M. C. R. 1103(52, 53), 1104(55), *1106*
Syper, L. 422(147a), 424(152c), *483*, 608(12), *656*
Syrjänen, K. 1216(126a, 126b), *1341*
Sytnyk, W. 1124(229), *1148*
Syvret, R. G. 687(216), *709*
Szady-Chelmieniecka, A. 1024(63), *1088*
Szafran, M. 143(564–566), 146(638), *192, 194*, 364(156), *389*, 580(205), 593(269, 281, 282), 594(289), 595(281, 282, 289), *601–603*
Szalontai, G. 1388(98), 1394, 1407(170b), *1439, 1442*
Szantay, C. 759(115), *833*
Szczodrowska, B. 353(96), *388*, 551(127), *599*
Szczubialka, K. 1086(377), *1096*
Szegstay, M. 1462(23), *1502*
Szele, I. 402(26a, 26b), *479*
Széll, T. 472(310b, 310c), *488*
Szemes, F. 1393, 1403(152c), 1432(453), *1441, 1453*
Szemik-Hojniak, A. 144(658), *194*
Szewczynska, M. 976(325), *1010*
Szmant, H. H. 15, 100(80), *181*

Taagepera, M. 101(388), *188*
Tabakovic, I. 1219(131), *1341*
Tabassum, N. 1129(282), 1133(282, 340), 1141(282), *1149, 1150*
Tabata, M. 170(740), *196*, 1352(57), *1364*
Tabatabai, M. 1374(41, 42a, 43), 1385(83), 1393(157b), *1436, 1437, 1439, 1442*
Tabatchikova, N. V. 977(342), *1010*
Tabet, J. C. 316(218–221), 322(249), *330, 331*
Tabul, A. 448(229a, 229b), *485*
Tachi, Y. 1209(106, 107), *1340*
Tada, M. 1180(63a, 63b), 1192(75a, 75b), 1202, 1205(95), *1339, 1340*
Taddei, H. 418(119), *482*
Tadros, S. P. 1482(88), *1504*
Tafi, A. 1383(75a, 75b, 75d, 75e), 1422(75e), *1439*

Taft, R. W. 5(19), 92(351), 99, 100(351, 377), 101(351, 388), 103(377), 109(413), *180, 187–189*,263, 296(16), 311(186), 312(193), *326, 329, 330,* 503(87–89), *526,* 531(20, 21), 534(23), 537, 539(38), 541(45), 543(49, 50), 551, 553(87), 559(38, 153), 564(170), 575(187), 586(20, 23, 153, 245), 587(23, 245), 591, 592(265), 593(20, 153), *597, 598, 600, 602,*877(187), 887(219), *906, 907*
Taga, T. 11(54), *180,* 232(36), *255*
Tagiev, D. B. 612(20), *656*
Taguchi, H. 472(310a), *488*
Taguchi, T. 460(288), *487,* 624(58), *657,* 769(146), *834*
Tai, Z. 365(158), *389*
Taing, M. 450(239j), *486*
Tairov, M. O. 1394(162), *1442*
Tait, B. 851, 852, 894(67), *903*
Tait, J. M. S. 274(39), *326*
Tajima, K. 865, 896(148), *905*
Tajima, M. 1076(234), *1092*
Tajima, S. 291(63, 64), 300(125), *327, 328*
Takabayashi, Y. 1360(171), *1367*
Takada, H. 1356(125), *1366*
Takada, T. 1246(172, 173), 1247(173), *1342*
Takahara, K. 774(191), *835*
Takahashi, A. 1066(196), *1091*
Takahashi, H. 1233(151a), *1341,* 1495(158), *1505*
Takahashi, K. 345(49), *387,* 1403(233b), *1445*
Takahashi, M. 110, 112, 121(444), *189,* 851, 852(71), 855(113), 856, 863, 864(126), 887(218), 890(236), 892(71), *903, 904, 907,* 1164, 1290(36), *1338*
Takahashi, S. 291(64), *327*
Takahashi, T. 766(142), *834*
Takahashi, T. T. 813(299), *837*
Takahira, K. 1196(89), *1340*
Takai, N. 953, 962(248), *1008*
Takai, Y. 93(362), *187*
Takaki, H. 1109(31), *1144*
Takaki, K. 414(86b), *481*
Takaki, K. S. 450(239d), *486*
Takakura, H. 1185(68, 69), 1186(71), *1339*
Takami, A. 773(160), *834*
Takamitsu, N. 1470(67), *1504*
Takamuku, T. 170(740), *196*
Takamura, M. 695(269), *710*
Takamura, Y. 1066(197), *1091*
Takanashi, K. 874(172), *905*
Takano, S. 463(276d), *487*
Takaoka, M. 774(194), *835*
Takasuka, M. 551(133), *599*
Takaya, Y. 1079(240), *1092*
Takayanagi, H. 1192(79, 81), *1339*
Takayanagi, M. 1318(303), *1345*
Takayanagi, R. 874(171), *905*
Takayanagi, T. 974(318), *1010*
Takazawa, K. 37, 46, 110, 111(182), *183*
Takeda, K. 1352(57), *1364*
Takeda, S. 367(171), *390,* 770(158), 773(158, 159), *834*
Takeda, Y. 448(226), *485*
Takehira, K. 1213(114, 117), *1340*
Takekoshi, T. 1495(154), *1505*
Takemura, H. 353(99), *388,* 1380(58a–e), *1437*
Takenaka, N. E. 826(338), *838*
Takenaka, S. 414(90k), *481*
Takesako, K. 1163(33a), *1338*
Takeshita, H. 1411(299), *1447*
Takeshita, M. 1392(139), 1403(231a), *1441, 1445*
Takeshita, R. 938(128), *1006*
Takeuchi, H. 614(25), *656*
Takeuchi, M. 697(284), *711*
Takeuchi, T. 1262(195), *1343,* 1417(333), *1448*
Takeya, H. 1380(61b, 62), *1438*
Takhi, M. 675(113), *707*
Takhistov, V. V. 301(131), *328*
Takiyama, M. M. K. 1362(212), *1368*
Talanov, V. S. 1392(126), *1441*
Talaska, G. 1361(188, 189), *1367*
Talipov, S. A. 204(10), *220*
Tamelen, E. E. van 448(232b), *486*
Tamminen, T. 998(451), *1013*
Tamura, H. 774(191), *835*
Tamura, K. 856, 863, 864(126), *904*
Tamura, T. 1419(345), *1450*
Tamura, Y. 447(221b, 222a, 222c–e), 448(227), *485,* 1227, 1228(144), 1230(148), *1341*
Tamura, Z. 992(430), *1013*
Tan, S. N. 975(323), *1010*
Tanabe, K. 177(768), *197,* 1022(54, 55), *1088*
Tanabe, S. 147(703, 705, 706), 148(705), 149(703, 705), 150, 157, 167, 168(705), *195*
Tanaka, H. 1066(197), *1091,* 1495(158, 159), 1498(159), *1505*
Tanaka, J. 219(36), *221*
Tanaka, K. 143, 179(546), *192,* 323(256), *331,* 1289(242), *1344,* 1380(56), *1437*
Tanaka, M. 972(303), *1010*
Tanaka, N. 1071(206), *1092*
Tanaka, S. 1086(379), 1087(381), *1096*
Tanaka, T. 409(58), *480,* 1361(187), *1367*
Tang, F. 1393(152b), *1441*
Tang, H. X. 1084(306), *1094*
Tang, W. Z. 1086(358, 364), *1096*
Tang, X.-S. 1108(16), *1144*
Tangermann, U. 145(634), *194*
Tanigaki, T. 865(149), *905*
Tanigaki, Y. 871, 874(160), *905*

Taniguchi, H. 414(86a), *481*, 662(11), *705*, 1054(162), *1091*, 1402, 1409(223), *1444*
Taniguchi, M. 447(221a, 221b), *485*
Taniguchi, S. 874(171), *905*
Tanikaga, R. 419(132), *482*
Tanimoto, N. 551, 554(110), *599*
Tanimoto, Y. 692(247), *710*
Tanimura, R. 853, 855, 856(90), 890(232), *904, 907*
Tannenbaum, H. P. 293(72), 314(203), *327, 330*
Tanzcos, I. 974(316), *1010*
Tao, X.-L. 1178, 1337(57a, 57b), *1339*
Taphanel, M. H. 300(123), *328*
Tapia, R. 813(301, 303), *837, 838*
Tappel, A. L. 852, 856(85), 890(229), *903, 907*
Tapramaz, R. 1104(56), *1106*
Tarakeshwar, P. 161(737), *196*
Taraschi, T. F. 888(226), *907*
Tarbell, D. S. 459(268), 472(311g), *487, 488*
Tarbin, J. A. 1273(206), *1343*
Tascón, M. L. 982(366), *1011*
Tashiro, M. 1290(245), *1344*, 1377(45a), *1437*
Tashiro, Y. 1024(58), *1088*
Taskinen, E. 237(61), *256*
Tata, M. 365(164), *389*
Tatarkovicova, V. 1354(105), *1365*
Tatchell, A. R. 626(61), *657*
Tateiwa, J. 265, 272(29b), *326*
Tatematsu, A. 275(44), 299(92), *326, 327*
Tatikolov, A. S. 1130(319), *1150*
Taton, D. 1388(100a), *1439*
Tatsukawa, R. 1351(46), *1364*
Tatsuma, T. 1362(211), *1368*
Tatsumi, K. 1086(376), *1096*, 1362(206), *1368*
Tatsumi, T. 414(90f), *481*
Tatsuta, K. 680(146), *708*
Taub, I. A. 1113(83), *1145*
Taurins, A. 420(139), *482*
Tauszik, G. 178(771), *197*
Tavani, C. 1073(225), *1092*
Taveras, A. G. 470(300d), *488*
Taverna, M. 19(99, 100), *182*
Tayima, Y. 1348(27), *1363*
Taylor, A. 78(278), 147, 148(710), *186, 195*, 375(242), *392*
Taylor, D. R. 807(271), *837*
Taylor, E. C. 416(108, 109a, 110b, 111), *481, 482*, 1312(284a, 284b), 1314(284a, 284b, 288, 289), 1315(292), *1345*
Taylor, G. A. 1123(212), *1148*
Taylor, J. K. 1227(146), *1341*
Taylor, J. R. N. 913(14), *1003*
Taylor, M. 665(44), *705*
Taylor, N. J. 795(234), *836*, 1408(282), *1447*
Taylor, P. 367(180), *390*
Taylor, P. J. 503(89), *526*, 533(22), 535(33), 536(33, 36, 37), 537(33), 539, 540(37),
557(33, 145), 560, 564(145), 588(33, 37, 257), 590(145), 591(37), *597, 599, 602*, 876–878(186), 879(196), 882(186, 196), *906*
Taylor, R. 535(31), *597*
Taylor, R. J. 666(45), *705*
Taylor, W. C. 807(279–281), *837*
Taylor, W. I. 1158, 1175(14a), *1337*
Tazaki, M. 677(134, 135), *707*
Tchaikovskaya, O. N. 1016(3), 1017(8), *1087*
Tchapla, A. 913(5), *1003*
Tchir, W. J. 1484, 1485(91), *1504*
Teat, S. J. 691(241), *710*
Tebben, A. 1180, 1183, 1186(62), *1339*
Tedder, J. M. 307(152), *329*
Tee, O. S. 38, 127(185), *184*, 716(11), *831*
Teitzel, H. 933, 937(101), *1005*
Teixeira, J. A. S. 227, 248(15), 252(94), *254, 257*
Teixeira-Dias, J. J. C. 561(164, 165), *600*
Tellgren, R. 144(664), *194*
Telliard, W. A. 1348(14), *1363*
Temple, R. G. 19(96, 98), *182*
Teng, C.-M. 851, 853, 857(61), 895(268), *903, 908*
Teng, H. 944(145), *1006*
Tenon, J. A. 727(37, 38), *831*
Tepper, A. W. J. W. 977(344), *1011*
Tepper, D. 493, 506(46), *526*
Terabe, S. 972(307), *1010*
Terada, M. 695(275), *711*
Terada, Y. 1158(19–21, 23, 24), 1159(24, 25a, 25b), 1172(25a, 25b), 1178(21), 1324(309), *1338, 1345*
Terao, J. 891(247), 894(263), *907, 908*
Terao, S. 624(56), *657*
Terao, Y. 1357(137), *1366*
Terashita, Z. 624(56), *657*
Terauchi, S. 874, 874(160), *905*
Terenghi, G. 685(200), *709*
Terenghi, M. G. 626(59), *657*
Terent'ev, P. B. 301(130), *328*
Tereshchenko, L. Ya. 1081(263, 264), *1093*
Terlouw, J. K. 110(451), *190*
Ternes, W. 739(59), *832*
Terpager, F. 373, 383, 384(238), *392*
Terriere, G. 1489(109, 115), *1504*
Terry, K. A. 930, 937(80), *1004*
Tersch, R. L. v. 57(243), *185*
Tersch, R. L. von 99, 101(378), *188*
Terzian, P. 1361(195, 196), *1367*
Terzian, R. 1084(313, 318), *1095*, 1100(26), *1105*, 1129(284, 286), *1149*
Tesfai, Z. 1177(56), *1339*
Testa, G. 1079(242), *1092*
Tetenbaum, M. T. 425(161), *483*
Teuber, H.-J. 1257, 1260(187a–c), *1342*
Tevesz, M. J. S. 314, 315(208), *330*

Texier, H. 928(61), 929, 954(62), *1004*
Texier, I. 1073(216), 1085(336, 353), 1086(336, 369), *1092, 1095, 1096*
Tezuka, T. 407(51), *479*
Thalladi, V. R. 219(35), *221*
Theilacker, W. 436(196), *484*
Theissen, H. 335(2), *386*
Theissling, C. B. 110(427), *189*, 299(99), *328*
Theobald, N. 1356(123), *1366*
Theriault, R. J. 412(74c), *480*
Thiebaud, S. 663(19), *705*
Thiel, W. 1025(84), *1089*
Thiele, J. 715(4), *830*
Thielking, G. 283(52), *327*
Thiéry, V. 680(150), *708*
Thigpen, K. 607(4), *656*
Thirring, K. 1244(171), *1342*
Thistlethwaite, P. J. 493(40), *525*
Thoden van Velzen, E. U. 1421(369), *1450*
Thoennessen, H. 1417(329a), *1448*
Thoer, A. 685(197), *709*
Thom, R. 68, 70(260), *185*
Thomas, B. 1482(88), *1504*
Thomas, C. B. 265, 319(28), *326*
Thomas, E. W. 1130(306), *1150*
Thomas, J. D. 19(101), *182*
Thomas, J. P. 855(107), *904*
Thomas, J. R. 15(79), *181*
Thomas, M. 241(78), *256*, 1330, 1332, 1333(326), *1346*
Thomas, M. J. 845(37), *902*, 1130(305), *1150*
Thomas, O. 985(385, 386), *1012*
Thomas, P. S. 19(101), *182*
Thomas, S. R. 102(394), *188*, 892(254), *907*
Thomas, T. D. 102(394), *188*
Thomas-Barberán, F. A. 973(309), *1010*
Thompson, A. 1136(358, 359), *1151*
Thompson, H. W. 170(742, 743), 171, 174(742), *196*
Thomsen, C. 931, 937(83), *1005*
Thomsen, M. 339(18), *386*
Thomson, B. 1461(16), *1502*
Thomson, R. H. 715, 718, 720(1b), *830*
Thondorf, I. 1374(34, 39), 1378(50a, 51), 1385(85), 1386(51, 85, 87), 1402, 1404, 1406(222), 1411(295, 298), 1419(358), 1420(359c, 361, 364), *1436, 1437, 1439, 1444, 1447, 1450*
Thönnessen, H. 1383, 1421(67k), 1424(395), *1438, 1451*
Thormann, Th. 375, 376(249), *392*
Thorpe, F. G. 416(112), *482*
Thozet, A. 200(1), 219(37), *220, 221*, 549(66), *598*
Thruston, A. D. Jr. 1354, 1355(100), *1365*
Thuéry, P. 1380(58c), 1415(313), 1435(465b), *1437, 1448, 1454*

Thulstrup, E. W. 368, 373(190, 196), 374(190), 383(282), *390, 393*
Thulstrup, P. W. 375, 376(248), *392*
Thurkauf, A. 429(173), *483*
Thurman, C. 1348(26), *1363*
Thurman, E. M. 1353(79), *1365*
Tiainen, E. 998(451), *1013*
Tian, B. M. 971(301), *1010*
Tian, Y. 697(283), *711*
Tiedemann, P. W. 83(291), *186*
Tieze, L.-F. 674(107), *707*
Tikannen, L. 1360(185), *1367*
Tikkanen, L. 1360(182), *1367*
Tillett, J. G. 299(91), *327*
Timmerman, P. 1370(4), 1391, 1393(110), 1403(242), 1406(262b, 263b, 263d), 1417(4), 1428(263d, 437b), 1432(454), 1433(456a), *1435, 1440, 1445, 1446, 1453*
Timmermann, B. 305(151), *329*
Ting, Y. 1401(220a), *1444*
Tinucci, L. 683(180), 686(210), *708, 709*, 1082(281), *1094*
Tipikin, D. S. 1018(23), *1087*
Tischenko, D. V. 396(1), *478*
Tishchenko, O. 47, 48, 56(220), 58(244), 83–86, 88(299), 89(220, 299, 317b), 90, 91(299), 143(551), 144(680), *184–186, 192, 195*
Tishchenko, V. 18(90), *181*
Tismaneanu, R. 619(40), *656*
Tisnes, P. 400(18b), *478*
Tissot, A. 1073(217), *1092*
Tisue, G. T. 801, 802(245), *836*
Tite, E. L. 1383, 1421(67a, 67b), *1438*
Titov, A. I. 638(120), *658*
Tius, M. A. 429(173), *483*
Tjessem, K. 170, 171(757), *197*
Tobe, Y. 1048(158, 159), *1090, 1091*
Tobiason, F. L. 1462(35, 36), *1503*
Tobinaga, S. 1216, 1283, 1287(122c), 1288(239, 240), 1289(240), 1297(122c), *1340, 1344*
Tobita, S. 291(63), *327*, 493(36, 37), *525*
Tochtermann, W. 826(339), *838*
Toda, F. 1289(242), *1344*
Todd, M. H. 459(269), *487*
Todorovic, M. 931, 935, 937(85), *1005*
Toffcl, P. 1109(46), *1144*
Togashi, H. 432(182a, 183c), *484*
Tohma, H. 1178(58), 1230(148), 1246(172, 173), 1247(173), *1339, 1341, 1342*
Tohma, S. 684(193), *709*
Toke, L. 1387, 1391(93), *1439*
Tokhadze, K. 143(557), *192*
Tokita, S. 1396(184), *1443*
Tokitoh, N. 1399(201a), *1443*
Tokiwa, H. 92(347), *187*
Tokuda, Y. 410(68), *480*

Tokumaru, K. 174, 177(766), *197*
Tokumaru, S. 855(104), *904*
Tokumura, K. 1030(104), *1089*
Tokunaga, A. 851, 868, 870(51), *903*
Tokunaga, T. 695(279), *711*
Tolbert, L. M. 492(8, 9, 17), 493(8, 9, 17, 29, 47, 48, 50), 498(9, 47), 511(47), 522(8, 9, 17), *525, 526*
Tolbert, P. E. 1359(155), *1366*
Toledo, M. A. 650(162), *659*
Tolkunov, S. V. 433(187a), *484*
Tolley, M. S. 761(117), *833*
Tollin, G. 1134(349), *1151*
Tolman, W. B. 1404(245), *1445*
Tomás, C. 1001(472), *1013*
Tomas, J. 813(298), *837*
Tomasallo, C. 12(60b), *181*
Tomás-Barberán, F. A. 973, 981(313), *1010*
Tomasi, J. 4, 100(2), *179*, 368, 372, 373(200), *390*
Tomassetti, M. 975(321), 977(334), *1010*
Tomaszewska, M. 1083(305), *1094*
Tomaszewski, J. E. 829(344), *838*
Tomer, K. B. 299(89, 93), *327*
Tominaga, H. 414(90f), *481*
Tomino, I. 692(247), *710*
Tomioka, H. 1059(178, 179), *1091*
Tomioka, K. 1278(220), *1343*
Tomisic, V. 726(35), *831*
Tomita, E. 1115(98), *1145*
Tomita, I. 913(28), *1003*
Tomita, K. 687(222, 223), *709*
Tomiyama, S. 143(546, 547), 179(546), *192*, 871, 874(160), *905*
Tomizawa, G. 647(152), *659*, 687(223–225), *709*
Tomizawa, K. 408(53), *479*
Tomkiewicz, M. 1117(115), 1120(115, 179), *1146, 1147*
Tomlin, C. 1347, 1350–1352(3), *1363*
Tommasi, I. 686(212), *709*
Tommos, C. 1108(16), *1144*
Tomooka, C. S. 450(246g), *486*, 702(328, 329, 331), *712*
Tompkins, J. A. 593, 595, 596(279), *602*
Tonachini, G. 18(83b), 138(526), *181, 191*
Tonegutti, C. A. 957(213), *1007*
Tong, H. 1082(280), *1094*
Tong, W. 1380(55, 58f), *1437*
Tong, Z. 1077(237), *1092*
Tonogai, S. 1470(68, 69), *1504*
Tonogai, Y. 953, 957(219), *1008*
Toppet, S. 583(231), 594–596(290), *601, 603*
Topping, C. M. 691(240), *710*, 782(220), *836*
Topsom, R. D. 100(380), 101(388), *188*, 311(186), *329*
Toribio, L. 953(209, 256), 955(209), 962(256), *1007, 1009*

Torii, S. 1156(3, 4, 6), 1158(3, 4, 6, 13), 1175(3, 4, 6), *1337*
Torijas, M. C. 981(360), *1011*
Tormos, R. 744(71, 72), *832*, 1038(122c), 1039(123), 1040(124–126), 1042(127–129), 1055(169), 1071(204), *1090–1092*
Torok, B. 613, 644(23), *656*
Toro-Labbe, A. 90(343), *187*
Toropin, N. V. 725(28), *831*
Torrado, J. J. 985(388), *1012*
Torrado, S. 985(388), *1012*
Torrecillas, R. 1495(138), *1505*
Torregrosa, J. 1362(207), *1368*
Torrent, M. 455(266, 267), *487*
Torres, G. 477(325), *489*
Torres-Martinez, C. L. 1084(323), *1095*
Torres-Pinedo, R. 1393(140), *1441*
Toscano, M. 89, 90(325), *187*, 898(276), *908*
Toscano, R. A. 551(93), *598*
Toscano, T. 84, 89–91(302), *186*
Toshima, K. 642(139), *659*, 680(146), *708*
Tosic, M. 871(165), 897(274), *905, 908*, 1139, 1142(370), *1151*
Toste, F. D. 460(287), *487*, 624(57), *657*, 665(42, 43), 681(162), *705, 708*, 766, 767(145), *834*
Toth, R. 933(105), 935, 936(113), *1005*
Totten, C. E. 813(297), *837*
Toullec, J. 714, 715(1a), *830*
Townsend, J. D. 730(47), *832*
Towson, J. C. 417(117, 118), *482*
Toyoda, A. 1083(304), *1094*
Toyoda, J. 219(36), *221*
Toyoda, K. 1186(71), 1187(72, 73), *1339*
Toyoda, S. 1489, 1491(110), *1504*
Toyoda, T. 686(209), *709*, 935(115), *1005*
Toyoda, Y. 1206(105), *1340*
Toyonari, M. 1233(151a), *1341*
Trahanovsky, W. S. 1216(119a), *1340*
Trainor, R. W. 762(127), *834*
Trainor, T. M. 315(214), *330*
Tramer, A. 143(588), 144(588, 600, 601), *193*, 493, 512, 513(35), *525*
Tramontini, M. 684(186), *709*
Tran, Q. 319(230), *330*
Tran-Thi, T.-H. 20, 23, 78, 93–96, 101, 106, 107, 109(115), *182*, 493(64–66), 500(78), 502, 507(66), 511(65, 66), 515, 522(66), *526*
Tratnyek, P. G. 1084(316), *1095*
Traven, V. F. 776(200), *835*
Travnikova, T. A. 985(389), *1012*
Trawczynski, J. 608(9), *656*
Tregub, N. G. 778(213), *836*
Trehan, I. R. 774(185, 186), *835*
Treinin, A. 1130(321), *1150*
Trejo, L. M. 546(58), *598*

Treleaven, W. D. 1383, 1421(67f), *1438*
Treloar, P. 983(379), *1011*
Tremblay, J.-F. 19(92), *182*, 1348(24), *1363*
Tremp, J. 1348(10), *1363*
Treutter, D. 913(1), *1002*, 1355(114), *1365*
Trevor, D. J. 83(291), *186*
Trifunac, A. D. 1103(48), *1106*
Trillas, M. 1086(367), *1096*
Trindle, C. 111(453), *190*
Trinoga, M. 868, 870(157), *905*, 1108(10), *1144*
Tripathi, G. N. R. 129(494, 495), 132(512), 133(494, 495, 512), *191*, 1115(109–111), 1120(187), 1123(196, 197), 1128(187, 197), 1129(197, 288, 291), 1130(320), 1132(187, 196, 197, 288, 291, 320), 1133(197), 1134(288), 1135(109, 352), 1136(352), *1146–1151*
Tripathi, H. B. 1047(148), *1090*
Trivedi, B. K. 851, 852, 894(67), *903*
Trivedi, G. K. 1311(283), *1345*
Trivedi, N. R. 650(166), *659*
Trivunac, K. V. 1001(476), *1014*
Troe, J. 1105(64, 65), *1106*
Troganis, A. 998(454), *1013*
Trojanowicz, M. 929(64), 976(325), 977(64), *1004, 1010*, 1100(29), *1106*
Tromelin, A. 472(314), *488*
Trommsdorf, H. P. 371(223), *391*
Trommsdorff, H. P. 371(228), *391*
Tromp, M. N. J. L. 871, 873, 898(169), *905*
Tropper, F. 673(97), *707*
Trost, B. M. 460(287), *487*, 624(57), *657*, 665(42, 43), 681(162), 691(236, 237), *705, 708, 710*, 766, 767(145), *834*
Trotter, B. W. 1318(302), *1345*
Trova, M. P. 450(239e, 246i), *486*
Trtanj, M. I. 931, 935, 937(85), *1005*
Truce, W. E. 467(295a), *487*, 685(194), *709*, 818(321), *838*
Trucks, G. W. 4, 100(2), *179*, 368, 372, 373(200), *390*
Trueblood, K. N. 651(169), *660*
Truesdell, J. W. 429(173), *483*
Truhlar, D. G. 20, 25(130), 57(242), *182, 185*
Truipathi, G. N. R. 1132(330–338), 1133(335), 1135, 1136(338), *1150*
Truscott, T. G. 1136(358, 359, 364), 1137(364), *1151*
Truter, M. R. 204(9), *220*
Trzeciak, A. M. 752(104), *833*
Tsai, A. L. 15(72), *181*
Tsai, F. C. S. 1425(403b), *1451*
Tsai, S. J. 1082(286), *1094*
Tsai, Y.-F. 1240(166), *1342*
Tsang, W. 5, 144(25), *180*, 1105(62), *1106*, 1111, 1118(60), *1145*
Tsank, S. 627(63), *657*

Tsantrizos, Y. S. 1401(211), *1444*
Tschaen, D. 670(78), *706*
Tschierske, C. 674(103), *707*
Tse, C. S. 1470(56), *1503*
Tseitlin, G. M. 235(51), *255*
Tseng, J. M. 1085(335), *1095*
Tsiliopoulos, E. 647(148), *659*
Tsimidou, M. 998(454), *1013*
Tsubaki, K. 1380(56), *1437*
Tsubaki, T. 633(94), 634(95), *658*
Tsuboi, H. 844(22), 856, 865(127), *902, 904*
Tsuchida, H. 913(13), *1003*
Tsuchiya, J. 853(90), 855(90, 113, 119), 856(90, 119), 882(204), 887(218, 221), 890(232), *904, 906, 907*
Tsuchiya, K. 763(135), *834*
Tsuchiya, S. 855(117), *904*
Tsuchiya, T. 674(108), *707*
Tsudera, T. 1392(124), *1441*
Tsue, H. 1374(38a, 38b), *1436*
Tsugawa, D. 1402(224a), *1444*
Tsugoshi, T. 447(222a, 222c), 448(227), *485*
Tsuji, J. 664(27), *705*, 1192(79, 81), *1339*
Tsuji, M. 309(172, 173), 310(174), *329*
Tsuji, S. 953, 957(219), *1008*
Tsuji, T. 448(231), *486*, 1062(186), *1091*
Tsujihara, K. 673(98), *707*
Tsukazaki, M. 786(231), *836*
Tsukube, H. 1194(83b), *1339*
Tsumura, K. 1084(322), *1095*
Tsunetsugu, J. 818(318–320), *838*
Tsunoda, S. 844, 891(21), *902*
Tsunoi, S. 972(303), *1010*
Tsuruta, Y. 951, 953(195, 196), *1007*
Tsutsui, T. 1359(146), *1366*
Tsutsumi, K. 493, 512(38, 39), *525*
Tsuzuki, S. 20, 44(120), 90(344), *182, 187*, 551, 556(117), *599*
Tsvetkov, V. F. 227(13), *254*
Tsymbal, I. F. 1417(326), 1418, 1420(341), *1448, 1450*
Tu, Y. P. 296(83), *327*
Tubul, A. 448(230), *485*
Tucci, F. C. 1424(398), 1430(441b, 441c), *1451, 1453*
Tuck, D. G. 580(207), *601*
Tucker, J. A. 1382, 1383, 1421(67c), *1438*
Tuckcr, W. P. 1290(249), *1344*
Tudos, F. 1462(23), *1502*
Tümmler, R. 310(175), *329*
Tundo, P. 663(18), *705*
Tung, C. 1077(237), *1092*
Tunikova, S. A. 1001(475), 1002(478), *1014*
Tunnel, R. L. 1357(133), *1366*
Tunstad, L. M. 1382, 1383, 1421(67c), *1438*
Tuomi, W. V. 638(121), *658*
Tupparinen, K. 349(74), *388*
Turchi, I. J. 1314(289), *1345*

Author Index

Tureček, F. 110(434, 448, 449), 114(459), 121(434), 129, 135(459), *189, 190,* 260, 264(2), 305(148), *325, 329*
Turek, P. 1427(408c), *1451*
Turner, D. T. 1103(51), *1106*
Turner, D. W. 110, 112(446), *189*
Turner, E. 503(89), *526*
Turner, P. 1432(447), *1453*
Turner, R. S. 1497, 1498(170), *1506*
Turnes, I. 1354(110), *1365*
Turnes, M. I. 971(294), *1009,* 1348(17), 1354(17, 101), *1363, 1365*
Turrión, C. 228, 229(16), 230(25), 232(38), 237(61), 241(80), *254-256*
Turro, N. J. 477(328), *489*
Turujman, S. A. 358(112), *388*
Tuszynski, W. 305(147), *329*
Tuxen, N. 1357(136), *1366*
Tyeklár, Z. 1194(87), *1340*
Tykarska, E. 143(565), *192,* 580(205), *601*
Tyl, R. W. 1348(30), *1363*
Tyle, Z. 432(182b, 182d), *484*
Tyman, J. H. P. 606(1), *656*
Typke, V. 171(762), *197*
Tyurin, V. A. 887(222, 225), *907*
Tzeng, S. C. 236(53), *255*

Uber, W. 1122(192), *1147*
Uccella, N. 898(276), *908*
Uceda, M. 913(6), *1003*
Uchida, A. 818(319), *838*
Uchimaru, T. 89(339), 90(339, 341, 344), *187,* 680(157), *708*
Uchiyama, K. 811(293), *837*
Ucun, F. 729(42), *831*
Udagawa, Y. 107(409), 143(409, 575), 144(575), *188, 192*
Udenfriend, S. 405(43a), *479*
Ue, M. 1048(158, 159), *1090, 1091*
Ueda, E. 424(155b), *483*
Ueda, I. 953, 970(292), *1009*
Ueda, J. 1388(100b), *1439*
Ueda, K. 447, 448(222b), *485*
Ueda, M. 1383(67h), 1409(289a), 1417(332, 333), 1421(67h), *1438, 1447, 1448*
Ueda, S. 697(284), *711,* 1380(61a), *1438*
Ueki, S. 770(158), 773(158, 159), *834*
Uemura, S. 265, 272(29b), *326,* 614(25), *656*
Ueno, M. 695(276, 277), *711*
Ueno, Y. 845(36), *902*
Ueyama, T. 1419(349), *1450*
Uff, B. C. 799(237), *836*
Ugozzoli, F. 1374(42a), 1387(90), 1391(118a, 121a), 1392(90), 1393(156), 1394, 1395(90), 1396(118a, 121a, 182a, 182b, 188), 1397, 1399, 1403(197b), 1424(389b, 397b), 1427(410a), 1435(467), *1436, 1439, 1440, 1442, 1443, 1450, 1451, 1454*
Ugrinov, A. 1024(65), *1088*
Ugrinov, U. 360(136), *389*
Ulatowski, T. G. 417(118), *482*
Uličný, J. 93, 95(361), *187,* 384(290), *393*
Ullrich, V. 13(67), *181*
Ulrich, C. 1120(177), *1147*
Ulrich, G. 1391(120b), 1427(408c), *1440, 1451*
Umamaheshwari, V. 617(34), *656*
Umeda, F. 874(171), *905*
Umemoto, T. 647(152), *659,* 687(222-225), *709*
Umezawa, B. 1065(195), 1066(196), *1091*
Umezawa, H. 447(222f), *485*
Umezawa, I. 1318(300a), *1345*
Umezawa, T. 1216(125), *1341*
Un, J. S. 1372(13), *1436*
Un, S. 129(490), *190*
Undenfriend, S. 405(44), *479*
Underberg, J. M. 78(277), *186*
Underberg, W. J. M. 498(72a, 72b), *526*
Underhill, A. E. 1495(148), *1505*
Uneyama, K. 811(294), *837*
Ungaretti, L. 216(27), *221*
Ungaro, R. 683(169), *708,* 1387(90), 1391(110, 118a, 121a-c), 1392(90, 136, 137), 1393(110, 141, 151, 156, 158), 1394(90, 171, 176), 1395(90, 177, 180), 1396(118a, 121a-c,182a, 182b, 185, 186a, 186c, 188), 1397, 1399(197b), 1403(197b, 232a, 234e, 235, 240b), 1404(244), 1405(232a, 235, 255), 1409(287, 288), 1427(410a), 1430(439), 1432(446), 1435(462b), *1439-1443, 1445-1447, 1451, 1453, 1454*
Unny, I. R. 639(131), *659*
Uno, H. 851, 868, 870(51), *903*
Uno, T. 669(64), *706*
Unterberg, C. 375(247), *392*
Unwin, P. R. 1086(371), *1096*
Uozumi, Y. 669(66, 67), *706*
Upham, R. A. 291(66), *327*
Uppu, R. M. 637(107, 108), *658*
Uptoon, C. G. 1485(95), *1504*
Urakawa, T. 1483(89), *1504*
Urano, S. 855(113), 891(244), *904, 907*
Urban, F. J. 663(22), *705*
Urban, J. J. 57(243), 99, 101(378), *185, 188*
Urban, W. 493, 495(32), *525*
Urbaniak, M. 1417(330a), *1448*
Urbano, A. 650(162), *659*
Urbano, F. J. 972(304), *1010*
Urbano, M. R. 618(36), *656*
Urbe, I. 953, 963(261), *1009*
Uri, N. 404(35), *479*
Urjasz, H. 144(676), *195*

Ursini, F. 890(234), *907*
Ursini, O. 274(41), *326*
Ushijima, H. 1411(299), *1447*
Usui, S. 1084(322), *1095*
Utley, J. H. P. 1282(229), *1343*
Utsumi, H. 855(117), *904*
Uyemura, S. A. 895(266), *908*
Uzawa, H. 766(142), *834*
Uzhinov, B. M. 492, 493, 501(13), *525*, 1020(35d), *1087*
Uznanski, P. 493, 511(65), *526*

Vaal, M. J. 663(24), *705*
Vaccari, A. 644(143), *659*
Vadgama, P. 983(379), *1011*
Vaes, J. 565(171), *600*
Vaid, R. K. 1224(140b), *1341*
Vainiotalo, A. 360(139), *389*
Vainiotalo, P. 322(250), *331*
Vairamani, M. 293(77), *327*
Vaish, S. P. 1134(349), *1151*
Vaissermann, J. 748(84–86), *832, 833*
Vajda, E. 238(73), *256*, 551, 555(111, 112), *599*
Valcárcel, M. 319(226, 227), *330*, 932, 937(92), 946(160), 986(392), *1005, 1006, 1012*
Valcic, S. 305(151), *329*
Valderrama, J. A. 813(301–303), *837, 838*, 1283(230), *1343*
Valdés-Martínez, J. 551(93), *598*
Valencia, G. 651(172), *660*
Valentao, P. 913(18, 20), 958(20), *1003*
Valentine, J. S. 1118(152), *1147*
Valero, E. 979(356), *1011*
Valgimigli, L. 45, 140(198, 201), *184*, 852, 868(86), 877(191, 193), 878(86, 191, 193), 879(191, 194, 199), 880(199), 881(191, 193), 882(86, 191, 194), 883(191, 194, 205), 884(193), 887(205), 900(280), *903, 906, 908*
Vallet, P. 110(433), *189*
Valoti, E. 683(180), 686(210), *708, 709*
Valoti, M. 953, 968(275), *1009*
Valpuesta, M. 477(325), *489*
Valters, R. E. 740(63), *832*
Van, P. 492, 493(16), *525*
Van Alsenoy, C. 170(756), *197*
Van Alsenoy, K. 4(8), *179*
Van Bekkum, H. 631, 640(74), *657*
Van Beylen, M. 93, 103(363), *187*
Van Damme, A. 990(424), *1012*
vanDam-Mieras, M. C. E. 852(78), *903*
Vandenborre, M. T. 549(67), 550(68), *598*
Vander Donckt, E. 492, 493, 501(10), *525*
Van der Greef, J. 110(430), *189*
Van der Wieln, V. W. M. 1074(231), *1092*

Van Dorsselaer, A. 1427(421a), *1452*
Vane, J. 11(50), *180*
Vane, J. R. 10(45), *180*
Van Gelder, J. M. 1411(298), *1447*
Van Haverbeke, A. Y. 110(431), *189*
Van Haverbeke, Y. 121(465, 466), *190*
Van Hulst, N. F. 1391(112), *1440*
vanKessel, J. 851, 885(46), *903*
Vankoten, M. A. 631, 640(74), *657*
Van Lerberghe, D. 534(26), *597*
Van Lieshout, M. 953, 968(277, 278), *1009*
Van Loon, J.-D. 1393(151, 156), 1401(220a), 1403(234e, 240b), 1405(248), 1406(258a), 1432(449b), *1441, 1442, 1444–1446, 1453*
Van Miller, J. P. 1348(30), *1363*
Van Mourik, T. 148(726), *196*
Vanneste, W. 1123(199), *1148*
Van Poucke, L. C. 83, 84(298), *186*, 311(187), *329*
Vanquickenborne, L. G. 89, 90(328), 119(463), 127, 144(472), *187, 190*
Vanwilligen, H. 1057(172), *1091*
Varadarajan, A. 1048(154), *1090*
Varadarajan, R. 634(97), *658*, 1402(227), *1445*
Varfolomeyev, S. D. 979(353), *1011*
Varghese, H. T. 378(271), *392*
Varma, A. 1085(352), *1095*
Varma, C. 1117(114), *1146*
Varma, C. A. G. O. 1047(150), *1090*
Varma, K. S. 1495(148), *1505*
Varma, R. S. 422(143), *483*
Varquez-Lopez, E. M. 580(207), *601*
Varsanyi, G. 35, 36, 66, 78(167), *183*, 368, 375(187), *390*
Vartanyan, L. S. 1108(20), 1118, 1119(20, 140), *1144, 1146*
Varvoglis, A. 1075(232b, 232c), *1092*, 1224(140c–e), *1341*
Vasak, M. 511(103), *527*
Vasilyev, A. N. 620(46), *657*
Vaz, R. J. 983(373), *1011*
Vaziri-Zand, F. 662(3), *705*
Vázquez, E. 680(154), *708*
Vázquez, M. D. 982(366), *1011*
Veen, N. J. van der 1403(238), *1445*
Veening, H. 951, 953(191), *1007*
Vega, J. R. de la 721(23), *831*
Veggel, F. C. J. M. van 1385(79b, 84), 1386(84), 1391(119), 1406(263d), 1427(407a), 1428(263d, 437b), 1432(455), 1433(456a), *1439, 1440, 1446, 1451, 1453*
Veggel, F. J. C. M. van 1403(238), *1445*
Veglia, A. V. 365(162), *389*, 477(324b, 326b), *489*
Veiga, L. A. de 233(39), *255*
Veijanen, A. 1363(220), *1368*
Veit, M. 913(19), *1003*
Velasco, A. 929, 987(70), *1004*

Velicky, R. W. 236(55), *255*
Velino, B. 551, 553(73), *598*
Velzen, P. N. T. van 299(101), *328*
Venceslada, J. L. B. 1362(216), *1368*
Venema, D. P. 913(23), 953, 961(242), *1003, 1008*
Vener, M. V. 95, 106, 107(372), *188*, 551, 553(101), *599*
Venezia, M. 1084(326), *1095*
Venkatachalapathy, C. 459(270), *487*, 774(183), *835*
Venkataraman, B. 1017(13), *1087*
Venkataraman, D. 670(77), *706*
Venkataramani, P. S. 1283(232), *1343*
Venkatesan, P. 868(158), *905*
Venkateswarlu, A. 1465(43), *1503*
Venkateswarlu, Y. 631(84), *658*
Ventura, F. 1355(119), *1365*
Venturi, M. 761(118), *833*
Venturo, V. A. 35(163), 143, 144, 147(164), *183*
Vepsäläinen, J. 360(139), *389*
Vera, S. 953, 961(238), *1008*
Vera, W. 471(305), *488*
Vera-Avila, L. E. 944(140), 945(150), *1006*
Verboom, A. 1391(112), *1440*
Verboom, W. 377(251), *392*, 635(101), *658*, 782(218), *836*, 1370(4), 1383(67d), 1387(90), 1391(110, 118a), 1392(90, 125b, 137), 1393(110, 151, 156), 1394(90), 1395(90, 177), 1396(118a, 189), 1401(220a), 1402(224b), 1403(125b, 234e, 240b, 242), 1405(224b, 248, 253, 254a, 254b), 1406(254a, 258a, 259, 263b, 263d), 1408(282), 1409(287), 1417(4), 1421(67d), 1422(378), 1423(380b), 1424(382a, 382b, 383), 1427(410b), 1428(263d, 437a, 437b), 1430(440), 1432(449b, 454, 455), 1433(456a, 456b), *1435, 1438–1447, 1450, 1451, 1453*
Vercruysse, G. 534(27), *597*
Verdú-Andrés, J. 942(135), 953, 954(198), *1006, 1007*
Verduyn, A. J. 941(133), *1006*
Verevkin, S. P. 229(19), 230(22), 231(22, 34), 250, 251(19), *254, 255*
Verevkin, S. R. 227(13), *254*
Vergnani, T. 403(29), *479*
Verhoeven, T. R. 413(82a), *480*
Verkade, J. G. 998(455, 456), *1013*
Verkruijsse, H. D. 677(138), *707*
Verma, B. S. 757(107), *833*
Verpenaux, J.-N. 414(94a), *481*
Versari, A. 972(308), *1010*
Vertianen, T. 1360(179), *1367*
Vervloet, M. 109(414), *189*
Vesely, J. A. 405(37a), *479*
Vesely, M. 1084(328), *1095*

Vestner, S. 429(175), *483*
Vetoshkin, E. V. 371(223, 228), *391*
Vetter, S. 323(266), *331*
Vialaton, D. 1086(357), *1096*
Vicens, J. 320(239), *331*, 1370(1b), 1372(15, 20, 24a), 1380(58c), 1391(114), 1415(313), 1427(423a, 423b), *1435–1437, 1440, 1448, 1452*
Vicente, F. 1353(72), *1364*
Victorov, A. V. 888(226), *907*
Vidal, Y. 405(39b), *479*
Vidsiunas, E. K. 978(348), *1011*
Vieira, A. J. S. C. 1115(106), *1146*
Vieira, I. D. 981(359), *1011*
Vieira, T. de O. 1312, 1315(285), *1345*
Vigne, B. 412(74a), *480*
Vijfhuizen, P. C. 290(60), *327*
Vijgenboom, E. 977(344), *1011*
Vijgh, W. J. F. van der 871, 873, 898(169), *905*
Vila, A. S. 1084(324), *1095*
Vilaplana, J. 932(96), *1005*
Vílchez, J. L. 993(435), *1013*
Vilchez, L. R. 1351(53), *1364*
Villaiain, J. 887(223), *907*
Villani, C. 1383(75d), *1439*
Villanueva-Camañas, R. M. 953, 961(237), *1008*
Villasenor, J. 1087(383), *1096*
Villieras, F. 944(144), *1006*
Vinader, V. 677(125), *707*
Viñas, P. 953, 960(234), *1008*
Vincent, M. 818(314), *838*
Vincent, W. R. 78(277), *186*, 498(72a, 72b), *526*
Vincent-Falquet-Berny, M. F. 229, 230(23), *254*
Vincenti, M. 322(250), *331*, 1086(365), *1096*
Vincow, G. 1124(228, 237), *1148*
Vineis, P. 1361(188, 189), *1367*
Vinogradov, M. G. 1086(378), *1096*
Vinogradov, S. 47, 143(207–209), 170(209), *184*
Vinogradov, S. N. 368(183), *390*, 530, 557(5), 592(268), *596, 602*
Vinqvist, M. R. 851(49–51, 54–56, 59, 60, 70), 852(60), 855(116), 856(50), 858(56), 862(49), 863(49, 55), 868, 870(51), 872(50), 877(116), 878(50), 882(55), 885(54, 55), 886(55), 887(55, 60, 220), 888(60), 889(56), 899, 901(279), *903, 904, 907, 908*
Vinson, J. A. 913, 920(27), *1003*, 1355(115, 116), *1365*
Virelezier, H. 322(249), *331*
Vishnoi, S. C. 941(134), *1006*
Vishurajjala, B. 1274(210a), *1343*
Vishwakarma, J. N. 668(57), *706*

Viswanadha Rao, G. K. 293(77), 294(78), *327*
Visy, J. 1227(141), *1341*
Vital, J. J. 1417, 1424(327), *1448*
Vitharana, D. 694(266), *710*
Vittal, J. J. 1424(391), *1450*
Vitullo, V. P. 806(255–257, 259), *837*
Vivekananda, S. 110(448), *189*
Vlach, J. 1415(316), *1448*
Vlasov, V. M. 312(193), *330*
Vliet, A. van der 636(104), 637(109), *658*
Vocanson, F. 1372(11, 19), 1378(52b), 1402(224d), 1415(312, 318a), *1436, 1437, 1445, 1448*
Voerckel, V. 405(48a), *479*
Voets, R. 83, 84(298), *186*, 311(187), *329*
Vogel, E. 824(336), *838*
Vogel, H.-P. 147(715), *195*, 992(427), *1012*
Vogt, A. 774(175), *835*
Vogt, A. H. G. 631, 640(74), *657*, 774(179), *835*
Vogt, W. 635(100), *658*, 1374(33b, 33c, 36c, 39, 40, 43), 1378(50a, 51), 1386(51), 1396(195a, 195b), 1403(239), 1406(257, 262a), 1418(342), 1419(356b, 358), 1420(342, 360, 361, 363, 364, 365b, 366), 1421(342), 1431(443a, 444), 1432(449b), 1435(464a), *1436, 1437, 1443, 1445, 1446, 1450, 1453, 1454*,1462(30), *1503*
Vögtle, F. 580(206), *601*, 1378(48a, 48b), 1432(446), *1437, 1453*
Volante, R. P. 632(89), *658*, 670(78), *706*
Vol'eva, V. B. 651, 654(167), *659*, 747(82, 83), *832*
Volgger, D. 974(315), *1010*
Völker, S. 493(52), *526*
Volkert, O. 1117(116), *1146*
Volkonskii, A. Yu. 312(193), *330*
Volksen, W. 1493(123), 1495(123, 135, 136), *1505*
Volland, R. 1123, 1124(216), *1148*
Vollbrecht, A. 1417(329a, 329b), 1418(339), 1424(395), *1448, 1450, 1451*
Vollmuth, S. 1360(164), *1367*
Volodin, A. M. 18(85), *181*
Volod'kin, A. A. 745(78), 811(292), *832, 837*, 1119(157, 158), *1147*
Volpe, P. L. Q. 957(213), *1007*
Voncina, E. 1360(161), *1367*
Von Laue, L. 371(223), *391*
Voogt, P. de 1349(35), *1363*
Vorlop, K. D. 976(327), *1010*
Vorogushin, A. 451(252), *486*
Vorontsov, E. 640(132), *659*
Vorozhtzov, N. N. Jr. 18(91), *182*
Vosejpka, P. C. 406(49c), *479*
Vossen, R. C. R. M. 852(78), *903*
Votoupal, K. L. 924, 925, 927(40), *1004*
Vouros, P. 315(214), *330*
Voyer, N. 477(326c), *489*
Vozhdaeva, V. N. 18(89), *181*
Vrana, R. P. 934, 937(106), *1005*
Vreekamp, R. H. 1405, 1406(254a), *1446*
Vrolix, E. 561(162), 584(233), *600, 601*
Vuano, B. M. 1381(65a), *1438*
Vuorela, H. 953, 968(273), 973(312), *1009, 1010*
Vuorinen, M. 1363(220), *1368*
Vuorinen, P. J. 1363(220), *1368*
Vyskocil, S. 691(239), *710*
Vysotsky, M. 1370, 1430(6), 1435(461), *1435, 1453*
Vysotsky, M. O. 1393(161), 1394(162, 166), *1442*

Wachter, P. 430(180a, 180b), *484*
Wachter-Jurcsak, N. 1028(96), *1089*
Wachter-Jurszak, N. 336(8), *386*
Wada, M. 951, 953(193, 194), *1007*
Wadgaonkar, P. P. 415(100), *481*
Wageningen, A. M. A. van 1406(263d), 1428(263d, 437a), 1433(456b), *1446, 1453*
Wagmen, N. 1347(8), *1363*
Wagner, E. R. 1461, 1463, 1470(18), *1502*
Wagnerova, D. M. 1080(250), *1093*
Waguepack, Y. Y. 365(164), *389*
Wai, C. M. 1394(167c), *1442*
Waiblinger, F. 1026(90), *1089*
Waite, T. D. 1100(33), *1106*
Wakefield, D. H. 776(200), *835*
Waki, H. 323(256), *331*
Wakisaka, A. 174, 177(766), *197*
Wakita, H. 170(740), *196*
Wakita, M. 1498(175), *1506*
Wakselman, C. 1272(205a, 205b), *1343*
Waldman, S. R. 764(137), *834*
Waldmann, H. 1216(119c), 1220(134), *1340, 1341*
Walia, S. 1050(161), *1091*
Walker, A. 1430(442b), *1453*
Walker, E. R. 1381(65b), *1438*
Walker, G. N. 433(188a), *484*
Walker, J. A. 5, 144(25), *180*
Walker, J. K. 1250(178), 1252(179), *1342*
Walker, J. R. L. 913(11, 12), *1003*
Walker, L. E. 1470(61), *1503*
Walker, P. P. 408(54h), *480*
Walker, R. W. 1105(64, 65), *1106*
Walker, S. 53, 56(238), *185*, 674(106), *707*
Walker-Simmons, M. K. 1237(164), *1342*
Wallace, G. G. 978(345), *1011*
Wallace, J. 19(94), *182*
Wallace, R. A. 1312(287), *1345*
Wallbank, P. 1227(143), *1341*
Wallenborg, S. R. 953, 955(208), *1007*

Walling, C. 405(41c, 41d), 406(50), *479*, 876, 892(185), *906*, 1119(159), *1147*
Wallington, T. J. 138(528), *191*, 1104(61), *1106*
Walsdock, M. J. 1356(127), *1366*
Walsh, E. B. 448(233), *486*
Walter, H. F. 1111(74), *1145*
Walters, M. R. 925(50), *1004*
Walton, J. C. 860(141), *905*
Waltos, A. M. 811(291), *837*
Waluk, J. 129, 132, 133(498), *191*, 368, 373(196), *390*
Wamer, R. E. 549(65), *598*
Wamme, N. 1372(14), *1436*
Wan, C. 1083(292), *1094*
Wan, C. C. 1082(279), 1083(303), *1094*
Wan, J. K. S. 1018(22), *1087*, 1120(181, 185), 1124(229, 230, 239), *1147, 1148*
Wan, P. 1024(56, 57), 1033(107–110), 1034(111), 1035(114, 115, 117), 1036(117–119), 1038(120), 1044(132), *1088–1090*
Wan, Z. 1045(138), *1090*
Wang, A. P. L. 281(51), *327*
Wang, B. 976(328), *1010*
Wang, C.-K. 944(141), *1006*
Wang, C. M. 1085(355), *1096*
Wang, D. 684(191, 192), *709*, 1129, 1134(289), *1149*
Wang, F. 1024(61), *1088*
Wang, G.-Q. 1403(236), *1445*
Wang, H. 982(364), *1011*
Wang, J. 367(177), *390*, 953, 955(204), 971(301), *1007, 1010*, 1086(380), *1096*, 1405(250), 1407(275), 1427(275, 414b), *1446, 1447, 1451*
Wang, J. K. 493(22), *525*
Wang, J.-X. 662(7), *705*
Wang, K.-H. 1085(338, 341, 342), *1095*
Wang, K. M. 994(436), *1013*
Wang, L. 929, 981(66), *1004*, 1361(198), *1367*
Wang, L.-J. 892(250), *907*
Wang, L. M. 1018(20), *1087*
Wang, L.-S. 987(401), *1012*
Wang, M.-X. 1432(452), *1453*
Wang, P.-S. 634(98), *658*, 1402(225), *1445*
Wang, Q.-E. 995(442), *1013*
Wang, R. 871, 898(164), *905*
Wang, R.-P. 671(83), *706*
Wang, R. X. 685(201), *709*
Wang, S. 986(396), *1012*, 1081(265), *1093*, 1180, 1183, 1186(62), 1240(167), *1339, 1342*
Wang, W. 795(233), *836*, 913(24), *1003*
Wang, W.-G. 1410(292), *1447*
Wang, X. 1471(71), *1504*
Wang, Y. 1082(280), *1094*
Wang, Y. Y. 1083(303), *1094*
Wang, Y. Z. 1084(306), *1094*
Wang, Z. 171(762), *197*
Wang, Z.-G. 447(222f), *485*
Wanhao, H. 702(335, 336), *712*
Wani, T. 414(89c), *481*
Warashina, T. 1124(231, 234), *1148*
Ward, B. 129, 131(502), *191*
Ward, R. 1224(140f), *1341*
Ward, R. S. 1158(17a), 1242(168), 1244(169), 1255(182), *1337, 1342*
Waring, A. J. 471(301), *488*, 807(283–285), 811(286–289), *837*
Warmuth, R. 1402(228a, 228b), 1409(288), 1428(426a, 434a, 434b), *1445, 1447, 1452*
Warnatz, J. 1105(64, 65), *1106*
Warncke, K. 1108(16), *1144*
Warner, J. M. 1025(73), *1088*
Warrener, R. N. 745, 746(80), *832*
Warshawsky, A. 472(311j), *488*
Warshel, A. 5(15), 179(778), *180, 198*
Wasikiewicz, W. 1427(412a, 412b), 1431(443b), *1451, 1453*
Wasserman, H. A. 1080(259b), *1093*
Wasylishen, R. E. 1470(57), *1503*
Watabe, T. 827(340), *838*
Watanabe, H. 148(729, 730), 149, 153, 156(729), 160(729, 730), 170(729), *196*
Watanabe, I. 1351(46), *1364*
Watanabe, K. 477(326a), *489*, 874(172), 894(260), *905, 908*, 986(394), *1012*
Watanabe, M. 403(31), *479*
Watanabe, S. 695(270, 279), *710, 711*, 776(206), *835*, 951, 953(195, 196), *1007*, 1428(435a, 435b, 436), *1452*
Watanabe, T. 144(598), 147–150, 157, 167, 168(705), *193, 195*, 773(160), *834*, 972(307), 987(405), *1010, 1012*
Watanabe, Y. 856(129), *904*, 1196(89), 1213(117), *1340*
Waterhouse, A. L. 920(31), 953(229), 960(228, 229), 965(31), 990(420), *1003, 1008, 1012*
Waterhouse, D. 397(8), *478*
Waterman, K. J. 1073(219), *1092*
Waters, W. A. 129, 135(488), *190*, 404(36), *479*, 1109(33, 34, 36, 37), 1120(34, 36, 37, 171, 172), 1124(34, 226, 227), *1144, 1147, 1148*
Waterson, A. G. 449(234c), *486*
Watkins, M. J. 129, 144(478), *190*
Watkinson, J. G. 53(235), *185*
Watson, C. G. 412(72), *480*
Watson, C. H. 325(278), *332*
Watson, J. T. 315(212), *330*
Watson, P. S. 1318(301b), *1345*
Watson, W. H. 447(222i), *485*, 580(206), *601*, 1390, 1392(107), 1393(147), 1394(147, 165), 1401(220c), 1409(285), 1427(414b), *1440–1442, 1444, 1447, 1451*

Watts, R. J. 1348, 1350, 1352(12), *1363*
Wawer, I. 366(165), *389*
Wayland, B. B. 548(62), *598*
Wayner, D. D. M. 45(191, 197), 139(191, 531), 179(197), *184, 191*, 851, 857(62), 897(273), 901(288), *903, 908*
Weatherhead, R. H. 1393(146), *1441*
Weaver, L. K. 1361(199, 200), *1368*
Webb, A. C. 141(539), *192*
Webb, G. A. 359(123), *389*
Webb, J. L. 823(328), *838*
Webb, K. S. 415(101), *481*
Webb, S. P. 493, 498(42), *525*
Webber, P. R. A. 1427(413), *1451*
Weber, B. A. 885(209, 210), *906*
Weber, E. 727(39), *831*
Weber, J. 1044(131), *1090*
Weber, P. M. 110(418, 419), 121(419), *189*, 1017(9, 10), *1087*
Weber, T. 12(60b), *181*
Webster, C. 937, 938(123), 953, 965(266), *1005, 1009*
Webster, G. B. 429, 430(179c, 179e), *483, 484*
Webster, G. R. B. 1073(215), 1074(231), *1092*
Wedemeyer, K.-F. 715, 724(3), 749(96), 750(101), 759(111, 112), 761(124), 799, 801, 814, 822(3), 830(348), *830, 833, 838*
Weeding, T. L. 293(71), *327*
Weeks, J. A. 1356(128), *1366*
Weenen, H. 885(209, 211), *906*
Wegelius, E. 1420(365b), *1450*
Wehrli, F. W. 349(64), *387*
Wehry, E. L. 498(71), *526*
Wei, C.-P. 1237(162e), *1342*
Wei, D. 1361(198), *1367*
Wei, H. 1082(283), *1094*
Wei, H. F. 608(7), *656*
Wei, J. 416, 417(116), *482*
Wei, T. Y. 1082(279), 1083(292, 303), *1094*
Weichert, A. 145, 156(636), 179(781), *194, 198*
Weickhardt, K. 701(315, 316), *711*
Weill, A. 1459(12), *1502*
Weinberg, D. V. 673(96), *707*
Weinberg, N. L. 410(66a), *480*, 1156, 1158(5, 8), 1172(8), 1175(5, 8), *1337*
Weinelt, F. 1417(330b), *1448*
Weiner, S. A. 1136(355), *1151*
Weinhold, F. 4(4a, 4b, 5), *179*
Weinraub, D. 1142(380), *1152*
Weinstein, B. 400(19), *478*
Weisburger, J. H. 1360(175), *1367*
Weiser, J. 1382, 1383, 1421(67c), *1438*
Weisgraber, K. H. 1159(29), *1338*
Weiss, H. C. 219(35), *221*
Weiss, J. 1103(50), *1106*, 1113(82), *1145*
Weiss, R. 413(81), *480*, 719(19), *831*
Weiss, U. 721(25), *831*

Wel, H. van der 302(138), *328*
Well, G. 1231(150), *1341*
Weller, A. 491, 492(5), 493, 495(5, 32, 33), 497, 498(69), 503(5), *525, 526*
Welling, L. 1363(220), *1368*
Wellings, I. 399(13a), *478*
Wellmann, J. 404(34), *479*
Wells, R. J. 319(223), *330*, 931, 937(84), *1005*
Welsh, H. L. 67(247–249), *185*
Welti, D. 35(162), *183*
Wemmer, D. 343(42), 347(42, 55), 348, 349(55), 356(42), *387*
Wemmes, D. 129(485), *190*
Wendel, V. 1403(237), *1445*
Weng, L. 1353(91), *1365*, 1387, 1388(96), *1439*
Wenglowsky, S. 457(263), *487*
Weniger, K. 300(113), *328*
Wenkert, E. 830(346), *838*
Wennerström, H. 348(57), 349(57, 58), *387*
Wentrup, C. 1478(85–87), 1486(87), *1504*
Werkhoven-Goewie, C. E. 1353(80), *1365*
Werner, A. 638(119), *658*
Werner, S. 461(290), *487*
Werstler, D. D. 1482(88), 1485(96), *1504*
Wertz, J. E. 1120(169), *1147*
Wessely, F. 1325(311a–c), *1345*
West, F. 1307(278), *1344*
West, P. R. 1120(176, 178), *1147*
West, R. 555(140), *599*
West, T. R. 671(84), *706*
Westin, G. 1267(202a), *1343*
Westlund, N. 786(232), *836*
West-Nielsen, M. 344(44), *387*
Westphal, A. 144(679), *195*, 377(253), *392*, 578(194), 581(216), *601*, 992(428), *1012*
Westra, J. G. 1194(83a), *1339*
Westrum, E. F. Jr. 238–240(66), *256*
Wettermarck, G. 558–560(150), *600*
Wey, S.-J. 702(325), *712*
Whang, C.-W. 971(299), 972(302), *1009, 1010*
Whangbo, M.-H. 669(61), *706*
Whatley, L. S. 555(140), *599*
Wheals, B. B. 925(53), *1004*
Wheatley, P. J. 11(53), *180*, 232(36), *255*
Wheeler, R. A. 31(138), 35(168), 111, 112(138, 168), 113(138), 114(168), 132, 133, 136(511, 513), *183, 191*
Whelan, B. A. 448(226), *485*
Whiffen, D. H. 36(176), *183*
Whipple, M. R. 511(103), *527*
Whitakker, J. M. 129(484), *190*
Whitakker, M. M. 129(484), *190*
White, A. H. 1382(73), 1388(97), 1413(304, 305), *1438, 1439, 1447*
White, A. J. P. 761(118), *833*
White, D. 1244(170), *1342*
White, D. M. 466(292), *487*

White, G. F. 1348(31), *1363*
White, J. D. 1244(171), *1342*
White, R. 925(49), *1004*
White, R. L. 370(213), *391*
White, S. C. 170, 171, 174(742), *196*
White, W. N. 460(271b), 463(276c), *487*
Whiting, A. 692(249), *710*
Whiting, D. A. 12(58), *181*, 806(252), *836*
Whitney, J. 305(151), *329*
Whitson, P. E. 1167(38), *1338*
Whittaker, A. K. 1470, 1471(59), *1503*
Wiberg, K. B. 87(308), *186*
Wicki, M. 786(231), *836*
Wickleder, C. 493(26, 27), 515(27), *525*
Widdowson, D. A. 1297(257), *1344*
Widén, C.-J. 349(63), *387*
Widmer, U. 466(291), *487*
Wieghardt, K. 129, 132, 133, 135(514), *191*
Wieland, S. 620(47), *657*
Wierzchowski, K. L. 1117(130, 132), *1146*
Wierzejewska, M. 144(604), *193*, 594, 595(285), *603*
Wiesen, P. 18(83a), *181*
Wiesli, U. 324(269), *331*
Wijmenga, S. S. 1405(248), *1445*
Wilde, H. 1424(392), *1450*
Wilderer, P. A. 303(144), *329*
Wilhoit, R. C. 241(75), *256*
Wilkins, J. P. G. 319(228), *330*
Wilkins, R. F. 684(188, 189), *709*
Wilkinson, F. 1057(173), *1091*
Willbrand, D. M. 493(21, 23), *525*
Williams, A. 1393(146), *1441*
Williams, A. R. 730(47), *832*
Williams, D. H. 260(1), 263(19), 290(59), *325–327*
Williams, D. J. 761(117, 118), *833*, 1432(449a), *1453*
Williams, D. L. H. 419(123b), *482*
Williams, D. R. 35(153, 156), 36, 37, 66, 67(153), 107(156), *183*
Williams, D. T. 319(230), *330*
Williams, E. L. 449(234b), *486*
Williams, P. 343(42), 347(42, 55), 348, 349(55), 356(42), *387*
Williams, R. J. J. 1459(7, 8), *1502*
Williamson, B. L. 307(157, 158), 308(158), *329*
Williamson, G. 895(267), *908*
Willigen, H. van 1016(4), *1087*
Willmore, W. G. 854(97), *904*
Willson, J. M. 1071(208), *1092*
Willson, R. L. 1110(54), 1111(66, 67), 1119, 1128(155), 1129(299), 1130(299, 308, 324), 1131(155), 1134(155, 347, 348), 1137, 1138(308), *1145, 1147, 1150, 1151*
Wilson, C. C. 178(770), *197*, 579(196), *601*
Wilson, D. 1494(129), *1505*

Wilson, E. B. 375(239), *392*
Wilson, F. X. 786(232), *836*
Wilson, G. 933, 937(101), *1005*
Wilson, H. W. 35(154, 160), 36(154), *183*
Wilson, J. D. 95(368), *188*
Wilson, K. 617(33), *656*
Wilson, K. J. 735(53), *832*
Wilson, P. 953, 965(266), *1009*
Wilson, R. L. 890(230), *907*
Wilson, S. R. 698(293), *711*
Wilson, W. D. 356(108), *388*
Wimalasena, K. 612(19), *656*
Winberg, H. 716(9), *831*
Winer, A. M. 1105(63), *1106*
Wingbermühle, D. 461(290), *487*
Winget, P. 57(242), *185*
Winiarski, J. 405(46a), *479*
Winkler, E. L. 1459(13), *1502*
Winkler, M. R. 319(228), *330*
Winnik, W. 314, 315(208), *330*
Winter, T. J. 438(199), *484*
Winter, W. 1169(43), *1338*
Winterle, J. 845, 846(31), *902*
Winters, M. P. 671(83), *706*
Wintgens, D. 229, 250(20), *254*
Wipf, P. 461(289), *487*, 677(120), *707*
Wipff, G. 1391(120b), *1440*
Wirz, J. 38(184), 127(184, 468), *184, 190*, 229, 250(20), *254*, 716(7), *831*, 1017(14), 1035(116), *1087, 1089*
Wiseman, A. 412(71b), *480*
Wiseman, G. H. 430(180a), *484*
Wiseman, J. R. 447(220), *485*
Wisted, E. E. 1352(66), *1364*
Wit, J. de 110, 112, 121(444), *189*, 265(24), *326*
Witanowski, M. 144(655), *194*
Witiak, D. T. 851, 852, 894(67), *903*
Witkop, B. 405(44), *479*
Wittek, H. 35(147), *183*
Wittek, P. J. 429(174), *483*
Witter, D. 141(539), *192*
Wittkowski, R. 315(215), *330*
Wittmann, H. 818(315), *838*
Witz, M. 647(150), *659*, 687(221), *709*
Wöhnert, J. 1378, 1386(51), *1437*
Wohrle, D. 1082(277), *1094*
Wojciechowski, G. 377(263, 264), 386(307), *392, 393*
Wojciechowski, W. 35, 47, 58(170), *183*
Wojcierchauski, K. 405(46b), *479*
Wojnarovits, L. 1100(30), 1102(47), *1106*, 1115(91), 1128(91, 275, 276), 1142(378), *1145, 1149, 1151*
Wolf, A. 977(336), *1010*
Wolf, C. 1420(365a), *1450*
Wolf, G. 977(336), *1010*
Wolf, J. F. 263, 296(16), *326*

Wolf, K. 383(284), *393*
Wolfe, J. P. 671(80), *706*
Wolfender, J. L. 953, 967(270), *1009*
Wolff, A. 1374(33b, 42a), *1436*
Wolff, C. 826(339), *838*, 1419(346), *1450*
Wolff, J. J. 211, 216(20), *221*
Wolff, L. 450(240), *486*
Wolff, S. P. 856, 857(131), *904*
Wollenberger, U. 976(331), 977(344), 978(346), 979(357), *1010, 1011*
Wollschann, P. 358(119, 120), *388*
Wolschann, P. 144(647, 656, 667), *194*, 370(215), 386(215, 305, 306), *391, 393*, 551, 556(118), *599*, 992(431), *1013*
Won, M. S. 982(367), *1011*
Wonchoba, E. R. 669(62), *706*
Wong, A. S. 1360(165), *1367*
Wong, C. S. 1351(51), *1364*
Wong, H. N. C. 448(229d), *485*
Wong, M. K. 946, 953(161), *1006*
Wong, M. W. 4, 100(2), *179*, 368(200, 203), 372, 373(200), *390, 391*, 1478, 1486(87), *1504*
Wong, S. K. 1124(229, 230), *1148*
Wong, T. S. 1470(56), *1503*
Woo, H.-G. 667(50), *705*
Wood, C. E. C. 230(29), *254*
Wood, J. L. 85(304), *186*, 548(61), *598*
Wood, K. V. 308(163, 165), 309(165), *329*
Wood, M. R. 1318(302), *1345*
Wood, P. M. 1118(153), *1147*
Woodbrey, J. C. 1462(22), *1502*
Woodgate, P. D. 416(114), *482*
Woodin, R. L. 291(65), *327*
Woodrow, D. 933, 937(101), *1005*
Woods, R. J. 1098(1), *1105*
Woodward, D. L. 687(218), *709*
Woodward, R. B. 717(12), *831*
Woodward, S. 691(240–242), *710*, 782(219, 220), *836*
Woolard, A. C. S. 856, 857(131), *904*
Woolfe, G. J. 1025(82), *1089*
Worms, P. 475(318), *488*
Worrall, D. 1057(173), *1091*
Worth, B. R. 1261(194), *1342*
Wozniak, K. 336, 343(40), 344(40, 44), 345, 354, 355, 366(40), *387*
Wray, J. W. 1349, 1360(33), *1363*
Wray, V. 350(78), *388*
Wright, A. J. 1383, 1421(67m), *1438*
Wright, F. G. 129, 131(499), *191*
Wright, F. J. 1127(268), *1149*
Wright, I. G. 429, 430(179e), *484*
Wright, J. D. 1000(469), *1013*
Wright, J. S. 20, 23(116), 31(140), 44(188), 45(140, 192), 110, 113(440), 139(116, 192), 140(116, 140, 192), *182–184, 189*, 867, 871, 895, 896(154), 897(154, 275), 898(154), 899(154, 278), *905, 908*
Wright, T. G. 30(136), 110, 112, 121(444), 129(476), 144(677), *183, 189, 190, 195*
Wronska, U. 891(241), *907*
Wrzyszcz, J. 608(8, 9, 12), *656*
Wu, C. 1361(198), *1367*
Wu, C.-L. 851, 853, 857(61), *903*
Wu, D. 231, 234, 241(33), *255*
Wu, H.-D. 367(175), *390*
Wu, J. 1100(31), *1106*
Wu, J. G. 639(124), *659*
Wu, J. M. 1361(197), *1367*
Wu, K. J. 323(260), *331*
Wu, M. 851, 852, 894(67), *903*
Wu, M. H. 698(291), *711*
Wu, N. C. 410(66a), *480*
Wu, S. 1380(58f), *1437*
Wu, S.-H. 1391–1393(122a), 1394(168), *1440, 1442*
Wu, S.-S. 1085(343), *1095*
Wu, T. 449(234c), *486*
Wu, Y.-D. 20, 45(125), 104(399), 140(125), *182, 188*
Wuest, J. D. 551, 553(82), *598*
Wujek, D. G. 852(80), *903*
Wulf, O. R. 35(174), 47(213–217), 48(217), *183, 184*, 555(141), *599*
Wulf, W. D. 450(246b), *486*
Wulff, W. D. 454(253, 255), 456(255, 260), 457(263), *486, 487*, 693(253, 254), *710*, 735(52, 53), *832*
Wunderlich, H. 200(5), *220*, 550(69), *598*
Würsch, J. 433(189c), *484*
Wuts, P. G. M. 662, 666(1), *704*
Wyatt, A. H. 498(75), *526*
Wyatt, P. A. H. 492, 493, 501, 507(12), *525*
Wylie, R. 1297(260b), *1344*
Wynberg, H. 685(195), *709*, 1292(251), *1344*
Wyness, P. 1084(308, 309), *1094*
Wysong, R. D. 650(163), *659*
Wyssbrod, H. R. 106, 148(404), *188*

Xi, B. 1087(382), *1096*
Xi, F. 843(283), 851(48), 867–869(152), 871(152, 159), 901(283), *903, 905, 908*
Xi, H. 1424(388a, 389a), *1450*
Xia, G. 1045(142), *1090*
Xia, Y. 1432(445), *1453*
Xiang, S. H. 620(41), *657*
Xiangge, Z. 702(335, 336), *712*
Xiao, G. 15(72), *181*
Xie, D. 1392(133), *1441*
Xie, J. 1432(452), *1453*
Xing, Y. D. 448(229d), *485*
Xing, Z. 662(7), *705*

Xu, B. 1374(42b), 1396(191a, 191b, 193), *1436, 1443*
Xu, D. 377(261), *392*, 620(41), *657*
Xu, D. K. 983(378), *1011*
Xu, F. 1362(209), *1368*
Xu, L. 925, 929(54), 990(422), *1004, 1012*
Xu, R. 305(148), *329*, 1318(304), *1345*
Xu, S. L. 450(239i), 452(243), *486*, 818(312, 313), *838*
Xu, T. 171(762), *197*
Xu, W. 1424(391), *1450*
Xu, X. 1360(184), *1367*
Xu, Y. 377(261), *392*

Yabunaka, Y. 855, 857(108), *904*
Yadav, G. D. 619(39), *656*
Yagi, A. 953, 966(268), *1009*
Yagi, H. 14(69), *181*, 829(344), *838*, 1159(29), *1338*
Yagi, K. 854, 855(103), *904*
Yagi, Y. 845(40), *902*
Yagoubi, N. 953, 955(202), *1007*
Yagupolskii, L. M. 312(193), *330*
Yagupolskii, Yu. L. 312(193), *330*
Yakimchenko, O. E. 1104(58), *1106*
Yakura, T. 1227, 1228(144), 1230(148), *1341*
Yakura, Y. 1178(58), *1339*
Yalpani, M. 430(181d), *484*
Yamabe, S. 762(125), *833*
Yamabe, T. 18(86), *181*
Yamada, A. 1057(170), *1091*, 1233(152), *1341*
Yamada, F. 143(546, 547), 179(546), *192*
Yamada, H. 211(19), *220*, 412(74b), *480*
Yamada, K. 1374, 1406(36b), *1436*
Yamada, M. 1201(93), *1340*
Yamada, S. 1278(220), *1343*, 1489, 1491(116), *1505*
Yamada, T. 855, 856(119), *904*
Yamada, Y. 1315(293, 294), *1345*, 1495(132, 133), *1505*
Yamagishi, T. 1462(25), *1503*
Yamaguchi, B. 1048(158), *1090*
Yamaguchi, F. 414(89b), *481*
Yamaguchi, H. 447(223), *485*, 1419(350b), *1450*
Yamaguchi, M. 170(740), *196*, 429(172), *483*, 674(100), 680(156–160), 681(161), 682(166), *707, 708*
Yamaguchi, T. 170(740), *196*
Yamaguchi, Y. 811(290), *837*
Yamakawa, A. 1459(10), *1502*
Yamakawa, Y. 1417(333), *1448*
Yamamoto, A. 972(307), *1010*
Yamamoto, H. 219(36), *221*, 447(222g), 460(283b), *485, 487*, 677(128–130), 681(164), 688(226–229), 689(230–235), 691(243, 244), 693(250–252, 255–261), *707–710*, 1396(186b), *1443*
Yamamoto, J. 774(191–194), *835*
Yamamoto, K. 692(248), *710*, 892(251), *907*
Yamamoto, N. 667(50), *705*
Yamamoto, O. 1100(20), *1105*
Yamamoto, S. 1057(170), *1091*
Yamamoto, T. 1352(56), *1364*, 1372(18), *1436*
Yamamoto, Y. 664(35), *705*, 763(134), *834*, 882(204), 887(221), 890(235), *906, 907*
Yamamura, S. 1080(257), *1093*, 1158(17b, 17c, 19–21, 23, 24), 1159(24, 25a, 25b, 26, 30), 1163(31, 32), 1164(35, 36), 1172(25a, 25b, 48–50), 1175(53), 1177(55), 1178(21, 57a, 57b, 59), 1180(53, 61a, 61b, 65, 66), 1183(67), 1185(68, 69), 1186(70, 71), 1187(72, 73), 1188(74), 1224(35), 1227(30), 1290(36), 1314(290), 1315(31, 32, 291, 295a, 295b, 296, 297), 1317(298), 1318(299, 306), 1324(307–309), 1325(310a, 310b), 1337(57a, 57b, 332), *1337–1339, 1345, 1346*
Yamamura, T. 871, 874(160), *905*, 1362(211), *1368*
Yamana, H. 774(191–193), *835*
Yamanaka, I. 414(90h–j), *481*
Yamane, T. 412(75c), *480*
Yamanoi, T. 674(105), *707*
Yamashita, A. 454(254), *486*
Yamashita, H. 1172(48), *1338*
Yamashita, M. 1204(98), *1340*
Yamashita, N. 1356(124), *1366*
Yamashita, Y. 699(312), *711*, 811(293), *837*, 1489(111, 112), 1491(111), *1504*
Yamaski, H. 1498(176), *1506*
Yamato, T. 1057(170), *1091*, 1377(45a, 45b, 46a–c), *1437*
Yamauchi, R. 845(36, 40), *902*
Yamauchi, S. 1021(36), *1088*
Yamauchi, Y. 272(37), *326*
Yamawaki, J. 662(6), *705*
Yamazaki, I. 1120(170), *1147*
Yamdagni, R. 313(197), *330*
Yan, C.-F. 441(205), *484*
Yan, D. 1495(155), *1505*
Yan, X. 1082(283), *1094*
Yanagi, A. 1394(172b), *1442*
Yanagisawa, T. 291(63), 300(125), *327, 328*
Yanase, M. 1403(240a), *1445*
Yanase, T. 874(171), *905*
Yáñez, M. 83(296b), *186*, 580, 581(212), *601*, 322, 325(244), *331*
Yáñez-Sedeno, P. 982(362), *1011*
Yang, C. 1033(109, 110), *1089*, 1361(186), *1367*
Yang, C. Q. 1057(172), *1091*
Yang, D.-D. H. 1073(218), *1092*

Yang, F. 1397(197c), 1398(198), 1403(197c), *1443*
Yang, G.-Y. 953, 969(283), *1009*
Yang, L. 990(417), *1012*, 1080(258), *1093*
Yang, L.-M. 1403(236), *1445*
Yang, N. C. 1073(218), *1092*
Yang, S. S. 110, 112, 139(442), *189*
Yang, W. 89, 90(320, 322), *187*
Yang, Y. 1305(269), *1344*
Yang-zhi, L. 813(296), *837*
Yano, M. 1372(18), *1436*
Yanovskii, A. I. 1290, 1292(247), *1344*
Yao, B. 609(18), *656*, 1402(224c), *1444*
Yao, Q. 894(263), *908*
Yaozhong, J. 702(335, 336), *712*
Yarboro, T. L. 414(97a), *481*
Yaremko, A. M. 371(229), *391*
Yaropolov, A. I. 977(333), *1010*
Yarwood, J. 170(754), 174(764), *197*
Yasuda, M. 1022(54, 55), *1088*
Yasuda, T. 551, 554(110), *599*
Yasuhara, A. 1352(56), *1364*
Yasumatsu, M. 1377(46b, 46c), *1437*
Yates, J. R.III 305(148), *329*
Yates, K. 1022(52, 53), *1088*
Yates, P. 1180(60a, 60b), 1327(60a, 60b, 313), *1339, 1345*
Yawalkar, A. A. 1084(311), *1094*
Ye, M. 1098(7), *1105*, 1135, 1136(354), *1151*
Yeddanapalli, L. M. 1461(17), *1502*
Yeh, M. 1393(152b), *1441*
Yenes, S. 636(103), *658*
Yenkie, M. K. N. 1002(481), *1014*
Yeo, A. M. H. 300(118), *328*
Yeoh, S. L. 535(31), *597*
Yi, M. 144(597), *193*
Yianni, P. 1495(150, 151), *1505*
Yildiz, M. 551, 557(122), *599*, 729(43), *831*
Yip, C.-W. 697(281), *711*
Yododzinski, L. H. 469(297a), *488*
Yohann, D. 1163(33c), *1338*
Yohannes, D. 670(73), *706*
Yohihara, K. 129, 131(505), *191*
Yokawa, A. 1029(102), *1089*
Yokosu, H. 951, 953(192), *1007*
Yokota, Y. 935(115), *1005*
Yoneda, N. 648(154), *659*
Yonehara, H. 1306(271), *1344*
Yoo, M. R. 1086(373), *1096*
Yoon, D. Y. 1495(135, 136), *1505*
Yoon, J. 1085(346), *1095*, 1362(205), *1368*, 1423(381), 1428(425b, 426b, 434a), 1429(426b, 430a, 430b), 1433(458), *1450, 1452, 1453*
Yoon, M. 1047(151), *1090*
Yoon, T. H. 1407(273a), *1446*
Yoon, T. Y. 447(224), *485*
Yorke, C. P. 319(228), *330*

Yoshida, H. 1124(231, 234), *1148*
Yoshida, K. 1175(52), *1338*
Yoshida, M. 665(40), 669(63), *705, 706*
Yoshida, T. 323(256), *331*, 855(117), *904*
Yoshida, Y. 323(256), *331*, 447, 448(222b), *485*, 953, 961(241), *1008*
Yoshihara, M. 1495(133), *1505*
Yoshikawa, N. 695(273, 274), *710, 711*
Yoshikawa, T. 913(22), *1003*
Yoshikawa, Y. 851, 852, 892(71), *903*
Yoshimi, N. 1361(187), *1367*
Yoshimura, K. 1380(58a, 58d), *1437*
Yoshimura, M. 1379(54d), 1404(246b), *1437, 1445*
Yoshimura, Y. 987(405), *1012*
Yoshino, A. 345(49), *387*
Yoshino, N. 1383, 1421(67l), *1438*
Yoshioka, H. 412(74b), *480*
Yoshiuchi, H. 143(582), *193*
Yoshiya, H. 1290(245), *1344*
Yoshizawa, I. 874(172), *905*
Yoshizawa, K. 18(86), *181*
You, X. Z. 685(201), *709*
Young, D. C. 749(99), *833*
Young, W. F. 1358(144), *1366*
Yperman, J. 83, 84(298), *186*, 311(187), *329*
Yrjonen, T. 953, 968(273), *1009*
Yu, A.-M. 983(378), *1011*
Yu, C.-M. 664(32), *705*
Yu, D. 702(344), *712*
Yu, H. 129, 132(508), *191*
Yu, I. Y. 1427(423b), *1452*
Yu, S. K.-T. 934, 937(106), *1005*
Yu, W. 1080(258), *1093*
Yu, Y.-H. 1432(445), *1453*
Yuan, H. 1378(50b), 1400(206), 1431(50b), *1437, 1444*
Yuan, H. L. 403(32), *479*
Yuan, Z.-H. 1082(291), *1094*
Yuasa, K. 414(90f), *481*
Yuasa, M. 1318(300a, 300b), *1345*
Yudilevich, I. A. 432(186), *484*
Yudin, A. K. 697(285), *711*
Yue, T.-Y. 670(71), *706*
Yufit, D. S. 824(331), *838*
Yuk, J. Y. 663(14), *705*
Yumanaka, I. 414(90e), *481*
Yumura, T. 18(86), *181*
Yun, S. 1428(427), *1452*
Yurchenko, V. N. 235(51), *255*
Yusubov, N. N. 640(134), *659*
Yutaka, Y. 447(222h), *485*

Zabel, F. 5, 137, 138(21), *180*
Zabik, M. J. 315(212), *330*
Zaborovska, M. 493, 512, 513(35), *525*
Zacharias, J. D. E. 375(250), *392*

Zacharov, A. I. 236(54), *255*
Zafra, A. 993(435), *1013*
Zagorevskii, D. V. 275, 277(42, 43), *326*
Zagrodzki, B. 1025(68), *1088*
Zahalka, H. 864, 865(144), *905*
Zahler, R. E. 397, 467(5), *478*
Zahn, G. 1417(330b), *1448*
Zahraa, O. 1080(249), *1093*
Zaidi, J. H. 807(283–285), *837*
Zaitsev, B. E. 730(48), *832*
Zaitsev, N. K. 515(96), *527*, 985(389), *1012*
Zaitseva, Yu. N. 730(48), *832*
Zajc, A. 1360(164), *1367*
Zak, H. 429(177), *483*
Zakrzewski, V. G. 4, 100(2), *179*, 368, 372, 373(200), *390*
Zanarotti, A. 1307(279), 1308(280), *1344*
Zang, W. 1045(136), *1090*
Zanoni, L. 1082(287), *1094*
Zapf, A. 936, 937(117), *1005*
Zapotoczny, S. 1086(377), *1096*
Zappia, G. 1422(375), *1450*
Zaragoza Dörwald, F. 450(245), *486*
Zare, R. N. 37(180), 110–112(426), *183, 189*
Zareba, M. 852(84), *903*
Zauer, K. 244(87), *256*
Zavodnik, V. E. 200(4), *220*
Zavoianu, D. 230(31), *255*
Zawadzki, J. 632(88), *658*
Zawadzki, M. 608(9, 12), *656*
Zaworotko, M. J. 207(12), 211(21), *220, 221*, 530(15), *597*
Zazulak, W. 724, 725(27), *831*
Zeegers-Huyskens, T. 313(196), *330*, 377(260, 265, 266), *392*, 997(449), *1013*
Zeegers-Huyskens, Th. 35(170), 37(181), 38(183), 47(170, 210), 58(170, 245), 67(245), 83(289), 143(561–563), 145(610, 611, 613–619), 146(610, 611, 613–615, 617–619), 147(614, 619), 170(755), *183–186, 192, 193, 197*, 371(219–221), 372(220, 221), 377(219, 220), *391*, 530(8), 534(24, 26–28), 559(158–161), 561(162), 562(166, 167), 563(168), 565(171–173), 566(174, 175), 568(159, 160, 176), 569(177, 178), 570(158), 571(158, 179), 573(180, 181), 574(28, 182), 575(158), 583(225, 227–232), 584(233–235), 585(236), 593(28, 272), 594(287, 290), 595, 596(28, 182, 272, 290, 296, 297), *597, 600–603*
Zefirov, N. S. 763(133), *834*, 1470(53), *1503*
Zehnacker-Rentien, A. 78(282), *186*
Zeijden, A. A. H. van der 684(185), *708*
Zelesko, M. J. 416(111), *482*
Zemann, A. 974(315), *1010*
Zemel, H. 1099(17), *1105*, 1126–1128, 1136, 1138(263), *1149*

Zen, J.-H. 953, 969(283), *1009*
Zen, S. 674(108), *707*
Zeng, J. 171(762), *197*
Zeng, X. 1387, 1388(96), *1439*
Zenk, M. H. 1216(124), 1246(172), *1341, 1342*
Zenobi, R. 260(4), 323(4, 265–267), 324(269, 275), 325(276), *325, 331, 332*
Zepp, C. M. 685(199), *709*
Zepp, R. G. 1082(270), *1093*
Zerbetto, F. 129, 133(497), *191*
Zeta, L. 680(148), *708*
Zetta, L. 683(176), *708*, 1374(33b), 1431(444), *1436, 1453*, 1462(27, 28), *1503*
Zewail, A. H. 143(584), *193*, 493(21–23), *525*
Zeya, M. 1048(152), *1090*
Zgierski, M. 371, 377(222), *391*
Zgierski, M. Z. 179(779), *198*, 377(254), *392*
Zhadanov, Yu. A. 743(69), *832*
Zhang, A. 982(364), *1011*
Zhang, A. Q. 935, 937(114), *1005*
Zhang, B. 1073(218), *1092*
Zhang, C. 448(229d), *485*
Zhang, F. 995(442), *1013*, 1374(31a), 1377(46c), *1436, 1437*
Zhang, F.-Y. 697(281, 282), *711*
Zhang, G. 1362(208), *1368*
Zhang, G.-Q. 982(372), *1011*
Zhang, H. 984(381), *1011*, 1233(151c), *1341*, 1393(145b), *1441*
Zhang, H. B. 620(41), *657*
Zhang, H. W. 637(107), *658*
Zhang, H.-Y. 17(81), *181*, 982(372), *1011*
Zhang, J. 1024(61), 1073(218), *1088, 1092*, 1360(184), 1361(186), *1367*
Zhang, J. Y. 281(51), *327*
Zhang, K. 620(41), *657*
Zhang, L. 990(417), *1012*
Zhang, L.-D. 1082(291), *1094*
Zhang, M. 450(246f), *486*, 662(7), *705*, 1045(142), *1090*
Zhang, P. 985(391), *1012*
Zhang, W. 349(60), *387*, 698(293–295), *711*
Zhang, W.-C. 634(99), *658*, 1410(292), *1447*
Zhang, X. 144(684), *195*, 1459(9), 1463(9, 40), 1470(59), 1471(59, 74–79), 1473(75, 76), 1474(77), 1475(78, 79), 1488(103), *1502–1504*
Zhang, X.-L. 1432(445), *1453*
Zhang, Y. 1387(88), 1400(206), *1439, 1444*
Zhang, Y.-L. 855(114), *904*
Zhang, Z.-Q. 984(381), *1011*
Zhang, Z.-Z. 1387, 1388(96), *1439*
Zhao, B. 997(448), *1013*
Zhao, B. P. 1038(120), *1089*
Zhao, C.-X. 1432(445), *1453*
Zhao, G. 1495(155), *1505*
Zhao, J. 1086(369), *1096*

Zhao, Q. 352(92, 94), 367(180), *388, 390*
Zhao, W. 1085(354), *1096*
Zhao, X. S. 1081(268), *1093*
Zhao, Z. W. 1109(51), *1144*
Zheglova, D. K. 351(87), *388*
Zhelev, Zh. Zh. 887(222), *907*
Zheng, D. K. 359(131), *389*
Zheng, Q. 1305(269), *1344*
Zheng, Q.-Y. 1409(286a), 1432(452), *1447, 1453*
Zheng, X.-F. 694(266), *710*
Zheng, Y. J. 105(402), *188*
Zheng, Y.-S. 634(99), *658*, 1432(448a, 452), *1453*
Zhi, L. 702(336), *712*
Zhi, Z. L. 986(392), 987(401), *1012*
Zhidomirov, G. M. 1123(200), *1148*
Zhigulev, A. V. 987(408), *1012*
Zhilinskaya, K. I. 930(81), *1004*
Zhong, M. 776(199), *835*
Zhong, Z. 1379(54e), 1427(420b), *1437, 1452*
Zhong, Z.-L. 1435(468), *1454*
Zhou, J. 359(131), *389*, 990(417), *1012*, 1400(206), *1444*
Zhou, M. 593, 595(278), *602*
Zhou, T. 986(396), *1012*
Zhou, X. 93, 94(360), 129(509, 510), 131(510), 132(360, 509, 510), 133(509, 510), *187, 191*
Zhou, Y. 1045(142), *1090*
Zhu, C. 695(271), *710*
Zhu, F. 1082(285), *1094*
Zhu, G.-D. 450(235), *486*
Zhu, H.-W. 593, 595(278), *602*
Zhu, J. 420(141), *483*, 670(69), *706*, 1045(135), *1090*, 1104(59), *1106*
Zhu, L. 1362(214), *1368*
Zhu, P. Y. 673(92), *706*
Zhu, S. Z. 312(193), *330*
Zhu, Y. 1034(113), *1089*
Zhu, Z. 666(49), *705*
Zhuang, H.-S. 995(442), *1013*
Zhuo, J.-C. 359(133), *389*
Ziat, K. 1374(33c), *1436*
Zia-Ul-Haq, M. 740(66), *832*
Ziegler, G. 1463(38), *1503*
Zieliński, H. 853, 857(89), *904*
Ziemer, B. 1403(237), *1445*
Zierkiewicz, W. 35, 47(170), 58(170, 245), 67(245), *183, 185*
Ziessel, R. 1391(120b), 1427(408c), 1435(469b), *1440, 1451, 1454*
Ziganshina, A. U. 1419(355), *1450*

Zigmond, M. J. 874(175), *905*
Zilberman, J. 397(7), *478*
Zimmer, D. 319(233, 234), *331*
Zimmer, H. 1260(190), *1342*
Zimmerman, H. E. 816(309), *838*
Zimmermann, J. 1408(279b), *1447*
Zinke, A. 1463(37, 38), *1503*
Zinke, A. J. 1484(93), *1504*
Ziogas, A. 1172(51a), *1338*
Ziolkowski, J. J. 752(104), *833*
Zoda, M. F. 1275(214), *1343*
Zolfigol, M. A. 632(87), 638(123), *658, 659*
Zollinger, H. 402(26a, 26b), *479*
Zollner, H. 854(100), *904*
Zolotov, Yu. A. 987(408), 988(413), *1012*
Zombeck, A. 1205(102), *1340*
Zong, H.-X. 634(98), *658*, 1402(225), *1445*
Zoran, A. 421(126), *482*
Zordan, F. 663(18), *705*
Zorkii, P. M. 200(4), *220*
Zotova, O. A. 730(48), *832*
Zou, H. 1360(184), 1361(186), *1367*
Zou, J. 365(158), *389*
Zouiten, N. 913(4), *1003*
Zrinski, I. 97, 101(374), *188*
Zubarev, V. 1101(38), *1106*, 1134(341), *1151*
Zubarev, V. E. 110(422), *189*
Zuberi, S. S. 687(221), *709*
Zubovics, Z. 818(315), *838*
Zucco, M. 1255(185), *1342*
Zuerker, J. 807(273), *837*
Zummack, W. 731(49), *832*
Zumwalt, L. R. 20(103), 48, 53, 56(231), *182, 185*
Zundel, G. 47, 53(226), 143(573), 144(607, 653, 671–676), 145(573, 620, 628–635, 637), 146(642), 147(646), *184, 192–195*,353(95), 355(106, 107), 364(157), 371(218, 233), 377(233, 257, 267), 386(304), *388, 389, 391–393*,530(6), 551(126), 593(270, 273–275), 594(126, 270, 286), 595(126, 273–275, 286, 292, 294, 295, 298, 299), 596(270, 292, 294, 295, 298), *596, 599, 602, 603*,742(68), *832*
Zupan, M. 655(190), *660*, 687(219), *709*
Zwaal, R. F. A. 852(78), *903*
Zweifel, G. 414(97c), *481*
Zweifel, H. 748(89), *833*
Zwicker, E. F. 1127(270), *1149*
Zwier, T. S. 148, 160(720), *196*
Zwolinski, B. J. 241(75), *256*
Zydowsky, T. M. 813(297), *837*
Zyubin, A. S. 129, 131(507), *191*

Index compiled by K. Raven

Subject Index

Abscisic acids, synthesis of 1237, 1242
Absorption spectroscopy, of phenoxyl radicals
  1127–1131
Accufluor, as fluorinating agent 655
Acenaphthenequinone monoxime, tautomerism
  in 360
Acenaphthoquinones—see
  2-Hydroxyacenaphthoquinone
Acetonitrile,
  complexes with phenol 170–177
  dielectric constant of 170
Acetophenones—see also
  Fluoroacetophenones,
  Hydroxyacetophenones
  ^{13}C chemical shifts for 338
Acetoxylation 1167
Acetylcholine esterase mimic 1406
o-Acetylhydroxy[2.2]paracyclophanes 639
Acetylnaphthols, ^{17}OH chemical shifts for 340
o-Acetylphenol, synthesis of 438
1-Acetylsugars 674
Acid–base dissociation 992
Acid–base equilibria, of phenoxyl radicals
  1132–1135
Acidity,
  gas-phase 5, 97–100
  increase upon electronic excitation 108
  in solution 101
  intrinsic 101–103
Acid strength 224
Acoradienes, electrosynthesis of 1177
Acourtia isocedrene, retrosynthetic pathways
  toward 1185, 1188
Acrylophenones, formation of 631
Acylation, of phenols 629–632, 933, 934
  fluorescent labels for 951
C-Acylation 685, 686
O-Acylation,
  of calix[n]arenes 1387–1389
  of calixresorcinols 1417, 1418

Acylbenzenediols, thermochemistry of 244
Acylphenols—see also o-Acetylphenol
  hydrogen transfer in 1020–1022
  thermochemistry of 232, 233
4-Acylphenols, photooxidation of 1084
1-Adamantyl halides, reactions of 677
p-(1-Adamantyl)phenols, synthesis of 606, 607
Aerophobin-1, biosynthesis of 1315, 1316
Aeroplysinin-1, synthesis of 1327, 1328
Aerothionin, biosynthesis of 1315, 1316
Ageing, free-radical theory of 140, 141
Agent Orange 12
Aggregation 365
Alcoholic beverages,
  analysis of 957, 960, 961, 972, 973, 985,
    1000
  phenolic compounds in 940
Aldehydes—see also Benzaldehydes
  addition of diethylzinc to 693
  α-amino—see α-Amino aldehydes
  asymmetric silylcyanation of 694, 695
Aldol condensation,
  intermolecular 433–437
  intramolecular 426–433
Aldol reaction, asymmetric 691
Alizarin, thermochemistry of 248
Alkanols, comparison of phenol with 228, 229
C-Alkenylation 680, 681
Alkenylphenols, addition reactions of
  1022–1024
Alkoxides, diastereomeric 1396
4-Alkoxycyclohexadienones, synthesis of 651,
  654
Alkoxyhydroindolenones, synthesis of 652
Alkoxyhydroquinolenones, synthesis of 652
Alkoxyphenols, thermochemistry of 236, 237
3-Alkoxyphenols, reactions with aldehydes
  1382
2-(Alkylaminomethyl)phenols, IR spectra of
  370

The Chemistry of Phenols. Edited by Z. Rappoport
© 2003 John Wiley & Sons, Ltd ISBN: 0-471-49737-1

Alkylation,
 of calixarenes 1403
 of phenols 606–629, 941
C-Alkylation 676–680
O-Alkylation 662–670
 of calix[n]arenes 1387–1389
 of calixresorcinols 1417–1419
Alkylbenzenediols, thermochemistry of 242, 243
Alkylphenol ions, metastable 268
Alkylphenols—see also Methylphenols, n-Pentadecylphenol
 antioxidant activity of 1349
  substituent effects on 860, 861
 fragmentation of 265–270
 in wastewater effluents 1356
 irradiation of 1104
 isomerization of 749
 mass spectra of 288, 298
 photooxidation of 1084
 thermochemistry of 229–231
Alkyl phenyl ethers, ionized 299
Alkyl propiolates, reactions with phenol, palladium-catalyzed 681
Alkyl radicals 1103
2-Alkylresorcinols 1382
(1-Alkynyl)carbene tungsten complexes, reactions of 669
Allyl aryl ethers, rearrangement of 761–778—see also Allyl phenyl ethers
Allylation, asymmetric 665
C-Allylation 677
O-Allylation 664, 665
Allyl carbonates, as allylating reagents 664
π-Allylpalladium compounds 664
2-Allylphenol, IR spectra of 370
Allylphenols, rearrangement of 750, 752–758
o-Allylphenols 301
 photocyclization of 1038–1047
  competition with dehalogenation 1042
 synthesis of 460–466, 613, 624
Allyl phenyl ethers 301, 302
 rearrangement of 460–466, 613
Aluminum tris(2,6-diphenylphenoxide) 688
Amaryllidaceae alkaloids 1289
 synthesis of 1246, 1249
Amberlyst catalysts 606, 619
AMD biosensors 974, 976 978
Amination, anodic 1169
Amines—see Arylamines, Catecholamines, Cepharamine
α-Amino aldehydes, N-protected 621, 622
4-Aminoantipyrine 950
Aminocalixarenes 1406, 1407, 1411
β-Amino-o-hydroxybenzyl alcohols 621
2-Amino-3-hydroxypyridine, as MALDI matrix 324
Aminomethylation, of calixresorcinols 1419

p-Aminophenol,
 as MALDI matrix 324
 ion/molecule adducts of 314
3-Aminophenol, ion/solvent adducts for 296
Aminophenols, thermochemistry of 234, 235
1-Aminopyrene, photoacidity of 503
Aminoquinones, formation of 1076
Aminyl radicals 1110
AM1 method 982
Amperometric detection (AMD) 983, 984
Analytical chemistry,
 acronyms used in 912
 computational methods used in 929
Anharmonic effects 371, 377, 382
Anharmonicities, of OH and OD stretching modes 82, 83
Aniba neolignans 1180
Aniline, ionized 299
Anilines—see Thiacalix[4]anilines
Anilinyl radicals 1109
Anion/molecule complexes 316, 318
Anions—see also Heteropolyanions
 cresolate 315
 hypericin 345
 naphtholate 506, 512, 515
 phenol 310
 phenolate 93–97, 110, 315, 336, 495–497, 638
 salicylate 371
Anisole,
 proton affinity of 88, 89
 protonated, mass spectrum of 123–125
Anisoles—see also 4-Bromoanisole, Fluoroanisoles, Hydroxyanisoles
 anodic oxidation of 1177
Anisotropic solvents 368
Anisotropy, caused by OH group 339
Anthocyanins 957, 960, 965
1,8,9-Anthracenetriol 380—see also 1,8,9-Trihydroxyanthracene
Anthracenones—see 1,8-Dihydroxy-9(10H)-anthracenone
Anthralin,
 chemical shifts for, isotope effects on 346
 IR spectra of 368, 380
 LD spectra of 380, 381
 Raman spectra of 380
 tautomerism in 357
Anthraquinones—see also Hydroxyanthraquinones
 ^{13}C chemical shifts for 338
 synthesis of 444, 446, 447, 1170, 1239, 1240, 1243
Anthrols, enthalpies of formation for 227–229
Anthropogenic compounds, non-estrogenic 925
Antibiotics, phenolic 995
Anticarcinogenicity 913
Antimutagenicity 913

Subject Index

Antioxidant activity 842–845
  practical limit to 899, 900
Antioxidants—*see also* Co-antioxidants
  analysis of 941, 947, 949, 955, 956, 960, 961, 981, 982
  as food preservatives 982
  biologically active 913
  calculations on 895–899
  chain-breaking 840, 874
  efficiency of 850–895, 900
    media effects on 876–895
    structural effects on 859–876
  future prospects for 899–901
  hydrogen atom donating ability of 865–867
  in aircraft fuel 990
  in alcoholic beverages 973
  in drinking water 963
  induction period for 843
  in foodstuffs 925
  inhibition rate constants of 992
  in lipid membranes 884–895
    electrostatic effects on 888, 889
    factors controlling 886–892
    synergistic effects between 889–891
  in liposomes 852, 876
  in low-density polyethylene 997
  in micelles 852, 876, 884, 894
  in plant extracts 966
  phenols as 15, 139–143, 839–901, 1349
    substituent effects on 860, 861
    thermochemistry of 179
  reaction products of 845–850
  stoichiometric factors for 842–845
    measurement of 859
  substituent effects on, calculation of 897, 898
  water-soluble 885
Antioxygenicity 913
Antipyrine 968
Antitumour agents 641
Arbuzov reaction 1404
Arene oxides, K-region 721
Arenepolyols, thermochemistry of 252, 253
Arenolquinones, thermochemistry of 248–250
Arenols—*see also* Arenepolyols, Hydroxyarenes
  definition of 225
  tautomeric 250–253
  thermochemistry of 227–240, 250–253
Aromaticity 95
Arylamines, as phenol precursors 399, 400
O-Arylation 670–673
Arylazonaphthols, thermochemistry of 252
Arylazophenols,
  tautomerism in 724–726
  thermochemistry of 252
Aryl benzoates, Fries rearrangement of 642, 643

Arylbismuth 671, 673
Arylboronic acid 671, 673
Arylbutenes, formation of 613
Aryl-2-cyclohexenones 653
Aryl ethers—*see also* Allyl aryl ethers, Diaryl ethers, Phenyl ethers, Propargyl aryl ethers
  formation from calixarenes 1387
Aryl halides, as phenol precursors 396, 397
N-Arylhydroxylamines,
  isomerization of 801–805
  oxidation of 419
3-Arylindoles, synthesis of 1236
Aryl ketones, oxidation of 424, 425
Aryloxylium cations 179
Asatone, synthesis of 1178, 1179
Ash, from incineration of municipal waste, phenolic compounds in 938
Aspersitin, synthesis of 1327, 1328
Aspirin 10, 11
Atmospheric pressure chemical ionization (APCI) techniques 319
Atomic emission detection (AED) 937
Atomic softness 90
Atropisomers 1390, 1391
  NMR spectra of 1391
Aurones 1297, 1301
Autofluorescence 994
Autoxidation 841, 842
  accelerated, of methyl linoleate 982
  classical rate law of 885
  inhibition by phenols 842–850
    deuterium isotope effects on 844
    efficiency of 850–895
  retarders of 843, 844
Aziridination, asymmetric 702
Azocines, formation of 1060
Azo dyes 950
Azoxy compounds, oxidation of 419

Badger–Bauer rule 555
Badger-type relations 133
Baeyer–Villiger oxidation 424, 647, 701, 799
Baker–Venkataraman rearrangement 740, 743, 778
Bamberger rearrangement 419, 801
Basicity, gas-phase 5, 261
Bastadin-6, synthesis of 1315, 1319
Batzellines 1337
BE-10988 1306
Benzaldehydes—*see also* Fluorobenzaldehydes, Hydroxybenzaldehydes
  oxidation of 419, 420, 422, 423
Benzamides—*see* 2-Hydroxybenzamide
Benzanilides—*see* 2-Hydroxybenzanilide
Benzannulation, in synthesis of phenols 450–459

Benzaurins, tautomerism in 357, 358
Benzene,
  exposure to 920
  oxidation of 17
Benzenediols—*see also* Acylbenzenediols, Alkylbenzenediols, Chlorobenzenediols, Dihydroxybenzenes, Nitrobenzenediols
  thermochemistry of 242–246
Benzimidazoles, hydrogen transfer in 1025
Benzoates—*see* Aryl benzoates, Heptabenzoates
1,3-Benzodioxanes, rearrangement of 824, 825
Benzofurans—*see also* Dibenzofurans, Dihydrobenzofurans
  formation of 1255, 1256
Benzoic acids—*see also* Hydroxybenzoic acids
  fluorination of 646
Benzonitriles—*see also* 4-Hydroxybenzonitrile
  complexes with phenol 177, 178
Benzophenones—*see also* Hydroxybenzophenones
  ^{13}C chemical shifts for 338
Benzopyrans, formation of 619
Benzoquinone O-oxide 1057
Benzoquinone oximes 235
  tautomerism in 251
Benzoquinones—*see* Dihydroxybenzoquinones
Benzothiazolinone hydrazones 980, 987
Benzoxazine, reactions with model phenols 1473, 1474
Benzoxazoles,
  hydrogen transfer in 1025
  rearrangement of 823
Benzoylation, of phenols 934
Benzoylhydroxy[2.2]paracyclophanes 639
Benzthiazoles, hydrogen transfer in 1025
Benzylamines—*see* p-Hydroxybenzylamine
Benzylic cleavage 272
Benzylisoquinoline alkaloids 1292, 1297
  synthesis of 1159, 1274, 1275, 1312–1314
Benzylphenols, synthesis of 459, 460
Benzyl phenyl ethers, rearrangement of 459, 460
BHT, pulse radiolysis of 1102, 1103
Bicyclocalix[4]arenes 1431
Bicyclo[4.1.0]hexenones, photoformation of 1048
Bicyclo[2.2.2]octenediones 299
  synthesis of 1220, 1222
Bicyclo[2.2.2]octenones, synthesis of 1327
BINAP 701
BINOL complexes 688, 691–696
Biodegradation 1362, 1363
Biological activity, non-toxic, of phenols 1361

Bioluminescence 981
Biomethylation 1362, 1363
Biosensors 974–981
  amplification processes 979
  colorimetric detection 980, 981
  electrochemical detection 974–980
  electroimmunological 977
  spectrophotometric detection 980, 981
Biosynthesis, of phenols 411–413
Biphenols, synthesis of 655
Biphenyls—*see also* Hydroxybiphenyls
  polychlorinated, determination of 933
Bis(3-alkoxybenzoyl) peroxides, reactions of 1257, 1260
Bis-anhydrides 1406
Bis-carplexes 1434
Bis-cavitands 1432
Bis-crowns, of calixarenes 1396
2,6-Bis(diethylaminomethyl)phenol di-N-oxide 355
Bishomooxacalix[4]arenes 1378
Bishydroxymethylated phenols, thermal dehydration of 1379, 1380
Bis(maleimides) 1495, 1497
Bismuth compounds, as oxidants 1330–1332
Bis(1-nitroso-2-naphtholato)metal(II) complexes, as oxidation catalysts 1203, 1204
Bis-phenol A 619, 620, 1348, 1357, 1360, 1487, 1488
Bis-phenols, condensation with formaldehyde 1377
Bis(pyridinium) iodotetrafluoroborate, as iodinating agent 651
Bis-resorcarenes 1420
N,N'-Bis(salicylidene)phenylene diamine, tautomerism in 359
Bis-spirodienones 1410, 1411
Bleaching mills, chlorinated phenolic compounds from 932
B3LYP/cc-pVTZ calculations 372–374
B3LYP/6–31G* calculations 378, 383, 385
Boc-protection 1406
Bond angles, in mono-/poly-hydroxybenzenes 200, 201, 203, 204
Bond dissociation, O–H 129–143
Bond dissociation energy 5, 224, 895
  O–H 139, 140, 982
Bond dissociation enthalpy 3
Bond lengths, in mono-/poly-hydroxybenzenes 200, 201, 203, 204
Bond orbitals, natural 3
Borins—*see* 1,3,2-Dioxaborins
Bovine serum albumin 978
Boyland–Sims oxidation 409
Boyland–Sims rearrangement 801
Bromination 649–651
  of calixarenes 1403

Subject Index 1633

of calixresorcinols 1419
para-selective 651
solid-phase 654
solvent effects 650
Bromine atoms, as radiolytic products 1101, 1102
Bromoalkanes, irradiation of 1101
4-Bromoanisole, protonated, mass spectrum of 121, 123
Bromocyclohexadienones,
  as intermediates in electrophilic bromination of phenols 650
  synthesis of 651
Bromonium tetrafluoroborate, as brominating agent 651
4-Bromophenol, protonated, mass spectrum of 121, 123
Bromophenols,
  electrooxidation of 1159, 1163
  fragmentation of 275
  photolysis of 1018
N-Bromosuccinimide, as brominating agent 649, 650, 654
Bromoxynil 1073, 1350
Brønsted acidity 498, 499, 507, 524
Brønsted acids,
  as catalysts 612, 642–645
  extremely strong neutral 312
Brønsted basicity 499
Brønsted relation 522
Brunauer–Emmett–Teller (BET) specific surface areas 944
Bucherer reaction 401, 402
Buckingham formula 170
Burechellin, synthesis of 1250, 1251
1-(*t*-Butylperoxy)-1,2-benziodoxol-3(1*H*)-one, as oxidant 1224, 1233
n-Butylphenol, mass spectrum of 268
p-*t*-Butylphenol,
  acid-catalyzed condensation of 1371, 1372
  bromination of 650
  formation of 616, 620
  sulfur-bridged dimer of 1380
t-Butylphenols,
  bromination of 654
  IR spectra of 369

Cacalal, formation of 1266, 1269
Cacalol, oxidation of 1266, 1269
Caesium carbonate, as catalyst for Williamson synthesis 663
Caffeic acid conjugates 913
Calixarenes—*see also* Aminocalixarenes, Calix[3]arenes, Calix[4]arenes, Calix[5]arenes, Calix[6]arenes, Calix[8]arenes, Homocalixarenes,

Iodocalixarenes, Multicalixarenes, Nitrocalixarenes, Thiacalixarenes
  as stoppers 1432
  chemical shifts for, ^{17}OH 341
  chirality of 1385, 1403, 1435
  conformation of 1384–1387
    1,2-alternate 1384, 1391
    1,3-alternate 1384, 1391, 1396, 1427
    boat 1385
    chair 1385, 1386
    cone 1384–1387, 1391
    cone-to-cone inversion 1413
    diamond 1386
    partial cone 1384, 1385, 1391
  definitions of 1370
  double 1425–1427
  hydroxy groups in,
    position of 1370, 1371
    replacement of 1401, 1402
  mass spectra of 320–322
  monodehydroxylated 1411
  nitration of 633–635
  spherand-type 1411
  synthesis of 1377–1384
    by fragment condensations 1374–1376
    by one-pot procedures 1371, 1372, 1378
    stepwise 1373, 1374, 1380
  use in chromatographic separations 1435
Calix[3]arenes—*see* Hexahomotrioxacalix[3]arenes
Calix[4]arenes—*see also* Bicyclocalix[4]arenes, Bishomooxacalix[4]arenes, Dihomoazacalix[4]arene, Mercaptocalix[4]arenes
  all-homo 1378
  annelated 1431
  conformation of 1384, 1385, 1389, 1391, 1392, 1427
  crown ethers of 1395–1398
    wide-rim 1405
  diesters of 1393, 1394
  diethers of 1393, 1394
  exo–*endo* 1378
  monoesters of 1392, 1393
  monoethers of 1392, 1393
  perhydroxanthene derivatives of 1411, 1413
  reactions with di-multi-functional reagents 1395, 1396
  synthesis of 1374–1376, 1378
  tetracyclohexanol derivatives of 1411, 1413
  tetraesters of 1392
  triesters of 1393, 1394
  triethers of 1393, 1394
  with substituted bridges 1378

Calix[5]arenes,
    conformation of 1390
    crown ethers of 1396
    synthesis of 1374, 1375
Calix[6]arenes,
    as catalysts, in phenol O-alkylation 662
    crown ethers of 1397–1400
    partial ethers/esters of 1394, 1395
    synthesis of 1374
Calix[8]arenes,
    crown ethers of 1400, 1401
    protected 1395
    reactions with isopropyl chloride 609, 611
Calix[4]-bis-crowns 1398
Calix[4]crowns, 1,3-dihydroxy- 1395
Calixcryptands 1396
Calix[4]cyclohexanols 1411, 1413
Calix[4]cyclohexanones 1413
Calixhydroquinones 1409, 1410
Calixnaphthols 1410
Calixphenols 1370
    conformational properties of 1384, 1385
    Cr(CO)^3 complexes of 1413
    hydrogenation of, face selectivity of 1413, 1414
    metalation of the π-electron system 1413, 1414
    modification of the methylene bridges 1408
    modification on the wide rim 1402–1407
        of substituents 1403–1407
        selectivity transfer from the narrow to the wide rim 1402, 1403
    oxidation of the aromatic systems 1409–1411
    reactions of the hydroxy groups 1387–1402
    rearrangement of 1407, 1409
    reduction of the aromatic systems 1411–1413
    synthesis of 1371, 1372
Calix[4]pyrogallol 1417
Calixpyrroles 1430
Calixquinones 1409, 1410—see also Calixhydroquinones
Calixresorcinols 1370—see also Resorcarenes
    electrophilic substitution reactions of 1419–1421
    reactions of the hydroxy groups 1417–1419
    reactions of the substituents at the bridges 1421, 1422
    synthesis of 1381–1384
    trans-cavity bridged 1420
Calix[4]resorcinols,
    conformation of 1385–1387
    synthesis of 1385–1387
Calix[4]spherands 1396
Calixsugars 1405

Calix[4]tubes 1427
Cannabinols—see Hexahydrocannabinol
Capacity factors 954, 970
Capacity ratios 954
Capillary chromatography 962
Capillary electrochromatography (CEC) 953
Capillary electrophoresis 983
Carbenes,
    formation of 1057–1059, 1072
    insertion into solvent molecules 1042, 1043
Carbon,
    activated 945, 946
    derived from novolac resins 1488
    derived from resole resins 1488–1491
Carbon fibre microelectrodes 982
Carbonization 1489, 1490
Carbon monoxide, expulsion of 275, 302, 314, 315
    from phenol radical cations 263–265, 289
Carbon paste electrodes (CPEs) 955, 981, 982
Carbon tetrachloride solutions, radiolysis of 1100, 1101
Carbonyl reduction, asymmetric 691, 692
Carboxamides—see N-Fluorocarboxamides
2-Carboxy-4'-hydroxyazobenzene, as MALDI matrix 323
C-Carboxylation 685–687
Carceplexes 1428—see also Bis-carceplexes, Hemicarceplexes
Carcerands 1427–1429—see also Hemicarcerands
Carceroisomerism 1430
Carpanone, synthesis of 1311
    electrochemical 1159, 1163, 1206
    using O_2–Co(II)(salen) complex 1206, 1207, 1209
Carvacrol complexes 365
Catalysis, phase-transfer 662
Catechins 934, 965
    as antioxidants 894
Catechol 236
    acylation of 631
    thermochemistry of 242
Catecholamines 871, 956, 978
Catechol monomethyl ether, mass spectrum of 289
Catechol oxidase 978, 981
Catechols 975, 976—see also (ω-Phenylalkyl)catechols
    as antioxidants 872, 895, 901
        calculations for 898, 899
        effect of intramolecular hydrogen bonding on 871
    cytotoxicity of 874
    identification of, in green tea 305
    mass spectra of 272–275
    synthesis of 432
Catenanes 1432

'Catherine wheel'-type structure 67
Cation exchange resins 620
Cations—see also Dications
  aryloxylium 179
  hydroxybenzyl 295
  hydroxyphenyl 273–277
  phenoxy 273–277, 289
Cavitands—see also Bis-cavitands, Oligocavitands
  resorcarene-derived 1421–1425
  self-folding 1430
Caviteins 1425
C_{18}-bonded silica 944
C_{18} cartridges 946
CEC 953
8,14-Cedranoxide, electrosynthesis of 1186, 1189
2-epi-Cedrene-isoprenologue, electrosynthesis of 1186, 1189
Cepharamine, synthesis of 1312
Cerium compounds, as oxidants 1333, 1335–1337
Chalcones,
  hydrogen transfer in 1027, 1028
  photochemical double-bond isomerism in 1030
  photochemical oxidative dimerization of 1032
  photochromic 1029
  photocyclization of 1028–1030
CHAOS-B program 991
Charcoal, activated, phenol adsorption on 944
Charge-stripping (CS) mass spectrometry 275
Charge-transfer (CT) complexes 1021
Chelation, in intramolecular hydrogen bonding 371, 377
Chelation intermediates 683
Chemical ionization (CI) mass spectrometry 291–298
Chemical shifts,
  ^{13}C 338, 339
    isotope effects on 358
    patterns of 339
    deuterium isotope effects on 342–349
    relay effect in 346
  ^{19}F, isotope effects on 358
    for phenolate anion 336
  ^{15}N, isotope effects on 358
  ^{17}O,
    calculation of 366
    in studies of tautomerism 356, 357, 359, 360
  OH 336–338
    relationship with temperature 336, 337
  ^{17}OH,
    hydrogen bonding effects on 340, 341
    solvent effects on 341
    substituent effects on 339, 340

Chemiluminescence 981
Chemiluminescence detection (CLD) 995
Chemometrics 929
Chlorination 649
Chlorine atoms, in one-electron oxidation 1100, 1101
Chlorine disinfection 925
Chlorins, reactions of 1082
Chlorobenzenediols, thermochemistry of 244–246
γ-Chlorobutyroyl chloride, as acylating agent 640
Chlorocyclohexadienones, synthesis of 651
4-Chloro-1,2/3-dihydroxybenzenes 1085
4-Chloro-1-hydroxynaphthalene, irradiation of 1073
Chloromethylation, of calixarenes 1403
2-Chloro-4-nitrophenol, gas-phase acidity of 312
p-Chlorophenol, NICI mass spectrum of 314
Chlorophenols—see also Pentachlorophenol, Polychlorophenols, Tetrachlorophenols
  determination of 933, 954, 964
  photooxidation of 1085–1087
Chlorous acid, as oxidant 1271, 1272
Chromans, formation of 677
2H-Chromenes, synthesis of 1242, 1246
Chromium compounds, as oxidants 1278, 1288
Chrysazin, IR spectrum of 383, 384
Chrysophanols, synthesis of 444
CI/CID mass spectrometry 296
CIDEP studies, of photooxidation 1017
CIDNP studies 637
Cinnamic acid derivatives 960—see also 2,4-Dimethoxycinnamic acid esters
Cinnamic acids—see Hydroxycinnamic acids, trans-4-Hydroxy-α-cyanocinnamic acid
Circular dichroism 992
Claisen condensation,
  intramolecular 426–433
  simultaneous Michael addition with 436, 437
Claisen rearrangement 613, 624, 761–769
  aromatic 460–466
  asymmetric 693
  tandem 1427
Clark oxygen electrode 977, 979
Claycop, use in nitration 633
Cluster formation 929
Clustering energies, gas-phase 224
CMPO derivatives 1406
Co-antioxidants 892
Cobalt–amine complexes, as oxidation catalysts 1199, 1201, 1205, 1206, 1209, 1211
Cobalt compounds, as oxidants 1287, 1288, 1302, 1303

Cobalt(II)–phthalocyaninetetrasulfonate, as oxidation catalyst 1202
Collision-induced dissociation (CID) mass spectrometry 275, 293
Complexes,
 formation in UV-visible detection methods 988–990
 hydrogen-bonded—*see* Hydrogen-bonded complexes
 proton-transfer 363, 364
 stereoisomeric 576
 weak 364–366
Concave bases 1399
Condensed phases 224, 225
Conformation maps, of dihydroxybenzenes 205–208, 210
Coniferyl alcohol 868
Copper–amine complexes, as oxidation catalysts 1194, 1197, 1200
Copper compounds, as oxidants 1286, 1288, 1305–1307
Copper(I) complexes, binuclear, as oxidants 1192, 1195, 1198
Copper(II) perchlorate, as allylation catalyst 677
Core valence bifurcation index 75, 76
Coreximine, formation of 1306
Cotitration studies 335
Coulometric arrays 960, 961
Coumarans—*see also* Phenylcoumarans
 formation of 753–758
Coumaric acid 1057
Coumarins—*see* 4-Methylcoumarin
Coumestans, synthesis of 1219, 1222
Coupled clusters 3
Coupling, with diazonium ions 987, 988
Coupling constants 349, 350
 use in description of tautomeric equilibria 358, 359
anti-Cram addition 688
Cram-model intermediates 683
Creosote oil 1350, 1351
o-Cresol,
 mass spectra of 265, 266, 280, 281
 reactions with cumene hydroperoxide 607
p-Cresol,
 mass spectra of 265, 266
 protonation of 263
 reactions with cumene hydroperoxide 607
4-Cresol, as dopand 365
Cresolate anions 315
Cresols—*see also* α,α,α-Trifluorocresols
 t-butylation of 319
 electrolysis of 1167
 radiolysis of 1103
m-Cresyl acetate, Fries rearrangement of 640
Crinine 1289
Crininone, synthesis of 1289

Critical points 31
Cross-coupling reactions 1216
 oxidative 1292, 1297, 1299
Crosslinking 1100
Crown ethers, of calixarenes 1396–1399—*see also* Calix[4]crowns
Cryptands—*see* Calixcryptands
Crystal effects, on LD IR spectroscopy 368
Crystal structure determination, computer-based 549
C−S bond fission, homolytic 1079
Cumene, as substrate for antioxidants 859
4-Cumyl-1-naphthol, formation of 607
Cumyloxyl radical 877
Cumylphenols, formation of 607
4-Cumylresorcinol, formation of 607
Curcumin 867, 868, 870
Curie-point carbon-isotope-ratio mass spectrometry 303
Curie-point Py/GC/MS 303
Curie-point pyrolyzer 938
p-Cyanophenol,
 gas-phase acidity of 312
 ion/molecule adducts of 314
4-Cyanophenol, photooxidation of 1084
Cyanophenols,
 photoacidity of 494, 498
 theoretical aspects of 78–81
 thermochemistry of 233
Cyanosilylation, asymmetric 691
Cyclic dipeptides, thermochemistry of 231
Cyclic voltammetry 975, 982
Cyclization reactions 425, 426
 radical 438
[3+2]Cycloaddition reactions 1167
[4+2]Cycloaddition reactions 1159
[5+2]Cycloaddition reactions, cationic 1167, 1180, 1183, 1185, 1187, 1250, 1252
Cycloalkenones—*see also* Cyclobutenones, Cyclohexenones
 rearrangement of 816, 818–821
Cycloaromatization 439, 440
Cyclobutenones, as dienylketene precursors 450–452
α-Cyclodextrin, inclusion complexes with phenols 951, 952
β-Cyclodextrin 971, 982, 990, 992
 complexes of 365
γ-Cyclodextrin 992
Cyclohexa-1,3-dien-5-one—*see o*-Isophenol
Cyclohexadienones 1035—*see also* 4-Alkoxycyclohexadienones, Halocyclohexadienones, *o*-Isophenol, Nitrocyclohexadienones
 heats of formation of 40
 ionized 264, 265
 synthesis of 1080, 1082

Cyclohexanes—see
  Hydroxyphenylcyclohexanes
Cyclohexanols—see Calix[4]cyclohexanols
Cyclohexanones—see
  Calix[4]cyclohexanones
Cyclohexenones—see Aryl-2-cyclohexenones
4-Cyclohexylphenol, formation of 613
Cyclopentadiene 5-carboxylic acid 1057
Cyclopropanation, asymmetric 697
Cytochrome P-450 linked microsomal *Papaver*
  enzyme 1216
Cytotoxicity 654
CZE 971–974

Dakin oxidation 419, 420
Dakin rearrangement 799, 800
Dehalogenation 1098–1100
Dehydrodiconiferyl alcohol, mass spectrum of
  303, 304
Dehydrodieugenol 302–306
  mass spectrum of 304–306
Dehydrodiisoeugenol, mass spectrum of 303,
  304
Dehydrodivanillin, mass spectrum of 304, 305
Dehydrozingerone 868, 870
Delphinium formosum 944
Dendrimers 1387
  aminosugar 1388
  of calixphenols 1387
  of resorcinols 1417
Dendrograms 929
Density functional theory 3, 368
Denudatin B, synthesis of 1240, 1245
Deoxybouvardin 1315, 1321
  synthesis of 1318
Deoxyschizandrin, synthesis of 1278, 1280
Deprotonation 92–105
  energy of 3, 97–100
Depsides 942–944
Derivative isotherm summation 944
Derivative spectrometry 929, 984
Derivatization, pre-analysis 985–991
DesMarteau sulfonimide, as electrophilic
  fluorinating agent 647
Desulfurization 1079
Detoxification, of phenolic compounds
  1361–1363
Deuterium labelling, exhaustive 314
Diacetylnickel, as catalyst for Williamson
  synthesis 663
Diacetyl-2,4,6-trihydroxybenzene, solid-state
  NMR spectra of 366
(Diacyloxyiodo)benzenes, as oxidants 1224,
  1226–1228, 1230–1235, 1237, 1239–1242
Diaryl ethers,
  formation of 670–673
  rearrangement of 466–470

2,4-Diaryl-6-(2-hydroxy-4-methoxyphenyl)-
  1,3,5-triazine 353
Diastereomers, formation of,
  in oxidation of thiacalixarenes 1415
  in synthesis of resorcarenes 1381
Diazo compounds, reactions of 666, 667
Diazonium compounds, as phenol precursors
  402–404
Diazonium coupling,
  of calixarenes 1403
  of calixresorcinols 1419
Diazotation 402–404
Diazotization 988
Dibenzoazonine 1297
  synthesis of 1275, 1276
Dibenzodioxepens, reactions with Lewis acids
  653
Dibenzo-*p*-dioxins 1086
Dibenzodioxocines 1308, 1309, 1311
Dibenzofurans, formation of 1289, 1291
Dibenzopyranones, formation of 1257
Dibenzoylmethane enol, IR spectrum of 382
2,6-Di-*t*-butyl-4-(4′-methoxyphenyl)phenoxyl
  radical 865
2,6-Di-*t*-butyl-4-methylphenol (BHT) 868
  ^{13}C chemical shifts for 338
2,6-Di-*t*-butylphenol 307
  complexes of 366
  formation of 620
  steric effects in 355
3,5-Di-*t*-butylphenol 307
2,4-Di-*t*-butylphenols,
  reactions with cinnamyl alcohol 619
  transalkylation of 616
Dications, O,C-diprotonated gitonic 653
2,3-Dichloro-5,6-dicyano-*p*-benzoquinone
  (DDQ), as oxidant 1247–1255
Dichlorophen 1350
2,3-Dichlorophenol, deprotonation of 335
2,4-Dichlorophenol,
  complex with triethylamine 364
  photooxidation of 1086
2,5-Dichlorophenol,
  complex with *N*-methylpiperidine 364
  deprotonation of 335
p-Dichlorophenoxyacetic acid, irradiation of
  1079
Dichlorovancomycin aglycone, synthesis of
  1318
Dichroic ratio 373
Dielectric constant, of acetonitrile 170
Diels–Alder reactions 692, 693, 1033, 1178,
  1192, 1219, 1230, 1237, 1327
  in regioselective synthesis of phenols
    440–450
  of quinone monoketals 1235, 1236
  retro- 299, 315
Dienone–phenol rearrangement 470–472

Dienones—see also Cyclohexadienones, Estradienones, Homoerysodienones, Spirodienones
  enolization constants of 716
  rearrangement of 806–818
Dienylketenes, reactions with alkynes 450–453
Diesel fuel, phenolic compounds in 948
N,N-Diethyl-p-phenylenediamine 987
Difference enthalpies 228, 229
Difference spectra 992
Differential pulse voltammetry (DPV) 981, 982
Differential refractometric detection (DRD) 953
Difluorocyclohexadienones, formation of 648
Dihalophenols,
  IR spectra of 370
  thermochemistry of 239, 240
Dihomoazacalix[4]arene, hydrogen bonding in 353
Dihydrobenzofurans—see also 2-Methyldihydrobenzofuran
  electrosynthesis of 1180, 1192
  formation of 1240, 1245, 1255, 1256
  oxidation of 1333, 1335
Dihydroerysodine 1297
Dihydroflavonols 965
Dihydrofurans—see also Dihydrobenzofurans, 5-Hydroxy-6,7-dimethyl-2,3-dihydrofurans
  formation of 1039
2,3-Dihydro-5-hydroxy-2,2,4-trimethyl-naphtho[1,2-b]furan, as antioxidant 863
Dihydropyrans, formation of 1039
Dihydroquercetin 872, 873
2,6-Dihydroxyacetophenones, isotopic perturbation in 339
Dihydroxyalkylbenzenes, fragmentation of 270, 272, 273
1,4-Dihydroxy-3-aminocyclohexa-2,5-dienyl, as reaction intermediate 1083
1,8-Dihydroxy-9(10H)-anthracenone 380—see also Anthralin, Dithranol
Dihydroxyanthraquinones, chemical shifts for, isotope effects on 345, 346
Dihydroxybenzenes—see also Benzenediols, 4-Chloro-1,2/3-dihydroxybenzenes, Dihydroxyalkylbenzenes, Hydroxyphenols
  methylation of 308
  synthesis of 408, 409, 411
1,2-Dihydroxybenzenes,
  antioxidant activity of 871
  conformation of 204–207
  ^{17}OH chemical shifts for 341
  structural chemistry of 204–207
1,3-Dihydroxybenzenes,
  conformation of 207, 208
  structural chemistry of 207, 208
1,4-Dihydroxybenzenes, structural chemistry of 209–211
2,4-Dihydroxybenzoic acid, photoreactions of 1080
2,5-Dihydroxybenzoic acid, as MALDI matrix 323
2,6-Dihydroxybenzoic acid, mass spectrum of 285
2,2′-Dihydroxybenzophenones,
  chemical shifts for, isotope effects on 346
  hydrogen bonding in 352
  IR/Raman spectra of 379
Dihydroxybenzopyrylium ions 285, 288
Dihydroxybenzoquinones, IR spectra of 382
3,4-Dihydroxybenzyl carboxylates, deprotonation of 316
Dihydroxybiphenyls 1071, 1072
  photooxidation of 1084
3,4-Dihydroxy-3-cyclobutene-1,2-dione, IR spectrum of 382
4,5-Dihydroxynaphthalene-2,7-disulfonates, hydrogen bonding in 352
1,4-Dihydroxy-5,8-naphthoquinones, chemical shifts for, isotope effects on 345
Dihydroxyphenols—see also Trihydroxybenzenes
  mass spectra of 305
  photooxidation of 1084, 1085
2,6-Dihydroxypyridine, reactions with aldehydes 1382
Dihydroxytoluenes, protonation of 309
Diiodotyrosine, reactions with solvated electrons 1099
2,6-Diisopropylphenol, IR spectra of 369
5,7-Diisopropyltocopheroxyl radical 894
Dilinoleoylphosphatidyl choline bilayers 885
Dimerization 547–550, 999
Dimers, alkylidene-linked, fragment condensation of 1382
Di-π-methane reactions 1042, 1049, 1050, 1053
2,4-Dimethoxycinnamic acid esters, tetramerization of 1383, 1384
$trans$-3,5-Dimethoxy-4-hydroxycinnamic acid, as MALDI matrix 323
4,8-Dimethoxy-1-naphthol, stereoselective hydroxyalkylation of 623
3-(Dimethylamino)phenol, ion/solvent adducts for 296
p-N,N-Dimethylaminosalicylic acid, hydrogen bonding in 1047
Dimethyl carbonate, as methylating reagent 663
2,2-Dimethylphenol, photooxidation of 1017
2,6-Dimethylphenol, reactions with 1-bromoadamantane 606, 607
Dinaphthyl ketones, electron transfer reactions of 1019
2,4-Dinitrophenol, gas-phase acidity of 312
2,5-Dinitrophenol, complexes of 366

2,6-Dinitrophenol, hydrogen bonding in 352
2,4-Dinitrosoresorcinol, synthesis of 639
Dinoterb 1350
Diode array-UVD (DA-UVD) 953
1,3,2-Dioxaborins, formation of 683
Dioxanes—see 1,3-Benzodioxanes
Dioxepens—see Dibenzodioxepens
Dioxins—see Dibenzo-p-dioxins
Dioxiranes, as oxidants 1255, 1257, 1258
Dioxocines—see Dibenzodioxocines
Dioxygen–copper complexes, as oxidants 1194, 1198
Dipeptides, cyclic—see Cyclic dipeptides
1,1′-Diphenylphosphinoferrocene ligand 671
1,3-Diphosphates, rearrangement of 1393, 1394
Dipole moment interactions 955
Diquinocyclobutene, synthesis of 1292
Directed *ortho* metalation (DoM) reaction 781, 792
Discorhabdins 1337
  electrosynthesis of 1178
Discriminating power, evaluation of 929
Disinfectants, analysis of 970
Disinfection by-products (DBPs) 1355
Dispersion forces 955
Distonic ions 285, 305
Dithranol,
  chemical shifts for, isotope effects on 346
  IR spectra of 368, 380
  LD spectra of 380, 381
  Raman spectra of 380
  tautomerism in 357
[Di(trifluoroacetoxy)iodo]benzene, as oxidant 1224, 1227–1229, 1237, 1240, 1244, 1247
Dityrosines, formation of 1163, 1164, 1166
  enzymatic 1224
Divinylbenzene–hydrophilic methacrylate copolymer 944
DNOC 1350
Domesticine 1064
Dötz benzannulation 454–459
Double Fourier transform filtering 984, 985
DRD 953
Drinking water,
  analysis of 946, 947, 962, 963, 983
  maximum admissible concentration of phenols in 928
  regulations for 1358, 1359
Drugs, phenolic compounds as 916, 917
Duff reaction 685
Dyes, phenolic compounds as 921–924
  analysis of 970
Dysazecine 1066

EDTA titration 988
Egg lecithin, Vitamin E inhibited oxidation of 884
Elbs reaction 408, 409
Electrochemical detection (ELD) 953, 985
Electrochemical methods,
  amperometric detection 983, 984
  polarography 984
  potentiometric titration 984
  voltammetric detection 981–983
Electron affinities 315
Electron-capture detection (ECD) 937
Electron-capture negative-ion (EC-NI) mass spectrometry 315
π-Electron forces 955
Electronic delocalization, in keto form of phenol 39
Electronic excitation 105–110
Electronic Localization Function (ELF) 68–78
  topology of 68–70
Electron proton resonance (EPR) spectroscopy 113
Electrons, hydrated 1098
Electron spin resonance (ESR) spectroscopy,
  in measurement of antioxidant efficiency 855, 856
  of phenoxyl radicals 1120–1127
Electron transfer,
  in antioxidant reactions, solvent effects on 880, 881
  photochemical 1019, 1020
Electrooxidation,
  cationic reactions 1165–1192
  compared with (diacetoxyiodo)benzene-promoted oxidation 1227
  radical coupling reactions 1156–1165
Electrophilic F^+ species 646
Electrophilic substitution positional indices 71
Electrophilic substitution reactions 1402
  of calixresorcinols 1419–1421
Electrophoresis,
  of biological samples 973
  of environmental samples 971, 972
  of foodstuffs 972, 973
  of industrial samples 974
Electrospray mass ionization (ESI) mass spectrometry 305, 320
Electrostatic potentials 102
Ellagic acid dihydrate, IR spectrum of 375, 376
Emulsion formation, in hydrophobic compounds 338
Ene reactions, asymmetric 693
Engelhard F-24 catalyst 650

Enolization constants 716
Enones—see Futoenones
Enthalpies of formation 5, 224
　of ancilliary deoxygenated compounds
　　related to arenols 226, 227
　of anthrol 227–229
　of arenediols 241–247
　of arenetriols 241, 247, 248
　of arenolquinones 248–250
　of naphthol 227–229
　of phenol 227, 228
　of phenols,
　　with carbon-bonded substituents
　　　229–233
　　with halogen substituents 237–240
　　with nitrogen-bonded substituents
　　　233–236
　　with oxygen-bonded substituents 236,
　　　237
　　with sulfur-bonded substituents 237
　of tautomeric arenols 250–253
　of trimethoxybenzenes 247, 248
Enthalpies of fusion 225
Enthalpies of sublimation 225
Enthalpies of vaporization 225
Enthalpy, excess 224
Entropy 224
Environmental effects, of substituted phenols
　1347–1363
Environmental Protection Agency (EPA) 925,
　927
　priority phenols 945, 962, 971, 976
　wastewater methods 932
Environmental samples, preconcentration of
　932
Enzymatic oxidation 1216–1224
Enzyme–glutathione binary complexes 105
Epi-deoxyschizandrins, formation of 1278
6a-Epipretazettine, synthesis of 1244, 1248
Epoxidation,
　asymmetric 695
　of alkenes 698, 699
meso-Epoxides, enantioselective cleavage of
　702
1,2-α-Epoxyglucose 674, 675
Epoxyquinomicin B, synthesis of 1262, 1265
Epoxy resins 1492, 1493
　hardeners for 941
Eremopetasidione, racemic, synthesis of 1237,
　1239
Erybidine, synthesis of 1278
Erysodienone, formation of 1297
Erysodines—see Dihydroerysodine
Escherichia coli 981
Estra-4,9-dien-3,7-dione 652
Estradienones—see
　β-Fluoro-1,4-estradien-3-ones
Estradiol, ionized 266

17β-Estradiol 925
17β-Estradiol fatty acid esters, deprotonation
　of 316
Estrogenic endocrine-disrupting chemicals 925
Ethenylation 680
Ethylene, expulsion of 302
C-Ethynylation 682
Eugenol,
　complexes of 365
　dimerization of 1285
　electrolysis of 1158, 1160
　ionized 305
　mass spectra of 302–306
Euglobals 1192
　electrosynthesis of 1193
European Community directives 927
Excitation energies 105
　lowest 108
Extrathermodynamic equation 559

Fats, analysis of 961
Fatty acids, determination of 936
Feces, phenolic compounds in, determination
　of 969
Fermi coupling 377
Fermi resonance 385
Ferrier reaction 675
Flame-ionization detection (FID) 937
Flame-retardants, in human plasma 931
Flash photolysis 1127
Flash vacuum pyrolysis 1478
Flavanols 960
Flavanones 965
Flavones 957, 965
Flavonoids 913, 934, 935, 965, 967, 973, 982,
　992, 994, 1297
　as antioxidants 871–873, 892, 901
　　calculations for 898, 899
　　in aqueous micelles 894
　　in lipid membranes 894, 895
　coronary disease and 920
Flavonols 934, 935, 965—see also
　Dihydroflavonols
Flavor compounds, detection of 938
Flavylium salts 1029
Floating peroxyl radical theory 889
Fluorescence/chemiluminescence techniques,
　in measurement of antioxidant efficiency
　853, 854
Fluorescence detection (FLD) 953, 985
Fluorescence enhancement 950–952, 993
Fluorescence quenching 993, 994
Fluorescence titration 497
Fluorescent labelling 950, 951, 995
Fluorination 646–649
　anodic 648
　ortho-selective 647

Subject Index

oxidative 648, 1227, 1229
   solvent effects 646, 647
C-Fluorination 687, 688
Fluoroacetophenones, Baeyer–Villiger oxidation of 647
Fluoroanisoles, proton affinities of 88, 89
Fluorobenzaldehydes, Baeyer–Villiger oxidation of 647
N-Fluorocarboxamides, as electrophilic fluorinating agents 647
Fluorocyclohexadienones—see also Difluorocyclohexadienones
   synthesis of 648, 651, 655
β-Fluoro-1,4-estradien-3-ones, synthesis of 655
Fluorohydroindolenones, synthesis of 652
Fluorohydroquinolenones, synthesis of 652
3-Fluoromethyl-4-nitrophenol 1351
Fluorophenol radical cations 291
N-Fluoropyridinium-2-sulfonates, as electrophilic fluorinating agents 647
Folin–Ciocalteu method 990, 1000, 1001
Folin–Denis reagent 990
Forensic identification, use of phenols in 913
Formylation,
   of calixarenes 1403
   of phenols 626–629
C-Formylation 685
N-Formylwilsonirine, synthesis of 1330
Förster cycle 493, 495–498, 501, 502, 515, 517, 518, 525
Fourier transform filter 929, 981
Fourier transform IR spectroscopy 547
Fractionation factors 367
Fragmentation, metastable 275
Franck–Condon simulation 107
FRAP (ferric reducing ability of plasma) assay 857
Free radical oxidation, inhibition of, kinetics and mechanism of 842–845
Free radicals, damage caused by 968
Free radical scavenging 982
Fremy's salt, as oxidant 1259–1265
Friedel–Crafts acylation 606, 629–632, 642, 676, 685, 686
Friedel–Crafts alkylation 606–629, 676
   Brønsted acid catalyzed 612
   Lewis acid catalyzed 607–611
   solid acid catalyzed 612–621
   stereoselective 621–626
   under supercritical conditions 621
Fries rearrangement 472–478, 631, 685, 773, 774, 776, 778
   Brønsted acid catalyzed 642–645
   homo-, anionic 783, 785, 786, 795, 797
   lateral 786, 798
   Lewis acid catalyzed 639–642
   metallo- 783, 794

ortho-, anionic 778, 781–786, 792–798, 1408, 1409
   remote anionic 795, 798
   solvent effects 644
   thia- 776
Fruit juices, analysis of 957–960
FT-ICR mass spectrometry 322
Fuchsones, tautomerism in 357, 358
Fukui function 90
Fulvic acids 946
Furans—see also Benzofurans, Dihydrofurans, Tetrahydrobenzonaphthofurans
   rearrangement of 823
Futoenones, formation of 1180, 1184
Futoquinols, formation of 1172

Gallic acid,
   complexes of 365
   ionized, mass spectrum of 285
   irradiation of 1079
   thermochemistry of 247, 248
Gallium lithium bis(naphthoxide), as catalyst, in asymmetric cleavage of *meso*-epoxides 667, 668
Gas chromatography,
   derivatization,
      acylation 933, 934
      alkylation 935–937
      bromination 937
      silylation 934, 935
   end analysis 937, 938
      of environmental samples 938
      of foodstuff samples 938–941
      of industrial samples 941, 942
   sample preparation 1352
      simultaneous distillation and extraction 931, 932
      solid-phase extraction vs. liquid–liquid extraction 930, 931
      solid-phase microextraction 932
      supercritical fluid extraction 932
Gasoline, phenolic compounds in 948
Gas-phase species, in thermochemistry 224
GC/MS analysis 319
GC/MS recognition 314
Gel permeation chromatography (GPC) 953
Genistein–piperazine complexes 367
Gentianaceae 967
Gibbs energy 224
Gibbs reagent 990
Gimnomitrol, synthesis of 1250, 1253
Glassy carbon electrodes (GCEs) 956, 982, 983
Glucopyranosylphenols, bromination of 650
β-Glucosidase 969

1642 Subject Index

o-β-C-Glucosylphenols 642
C-Glycosidation 679, 680
O-Glycosidation 673–675
Glycosides 958, 965, 967
Glycosyl fluorides 673, 674, 679, 680
Glyoxylates, chiral 623
Gossypol, ^{13}C chemical shifts for, deuterium isotope effects on 345
GPC 953
Grapes, resveratrol in 920
Graphitized carbon black (GCB) 947
Groundnuts, resveratrol in 920
Ground water, analysis of 946

Halocyclohexadienones—see also Bromocyclohexadienones, Chlorocyclohexadienones, Fluorocyclohexadienones
 formation of 648, 650, 652, 654, 655
Halogenation 985—see also Bromination, Chlorination, Fluorination, Iodination
 anodic 1168
 electrophilic 606, 645–651
Halomagnesium phenolates, crystal structure of 622
Halophenols—see also Bromophenols, Chlorophenols, Dihalophenols, Monohalophenols, Pentafluorophenol, Polyhalophenols
 elimination of hydrogen halides from 1057, 1058, 1072–1075
 proton affinities of 86–88
 structural chemistry of 219, 220
o-Halophenols, intramolecular hydrogen bonding in 47–57
m-Halophenols, theoretical aspects of 57–64, 66–68
p-Halophenols, theoretical aspects of 57, 58, 65–68
4-Halophenols, photooxidation of 1084
Hammett constants 71
Hammett parameters 350
Hammett plots 999
Hammett sigma values 522
Hammett-type structure–acidity correlations 506
Hartree Fock method 3, 68
Heat capacity 224
Heck reaction 1406
Helminthosporal, electrosynthesis of 1180, 1185
Hemicarceplexes 1428
Hemicarcerands 1428
Hemodialysis 969
Henry's law 984
Heptabenzoates 1395
n-Heptylresorcinol, mass spectrum of 272

Herbicides,
 analysis of 964
 phenolic 936
Heterocyclic compounds, hydrogen transfer in 1025–1027
Heteropolyacids, as oxidation catalysts 1213
Heteropolyanions, as oxidation catalysts 1204
Heterotropanone, synthesis of 1178, 1179
Heterotropatrione, synthesis of 1178, 1179
Hexahomotrioxacalix[3]arenes 1379
Hexahydrocannabinol 1034
Hexamethylenetetramine–trifluoroacetic acid system 627
2-Hexenyl phenyl ether 624
n-Hexylresorcinol, mass spectrum of 272
Histograms,
 of bond lengths,
  in 1,2-dihydroxybenzenes 204, 205
  in 1,4-dihydroxybenzenes 209
 of bond lengths/angles,
  in ortho-substituted phenols 214
  in para-substituted phenols 215
Hock–Sergeev reaction 799
Holands 1433
Homoaerothionin, biosynthesis of 1315, 1316
Homocalixarenes, synthesis of 1376–1381
Homoerysodienones, formation of 1297
B-Homoerythrina alkaloids 1297
 synthesis of 1300
Honey, analysis of 961, 973
Horseradish peroxidase 974, 977, 978, 981
 as oxidation catalyst 1216, 1217, 1219–1221, 1224, 1225
HPLC, normal phase 953
H-point standard additions method 954
HSZ-360 catalyst 680
Hydrocarbons,
 aromatic,
  determination of 933
  exposure to 930, 931
  monohydroxylation of 410, 411
  reactions with hydrogen peroxide 404, 405
 condensed polycyclic, exposure to 920
Hydrogen acceptors 201
Hydrogenation, catalytic 1406
Hydrogen atom abstraction 1108, 1109
Hydrogen atoms,
 as radiolytic products 1098
 reactions of 1102
Hydrogen atom transfer (HAT), in antioxidant reactions 896, 897
 solvent effects on 881–884
Hydrogen-bond acidity 535–546
 from partition coefficients 543–546
 log K scales for 535–540
Hydrogen-bond basicity 531–535, 876, 877
 spectroscopic scales of 589–593

thermodynamic scales of 586–589
Hydrogen-bonded complexes 312
  between phenols and
    bis-1,8-(dimethylamino)naphthalene
    997
  binding energies for 577, 578
  dissociation energies for 577, 578
  geometry of 579–582
  IR spectra of 375–377
  thermodynamic properties of 557–577
Hydrogen bonding 224
  antioxidant activity, effect on 892–894
  between phenol and acetonitrile 170–177
  between phenolic hydrogen and solvent
    881–883
  bifurcated 352
  in hydroxyarenes 492, 503, 505, 506, 519,
    520, 524
  in 2-hydroxybenzoyl compounds 553, 554
  in mono-/poly-hydroxybenzenes 200–203
  in nitrophenols 216–218
  intramolecular 551–557
    in antioxidants 867–870, 893, 896, 900
    in o-halophenols 47–57
    in 2-halophenols 554–556
    in 2-hydroxybenzoyl compounds
      377–382
    in 2-hydroxyphenols 556
    in hydroxyquinones 382–384
    in $ortho$-Mannich bases 385, 386
    in 2-methoxyphenols 556
    in 2-methylaminophenols 556, 557
    in 2-methyliminophenols 556, 557
    in 2-nitrophenols 384, 385, 553
    in 2-nitrosophenols 385
    in Schiff bases 385
    in stabilization of deepened cavities in
      cavitands 1425
  resonance-enhanced 370, 371
  IR spectra, effect on 370, 371, 375–386
  multiple, to the same acceptor 351
  protonation and 593–596
  ranking of substituents as partners 350, 351
  resonance-assisted 342, 344
  strength of 350, 378
  strong 352, 353
Hydrogen-bonding ability 143–178
Hydrogen-bonding sites 582–585
Hydrogen bonds,
  O−H···N, short 178
  three-centered 580
Hydrogen/deuterium exchange, in mass
  spectrometry 277
Hydrogen donors 201
Hydrogen peroxide, in phenolic samples 941
Hydrogen transfer reactions 280, 1108
  photochemical 1020–1028
Hydrophobocity 970

Hydroquinone monomethyl ether, mass
  spectrum of 289
Hydroquinones 236
  acylation of 631
  thermochemistry of 242
p-Hydroquinones, synthesis of 449
1,4-Hydroquinones,
  antioxidant activity of 874, 875
  cytotoxicity of 874
2-Hydroxyacenaphthophenone, steric effects in
  355
o-Hydroxyacetophenone 1021
2-Hydroxyacetophenone, IR spectrum of 377,
  378
4-Hydroxyacetophenone 639
Hydroxyacetophenones 631—see also
  2,6-Dihydroxyacetophenones,
  2,4,6-Trihydroxyacetophenone
  ion/molecule reactions of 308
  mass spectra of 296
  synthesis of 434, 644
o-Hydroxyacyl derivatives,
  hydrogen bonding in 350
OH chemical shifts for 338
C-Hydroxyalkylation 682–685
Hydroxyanisoles, protonation of 309
Hydroxyanthraquinones—see also
  Dihydroxyanthraquinones,
  1,3,8-Trihydroxyanthraquinones
  photoreactions of 1075, 1076
Hydroxyarenes—see also Arenols
  photoacidity of 491–525
3-(2-Hydroxyaryl)-3-hydroxyindolones 608,
  610
o-Hydroxyaryl ketones 640
o-Hydroxyazo aromatics, tautomerism in 356,
  358
2-Hydroxybenzalaniline N-oxide,
  thermochemistry of 233
Hydroxybenzaldehydes—see also
  4-Hydroxy-3-methoxybenzaldehyde
  mass spectra of 296
  thermochemistry of 232, 233
2-Hydroxybenzamide, thermochemistry of 232
2-Hydroxybenzanilide, thermochemistry of 232
Hydroxybenzenes—see also
  Dihydroxybenzenes, Phenols,
  Polyhydroxybenzenes,
  Trihydroxybenzenes
  structural chemistry of 200–203
2-Hydroxy-1,3,5-benzenetricarbaldehyde,
  synthesis of 627
p-Hydroxybenzoic acid, as MALDI
  matrix 324
Hydroxybenzoic acids—see also
  Dihydroxybenzoic acids,
  2-Hydroxy-4-trifluoromethylbenzoic acid,
  3,4,5-Trihydroxybenzoic acid

Hydroxybenzoic acids—see also
　　Dihydroxybenzoic acids,
　　2-Hydroxy-4-trifluoromethylbenzoic
　　acid, 3,4,5-Trihydroxybenzoic acid
　　(continued)
　mass spectra of 285
　thermochemistry of 232
4-Hydroxybenzoisonitrile, photoformation of 1054
4-Hydroxybenzonitrile, photorearrangement of 1054
Hydroxybenzophenones—see also
　　2,2′-Dihydroxybenzophenones
　IR spectra of 380
(2-Hydroxybenzoyl)benzoylmethane, IR spectrum of 382
o-Hydroxybenzoyl compounds,
　coupling constants for 349
　IR spectra of 371
2-Hydroxybenzoyl compounds, IR spectra of 377
Hydroxybenzyl alcohols,
　CI mass spectra of 292, 293
　coupling constants for 349
p-Hydroxybenzylamine, reactions with model phenols 1474, 1475
Hydroxybenzyl cations 265, 295
2-(Hydroxybenzyl)indanes, mass spectra of 277–279
Hydroxybiphenyls 277—see also
　　Octachlorodihydroxybiphenyls
β-Hydroxycarbonyl compounds, chelated hydrogen bonding in 377
Hydroxycarboxylic acids,
　decarboxylation of 1079
　reactions in presence of titanium oxide 1080
Hydroxycinnamic acids 1057—see also trans-3,5-Dimethoxy-4-hydroxycinnamic acid, trans-3-Methoxy-4-hydroxycinnamic acid
　mass spectra of 285
trans-4-Hydroxy-α-cyanocinnamic acid, as MALDI matrix 323
Hydroxycyclohexadienyl radicals 1098, 1100
5-Hydroxy-6,7-dimethyl-2,3-dihydrofurans, as antioxidants 863
8-Hydroxy-N,N-dimethyl-1-naphthylamine, hydrogen bonding in 352
6-Hydroxy-2,5-dimethyl-2-phytyl-7,8-benzochroman, as antioxidant 863
3-Hydroxyhomophthalates, synthesis of 439, 440
Hydroxylamines—see N-Arylhydroxylamines
Hydroxylation 1167
Hydroxyl radicals 1098
2-Hydroxymandelic acids, formation of 623
4-Hydroxy-3-methoxybenzaldehyde, nitration of 632

1-Hydroxy-2-methoxybenzene, ^{17}OH chemical shift for 341
(Hydroxymethyl)phenols,
　IR spectra of 370
　mass spectra of 281, 282
2-Hydroxy-6-methyl-m-phthalic acid, hydrogen transfer in 1047
2-Hydroxy-1-naphthaldehyde, ^{17}OH chemical shift for 340
Hydroxynaphthalenes—see also
　　4-Chloro-1-hydroxynaphthalene
　^{13}C chemical shifts for 338
　coupling constants for 350
N-(2-Hydroxy-1-naphthalenylmethylene)amine, tautomerism for 359
2-Hydroxynaphthoquinone, IR spectrum of 382
o-Hydroxynitroso compounds, tautomerism for 356, 360, 361
4-Hydroxy[2.2]paracyclophane 640
9-Hydroxyphenalen-1-one, tautomerism in 357
1-(2-Hydroxyphenol)-3-naphthyl-1,3-propanediones, tautomerism in 357
Hydroxyphenols—see also
　　Dihydroxybenzenes, Dihydroxyphenols,
　　Polyhydroxybenzenes,
　　Trihydroxybenzenes
　autoxidation of 1118, 1119
p-Hydroxyphenylacetic acid, formation of 1044
2-Hydroxyphenylacetylene, photohydration of 1022
2-Hydroxyphenylbenzoxazole, hydrogen transfer in 1026
Hydroxyphenyl cations 273–277
Hydroxyphenylcyclohexanes, mass spectra of 265
2-Hydroxyphenyllapazole, hydrogen transfer in 1026
N-(Hydroxyphenyl)maleimide, polymerization of 1498
Hydroxyphenyl radicals 1098, 1099
4-Hydroxyphenyl-5-thio-D-xylopyranoside 609, 610
Hydroxyphenyltriazines, hydrogen bonding in 1026
2-[4-(2-Hydroxyphenyl)tricyclo[5.2.1.0(2,6)]dec-8-yl]phenol 608, 609
3-Hydroxypicolinic acid, as MALDI matrix 323
Hydroxypropiophenones, synthesis of 645
Hydroxypyrenes—see also
　　8-Hydroxy-1,3,6-tris(N,N-dimethylsulfonamido)pyrene
　^{13}C chemical shifts for 338
o-Hydroxypyridines, ^{17}OH chemical shifts for 341

Subject Index 1645

8-Hydroxyquinoline N-oxides,
  chemical shifts for, isotope effects on 348
  coupling constants for 349
  hydrogen bonding in 352, 353
Hydroxyquinones, IR spectra of 382–384
Hydroxystyrenes,
  ionized 265
  photohydration of 1022
Hydroxytetrachlorophenoxyl radical anions 1086
Hydroxytetralins, reactions of 652
2-Hydroxy-4-trifluoromethylbenzoic acid, irradiation of 1079
8-Hydroxy-1,3,6-tris(N,N-dimethylsulfonamido)pyrene (HPTA),
  absorption spectra of 495, 496, 511
  fluorescence spectra of 497, 504, 511
  hydrogen bonding in 503
  photoacidity of 493, 495, 496, 498
Hyperfine coupling constants 113
Hypericin,
  anion of 345
  irradiation of 1045
  vibrational structure of 384
Hypertension, treatment of 913
Hyperthermal negative surface ionization detection 937

IEC 953
Imidazoles—see Benzimidazoles
Imides—see Polyimides
Imine N-oxides, thermochemistry of 233
Imino dyes 950
Indanes—see 2-(Hydroxybenzyl)indanes
Indolenones—see Alkoxyhydroindolenones, Fluorohydroindolenones
Indoles—see 3-Arylindoles
Indolones—see 3-(2-Hydroxyaryl)-3-hydroxyindolones
Inductive effects 102
Infrared polarization spectroscopy 368
Infrared spectroscopy 367–386
  density functional theory and 368
  hydrogen bonding and 368, 370, 371, 375–386
  in analysis 997
  linear dichroism (LD) 368
  near-, overtone and combination bands in 371
  proton transfer and 371
  tunnelling effects and 371
Inhibited oxygen uptake (IOU) measurement techniques 850, 851
Inhibition rate constants 992
Insecticides 948
Iodination,
  of calixarenes 1403
  solid-phase 651
Iodine, as oxidant 1266, 1269
Iodine disinfection 928
Iodobenzenes, hypervalent, as oxidants 1224–1247
Iodocalixarenes 1406
Iodonium phenolate, photoreactions of 1075
Ion chromatography 953, 957
Ion exchange chromatography (IEC) 953
Ionic liquids, immobilized 620
Ionization 110–129
Ionization energy 5, 110, 261
Ionization potentials 895
  calculation of 954
Ion/molecule complexes 314
  formed during fragmentation of phenolic anions 316–318
Ion/molecule reactions 307–310
Ion/neutral complexes 300, 301
Ion spray techniques 319
Ioxynil 1350
Ipso nitration 1402, 1403
Ipso substitution 1402
Iridium compounds, as oxidants 1294, 1298
Iron compounds, as oxidants 1287–1301
Iron tetrasulphophthalocyanine, as oxidation catalyst 1211–1213
Isatins, reactions with aryloxymagnesium bromides 608–610
o-Isoaniline 299
Isoasatone, synthesis of 1178, 1179
Isobatzelline C, synthesis of 1336, 1337
Isoboldine 1070
Isobutylene, as alkylating agent 941
Isocomene, synthesis of 1252
Isodihydrofutoquinol A,
  DDQ oxidation of 1172
  formation of 1171
Isodityrosines 1315
  formation of 1163, 1164, 1166
Isoentropic relationships 559
Isoeugenols 868
  electrolysis of 1158, 1161
  ionized 305
  mass spectra of 303, 304
Isoflavones 965
Isoflavonoids 913
Isofutoquinols—see also Isodihydrofutoquinol A
  formation of 1172
Isoitalicene, electrosynthesis of 1188, 1191
Isonitriles—see 4-Hydroxybenzoisonitrile
o-Isophenol 298–300
2-Isopropyl-5-methylphenol,
  photorearrangement of 1048
Isoquinolines—see Tetrahydroisoquinolines
Isosilybin, synthesis of 1308, 1310
Isosteganacins, synthesis of 1300, 1302

Isostegane,
    formation of 1242
    synthesis of 1278
Isothebaine 1070
o-Isotoluene 298
Isotope effects,
    equilibrium 361–363
    on chemical shifts 342–349
    primary 347–349
    solvent 346
    vs. bond order 342–344

K-13 1315, 1321
    synthesis of 1318
Kalman filter 929, 987
Katsurenone, synthesis of 1240, 1245
Kerosene, phenolic compounds in 948
Ketenes—see Dienylketenes
Ketides—see Polyketides
Keto-enols, formation of 1028
Ketones,
    aryl 424, 425
    dinaphthyl 1019
    o-hydroxyaryl 640
    unsaturated diazo- 450, 451, 453
Ketosteroid isomerase, hydrogen bonding in 352
Kinetic energy release 263
Kinetic shifts 275
Kinetic solvent effect (KSE), on antioxidant activity 877–881
Knoevenagel condensation 1405
Kohn–Sham method 68
Koilands 1427
Koilates 1427
Kolbe–Schmidt reaction 685–687
Konig–Knorr method 673
K–T analysis 503, 504, 513, 514

Lactases 976
Lamiaceae 967
Langmuir isotherm 944
Lariciresinol, enzymatic formation of 1216
Lead compounds, as oxidants 1292, 1325–1330
Leguminosae 967
Lewis acid catalysis 607–611, 631
Lignans, synthesis of 1158, 1242, 1274
Lignin model compounds 302–306, 867, 868
    phosphitylation of 999
Lignins 962, 997, 998
    addition to phenolic resins 1002
    characterization of 941
    degradation of 928
Lipids, peroxidation of,
    inhibition by phenols 884–895
    initiators for 885, 900, 901

Lipoproteins, low-density 913
Liquid chromatography,
    derivatization 949
        fluorescent tags and fluorescence enhancement 950–952
        formation of azo dyes 950
        formation of imino dyes 950
    end analysis 952–957
        of biological samples 965–969
        of environmental samples 962–965
        of foodstuffs 957–961
        of industrial products 969, 970
    sample preparation 942–944, 1352
        of biological samples 948, 949
        of environmental samples 945–947
        of foodstuffs 947–949
        of industrial products 948
        solid-phase extraction vs. liquid–liquid extraction 944–949
Liquid fuels, analysis of 970
Liquid–liquid extraction (LLE) 930, 931, 936, 942, 944–950, 960, 962, 968, 984, 990, 993
    pressurized 933, 935
Localization domains 70–72, 76, 77, 79
Low-density lipoprotein (LDL) 892
Lowenthal's method 1000
Lucerne cell cultures 978
Luteolin 872, 873
Lüttringhaus rearrangement 759

Magnetic field effects, in cobalt(II)-catalyzed oxidations 1202
Maize kernels, phenolic compounds in 994
Makaluvamines 1337
    synthesis of 1264
MALDI matrices, phenolic compounds as 323–325
Maleimides—see Bis(maleimides),
    N-(Hydroxyphenyl)maleimide,
    Poly[N-(substituted phenyl)maleimides]
Mandelic acids—see Hydroxymandelic acids
Manganese compounds, as oxidants 1278, 1279, 1282–1288
Mannich bases 353
    ^{13}C chemical shifts for, isotope effects on 361
    elimination of amines from 1034
    tautomerism in 356, 358
ortho-Mannich bases, IR spectra of 385, 386
Mannich reaction 684, 1406, 1419, 1420
o-β-C-Mannosylphenols 642
Marcus' bond-energy–bond-order (MBEBO) theory 522
Maritidines, synthesis of 1246, 1249, 1278, 1279
Mass scale calibrants 320

Subject Index 1647

Mass spectrometry 121–127, 937, 953
  as analytical method 999, 1000
  CA 123–126
    charge-stripping (CS) 275
    chemical ionization (CI) 291–298
    collision-induced dissociation (CID) 275, 293
    Curie-point carbon-isotope-ratio 303
    electron capture negative-ion (EC-NI) 315
    electrospray mass ionization (ESI) 305, 320
  FT-ICR 322
  MALDI 305, 320, 323–325, 1000
  MIKE 124–126
    negative-ion chemical ionization (NICI) 310, 314–316
    neutralization/reionization (NR) 275
McConnell relation 113, 136
McLafferty reaction 266, 268, 269, 272, 277, 280, 298
Megaphones, synthesis of 1250, 1252
Mercaptocalix[4]arenes, synthesis of 1402
[1.1.1.1]Metacyclophane framework 322
[1$_n$]Metacyclophanes 1370
Metalation, directed 677
Metathesis reactions 1432
trans-3-Methoxy-4-hydroxycinnamic acid, as MALDI matrix 323
Methoxylation 1167
2-Methoxynaphthalene, acylation of 630
Methoxynaphthols, K–T analysis of 504
4-Methoxyphenol,
  electron-transfer oxidation of 1019
  nitration of 636
  photooxidation of 1084
  reactions with ethylmagnesium bromide 622
Methoxyphenols,
  formylation of 626
  mass spectra of 296
o-Methoxyphenols, antioxidant activity of 867–870
p-Methoxyphenols,
  antioxidant activity of 860, 861
  nitration of 633
8-Methoxy-1,3,6-tris(*N*,*N*-dimethylsulfonamido)pyrene (MPTA), fluorescence spectra of 504, 505
Methylaluminium bis[2,4,6-tri(*t*-butyl)phenoxide] (MAT) 688
3-(Methylamino)phenol, ion/solvent adducts for 296
Methylation, gas-phase 309
3-Methyl-2-benzothiazolidinone hydrazone 963
4-Methylcoumarin, formation of 630
2-Methyl-4-cumylphenol, formation of 607
4-Methyl-2,6-diacetylphenol, hydrogen transfer in 1021

4-Methyl-2,6-diamidophenol, hydrogen transfer in 1022
2-Methyldihydrobenzofuran 613
6,7-Methylenedioxyflav-3-ene, synthesis of 1255
Methylnaphthols, MS/MS study of 319
Methylphenols—*see also* Polymethylphenols
  MS/MS study of 319
  reactions with 1-bromoadamantane 606, 607
Methyne transfer 308
Micellar electrokinetic chromatography (MEC) 972
Micellar liquid chromatography (MLC) 953
Michael addition/aldol condensation, tandem 434
Microbore columns 953, 955
Microcolumn chromatography 963
  reversed phase 953
Microorganisms, incorporated in biosensors 978, 979
Microspectrofluorometry 994
Microwave-assisted processes (MAP), during analytical sample preparation 930, 932, 946, 947
Miltirone, synthesis of 1261, 1264
Mitsunobu reaction 663, 674, 1418
MLC 953
Mobile phase optimization 954
Molecular geometry, calculations of 895, 896
Molecular graphs 32
Molecular orbitals (MOs) 3
Molecular symmetry, study of 368
Molybdenum compounds,
  as allylation catalysts 677
  as oxidants 1273, 1275
Monobenzylation 1395
Monocyclic phenols, keto–enol tautomerism in 715–719
Monohalophenols, thermochemistry of 237–239
Monolayers, self-assembled 997
  on gold surfaces 1421
Monte Carlo simulation 107
Montmorillonite catalysis 613–617, 627, 632
MOPAC program 954
Morphine 237
  enzymatic formation of 1216
Morphine alkaloids, synthesis of 1312
Mukaiyama asymmetric aldol reaction 702
Müller–Röscheisen cyclization 1378
Multicalixarenes 1432–1434
Multidimensional chromatography 955
Multilinear regression 929, 984
Mushroom tyrosinase, as oxidation catalyst 1218–1223

β-NADH 979
Nafion-H catalysis 606, 613
  for Fries rearrangements 642–644
Naphthaldehydes—see
  2-Hydroxy-1-naphthaldehyde
Naphthalenediols,
  as antioxidants 899
  thermochemistry of 241, 246, 247
Naphthalenes—see Hydroxynaphthalenes,
  2-Methoxynaphthalene
Naphthazarin, IR spectrum of 383
2-Naphthoates, calix[4]arene-based,
  photocyclization of 1077
Naphtholate anions 506, 512, 515
1-Naphthol/2-naphthol photoacids 510–520,
  523, 524
Naphthols—see also Acetylnaphthols,
  Arylazonaphthols, Calixnaphthols,
  4-Cumyl-1-naphthol, Methoxynaphthols,
Naphthols—see also (continued)
  Methylnaphthols, Nitronaphthols,
  Nitrosonaphthols,
  Tetrahydropyranylnaphthols
  absorption spectra of 510, 516–518, 520,
    521
  bromination of 650
  enthalpies of formation for 227–229
  fluorescence spectra of 512, 513, 516–518
  fluorination of 647
  glycopyranosides of, rearrangement of 641
  hydrogen bonding in 503, 519
  nitration of 633
  nitrosation of 636
  photoacidity of 494, 498, 501, 507, 512,
    515, 523
  reactions of,
    with 4-alkylcyclohexanones 614, 615
    with cumene hydroperoxide 607
  synthesis of 401, 402
Naphthoquinone oximes, tautomerism in 251,
  252
Naphthoquinones—see also
  Acenaphthoquinones,
  1,4-Dihydroxy-5,8-naphthoquinones,
  2-Hydroxynaphthoquinone
  synthesis of 444
Naphthyl acetates, Fries rearrangement of 639
Naphthylamines—see also
  8-Hydroxy-N,N-dimethyl-
    1-naphthylamine
  as phenol precursors 401, 402
Naringin, complexes of 365
National Institute of Health (NIH) shift 14
Negative ion chemical ionization (NICI) mass
  spectrometry 310, 314–316
Negishi reaction 1406
Nematic liquid crystals 372–375
Neolignans 1278

electrolysis of 1171
synthesis of 1158, 1242, 1244, 1274, 1314
  by enzymatic oxidation 1217, 1218
  electrochemical 1178–1180, 1184
Neural networks 925, 929
  artificial 977
Neutralization/reionization (NR) mass
  spectrometry 275
Newman–Kwart rearrangement 1401, 1415
Nickel compounds, as oxidants 1287, 1302,
  1304
NIOSH/OSHA 925, 927
Nishiguchi method 410, 411
Nitrate ion, photoreactions with 1071
Nitration 606
  by hexanitroethane 638
  by metal nitrates 638, 639
  by tetranitromethane 638
  ipso 634, 635
  of calixarenes 1402, 1403
  ortho 633, 634
  peroxynitrite-induced 636, 637
  regioselectivity of 632–635
Nitric oxide radical 638
Nitriles—see Benzonitriles
Nitroarenes, reactions with peroxides 405, 406
Nitrobenzenediols, thermochemistry of 243,
  244
Nitrobenzenes, oxidation of 419
2-Nitro-4-t-butylphenol, synthesis of 639
Nitrocalixarenes, synthesis of 633–635
Nitrocyclohexadienones, formation of 1403
Nitronaphthols, synthesis of 406
Nitronium radical 638
p-Nitrophenol,
  gas-phase acidity of 312
  ion/molecule adducts of 314
  mass spectra of 290, 314
Nitrophenols—see also Dinitrophenols,
  3-Fluoromethyl-4-nitrophenol,
  2,4,6-Trinitrophenol
  CI mass spectra of 295
  determination of 954, 964
  photoacidity of 506
  structural chemistry of 216–219
  synthesis of 405, 406, 632, 633
  theoretical aspects of 78, 81, 82
  thermochemistry of 234–236
2-Nitrophenols, IR spectra of 384, 385
3-Nitrophenols, irradiation of 1073
5-Nitrophenols, synthesis of 639
2-Nitroresorcinol, IR LD spectrum of 384, 385
5-Nitrosalicylaldehyde, NMR spectra of 336
Nitrosation 606
  by metal nitrites 639
  by nitrous acid 638
  peroxynitrite-induced 636, 637
Nitrosobenzenes 253

Subject Index

Nitrosonaphthols 385
 tautomerism in 251, 252, 360, 361
Nitrosonium ions 637
Nitrosophenols,
 tautomerism in 251, 360, 361, 721–724
 thermochemistry of 235
2-Nitrosophenols, IR spectra of 385
4-Nitrosophenols, formation of 637
NO_2 radical 639
Noroxycodeine, synthesis of 1275, 1276
Novolac resins 367, 1457–1460
 model compounds of 1463–1469
 reactions with HMTA 1470–1475
 structure of 1469
 synthesis of 1460–1463
Nuclear magnetic resonance spectroscopy 335–367
 calculations for 366
 chemical shifts 336–341
  isotope effects on 342–349
 coupling constants 349, 350
 fractionation factors and 367
 hydrogen bonding and 350–354
 in quantitative analysis 997, 998
 in structural and functional characterization 998, 999
 of atropisomers of calix[4]arene tetraethers 1391
 of phenol complexes 363–366
 OH exchange and 336
  steric effects on 354, 355
 proton transfer and 355
 solid-state 366, 367
 tautomerism and 356–363
Nuclear waste, treatment of 1435
Nucleophilic aromatic substitution, chelation-assisted 1415
Nucleus-independent chemical shift (NICS) 95
Nylon 6 178
Nylon 66 178

Octachlorodihydroxybiphenyls 1074
Octachlorodioxin 1074
Octaphosphates, of calixresorcinols 1417
Octaphosphinites, of calixresorcinols 1417
Octasulfonates, of calixresorcinols 1417
Octatrimethylsilyl derivatives, of calixresorcinols 1417
sec-Octylphenols 265
OF 4949-I 1315, 1317, 1321, 1322
OH exchange 336
OH/H increment exchange energies 228, 229
Oils,
 analysis of 961, 986
 phenolic compounds in 948
Olah's nitronium/nitrosonium salts 606
Oligocavitands 1434

Oligomerization 998
Olive oil, phenolic compounds in 999, 1000
Optical fibres 981
Organoleptic activity 925
Organometallic compounds, oxidation of 414–419
Orientalinone 1070
Orientation factors 373, 374
Ortho attack, in alkylation of phenols 309
9-Oxaisotwist-8-en-2-one, formation of 1180, 1181
Oxazines—see Benzoxazine
Oxazoles—see Benzoxazoles
Oxazolidines, formation of 1420
Oxepines, rearrangement of 824, 826–830
Oxetenes, formation of 1059
Oxidants, biologically active 913
Oxidation,
 anodic 1158–1165, 1169, 1171, 1172, 1174–1177
 enzymatic 1216–1224
Oxidative coupling 985–987
Oxidative stress 968
Oxidizability 842
Oxiranes—see Dioxiranes
4-Oxocyclohexa-2,5-dienylidene 1057
Oxodecalines, synthesis of 652
μ-Oxodicopper(III), as oxidant 1194
Oxomethylene ylides 653
Oxygen electrodes 851

Pallescensin B, synthesis of 1237, 1241
Paracyclophanes—see o-
 Acetylhydroxy[2.2]paracyclophanes,
 Benzoylhydroxy[2.2]paracyclophanes,
 4-Hydroxy[2.2]paracyclophane
Partial least squares (PLS) method 929, 981, 987
Partition properties 970
Peanuts, as source of peroxidase 977
Pechman–Duisberg condensation 995
Pechmann reaction 680
Pedaliaceae 967
Pentachlorophenol 1350
 complex with 2-methylpyridine 178
 photolysis of 1086
 theoretical aspects of 81–83
 thermochemistry of 240
Pentachlorophenolates, photoreactions of 1086, 1087
n-Pentadecylphenol, mass spectrum of 268
n-Pentadecylresorcinol, mass spectrum of 272
Pentafluorobenzoylation, of phenols 934
Pentafluorobenzylation, of phenols 937
Pentafluorophenol,
 reactions with solvated electrons 1099

Pentafluorophenol, (continued)
  thermochemistry of 240
Pentalenene, electrosynthesis of 1183, 1187
Peptides 1117—see also Dipeptides
Perborates, as oxidants 1272, 1273
Perfluoro compounds, as fluorinating agents 647
Periodates, as oxidants 1267, 1270
Periodic acid, as oxidant 1267, 1270, 1271
Permethylation 941
Peroxidases 974
  biosensors based on 977, 978
Peroxides,
  as oxidants 1255, 1257, 1259
  rearrangement of 799–801
Peroxy acids, as oxidizing agents 408–410
Peroxyl radicals 841–845
  in one-electron oxidation 1101
  reactions with phenolic antioxidants 850–855
Peroxyoxalates, as oxidants 1255, 1257, 1259
Peroxy radical scavenging 992
Pervaporation 983, 984
Perylenequinones, chemical shifts for, isotope effects on 345
Pesticides,
  determination of 933
  phenolic 936, 1350
Petroleum cracking 7
Pharmaceuticals 916, 917
  analysis of 987, 988, 995
Phase-transfer methods 673
Phellinus igniarius 977
Phenanthrenediols, thermochemistry of 241, 247
Phenol,
  absorption spectra of 495–497
  anions of, gaseous 310
  atom-in-molecule analysis of 31–34
  complexes of 145–178, 364, 377
  enthalpy of formation for 227, 228
  equilibrium structure of 20–25
  functionalization of,
    nucleophilic/electrophilic 748
  geometry of 20, 22, 23
  history of the discovery of 6, 7
  hydrogen bonding in 143–178, 519
  ionized 261, 298–306
    ortho tautomer of 265
    potential energy surface of 31
  IR polarization spectra of 372–375
  keto–enol tautomerism in 38–40, 46, 127–130
  molecular bonding patterns in 21, 26–31
  nitration of 632, 633
  OH group, characteristic vibrations of 369–371
  PE spectra of 112
  phosphorescent 109, 110
  photoacidity of 494, 497, 498, 500, 502
  photooxidation of 1080–1084
  physicochemical properties of 4–6
  protonated 83–86, 261, 282, 285
  radical cation of,
    isomers of 114–121
    molecular and electronic structure of 111–114
    potential energy surface of 116, 119–121
  reactions of 608–610
    with acetone 620
    with bromoadamantane 606, 607
    with *exo*-2-bromonorbornane 607
    with 2-buten-1-ol 613
    with *t*-butyl alcohol 616, 620
    with cyclohexanone 613
    with cyclohexyl bromide 607
    with dicyclopentadiene 608
    with β,β-dimethylacrylic acid 631
    with hydroperoxides 607
    with isobutene 620
    with methanol 618
  theoretical structures of 38–41, 44–46
  usage and production of 7–20
    industrial synthesis 18, 19, 22
  vibrational modes in 34–46
Phenol—acetonitrile complexes 170–177
Phenolacetyl nitrate–silica complex 632
Phenol adsorption 1002
Phenol–amine complexes 377
Phenol–ammonia complexes 377
Phenolate anion 93–97
  absorption spectra of 495–497
  chemical shift for 336
  Dewar benzene-type isomer of 315
  reactions with alkyl nitrites 638
  triplet states of 110
Phenolates,
  alkali metal 103–105
  hydrogen-bonded, ^{17}O chemical shifts for 360
Phenol–azaaromate complexes 377
Phenol–benzonitrile complex 177, 178
Phenol–dienone rearrangements 651–655
Phenol–formaldehyde ratio 1457
Phenol–formaldehyde resins 606, 608, 626–628, 1457
  functionality in 1457–1460
Phenolic acids 965–967
  antioxidant efficiency of 982
Phenolic aldehydes 965
Phenolic anion/molecule adducts 312–314
Phenolic esters, rearrangement of 472–478, 773, 774, 776, 778, 781–798
Phenolic ethers, rearrangement of 759–761
Phenol index number 1352

Phenol–ketone novolacs 1487, 1488
Phenol–nitrile complexes 377
Phenol radical cations 1101
  fragmentation of 289–291
Phenols—*see also* Biphenols, Bis-phenols, Hydroxybenzenes, Polyphenols
  acidities of, gas-phase 310–312
  acylation of 629–632, 933, 934
    Lewis acid catalyzed 631
    montmorillonite-catalyzed 632
    pyridine-catalyzed 631
  adsorption of 944
  alkylation of 606–629, 941
    Brønsted acid catalyzed 612
    Lewis acid catalyzed 607–611
    solid acid catalyzed 612–621
    stereoselective 621–626
    under supercritical conditions 621
  as antioxidants 139–143, 840–901
    ortho-substituted 845
    thermochemistry of 139, 140, 179
  autoxidation of 1118, 1119
  bromination of 649–651
  π-cation interaction of 322
  chlorination of 649
  comparison with isoelectronic methyl, amino and fluoro aromatic derivatives 226
  2,6-disubstituted, hydrogen bonding in 378, 379
  fluorination of 646–649
  hindered, photosensitization of 1018
  hydrogen atom abstraction from 141–143
  hydrogen-bonding abilities of 143–178
  hydrogen bonding in 200–203, 216–218, 503, 881–883
    intramolecular 47–57, 384, 385, 553–557
  ionized, reversible intramolecular hydrogen transfer in 277–280
  mass spectra of 291–298
  metal-coordinated, tautomerism in 731, 732, 735
  monocyclic—*see* Monocyclic phenols
  nitration of 632–639
  nitrosation of 636–638
  octylation of 319
  oxidation of 408–410, 1108
    electrochemical 1154–1192
    enzymatic 1216–1224
    with ammonium nitrate–acetic anhydride 1259, 1264, 1265, 1267
    with *p*-benzoquinones 1247–1255
    with chlorous acid 1271, 1272
    with di-*t*-butyl peroxides 1255, 1257, 1259
    with dioxiranes 1255, 1257, 1258

    with dioxygen–*t*-butoxide 1265, 1266, 1268
    with dioxygen–metal complexes 1192, 1194–1207
    with free radicals 1110, 1112
    with Fremy's salt 1259–1265
    with hydrogen peroxide–metal complexes 1207–1215
    with hypervalent iodobenzenes 1224–1247
    with iodine 1266, 1269
    with metal compounds/ions 1109, 1110, 1273–1337
    with perborates 1272, 1273
    with periodic acids/periodates 1267, 1270, 1271
    with radical cations 1113
  polycyclic—*see* Polycyclic phenols
  reactions of,
    with aryloxymagnesium bromides 608, 609
    with fluorophenyl-7-hydroxyheptanoates 623, 624
    with glycosides 642
    with hydroxyl radicals 1113–1116
  self-association of 377
  structural chemistry of 213–216
  substituted, *ortho* effects in 280–288, 295
  synthesis of,
    biochemical 411–413
    by benzannulation 450–459
    by condensation 425–438
    by cycloaddition 439–459
    by hydroxylation 404–414
    by oxidation of carbonyl groups 419–425
    by oxidation of nitrogen derivatives 419–423
    by oxidation of organometallics 414–419
    by rearrangement 459–478
    from aryl halides 396, 397
    from nitrogen derivatives 399–404
    from sulfonic acids 397–399
Phenol–Schiff base complexes 377
Phenol–N,N,N-trialkylamine oxide complexes 364
Phenol–triazabicyclodecene base complexes 377
Phenol trimethylsilyl ether, fluorination of 647
Phenol–water complexes 145–169
PhOH(H$_2$O)$_1$ 149–156
PhOH(H$_2$O)$_2$ 149, 150, 152–155
PhOH(H$_2$O)$_3$ 156–160
PhOH(H$_2$O)$_4$ 160–169
Phenones—*see* Acenaphthoquinones, Acetophenones, Acrylophenones,

Subject Index

Phenones—*see* Acenaphthoquinones, Acetophenones, Acrylophenones, Benzophenones, Hydroxypropiophenones, Pivalophenones (*continued*)
  Benzophenones, Hydroxypropiophenones, Pivalophenones
Phenoxenium ion—*see* Phenoxy cation
Phenoxy cation 273–277, 289
Phenoxyl radicals 15, 129, 131–139, 310, 638, 865, 1017–1019, 1098, 1100
  acid–base equilibria of 1132–1135
  comparison with isoelectronic radicals 1127
  decomposition of 137–139
  ESR spectra of 1120–1127
  fast detection techniques for 1108
  formation of 1108–1119
    by autoxidation of phenols and hydroxyphenols 1118, 1119
    by intramolecular electron transfer 1117
    by oxidation of phenols 1109–1113
    by oxidative replacement of substituents 1116, 1117
    by reaction of hydroxyl radicals with phenols 1113–1116
  optical spectra of 1127–1132
  reactions of 1135–1140
  reduction potentials of 1138, 1140–1143
Phenoxymagnesium halides, alkylation of 622
Phenoxyselenation 669
Phenyl acetates, Fries rearrangement of 639, 644
Phenylalanine 231
(ω-Phenylalkyl)catechols, mass spectra of 272, 273, 275
Phenylalkyl ethers, CI mass spectra of 293
C-Phenylation 680–682
Phenylcoumarans, enzymatic formation of 1216
Phenyl ethers—*see* Alkyl phenyl ethers, Allyl phenyl ethers, Benzyl phenyl ethers, 2-Hexenyl phenyl ether
Phenylhydrazones, thermochemistry of 252
Phenyliminobenzoxathiole, photoreactions of 1075
Phenyl isocyanate, photoreactions of 1075
o-Phenylphenol, photodimerization of 1017
2-Phenylphenol 1350
  photooxidation of 1084
Phenylpropanoids, CI mass spectra of 293
Phenyl radicals 1100
Phloroglucinol,
  acylation of 631
  thermochemistry of 247
Phosphates—*see* 1,3-Diphosphates
Phosphatidylcholine liposomes 851
Photoacidity 491–494

current understanding of 523–525
free-energy correlations with reactivity 522, 523
origin of 498–507
thermodynamic aspects of 493, 495–498
Photoacids, electronic structure of 507–522
Photoacoustic methods 897
Photoaddition reactions 1022–1024
Photobasicity 492, 501–503
Photocyclization reactions 1067–1070
  involving aryl radicals 1063–1067
  of chalcones 1028–1030
  of naphthoates 1077
Photocycloaddition reactions,
  intermolecular 1059–1061
  intramolecular 1061–1063
Photodecomposition, solar-induced, of chlorophenols 1085
Photodissociative states 1017
Photoelectron spectroscopy, of phenol 112
Photo-Fenton oxidation 1081
Photoformylation 1071
Photo-Fries reactions 1055
  intermolecular 1057
Photoionization 1016
  two-photon processes in 1017
Photolabile protecting groups 1043–1045
Photooxidation 1017–1019
  of alkylphenols 1084
  of chlorophenols 1085–1087
  of dihydroxyphenols 1084, 1085
  of phenol 1080–1084
Photorearrangements,
  side-chain 1048–1055
  skeletal 1048
  with side-chain migration 1055–1057
Physicochemical properties, of phenol 4–6
Physiological monitoring 968, 969
Picolinic acids—*see* 3-Hydroxypicolinic acid
Picric acid 236
Piezoelectric detection 1001, 1002
Pimentel model 503
Pinoresinol, enzymatic formation of 1216
Piperazinomycin 1315, 1321
  synthesis of 1318
Pivalophenones, synthesis of 642
pK_a values,
  determination of 954
  of phenolphthalein 974
  of polychlorophenols 970
  of radical species 1133–1135
Plant extracts,
  analysis of 957, 965–968, 973
  phenolic compounds in 942
Plasma, phenolic compounds in 969
Plastics, phenolic compounds in 948
Platt free-electron model 508
Platt notation 495, 508, 509

PLS algorithm 985, 993
Polar solvents 878, 879
Pollutants, phenolic compounds as 927, 976
  analysis of,
    in environmental and microbial samples 935, 936
    in water 938, 945–947, 954, 955, 962–965, 971, 972, 985, 986, 993
  estrogenic activity of 925
Polychlorophenols 954
Polycyclic phenols, keto–enol tautomerism in 719–721
Polygalaceae 967
Polyhalophenols—*see also* Polychlorophenols
  thermochemistry of 240
Polyhydroxybenzenes, structural chemistry of 200–204
Polyimides 1493–1502
  linear 1494–1496
  origins of 1494
Polyketides, synthesis of 426
Polymethylphenols 954
Polyphenol oxidases 978, 980, 981
Polyphenols,
  acylation of 631
  determination of 1354
Poly(phenylene oxides) 1491, 1492
Polystyrene, free-radical polymerization of 629
Polystyrene–divinylbenzene copolymer 944, 947
Poly[*N*-(substituted phenyl)maleimides] 1495, 1497, 1498
Porphyrin complexes, as oxidation catalysts 1201, 1205, 1211
Porphyrins—*see also meso-*
  Tetraphenylporphyrin
  reactions of 1082
Potential energy surface 3
Potentials, double-minima 371, 377
  symmetrical OH stretching 382
Potential well, shape of 349
Preconcentration, of analytical samples 930, 932, 934, 936, 942, 944–949, 955, 962–964, 971, 972, 983–987, 995
Prenylphenols, ^{13}C chemical shifts for 338
Pressure transducers 851
Prestegane A 1244
  oxidative coupling of 1286
Principal components regression (PCR) method 929
Proanthocyanidins 948
Prooxidant effect 845, 849, 891, 892, 895, 899
Propargyl aryl ethers, rearrangement of 770, 773, 779–782
o-Propenylphenols 301
5-Propylguaiacol, mass spectrum of 303, 304
Protocatechuic acid, irradiation of 1079

Proton affinities 3, 5, 86–89, 261
  local 261, 262, 296, 308
Protonation 83–92
  hydrogen bonding and 593–596
  regioselectivity of 89–92
Proton polarizabilities 593
Proton sponges 997
Proton transfer 316, 355, 356, 1108
  excited-state intramolecular 729, 730
  in hydroxyarenes 492, 523, 524
  in photoacids 502
Pseudomonas cepacia 638
Pseudomonas putida 979
PTFE membranes 945
Pulse radiolysis 1099, 1109, 1113
Pummerer rearrangement 671, 674
Pummerer's ketone 1158, 1288
Pyranones—*see* Dibenzopyranones
Pyrans—*see also* Benzopyrans, Dihydropyrans
  rearrangement of 823, 824
Pyrenes—*see* 1-Aminopyrene, Hydroxypyrenes
Pyrenols,
  hydrogen bonding in 503
  photoacidity of 493, 498, 502, 523
Pyridine methoxy cupric chloride complex, as oxidant 1192
Pyridines—*see also*
  2-Amino-3-hydroxypyridine,2,6-Dihydroxypyridine, *o*-Hydroxypyridines
  photoreactions of 1075
Pyrimidinopteridine *N*-oxide 1019
Pyrogallo[4]arenes 320, 322
Pyrogallols 1382—*see also*
  Calix[4]pyrogallol
  acylation of 631
  thermochemistry of 247
Pyrolysis-gas chromatography 941
α-Pyrones, condensation of 430
Pyrroles—*see* Calixpyrroles
Pyrroloiminoquinone alkaloids 1178, 1264
Pyrroloquinoline quinone-dependent glucose dehydrogenase 979
Pyrylium salts, rearrangement of 823, 825

Quantitative structure–activity relationships (QSAR) 970, 982
Quantitative structure–affinity relationships (QSAR) 992
Quantum Chemistry Library Data Base 4
Quercetin 872, 873
  as antioxidant 894
Quinizarin, IR spectrum of 383

Quinolenones—see
    Alkoxyhydroquinolenones,
    Fluorohydroquinolenones
Quinolines—see Tetrahydroquinoline
Quinols—see Futoquinols, Isofutoquinols,
    Ubiquinols
Quinone ketals, formation of 1170, 1171
p-Quinone methide radical 1158
Quinone methides 1217, 1282
    dimerization of 1255, 1256
    enzymatic formation of 1223
    from biphenyl derivatives 1035–1038
    from fluorenols 1038
    intramolecular cyclization of 1220
    reactions of 1308
    synthesis of 1253, 1255
m-Quinone methides 1024
    formation from phenol derivatives 1035
o-Quinone methides,
    formation from phenol derivatives
        1032–1035, 1192, 1309, 1311,
        1312
    in resole formation 1478–1483
p-Quinone methides 1196
    dimeric 1159
    formation from phenol derivatives 1035,
        1307, 1309
Quinone monoketals,
    Diels–Alder reactions of 1235
    synthesis of 1228, 1231, 1250, 1335
Quinone oxime tautomers 251, 252
Quinones—see also Aminoquinones,
    Anthraquinones, Arenolquinones,
    Benzoquinones, Calixquinones,
    Hydroquinones, Hydroxyquinones,
    Naphthoquinones, Perylenequinones,
    Scabequinone
    reduction of 1108
Quinone units 1396
Quinoprotein glucose dehydrogenase 977,
    978

Radiation chemistry 1097
    gas-phase 1104, 1105
    in alkane solvents 1102, 1103
    in aqueous solution 1098–1100
    in organic halogenated solvents 1100–1102
    in solid matrices 1103, 1104
    of aqueous phenol solutions 1113
    of neat phenols 1103
Radical anions,
    formation of 1098
    hydroxytetrachlorophenoxyl 1086
Radical cations,
    fluorophenol 291
    gaseous 261–263
    in oxidation of phenols 1113

phenol 111–121, 289–291, 1101
Radical coupling reactions 1156–1165
Radicals,
    alkyl 1103
    aminyl 1110
    anilinyl 1109
    cumyloxyl 877
    5, 7-diisopropyltocopheroxyl 894
    hydroxycyclohexadienyl 1098, 1100
    hydroxyl 1098
    hydroxyphenyl 1098, 1099
    nitric oxide 638
    nitronium 638
    NO_2 639
    peroxyl 841–845
    phenoxyl 15, 129, 131–139, 310, 638, 865,
        1017–1019, 1098, 1100,
        1108–1143
    phenyl 1100
    p-quinone methide 1158
    semiquinone 871, 874, 893, 1109
TEMPO 366
    α-tocopherol 846, 849
    tyrosyl 129, 1130
Radical stabilization energies 140
Radiolysis, gas-phase 307
γ-Radiolysis 1099
Raman spectroscopy 368, 379, 380, 997, 1132
Raschig–Dow process 7
Reactivity indices 89–92
Rearrangements,
    hydroxy group migrations 748, 749
    isomerizations,
        cis–trans 743–745
    of alkylphenols 749
    of phenols containing unsaturated
        side-chains 750, 752–758
    of functionalized arenes 799–805
    of heterocyclic compounds 823–830
    of non-aromatic carbocycles 806–822
    of *O*-substituted phenol derivatives
        759–798
    phenol–dienone 745–748
Reduction potentials,
    calculation of 897, 898
    of phenoxyl radicals 1138, 1140–1143
    one-electron 895, 896
Rcichardt's betaine dye, solvatochromic shifts
    of 540–543
Reimer–Tiemann reaction 685, 1071
Resole resins 1457, 1458, 1460
    model compounds of 1476–1487
    synthesis of 1475, 1476
Resol prepolymers 969
Resonance Assisted Hydrogen Bonding
    (RAHB) 552
Resonance energies, in keto/enol forms of
    phenol 39

Resonance-enhanced multiphoton ionization 1000
Resorcarenes 1370—*see also* Bis-resorcarenes, Calixresorcinols
 methylene-bridged 1383
 synthesis of 1381–1384
Resorc[4]arenes 320, 322
Resorcarene–sugar clusters 1418
Resorcinol benzoates, Fries rearrangement of 645
Resorcinol monomethyl ether, mass spectrum of 289
Resorcinols 236—*see also* 2-Alkylresorcinols, Calixresorcinols, 4-Cumylresorcinol, *n*-Heptylresorcinol, *n*-Hexylresorcinol, 2-Nitroresorcinol, *n*-Pentadecylresorcinol
 acylation of 631
 mass spectra of 272
 reactions with cumene hydroperoxide 607
 synthesis of 428, 429, 436, 437
 thermochemistry of 242
Resorcylic acids 426, 427, 430, 431
 ionized, mass spectra of 285
trans-Resveratrol, determination of 319
Retention times, correlation with boiling points 938
Rhenium compounds, as oxidants 1278, 1286, 1288
Rhodium compounds, as oxidants 1287, 1302, 1304
Rhodium tetraacetate, as catalyst, in *O*-alkylation 666
Rhodococcus 978
Ring current effects 336
Ring inversion, cone-to-cone 1385, 1386
Ring-walk isomerization 283
Ripening, state of, effect on organoleptic properties of foodstuffs 913
River water, analysis of 946, 947, 962, 977
Road particulate matter, phenolic compounds in 938
Rotation, around C–O bond 353, 354
Rotaxanes 1432
Rubazoic acids 349
 tautomerism in 361–363
Ruthenium chlorobenzene complexes 673
Ruthenium compounds,
 as oxidants 1286, 1288, 1300, 1302
 as oxidation catalysts 1213–1215

Salen complexes 688, 697–701
 as catalysts,
  in kinetic resolution of racemic epoxides 667, 668
  in oxidation 1199, 1201, 1202, 1205, 1209, 1211
Salicylaldehyde imines, complexes of 697–704

Salicylaldehydes—*see also* 5-Nitrosalicylaldehyde
 chemical shifts for, isotope effects on 348
 formation of 626, 627
 ionized, mass spectra of 124, 126
 IR spectra of 377, 378
Salicylamide, IR spectrum of 377, 378
Salicylate anion, IR spectrum of 371
Salicylates 1021
 IR spectra of 377, 378
 synthesis of 438
Salicylic acids 232—*see also* Sulfosalicylic acid
 mass spectra of 285
 photoreactions of 1080
N-Salicylideneaniline, photo-induced transformation of 1024
Salicylidene derivatives, hydrogen transfer in 1024, 1025
Salting out 983
Sample preservation 945, 946
Scabequinone, synthesis of 1261
Schelhammeridine 1297
Schiarisanrin natural products 653, 654
Schiff bases,
 coupling constants for 349
 hydrogen transfer in 1024
 solid-state 366, 367
 synthesis of 1405, 1406
 tautomerism in 356, 359, 360, 726–734
Schizandrin 1278
β-Scission reactions 877
Scoulerine, formation of 1306
Seawater, analysis of 982
Secoaglucovancomycin, synthesis of 1324, 1325
Secoisolariciresinol, enzymatic formation of 1216
Selectfluor, as fluorinating agent 646, 647, 655
Selenium compounds, as oxidants 1330, 1332–1334
Self-association 546–550
Semiquinone radical 871, 874, 893, 1109
Sensor techniques, use of calixarenes in 1435
Sequential injection analysis (SIA) 984, 986
Serum, phenolic compounds in 968
Sewage, phenolic pollutants in 931
SFC 953
[2,3]-Sigmatropic rearrangement 677, 697
Silica gel membranes 1082
Silphinene, electrosynthesis of 1183, 1187
Silver compounds, as oxidants 1307–1311
Silybin, synthesis of 1308, 1310
Silydianin 1180
Silylation, of phenols 934, 935
Silylcyanation, asymmetric 694, 695, 702
Simmons–Smith cyclopropanation, asymmetric 694

Simmons–Smith reagent 677, 678
Simultaneous distillation and extraction (SDE) 949
Sinapic acid, electrolysis of 1158, 1159, 1162
Single electron transfer (SET), as antioxidant mechanism 844, 896, 897
Size exclusion chromatography 953
Slash pine bark, phenolic compounds in 944
Smiles rearrangement 466–470, 759
S_NAr reactions 673
Soil samples, phenolic compounds in,
   analysis of 932, 965, 972, 985
   field screening of 938
Sol-gel technique 1082
Solid acid catalysis 612–621
Solid-phase extraction (SPE) 930–933, 936, 942, 944–950, 955, 958, 960, 962–964, 969, 972, 985, 986, 995, 1354
Solvation energy 500, 992
Solvation free energy 5
Solvatochromic comparison method, for solvent hydrogen-bond basicity 591
Solvent effects,
   on antioxidant activity of phenols 876–884
   on Raman spectra of phenoxyl radicals 1132
   on reactivity of phenols 179
Solvent interactions 877–880
   with phenolic antioxidants 880–884
Solvent parameters 876
Solvent viscosity 883, 884
Sonication methods 932
Soya extracts, analysis of 957
Specific heats 5
Spin density 135–137
   substituent effects on 1123
Spin–orbit coupling 110
Spirocyclic cyclohexadienone ring system, synthesis of 653
Spirodienones 1410, 1411—see also Bis-spirodienones, Tris-spirodienones
Spiro-oxazoles, formation of 1172
Square-wave voltammetry 982
$S_{RN}1$ reactions 1073
Staphylococcus aureus, methicillin-resistant 1318
Steam distillation 931
Steganacin 1278
   synthesis of 1281
Steganes 1278
   formation of 1242
Stereoelectronic effects, on antioxidant activity of phenols 861
Stereoisomeric complexes 576
Steric effects 354, 355
Stilbenes,
   irradiation of 1079

synthesis of 1405
Stille reaction 1406
Streptomyces antibioticus 977
Stretching potential, OH 371
Strong anion exchanger (SAX) 947
Structural chemistry 199–220
   steric and electronic effects on 212–216
Structure–activity models 925
Styrene, as substrate for antioxidants 858
Styrene–divinylbenzene copolymer 945, 946, 962
Styrenes—*see* Hydroxystyrenes
Substituent effects, additivity of 263, 899
Sugars—*see* 1-Acetylsugars, Calixsugars
Sulfides, asymmetric oxidation of 699
Sulfinyl derivatives, formation from thiacalixarenes 1415
Sulfonic acids, as phenol precursors 397–399
Sulfosalicylic acid, thermochemistry of 237
Sulfuryl chloride, as chlorinating agent 649
Supercritical fluid chromatography (SFC) 953, 955
Surface plasmon resonance 1000
Surfactants 948, 972
Suzuki coupling 1406, 1424
Sympathomimetic drugs 995
Synergism 889–891
Syringic acid, irradiation of 1079
Syringyl alcohol, catalytic oxidation of 1205, 1208

Tannic acids, titration of 1000
Tannins 940, 948
   use in embalming 913
Tautomerism 356–363
   arylazophenol–quinone arylhydrazone 724–726
   enol–enol 721
   in *o*-hydroxyazo aromatics 356, 358
   in Mannich bases 356, 358
   in nitrosophenols 360, 361
   in Schiff bases 356, 359, 360
   in side-chains 740, 742, 743
   interconversion barriers in 359
   keto–enol 38–40, 46, 127–130, 714–737
   nitrosophenol–quinone oxime 721–724
   ring–chain 737–741
Tea catechins,
   analysis of 961
   biological activity of 913
TEAC (Trolox equivalent antioxidant capacity) 894
Tellurium compounds, as oxidants 1288
Template effect 1428
   in formation of calixarenes 1377, 1378
Tetrabenzoxazines,

as precursors of other chiral derivatives 1420
 1,2; 3,4-bis-bridged 1420
 formation of 1419, 1420
Tetrabromobisphenol A 1351
2,3,5,6-Tetrachlorophenol—N,N-dimethylaniline complex 364
Tetrachlorophenols, photoreactions of 1074
Tetrafluoroethylene, addition to sodium phenoxides 669, 670
Tetrahydrobenzonaphthofurans 614, 615
Tetrahydroisoquinolines,
 electrolysis of 1159
 oxidation of 1329
Tetrahydropyranylphenols, rearrangement of 641
Tetrahydroquinoline, as arylating agent 653
Tetralins—see Hydroxytetralins
2,2,6,6-Tetramethyl-1-piperidinyloxy radical (TEMPO), complexes of 366
Tetranitromethane, photoreactions with 1071
meso-Tetraphenylporphyrin, as electron acceptor 1019
Tetrasulfones, formation from thiacalixarenes 1415
Thallium compounds, as oxidants 1278, 1281, 1286, 1288, 1311–1325
Thebaine, enzymatic formation of 1216
Thermal analysis 1002
Thermal desorption 938
Thermospray techniques 319
Thiacalix[4]anilines, synthesis of 1416
Thiacalixarenes,
 reactions of 1415, 1416
 sulfur bridges in, oxidation of 1415
 synthesis of 1380, 1381
Thiacalix[4]arenes, conformation of 1415, 1416
Thiazoles—see Benzthiazoles
Thin layer chromatography (TLC) 953
Thioacyl derivatives, hydrogen bonding in 350
Thiocarbamates, rearrangement of 1401
Thiomethylation, of calixresorcinols 1419
Thiophenols 963
Thiourea, photoreactions of 1075
Thymol complexes 365
Tissue biosensors 978
Titanium phenoxide 691
Titration, thermometric 1000
Titration studies,
 of antioxidant efficiency 854, 855
 of deprotonation of phenols 335
TLC 953
Tobacco extracts, analysis of 957
Tobacco smoke, phenolic compounds in 932
α-Tocopherol 844–850, 861, 872, 883–888, 892–895, 1111, 1335
 oxidation of 1080

α-Tocopherol quinone 846
α-Tocopherol radical 846, 849
Toluenes—see Dihydroxytoluenes
Toxicity, of phenols 925, 1359–1361
Toxicological monitoring 968, 969
Transbutylation 1402, 1415
 selective 1403
Transition moments 105, 106
 direction of 368, 381
Transmembrane diffusion, of α-tocopherol 887, 888
TRAP (total radical antioxidant parameter) assay 857
Triazacyclononane complexes, as oxidation catalysts 1209, 1212
Triazines—see Hydroxyphenyltriazines
2,3,5-Trichlorophenol, deprotonation of 335
2,4,6-Trichlorophenol, irradiation of 1086, 1087
Trichosporon beigelii 978
Triflates,
 in *ortho*-acylation 629, 630
 in Fries rearrangements 639, 642
Trifluoroacetylation, of phenols 934
1-Trifluoroacetyl sugars 674
Trifluoroacetylwilsonirine, synthesis of 1275
α,α,α-Trifluorocresols, mass spectra of 290, 291
4-(2,2,2-Trifluoro-1-hydroxyethyl)phenol 608
Trifluoromethylphenols—see also 2,4,6-Tris(trifluoromethyl)phenol
 gas-phase acidity of 312
 photooxidation of 1084
4-Trifluorosulfonylphenol, gas-phase acidity of 312
2,4,6-Trihydroxyacetophenone, as MALDI matrix 323, 324
1,8,9-Trihydroxyanthracene, as MALDI matrix 323
1,3,8-Trihydroxyanthraquinones, chemical shifts for, isotope effects on 345
Trihydroxybenzenes—see also Dihydroxyphenols
 structural chemistry of 204, 211, 212
3,4,5-Trihydroxybenzoic acid, mass spectrum of 285
4,4'-(Trimethylene)bis(2,6-di-*t*-butylphenoxy) diradical 1018
Trimethylolphenol 627
Trimethylsilyloxydienes, in synthesis of phenols 441–443
2,4,6-Trinitrophenol,
 complexes of 366
 hydrogen bonding in 352
Triphenylphosphine oxide, complexes of 367
Triphenylpyrylium tetrafluoroborate, as electron-accepting sensitizer 1040

Triquinanes,
  electrosynthesis of 1188, 1191
  synthesis of 1237, 1238, 1252
Tris-spirodienones 1411
2,4,6-Tris(trifluoromethyl)phenol, gas-phase acidity of 312
Trolox 862, 889
Trolox C 1111
Tropolones,
  IR spectra of 382, 383
  rearrangement of 818, 822
1,2-Tungstophosphoric acid 619
Tyrosinase 955, 969, 978, 979, 981
  biosensors based on 974–977
  recombinant 977
Tyrosine protein residues 105
Tyrosines 15—see also Diiodotyrosine, Dityrosines
  irradiation of 1100
  thermochemistry of 231, 232
Tyrosine spirolactones, as peptide synthons 1172, 1174
Tyrosyl radicals 129, 1130

Ubiquinols 867, 874, 875
  as antioxidants, in lipid membranes 892–894
Ullmann reaction 670, 1406
Ultraviolet multiwavelengths absorptiometry (UVMA) 985
Umbellulone, formation of 1048
α,β-Unsaturated α'-diazoketones, as dienylketene precursors 450, 451, 453
UNSCRAMBLER program 987
Uremia 969
Urine, phenolic compounds in 920, 930, 931, 934, 935
  determination of 968, 969
Usnic acid, tautomerism in 357
UV-visible detection (UVD) methods 937, 953
  chemiluminescence 853, 995
  colorimetry 990, 991
  fluorescence 853, 993–995
  in measurement of antioxidant efficiencies 851–854, 856–858
  spectrophotometry 984–990

Valence basin synaptic order 69
Valence shell charge 32
Vanadium compounds, as oxidants 1273–1278, 1286, 1288, 1312, 1313
Vancomycin 1315, 1321, 1324
  resistance 1318

Vanillic acid,
  irradiation of 1079
  photooxidation of 1080
Vanillin,
  complexes of 365
ion/molecule reactions of 308
  nitration of 632
Veratric acid, irradiation of 1079
Vibrational modes, of phenol 34–44
Vibrational spectroscopy 367
Vibrational transitions, polarization directions of 368
Vicarious nucleophilic substitution (VNS) 405
Vitamin C 889–891
Vitamin E 129, 840, 884, 1111
  synergistic interactions with Vitamin C 889–891
(−)-Vitisin B, synthesis of 1218
(+)-Vitisin C, synthesis of 1218

Wacker-type reaction 669
Wallach rearrangement 419, 801
Wastewater,
  analysis of 962, 971, 972, 979, 985, 986
  phenolic pollutants in 931–933
Wessely–Moser reaction 748, 751
Wessely oxidation 1325–1328
Wheland intermediates 639
Williamson synthesis 662
Wilson's notation 375
Wine,
  analysis of 947, 960, 961, 972, 973, 1000
  phenolic acids in 942, 943
  polyphenols in 935
Wittig–Horner-type olefination 1405
Wolff–Kishner reduction 639
Wolff rearrangement 1057

Xenoestrogens 925, 936
Xylenols, irradiation of 1084

Ylides, oxomethylene 653

Zeolite catalysis 616–620, 640, 644, 645
Zero-crossing technique 984
Zinc chloride, as allylation catalyst 677
Zinc phenoxide 691
Zirconium cyclopentadienyl complexes 684
Zwitterionic/dipolar resonance structures, for phenol 226
Zwitterionic intermediates 1062

Index compiled by P. Raven